BURGER'S MEDICINAL CHEMISTRY AND DRUG DISCOVERY

BURGER'S MEDICINAL CHEMISTRY AND DRUG DISCOVERY

BURGER'S MEDICINAL CHEMISTRY AND DRUG DISCOVERY

Sixth Edition

Volume 1: Drug Discovery

Edited by

Donald J. Abraham
Department of Medicinal Chemistry
School of Pharmacy
Virginia Commonwealth University
Richmond, Virginia

Burger's Medicinal Chemistry and Drug Discovery
is available Online in full color at
www.mrw.interscience.wiley.com/bmcdd.

WILEY-INTERSCIENCE

A John Wiley and Sons, Inc., Publication

Cover Description   The molecule on the cover is hemoglobin with the allosteric effector RSR 13 attached. Three groups initiated structure-based drug design in the middle to late 1970s. Two of the groups, Peter Goodford's at the Burroughs Wellcome Laboratories in London and the editors' at the School of Pharmacy at the University of Pittsburgh, worked on hemoglobin while David Matthews at Agouron Pharmaceuticals worked on dihydrofolate reductase. Max Perutz and his coworkers' solution of the phase problem produced the first three-dimensional structure of a protein whose coordinates were of interest for drug design. The Editor worked with Max Perutz from 1980 to 1988, attempting to design antisickling agents. One of the active antisickling molecules, clofibric acid, would not lead to a sickle cell drug but to an allosteric effector RSR 13 designed and synthesized at Virginia Commonwealth University, which has been studied clinically for treatment of metastatic brain cancer. Max Perutz, whose work would provide the underpinnings for structure-based drug design, passed away in February 2002. His spirit and love for science that transferred to his students, postdoctoral fellows, visiting scientists, and colleagues worldwide, like Professor Burger's, is the leaven and inspiration for new discoveries.

Library of Congress Cataloging-in-Publication Data:

Burger's medicinal chemistry and drug discovery.—6th ed., Volume 1: drug discovery/
Donald J. Abraham, editor

ISBN 0-471-27090-3 (v. 1: acid-free paper)

Printed in the United States of America.

10 9 8 7 6 5 4 3 2 1

*To Alfred Burger and Max Perutz for their mentorship
and life-long passion for science.*

BURGER MEMORIAL EDITION

The Sixth Edition of Burger's Medicinal Chemistry and Drug Discovery is being designated as a Memorial Edition. Professor Alfred Burger was born in Vienna, Austria on September 6, 1905 and died on December 30, 2000. Dr. Burger received his Ph.D. from the University of Vienna in 1928 and joined the Drug Addiction Laboratory in the Department of Chemistry at the University of Virginia in 1929. During his early years at UVA, he synthesized fragments of the morphine molecule in an attempt to find the analgesic pharmacophore. He joined the UVA chemistry faculty in 1938 and served the department until his retirement in 1970. The chemistry department at UVA became the major academic training ground for medicinal chemists because of Professor Burger.

Dr. Burger's research focused on analgesics, antidepressants, and chemotherapeutic agents. He is one of the few academicians to have a drug, designed and synthesized in his laboratories, brought to market [Parnate, which is the brand name for tranylcypromine, a monoamine oxidase (MAO) inhibitor]. Dr. Burger was a visiting Professor at the University of Hawaii and lectured throughout the world. He founded the *Journal of Medicinal Chemistry, Medicinal Chemistry Research,* and published the first major reference work *"Medicinal Chemistry"* in two volumes in 1951. His last published work, a book, was written at age 90 (*Understanding Medications: What the Label Doesn't Tell You,* June 1995). Dr. Burger received the Louis Pasteur Medal of the Pasteur Institute and the American Chemical Society Smissman Award. Dr. Burger played the violin and loved classical music. He was married for 65 years to Frances Page Burger, a genteel Virginia lady who always had a smile and an open house for the Professor's graduate students and postdoctoral fellows.

PREFACE

The Editors, Editorial Board Members, and John Wiley and Sons have worked for three and a half years to update the fifth edition of Burger's Medicinal Chemistry and Drug Discovery. The sixth edition has several new and unique features. For the first time, there will be an online version of this major reference work. The online version will permit updating and easy access. For the first time, all volumes are structured entirely according to content and published simultaneously. Our intention was to provide a spectrum of fields that would provide new or experienced medicinal chemists, biologists, pharmacologists and molecular biologists entry to their subjects of interest as well as provide a current and global perspective of drug design, and drug development.

Our hope was to make this edition of Burger the most comprehensive and useful published to date. To accomplish this goal, we expanded the content from 69 chapters (5 volumes) by approximately 50% (to over 100 chapters in 6 volumes). We are greatly in debt to the authors and editorial board members participating in this revision of the major reference work in our field. Several new subject areas have emerged since the fifth edition appeared. Proteomics, genomics, bioinformatics, combinatorial chemistry, high-throughput screening, blood substitutes, allosteric effectors as potential drugs, COX inhibitors, the statins, and high-throughput pharmacology are only a few. In addition to the new areas, we have filled in gaps in the fifth edition by including topics that were not covered. In the sixth edition, we devote an entire subsection of Volume 4 to cancer research; we have also reviewed the major published Medicinal Chemistry and Pharmacology texts to ensure that we did not omit any major therapeutic classes of drugs. An editorial board was constituted for the first time to also review and suggest topics for inclusion. Their help was greatly appreciated. The newest innovation in this series will be the publication of an academic, "textbook-like" version titled, "Burger's Fundamentals of Medicinal Chemistry." The academic text is to be published about a year after this reference work appears. It will also appear with soft cover. Appropriate and key information will be extracted from the major reference.

There are numerous colleagues, friends, and associates to thank for their assistance. First and foremost is Assistant Editor Dr. John Andrako, Professor emeritus, Virginia Commonwealth University, School of Pharmacy. John and I met almost every Tuesday for over three years to map out and execute the game plan for the sixth edition. His contribution to the sixth edition cannot be understated. Ms. Susanne Steitz, Editorial Program Coordinator at Wiley, tirelessly and meticulously kept us on schedule. Her contribution was also key in helping encourage authors to return manuscripts and revisions so we could publish the entire set at once. I would also like to especially thank colleagues who attended the QSAR Gordon Conference in 1999 for very helpful suggestions, especially Roy Vaz, John Mason, Yvonne Martin, John Block, and Hugo

Kubinyi. The editors are greatly indebted to Professor Peter Ruenitz for preparing a template chapter as a guide for all authors. My secretary, Michelle Craighead, deserves special thanks for helping contact authors and reading the several thousand e-mails generated during the project. I also thank the computer center at Virginia Commonwealth University for suspending rules on storage and e-mail so that we might safely store all the versions of the author's manuscripts where they could be backed up daily. Last and not least, I want to thank each and every author, some of whom tackled two chapters. Their contributions have provided our field with a sound foundation of information to build for the future. We thank the many reviewers of manuscripts whose critiques have greatly enhanced the presentation and content for the sixth edition. Special thanks to Professors Richard Glennon, William Soine, Richard Westkaemper, Umesh Desai, Glen Kellogg, Brad Windle, Lemont Kier, Malgorzata

Dukat, Martin Safo, Jason Rife, Kevin Reynolds, and John Andrako in our Department of Medicinal Chemistry, School of Pharmacy, Virginia Commonwealth University for suggestions and special assistance in reviewing manuscripts and text. Graduate student Derek Cashman took able charge of our web site, *http://www.burgersmedchem.com,* another first for this reference work. I would especially like to thank my dean, Victor Yanchick, and Virginia Commonwealth University for their support and encouragement. Finally, I thank my wife Nancy who understood the magnitude of this project and provided insight on how to set up our home office as well as provide John Andrako and me lunchtime menus where we often dreamed of getting chapters completed in all areas we selected. To everyone involved, many, many thanks.

DONALD J. ABRAHAM
Midlothian, Virginia

Dr. Alfred Burger

Photograph of Professor Burger followed by his comments to the American Chemical Society 26th Medicinal Chemistry Symposium on June 14, 1998. This was his last public appearance at a meeting of medicinal chemists. As general chair of the 1998 ACS Medicinal Chemistry Symposium, the editor invited Professor Burger to open the meeting. He was concerned that the young chemists would not know who he was and he might have an attack due to his battle with Parkinson's disease. These fears never were realized and his comments to the more than five hundred attendees drew a sustained standing ovation. The Professor was 93, and it was Mrs. Burger's 91st birthday.

Opening Remarks

ACS 26th Medicinal Chemistry Symposium

June 14, 1998
Alfred Burger
University of Virginia

It has been 46 years since the third Medicinal Chemistry Symposium met at the University of Virginia in Charlottesville in 1952. Today, the Virginia Commonwealth University welcomes you and joins all of you in looking forward to an exciting program.

So many aspects of medicinal chemistry have changed in that half century that most of the new data to be presented this week would have been unexpected and unbelievable had they been mentioned in 1952. The upsurge in biochemical understandings of drug transport and drug action has made rational drug design a reality in many therapeutic areas and has made medicinal chemistry an independent science. We have our own journal, the best in the world, whose articles comprise all the innovations of medicinal researches. And if you look at the announcements of job opportunities in the pharmaceutical industry as they appear in *Chemical & Engineering News,* you will find in every issue more openings in medicinal chemistry than in other fields of chemistry. Thus, we can feel the excitement of being part of this medicinal tidal wave, which has also been fed by the expansion of the needed research training provided by increasing numbers of universities.

The ultimate beneficiary of scientific advances in discovering new and better therapeutic agents and understanding their modes of action is the patient. Physicians now can safely look forward to new methods of treatment of hitherto untreatable conditions. To the medicinal scientist all this has increased the pride of belonging to a profession which can offer predictable intellectual rewards. Our symposium will be an integral part of these developments.

CONTENTS

BURGER'S MEDICINAL CHEMISTRY AND DRUG DISCOVERY

CHAPTER ONE

History of Quantitative Structure-Activity Relationships

C. D. Selassie
Chemistry Department
Pomona College
Claremont, California

Contents

Burger's Medicinal Chemistry and Drug Discovery
Sixth Edition, Volume 1: Drug Discovery
Edited by Donald J. Abraham
ISBN 0-471-27090-3   © 2003 John Wiley & Sons, Inc.

1 INTRODUCTION

It has been nearly 40 years since the quantitative structure-activity relationship (QSAR) paradigm first found its way into the practice of agrochemistry, pharmaceutical chemistry, toxicology, and eventually most facets of chemistry (1). Its staying power may be attributed to the strength of its initial postulate that activity was a function of structure as described by electronic attributes, hydrophobicity, and steric properties as well as the rapid and extensive development in methodologies and computational techniques that have ensued to delineate and refine the many variables and approaches that define the paradigm. The overall goals of QSAR retain their original essence and remain focused on the predictive ability of the approach and its receptiveness to mechanistic interpretation.

Rigorous analysis and fine-tuning of independent variables has led to an expansion in development of molecular and atom-based descriptors, as well as descriptors derived from quantum chemical calculations and spectroscopy (2). The improvement in high-throughput screening procedures allows for rapid screening of large numbers of compounds under similar test conditions and thus minimizes the risk of combining variable test data from many sources.

The formulation of thousands of equations using QSAR methodology attests to a validation of its concepts and its utility in the elucidation of the mechanism of action of drugs at the molecular level and a more complete understanding of physicochemical phenomena such as hydrophobicity. It is now possible not only to develop a model for a system but also to compare models from a biological database and to draw analogies with models from a physical organic database (3). This process is dubbed *model mining* and it provides a sophisticated approach to the study of chemical-biological interactions. QSAR has clearly matured, although it still has a way to go. The previous review by Kubinyi has relevant sections covering portions of this chapter as well as an extensive bibliography recommended for a more complete overview (4).

1.1 Historical Development of QSAR

More than a century ago, Crum-Brown and Fraser expressed the idea that the physiological action of a substance was a function of its chemical composition and constitution (5). A few decades later, in 1893, Richet showed that the cytotoxicities of a diverse set of simple organic molecules were inversely related to their corresponding water solubilities (6). At the turn of the 20th century, Meyer and Overton independently suggested that the narcotic (depressant) action of a group of organic compounds paralleled their olive oil/water partition coefficients (7, 8). In 1939 Ferguson introduced a thermodynamic generalization to the correlation of depressant action with the relative saturation of volatile compounds in the vehicle in which they were administered (9). The extensive work of Albert, and Bell and Roblin established the importance of ionization of bases and weak acids in bacteriostatic activity (10–12). Meanwhile on the physical organic front, great strides were being made in the delineation of substituent effects on organic reactions, led by the seminal work of Hammett, which gave rise to the "sigma-rho" culture (13, 14). Taft devised a way for separating polar, steric, and resonance effects and introducing the first steric parameter, E_S (15). The contributions of Hammett and Taft together laid the mechanistic basis for the development of the QSAR paradigm by Hansch and Fujita. In 1962 Hansch and Muir published their brilliant study on the structure-activity relationships of plant growth regulators and their dependency on Hammett constants and hydrophobicity (16). Using the octanol/water system, a whole series of partition coefficients were measured, and thus a new hydrophobic scale was introduced (17). The parameter π, which is the relative hydrophobicity of a substituent, was defined in a manner analogous to the definition of sigma (18).

$$\pi_X = \log P_X - \log P_H \qquad (1.1)$$

P_X and P_H represent the partition coefficients of a derivative and the parent molecule, respectively. Fujita and Hansch then combined these hydrophobic constants with Hammett's electronic constants to yield the linear Hansch equation and its many extended forms (19).

$$\text{Log } 1/C = a\sigma + b\pi + ck \qquad (1.2)$$

Hundreds of equations later, the failure of linear equations in cases with extended hydrophobicity ranges led to the development of the Hansch parabolic equation (20):

$$\begin{aligned} \text{Log } 1/C = {} & a \cdot \log P \\ & - b(\log P)^2 + c\sigma + k \end{aligned} \qquad (1.3)$$

The delineation of these models led to explosive development in QSAR analysis and related approaches. The Kubinyi bilinear model is a refinement of the parabolic model and, in many cases, it has proved to be superior (21).

$$\text{Log } 1/C = a \cdot \log P$$
$$- b \cdot \log(\beta \cdot P + 1) + k \qquad (1.4)$$

Besides the Hansch approach, other methodologies were also developed to tackle structure-activity questions. The Free-Wilson approach addresses structure-activity studies in a congeneric series as described in Equation 1.5 (22).

$$\text{BA} = \sum a_i x_i + u \qquad (1.5)$$

BA is the biological activity, u is the average contribution of the parent molecule, and a_i is the contribution of each structural feature; x_i denotes the presence $X_i = 1$ or absence $X_i = 0$ of a particular structural fragment. Limitations in this approach led to the more sophisticated Fujita-Ban equation that used the logarithm of activity, which brought the activity parameter in line with other free energy-related terms (23).

$$\text{Log } BA = \sum G_i X_i + u \qquad (1.6)$$

In Equation 1.6, u is defined as the calculated biological activity value of the unsubstituted parent compound of a particular series. G_i represents the biological activity contribution of the substituents, whereas X_i is ascribed with a value of one when the substituent is present or zero when it is absent. Variations on this activity-based approach have been extended by Klopman et al. (24) and Enslein et al. (25). Topological methods have also been used to address the relationships between molecular structure and physical/biological activity. The minimum topological difference (MTD) method of Simon and the extensive studies on molecular connectivity by Kier and Hall have contributed to the development of quantitative structure property/activity relationships (26, 27). Connectivity indices based on hydrogen-suppressed molecular structures are rich in information on branching, 3-atom fragments, the degree of substitution, proximity of substituents and length, and heteroatom of substituted rings. A method in its embryonic state of development uses both graph bond

distances and Euclidean distances among atoms to calculate E-state values for each atom in a molecule that is sensitive to conformational structure. Recently, these electrotopological indices that encode significant structured information on the topological state of atoms and fragments as well as their valence electron content have been applied to biological and toxicity data (28). Other recent developments in QSAR include approaches such as HQSAR, Inverse QSAR, and Binary QSAR (29–32). Improved statistical tools such as partial least square (PLS) can handle situations where the number of variables overwhelms the number of molecules in a data set, which may have collinear X-variables (33).

1.2   Development of Receptor Theory

The central theme of molecular pharmacology, and the underlying basis of SAR studies, has focused on the elucidation of the structure and function of drug receptors. It is an endeavor that proceeds with unparalleled vigor, fueled by the developments in genomics. It is generally accepted that endogenous and exogenous chemicals interact with a binding site on a specific macromolecular receptor. This interaction, which is determined by intermolecular forces, may or may not elicit a pharmacological response depending on its eventual site of action.

The idea that drugs interacted with specific receptors began with Langley, who studied the mutually antagonistic action of the alkaloids, pilocorpine and atropine. He realized that both these chemicals interacted with some receptive substance in the nerve endings of the gland cells (34). Paul Ehrlich defined the receptor as the "binding group of the protoplasmic molecule to which a foreign newly introduced group binds" (35). In 1905 Langley's studies on the effects of curare on muscular contraction led to the first delineation of critical characteristics of a receptor: recognition capacity for certain ligands and an amplification component that results in a pharmacological response (36).

Receptors are mostly integral proteins embedded in the phospholipid bilayer of cell membranes. Rigorous treatment with detergents is needed to dissociate the proteins from the membrane, which often results in loss of integrity and activity. Pure proteins such as enzymes also act as drug receptors. Their relative ease of isolation and amplification have made enzymes desirable targets in structure-based ligand design and QSAR studies. Nucleic acids comprise an important category of drug receptors. Nucleic acid receptors (aptamers), which interact with a diverse number of small organic molecules, have been isolated by *in vitro* selection techniques and studied (37). Recent binary complexes provide insight into the molecular recognition process in these biopolymers and also establish the importance of the architecture of tertiary motifs in nucleic acid folding (38). Groove-binding ligands such as lexitropsins hold promise as potential drugs and are thus suitable subjects for focused QSAR studies (39).

Over the last 20 years, extensive QSAR studies on ligand-receptor interactions have been carried out with most of them focusing on enzymes. Two recent developments have augmented QSAR studies and established an attractive approach to the elucidation of the mechanistic underpinnings of ligand-receptor interactions: the advent of molecular graphics and the ready availability of X-ray crystallography coordinates of various binary and ternary complexes of enzymes with diverse ligands and cofactors. Early studies with serine and thiol proteases (chymotrypsin, trypsin, and papain), alcohol dehydrogenase, and numerous dihydrofolate reductases (DHFR) not only established molecular modeling as a powerful tool, but also helped clarify the extent of the role of hydrophobicity in enzyme-ligand interactions (40–44). Empirical evidence indicated that the coefficients with the hydrophobic term could be related to the degree of desolvation of the ligand by critical amino acid residues in the binding site of an enzyme. Total desolvation, as characterized by binding in a deep crevice/pocket, resulted in coefficients of approximately 1.0 (0.9–1.1) (44). An extension of this agreement between the mathematical expression and structure as determined by X-ray crystallography led to the expectation that the binding of a set of substituents on the surface of an enzyme would yield a coefficient of about 0.5 (0.4–0.6) in the regression equation, indicative of partial desolvation.

Probing of various enzymes by different ligands also aided in dispelling the notion of Fischer's rigid lock-and-key concept, in which the ligand (key) fits precisely into a receptor (lock). Thus, a "negative" impression of the substrate was considered to exist on the enzyme surface (geometric complementarity). Unfortunately, this rigid model fails to account for the effects of allosteric ligands, and this encouraged the evolution of the induced-fit model. Thus, "deformable" lock-and-key models have gained acceptance on the basis of structural studies, especially NMR (45).

It is now possible to isolate membrane-bound receptors, although it is still a challenge to delineate their chemistry, given that separation from the membrane usually ensures loss of reactivity. Nevertheless, great advances have been made in this arena, and the three-dimensional structures of some membrane-bound proteins have recently been elucidated. To gain an appreciation for mechanisms of ligand-receptor interactions, it is necessary to consider the intermolecular forces at play. Considering the low concentration of drugs and receptors in the human body, the law of mass action cannot account for the ability of a minute amount of a drug to elicit a pronounced pharmacological effect. The driving force for such an interaction may be attributed to the low energy state of the drug-receptor complex: $K_D = [Drug][Receptor]/[Drug\text{-}Receptor\ Complex]$. Thus, the biological activity of a drug is determined by its affinity for the receptor, which is measured by its K_D, the dissociation constant at equilibrium. A smaller K_D implies a large concentration of the drug-receptor complex and thus a greater affinity of the drug for the receptor. The latter property is promoted and stabilized by mostly noncovalent interactions sometimes augmented by a few covalent bonds. The spontaneous formation of a bond between atoms results in a decrease in free energy; that is, ΔG is negative. The change in free energy ΔG is related to the equilibrium constant K_{eq}.

$$\Delta G° = -RT \ln K_{eq} \qquad (1.7)$$

Thus, small changes in $\Delta G°$ can have a profound effect on equilibrium constants.

Table 1.1 Types of Intermolecular Forces

Bond Type	Bond Strength (kcal/mol)	Example
1. Covalent	40–140	CH_3CH_2O—H
2. Ionic (Electrostatic)	5	$R_4\overset{+}{N}$⁗⁗⁗$\overset{-}{O}$—C—
3. Hydrogen	1–10	—NH⁗⁗O—H
4. Dipole–dipole	1	R_3N:⁗⁗C=O
5. van der Waals	0.5–1	C⁗⁗⁗C
6. Hydrophobic	1	

In the broadest sense, these "bonds" would include covalent, ionic, hydrogen, dipole-dipole, van der Waals, and hydrophobic interactions. Most drug-receptor interactions constitute a combination of the bond types listed in Table 1.1, most of which are reversible under physiological conditions.

Covalent bonds are not as important in drug-receptor binding as noncovalent interactions. Alkylating agents in chemotherapy tend to react and form an immonium ion, which then alkylates proteins, preventing their normal participation in cell divisions. Baker's concept of active site directed irreversible inhibitors was well established by covalent formation of Baker's antifolate and dihydrofolate reductase (46).

Ionic (electrostatic) interactions are formed between ions of opposite charge with energies that are nominal and that tend to fall off with distance. They are ubiquitous and because they act across long distances, they play a prominent role in the actions of ionizable drugs. The strength of an electrostatic force is directly dependent on the charge of each ion and inversely dependent on the dielectric constant of the solvent and the distance between the charges.

Hydrogen bonds are ubiquitous in nature: their multiple presence contributes to the sta-

bility of the (αhelix and base-pairing in DNA. Hydrogen bonding is based on an electrostatic interaction between the nonbonding electrons of a heteroatom (e.g., N, O, S) and the electron-deficient hydrogen atom of an —OH, SH, or NH group. Hydrogen bonds are strongly directional, highly dependent on the net degree of solvation, and rather weak, having energies ranging from 1 to 10 kcal/mol (47, 48). Bonds with this type of strength are of critical importance because they are stable enough to provide significant binding energy but weak enough to allow for quick dissociation. The greater electronegativity of atoms such as oxygen, nitrogen, sulfur, and halogen, compared to that of carbon, causes bonds between these atoms to have an asymmetric distribution of electrons, which results in the generation of electronic dipoles. Given that so many functional groups have dipole moments, ion-dipole and dipole-dipole interactions are frequent. The energy of dipole-dipole interactions can be described by Equation 1.8, where μ is the dipole moment, θ is the angle between the two poles of the dipole, D is the dielectric constant of the medium and r is the distance between the charges involved in the dipole.

$$E = 2\mu_1\mu_2\cos\,\theta_1\cos\,\theta_2/Dr^3 \qquad (1.8)$$

Although electrostatic interactions are generally restricted to polar molecules, there are also strong interactions between nonpolar molecules over small intermolecular distances. Dispersion or London/van der Waals forces are the universal attractive forces between atoms that hold nonpolar molecules together in the liquid phase. They are based on polarizability and these fluctuating dipoles or shifts in electron clouds of the atoms tend to induce opposite dipoles in adjacent molecules, resulting in a net overall attraction. The energy of this interaction decreases very rapidly in proportion to $1/r^6$, where r is the distance separating the two molecules. These van der Waals forces operate at a distance of about $0.4-0.6$ nm and exert an attraction force of less than 0.5 kcal/mol. Yet, although individual van der Waals forces make a low energy contribution to an event, they become significant and additive when summed up over a large area with close surface contact of the atoms.

Hydrophobicity refers to the tendency of nonpolar compounds to transfer from an aqueous phase to an organic phase (49, 50). When a nonpolar molecule is placed in water, it gets solvated by a "sweater" of water molecules ordered in a somewhat icelike manner. This increased order in the water molecules surrounding the solute results in a loss of entropy. Association of hydrocarbon molecules leads to a "squeezing out" of the structured water molecules. The displaced water becomes bulk water, less ordered, resulting in a gain in entropy, which provides the driving force for what has been referred to as a hydrophobic bond. Although this is a generally accepted view of hydrophobicity, the hydration of apolar molecules and the noncovalent interactions between these molecules in water are still poorly understood and thus the source of continued examination (51–53).

Because noncovalent interactions are generally weak, cooperativity by several types of interactions is essential for overall activity. Enthalpy terms will be additive, but once the first interaction occurs, translational entropy is lost. This results in a reduced entropy loss in the second interaction. The net result is that eventually several weak interactions combine to produce a strong interaction. One can safely state that it is the involvement of myriad interactions that contribute to the overall selectivity of drug-receptor interactions.

2 TOOLS AND TECHNIQUES OF QSAR

2.1 Biological Parameters

In QSAR analysis, it is imperative that the biological data be both accurate and precise to develop a meaningful model. It must be realized that any resulting QSAR model that is developed is only as valid statistically as the data that led to its development. The equilibrium constants and rate constants that are used extensively in physical organic chemistry and medicinal chemistry are related to free energy values ΔG. Thus for use in QSAR, standard biological equilibrium constants such as K_i or K_m should be used in QSAR studies. Likewise only standard rate constants should be deemed appropriate for a QSAR analysis. Percentage activities (e.g., % inhibition of growth at certain concentrations) are *not* appropriate biological endpoints because of the nonlinear characteristic of dose-response relationships. These types of endpoints may be transformed to equieffective molar doses. Only equilibrium and rate constants pass muster in terms of the free-energy relationships or influence on QSAR studies. Biological data are usually expressed on a logarithmic scale because of the linear relationship between response and log dose in the midregion of the log dose-response curve. Inverse logarithms for activity ($\log 1/C$) are used so that higher values are obtained for more effective analogs. Various types of biological data have been used in QSAR analysis. A few common endpoints are outlined in Table 1.2.

Biological data should pertain to an aspect of biological/biochemical function that can be measured. The events could be occurring in enzymes, isolated or bound receptors, in cellular systems, or whole animals. Because there is considerable variation in biological responses, test samples should be run in duplicate or preferably triplicate, except in whole animal studies where assay conditions (e.g., plasma concentrations of a drug) preclude such measurements.

Table 1.2 Types of Biological Data Utilized in QSAR Analysis

Source of Activity	Biological Parameters
1. Isolated receptors	
Rate constants	Log k_{cat}; Log k_{uncat}; Log k
Michaelis–Menten constants	Log $1/K_m$
Inhibition constants	Log $1/K_i$
Affinity data	pA_2; pA_1
2. Cellular systems	
Inhibition constants	Log $1/IC_{50}$
Cross resistance	Log CR
In vitro biological data	Log $1/C$
Mutagenicity states	Log TA_{98}
3. "*In vivo*" systems	
Biocencentration factor	Log BCF
In vivo reaction rates	Log I (Induction)
Pharmacodynamic rates	Log T (total clearance)

It is also important to design a set of molecules that will yield a range of values in terms of biological activities. It is understandable that most medicinal chemists are reluctant to synthesize molecules with poor activity, even though these data points are important in developing a meaningful QSAR. Generally, the larger the range (>2 log units) in activity, the easier it is to generate a predictive QSAR. This kind of equation is more forgiving in terms of errors of measurement. A narrow range in biological activity is less forgiving in terms of accuracy of data. Another factor that merits consideration is the time structure. Should a particular reading be taken after 48 or 72 h? Knowledge of cell cycles in cellular systems or biorhythms in animals would be advantageous.

Each single step of drug transport, binding, and metabolism involves some form of partitioning between an aqueous compartment and a nonaqueous phase, which could be a membrane, serum protein, receptor, or enzyme. In the case of isolated receptors, the endpoint is clear-cut and the critical step is evident. But in more complex systems, such as cellular systems or whole animals, many localized steps could be involved in the random-walk process and the eventual interaction with a target.

Usually the observed biological activity is reflective of the slow step or the rate-determining step.

To determine a defined biological response (e.g., IC_{50}), a dose-response curve is first established. Usually six to eight concentrations are tested to yield percentages of activity or inhibition between 20 and 80%, the linear portion of the curve. Using the curves, the dose responsible for an established effect can easily be determined. This procedure is meaningful if, at the time the response is measured, the system is at equilibrium, or at least under steady-state conditions.

Other approaches have been used to apply the additivity concept and ascertain the binding energy contributions of various substituent (R) groups. Fersht et al. have measured the binding energies of various alkyl groups to aminoacyl-tRNA synthetases (54). Thus the ΔG values for methyl, ethyl, isopropyl, and thio substituents were determined to be 3.2, 6.5, 9.6, and 5.4 kcal/mol, respectively.

An alternative, generalized approach to determining the energies of various drug-receptor interactions was developed by Andrews et al. (55), who statistically examined the drug-receptor interactions of a diverse set of molecules in aqueous solution. Using Equation 1.9, a relationship was established between ΔG and E_X (intrinsic binding energy), E_{DOF} (energy of average entropy loss), and the $\Delta S_{r,t}$ (energy of rotational and translational entropy loss).

$$\Delta G = T\,\Delta S_{r,t} + n_{DOF}E_{DOF} + n_X E_X \quad (1.9)$$

E_X denotes the sum of the intrinsic binding energy of each functional group of which n_X are present in each drug in the set. Using Equation 1.9, the average binding energies for various functional groups were calculated. These energies followed a particular trend with charged groups showing stronger interactions and nonpolar entities, such as sp^2, sp^3 carbons, contributing very little. The applicability of this approach to specific drug-receptor interactions remains to be seen.

2.2 Statistical Methods: Linear Regression Analysis

The most widely used mathematical technique in QSAR analysis is multiple regression analysis (MRA). We will consider some of the basic tenets of this approach to gain a firm understanding of the statistical procedures that define a QSAR. Regression analysis is a powerful means for establishing a correlation between independent variables and a dependent variable such as biological activity (56).

$$Y_i = b + aX_i + E_i \qquad (1.10)$$

Certain assumptions are made with regard to this procedure (57):

1. The independent variables, which in this case usually include the physicochemical parameters, are measured without error. Unfortunately, this is not always the case, although the error in these variables is small compared to that in the dependent variable.

2. For any given value of X, the Y values are independent and follow a normal distribution. The error term E_i possesses a normal distribution with a mean of zero.

3. The expected mean value for the variable Y, for all values of X, lies on a straight line.

4. The variance around the regression line is constant. The "best" straight line for model $Y_i = b + aZ_i + E$ is drawn through the data points, such that the sum of the squares of the vertical distances from the points to the line is minimized. Y represents the value of the observed data point and Y_{calc} is the predicted value on the line. The sum of squares SS $= \Sigma\,(Y_{obs} - Y_{calc})^2$.

$$Y_{obs} = aX_i + b + E_i \qquad (1.11)$$

$$Y_{calc} = aX_i + b \qquad (1.12)$$

$$E = Y_{obs} - aX_i - b \qquad (1.13)$$

$$\sum_{i=1}^{n} E_i^{2} = \sum \Delta^2 = SS$$
$$= \sum\,(Y_{obs} - Y_{calc})^2 \qquad (1.14)$$

Thus, SS $= \sum_{i=1}^{n}\,(Y_{obs} - aX_i - b)^2 \qquad (1.15)$

Expanding Equation 1.15, we obtain

$$SS = \sum_{i=1}^{n}\,(Y_{obs}^{\,2} - Y_{obs}aX_i - Y_{obs}b$$
$$- Y_{obs}aX_i + a^2X_i^{\,2} + aX_ib \qquad (1.16)$$
$$- bY_{obs} + abX_i + b^2)$$

Taking the partial derivative of Equation 1.14 with respect to b and then with respect to a, results in Equations 1.17 and 1.18.

$$\frac{dSS}{db} = \sum_{i=1}^{n} -2(Y_{obs} - b - aX_i) \qquad (1.17)$$

$$\frac{dSS}{da} = \sum_{i=1}^{n} -2X_i(Y_{obs} - b - aX_i) \qquad (1.18)$$

SS can be minimized with respect to b and a and divided by -2 to yield the normal Equations 1.19 and 1.20.

$$\sum_{i=1}^{n}\,(Y_{obs} - b - aX_i) = 0 \qquad (1.19)$$

$$\sum_{i=1}^{n} X_i(Y_{obs} - b - aX_i) = 0 \qquad (1.20)$$

These "normal equations" can be rewritten as follows:

$$b \sum_{i=1}^{n} X_i + a \sum_{i=1}^{n} X_i^{\,2} = \sum_{i=1}^{n} X_iY_{obs} \qquad (1.21)$$

$$b + a \sum_{i=1}^{n} X_i = \sum_{i=1}^{n} Y_{obs} \qquad (1.22)$$

The solution of these simultaneous equations yields a and b. More thorough analyses of these procedures have been examined in detail (19, 58–60). The following simple example, illustrated by Table 1.3, will illustrate the nuances of a linear regression analysis.

Table 1.3 Antibacterial Activity of N'-(R-phenyl)sulfanilamides

Compound	$\sigma(X)$	Observed BA (Y)
1. 4-CH$_3$	−0.17	4.66
2. 4-H	0	4.80
3. 4-Cl	0.23	4.89
4. 2-Cl	0.23	5.55
5. 2-NO$_2$	0.78	6.00
6. 4-NO$_2$	0.78	6.00

k = no. of variables = 1
n = no. of data points = 6
$\Sigma\,X$ = 1.85
$\Sigma\,Y$ = 31.90
$\Sigma\,X^2$ = 1.352
$\Sigma\,Y^2$ = 171.45
$\Sigma\,XY$ = 10.968

For linear regression analysis, $Y = ax + b$

$$a = (n \cdot \sum xy - \sum x \cdot \sum y)/n \cdot \sum x^2$$

$$- \left(\sum x\right)^2 = 1.45$$

$$b = \left(\sum y - a \cdot \sum x\right)/n = 4.869$$

$$r^2 = \sum xy - \sum x \cdot \sum y/n)^2 / (\sum x^2$$

$$- \left(\sum x\right)^2/n) \cdot \left(\sum y^2 - \left(\sum y\right)^2/n\right)$$

$$= 0.875 \quad \therefore r = 0.935$$

$$s^2 = (1 - r^2)$$

$$\times \left(\sum y^2 - \left(\sum y\right)^2/n\right)/(n - k - 1)$$

$$= 0.058 \quad \therefore s = 0.240$$

$$F = r^2 \cdot (n - k - 1)/k(1 - r^2) = 28.52$$

The correlation coefficient r, the total variance SS_T, the unexplained variance SSQ, and the standard deviation, are defined as follows:

$$r^2 = 1 - \frac{\Sigma\,\Delta^2}{SS_T} \tag{1.23}$$

$$SS_T = \sum (Y_{obs} - Y_{mean})^2 \tag{1.24}$$

$$= \sum y^2 - \left(\sum y\right)^2/n$$

$$\sum \Delta^2 = SSQ = \sum (Y_{obs} - Y_{calc})^2 \tag{1.25}$$

$$s = \sqrt{\frac{\Sigma\,\Delta^2}{n - k - 1}} = \sqrt{\frac{SSQ}{n - k - 1}} \tag{1.26}$$

The correlation coefficient r is a measure of quality of fit of the model. It constitutes the variance in the data. In an ideal situation one would want the correlation coefficient to be equal to or approach 1, but in reality because of the complexity of biological data, any value above 0.90 is adequate. The standard deviation is an absolute measure of the quality of fit. Ideally s should approach zero, but in experimental situations, this is not so. It should be small but it cannot have a value lower than the standard deviation of the experimental data. The magnitude of s may be attributed to some experimental error in the data as well as imperfections in the biological model. A larger data set and a smaller number of variables generally lead to lower values of s. The F value is often used as a measure of the level of statistical significance of the regression model. It is defined as denoted in Equation 1.27.

$$F_{k_2 - k_1, n - k_2} = \frac{(SS_1 - SS_2)}{SS_2} \frac{(n - k_2 - 1)}{k_2 - k_1} \tag{1.27}$$

A larger value of F implies a more significant correlation has been reached. The confidence intervals of the coefficients in the equation reveal the significance of each regression term in the equation.

To obtain a statistically sound QSAR, it is important that certain caveats be kept in mind. One needs to be cognizant about collinearity between variables and chance correlations. Use of a correlation matrix ensures that variables of significance and/or interest are orthogonal to each other. With the rapid proliferation of parameters, caution must be exercised in amassing too many variables for a QSAR analysis. Topliss has elegantly demonstrated that there is a high risk of ending up with a chance correlation when too many variables are tested (62).

Outliers in QSAR model generation present their own problems. If they are badly fit by the model (off by more than 2 standard deviations), they should be dropped from the data set, although their elimination should be noted and addressed. Their aberrant behavior may be attributed to inaccuracies in the testing procedure (usually dilution errors) or unusual behavior. They often provide valuable information in terms of the mechanistic interpretation of a QSAR model. They could be participating in some intermolecular interaction that is not available to other members of the data set or have a drastic change in mechanism.

2.3 Compound Selection

In setting up to run a QSAR analysis, compound selection is an important angle that needs to be addressed. One of the earliest manual methods was an approach devised by Craig, which involves two-dimensional plots of important physicochemical properties. Care is taken to select substituents from all four quadrants of the plot (63). The Topliss operational scheme allows one to start with two compounds and construct a potency tree that grows branches as the substituent set is expanded in a stepwise fashion (64). Topliss later proposed a batchwise scheme including certain substituents such as the 3,4-Cl_2, 4-Cl, 4-CH_3, 4-OCH_3, and 4-H analogs (65). Other methods of manual substituent selection include the Fibonacci search method, sequential simplex strategy, and parameter focusing by Magee (66–68).

One of the earliest computer-based and statistical selection methods, cluster analysis was devised by Hansch to accelerate the process and diversity of the substituents (1). Newer methodologies include D-optimal designs, which focus on the use of det $(X'X)$, the variance-covariance matrix. The determinant of this matrix yields a single number, which is maximized for compounds expressing maximum variance and minimum covariance (69–71). A combination of fractional factorial design in tandem with a principal property approach has proven useful in QSAR (72). Extensions of this approach using multivariate design have shown promise in environmental QSAR with nonspecific responses, where the clusters overlap and a cluster-based design approach has to be used (73). With strongly clustered data containing several classes of compounds, a new strategy involving local multivariate designs within each cluster is described. The chosen compounds from the local

designs are grouped together in the overall training set that is representative of all clusters (74).

3 PARAMETERS USED IN QSAR

3.1 Electronic Parameters

Parameters are of critical importance in determining the types of intermolecular forces that underly drug-receptor interactions. The three major types of parameters that were initially suggested and still hold sway are electronic, hydrophobic, and steric in nature (20, 75). Extensive studies using electronic parameters reveal that electronic attributes of molecules are intimately related to their chemical reactivities and biological activities. A search of a computerized QSAR database reveals the following: the common Hammett constants (σ, σ^+, σ^-) account for 7000/8500 equations in the Physical organic chemistry (PHYS) database and nearly 1600/8000 in the Biology (BIO) database, whereas quantum chemical indices such as HOMO, LUMO, BDE, and polarizability appear in 100 equations in the BIO database (76).

The extent to which a given reaction responds to electronic perturbation constitutes a measure of the electronic demands of that reaction, which is determined by its mechanism. The introduction of substituent groups into the framework and the subsequent alteration of reaction rates helps delineate the overall mechanism of reaction. Early work examining the electronic role of substituents on rate constants was first tackled by Burckhardt and firmly established by Hammett (13, 14, 77, 78). Hammett employed, as a model reaction, the ionization in water of substituted benzoic acids and determined their equilibrium constants K_a. See Equation 1.28. This led to an operational definition of σ, the substituent constant. It is a measure of the size of the electronic effect for a given substituent and represents a measure of electronic charge distribution in the benzene nucleus.

$$\sigma_X = \log K_X - \log K_H \quad or$$
$$\log(K_X/K_H) = -pK_X + pK_H \tag{1.29}$$

Electron-withdrawing substituents are thus

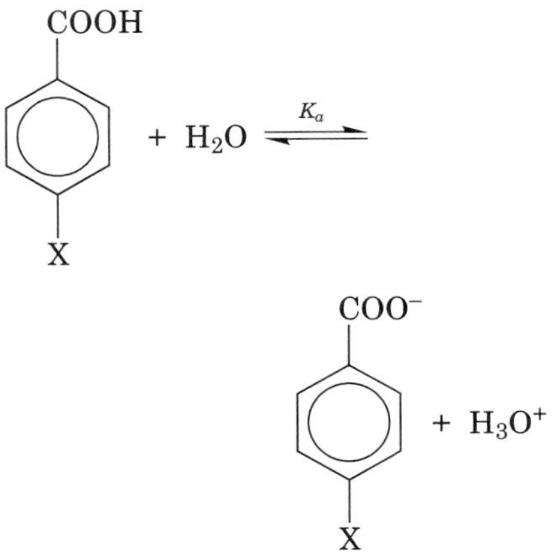

COOH

$+ H_2O \xrightleftharpoons{K_a}$

X

(1.28)

COO$^-$

$+ H_3O^+$

X

characterized by positive values, whereas electron-donating ones have negative values. In an extension of this approach, the ionization of substituted phenylacetic acids was measured.

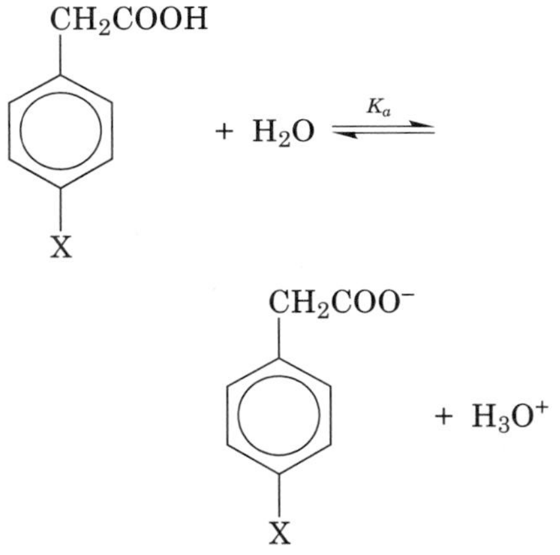

CH$_2$COOH

$+ H_2O \xrightleftharpoons{K_a}$

X

(1.30)

CH$_2$COO$^-$

$+ H_3O^+$

X

The effect of the 4-Cl substituent on the ionization of 4-Cl phenylacetic acid (PA) was found to be proportional to its effect on the ionization of 4-Cl benzoic acid (BA).

$$\log(K'_{Cl(PA)}/K'_{H(PA)}) \propto \log(K_{Cl(BA)}/K_{H(BA)})$$

Since $\log(K_{Cl(BA)}/K_{H(BA)}) = \sigma$,

then $\log \dfrac{K'_{Cl}}{K'_H} = \rho \cdot \sigma$　　(1.31)

ρ (rho) is defined as a proportionality or reaction constant, which is a measure of the sus-

ceptibility of a reaction to substituent effects. A positive rho value suggests that a reaction is aided by electron withdrawal from the reaction site, whereas a negative rho value implies that the reaction is assisted by electron donation at the reaction site. Hammett also drew attention to the fact that a plot of log K_A for benzoic acids versus log k for ester hydrolysis of a series of molecules is linear, which suggests that substituents exert a similar effect in dissimilar reactions.

$$\log \frac{k_X}{k_H} \propto \log \frac{K_X}{K_H} = \rho \cdot \sigma \qquad (1.32)$$

Although this expression is empirical in nature, it has been validated by the sheer volume of positive results. It is remarkable because four different energy states must be related.

A correlation of this type is clearly meaningful; it suggests that changes in structure produce proportional changes in the activation energy $\Delta G^{\ddagger}$ for such reactions. Hence, the derivation of the name for which the Hammett equation is universally known: linear free energy relationship (LFER). Equation 1.32 has become known as the Hammett equation and has been applied to thousands of reactions that take place at or near the benzene ring bearing substituents at the meta and para positions. Because of proximity and steric effects, ortho-substituted molecules do not always follow this maxim and are subject to different parameterizations. Thus, an expanded approach was established by Charton (79) and Fujita and Nishioka (80). Charton partitioned the ortho electronic effect into its inductive, resonance, and steric contributions; the factors α, β, and X are susceptibility or reaction constants and h is the intercept.

$$\text{Log } k = \alpha\sigma_I + \beta\sigma_R + Xr_v + h \qquad (1.33)$$

Fujita and Nishioka used an integrated approach to deal with ortho substituents in data sets including meta and para substituents.

$$\text{Log } k = \rho\sigma + \delta E_S^{\text{ortho}} + fF_{\text{ortho}} + C \qquad (1.34)$$

For ortho substituents, para sigma values

were used in addition to Taft's E_S values and Swain-Lupton field constants F_{ortho}.

The reason for employing alternative treatments to ortho-substituted aromatic molecules is that changes in rate or ionization constants mediated by meta or para substituents are mostly changes in ($H^{\ddagger}$ or ΔH° because substitution does not affect $\Delta S^{\ddagger}$ or ΔS°. Ortho substituents affect both enthalpy and entropy; the effect on entropy is noteworthy because entropy is highly sensitive to changes in the size of reagents and substituents as well as degree of solvation. Bolton et al. examined the ionization of substituted benzoic acids and measured accurate values for ΔG, ΔH, and ΔS (81). A hierarchy of different scenarios, under which an LFER operates, was established:

1. ΔH° is constant and ΔS varies for a series.
2. ΔS° is constant and ΔH varies.
3. ΔH° and ΔS° vary and are shown to be linearly related.
4. Precise measurements indicated that category 3 was the prevalent behavior in benzoic acids.

Despite the extensive and successful use in QSAR studies, there are some limitations to the Hammett equation.

1. Primary σ values are obtained from the thermodynamic ionizations of the appropriate benzoic acids at 25°C; these are reliable and easily available. Secondary values are obtained by comparison with another series of compounds and are thus subject to error because they are dependent on the accuracy of a measured series and the development of a regression line using statistical methods.

2. In some multisubstituted compounds, the lack of additivity needs to be noted. Proximal effects are operative and tend to distort electronic contributions. For example,

$$\sum \sigma_{calc}(3,4,5\text{-trichlorobenzoic acid})$$
$$= 0.97;$$

that is, $2\sigma_M + \sigma_P$ *or* $2(0.37) + 0.23$

$$\sum \sigma_{obs}(3,4,5\text{-trichlorobenzoic acid}) = 0.95$$

Sigma values for smaller substituents are more likely to be additive. However, in the case of 3-methyl, 4-dimethylaminobenzoic acid, the discrepancy is high. For example,

$$\sum \sigma_{calc}(3\text{-CH}_3, 4\text{-N(CH}_3)_2 \text{ benzoic acid})$$
$$= -0.90$$
$$\sum \sigma_{obs}(3\text{-CH}_3, 4\text{-N(CH}_3)_2 \text{ benzoic acid})$$
$$= -0.30$$

The large discrepancy may be attributed to the twisting of the dimethylamino substituent out of the plane of the benzene ring, resulting in a decrease in resonance. Exner and his colleagues have critically examined the use of additivity in the determination of σ constants (82).

3. Changes in mechanism or transition state cause discontinuities in Hammett plots. Nonlinear plots are often found in reactions that proceed by two concurrent pathways (83, 84).

4. Changes in solvent may lead to dissimilarities in reaction mechanisms. Thus extrapolation of σ values from a polar solvent (e.g., CH_3CN) to a nonpolar solvent such as benzene has to be approached cautiously. Solvation properties will differ considerably, particularly if the transition state is polar and/or the substituents are able to interact with the solvent.

5. A strong positional dependency of sigma makes it imperative to use appropriate values for positional, isomeric substituents. Substituents ortho to the reaction center are difficult to describe and thus one must resort to a Fujita-Nishioka analysis (80).

6. Thorough resonance or direct conjugation effects cause a breakdown in the Hammett equation. When coupling occurs between the substituent and the reaction center through the pi-electron system, reactivity is enhanced, diminished, or mitigated by separation. In a study of X-cumyl chlorides, Brown and Okamoto noticed the strong conjugative interaction between lone-pair,

para substituents and the vacant ρ-orbital in the transition state, which led to deviations in the Hammett plot (85). They defined a modified LFER applicable to this situation.

$$\text{Log} \frac{k_Y}{k_H} = (\rho^+)(\sigma^+) \qquad (1.35)$$

σ^+ was a new substituent constant that expressed enhanced resonance attributes. A similar situation was noticed when a strong donor center was present as a reactant or formed as a product (e.g., phenols and anilines). In this case, strong resonance interactions were possible with electron-withdrawing groups (e.g., NO_2 or CN). A scale for such substituents was constructed such that

$$\text{Log} \frac{k_Y}{k_H} = (\rho^-)(\sigma^-) \qquad (1.36)$$

One shortcoming of the benzoic acid system is the extent of coupling between the carboxyl group and certain lone-pair donors. Insertion of a methylene group between the core (benzene ring) and the functional group (COOH moiety) leads to phenylacetic acids and the establishment of σ^0 scale from the ionization of X-phenylacetic acids. A flexible method of dealing with the variability of the resonance contribution to the overall electronic demand of a reaction is embodied in the Yukawa-Tsuno equation (86). It includes normal and enhanced resonance contributions to an LFER.

$$\text{Log} \frac{k_Y}{k_H} = \rho[\sigma + r(\sigma^+ - \sigma)] \qquad (1.37)$$

where r is a measure of the degree of enhanced resonance interaction in relation to benzoic acid dissociations ($r = 0$) and cumyl chloride hydrolysis ($r = 1$).

Most of the Hammett-type constants pertain to aromatic systems. In evaluating an electronic parameter for use in aliphatic systems, Taft used the relative acid and base hydrolysis rates for esters. He developed equation 1.38 as a measure of the inductive effect

(σ^*) of a substituent R' in the ester R' COOR, where B and A refer to basic and acidic hydrolysis, respectively.

$$\sigma^* = \frac{1}{2.48} \left[\log(k/k_O)_B - \log(k/k_O)_A\right] \qquad (1.38)$$

The factor of 2.48 was used to make σ^* equiscalar with Hammett σ values. Later, a σ_I scale derived from the ionization of 4-X-bicyclo[2.2.2]octane-1-carboxylic acids was shown to be related to σ^* (87, 88). It is now more widely used than σ^*.

$$\sigma_I(X) = 0.45\sigma^*(CH_2X) \qquad (1.39)$$

Ionization is a function of the electronic structure of an organic drug molecule. Albert was the first to clearly delineate the relationship between ionization and biological activity (89). Now, pK_a values are widely used as the independent variable in physical organic reactions and in biological systems, particularly when dealing with transport phenomena. However, caution must be exercised in interpreting the dependency of biological activity on pK_a values because pK_a values are inherently composites of electronic factors that are used directly in QSAR analysis.

In recent years, there has been a rapid growth in the application of quantum chemical methodology to QSAR, by direct derivation of electronic descriptors from the molecular wave functions (90). The two most popular methods used for the calculation of quantum chemical descriptors are *ab initio* (Hartree-Fock) and semiempirical methods. As in other electronic parameters, QSAR models incorporating quantum chemical descriptors will include information on the nature of the intermolecular forces involved in the biological response. Unlike other electronic descriptors, there is no statistical error in quantum chemical computations. The errors are usually made in the assumptions that are established to facilitate calculation (91). Quantum chemical descriptors such as net atomic changes, highest occupied molecular orbital/lowest unoccupied molecular orbital (HOMO-LUMO) energies, frontier orbital electron densities, and superdelocalizabilities have been shown to correlate well with various biological activities (92). A mixed approach using frontier orbital theory and topological parameters have been used to calculate Hammett-like substituent constants (93).

$$\sigma = -2.480\Delta N - 7.894\Delta E$$
$$- 0.605 D_X/D_H \cdot (EA_H/EA_X) \qquad (1.40)$$
$$+ 0.009 \, \text{Ø} S_X + 0.028 \sum \pi + 0.279$$
$$n = 150, \quad r^2 = 0.947,$$
$$s = 0.079, \quad F = 789.9$$

In Equation 1.40, ΔN represents the extent of electron transfer between interacting acid-base systems; ΔE is the energy decrease in bimolecular systems underlying electron transfer; $D_X/D_H \cdot (EA_H/EA_X)$ corresponds to electron affinity and distance terms; and $\text{Ø} S_X$ factors the electrotopological state index, whereas $\sum \pi$ is the number of all π-electrons in the functional group. Observed principal component analysis (PCA) clustering of 66 descriptors derived from AM1 calculations was similar to that previously reported for monosubstituted benzenes (94, 95). The advantages of quantum chemical descriptors are that they have definite meaning and are useful in the elucidation of intra- and intermolecular interactions and can easily be derived from the theoretical structure of the molecule.

3.2 Hydrophobicity Parameters

More than a hundred years ago, Meyer and Overton made their seminal discovery on the correlation between oil/water partition coefficients and the narcotic potencies of small organic molecules (7, 8). Ferguson extended this analysis by placing the relationship between depressant action and hydrophobicity in a thermodynamic context; the relative saturation of the depressant in the biophase was a critical determinant of its narcotic potency (9). At this time, the success of the Hammett equation began to permeate structure-activity studies and hydrophobicity as a determinant was relegated to the background. In a landmark study, Hansch and his colleagues de-

vised and used a multiparameter approach that included both electronic and hydrophobic terms, to establish a QSAR for a series of plant growth regulators (16). This study laid the basis for the development of the QSAR paradigm and also firmly established the importance of lipophilicity in biosystems. Over the last 40 years, no other parameter used in QSAR has generated more interest, excitement, and controversy than hydrophobicity (96). Hydrophobic interactions are of critical importance in many areas of chemistry. These include enzyme-ligand interactions, the assembly of lipids in biomembranes, aggregation of surfactants, coagulation, and detergency (97–100). The integrity of biomembranes and the tertiary structure of proteins in solution are determined by apolar-type interactions.

Molecular recognition depends strongly on hydrophobic interactions between ligands and receptors. Excellent treatises on this subject have been written by Taylor (101) and Blokzijl and Engerts (51). Despite extensive usage of the term *hydrophobic bond*, it is well known that there is no strong attractive force between apolar molecules (102). Frank and Evans were the first to apply a thermodynamic treatment to the solvation of apolar molecules in water at room temperature (103). Their "iceberg" model suggested that a large entropic loss ensued after the dissolution of apolar compounds and the increased structure of water molecules in the surrounding apolar solute. The quantitation of this model led to the development of the "flickering" cluster model of Némethy and Scheraga, which emphasized the formation of hydrogen bonds in liquid water (104). The classical model for hydrophobic interactions was delineated by Kauzmann to describe the van der Waals attractions between the nonpolar parts of two molecules immersed in water. Given that van der Waals forces operate over short distances, the water molecules are squeezed out in the vicinity of the mutually bound apolar surfaces (49). The driving force for this behavior is not that alkanes "hate" water, but rather water that "hates" alkanes (105, 106). Thus, the gain in entropy appears as the critical driving force for hydrophobic interactions that are primarily governed by the repulsion of hydrophobic solutes from the solvent water and the limited but important capacity of water to maintain its network of hydrogen bonds.

Hydrophobicities of solutes can readily be determined by measuring partition coefficients designated as P. Partition coefficients deal with neutral species, whereas distribution ratios incorporate concentrations of charged and/or polymeric species as well. By convention, P is defined as the ratio of concentration of the solute in octanol to its concentration in water.

$$P = [\text{conc}]_{\text{octanol}}/[\text{conc}]_{\text{aqueous}} \quad (1.41)$$

It was fortuitous that octanol was chosen as the solvent most likely to mimic the biomembrane. Extensive studies over the last 35 years (40,000 experimental P-values in 400 different solvent systems) have failed to dislodge octanol from its secure perch (107, 108).

Octanol is a suitable solvent for the measurement of partition coefficients for many reasons (109, 110). It is cheap, relatively nontoxic, and chemically unreactive. The hydroxyl group has both hydrogen bond acceptor and hydrogen bond donor features capable of interacting with a large variety of polar groups. Despite its hydrophobic attributes, it is able to dissolve many more organic compounds than can alkanes, cycloalkanes, or aromatic hydrocarbons. It is UV transparent over a large range and has a vapor pressure low enough to allow for reproducible measurements. It is also elevated enough to allow for its removal under mild conditions. In addition, water saturated with octanol contains only 10^{-3} M octanol at equilibrium, whereas octanol saturated with water contains 2.3 M of water. Thus, polar groups need not be totally dehydrated in transfer from the aqueous phase to the organic phase. Likewise, hydrophobic solutes are not appreciably solvated by the 10^{-3} M octanol in the water phase unless their intrinsic log P is above 6.0. Octanol begins to absorb light below 220 nm and thus solute concentration determinations can be monitored by UV spectroscopy. More important, octanol acts as an excellent mimic for biomembranes because it shares the traits of

amphiphilicity and hydrogen-bonding capability with phospholipids and proteins found in biological membranes.

The choice of the octanol/water partitioning system as a standard reference for assessing the compartmental distribution of molecules of biological interest was recently investigated by molecular dynamics simulations (111). It was determined that pure 1-octanol contains a mix of hydrogen-bonded "polymeric" species, mostly four-, five-, and six-membered ring clusters at 40°C. These small ring clusters form a central hydroxyl core from which their corresponding alkyl chains radiate outward. On the other hand, water-saturated octanol tends to form well-defined, inverted, micellar aggregates. Long hydrogen-bonded chains are absent and water molecules congregate around the octanol hydroxyls. "Hydrophilic channels" are formed by cylindrical formation of water and octanol hydroxyls with the alkyl chains extending outward. Thus, water-saturated octanol has centralized polar cores where polar solutes can localize. Hydrophobic solutes would migrate to the alkyl-rich regions. This is an elegant study that provides insight into the partitioning of benzene and phenol by analyzing the structure of the octanol/water solvation shell and delineating octanol's capability to serve as a surrogate for biomembranes.

The shake-flask method, so-called, is most commonly used to measure partition coefficients with great accuracy and precision and with a log P range that extends from -3 to $+6$ (112, 113). The procedure calls for the use of pure, distilled, deionized water, high-purity octanol, and pure solutes. At least three concentration levels of solute should be analyzed and the volumes of octanol and water should be varied according to a rough estimate of the log P value. Care should be exercised to ensure that the eventual *amounts* of the solute in each phase are about the same after equilibrium. Standard concentration curves using three to four known concentrations in water saturated with octanol are usually established. Generally, most methods employ a UV-based procedure, although GC and HPLC may also be used to quantitate the concentration of the solute.

Generally, 10-mL stopped centrifuge tubes or 200-mL centrifuge bottles are used. They are inverted gently for 2–3 min and then centrifuged at 1000–2000 g for 20 min before the phases are analyzed. Analysis of both phases is highly recommended, to minimize errors incurred by adsorption to glass walls at low solute concentration. For highly hydrophobic compounds, the slow stirring procedure of de Bruijn and Hermens is recommended (114). The filler probe extractor system of Tomlinson et al. is a modified, automated, shake-flask method, which is efficient, fast, reliable, and flexible (115).

Partition coefficients from different solvent systems can also be compared and converted to the octanol/water scale, as was suggested by Collander (116). He stressed the importance of the following linear relationship: $\log P_2 = a \log P_1 + b$. This type of relationship works well when the two solvents are both alkanols. However, when two solvent systems have varying hydrogen bond donor and acceptor capabilities, the relationship tends to fray. A classical example involves the relationship between $\log P$ values in chloroform and octanol (117, 118).

$$\text{Log } P_{\text{CHCl}_3} = 1.012 \log P_{\text{oct}} - 0.513 \quad (1.42)$$

$$n = 72, \quad r^2 = 0.811, \quad s = 0.733$$

Only 66% of the variance in the data is explained by this equation. However, a separation of the various solutes into OH bond donors, acceptors, and neutrals helped account for 94% of the variance in the data. These restrictions led Seiler to extend the Collander equation by incorporating a corrective term for H-bonding in the cyclohexane system (119). Fujita generalized this approach and formulated Equation 1.43 as shown below (120).

$$\log P_2 = a \log P_1 + \sum b_i \cdot \text{HB}_i + C \quad (1.43)$$

P_1 is the reference solvent and HB_i is an H-bonding parameter. Leahy et al. suggested that a more sophisticated approach incorporating four model systems would be needed to adequately address issues of solute partitioning in membranes (121). Thus, four distinct solvent types were chosen—apolar, amphiprotic, proton

donor, and proton acceptor—and they were represented by alkanes, octanol, chloroform, and propyleneglycol dipelargonate (PGDP), respectively. The demands of measuring four partition coefficients for each solute has slowed progress in this particular area.

3.2.1 Determination of Hydrophobicity by Chromatography. Chromatography provides an alternate tool for the estimation of hydrophobicity parameters. R_m values derived from thin-layer chromatography provide a simple, rapid, and easy way to ascertain approximate values of hydrophobicity (122, 123).

$$R_m = \log(1/R_f - 1) \quad (1.44)$$

Other recent developments in chromatography techniques have led to the development of powerful tools to rapidly and accurately measure octanol/water partition coefficients. Countercurrent chromatography is one of these methods. The stationary and mobile phases include two nonmiscible solvents (water and octanol) and the total volume of the liquid stationary phase is used for solute partitioning (124, 125). Log P_{app} values of several diuretics including ionizable drugs have been measured at different pH values using countercurrent chromatography; the log P values ranged from -1.3 to 2.7 and were consistent with literature values (126).

Recently, a rapid method for the determination of partition coefficients using gradient reversed phase/high pressure liquid chromatography (RP-HPLC) was developed. This method is touted as a high-throughput hydrophobicity screen for combinatorial libraries (127, 128). A chromatography hydrophobicity index (CHI) was established for a diverse set of compounds. Acetonitrile was used as the modifier and 50 mm ammonium acetate as the mobile phase (127). A linear relationship was established between Clog P and CHIN for neutral molecules.

$$\text{Clog } P = 0.057 \text{ CHIN} - 1.107 \quad (1.45)$$

$$n = 52, \quad r^2 = 0.724, \quad s = 0.82, \quad F = 131$$

A more recent study using RP-HPLC for the determination of log P (octanol) values for neutral and weakly acidic and basic drugs, revealed an excellent correlation between log P_{oct} and log K_W values (129). Log P_{oct} values determined in this system are referred to as Elog P_{oct}. They were expressed in terms of solvation parameters.

$$\text{Elog } P_{oct} = 0.204 + 0.452 R_2$$
$$- 1.053 \pi_2^H - 0.041 \sum \alpha_2^H \quad (1.46)$$
$$- 3.410 \sum \beta_2^O + 3.842 V_X$$
$$n = 35, \quad r^2 = 0.960, \quad s = 0.244$$

In this equation, R_2 is the excess molar refraction; π_2^H is the dipolarity/polarizability; $\sum \alpha_2^H$ and $\sum \beta_2^O$ are the summation of hydrogen bond acidity and basicity values, respectively; and V_X is McGowan's volume.

3.2.2 Calculation Methods. Partition coefficients are additive-constitutive, free energy-related properties. Log P represents the overall hydrophobicity of a molecule, which includes the sum of the hydrophobic contributions of the "parent" molecule and its substituent. Thus, the π value for a substituent may be defined as

$$\pi_X = \log P_{R-X} - \log P_{R-H} \quad (1.47)$$

π_H is set to zero. The π-value for a nitro substituent is calculated from the log P of nitrobenzene and benzene.

$$\pi_{NO_2} = \log P_{\text{nitrobenzene}} - \log P_{\text{benzene}}$$
$$= 1.85 - 2.13 = -0.28$$

An extensive list of π-values for aromatic substituents appears in Table 1.4. Pi values for side chains of amino acids in peptides have been well characterized and are easily available (130–132). Aliphatic fragments values were developed a few years later. For a more extensive list of substituent value constants, refer to the extensive compilation by Hansch et al. (133). Initially, the π-system was applied only to substitution on aromatic rings and when the hydrogen being replaced was of innocuous character. It was apparent from the

beginning that not all hydrogens on aromatic systems could be substituted without correction factors because of strong electronic interactions. It became necessary to determine π values in various electron-rich and -deficient systems (e.g., X-phenols and X-nitrobenzenes). Correction factors were introduced for special features such as unsaturation, branching, and ring fusion. The proliferation of π-scales made it difficult to ascertain which system was more appropriate for usage, particularly with complex structures.

The shortcomings of this approach provided the impetus for Nys and Rekker to design the fragmental method, a "reductionist" approach, which was based on the statistical analysis of a large number of measured partition coefficients and the subsequent assignment of appropriate values for particular molecular fragments (118, 134). Hansch and Leo took a "constructionist" approach and developed a fragmental system that included correction factors for bonds and proximity effects (1, 135). Labor-intensive efforts and inconsistency in manual calculations were eliminated with the debut of the automated system CLOGP and its powerful SMILES notation (136–138). Recent analysis of the accuracy of CLOGP yielded Equation 1.48 (139).

$$\text{MLOGP} = 0.959 \text{ CLOGP} + 0.08 \quad (1.48)$$
$$n = 12,107, \quad r^2 = 0.973, \quad s = 0.299$$

The Clog P values of 228 structures (1.8% of the data set) were not well predicted. It must be noted that Starlist (most accurate values in the database) contains almost 300 charged nitrogen solutes (ammonium, pyridinium, imidazolium, etc.) and over 2200 in all, which amounts to 5% of Masterfile (database of measured values). CLOGP adequately handles these molecules within the 0.30 standard deviation limit. Most other programs make no attempt to calculate them. For more details on calculating log P_{oct} from structures, see excellent reviews by Leo (140, 141).

The proliferation of methodologies and programs to calculate partition coefficients continues unabated. These programs are based on substructure approaches or whole-molecule approaches (142, 143). Substructure

Table 1.4   Substituent Constants for QSAR Analysis

No.	Substituent	Pi	MR	L	B1	B5	S-P	S-M
1	$+N(CH_3)_3$	-5.96	1.94	4.02	2.57	3.11	0.82	0.88
2	$EtN(CH_3)_3+$	-5.44	2.87	5.58	1.52	4.53	0.13	0.16
3	$CH_2N(CH_3)_3+$	-4.57	2.40	4.83	1.52	4.08	0.44	0.40
4	CO_2-	-4.36	0.61	3.53	1.60	2.66	0.00	-0.10
5	$+NH_3$	-4.19	0.55	2.78	1.49	1.97	0.60	0.86
6	$PR-N(CH_3)_3+$	-4.15	3.33	6.88	1.52	5.49	-0.01	0.06
7	CH_2NH_3+	-4.09	1.01	4.02	1.52	3.05	0.29	0.32
8	IO_2	-3.46	6.35	4.25	2.15	3.66	0.78	0.68
9	$C(CN)_3$	-2.33	1.86	3.99	2.87	4.12	0.96	0.97
10	$NHNO_2$	-2.27	1.07	4.50	1.35	3.66	0.57	0.91
11	$C(NO_2)_3$	-2.01	2.27	4.59	2.55	3.72	0.82	0.72
12	$SO_2(NH_2)$	-1.82	1.23	4.02	2.04	3.05	0.60	0.53
13	$C(CN)=C(CN)_2$	-1.77	2.58	6.46	1.61	5.17	0.98	0.77
14	$CH_2C=O(NH_2)$	-1.68	1.44	4.58	1.52	4.37	0.07	0.06
15	$N(COCH_3)_2$	-1.68	2.48	4.45	1.35	4.33	0.33	0.35
16	SO_2CH_3	-1.63	1.35	4.11	2.03	3.17	0.72	0.60
17	$P(O)(OH)_2$	-1.59	1.26	4.22	2.12	2.88	0.42	0.36
18	$S=O(CH_3)$	-1.58	1.37	4.11	1.40	3.17	0.49	0.52
19	$N(SO_2CH_3)_2$	-1.51	3.12	4.83	1.36	3.72	0.49	0.47
20	$C=O(NH_2)$	-1.49	0.98	4.06	1.50	3.07	0.36	0.28
21	$CH(CN)_2$	-1.45	1.43	3.99	1.85	4.12	0.52	0.53
22	$CH_2NHCOCH_3$	-1.43	1.96	5.67	1.52	4.75	-0.05	0.05
23	$NHC=S(NH_2)$	-1.40	2.22	5.06	1.35	4.18	0.16	0.22
24	$NH(OH)$	-1.34	0.72	3.87	1.35	2.63	-0.34	-0.04
25	$CH=NNHCONHNH_2$	-1.32	2.42	7.57	1.60	4.55	0.16	0.22
26	$NHC=O(NH_2)$	-1.30	1.37	5.06	1.35	3.61	-0.24	-0.03
27	$C=O(NHCH_3)$	-1.27	1.46	5.00	1.54	3.16	0.36	0.35
28	2-Aziridinyl	-1.23	1.19	4.14	1.55	3.24	-0.10	-0.06
29	NH_2	-1.23	0.54	2.78	1.35	1.97	-0.66	-0.16
30	$NHSO_2CH_3$	-1.18	1.82	4.70	1.35	4.13	0.03	0.20
31	$P(O)(OCH_3)_2$	-1.18	2.19	5.04	2.42	3.25	0.53	0.42
32	$C(CH_3)(CN)_2$	-1.14	1.90	4.11	2.81	4.12	0.57	0.60
33	$N(CH_3)SO_2CH_3$	-1.11	2.34	4.83	1.35	3.72	0.24	0.21
34	SO_2Et	-1.10	1.81	4.92	2.03	3.49	0.77	0.66
35	CH_2NH_2	-1.04	0.91	4.02	1.52	3.05	-0.11	-0.03
36	1-Tetrazolyl	-1.04	1.83	5.28	1.71	3.12	0.50	0.52
37	CH_2OH	-1.03	0.72	3.97	1.52	2.70	0.00	0.00
38	$N(CH_3)COCH_3$	-1.02	1.96	4.77	1.35	3.71	0.26	0.31
39	$NHCHO$	-0.98	1.03	4.22	1.35	3.61	0.00	0.19
40	$NHC(=O)CH_3$	-0.97	1.49	5.09	1.35	3.61	0.00	0.21
41	$C(CH_3)(NO_2)_2$	-0.88	2.17	4.59	2.55	3.72	0.61	0.54
42	$NHNH_2$	-0.88	0.84	3.47	1.35	2.97	-0.55	-0.02
43	OSO_2CH_3	-0.88	1.70	4.66	1.35	4.10	0.36	0.39
44	$SO_2N(CH_3)_2$	-0.78	2.19	4.83	2.03	4.08	0.65	0.51
45	$NHC=S(NHC_2H_5)$	-0.71	3.17	7.22	1.45	4.38	0.07	0.30
46	$SO_2(CHF_2)$	-0.68	1.31	4.11	2.03	3.70	0.86	0.75
47	OH	-0.67	0.29	2.74	1.35	1.93	-0.37	0.12
48	CHO	-0.65	0.69	3.53	1.60	2.36	0.42	0.35
49	$CH_2CHOHCH_3$	-0.64	1.64	4.92	1.52	3.78	-0.17	-0.12
50	$CS(NH_2)$	-0.64	1.81	4.10	1.64	3.18	0.30	0.25
51	$OC=O(CH_3)$	-0.64	1.25	4.74	1.35	3.67	0.31	0.39
52	$SOCHF_2$	-0.63	1.33	4.70	1.40	3.70	0.58	0.54
53	4-Pyrimidinyl	-0.61	2.18	5.29	1.71	3.11	0.63	0.30
54	2-Pyrimidinyl	-0.61	2.18	6.28	1.71	3.11	0.53	0.23

Table 1.4 *(Continued)*

No.	Substituent	Pi	MR	L	B1	B5	S-P	S-M
55	P(CF$_3$)$_2$	−0.59	1.99	4.96	1.40	3.86	0.69	0.60
56	CH$_2$CN	−0.57	1.01	3.99	1.52	4.12	0.18	0.16
57	CN	−0.57	0.63	4.23	1.60	1.60	0.66	0.56
58	COCH$_3$	−0.55	1.12	4.06	1.60	3.13	0.50	0.38
59	CH$_2$P=O(OEt)$_2$	−0.54	3.58	7.10	1.52	5.73	0.06	0.12
60	P=O(OEt)$_2$	−0.52	3.12	6.26	2.52	5.58	0.60	0.55
61	NHCOOMe	−0.52	1.57	5.84	1.45	3.99	−0.17	−0.02
62	NHC=O(NHC$_2$H$_5$)	−0.50	2.32	7.29	1.45	3.98	−0.26	0.04
63	NHC=O(CH$_2$Cl)	−0.50	1.98	6.26	1.55	4.26	−0.03	0.17
64	NHCH$_3$	−0.47	1.03	3.53	1.35	3.08	−0.70	−0.21
65	N(CH$_3$)COCF$_3$	−0.46	1.95	5.20	1.56	3.96	0.39	0.41
66	C=S(NHCH$_3$)	−0.46	2.23	5.00	1.88	3.18	0.34	0.30
67	NHC=S(CH$_3$)	−0.42	2.34	5.09	1.45	4.38	0.12	0.24
68	C(Et)(NO$_2$)$_2$	−0.35	3.66	4.92	2.55	3.72	0.64	0.56
69	CO$_2$H	−0.32	0.69	3.91	1.60	2.66	0.45	0.37
70	C(OH)(CH$_3$)$_2$	−0.32	1.64	4.11	2.40	3.17	0.60	0.47
71	EtCO$_2$H	−0.29	1.65	5.97	1.52	3.31	−0.07	−0.03
72	NO$_2$	−0.28	0.74	3.44	1.70	2.44	0.78	0.71
73	CH=NNHCSNH$_2$	−0.27	2.96	7.16	1.60	5.41	0.40	0.45
74	NHCN	−0.26	1.01	3.90	1.35	4.05	0.06	0.21
75	CH$_2$C(OH)(CH$_3$)$_2$	−0.24	2.11	4.92	1.52	4.19	−0.17	−0.16
76	CH=CHCHO	−0.23	1.69	5.76	1.60	3.46	0.13	0.24
77	NHCH$_2$CO$_2$Et	−0.21	2.69	7.91	1.35	5.77	−0.68	−0.10
78	CH$_2$OCH$_3$	−0.21	1.21	4.78	1.52	3.40	0.01	0.08
79	NHC=OCH(CH$_3$)$_2$	−0.18	2.43	5.53	1.35	4.09	−0.10	0.11
80	CH$_2$OC=O(CH$_3$)	−0.17	1.65	5.46	1.52	4.46	0.05	0.04
81	CH$_2$N(CH$_3$)$_2$	−0.15	1.87	4.83	1.52	4.08	0.01	0.00
82	CH$_2$SCN	−0.14	1.81	6.63	1.52	3.41	0.14	0.12
83	1-Aziridinyl	−0.12	1.35	4.14	1.35	3.24	−0.22	−0.07
84	NO	−0.12	0.52	3.44	1.70	2.44	0.91	0.62
85	ONO$_2$	−0.12	0.85	4.46	1.35	3.62	0.70	0.55
86	S=O(C$_6$H$_5$)	−0.07	3.34	4.62	1.40	6.02	0.44	0.50
87	CH$_2$SO$_2$C$_6$H$_5$	−0.06	3.79	8.33	1.52	3.78	0.16	0.15
88	OCH$_3$	−0.02	0.79	3.98	1.35	3.07	−0.27	0.12
89	C=O(OCH$_3$)	−0.01	1.29	4.73	1.64	3.36	0.45	0.36
90	H	0.00	0.10	2.06	1.00	1.00	0.00	0.00
91	C=O(CF$_3$)	0.02	1.12	4.65	1.70	3.67	0.80	0.63
92	CH=C(CN)$_2$	0.05	1.97	6.46	1.60	5.17	0.84	0.66
93	SO$_2$(F)	0.05	0.87	3.33	2.01	2.70	0.91	0.80
94	COEt	0.06	1.58	4.87	1.63	3.45	0.48	0.38
95	C(CF$_3$)$_3$	0.07	2.08	4.11	3.13	3.64	0.55	0.55
96	NH—Et	0.08	1.50	4.83	1.35	3.42	−0.61	−0.24
97	NHC=O(CF$_3$)	0.08	1.43	5.62	1.79	3.61	0.12	0.30
98	SC=O(CH$_3$)	0.10	1.84	5.11	1.70	4.01	0.44	0.39
99	CF$_3$	0.10	0.50	3.30	1.99	2.61	0.54	0.43
100	OCH$_2$F	0.10	0.72	4.57	1.35	3.07	0.02	0.20
101	CH=CHNO$_2$(TR)	0.11	1.64	4.29	1.60	4.78	0.26	0.32
102	CH$_2$F	0.13	0.54	3.30	1.52	2.61	0.11	0.12
103	F	0.14	0.09	2.65	1.35	1.35	0.06	0.34
104	C(OMe)$_3$	0.14	2.48	4.78	2.56	4.29	−0.04	−0.03
105	SECF$_3$	0.15	1.63	4.50	1.85	4.09	0.45	0.44
106	NHC=O(OEt)	0.17	2.12	7.25	1.35	3.92	−0.15	0.11
107	CH$_2$Cl	0.17	1.05	3.89	1.52	3.46	0.12	0.11
108	N(CH$_3$)$_2$	0.18	1.56	3.53	1.35	3.08	−0.83	−0.16

Table 1.4 (*Continued*)

No.	Substituent	Pi	MR	L	B1	B5	S-P	S-M
109	CHF_2	0.21	0.52	3.30	1.71	2.61	0.32	0.29
110	$CCCF_3$	0.22	1.41	5.90	1.99	2.61	0.51	0.41
111	$SO_2C_6H_5$	0.27	3.32	5.86	2.03	6.02	0.68	0.62
112	$COCH(CH_3)_2$	0.29	1.98	4.84	1.99	4.08	0.47	0.38
113	$OCHF_2$	0.31	0.79	3.98	1.35	3.61	0.18	0.31
114	$CH_2SO_2CF_3$	0.33	1.75	5.35	1.52	4.07	0.31	0.29
115	$C(NO_2)(CH_3)_2$	0.33	2.06	4.59	2.58	3.72	0.20	0.18
116	$P(O)(OPR)_2$	0.35	4.05	7.07	2.52	6.90	0.50	0.38
117	$CH_2S{=}O(CF_3)$	0.37	1.90	5.35	1.52	4.07	0.24	0.25
118	OCH_2CH_3	0.38	1.25	4.80	1.35	3.36	-0.24	0.10
119	SH	0.39	0.92	3.47	1.70	2.33	0.15	0.25
120	$N{=}NCF_3$	0.40	1.39	5.45	1.70	3.48	0.68	0.56
121	CCH	0.40	0.96	4.66	1.60	1.60	0.23	0.21
122	$N{=}CCl_2$	0.41	1.84	5.65	1.70	4.54	0.13	0.21
123	SCCH	0.41	1.62	4.08	1.70	4.85	0.19	0.26
124	SCN	0.41	1.34	4.08	1.70	4.45	0.52	0.51
125	$P(CH_3)_2$	0.44	2.12	3.88	2.00	3.32	0.06	0.03
126	$NHSO_2C_6H_5$	0.45	3.79	8.24	1.35	3.72	0.01	0.16
127	$SO_2NHC_6H_5$	0.45	3.78	8.24	2.03	4.50	0.65	0.56
128	CH_2CF_3	0.45	0.97	4.70	1.52	3.70	0.09	0.12
129	NNN	0.46	1.02	4.62	1.50	4.18	0.08	0.37
130	NNN	0.46	1.02	4.62	1.50	4.18	0.08	0.37
131	4-Pyridyl	0.46	2.30	5.92	1.71	3.11	0.44	0.27
132	$N{=}NN(CH_3)_2$	0.46	2.09	5.68	1.77	3.90	0.44	0.27
133	$C{=}O(NHC_6H_5)$	0.49	3.54	8.24	1.63	4.85	-0.03	-0.05
134	2-Pyridyl	0.50	2.30	6.28	1.71	3.11	0.41	0.23
135	$OCH_2CH{=}CH_2$	0.51	1.61	6.22	1.35	4.42	0.17	0.33
136	$C{=}O(OEt)$	0.51	1.75	5.95	1.64	4.41	-0.25	0.09
137	$S{=}O(CF_3)$	0.53	1.31	4.70	1.40	3.70	0.45	0.37
138	$CHOHC_6H_5$	0.54	3.15	4.62	1.73	6.02	0.69	0.63
139	OCH_2Cl	0.54	1.20	5.44	1.35	3.13	-0.03	0.00
140	$SO_2(CF_3)$	0.55	1.29	4.70	2.03	3.70	0.08	0.25
141	CH_3	0.56	0.57	2.87	1.52	2.04	0.96	0.83
142	SCH_3	0.61	1.38	4.30	1.70	3.26	-0.17	-0.07
143	$SC{=}O(CF_3)$	0.66	1.82	5.55	1.70	4.51	0.00	0.15
144	$COC(CH_3)_3$	0.69	2.44	4.87	1.87	4.42	0.46	0.48
145	$CH{=}NC_6H_5$	0.69	3.30	8.50	1.70	4.07	0.32	0.27
146	$P{=}O(C_6H_5)_2$	0.70	5.93	5.40	2.68	6.19	0.42	0.35
147	Cl	0.71	0.60	3.52	1.80	1.80	.530	.380
148	$N{=}CHC_6H_5$	0.72	3.30	8.40	1.70	4.65	0.23	0.37
149	$SeCH_3$	0.74	1.70	4.52	1.85	3.63	-0.55	-0.08
150	SCH_2F	0.74	1.34	4.89	1.70	3.41	0.00	0.10
151	$OCH{=}CH_2$	0.75	1.14	4.98	1.35	3.65	0.20	0.23
152	CH_2Br	0.79	1.34	4.09	1.52	3.75	-0.09	0.21
153	$CCCH_3$	0.81	1.41	5.47	1.60	2.04	0.14	0.12
154	$CH{=}CH2$	0.82	1.10	4.29	1.60	3.09	0.03	0.21
155	Br	0.86	0.89	3.82	1.95	1.95	-0.16	-0.08
156	$NHSO_2CF_3$	0.93	1.75	5.26	1.35	4.00	0.23	0.39
157	$OSO_2C_6H_5$	0.93	3.67	8.20	1.35	3.64	0.39	0.44
158	1-Pyrryl	0.95	1.95	5.44	1.71	3.12	0.33	0.36
159	$N(CH_3)SO_2CF_3$	1.00	2.28	5.26	1.54	4.00	0.37	0.47
160	$SCHF_2$	1.02	1.38	4.30	1.70	3.94	0.44	0.46
161	CH_2CH_3	1.02	1.03	4.11	1.52	3.17	0.37	0.33
162	OCF_3	1.04	0.79	4.57	1.35	3.61	-0.15	-0.07

Table 1.4 (*Continued*)

No.	Substituent	Pi	MR	L	B1	B5	S-P	S-M
163	OCH$_2$CH$_2$CH$_3$	1.05	1.71	6.05	1.35	4.42	0.35	0.38
164	C=O(C$_6$H$_5$)	1.05	3.03	4.57	1.92	5.98	−0.25	0.10
165	NHCO$_2$C$_4$H$_9$	1.07	3.05	9.50	1.45	5.05	0.43	0.34
166	S—Et	1.07	1.84	5.16	1.70	3.97	−0.05	0.06
167	N(CF$_3$)$_2$	1.08	1.43	4.01	1.52	3.58	0.03	0.18
168	CHCl$_2$	1.09	1.53	3.89	1.88	3.46	0.53	0.40
169	CH$_2$CH=CH$_2$	1.10	1.45	5.11	1.52	3.78	0.32	0.31
170	CH$_2$I	1.10	1.86	4.36	1.52	4.15	−0.14	−0.11
171	NH—Bu	1.10	2.43	6.88	1.35	4.87	0.11	0.10
172	CClF$_2$	1.11	1.07	3.89	1.99	3.46	−0.51	−0.34
173	I	1.12	1.39	4.23	2.15	2.15	0.46	0.42
174	Cyclopropyl	1.14	1.35	4.14	1.55	3.24	0.18	0.35
175	C(CH$_3$)=CH$_2$	1.14	1.56	4.29	1.73	3.11	−0.21	−0.07
176	NCS	1.15	1.72	4.29	1.50	4.24	0.05	0.09
177	SCH$_2$CH=CH$_2$	1.15	2.26	6.42	1.70	5.02	0.38	0.48
178	N(Et)$_2$	1.18	2.49	4.83	1.35	4.39	0.12	0.19
179	OSO$_2$CF$_3$	1.23	1.45	5.23	1.35	3.24	−0.72	−0.23
180	SF$_5$	1.23	0.99	4.65	2.47	2.92	0.53	0.56
181	OCHCl$_2$	1.26	1.69	3.98	1.35	4.41	0.68	0.61
182	CF$_2$CF$_3$	1.26	0.92	4.11	1.99	3.64	0.26	0.38
183	C(OH)(CF$_3$)$_2$	1.28	1.52	4.11	2.61	3.64	0.52	0.47
184	SCH=CH$_2$	1.29	1.77	5.33	1.70	4.23	0.30	0.29
185	NHC$_6$H$_5$	1.37	3.00	4.53	1.35	5.95	0.20	0.26
186	SCH(CH$_3$)$_2$	1.41	2.41	5.16	1.70	4.41	−0.56	−0.02
187	SCF$_3$	1.44	1.38	4.89	1.70	3.94	0.07	0.23
188	OC=O(C$_6$H$_5$)	1.46	3.23	8.15	1.64	4.40	0.50	0.40
189	COOC$_6$H$_5$	1.46	3.02	8.13	1.94	3.50	0.13	0.21
190	Cyclobutyl	1.51	1.79	4.77	1.77	3.82	0.44	0.37
191	O—Bu	1.52	2.17	6.86	1.35	4.79	−0.14	−0.05
192	CH(CH$_3$)$_2$	1.53	1.50	4.11	1.90	3.17	−0.32	0.10
193	CHBr$_2$	1.53	1.68	4.09	1.92	3.75	−0.15	−0.04
194	Pr	1.55	1.50	4.92	1.52	3.49	0.32	0.31
195	C(F)(CF$_3$)$_2$	1.56	1.34	4.11	2.45	3.64	−0.13	−0.06
196	C$_6$H$_4$(NO$_2$)-p	1.64	3.17	7.66	1.71	3.11	0.05	0.09
197	CH$_2$OC$_6$H$_5$	1.66	3.22	8.19	1.52	3.53	0.13	0.10
198	N=NC$_6$H$_5$	1.69	3.13	8.43	1.70	4.31	0.07	0.06
199	SO$_2$CF$_2$CF$_3$	1.73	1.97	5.35	2.03	4.07	0.39	0.32
200	CF$_2$CF$_2$CF$_2$CF$_3$	1.74	1.77	6.76	1.99	5.05	1.08	0.92
201	1-Cyclopentenyl	1.77	2.21	5.24	1.91	3.08	0.52	0.47
202	OCF$_2$CHF$_2$	1.79	1.08	5.23	1.35	3.94	−0.05	−0.06
203	C$_6$H$_4$(OCH$_3$)-p	1.82	3.17	7.71	1.80	3.11	0.25	0.34
204	CH$_2$SCF$_3$	1.83	1.76	5.82	1.52	4.10	−0.08	0.05
205	C$_6$H$_5$	1.96	2.54	6.28	1.71	3.11	0.04	0.01
206	C(CH$_3$)$_3$	1.98	1.96	4.11	2.60	3.17	−0.01	0.06
207	CCl$_3$	1.99	2.01	3.89	2.64	3.46	−0.20	−0.10
208	CH$_2$Si(CH$_3$)$_3$	2.00	2.96	5.39	1.52	4.75	0.46	0.40
209	CH$_2$C$_6$H$_5$	2.01	3.00	4.62	1.52	6.02	−0.21	−0.16
210	CH(CH$_3$)(Et)	2.04	1.96	4.92	1.90	3.49	−0.09	−0.08
211	C$_6$H$_4$F-p	2.04	2.53	6.87	1.71	3.11	−0.12	−0.08
212	OC$_5$H$_{11}$	2.05	2.63	8.11	1.35	5.81	0.06	0.12
213	N(C$_3$H$_7$)$_2$	2.08	3.24	6.07	1.35	5.50	−0.34	0.10
214	OC$_6$H$_5$	2.08	2.77	4.51	1.35	5.89	−0.93	−0.26
215	C$_6$H$_4$N(CH$_3$)$_2$-p	2.10	3.99	7.75	1.79	3.11	−0.03	0.25
216	Bu	2.13	1.96	6.17	1.52	4.54	−0.56	−0.06

Table 1.4 (*Continued*)

No.	Substituent	Pi	MR	L	B1	B5	S-P	S-M
217	Cyclopentyl	2.14	2.20	4.90	1.90	4.09	0.28	0.35
218	CHI_2	2.15	3.15	4.36	1.95	4.15	-0.14	-0.05
219	SC_6H_5	2.32	3.43	4.57	1.70	6.42	0.26	0.26
220	1-Cyclohexenyl	2.33	2.67	6.16	2.23	3.30	0.07	0.23
221	$OCCl_3$	2.36	2.18	5.44	1.35	4.41	-0.08	-0.10
222	$C(Et)(CH_3)_2$	2.37	2.42	4.92	2.60	3.49	0.35	0.43
223	$CH_2C(CH_3)_3$	2.37	2.42	4.89	1.52	4.18	-0.18	-0.06
224	$SC_6H_4NO_2$-p	2.39	4.11	4.92	1.70	7.86	-0.17	-0.05
225	SCF_2CHF_2	2.43	1.84	5.60	1.70	4.55	0.24	0.32
226	C_6H_4Cl-p	2.61	3.04	7.74	1.80	3.11	-0.07	-0.04
227	C_6F_5	2.62	2.40	6.87	1.71	3.67	0.12	0.15
228	C_5H_{11}	2.63	2.42	6.97	1.52	4.94	0.27	0.26
229	CCC_6H_5	2.65	3.32	8.88	1.71	3.11	-0.15	-0.08
230	CBr_3	2.65	2.88	4.09	2.86	3.75	0.16	0.14
231	EtC_6H_5	2.66	3.47	8.33	1.52	3.58	0.29	0.28
232	$C_6H_4(CH_3)$-p	2.69	3.00	7.09	1.84	3.11	-0.12	-0.07
233	C_6H_4I-p	3.02	3.91	8.45	2.15	3.11	-0.03	0.06
234	C_6H_4I-m	3.02	3.91	6.72	1.84	5.15	0.12	0.15
235	1-Adamantyl	3.37	4.03	6.17	3.16	3.49	-0.15	-0.05
236	$C(Et)_3$	3.42	3.36	4.92	2.94	4.18	0.10	0.14
237	$CH(C_6H_5)_2$	3.52	5.43	5.15	2.01	6.02	0.06	0.13
238	$N(C_6H_5)_2$	3.61	5.50	5.77	1.35	5.95	0.01	0.08
239	Heptyl	3.69	3.36	9.03	1.52	6.39	-0.13	-0.12
240	$C(SCF_3)_3$	4.17	4.40	5.82	3.32	5.00	-0.20	-0.07
241	C_6Cl_5	4.96	4.95	7.74	1.81	4.48	-0.05	-0.03

methods are based on molecular fragments, atomic contributions, or computer-identified fragments (1, 106, 107, 144–147). Whole-molecule approaches use molecular properties or spatial properties to predict $\log P$ values (148–150). They run on different platforms (e.g., Mac, PC, Unix, VAX, etc.) and use different calculation procedures. An extensive, recent review by Mannhold and van de Waterbeemd addresses the advantages and limitations of the various approaches (143). Statistical parameters yield some insight as to the effectiveness of such programs.

Recent attempts to compute $\log P$ calculations have resulted in the development of solvatochromic parameters (151, 152). This approach was proposed by Kamlet et al. and focused on molecular properties. In its simplest form it can be expressed as follows:

$$\text{Log } P_{\text{oct}} = aV + b\pi^* + c\beta_H + d\alpha_H + e \quad (1.49)$$

V is a solute volume term; π^* represents the solute polarizability; β_H and α_H are measures of hydrogen bond acceptor strength and hydrogen bond donor strength, respectively; and e is the intercept. An extension of this model has been formulated by Abraham and used by researchers to refine molecular descriptors and characterize hydrophobicity scales (153–156).

3.3 Steric Parameters

The quantitation of steric effects is complex at best and challenging in all other situations, particularly at the molecular level. An added level of confusion comes into play when attempts are made to delineate size and shape. Nevertheless, sterics are of overwhelming importance in ligand-receptor interactions as well as in transport phenomena in cellular systems. The first steric parameter to be quantified and used in QSAR studies was Taft's E_S constant (157). E_S is defined as

$$E_S = \log(k_X/k_H)_A \quad (1.50)$$

where k_X and k_H represent the rates of acid hydrolysis of esters, XCH_2COOR and CH_3COOR,

respectively. To correct for hyperconjuga-
tion in the α-hydrogens of the acetate moiety, Hancock devised a correction on E_S such
that

$$E_S{}^C = E_S + 0.306(n - 3) \qquad (1.51)$$

In Equation 1.51, n represents the number of α-hydrogens and 0.306 is a constant
derived from molecular orbital calculations
(158). Unfortunately, the limited availability of E_S and $E_S{}^C$ values for a great number
of substituents precludes their usage in
QSAR studies. Charton demonstrated a
strong correlation between E_S and van der
Waals radii, which led to his development of
the upsilon parameter v_X (159).

$$v_X = r_X - r_H = r_X - 1.20 \qquad (1.52)$$

where r_X and r_H are the minimum van der
Waals radii of the substituent and hydrogen,
respectively. Extension of this approach
from symmetrical substituents to nonsymmetrical substituents must be handled with
caution.

One of the most widely used steric parameters is molar refraction (MR), which has
been aptly described as a "chameleon" parameter by Tute (160). Although it is generally considered to be a crude measure of
overall bulk, it does incorporate a polarizability component that may describe cohesion and is related to London dispersion
forces as follows: MR $= 4\pi N\alpha/3$, where N is
Avogadro's number and α is the polarizability of the molecule. It contains no information on shape. MR is also defined by the
Lorentz-Lorenz equation:

$$MR = [(n^2 - 1)/(n^2 + 2)] \\ \times (MW/density) \qquad (1.53)$$

MR is generally scaled by 0.1 and used in biological QSAR, where intermolecular effects
are of primary importance. The refractive index of the molecule is represented by n. With
alkyl substituents, there is a high degree of
collinearity with hydrophobicity; hence, care

must be taken in the QSAR analysis of such
derivatives. The MR descriptor does not distinguish shape; thus the MR value for amyl
($-CH_2CH_2CH_2CH_2CH_3$) is the same as that
for $[-C(Et)(CH_3)_2]$: 2.42. The coefficients
with MR terms challenge interpretation, although extensive experience with this parameter suggests that a negative coefficient implies steric hindrance at that site and a
positive coefficient attests to either dipolar interactions in that vicinity or anchoring of a
ligand in an opportune position for interaction
(161).

The failure of the MR descriptor to adequately address three-dimensional shape issues led to Verloop's development of STERI-
MOL parameters (162), which define the
steric constraints of a given substituent along
several fixed axes. Five parameters were
deemed necessary to define shape: L, B1, B2,
B3, and B4. L represents the length of a substituent along the axis of a bond between the
parent molecule and the substituent; B1 to B4
represent four different width parameters.
However, the high degree of collinearity between B1, B2, and B3 and the large number of
training set members needed to establish the
statistical validity of this group of parameters
led to their demise in QSAR studies. Verloop
subsequently established the adequacy of just
three parameters for QSAR analysis: a slightly
modified length L, a minimum width B1, and a
maximum width B5 that is orthogonal to L
(163). The use of these insightful parameters
have done much to enhance correlations with
biological activities. Recent analysis in our
laboratory has established that in many cases,
B1 alone is superior to Taft's E_S and a combination of B1 and B5 can adequately replace E_S
(164).

Molecular weight (MW) terms have also
been used as descriptors, particularly in cellular systems, or in distribution/transport studies where diffusion is the mode of operation.
According to the Einstein-Sutherland equation, molecular weight affects the diffusion
rate. The Log MW term has been used extensively in some studies (159–161) and an example of such usage is given below. In correlating
permeability (Perm) of nonelectrolytes through
chara cells, Lien et al. obtained the following
QSAR (168):

Log Perm

$$= 0.889 \log P^* - 1.544 \log \mathrm{MW} \quad (1.54)$$

$$- 0.144 H_b + 4.653$$

$$n = 30, \quad r^2 = 0.899,$$

$$s = 0.322, \quad F = 77.39$$

In QSAR 54, Log P^* represents the olive oil/water partition coefficient, MW is the molecular weight of the solute and defines its size, and H_b is a crude approximation of the total number of hydrogen bonds for each molecule. The molecular weight descriptor has also been an omnipresent variable in QSAR studies pertaining to cross-resistance of various drugs in multidrug-resistant cell lines (169). $\sqrt[3]{\mathrm{MW}}$ was used because it most closely approximates the size (radii) of the drugs involved in the study and their interactions with GP-170. See QSAR 1.55.

Log CR $= 0.70 \sqrt[3]{\mathrm{MW}}$

$$- 1.01 \log(\beta \cdot 10^{\sqrt[3]{\mathrm{MW}}} + 1)$$

$$- 0.10 \log P + 0.38 I \qquad (1.55)$$

$$- 3.08$$

$$n = 40, \quad r^2 = 0.794, \quad s = 0.344$$

$$\log \beta = -6.851 \quad \text{optimum } \sqrt[3]{\mathrm{MW}} = 7.21$$

3.4 Other Variables and Variable Selection

Indicator variables (I) are often used to highlight a structural feature present in some of the molecules in a data set that confers unusual activity or lack of it to these particular members. Their use could be beneficial in cases where the data set is heterogeneous and includes large numbers of members with unusual features that may or may not impact a biological response. QSAR for the inhibition of trypsin by X-benzamidines used indicator variables to denote the presence of unusual features such as positional isomers and vinyl/carbonyl-containing substituents (170). A recent study on the inhibition of lipoxygenase catalyzed production of leukotriene B4 and 5-hydroxyeicosatetraenoic from arachidonic

acid in guinea pig leukocytes by X-vinyl catechols led to the development of the following QSAR (171):

Log $1/C$

$$= 0.49(\pm 0.11)\log P$$

$$- 0.75(\pm 0.22)\log(\beta \cdot 10^{\log P} + 1) \quad (1.56)$$

$$- 0.62(\pm 0.18)\mathrm{D2}$$

$$- 1.13(\pm 0.20)\mathrm{D3} + 5.50(\pm 0.33)$$

$$n = 51, \quad r^2 = 0.801, \quad s = 0.269,$$

$$\text{Log } P_O = 4.61(\pm 0.49) \quad \text{Log } \beta = -4.33$$

The indicator variables are D2 and D3; for simple X-catechols, D2 = 1 and for X-naphthalene diols, D3 = 1. The negative coefficients with both terms (D2 and D3) underscore the detrimental effects of these structural features in these inhibitors. Thus, discontinuities in the structural features of the molecules of this data set are accounted for by the use of indicator variables. An indicator variable may be visualized graphically as a constant that adjusts two parallel lines so that they are superimposable. The use of indicator variables in QSAR analysis is also described in the following example. An analysis of a comprehensive set of nitroaromatic and heteroaromatic compounds that induced mutagenesis in TA98 cells was conducted by Debnath et al., and QSAR 1.57 was formulated (172).

Log TA98

$$= 0.65(\pm 0.16)\log P$$

$$- 2.90(\pm 0.59)\log(\beta \cdot 10^{\log P} + 1)$$

$$- 1.38(\pm 0.25) E_{\mathrm{LUMO}} \qquad (1.57)$$

$$+ 1.88(\pm 0.39)I_1 - 2.89(\pm 0.81)I_a$$

$$- 4.15(\pm 0.58)$$

$$n = 188, \quad r^2 = 0.810, \quad s = 0.886,$$

$$\text{Log } P_O = 4.93(\pm 0.35) \quad \text{Log } \beta = -5.48$$

TA98 represents the number of revertants per nanomole of nitro compound. E_{LUMO} is the energy of the lowest unoccupied molecular orbital and I_a is an indicator variable that signifies the presence of an acenthrylene ring in the mutagens. I_1 is also an indicator variable that pertains to the number of fused rings in the data set. It acquires a value of 1 for all congeners containing three or more fused rings and a value of zero for those containing one or two fused rings (e.g., naphthalene, benzene). Thus, the greater the number of fused rings, the greater the mutagenicity of the nitro congeners. The E_{LUMO} term indicates that the lower the energy of the LUMO, the more potent the mutagen. In this QSAR the combination of indicator variables affords a mixed blessing. One variable helps to enhance activity, whereas the other leads to a decrease in mutagenicity of the acenthrylene congeners. In both these QSAR, Kubinyi's bilinear model is used (21). See Section 4.2 for a description of this approach.

3.5   Molecular Structure Descriptors

These are truly structural descriptors because they are based only on the two-dimensional representation of a chemical structure. The most widely known descriptors are those that were originally proposed by Randic (173) and extensively developed by Kier and Hall (27). The strength of this approach is that the required information is embedded in the hydrogen-suppressed framework and thus no experimental measurements are needed to define molecular connectivity indices. For each bond the C_k term is calculated. The summation of these terms then leads to the derivation of X, the molecular connectivity index for the molecule.

$$C_k = (\delta_i \delta_j)^{-0.5} \quad \text{where} \quad \delta = \sigma - h \quad (1.58)$$

δ is the count of formally bonded carbons and h is the number of bonds to hydrogen atoms.

$$^1X = \sum C_k = \sum (\delta_i \delta_j)_k^{-0.5} \quad (1.59)$$

^1X is the first bond order because it considers only individual bonds. Higher molecular connectivity indices encode more complex attributes of molecular structure by considering longer paths. Thus, ^2X and ^3X account for all two-bond paths and three-bond paths, respec-

tively, in a molecule. To correct for differences in valence, Kier and Hall proposed a valence delta (δ^v) term to calculate valence connectivity indices (175).

Molecular connectivity indices have been shown to be closely related to many physicochemical parameters such as boiling points, molar refraction, polarizability, and partition coefficients (174, 176). Ten years ago, the E-State index was developed to define an atom- or group-centered numerical code to represent molecular structure (28). The E-State was established as a composite index encoding both electronic and steric properties of atoms in molecules. It reflects an atom's electronegativity, the electronegativity of proximal and distal atoms, and topological state. Extensions of this method include the HE-State, atom-type E-State, and the polarity index Q. Log P showed a strong correlation with the Q index of a small set ($n = 21$) of miscellaneous compounds (28). Various models using electrotopological indices have been developed to delineate a variety of biological responses (177–179). Some criticism has been leveled at this approach (180, 181). Chance correlations are always a problem when dealing with such a wide array of descriptors. The physicochemical interpretation of the meaning of these descriptors is not transparent, although attempts have been made to address this issue (27).

4   QUANTITATIVE MODELS

4.1   Linear Models

The correlation of biological activity with physicochemical properties is often termed an *extrathermodynamic relationship*. Because it follows in the line of Hammett and Taft equations that correlate thermodynamic and related parameters, it is appropriately labeled. The Hammett equation represents relationships between the logarithms of rate or equilibrium constants and substituent constants. The linearity of many of these relationships led to their designation as linear free energy relationships. The Hansch approach represents an extension of the Hammett equation from physical organic systems to a biological milieu. It should be noted that the simplicity of the approach belies the tremendous complexity of the intermolecular interactions at play in the overall biological response.

Biological systems are a complex mix of heterogeneous phases. Drug molecules usually traverse many of these phases to get from the site of administration to the eventual site of action. Along this random-walk process, they perturb many other cellular components such as organelles, lipids, proteins, and so forth. These interactions are complex and vastly different from organic reactions in test tubes, even though the eventual interaction with a receptor may be chemical or physicochemical in nature. Thus, depending on the biological system involved—isolated receptor, cell, or whole animal—one expects the response to be multifactorial and complex. The overall process, particularly *in vitro* or *in vivo*, studies a mix of equilibrium and rate processes, a situation that defies easy separation and delineation.

Meyer and Overton were the first to attempt to get a grasp on biological responses by noting the relationship between oil/water partition coefficients and their narcotic activity. Ferguson recognized that equitoxic concentrations of small organic molecules was markedly influenced by their phase distribution between the biophase and exobiophase. This concept was generalized in the form of Equation 1.60 and extended by Fujita to Equation 1.61 (182, 183).

$$C = kA^m \qquad (1.60)$$

$$\text{Log } 1/C = m \text{ Log}(1/A) + \text{constant} \qquad (1.61)$$

C represents the equipotent concentration, k and m are constants for a particular system, and A is a physicochemical constant representative of phase distribution equilibria such as aqueous solubility, oil/water partition coefficient, and vapor pressure. In examining a large and diverse number of biological systems, Hansch and coworkers defined a relationship (Equation 1.62) that expressed biological activity as a function of physicochemical parameters (e.g., partition coefficients of organic molecules) (19).

$$\text{Log } 1/C = a \log P + b \qquad (1.62)$$

Model systems have been devised to elucidate

the mode of interactions of chemicals with biological entities. Examples of linear models pertaining to nonspecific toxicity are described. The effects of a series of alcohols (ROH) have been routinely studied in many model and biological systems. See QSAR 1.63–1.67.

4.1.1 Penetration of ROH into Phosphatidylcholine Monolayers (184)

$$\text{Log } 1/C = 0.87(\pm 0.01)\log P$$
$$+ 0.66(\pm 0.01) \qquad (1.63)$$
$$n = 4, \quad r^2 = 0.998, \quad s = 0.002$$

4.1.2 Changes in EPR Signal of Labeled Ghost Membranes by ROH (185)

$$\text{Log } 1/C = 0.93(\pm 0.09)\log P$$
$$- 0.41(\pm 0.16) \qquad (1.64)$$
$$n = 6, \quad r^2 = 0.996, \quad s = 0.092$$

4.1.3 Induction of Narcosis in Rabbits by ROH (184)

$$\text{Log } 1/C = 0.72(\pm 0.16)\log P$$
$$+ 1.35(\pm 0.12) \qquad (1.65)$$
$$n = 11, \quad r^2 = 0.924, \quad s = 0.142$$

4.1.4 Inhibition of Bacterial Luminescence by ROH (185)

$$\text{Log } 1/C = 1.10(\pm 0.07)\log P$$
$$+ 0.16(\pm 0.12) \qquad (1.66)$$
$$n = 8, \quad r^2 = 0.996, \quad s = 0.103$$

4.1.5 Inhibition of Growth of Tetrahymena pyriformis by ROH (76, 186)

$$\text{Log } 1/C = 0.82(\pm 0.04)C\log P$$
$$+ 0.89(\pm 0.10) \qquad (1.67)$$
$$n = 34, \quad r^2 = 0.982, \quad s = 0.173$$

In all cases, there is a strong dependency on

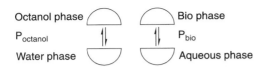

Figure 1.1. Log P_{octanol} mirrors Log P_{bio}.

log P_{oct} because all these processes involve transport of alcohols through membranes. The low intercepts speak to the nonspecific nature of the alcohol-mediated toxic interaction. An equilibrium-pseudoequilibrium modeled by log P can be defined as shown in Fig. 1.1.

The Hammett-type relationship for this conceptual idea of distribution is

$$\text{Log } P_{\text{bio}} = a \cdot \log P_{\text{octanol}} + b \quad (1.68)$$

This postulate assumes that steric, hydrophobic, electronic, and hydrogen bonding factors that affect partitioning in the biophase are handled by the octanol/water system. Given that the biological response (log $1/C$) is proportional to log P_{bio}, then it follows that

$$\text{Log } 1/C = a \cdot \log P_{\text{octanol}} + \text{constant} \quad (1.69)$$

Hansch and coworkers have amply demonstrated that Equation 1.69 applies not only to systems at or near phase distribution equilibrium but also to systems removed from equilibrium (184, 185).

4.2 Nonlinear Models

Extensive studies on development of linear models led Hansch and coworkers to note that a breakdown in the linear relationship occurred when a greater range in hydrophobicity was assessed with particular emphasis placed on test molecules at extreme ends of the hydrophobicity range. Thus, Hansch et al. suggested that the compounds could be involved in a "random-walk" process: low hydrophobic molecules had a tendency to remain in the first aqueous compartment, whereas highly hydrophobic analogs sequestered in the first lipoidal phase that they encountered. This led to the formulation of a parabolic equation, relating biological activity and hydrophobicity (187).

$$\text{Log } 1/C = -a(\log P)^2 + b \cdot \log P$$
$$+ \text{constant} \quad (1.70)$$

In the random-walk process, the compounds partition in and out of various compartments and interact with myriad biological components in the process. To deal with this conundrum, Hansch proposed a general, comprehensive equation for QSAR 1.71 (188).

$$\text{Log } 1/C = -a(\log P)^2 + b \cdot \log P$$
$$+ \rho\sigma + \delta E_{\text{S}} + \text{constant} \quad (1.71)$$

The optimum value of log P for a given system is log P_{O} and it is highly influenced by the number of hydrophobic barriers a drug encounters in its walk to its site of action. Hansch and Clayton formulated the following parabolic model to elucidate the narcotic action of alcohols on tadpoles (189).

4.2.1 Narcotic Action of ROH on Tadpoles

$$\text{Log} 1/C = 1.38(\pm 0.34)\log P$$
$$- 0.08(\pm 0.07)(\text{Log} P)^2 \quad (1.72)$$
$$+ 0.52(\pm 0.34)$$
$$n = 10, \quad r^2 = 0.990, \quad s = 0.210,$$
$$\text{Log } P_{\text{O}} = 8.69(5.78 - 43.43)$$

This is an example of nonspecific toxicity where the last step probably involves partitioning into a hydrophobic membrane. Log P_{O} represents the optimal hydrophobicity (as defined by log P) that elicits a maximal biological response.

Despite the success of the parabolic equation, there are a number of worrisome limitations. This approach forces the data into a symmetrical parabola, with the result that there are usually deviations between the experimental and parabola-calculated data. Second, the ascending slope is curved and inconsistent with the observed linear data. Thus, the slope of a linear model cannot be compared to the curved slope of the parabola. In 1973 Franke devised a sophisticated, empirical model consisting of a linear ascending part and a parabolic part (190). See Equations 1.73 and 1.74.

$$\text{Log } 1/C = a \cdot \log P + c$$
$$\text{(if } \log P < \log P_X)$$ (1.73)

$$\text{Log } 1/C = -a(\log P)^2 + b \cdot \log P + c$$
$$\text{(if } \log P > \log P_X)$$ (1.74)

The binding of drugs to proteins is linearly dependent on hydrophobicity up to a limited value, log P_X, after which steric hindrance causes the linear dependency to alter to a nonlinear one. The major limitation of this approach involves the inclusion of highly hydrophobic congeners that tend to cause systematic deviations between experimental and predicted values.

Another cutoff model, which deals with nonlinearity in biological systems, is one defined by McFarland (191). It attempts to elucidate the dependency of drug transport on hydrophobicity in multicompartment models. McFarland addressed the probability of drug molecules traversing several aqueous lipid barriers from the first aqueous compartment to a distant, final aqueous compartment. The probability $P_{0,n}$ of a drug molecule to access the final compartment n of a biological system was used to define the drug concentration in this compartment.

$$\text{Log } C_R = a \cdot \log P - 2a \cdot \log(P + 1)$$
$$+ \text{ constant}$$ (1.75)

The ascending and descending slopes are equal ($=1$) and linear. However, a major drawback of this model is that it forces the activity curves to maximize at $\log P = 0$. These studies were extended by Kubinyi, who developed the elegant and powerful bilinear model, which is superior to the parabolic model and is extensively used in QSAR studies (192).

$$\text{Log } 1/C = a \cdot \log P - b \cdot \log(\beta \cdot P + 1)$$
$$+ \text{ constant}$$ (1.76)

where β is the ratio of the volumes of the organic phase and the aqueous phase. An important feature of this model lies in the symmetry of the curves. For aqueous phases of this model system, symmetrical curves with linear ascending and descending sides (like a teepee) and a limited parabolic section around the hydrophobicity optimum are generated. Unsymmetrical curves arise for the lipid phases. It is highly compatible with the linear model and allows for quick comparisons of the ascending slopes. It can also be used with other parameters such as MR and σ, where it appears to pinpoint a change in mechanism similar to the breaks in linearity of the Hammett equation. The following example of the bilinear model reveals the symmetrical nature of the curve.

4.2.2 Induction of Ataxia in Rats by ROH

$$\text{Log } 1/C = 0.77(\pm 0.10)\log P$$
$$- 1.53(\pm 0.12)\log(\beta \cdot P + 1)$$ (1.77)
$$+ 1.68(\pm 0.12)$$
$$n = 35, \quad r^2 = 0.887,$$
$$s = 0.165, \quad \log P_O = 2.0$$

The bilinear model has been used to model biological interactions in isolated receptor systems and in adsorption, metabolism, elimination, and toxicity studies, although it has a few limitations. These include the need for at least 15 data points (because of the presence of the additional disposable parameter β and data points beyond optimum Log P. If the range in values for the dependent variable is limited, unreasonable slopes are obtained.

4.3 Free-Wilson Approach

The Free-Wilson approach is truly a structure-activity-based methodology because it incorporates the contributions made by various structural fragments to the overall biological activity (22, 193, 194). It is represented by Equation 1.78.

$$\text{BA}_i = \sum_j a_j X_{ij} + \mu$$ (1.78)

Indicator variables are used to denote the presence or absence of a particular structure feature.

Like classical QSAR, this *de novo* approach assumes that substituent effects are additive and constant. BA is the biological activity; X_j is the *j*th substituent, which carries a value 1 if present, 0 if absent. The term a_j represents the contribution of the *j*th substituent to biological activity and μ is the overall average activity. The summation of all activity contributions at each position must equal zero. The series of linear equations that are formulated are solved by linear regression analysis. It is necessary for each substituent to appear more than once at a position in different combinations with substituents at other positions.

There are certain advantages to the Free-Wilson method that have been addressed (193–195). Any type of quantitative biological data can be subject to such analysis. There is no need for any physicochemical constants. The molecules of a series may be structurally dissected in any way and multiple sites of substitution are necessary and easily accommodated (196). Limitations include the large number of molecules with varying substituent combinations that are needed for this analysis and the inability of the system to handle nonlinearity of the dependency of activity on substituent properties. Intramolecular interactions between the substituent are not handled very well, although special treatments can be used to accommodate proximal effects. Extrapolation outside of the substituents used in the study is not feasible. Another problem inherent with this approach is that usually a large number of variables is required to describe a smaller number of compounds, which creates a statistical faux pas. Fujita and Ban modified this approach in two important ways (23). They expressed the biological activity on a logarithmic scale, to bring it into line with the extrathermodynamic approach, as seen in the following equation:

$$\text{Log } X_C = \sum a_i X_i + \mu \qquad (1.79)$$

This allowed the derived substituent constants to be compared with other free energy-related parameters. The overall average intercept u took on a new look, as it were, akin to an intercept in other QSAR analyses.

Recent analyses of a Free-Wilson type have included the *in vitro* inhibitory activity of a series of heterocyclic compounds against *K. pneumonia* (197). Other applications of the Free-Wilson approach have included studies on the antimycobacterial activity of 4-alkyl-thiobenzanilides, the antibacterial activity of fluoronapthyridines, and the benzodiazepine receptor-binding ability of some non-benzodiapzepine compounds such as 3-X-imidazo-[1,2-*b*]pyridazines, 2-phenylimidazo[1,2-*α*]pyridines, 2-(alkoxycarbony)imidazo[2,1-*p*]benzothiazoles, and 2-arylquinolones (198–200).

4.4 Other QSAR Approaches

The similarity in approaches of Hansch analysis and Free-Wilson analysis allows them to be used within the same framework. This is based on their theoretical consistency and the numerical equivalencies of activity contributions. This development has been called the mixed approach and can be represented by the following equation:

$$\text{Log } 1/C = \sum a_i + \sum c_j \varnothing_j + \text{constant} \quad (1.80)$$

The term a_i denotes the contribution for each *i*th substituent, whereas $\varnothing_j$ is any physicochemical property of a substituent X_j. For a thorough review of the relationship between Hansch and Free-Wilson analyses, see the excellent reviews by Kubinyi (58, 195). A recent study of the P-glycoprotein inhibitory activity of 48 propafenone-type modulators of multidrug resistance, using a combined Hansch/Free-Wilson approach was deemed to have higher predictive ability than that of a stand-alone Free-Wilson analysis (201). Molar refractivity, which has a high collinearity with molecular weight, was a significant determinant of modulating ability. It is of interest to note that molecular weight has been shown to be an omnipresent parameter in cross-resistance profiles in multidrug-resistance phenomena (167).

5 APPLICATIONS OF QSAR

Over the last 40 years, the glut in scientific information has resulted in the development of thousands of equations pertaining to struc-

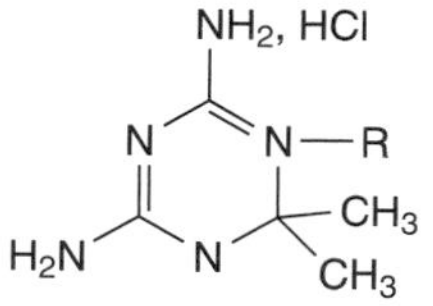

Figure 1.2. 4,6-Diamino-1,2-dihydro-2,2-dimethyl-1R-s-triazines.

ture-activity relationships in biological systems. In its original definition, the Hansch equation was defined to model drug-receptor interactions involving electronic, steric, and hydrophobic contributions. Nonlinear relationships helped refine this approach in cellular systems and organisms where pharmacokinetic constraints had to be considered and tackled. They have also found increased utility in addressing the complex QSAR of some receptor–ligand interactions. In many cases the Kubinyi bilinear model has provided a sophisticated approach to delineation of steric effects in such interactions. Examples of ligand–receptor interactions will be drawn from receptors such as the much-studied dihydrofolate reductases (DHFR), α-chymotrypsin and 5α-reductase (202–204).

5.1 Isolated Receptor Interactions

The critical role of DHFR in protein, purine, and pyrimidine synthesis; the availability of crystal structures of binary and ternary complexes of the enzyme; and the advent of molecular graphics combined to make DHFR an attractive target for well-designed heterocyclic ligands generally incorporating a 2,4-diamino-1,-3-diazapharmacophore (205). The earliest study focused on the inhibition of DHFR by 4,6-diamino-1,2-dihydro-2,2-dimethyl-1R-s-triazines, the structure of which is shown in Fig. 1.2 (202).

5.1.1 Inhibition of Crude Pigeon Liver DHFR by Triazines (202)

$$\text{Log } 1/\text{IC}_{50} = 2.21(\pm 1.00)\pi$$
$$- 0.28(\pm 0.17)\pi^2$$
$$+ 0.84(\pm 0.76)D$$
$$+ 2.58(\pm 1.30)$$

(1.81)

$$n = 15, \quad r^2 = 0.861,$$
$$s = 0.553, \quad \pi_{\text{O}} = 4(3.6 - 6.0)$$

In all equations, n is the number of data points, r^2 is the square of the correlation coefficient, s represents the standard deviation, and the figures in parentheses are for construction of the 95% confidence intervals. π represents the hydrophobicity of the substituent R and π_{O} is the optimum hydrophobic contribution of the R substituent. D is an indicator variable that acquires a value of 1.0 when a phenyl ring is present on the nitrogen and a value of zero for all other R. This is an example of a Hansch-Fujita-Ban analysis, where the indicator variable D establishes the contribution and thus the importance of a phenyl ring in DHFR inhibition. This equation has some limitations. Improper choice of N-substituents led to a high degree of collinearity between size and hydrophobicity and in terms of electronic contributions, spanned space was limited and thus inadequate. A subsequent study on the binding of these compounds to DHFR isolated from chicken liver was more revealing.

5.1.2 Inhibition of Chicken Liver DHFR by 3-X-Triazines (207)

$\text{Log } 1/K_{\text{i}}$

$$= 1.01(\pm 0.14)\pi'$$
$$- 1.16(\pm 0.19)\log(\beta \cdot 10^{\pi'} + 1)$$
$$+ 0.86(\pm 0.57)\sigma + 6.33(\pm 0.14)$$

(1.82)

$$n = 59, \quad r^2 = 0.821, \quad s = 0.906,$$
$$\pi'_{\text{O}} = 1.89(\pm 0.36) \quad \log \beta = -1.08$$

In this example, the R group on the 2-nitrogen was restricted to an (3-X-phenyl) aromatic ring (205). Accurate K_{i} values were obtained from highly purified DHFR isolated from chicken liver. In most cases, π' represented the hydrophobicity of the substituent except in certain instances where X = $-$OR or $-$CH$_2$ZC$_6$H$_4$-Y. It was ascertained that alkoxy substituents were not making direct hydrophobic contact with the enzyme, given that their inhibitory activities were essentially constant from the methoxy to the nonyloxy substituent. In the bridged substituents where Z = O, NH, S, Se, the Y substituent again did not contact the enzyme surface. Variation in Y led to the same, constant biological activity. The coefficient with π' suggests that the substituent is engulfed in a hydrophobic pocket that has an optimal π'_O of 2. This value is consistent with that seen in the crude pigeon liver DHFR corrected for the presence of the phenyl group $(4.0 - 2.0 = 2)$. The 0.86 ρ value (coefficient with σ) suggests that there could be a dipolar interaction between the electron deficient phenyl ring and a region of positively charged electrostatic potential in the enzyme, perhaps an arginine, lysine, or histidine residue. Hathaway et al. developed a QSAR for the inhibition of human DHFR by 3-X-triazines and obtained Equation 1.83 (208).

5.1.3 Inhibition of Human DHFR by 3-X-Triazines (208)

Log $1/K_i$

$$= 1.07(\pm 0.23)\pi'$$
$$- 1.10(\pm 0.26)\log(\beta \cdot 10^{\pi'} + 1) \quad (1.83)$$
$$+ 0.50(\pm 0.19)I + 0.82(\pm 0.66)\sigma$$
$$+ 6.07(\pm 0.21)$$
$$n = 60, \quad r^2 = 0.792, \quad s = 0.308,$$
$$\pi'_O = 2.0(\pm 0.87) \quad \log \beta = -0.577$$

The enhanced activity of the "bridged" substituents was corrected by the indicator variable I. Note that triazines bearing the bridge moieties $-CH_2NHC_6H_4Y$, $-CH_2OC_6H_4Y$, and $-CH_2SC_6H_4Y$ had unusually high enzyme binding activity. Note that the $-CH_2NHC_6H_5$ bridge is present in the endogenous substrate, folic acid. The bilinear dependency on hydrophobicity of the substituents parallels that seen in the case of chicken liver DHFR. A similar QSAR was obtained for DHFR isolated from L1210 murine leukemia cells (209).

5.1.4 Inhibition of L1210 DHFR by 3-X-Triazines (209)

Log $1/K_i$

$$= 0.98(\pm 0.14)\pi' \quad (1.84)$$
$$- 1.14(\pm 0.20)\log(\beta \cdot 10^{\pi'} + 1)$$
$$+ 0.79(\pm 0.57)\sigma + 6.12(\pm 0.14)$$
$$n = 58, \quad r^2 = 0.810, \quad s = 0.264,$$
$$\pi'_O = 1.76(\pm 0.28) \quad \log \beta = -0.979$$

The consistency in these models versus prokaryotic DHFR is established by the coefficient with the hydrophobic term, the optimum π' value, and the rho value. These numerical coefficients can be contrasted sharply with those obtained from fungal and protozoal DHFR. Inhibition constants were determined for 3-X-triazines versus *Pneumocystis carinii* DHFR (210).

5.1.5 Inhibition of *P. carinii* DHFR by 3-X-Triazines (210)

Log $1/K_i$

$$= 0.73(\pm 0.12)\pi'$$
$$- 1.36(\pm 0.35)\log(\beta \cdot 10^{\pi'} + 1)$$
$$- 0.78(\pm 0.42)I_{OR} \quad (1.85)$$
$$+ 0.28(\pm 0.21)MR_Y$$
$$+ 6.48(\pm 0.23)$$
$$n = 43, \quad r^2 = 0.840, \quad s = 0.435,$$
$$\pi'_O = 3.99(\pm 0.68) \quad \log \beta = -3.925$$

In Equation 1.85, I_{OR} is an indicator variable that assumes a value of 1 when an alkoxy substituent is present and 0 for all other substituents. It is of interest to note that the Y substituent on the second phenyl ring now contributes to activity. The MR_Y term suggests that it most probably accesses a polar region of the active site of the enzyme. The positive coefficient with MR_Y suggests that an increase in bulk and/or polarizability enhances binding. The descending slope of the bilinear equation is much steeper $(1.36 - 0.73 = 0.63)$ than that seen with the mammalian and avian enzymes.

A similar model is obtained vs. the bifunctional protozoal DHFR from *Leishmania major*, which is coupled to thymidylate synthase (211).

5.1.6 Inhibition of *L. major* DHFR by 3-X-Triazines (211)

$$\text{Log } 1/K_i$$

$$
\begin{aligned}
= \; & 0.65(\pm 0.08)\pi' \\
& - 1.22(\pm 0.29)\log(\beta \cdot 10^{\pi'} + 1) \\
& - 1.12(\pm 0.29)I_{\text{OR}} \\
& + 0.58(\pm 0.16)\text{MR}_{\text{Y}} \\
& + 5.05(\pm 0.16)
\end{aligned}
\tag{1.86}
$$

$$n = 41, \quad r^2 = 0.931, \quad s = 0.298,$$

$$\pi'_{\text{O}} = 4.54 \quad \log \beta = -4.491$$

QSAR analysis on a limited set of 3-X-triazines assayed by Chio and Queener versus *Toxoplasmosis gondii* led to the formulation of Equation 1.87 (202, 212).

5.1.7 Inhibition of *T. gondii* DHFR by 3-X-Triazines

$$\text{Log } 1/\text{IC}_{50} = 0.39(\pm 0.20)\pi'$$
$$- 0.43(\pm 0.19)\text{MR}_{\text{Y}} + 6.65(\pm 0.30) \tag{1.87}$$

$$n = 17, \quad r^2 = 0.810, \quad s = 0.289$$

A quick comparison of QSAR 1.82–1.84 reveals the strong similarity between the avian and mammalian models. In fact because of its increased stability, chicken liver DHFR has often been used as a surrogate for human DHFR in enzyme-inhibition studies. The intercepts, coefficients with π'_3 and optimum π'_{O} for avian (6.33, 1.01, 1.9), human (6.07, 1.07, 2.0), and mouse leukemia (6.12, 0.98, 1.76) can be compared to the corresponding values for *P. carinii* (6.48, 0.73, 3.99) and *Leishmania major* (5.05, 0.65, 4.54). QSAR 1.81 and 1.87 are not included in the comparison because crude pigeon enzyme was used in

the former and the testing for QSAR 1.87 was conducted under different assay conditions; K_i values were not determined. A noteworthy difference between these models is the wide disparity in π_{O} values. The binding site of the protozoal and fungal species comprises an extensive hydrophobic surface unlike the abbreviated pockets in the mammalian and avian enzymes. The positive coefficients with the MR_{Y} terms suggests that added bulk on the bridged phenyl ring enhances inhibitory potency. The study versus *T. gondii* DHFR (QSAR 1.87) included a number of mostly small, polar substituents (NH_2, NO_2, CONMe_2) on the bridged phenyl and their activities were considerably lower than the unsubstituted analog. Comparative QSAR can be useful, particularly if the biological data are consistent (tested under the same assay conditions, excellent purity of enzymes, substrates, inhibitors, buffers), and the choice of substituents is appropriate.

One of the major problems that arises with some QSAR studies is extrapolation from beyond spanned space. Predictive ability is sound when one has probed an adequate range in electronic, hydrophobic, and steric space. At the onset of the study, the training set should address these concerns. Lack of adequate attention to such issues can result in QSAR models that are misleading. When examined on its own, such a model may appear to withstand statistical rigor and apparent transparency but, on being subjected to lateral validation, loopholes emerge. A brief study to illustrate this phenomenon is outlined below.

Four different QSAR were derived for the inhibition of DHFR from rat liver, human leukemia, mouse L1210, and bovine liver by 2,4-diamino, 5-Y, 6-Z-quinazolines (Fig. 1.3) (202, 213–215). A comparison of their QSAR presents an interesting study on the importance of spanned space in delineating enzyme-receptor interactions.

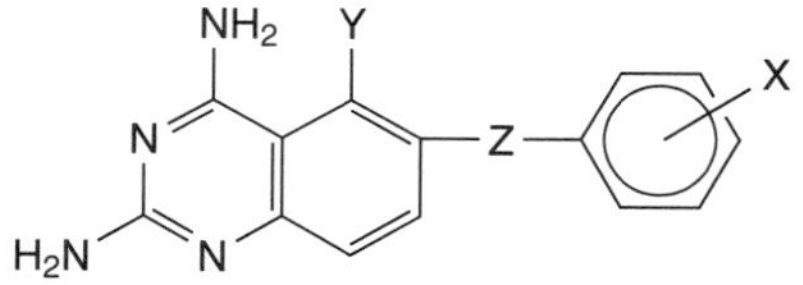

Figure 1.3. 2,4-Diamino, 5-Y, 6-Z-quinazolines.

5.1.8 Inhibition of Rat Liver DHFR by 2,4-Diamino, 5-Y, 6-Z-quinazolines (213)

Log $1/IC_{50}$

$$= 0.78(\pm 0.12)\pi_5$$

$$+ 0.81(\pm 0.12)MR_6$$

$$- 0.06(\pm 0.02)MR_6{}^2 \qquad (1.88)$$

$$- 0.73(\pm 0.49)I_1 - 2.15(\pm 0.38)I_2$$

$$- 0.54(\pm 0.21)I_3 - 1.40(\pm 0.41)I_4$$

$$+ 0.78(\pm 0.37)I_6$$

$$- 0.20(\pm 0.12)MR_6 \cdot I$$

$$+ 4.92(\pm 0.23)$$

$$n = 101, \quad r^2 = 0.924, \quad s = 0.441,$$

$$MR_{6,O} = 6.4(\pm 0.8)$$

5.1.9 Inhibition of Human Liver DHFR by 2,4-Diamino, 5-Y, 6-Z-quinazolines (214)

Log $1/K_i$

$$= -2.87(\pm 0.16)I_1$$

$$+ 0.29(\pm 0.14)I_2$$

$$\qquad\qquad (1.89)$$

$$- 0.38(\pm 0.11)MR_6$$

$$- 0.29(\pm 0.06)\pi_R$$

$$- 0.19(\pm 0.07)MR_R + 10.12(\pm 0.45)$$

$$n = 47, \quad r^2 = 0.914, \quad s = 0.420$$

5.1.10 Inhibition of Murine L1210 DHFR by 2,4-Diamino, 5-Y, 6-Z-quinazolines (214)

Log $1/IC_{50}$

$$= 0.49(\pm 0.11)I_2$$

$$\qquad\qquad (1.90)$$

$$- 1.23(\pm 0.25)I_3$$

$$- 0.30(\pm 0.07)MR_6$$

$$- 0.12(\pm 0.04)\pi_R + 9.36(\pm 0.27)$$

$$n = 24, \quad r^2 = 0.817, \quad s = 0.235$$

5.1.11 Inhibition of Bovine Liver DHFR by 2,4-Diamino, 5-Y, 6-Z-quinazolines (215)

$$Log\ 1/IC_{50} = 0.70(\pm 0.24)MR_6$$

$$+ 4.72(\pm 0.59) \qquad (1.91)$$

$$n = 11, \quad r^2 = 0.823, \quad s = 0.420$$

These QSAR vary in size and the number of variables used to define inhibitory activity. Selassie and Klein have described a more thorough comparative analysis of these QSAR (202). A brief focus on the MR_6 term reveals that its coefficients vary remarkably in all four sets. QSAR 1.88 is a parabola with an optimum of 6.4. Because it is parabolic in nature, the coefficient of the ascending slope cannot be compared with the linear slopes in QSAR 1.89–1.91. Figure 1.4 illustrates the problems with QSAR 1.89–1.91, which failed to test analogs across the available space.

Figure 1.4 reveals that QSAR 1.89 and 1.90 were sampled in the suboptimal MR_6 range; thus, the negative dependency on MR_6. On the other hand, QSAR 1.91 was focused on the ascending portion of the curve and thus only molecules in the 0.1–3.4 range were tested. Thus, with a limited set of compounds, one gets a misleading picture of the biological interactions.

Enzymatic reactions in nonaqueous solvents have generated a great deal of interest, fueled in part by the commercial application of enzymes as catalysts in specialty synthesis. The increasing demand for enantiopure pharmaceuticals has accelerated the study of enzymatic reactions in organic solvents containing

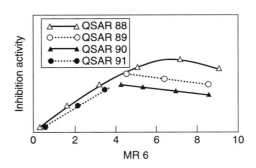

Figure 1.4. Gaps in spanned space of MR6 for 2,4-diamino-quinazolines.

little or no water (216). To investigate the substrate specificity of α-chymotrypsin in pentanol, a series of X-phenyl esters of N-benzoyl-L-alanine (Fig. 1.5) were synthesized and their binding constants were evaluated in buffer and in pentanol (203). The following QSAR 1.92 and 1.93 were derived in phosphate buffer and pentanol.

5.1.12 Binding of X-Phenyl, N-Benzoyl-L-alaninates to α-Chymotrypsin in Phosphate Buffer, pH 7.4 (203)

Log $1/K_M$

$$= 0.28(\pm 0.11)\pi + 0.51(\pm 0.24)\sigma^- \quad (1.92)$$

$$+ 0.38(\pm 0.23)MR + 3.70(\pm 0.24)$$

$$n = 16, \quad r^2 = 0.834, \quad s = 0.198$$

5.1.13 Binding of X-Phenyl, N-Benzoyl-L-alaninates to α-Chymotrypsin in Pentanol (203)

Log $1/K_M = 0.25(\pm 0.09)\pi$

$$+ 0.24(\pm 0.18)\sigma^- \quad (1.93)$$

$$+ 4.10(\pm 0.09)$$

$$n = 17, \quad r^2 = 0.762, \quad s = 0.156$$

Outliers in QSAR 1.92 included the 4-t-butyl and 4-OH analogs, whereas the 4-CONH$_2$ analog was an outlier in QSAR 1.93. These results were recently reanalyzed by Kim (217, 218) with respect to the role of enthalpic and entropic contributions to ligand binding with α-chymotrypsin. Use of the Fujiwara hydrophobic enthalpy parameter π_H and the hydrophobic entropy parameter π_S led to the development of QSAR 1.94 and 1.95 (219).

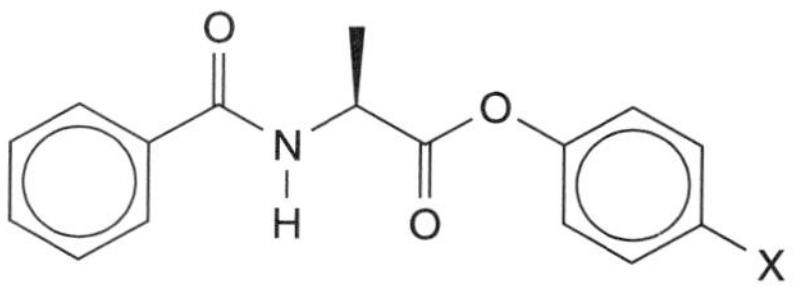

Figure 1.5. X-Phenyl, N-benzoyl-L-alaninates.

5.1.14 Binding of X-Phenyl, N-Benzoyl-L-alaninates in Aqueous Phosphate Buffer (218)

Log $1/K_M$

$$= 0.38(\pm 0.11)\pi_H + 0.19(\pm 0.07)\pi_S$$
$$+ 0.53(\pm 0.11)\sigma^- \quad (1.94)$$

$$+ 0.26(\pm 0.10)MR + 3.77(\pm 0.11)$$

$$n = 15, \quad r^2 = 0.806, \quad s = 0.200$$

5.1.15 Binding of X-Phenyl, N-Benzoyl-L-alaninates in Pentanol (218)

Log $1/K_M$

$$= 0.21(\pm 0.08)\pi_H + 0.31(\pm 0.05)\pi_S \quad (1.95)$$

$$+ 0.20(\pm 0.08)\sigma^- + 4.16(\pm 0.04)$$

$$n = 15, \quad r^2 = 0.787, \quad s = 0.160$$

The disappearance of the MR term in QSAR 1.93 and 1.95 is significant. The MR term usually relates to nonspecific, dispersive interactions in polar space. Thus, its presence in QSAR 1.92 and 1.94 suggests that substrates bearing polarizable substituents may displace the ordered-category II water molecules. In pentanol, the substrate may be faced with the task of displacing pentanol, not water, from the enzyme and thus the MR term is no longer of consequence. QSAR 1.94 also indicates that the enthalpy term π_H plays a more critical role in binding than the entropy term π_S. Note that these roles are reversed in QSAR 1.95, suggesting that binding in pentanol is largely an entropic-driven process. Similar results were obtained by Compadre et al. in a study on the hydrolysis of X-phenyl-N-benzoyl-glycinates by cathepsin B in aqueous buffer and acetonitrile (220). Kim's analysis provides an excellent example of a study that focuses on mechanistic interpretation and clearly demonstrates that a thermodynamic approach in QSAR can provide pertinent information about the energetics of the ligand binding process.

5α-Reductase, a critical enzyme in male sexual development, mediates the reduction of testosterone to dihydrotestosterone (DHT). Elevated levels of DHT in certain disease states such as benign prostatic hypertrophy and prostatic cancer drives the need for effective inhibitors of 5α-reductase. A recent QSAR study on inhibition of human 5α-reductase, type 1 by various steroid classes was carried out by Kurup et al. (204, 221, 222). A few of the models will be examined to demonstrate the importance and power of lateral validation. The three classes of steroidal inhibitors are depicted in Fig. 1.6.

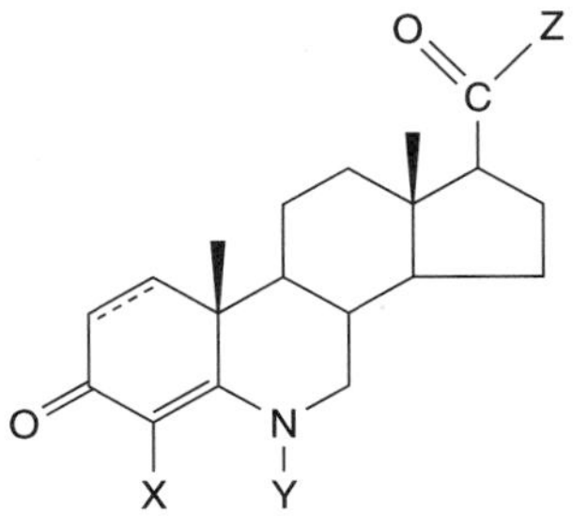

(I)

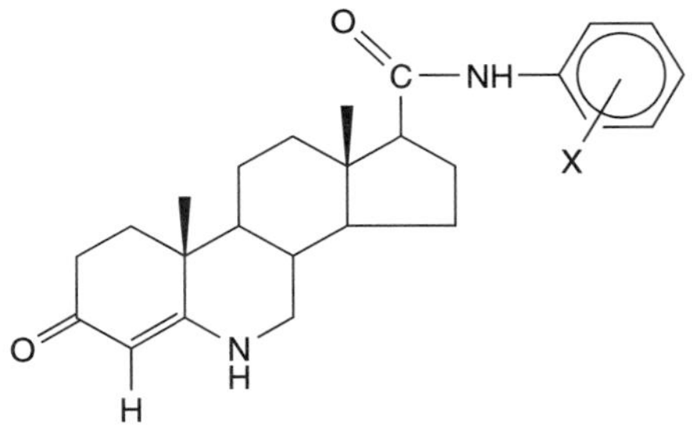

(II)

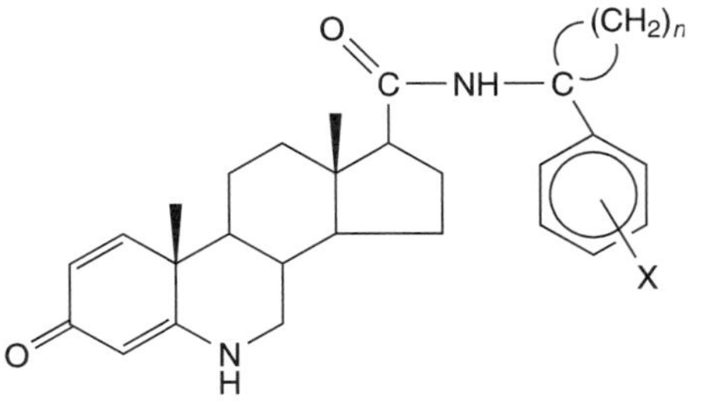

(III)

Figure 1.6. Steroidal inhibitors of 5α-reductase.

5.1.16 Inhibition of 5-α-Reductase by 4-X, N-Y-6-azaandrost-17-CO-Z-4-ene-3-ones, I

Log $1/K_i$

$$= 0.42(\pm 0.22)\mathrm{Clog}\,P$$
$$- 1.47(\pm 0.43)I - 0.32(\pm 0.30)L_Y \quad (1.96)$$
$$+ 6.88(\pm 0.13)$$

$$n = 21, \quad r^2 = 0.829, \quad s = 0.406$$

outliers: X = Y = H, Z = $NHCMe_3$;

X = Me, Y = H, Z = CH_2CHMe_2

5.1.17 Inhibition of 5-α-Reductase by 17β-(N-(X-phenyl)carbamoyl)-6-azaandrost-4-ene-3-ones, II

$$\mathrm{Log}\,1/K_i = 0.35(\pm 0.09)\mathrm{Clog}\,P$$
$$+ 0.26(\pm 0.11)\mathrm{B5}_{ortho} \quad (1.97)$$
$$+ 5.08(\pm 0.58)$$

$$n = 12, \quad r^2 = 0.942, \quad s = 0.154$$

outlier: 2,5-$(CF_3)_2$

5.1.18 Inhibition of 5-α-Reductase by 17β-(N-(1-X-phenyl-cycloalkyl)carbamoyl)-6-azaandrost-4-ene-3-ones, III

$$\mathrm{Log}\,1/K_i = 0.32(\pm 0.17)\mathrm{Clog}\,P$$
$$+ 6.34(\pm 1.15) \quad (1.98)$$
$$n = 5, \quad r^2 = 0.920, \quad s = 0.090$$

outlier: $n = 5$, X = 4-t-Bu

In all these equations, the coefficients with hydrophobicity as represented by Clog P, suggest that binding of these azaandrostene-ones occurs on the surface of the binding site where partial desolvation can occur. I is an indicator variable that pinpoints the negative effect of a double bond at C-1. A bulky substituent on N-6 is detrimental to activity, whereas a large substituent in the ortho position on the aromatic ring enhances activity (QSAR 1.97). The bulky ortho substituents (mostly t-Bu) may destroy coplanarity with the amide bridge by perhaps twisting of the phenyl ring and enhancing its hydrophobic contact with the binding site on the enzyme. Note that the larger intercept in QSAR 1.98 versus QSAR 1.97 suggests that hydrophobicity is more important in this area.

5.2 Interactions at the Cellular Level

QSAR analysis of studies at the cellular level allows us to get a handle on the physicochemical parameters critical to pharmacokinetics processes, mostly transport. Cell culture systems offer an ideal way to determine the optimum hydrophobicity of a system that is more complex than an isolated receptor. Extensive QSAR have been developed on the toxicity of 3-X-triazines to many mammalian and bacterial cell lines (202, 209). A comparison of the cytotoxicities of these analogs vs. sensitive murine leukemia cells (L1210/S) and methotrexate-resistant murine leukemia cells (L1210/R) reveals some startling differences.

5.2.1 Inhibition of Growth of L1210/S by 3-X-Triazines (209)

$\text{Log } 1/\text{IC}_{50}$

$$= 1.13(\pm 0.18)\pi$$
$$- 1.20(\pm 0.21)\log(\beta \cdot 10^{\pi} + 1)$$
$$+ 0.66(\pm 0.23)I_R \qquad (1.99)$$
$$- 0.32(\pm 0.17)I_{OR}$$
$$+ 0.94(\pm 0.37)\sigma + 6.72(\pm 0.13)$$

$$n = 61, \quad r^2 = 0.792, \quad s = 0.241,$$
$$\pi_O = 1.45(\pm 0.93) \quad \log \beta = -0.274$$

5.2.2 Inhibition of Growth of L1210/R by 3-X-Triazines (209)

$\text{Log } 1/\text{IC}_{50}$

$$= 0.42(\pm 0.05)\pi - 0.15(\pm 0.05)\text{MR} \quad (1.100)$$
$$+ 4.83(\pm 0.11)$$

$$n = 62, \quad r^2 = 0.885, \quad s = 0.220$$

There is a radical difference between these two QSAR. QSAR 1.99 is very similar to the one (QSAR 1.84) obtained versus the L1210

DHFR and it can be posited that the cytotoxicity in the sensitive cell line results from the inhibition of the enzyme. The intercepts suggest that slight interference with folate metabolism significantly affects growth. A comparison of the sensitive and resistant QSAR reveals a substantial difference in the coefficients with π. The lack of many variables in QSAR 1.100 and its overall simplicity suggests that inhibition of the enzyme is not the critical step, but rather transport to the site of action in these resistant cells may be of utmost importance. This particular cell line was resistant to methotrexate by virtue of elevated levels of DHFR and also overexpression of glycoprotein, GP-170 (209). Thus, modified transport through the dysfunctional membrane would severely curtail the partitioning process, resulting in a coefficient with π that is only one-half (0.42) of what is normally seen. The negative coefficient with the MR term indicates that size plays a role, albeit a negative one, in passage through the GP-170-fortified membrane and to the site of action.

The QSAR paradigm has been shown to be particularly useful in environmental toxicology, especially in acute toxicity determinations of xenobiotics (223). There has recently been an emphasis on "transparent, mechanistically comprehensive QSAR for toxicity," a move that is welcomed by many researchers in the field (224, 225). Cronin and Schultz developed QSAR 1.101 to describe the polar, narcotic toxicity of a large set of substituted phenols. A number of phenols with ionizable or reactive groups (e.g., —COOH, —NO$_2$, —NO, —NH$_2$, or —NHCOCH$_3$) were omitted from the final analysis (226).

5.2.3 Inhibition of Growth of *Tetrahymena pyriformis* (40 h)

$\text{Log } 1/C$

$$= 0.67(\pm 0.02)\text{Clog } P \qquad (1.101)$$
$$- 0.67(\pm 0.55)E_{\text{LUMO}} - 1.12$$
$$n = 120, \quad r^2 = 0.893, \quad s = 0.271$$

Using Hammett σ constants, Garg et al. rederived QSAR 1.102 for the same set and QSAR 1.103 and 1.104 for the diverse set of multi-, di-, and monophenols, which were se-

questered into two subsets containing electron-releasing and electron-attracting substituents, respectively (227).

5.2.4 Inhibition of Growth of *T. pyriformis* by Phenols (using σ) (227)

Log $1/C$

$$= 0.64(\pm 0.04)\text{Clog}\,P \qquad (1.102)$$

$$+ 0.61(\pm 0.12)\sigma + 1.84(\pm 0.13)$$

$$n = 119, \quad r^2 = 0.896, \quad s = 0.265$$

5.2.5 Inhibition of Growth of *T. pyriformis* by Electron-Releasing Phenols (227)

$$\text{Log}\,1/C = 0.66(\pm 0.05)\text{Clog}\,P$$
$$+ 1.63(\pm 0.15) \qquad (1.103)$$

$$n = 44, \quad r^2 = 0.946, \quad s = 0.182$$

5.2.6 Inhibition of Growth of *T. pyriformis* by Electron-Attracting Phenols (227)

$$\text{Log}\,1/C = 0.63(\pm 0.07)\text{Clog}\,P$$
$$+ 0.54(\pm 0.16)\sum\sigma \qquad (1.104)$$
$$+ 1.92(\pm 0.18)$$

$$n = 100, \quad r^2 = 0.836, \quad s = 0.327$$

There is excellent agreement between QSAR
1.101 and QSAR 1.104, in terms of the importance of hydrophobicity and electron demand of
the substituents: the coefficients with Clog$P$ are
similar and there is a good correspondence be-
tween E_{LUMO} and σ. Nevertheless, separation of
the phenols into subsets, based on their elec-
tronic attributes, indicates that different mechanisms of toxicity might be operative in this organism, a phenomenon that has been duplicated
in mammalian cells (228). In a recent extension
of toxicity studies on aromatics, Cronin and
Schultz used a two-parameter or response-sur-
face approach to define toxicity (229). In addition, indicator variables and group counts were
included to broaden the applicability of the approach. An excellent comparison of the different
modeling approaches (MLR, PLS, and Bayesian-
regularized neural networks) in QSAR is also
made (229).

5.2.7 Inhibition of Growth of *T. pyriformis* by Aromatic Compounds (229)

Log $1/\text{IgC}_{50}$

$$= 0.633\log P - 0.526E_{\text{LUMO}} \qquad (1.105)$$

$$+ 0.721I_{2,4\,\text{AP}} - 1.61I_{\text{strong acid}}$$

$$+ 0.314\sum\text{H-donor} - 1.39$$

$$n = 268, \quad r^2 = 0.780, \quad s = 0.393$$

The indicator variables $I_{2,4\,\text{AP}}$ and $I_{\text{strong acid}}$
suggest that 2- and 4-amino-substituted phe-
nols enhance toxicity, whereas strong acids
decrease toxicity, respectively. The H-bond
donor parameter may be correcting for the
added potency of amino phenols. The low r^2
may be attributed to inherent variability in
biological data and to the commingling of data
from four different studies. The wide variety
of compounds with different toxicity mechanisms, present in this combined study, would
also be a contributing factor to the low r^2.
Overall, this regression-based approach shows
adequate predictability and is transparent,
thus aiding in mechanistic interpretation.

5.3 Interactions *In Vivo*

The paucity of QSAR studies in whole animals
is understandable in terms of the costs, the
heterogeneity of the biological data, and the
complexity of the results. Nevertheless, in the
few studies that have been done, excellent
QSAR have been obtained, despite the small
number of subjects in the data set (164). One
particular example is insightful. The renal and
nonrenal clearance rates of a series of 11
β-blockers, including bufuralol, tolamolol,
propranolol, alprenolol, oxprenolol, acebutol,
timolol, metoprolol, prindolol, atenolol, and
nadolol were measured (230). The following
QSAR were formulated using those data (164).

5.3.1 Renal Clearance of β-Adrenoreceptor Antagonists

$$\text{Log}\,k = -0.42(\pm 0.12)\text{Clog}\,P$$
$$+ 2.35(\pm 0.24) \qquad (1.106)$$

$$n = 10, \quad r^2 = 0.888, \quad s = 0.185$$

5.3.2 Nonrenal Clearance of β-Adrenoreceptor Antagonists

$$\text{Log } k = 1.94(\pm 0.61)\text{Clog } P$$

$$- 2.00(\pm 0.80)\log(\beta \cdot P + 1) \quad (1.107)$$

$$+ 1.29(\pm 0.30)$$

$$n = 10, \quad r^2 = 0.950, \quad s = 0.168,$$

$$\text{Clog } P_O = 2.6 \pm 1.5 \quad \log \beta = -0.813$$

outlier: oxprenolol

It is apparent from QSAR 1.106 and 1.107, that the hydrophobic requirements of the substrates vary considerably. As expected, renal clearance is enhanced in the case of hydrophilic drugs, whereas nonrenal clearance shows a strong dependency on hydrophobicity. Note that QSAR 1.107 is stretching the limits of the bilinear model with only 10 data points! The 95% confidence intervals are also large but, nevertheless, the equations serve to emphasize the difference in clearance mechanisms that are clearly linked to hydrophobicity.

In formulating QSAR, it is useful to use a well-designed series to optimize a particular biological activity. It is also important to ensure that the ratio of compounds to parameters is 5, so that collinearity is minimized while spanned space is maximized. A normal distribution of biological data is necessary. A violation of these guidelines usually leads to statistically insignificant QSAR or models that defy predictability. One of our earliest works on the inhibition of *E. coli* DHFR by 2,4-diamino-5-X-benzyl pyrimidines led to the derivation of the following equation (231):

$$\text{Log } 1/K_i = -1.13\sigma_R + 5.54 \quad (1.108)$$

$$n = 10, \quad r^2 = 0.972, \quad s = 0.182$$

Most of the variance in these data was explained by the Hammett through-resonance constant (σ_R). It implied that electron-releasing substituents enhanced inhibitory potency. Later, expanded and extensive studies on this system revealed that inhibition of the bacterial enzyme was related to mostly

steric effects and there was no dependency on electronic terms. Careful analysis of the initial data revealed that it had a limited range in hydrophobicity and steric attributes. The lack of other QSAR to validate the findings in QSAR 1.108 made it statistically significant, at that time, but mechanistically weak. Most weaknesses in QSAR formulations usually violate the compound-to-parameter ratio rule (232, 233).

6 COMPARATIVE QSAR

6.1 Database Development

There are literally dozens of databases containing information about chemical structures, synthetic methods, and reaction mechanisms. The C-QSAR database is a database for QSAR models (164, 234). It was designed to organize QSAR data on physical (PHYS) organic reactions as well as chemical-biological (BIO) interactions, in numerical terms, to bring cohesion and understanding to mechanisms of chemical-biodynamics. The two databases are organized on a similar format, with the emphasis on reaction types in the PHYS database. The entries in the BIO database are sequestered into six main groups: macromolecules, enzymes, organelles, single-cell organisms, organs/tissues, and multicellular organisms (e.g., insects). The combined databases or the separate PHYS or BIO databases can be searched independently by a string search or searching using the SMILES notation. A SMILES search can be approached in three ways: one can identify every QSAR that contains a specific molecule, one can use a MERLIN search that locates all derivatives of a given structure, or one can search on single or multiple parameters. For a more thorough description of the C-QSAR database and ways to search it, see Hansch et al. (234) and Hansch et al. (164). The net result of searching the QSAR database is to "mine" for models; one could thus call it model-mining.

6.2 Database: Mining for Models

To enhance our understanding of ligand-receptor interactions and bring coherence to these relationships, there needs to be a con-

Table 1.5 Rho Values for Chemical and Biochemical Reactions

	Solvent	Radical Reagent	n	$\rho^+ (\sigma^+)$
	Hydrogen Abstraction from Unhindered Phenols			
1	CCl_4	$(CH_3)_3CO \cdot$	14	$-1.81 (\pm 0.77)$
2	Benzene	$(CH_3)_3CO \cdot$	12	$-0.82 (\pm 0.08)$
3	CCl_4	$(CH_3)_3CO \cdot$	5	$-0.82 (\pm 0.16)$
	X-phenols–Enzyme Systems			
1	Horseradish peroxidase	—	13	$-2.68 (\pm 0.78)$
2	Lactoperoxidase	—	11	$-1.34 (\pm 0.55)$

certed effort not only to develop high-quality regressions but also to create models that resonate with those drawn from mechanistic organic chemistry. A comprehensive, integrated database C-QSAR allows us to do so; it contains over 16,000 examples drawn from all facets of chemistry and biology. An example on the toxicity of X-phenols will illustrate the usefulness of this database (164, 228, 235–238). Recently, increasing numbers of QSAR for phenols have been based on Brown's σ^+ term, an electronic term that was first designed to rationalize electronic effects of substituents on electrophilic aromatic substitution. Studies conducted at EPA gave early indications that embryologic defects of rat embryos *in vitro* could be correlated by σ^+, as seen in QSAR 1.109109 (239).

6.2.1 Incidence of Tail Defects of Embryos (235)

$$Log\ 1/C = -0.58(\pm 0.21)\sigma^+$$
$$+ 3.51(\pm 0.14) \tag{1.109}$$
$$n = 10, \quad r^2 = 0.832, \quad s = 0.189$$

Soon, this parameter was shown to correlate radical reactions in chemistry as well as chemical-biological interactions in an extensive compilation (240). Another older study by Richard et al. on the inhibition of replicative DNA synthesis in Chinese hamster ovary cells was examined and led to the development of Equation 1.110 (241). Again, there was a dependency on σ^+.

6.2.2 Inhibition of DNA Synthesis in CHO Cells by X-Phenols (236)

$$Log\ 1/C = -0.74(\pm 0.34)\sigma^+$$
$$- 1.02(\pm 0.41)CMR \tag{1.110}$$
$$+ 6.97(\pm 1.16)$$
$$n = 9, \quad r^2 = 0.915, \quad s = 0.305$$

These Brown ρ^+ values were in line with those obtained from chemical and biological systems (228) see Table 1.5.

Cytotoxicity studies of X-phenols versus L1210 cells in culture led to an unusual result, which was baffling but reminiscent of Hammett plots related to changes in mechanism (228).

6.2.3 Inhibition of Growth of L1210 by X-Phenols

$$Log\ 1/IC_{50}$$
$$= -0.83(\pm 0.18)\sigma^+$$
$$+ 0.74(\pm 0.28)\sigma^{+2}$$
$$+ 0.56(\pm 0.15)\log P \tag{1.111}$$
$$- 0.45(\pm 0.21)\log(\beta \cdot P + 1)$$
$$+ 2.70(\pm 0.26)$$
$$n = 39, \quad r^2 = 0.913, \quad s = 0.229,$$
$$Log\ P_O = -0.18 \quad Log\ \beta = -2.28$$
$$\text{outliers: } 4\text{-}C_2H_5, 3\text{-}NH_2$$

Sequestering of the data into two subsets with

varying electronic attributes ($\sigma^+ > 0$ and $\sigma^+ < 0$) led to the derivation of the following equations.

6.2.4 Inhibition of Growth of L1210 by Electron-Withdrawing Substituents ($\sigma^+ > 0$)

$$\text{Log } 1/\text{IC}_{50} = 0.62(\pm0.16)\text{Log } P$$
$$+ 2.35(\pm0.31) \tag{1.112}$$

$$n = 15, \quad r^2 = 0.845, \quad s = 0.232,$$

outlier: 3-OH

6.2.5 Inhibition of Growth of L1210 by Electron-Donating Substituents ($\sigma^+ < 0$)

$$\text{Log } 1/\text{IC}_{50} = -1.58(\pm0.26)\sigma^+$$
$$+ 0.21(\pm0.06)\text{Log } P \tag{1.113}$$
$$+ 3.10(\pm0.24)$$

$$n = 23, \quad r^2 = 0.898, \quad s = 0.191,$$

outliers: 3-NH$_2$,4-NHAc

In QSAR 1.113, 62% of the variance is accounted for by σ^+ and 28% is explained by log P. It appears that free-radical-mediated toxicity is responsible for the growth-inhibitory effects of the phenols. Homolytic bond dissociation energies related to the homolytic cleavage of the OH bond in the following reaction: ($\text{X—C}_6\text{H}_4\text{OH} + \text{C}_6\text{H}_5\text{O} \cdot \rightarrow \text{X—C}_6\text{H}_4\text{O} \cdot + \text{C}_6\text{H}_5\text{OH}$) have been used in lieu of σ^+ values. The net result is similar, as seen in QSAR 1.114 (242).

$$\text{Log } 1/\text{IC}_{50} = -0.21(\pm0.03)\text{BDE}$$
$$+ 0.21(\pm0.04)\text{Log } P \tag{1.114}$$
$$+ 3.11(\pm0.17)$$

$$n = 52, \quad r^2 = 0.920, \quad s = 0.202,$$

outliers: 4-NHAc, 3-NH$_2$, 3-NMe$_2$

This data set contains a wide diversity of phenolic inhibitors, including a large number of ortho-substituted compounds, estrogenic phenols (β-estradiol, DES, nonyl phenol), and other antioxidants whose activities are well predicted by this model. The model suggests that cytotoxicity is an outcome of phenoxy radical formation and subsequent interaction with a relatively nonpolar receptor. The small hydrophobic coefficient suggests that DNA could be a likely target.

The appearance of the σ^+ parameter in a large number of reactions and interactions involving X-phenols indicates that the phenoxy radical can be a potent, reactive intermediate in myriad reactions. The availability of a fast, easily retrievable computerized database to corroborate this phenomenon was useful. This approach of lateral validation was crucial in establishing a QSAR model that was not only statistically significant but also mechanistically interpretable.

6.3 Progress in QSAR

The last four decades have seen major changes in the QSAR paradigm. In tandem with developments in molecular modeling and X-ray crystallography, it has impacted drug design and development in many ways. It has also spawned 3D QSAR approaches that are routinely used in computer-assisted molecular design. In terms of ligand design, it shares center stage with other approaches such as structure-based ligand design and other rational drug design approaches including docking methods and genetic algorithms (243). Success stories in QSAR have been recently reviewed (244, 245). Bioactive compounds have emerged in agrochemistry, pesticide chemistry, and medicinal chemistry.

Bifenthrin, a pesticide, was the product of a design strategy that used cluster analysis (244) (Fig. 1.7). Guided by QSAR analysis, the chemists at Kyorin Pharmaceutical Company designed and developed Norfloxacin, a 6-fluoro quinolone, which heralded the arrival of a new class of antibacterial agents (246) (Fig. 1.7). Two azole-containing fungicides, metconazole (Fig. 1.8) and ipconazole were launched in 1994 in France and Japan, respectively (247). Lomerizine, a 4-F-benzhydryl-4-(2,3,4-trimethoxy benzyl) piperazine, was introduced into the market in 1999 after extensive design strategies using QSAR (248) (Fig. 1.8). Flobufen, an anti-inflammatory agent was designed by Kuchar et al. as a long-acting agent without the usual gastric toxicity

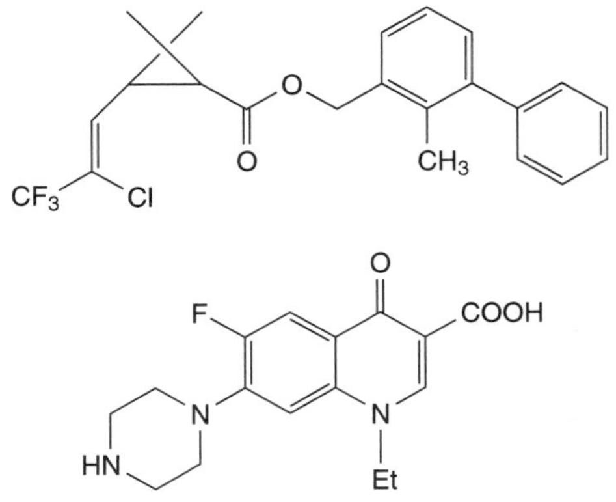

Figure 1.7. Bifenthrin and Norfloxacin.

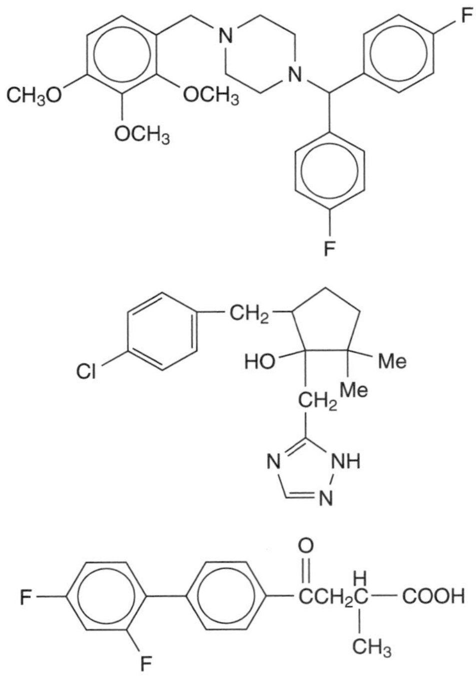

Figure 1.8. Lomerizine, Metconazole, and Flobufen.

(249) (Fig. 1.8). It is currently in clinical trial. Other examples of the commercial utility of QSAR include the development of metamitron and bromobutide (250). In most of these examples, QSAR was used in combination with other rational drug-design strategies, which is a useful and generally fruitful approach.

In addition to these commercial successes, the QSAR paradigm has steadily evolved into a science. It is empirical in nature and it seeks to bring coherence and rigor to the QSAR models that are developed. By comparing models one is able to more fully comprehend scientific phenomena with a "global" perspective; trends in patterns of reactivity or biological activity become self-evident.

7 SUMMARY

QSAR has done much to enhance our understanding of fundamental processes and phenomena in medicinal chemistry and drug design (251). The concept of hydrophobicity and its calculation has generated much knowledge and discussion as well as spawned a mini-industry. QSAR has refined our thinking on selectivity at the molecular and cellular level. Hydrophobic requirements vary considerably between tumor-sensitive cells and resistant ones. It has allowed us to design more selectivity into antibacterial agents that bind to dihydrofolate reductase. QSAR studies in the pharmacokinetic arena have established different hydrophobic requirements for renal/nonrenal clearance, whereas the optimum hy-

drophobicity for CNS penetration has been determined by Hansch et al. (252). QSAR has helped delineate allosteric effects in enzymes such as cyclooxygenase, trypsin, and in the well-defined and complex hemoglobin system (253, 254).

QSAR has matured over the last few decades in terms of the descriptors, models, methods of analysis, and choice of substituents and compounds. Embarking on a QSAR project may be a daunting and confusing task to a novice. However, there are many excellent reviews and tomes (1, 4, 19, 58–60) on this subject that can aid in the elucidation of the paradigm. Dealing with biological systems is not a simple problem and in attempting to develop a QSAR, one must always be cognizant of the biochemistry of the system analyzed and the limitations of the approach used.

REFERENCES

1. C. Hansch and A. Leo, *Substituent Constants for Correlation Analysis in Chemistry and Biology*, John Wiley & Sons, New York, 1979.

2. D. J. Livingstone, *J. Chem. Inf. Comput. Sci.*, **40**, 195 (2000).

3. C. Hansch, A. Kurup, R. Garg, and H. Gao, *Chem. Rev.*, **101**, 619 (2001).

4. H. Kubinyi in M. Wolff, Ed., *Burger's Medicinal Chemistry and Drug Discovery, Volume 1: Principles and Practice*, John Wiley & Sons, New York, 1995, p. 497.

5. A. Crum-Brown and T. R. Fraser, *Trans. R. Soc. Edinburgh*, **25**, 151 (1868).

6. C. Richet and C. R. Seancs, *Soc. Biol. Ses. Fil.*, **9**, 775 (1893).

7. H. Meyer, *Arch. Exp. Pathol. Pharmakol.*, **42**, 109 (1899).

8. E. Overton, *Studien Uber die Narkose*, Fischer, Jena, Germany, 1901.

9. J. Ferguson, *Proc. R. Soc. London Ser. B*, **127**, 387 (1939).

10. A. Albert, S. Rubbo, R. Goldacre, M. Darcy, and J. Stove, *Br. J. Exp. Pathol.*, **26**, 160 (1945).

11. A. Albert, *Selective Toxicity: The Physicochemical Bases of Therapy*, 7th ed., Chapman and Hall, London, 1985, p. 33.

12. P. H. Bell and R. O. Roblin, Jr. *J. Am. Chem. Soc.*, **64**, 2905 (1942).

13. L. P. Hammett, *Chem. Rev.*, **17**, 125 (1935).

14. L. P. Hammett, *Physical Organic Chemistry*, 2nd ed., McGraw-Hill, New York, 1970.

15. R. W. Taft, *J. Am. Chem. Soc.*, **74**, 3120 (1952).

16. C. Hansch, P. P. Maloney, T. Fujita, and R. M. Muir, *Nature*, **194**, 178 (1962).

17. R. Nelson Smith, C. Hansch, and M. M. Ames, *J. Pharm. Sci.*, **64**, 599 (1975).

18. T. Fujita, J. Iwasa, and C. Hansch, *J. Am. Chem. Soc.*, **86**, 5175 (1964).

19. C. Hansch and A. Leo in S. R. Heller, Ed., *Exploring QSAR. Fundamentals and Applications in Chemistry and Biology*, American Chemical Society, Washington, DC, 1995.

20. C. Hansch, *Acc. Chem. Res.*, **2**, 232 (1969).

21. H. Kubinyi, *Arzneim.-Forsch.*, **26**, 1991 (1976).

22. S. M. Free and J. W. Wilson, *J. Med. Chem.*, **7**, 395 (1964).

23. T. Fujita and T. Ban, *J. Med. Chem.*, **14**, 148 (1971).

24. G. Klopman, *J. Am. Chem. Soc.*, **106**, 7315 (1984).

25. B. W. Blake, K. Enslein, V. K. Gombar, and H. H. Borgstedt, *Mutat. Res.*, **241**, 261 (1990).

26. Z. Simon, *Angew. Chem. Int. Ed. Eng.*, **13**, 719 (1974).

27. L. H. Hall and L. B. Kier, *J. Pharm. Sci.*, **66**, 642 (1977).

28. L. B. Kier and L. H. Hall, *Molecular Structure Description. The Electrotopological State*, Academic Press, San Diego, CA, 1999.

29. W. Tong, D. R. Lowis, R. Perkins, Y. Chen, W. J. Welsh, D. W. Goddette, T. W. Heritage, and D. M. Sleehan, *J. Chem. Inf. Comput Sci.*, **38**, 669 (1998).

30. S. J. Cho, W. Zheng, and A. Tropsha, *Pac. Symp. Biocomput.*, 305 (1998).

31. H. Gao and J. Bajorath, *J. Mol. Diversity*, **4**, 115 (1999).

32. H. Gao, C. Williams, P. Labute, and J. Bajorath, *J. Chem. Inf. Comput. Sci.*, **39**, 164 (1999).

33. W. J. DunnIII, S. Wold, U. Edlund, S. Hellberg, and J. Gasteeger, *Quant. Struct.-Act. Relat.*, **3**, 131 (1984).

34. J. Langley, *J. Physiol.*, **1**, 367 (1878).

35. P. Ehrlich, *Klin. Jahr.*, **6**, 299 (1897).

36. J. N. Langley, *J. Physiol.*, **33**, 374 (1905).

37. M. Famulok, *Curr. Opin. Struct. Biol.*, **9**, 324 (1999).

38. K. Y. Wang, S. Swaminathan, and P. H. Bolton, *Biochemistry*, **33**, 7617 (1994).

39. J. W. Lown in S. Neidle and M.-J. Waring, Eds., *Molecular Aspects of Anticancer Drug-DNA Interactions*, Macmillan, Basinstoke, UK, 1993, p. 322.

40. L. Morgenstern, M. Recanatini, T. E. Klein, W. Steinmetz, C. Z. Yang, R. Langridge, and C. Hansch, *J. Biol. Chem.*, **262**, 10767 (1987).

41. R. N. Smith, C. Hansch, K. H. Kim, B. Omiya, G. Fukumura, C. D. Selassie, P. Y. C. Jow, J. M. Blaney, and R. Langridge, *Arch. Biochem. Biophys.*, **215**, 319 (1982).

42. C. Hansch, T. Klein, J. McClarin, R. Langridge, and N. W. Cornell, *J. Med. Chem.*, **29**, 615 (1986).

43. C. D. Selassie, Z. X. Fang, R. Li, C. Hansch, T. Klein, R. Langridge, and B. T. Kaufman, *J. Med. Chem.*, **29**, 621 (1986).

44. J. M. Blaney and C. Hansch in C. A. Ramsden, Ed., *Comprehensive Medicinal Chemistry. The Rational Design, Mechanistic Study and Therapeutic Application of Chemical Compounds, Vol. 4, Quantitative Drug Design*, Pergamon, Elmsford, NY, 1990, p. 459.

45. G. C. K. Roberts, *Pharmacochem. Libr.*, **6**, 91 (1983).

46. A. A. Kumar, J. H. Mangum, D. T. Blankenship, and J. H. Freisheim, *J. Biol. Chem.*, **256**, 8970 (1981).

47. G. D. Rose and R. Wolfenden, *Annu. Rev. Biophys. Biomol. Struct.*, **22**, 381 (1993).

48. A. T. Hagler, P. Dauber, and S. Lifson, *J. Am. Chem. Soc.*, **101**, 5131 (1979).

49. W. Kauzmann, *Adv. Protein Chem.*, **14**, 1 (1959).

50. A. Ben-Naim, *Pure Appl. Chem.*, **69**, 2239 (1997).

51. W. Blokzijl and J. B. F. N. Engberts, *Angew. Chem. Int. Ed. Engl.*, **32**, 1545 (1993).

52. N. Muller, *Acc. Chem. Res.*, **23**, 23 (1990).

53. F. Eisenhaber, *Perspect. Drug Discov. Des.*, **17**, 27 (1999).

54. A. R. Fersht, J. S. Shindler, and W. C. Tsui, *Biochemistry*, **19**, 5520 (1980).

55. P. R. Andrews, D. J. Craik, and J. L. Matin, *J. Med. Chem.*, **27**, 1648 (1984).

56. N. R. Draper and H. Smith, *Applied Regression Analysis*, 2nd ed., John Wiley & Sons, New York, 1981.

57. Y. Martin in G. Grunewald, Ed., *Quantitative Drug Design*, Marcel Dekker, New York, 1978, p. 167.

58. H. Kubinyi in R. Mannhold, P. Krogsgaard-Larsen, and H. Timmerman, Eds., *QSAR: Hansch Analysis and Related Approaches*, VCH, New York, 1993, p. 91.

59. R. Franke in W. Th. Nauta and R. F. Rekker, Eds., *Theoretical Drug Design Methods*, Elsevier Science, Amsterdam/New York, 1983, p. 395.

60. C. Hansch in C. J. Cavallito, Ed., *Structure Activity Relationships*, Vol. **1**, Pergamon, Oxford, UK, 1973, p. 75.

61. J. K. Seydel, *Int. J. Quantum Chem.*, **20**, 131 (1981).

62. J. G. Topliss and R. P. Edwards, *J. Med. Chem.*, **22**, 1238 (1979).

63. P. N. Craig, *J. Med. Chem.*, **14**, 680 (1971).

64. J. G. Topliss, *J. Med. Chem.*, **15**, 1006 (1972).

65. J. G. Topliss, *J. Med. Chem.*, **20**, 463 (1977).

66. T. M. Bustard, *J. Med. Chem.*, **17**, 777 (1974).

67. F. Darvas, *J. Med. Chem.*, **17**, 799 (1974).

68. P. S. Magee in J. Miyamoto and P. C. Kearney, Eds., *Pesticide Chemistry: Human Welfare and Environment, Proceedings of the International Congress on Pesticide Chemistry*, Vol. **1**, Pergamon, Oxford, UK, 1983, p. 251.

69. T. J. Mitchell, *Technometrics*, **16**, 203 (1974).

70. T. Moon, M. H. Chi, D. H. Kim, C. N. Yoon, and Y. S. Choi, *Quant. Struct.-Act. Relat.*, **19**, 257 (2000).

71. M. Baroni, S. Clementi, G. Cruciani, N. Kettaneh-Wold, and S. Wold, *Quant. Struct.-Act. Relat.*, **12**, 225 (1993).

72. M. Sjostrom and L. Eriksson in H. van de Waterbeemd, Ed., *Chemometric Methods in Molecular Design*, VCH, Weinheim, Germany, 1995, p. 63.

73. L. Eriksson, E. Johansson, M. Muller, and S. Wold, *Quant. Struct.-Act. Relat.*, **16**, 383 (1997).

74. L. Eriksson, E. Johansson, M. Muller, and S. Wold, *J. Chemom.*, **14**, 599 (2000).

75. C. Hansch and T. Fujita, *J. Am. Chem. Soc.*, **86**, 1616 (1964).

76. C-QSAR Database, BioByte Corp., Claremont, CA.

77. G. N. Burckhardt, W. G. K. Ford, and E. Singelton, *J. Chem. Soc.*, **17** (1936).

78. L. P. Hammett, *J. Chem. Ed.*, **43**, 464 (1966).

79. M. Charton, *Prog. Phys. Org. Chem.*, **8**, 235 (1971).

80. T. Fujita and T. Nishioka, *Prog. Phys. Org. Chem.*, **12**, 49 (1976).

81. P. D. Bolton, K. A. Fleming, and F. M. Hall, *J. Am. Chem. Soc.*, **94**, 1033 (1972).

82. K. Kalfus, J. Kroupa, M. Vecera, and O. Exner, *Collect. Czech. Chem. Commun.*, **40**, 3009 (1975).

83. M. Bergon and J. P. Calmon, *Tetrahedron Lett.*, **22**, 937 (1981).

84. J. Schreck, *J. Chem. Ed.*, **48**, 103 (1971).

85. H. C. Brown and Y. Okamoto, *J. Am. Chem. Soc.*, **80**, 4979 (1958).

86. Y. Tsuno, T. Ibata, and Y. Yukawa, *Bull. Chem. Soc. Jpn.*, **32**, 960, 965, 971 (1959).

87. J. D. Roberts and W. T. Moreland, *J. Am. Chem. Soc.*, **75**, 2167 (1953).

88. K. Bowden in C. A. Ramsden, Ed., *Comprehensive Medicinal Chemistry. The Rational Design, Mechanistic Study and Therapeutic Application of Chemical Compounds*, Vol. 4: *Quantitative Drug Design*, Pergamon, Elmsford, NY, 1990, p. 212.

89. A. Albert, *Selective Toxicity: The Physicochemical Bases of Therapy*, 7th ed., Chapman and Hall, London, 1985, p. 379.

90. M. Karelson, V. S. Lobanov, and A. R. Katritzky, *Chem. Rev.*, **96**, 1027 (1996).

91. P. S. Magee in *ACS Symposium Series 37*, American Chemical Society, Washington, DC, 1980.

92. S. P. Gupta, *Chem. Rev.*, **91**, 1109 (1991).

93. J. J. Sullivan, A. D. Jones, and K. K. Tangi, *J. Chem. Int. Comput. Sci.*, **40**, 1113 (2000).

94. M. Cocchi, M. C. Menziani, F. Fanelli, P. G. Debenedetti, *J. Mol. Struct.*, **331**, 79 (1995).

95. M. Cocchi, M. Menziani, P. G. Debenedetti, A. Cruciani, *Chemom. Intell. Lab. Sys.*, **14**, 209 (1992).

96. J. H. Hildebrand, *Proc. Natl. Acad. Sci. USA*, **76**, 194 (1979).

97. G. D. Rose, A. R. Geselowitz, G. J. Lesser, R. H. Lee, and M. H. Zehfus, *Science*, **229**, 834 (1985).

98. H. J. Schneider, *Angew. Chem. Int. Ed. Engl.*, **30**, 1417 (1991).

99. J. N. Israelachvili and H. Wennerstrom, *J. Phys. Chem.*, **96**, 520 (1992).

100. J. J. H. Nusselder and J. B. F. N. Engberts, *Langmuir*, **7**, 2089 (1991).

101. P. J. Taylor in C. A. Ramsden, Ed., *Comprehensive Medicinal Chemistry. The Rational Design, Mechanistic Study and Therapeutic Application of Chemical Compounds, Vol. 4, Quantitative Drug Design*, Pergamon, Elmsford, NY, 1990, p. 241.

102. J. H. Hildebrand, *J. Phys. Chem.*, **72**, 1841 (1969).

103. H. S. Frank and M. W. Evans, *J. Chem. Phys.*, **13**, 507 (1945).

104. G. Nemethy and H. A. Scheraga, *J. Chem. Phys.*, **36**, 3382 (1962).

105. A. D. J. Haymet, K. A. T. Silverstin, and K. A. Dill, *Faraday Discuss.*, **103**, 117 (1996).

106. K. A. T. Silverstein, K. A. Dill, and A. D. J. Haymet, *J. Chem. Phys.*, **114**, 6303 (2001).

107. A. J. Leo and C. Hansch, *Perspect. Drug Discov. Des.*, **17**, 1 (1999).

108. R. N. Smith, C. Hansch, and M. A. Ames, *J. Pharm. Sci.*, **64**, 599 (1975).

109. A. Leo and C. Hansch, *J. Org. Chem.*, **36**, 1539 (1971).

110. B. C. Lippold and M. S. Adel, *Arch. Pharm.*, **305**, 417 (1972).

111. S. E. Debolt and P. A. Kollman, *J. Am. Chem. Soc.*, **117**, 5316 (1995).

112. A. Leo, *J. Pharm. Sci.*, **76**, 166 (1987).

113. A. Leo, *Methods Enzymol.*, **202**, 544 (1991).

114. J. de Bruijn and J. Hermens, *Quant. Struct.-Act. Relat.*, **9**, 11 (1990).

115. E. Tomlinson, S. S. David, G. D. Parr, M. James, N. Farraj, J. F. M. Kinkel, D. Gaisser, and H. J. Wynn in W. J. Dunn III, J. H. Block, and R. S. Pearlman, Eds., *Partition Coefficient,* *Determination and Estimation*, Pergamon, Oxford, UK, 1986, p. 83.

116. R. Collander, *Acta Chem. Scand.*, **5**, 774 (1951).

117. A. Leo, C. Hansch, and D. Elkins, *Chem. Rev.*, **71**, 525 (1971).

118. R. F. Rekker, *The Hydrophobic Fragmented Constant. Its Derivation and Application: A Means of Characterizing Membrane Systems*, Elsevier, Amsterdam, 1977, p. 131.

119. P. Seiler, *Eur. J. Med. Chem.*, **9**, 473 (1974).

120. T. Fujita, T. Nishioka, and M. Nakajima, *J. Med. Chem.*, **20**, 1071 (1977).

121. D. E. Leahy, P. J. Taylor, and A. R. Wait, *Quant. Struct.-Act. Relat.*, **8**, 17 (1989).

122. J. C. Dearden, A. M. Patel, and J. M. Thubby, *J. Pharm. Pharmacol.*, **26** (Suppl.), 75P (1974).

123. W. Draber, K. H. Buchel, and K. Dickore, *Proc. Int. Congr. Pest. Chem., 2nd ed., 1971*, **5**, 153 (1972).

124. P. Vallat, N. El Tayar, B. Testa, I. Slacanin, A. Martson, and K. Hostettmann, *J. Chromatogr.*, **504**, 411 (1990).

125. A. Berthod, Y. I. Han, and D. W. Armstrong, *J. Liq. Chromatogr.*, **11**, 1441 (1988).

126. A. Berthod, S. Carola-Broch, and M. C. Garcia-Alvarex-Cogne, *Anal. Chem.*, **71**, 879 (1999).

127. K. Valko, C. Beran, and D. Reynolds, *Anal. Chem.*, **69**, 2022 (1997).

128. K. Valko, C. M. Du, C. Bevan, D. P. Reynolds, and M. H. Abraham, *Curr. Med. Chem.*, **8**, 1137 (2001).

129. F. Lombardo, M. Y. Shalaeva, K. A. Tupper, F. Gao, and M. H. Abraham, *J. Med. Chem.*, **43**, 2922 (2000).

130. J. L. Fauchére and V. Pliska, *Eur. J. Med. Chem.*, **18**, 369 (1983).

131. J. L. Fauchére in B. Testa, Ed., *Advances in Drug Research*, Vol. **15**, Academic Press, London/New York, 1986, p. 29.

132. M. Akamatsu, Y. Yoshida, H. Nakamura, M. Asao, H. Iwamura, and T. Fujita, *Quant. Struct.-Act. Relat.*, **8**, 195 (1989).

133. C. Hansch, A. Leo, and D. Hoekman in S. R. Heller, Ed., *Exploring QSAR: Hydrophobic, Electronic and Steric Constants*, Vol. **2**, American Chemical Society Professional Reference Book, Washington, DC, 1995.

134. G. G. Nys and R. F. Rekker, *Chim. Ther.*, **8**, 521 (1973).

135. A. Leo, P. Y. C. Jow, C. Silipo, and C. Hansch, *J. Med. Chem.*, **14**, 865 (1979).

136. D. Weininger, *J. Chem. Int. Comput. Sci.*, **28**, 31 (1988).

137. D. Weininger, A. Weininger, and J. L. Weininger, *J. Chem. Int. Comput. Sci.*, **29**, 97 (1989).

138. A. Leo in C. A. Ramsden, Ed., *Comprehensive Medicinal Chemistry. The Rational Design, Mechanistic Study and Therapeutic Application of Chemical Compounds, Vol. 4, Quantitative Drug Design*, Pergamon, Elmsford, NY, 1990, p. 315.

139. A. Leo, personal communication.

140. A. Leo, *Chem. Rev.*, **93**, 1281 (1993).

141. A. J. Leo and D. Hoekman, *Perspect. Drug Discov. Des.*, **18**, 19 (2000).

142. H. van de Waterbeemd and R. Mannhold, *Quant. Struct.-Act. Relat.*, **15**, 410 (1996).

143. R. Mannhold and H. van de Waterbeemd, *J. Comput.-Aided Mol. Des.*, **15**, 337 (2001).

144. R. F. Rekker and H. M. DeKort, *Eur. J. Med. Chem.*, **14**, 479 (1979).

145. G. Klopman, J. W. Li, S. Wang, and M. Dimayuga, *J. Chem. Inf. Comput. Sci.*, **34**, 752 (1994).

146. A. K. Ghose and G. M. Crippen, *J. Med. Chem.*, **28**, 333 (1985).

147. T. Suzuki and Y. Kudo, *J. Comput.-Aided Mol. Des.*, **4**, 155 (1990).

148. I. Moriguchi, S. Hirono, Q. Liu, I. Nakagome, and Y. Matsushita, *Chem. Pharm. Bull.*, **40**, 127 (1992).

149. G. E. Kellogg, G. J. Joshi, and D. J. Abraham, *J. Med. Chem. Res.*, **1**, 444 (1992).

150. J. Devillers, D. Domine, C. Guillon, and W. J. Karcher, *J. Pharm. Sci.*, **87**, 1086 (1998).

151. M. J. Kamlet, P. W. Carr, R. W. Taft, and M. H. Abraham, *J. Am. Chem. Soc.*, **103**, 6062 (1981).

152. M. J. Kamlet, J. L. Abboud, M. Abraham, and R. Taft, *J. Org. Chem.*, **48**, 2877 (1983).

153. J. A. Platts, D. Butina, M. H. Abraham, and A. Hersey, *J. Chem. Inf. Comput Sci.*, **39**, 835 (1999).

154. Y. Ishihama and N. Asakawa, *J. Pharm. Sci.*, **88**, 1305 (1999).

155. J. A. Platts, M. H. Abraham, D. Butina, and A. Hersey, *J. Chem. Inf. Comput. Sci.*, **40**, 71 (2000).

156. A. J. Leo, *J. Pharm. Sci.*, **89**, 1567 (2000).

157. R. W. Taft in M. S. Newman, Ed., *Steric Effects in Organic Chemistry*, John Wiley & Sons, New York, 1956, p. 556.

158. K. Hancock, E. A. Meyers, and B. J. Yager, *J. Am. Chem. Soc.*, **83**, 4211 (1961).

159. M. Charton in M. Charton and I. Motoc, Eds., *Steric Effects in Drug Design*, Springer, Berlin, 1983, p. 57.

160. M. S. Tute in C. A. Ramsden, Ed., *Comprehensive Medicinal Chemistry. The Rational Design, Mechanistic Study and Therapeutic Application of Chemical Compounds, Vol. 4, Quantitative Drug Design*, Pergamon, Elmsford, NY, 1990, p. 18.

161. C. Hansch and T. Klein, *Acc. Chem. Res.*, **19**, 392 (1986).

162. A. Verloop, W. Hoogenstraaten, and J. Tipker in E. J. Ariens, Ed., *Drug Design*, Vol. **VII**, Academic Press, New York/London, 1976, p. 165.

163. A. Verloop, *The STERIMOL Approach to Drug Design*, Marcel Dekker, New York, 1987.

164. C. Hansch, D. Hoekman, A. Leo, D. Weininger, and C. D. Selassie, unpublished results.

165. V. A. Levin, *J. Med. Chem.*, **23**, 682 (1980).

166. E. J. Lien and P. H. Wang, *J. Pharm. Sci.*, **69**, 648 (1980).

167. C. D. Selassie, C. Hansch, and T. Khwaja, *J. Med. Chem.*, **33**, 1914 (1990).

168. E. J. Lien, L. L. Lien, and H. Gao in F. Sanz, J. Guiraldo, and F. Manaut, Eds., *QSAR and Molecular Modelling: Concepts, Computational Tools and Biological Applications*, Prous Science, Barcelona/Philadelphia, 1995, p. 94.

169. C. Selassie, unpublished results.

170. M. Recanatini, T. Klein, C. Z. Yang, J. McClarin, R. Langridge, and C. Hansch, *Mol. Pharmacol.*, **29**, 436 (1986).

171. Y. Naito, M. Sugiura, Y. Yamamura, C. Fukaya, K. Yokoyama, Y. Nakagawa, T. Ikeda, M. Senda, and T. Fujita, *Chem. Pharm. Bull.*, **39**, 1736 (1991).

172. A. K. Debnath, R. L. L. de Compadre, G. Debnath, A. J. Shusterman, and C. Hansch, *J. Med. Chem.*, **34**, 786 (1991).

173. M. Randic, *J. Am. Chem. Soc.*, **97**, 6609 (1975).

174. L. B. Kier and L. H. Hall, *Molecular Connectivity in Chemistry and Drug Research*, Academic Press, New York/London, 1976.

175. L. B. Kier and M. H. Hall, *J. Pharm. Sci.*, **72**, 1170 (1983).

176. L. H. Hall and L. B. Kier, *J. Pharm. Sci.*, **64**, 1978 (1975).

177. J. Gough and L. H. Hall, *J. Chem. Inf. Comput. Sci.*, **39**, 356 (1999).

178. J. K. Boulamwini, K. Raghavan, M. Fresen, Y. Pommier, K. Kohn, and J. Weinstein, *Pharm. Res.*, **13**, 1892 (1995).

179. V. E. F. Heinzen, V. Cechinel, and R. A. Yunes, *Farmaco*, **54**, 125 (1999).

180. R. L. Lopez de Compadre, C. M. Compadre, R. Castillo, and W. J. DunnIII, *Eur. J. Med. Chem.*, **18**, 569 (1983).

181. H. Kubinyi, *Quant. Struct.-Act. Relat.*, **14**, 149 (1995).

182. P. A. J. Janssen and N. B. Eddy, *J. Med. Pharm. Chem.*, **2**, 31 (1960).

183. T. Fujita in C. A. Ramsden, Ed., *Comprehensive Medicinal Chemistry. The Rational Design, Mechanistic Study and Therapeutic Application of Chemical Compounds, Vol. 4, Quantitative Drug Design*, Pergamon, Elmsford, NY, 1990, p. 503.

184. C. Hansch, D. Kim, A. J. Leo, E. Novellino, C. Silipo, and A. Vittoria, *CRC Crit. Rev. Toxicol.*, **19**, 185 (1989).

185. C. Hansch and W. J. DunnIII, *J. Pharm. Sci.*, **61**, 1 (1972).

186. T. W. Schultz and M. Tichy, *Bull. Environ. Contam. Toxicol.*, **51**, 681 (1993).

187. J. T. Penniston, L. Beckett, D. L. Bentley, and C. Hansch, *Mol. Pharmacol.*, **5**, 333 (1969).

188. C. Hansch, *Adv. Chem. Ser.*, **114**, 20 (1972).

189. C. Hansch and J. M. Clayton, *J. Pharm. Sci.*, **62**, 1 (1973).

190. R. Franke and W. Schmidt, *Acta Biol. Med. Germ.*, **31**, 273 (1973).

191. J. McFarland, *J. Med. Chem.*, **13**, 1192 (1970).

192. H. Kubinyi and O. H. Kehrhahn, *Arzneim.-Forsch.*, **28**, 598 (1978).

193. H. Kubinyi, *Arzneim.-Forsch.*, **29**, 1067 (1979).

194. R. Franke in W. Th. Nauta and R. F. Rekker, Eds., *Theoretical Drug Design Methods*, Elsevier, New York, 1984, p. 256.

195. H. Kubinyi in C. A. Ramsden, Ed., *Comprehensive Medicinal Chemistry. The Rational Design, Mechanistic Study and Therapeutic Application of Chemical Compounds, Vol. 4, Quantitative Drug Design*, Pergamon, Elmsford, NY, 1990, p. 539.

196. C. John Blankley in J. G. Topliss, Ed., *Quantitative Structure Activity Relationships of Drugs*, Academic Press, New York, 1983, p. 5.

197. E. Yalcin, S. E. Sener, I. Owren, and O. Temiz in E. Sanz, J. Giraldo, and F. Manaut, Eds., *QSAR and Molecular Modelling: Concepts, Computational Tools and Biological Applications*, Prous Science, Barcelona/Philadelphia, 1995, p. 147.

198. J. Kunes, J. Jachym, P. Tirasko, Z. Odlerova, and K. Waisser, *Collect. Czech. Chem. Commun.*, **62**, 1503 (1997).

199. Y. Terada and K. Naya, *Pharmazie*, **55**, 133 (2000).

200. S. P. Gupta and A. Paleti, *Bioorg. Med. Chem.*, **6**, 2213 (1998).

201. C. Tmej, P. Chiba, M. Huber, E. Richter, M. Hitzler, K. J. Schaper, and G. Ecker, *Arch. Pharm.*, **331**, 233 (1998).

202. C. Selassie and T. E. Klein in J. Devillers, Ed., *Comparative QSAR*, Taylor & Francis, Washington, DC, 1998, p. 235.

203. C. D. Selassie, W. X. Gan, M. Fung, and R. Shortle in F. Sanz, J. Giraldo, and F. Manaut, Eds., *QSAR and Molecular Modelling: Concepts, Computational Tools and Biological Applications*, Prous Science, Barcelona/Philadelphia, 1995, p. 128.

204. A. Kurup, R. Garg, and C. Hansch, *Chem. Rev.*, **100**, 909 (2000).

205. J. M. Blaney, C. Hansch, C. Silipo, and A. Vittorio, *Chem. Rev.*, **84**, 333 (2000).

206. C. Hansch, *Ann. N. Y. Acad. Sci.*, **186**, 235 (1971).

207. C. Hansch, B. A. Hathaway, Z. R. Guo, C. D. Selassie, S. W. Dietrich, J. M. Blaney, R. Langridge, K. W. Volz, and B. T. Kaufman, *J. Med. Chem.*, **27**, 129 (1984).

208. B. A. Hathaway, Z. R. Guo, C. Hansch, T. J. Delcamp, S. S. Susten, and J. H. Freisheim, *J. Med. Chem.*, **27**, 144 (1984).

209. C. D. Selassie, C. D. Strong, C. Hansch, T. Delcamp, J. H. Freisheim, and T. A. Khwaja, *Cancer Res.*, **46**, 744 (1986).

210. C. K. Marlowe, C. D. Selassie, and D. V. Santi, *J. Med. Chem.*, **38**, 967 (1995).

211. R. G. Booth, C. D. Selassie, C. Hansch, and D. V. Santi, *J. Med. Chem.*, **30**, 1218 (1987).

212. L. C. Chio and S. F. Queener, *Antimicrob. Agents Chemother.*, **37**, 1916 (1993).

213. J. Y. Fukunaga, C. Hansch, and E. E. Stellar, *J. Med. Chem.*, **19**, 605 (1976).

214. B. K. Chen, C. Horvath, and J. R. Bertino, *J. Med. Chem.*, **22**, 483 (1979).

215. N. V. Harris, C. Smith, and K. Bowden, *Eur. J. Med. Chem.*, **27**, 7 (1992).

216. A. M. Klibanov, *Nature*, **409**, 241 (2001).

217. K. H. Kim, *J. Comput.-Aided Mol. Des.*, **15**, 367 (2001).

218. K. H. Kim, *Bioorg. Med. Chem.*, **9**, 1951 (2001).

219. K. Nakamura, K. Hayashi, I. Ueda, and H. Fujiwara, *Chem. Pharm. Bull.*, **43**, 369 (1995).

220. C. M. Compadre, R. J. Sanchez, C. Bhurane-swaran, R. L. Compadre, D. Plunkett, and S. G. Novick in C. G. Wermuth, Ed., *Trends in QSAR and Molecular Modelling*, Escom, Strasbourg, France, 1993, p. 112.

221. S. V. Frye, C. D. Haffner, P. R. Maloney, R. A. Mook, Jr., G. F. Dorsey, R. N. Hiner, C. M. Cribbs, T. N. Wheeler, J. A. Ray, R. C. Andrews, K. W. Batchelor, H. N. Branson, J. D. Stuart, S. L. Schwiker, J. Van Arnold, S. Croom, D. M. Bickett, M. L. Moss, G. Tian, R. J. Unwalla, F. W. Lee, T. K. Tippin, M. K. James, M. K. Grizzle, J. E. Long, and S. V. Schuster, *J. Med. Chem.*, **37**, 2352 (1994).

222. S. V. Frye, C. D. Haffner, P. R. Maloney, R. N. Hiner, G. F. Dorsey, R. A. Roe, R. J. Unwalla, K. W. Batchelor, H. N. Branson, J. D. Stuart, S. L. Schwiker, J. Van Arnold, D. M. Bickett, M. L. Moss, G. Tian, F. W. Lee, T. K. Tippin, M. K. James, M. K. Grizzle, J. E. Long, and D. K. Croom, *J. Med. Chem.*, **38**, 2621 (1995).

223. M. T. D. Cronin and J. C. Dearden, *Quant. Struct.-Act. Relat.*, **14**, 518 (1995).

224. M. T. D. Cronin, B. W. Gregory, and T. W. Schultz, *Chem. Res. Toxicol.*, **11**, 902 (1998).

225. T. W. Schultz, *Chem. Res. Toxicol.*, **12**, 1262 (1999).

226. M. T. D. Cronin and T. W. Schultz, *Chemosphere*, **32**, 1453 (1996).

227. R. Garg, A. Kurup, and C. Hansch, *Crit. Rev. Toxicol.*, **31**, 223 (2001).

228. C. D. Selassie, T. V. DeSoyza, M. Rosario, H. Gao, and C. Hansch, *Chem.-Biol. Interact.*, **113**, 175 (1998).

229. M. T. D. Cronin and T. W. Schultz, *Chem. Res. Toxicol.*, **14**, 1284 (2001).

230. P. H. Hinderling, O. Schmidlin, and J. K. Seydel, *J. Pharmacokinet. Biopharm.*, **12**, 263 (1984).

231. C. Selassie and T. E. Klein in H. Kubinyi, Ed., *3D QSAR in Drug Design. Theory, Methods and Applications*, Escom Science, Leiden, The Netherlands, 1993, p. 257.

232. O. Geban, H. Ertepinar, M. Yurtsever, S. Ozden, and F. Gumus, *Eur. J. Med. Chem.*, **34**, 753 (1999).

233. S. Daunes, C. D'Silva, H. Kendrick, V. Yardley, and S. L. Croft, *J. Med. Chem.*, **44**, 2976 (2001).

234. C. Hansch, H. Gao, and D. Hoekman in J. Devillers, Ed., *Comparative QSAR*, Taylor & Francis, Washington, DC, 1998, p. 285.

235. C. Hansch, B. R. Telzer, and L. Zhang, *Crit. Rev. Toxicol.*, **25**, 67 (1995).

236. R. Garg, S. Kapur, and C. Hansch, *Med. Res. Rev.*, **21**, 73 (2000).

237. L. Zhang, H. Gao, C. Hansch, and C. Selassie, *J. Chem. Soc. Perkin Trans. 2*, 2553 (1998).

238. C. Hansch, S. McKarns, C. J. Smith, and D. J. Doolittle, *Chem.-Biol. Interact.*, **127**, 61 (2000).

239. L. A. Oglesby, M. T. Ebon-McCoy, T. R. Logsdon, F. Copeland, P. E. Beyer, and R. J. Kavlock, *Teratology*, **45**, 11 (1992).

240. C. Hansch and H. Gao, *Chem. Rev.*, **97**, 2995 (1997).

241. A. M. Richard, J. K. Hongslo, P. F. Boone, and J. A. Holme, *Chem. Res. Toxicol.*, **4**, 151 (1991).

242. C. D. Selassie, A. J. Shusterman, S. Kapur, R. P. Verma, L. Zhang, and C. Hansch, *J. Chem. Soc. Perkin Trans. 2*, 2729 (1999).

243. D. Boyd in A. L. Parrill and M. Rami-Reddy, Eds., *Rational Drug Design, ACS Symposium Series 719*, American Chemical Society, Washington, DC, 1999, p. 346.

244. E. Plummer in C. Hansch and T. Fujita, Eds., *Classical and Three-Dimensional QSAR in Agrochemistry, ACS Symposium Series 606*, American Chemical Society, Washington, DC, 1995, p. 241.

245. T. Fujita, *Quant. Struct.-Act. Relat.*, **16**, 107 (1997).

246. H. Koga, A. Itoh, S. Murayama, S. Suzue, and T. Irikura, *J. Med. Chem.*, **23**, 1358 (1980).

247. H. Chuman, A. Ito, T. Shaishoji, and S. Kumazawa in C. Hansch and T. Fujita, Eds., *Classical and Three-Dimensional QSAR in Agrochemistry, ACS Symposium Series 606*, American Chemical Society, Washington, DC, 1995, p. 171.

248. J. Ohtaka and G. Tsukamoto, *Chem. Pharm. Bull.*, **35**, 4117 (1987).

249. M. Kuchar, E. Maturova, B. Brunova, J. Grimova, H. Tomkova, and K. J. Holubek, *Collect. Czech. Chem. Commun.*, **53**, 1862 (1988).

250. T. Fujita in G. Jolles and K. R. H. Wooldridge, Eds., *Drug Design: Fact or Fantasy*, Academic Press, London, 1984, p. 19.

251. J. G. Topliss, *Perspect. Drug Discov. Des.*, **1**, 253 (1993).

252. C. Hansch, J. P. Bjorkroth, and A. Leo, *J. Pharm. Sci.*, **76**, 663 (1987).

253. C. Hansch, R. Garg, and A. Kurup, *Bioorg. Med. Chem.*, **9**, 283 (2001).

254. R. Garg, A. Kurup, S. B. Mekapati, and C. Hansch, *Bioorg. Med. Chem.*, in press (2002).

CHAPTER TWO

Recent Trends in Quantitative Structure-Activity Relationships

A. Tropsha
Laboratory for Molecular Modeling
School of Pharmacy
University of North Carolina
Chapel Hill, North Carolina

Contents

Burger's Medicinal Chemistry and Drug Discovery
Sixth Edition, Volume 1: Drug Discovery
Edited by Donald J. Abraham
ISBN 0-471-27090-3　© 2003 John Wiley & Sons, Inc.

1 INTRODUCTION

Quantitative structure-activity relationship (QSAR) methodology was introduced by Hansch et al. in the early 1960s (1, 2). The approach stemmed from linear free-energy relationships in general and the Hammett equation in particular (3). It is based on the assumption that the difference in structural properties accounts for the difference in biological activities of compounds. According to this approach, the structural changes that affect the biological activities of a set of congeners are of three major types: electronic, steric, and hydrophobic (4). These structural properties are often described by Hammett electronic constants (5), Verloop STERIMOL parameters (6), hydrophobic constants (5), to name but a few. The relationship between a biological activity (or chemical property) and the structural parameters is obtained through the use of linear or multiple linear regression (MLR) analysis. The fundamentals and applications of this method in chemistry and biology have been summarized by Hansch and Leo (4) and an account of the most recent developments in this area of traditional QSAR appears in the chapter by Celassie in this series (7). As discussed in that chapter, the history of modern QSAR counts over 40 years of active research in method development and its applications. It is practically impossible to review all, even relatively recent, developments in the field in a single chapter. Several reviews and monographs on QSAR and its applications have been published in recent years (4, 8–12) and the reader is referred to this collection of general references and publications cited therein for additional in-depth information.

One of the most characteristic features of the modern age QSAR as an integral part of drug design and discovery is an unprecedented growth of biomolecular databases, which contain data on chemical structure and, in some cases, biological activity (or other relevant drug properties such as toxicity or mutagenicity) of chemicals. Figure 2.1 illustrates the fast growth of one of such databases, the Chemical Abstract Service (CAS) registry file (13). The

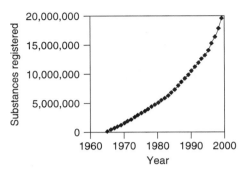

Figure 2.1. Growth in the number of chemical compounds, excluding biopolymers, registered by the Chemical Abstract Service (CAS).

growth has been phenomenal: CAS currently contains more than 39 million compounds, including biological sequences [and it does not include chemical libraries, which literally include billions of compounds (14)]. Naturally, the growth of molecular databases has been concurrent with the acceleration of the drug discovery process. According to an excellent, recent historical account of drug discovery (15), as the result of high throughput screening (HTS) technologies, the amount of raw data points obtained by a large pharmaceutical company per year has increased from approximately 200,000 at the beginning of last decade to around 50 million today. The total number of drugs used worldwide is approximately 80,000, which reportedly act at less than 500 confirmed molecular targets (15). Recent estimates suggest that the number of potential targets lies between 5000 and 10,000, approximately 10-fold greater than the number of targets currently pursued (15). Although traditional QSAR modeling has been typically limited to deal with a maximum of several dozen compounds at a time, rapid generation of large quantities of data requires new methodologies for data analysis. New approaches need to be developed to establish QSAR models for hundreds, if not thousands, of molecules. These new methods should be robust, yet computationally efficient, to compete with the experimental methods of drug discovery, such as combinatorial chemistry and HTS.

This chapter concentrates on recent trends and developments in QSAR methodology, which are characterized by the growing *size of the data sets* subjected to the QSAR analysis, use of *multiple descriptors* of chemical structure, application of both *linear and, especially, nonlinear optimization algorithms* applicable to multidimensional data modeling, growing emphasis on the rigorous model *validation*, and application of QSAR models as *virtual screening* tools in database mining and chemical library design. We begin by presenting a unified concept of QSAR, emphasizing common aspects of different QSAR methodologies. We then consider some popular approaches to the derivation of molecular descriptors and optimization algorithms in the context of three important components of any QSAR investigation: model development, model validation, and model utility. We conclude with several remarks on present status and future developments in this exciting research discipline.

1.1 A Unified Concept of QSAR

An inexperienced user or sometimes even an avid practitioner of QSAR could be easily confused by the multitude of methodologies and naming conventions used in QSAR studies. Two-dimensional (2D) and three-dimensional (3D) QSAR, variable selection and artificial neural network methods, comparative molecular field analysis (CoMFA), and binary QSAR present examples of various terms that may appear to describe totally independent approaches, which cannot be even compared to each other. In fact, any QSAR method can be generally defined as the application of mathematical and statistical methods to the problem of finding empirical relationships (QSAR models) of the form $P_i = \hat{k}(D_1, D_2, \cdots D_n)$, where P_i are biological activities (or other properties of interest) of molecules, $D_1, D_2, \cdots, D_n$ are calculated (or, sometimes, experimentally measured) structural properties (molecular descriptors) of compounds, and $\hat{k}$ is some empirically established mathematical transformation that should be applied to descriptors to calculate the property values for all molecules. The relationship between values of descriptors D and target properties P can be linear [e.g., multiple linear regression (MLR)

Structure Id	Target Property (EC_{50}, K_i, etc.)	Structural Properties (descriptors)		
Comp. 1	P1	D11	D12 $\cdots$	D1n
Comp. 2	P2	D21	D22 $\cdots$	D2n
...	...	"	" "	"
Comp. m	Pm	Dm1	Dm2 $\cdots$	Dmn

$$\{P\} = \hat{K}\{D\}$$

Figure 2.2. Standard QSAR table is a general starting point of any QSAR approach.

as in the Hansch QSAR approach], where target property can be calculated directly from the descriptor values, or nonlinear (such as artificial neural networks or classification QSAR methods), where descriptor values are used in characterizing chemical similarity between molecules, which in turn is used to predict compound activity. In general, each compound can be represented by a point in a multidimensional space, in which descriptors $D_1, D_2, \cdots, D_n$ serve as independent coordinates of the compound. The goal of QSAR modeling is to establish a trend in the descriptor values, which correlates, in a linear or nonlinear fashion, with the trend in biological activity. All QSAR approaches imply, directly or indirectly, a simple similarity principle, which for a long time has provided a foundation for experimental medicinal chemistry: compounds with similar structures are expected to have similar biological activities. This implies that points representing compounds with similar activities in multidimentional descriptor space should be geometrically close to each other, and vice versa.

Despite formal differences between various methodologies, any QSAR method is based on a QSAR table, which can be generalized, as shown in Fig. 2.2. To initiate a QSAR study, this table must include some identifiers of chemical structures (e.g., company's ID numbers, first column of the table in Fig. 2.2), reliably measured values of biological activity [or any other target property of interest (e.g., solubility, metabolic transformation rate, etc.;

second column)], and calculated values of molecular descriptors in all remaining columns (sometimes, experimentally determined physical properties of compounds can be used as descriptors as well).

The differences in various QSAR methodologies can be understood in terms of types of target property values, types of descriptors, and differences in optimization algorithms used to relate descriptors to the target properties. The target property values can be defined as activity classes [i.e., active or inactive, frequently encoded numerically for the purpose of the subsequent analysis as one (for active) or zero (for inactive)] or as a continuous range of values; the corresponding methods of data analysis are referred to as classification or continuous property QSAR, respectively. Descriptors can be generated from various representations of molecules (e.g., 2D chemical graphs or 3D molecular geometries), giving rise to the terms of 2D- or 3D-QSAR, respectively. Finally, the types of optimization algorithms used in the QSAR model development lead to the definitions of linear versus nonlinear QSAR methods.

In some cases, the types of biological data, the choice of descriptors, and the class of optimization methods are closely related and mutually inclusive. For instance, multiple linear regression can be applied only when a relatively small number of molecular descriptors are used (at least five to six times smaller than the total number of compounds) and the target property is characterized by a continuous range of values. The use of multiple descriptors makes it impossible to use MLR because of a high chance of spurious correlation (16) and requires the use of partial least squares or nonlinear optimization techniques. However, in general, for any given data set a user could choose between various types of descriptors and various optimization schemes, combining them in a practically mix-and-match mode, to arrive at statistically significant QSAR models in a variety of ways. This situation is in essence analogous to molecular mechanics calculations (17), where different force fields and differently derived parameters are developed by different groups, although the common goal is to compute (unique) optimized geometries of molecules from their chemical composition and coordinates of all atoms. Thus, in general, all QSAR models can be universally compared in terms of their statistical significance and, most important, their ability to predict accurately biological activities (or other target properties) of molecules not included in the training set (cf. molecular mechanics, where different methods are ultimately compared by their ability to reproduce experimental molecular geometries). This concept of statistical robustness and the predictive ability as universal characteristics of any QSAR model independent of the particulars of individual approaches should be kept in mind as we consider examples of QSAR tools, their applications, and pitfalls in the subsequent sections of this chapter.

1.2 The Taxonomy of QSAR Approaches

Many different approaches to QSAR have been developed since Hansch's seminal work. As briefly discussed above, the major differences between these methods can be analyzed from two viewpoints: (1) the types of structural parameters that are used to characterize molecular identities, starting from different representation of molecules, from simple chemical formulas to three-dimensional conformations; and (2) the mathematical procedure that is employed to obtain the quantitative relationship between these structural parameters and biological activity.

On the basis of the origin of molecular descriptors used in calculations, QSAR methods can be divided into three groups. One group is based on a relatively small number (usually many times smaller than the number of compounds in a data set) of physicochemical properties and parameters describing, for example, hydrophobic, steric, and electrostatic effects. Usually, these descriptors are used as independent variables in multiple regression approaches (18). In the literature, these methods are typically referred to as Hansch analysis (8). These types of descriptors and corresponding linear optimization methods used in traditional QSAR analyses are discussed extensively in the chapter by Celassie (7) and therefore is not reviewed here.

More recent methods are based on quantitative characteristics of molecular graphs (molecular topological descriptors). Because molecular graphs or structural formulas are "two-dimensional," these methods are referred to as 2D-QSAR. Most of the 2D-QSAR methods are based on graph theoretical indices, which have been extensively studied by Randíc (19) and Kier and Hall (20–22). They include, for example, molecular connectivity indices (19, 20), molecular shape indices (23, 24), topological (25) and electrotopological state indices (26–29), and atom-pair descriptors (30, 31). Sometimes, topological descriptors are also combined with physicochemical properties of molecules. Although these structural indices represent different aspects of molecular structures, and, what is important for QSAR, different structures provide numerically different values of indices, their physicochemical meaning is frequently unclear. The successful applications of topological indices combined with multiple linear regression (MLR) analysis have been summarized by Kier and Hall (20, 21, 28).

The third group of methods is based on descriptors derived from spatial (three-dimensional) representation of molecular structures. Correspondingly, these methods are referred to as three-dimensional or 3D-QSAR; they have become increasingly popular with the development of fast and accurate computational methods for generating 3D conformations and alignments of chemical structures. The early examples of 3D-QSAR include molecular shape analysis (MSA) (32), distance geometry (33, 34), and Voronoi techniques (35). The first method uses shape descriptors and multiple linear regression analysis, whereas the latter methods apply atomic refractivity as structural descriptors and the solution of mathematical inequalities to obtain the quantitative relationships. These two methods have been applied to the study of structure-activity relationships of many data sets by Hopfinger (e.g., Refs. 36, 37) and Crippen (e.g., Refs. 38, 39), respectively.

Perhaps the most popular example of 3D-QSAR is the comparative molecular field analysis (CoMFA), developed by Cramer et al. (40), which has elegantly combined the power of 3D molecular modeling and partial least-square (PLS) optimization technique (41, 42) and found wide applications in medicinal chemistry and toxicity analysis (see below). Most of

3D-QSAR methods require 3D alignment of all molecules according to a pharmacophore model or based on ligand docking to a receptor-binding site. Descriptors in the case of CoMFA (40, 43) and CoMFA-like methods such as COMBINE (44), COMSiA (45), and QsiAR (46) represent electrostatic, steric, and hydrophobic field values (to name but a few examples) in the grid points surrounding molecules.

Finally, QSAR methods can also be classified by the type of the correlation methods used in model development. Linear methods include linear regression or MLR, PLS (41, 42, 47), or principal component regression (PCR), whereas nonlinear methods can be exemplified, for example, by k-Nearest Neighbors (kNN) (48, 49) and artificial neural networks (50) methods. An example of the linear methods is provided by the ADAPT system, which employs topological indices as well as other calculable structural parameters (e.g., steric and quantum mechanical parameters), and the MLR method for QSAR analysis. It has been extensively applied to QSAR/QSPR studies in analytical chemistry, toxicity analysis, and other biological activity prediction (51–54). Parameters derived from various experiments through chemometric methods have also been used in the study of peptide QSAR (55), where PLS analysis was employed. The latter technique has been used almost exclusively in 3D-QSAR, where the number of descriptors characterizing molecular fields may exceed the number of compounds by orders of magnitude.

There has been a great deal of interest, especially more recently, in the use of data mining methods to extract the information from large and/or chemically inhomogeneous data sets. Examples of these methods include pattern recognition (56, 57), automated structure evaluation (58, 59), neural network (60–62), and machine learning (63–65). Recent trends in QSAR studies also include developing optimal QSAR models through variable selection, that is, by selecting a subset of available descriptors in either MLR, PLS, or nonlinear classification or artificial neural networks (ANN) analysis as applied either in 2D- (66–72) or in 3D-QSAR (73). These methods employ either generalized simulated annealing

(67), or genetic algorithms (68), or evolutionary algorithms (69–72) as optimization tools. The effectiveness and convergence of these algorithms are strongly affected by the choice of a fitting function, which drives the optimization process (70–72). It has been demonstrated that optimization combined with variable selection effectively improves QSAR models as compared to those without variable selection. For example, GOLPE (74) was developed through the use of chemometric principles and q^2-GRS (75) was developed on the basis of independent CoMFA analysis of small areas of CoMFA descriptor space, to address the issue of region selection. Both of these methods have been shown to improve QSAR models compared to the original CoMFA technique.

Different QSAR methods have their own strengths and weaknesses. For example, 3D-QSAR methods generally result in the diagrams of important molecular fields that can be easily interpreted in terms of specific steric and electrostatic interactions important for the ligand binding to their receptor. However, the need to align structures in 3D, which is time-consuming and subjective, precludes the use of 3D-QSAR techniques for the analysis of large data sets. On the other hand, 2D-QSAR methods are much faster and more amenable to automation because they require no conformational search and structural alignment. Thus, 2D methods are best suited for the analysis of large numbers of compounds and computational screening of molecular databases; however, the interpretation of the resulting models in familiar chemical terms is frequently difficult, if not impossible.

The generality of the QSAR modeling approach as a drug discovery tool, irrespective of descriptor types or optimization algorithms, can be best demonstrated in the context of inverse QSAR, which can be defined as designing or discovering molecular structures with a desired property on the basis of QSAR models (76–78). In practical terms, inverse QSAR also includes searching for molecules with a desired target property in chemical databases or virtual chemical libraries. These considerations emphasize the universal importance of establishing QSAR model robustness and predictive ability as opposed to concentrating on

explanatory power, which has been a characteristic feature of many traditional QSAR approaches.

2 MULTIPLE DESCRIPTORS OF MOLECULAR STRUCTURE

It has been said frequently that there are three keys to the success of any QSAR model building exercise: descriptors, descriptors, and descriptors. Many different molecular representations have been proposed, exemplified by Hansch-type parameters (2), topological indices (19, 79), quantum mechanical descriptors (80), molecular shapes (32, 81), molecular fields (40), atomic counts (82), 2D fragments (83–85), 3D fragments (86–88), molecular eigenvalues (89), molecular multipole moments (90), E-state fields (28), molecular fragment-based hash codes (91, 92), and molecular holograms (93). A recent review by Livingstone provides an excellent survey of various 2D and 3D descriptors, along with some associated diversity and similarity functions (9). Various physicochemical parameters such as the partition coefficient, molar refractivity, and quantum mechanical quantities such as highest occupied molecular orbital (HOMO) and lowest occupied molecular orbital (LUMO) energies have been used to represent molecular identities in early QSAR studies by the use of linear and multiple linear regression. However, these descriptors are not suited for the analysis of large numbers of molecules, either because of the lack of physicochemical parameters for compounds yet to be synthesized or because of the computational expenses required by quantum mechanical methods. Recent years have seen the application of various topological descriptors that are usually derived from either 2D or 3D molecular structural information based on the graph theory or molecular topology (20–22, 94). These descriptors are generated on the basis of the molecular connectivity, 3D molecular topography, and molecular field properties.

2.1 Topological Descriptors

Two widely applied examples of 2D molecular descriptors are molecular connectivity indices

(MCI) and atom-pair (AP) descriptors. Molecular connectivity indices, χ, were first formulated by Randíc (19) and subsequently generalized and extended by Kier and Hall (20–22). The fundamentals and applications of molecular connectivity indices have been thoroughly reviewed (22, 28). A popular MolConnZ software (95) affords the computation of a wide range of topological indices of molecular structure. These indices include (but are not limited to) the following descriptors: simple and valence path, cluster, path/cluster and chain molecular connectivity indices, kappa molecular shape indices, topological and electrotopological state indices, differential connectivity indices, the graph's radius and diameter, Wiener and Platt indices, Shannon and Bonchev-Trinajstíc information indices, counts of different vertices, and counts of paths and edges between different kinds of vertices (19, 20, 96–100).

Overall, MolConnZ (95) produces over 400 different descriptors. Most of these descriptors characterize chemical structure, but several depend on the arbitrary numbering of atoms in a molecule and are introduced solely for bookkeeping purposes. In a typical QSAR study, only about one-half of all possible MolConnZ descriptors are eventually used, after deleting descriptors with zero value or zero variance. Figure 2.3 provides a summary of these molecular descriptors and presents some algorithms used in their derivation.

The idea of using atom pairs as molecular features in structure-activity studies was first proposed by Carhart et al. (84). AP descriptors are defined by their atom types and topological distance bins. An AP is a substructure defined by two atom types and the shortest path separation (or graph distance) between the atoms. The graph distance is defined as the smallest number of atoms along the path connecting two atoms in a molecular structure. The general form of an atom-pair descriptor is as follows:

atom type i —— (distance) —— atom type j

where atom chemical types are typically defined by the user. For example, 15 atom types can be defined by use of the SYBYL mol2 format (101) as follows: (1) negative charge center (NCC); (2) positive charge center (PCC); (3) hydrogen bond acceptor (HA); (4) hydrogen bond donor (HD); (5) aromatic ring center (ARC); (6) nitrogen atoms (N); (7) oxygen atoms (O); (8) sulfur atoms (S); (9) phosphorous atoms (P); (10) fluorine atoms (FL); (11) chlorine, bromine, iodine atoms (HAL); (12) carbon atoms (C); (13) all other elements (OE); (14) triple bond center (TBC); and (15) double bond center (DBC). Apparently, the total number of pairwise combinations of all 15 atom types is 120. Furthermore, distance bins should be defined to discriminate between identical atom pairs separated by different graph distances and therefore representing different molecular substructures. Thus, 15 distance bins can be introduced in the interval between graph distance zero (i.e., zero atoms separating an atom pair) to 14 and greater. Thus, in this format a total of 1800 (120 × 15) AP descriptors can be generated for any molecular structure. An example of an atom-pair descriptor is shown on Fig. 2.4. Frequently, as applied to particular data sets, many of the theoretically possible AP descriptors have zero value (implying that certain atom types or atom pairs are absent in molecular structures). For instance, in our recent studies of 48 anticonvulsant agents, only 273 descriptors with nonzero value and nonzero variance were generated (102).

2.2 3D Descriptors

The rapid increase in structural three-dimensional (3D) information of bioorganic molecules (103, 104), coupled with the development of fast methods for 3D structure generation [e.g., CONCORD (105, 106) and CORINA (107)] and alignment [e.g., Active Analog Approach (43, 108)], have led to the development of 3D structural descriptors and associated 3D-QSAR methods. Many 3D-QSAR methods (considered below) make use of so-called molecular field descriptors. To calculate these descriptors, steric and electrostatic fields of all molecules are sampled with a probe atom, usually carbon sp^3 bearing a +1 charge, on a rectangular grid that encompasses structurally aligned molecules. The values of both van der Waals and electrostatic interactions between the probe atom and all

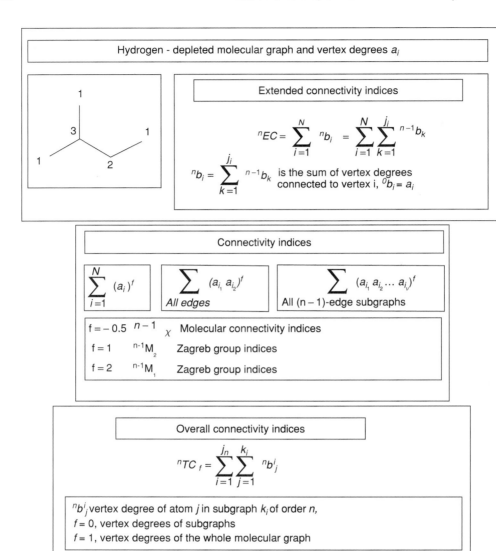

Figure 2.3. Examples of topological descriptors frequently used in QSAR studies.

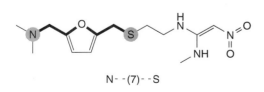

N- -(7)- -S

Figure 2.4. Example of an AP descriptor: two atom types, aliphatic nitrogen and aliphatic sulfur, separated by the shortest chemical graph path of seven.

atoms of each molecule are calculated in every lattice point by use of the force field equation described above and entered into the CoMFA QSAR table (Fig. 2.5), which typically contains thousands of columns. Additional molecular field descriptors such as HINT (Hydropathic INTeraction) descriptors (109) could improve the CoMFA model. PLS algorithms coupled with leave-one-out (LOO) cross-validation is typically used to arrive at statistically significant CoMFA models.

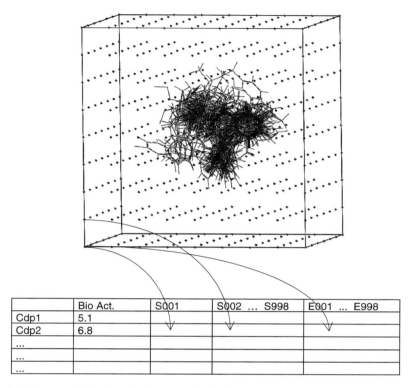

	Bio Act.	S001	S002 ... S998	E001 ... E998
Cdp1	5.1			
Cdp2	6.8			
...				
...				
...				

Figure 2.5. Process of steric and electrostatic descriptor generation in CoMFA. Note that this process results in a familiar QSAR table (cf. Fig. 2.2). PLS is used as a standard analytical technique in CoMFA.

One of the most attractive features of the CoMFA and CoMFA-like methods is that, because of the nature of molecular field descriptors, these approaches yield models that are relatively easy to interpret in chemical terms. Famous CoMFA contour plots, which are obtained as a result of any successful CoMFA study, tell chemists in rather plain terms how the change in the compounds' size or charge distribution as a result of chemical modification correlate with the binding constant or activity. These observations may immediately suggest to a chemist possible ways to modify molecules to increase their potencies. However, as demonstrated in the next section, these predictions should be taken with caution only after sufficient work has been done to prove the statistical significance and predictive ability of the models.

By analogy with 2D atom-pair descriptors (Fig. 2.4), 3D AP descriptors can also be de-fined through the use of similar atom types and atom pairs and 3D molecular topography; in this case, a physical distance between atom types is used in place of chemical graph distance. The distance between two "atoms" is measured and then assigned into one or two distance bins. Typically, the width of each distance bin is chosen as 1.0 Å. Because it is also designed to let the adjacent bins have 10% overlap with each other, the actual length of each distance bin is 1.2 Å. Any distance located in the overlap region is assigned to both bins. This "fuzzy distance" concept is adopted to alleviate the possible unfavorable boundary effects of the distance bins. For example, with strict boundary conditions, a distance of 2.05 Å will be assigned only to bin No. 2, but it can be reasonably argued that it is almost as close to the upper half of bin No. 1 as to bin No. 2. With fuzzy boundary conditions, 2.05 Å belongs to both bin No. 1 and bin No. 2, allowing a possible match to either. All the distances greater than 20 Å are assigned into the last bin.

3   QSAR MODELING APPROACHES

3.1   3D-QSAR

Two original 3D-QSAR methods, CoMFA (40) and GRID (110), were developed almost simultaneously in the mid- to late-1980s (9). Since its introduction, the CoMFA approach has rapidly become one of the most popular methods of QSAR. Over the years, this approach has been applied to a wide variety of receptor and enzyme ligands [many reviews appeared in a recent monograph (10)]. Undoubtedly, the further development of this and related methods is of great importance and interest to many scientists working in the area of rational drug design.

CoMFA methodology is based on the assumption that because, in most cases, the drug-receptor interactions are noncovalent, the changes in the biological activities or binding affinities of sample compounds correlate with changes in the steric and electrostatic fields of these molecules. In a standard CoMFA procedure, all molecules under investigation are first structurally aligned, and the steric and electrostatic fields around them are sampled with probe atoms, usually sp^3 carbon with a +1 charge, on a rectangular grid that encompasses aligned molecules. The results of the field evaluation in every grid point for every molecule in the data set are placed in the CoMFA QSAR table, which therefore contains thousands of columns (Fig. 2.5). The analysis of this table by the means of standard multiple regression is practically impossible; however, the application of special multivariate statistical analysis routines, such as PLS analysis and LOO cross-validation ensures the statistical significance of the final CoMFA equation. The outcome from this procedure is a cross-validated correlation coefficient R^2 (q^2), which is calculated according to the formula

$$q^2 = 1 - \frac{\Sigma\ (y_i - \hat{y}_i)^2}{\Sigma\ (y_i - \bar{y})^2}, \qquad (2.1)$$

where y_i, $\hat{y}_i$, and $\bar{y}$ are the actual, estimated, and averaged (over the entire data set) activities, respectively. The summations in Equation 2.1 are performed over all compounds, which are used to build a model for the training set. The statistical meaning of the q^2 is different from that of the conventional r^2: a q^2 value greater than 0.3 is often considered significant (111).

Despite obviously successful and growing application of CoMFA in molecular design, several problems intrinsic to this methodology have persisted. Studies revealed that CoMFA results can be extremely sensitive to a number of factors, such as alignment rules, overall orientation, lattice placement, step size, and probe atom type (40, 75, 112–114). The problem of three-dimensional alignment has been the most notorious among others. Even with the development of automated or semiautomated alignment protocols such as the Active Analog Approach (108, 115) or DISCO (116) and the opportunity to use, in some cases, the structural information about the target receptor (112, 117) to align molecules, in general there is no standard recipe as to how to align all molecules under consideration in a unique and unambiguous fashion. A QSAR analysis of 60 acetylcholinesterase inhibitors (117) is particularly illustrative with respect to this point. In that study, the combination of structure-based alignment and CoMFA was employed to obtain a QSAR model for 60 chemically diverse inhibitors of acetylcholinesterase (AChE). The great structural diversity of the AChE inhibitors, ranging from choline to decamethonium, made it practically impossible to structurally align all the inhibitors in any unbiased way and generate a unique three-dimensional pharmacophore. X-ray crystallographic analysis of AChE from *Torpedo californica* (EC 3.1.1.7) (118), followed by X-ray determination of the complexes of the enzyme with three structurally diverse inhibitors, tacrine, edrophonium, and decamethonium (119), provided crucial information with respect to the orientation of these inhibitors in the active site of the enzyme. The crystallographic data indicated that each of the three inhibitors had a unique binding orientation in the active site of the enzyme (Fig. 2.6). Their natural structural alignment would probably never have been predicted by any of the existing automated algorithms for ligand align-

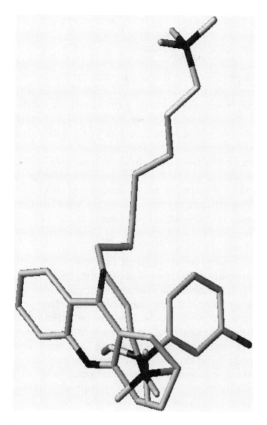

Figure 2.6. Superposition of three inhibitors of AChE in the active site of the enzyme based on crystallographic structures of enzyme-inhibitor complexes. Obviously, no common pharmacophore can be found for these molecules.

ment or even by the researcher's imagination on the basis of the ligand chemical structure alone. This consideration demonstrates the general difficulty of generating a unique and meaningful alignment in 3D-QSAR studies that leads to interpretable and predictive models.

The 3D alignment problem is the main source of ambiguity in obtaining and analyzing CoMFA results, especially in the case of structurally diverse compounds. However, it was also shown that, even if the structural alignment is fixed, the resulting q^2 value could also be sensitive to the orientation of rigidly aligned molecules on the user terminals (75), which can be explained as follows.

The grid orientation in CoMFA is fixed in the coordinate system of the computer; thus, every time when the orientation of the molecular aggregate is changed, the size of the grid may change but not its orientation. The orientation of the assembled molecules therefore affects the placement of probe atoms, which, in turn, influences the field sampling process. This leads to the variability of the q^2 values, mostly attributable to the reasons outlined earlier. The effect of variability of q^2 as a function of molecular aggregate orientation was more pronounced in the case of structurally diverse molecules (e.g., cephalotaxine esters and 5-HT$_{1A}$ receptor ligands) than in the case of much less structurally diverse molecules (e.g., HIV protease inhibitors) (75). This effect may be attributed to the fact that the pattern of probe atom placement with respect to the aligned molecules changes more dramatically when one changes the orientation of more structurally diverse molecules than it does when the data set is composed of structurally similar molecules.

In the conventional CoMFA implementation, the steric and electrostatic fields, which theoretically form a continuum, are sampled on a fairly coarse grid. As a result, these fields are represented inadequately, and the results are not strictly reproducible. Intuitively, decreasing the grid spacing may increase the adequacy of sampling, as was suggested by Cramer et al. (120). Indeed, it was shown that decreasing the grid spacing from 2.0 to 1.0 Å minimized the fluctuation in the observed q^2 values (75). Most probably, the reason for this phenomenon is that the decrease in grid spacing increases the number of probe atoms, which in turn should raise the probability of placing the probe atoms in a region where the steric and electrostatic field changes can be best correlated with biological activity. However, as was noticed by Cramer et al. (120), the increase in the number of probe atoms also increases the noise in PLS analysis and leads to a less statistically significant q^2 (121).

An important feature of conventional CoMFA routine is that it assumes equal sampling and *a priori* equal importance of all lattice points for PLS analysis, whereas the final CoMFA result actually emphasizes the limited areas of three-dimensional space as important for biological activity. Indeed, the deficiencies of conventional CoMFA routine mentioned earlier may be effectively dealt with by eliminating from the analyses those areas of three-dimensional space where changes in steric and electrostatic fields do not correlate with changes in biological activity. The q^2-GRS routine was devised (75) to eliminate those areas from the analysis based on the (low) value of the q^2 obtained for such regions individually. The major feature of this routine is that it optimizes the region selection for the final PLS analysis. In this regard, it is intellectually analogous to the GOLPE approach (74).

3D-QSAR remains an active area of research and method development. Several recent approaches such as COMSiA (45), QSiAR (46), and GRIND (122) address the most notorious CoMFA problems dealing with the grid artifacts. However, it should be kept in mind that 3D-QSAR modeling is a difficult process. It is reasonably successful when underlying molecules are relatively rigid and similar, so that the identification of the 3D pharmacophore is straightforward. With the increased complexity and flexibility of molecules and a possibility of multiple mechanisms of binding with the receptor, the derivation of unambiguous pharmacophore and unique alignment is sometimes practically impossible (as shown above in the case of AchE inhibitors), and extreme care is important in trying to obtain reproducible and validated QSAR models.

3.2 The Descriptor Pharmacophore Concept and Variable Selection QSAR

The term *pharmacophore*, introduced by Ehrlich in the early 1900s (123), was originally referred to the molecular framework that carries (*phoros*) the essential features responsible for a drug's (*pharmacon*) activity. Nowadays, this term has almost the opposite meaning as applied to three-dimensional (3D) molecular structure. A 3D pharmacophore is defined as a collection of particular chemical features (functional groups) and their spatial arrangement, which define pharmacological specificity of a series of compounds (124). The pharmacophore concept assumes that structurally diverse molecules bind to their receptor site in

a similar way, with their pharmacophoric elements interacting with the same functional groups of the receptor.

The pharmacophore concept plays a very important role in guiding the drug discovery process. Pharmacophore models help medicinal chemists gain an insight into the key interactions between ligand and receptor when the receptor structure has not been determined experimentally. A pharmacophore can be used as a basis for the alignment rules in 3D-QSAR analysis for the lead compound optimization (125). Furthermore, a pharmacophore can be directly used as the search query for 3D database mining, which is a common and efficient approach for discovery of lead compounds (126).

Pharmacophore identification refers to the computational way of identifying the essential 3D structural features and configurations that are responsible for the biological activity of a series of compounds. It is computationally intensive, requiring searching two huge spaces: the available conformations for each compound and the possible correspondence (alignment) between different compounds. A number of approaches and computer programs have been specifically developed for pharmacophore identification including, for example, Active Analog Approach, AAA (108, 127, 128), Ensemble distance geometry (129), DISCO (116), Chem-X (130), Catalyst/Hypo (131, 132), Catalyst/HipHop (133, 134), and Apex-3D (135).

An obvious parallel can be established between the identification of descriptors contributing the most to the correlation with biological activity, and search for pharmacophoric elements, which are mainly responsible for the specificity of drug action. Indeed, individual pharmacophoric elements are typically identified in the course of experimental structure-activity studies. Considering molecules as a collection of substructures, pharmacophoric elements can also be viewed as specific chemical features selected from all chemical fragments present in a molecular data set. Thus, the selection of specific pharmacophoric features responsible for biological activity is directly analogous to the selection of specific chemical descriptors contributing to the most explanatory QSAR model. Frequently, the

QSAR modeling that involves descriptor (feature) selection is referred to as variable selection QSAR.

This consideration emphasizes the analogy between pharmacophore identification and variable selection QSAR. On the basis of this analogy, we now expand the notion of chemical pharmacophore to that of the more general descriptor pharmacophore. We shall define descriptor pharmacophore as a special subset of molecular descriptors (of any nature, not only chemical functional groups) optimized in the process of variable selection QSAR, to achieve the most significant correlation between descriptor values and biological activity.

Similar to the common areas of application of chemical pharmacophores, descriptor pharmacophores can be applied for database mining. First, a preconstructed QSAR model can be used as a means of screening compounds from existing databases (or virtual libraries) for high predicted biological activity. Alternatively, variables selected by QSAR optimization can be used for similarity searches to improve the performance of the rational library design or database mining methods. The advantage of this approach for database mining is that it affords not only the compound selection but also the quantitative prediction of their activity.

3.2.1 Linear Models. Variable selection approaches can be applied in combination with both linear and nonlinear optimization algorithms. Exhaustive analysis of all possible combinations of descriptor subsets to find a specific subset of variables that affords the best correlation with the target property is practically impossible because of the combinatorial nature of this problem. Thus, stochastic sampling approaches such as genetic or evolutionary algorithms (GA or EA) or simulated annealing (SA) are employed. To illustrate one such application we shall consider the GA-PLS method, which was implemented as follows (136).

Step 1. Multiple descriptors such as molecular connectivity indices or atom pair descriptors (cf. Section 2.1) are generated initially for every compound in a data set.

Step 2. An initial population of 100 different random combinations of subsets of these

descriptors (parents) is generated as follows. Each parent is described by a string of random binary numbers (i.e., one or zero), with the length (total number of digits) equal to the total number of descriptors selected for each data set. The value of one in each string implied that the corresponding descriptor is included for the parent, and the value of zero implies that the descriptor is excluded.

Step 3. For every random combination of descriptors (i.e., every parent), a QSAR equation is generated for the training data set by use of the PLS algorithm (41). Thus, for each parent a q^2 value is obtained, and some function of q^2 is used as a fitness function to guide GA.

Step 4. Two parents are selected randomly and subjected to a crossover (i.e., the exchange of the equal length substrings), which produces two offspring. Each offspring is subjected to a random single-point mutation, that is, a randomly selected one (or zero) is changed to zero (or one) and the fitness of each offspring is evaluated as described above (cf. Step 3).

Step 5. If the resulting offspring are characterized by a higher value of the fitness function, then they replaced parents; otherwise, the parents are kept.

Step 6. Steps 3–5 are repeated until a predefined convergence criterion is achieved. For the convergence criterion one can use the difference between the maximum and minimum values of the fitness function. Calculations are terminated when this difference falls below a certain threshold (e.g., 0.02).

In summary, each parent in this method represents a QSAR equation with randomly chosen variables, and the purpose of the calculation is to evolve from the initial population of the QSAR equations to the population with the highest average value of the fitness function. In the course of the GA-PLS process, the initial number of members of the population (100) is maintained while the average value of the fitness function for the whole population converges to a high number. The best model is characterized by the highest value of the fitness function as well as by specific descriptor selection (descriptor pharmacophore) that affords such a model.

3.2.2 Nonlinear Models.

Most of the QSAR approaches assume the existence of a linear relationship between a biological activity and molecular descriptors. However, the fast collection of structural and biological data, as a consequence of the recent development of combinatorial chemistry and high throughput screening technologies, has challenged traditional QSAR techniques. First, 3D methods may be computationally too expensive for the analysis of a large volume of data; and in some cases, an automated and unambiguous alignment of molecular structures is not achievable. Second, although existing 2D techniques are computationally efficient, the assumption of linearity in the SAR may not hold true, especially when a large number of structurally diverse molecules are included in the analysis.

These considerations provide an impetus for the development of fast, nonlinear, variable selection QSAR methods that can avoid the aforementioned problems of linear QSAR. Several nonlinear QSAR methods have been proposed in recent years. Most of these methods are based on either artificial neural network (ANN) (50, 61, 137–142) or machine learning techniques (65, 143–145). Given that optimization of many parameters is involved in these techniques, the speed of the analysis is relatively slow. More recently, Hirst reported a simple and fast nonlinear QSAR method (146), in which the activity surface was generated from the activities of training set compounds based on some predefined mathematical function.

For illustration, we shall consider here one of the nonlinear variable selection methods that adopts a k-Nearest Neighbor (kNN) principle to QSAR [kNN-QSAR (49)]. Formally, this method implements the active analog principle that lies in the foundation of the modern medicinal chemistry. The kNN-QSAR method employs multiple topological (2D) or topographical (3D) descriptors of chemical structures and predicts biological activity of any compound as the average activity of k most similar molecules. This method can be used to analyze the structure-activity relationships (SAR) of a large number of compounds where a nonlinear SAR may predominate.

In principle, the kNN technique is a conceptually simple, nonlinear approach to pattern-recognition problems (147). In this method, an unknown pattern is classified according to the majority of the class labels of its k nearest neighbors of the training set in the descriptor space. Many variations of the kNN method have been proposed in the past and new and fast algorithms have continued to appear in recent years (148, 149). The applications of the kNN principle in chemistry have been summarized by Strouf (150). In the area of biology, Raymer et al. have successfully applied a kNN pattern-recognition technique with simultaneous feature selection and classification in the analysis of water distribution in protein structures (151). In the area of QSPR, Basak et al. have applied this principle, combined with principal component analysis and graph theoretical indices, in the estimation of physicochemical properties of organic compounds (152–155).

The assumptions underlying the kNN-QSAR method are as follows. First, structurally similar compounds should have similar biological activities, and the activity of a compound can be predicted (or estimated) simply as the average of the activities of similar compounds. Second, the perception of structural similarity is relative and should always be considered in the context of a particular biological target. Given that the physicochemical characteristics of the receptor-binding site vary from one target to another, the structural features that can best explain the observed biological similarities between compounds are different for different biological endpoints. These critical structural features can be defined as the descriptor pharmacophore (DP) for the underlying biological activity. Thus, one of the tasks of building a kNN-QSAR model is to identify the best DP. This is achieved by the "bioactivity-driven" variable selection, that is, by selecting a subset of molecular descriptors that afford a highly predictive kNN-QSAR model. Because the number of all possible combinations of descriptors is huge, an exhaustive search of these combinations is not possible. Thus, a stochastic optimization algorithm (i.e., simulated annealing) has been adopted for an efficient sampling of the combinatorial space. Fig-

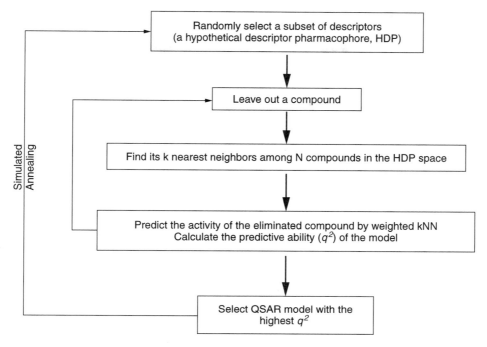

Figure 2.7. Flowchart of the kNN method (49).

ure 2.7 shows the overall flowchart of the kNN-QSAR method, which involves the following steps.

1. Select a subset of n descriptors randomly (n is a number between 1 and the total number of available descriptors) as a hypothetical descriptor pharmacophore (HDP).

2. Validate this HDP by a standard cross-validation procedure, which generates the cross-validated R^2 (or q^2) value for the kNN-QSAR model built by use of this HDP. The standard leave-one-out procedure has been implemented as follows: (i) Eliminate a compound from the training set. (ii) Calculate the activity of the eliminated compound, which is treated as an unknown, as the average activity of the k most similar compounds found in the remaining molecules (k is set to 1 initially). The similarities between compounds are calculated using only the selected descriptors (i.e., the current trial HDP) instead of the whole set of descriptors. (iii) Repeat this procedure until every compound in the training set has been eliminated and predicted once. (iv)

Calculate the cross-validated R^2 (or q^2) value (cf. Equation 2.1). (v) Repeat calculations for $k = 2, 3, 4, \ldots, n$. The upper limit of k is the total number of compounds in the data set; however, the best value is found empirically between 1 and 5. The k that leads to the best q^2 value is chosen for the current kNN-QSAR model.

3. Repeat steps 1 and 2, the procedure of generating trial HTPs and calculating corresponding q^2 values. The goal is to find the best HTP that maximizes the q^2 value of the corresponding kNN-QSAR model. This process is driven by a generalized simulated annealing by use of q^2 as the objective function.

4   VALIDATION OF QSAR MODELS

One of the most important characteristics of QSAR models is their predictive power. The latter can be defined as the ability of a model to predict accurately the target property (e.g., biological activity) of compounds that were not used for model development. The typical problem of QSAR modeling is that at the time of

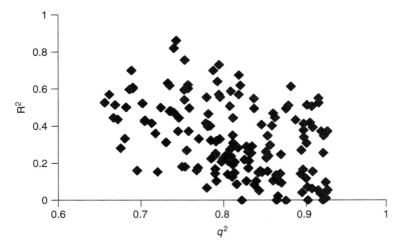

Figure 2.8. Beware of q^2! External R^2 (for the test set) presents no correlation with the "predictive" LOO $q2$ (for the training set). (Adopted from Ref. 163.)

model development a researcher has, essentially, only training set molecules, so predictive ability can be characterized only by statistical characteristics of the training set model, and not by true external validation. Recent research demonstrates that external validation must be made, indeed, a mandatory part of model development. This goal can be achieved by a division of an experimental SAR data set into the training and test sets, which are used for model development and validation, respectively.

It has been shown that the more independent variables are involved in MLR QSAR analysis, the higher the probability of a chance correlation between predicted and observed activities, even if only a small portion of variables is included in the final QSAR equation (16). This conclusion is true not only for MLR QSAR, but also for any QSAR approach when the number of variables (descriptors) is comparable to or higher than the number of compounds in a data set. Thus, model validation is one of the most important aspects of QSAR analysis.

4.1 Beware of q^2

To validate a QSAR model, most of researchers apply the leave-one-out (LOO) or leave-some-out (LSO) cross-validation procedures. The outcome from this procedure is a cross-validated correlation coefficient R^2 (q^2) (Equation 2.1). Frequently, q^2 is used as a criterion of both robustness and predictive ability of the model. Many authors consider high q^2 (for instance, $q^2 > 0.5$) as an indicator or even as the ultimate proof of the high predictive power of the QSAR model. They do not test the models for their ability to predict the activity of compounds of an external test set (i.e., compounds that have not been used in the QSAR model development). There are several examples of recent publications, in which the authors claim that their models have high predictive ability without validating them by use of an external test set (156–160). Some authors validate their models by the use of only one or two compounds that were not used in QSAR model development (161, 162) and still claim that their models are highly predictive. In contrast with such expectations, it has been shown that if a test set with known values of biological activities is available for prediction, there exists no correlation between LOO cross-validated q^2 and correlation coefficient R^2 between the predicted and observed activities for the test set [Fig. 2.8; (46, 163)].

4.2 Rational Selection of Training and Test Sets

As discussed earlier, to obtain a reliable (validated) QSAR model, an available data set should be divided into the training and test sets. Ideally, this division must be performed such that points representing both training and test set are distributed within the whole descriptor space occupied by the entire data set, and each point of the test set is close to at least one point of the training set. This approach ensures that the similarity principle can be employed for the activity prediction of the test set. Unfortunately, as we shall see below, this condition cannot always be satisfied.

Many authors use external test sets for validation of QSAR models, but do not provide any rationale as to how and why certain compounds were chosen for the test set (164, 165). One of the most widely used methods for dividing a data set into training and test sets is a mere random selection (166, 167). Some authors assign whole structural subgroups of molecules to the training set or the test set (168, 169). Another frequently used approach is based on the activity sampling. The whole range of activities is divided into bins, and compounds belonging to each bin are randomly (or in some regular way) assigned to the training set or test set (170, 171). These methods (166, 170, 171) cannot guarantee that the training set compounds represent the entire descriptor space of the original data set, and that each compound point of the test set is close to at least one point of the training set.

In several publications, the division of a data set into training and test sets is performed by use of the Kohonen's Self-Organizing Map (SOM) (172). Representative points falling into the same areas of the SOM are randomly selected for the training and test sets (173, 174). SOM preserves the closeness between points (points that are close to each other in the multidimensional descriptor space are close to each other on the map). Therefore, it is anticipated that the training and test sets must be scattered within the whole area occupied by representative points in the original descriptor space, and that each point of the test set is close to at least one point of the training set. The drawback of this method is that the quantitative methods of prediction use exact values of distances between representative points; because SOM is a nonlinear projection method, the distances between points in the map are distorted.

The division of a data set into the training and test sets can be performed by the use of various clustering techniques. In Burden and Winkler (175) and Burden et al. (176) the K-means clustering algorithm (177) was used, and from each cluster one compound for the training set was randomly selected. In Potter and Matter (178), to select a representative subset from a data set, hierarchical clustering and the maximum dissimilarity method (179–181) were used. The authors showed that both methods choose representative subsets of compounds much better than the random selection. Compounds selected through use of the maximum dissimilarity method were used as training sets in 3D-QSAR studies, with all remaining compounds composing the test set. In Wu et al. (166) the Kennard-Stone (182–184) method, which is similar to the maximum dissimilarity method, was applied to the classification of NIR spectra and QSAR analysis. The drawbacks of clustering methods are that different clusters contain different numbers of points and have different densities of representative points. Therefore, the closeness of each point of the test set to at least one point of the training set is not guaranteed. The maximum dissimilarity and Kennard-Stone methods guarantee that the points of the training set are distributed more or less evenly within the whole area occupied by representative points, and the condition of closeness of the test set points to the training set points is satisfied. The maximum distance between training and test set points in these methods does not exceed the radius of the probe sphere.

To select a representative subset of samples from the whole data set, factorial designs (185, 186) and D-optimal designs (187) were used (166, 173, 188). Factorial designs presume that different sample properties (such as substituent groups at certain positions) are divided into groups. The training set includes one representative for each combination of properties. For a diverse data set this approach is impractical, and fractional factorial designs are used, in which only a part of all combinations is included into the training set. Generally, this approach does not guarantee the closeness of the test set points to the training set points in the descriptor space. D-optimal design algorithms select samples that maximize the $|X'X|$ determinant, where X is the information (variance-covariance) matrix of independent variables (descriptors) (189, 190). The points maximizing the $|X'X|$ determinant are spanned across the whole area occupied by representative points. They can be used as a training set, and the points not selected then are used as the test set (166, 173).

In Wu et al. (166) four methods of sample selection (random, SOM, Kennard-Stone design, and D-optimal design) were compared. The best models were built when Kennard-Stone and D-optimal designs were used. SOM was better than random selection, and D-optimal design was slightly better than the random selection.

4.3 Guiding Principles of Safe QSAR

A widely used approach to establish the model robustness is so-called y-randomization (randomization of response, i.e., in our case, activities) (191). It consists of repeating the calculation procedure with randomized activities and subsequent probability assessment of the resultant statistics. Frequently, it is used along with cross-validation. It is expected that models obtained for the data set with randomized activity should have low values of q^2; otherwise, the original model should be considered insignificant. We suggest that the y-randomization test is a mandatory component of model validation.

Several authors have suggested that the only way to estimate the true predictive power of a QSAR model is to compare the predicted and observed activities of an (sufficiently large) external test set of compounds that were not used in the model development (46, 163, 192–194). To estimate the predictive power of a QSAR model, we recommended use of the following statistical characteristics of the test set (163): (i) correlation coefficient R between the predicted and observed activities; (ii) coefficients of determination (195) (predicted vs. observed activities R_0^2, and observed vs. predicted activities $R_0'^2$); (iii) slopes k and k' of the regression lines through the origin. We consider a QSAR model predictive, if the following conditions are satisfied (163):

$$q^2 > 0.5 \qquad (2.2)$$

$$R^2 > 0.6 \qquad (2.3)$$

$$\frac{(R^2 - R_0^2)}{R^2} < 0.1 \ \text{ or } \ \frac{(R^2 - R_0'^2)}{R^2} < 0.1 \quad (2.4)$$

$$0.85 \le k \le 1.15 \ \text{ or } \ 0.85 \le k' \le 1.15 \quad (2.5)$$

The lack of the correlation between q^2 and R^2 was noted in Kubinyi et al. (46), Novellino et al. (192), Norinder (193), and in our recent publication (163), where we demonstrated that all of the above-mentioned criteria are necessary to adequately assess the predictive ability of a QSAR model. We suggest (163) that the external test set must contain at least five compounds, representing the whole range of both descriptor and activities of compounds included into the training set.

5 QSAR MODELS AS VIRTUAL SCREENING TOOLS

5.1 Data Mining and SAR Analysis

Data mining has been of interest to researchers in machine learning, pattern recognition, artificial intelligence, database statistics, and so forth for many years, and widely applied in science, business, and government. Now, chemoinformatitians have also started to plunge into this field because of the increased quantity of data in the drug discovery process. *Data mining* can be defined as the process of discovering valid, novel, understandable, and potentially useful patterns in data (196, 197). Data mining is an interactive and iterative, multiple-step process, involving the decisions made by the user. It may include data collection, data cleaning, data engineering, algorithm engineering, algorithm running, result evaluation, and knowledge utilization (198, 199).

Data mining methods can be generally divided into two types, unsupervised and supervised. Whereas *unsupervised* methods seek *informative* patterns, which directly display the interesting relationship among the data, *supervised* methods discover *predictive* patterns, which can be used later to predict one or more attributes from the rest.

A wide variety of supervised data mining methods have been applied for analyzing structure-activity data sets, besides the traditional linear regression methods. Most of them are nonlinear and nonparametric and need no statistical assumptions to apply them. Decision tree and rule induction methods, such as ID3 (200), CART (201), and FIRM (202–204) usually use univariate splits to generate a model in the form of a tree or propositional logic. The inferred model is easy to comprehend, but the approximation power may be significantly restricted by a particular tree or rule representation. Inductive logic programming methods, such as GOLEM (64) and PROGOL (65), are designed to induce a model from the more flexible representation of first-order predicate logic. However, this generality comes at the price of significant computational demands. *Nonlinear regression and classification* methods, such as various neural networks (60–62), train a model by fitting linear and nonlinear combinations of basis functions to the combinations of the input variables. They may be powerful in terms of approximation, but they are statistically poorly characterized, slow (205), and difficult to interpret in chemical terms. *Example-based* methods, such as nearest-neighbor methods (147), use representative examples from the database as an approximate model and predicate new samples on the basis of the properties of the most similar examples in the model. They are asymptotically powerful for approximating properties, but also difficult to interpret. Furthermore, their performance is strongly dependent on a well-defined distance metric to evaluate distances between data points.

Data mining of chemical databases is still at its very early stage. Nevertheless, as a result of the data explosion in pharmaceutical industry, it is expected that data mining techniques will play an increasingly important role in the drug discovery process. Future studies may include, for example, the definition of chemical space, the validation of various algorithms (206), and the representation of extremely large virtual databases (207).

5.2 Virtual Screening

Although combinatorial chemistry and HTS have offered medicinal chemists a much broader range of possibilities for lead discovery and optimization, the number of chemical compounds that can be reasonably synthesized, which is sometimes called "virtual chemistry space," is still far beyond today's capability of chemical synthesis and biological assay. Therefore, medicinal chemists continue to face the same problem as before: Which compounds should be chosen for the next round of synthesis and testing? For chemoinformatitians, the task is to develop and utilize various computer programs to evaluate a very large number of chemical compounds and recommend the most promising ones for bench medicinal chemists. This process can be called *virtual screening* (208) or chemical database searching. A large number of computational methods exist for virtual screening, but which one is chosen will depend on the information available and the task at hand in practice.

A substructure search will typically be undertaken if a lead compound has been found. The search query will retrieve all the structures in a database that contain the substructures present in the lead compound that are believed to be important for activity (209). According to graph theory, it is equivalent to searching a series of topological graphs for the existence of a subgraph isomorphism with a specified query graph. Subgraph isomorphism is an NP-complete problem (210), which means that for it, there are no algorithms whose worst-case time requirements do not rise exponentially with the size of the input. However, various backtracking algorithms (211–213) and partitioning algorithms (214–217) have been developed since the 1950s, to reduce the *average* time required for chemical substructure searching. Today, almost all the chemical database software includes the function of substructure searching.

A similarity search provides a way forward by retrieving the structures that are similar, but not identical, to a lead compound (94). Therefore, it overcomes some limitations of substructure search, for example, not requiring specific knowledge about the substructures responsible for activity, and being able to rank the output structures according to the overall similarity. The search query usually involves a set of descriptors that collectively specify the whole structure of the lead compound. This set of descriptors is compared with the corresponding set of descriptors for each compound in the database, and then a measure of similarity is calculated between them. There are a wide variety of molecular descriptors for similarity searching (cf. Section 2). Not a single set of molecular descriptors has been found as the best choice in all the cases. The present trend in descriptor selection is to use combined descriptors with many different types. The similarity coefficients that are often used for measuring the similarity between two structures includes Manhattan distance, Euclidean distance, Soergel distance, Tanimoto coefficient, Dice coefficient, Cosine coefficient, and so forth (218), and again no clear-cut winner has been found among them (219). Virtual screening based on QSAR models can serve as a powerful approach to the design of targeted chemical libraries, as illustrated in the following section.

5.3 Rational Library Design by use of QSAR

As discussed earlier, combinatorial chemical synthesis and high throughput screening have significantly increased the speed of the drug discovery process (220–222). However, it remains impossible to synthesize all of the library compounds in a reasonably short period of time. For instance, 3000^3 (2.7×10^{10}) compounds can be synthesized from a molecular scaffold with three different substitution positions when each of the positions has 3000 different substituents. If a chemist could synthesize 1000 compounds per week, 27 million weeks (~0.5 million years) would be required to synthesize all these compounds. Furthermore, many of these compounds can be structurally similar to each other, thus making redundant the chemical information contained in the library. There is a need for rational library design (i.e., rational selection of a subset of available building blocks for combinatorial chemical synthesis), so that a maximum amount of information can be obtained while a minimum number of compounds are synthesized and tested. Similarly, there is a closely related task in computational database mining, that is, rational selection of a subset of compounds from commercially available or proprietary databases for biological testing.

Thus, in many practical cases, the exhaustive synthesis and evaluation of combinatorial libraries is prohibitively expensive, time-con-

suming, or redundant (223). Modern rational approaches to the design of combinatorial libraries have been explored in a recent monograph (224). Theoretical analysis of available experimental information about the biological target or pharmacological compounds capable of interacting with the target can significantly enhance the rational design of targeted chemical libraries. In many cases, the number of compounds with known biological activity is sufficiently large to develop viable QSAR models for such data sets. These models can be used as a means of selecting virtual library compounds (or actual compounds from existing databases) with (high) predicted biological activity. Alternatively, if a variable selection method has been employed in developing a QSAR model, the use of only selected variables can improve the performance of the rational library design or database mining methods on the basis of the similarity to a probe. This procedure of use of only selected variables in a similarity search in the descriptor space is analogous to more traditional use of conventional chemical pharmacophores in database mining.

QSAR models can be employed for rational design of targeted chemical libraries and database mining by predicting biologically active structures in virtual or actual chemical libraries (225–227). To illustrate this approach, we consider the design of a pentapeptide combinatorial library with the bradykinin activity by use of a QSAR model derived for a small bradykinin peptide data set. Figure 2.9 shows the schematic diagram illustrating the targeted pentapeptide combinatorial library design by use of the FOCUS-2D method (225, 226). The algorithm includes the description, evaluation, and optimization steps.

To identify potentially active compounds in the virtual library, FOCUS-2D employs stochastic optimization methods such as SA (228, 229) and GA (230–232). The latter algorithm was used for targeted pentapeptide library design as follows. Initially, a population of 100 peptides is randomly generated and encoded by use of topological indices or amino acid-dependent physicochemical descriptors. The fitness of each peptide is evaluated by its biological activity predicted from a preconstructed QSAR equation (see below). Two par-

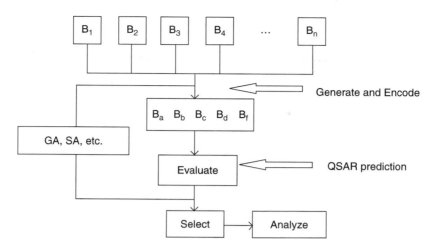

Figure 2.9. Flowchart of the library design approach by FOCUS-2D.

ent peptides are chosen by use of the roulette wheel selection method (i.e., high fitting parents are more likely to be selected). Two offspring peptides are generated by a crossover (i.e., two randomly chosen peptides exchange their fragments) and mutations (i.e., a randomly chosen amino acid in an offspring is changed to any of 19 remaining amino acids). The fitness of the offspring peptides is then evaluated and compared with that of the parent peptides, and the two lowest scoring peptides are eliminated. This process is repeated for 2000 times to evolve the population.

Design of a Targeted Library with Bradykinin (BK) Potentiating Activity. The results obtained with the FOCUS-2D and a QSAR-based prediction are shown in Figure 2.10. The position-dependent frequency distributions of amino acids in the highest scoring pentapepeptides are shown before (Fig. 2.10a) and after (Fig. 2.10b,c) FOCUS-2D. To evaluate the efficiency of stochastic sampling, the entire pentapeptide library (which includes as many as 3.2 million molecules) was also generated and subjected to evaluation by use of the same QSAR model, and the results are shown in Fig. 2.10c. Apparently, the results after FO-CUS-2D and the exhaustive search were very similar to each other. FOCUS-2D selected the following amino acids: E, I, K, L, M, Q, R, V, and W. Interestingly, these selected amino acids included most of those found in the two experimentally most active pentapeptides,

VEWAK and VKWAP (excluded from the training set for the QSAR model development). Furthermore, the actual spatial positions of these amino acids were correctly identified: the first and fourth positions for V; the second and fifth positions for E; the third position for W; and the second and fifth positions for K. More detailed analysis of these results (cf. Fig. 2.10b,c) may suggest which residues should be preferably chosen for each position in the pentapeptide to achieve a limited size library with high predicted bradykinin activity.

6 CONCLUSIONS

In this chapter, we have reviewed recent and developing trends in the field of QSAR. We have provided common terminology and presented a unified concept of the QSAR approach. We have emphasized that, regardless of the origin of molecular descriptors, any QSAR modeling exercise starts from constructing a two-dimensional data array (Fig. 2.2), which lists molecular IDs, values of the target (or dependent) property of each compound, and values of descriptors (independent variables) for each compound. We have considered various protocols employed by QSAR practitioners to develop quantitative models of biological activity by the use of chemical descriptors and linear or nonlinear optimiza-

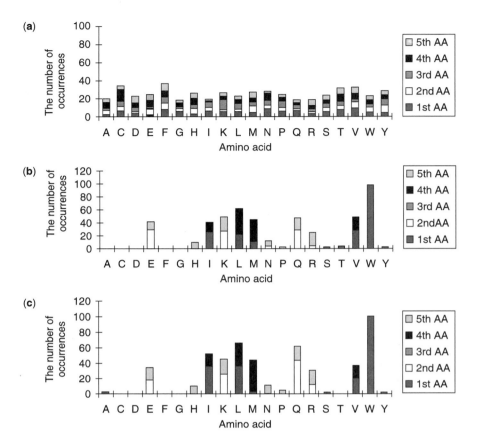

Figure 2.10. Ratonal selection of building blocks for library design by use of FOCUS-2D and a QSAR model for activity prediction: (a) initial population; (b) final population after FOCUS-2D; and (c) final population after the exhaustive search.

tion techniques. We have particularly empha-sized that the true power of any QSAR model comes from its statistical significance and the model's ability to predict accurately biological properties of chemical compounds both in the training and, most important, in the test sets. One of the important research challenges in the QSPR modeling remains finding descrip-tor types, correlation approaches, and ade-quate statistical characteristics of the training set only, which may ensure high predictive power of the models.

In conclusion, we strongly advocate rigor-ous validation of QSAR models before their practical application or interpretation. The practical guidelines for the development of statistically robust and predictive QSAR mod-els can be summarized as follows:

1. Establish an SAR database through the use of reliable quantitative measurements of the target property and a preferred set of molecular descriptors.

2. Divide the underlying data set into training and test sets through the use of diversity sampling algorithms.

3. Develop training set models through the use of available QSAR methods or commer-cial software. Characterize these models with internal validation parameters, as dis-cussed in this chapter, and define the appli-cability domain for each model.

4. Validate training set models through the use of an external test set and calculate the external validation parameters, as dis-cussed in this chapter. Ideally, repeat the

procedure of training and test selection and external validation several times to identify the QSAR model for the smallest training set that affords adequate prediction power for the biggest test set.

5. Finally, explore and exploit validated QSAR models for possible mechanistic interpretation and prediction.

In the modern age of medicinal chemistry, QSAR modeling remains one of the most important instruments of computer-aided drug design. Skillful application of various methodologies discussed in this chapter will afford validated QSAR models, which should continue to enrich and facilitate the experimental process of drug discovery and development.

REFERENCES

1. C. Hansch, R. M. Muir, T. Fujita, P. P. Maloney, E. Geiger, and M. Streich, *J. Am. Chem. Soc.*, **85**, 2817 (1963).

2. T. Fujita, J. Iwasa, and C. Hansch. *J. Am. Chem. Soc.*, **86**, 5175 (1964).

3. L. P. Hammett, *Chem. Rev.*, **17**, 125 (1935).

4. C. Hansch and A. Leo in S. R. Heller, Ed., *Exploring QSAR: Fundamentals and Applications in Chemistry and Biology*, American Chemical Society, Washington, DC, 1995.

5. C. Hansch, A. Leo, and D. Hoekman in S. R. Heller, Ed., *Exploring QSAR: Hydrophobic, Electronic, and Steric Constants*. American Chemical Society, Washington, DC, 1995.

6. A. Verloop, W. Hoogenstraaten, and J. Tipker in E. J. Ariens, Ed., Drug Design, Vol. **VII**, Academic Press, New York, 1976, 165 pp.

7. C. Selassie, this volume, Chapter 1.

8. H. Kubinyi in R. Mannhold, P. Krogsgaard-Larsen, and H. Timmerman, Eds., *Methods and Principles in Medicinal Chemistry*, Vol. **1**, VCH, New York, 1993.

9. D. Livingstone, *Data Analysis for Chemists: Applications to QSAR and Chemical Product Design*, Oxford University Press, Oxford, UK, 1995.

10. H. Kubinyi, G. Folkers, and Y. Martyn, Eds., 3D QSAR in Drug Design, Vols. **2** and **3**, Kluwer/ESCOM, Dordrecht, The Netherlands, 1998.

11. M. Karelson, *Molecular Descriptors in QSAR/QSPR*, Wiley-Interscience, New York, 2000.

12. D. J. Livingstone, *J. Chem. Inf. Comput. Sci.*, **40**, 195–209 (2000).

13. Chemical Abstracts Service (CAS), Columbus, OH. May be accessed at http://www.cas.org

14. D. S. Tan, M. A. Foley, M. D. Shair, and S. L. Schreiber, *J. Am. Chem. Soc.*, **120**, 8565–8566 (1998).

15. J. Drews, *Science*, **287**, 1960–1964 (2000).

16. J. G. Topliss and R. P. Edwards, *J. Med. Chem.*, **22**, 1238 (1979).

17. U. Burkert and N. L. Allinger, *Molecular Mechanics*, American Chemical Society, Washington, DC, 1982.

18. C. Hansch and T. Fujita, *J. Am. Chem. Soc.*, **86**, 1616–1626 (1964).

19. M. Randíc, *J. Am. Chem. Soc.*, **97**, 6609–6615 (1975).

20. L. B. Kier and L. H. Hall, *Molecular Connectivity in Chemistry and Drug Research*, Academic Press, New York, 1976.

21. L. B. Kier and L. H. Hall, *Molecular Connectivity in Structure-Activity Analysis*, Research Studies Press, Chichester, UK, 1986.

22. L. B. Kier and L. H. Hall in K. B. Lipkowitz and D. B. Boyd, Eds., *Reviews in Computational Chemistry II*, VCH, Weinheim/New York, 1991, pp. 367–422.

23. L. B. Kier, *Quant. Struct.-Act. Relat.*, **4**, 109–116 (1985).

24. L. B. Kier, *Quant. Struct-Act. Relat.*, **6**, 8–12 (1987).

25. L. H. Hall and L. B. Kier, *Quant. Struct.-Act. Relat.*, **9**, 115–131 (1990).

26. L. H. Hall, B. K. Mohney, and L. B. Kier, *Quant. Struct.-Act. Relat.*, **10**, 43–51 (1991).

27. L. H. Hall, B. K. Mohney, and L. B. Kier, *J. Chem. Inf. Comput. Sci.*, **31**, 76–82 (1991).

28. L. B. Kier and L. H. Hall, *Molecular Structure Description: The Electrotopological State*, Academic Press, Orlando, FL, 1999.

29. G. E. Kellogg, L. B. Kier, P. Gaillard, and L. H. Hall, *J. Comput.-Aided Mol. Des.*, **10**, 513–520 (1996).

30. R. P. Sheridan, R. B. Nachbar, and B. L. Bush, *J. Comput.-Aided Mol. Des.*, **8**, 323–340 (1994).

31. H. Matter, *J. Med. Chem.*, **40**, 1219–1229 (1997).

32. A. J. Hopfinger, *J. Am. Chem. Soc.*, **102**, 7196 (1980).

33. G. M. Crippen, *J. Med. Chem.*, **22**, 988–997 (1979).

34. G. M. Crippen, *J. Med. Chem.*, **23**, 599–606 (1980).

35. L. G. Boulu and G. M. Crippen, *J. Comput. Chem.*, **10**, 673 (1989).

36. U. Holzbrabe and A. J. Hopfinger, *J. Chem. Inf. Comput. Sci.*, **36**, 1018 (1996).

37. A. J. Hopfinger, B. J. Burke, and W. J. Dunn, *J. Med. Chem.*, **37**, 3768 (1994).

38. S. Srivastava and G. M. Crippen, *J. Med. Chem.*, **36**, 3572 (1993).

39. M. P. Bradley and G. M. Crippen, *J. Med. Chem.*, **36**, 3171 (1993).

40. R. D. Cramer III, D. E. Patterson, and J. D. Bunce, *J. Am. Chem. Soc.*, **110**, 5959–5967 (1988).

41. S. Wold, A. Ruhe, H. Wold, and W. J. Dunn III, *SIAM J. Sci. Stat. Comput.*, **5**, 735–743 (1984).

42. P. Geladi and B. R. Kowalski, *Anal. Chim. Acta*, **185**, 1–17 (1986).

43. G. R. Marshall and R. D. Cramer III, *Trends Pharmacol. Sci.*, **9**, 285–289 (1988).

44. C. Pérez, M. Pastor, A. R. Ortiz, and F. Gago, *J. Med. Chem.*, **41**, 836–852 (1998).

45. G. Klebe in H. Kubinyi, G. Folkers, and Y. C. Martin, Eds., 3D QSAR in Drug Design, Vol. **3**, Kluwer/ESCOM, Dordrecht, The Netherlands, 1998, pp. 87–104.

46. H. Kubinyi, F. A. Hamprecht, T. Mietzner, *J. Med. Chem.*, **41**, 2553–2564 (1998).

47. S. Wold in H. van de Waterbeemd, Ed., *Chemometrics Methods in Molecule Design*, VCH, Weinheim/New York, 1995, pp. 195–218.

48. B. Hoffman, S. J. Cho, W. Zheng, S. Wyrick, D. E. Nichols, R. B. Mailman, and A. Tropsha, *J. Med. Chem.*, **42**, 3217–3226 (1999).

49. W. Zheng and A. Tropsha, *J. Chem. Inf. Comput. Sci.*, **40**, 185–194 (2000).

50. Ajay, *J. Med. Chem.*, **36**, 3565–3571 (1993).

51. L. S. Anker and P. C. Jurs, *Anal. Chem.*, **62**, 2676 (1990).

52. P. C. Jurs, J. W. Ball, and L. S. Anker, *J. Chem. Inf. Comput. Sci.*, **32**, 272 (1992).

53. T. M. Nelson and P. C. Jurs, *J. Chem. Inf. Comput. Sci.*, **34**, 601 (1994).

54. D. T. Stanton and P. C. Jurs, *J. Chem. Inf. Comput. Sci.*, **32**, 109 (1992).

55. S. Hellberg, M. Sjostrom, B. Skagerberg, and S. Wold, *J. Med. Chem.*, **30**, 1126–1135 (1987).

56. B. R. Kowalski and C. F. Bender, *J. Am. Chem. Soc.*, **96**, 916–918 (1974).

57. K. C. Chu, R. J. Feldmann, N. B. Shapiro, G. F. Harard, and R. I. Geran, *J. Med. Chem.*, **18**, 539–545 (1975).

58. G. Klopman, *J. Am. Chem. Soc.*, **106**, 7315–7321 (1984).

59. G. Klopman, *Quant. Struct.-Act. Relat.*, **11**, 176–184 (1992).

60. T. Aoyama, Y. Suzuki, and H. Ichikawa, *J. Med. Chem.*, **33**, 2583–2590 (1990).

61. S.-S. So and W. G. Richards, *J. Med. Chem.*, **35**, 3201–3207 (1992).

62. F. R. Burden, B. S. Rosewarne, and D. A. Winkler, *Chemom. Intel. Lab. Syst.*, **38**, 127–137 (1997).

63. G. Bolis, L. Di Pace, and F. Fabrocini, *J. Comput.-Aided Mol. Des.*, **5**, 617–628 (1991).

64. R. D. King, S. H. Mugglfton, R. A. Lewis, and M. J. E. Sternberg, *Proc. Natl. Acad. Sci. USA*, **89**, 11322–11326 (1992).

65. R. D. King, S. H. Muggleton, A. Srinivasan, and M. J. E. Sternberg, *Proc. Natl. Acad. Sci. USA*, **93**, 438–442 (1996).

66. S. Clementi and S. Wold in H. van de Waterbeemd, Ed., *Chemometrics Methods in Molecular Design*, VCH, Weinheim/New York, 1995, pp. 319–338.

67. J. M. Sutter, S. L. Dixon, and P. C. Jurs, *J. Chem. Inf. Comput. Sci.*, **35**, 77 (1995).

68. D. Rogers and A. J. Hopfinger, *J. Chem. Inf. Comput. Sci.*, **34**, 854–866 (1994).

69. H. Kubinyi, *Quant. Struct.-Act. Relat.*, **13**, 285–294 (1994).

70. H. Kubinyi, *Quant. Struct.-Act. Relat.*, **13**, 393–401 (1994).

71. B. T. Luke, *J. Chem. Inf. Comput. Sci.*, **34**, 1279–1287 (1994).

72. S.-S. So and M. Karplus, *J. Med. Chem.*, **39**, 1521–1530 (1996).

73. K. Hasegawa, T. Kimura, and K. Funatsu, *J. Chem. Inf. Comput. Sci.*, **39**, 112–120 (1999).

74. M. Baroni, G. Costantino, G. Cruciani, D. Riganelli, R. Valigi, and S. Clementi, *Quant. Struct.-Act. Relat.*, **12**, 9–20 (1993).

75. S. J. Cho and A. Tropsha, *J. Med. Chem.*, **38**, 1060–1066 (1995).

76. L. B. Kier, L. H. Hall, and J. W. Frazer, *J. Chem. Inf. Comput. Sci.*, **33**, 143 (1993).

77. L. H. Hall, L. B. Kier, and J. W. Frazer, *J. Chem. Inf. Comput. Sci.*, **33**, 148 (1993).

78. L. H. Hall, R. S. Dailey, and L. B. Kier, *J. Chem. Inf. Comput. Sci.*, **33**, 598 (1993).

79. L. H. Hall and L. B. Kier in K. B. Lipkowitz and D. B. Boyd, Eds., *Reviews in Computational Chemistry II*, VCH, Weinheim/New York, 1991, pp. 367–422.

80. A. K. Debnath, R. L. Lopez de Compadre, G. Debnath, A. J. Shusterman, and C. Hansch, *J. Med. Chem.*, **34**, 786–797 (1991).

81. A. N. Jain, K. Koile, and D. Chapman, *J. Med. Chem.*, **37**, 2315–2327 (1994).

82. F. R. Burden, *Quant. Struct.-Act. Relat.*, **15**, 7–11 (1996).

83. P. G. Dittmar, N. A. Farmer, W. Fisanick, R. C. Haines, and J. Mockus, *J. Chem. Inf. Comput. Sci.*, **23**, 93–102 (1983).

84. R. E. Carhart, D. H. Smith, and R. Venkataraghavan, *J. Chem. Inf. Comput. Sci.*, **25**, 64–73 (1985).

85. R. Nilakantan, N. Bauman, J. S. Dixon, and R. Venkataraghavan, *J. Chem. Inf. Comput. Sci.*, **27**, 82–85 (1987).

86. C. A. Pepperrell and P. Willett, *J. Comput.-Aided Mol. Des.*, **5**, 455–474 (1991).

87. R. Nilakantan, N. Bauman, and R. Venkataraghavan, *J. Chem. Inf. Comput. Sci.*, **33**, 79–85 (1993).

88. R. P. Sheridan, M. D. Miller, D. J. Underwood, and S. K. Kearsley, *J. Chem. Inf. Comput. Sci.*, **36**, 128–136 (1996).

89. F. R. Burden, *Quant. Struct.-Act. Relat.*, **16**, 309–314 (1997).

90. B. D. Silverman and D. E. Platt, *J. Med. Chem.*, **39**, 2129–2140 (1996).

91. D. A. Winkler, F. R. Burden, and A. Watkins, *Quant. Struct.-Act. Relat.*, **17**, 14–19 (1998).

92. R. D. Brown and Y. C. Martin, *J. Chem. Inf. Comput. Sci.*, **37**, 1–9 (1997).

93. D. A. Winkler and F. R. Burden, *Quant. Struct.-Act. Relat.*, **17**, 224–231 (1998).

94. G. M. Downs and P. Willett in K. B. Lipkowitz and D. B. Boyd, Eds., Reviews in Computational Chemistry, Vol. **7**, VCH, Weinheim/New York, 1996, pp. 1–65.

95. Molconn-Z version 3.5, Hall Associates Consulting, Quincy, MA.

96. M. Petitjean, *J. Chem. Inf. Comput. Sci.*, **32**, 331–337 (1992).

97. H. Wiener, *J. Am. Chem. Soc.*, **69**, 17 (1947).

98. J. R. Platt, *J. Phys. Chem.*, **56**, 328 (1952).

99. C. Shannon and W. Weaver, *Mathematical Theory of Communication*, University of Illinois, Urbana, 1949.

100. D. Bonchev, O. Mekenyan, and N. Trinajstíc, *J. Comput. Chem.*, **2**, 127–148 (1981).

101. The program Sybyl is available from Tripos Associates, St. Louis, MO.

102. M. Shen, A. LeTiran, Y. Xiao, H. Kohn, and A. Tropsha, *J. Med. Chem.*, **45**, 2811–2823 (2002).

103. F. H. Allen, J. E. Davies, J. J. Galloy, O. Johnson, O. Kennard, C. F. Macrae, E. M. Mitchell, G. F. Mitchell, J. M. Smith, and D. G. Watson, *J. Chem. Inf. Comput. Sci.*, **31**, 187–204 (1991).

104. F. H. Allen, S. Bellard, M. D. Brice, B. A. Cartwright, A. Doubleday, H. Higgs, T. Hummelink, B. G. Hummelink-Peters, O. Kennard, W. D. S. Motherwell, J. R. Rodgers, and D. G. Watson, *Acta Crystallogr. Sect. B*, **B35**, 2331–2339 (1979).

105. A. Rusinko III, J. M. Skell, R. Balducci, C. M. McGarity, and R. S. Pearlman, *Concord, A Program for the Rapid Generation of High Quality Approximate 3-Dimensional Molecular Structures*, The University of Texas at Austin and Tripos Associates, St. Louis, MO, 1988.

106. R. S. Pearlman, *Chem. Des. Aut. News*, **2**, 1–6 (1987).

107. J. Gasteiger, C. Rudolph, and J. Sadowski, *Tetrahedron Comput. Methodol.*, **3**, 537–547 (1990).

108. G. R. Marshall, C. D. Barry, H. E. Bosshard, R. A. Dammkoehler, and D. A. Dunn in E. C. Olson and R. E. Christoffersen, Eds., Computer-Assisted Drug Design, Vol. **112**, American Chemical Society, Washington DC, 1979, pp. 205–226.

109. G. E. Kellogg, S. F. Semus, and D. J. Abraham, *J. Comput.-Aided Mol. Des.*, **5**, 545–552 (1991).

110. P. J. Goodford, *J. Med. Chem.*, **28**, 849–857 (1985).

111. A. Agarwal, P. P. Pearson, E. W. Taylor, H. B. Li, T. Dahlgren, M. Herslof, Y. Yang, G. Lambert, D. L. Nelson, J. W. Regan, and A. R. Martin, *J. Med. Chem.*, **36**, 4006–4014 (1993).

112. C. L. Waller, T. I. Oprea, A. Giolitti, and G. R. Marshall, *J. Med. Chem.*, **36**, 4152–4160 (1993).

113. A. K. Debnath, C. Hansch, K. H. Kim, and Y. C. Martin, *J. Med. Chem.*, **36**, 1007–1016 (1993).

114. M. Y. Brusniak, R. S. Pearlman, K. A. Neve, and R. E. Wilcox, *J. Med. Chem.*, **39**, 850–859 (1996).

115. Y. C. Martin, *Methods Enzymol.*, **203**, 587–613 (1991).

116. Y. C. Martin, M. G. Bures, E. A. Danaher, J. DeLazzer, I. Lico, and P. A. Pavlik, *J. Comput.-Aided Mol. Des.*, **7**, 83–102 (1993).

117. S. J. Cho, M. G. Serrano, J. Bier, and A. Tropsha, *J. Med. Chem.*, **39**, 5064–5071 (1996).

118. J. L. Sussman, M. Harel, F. Frolow, C. Oefner, A. Goldman, L. Toker, and I. Silman, *Science*, **253**, 8872–8879 (1991).

119. M. Harel, I. Schalk, L. Ehret-Sabatier, F. Bouet, M. Goeldner, C. Hirth, P. H. Axelsen, I. Silman, and J. L. Sussman, *Proc. Natl. Acad. Sci. USA*, **90**, 9031–9035 (1993).

120. R. D. Cramer III, S. A. DePriest, D. E. Patterson, and P. Hecht in H. Kubinyi, Ed., *3D QSAR in Drug Design: Theory, Methods, and Applications*, ESCOM Scientific, Leiden, The Netherlands, 1993, pp. 443–485.

121. M. Baroni, G. Costantino, G. Cruciani, D. Riganelli, R. Valigi, and S. Clementi, *Quant. Struct.-Act. Relat.*, **12**, 9–20 (1993).

122. M. Pastor, G. Cruciani, I. McLay, S. Pickett, and S. Clementi, *J. Med. Chem.*, **43**, 3233–3243 (2000).

123. P. Ehrlich, *Dtsch. Chem. Ges.*, **42**, 17 (1909).

124. C. Humblet and G. R. Marshall, *Annu. Rep. Med. Chem.*, **15**, 267–276 (1980).

125. S. A. DePriest, D. Mayer, C. B. Naylor, and G. R. Marshall, *J. Am. Chem. Soc.*, **115**, 5372–5384 (1993).

126. S. Wang, D. W. Zaharevitz, R. Sharma, V. E. Marquez, N. E. Lewin, L. Du, P. M. Blumberg, and G. W. A. Milne, *J. Med. Chem.*, **37**, 4479–4489 (1994).

127. I. Motoc, R. A. Dammkoehler, and G. R. Marshall, *Mathematics and Computational Concepts in Chemistry*, Ellis Horwood, Chichester, UK, 1985, pp. 222–251.

128. D. Mayer, C. B. Naylor, I. Motoc, and G. R. Marshall, *J. Comput.-Aided Mol. Des.*, **1**, 3–16 (1987).

129. R. P. Sheridan, R. Nilakantan, J. S. Dixon, and R. Venkataraghavan, *J. Med. Chem.*, **29**, 899–906 (1986).

130. G. W. A. Milne, M. C. Nicklaus, J. S. Driscoll, S. Wang, and D. Zaharevitz, *J. Chem. Inf. Comput. Sci.*, **34**, 1219–1224 (1994).

131. Catalyst/Hypo Tutorial, version 2.0, BioCAD Corp., Mountain View, CA, 1993.

132. P. W. Sprague, *Perspect. Drug Discov. Des.*, **3**, 1–20 (1995).

133. D. Barnum, J. Greene, A. Smellie, and P. Sprague, *J. Chem. Inf. Comput. Sci.*, **36**, 563–571 (1996).

134. HipHop Tutorial, version 2.3, Molecular Simulation Inc., Sunnyvale, CA, 1995.

135. V. Golender and B. Vesterman, *Network Science* (http://www.netsci.org/Science/Compchem/feature09. html).

136. (a) Available from the author's WWW home page at http://mmlin1.pha.unc.edu/~jin/QSAR/ (b) A. Tropsha, S. J. Cho, and W. Zheng in A. L. Parrill and M. R. Reddy, Eds., *Rational Drug Design: Novel Methodology and Practical Applications*, ACS Symposium Series 719, 1999, pp. 198–211.

137. T. A. Andrea and H. Kalayeh, *J. Med. Chem.*, **34**, 2824–2836 (1991).

138. J. D. Hirst, R. D. King, and M. J. Sternberg, *J. Comput.-Aided Mol. Des.*, **8**, 405–420 (1994).

139. J. D. Hirst, R. D. King, and M. J. Sternberg, *J. Comput.-Aided Mol. Des.*, **8**, 421–432 (1994).

140. I. V. Tetko, V. Yu. Tanchuk, N. P. Chentsova, S. V. Antonenko, G. I. Poda, V. P. Kukhar, and A. I. Luik, *J. Med. Chem.*, **37**, 2520–2526 (1994).

141. D. T. Manallack, D. D. Ellis, and D. J. Livingstone, *J. Med. Chem.*, **37**, 3758–3767 (1994).

142. D. J. Maddalena and G. A. Johnston, *J. Med. Chem.*, **38**, 715–724 (1995).

143. G. Bolis, L. Pace, and F. A. Fabrocini, *J. Comput.-Aided Mol. Des.*, **5**, 617–628 (1991).

144. R. D. King, S. Muggleton, R. A. Lewis, and M. J. Sternberg, *Proc. Natl. Acad. Sci. USA*, **89**, 11322–11326 (1992).

145. A. N. Jain, T. G. Dietterich, R. H. Lathrop, D. Chapman, R. E. Critchlow Jr., B. E. Bauer, T. A. Webster, and T. Lozano-Perez, *J. Comput.-Aided Mol. Des.*, **8**, 635–652 (1994).

146. J. D. Hirst, *J. Med. Chem.*, **39**, 3526–3532 (1996).

147. V. S. Rose, J. Wood, and H. J. H. MacFie in H. van de Waterbeemd, Ed., *Advanced Computer-Assisted Techniques in Drug Discovery*, VCH, Weinheim/New York, 1995, pp. 228–242.

148. Y. Hamamoto, S. Uchimura, and S. Tomita, *IEEE Trans. Pattern Anal. Machine Intell.*, **19**, 73–79 (1997).

149. A. Djouadi and E. Bouktache, *IEEE Trans. Pattern Anal. Machine Intell.*, **19**, 277–282 (1997).

150. O. Strouf, *Chemical Pattern Recognition*, Research Studies Press, Chichester, UK, 1986.

151. M. L. Raymer, P. C. Sanschagrin, W. F. Punch, S. Venkataraman, E. D. Goodman, and L. A. Kuhn, *J. Mol. Biol.*, **265**, 445–464 (1997).

152. S. C. Basak and G. D. Grunwald, *SAR QSAR Environ. Res.*, **3**, 265–277 (1995).

153. S. C. Basak, S. Bertelsen, and G. D. Grunwald, *Toxicol. Lett.*, **79**, 239–250 (1995).

154. S. C. Basak and G. D. Grunwald, *Chemosphere*, **31**, 2529–2546 (1995).

155. S. C. Basak and G. D. Grunwald, *New J. Chem.*, **19**, 231 (1995).

156. X. Gironés, A. Gallegos, and C.-D. Ramon, *J. Chem Inf. Comput. Sci.*, **46**, 1400–1407 (2000).

157. B. Bordás, T. Kömíves, Z. Szántó, and A. Lopata, *J. Agric. Food Chem.*, **48**, 926–931 (2000).

158. Y. Fan, L. M. Shi, K. W. Kohn, Y. Pommier, and J. N. Weinstein, *J. Med. Chem.*, **44**, 3254–3263 (2001).

159. M. Randíc and S. C. Basak, *J. Chem. Inf. Comput. Sci.*, **40**, 899–905 (2000).

160. T. Suzuki, K. Ide, M. Ishida, and S. Shapiro, *J. Chem. Inf. Comput. Sci.*, **41**, 718–726 (2001).

161. M. Recanatini, A. Cavalli, F. Belluti, L. Piazzi, A. Rampa, A. Bisi, S. Gobbi, P. Valenti, V. Andrisano, M. Bartolini, and V. Cavrini, *J. Med. Chem.*, **43**, 2007–2018 (2000).

162. J. A. Morón, M. Campillo, V. Perez, M. Unzeta, and L. Pardo, *J. Med. Chem.*, **43**, 1684–1691 (2000).

163. A. Golbraikh and A. Tropsha, *J. Mol. Graphics Model.*, **20**, 269–276 (2002).

164. J. Huuskonen, *J. Chem. Inf. Comput. Sci.*, **41**, 425–429 (2001).

165. I. V. Tetko, V. V. Kovalishyn, and D. J. Livingstone, *J. Med. Chem.*, **44**, 2411–2420 (2001).

166. W. Wu, B. Walczak, D. L. Massart, S. Heuerding, F. Erni, I. R. Last, and K. A. Prebble, *Chemom. Intell. Lab. Syst.*, **33**, 35–46 (1996).

167. A. Yasri and D. Hartsough, *J. Chem. Inf. Comput. Sci.*, **41**, 1218–1227 (2001).

168. P. Bernard, D. B. Kireev, J. R. Chretien, P. L. Fortier, and L. Coppet, *J. Comput.-Aided Mol. Des.*, **13**, 355–371 (1999).

169. Y. Takeuchi, E. F. B. Shands, D. D. Beusen, and G. R. Marshall, *J. Med. Chem.*, **41**, 3609–3623 (1998).

170. G. V. Kauffman and P. C. Jurs, *J. Chem. Inf. Comput. Sci.*, **41**, 1553–1560 (2001).

171. B. E. Mattioni and P. C. Jurs, *J. Chem. Inf. Comput. Sci.*, **42**, 94–102 (2002).

172. J. Gasteiger and J. Zupan, *Angew. Chem.*, **32**, 503 (1993).

173. Y. L. Loukas, *J. Med. Chem.*, **44**, 2772–2783 (2001).

174. P. Bernard, M. Pintore, J. Y. Berthon, and J. R. Chretien, *Eur. J. Med. Chem.*, **36**, 1–19 (2001).

175. F. R. Burden and D. A. Winkler, *J. Med. Chem.*, **42**, 3183–3187 (1999).

176. F. R. Burden, M. G. Ford, D. C. Whitley, and D. A. Winkler, *J. Chem. Inf. Comput. Sci.*, **40**, 1423–1430 (2000).

177. M. J. Adams, *Chemometrics in Analytical Spectroscopy*, The Royal Society of Chemistry, London, 1995.

178. T. Potter and H. Matter, *J. Med. Chem.*, **41**, 478–488 (1998).

179. M. Lajiness, M. A. Johnson, and G. M. Maggiora in J. L. Fauchere, Ed., *Quantitative Structure-Activity Relationships in Drug Design*, Alan R. Liss, New York, 1989, pp. 173–176.

180. R. Taylor, *J. Chem. Inf. Comput. Sci.*, **35**, 59–67 (1995).

181. M. Snarey, N. K. Terrett, P. Willett, and D. J. Wilton, *J. Mol. Graphics Model.*, **15**, 372–385 (1997).

182. R. W. Kennard and L. A. Stone, *Technometrics*, **11**, 137–148 (1969).

183. B. Bourguignon, P. F. Deaguiar, K. Thorre, and D. L. Massart, *J. Chromatogr. Sci.*, **32**, 144–152 (1994).

184. B. Bourguignon, P. F. Deaguiar, M. S. Khots, and D. L. Massart, *Anal. Chem.*, **66**, 893–904 (1994).

185. S. Hellberg, L. Eriksson, J. Jonsson, F. Lindgren, M. Sjostrom, B. Skagerberg, S. Wold, and P. Andrews, *Int. J. Pept. Protein Res.*, **37**, 414–424 (1991).

186. L. Eriksson and E. Johansson, *Chemom. Intell. Lab. Syst.*, **34**, 1–19 (1996).

187. R. Carlson, *Design and Optimization in Organic Synthesis*, Elsevier, Amsterdam/New York, 1992.

188. E. J. Martin and R. E. Critchlow, *J. Comb. Chem.*, **1**, 32–45 (1999).

189. A. Miller and N.-K. Nguyen, *Appl. Stat.*, **43**, 669–678 (1994).

190. T. J. Mitchell, *Technometrics*, **42**, 48–54 (2000).

191. S. Wold and L. Eriksson in H. van de Waterbeemd, Ed., *Chemometrics Methods in Molecular Design*, VCH, Weinheim/New York, 1995, pp. 309–318.

192. E. Novellino, C. Fattorusso, and G. Greco, *Pharm. Acta Helv.*, **70**, 149–154 (1995).

193. U. Norinder, *J. Chemom.*, **10**, 95–105 (1996).

194. N. S. Zefirov and V. A. Palyulin, *J. Chem. Inf. Comput. Sci.*, **41**, 1022–1027 (2001).

195. L. Sachs, *Applied Statistics: A Handbook of Techniques*, Springer-Verlag, Berlin/New York, 1984.

196. U. M. Fayyad, G. Piatetsky-Shapiro, P. Smyth, and R. Uthurusamy, *Adavnces in Knowledge Discovery and Data Mining*, AAAI Press/The MIT Press, Cambridge, MA, 1996.

197. G. H. John, *Enhancements to the Data Mining Process*, Ph.D thesis, Stanford University, 1997.

198. U. M. Fayyad, G. Piatetsky-Shapiro, and P. Smyth, *From Data Mining to Knowledge Discovery*, AAAI Press/The MIT Press, Cambridge, MA, 1995.

199. E. Simoudis, *IEEE Expert.*, **11**, 26–33 (1996).

200. M. A. Razzak and R. C. Glen, *J. Comput.-Aided Mol. Des.*, **6**, 349–383 (1992).

201. R. D. King, J. D. Hirst, and M. J. E. Sternberg, *Appl. Artif. Intell.*, **9**, 213–234 (1994).

202. S. S. Young and D. M. Hawkins, *J. Med. Chem.*, **38**, 2784–2788 (1995).

203. S. S. Young and D. M. Hawkins, *SAR QSAR Environ. Res.*, **8**, 183–193 (1998).

204. D. M. Hawkins, S. S. Young, and A. Rusinko, *Quant. Struct.-Act. Relat.*, **16**, 1–7 (1997).

205. R. King, R. Henery, C. Feng, and A. Sutherland in D. Michie, S. Muggleton, and F. Furukawa, Eds., Machine Intelligence and Inductive Learning, Vol. **13**, Oxford University Press, Oxford, UK, 1994.

206. S. S. Young, M. Farmen, and A. Rusinko, *Network Science* (http://www.netsci.org/Science/Screening/feature09.html)

207. J. M. Barnard and G. M. Downs, *Perspect. Drug Discov. Des.*, **7/8**, 13–30 (1997).

208. W. P. Walters, M. T. Stahl, and M. A. Murcko, *Drug Discov. Today*, **3**, 160–178 (1998).

209. J. M. Barnard, *J. Chem. Inf. Comput. Sci.*, **33**, 532–538 (1993).

210. S. A. Cook, Proceedings of the Third Annual ACM Symposium on the Theory of Computing, ACM, New York, 1971, pp. 151–158.

211. L. C. Ray and R. A. Kirsch, *Science*, **126**, 814–819 (1957).

212. X. Jun and Z. Maosen, *Tetrahedron Comput. Methodol.*, **2**, 75–83 (1989).

213. A. Dengler and I. Ugi, *Comput. Chem.*, **15**, 103–107 (1991).

214. E. H. Sussenguth, *J. Chem. Doc.*, **5**, 36–43 (1965).

215. J. Figueras, *J. Chem. Doc.*, **12**, 237–244 (1972).

216. J. R. Ullmann, *J. Assoc. Comput. Mach.*, **23**, 31–42 (1976).

217. A. Von Scholley, *J. Chem. Inf. Comput. Sci.*, **24**, 235–241 (1984).

218. P. H. A. Sneath and R. R. Sokal, *Numerical Taxonomy*, Freeman, San Francisco, 1973.

219. P. Willett and V. A. Winterman, *Quant. Struct.-Act. Relat.*, **5**, 18–25 (1986).

220. M. A. Gallop, R. W. Barret, W. J. Dower, S. P. A. Fodor, and E. M. Gordon, *J. Med. Chem.*, **37**, 1233–1251 (1994).

221. E. M. Gordon, R. W. Barret, W. J. Dower, S. P. A. Fodor, and M. A. Gallop, *J. Med. Chem.*, **37**, 1385–1401 (1994).

222. W. A. Warr, *J. Chem. Inf. Comput. Sci.*, **37**, 134–140 (1997).

223. R. P. Sheridan and S. K. Kearsley, *J. Chem. Inf. Comput. Sci.*, **35**, 310–320 (1995).

224. A. K. Ghose and V. N. Viswanadhan, Eds., *Combinatorial Library Design and Evaluation for Drug Discovery: Principles, Methods, Software Tools and Applications*, Marcel Dekker, New York, 2001.

225. W. Zheng, S. J. Cho, and A. Tropsha, *J. Chem. Inf. Comp. Sci.*, **38**, 251–258 (1998).

226. S. J. Cho, W. Zheng, and A. Tropsha, *J. Chem. Inf. Comp. Sci.*, **38**, 259–268 (1998).

227. A. Tropsha, S. J. Cho, and W. Zheng in A. L. Parrill and M. R. Reddy, Eds., *Rational Drug Design: Novel Methodology and Practical Applications*, ACS Symposium Series 719, American Chemical Society, Washington, DC, 1999, pp. 198–211.

228. I. O. Bohachevsky, M. E. Johnson, and M. L. Stein, *Technometrics*, **28**, 209–217 (1986).

229. J. H. Kalivas, J. M. Sutter, and N. Roberts, *Anal. Chem.*, **61**, 2024–2030 (1989).

230. D. E. Goldberg, *Genetic Algorithm in Search, Optimization, and Machine Learning*, Addison-Wesley, Reading, MA, 1989.

231. J. H. Holland, *Sci. Am.*, **267**, 66–72 (1992).

232. S. Forrest, *Science*, **261**, 872–878 (1993).

CHAPTER THREE

Molecular Modeling in Drug Design

GARLAND R. MARSHALL
Washington University
Center for Computational Biology
St. Louis, Missouri

DENISE D. BEUSEN
Tripos, Inc.
St. Louis, Missouri

Contents

Burger's Medicinal Chemistry and Drug Discovery
Sixth Edition, Volume 1: Drug Discovery
Edited by Donald J. Abraham
ISBN 0-471-27090-3 © 2003 John Wiley & Sons, Inc.

1   INTRODUCTION

By historical imperative, the role of molecular modeling in drug design has been divided into two separate paradigms, one centered on the structure-activity problem that attempts to rationalize biological activity in the absence of detailed, three-dimensional structural information about the receptor, and the other focused on understanding the interactions seen in receptor-ligand complexes and using the known three-dimensional structure of the therapeutic target to design novel drugs. The

rapid increase in relevant structural information, attributed to advances in molecular biology to generate the target proteins in adequate quantities for study, and the equally impressive gains in NMR (1–9) and crystallography (10, 11) to provide three-dimensional structures as well as identify leads, have stimulated the need for design tools and the molecular modeling community is rapidly evolving useful approaches. The more common problem, however, is one in which the receptor can only be inferred from pharmacological studies and little, if any, structural information is available to guide in modeling. Nevertheless, useful information to guide the design and synthesis of potential novel therapeutics can be developed from an analysis of structure-activity data in the three-dimensional framework provided by current molecular modeling techniques. Although most of the techniques and approaches described have broader application than shown, the examples chosen should be sufficient to illustrate their use. A number of reviews (12–18) of computer-aided drug design have relevant sections covering portions of this chapter with different perspectives and are recommended for a more complete overview.

2 BACKGROUND AND METHODS

2.1 Molecular Mechanics

Molecular mechanics (19) treats a molecule as a collection of atoms whose interactions can be described by Newtonian mechanics. Because the mass of the nuclei is much greater than the mass of the electrons, one can separate (the Born-Oppenheimer approximation) the Schrödinger equation into a product of two functions: one for electrons, one for nuclei. For the purposes of molecular mechanics, the electronic function, initially developed to interpret spectroscopic data, is ignored; that is, the charge distribution is assumed to remain constant during changes in the position of the nuclei. Because molecular mechanics is based on classical physics, it cannot provide information about the electronic properties of molecules under study that are generally assumed fixed during the parameterization of the force field with experimental data.

A few words about the basics of molecular mechanics (19, 20) may provide the elements of understanding for what follows. This is not meant to be comprehensive, but rather a simple overview, to remind the reader of a few crucial points. For a comprehensive overview of molecular modeling, the reader is referred to the excellent text by Leach (21). The interactions between atoms are divided into bonded and nonbonded classes. Nonbonded forces between atoms are based on an attractive interaction that has a firm theoretical basis and varies as the inverse of the 6th power of

the distance between the atoms. It is balanced by a repulsion between the electronic clouds as the atoms come close and this interaction has been represented empirically by a variety of functional forms: exponential, 12th power, or 9th power of the distance between the atoms. The coefficients for these two interactions are parameterized for atom types, usually by element, so that the minimum of the combined functions corresponds to the sum of the experimental van der Waals radii for the two atoms.

In addition, bonded atoms are considered as a special case, with a "spring constant" determining the energy of deformation from experimental bond lengths. Atoms directly bonded to the same atom (one-three interactions) are eliminated from the van der Waals list and have a special energetic term relating the deviation from an ideal bond angle. Atoms having a one-four interaction define a torsional relation that is usually parameterized based on the types of the four connected atoms defining the torsion angle. The numerous combinations of atom types require an enormous number of parameters to be determined from either theoretical (quantum mechanics) and/or experimental data. Simplified force fields in which the torsional parameters depend only on the atoms at the end of a bond have been developed, to give approximate geometries for further refinement by quantum mechanics.

2.1.1 Force Fields. The basic assumption underlying molecular mechanics is that classical physical concepts can be used to represent the forces between atoms. In other words, one can approximate the potential energy surface by the summation of a set of equations representing pairwise and multibody interactions. These equations represent forces between atoms related to bonded and nonbonded interactions. Pairwise interactions are often represented by a harmonic potential $[\frac{1}{2}K_b(b - b_0)^2]$ that obeys Hooke's law (derived for a spring) for bonded atoms, restoring the bond distance to an equilibrium value b_0, and a van der Waals potential $[C_{12}(i,j)/r_{ij}^{12} - C_6(i,j)/r_{ij}^{6}]$ for nonbonded atoms. Similarly, distortion from an equilibrium valence angle (θ_0) describing the angle between three bonded atoms sharing a common atom is also penalized $[\frac{1}{2}K_\theta(\theta -$

$\theta_0)^2]$. A third class of interaction dependent on the dihedral angle ϕ between four bonded atoms is the torsional potential $\{K_\phi[1 + \cos(\phi - \delta)]\}$ used to account for orbital delocalization and to compensate for other deficiencies in the force field. A harmonic term $[\tfrac{1}{2}K_\xi(\xi - \xi_0)^2]$ is often introduced for dihedral angles ξ that are relatively fixed, such as those in aromatic rings. Coulomb's law $[q_iq_j/(4\pi\varepsilon_0\varepsilon_r r_{ij})]$ is the simplest approach to the contribution of electrostatics to the potential V:

$$V = \sum \frac{1}{2}K_b(b - b_0)^2 + \sum \frac{1}{2}K_\theta(\theta - \theta_0)^2$$
$$+ \sum \frac{1}{2}K_\xi(\xi - \xi_0)^2 + \sum K_\phi[1 + \cos(\phi - \delta)]$$
$$+ \sum [C_{12}(i,j)/r_{ij}^{12} - C_6(i,j)/r_{ij}^6]$$
$$+ \sum q_iq_j/(4\pi\varepsilon_0\varepsilon_r r_{ij}).$$

A central issue is the number of different atom types that are used in a particular force field. There is always a compromise between increasing the number to allow for the inclusion of more environmental effects (i.e., local electronic interactions) vs. the increase in the number of parameters to be determined to adequately represent a new atom type. In general, the more subtypes of atoms (how many different kinds of nitrogen, for example), the less likely that the parameters for a particular application will be available in the force field. The extreme, of course, would be a special atom type for each kind of atomic environment in which the parameters were chosen, so that the calculated properties of each molecule would simply reproduce the experimental observations. One major assumption, therefore, is that the force constants (parameters) and equilibrium values of the equations are functions of a limited number of atom types and can be transferred from one molecular environment to another. This assumption holds reasonably well where one may be primarily interested in geometric issues, but is not so valid in molecular spectroscopy. This had led to the introduction of additional equations, the so-called "cross-terms" which allow additional parameters to account for correlations between bond lengths and bond angles $[K_{b\theta}(b$

$- b_0)(\theta - \theta_0)]$, dihedral angles and bond angles, and so forth. Because of the lack of adequate parameterization of the more complex force fields that are usually specialized to one kind of molecule (e.g., proteins or nucleic acids), more simplified force fields have gained some popularity because of their general applicability, despite limited accuracy.

Examples are the Tripos force field (22), the COSMIC force field (23), and that of White and Bovill (24), which uses only two atom types, those at the end of the bond to parameterize the torsional potential rather than the four types of the atoms used to define the torsional angle. One has only to consider the number of combinations of 20 atom subtypes taken four at time (160,000) versus two at a time (400) to understand the explosion of parameters that occurs with increased atom subtypes. The simplifying assumption in parameterization of the torsional potential reduces to some extent the quality of the results (25), but allows the use of the simplified force fields (22) in many situations where other force fields would lack appropriate parameters. The situation can become complicated, however. For example, the amide bond is normally represented by one set of parameters, whether the configuration is *cis* or *trans*. Experimental data are quite compelling that the electronic state is different between the two configurations, and different parameter sets should be used for accurate results (Fig. 3.1). Only AMBER/OPLS currently distinguishes between these two conformational states (26). Certainly, the limited parameterization of simplified force fields would not allow accurate prediction of spectra that is more reflective of the dynamic behavior of the molecule.

Accurate estimates of energy may require accurate representation of the dynamics of molecules and justify derivation of the larger number of parameters. The new version (27) of the Allinger force field, MM3, has the objective of reproducing spectral data more accurately than MM2. Much of the chemistry remains to be incorporated into appropriate force fields. Only recently have adequate modifications been made to the force fields developed for organic molecules to include some metals (28–31). Carlsson (32, 33) recently de-

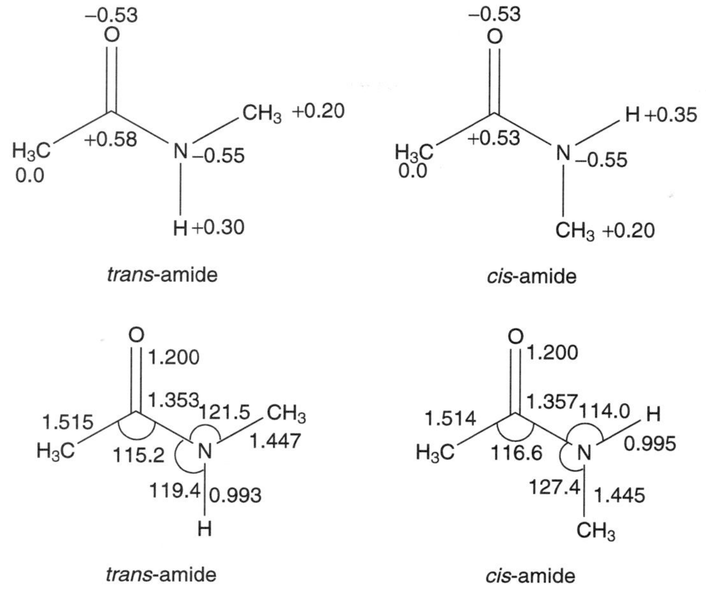

Figure 3.1. Differences in OPLS charge distribution (top) between *cis*- and *trans*-isomers of amide bond and geometries (bottom) as calculated by *ab initio* parameterization (26).

veloped a functional form that allows electronic *d*-orbitals of metals to be reasonably represented within molecular mechanics.

Because different force fields may use different mathematical representations of the forces between atoms and the details of their parameterization will in general differ also, it is unwise to use parameters derived for one force field to replace missing parameters in another. One often hears of a "balanced" parameter set that reproduces well the phenomena under consideration, but which is inadequate for other applications. A comparison by Burkert and Allinger (19) shows the different van der Waals (VDW) potentials used in several of the popular force fields, and the situation has not improved significantly in the intervening years. Because of other differences in parameters and functional forms of the equations used in the rest of the individual force fields, these quite different approaches to the VDW potential give excellent results when used in the correct combination. Indiscriminant combination of one part of a force field with another derived independently would lead to considerable divergence in the calculated results from those obtained by experimental observation.

The most extreme difference between force fields arises in the method by which the hydrogen bond is included. Because atoms involved in a hydrogen bond are often closer than the sum of their VDW radii, they must be handled in a special manner. Several force fields have special functional forms with angular dependency that not only have special VDW parameters, to ensure that the close approach of the atoms involved is calculated correctly, but that the angular distribution observed for hydrogen bonds is also reproduced. Hagler et al. (34) used an amide hydrogen with a zero VDW radius for hydrogen bonding and a slightly greater nitrogen radius to give a correct amide hydrogen bond distance. The charges on the atoms involved (including the amide hydrogen) are adjusted to give an appropriate balance of VDW repulsion and dipole attraction. Clearly, the method for handling the electrostatic interaction is an integral part of each force field and cannot be modified independently.

2.1.2 Electrostatics. The most difficult aspect of molecular mechanics is electrostatics (35–38). In most force fields, the electronic distribution surrounding each atom is treated as a monopole with a simple coulombic term for the interaction. The effect of the surrounding medium is generally treated with a continuum model by use of a dielectric constant. More detailed approaches with distributed multipole representations of the electron distribution (39, 40) and/or efforts to deal with dielectric inhomogeneity through solution of the Poisson equation are clear improvements and have become routine in many studies. Other difficulties arise in dealing with macromolecular systems, given that the electrostatic interaction is long ranged ($1/r$) and the interactions cannot be arbitrarily terminated with distance. Electrostatic interactions range from those operating only at very short distances that are nonspecific (dispersive interactions, r^{-6} dependency) to those operating at very long distances with a high degree of specificity (charge-charge interactions, r^{-1} dependency).

Dispersive Interactions (r^{-6}). These are attributed to interaction of induced dipoles within the electron clouds as molecules come in proximity and are responsible for the attractive part of the nonbonded van der Waals interaction.

Dipole–Dipole Interactions (r^{-3}). Because of the nonsymmetrical distribution of electrons between atoms of different size and electronegativity, bonds have associated permanent dipoles. The interaction energy between two of these dipoles depends on their relative orientation. This is basically the interaction underlying the phenomenon of the hydrogen bond. Although some force field authors use a special hydrogen bonding potential with an orientation dependency, simple partial charge representations combined with appropriate VDW parameters can reproduce the effect as well (34).

Charge–Dipole Interactions (r^{-2}). A charge interacting with a permanent dipole can be handled simply by considering the charge interacting with the two charges at the poles of the dipole. Alternatively, if the distance between the poles of the dipole is small compared with that between the centers of the ion and the dipole, then the potential energy Φ can be approximated as

$$\Phi = e\mu \cos \Theta / r^2$$

where e is the charge of ion, μ is the dipole moment, Θ is the angle between the vector connecting the center of the dipole with charge and dipole orientation, and r is the distance between the center of the ion and the center of the dipole.

Charge-Charge Interactions (r^{-1}). The energy of interaction between two charges q_1 and q_2 is given by Coulomb's law:

$$E = \frac{q_1 q_2}{4\pi\varepsilon r_{12}}$$

where r_{12} is the distance separating charges and ε is the dielectric constant of the medium.

To evaluate atom-atom interactions using Coulomb's law, the concept of net atomic charge is invoked. This amounts to representing charge as a point, a monopole, and is an artificial construct. Nevertheless, this is the common method. Recent improvements in calculating an appropriate set of point charges, to accurately reproduce the molecular electrostatic potential derived by quantum calculations, have been reported (41).

In an effort to increase the quality of electrostatic representations, dipole and higher multipole moments have been used. There are advantages in these more accurate representations, with a relatively small computational increase attributed to the reductions in distances over which the higher moments have to be summed, although they do require additional effort in the derivation of the parameters for the higher moments themselves. A good example is the distributed multipole model of electrostatics derived for peptides. A review by Williams (42) discusses the problems of deriving a distributed multipole expansion of charge representation that accurately reproduces the molecular electrostatic potential derived from quantum calculations. Comparisons were made between atomic multipoles, bond dipole, and restricted bond dipole models. Williams finds that a model for the electrostatic potential based on bond dipoles supplemented with monopoles (for ions) and atomic dipoles (for lone pairs) is most useful. Dipole-dipole energy converges much faster than monopole-monopole energy. Molecular charge at any desired position in a molecule is not a physically measurable quantity; one can only calculate a delocalized electron probability distribution from quantum theory. Clearly, the more complex the representation, the more accurately one can approximate the quantum mechanical results, and the more realistic should be the results obtained. One complexity of electrostatics is the long distances over which interactions occur. Appropriate means of truncating the long-range forces to maintain the accuracy of simulations are necessary (43–45) and progress in better approximations has been reported (46). The difficulties with cutoff schemes were demonstrated (47, 48) by significant variations in the behavior of a 17-residue helical peptide simulated with explicit waters, using various electrostatic schemes and by studies (49) of a pentapeptide in aqueous ionic solution (50). In both cases, the Ewald approximation in which periodicity is assumed (which allows summation over much longer distances) gave superior results (47–49).

2.1.2.1 The Dielectric Problem and Solvation. Although methods of localizing charge just described may give reasonable results, the use of Coulomb's law with a dielectric constant, a scaling factor related to the polarizability of the medium between the charges, is clearly of concern. The dielectric at the molecular level is neither homogeneous nor continuous, nor even well defined, and thus violates the basic assumption of Coulomb's law. Although the use of a low, uniform dielectric is more nearly correct in dynamical simulations where all solute and solvent atoms are explicitly included, a variety of comparisons of experimental data with the results of calculation by use of a simplified solvent model have led to the realization that much better approaches are needed. Initial efforts (51) led to the proposal of a variable dielectric ($1/R$ or $1/4R$). More recently, the use of approaches that model the inhomogeneity of the dielectric at the interface between the solute and solvent by use of the Poisson-Boltzman equation have shown considerable promise (52, 53). An alternative approach that uses the mirror charge approximation has been described by Schaefer and Froemmel (54). Excellent reviews (35–38) of the electrostatic problem have appeared, to which the reader is referred.

Much effort has been given to simple continuum models of solvation to explain the origin of solvent effects on conformational equilibria and reaction rates. The current status of such efforts, as well as simulations to rationalize solvation effects, has been reviewed by Richards et al. (55). There are two general approaches to the continuum models. The first is reaction field theory (Bell, Kirkwood, Onsager) that follows the classical treatment of Debye-Huckel. The solvent is considered in terms of charge distribution, polarizability, and dielectric constant. The solvation energy is determined simply by considering the solute as a point dipole that interacts with the induced charge distribution in the solvent (Onsager reaction field). An extension by Sinangolou in the 1960s partitioned solvation energy into cavity formation, solvent-solute interaction, and the "free volume" of the solute. The logical extension of this approach is scaled-particle theory (56), in which the free energy of formation of a hard-sphere cavity of diameter $a2$ in a hard-sphere solvent of diameter a and number density p is scaled to the exact solution for small cavity sizes. Alternatively, the virtual charge approach used a system of effective and virtual charges interacting in the gas phase. The Hamiltonian of the system is modified to include an imaginary particle, a "solvaton" with an opposite charge for each of the solute atoms and solved by the SCF procedure. These continuum models have met with limited success (trends and relative effects of solvation can be predicted), although highly specific molecular interactions, such as those involving hydrogen-bonding groups, cannot be accommodated.

In the equation for calculating affinity of a drug for a receptor, the ligand is solvated either by the receptor or by the solvent. This competition means that accurate determination of the free energy of solvation is important in understanding differences in affinities. Solvation free energy (G_{sol}) can be approximated by three terms: G_{cav}, the formation of a cavity in the solvent to hold the solute; G_{vdW} and G_{pol}, the interaction between solute and solvent divided between VDW and electrostatic forces, respectively:

$$G_{sol} = G_{cav} + G_{vdW} + G_{pol}.$$

There are four theoretical approaches to the problem:

1. Scaled Particle Theory (56)

   The essence of the scaled particle theory is that formation of a cavity in a fluid requires work. The theory for hard spheres has been well developed from statistical mechanics, and the work, $W(R, \rho)$, can be calculated as follows:

$$W(R,\rho)/kT = -\ln(1 - y) + (3y/1 - y)R$$
$$+ [(3y/1 - y) + 9/2*(y/1 - y)^2]R^2$$
$$+ yPR^3/rkT$$

   where $y = \pi\rho\sigma_1{}^3/6$, $R = \sigma_2/\sigma_1$, σ_2 is the diameter of the hard-sphere solute, σ_1 is the diameter of the solvent, and ρ is the number density of the fluid (N/V).

   Because this theory includes no interaction between solvent and solute (i.e., only G_{cav} is calculated), effective volumes for nonspherical compounds with interactive groups are normally calibrated from experiment. This is one way to deal with the energy of interaction between solvent and solute. For further discussion, see Pollack (57).

2. Charge Image (or Virtual Charge) Method (54) (Method for G_{pol} calculation)

   This model replaces the solute-continuum model with one in which a system of charges derived from the solute and virtual charges in the adjacent space interact in the gas phase. A set of mirror charges reflected at the dielectric boundary are created and used in the calculation of the electrostatics.

3. Boundary Element Method (58) (Method for G_{pol} calculation)

   In this approximation, the system is modeled by calculating the appropriate surface charges at the dielectric boundary. This is similar to fitting charges at atomic centers to reproduce the molecular electrostatic potential. For a quantum-mechanical equivalent, Tomasi et al. (59) introduced a charge distribution on the surface of a cavity of realistic shape to introduce the solvation term in the Hamiltonian of the solute. The charge distribution on the surface of the cavity depends on the solute's electric field, which is affected in turn by polarization from the cavity's surface. An iterative QM procedure is used to obtain the perturbation term. Cramer and Trular have developed AMSOL to include a solvent approximation in calculations of molecular systems. The approach has been calibrated by comparison of theoretical and experimental solvation free energies for numerous molecular species (60).

4. Poisson-Boltzmann Equation (53) (Method for G_{pol} calculation)

   Generalization of the Debye-Huckel theory leads directly to the Poisson-Boltzmann equation that describes the electrostatic potential of a field of charges with dielectric discontinuities. This equation has been solved analytically for spherical and elliptical cavities, but must be solved by finite-difference methods on a grid for more complicated systems. One exciting advance in this area is the development of an approximate equation for the reaction field acting on a macromolecular solute, attributed to the surrounding water and ions (61). By combining these equations with conventional molecular dynamics, solvation free energies were obtained similar to those with explicit solvent molecules, at little computational cost over vacuum simulations. This implies that a more nearly correct solution to the electrostatics problem might minimize the solvation problem. Other approaches to evaluations of G_{sol} have recently appeared in the literature. Still et al. (62) estimated $G_{cav} + G_{vdW}$ by the solvent-accessible surface area times 7.2 cal/mol/Å^2. G_{pol} is estimated from the generalized Born equation. Effective solvation terms have been added (63, 64) to molecular mechanics force fields to improve molecular dynamics simulations without the cost of modeling explicit solvent. Zauhar (65) combined the polarization-charge technique with molecular mechanics to effectively minimize a tripeptide in solvent.

One final refinement may be necessary in some situations: the inclusion of electric polarizability, for example, by inclusion of induced dipoles, or distributed polarizability (66) in the electrostatic representation of the model. Kuwajima and Warshel (67) recently examined the effects of this refinement in modeling crystal structures of polymorphs of ice. Such models including polarizability have been previously shown useful for predicting the properties of crystalline polymorphs of polymers by Sorensen et al. (68). Caldwell et al. (69) included implicit nonadditive polarization energies in water-ion outcomes, resulting in improved accuracy. At the semiempirical level of quantum theory, Cramer and Truhlar (70–73) added solvation and solvent effects on polarizability to AM1, with impressive agreement between experimental and calculated solvation energies (60). Rauhut et al. (74) also introduced an arbitrarily shaped cavity model by use of standard AM1 theory.

2.1.2.2 The "Hydrophobic" Effect. Water has been the nemesis of solvation modeling because of its rather unique thermodynamic properties, as reviewed by Frank (75) and Stillinger (76). The biochemical literature discusses at length "hydrophobic effects" (77). This effect is not "hydrophobic" at all because the enthalpic interaction of nonpolar solutes with water is favorable. This, however, is counterbalanced by an unfavorable entropic interaction that is interpreted as an induced structuring of the water by the nonpolar solute. Water interacts less well with the nonpolar solute than it does with itself because of the lack of hydrogen-bonding groups on the solute. This creates an interface similar to the air-water interface, with a resulting surface tension attributed to the organization of the hydrogen-bonded patterns available. This is the so-called iceberg formation around nonpolar solutes in water, first suggested by Frank and Evans. Studies by both molecular dynamics (78–80) and Monte Carlo simulations (81) support this interpretation (76), although there is still considerable controversy in interpretation of experimental data (82).

2.1.2.3 Polarizability. The traditional approaches in molecular mechanics have excluded the effects of charge on induced dipoles and multibody effects. This approximation becomes a serious limitation when dealing with charged systems and molecules like water that are highly polar. A recent paper (83) from the Kollman group described nonadditive many-body potential models to calculate ion solvation in polarizable water with good agreement with experimental observation. It was necessary to include a three-body potential (ion-water-water) in the molecular dynamics simulation of the ionic solution to obtain quantitative agreement with solvation enthalpies and coordination numbers. Inclusion of a bond-dipole model with polarizability in molecular dynamics simulations has given excellent agreement in predicting physical properties of polymers by Sorensen et al. (68).

A novel approach based on the concept of charge equilibration has been suggested by Rappe and Goddard (84) that allows the inclusion of polarizabilities in molecular dynamics calculations.

2.1.3 The Potential Surface. The set of equations that describe the sum of interactions between the ensemble of atoms under consideration is an analytical representation of the Born-Oppenheimer surface, which describes the energy of the molecule as a function of the atomic positions. Many important properties of the molecule can be derived by evaluation of this function and its derivatives. For example, setting the value of the first derivative to zero and solving for the coordinates of the atoms leads one to minima, maxima, and saddlepoints. Evaluation of the sign of the second derivative can determine which of the above have been found. It is a straightforward procedure to calculate the vibrational frequencies from the force constants by evaluation of the eigenvalues of the secular determinant (the mass-weighted matrix; see textbook on vibrational spectroscopy). Gradient methods for the location of energy minima and transition states are an essential part of any molecular modeling package. It is essential to remember, however, that minimization is an iterative method of geometrical optimization that is dependent on starting geometry, unless the potential surface contains only one minimum (a condition not found for any system of sufficient complexity to be of real interest).

The ability to locate both minima and transition points enables one to determine the minimum energy reaction path between any two minima. In the case of flexible molecules, these minima could correspond to conformers and the reaction path would correspond to the most likely reaction coordinate. One could estimate the rate of transition by determination of the height of the transition states (the activation energy) between the minima. Elbers (85) developed a new protocol for the location of minima and transition states and applied it to the determination of reaction paths for the conformational transition of a tetrapeptide (86). Huston and Marshall (87) used this approach to map the reaction coordinates of the α- to 3_{10}-helical transition in model peptides.

Despite the limitations that curtail exact quantitative applications, molecular mechanics can provide three-dimensional insight as the geometric relations between molecules are adequately represented. Electrical field potentials can be calculated and compared to give a qualitative basis for rationalizing differences in activity. Molecular modeling and its graphical representation allow the medicinal chemist to explore the three-dimensional aspects of molecular recognition and to generate hypotheses that lead to design and synthesis of new ligands. The more accurate the representation of the potential surface of the molecular system under investigation, the more likely that the modeling studies will provide qualitatively correct solutions.

2.1.3.1 Optimization.
The search for the optimal solution to a complex problem is common to many areas in science and engineering and does not have a general solution. Numerous approaches to this problem, which is generally referred to as optimization, have been used in chemistry: most commonly, distance geometry, molecular dynamics, stochastic methods such as Monte Carlo sampling, and systematic, or grid, search. Most rely on minimization, often combined with a stochastic search. Minimization algorithms have been thoroughly characterized with regard to their convergence properties, but, in general only locate the closest local minima to the starting geometry of the system. A stochastic approach to starting geometries can be combined with minimization to find a subset of minima in the hope that the global minimal is contained

within the subset and can readily be identified by its potential value compared with that of the other minima.

2.1.3.2 Potential Smoothing.
One approach to global optimization that has shown promise is potential smoothing (88). This approach uses a mathematical transformation to smooth the multidimensional potential energy surface of a molecule, reducing the high frequency complexity of the surface and making it much easier to search for minimum energy conformations. This concept was first used to deform the conformational potential energy surface in the diffusion equation method (DEM) of Piela and coworkers (89). Search procedures will not confront multiple local minima on the deformed surface. If the procedure is reversed iteratively, then one can trace the path back into a region that lies near the global minimum of the undeformed potential surface. Ponder et al. (88, 90) improved the procedure for tracing back from one partially deformed surface to the next by including a local search procedure to limit detection of false minima.

One of the best known benchmark problems for conformational search involves the determination of the low energy conformations of the highly flexible cycloheptadecane (91, 92). This system continues to serve as a test for newly developed search methods (93). Although not a particularly large molecule, this system is a challenge because of its flexibility and the close energy spacing of the lower lying minima. Extensive analysis through a variety of search methods has located exactly 263 minima within 3.0 kcal/mole of the purported global minimum. The potential smoothing search (PSS) (88) was dramatically effective at locating many of the lowest energy structures for cycloheptadecane. Although the global minimum for cycloheptadecane was not located, the second lowest energy structure was located and differed by only 0.01 kcal/mole. Based on its MM2 vibrational frequencies, the global minimum is entropically disfavored relative to all of the minima located by the smoothing procedure. The PSS method was also applied to obtain the minimum energy conformation of the TM helix dimer of glycophorin A (GpA) (94), previously solved by solution NMR spectroscopy (95).

2.1.3.3 Genetic Algorithm. Another approach to global optimization is the genetic algorithm. This approach is based on biological evolution and is analogous to natural selection (96–98). In applications to computational chemistry, evolution on the computer has been shown to be an efficient approach to global optimization, although because of sampling issues, there is no guarantee that the global optimum has been found in any particular application (99).

2.1.3.3.1 Characteristics of the Genetic Algorithm. In analogy to natural selection, the parameters to be optimized are encoded in a bit string and strung together in a "chromosome." Each chromosome in the population represents a particular genotype or solution to the problem under consideration (i.e., a specific set of values for the parameters that determine the configuration of the system under study). The values of the parameters have to be decoded for the "fitness" of a particular genotype to be evaluated. Once the fitness of each chromosome in the population has been evaluated, then the more "fit" members are allowed to reproduce, mutate, or cross over with other members of the parent population to generate a new daughter population. This process is repeated until the fitness of the population converges, or until the available computer cycles are consumed.

2.1.3.3.2 Example of Conformational Analysis. The simplifying assumption of rigid geometry is used to reduce the computational complexity of the model problem of conformational analysis. The elimination of variables is rationalized based on the high energy cost associated with bond length distortions and the ability to accommodate bond angle deformations by a reduced set of van der Waals radii. To represent the conformation of a molecule, one needs only to specify the values of the torsional angles associated with rotatable bonds. One can assign a set number N of bits, 6 for example, to represent 2^N values for the torsional angles. Each set of 6 bits can be considered a "gene" and crossover allowed only at gene boundaries, if desired. Thus, the conformation of a molecule can be encoded as a set of torsional genes. The actual coordinates of the molecule corresponding to each genotype must be generated for the fitness function F,

in this case internal energy, to be numerically evaluated by molecular mechanics. Each chromosome in the population is evaluated for its internal energy and a subset of the more fit selected for reproduction. The degree of limitation on reproductive fitness is analogous to the selective pressure brought to bear on a population (i.e., selection of the fittest). This is a parameter that can be varied in most GA programs and one must balance selective pressure against maintaining some variation in the population for evolution to occur (to avoid being trapped in a local minimum). The set of chromosomes to be reproduced can be based on some arbitrary criteria (the top 50%), all those with fitness at least half that of the most fit chromosome detected, or the fitness scaled in some way and chromosomes reproduced in proportion to their scaled fitness.

Given a subset of chromosomes to reproduce, several operations analogous to evolution are invoked. First is mutation, where a certain number of randomly selected bits are mutated from 0 to 1 or vice versa in the daughter chromosome. This would allow for changes in the settings of one or more torsional angles. A certain number of pairs of chromosomes are also selected for crossover and one or more locations between genes (if specified) are randomly selected and the two pieces derived from each parent chromosome swapped, to generate two or more novel chromosomes. This would allow for different subsets of conformations to be combined; this provides a mechanism for concerted changes or jumps over barriers to find minima that would be difficult to sample by mutation alone. This would appear to be the feature that provides the analogous behavior to simulated annealing in efficient searching of parameter space. In this case, however, the search is more directed by the selective pressure of increasing the "fitness" or facing elimination from the population. In other words, each new generation should have eliminated a significant portion of the less fit members of the previous generation and propagated those torsional values that generate good local conformational states.

2.1.3.3.3 Schema and the Building Block Hypothesis. Once a population of good local substates has been established, then crossover can probe the combination of these subconformations that have positive interactions leading to more fit progeny. In the jargon of computer science, the subpattern of 1's and 0's giving a preferred subconformation would be a schema (or building block). According to the most accepted theory, the building block hypothesis, the genetic algorithm initially detects biases toward fitness in lower order (fewer identical bits) schemas and converges on this part of search space (the entire set of bit strings). By combining information from lower order schema through crossovers, biases in higher order schemas are detected and propagated.

The strong convergence property of the genetic algorithm is a major attraction. Given sufficient members of the population and sufficient evolutionary time (number of generations), then one can expect convergence if the fitness function is based on the optimal combination of locally optimized substructures. Some fitness functions are termed "deceptive," in that low order schemas are not present in higher order schemas and their propagation slows detection of the more fit higher order schemas. Another problem arises when the population size is too small or the selection factor too high. Then, the genetic algorithm can magnify a small sampling error and prematurely converge in a local optimum.

2.1.3.3.4 Mutations and Encoding. There are different ways to encode binary numbers by bit strings and these can have some influence on the impact of mutation. Traditional binary encoding requires that all bits be changed for some cases if the digital value is to be simply incremented. This causes erratic behavior near an optimum, with mutation and mutations in higher order bits having more effect than in lower order bits.

2.1.3.3.5 Crossovers and Encoding. In our example, we indicated that one might want to separate the bit string into genes corresponding to torsional angles because the gene has a coherent meaning in the context of the problem. If one restricts crossovers to the junctions between genes, then the coherence of the conformation of molecular fragments is preserved and one is more likely to make a successful crossover producing more fit offspring. There are methods such as random-key encoding

(97) to generalize the process of crossovers without requiring customized crossover operators that are problem specific, although this is beyond the scope of this chapter.

2.1.3.3.6 Examples of Applications to Biochemical Problems. McGarrah and Judson (100) explored the impact of different parameters setting on the ability of the GA to explore the conformational space of cyclo(Gly_6). Each residue was represented by four angles, each with a string of four bits (1/16 of range). A selection fraction of 50% was used, which eliminated the lower half in fitness from reproduction. Population sizes of 10, 50, and 100 were tested. Each group was divided into four niche populations with communication between groups. Local minimization was performed for each chromosome before evaluation. They concluded that it was of little use to examine a population size of less than 100 members for the 24 variables examined. As soon as convergence in the average is detected in a population, it should be cross-fertilized from another niche or GA evolution should terminate. It is a clear example of a hybrid approach, in which GA does a rough search for minima and local minimization to find the closest local minimum.

Judson et al. (101) examined the use of a genetic algorithm to find low energy conformers of 72 small to medium organic molecules (1–12 rotatable bonds) whose crystal structures were known. They used the elitist strategy, in which the best individual from each generation is propagated without modification. A population size of 10 times the number of the nonring dihedral angles being varied was chosen. Each molecule was allowed to run for 10,000 energy evaluations, or until the population was bit converged. In a few cases, conformers with lower energies than those observed in the crystal structure were found. A comparison with CSEARCH in SYBYL (Tripos, Inc.) was made, but the differences in efficiencies found were not compelling. In only 9 of the 72 cases examined, did the GA find its best conformer had energy greater than the crystal structure, with the largest deviation being only 0.8 kcal/mol.

The GA approach has also been applied to the docking problem with dihydrofolate reductase, arabinose binding protein, and sialidase

(98). A typical run took minutes on a workstation and the predicted conformations agreed with those observed crystallographically in all cases. Meadows and Hajduk (102) used experimental constraints with a GA algorithm to dock biotin to stepavidin. Judson et al. (101) also reported docking of flexible molecules into the active sites of thermolysin, carboxypeptidase, and dihydrofolate reductase. In 9 of the 10 cases examined, the GA found conformations within 1.6 Å root-mean-square (rms) of the relaxed crystal conformation.

This approach has also been used in the PRO_LIGAND *de novo* design program (103) to optimize the structure of ligands for a binding site. A set of candidate structures was generated and then crossover between molecular fragments used to optimize the predicted binding mode. This is similar to the SPLICE program of Ho and Marshall (104) that evolves ligands with more favorable interactions with a given site.

Payne and Glen (105) studied several different aspects of molecular recognition with genetic algorithms. Conformations and orientations were determined which best-fit constraints such as inter- or intramolecular distances, electrostatic surface potentials, or volume overlaps with up to 30 degrees of freedom.

2.1.4 Systematic Search and Conformational Analysis.

Because of the convoluted nature of the potential energy surface of molecules, minimization usually leads to the nearest local minimum (106, 107) and not the global minimum. In addition, many problems in structure-activity studies require geometric solutions that may not be at the global minimum of the isolated molecule. To scan the potential surface with some surety of completeness, systematic, or grid, search procedures have been developed. To understand the strengths and limitations of this approach, some of the algorithmic details must be considered. These are discussed in depth in a review by Beusen et al. (108).

2.1.4.1 Rigid Geometry Approximation. A simplifying assumption that is usually invoked to reduce the computational complexity of the problem through elimination of variables is that of rigid geometry. The rationale is

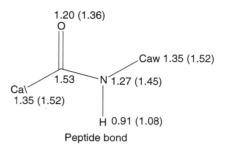

Figure 3.2. Calibrated set of van der Waals radii for peptide backbone for use with rigid geometry approximation (109). Usual radii shown in parentheses. Carbonyl carbon not modified.

based on the high energy cost associated with bond-length distortions and the ability to accommodate bond-angle deformations by a reduced set of VDW radii. This approach is compatible with problems where one is most interested in eliminating conformations that are energetically unlikely (i.e., sterically disallowed) because of VDW interactions, which cannot be relieved by bond-angle deformation. A successful application requires that one calibrate an appropriate set of VDW radii for the particular application area. Iijima et al. (109) calibrated such a set (Fig. 3.2) for peptide application by comparison with experimental crystallographic data from proteins and peptides.

2.1.4.2 Combinatorial Nature of the Problem. Using the rigid geometry assumption, one can analyze the combinatorial complexity of a simplified approach to the problem with some ease. Let us assume a molecule (Fig. 3.3) of N atoms with T torsional degrees of freedom (i.e., rotatable bonds). For each torsional degree of freedom T, explored at a given angular increment in degrees A, there are $360/A$ values to be examined for each T. This means that $(360/A)^T$ sets of angles, each describing a unique conformation, need to be examined for steric conflict. For each conformer, the starting geometry will have to be modified by applying the appropriate transformation matrices to different subsets of atoms to generate the coordinates of the conformation. For each conformation, $N(N-1)/2$ distance determinations will have to be calculated to a first approximation (this does not exclude bonded at-

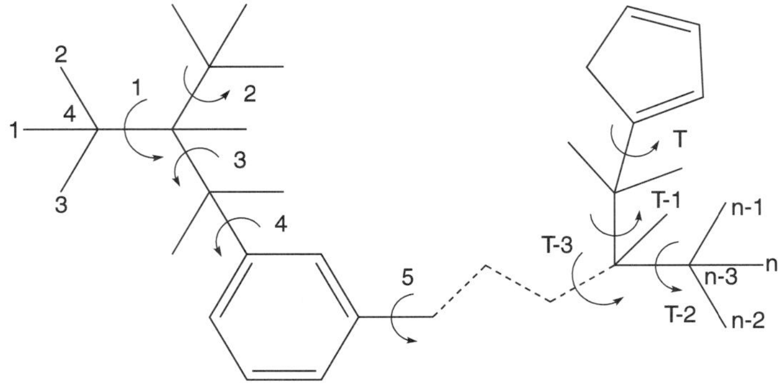

Figure 3.3. Schematic diagram of molecule with N atoms and T rotatable bonds.

oms and atoms bonded to the same atom from the check, which is necessary) and checked against the allowed sum of VDW radii for the two atoms involved. The number of VDW comparisons V is given by

$$V = (360/A)^T \times N(N - 1)/2$$

It should be clear that the VDW comparisons are the rate-limiting step by their sheer number, and any algorithmic improvement that reduces the number of such checks or enhances the efficiency of performing such checks is of value.

2.1.4.3 Pruning the Combinatorial Tree. From this simplified analysis, a systematic search of other than the smallest molecules at a coarse increment would appear daunting. A hybrid approach with a coarse grid search followed by minimization has been successfully used to locate minima. There are a number of algorithmic improvements over the "brute force" approach that enhances the applicability of the systematic search itself. To understand these improvements, some concepts need to be defined. First is the concept (110) of *aggregate*, a set of atoms whose relative positions are invariant to rotation of the T rotational degrees of freedom. *n*-Butane is divided into aggregates as an illustration (Fig. 3.4).

In this simple example, the atoms in an aggregate are all either directly bonded or have a 1–3 relationship (i.e., are related by a bond angle). Because of the rigid geometry approximation, their relative positions are fixed. Atoms contained within the same aggregate do not, therefore, have to be included in the set of those that undergo VDW checks for each con-

formation. For linear molecules, there are $n - 1$ bonds and the number of 1–3 interactions depends on the valence of the atom. This simplification leads to a reduction of the number of VDW checks by the factor $N(N - 1)/2$, which is multiplied by the number of conformations.

How can one reduce the number of conformations that have to be checked? Here the concept of construction becomes useful. One constructs the conformations in a stepwise fashion, starting with an initial aggregate and adding a second aggregate at a given torsional increment for the torsional variable T that is applied to the rotatable bond connecting the two. If any pair of atoms overlaps for that increment, then one can terminate the construction because no addition operation will relieve that steric overlap. In effect, one has truncated the combinatorial possibilities that would have included that subconformation; that is, one has pruned the combinatorial tree.

2.1.4.4 Rigid Body Rotations. If one constructs the molecule stepwise by the addition of aggregates, then one has two sets of atoms to consider. First are those in the partial mol-

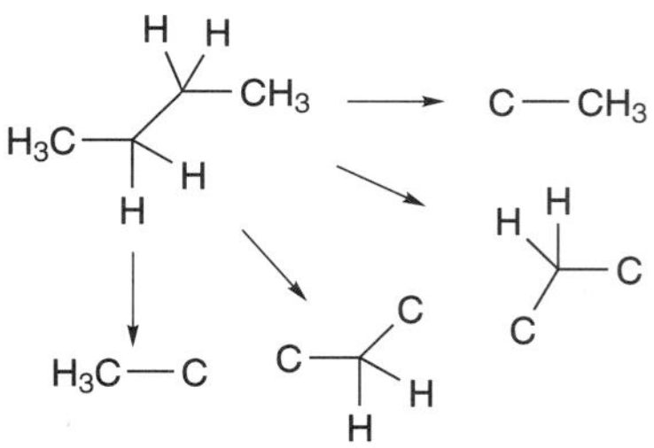

Figure 3.4. Decomposition of *n*-butane molecule into aggregates.

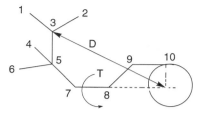

Figure 3.5. Distance between atoms (1–7) and atom 10 separated by a single rotatable bond T can be described with a transformation of the equation of a circle describing the locus of atom 10 as bond T is rotated. Notice that distance D between any atom (1–7) and center of circle of rotation of atom 10 that is on axis of rotation is fixed, regardless of value of T.

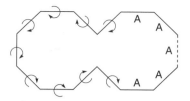

Figure 3.6. Scheme for combining systematic search with analytical solution for closure. Bonds indicated by arrows were systematically scanned, whereas those indicated by A were analytically determined. Dotted bond can represent either chemical bond or experimental distance determination (NOE, etc.).

ecule (set A), previously constructed, that have been found to be in a sterically allowed partial conformation. For each possible addition of the aggregate, the atoms of the aggregate (set B) must be checked against those in the partial molecule. If one uses the concept of a rigid body rotation, then one can describe the locus of possible positions of any atom in set B as a circle whose center lies on the axis of rotation T_i (the interconnecting bond) at a distance along the axis that can be calculated. The formula for a circle can be transformed to represent the possible distances between the atom b in set B and any atom a in set A, as shown in Fig. 3.5. An equation with scalar coefficients that describes the variable distance between two atoms as a function of a single torsional variable was derived (111), which has a discriminant whose evaluation can be used to determine whether atom a and atom b will:

- be in contact, despite changes in the value of the torsional rotation of the aggregate, which implies that the current partial conformation has to be discarded, given that there is no possible way to add the aggregate that is sterically allowed;
- never come in contact for any value of the torsional rotation, so that this pair of atoms can be removed from consideration regarding this aggregate; or
- come in contact for some values of the torsional rotation that can be calculated for that pair and that removes a segment of the

torsional circle from consideration for other atom pairs. If all segments of the torsional circle are disallowed by combinations of the angular requirements of different atom pairs, then the partial conformation of the molecule is disallowed because further construction is not feasible. As a first approximation, this removes a degree of torsional freedom from the problem, reducing T to $T - 1$ torsional degrees of freedom. At a 10° torsional scan, an approximate reduction in computational complexity of a factor of 36 results.

2.1.4.5 The Concept and Exploitation of Rings. Realization that many of the relevant constraints in chemistry can be expressed as interatomic distances, VDW interactions, nuclear Overhauser effect constraints, and so forth allows use of the concept of a virtual ring in which the constraint forms the closure bond. Small rings up to six members can be solved analytically (112), so that one can search the torsional degrees of freedom associated with a constraint until only five remain and then solve the problem analytically (Fig. 3.6). The torsional angles for those degrees of freedom are no longer sampled on a grid, thus removing the problem of grid tyranny, in which valid conformations are missed by the choice of increment and starting conformation. This approach is then a hybrid because only part of the conformational space is searched with regular torsional increments. It is, however, much more efficient to solve a set of equations than search 5 torsional degrees of freedom.

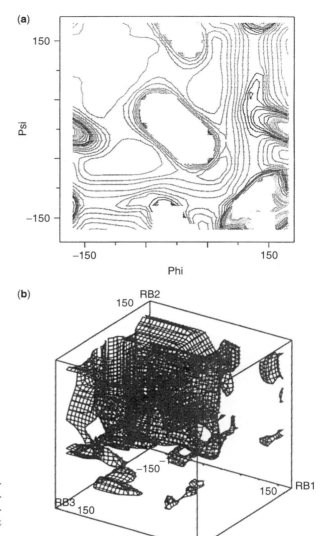

Figure 3.7. (a) Two-dimensional (Ramachandran) plot of energy vs. backbone torsional angles, Φ and Ψ, for *N*-acetyl-valine-methylamide. (b) Three-dimensional plot of energy vs. torsional angles, Φ, Ψ, and $\chi 1$, for *N*-acetyl-valine-methylamide.

2.1.4.6 Conformational Clustering and Families. In a congeneric series, the correspondence between torsional rotation variables is maintained as one compares molecules, and a direct comparison of the values allowed for one molecule with those allowed for another is meaningful. Two- (2D) or three-dimensional (3D) plots (Fig. 3.7) of torsional variables against energy often provide considerable insight into the difference in conformational flexibility between two molecules. Such a plot of the peptide backbone torsional angles

Φ, Ψ is known as a Ramachandran plot. When more than three torsional variables become necessary to define the conformation of the molecule under consideration, then multiple plots become necessary to represent the variables. Unless special graphical functions are included in the software, then correlations between plots become difficult, given that each plot is a projection of a multidimensional space. One approach to this problem is to use cluster analysis programs to identify those values of the multidimensional variables that

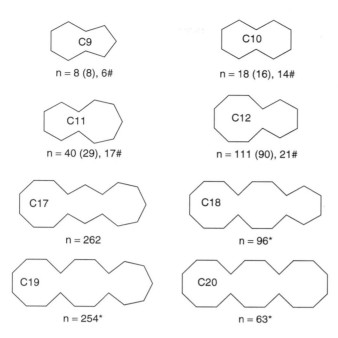

Figure 3.8. Cycloalkane rings and number of local minima found by various search strategies. n, number of conformers with MM2 (117); parentheses, number of conformers with MM3 (117); #, number of conformers within 25 kJ/mol of global minima [MM2 (92)]; *, number of minima found within 3 kcal/mol of global minima (115).

are adjacent in *N*-space. The clusters of conformers that result have been referred to as *families*. A member of a family is capable of being transformed into another conformer belonging to the same family without having to pass over an energy barrier; that is, the members of a family exist within the same energy valley.

Because of the combinatorial nature of systematic search, one is often faced with large numbers of conformers that have to be analyzed. For some problems, energetic considerations are appropriate and conformers can be clustered with the closest local minimum, providing to a first approximation an estimate of the entropy associated with each minima by the number of conformers associated, in that they can come from a grid search that approximates the volume of the potential well. A single conformer, perhaps the one of lowest energy, can be used with appropriately adjusted error limits in further analyses as representative of the family.

2.1.4.7 Conformational Analysis. Although interaction with a receptor will certainly perturb the conformational energy surface of a flexible ligand, high affinity would suggest that the ligand binds in a conformation that is not exceptionally different from one of its low

energy minima. Mapping the energy surface of the ligand in isolation to determine the low energy minima will, at the very least, provide a set of candidate conformations for consideration, or as starting points for further analyses. The problem of finding the global minimum on a complicated potential surface is common to many areas, and lacks a general solution. Minimization procedures locate the closest local minimum depending on the starting conformation. Several strategies have developed to map the potential surface and locate minima. For an excellent overview of the different approaches, the reader is referred to the surveys by Leach (113) and by Burt and Greer (114). Stochastic methods such as Monte Carlo have been advocated (115) for conformational analysis and their usefulness demonstrated on carbocyclic ring systems (91, 115–121) (Fig. 3.8). Molecular dynamics can be used to explore the potential energy surface, often with simulated annealing to help overcome activation-energy barriers, but exploration is concentrated in local minima and duplication of the surface explored is controlled by Boltzmann's law. A systematic, or grid, search samples conformations in a regular fashion, at least in the parameter space (usually torsional space) that is incremented.

Comparisons of a variety of methods were made on cycloheptadecane by Saunders et al. (91) and it was concluded that the stochastic method was most efficient. In one of the few independent comparisons of the effectiveness of these procedures, Boehm et al. (122) studied the sampling properties on the model system caprylolactam, a nine-membered ring, and concluded that systematic search was both inefficient and ineffective at finding the minima found by the other methods when the number of conformers examined was limited.

2.1.4.8 Other Implementations of Systematic Search. Numerous other implementations of systematic, or grid, search programs exist in the literature and those with protein applications have been reviewed by Howard and Kollman (123), whereas those for small or medium sized molecules are included in the reviews by Burt and Greer (114) and by Leach (113). One of the more widely used programs in organic chemistry, MACROMODEL, has a search module (124) coupled to energy minimization for conformational analysis. MACROSEARCH has been developed by Beusen et al. (125) to generate the set of conformers consistent with experimental NMR data and used to determine the conformation of a 15-residue peptide antibiotic.

2.1.5 Statistical Mechanics Foundation (126).

To understand the relationships between the simulation methods and the desired thermodynamic quantities, a short review of the major concepts of statistical mechanics may be in order. This is not meant to be comprehensive, but rather to remind the reader of the relevant ideas.

The set of configurations generated by the Monte Carlo simulation generates what J. Willard Gibbs would call an "ensemble," assuming that the number of molecules in the simulation was large and the number of configurations was also large. This ensures that the possible arrangements of molecules that are energetically reasonable have been adequately sampled. One is often interested in the statistical weight W of a particular observable. For example, a particular conformation of a solute molecule, say, the staggered rotamer of ethane, could be compared with another conformer, the eclipsed rotamer, in a simulation with solvent. If more configurations of the surrounding solvent molecules of equivalent energy were available to the staggered than to the eclipsed, then the staggered would have a higher statistical weight. From the inscription on Boltzmann's tomb, we all recall that $S = k \ln W$, where S is the entropy and k is Boltzmann's constant. Thus, we have a link between statistics and thermodynamics. W in this case would be the number of configurations associated with the particular conformation of ethane under consideration divided by the total number of configurations sampled. This would have to be weighted by their energy, of course, unless the distribution was already Boltzmann weighted, as happens when one uses the Metropolis algorithm (127).

Another way of stating this is that the probability P_i of a particular configuration N_i is proportional to its Boltzmann probability divided by the Boltzmann probability of all the other configurations or states:

$$P_i = \exp(-E_i/kT) \Big/ \sum_{i=1}^{N} \exp(-E_i/kT)$$

The denominator in this equation has been given a special name, *partition function*, often symbolized by Z, which is derived from the German *Zustandsumme* (sum over states). The successive terms in the partition function describe the partition of the configurations among the respectives states available. One can express the thermodynamic state functions of an ideal gas in terms of the molecular partition function Z as follows:

$$S = k \ln W = kN \ln Z/N + U/T + kN$$

where N is the number of molecules and U is the internal energy. From this and the assumption of an ideal gas $pV = NkT$, the Gibbs free energy $G = U - TS + pV$ leads to

$$G = -NkT \ln Z/N$$

and similarly, the Helmholtz free energy $A = U - TS$ leads to the expression

$$A = -kT \ln Z^N/N!$$

all of which may be more familiar if expressed in terms of enthalpy, $H = U + pV$.

In summary, by simulating a relevant statistical sample of the possible arrangements of molecules when interacting, one can derive the macroscopic thermodynamic properties by statistical analysis of the results. In this case, one is deriving the partition function not by theoretical analysis of the quantum states available to the molecule, but through simulation. In other words, the average properties are valid if the Monte Carlo or molecular dynamics trajectories are ergodic, that is, constructed such that the Boltzman distribution law is in accord with the relative frequencies with which the different configurations are sampled. (An ergodic system is by definition one in which the time average of the system is the same as the ensemble average.) A basic concept in statistical mechanics is that the system will eventually sample all configurations, or microscopic states, consistent with the conditions (temperature, pressure, volume, other constraints) given sufficient time. In other words, a trajectory of sufficient length (in time) would sample configuration space.

2.1.6 Molecular Dynamics (37, 126, 128).
Molecular dynamics is a deterministic process based on the simulation of molecular motion by solving Newton's equations of motion for each atom and incrementing the position and velocity of each atom by use of a small time increment. If a molecular mechanics force field of adequate parameterization is available for the molecular system of interest and the phenomenon under study occurs within the time scale of simulation, this technique offers an extremely powerful tool for dissecting the molecular nature of the phenomenon and the details of the forces contributing to the behavior of the system.

In this paradigm, atoms are essentially a collection of billiard balls, with classical mechanics determining their positions and velocities at any moment in time. As the position of one atom changes with respect to the others, the forces that it experiences also change. The forces on any particular atom can be calculated

by evaluation of the energy of the system using the appropriate force field. From physics,

$$F = ma = -\delta V/\delta r = m\,\delta^2 r/\delta t^2$$

where F is the force on the atom, m is the mass of the atom, a is the acceleration, V is the potential energy function, and r represents the cartesian coordinates of the atom. Using the first derivative of the analytical expression for the force field allows the calculation of the force felt on any atom as a function of the position of the other atoms.

2.1.6.1 Integration. In this simulation, we use numerical integration; that is, we choose a small time step (smaller than the period of fastest local motion in the system) such that our simulation moves atoms in sufficiently small increments, so that the position of surrounding atoms does not change significantly per incremental move. In general, this means that the time increment is on the order of 10^{-15} s (1 femtosecond). This reflects the need to adequately represent atomic vibrations that have a time scale of 10^{-15} to 10^{-11} s. For each picosecond of simulation, we need to do 1000 iterations of the simulation. For each iteration, the force on each atom must be evaluated and its next position calculated. For simulations involving molecules in solvent, sufficient solvent molecules must be included, so that the distance from any atom in the solute to the boundary of the solvent is larger than the decay of the intermolecular interaction between the solute and solvent molecules. This requires several hundred solvent molecules for even small solutes, and the computations to do a single iteration are sufficiently large that simulations of more than several hundred picoseconds for proteins with explicit solvent are still rare. Efforts to increase the time step and thus allow for longer simulations without sacrificing the accuracy of the methodology are under investigation. Combination of normal mode calculations with explicit numerical integration allows time steps up to 50 ps for model systems (129). A similar approach has been shown effective by Schlick and Olson (130) in modeling supercoiling of DNA.

Let us attempt a rough trajectory through molecular dynamics. We have a system of N atoms obeying classical Newtonian mechanics. In such a system, we can represent the total energy E_{tot} as the sum of kinetic energy E_{kin} and potential energy V_{pot}:

$$E_{tot}(t) = E_{kin}(t) + V_{pot}(t)$$

where the potential energy is a function of the coordinates, $V_i = f(r_i)$ for atoms i to N and r_i represents cartesian coordinates of atom i; and the kinetic energy depends on the motion of the atoms:

$$E_{kin}(t) = \sum \frac{1}{2} M_i V_i^2(t)$$

where M_i is the mass of atom i and V_i is the velocity of atom i.

The energy undergoes constant redistribution because of the movements of the atoms, resulting in changes in their positions on the potential surface and in their velocities. At each iteration $(t \rightarrow t + \Delta t)$, an atom i moves to a new position $[r_i(t) \rightarrow r_i(t + \Delta t)]$, and it experiences a new set of forces. The basic assumption is that the time step Δt is sufficiently small that the position of atom i at $t + \Delta t$ can be linearly extrapolated from its velocity at time t and the acceleration resulting from the forces felt by atom i at time t. If Δt is long enough for the atoms surrounding atom i to change their position so that the forces felt by atom i will change during Δt, then the approximation is not valid and the simulation will deviate from that observed with a shorter Δt. After each atom is moved, the forces on the first atom based on the new positions of the other $N - 1$ atoms can be recalculated and a new iteration begun. Several algorithms exist for numerical integration. The ones by Verlet and Gear are in common use, with the one by Verlet being computationally more efficient (126). A variant of the Verlet algorithm in common use is called the leapfrog algorithm. The calculation of the velocity is done at $t - \Delta t/2$, whereas the calculation of the force occurs at t to derive the new velocity at $t = \Delta t/2$. In other words,

$$V_i(t + \Delta t/2) = V_i(t - \Delta t/2) + F_i(t)\Delta t/M_i.$$

The atomic position of atom i is calculated by adding the incremental change in position,

$V_i(t + \Delta t/2) \cdot \Delta T$, to the original position $V_i(t)$. By staggering the evaluation of the velocity and force calculations by $\Delta t/2$, an improvement in the simulation performance is obtained.

2.1.6.2 Temperature. For simulations that can be compared with experimental results, one must be able to control the temperature of the simulation. The temperature of a system is a function of the kinetic energy, $E_{kin}(t)$:

$$T(t) = E_{kin}(t)/\frac{3}{2}Nk$$

where k is Boltzmann's constant.

One can perform molecular dynamics simulations, at a constant temperature T_c, by scaling all atomic velocities $V_i(t)$ at each step by a factor t derived from

$$\delta T(t)/\delta t = [T_c - T(t)]/t$$

where T_c is the desired temperature.

2.1.6.3 Pressure and Volume. Depending on the simulation that one desires to accomplish, either the pressure or volume must be maintained constant. Constant volume is the easiest to perform because the boundaries of the system are maintained with all molecules confined within those boundaries and the pressure allowed to change during the simulation.

2.1.7 Monte Carlo Simulations. The Monte Carlo method (126) is based on statistical mechanics and generates sufficient different configurations of a system by computer simulation to allow the desired structural, statistical, and thermodynamic properties to be calculated as a weighted average of these properties over these configurations. The average value $\langle X \rangle$ of the property X can be calculated by the following formula:

$$\langle X \rangle = \sum_{i=1}^{N} X_i \exp(-E_i/kT)V'$$

$$\sum_{i=1}^{N} \exp(-E_i/kT)$$

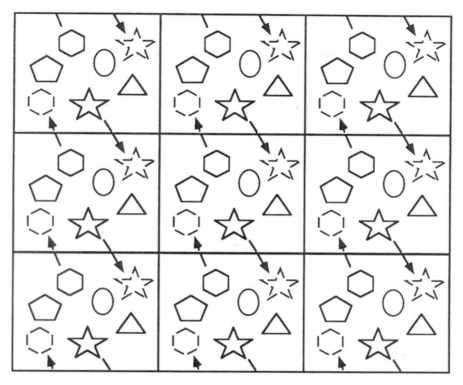

Figure 3.9. Schematic diagram of simulation with periodic boundary conditions in which adjacent cells are generated by simple translations of coordinates.

where N is the number of configurations, E_i is the energy of configuration i, k is Boltzmann's constant, and T is temperature.

If we have sufficiently sampled the possible arrangements of molecules in the simulation and have an accurate method to calculate their energy E, then the above formula will give a Boltzmann weighted average of the property X.

In practice, one must compromise the number of molecules in the simulation and/or the number of configurations calculated to conserve computer cycles. Two essential techniques that are utilized are periodic boundary conditions and sampling algorithms, which we discuss separately.

Although it is important to minimize the number of molecules in either Monte Carlo or molecular dynamics simulations for computational convenience, surface effects at the interface between the simulated solvent and the surrounding vacuum could seriously distort the results. To approximate an "infinite" liquid, one can surround the box of molecules by simple translations to generate periodic images. Each atom in the central box has a set of related molecules in the virtual boxes surrounding the central one (Fig. 3.9). The energy calculations for pairwise interactions consider only the interaction of a molecule, or its "ghost," with any other molecule, but not both. In practice, this is accomplished by limiting pairwise interactions to distances less than one-half the length of the side of the box. Real concerns often arise regarding convergence of electrostatic terms because of the linear dependency on distance.

For any large nontrivial system, the total number of possible configurations is beyond comprehension. Consider a set of protons in a magnetic field: the magnetic moments can be either aligned with or opposed to the magnetic field. For only 50 protons, there are 2^{50} combinations, which is a large number. For a small cyclic pentapeptide, there are potentially 36^{10} conformations if one considers a $10°$ scan of the torsional variables Φ, Ψ. Clearly, some of these are energetically unreasonable because the conformation requires overlap of two or more atoms in the structure. Monte Carlo simulations are successfully performed by sampling only a limited set of the energetically feasible conformations, say, 10^6 out of 10^{100} theoretical possibilities. The reason for this success is that the Monte Carlo schemes sample those states that are statistically most important. One could sample all states, calculate the energy of each, and then Boltzmann-weight its contribution to the average. Alternatively, one can ignore those states that are energetically high so that they contribute little, if any, weight to the average, and concentrate on those of low energy. In other words, we look only where there are reasonable answers energetically. This is called *importance sampling*, which is the key to the Monte Carlo procedure.

One aspect shared by Monte Carlo methods and molecular dynamics is the ability to cross barriers. In the case of Monte Carlo, barrier crossing occurs both by random selection of variables and by acceptance of higher energy states on occasion. Both methods require an equilibration period to eliminate bias associated with the starting configuration. When one considers randomly filling a box with molecules with arbitrary choices for position and orientation, it should be obvious that most examples would result in high energy, especially if the density of such a simulation is made to resemble that of a liquid in which adjacent molecules are often in VDW contact. High energy configurations contribute very little to the properties we are trying to evaluate because they are Boltzmann weighted. It is, therefore, extremely inefficient to randomly calculate configurations. One needs procedures, often referred to as importance sampling, that selectively calculate configurations that will be representative of allowed states. In fact, if one can guarantee that the energy of the configurations actually has a Boltzmann distribution, then one can simply average the properties. In practice, this has been accomplished by an algorithm suggested by Me-

tropolis et al. (127). One essentially uses a Markov process in which the current configuration becomes the basis for generating the next.

1. A molecule in the current configuration is chosen at random and its degrees of freedom randomly varied by small increments.
2. The energy of the new configuration is evaluated and compared with that of the starting configuration.
3. If the new energy is lower, the new configuration is accepted and becomes the basis for the next random perturbation.
4. If the energy is higher, $E(\text{new}) > E(\text{old})$, then a random number between 0 and 1 is generated and compared with $\exp\{-[E(\text{new}) - E(\text{old})]\}/kT$. If the number is less, then the configuration is accepted and the process continues by generating a new configuration. If the number is greater, then the configuration is rejected and the process resumes with the old configuration.

In this way, configurations of lower energy are accepted and the system eventually "minimizes" to sample the higher populated lower energy configurations; at the same time, higher energy configurations are included but only in proportion to their Boltzmann distribution, which is clearly a function of temperature of the simulation. Because the configurations occur with a probability depending on their energy and proportional to the Boltzmann distribution, one can simply average thermodynamic properties over this distribution of configurations,

$$<X> = 1/N \sum_{i=1}^{N} X_i$$

where the sum covers the N configurations generated. Because one often does not know an appropriate starting configuration, the initial part of the run may be used to "minimize," or equilibrate the system, and only

(a)

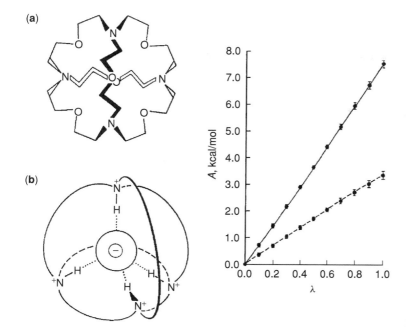

(b)

A, kcal/mol

Figure 3.10. Estimation of difference in affinity ($\Delta\Delta G$) of the two anions Cl⁻ and Br⁻ for the cryptand SC24 [(a) structural formula; (b) schematic of complex formed with halide ion] as the parameters for Cl⁻ are slowly mutated into those for Br⁻ in water (- - -) as well as in the complex (—). Used with permission (138).

the latter part of the simulation analyzed once the configurational energy has stabilized.

A useful application has combined Monte Carlo sampling with variable temperatures (simulated annealing) to encourage barrier crossing to optimize the docking of ligands into active sites. Random displacements of rigid body translation and rotation (6 degrees of freedom) and of internal torsional rotations in a substrate within the binding site cavity were performed with Metropolis sampling and a temperature program. This procedure reproduced the crystallographically observed structure of the complex for several test cases (131).

2.1.8 Thermodynamic Cycle Integration (132–134). Thermodynamic cycle integration is an approach that allows calculation of the free energy difference between two states. In this method, one takes advantage of the state-function nature of a thermodynamic cycle and eliminates the paths of the simulation with long time constants (e.g., formation of a complex requiring diffusion). As an example, the difference in affinity of two ligands (L and M) for the same enzyme or receptor R is described by the following thermodynamic cycle:

$$
\begin{array}{ccc}
R + L & \xrightarrow{\Delta A1} & RL \\
\Delta A3 \downarrow & & \Delta A4 \uparrow \\
R + M & \xrightarrow{\Delta A2} & RM
\end{array}
$$

Because the thermodynamic values of the two states do not depend on the path between the states, one can write the following equation:

$\Delta\Delta A$ = difference in affinity of L and M for R

$$= \Delta A2 - \Delta A1 = \Delta A4 - \Delta A3$$

By simulating the mutation of L into M, paths A3 and A4, one can avoid the long simulation required for diffusion of the ligands, paths A1 and A2, into the receptor. One simply incrementally modifies the potential functions representing ligand L to those representing ligand M during the course of the simulation, making sure that the perturbations are introduced gradually and that the surrounding atoms have time to relax from the perturbation (Fig. 3.10). Either Monte Carlo (135) or molecular dynamics simulations can utilize this technique. Many interesting applications have appeared in the literature (132, 134, 136, 137). Their success appears directly related to sampling problems and minimal perturbation of the ligand to ensure equilibration.

2.1.9 Non-Boltzmann Sampling. There are equivalent molecular dynamics and Monte Carlo procedures that allow one to sample regions of configuration space that are not minima, transition states, for example. One can generate a Monte Carlo trajectory for a system E_V that has energetics similar to that of the Boltzmann system E_0, with sampling in the region associated with a transition barrier by subtracting a potential V, to reduce the barrier:

$$E_0 = E_V - V.$$

Alternatively, one may want to obtain meaningful statistics for a rare event without oversampling the lower energy states. This can be accomplished by adding a potential W, which is zero for the the interesting class of configurations and very large for all others (Fig. 3.11):

$$E_0 = E_V + W.$$

The details of these sampling procedures that allow one to focus on the aspect of the problem of interest are the subject of a review by Beveridge (133). Application of this approach to determining conformational transitions in model peptides (137, 139, 140) are exemplified in the work of Elber's group on helix-coil (85, 86, 141), the Brooks group on turn-coil (142–146), and Huston and Marshall and Smythe et al. (147, 148) on helical transitions in peptides.

2.2 Quantum Mechanics: Applications in Molecular Mechanics

Detailed discussion of quantum mechanics (149) is clearly beyond the scope of this review, and its applications to molecular mechanics and modeling will be briefly summarized. Molecular mechanics is based on the laws of classical physics and deals with electronic interactions by highly simplified approximations such as Coulomb's law. All forces operating in intermolecular interactions are essentially electronic in nature. Any effort to quantitate those forces requires detailed information

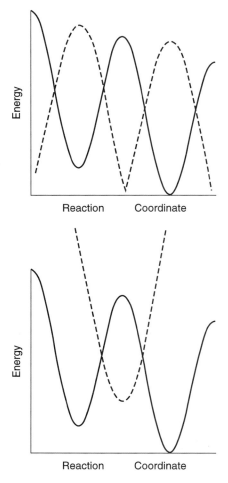

Figure 3.11. Schematic diagrams of methods for modifying the potential surface to allow adequate sampling during simulations.

about the nuclear positions and the electron distribution of the molecules involved. At considerable computational cost, quantum mechanics provides information about both nuclear position and electronic distribution. Molecular mechanics is built on the assumption that electronic interactions can be adequately accounted for by parameterization. Although most of the systems of interest in biology are too large for the direct application of quantum mechanics, quantum mechanics has at least three essential roles to play in drug design (149): (1) charge approximations, (2)

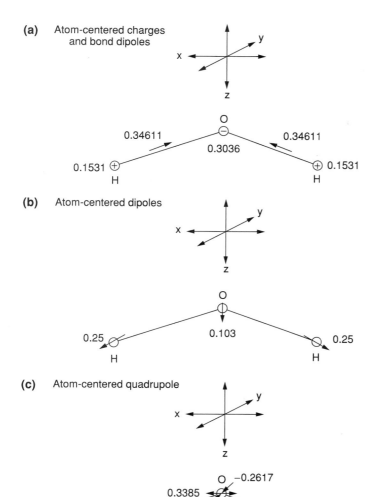

Figure 3.12. Different approaches to localization of charge used in electrostatic models. (a) Atom-centered monopole; (b) atom-centered dipole; and (c) atom-centered quadrapole.

characterization of molecular electrostatic potentials, and (3) parameter development for molecular mechanics.

2.2.1 Parameterization of Charge. Estimates of charges in molecular mechanics can be derived, in general, by application of one of the many different quantum chemical approaches, either *ab initio* or semiempirically. Quantum mechanical methods are available for calculating the electron probability distri-

butions for all the electrons in a molecule and then partitioning those distributions to yield representations for the net atomic charges of atoms in the molecule, either as atom-centered charges or as more complex distributed multipole models (39, 42) (Fig. 3.12).

2.2.1.1 Atom-Centered Point Charges. In the Mulliken population analysis, all the one-center charge on an atom is assigned to that atom, whereas the two-center charge is divided equally between the two atoms in the overlap (even if the electronegativities of the two atoms are quite dissimilar). The sum is the gross atomic population, and the net atomic charge is simply this plus the nuclear charge. The result is very sensitive to the basis set (the number of atomic orbitals) used. Despite poor fit of the molecular electrostatic potential derived with point charges to the *ab initio* electrostatic potential, or that derived from a distributed multipole analysis (150), widespread use continues because they do reflect chemical trends and are reportedly compatible with known electronegativities. In addition, this option is commonly available in software packages. Unfortunately, poor representation of the electric field surrounding the molecule results from use of atom-centered monopole models (42), even when more careful methods are used to distribute the charge.

2.2.1.2 Methods to Reproduce the Molecular Electrostatic Potential (MEP).

The electrostatic potential surrounding the molecule that is created by the nuclear and electronic charge distribution of the molecule is a dominant feature in molecular recognition. Williams reviews (42) methods to calculate charge models to accurately represent the MEP as calculated by *ab initio* methods by use of large basis sets. The choice between models (monopole, dipole, quadrapole, bond dipole, etc., Fig. 3.12) depends on the accuracy with which one desires to reproduce the MEP. This desire has to be balanced by the increased complexity of the model and its resulting computational costs when implemented in molecular mechanics.

The first problem is to select points where the MEP is to be evaluated and eventually fitted, the position of the shell outside the VDW radii of the atoms in the molecule, and the spacing of grid points on that shell. Sampling too close to the nuclei gives rise to anomalies because the potential around nuclei is always positive. Singh and Kollman (151) report the use of four surfaces at 1.4, 1.6, 1.8, and 2.0 times the VDW radii, with a density of one to five points per $\mathring{A}^2$. This paradigm was reported to give an adequate sampling to which the fitted charges were fairly insensitive, at least at the higher values. An improved procedure, the restrained electrostatic potential fit (RESP), was developed by Bayly et al. (41) to enhance transferability of the resulting point charges.

Williams (42) derived a procedure to derive the best fit to a given MEP with a defined set of monopoles, dipoles, and so forth.

Typically, fragments of molecules of interest are analyzed by *ab initio* techniques to generate their MEPs that are the reference for parameterization of charge. Besler et al. (152) reported fitting of atomic charges to the electrostatic potentials calculated by the semiempirical methods AM1 and MINDO. The MINDO charges derived by fitting the MEP can be linearly scaled to agree with results derived from *ab initio* calculations. Among the motivations for semiempirical methods are the facts that semiempirical methods using high quality basis sets often yield better results than *ab initio* techniques employing minimal basis sets, and the significant reduction in computational time in moving from *ab initio* to semiempirical calculations. Rauhut and Clark (153) used the AM1 wave function to develop a multicenter point-charge model in which each hybrid natural atomic orbital is represented by two charges located at the centroid of each lobe. Thus, up to nine charges (4 orbitals and 1 core charge) are used to represent heavy atoms. Results using this approach affirm the observations that distributed charges are more successful than atom-centered charges in reproducing intermolecular interactions (154, 155).

2.2.2 Parameter Derivation for Force Fields.

Because molecular mechanics is empirical, parameters are derived by iterative evaluation of computational results, such as molecular geometry (bond lengths, bond angles, dihedrals) and heats of formation, compared with experimental values (20). Lifson has coined the expression "consistent" for force fields in which structures, energies of formation, and vibrational spectra have all been used in parameterization by least-squares optimization. In the case of bond lengths, bond angles, and VDW parameters, crystallography has provided most of the essential experimental database. Major efforts (156) to derive general sets of parameters from quantum mechanical calculation have been made, especially for systems for which adequate experimental data are unavailable. Although quantum mechanics is certainly adequate for initial approximations of parameters and essential for charge approximations, a detailed analysis indicates that *in vacuo* calculations neglect many-body effects and can be misleading. A major effort by Hehre (personal communication) to derive parameters for water from extensive *ab initio* calculations with large basis sets failed even to give a parameter set that reproduced the radial distribution for bulk water. Parameters derived from relevant experimental data in condensed phase (especially if available in the solvent of theoretical interest) are generally more capable of accurately predicting results because the many-body effects are implicitly included in the parameterization. The basic assumption is that these "effective" two-body potentials implicitly incorporate many-body interaction energies.

Jorgensen has parameterized by fitting properties of bulk liquids to Monte Carlo simulations to give the AMBER/OPLS force field (26, 157, 158). Conceptually, one is attracted by the use of liquids and their observable properties as constraints during the derivation of a force field that is destined to study the properties of solvated molecules.

2.2.3 Modeling Chemical Reactions and Design of Transition-State Inhibitors.

In cases, such as enzyme reactions, where chemical transformations occur, quantum chemical methods must be used to deal with electronic changes in hybridization and bond cleavage (159, 160). Hybrid applications (161–163) in which the reaction core is modeled quantum mechanically and the rest by molecular mechanics would appear a viable option. Alternatively, the geometry of the transition state has been modeled by molecular mechanics, with force constants derived from *ab initio* calculations that predict with amazing accuracy the relative selectivity of reactions. Andrews and coworkers (164) pioneered modeling of transition states (165) of enzymatic reactions to design transition-state inhibitors.

3   KNOWN RECEPTORS

A significant challenge is the design of novel ligands for therapeutic targets in which the three-dimensional structure has been deter-

mined by either X-ray crystallography or NMR (12, 13, 166). The availability of the coordinates of all the atoms of the target suggests use of modeling of the site and interaction with prospective ligands. Qualitative information can be discerned by simple examination of complexes by the use of molecular graphics and improvement of known ligands made by searching for accessory binding interactions through ligand modification. This approach was pioneered by groups at Wellcome Research Laboratories (167–169) in designing analogs of 2,3-diphosphorylglycerate (Fig. 3.13), to modulate oxygen binding to hemoglobin, and at Burroughs-Wellcome (170), to enhance affinity of dihydrofolate reductase (DHFR) antagonists. When used in an iterative fashion, novel compounds with improved affinity result (166, 171, 172). Quantification of interactions and design of novel ligands require application of molecular and statistical mechanics to quantify the enthalpy and entropy of binding. In other words, experimental measurements reflect free energies of binding and both enthalpic and entropic contributions must be estimated for prediction of affinities as part of the design process. When combined with combinatorial chemistry and high throughput screening, rapid identification of therapeutic candidates is feasible, as witnessed in the case of factor Xa antagonists (173) or TAR RNA inhibitors as possible HIV drugs (174).

3.1   Definition of Site

The availability of three-dimensional structural information on a potential therapeutic target does not guarantee identification of the site of action of the substrate, or inhibitor, unless the structure of a relevant complex has been determined. In fact, conformational changes often occur during binding of ligands to enzymes that are not reflected in the three-dimensional structure of the enzyme alone. Illustrative examples are the major conformational changes seen (175, 176) in HIV protease on binding the inhibitor MVT-101 (Fig. 3.14) and the changes in domain orientation observed (177) in the complex of an anti-HIV peptide antibody with the peptide. Until the two β-strand flaps have been folded in, to complete the active site of HIV protease, many of

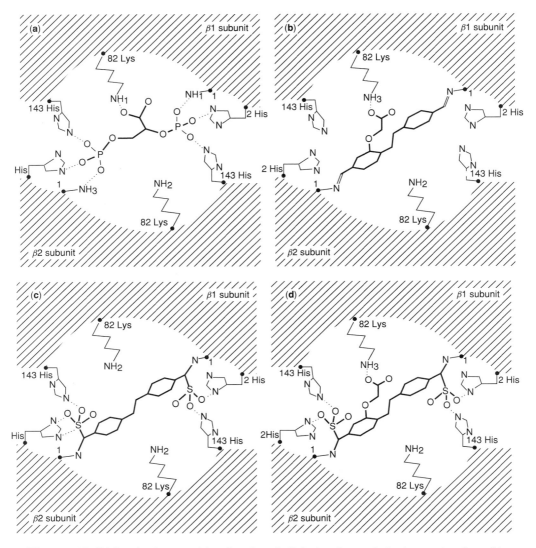

Figure 3.13. Diphosphoglycerate (a) and analogs (b-d) designed to optimize interactions bound in schematic model of hemoglobin. Used with permission (169).

the important interactions for recognition in this proteolytic system have not been defined. In other cases of therapeutic targets, allosteric sites are involved in regulation of binding and cannot clearly be discerned from the crystal structure available. Here NMR offers a highly complementary approach where transfer and isotope-edited NOEs as well as magic angle spinning NMR on solid samples can help identify those residues of the therapeutic target (Fig. 3.15) involved in receptor interaction (178–180).

One significant concern of structure-based design is the dynamics of the target itself. How stable is the active site to modifications in the ligand? Are there alternative potential binding sites that could compete for the ligand? The geometrical identity of serine protease catalytic residues, for example, argues that the specificity essential for biological utility ensures a relatively rigid three-dimensional arrangement of functionality in the active site that determines molecular recognition and

(a)

(b)

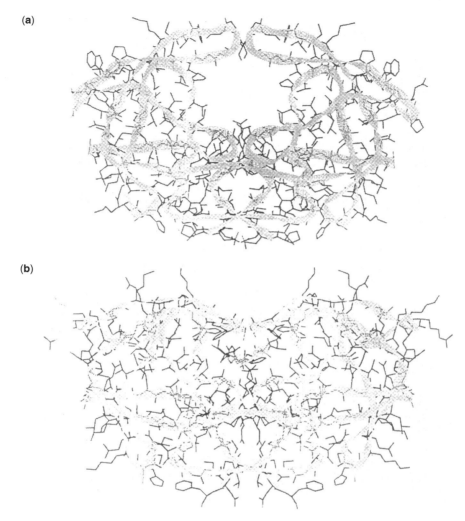

Figure 3.14. Ribbon diagram of HIV-1 protease in the absence of inhibitor (a) and when bound to the inhibitor MVT-101 (b). Diagrams based on crystal structures as reported by Miller et al. (175, 176).

discrimination. The active site has had no evolutionary pressure to optimize binding *per se*, but rather rates of interaction and discrimination among the limited repertoire of the biological milieu. One classic example (181) of difficulty in interpretation of binding as a result of ligand modification occurred when an analog designed to bind to a specific site on hemoglobin actually found a more appropriate site within the packed side chains of the protein molecule (Fig. 3.16). This example emphasizes the importance of protein dynamics. Alternate conformations of the protein that are easily

accessible at room temperature may be difficult to characterize experimentally because of relatively low abundance and/or lack of resolution of the experimental techniques used. Computationally, they are problematic as well because of the complexity of the energy surface for a macromolecule.

3.2 Characterization of Site

3.2.1 Volume and Shape. Most substrate-enzyme or receptor-ligand interactions occur within pockets, or cavities, buried within pro-

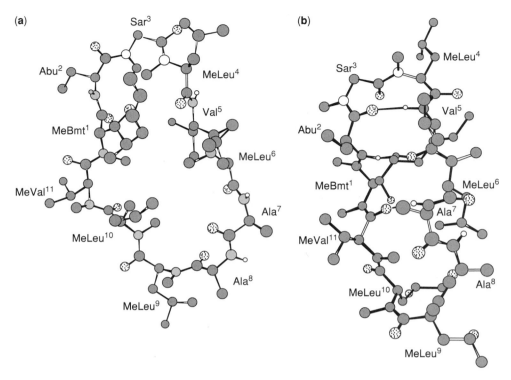

Figure 3.15. Bound conformation of cyclosporin (a) as determined by NMR compared with solution conformation (b) (178). Residues involved with interaction with cyclophilin are indicated on (a) in bold.

teins. Inside these invaginations, a microenvironment is established that favors desolvation and binding of the ligand, despite the entropic cost of fixing the relative geometries of the two molecules. Knowledge of the three-dimensional structure of such cavities can assist the study of binding interactions and the design of novel ligands as potential therapeutics. Several algorithms to find, display, and characterize cavity-like regions of proteins as potential binding sites have been developed. Kuntz et al. (13, 183) described a program, DOCK, to explore the steric complementarity between ligands and receptors of known three-dimensional structure. Using the molecular surface of a receptor, a volumetric representation of the chosen binding cavity is approximated by use of a set of spheres of various sizes that have been mathematically "packed" within it (Fig. 3.17). The set of distances between the centers of the spheres serves as a compact representation of the shape of the cavity. The use

of the relative distance paradigm allows comparison without the need for orientation of one shape with respect to the other. Potential ligands are characterized in a similar fashion by generating a set of spheres that mimic the shape of the ligand. Matching the distance matrix of the cavity with that of a potential ligand provides an efficient screen for selection of complementary shapes. Voorintholt et al. (184) used three-dimensional lattices to calculate density maps of proteins. In these maps, lattice points were assigned as a function of the distance to the nearest atom. This technique is effective in delineating regions of low density where channels and cavities exist. Ho and Marshall (185) implemented a search function in CAVITY to allow the investigator to isolate a single cavity of interest by specifying a seed point. From this seed point, the algorithm systematically explored the entire volume of the cavity, following its borders and effectively filling every crevice within it; that

(a)

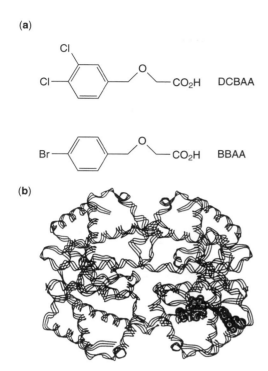

(b)

Figure 3.16. Diagram of crystal structure (181, 182) of hemoglobin in which [(*p*-bromobenzyl)oxy] acetic acid (BBAA) and [(3,4-dichlorobenzyl)oxy] acetic acid (DCBAA) bind to different sites. BBAA shown at lower right with edge-on view of phenyl ring. Coordinates of complexes supplied by Prof. Donald J. Abraham, Medical College of Virginia.

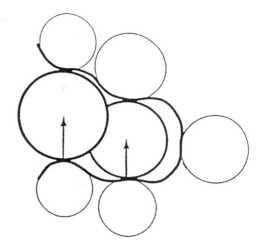

Figure 3.17. Schematic diagram of small cavity (formed by five atoms with molecular surface indicated by thick line) in which two spheres of different size have been packed according to DOCK program (183). Used with permission.

is, a three-dimensional cast of the internal volume was produced using techniques of solid modeling. This cast, or cavity, can be used as a simplified representation of the VDW surface of the receptor for drug design applications (Fig. 3.18). This and many of the techniques that follow allow a simplified display that focuses on the aspect of the problem under consideration.

3.2.2 Hydrogen-Bonding and Other Group Binding Sites. In evaluating potential ligands, knowledge of the optimal position of particular atoms, or functional groups, within the site can provide valuable insight. A popular program, GRID (186), allows a probe atom, or group, to explore the receptor site cavity on a lattice, or grid, while estimating the enthalpy of interaction. A 3D contour map can be gen-

erated from the lattice of interaction energies that gives a graphical representation of the optimal positions for the atom, or group, in question. Similar ideas are embodied in the field mapping used in the comparative molecular field analysis (CoMFA) paradigm of Cramer et al. (187), and were first presented in its predecessor, DYLOMMS (188). Ideal positions for hydrogen-bond donors or acceptors can be mapped in this fashion as a preface to ligand design.

To eliminate the grid that limits the resolution of the local mimina found and orient the functional groups of ligands for optimal connection in generating novel structures, multiple copies of ligands can be distributed in the active site by simulation and their relative distributions examined (189, 190). Miranker and Karplus (189, 190) used molecular dynamics to investigate the interaction of multiple copies of small ligands, such as MeOH, acetonitrile, and methylacetamide, with active-site cavities. After simulated annealing, the quenched populations of ligands concentrated in various orientations at points within the receptor where optimal binding would occur. By connecting the ligands with the most energetically favored binding with fragments from a library of molecular fragments, when the se-

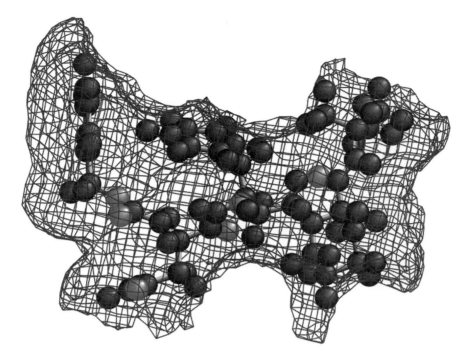

Figure 3.18. Cast of active site of HIV-1 protease as generated by CAVITY program (185).

lection is based on overlap with the carbon-carbon bonds of small ligands with fragments, novel compounds could be designed for possible synthesis (Fig. 3.19). This may be an effective means of designing compounds with high affinity, given that designs based on natural ligands may be biased toward less optimal interactions.

3.2.3 Electrostatic and Hydrophobic Fields. Because the concept of complementarity underlies much of our design of ligands, surface display of properties such as hydrophobicity and electrostatic potential offer a synopsis of the properties of the active site. Molecular surface displays may be color-coded to depict electrostatic potentials (191) and other properties. This can be done with various surface displays (dots, contour, rendered) as well as the cavity display. In the case of CAVITY (185), the loci of filler atoms necessary to pack the cavity is computed and those present within the outermost layer of the filler solid are identified. They essentially form the lining of the cavity. In other words, these points lie along the cav-

ity-pocket interface and are positioned where electrostatic interactions between the pocket and the binding ligand may be represented. At each of these positions, the electrostatic potential of the atoms forming the cavity is calculated by use of the method described by Weiner et al. (192, 193), to compute the electrostatic potential at a specific point in space within a system of charges. These values are then normalized, assigned a color, and displayed.

Researchers interested in studying the electrostatic interactions between binding molecular entities usually do so by color-coding the molecular surface of each molecule by electrostatic potential. These surfaces are then docked and visually inspected to note regions of electrostatic complementarity and disparity. Although this method is quite effective, it requires the viewer to scrutinize both electrostatic color surfaces and mentally estimate the degree of electrostatic attraction and repulsion. An effective way to view the electrostatic interaction is to color-code the electrostatic complementarity between the ligand

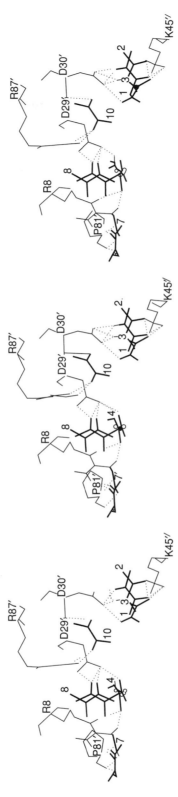

Figure 3.19. Stereoview of 10 minima of *N*-methyl-acetamide bound to open end (S4′) of the active site region of HIV protease by simulation with multiple fragments (189, 190). Left pair for cross-eyed and right pair for wall-eyed viewing. Used with permission.

and receptor (185). At every cavity-pocket interface point, the electrostatic potential of both the atoms forming the cavity and those of the binding ligand are calculated. A rough approximation of complementarity is computed by multiplying these potentials together. A favorable electrostatic interaction is produced when the electrostatic potentials are opposite in sign. Therefore, favorable interactions are indicated when the product of these values is a negative number. Likewise, unfavorable interactions are indicated when the product of these values is a positive number and the potential of the cavity and that of the binding ligand have the same sign. These products are then normalized, assigned a color, and displayed.

In a similar way, an estimate of the hydrophobic character of a segment of the surface can be quantitated and indicated through color coding. The ability to rapidly switch between these hydrophobic and electrostatic surface representations, to visually integrate the optimal complementarity between site and potential ligand to be designed, is helpful.

3.3 Design of Ligands

3.3.1 Visually Assisted Design. In the process of optimization of a lead, one needs to ascertain where modification is feasible. Although visualization of the excess space available in the active-site cavity by directly examining ligands is useful for locating selected regions where ligand modifications may be made, it is not well suited for fully characterizing the void that exists between the ligand and the receptor, the ligand-receptor gap region; information concerning the relative dimensions of free space is difficult to discern. To facilitate the display of this information, Ho and Marshall (185) developed another algorithm to color-code the cavity display by the ligand-receptor nearest atom gap distance. The actual VDW, surface-to-surface distance (not center to center) between the ligand and enzyme atoms is calculated. When the ligand-receptor distances have been calculated at all cavity-pocket interface lattice points, a user-defined color-coding scale is implemented to generate the displays. This highlights those

areas that are less well packed and available for ligand modification.

3.3.2 Three-Dimensional Databases. Medicinal chemists have recognized the potential of searching three-dimensional chemical databases to aid in the process of designing drugs for known, or hypothetical, receptor sites. Several databases are well known, such as the Cambridge Crystallographic Database (194) (CSD). The crystal coordinates of proteins and other large macromolecules are deposited into the Brookhaven Protein Databank (195). The conformations present in crystallographic databases reflect low energy conformers that should be readily attainable in solution and in the receptor complex. The three-dimensional orientation of the key regions of the drug that are crucial for molecular recognition and binding are termed the *pharmacophore*. The investigator searches the three-dimensional database through a query for fragments that contain the pharmacophoric functional groups in the proper three-dimensional orientation. Using these fragments as "building blocks," completely novel structures may be constructed through assembly and pruning (196). Receptor sites are complex both in geometrical features and in their potential energy fields, and many diverse compounds can bind to the same protein by occupying various combinations of subsites. Noncrystallographic databases have been developed as well. One example is the three-dimensional database of structures from Chemical Abstracts generated through CONCORD (197–199) that contains over 700,000 entries. The use of such databases is most applicable when the binding of a particular ligand and its receptor is well understood in terms of functional group recognition, and a crystal structure of the complex is known (200). One approach to ligand design is to develop novel chemical architectures (i.e., scaffolds) that position the pharmacophoric groups, or their bioisosteres, in the correct three-dimensional arrangement.

Gund conceived the first prototypic program designed to search for molecules that match three-dimensional pharmacophoric patterns (201, 202). This program, MOLPAT, performed atom-by-atom searches to verify comparable interatomic distances between pattern and candidate structures. Although rigorous, this approach was tedious and required optimization. Lesk (203) devised a method that used the geometric attributes of the query to screen potential candidates. Similarly, Jakes and Willett (204) proposed that screens based on interatomic distances and atom types could considerably augment search efficiency. Furthermore, Jakes et al. (205) showed that methods widely used in two-dimensional structure retrieval could be applied to three-dimensional searches, to remove the vast majority of compounds before more rigorous comparisons. This was validated in test searches against a subset of the CSD. This concept was furthered by Sheridan et al. (200), who included screens based on aromaticity, hybridization, connectivity, charge, position of lone pairs, and centers of mass of rings. To contain this wealth of information, an inverted bit map [the presence or absence of a feature is encoded as a 1 or 0 (bit) at a particular location in a "keyword"] was employed for highly efficient screening, hundreds of thousands of compounds in minutes.

Similar database searching methods have been incorporated into a number of current database searching systems. Programs such as CAVEAT (206), ALADDIN (Abbott) (207), 3DSEARCH (Lederle) (208), MACCS-3D (209), CHEM-X (210), UNITY (211), and others contain considerable functionality useful for such an approach. CAVEAT (206) is designed to assist a chemist in identifying cyclic structures that could serve as the foundation for novel compounds. In particular, it allows an investigator to rapidly search structural databases for compounds containing substituent bonds that satisfy a specific geometric relationship. ALADDIN (207), 3DSEARCH (208), MACCS-3D (209), and CHEM-X (210) are similar, in that geometric relationships between various user-defined atomic components can be used as a query to retrieve matching structures. Features have been included to allow the user to delineate molecular characteristics (atom type, bond angles, torsional constraints, etc.) to ensure the retrieval of relevant compounds. Additional constraints have been incorporated into 3DSEARCH (208) and ALADDIN (207), including the consideration of retrieved ligand-receptor volume comple-

mentarity. Furthermore, CHEM-X (210) performs a rule-based conformational search on each structure in the database to account for conformational flexibility. For a comprehensive review of three-dimensional chemical database searching, see Martin et al. (212, 213).

Pharmaceutical companies have developed three-dimensional databases for their compound files to help prioritize candidates for screening (210, 214). An essential component in such a system is a method for assessing similarity (212, 215). Because most compound databases were entered as two-dimensional structures, this has required conversion to a three-dimensional format. Programs have proved (197–199, 216) useful in generating plausible three-dimensional structures from the connectivity data, as reviewed by Sadowski and Gasteiger (217). Because of the inherent flexibility in most compounds, the use of a single conformation to represent the three-dimensional potential for interaction of a molecule is a clear limitation. Development of three-dimensional databases with a compact, coded representation of the conformational states available to each compound is a logical next step. Efficient use of such a database requires methods for evaluating three-dimensional similarities. In addition to identification of compounds that can present an appropriate three-dimensional pattern, compounds must also fit within the receptor cavity. Based on a shape-matching algorithm, Sheridan et al. (200) screened candidate compounds to select those whose volumes would fit within the combined volumes of known active compounds. Previously, this group used (218) the same algorithm to help identify potential ligands for papain and carbonic anhydrase, by screening compounds from the CSD. Screening of the active site of HIV protease identified (219) haloperidol (Fig. 3.20) as an inhibitor of the enzyme and provided a novel chemical lead for further investigation. Burt and Richards (220) introduced flexible fitting of molecules to a target structure, with assessment of molecular similarity as a means of dealing with the conformational problem.

The use of preliminary screens can eliminate the vast majority of compounds before more rigorous, and computationally demanding, pattern-matching comparisons (212, 213).

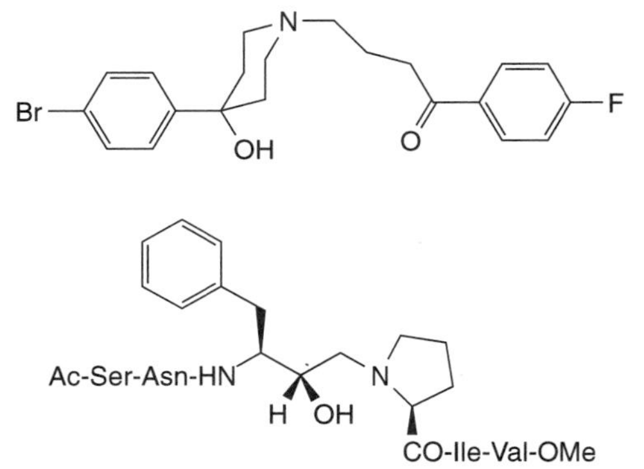

Figure 3.20. Structure of bromperidol (top) found by DOCK program when used on active site of HIV-1 protease (219) compared with structure of JG-365 (bottom), a typical substrate-derived inhibitor.

This search strategy is indeed very quick and efficient; however, all retrieved compounds must contain every query component as defined in the preliminary screens. As the number and complexity of the query elements increase, one would anticipate fewer true hits, but a corresponding rise in the number of near-misses. If such near-misses could be recovered, effective ligands may simply arise from slight conformational modification to maximize receptor interactions. Furthermore, the retrieval and combinatorial assembly of numerous pharmacophore subcomponents would intuitively produce many more diverse structures than the quest for a single compound in the database incorporating the entire pharmacophore, that is, all requirements of the query. This suggests an approach that would retrieve compounds containing any combination of a minimum number of matching pharmacophoric elements.

Methods have been developed that employ this "divide-and-conquer" approach to ligand development. The active site is partitioned into subsites, each containing several pharmacophoric elements. Chemical fragments complementary to each subsite are then designed or retrieved from databases. Finally, fragments are linked to form aggregate ligands. The advantage of this approach is that ligand diversity can be tremendously augmented through the combinatorial assembly of numerous subcomponents. DeJarlais was per-

haps the first to employ this philosophy in a novel application (220) of the program DOCK. This well-known program searches three-dimensional databases of ligands and determines potential binding modes of any that will fit within a target receptor (183). However, only a single, static conformation of each database structure is maintained, disregarding ligand flexibility. In DeJarlais' method, conformational flexibility was later introduced by dividing individual ligands into fragments overlapping at rotatable bonds. Each fragment was first docked separately into various receptor regions. Attempts were then made to reassemble the component parts into a legitimate structure. A current example of this approach is the program LUDI, written by Böhm (221, 222). In this program, a receptor volume of interest is scanned to determine subsites where hydrogen bonding or hydrophobic contact can occur. Small complementary molecules are then chosen from a database and positioned within these subsites to optimize binding energy. The process concludes with the selection of various bridging fragments to link subsets of small molecules.

Chau and Dean published a series of articles addressing whether small molecular fragments, with transferable properties, could be generated for further use in automated site-directed drug design (223–225). A program was developed to combinatorially generate all three-, four-, and five-atom fragments containing any geometrically allowed combination of H, C, N, O, F, and Cl. Aromatic fragments were produced as well. Searches of the Cambridge Structural Database (194) were performed to determine the most frequently occurring fragments. To utilize these fragments as components for ligand assembly, more data were necessary to better characterize them. They were analyzed, therefore, to statistically ascertain bond lengths from the CSD to provide some geometrical constraints for structure assembly. Finally, the transferability of atomic residual charges was studied by comparing charges generated for the atoms in each fragment with charges calculated for whole molecules containing the fragment.

Another approach, FOUNDATION (226), searches three-dimensional databases of chemical structures for a user-defined query consisting of the coordinates of atoms and/or bonds. All possible structures that contain any combination of a user-specified minimum number of matching atoms and/or bonds are retrieved. Combinations of hits can be generated automatically by a companion program (104), SPLICE, which trims molecules found from the database to fit within the active site and then logically combines them by overlapping bonds to maximize their interactions with the site (Fig. 3.21). The addition of bridging fragments to those recovered from the database allows generation of many novel ligands for further evaluation.

3.3.3 De Novo Design. Design of novel chemical structures that are capable of interacting with a receptor of known structure uses methodology that is much more robust, given that the geometric foundations of molecular sciences are much firmer than the thermodynamic ones. Techniques for the design of novel structures to interact with a known receptor site are becoming more available and show promise (227–229). It has become quite evident that much of a molecule acts simply as a scaffold to align the appropriate groups in the three-dimensional arrangement that is crucial for molecular recognition. By understanding the pattern for a particular receptor, one can transcend a given chemical series by replacing one scaffold with another of geometric equivalence. This offers a logical way to dramatically change the side-effect profile of the drug as well as its physical and metabolic attributes. Various software tools are already under development to assist the chemist in this design objective. Lewis and Dean described their approaches to molecular templates in a series of papers (230, 231). An alternative approach, BRIDGE (Dammkoehler et al., unpublished), is based on geometric generation of possible cyclic compounds as scaffolds, given constraints derived from the types of chemistry the chemist is willing to consider. Nishibata and Itai (232, 233) published a Monte Carlo approach to generating novel structures that fit a receptor cavity. Pearlman and Murko (234) combined a similar approach with molecular dynamics with illustrative applications to HIV protease and FK506 binding protein. CAVEAT is a program developed by

Bartlett to find cyclic scaffolds (207) by searching the CSD (195) for the correct vectorial arrangement of appended groups.

All of these approaches attempt to help the chemist discover novel compounds that will be recognized at a given receptor. Van Drie et al. (207) described a program, ALADDIN, for the design or recognition of compounds that meet geometric, steric, or substructural criteria, and Bures et al. (235) described its successful application to the discovery of novel auxin transport inhibitors. As our knowledge base of receptors grows, such tools will prove increasingly useful. The ability to transcend the chemical structure of lead compounds, while retaining the desired activity, should dramatically improve the ability to design away undesirable side effects. Böhm developed the program LUDI (221, 222) to construct ligands for active sites with an empirical scoring function to evaluate their construction.

3.3.4 Docking. The search for the global minimum, or the complete set of low energy minima, on the free energy surface when two molecules come in contact is commonly referred to as the "docking" problem [(236); see also Leach (21)]. Any useful molecular docking program must be computationally efficient in determining the most favorable binding mode, sufficiently sensitive in its scoring function to discriminate between alternate binding modes and the correct mode, and robust enough to allow various ligand-receptor systems to be studied.

3.3.4.1 Docking Methods. In the case of two proteins of known structure that can be approximated as rigid bodies, there are 6 degrees of freedom, the relative position (x, y, and z coordinates), and relative orientation (roll, pitch, and yaw to use the aeronautical expressions) to be explored. Several very intelligent approaches to this problem have been developed. The first and most well known approach is the DOCK program (http://www.cmpharm.ucsf.edu/kuntz/dock.html) (183) that was developed to solve the ligand-receptor problem. This program uses abstract representations (a set of spheres) of the convex shape on the receptor to be filled and the concave ligand and matches them to generate plausible binding modes with complementary

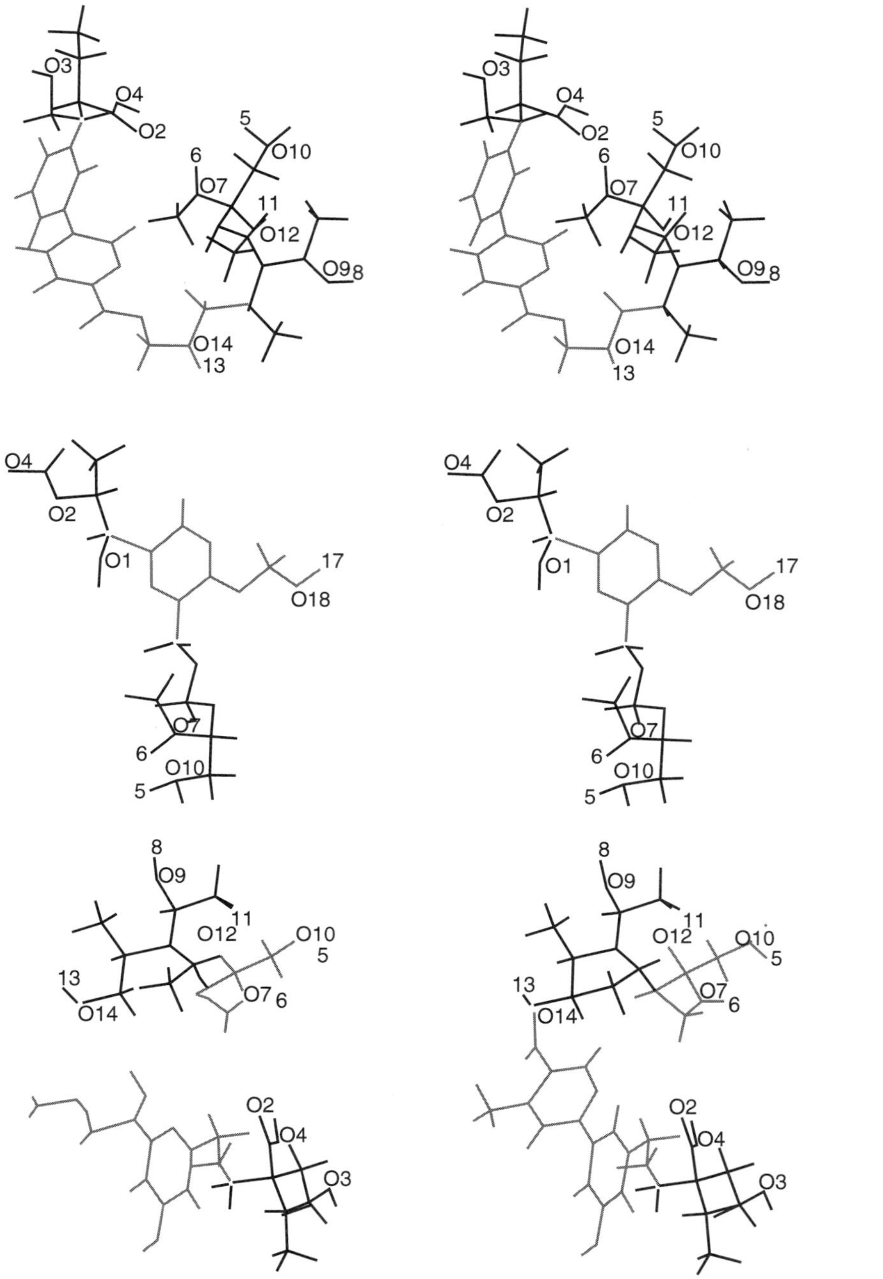

Figure 3.21. Combination by SPLICE (104) of fragments that bind to different subsites of NADP binding site of DHFR to generate a more optimal ligand.

surfaces. An example of the successful use of DOCK was the identification of 13 inhibitors of DHFR from *P. carinii* selected from the Fine Chemicals Directory. Of 40 compounds predicted to be active, these 13 showed IC_{50} values less than 150 micromolar. DOCK (13, 183) has been quite successful in finding non-congeneric molecules of the correct shape to interact with a receptor cavity (237–239). An overview of docking and scoring functions is available (240).

Another approach focusing on complementary surface maximization uses a grid representation of the surface in a series of slices. The slices from the target molecule are processed against the slices from the other molecules by use of a variant of the fast-Fourier transform (241–244) to identify those sections with the greatest complementarity. This approach has been incorporated and extended to electrostatic complementarity in FTDOCK (http://www.bmm.icnet.uk/ftdock/ftdock.html) by Gabb et al. (245). This approach is a relatively fast method for searching the 6 degrees of freedom and has reproduced the binding mode of several macromolecular complexes and is available in GRAMM (Global Range Molecular Matching, http://reco3.musc.edu/gramm/) that was judged the best when applied to identify the binding modes in a set of macromolecular complexes at the second (Fall, 1996) CASP evaluation of prediction methods.

Obviously, other degrees of freedom should be included to allow both molecules to undergo conformational changes (side-chain relaxation, at the very least, in the case of proteins). In many cases, the active site of the receptor is assumed to be rigid (rationalized on the basis of the specificity and affinity of the system) and a flexible ligand is docked. This limits the number of degrees of freedom to be explored. By simply generating a set of low energy conformers of the ligand and processing them sequentially with DOCK (220), one can sample on a low resolution scale; the flexible ligand problem can be addressed on the basis of shape complementarity.

FlexX is a program for flexibly docking ligands into binding sites, by use of an incremental construction algorithm that builds the ligands in the binding site (246). It starts by extracting a core fragment from the ligand.

The algorithm is dependent on selection of an appropriate base fragment, requiring one that makes enough specific contacts with the protein that a definite preference for binding orientation can be determined. FlexX holds bond lengths and angles invariant, using the values of the input ligand. The core is used as the base to which low energy fragment conformers are added, with these conformers based on a statistical evaluation of fragments in the Cambridge Structural Database.

3.3.4.2 Scoring Functions (247–260). Three-dimensional qualitative structure-activity relationship (3D QSAR) approaches based on the use of training sets of structures with measured affinities are often used to generate a model with predictive powers. The limitation in such methodologies is the necessity for a robust training set of diverse chemical structures to encompass the domain of possible interactions with the therapeutic target. At the beginning of a project, or when three-dimensional information on a novel target first becomes available, such data on a diverse set of chemical ligands are usually not available. For this reason, one would like to capitalize as much as possible on the physical chemistry of the possible interactions between the ligand and its receptor when the structure of the receptor is available. Because of the need to prioritize synthesis in structure-based design efforts and prioritize compounds in combinatorial libraries for screening as well as predict the structure of protein complexes, an increased interest in scoring functions (i.e., empirical approaches to predict affinities) have emerged. Several early attempts and their reported predictive ability are cited next.

1. Böhm (221, 222) analyzed 45 protein-ligand complexes (affinity range = -9 to -76 kJ/mol) and found the following equation by multiple regression analysis:

$$\Delta G_{\text{binding (kJ/mol)}} = 5.4 \Delta G_0 - 4.7 \Delta G_{\text{hb}}$$
$$- 8.3 \Delta G_{\text{ionic}} - 0.17 \Delta G_{\text{lipo}} + 1.4 \Delta G_{\text{rot}}$$

$$r^2 = 0.76, \ S = 7.9, \ q^2 = 0.696,$$
$$S \ (\text{press}) = 9.3 \ (2.2 \ \text{kcal/mol})$$

2. Krystek et al. (261) analyzed 19 protein-ligand complexes in an update of the Novotny approach (262).

$$\Delta G_{\text{binding (kcal/mol)}} = 11 - 0.025\Delta G_{\text{CSA}}$$
$$- \Delta G_{\text{EL}} + 0.6\ T_{\text{SC}}$$
$$r^2 = 0.69, S = 4.0$$

3. VALIDATE is a hybrid approach to predict the binding affinity of novel ligands for a receptor of known three-dimensional structure based on the calculation of several physicochemical properties of the ligand itself as well as a molecular mechanics analysis of the receptor-ligand complex (263). The properties of a diverse training set ($-\log K_i$, range = 2.47–14.00) of 51 crystalline complexes were analyzed by partial least squares (PLS) statistical methodology and neural network analysis to select a statistical model from a variety of parameters with the following properties:

$$r^2 = 0.81, S = 1.15, q^2 = 0.72,$$
$$S\ \text{(press)} = 1.29\ (1.75\ \text{kcal/mol})$$

The true measure of any model rests in its ability to predict the affinity of new compounds. This would include the prediction of unique ligands bound to receptors that exist in the base set as well as the affinities of unique ligand/receptor complexes. Three separate test sets were compiled for this purpose. The first set consisted of 14 inhibitors that were obtained from crystalline receptor/ligand complexes. Neither ligands nor their receptor classes were included in this training set. Included were 2 DHFR, 2 penicillipepsin, 3 carboxypeptidase, 2 alpha-thrombin, and 2 trypsinogen inhibitors as well as 3 DNA-binding molecules. Prediction of binding affinities gave a PLS predictive $r^2 = 0.786$, with an absolute average error of 0.693 log units. The second test set consisted of 13 HIV protease inhibitors whose initial conformation and alignment were derived from the CoMFA analysis done by Waller et al. (264). The selection of the inhibitors was based on maintaining a good range of activity as well as using several inhibitors from the published test set. The PLS predictive r^2 value was 0.565, with an absolute average error of 0.694. The predictive r^2 value is considerably lower than that of the first test set, although this is attributed to the smaller range and distribution of activity in this set. The absolute average error is almost identical.

Although shape complementarity is an important consideration and shows correlation with the energy of interaction, it does not consider the electrostatics of the system (the relative positioning of hydrogen-bond donors and acceptors, etc.). More sophisticated energetic functions are often used to refine the candidate binding modes found by DOCK, or in the docking process itself. The assumption of rigid geometry for the receptor allows a preprocessing of the energetic contribution of the receptor to each grid point of a lattice constructed within the active site cavity (131, 265, 266). This allows a simple estimation of the energy of interaction of each atom in the ligand by finding the energy of the lattice points that are closest followed by interpolation. By increasing the efficiency of the scoring function, more candidate binding modes can be evaluated and, thus, one resembling the global minimum is more likely to be found. This assumes that the scoring function used is sufficiently accurate to discriminate between the correct binding mode and others, and the problem is simply one of sampling. Most scoring functions used, however, deal almost essentially with the enthalpy of binding and ignore the entropy of binding. It should not be surprising, therefore, that the agreement between the predicted binding modes and those observed experimentally are not always perfect. As one is attempting to discriminate between alternate binding modes of the same complex, difficulties in estimating entropy and desolvation are minimal because many of the terms (solvation and entropy of isolated ligand and receptor) in the comparison cancel.

3.3.4.3 Search for the Correct Binding Mode (267–283). Just as there are many different approaches to the global minimization problem, most, if not all, have been applied to the docking problem. These include molecular dynamics, Monte Carlo sampling, systematic search (284), the genetic algorithm (101, 102, 105, 285, 286), and straight derivative optimization with multiple starting geometries. A combination of MD/MC has been shown (287, 288) to be a fairly efficient method for determining the free energy surface in smaller host-guest systems (289). The combination of molecular dynamics to locally sample with Monte Carlo that allows for conformational transitions provides adequate sampling if sufficient computational resources are available.

Wasserman and Hodge (290) used molecular dynamics to dock thermolysin inhibitors to an approximate model of the enzyme, with flexibility in the active site (38 of 314 residues) and ligand and with the rest of the enzyme represented by a grid approximation. A solvation model was used to compensate for desolvation in complex formation. To get 22 of 25 runs to orient the hydroxamate function correctly, the hydroxamate oxygens of the starting conformation were initialized within 4 Å of the zinc. If they were allowed to vary to 8 Å, then only 3 of 24 runs placed the ligand correctly. Obviously, there is a serious sampling problem.

Desmet et al. (291) used a truncated (dead-end elimination) search procedure to bind flexible peptides to the MHC I receptor. The translation/rotational space covered 6636 relative orientations and each nonglycine/proline residue of the peptide had 47 main-chain conformers. Side chains had threefold rotations about their chi angles and 28 side chains of the receptor were allowed to rotate. Seventy-four low energy structures were obtained with an average rmsd of 1 Å. The lowest energy structure had an rmsd of 0.56 Å. Peptides up to 20 residues were docked with this procedure.

King et al. (292) used an empirical binding free-energy function when docking MVT-101 to HIV protease. Forty-nine translation/rotations were examined with the Ponder/Richard rotamer library. Only a limited number of rotamers for each amino acid were examined: Thr(2), Ile(3), Nle(3), Nle(3), Gln(6), and Arg(5). According to the authors, 2.24×10^{10} discrete states were examined. Sixty-four low energy structures with an average rmsd of 1.36 Å were found. If the CHARMM potential was used with the same protocol, then the average rmsd was increased to 1.68 Å.

The genetic algorithm has been used by several groups (101, 102, 105, 285, 286, 293) to optimize the scoring function used. Encoding of the conformation of the ligand by torsional degrees of freedom and generating increasingly more fit sets of progeny by mutation and crossover have proved to be an effective search strategy. In one example (285), a Gray-coded binary string was used for the three translations, three rotations, and bond rotations that specified the binding mode, and a two-point crossover operator was used in the GA algorithm. In the four examples of complexes with known crystal structures, the results of rigid-body docking with a straightforward application of the GA were not encouraging, in that the correct binding mode was identified in only two of the four test cases. Restraining the GA to search subdomains (different binding hypotheses) in a systematic manner corrected this problem. Only the ligand was allowed flexibility and the GA procedure was repeated. Several binding modes similar to that seen in the experimental complex were found in each example, but ones with the lowest energy did not necessarily have the lowest rms from the experimental, pointing out deficiencies in the AMBER-like scoring function used.

Generally, no single scoring function can accurately predict the binding affinities for all types of ligands with all types of receptors. Consensus scoring (294, 295) is the simultaneous use of multiple different scoring functions to make virtual screening more predictive. CScore (Tripos, Inc.) is a consensus-scoring program that integrates several well-known scoring functions from the scientific literature. Each individual scoring function is used to predict the affinity of ligands in candidate complexes. CScore also creates a consensus column, containing integers that range from 0 to the total number of scoring functions. Each complex whose score exceeds the threshold for a particular function adds 1 to the value of the consensus; configurations below the threshold contribute a zero. Consensus columns can also be calculated from any combination of externally supplied indicators, so that key aspects of binding (e.g., the presence of a specific hydrogen bond) can be used to discriminate good configurations from bad ones. CScore can be used to rank multiple configurations of the same ligand docked with a receptor, or to rank selected configurations of different ligands docked to the same receptor.

These approaches implicitly assume that the observed receptor cavity has some physical stability (i.e., a static view) and is not significantly altered by binding of different ligands. Although there is no guarantee that this is true for any particular case under study, the specificity seen in biological systems argues that a receptor site has some functional significance in imposing its specific steric and electrostatic characteristics in the molecular recognition and selection process. One must always be prepared, however, for binding to sites other than that targeted, and possible exposure of cryptic sites that are not observed in the absence of the ligand (181). The current computational limits in molecular dynamics simulations restrict the chance of uncovering such alternative binding modes in our studies. If we can assume the binding mode of our candidate drug is nearly identical to that of a known compound, however, then we have a legitimate basis for thermodynamic perturbation calculations. Multiple or alternate binding modes remain a major fly in the ointment. Naruto et al. (284) demonstrated a systematic approach to the determination of productive binding modes for mechanism-based inhibitors (Fig. 3.22) that could select starting structures for complexes for molecular dynamics simulations. Combinations of methods, such as Monte Carlo or systematic search, to generate multiple starting configurations for simulations to improve sampling and thermodynamic reliability will increase as adequate computational power to support these hybrid approaches becomes more readily available.

Many technical limitations remain to be overcome before ligand design becomes reliable and routine. Many deficiencies in molecular mechanics previously cited remain that limit reliability. Adequate modeling of electrostatics remains elusive in many experimental systems of interest such as membranes. Newer derivations of force fields, such as MM3 (27, 296 and references therein), CHARMM (297, 298), AMBER/OPLS (157), ECEPP (299), and others (156, 300), are attempting to more accurately represent the experimental data, whereas others include a broader spec-

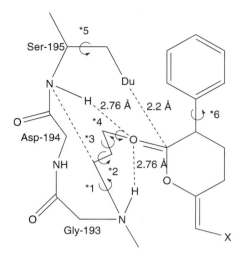

Figure 3.22. Use of systematic search to explore possible binding modes of mechanism-based inhibitors of chymotrypsin (284) by rotation of six bonds (*), which orient carbonyl of substrate relative to hydroxyl (Du) of Ser-195.

trum of chemistry such as metals (29–31, 301–305). Combinations of molecular mechanics with quantum chemistry (159, 160, 162, 306) are clearly necessary for problems in which chemical transformations are involved. Rather amazing agreement between calculation and experiment has been reported (165, 307) on the relative stabilities of transition-state structures, although there is some controversy (308) regarding this approach. In any case, this is another area of rapid growth as adequate computational resources become available. Riley et al. (309, 310) found an excellent correlation between the relative stabilities of conformers in manganese complexes of pentaazacrowns and their ability to catalyze the dismutation of superoxide.

3.4 Calculation of Affinity (260)

3.4.1 Components of Binding Affinity (255). The ability to calculate the affinity of prospective ligands based on the known three-dimensional structure of the therapeutic target would allow prioritization of synthetic targets. It would bring quantitation to the qualitative visualization of a potential ligand in the receptor site. Although this problem has been solved in principle, in practice, direct application of molecular mechanics has not yet

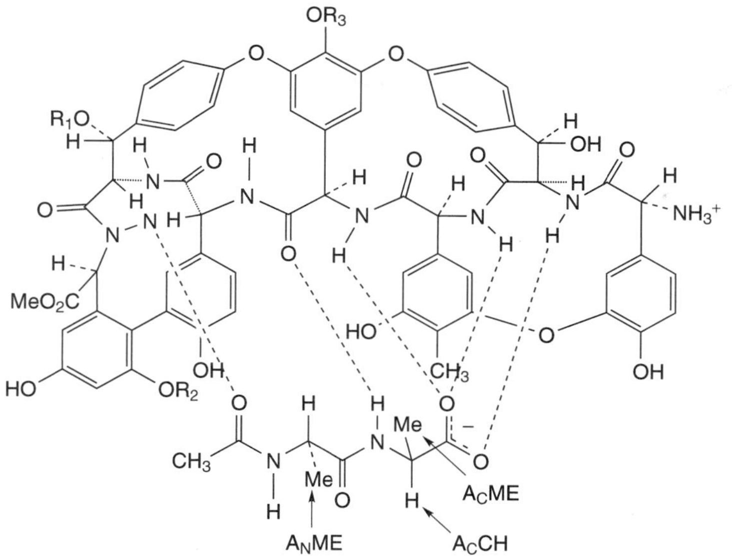

Figure 3.23. Vancomycin-peptide complex used by Williams et al. (311–315) to investigate components of free energy of binding.

proved to be a reliable indicator. The reasons behind this difficulty become more obvious if one dichotomizes the free energy of binding into a logical set of components.

For example, Williams (311–314) used a vancomycin-peptide complex (Fig. 3.23) as an experimental system in which to evaluate the various contributions to binding affinity. A similar analysis for antibody mutants was attempted by Novotny (262).

$$\Delta G = \Delta G_{(trans + rot)} + \Delta G_{rotors} + \Delta H_{conform}$$
$$+ \sum \Delta G_i + \Delta G_{vdW} + \Delta G_H$$

where $\Delta G_{(trans + rot)}$ is the free energy associated with translational and rotational freedom of the ligand. This has an adverse effect on binding of 50–70 kJ/mol (12–17 kcal/mol) at room temperature for ligands of 100–300 Da, assuming complete loss of relative translational and rotational freedom. ΔG_{rotors} is the free energy associated with the number of rotational degrees of freedom frozen. This is 5–6 kJ/mol (1.2–1.6 kcal/mol) per rotatable bond, assuming complete loss of rotational freedom. $\Delta H_{conform}$ is the strain energy introduced by complex formation (deformation in bond lengths, bond angles, torsional angles, etc. from solution states); $\sum \Delta G_i$ is the sum of interaction free energies between polar groups;

ΔG_{vdW} is the energy derived from enhanced van der Waals interactions in complex; and ΔG_H is the free energy attributed to the hydrophobic effect (0.125 kJ/mol per Å^2 of hydrocarbon surface removed from solvent by complex formation).

Through use of this analysis on the dipeptide-vancomycin system, estimates of the contribution of the hydrogen bonds to binding were made (312) that were considerably higher (−24 kJ/mol, −6 kcal/mol) than those derived experimentally. The most likely source of error is the assumption of complete loss of relative and internal entropy upon binding. In retrospect, Searle and Williams (313) examined the thermodynamics of sublimation of organic compounds without internal rotors, and showed that only 40–70% of theoretical entropy loss occurs on crystallization. This provides an estimate of the entropy loss to be expected on drug-ligand interaction. Applying this correction to the peptide-vancomycin system led (314) to a more conventional view of the hydrogen bond of between −2 and −8 kJ/mol (0.5–2.0 kcal/mol). Because several of the components in the binding energy estimate are directly related to the degree of order of the system (entropy), simulations in solvent may be necessary to quantitate the degree by which the relative motions of the ligand and protein are quenched and the restriction on rotational degrees of freedom upon complexation. Aqvist (316, 317) developed the linear interaction energy (LIE) method for calculating the ligand-binding free energies from molecular dynamics simulations. Verkhivker et al. (318) developed a hierarchical computational approach to structure and affinity prediction in which dynamics is combined with a simplified, knowledge-based energy function. Despite the focus on short peptides interacting with the SH2 domain with exhaustive calorimetric determination of binding entropy, enthalpy, and heat capacity changes, the overall correlation between computed and experimental binding affinity remained rather modest.

3.4.2 Binding Energetics and Comparisons.
Because of the difficulties in calculating binding free energies (see below), attempts to use ΔH as a means of correlation with binding affinities have often appeared in the literature, sometimes meeting with considerable success. These successes, however, are fortuitous and depend on simplifying assumptions as well as the well-known correlation (319) between ΔH and ΔG, which has been suggested as an unusual property of the solvent water. A similar correlation has been observed in nonaqueous systems and relates to higher entropy loss associated with stronger enthalpic interactions (313). It is a common assumption with congeneric series that the desolvation energies and entropic effects will be approximately the same across members of the series. This, often tacit, assumption may hold for most of the series, but complex formation is dependent on the total energetics of the complex, and what may appear a relatively innocuous change in a substituent may trigger a different binding mode in which the ligand has reoriented. This will likely have an impact on desolvation as well as entropic effects, in that the interactions of the majority of the ligand have changed environment.

3.4.3 Atom-Pair Interaction Potentials.
Affinities can be calculated based on ligand-receptor atom-pair interaction potentials that are statistical in nature rather than empirical. Muegge and Martin (320) derived these potentials from crystallographic data in the Protein Data Bank, drawing on hundreds or thousands of examples of each interaction type. Grzybowski et al. (321) combined a knowledge-based potential with a Monte Carlo growth algorithm that generated a very potent inhibitor of human carbonic anhydrase (322). The resulting equation for all the atom-pair interactions in a protein-ligand complex can yield free energies directly, given that solvation and entropic terms are treated implicitly.

3.4.4 Simulations and the Thermodynamic Cycle.
Given a known structure of a drug-receptor complex with a measured affinity of the ligand, the thermodynamic cycle paradigm allows calculation of the difference in affinity ($\Delta\Delta G$) with a novel ligand. Bash et al. (136) successfully calculated the effect of changing a phosphoramidate group (P-NH) to a phosphate ester (P-O) in transition-state analog inhibitors of thermolysin (Fig. 3.24). The difference in free energy between a benzenesulfonamide and its p-chloro derivative as an inhibitor of carbonic anhydrase has been calculated (323) as well. This is similar to the original application to enzyme-ligand work on benzamidine inhibitors of trypsin, in which the mutation of a proton to a fluorine was calculated (324). Hansen and Kollman (325) calculated differences in the free energy of binding of an inhibitor of adenosine deaminase as one changes a proton to a hydroxyl group by use of a model of the active site. Other examples (326–328) looked at the difference in binding of two stereoisomers of a transition-state inhibitor of HIV protease (Fig. 3.25) and the affinity of DHFR for methotrexate analogs (329). One obvious conclusion can be drawn: successful applications in the literature deal with relatively minor perturbations to a structure where there is less chance that the binding mode might be altered.

There is at least one example in the literature (330) in which the calculated affinity difference did not agree with the experimental date [binding of an antiviral agent to human rhinovirus HRV-14 and to a mutant virus in which a valine was mutated to a leucine (Fig. 3.26)]. Here a β-branched amino acid (Val) was converted into Leu, which lacks the isopropyl side chain adjacent to the peptide backbone besides the addition of a methyl group.

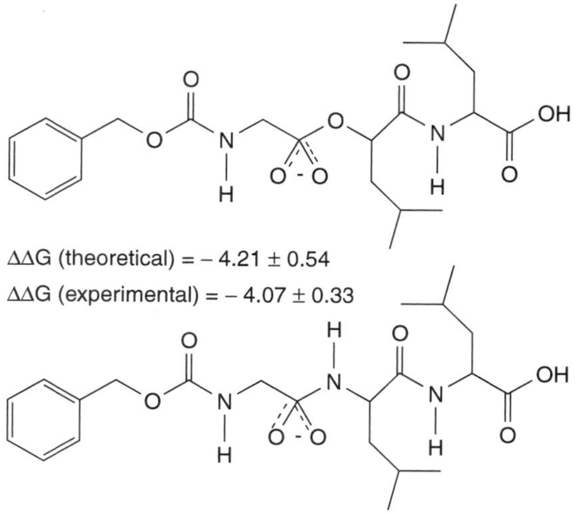

$\Delta\Delta G$ (theoretical) $= -4.21 \pm 0.54$

$\Delta\Delta G$ (experimental) $= -4.07 \pm 0.33$

Figure 3.24. Calculated (136) difference in affinity ($\Delta\Delta G$) compared with experimental value for two inhibitors of thermolysin.

The differences between calculation and experimental data may be related to rotational isomerism of the side chains that can be explicitly included (331). Despite the successful examples of this approach that appear in the literature, there exists a growing healthy skepticism regarding its general application. In a discussion (332) of the application of simulations to prediction of the changes in protein stability attributed to amino acid mutation, problems in adequate sampling, particularly of the unfolded state, as well as difficulties

with electrostatics were cited. A review of applications by Kollman (134) cites numerous other examples.

3.4.5 Multiple Binding Modes. Realistically, congeneric series that can be a useful construct exist only in the mind of the medicinal chemist. The orientation of the drug in the active site depends on a multitude of interactions and a minor perturbation in structure can destabilize the predominant binding mode in favor of another. As examples, detailed

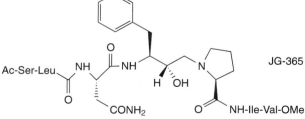

Figure 3.25. Structures of JG-365 and Ro 31–8959 in which chirality at crucial transition-state hydroxyl is reversed for optimal binding in the two analogs. An alteration in binding mode was predicted (333) to explain this observation that was subsequently confirmed by crystallography.

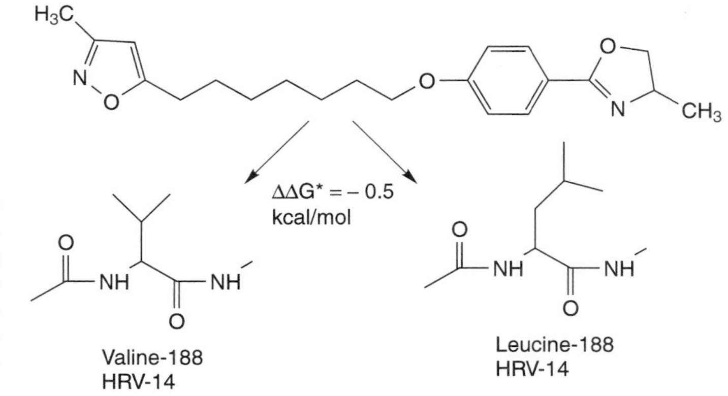

Figure 3.26. Calculated (330) relative affinity of a Sterling-Winthrop antiviral that binds to rhinovirus coat protein (HRV-14) and to the V188L mutant. Biological data indicate that V188L mutation drastically diminishes activity of the antiviral.

$\Delta\Delta G^* = -0.5$ kcal/mol

Valine-188
HRV-14

Leucine-188
HRV-14

analyses of the multiple binding modes shown with thyroxine analogs (334) by transthyretin, a transport protein, and enkephalin analogs (335) by an FAB fragment have been made through crystallography. For this reason, the probability of correct answers with thermodynamic integration studies is directly related to the similarity in structure between the ligand of interest and the reference compound. All three-dimensional methods for predicting affinity require a fundamental assumption about the binding mode (in other words, an orientation rule for aligning compounds in the model). Examination of series of ligands binding to the same site usually includes examples of similar compounds that have different binding modes [e.g., the change in orientation (Fig. 3.25) of the C-terminal portion of the Roche HIV protease inhibitor compared with JG-365] (333). Molecular modeling is currently capable of distinguishing correctly in many cases between alternate binding modes of the same ligand. Many components (desolvation, entropy of binding, etc. of the ligand), which cloud the issue of direct calculation of affinities are constant when comparing binding modes of the same compound and, therefore, do not have to be evaluated. The computational costs of exploring possible binding modes within the active site is nontrivial, however, especially when the protein is capable of reorganizing to expose alternative sites, as was the case for a series of ligands for hemoglobin (181).

In a similar fashion, it is generally assumed from the competitive behavior for binding shown by many agonists and antagonists that they bind at the same site on the receptor (certainly, the simplest hypothesis). Recent studies on G-protein-coupled receptors indicates that agonists and antagonists often have different binding sites, given that mutations in the receptor can affect the binding of one and not the other. An example of such a study on the angiotensin II receptor has been published (336). This story is only beginning to unfold, but appears to be a general phenomenon in G-protein receptors (337, 338). Examples of this phenomenon have been reported with antagonists derived from screening where the structure of antagonist and agonist differ dramatically, but also where the antagonists were obtained by minor structural modification of the natural agonist.

3.5 Protein Structure Prediction

Prediction methods for generating the 3D structure of a protein based on its sequence alone fall into several categories. There are hierarchical methods that predict secondary structures and then attempt to fold those elements together. There are simulation methods that attempt to fold the protein through the use of models of reduced complexity and then refine the prediction by using them to constrain all-atom models. Additionally, there are hybrids of these approaches that rely heavily on heuristics. These methods have been successful in limited cases in the hands of their authors, but have generally been found lacking when tested by others in a more thorough and objective manner. Nevertheless, partial successes indicate that signal has begun to emerge from the smoke and mirrors.

3.5.1 Homology Modeling. Often, the crystal structure of the therapeutic target is not available, but the three-dimensional structure of a homologous protein will have been determined. Depending on the degree of homology between the two proteins, it may be useful to model-build the structure of the unknown protein based on the known structure. Many models (339–341) of the various G-protein-coupled receptors have been built based on homology with bacterial rhodopsin. Models of the three-dimensional structures of human rennin (342) and HIV protease (343, 344) were built from crystal structures of homologous aspartyl proteinases as aids to drug design. The known structures of serine proteases have served as templates for models of phospholipase A2 (345) and convertases or subtilases (346). The crystal structure of the MHC class I receptor served to generate a hypothetical model of the foreign antigen-binding site of Class II histocompatibility molecules (347). Models of human cytochrome P450s have been built by homology as well (348).

One of the major difficulties facing construction of such models is the alignment problem that is compounded by multiple insertions and/or deletions. As the number of known homologous sequences increases, the alignment problem is lessened by consensus criteria. Although the interior core of the proteins is often quite similar, significant alterations can occur on surface loops, and much effort has been expended to fold these loops (123, 349). With regard to the utility of such models in drug design, one can expect that they will prove useful conceptually, but that the molecular details required for optimizing specificity, for example, would be deficient. One tries to exploit the often subtle differences that arise from sequence changes, which are reflected in the three-dimensional structure. Models built by homology would be expected to be weakest in those areas in which sequence differences were greatest.

3.5.2 Inverse Folding and Threading (350–353). This is the ultimate in motif recognition. One makes use of the ever-increasing database of known three-dimensional structures to generate a set of 3D folding motifs for proteins. The sequence of an unknown structure is systematically forced to adopt the coordinates of overlapping segments of the 3D motif and its energy evaluated. In essence, the local multibodied interactions induced by the 3D constraints are evaluted with an empirical pseudopotential that has been calibrated on the PDB database (354, 355) and that is capable of returning a low energy for native sequences compared with scrambled sequences or protein with other 3D structures. If one cannot discriminate native structures from other folding motifs, then there is little chance that an unknown sequence, which folds in a similar 3D pattern, would be discriminated. The basic assumption is that 3D homology exists between the test sequence and some sequence represented in the motif database. This is not necessarily true, inasmuch as many as 40% of the new structures by crystallography determined have no known 3D homologs. In fact, in an analysis of the genomes of several sequenced microorganisms (356), no more than 12% of the deduced proteins had detectable homology with proteins of known structure. In the CASP competition, however, the most predictive success has been with this approach when a 3D homology existed.

One interesting question that arises is an estimate of the number of protein motifs that exist. One way to approximate this is to assume random sampling of protein motif space and then analyze the frequency of new motifs in new crystal structures that leads to a number of approximately 1500 folds (357). Of course, such an estimate is always biased by size of protein, ease of crystallization, abundance, and so forth. Lattice approaches give a maximal estimate of 4000 folds (358). Over 1000 protein structures are known with approximately 120 folds (351).

At a more local level, proteins are generated from a set of architectural building blocks, helices, sheets, turns, and so forth. If one can accurately determine the location of these structural elements within a sequence, then the difficulty of assembly of these components is significantly easier because the degrees of freedom have been drastically reduced. Unfortunately, our ability to accurately determine these elements of secondary structure seems to have peaked at the 75% accuracy level (359, 360).

LINUS. LINUS (Local Independent Nucleating Units of Structure) (361) is an implementation of a hierarchical folding model in which protein sequences are subdivided into overlapping 50-residue fragments to assess the algorithm effectiveness in predicting short- and medium-range interaction as well as to limit computational complexity. The algorithms accumulate favorable structures within a sequence window, and repeat the process as the window is allowed to grow over the sequence. Obviously, this is an embodiment of the principle of hierarchical condensation of local initiation of folding. At the beginning, the segment length is six and the starting conformation set to all extended backbone. Starting at the *N*-terminus of the segment, three-residue subfragments are perturbed with backbone torsional values from a library to give a trial conformation. If two atoms overlap, the trial conformation is rejected. Otherwise, the energy is evaluated and selection depends on the Metropolis criterion. For each interaction cycle, 6000 iterations of this procedure are performed, 1000 iterations for equilibrium and 5000 samples. Conformations of chain segments that give a high frequency in the sample are frozen and the segment size increased. Backbone atoms and highly simplified side chains are used in the simulations. The simplified energy function has a vdW term, a hydrogen-bonding term, and a backbone torsional term.

Given the arbitrary fragmentation of the protein for computational efficiency, the predicted secondary structures were surprisingly accurate for the five cases examined, with helical and sheet boundaries within two residues of their corresponding native structures. Nevertheless, the rms differences were rather large, from 3 to 9 Å. Certainly, these results are quite encouraging and confirm the ideas from studies on lattices by Dill (362) and others that much of the secondary structure is encoded into local patterns of hydrophobic and polar residues.

GEOCORE (363). Amino acids are represented at the united atom level with explicit polar hydrogens with slightly reduced vdW radii. The approach uses a discrete set of Φ, Ψ values for each residue type: Gly has six, Pro has three, and most others have four or five

values. A contact between nonpolar atoms (carbon or sulfur) is worth -0.7 kcal/mol at closest contact and scaled down from there. Buried non-hydrogen-bonding groups get a penalty of 1.5 kcal/mol. Polar conflicts in which two donors or two acceptors are in contact are given a similar penalty. Constraint-based exhaustive search is used (systematic search with limits such that no steric overlap is allowed and that a compact structure is generated), a branch-and-bound method that guarantees that all globally or near-globally optimal conformations will be found, while neglecting less important conformations. The compact structure is guaranteed by a volume constraint about 60% higher than the volume of a native protein of the same size. Side chains are introduced in their most populated rotameric state from the PBD and only changed to an alternate rotamer to avoid a vdW contact. Four proteins were used to test the approach, avian pancreatic polypeptide (1PPT), crambin (1CRN), melittin (2MLT), and apamin (18 residues). Some 190 million conformations were generated for 1PPT, with 8217 having an energy not more than 16 kcal/mol above the optimum found. The conformation with the lowest rms to the native structure was within the 100 lowest energy conformations found, but the true native structure had a lower energy by use of the same energy function than that of any conformer found by 3–10%. This implies that the major problem was conformational sampling, not just an oversimplified potential function.

Genetic Algorithm. Le Grand and Merz (364) applied the genetic algorithm to a model of proteins using a rotamer library and the AMBER potential function. In a second study, they used a fragment library and a knowledge-based potential function. Sun (365) used a fragment library consisting of di- to pentapeptides and the Sippl potential. He predicted the structures of mellitin, avian pancreatic polypeptide, and apamin (both fragments from apamin and APP were included in the library, so it is not so surprising that the rms agreement for these two was around 1.5 Å). Bowie and Eisenberg (366) used the genetic algorithm with a fragment library of from 9 to 25 residues and their own knowledge-based potential. The fragment most similar to that of the sequence based on 3D profiles (367) was chosen. They were able to fold 50-residue fragments to within 4.0 Å based on the error in the distance matrix. This avoids the problem of embedding and generating the wrong chirality, which reduces the error estimate.

3.5.3 Contact Matrix.

Instead of searching the three-dimensional coordinate space, one can reduce dimensionality by focusing on generating an optimal contact map in 2D (368). The 3D coordinates of a correct contact map can be generated within 1 Å rms for the carbon alphas by distance geometry (369) or other methods (370). By use of the powers of the contact matrix as constraints that limit the contact matrices to compact structures, exploration of various potential interactions between secondary structural elements can be done efficiently. Because of the limited predictive ability of current secondary structure prediction paradigms, a set of plausible inputs to this procedure need to be generated, and the best structures that are derived evaluated further. This may be an efficient low resolution model builder and have some of the computational advantages of the hydrophobic core constraints used by Dill and coworkers. This approach based on geometrical constraints was originally proposed by Kuntz et al. in 1976 (371). The matrices of residue-residue contacts provide, at the very least, a significant partial solution to the prediction of long-range intersegmental contacts through a formalism explicitly describing the structure and some structure-related properties of a protein globule in terms of matrices of residue-residue contacts without explicit knowledge of secondary structure predictions, although they can be a useful source of constraints. In many ways, the success of this approach verifies the conclusions based on lattice models that secondary structures are implicit in the pattern of hydrophobic and hydrophilic residues and the requirements of compactness. The residue-residue contact matrices have some special properties as mathematical objects that can encode the geometrical requirements of compactness; the knowledge of these allows their treatment, starting with the sequence to generate a contact matrix that is consistent

with a compact structure. This is done within the framework of a simple and readily formalized geometric model.

The system of intraglobular residue-residue contacts of a protein of N residues may be represented as an $\mathbf{N} \times \mathbf{N}$ matrix of the carbon-alphas, whose elements are ones (contact) or zeros (lack of contact). Any reasonable definition of contact provides ones in the positions $(i, i + 1)$ that correspond to a peptide bond between two adjacent residues in the sequence. The same is true for the residues corresponding to the pair of cysteines forming a disulfide bond (these data may not be available as input and may be used as a test of correct prediction). This set of contacts describes the sequential covalent topology and is a constant part of the contact matrix which does not depend on the spatial structure of the polypeptide chain; however, any additional information on existing intraglobular contacts (e.g., from NMR data or disulfide linkage) can easily be introduced in the constant part A^c of the contact matrix $\mathbf{A}$:

$$A^c = \text{const.} \tag{3.1}$$

The number of contacts involving a given residue n_i (the coordination number of the ith residue)

$$n_i = \sum_j a_{ij} \tag{3.2}$$

are assumed to be approximate constants (coordination number) and are determined by a separate algorithm based on residue type and position in the sequence as well as predicted secondary structure.

A very important condition of spatial consistency of any given contact system is defined by the relation

$$b_{ij} = \sum_k a_{ik} a_{kj} \geq c, \quad \text{if } a_{ij} = 1 \tag{3.3}$$

In other words, the squared matrix of $\mathbf{A}$ should have its elements not less than c at any position where there is a nonzero element in matrix $\mathbf{A}$. More generally, there exists a set of specific constraints regulating the relationships of $\mathbf{A}$ with its powers A^2, A^3, and so forth. These relations are entirely analogous with those known from graph theory for connectivity (adjancency) matrices. The elements of the squared matrix represent the number of paths of length two, the cubed matrix, the number of paths of length three, and so forth. Finally, an obvious property of matrix $\mathbf{A}$ is its symmetry (for all contact definitions considered so far, if the ith residue is in contact with the jth, the jth residue is in contact with the ith, also).

$$a_{ij} = a_{ji} \qquad (3.4)$$

Thus, conditions 3.1–3.4 define the set of matrices $\mathbf{A_k}$ that correspond to spatially consistent, compact structures of protein chains. Besides these general conditions, mainly of geometrical origin, any matrix $\mathbf{A}$ describing the structure of a real protein molecule should also possess several more specific properties that may be derived from studies of the general properties of protein structures as exemplified in the Brookhaven Protein Databank. The central idea of the approach is to use both the general and specific properties of the contact matrix and its powers for the design of a gain (energy, penalty) function, $\Phi(A)$, so that the task of determining an appropriate intraglobular contact matrix might be formulated as a problem of maximization of $\Phi(A)$,

$$\Phi(A) \rightarrow \max_{A} \qquad (3.5)$$

with respect to A under conditions 3.1–3.4. In the simplest and clearest form, $\Phi(A)$ may be expressed in terms of the probabilities of contact between the residues of different types (or groups), q_{ij}. The solution of the problem provides the most probable residue-residue contact matrix $\mathbf{A}$ in

$$\Phi(A) = \prod_{\text{all contacts}} q_{ij} \rightarrow \max_{A}, \qquad (3.6)$$

which is the sense of the maximum likelihood principle. This condition may be rewritten in the form

$$\Phi'(A) = \sum_{\text{all contacts}} \ln q_{ij}$$

$$= \sum_{i,j} a_{ij} \ln q_{ij} \rightarrow \max_{A} \qquad (3.7)$$

It is clear that proper formulation and parameterization of this problem need the analysis of the voluminous experimental data on protein structure to derive the specific properties to be emulated.

This methodology has been used to predict the structure of loops of helical-bundle proteins, given the positions of the connection to the helices (372). Because of the uncertainties in secondary structure predictions that are used as inputs to constrain the search, any single prediction of the method must be viewed with skepticism. Development of scoring functions that discriminate between alternative models at the Cα level of resolution would complement this approach.

Distance Geometry. Aszodi et al. (373–375) explored the use of distance geometry as the metric for comparative modeling of structures. In the CASP2 target set, the methods generated an overall Cα rmsd of 1.85 Å for glutathione transferase based on close homologs with known structure. It had more difficulty with PNS1 and built models based on two different proteins. The correct fold was not obvious based on the CHARMM energy values for the two models.

Neural Networks. PROBE (376) is an integrated suite of neural network modules that predicts folding motif, secondary structure per residue, location of disulfide bonds, and surface accessibility of each residue. No critical assessment of the accuracy of the results from this package was given in the description, but is available for evaluation.

Discrimination Between Folds. Because of the inherent error in potential functions, secondary structure prediction methods, limited sampling, and so forth, one can anticipate that prediction of a variety of alternative structures (perhaps, by several methods) would be more likely to generate a correctly folded structure than any single prediction. The problem then becomes one of discriminating between the correct structure and alternatives that may be very similar in overall quality of fold. Park et al. (377) evaluated the ability of 18 low and medium resolution energy functions to discriminate correct from incorrect folds. Functions that were effective in protein threading were not competitive in discriminating the X-ray structure from ensembles of plausible structures, and vice versa. Obviously, these empirical functions have been derived to optimize their discriminate abilities for a given problem class and the training (selection) sets were different. In other words, the true physics has not been captured by any of the methods. Crippen (378) also raised serious doubts concerning the ability of "empirical" energy functions to identify correctly folded structures based on studies with simple lattice models. Thomas and Dill (379) described an iterative approach EN-ERGI to generate pairwise residue "energy" scores from the PDB. This is one alternative to the Boltzmann-based pairing frequency analysis used by others (380). The assumption that pairing frequencies are independent is not true based on lattice simulation and, therefore, the underlying assumption of the Boltzmann approach is flawed. The study that used two different sets of proteins to thread was able to classify 88% of 121 proteins having less than 25% homology and no homologs in the training set. The method appears to separate interactive free energies from chain configurational entropies and thus give a more realistic estimate.

4 UNKNOWN RECEPTORS

Until recently, receptors were hypothetical macromolecules whose existence was postulated on the basis of pharmacological experiments. Although recent advances in molecular biology have led to cloning and expression of many of those receptors whose existence was postulated as well as a plethora of subtypes, progress in most cases in defining their three-dimensional structure has yet to provide the medicinal chemist with the necessary atomic detail to design novel compounds. Without detailed information about the three-dimensional nature of the receptor, conventional computationally based approaches, such as

molecular dynamics and the Monte Carlo method, are not possible. One can only attempt to deduce an operational model of the receptor that gives a consistent explanation of the known data and, ideally, provides predictive value when considering new compounds for synthesis and biological testing. The utility of such an approach has been demonstrated by Bures et al. (235), who used the pharmacophoric pattern derived for the plant hormone auxin, to find four novel classes of active compounds by searching a corporate three-dimensional database of structures. In many ways, the approach that has evolved is analogous to the American parlor game of 20 questions, in which the medicinal chemist poses the questions in terms of novel three-dimensional chemical structures and attempts to interpret the response of the receptor in a consistent manner. The underlying hypothesis is a structural complementarity between the receptor and compounds that bind. In the same way that the receptor's existence could be deduced based on pharmacological data, some low resolution three-dimensional schematic of the receptor, at least with regard to the active site or binding pocket, can be deduced by analysis of structure-activity data. It is the purpose of this section to summarize the current approaches in use for receptors of unknown three-dimensional structure and evaluate their utility. For purposes of this section, *receptor* is often used in a completely generic sense, including enzymes and DNA, for example, as the macromolecular component (i.e., binding site) of recognition of biologically active small molecules.

4.1 Pharmacophore versus Binding-Site Models

4.1.1 Pharmacophore Models. It is often useful to assume that the receptor site is rigid and that structurally different drugs bind in conformations that present a similar steric and electronic pattern, the *pharmacophore*. Most drugs, because of inherent conformational freedom, are capable of presenting a multitude of three-dimensional patterns to a receptor. The pharmacophoric assumption led to a problem statement that logically is composed of two processes. First is the determina-

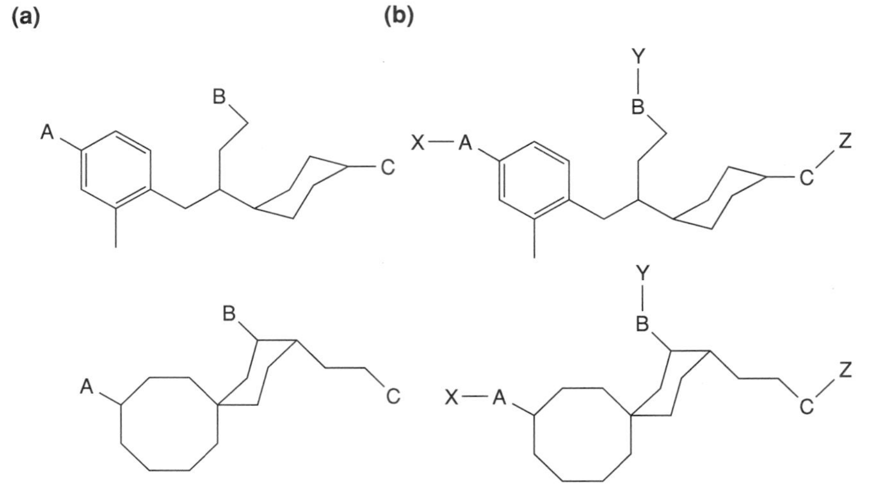

(a)   (b)

Figure 3.27. (a) Pharmacophore hypothesis with correspondence of functional groups in drugs, A = A', B = B', C = C'. (b) Binding-site hypothesis by use of drugs with hypothetical binding sites attached (X, Y, and Z overlap).

tion, by chemical modification and biological testing, of the relative importance of different functional groups in the drug to receptor recognition. This can give some indication of the nature of the functional groups in the receptor that are responsible for binding of the set of drugs. Second, a hypothesis is proposed (Fig. 3.27) concerning correspondence, either between functional groups (pharmacophore) in different congeneric series of the drug or between recognition site points postulated to exist within the receptor (binding-site model).

The intellectual framework for use of structure-activity data to extrapolate information regarding the ligand's partner, the receptor, is the concept of the pharmacophore. The pharmacophore, a concept introduced by Ehrlich at the turn of the 20th century, is the critical three-dimensional arrangement of molecular fragments (or distribution of electron density) that is recognized by the receptor and, in the case of agonists, that causes subsequent activation of the receptor upon binding. In other words, some parts of the molecule are essential for interaction, and they must be capable of assuming a particular three-dimensional pattern that is complementary to the receptor to interact favorably. One corollary of the pharmacophoric concept is the ability to replace the chemical scaffold holding the phar-

macophoric groups with retention of activity. This is the basis of the current activity (381, 382) in peptidomimetics, in which the amide backbone of peptides has been replaced by sugar rings, steroids (383, 384), benzodiazepines (385), or carbocycles (386, 387) (Fig. 3.28). In the pharmacophoric hypothesis, physical overlap of similar functional groups is assumed; that is, the carboxyl group from compound A physically overlaps with the corresponding carboxyl group from compound B and with the bioisosteric tetrazole ring of compound C.

One caveat that must be remembered is the probability of alternate, or multiple, binding modes. The interaction of a ligand with a binding site depends on the free energy of binding, a complex interaction with both entropic and enthalpic components. Simple modifications in structure may favor one of several nearly energetically equivalent modes of interaction with the receptor, and change the correspondence between functional groups that has previously been assumed and supported by experimental data. Changes in binding mode of an antibody FAB fragment to progesterone and its analogs have been shown by crystallography (390, 391) of the complexes. For this reason, analysis of agonists as a class is usually preferred, given that the necessity to both

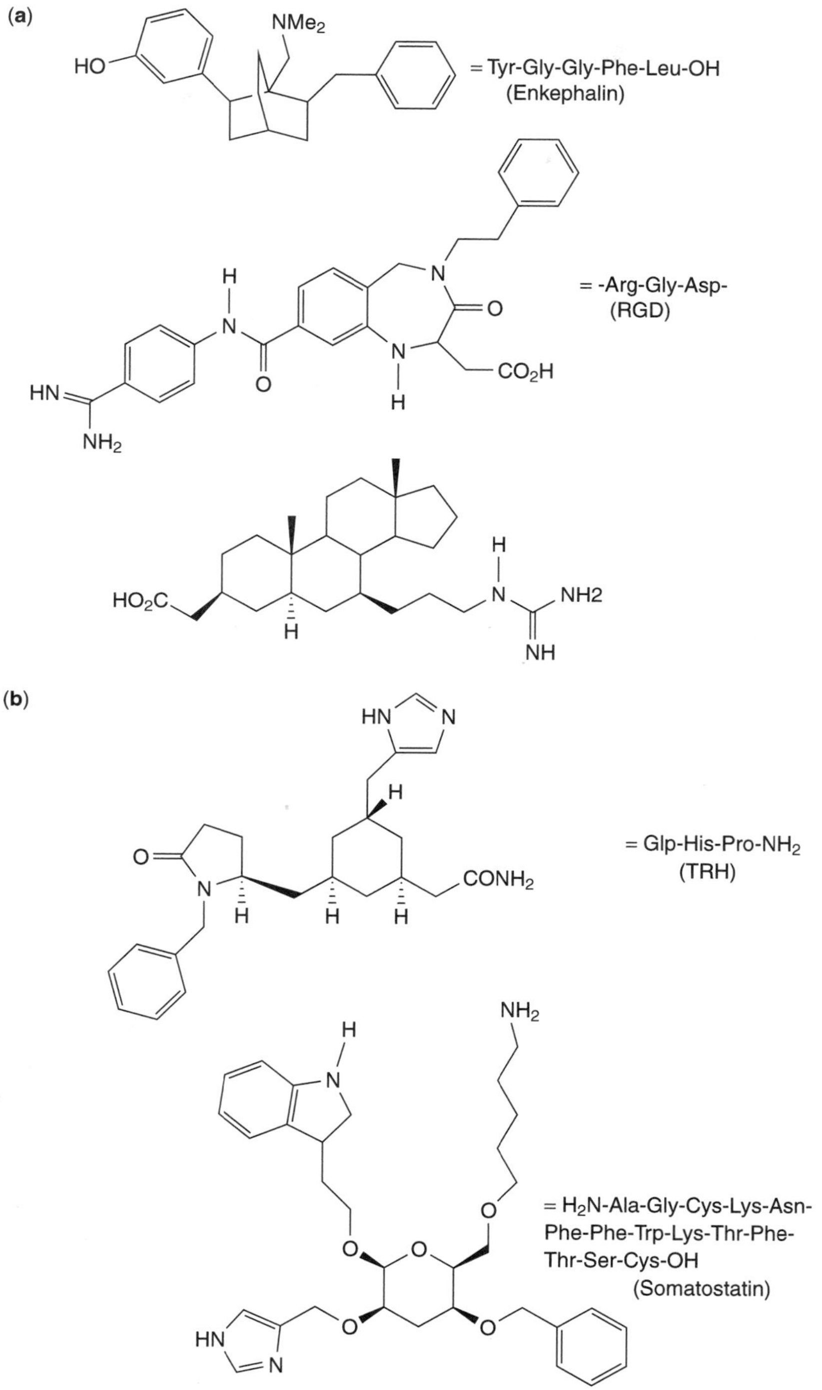

Figure 3.28. Peptidomimetics that have been designed based on iterative introduction of constraints into parent peptide and hypotheses concerning receptor-bound conformation. Enkephalin mimetic (388), RGD platelet GPIIb/IIIa receptor antagonists (384, 385), thyroliberin [TRH (387)], and somatostatin (383, 389). For an overview of recent approaches to peptidomimetic design, see the review by Bursavich and Rich (382).

bind and trigger a subsequent transduction event is more restrictive than the simple requirement for binding shared by antagonists (336). Compounds that clearly are inconsistent with models derived from large amounts of structure-activity data may be indicative of such changes in binding mode, and may require a separate structure-activity study to characterize their interaction. Despite its limitations, the pharmacophore approach is often the most appropriate because of lack of detailed information regarding the receptor and can yield useful insights, as seen in the case of clinical success with tyrosine kinase inhibitors (392, 393) and other recent examples (394).

4.1.2 Binding-Site Models. One major deficiency in the approach described above is the requirement for overlap of functional groups in accord with the pharmacophoric hypothesis. Although it is true that molecules having functional groups that show three-dimensional correspondence can interact with the same site, it is also true that a particular geometry associated with one site is capable of interacting with equal affinity with a variety of orientations of the same functional groups. One has only to consider the cone of nearly equal energetic arrangements of a hydrogen-bond donor and acceptor to realize the problem. Sufficient examples from crystal structures of drug-enzyme complexes and from theoretical simulation of binding compel the realization that the pharmacophore is a limiting assumption. Clearly, the observed binding mode in a complex represents the optimal position of the ligand in an asymmetric force field created by the receptor that is subject to perturbation from solvation and entropic considerations. Less restrictive is the assumption that the receptor-binding site remains relatively fixed in geometry when binding the series of compounds under study. Experimental support for such a hypothesis can be found in crystal structures of enzyme-inhibitor complexes, where the enzyme presents essentially the same conformation, despite large variations in inhibitor structures; studies of HIV-1 protease complexed with diverse inhibitors support this view (171).

In recent years, therefore, there has been an increasing effort to focus on the groups of the receptor that interact with ligands as being the common features for recognition of a set of analogs. When pharmacophore and binding-site hypotheses are compared, the binding-site model is physicochemically more plausible, in that overlap of functional groups in binding to a receptor is more restrictive than assuming the site remains relatively fixed when binding different ligands. However, the number of degrees of freedom in binding-site hypotheses, represented by the necessary addition of virtual bonds between groups A and X, B and Y, and C and Z in Fig. 3.27, is greater. Additional degrees of freedom complicate subsequent conformational analyses and may preclude any conclusions unless a sufficiently diverse set of compounds is available.

Other approaches to this problem have emphasized comparison of molecular properties rather than atom correspondences. Kato et al. (395) developed a program that allows construction of a receptor cavity around a molecule emphasizing the electrostatic and hydrogen-bonding capabilities. Other molecules can then be fit within the cavity to align them. This is similar in concept to the field-fit techniques available in the CoMFA module of SYBYL, in which the molecular field (electrostatic and steric) surrounding a selected molecule becomes the objective criterion for alignment of subsequent molecules for analysis. An example emphasizing molecular properties in pharmacophoric analysis was given by Moos et al. (396) on inhibitors of cAMP phosphodiesterase II.

4.1.3 Molecular Extensions. If we assume the binding-site points remain fixed and can augment our drug with appropriate molecular extensions that include the binding site (i.e., a hydrogen-bond donor correctly positioned next to an acceptor), we can then examine the set of possible geometrical orientations of site points to see whether one is capable of binding all the ligands. Here, the basic assumption of rigid site points is more reasonable, at least for enzymes that have evolved to catalyze reactions and must, therefore, position critical groups in a specific three-dimensional arrangement to create the correct electronic environment for catalysis. The program checks

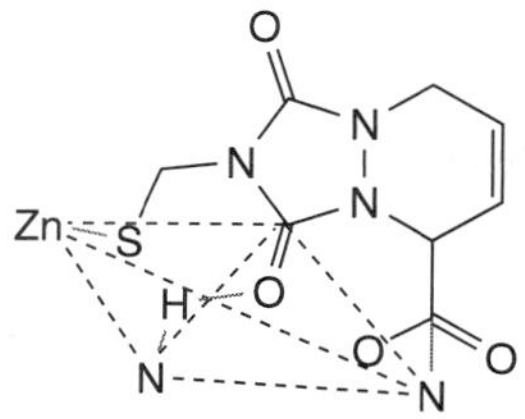

Figure 3.29. The use of active-site models in the Active Analog Approach. The structure shown is one of a series of ACE inhibitors analyzed. The thick gray lines are noncovalent interactions between the inhibitor and active-site points in the enzyme. The dashed lines correspond to the six interatomic distances monitored for each of the inhibitors analyzed.

this hypothesis by determining whether one or more geometrical arrangements of the postulated groups of site points is common to the set of active compounds. Such a geometrical arrangement of receptor groups becomes a candidate binding-site model, which can be evaluated for predictive merit.

In the study of the active site of angiotensin converting enzyme (ACE) by Mayer et al. (397), a binding site model (Fig. 3.29) was used by incorporating the active-site components as parts of each compound undergoing analysis. As an example, the sulfhydryl portion of captopril was extended to include a zinc bound at the experimentally optimal bond length and bond angle for zinc-sulfur complexes (Fig. 3.29). The orientation map (OMAP) (398), which is a multidimensional representation of the interatomic distances between pharmacophoric groups (Fig. 3.30), was based on the distances between binding-site points such as the zinc atom with the introduction of more degrees of torsional freedom to accommodate

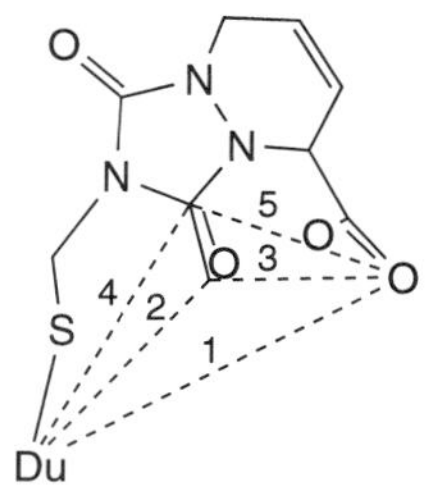

Figure 3.30. Distances used in five-dimensional OMAP used in analysis of ACE inhibitors.

the possible positioning of the zinc relative to ACE inhibitors such as captopril. Analyses of nearly 30 different chemical classes (Fig. 3.31) of ACE inhibitors led to a unique arrangement of the components of the active site postulated to be responsible for binding of the inhibitors. The displacement of the zinc atom in ACE to a location more distant from the carboxyl-binding Arg seen in carboxypeptidase A is compatible with the fact that ACE cleaves dipeptides from the C-terminus of peptides, whereas carboxypeptidase A cleaves single amino acid residues.

Visualization of the OMAP is useful to judge the additional information introduced as each new compound is added (Fig. 3.32). Computationally, it is much more efficient to treat the set of noncongeneric compounds simultaneously (111, 399), as we shall see, but reassuring when identical results are obtained if one uses the sequential procedure introducing each molecule in turn, where intermediate results may be visually verified. The use of computer graphics to confirm intermediate processing of data in convenient display modes becomes increasingly more important as the individual computations and numbers of molecules under consideration increase.

4.1.4 Activity versus Affinity. Given a consistent model of either type, a limitation is that one can only ask whether the compound under consideration can present the three-dimensional electronic pattern (pharmacophore) that is the current candidate. In other words, one is limited to predicting the presence or absence of activity, a binary choice. Even the presence of the appropriate pattern is insufficient to ensure biological activity. For example, competition with the receptor for occupied space by other parts of the molecule can inhibit binding and preclude activity. We can thus postulate the following conditions for activity:

1. The compound must be metabolically stable and capable of transport to the site for receptor interaction (interpretation of inactive compounds may be flawed by problems with bioavailability).

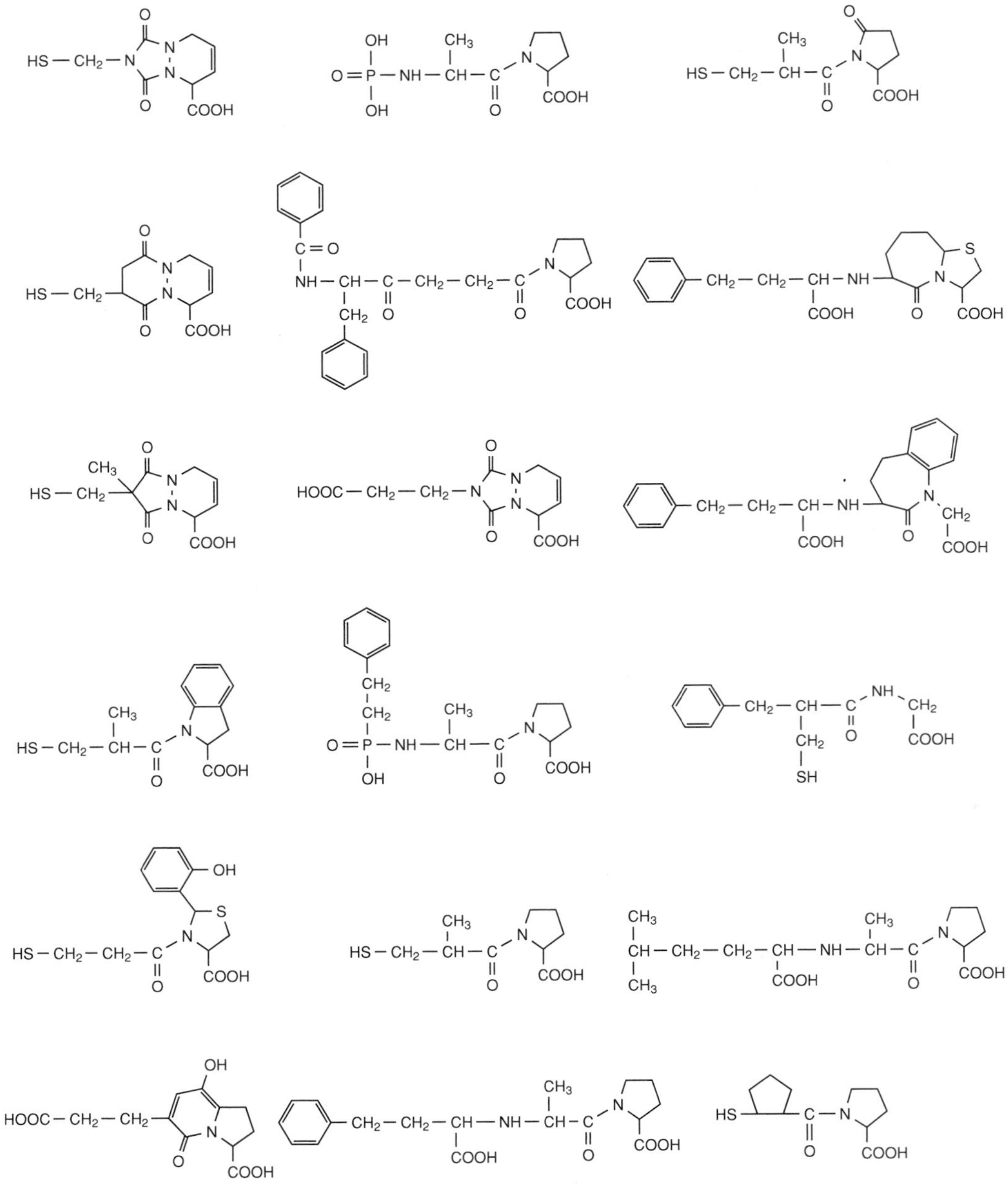

Figure 3.31. Compounds from different chemical classes of ACE inhibitors used in active-site analysis. Used with permission (397).

2. The compound must be capable of assuming a conformation that will present the pharmacophoric or binding-site pattern complementary to that of the receptor.

3. The compound must not compete with the receptor for space while presenting the pharmacophoric or binding-site pattern.

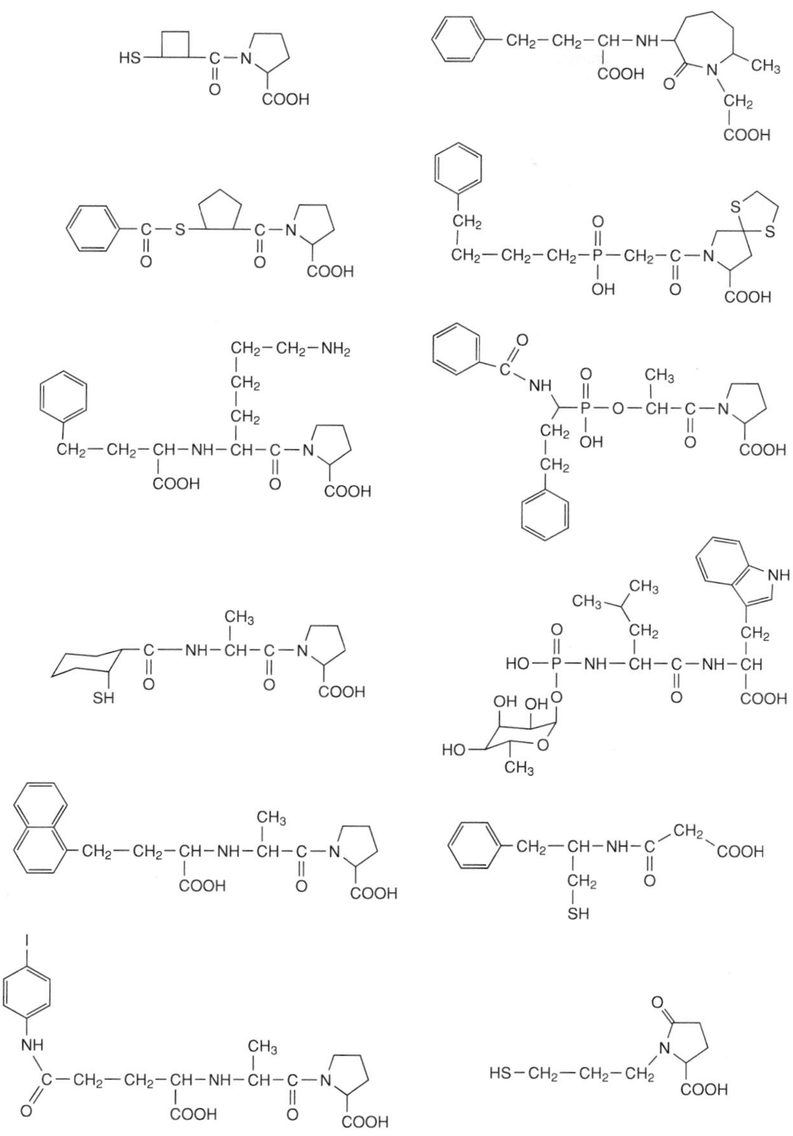

Figure 3.31. (Continued.)

Once these conditions are met, we can attempt to deal with the potency, or binding affinity. This belongs to the domain of three-dimensional quantitative structure-activity relationships (3D-QSARs) (400) and we illustrate the use of a particular variant, CoMFA (187, 401), on ACE inhibitors at the end of this chapter. Condition 3.3 allows us to utilize compounds capable of presenting the pharma-cophoric pattern, but incapable of binding, to help determine the location of receptor-occupied space in relation to the pharmacophore (receptor-mapping) (402). This allows a crude, low resolution map of the position of the receptor relative to the pharmacophoric elements and indicates in which directions chemical modifications may be productive.

The number and diversity of compounds

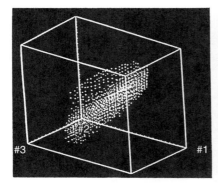 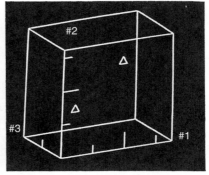

Figure 3.32. Change in OMAP (projection of three of the five dimensions) as new compounds were introduced to analysis of ACE inhibitors (397). Left is original OMAP of compound 1 (Fig. 3.30). Right is OMAP after completion of analysis.

available for analysis determine the methodology to be used. If there is a limited data set, then the pharmacophoric approach should be assessed first because of its fewer degrees of freedom. If no pharmacophoric patterns are consistent with the set of analogs, then introduction of logical molecular extensions to enable the active-site approach is warranted. Operationally, one first determines the set of potential pharmacophoric patterns consistent with the set of active analogs [leading to its name of Active Analog Approach (398)]. If there are sufficient data, then a unique pharmacophore, or active-site model, may be identifiable. The basic assumption behind efforts to infer properties of the receptor from a study of structure-activity relations of drugs that bind is the idea of *complementarity*. It follows that the stronger the binding affinity, the more likely that the drug fits the receptor cavity and aligns those functional groups that have specific interactions in a way complementary to those of the receptor itself. Certainly, our understanding of intermolecular interactions from studies of known complexes does not dissuade us of this notion, but may make us somewhat skeptical of the naive models that often result from such efforts. Andrews et al. (403) reviewed efforts of this type with regard to CNS drugs.

Clearly, the key to insight relies on chemical modification to determine the relative importance of functional groups for molecular recognition. Often more subtle effects than the simple presence or absence of a group are

important and then comparison of molecular properties becomes of interest. A major impediment to analysis is the definition of a common frame of reference by which to align molecules for comparison. This is equivalent to solving the three-dimensional pharmacophoric pattern, and implies that one has distinguished those properties of the molecules under consideration in a manner similar to the receptor. Initial efforts to rationalize structure-activity relationships (SARs) among noncongeneric systems was hampered by an "RMS mentality." That is, a point of view that required atomic centers to align rather than overlap of steric and electronically similar grouping of atoms. An example would be requiring the six atoms of aromatic benzene rings to overlap at each of the six atoms of the ring vertices rather than simple requirements for coincidence and coplanarity that would recognize the torus of electron density that the rings share in common (Fig. 3.33). In congeneric series, the difficulty in assignment of correspondence is less (nonexistent by definition). This allows a variety of approaches, including those based on molecular graph theory (404–407), to detect similarities between molecules that can form the basis of a correlation analysis. Extrapolation outside of the group of congenerically related compounds on which the analysis was based would appear difficult, if not impossible.

Although it is simpler to start an analysis with a congeneric series to identify the recognition elements, diversity in chemical struc-

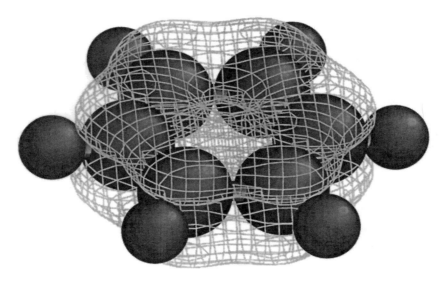

Figure 3.33. Torus of electron density representing benzene ring. Atom-to-atom correspondences of ring atoms used in normal fitting routines lead to overconstrained fits.

tures implies more information regarding the conformational requirements of the system. A congeneric series requires that the basic chemical framework of the molecule remains constant and that groups on the periphery are either modified (e.g., aromatic substitution) or substituted (e.g., tetrazole for carboxyl functional group). Implicit in this concept is the notion that the compounds bind to the receptor in a similar fashion and, therefore, the changes are localized and comparable for each position of modification. Introduction of degrees of freedom in the substituents as well as consideration of differences in properties that are conformationally dependent, such as the electric field, require conformational analysis in an effort to determine the relevant conformation for comparison.

The problem can be divided into two: what are the aspects of the molecules that are in common and that may provide the basis for molecular recognition, and which conformation for each molecule is appropriate to consider. For the first problem, studies on a congeneric series can often yield valuable insight. For determination of the three-dimensional arrangement of the crucial recognition elements, diversity in the chemical scaffolds imposes different constraints on possible three-

dimensional patterns and generates an opportunity for determining a unique solution.

4.2 Searching for Similarity

4.2.1 Simple Comparisons. To gain insight into molecular recognition, subtle differences in molecules must be perceived. Comparisons can be divided into two categories: those that are independent of the orientation and position of the molecule and those that depend on a known frame of reference. Simple comparisons deal with properties independent of a reference frame. For example, the magnitude of the dipole moment is frame independent, but the dipole itself is a vectorial quantity dependent on the orientation and conformation of the molecule. Similarly, the bond lengths, valence angles and torsion angles, and interatomic distances are independent of orientation. The *distance matrix*, composed of the set of interatomic distances (Fig. 3.34), is a convenient representation of molecular structure that is invariant to rotation and translation of the molecule, but which reflects changes in internal degrees of freedom. The *distance range matrix* is an extension (Fig. 3.34) that has two values for each interatomic distance

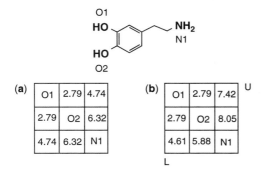

(a)

O1	2.79	4.74
2.79	O2	6.32
4.74	6.32	N1

(b)

O1	2.79	7.42	U
2.79	O2	8.05	
4.61	5.88	N1	

L

Figure 3.34. Distance matrix (a) in which unique interatomic distances for a particular conformation of a molecule are stored. Distance range matrix (b) in which ranges of interatomic distances representing conformational flexibilty of molecule are stored. U = upper bound, L = lower bound.

representing the upper and lower limits, or range, allowed for a given interatomic distance arising from the conformational flexibility of the molecule. Crippen (408) developed a procedure that will generate conformations that conform to the constraints represented by such a distance range matrix. This approach is used to generate structures from experimental measurements such as nuclear Overhauser effects in NMR experiments. The use of distance range matrices in the identification of pharmacophoric patterns was initially illustrated by Marshall et al. (398) (Fig. 3.35), and has recently been used by Clark et al. (409) in three-dimensional databases for representing the conformational flexibility of molecules. Pepperrell and Willett (410) examined several techniques for comparing molecules by use of distance matrices. Other descriptors for comparison of pharmacophoric

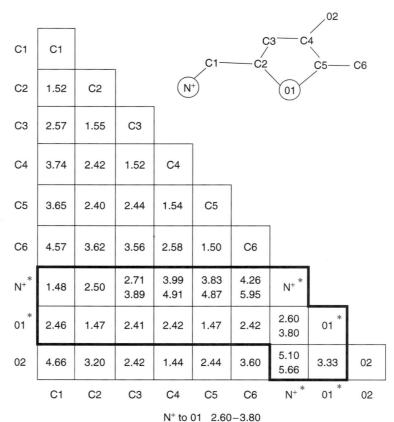

Figure 3.35. Distance range matrices used for illustration of analysis of muscarinic receptors (398). Used with permission.

	C1	C2	C3	C4	C5	C6	N+ *	01 *	02
C1	C1								
C2	1.52	C2							
C3	2.57	1.55	C3						
C4	3.74	2.42	1.52	C4					
C5	3.65	2.40	2.44	1.54	C5				
C6	4.57	3.62	3.56	2.58	1.50	C6			
N+ *	1.48	2.50	2.71 3.89	3.99 4.91	3.83 4.87	4.26 5.95	N+ *		
01 *	2.46	1.47	2.41	2.42	1.47	2.42	2.60 3.80	01 *	
02	4.66	3.20	2.42	1.44	2.44	3.60	5.10 5.66	3.33	02

N+ to 01   2.60–3.80

patterns and retrieval of similar substructures are under active investigation (411).

4.2.2 Visualization of Molecular Properties (412).

Although straightforward displays of molecular structure have proved to be extremely useful tools that enable medicinal chemists to visualize molecules and to compare their structural properties in three dimensions, of even greater potential utility is the display of the various chemical and physical properties of molecules in addition to their structures. Such displays allow the comparison not only of molecular shapes and three-dimensional structures, but also of molecular properties such as internal energy, electronic charge distribution, and hydrophobic character. A number of different properties have been displayed (412) in this manner in an effort to gain insight into molecular recognition in a series of compounds.

Among the more useful properties is the electrostatic potential. Any distribution of electrostatic charge, such as the electrons and nuclei of a molecule, creates an electrostatic potential in the surrounding space that at any given point represents the potential of the molecule for interacting with an electrostatic charge at that point. This potential is a very useful property for analyzing and predicting molecular reactive behavior. In particular, it has been shown to be an indicator of the sites or regions of a molecule to which an approaching electrophile or nucleophile is initially attracted or from which it is repelled (Fig. 3.36).

The major obstacle to use of electrostatic potentials in the comparison of different molecules has been the sheer volume of information produced. The traditional means of displaying such large amounts of data has been to display the electrostatic potential around a molecule as a two-dimensional contour map. The advent of computer graphics techniques have improved the situation by allowing three-dimensional contour maps to be displayed in color on the graphics screen and manipulated in real time along with a display of the molecule itself. An alternative mode for displaying molecular electrostatic potentials is to employ a dotted surface representation, with the dots taking on an appropriate color according to the electrostatic potential value at the relevant location. Such techniques were

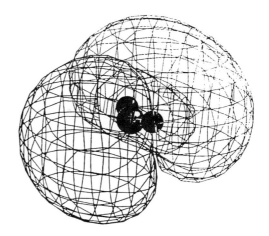

Figure 3.36. Molecular electrostatic potential for water. Positive potential superimposed on right surrounding hydrogens. Negative potential on left surrounding oxygen.

initially derived to display empirically determined potentials on the surface of proteins, but have since been used widely to display the electrostatic potentials on sets of small molecules for comparative purposes.

Other graphical uses of the electrostatic potential have been developed by Davis et al. (413), who were able to graphically align cyclic AMP and cyclic GMP, based on the superimposition of their respective electrostatic potential minima, and by Weinstein et al. (414), who oriented 5-hydroxytryptamine and 6-hydroxytryptamine based on the alignment of an electrostatically derived "orientation vector."

In a similar procedure to that described for the display of electrostatic potential, Cohen and colleagues developed a technique whereby the steric field surrounding a molecule can be displayed on a graphics screen as a three-dimensional isopotential contour map (415). The map is generated by calculating the VDW interaction energy between the molecule and a probe atom or molecule placed at varying points around the molecule of interest. This interaction energy is then contoured at specific levels to give the most stable VDW contour lines around the molecule, that is, the contour that represents the most favorable steric position for the probe as it is moved around the target.

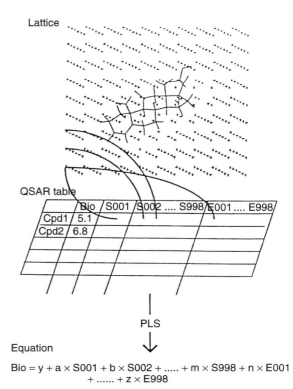

Figure 3.37. Calculation of electrostatic and VDW fields surrounding a series of molecules in defined orientations are used as a basis for 3D QSAR correlations in CoMFA (187, 401). Used with permission.

Lattice

QSAR table

	Bio	S001	S002	.... S998	E001	.... E998
Cpd1	5.1					
Cpd2	6.8					

PLS

Equation

$Bio = y + a \times S001 + b \times S002 + ... + m \times S998 + n \times E001 + ... + z \times E998$

A similar three-dimensional contour representation of a molecule can be obtained for both the electrostatic and steric fields of a molecule within the comparative molecular field analysis (CoMFA) methodology that has been developed by Cramer (187) to investigate 3D-QSARs (400). In this procedure, the molecule is surrounded by a regular lattice of points, at each point of which a van der Waals and an electrostatic interaction energy between the molecule and a probe atom is computed (Fig. 3.37). Isocontours can then be generated around individual molecules, displayed graphically, and they can be statistically compared throughout a series of molecules in an attempt to generate 3D-QSARs and hence to rationalize activity data. This is very similar to the GRID program (186), which uses various probe groups (416) to map potential interactions around a molecule. Inductive logic programming has been combined with CoMFA to develop a new approach (417) to pharmacophore mapping that does not require explicit superimposition of compounds.

In situations where, either from previous QSAR work or from experimental evidence, it is known or suspected that differences in the reactivity of a set of molecules are attributed primarily to their hydrophobic rather than their electrostatic properties; it is probably of more use to compare molecular surfaces that display hydrophobicity or polarity information. Indeed, dotted molecular surfaces color-coded by hydrophobic character have been used very successfully by Hansch and coworkers to rationalize QSARs from several different systems (418, 419). This concept has been extended to calculate the hydrophobic field surrounding a molecule by Kellogg and Abraham (420, 421) and utilized in CoMFA studies.

4.3 Molecular Comparisons

To compare molecules in a general way, a means of superposition, or correctly orienting the molecules in the same reference frame, must be available. A procedure for positioning an atom in the molecule at the center of the coordinate frame with other atoms positioned

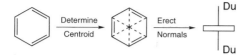

Figure 3.38. Construction of dummy vector perpendicular to plane of aromatic ring at centroid that allows superposition and coincidence of aromatic rings by fitting endpoints (Du) of dummy vector without requiring superposition of ring atoms.

Atom	Temperature factor	Atomic number
Carbon	60	25
Nitrogen	55	25
Oxygen	50	25
Sulfur	67	35
Phosphorus	70	35
Hydrogen	40	15
Bromine	65	50
Chlorine	60	35
Fluorine	40	40
Iodine	80	53
Sodium	87	85
Potassium	130	85
Calcium	130	85
Lithium	65	50
Aluminum	80	53
Silicon	80	55

Figure 3.39. Set of parameters to generate pseudo-electron density maps of molecules that can be contoured to approximately represent VDW surface (Ho and Marshall, unpublished).

along coordinate axes can be used, or the molecules can be successively fit to one that is used as the standard orientation. Danziger and Dean (422) described an approach that will find geometric similarities in positions of hydrogen-bonded atoms between two molecules. Least-squares-fitting procedures for designated atoms allow selectivity in orienting the molecules with predetermined conformations in the most appropriate manner. Kearsley (423) described an efficient method for fitting a series of molecules when atom-atom associations have been previously defined between members of the series. In some cases, the use of dummy atoms allows geometric superposition of groups such as aromatic rings without requiring superposition of the atoms composing the ring. By defining the centroid of the ring and erecting a normal to the plane of the ring, the dummy atom at the end of the normal and the centroid dummy atom can be used to superimpose the ring on another ring with similar dummy atoms (Fig. 3.38). This method leads to coincidence and coplanarity of the two ring systems without requiring the atoms composing the rings to be coincident. In other words, the rings can be viewed as two toruses of electron density without overemphasizing the positions of the atomic nuclei. In numerous studies [see review by Andrews et al. (403)] of biogenic amine ligands, this method of comparison of the aromatic ring components is essential to allow alignment of the nitrogens.

4.3.1 Volume Mapping. One method of displaying molecular surfaces that retains the ability to transform the display interactively has been developed by Marshall and Barry (424). The procedure involves computing a molecular pseudo-electron density map on a three-dimensional grid that surrounds the molecule whose atoms are replaced by dummy Gaussian atoms. Atom types are characterized by a half-width and an integrated density, chosen so that the Gaussians have a fixed value at a distance equal to the VDW radius (Fig. 3.39). Such density maps may be contoured in three dimensions to provide a chicken wire-like envelope around the molecule that corresponds to the van der Waals surface.

A concomitant benefit of this technique is that estimates of the molecular surface area and volume are generated as by-products of the contouring routines, whether the surface is being drawn around one or several molecules. Additionally, the generated surfaces and volumes are readily susceptible to logical operations, such as union, intersection, or subtraction, enabling the rapid determination of, for example, union or difference volumes among a series of molecules.

Once one has fixed the molecules in a common frame of reference, then comparison by a variety of techniques becomes feasible. As an example, difference in volume may be important in understanding the lack of seen activity in compounds that appear to possess all the prerequisites for activity seen in others in the series. In a congeneric series, a significant portion of the molecular structure is common to the molecules under comparison. This common volume that is shared logically should not contribute to differences in activity. By subtraction of the volume shared by two molecules, one obtains a difference map in which the volume occupied by one molecule and not the other remains (398). Correlations between the shared volume and the biological activity of a congeneric series of inhibitors of DHFR have been shown by Hopfinger (425). Simon and his colleagues (426) emphasized the use of both overlapping volume and nonoverlapping volume in QSAR studies in a quantitative methodology, the minimal steric difference, or MTD method. This approach has been enhanced to allow comparison of low energy conformers of each molecule and use of those that are sterically most similar. An application to substrates of acetylcholinesterase illustrates this facility (427).

4.3.2 Field Effects. Once the frame of reference has been established, other properties of molecules, such as the electrostatic field, can be compared as well. Because the electrostatic properties can be sampled on a grid, differences between the values of two molecules can be calculated and a difference map contoured. Such difference maps (428) highlight more clearly the similarities and differences between molecules. Hopfinger (429) integrated the difference between potential fields and showed this parameter to be useful in QSAR studies.

An approach to statistically quantifying the similarity between two molecular electrostatic potential surfaces was developed by Dean and coworkers (430, 431) and by Richards and coworkers (215). Here, the previously determined molecular electrostatic potential surfaces are projected outward onto surrounding spheres that provide a common surface of reference, and then statistical analyses are performed over the points on this common surface in an attempt to quantify the similarities or differences between the two molecules under consideration. Burt and Richards (432) introduced flexibility in the comparison of molecules based on their electrostatic potential fields.

4.3.3 Directionality. If one is comparing molecules that share interaction at a common site on a biological macromolecule, it is logical to assume that they may do so by interacting with similar sites in the receptor with optimal interaction shown by molecules with correctly oriented functional groups. If one does not have a three-dimensional model of the receptor from which to deduce potential interactive sites, then one can only attempt to deduce the potential interactive receptor-subsites by examination of the molecules that interact with them. Systematically, one can vary the conformation of a molecule and record the relative orientation of groups postulated, or shown experimentally, to play a dominant role in intermolecular interactions. In this way, one can map out the directionality of interactions of each functional group of the ligand in a common frame of reference. Comparison of these maps can often lead to hypotheses regarding pharmacophoric groups and their correspondence between molecules.

4.3.4 Locus Maps. One can generate a *locus* plot in coordinate space showing all the potential locations of one group relative to another by fixing one group in a particular orientation as a frame of reference and recording all possible coordinates of the other. An example would be the relative positions of the basic nitrogen to the aromatic ring in compounds such as dopamine interacting with biogenic amine receptors. One must choose the common fragment (in the example, the aromatic ring) of each molecule and its orientation to generate a similar frame of reference, so that the locus of positions of the atom (the basic nitrogen) leads to a meaningful comparison across a series of molecules (Fig. 3.40).

4.3.5 Vector Maps and Conformational Mimicry. Often, one is more interested in accessing the directionality of potential interaction rather than simply looking for overlap of atoms such as the basic nitrogen. In this case, for example, one is interested in determining both the locus of the lone pair of the nitrogen

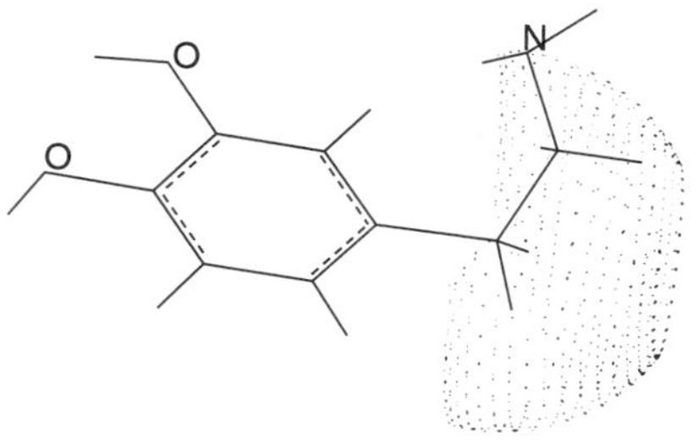

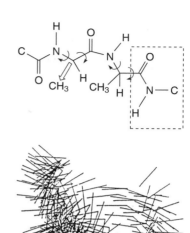

Figure 3.40. Locus of sterically allowed positions of nitrogen atom in dopamine relative to aromatic ring.

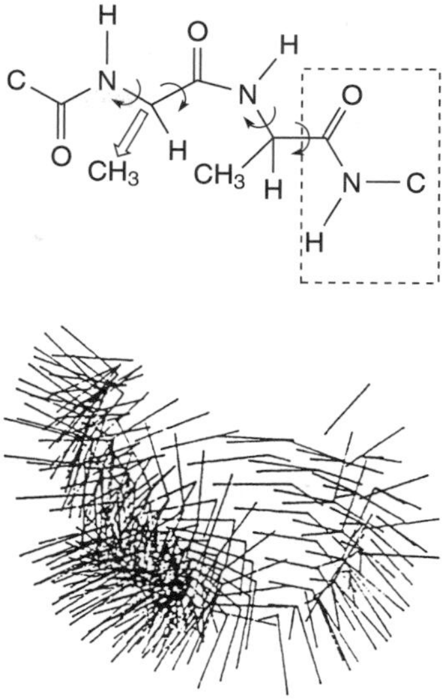

and the nitrogen as the ordered pair of coordinates determines a vector in the chosen frame of reference. The resulting plot of the locus of all possible vectors of the nitrogen lone pair constitutes a *vector map*. The combination of positional information with relative orientation offers considerable insight into potential interactions with a hypothetical receptor. The work of Lloyd and Andrews (233) postulating a common theme in CNS receptors based on an underlying biogenic amine pattern can be rationalized using the vector-map approach.

The use of vector maps is essential to the assessment of *conformational mimicry*, in that one attempts to determine the statistical probability that the conformation essential for activity will be preserved with a given chemical modification. An example will serve to illustrate this concept and its application. Modification of amide bonds (introduction of amide isosteres) in peptide drugs to increase metabolic stability may alter the potential accessible conformations. This may preclude the compound containing the isostere from adopting the correct orientation for receptor recognition and activation. In the general case, one has no specific information regarding which particular conformation is biologically relevant and can only assess whether the chemical modification mimics the amide bond in its conformational effects. This can be quantitatively assessed by the comparison of the percentage of vectors of the vector map of the parent amide bond that can be found in a comparable vector map of the analog.

Work by Zabrocki et al. (433) on the use of 1,5-disubstituted tetrazole rings as surrogates for the *cis*-amide bond illustrates this applica-

Figure 3.41. Vector map of the orientations of the C^{α}-C^{β} bond of Ala^1, with the methylamide fixed as a frame of reference of the dipeptide Ac-Ala-Ala-NH-CH$_3$ in which the central amide bond was *cis* (433). Used with permission.

tion. The linear dipeptide, acetyl-Ala-Ala-methylamide, with the amide bond between the two alanine residues in the *cis*-conformation, and the tetrazole analog, acetyl-AlaΨ[CN$_4$]Ala-methylamide, were modeled using the coordinates derived from diketopiperazines for the *cis*-amide bond or from the crystal structure of the cyclic tetrazole dipeptide. A systematic, or grid, search, which determines the sterically allowed conformations by systematically varying the torsional degrees of freedom, was used to generate a Ramachandran plot for each of the pairs of backbone torsional angles (Φ, Ψ) associated with each amino acid residue. The rigid geometry approximation was used with the set of scaled VDW radii, shown by Iijima et al. (109) to reproduce the experimental crystal data for proteins and peptides. When the *cis*-amide dipeptide model was calculated, the orientations of the C^{α}-C^{β} bond of Ala-1 with the methylamide fixed as a frame of reference were recorded for each sterically allowed conformation (Fig. 3.41). Use of the same orientation of the methylamide in the tetrazole allowed the program to determine which vectors, or orientations of the Ala-1 side chain relative to the methylamide, were common to both dipeptides. Alternatively, the acetyl group was used as the fixed frame of reference and the side-chain orientation of Ala-2 was used to monitor conformational mimicry. Because the quantitative results were essentially the same, the measurement of mimicry was shown to be independent of the chosen frame of reference. A torsional increment of 10 degrees was used, and a side-chain vector was assumed to correspond if both the carbon-α and carbon-β were within 0.2 Å of the coordinates of another vector. The percentage of orientations available to the analog that are available to the parent is referred to as the *conformational mimicry index*. For the tetrazole surrogate of the *cis*-amide bond, the conformational mimicry index is 88% [the number of vectors (747) common to both the tetrazole and *cis*-amide divided by the total number of vectors (849) allowed for the *cis*-amide]. The tetrazole analog has more conformational freedom than the *cis*-amide model with 33,359 conformers allowed compared to 14,912 allowed for the *cis*-amide of the 36^4 (or 1,679,616) possible conformations. This difference was easily visualized in plots of the vector maps for the two dipeptides.

A more recent example of the use of vector maps to evaluate conformational similarity is an application to β-turn mimetics by Ball et al. (434, 435). This led to a recognition that many of the various turn types described in peptides based on their backbone dihedral angles lead to quite similar topographical arrangements of the side chains. A new parameter, β [the dihedral angle formed by the backbone atoms $C_{(1)}$-$\alpha C_{(2)}$-$\alpha C_{(3)}$-$N_{(4)}$], was described (Fig. 3.42) that more readily facilitated comparison of the topography of the system.

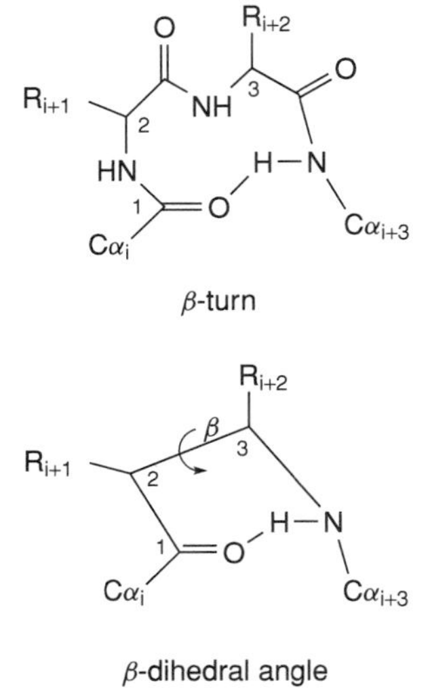

4.4 Finding the Common Pattern

If one assumes that a common binding mode exists for two or more compounds, then one can use the computer to verify the geometric feasibility of the assumption. One needs to determine whether it is possible for the two molecules to present a common geometric arrangement of the designated "important" functional groups for recognition. There are

Figure 3.42. Definition of new parameter β, the dihedral angle between the backbone atoms $C_{(1)}$-$\alpha C_{(2)}$-$\alpha C_{(3)}$-$N_{(4)}$ of peptides, used to describe the topography of reverse turns (434, 435).

two distinct approaches to this problem. The first that is associated with minimization methodology focuses on the existence issue. Is there a conformation that is energetically accessible to each of the molecules under consideration that will place the designated functional groups in a similar orientation? The second approach attempts to systematically enumerate all possible conformations and thereby derive all possible orientations or patterns to determine the set of patterns shared by the compounds under study. The latter approach, when it can be applied, can directly address the question of uniqueness of the common pattern.

The search for the global minimum, or complete set of low energy minima, on a potential surface is a common problem in science and engineering that does not have a general solution. Numerous approaches in chemistry have been used: most commonly stochastic methods such as distance geometry (408), molecular dynamics, and Monte Carlo sampling. Although distance geometry and molecular dynamics are widely used in the elucidation of solution conformations from NMR data, they have problems in conformational sampling and homogeneous treatment of data from

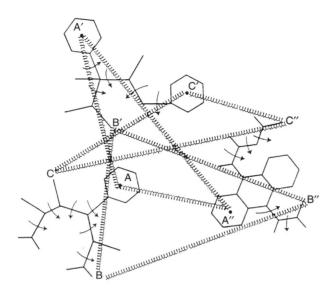

Figure 3.43. Simultaneous minimization of molecules to force overlap of pharmacophoric groups A, B, and C. Springs represent constraints between groups and only interatomic forces evaluated.

rigid and mobile domains. In general, the difficulties with most methods are similar to those seen with minimization procedures. If one is in the area of the global minimum, then one is likely to converge to that solution. Otherwise, one will be trapped in some local minimum. In contrast, systematic search methods are algorithmic, so that all sterically allowed conformations are generated at the selected torsional grid parameters. Systematic search methods, therefore, do not have problems in sampling and are path independent, but are combinatorial in complexity, which may limit the fineness of the sample grid and thus compromise the results. Only in small systems such as cycloalkane rings (121) and small peptides (90, 436) have the potential energy hypersurfaces been mapped.

4.4.1 Constrained Minimization. In cases where one has internal degrees of freedom, besides the six associated with position and orientation, the use of constrained minimization procedures becomes a useful technique. Often the standard molecule for comparison has a fixed conformation and the molecule to be fitted has internal degrees of freedom. Several groups have published methods for dealing with this problem. In case one has simultaneous degrees of freedom in both the molecule to be fitted and the target, a different

approach with simultaneous minimization of all variables is recommended (Fig. 3.43).

The combination of molecular mechanics with flexible minimization routines allows penalty functions to be assigned to force geometrical correspondence of groups, whereas individual molecules have their internal energy evaluated, but are invisible to the other molecules under consideration. A program has been described (437) with this capability and its use illustrated on histamine antagonists by Naruto et al. (438). Template forcing allows one molecule to be set up as a template and another molecule to be constrained to overlap in a specified manner. The strain energy involved in forcing correspondence gives an upper-bound estimate of the distortion energy required, given that the results depend on the initial-problem definition.

An alternative approach uses the distance geometry paradigm, in which all the constraints are combined to form the distance matrix from which energetically feasible conformations of the set of molecules are sought mathematically. Sheridan et al. (439) demonstrated this approach on acetylcholine analogs that are muscarinic agonists. Both of these approaches ask the same question and suffer from the same limitations, and differ only in computational technique. Each suffers from the local minima problem, in that each uses a minimization technique, and the results will be dependent on the starting geometries of the initial set of molecules. Both have the advantage that the unique constraints imposed by particular molecules enter consideration at an early stage and minimize comparison of conformations.

Another variant recently reported by Hodgkin et al. (440) uses a Monte Carlo search procedure to generate candidate pharmacophoric patterns. A reduced force-field parameter set is used initially to lower energy barriers between conformations to ensure greater configurational sampling. Candidate pharmacophores are then refined to produce low energy conformations of molecules overlaid in a common binding mode. Application to antagonists of the human platelet-activating factor led to a consistent binding model for a set of five diverse structures when active-site hydrogen-bonding groups were postulated. Barakat and Dean (441, 442) utilized simulated annealing to optimize structure matching by minimizing the difference matrix between the two molecules. A somewhat similar approach is that of Perkins and Dean (443), who used simulated annealing to search conformational space followed by cluster analysis for each molecule, with subsequent comparison of a small number of diverse conformers between different molecules.

4.4.2 Systematic Search and the Active Analog Approach. Once the existence of a common pattern has been determined, then the issue of uniqueness needs to be addressed. The Active Analog Approach (398) uses a systematic search to generate the set of sterically allowed conformations based on a grid search of the torsional variables at a given angular increment. For each sterically allowed conformation, a set of distances between the postulated pharmacophoric groups are measured. The set of distances, each of which represents a unique pharmacophoric pattern, constitutes an OMAP. Each point of the OMAP is simply a submatrix of the distance matrix and, as such, is invariant to global translation and rotation of the molecule. If the initial assumption is valid, that the same binding mode of interaction, or pharmacophoric pattern, is common to the set of molecules under consideration,

then the OMAP for each active molecule must contain the pattern encrypted in the set of distances. By logically intersecting the set of OMAPs, one can determine which patterns are common to all molecules (444). In other words, all potential pharmacophoric patterns consistent with the activity of the set of molecules can be found by this simple manipulation of OMAPs, and the question of uniqueness addressed directly (Fig. 3.44).

A good example is the work of Nelson et al. (445) on the receptor-bound conformation of morphiceptin. Based on structure-activity data, the tyramine portion and phenyl ring of residue three of morphiceptin, Tyr-Pro-Phe-Pro-NH_2, were postulated to be the pharmacophoric groups responsible for recognition and activation of the opioid μ-receptor. It was assumed further that the aromatic rings bound to the receptor in the different analogs were coincident and coplanar. A series of active analogs with a variety of conformationally constrained amino acid analogs in positions two and three were analyzed. A unique conformation was found for the two most constrained analogs that allowed overlap of the Phe and Tyr portions of the molecules (Fig. 3.45). In this case, a five-dimensional orientation map with distances between the nitrogen and normals to the two aromatic rings was used in the analysis.

The Active Analog Approach (Fig. 3.46) is appropriate for the unknown receptor problem, given that no objective criteria function, such as potential energy, can be used *a priori* in the absence of information regarding the receptor. Adequate sampling of the potential surface to ensure that the complete set of local minima is found is still problematic because of the phenomenon known as "grid tyranny." This relates to the fact that the combinatorial explosion that results by decreasing the increment of the torsion angles scanned limits one to a finite increment for a given problem, say, 10° for a seven-rotatable bond problem. Because the energetics of the system is very sensitive to interatomic distances, a conformation generated at the 10° increment may be sterically disallowed, but very close to a minimum. Relaxation of the structure might find the relevant conformation, for example, by allowing a torsional angle to vary by 1°. Im-

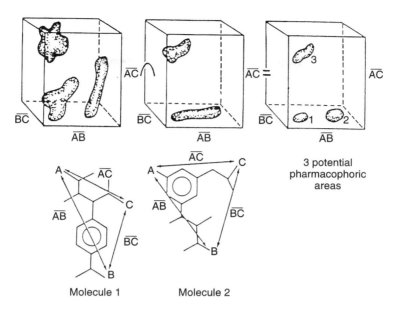

Figure 3.44. OMAPs generated for two molecules can be logically intersected to determine which three-dimensional patterns are common.

provements in algorithms described in the following section have helped to overcome this problem.

4.4.3 Strategic Reductions of Computational Complexity. Logically, the Active Analog Approach can be conceived as sequentially determining all the sterically allowed conformations for each molecule under consider-

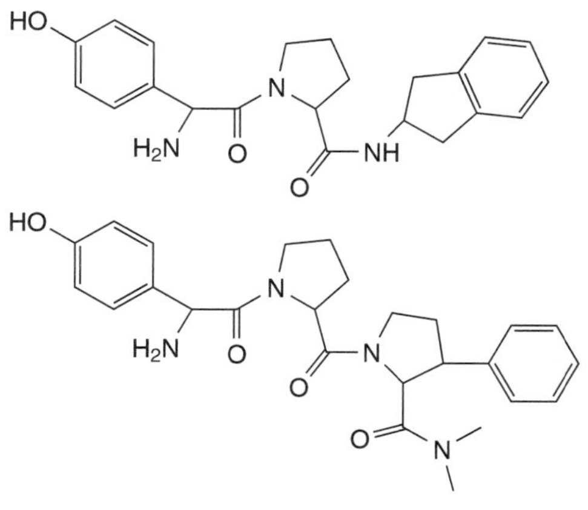

Figure 3.45. Conformations of two constrained analogs of morphiceptin in which aromatic rings of Tyr[1] and Phe[3] are overlapped (445).

ation, generation of an OMAP from those conformations, and logical intersection of the OMAPs to determine the common pharmacophoric patterns. A simple analysis will easily convince one that this is not feasible because of the computational complexity of the problem. For example, the set of 28 ACE inhibitors (Fig. 3.31), analyzed by Mayer et al. (397), have a total of 163 torsional degrees of freedom that have to be explored to find a common pattern, as seen in Table 3.1. If we were to determine all possible conformations for each molecule at 10° torsional scan, the scan parameter $(s) = 10°$ and the number of torsional increments $r = 360°/s$, or 36. For each molecule, there are r^n possibilities to be examined. For the set of molecules there are $(6 \times 36^3) + (7 \times 36^5) + (3 \times 36^6) + (5 \times 36^7) + (6 \times 36^8) + (1 \times 36^9)$ possible conformations to be generated and examined. If one compares each conformation of each molecule with all the conformations of the other molecules to find possible correspondences, the combinatorials of the problem explode and one reaches the same level of complexity as a complete conformational search of a peptide of 30 residues at a 10° scan (not currently feasible).

One is not interested in the conformational

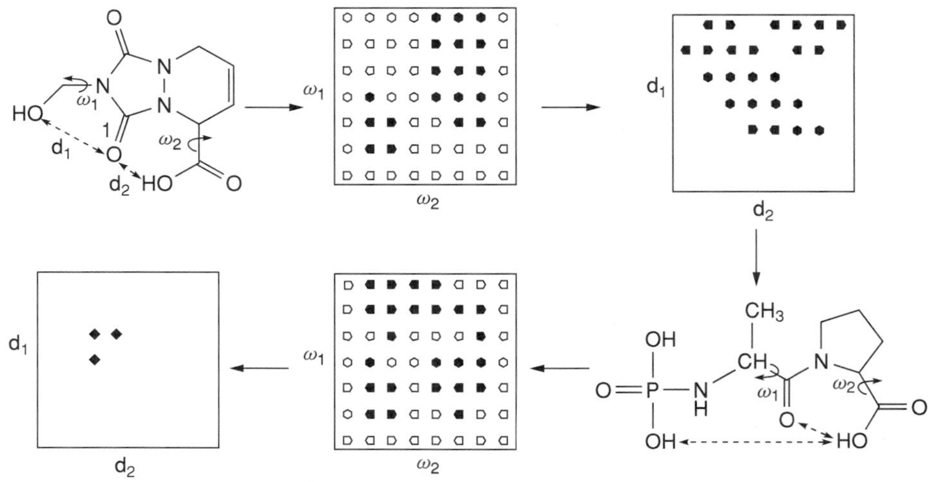

Figure 3.46. The flow of information in the Active Analog Approach (111, 399). Sterically allowed conformations (represented by filled circles on the ω_1, ω_2 torsional grid) of a molecule are determined and the distances (d_1, d_2, etc.) between pharmacophore elements are recorded for each. The resulting OMAP is used to constrain the next molecule in the series. Ideally, once all of the molecules have been evaluated, only a single point or cluster of points remains in the OMAP.

hyperspace of the set of the inhibitors, but rather the three-dimensional patterns common to the total set of inhibitors. Many conformations of a molecule often map into one three-dimensional pattern. Transformation of the multidimensional conformational hyperspace in a smaller-dimensioned OMAP space reduces the number of objects for comparison. If one starts with the most constrained inhibitor (fewest torsional degrees of freedom) and determined an OMAP for it, then one can use the upper and lower distance bounds as constraints for searches for the next molecule. In other words, one looks only where there are possible solutions to the problem. A more advanced approach simply examines each candidate solution from the initial OMAP to see whether all the other molecules are capable of presenting the same pattern. By changing the focus to the hypothesis of a common three-dimensional pattern, a more efficient approach has been devised (Fig. 3.46) (399). Clearly, the algorithms that one chooses to do the problem are important.

4.4.4 Alternative Approaches. A conceptually similar approach to receptor mapping has been taken by Ghose and Crippen (446–449), who used the distance geometry method to analyze site points and drug interactions. A site model was postulated with some initial estimates of force constants between the appropriate portion of the ligand and the site point. The binding energy for a particular binding mode can be calculated:

$$E_{\text{calcd}} = cE_c + Sx_{ti,tm}$$

where E_c is the conformational energy, c is a coefficient to be fit, x is the interaction of a site point i with the bound ligand point m, which depends on their types. The novel aspect of

Table 3.1   Degrees of Torsional Freedom to Specify ACE Active Site Geometry

Degrees of Freedom (n)	Number of Molecules	Total
3	6	18
5	7	35
6	3	18
7	5	35
8	6	48
9	1	9
Totals	28	163

this approach was the use of distance geometry to generate a variety of conformers binding within the postulated site and then finding a set of force constants between the postulated site points and ligand points that will predict the affinities of the compounds in the data set when bound in their optimal manner. With a site model of 11 attractive site points and 5 repulsive ones for DHFR, Ghose and Crippen (447) were able to derive force constants that fit 62 molecules, with an $R^2 = 0.90$, and predict the activity of 33 molecules, with an $R^2 = 0.71$. The compounds, however, are essentially an extended congeneric series because the core recognition portion of the inhibitor, the pyrimidine ring, is common to all the compounds.

Linschoten et al. (450) extended Crippen's method by use of lipophilicity to describe the binding of parts of the ligand to lipophilic areas of the receptor. Through the use of only a nine-point model of the turkey erythrocyte β-receptor and six energy parameters, they successfully modeled 58 compounds. Distance geometry approaches to receptor-site modeling have been reviewed (449, 451).

Simon and his coworkers have developed (426) a quantitative 3D-QSAR approach, the minimal steric (topologic) difference (MTD) approach. Oprea et al. (452) compared MTD and CoMFA on affinity of steroids for their binding proteins and found similar results. Snyder and colleagues (453) developed an automated method for pharmacophore extraction that can provide a clear-cut distinction between agonist and antagonist pharmacophores. Klopman (404, 454) developed a procedure for the automatic detection of common molecular structural features present in a training set of compounds. This has been used to produce candidate pharmacophores for a set of antiulcer compounds (404). Extensions (454) of this approach allow differentiation between substructures responsible for activity and those that modulate the activity.

Bersuker and Dimoglo (455) described a matrix-based approach that combines geometric and electronic features of a molecule, the electron-topological approach. For each molecule, an electron-topological matrix of congruity (ETMC) is constructed based on a conformer selected by conformational analysis.

The ETMC is essentially an interatomic distance matrix (Fig. 3.47), with the diagonal elements containing an electronic structural parameter (atomic charge, polarizability, HOMO energy, etc.). Off-diagonal elements for two atoms that are chemically bonded are used to store information regarding the bond (bond order, polarizability, etc.). Matrices for active compounds in a series are then searched for common features that are not shared by inactive compounds. The successful examples cited are predominately for small, relatively rigid structures where the conformational parameter does not confuse the analysis.

Martin et al. (456) developed a strategy for determining both the bioactive conformation and a superposition rule for each active molecule in a data set. In DISCO, a set of low energy conformers for each molecule is processed to locate atoms within the molecule and extensions for binding-site points for superposition. A clique-finding algorithm then finds superpositions containing at least one conformation of each molecule and a user-specified minimum number of site points.

Unlike methods that are limited to a precomputed set of rigid conformers, GASP (Genetic Algorithm Similarity Program) (457) allows full conformational flexibility of ligands. GASP employs a genetic algorithm for determining the correspondence between functional groups in different molecules and the alignment of these groups in a common geometry for receptor binding. For a set of ligands, GASP automatically identifies rotatable bonds and pharmacophore elements such as rings and potential hydrogen-bonding sites. A population of chromosomes is randomly constructed, where each chromosome represents a possible alignment of all the molecules. Chromosomes encode the torsion settings for rotatable bonds as well as the intermolecular mapping of elements. The fitness score of a particular alignment is the weighted sum of three terms: the number and similarity of overlaid elements, the common volume of all the molecules, and the internal van der Waals energy of each molecule. Using a mutation or crossover operator, child chromosomes are produced. Those with improved fitness scores replace the least-fit members of the existing population. The calculation terminates when

O₁	C₂	C₃	C₄	C₅	C₆	C₇	C₈	C₉	C₁₀	C₁₁	C₁₂	C₁₃	C₁₄	C₁₅	C₁₆	H₁₇
-0.23	5.09	5.22	5.69	6.59	4.22	2.82	2.30	1.75	3.54	4.74	4.96	6.11	6.98	7.09	6.84	1.89
	-0.01	1.00	2.52	2.48	2.54	2.97	4.33	5.14	5.15	4.96	3.79	6.29	6.45	6.79	7.50	6.23
		0.05	0.99	0.99	0.97	2.56	3.86	4.95	4.37	3.89	2.57	5.11	5.38	5.41	6.41	5.95
			-0.01	2.52	2.48	3.24	4.48	5.46	5.04	4.62	3.40	5.79	6.31	5.72	5.09	6.41
				-0.01	2.56	3.85	4.99	6.20	5.17	4.33	2.94	5.21	5.09	5.40	6.68	7.44
					0.05	1.42	2.46	3.69	2.84	2.49	1.42	3.87	4.37	4.38	5.06	4.61
						-0.03	1.40	2.39	2.43	2.83	2.40	4.36	5.07	5.05	5.20	3.40
							0.01	1.09	1.40	2.45	2.78	3.87	4.79	4.75	4.33	2.17
								0.18	2.40	3.69	4.13	4.98	5.94	5.89	5.12	0.93
									-0.03	1.41	2.40	2.58	3.64	3.60	2.92	2.70
										0.05	1.42	0.97	2.51	2.50	2.57	4.10
											-0.04	2.55	3.05	3.07	3.88	4.83
												0.05	0.99	1.00	0.99	5.16
													-0.01	2.53	2.49	6.22
														-0.01	2.49	6.07
															-0.01	4.97
																0.03

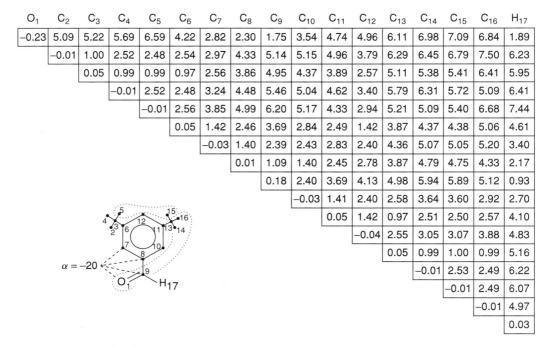

Figure 3.47. The electron-topological matrix of congruity (ETMC) for a 17-atom fragment proposed by Bersuker and Dimoglo (455) to encode geometrical and electronic features of molecules.

the fitness of the population fails to improve by a specified amount, or when the preset number of genetic operations is completed. GASP produces several sets of alignments and their associated pharmacophore elements.

4.4.5 Receptor Mapping. One can attempt to decipher physical properties of the receptor by use of data from both active and inactive analogs. Interpretation of results requires some understanding of the interactions between ligand and receptor that underlie molecular recognition. Oprea and Kurunczi (458) reviewed these interactions in the context of receptor mapping. A basic assumption is that a compound that contains the correct pharmacophoric elements and has the capability of positioning them correctly should be active. Compounds with these attributes that are inactive must be incapable of binding to the receptor in the correct orientation; that is, steric overlap with the receptor must occur. By calculating the combined volume of the active analogs superimposed in the correct orientation, one has mapped space that cannot be occupied

by the receptor and that must be available for binding. Inactive compounds mentioned above should possess novel volume requirements, some portion of which is likely to overlap with that occupied by the receptor. As an example of receptor mapping, Sufrin et al. (402) showed with amino acid analogs of methionine, which inhibited the enzyme, methionine:adenosyl transferase, that the data for a set of rigid amino acid inhibitors required the postulation of competition between the inactive analogs and the enzyme for a particular volume of space (Fig. 3.48). Summation of the volume requirements for the set of compounds, when oriented on the amino acid framework, yielded a minimum space from which the receptor could be excluded. Each amino acid had the necessary binding elements, but several were inactive. Each of the inactive analogs required extra volume not required by the active analogs and shared a small common unique volume whose occupancy by the enzyme would be sufficient to rationalize their inactivity.

Klunk et al. (459) used separate receptor

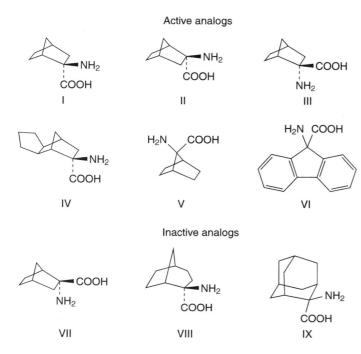

Active analogs

I   II   III

IV   V   VI

Inactive analogs

VII   VIII   IX

Figure 3.48. Example of receptor mapping of set of enzyme inhibitors that can be aligned on common amino acid framework. Set of inactive compounds all require common novel volume when compared with active compounds (402). Used with permission.

mapping of two different chemical classes of ligands to support the hypothesis that they bound to the same site. Calder et al. (460) argued that a successful correlative CoMFA model for 36 compounds of six chemical classes of GABA inhibitors indicated that the alignments used were significant. In some cases, comparison of volume maps for two receptors have allowed optimization of activity at one receptor with respect to the other. The work of Hibert et al. (461, 462), through the use of receptor mapping to increase the selectivity of a lead compound for the 5-HT$_{1A}$ receptor over the α_1-adrenoreceptor, has resulted in clinical trials for a novel chemical class. This steric-mapping approach has become relatively popular, and numerous examples appear in current journals (463) on a regular basis.

Although there are several feasible algorithms to deal with unions of molecular volumes, the use of pseudoelectron density functions calibrated to reproduce VDW radii (424) with three-dimensional contouring to represent the surface has allowed mathematical manipulation of the density associated with each lattice point to allow for union, intersec-

tion, and subtraction of volumes. Analytical representation of molecular volumes by Connolly (464, 465) and solvent-accessible surfaces by Kundrot et al. (466) may be an alternative that would allow optimization of volume overlap, for example, by minimizing the difference in volume between two structures. The solvent-accessible surface area can be used to approximate the free energy of hydration and a rapid, numerical procedure for its calculation has been reported (467).

4.4.6 Model Receptor Sites. One of the first visualizations of a receptor model is that of Beckett and Casey (468) for the opiate receptor published in 1954. Because morphine and many other compounds active at this receptor are essentially rigid, the model did not have to address the interaction of myriad numbers of flexible, naturally occurring opioid ligands, such as endorphins and enkephalin, which were only subsequently discovered. The model receptor had an anionic site to bind the charged nitrogen, a hydrophobic flat surface with a cleft to bind the phenyl ring, and a hydrophobic hydrocarbon bridge seen in morphine. Kier (469) published a number of pa-

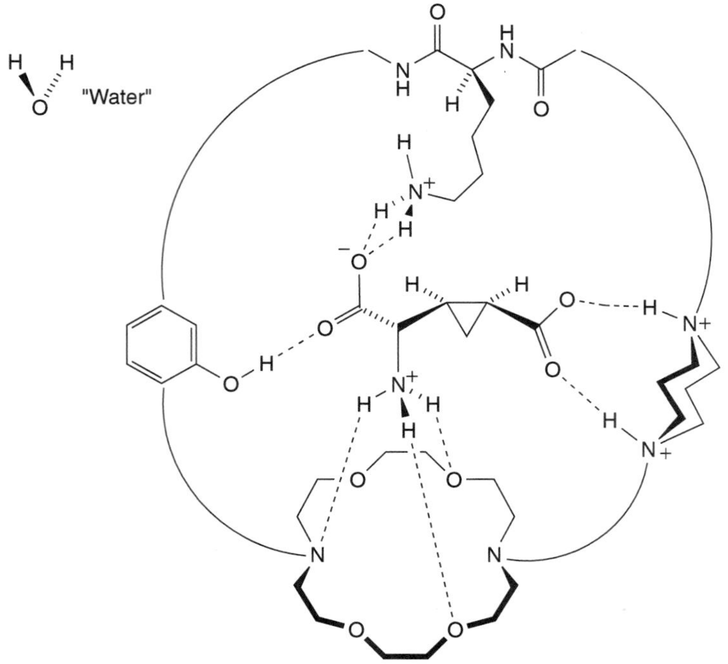

Figure 3.49. Peptidic pseudo-receptor used to calculate affinity of NMDA agonists and antagonists (453). Used with permission.

pers attempting to define the pharmacophore based on semiempirical molecular orbital calculations of *in vacuo* minimum-energy conformations. Although his basic concepts were valid, his emphasis on the global minima *in vacuo* limited his scope of applicability.

Humber et al. (470) used semirigid antipsychotic drugs, the so-called neuroleptics, which antagonize CNS dopamine transmission and displace dopamine from its receptor, to formulate a geometrical arrangement of receptor groups to rationalize their activity. Olson et al. (471) used this model to design a novel stereospecific dopamine antagonist and successfully predicted its stereochemistry.

Because we are reasonably convinced the receptor is a protein, construction of hypothetical sites from amino acid fragments and calculation of affinity for these sites should correlate with observed affinity, assuming that the type of interactions and their geometry is represented by the site in some reasonable manner. An individual fragment such as an indole ring from tryptophan does a good job of simulating a flat hydrophobic surface. Holtje and Tintelnot (472) constructed a site for chloramphenicol from arginine and histidine

by varying the distances of the amino acid from its postulated binding position and finding the optimal distance for correlation with observed affinity for the ribosome. Peptidic pseudoreceptors have been constructed (453) that correctly rank-order glutamate NMDA agonists and antagonists (Fig. 3.49).

An intermediate between unknown receptors and ones where the three-dimensional structure is known are models based on homology. For the medicinal chemist, the G-protein receptors have been of intense interest and numerous models (339, 340, 461, 473) of the various receptor types have been developed based on their presumed three-dimensional homology with bacteriorhodopsin (474). Mechanisms of signal transduction (475) and differences between agonists and antagonists (476) have even been rationalized based on such models. Nordvall and Hacksell (341) recently combined the construction of such a model for the muscarinic m1 receptor with constraints derived from steric mapping of muscarinic agonists. By adding the experimental constraints from ligand binding, a qualitative model was derived that was able to reproduce experimentally derived stereoselectivities.

4.4.7 Assessment of Model Predictability.

Because it is unlikely that there will be sufficient structure-activity data to uniquely define a model at atomic resolution in competition with crystallography, justification for model building must come from its potential predictive power and possible insight into the receptor-drug interaction before detailed three-dimensional information from either crystal structure or NMR studies. Certainly, the questions regarding the ability of a proposed drug to bind to the active site without steric conflict with the receptor can be addressed by the methods outlined above in a qualitative manner. The resolution of our receptor models is too crude, however, to subject them to molecular mechanics estimates of affinities. There are alternative paradigms, however, based on pattern recognition techniques in which a set of analogs and their activities are used, along with their physicochemical parameters, to generate a mathematical model that relates the values of the physicochemical parameters for a given analog with its activity. One such paradigm is comparative molecular field analysis (CoMFA), which combines the three-dimensional electrostatic and steric fields surrounding the analogs with powerful statistical techniques, partial least squares (PLS) (477) and cross-validation, to generate predictive models if a set of orientation rules are available for aligning the molecules for comparison and prediction. Alternative methods for assessing similarity and their use in QSAR schemes have been compared (215) with CoMFA. Another approach is the use of neural nets that learn to "see" patterns in much the same way as our own nervous system processes information. Examples of the use of this pattern-recognition approach include classification of mechanism of action for cancer chemotherapy (478) and QSAR studies of DHFR inhibitors (479, 480) and carboquinones (481). Machine learning has also been applied (482) to the QSAR problem. Trimethoprim analogs were successfully analyzed for their inhibition of DHFR and similar results to the original Hansch results were obtained. It is not clear that this paradigm could be applied to noncongeneric series, at least as outlined.

What appears crucial to such studies is the choice of training set, which encompasses as much of parameter space as one is likely to use in the predictive mode as well as tests of the predictive ability of resulting models. Given that one is dealing with a situation in which the number of variables is larger (often several times) than the number of observations, linear regression models are not applicable because chance correlations are highly probable. The use of cross-validation allows selection of correlations that are predictive in a self-consistent manner within the training set. This does not mean to imply that such internally self-consistent models have predictive power outside of the training set, or extremely close congeners.

DePriest et al. (483, 484) applied the CoMFA methodology to a series of 68 ACE (angiotensin-converting enzyme) inhibitors representing 28 different chemical classes. Through use of the binding-site geometry determined by Mayer et al. (397), a CoMFA model with a statistically significant cross-validated R^2 and considerable predictive ability for inhibitors outside of the training set was derived. Because the geometry of the ACE inhibitors was determined computationally by an active-site analysis rather than experimentally, a comparison of the results of the ACE series against thermolysin inhibitors, for which there were crystallographic data to explicitly define the binding-site geometry and the resulting alignment rules, was made, given that thermolysin is also a zinc-containing metallopeptidase with numerous similarities between ACE and thermolysin. Their results give strong support to both the Active Analog Approach (398) used to define the alignment rule for the ACE series and the CoMFA methodology itself. In the absence of an experimentally known active-site geometry, correlations were derived that explain as much as 84% of the variance in activities among a set of 68 diverse ACE inhibitors by use of CoMFA steric and electrostatic potentials plus a zinc indicator variable (Fig. 3.50). If the set of 68 ACE inhibitors was divided into three classes and correlations are derived for each class, CoMFA parameters alone explain 79–99% of the variance in activities. It was notable that statistically significant correla-

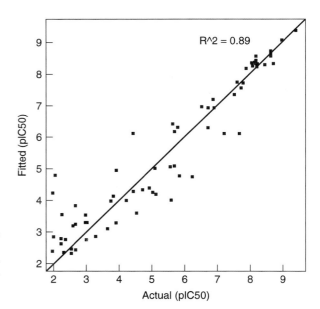

Figure 3.50. Plot of experimental versus predicted inhibition constants for 68 ACE inhibitors used in derivation of CoMFA model for the ACE active site (484). This plot shows the self-consistency of the model. Used with permission.

tions were found, in spite of the fact that CoMFA does not explicitly consider hydrophobicity or solvation. In further support of the active-site paradigm, the cross-validated results of the ACE series were equivalent to those of the thermolysin series (cross-validated R^2 = 0.65 to 0.70), for which the alignment rule was defined by crystallographic data.

The predictions for molecules outside the training sets are a valid test of the predictive ability of the model, rather than just a confirmation of self-consistency of the derived model. In other words, statistical analysis alone does not answer the question of a chance correlation (485) for the training set. One must investigate lateral correlations such as predictability. The predictive correlations presented by DePriest et al. (483, 484) represent a total of 66 diverse inhibitors that were not chosen as analogs of compounds present in the training set, but by selecting published papers on three different chemical classes and testing all compounds in the papers [predictive r^2 = 0.46 for the set of 66 compounds predicted, which had not been included in the training set for the ACE model with a zinc indicator of 10 (Fig. 3.51)]. The "predictive" r^2 was based only on molecules not included in the training set and was defined as

$$\text{predictive } r^2 = (\text{SD} - \text{``press''})/\text{SD}$$

where SD is the sum of the squared deviations between the affinities of molecules in the test set and the mean affinity of the training set molecules, and "press" is the sum of the squared deviations between predicted and actual affinity values for every molecule in the test set. It should be obvious from the equation that prediction of the mean value of the training set for each member of the test set would yield a predictive r^2 = 0. 35 out of the 66 predicted molecules had residuals less than one log value with a predictive r^2 value for the collective set of these 35 test molecules of 0.90. Of the 31 inhibitors with residuals greater than 1.0, 8 were carboxylates, 12 were phosphates, and 11 were thiols. Clearly, no single class of inhibitors dominated the distribution of residuals. Considering both the composition and the method of selection of the test data sets (range of activities over 7 log units), the fact that more than 50% of the molecules were predicted with correlations greater than r^2 = 0.90 lends strong support to the use of CoMFA as a tool for QSAR development.

Use of CoMFA as a predictive tool for receptors of known three-dimensional structure has also been explored. Klebe and Abraham

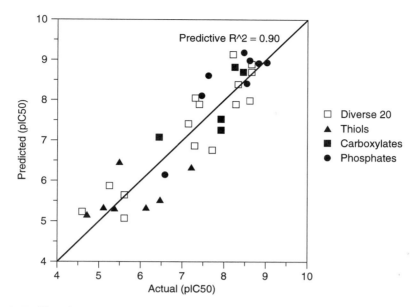

Figure 3.51. Plot of experimental versus predicted inhibition constants for 35 ACE inhibitors not used in derivation of CoMFA model (484). This plot indicates the predictability of the model. Used with permission.

(486) used two enzymes (thermolysin and renin) as well as antiviral activity against human rhinovirus, where the coat-protein receptor is known, to calibrate CoMFA methodology. They concluded that only enthalpies of binding and not binding affinities were predicted by CoMFA. Waller et al. (264) developed a predictive CoMFA model for the binding affinities of HIV-protease inhibitors based on crystal structures of complexes. Initial analysis of the 59 molecules in the training set representing five structurally diverse classes (hydroxyethylamine, statine, norstatine, ketoamide, and dihydroxyethylene) of transition-state protease inhibitors yielded a correlation with a cross-validated r^2 value of 0.786. To evaluate the predictive ability of this model, a test set of 18 additional inhibitors (487) was used that represented another class of transition-state isostere, hydroxyethylurea. The model expressed good predictive ability for the test set of hydroxyethylurea compounds ($r^2_{pred} = 0.624$) with all compounds predicted within 1.06 log unit (1.4 kcal/mol in binding affinity) of their actual activities, with an average absolute error of 0.58 log units (0.8 kcal/mol) across a range of 3.03 log units (Fig. 3.52). Pre-

dictions from this CoMFA model of HIV protease are being used to prioritize synthesis of *de novo*-designed HIV-protease inhibitors not included in development of the model.

Crippen developed a method (488) to objectively model the binding of small ligands to receptors, given the experimentally determined affinities of a set of ligands. The procedure, Vorom, used Voronoi polyhedra to generate the simplest geometrical model of the binding site. In a recent application to DHFR inhibitors (489), only eight analogs were used in the training set to derive the model and the affinities of 23/39 of the test set molecules were correctly predicted, with an average relative error of 0.83 kcal/mol for the remaining compounds.

5 CONCLUSIONS

Rapid advances in molecular and structural biology have provided ample therapeutic targets characterized in three dimensions. Tools to exploit this information are being rapidly developed and several strategies for *de novo* design of ligands, given an active site, are un-

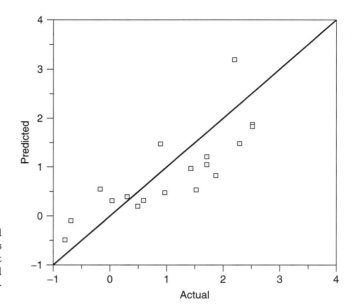

Figure 3.52. Plot of experimental versus predicted inhibition constants for 18 HIV-1 protease inhibitors not used in derivation of CoMFA model (264). This plot indicates the predictability of the model.

der investigation. It is already clear, however, that iterative approaches are necessary because of the lack of precision in predicting affinities for bound ligands. Molecular mechanics and computer graphics are essential components for design of novel ligands, and rapid progress in evolving a useful set of tools is apparent.

The ultimate goal in comparison of molecules with respect to their biological activity is insight into the receptor and its requirements for recognition and activation. Conjecture regarding the receptor is often a necessary part of rationalizing a set of structure-activity data. Although the problem of characterizing the active site of an unknown macromolecule indirectly is certainly challenging, the analysis of structure-activity data of a set of ligands, especially if their structural variety is wide, allows useful models of active sites to be developed. There are numerous caveats that must be acknowledged, however, such as flexibility of the receptor, multiple binding modes for ligands, and lack of uniqueness of most models because of limited experimental observations. Success in using these methods would appear to be increasing. This reflects both technological advances as well as insight into the problem and algorithmic improvements in our analytical approaches.

The game of 20 questions with receptors has progressed with experience. Ambiguity in interpretation of results and multiple models clearly reflect the uncertainties inherent in this indirect approach. Nevertheless, the absence of direct experimental data in many biological systems of intense therapeutic interest make this the only game available for many. It is hoped that the next decade will see further progress in our ability to extract three-dimensional information from structure-activity studies on unknown receptors.

This perspective has examined the approaches to molecular modeling and drug design and emphasized their limitations. The reader should be aware, however, that these tools are daily used on many problems of therapeutic interest with increasing success. This is clearly witnessed by publications of such studies in almost every issue of current major journals. For specific application areas, such as RNA (490, 491), DNA (492–496), membrane (497–507), or peptidomimetic modeling (382, 508–513), the reader is referred to the literature. The prediction of molecular properties, such as log P and correlation between substructures and metabolism, has led to a dramatic increase in efforts to correlate adsorption, distribution (514), metabolism (515–517), and elimination (ADME) with chemical structure (518–522). In addition, the advent of combinatorial chemistry has focused modeling efforts on prioritizing compounds (523–528) for high throughput screening based on chemical diversity (529–531), druglike properties (532, 533), predicted oral bioavailability (534, 535), and so forth.

6 ACKNOWLEDGMENTS

The work and influence of many talented collaborators as well as the National Institute of Health for grant support are gratefully acknowledged. Although my former colleagues' names are prominent in the references cited, their contributions are numerous and individual citations are avoided because of probable omissions. The author apologizes to many contributors to the field whose efforts have not been adequately recognized in this overview, the result of a somewhat arbitrary citation of references. Space and time limitations preclude a more thorough discussion of many important aspects.

REFERENCES

1. M. Salzmann, K. Pervushin, G. Wider, H. Senn, and K. Wuthrich, *J. Biomol. NMR*, **14**, 85–88 (1999).

2. P. J. Hajduk, R. P. Meadows, and S. W. Fesik, *Q. Rev. Biophys.*, **32**, 211–240 (1999).

3. P. J. Hajduk, G. Sheppard, D. G. Nettesheim, E. T. Olejniczak, S. B. Shuker, R. P. Meadows, D. H. Steinman, G. M. Carrera, P. A. Marcotte, J. Severin, K. Walter, H. Smith, E. Gubbins, R. Simmer, T. F. Holzman, D. W. Morgan, S. K. Davidsen, J. B. Summers, and S. W. Fesik, *J. Am. Chem. Soc.*, **119**, 5818–5827 (1997).

4. L. M. McDowell and J. Schaefer, *Curr. Opin. Struct. Biol.*, **6**, 624–629 (1996).

5. R. Ishima and D. A. Torchia, *Nat. Struct. Biol.*, **7**, 740–743 (2000).

6. L. M. McDowell, M. A. McCarrick, D. R. Studelska, W. J. Guilford, D. Arnaiz, J. L. Dallas, D. R. Light, M. Whitlow, and J. Schaefer, *J. Med. Chem.*, **42**, 3910–3918 (1999).

7. J. M. Moore, *Biopolymers*, **51**, 221–243 (1999).

8. J. Fejzo, C. A. Lepre, J. W. Peng, G. W. Bemis, Ajay, M. A. Murcko, and J. M. Moore, *Chem. Biol.*, **6**, 755–769 (1999).

9. C. J. Dinsmore, M. J. Bogusky, J. C. Culberson, J. M. Bergman, C. F. Homnick, C. B. Zartman, S. D. Mosser, M. D. Schaber, R. G. Robinson, K. S. Koblan, H. E. Huber, S. L. Graham, G. D. Hartman, J. R. Huff, and T. M. Williams, *J. Am. Chem. Soc.*, **123**, 2107–2108 (2001).

10. J. Hajdu, R. Neutze, T. Sjogren, K. Edman, A. Szoke, R. Wilmouth, and C. M. Wilmot, *Nat. Struct. Biol.*, **7**, 1006–1012 (2000).

11. A. Perrakis, R. Morris, and V. S. Lamzin, *Nat. Struct. Biol.*, **6**, 458–463 (1999).

12. C. R. Beddell, Ed., *The Design of Drugs to Macromolecular Targets*, John Wiley & Sons, New York, 1992.

13. I. D. Kuntz, *Science*, **257**, 1078–1082 (1992).

14. P. W. Finn and L. E. Kavraki, *Algorithmica*, **25**, 347–371 (1999).

15. D. Joseph-McCarthy, *Pharmacol. Ther.*, **84**, 179–191 (1999).

16. P. G. Mezey, *J. Mol. Model.*, **6**, 150–157 (2000).

17. E. F. Meyer, S. M. Swanson, and J. A. Williams, *Pharmacol. Ther.*, **85**, 113–121 (2000).

18. V. Schnecke and L. A. Kuhn, *Perspect. Drug Discov. Des.*, **20**, 171–190 (2000).

19. U. Burkert and N. L. Allinger in M. C. Caserio, Ed., *Molecular Mechanics, ACS Monograph 339*, Vol. **177**, American Chemical Society, Washington, DC, 1982.

20. J. P. Bowen and N. L. Allinger in K. B. Lipkowitz and D. B. Boyd, Eds., *Revisions in Computational Chemistry*, VCH, New York, 1991, pp. 81–98.

21. A. R. Leach, *Molecular Modelling: Principles and Applications*, 2nd ed., Prentice Hall, New York, 2001, 744 pp.

22. M. Clark, R. Cramer, and N. Van Opdenbosch, *J. Comput. Chem.*, **10**, 982–1012 (1989).

23. J. G. Vinter, A. Davis, and M. R. Saunders, *J. Comput.-Aided. Mol. Des.*, **1**, 31–51 (1987).

24. D. N. J. White and M. J. Bovill, *J. Chem. Soc. Perkin Trans. 2*, **12**, 1610–1623 (1977).

25. K. Gundertofte, J. Palm, I. Pettersson, and A. Stamvik, *J. Comput. Chem.*, **12**, 200–208 (1991).

26. W. L. Jorgensen and J. Gao, *J. Am. Chem. Soc.*, **110**, 4212–4216 (1988).

27. J.-H. Lii and N. L. Allinger, *J. Comput. Chem.*, **12**, 186–199 (1991).

28. J. Aqvist and A. Warshel, *J. Am. Chem. Soc.*, **112**, 2860 (1990).

29. R. D. Hancock, *Acc. Chem. Res.*, **23**, 253–257 (1990).

30. A. Vedani and D. W. Huhta, *J. Am. Chem. Soc.*, **112**, 4759–4767 (1990).

31. V. S. Allured, C. M. Kelly, and C. R. Landis, *J. Am Chem. Soc.*, **113**, 1–12 (1991).

32. A. E. Carlsson, *Phys. Rev. Lett.*, **81**, 477–480 (1998).

33. A. E. Carlsson and S. Zapata, *Biophys. J.*, **81**, 1–10 (2001).

34. A. T. Hagler, E. Hugler, and S. Lifson, *J. Am. Chem. Soc.*, **96**, 5319 (1974).

35. S. C. Harvey, *Proteins*, **5**, 78–92 (1989).

36. M. E. Davis and J. A. McCammon, *Chem. Rev.*, **90**, 509–521 (1990).

37. W. F. van Gunsteren and H. J. C. Berendsen, *Angew. Chem. Int. Ed. Engl.*, **29**, 992–1023 (1990).

38. C. E. Dykstra, *Chem. Rev.*, **93**, 2339–2353 (1993).

39. A. J. Stone and M. Alderton, *Mol. Phys.*, **56**, 1047–1064 (1985).

40. M. J. Dudek and J. W. Ponder, *J. Comput. Chem.*, **16**, 791–816 (1995).

41. C. I. Bayly, P. Cieplak, W. D. Cornell, and P. A. Kollman, *J. Phys. Chem.*, **97**, 10269–10280 (1993).

42. D. E. Williams in K. B. Lipkowitz and D. B. Boyd, Eds., *Revisions in Computational Chemistry*, VCH, New York, 1991, pp. 219–271.

43. R. J. Loncharich and B. R. Brooks, *Proteins*, **6**, 32–45 (1989).

44. J. Guenot and P. A. Kollman, *J. Comput. Chem.*, **14**, 295–311 (1993).

45. K. Tasaki, S. McDonald, and J. W. Brady, *J. Comput. Chem.*, **14**, 278–284 (1993).

46. J. Shimada, H. Kaneko, and T. Takada, *J. Comput. Chem.*, **14**, 867–878 (1993).

47. H. Schreiber and O. Steinhauser, *Chem. Phys.*, **168**, 75–89 (1992).

48. H. Schreiber and O. Steinhauser, *Biochemistry*, **31**, 5856–5860 (1992).

49. P. E. Smith and B. M. Pettit, *J. Chem. Phys.*, **95**, 8430–8441 (1991).

50. G. E. Marlow, J. S. Perkyns, and B. M. Pettit, *Chem. Rev.*, **93**, 2503–2521 (1993).

51. M. Whitlow and M. M. Teeter, *J. Am. Chem. Soc.*, **108**, 7163–7172 (1986).

52. M. K. Gilson, K. A. Sharp, and B. H. Honig, *J. Comput. Chem.*, **9**, 327–335 (1987).

53. A. Nicholls and B. Honig, *J. Comput. Chem.*, **12**, 435–445 (1991).

54. M. Schaefer and C. Froemmel, *J. Mol. Biol.*, **216**, 1045–1066 (1990).

55. W. G. Richards, P. M. King, and C. A. Reynolds, *Protein Eng.*, **2**, 319–327 (1987).

56. R. A. Pierotti, *Chem. Rev.*, **76**, 717–726 (1976).

57. G. L. Pollack, *Science*, **251**, 1323–1330 (1991).

58. R. J. Zauhar and R. S. Morgan, *J. Comput. Chem.*, **9**, 171–187 (1988).

59. J. Tomasi, R. Bonaccorsi, R. Cammi, et al., *Theochem. J. Mol. Struct.*, **80**, 401–424 (1991).

60. D. A. Liotard, G. D. Hawkins, G. C. Lynch, C. J. Cramer, and D. G. Truhlar, *J. Comput. Chem.*, **16**, 422–440 (1995).

61. K. Sharp, *J. Comput. Chem.*, **12**, 454–468 (1991).

62. W. C. Still, A. Tempczyk, R. C. Hawley, and T. Hendrickson, *Chem. Soc.*, **112**, 6127–6129 (1990).

63. C. A. Schiffer, J. W. Caldwell, P. A. Kollman, and R. M. Stroud, *Mol. Simul.*, **10**, 121–149 (1993).

64. P. F. W. Stouten, C. Frommel, H. Nakamura, and C. Sander, *Mol. Simul.*, **10**, 97–120 (1993).

65. R. J. Zauhar, *J. Comput. Chem.*, **12**, 575–583 (1991).

66. A. J. Stone, *Mol. Phys.*, **56**, 1065–1082 (1985).

67. S. Kuwajima and A. Warshel, *J. Phys. Chem.*, **94**, 460–466 (1990).

68. R. A. Sorensen, W. B. Liau, L. Kesner, and R. H. Boyd, *Macromolecules*, **21**, 200–208 (1988).

69. J. Caldwell, L. X. Dang, and P. A. Kollman, *J. Am. Chem. Soc.*, **112**, 9144–9147 (1990).

70. C. J. Cramer and D. G. Truhlar, *J. Am. Chem. Soc.*, **113**, 8305–8311 (1991).

71. C. J. Cramer, *J. Am. Chem. Soc.*, **113**, 8552–8554 (1991).

72. C. J. Cramer and D. G. Truhlar, *Science*, **256**, 213–217 (1992).

73. C. J. Cramer and D. G. Truhlar, *J. Comput. Chem.*, **13**, 1089–1097 (1992).

74. G. Rauhut, T. Clark, and T. Steinke, *J. Am. Chem. Soc.*, **115**, 9174–9181 (1993).

75. F. Franks in F. Franks, Ed., *Water, A Comprehensive Treatise*, Vol. **1**, Plenum Press, New York, 1975.

76. F. H. Stillinger, *Science*, **209**, 451–457 (1980).

77. L. R. Pratt, *Ann. Rev. Phys. Chem.*, **36**, 433–449 (1985).

78. J. P. M. Postma, H. J. C. Berendsen, and J. R. Haak, *Faraday Symp. Chem. Soc.*, **17**, 55–67 (1982).

79. B. G. Rao and U. C. Singh, *J. Am. Chem. Soc.*, **111**, 3125–3133 (1989).

80. I. Ohmine and H. Tanaka, *Chem. Rev.*, **93**, 2545–2566 (1993).

81. W. L. Jorgensen, J. Gao, and C. Ravimohan, *J. Phys. Chem.*, **89**, 3470–3473 (1985).

82. N. Muller, *Trends Biochem. Sci.*, **17**, 459–463 (1992).

83. L. X. Dang, J. E. Rice, J. Caldwell, and P. A. Kollman, *J. Am. Chem. Soc.*, **113**, 2481–2486 (1991).

84. A. K. Rappe and W. A. Goddard III, *J. Phys. Chem.*, **95**, 3358–3363 (1991).

85. R. Czerminski and R. Elber, *Int. J. Quantum Chem. Quantum Chem. Symp.*, **24**, 167–186 (1990).

86. C. Choi and R. Elber, *J. Chem. Phys.*, **94**, 751–760 (1991).

87. S. E. Huston and G. R. Marshall, *Biopolymers*, **34**, 74–90 (1994).

88. R. V. Pappu, R. K. Hart, and J. W. Ponder, *J. Phys. Chem. B*, **102**, 9725–9742 (1998).

89. L. Piela, *Collect. Czech. Chem. Commun.*, **63**, 1368–1380 (1998).

90. R. K. Hart, R. V. Pappu, and J. W. Ponder, *J. Comput. Chem.*, **21**, 531–552 (2000).

91. M. Saunders, K. N. Houk, Y.-D. Wu, W. C. Still, M. Lipton, G. Chang, and W. C. Guida, *J. Am. Chem. Soc.*, **112**, 1419–1427 (1990).

92. M. Saunders, *J. Am. Chem. Soc.*, **109**, 3150–3152 (1987).

93. J. T. Ngo and M. Karplus, *J. Am. Chem. Soc.*, **119**, 5657–5667 (1997).

94. R. V. Pappu, G. R. Marshall, and J. W. Ponder, *Nat. Struct. Biol.*, **6**, 50–55 (1999).

95. K. R. Mackenzie, J. H. Prestegard, and D. M. Engelman, *Science*, **276**, 131–133 (1997).

96. J. H. Holland, *Sci. Am.*, **July**, 66–72 (1992).

97. S. Forrest, *Science*, **261**, 872–878 (1993).

98. P. Willett, *Trends Biotechnol.*, **13**, 516–521 (1995).

99. J. E. Devillers, *Genetic Algorithms in Molecular Modeling*, Academic Press, New York, 1996.

100. D. B. McGarrah and R. S. Judson, *J. Comput. Chem.*, **14**, 1385–1395 (1993).

101. R. S. Judson, Y. T. Tan, E. Mori, C. Melius, E. P. Jaeger, A. M. Treasurywala, and A. Mathiowetz, *J. Comput. Chem.*, **16**, 1405–1419 (1995).

102. R. P. Meadows and P. J. Hajduk, *J. Biomol. NMR*, **5**, 41–47 (1995).

103. B. Waszkowycz, D. E. Clark, D. Frenkel, J. Li, C. W. Murray, B. Robson, and D. R. Westhead, *J. Med. Chem.*, **37**, 3994–4002 (1994).

104. C. M. W. Ho and G. R. Marshall, *J. Comput.-Aided Mol. Des.*, **7**, 623–647 (1993).

105. A. W. R. Payne and R. C. Glen, *J. Mol. Graphics*, **11**, 74–91 (1993).

106. H. A. Scheraga in K. B. Lipkowitz and D. B. Boyd, Eds., *Revisions in Computational Chemistry*, VCH, New York, 1992, pp. 73–142.

107. T. Schlick in K. B. Lipkowitz and D. B. Boyd, Eds., *Revisions in Computational Chemistry*, VCH, New York, 1992, pp. 1–71.

108. D. D. Beusen, E. F. B. Shands, S. F. Karasek, G. R. Marshall, and R. A. Dammkoehler, *THEOCHEM*, **370**, 157–171 (1996).

109. H. Iijima, J. B. Dunbar, Jr., and G. R. Marshall, *Proteins*, **2**, 330–339 (1987).

110. I. Motoc, R. A. Dammkoehler, and G. R. Marshall in N. Trinajstic, Ed., *Mathematic and Computational Concepts in Chemistry*, Ellis Horwood, Chichester, UK, 1986, pp. 222–251.

111. R. A. Dammkoehler, S. F. Karasek, E. F. B. Shands, and G. R. Marshall, *J. Comput.-Aided Mol. Des.*, **3**, 3–21 (1989).

112. N. Go and H. A. Scheraga, *Macromolecules*, **3**, 178–187 (1970).

113. A. R. Leach in K. B. Lipkowitz and D. B. Boyd, Eds., *Revisions in Computational Chemistry*, VCH, New York, 1991, pp. 1–55.

114. S. K. Burt and J. Greer, *Ann. Rep. Med. Chem.*, **23**, 285–294 (1988).

115. D. M. Ferguson and D. J. Raber, *J. Am. Chem. Soc.*, **111**, 4371–4378 (1989).

116. M. Saunders, *J. Comput. Chem.*, **10**, 203–208 (1989).

117. M. Saunders, *J. Comput. Chem.*, **12**, 645–663 (1991).

118. M. Saunders and H. A. Jimenez-Vazquez, *J. Comput. Chem.*, **14**, 330–348 (1993).

119. M. Saunders and N. Krause, *J. Am. Chem. Soc.*, **112**, 1791–1795 (1990).

120. A. V. Shah and D. P. Dolata, *J. Comput.-Aided Mol. Des.*, **7**, 103–124 (1993).

121. I. Kolossvary and W. C. Guida, *J. Am. Chem. Soc.*, **115**, 2107–2119 (1993).

122. H.-J. Boehm, G. Klebe, T. Lorenz, T. Mietzner, and L. Siggel, *J. Comput. Chem.*, **11**, 1021–1028 (1990).

123. A. E. Howard and P. A. Kollman, *J. Med. Chem.*, **31**, 1669–1675 (1988).

124. M. Lipton and W. C. Still, *J. Comput. Chem.*, **9**, 343–355 (1988).

125. D. D. Beusen, R. D. Head, J. D. Clark, W. C. Hutton, U. Slomczynska, J. Zabrocki, M. T. Leplawy, and G. R. Marshall in C. H. Schnei-

der and A. N. Eberle, Eds., *The Solution NMR Structures of Emerimicins III and IV Determined Using the New Program, MACROSE-ARCH*, ESCOM Scientific, Leiden, Netherlands, 1993, pp. 79–80.

126. M. P. Allen and D. J. Tildesley, *Computer Simulation of Liquids*, Oxford Science Publications, Oxford, UK, 1989, p. 385.

127. N. Metropolis, A. W. Rosenbluth, M. N. Rosenbluth, A. H. Teller, and E. Teller, *J. Chem. Phys.*, **21**, 1087 (1953).

128. J. A. McCammon and S. C. Harvey, *Dynamics of Protein and Nucleic Acids*, Cambridge University Press, Cambridge, UK, 1987, p. 234.

129. G. Zhang and T. Schlick, *J. Comput. Chem.*, **14**, 1212–1233 (1993).

130. T. Schlick and W. K. Olson, *Science*, **257**, 1110–1115 (1992).

131. D. S. Goodsell and A. J. Olson, *Proteins*, **8**, 195–202 (1990).

132. W. L. Jorgensen, *Acc. Chem. Res.*, **22**, 184–189 (1989).

133. D. L. Beveridge and F. M. DiCapua in W. van Gunsteren and P. K. Weiner, Eds., *Computer Simulation of Biomolecular Systems*, ESCOM Science, Leiden, Netherlands, 1989, pp. 1–26.

134. P. Kollman, *Chem. Rev.*, **93**, 2395–2417 (1993).

135. W. L. Jorgensen, *J. Phys. Chem.*, **87**, 5304–5314 (1983).

136. P. A. Bash, U. C. Singh, F. K. Brown, R. Langridge, and P. A. Kollman, *Science*, **235**, 574–576 (1987).

137. P. A. Kollman and K. M. Merz, *Acc. Chem. Res.*, **23**, 246–252 (1990).

138. T. P. Lybrand, J. A. McCammon, and G. Wipff, *Proc. Natl. Acad. Sci. USA*, **83**, 833–835 (1986).

139. J. Hermans, R. H. Yun, and A. G. Anderson, *J. Comput. Chem.*, **13**, 429–442 (1992).

140. J. Hermans, *Curr. Opin. Struct. Biol.*, **3**, 270–276 (1993).

141. R. Elber and M. Karplus, *J. Am. Chem. Soc.*, **112**, 9161–9175 (1990).

142. D. J. Tobias, J. E. Mertz, and C. L. Brooks III, *Biochemistry*, **30**, 6054–6058 (1991).

143. D. J. Tobias and C. L. Brooks III, *Biochemistry*, **30**, 6059–6070 (1991).

144. D. J. Tobias, S. F. Sneddon, and C. L. Brooks III, *J. Mol. Biol.*, **216**, 783–796 (1990).

145. S. F. Sneddon, D. J. Tobias, and C. L. Brooks III, *J. Mol. Biol.*, **209**, 817–820 (1989).

146. D. J. Tobias, S. F. Sneddon, and C. L. Brooks III in R. Lavery, J.-L. Rivail, and J. Smith,

Eds., Advances in Biomolecular Simulations, American Institute of Physics Conference Proceedings No. 239, Obernai, France, 1991, pp. 174–199.

147. M. L. Smythe, S. E. Huston, and G. R. Marshall, *J. Am. Chem. Soc.*, **115**, 11594–11595 (1993).

148. M. L. Smythe, S. E. Huston, and G. R. Marshall, *J. Am. Chem. Soc.*, **117**, 5445–5452 (1995).

149. G. H. Loew and S. K. Burt in C. A. Ramsden, Ed., *Quantitative Drug Design*, Pergamon Press, Oxford, UK, 1990, pp. 105–123.

150. S. L. Price and N. G. J. Richards, *J. Comput.-Aided Drug Des.*, **5**, 41–54 (1991).

151. U. C. Singh and P. A. Kollman, *J. Comput. Chem.*, **5**, 129 (1984).

152. B. H. Besler, K. M. Merz, Jr., and P. A. Kollman, *J. Comput. Chem.*, **11**, 431–439 (1990).

153. G. Rauhut and T. Clark, *J. Comput. Chem.*, **14**, 503–509 (1993).

154. J. G. Vinter and M. R. Saunders in D. J. Chadwick and K. Widdows, Eds., *Host-Guest Molecular Interactions: From Chemistry to Biology*, John Wiley & Sons, Chichester, UK, 1991, pp. 249–265.

155. C. A. Hunter and J. K. M. Sanders, *J. Am. Chem. Soc.*, **112**, 5525–5534 (1990).

156. U. Dinur and A. T. Hagler in K. B. Lipkowitz and D. B. Boyd, Eds., *Revisions in Computational Chemistry*, VCH, New York, 1991, pp. 99–164.

157. J. Pranata, S. G. Wierschke, and W. I. Jorgensen, *J. Am. Chem. Soc.*, **113**, 2810–2819 (1991).

158. J. Tirado-Rives and W. L. Jorgensen, *J. Am. Chem. Soc.*, **112**, 2773–2781 (1990).

159. A. Alex and T. Clark, *J. Comput. Chem.*, **13**, 704–717 (1992).

160. J. Aqvist and A. Warshel, *Chem. Rev.*, **93**, 2523–2544 (1993).

161. M. J. Field, P. A. Bash, and M. Karplus, *J. Comput. Chem.*, **11**, 700–783 (1990).

162. A. Warshel, *Computer Modeling of Chemical Reactions in Enzymes and Solutions*, John Wiley & Sons, New York, 1991, p. 236.

163. V. Daggett, S. Schroder, and P. Kollman, *J. Am. Chem. Soc.*, **113**, 8926–8935 (1991).

164. P. R. Andrews and D. A. Winkler in G. Jolles and K. R. H. Wooldridge, Eds., *Drug Design: Fact or Fantasy?*, Academic Press, New York, 1984, pp. 145–174.

165. J. E. Eksterowicz and K. N. Houk, *Chem. Rev.*, **93**, 2439–2461 (1993).

166. K. Appelt, R. J. Bacquet, C. A. Bartlett, C. L. J. Booth, S. T. Freer, M. A. M. Fuhry, M. R. Gehring, S. H. Herrmann, E. F. Howland, C. A. Janson, T. R. Jones, C.-C. Kan, V. Kathardekar, K. K. Lewis, G. P. Marzoni, D. A. Mathews, C. Mohr, E. W. Moomaw, C. A. Morse, S. J. Oatley, R. C. Ogden, M. R. Reddy, S. H. Reich, W. S. Schoettin, W. W. Smith, M. D. Varney, J. E. Villafranca, R. W. Ward, S. Webber, S. E. Webber, K. M. Welsh, and J. White, *J. Med. Chem.*, **34**, 1925–1934 (1991).

167. P. J. Goodford, *J. Med. Chem.*, **27**, 557–564 (1984).

168. C. R. Beddell, *Chem. Soc. Rev.*, **13**, 279–319 (1984).

169. R. Wootton in C. R. Beddell, Ed., *The Design of Drugs to Macromolecular Targets*, John Wiley & Sons, New York, 1992, pp. 49–83.

170. L. F. Kuyper, B. Roth, D. P. Baccanari, R. Ferone, C. R. Beddell, J. N. Champness, D. K. Stammers, J. G. Dann, F. E. Norrington, D. J. Baker, and P. J. Goodford, *J. Med. Chem.*, **28**, 303–311 (1985).

171. K. Appelt, *J. Comput.-Aided Mol. Des.*, **1**, 23–48 (1993).

172. M. von Itzstein, W.-Y. Wu, G. B. Kok, M. S. Pegg, J. C. Dyason, B. Jin, T. V. Phan, M. L. Smythe, H. E. White, S. W. Oliver, P. M. Colman, J. N. Varghese, D. M. Ryan, J. M. Woods, R. C. Bethell, V. J. Hotham, J. M. Cameron, and C. R. Penn, *Nature*, **363**, 418–423 (1993).

173. J. W. Liebeschuetz, S. D. Jones, P. J. Morgan, C. W. Murray, A. D. Rimmer, J. M. Roscoe, B. Waszkowycz, P. M. Welsh, W. A. Wylie, S. C. Young, H. Martin, J. Mahler, L. Brady, and K. Wilkinson, *J. Med. Chem.*, **45**, 1221–1232 (2002).

174. K. E. Lind, Z. Du, K. Fujinaga, B. M. Peterlin, and T. L. James, *Chem. Biol.*, **9**, 185–193 (2002).

175. M. Miller, M. Jaskolski, J. K. M. Rao, J. Leis, and A. Wlodawer, *Nature*, **337**, 576–579 (1989).

176. M. Miller, B. K. Sathyanarayana, A. Wlodawer, M. V. Toth, G. R. Marshall, L. Clawson, L. Selk, J. Schneider, and S. B. H. Kent, *Science*, **246**, 1149–1152 (1989).

177. R. L. Stanfield, M. Takimoto-Kamimura, J. M. Rini, A. T. Profy, and I. A. Wilson, *Structure*, **1**, 83–93 (1993).

178. S. W. Fesik, *J. Med. Chem.*, **34**, 2938–2945 (1991).

179. G. Otting, *Curr. Opin. Struct. Biol.*, **3**, 760–768 (1993).

180. S. O. Smith, *Curr. Opin. Struct. Biol.*, **3**, 755–759 (1993).

181. M. F. Perutz, G. Fermi, D. J. Abraham, C. Poyart, and E. Bursaux, *J. Am. Chem. Soc.*, **108**, 1064–1078 (1986).

182. A. S. Mehanna and D. J. Abraham, *Biochemistry*, **29**, 3944–3954 (1990).

183. I. D. Kuntz, J. M. Blaney, S. J. Oatley, R. Langridge, and T. E. Ferrin, *J. Mol. Biol.*, **161**, 269 (1982).

184. R. Voorintholt, M. T. Kosters, G. Vegter, G. Vriend, and W. G. J. Hol, *J. Mol. Graphics*, **7**, 243–245 (1989).

185. C. M. W. Ho and G. R. Marshall, *J. Comput.-Aided Mol. Des.*, **4**, 337–354 (1990).

186. P. J. Goodford, *J. Am. Chem. Soc.*, **28**, 849–856 (1985).

187. R. D. Cramer III, D. E. Patterson, and J. D. Bunce, *J. Am. Chem. Soc.*, **110**, 5959–5967 (1988).

188. R. D. Cramer III and M. Milne, The Lattice Model: A General Paradigm for Shape-Related Structure/Activity Correlation, in Proceedings of the 19th National Meeting of the American Chemical Society, American Chemical Society, Washington, DC, 1979.

189. A. Miranker and M. Karplus, *Proteins*, **11**, 29–34 (1991).

190. A. Caflisch, A. Miranker, and M. Karplus, *J. Med. Chem.*, **36**, 2142–2167 (1993).

191. P. K. Weiner, C. Landridge, J. M. Blaney, R. Schaefer, and P. A. Kollman, *Proc. Natl. Acad. Sci. USA*, **79**, 3754–3758 (1982).

192. S. J. Weiner, P. A. Kollman, D. A. Case, U. C. Singh, C. Ghio, G. Alagona, J. S. Profeta, and P. Weiner, *J. Am. Chem. Soc.*, **106**, 765–784 (1984).

193. S. J. Weiner, P. A. Kollman, D. T. Nguyen, and D. A. Case, *J. Comput. Chem.*, **7**, 230–252 (1986).

194. F. H. Allen, J. E. Davies, J. J. Galloy, O. Johnson, O. Kennard, C. F. Macrea, E. M. Mitchell, G. F. Mitchell, J. M. Smith, and D. G. Watson, *J. Chem. Inf. Comput. Sci.*, **31**, 187–204 (1991).

195. E. E. Abola, F. C. Bernstein, and T. F. Koetzle in P. S. Glaeser, Ed., *The Role of Data in Scientific Progress*, Elsevier, New York, 1985.

196. P. R. Andrews, E. J. Lloyd, J. L. Martin, and S. L. A. Munro, *J. Mol. Graphics*, **4**, 41–45 (1986).

197. R. S. Pearlman, *Chem. Des. Auto. News*, **2**, 1 (1987).

198. R. S. Pearlman, *CONCORD User's Manual*, Tripos Associates, St. Louis, MO, 1992.

199. R. S. Pearlman, *Chem. Des. Auto. News*, **8**, 3–15 (1993).

200. R. P. Sheridan, A. Rusinko III, R. Nilakantan, and R. Venkataraghavan, *Proc. Natl. Acad. Sci. USA*, **86**, 8165–8169 (1989).

201. P. Gund, W. T. Wipke, and R. Langridge, *Comput. Chem. Res. Ed. Technol.*, **3**, 5–21 (1974).

202. P. Gund, *Prog. Mol. Subcell. Biol.*, **11**, 117–143 (1977).

203. A. M. Lesk, *Commun. ACM*, **22**, 221–224 (1979).

204. S. E. Jakes and P. Willett, *J. Mol. Graphics*, **4**, 12–20 (1986).

205. S. E. Jakes, N. Watts, P. Willett, D. Bawden, and J. D. Fisher, *J. Mol. Graphics*, **5**, 41–48 (1997).

206. P. A. Bartlett, G. T. Shea, S. J. Telfer, and S. Waterman in S. M. Roberts, Ed., *Molecular Recognition: Chemical and Biological Problems*, Royal Society of Chemistry, London, 1989, pp. 182–196.

207. J. H. Van Drie, D. Weininger, and Y. C. Martin, *J. Comput.-Aided Mol. Des.*, **3**, 225–251 (1989).

208. R. P. Sheridan, R. Nilakantan, A. I. Rusinko, N. Bauman, K. S. Haraki, and R. Venkataraghavan, *J. Chem. Inf. Comput. Sci.*, **29**, 255–260 (1989).

209. Molecular Design, *MACCS-3D*, Molecular Design Ltd., San Leandro, CA, 1993.

210. Chemical Design, *CHEM-X*, Chemical Design Ltd., Oxford OX2 0JB, UK, 1993.

211. *UNITY User's Manual*, Tripos, Inc., St. Louis, MO, 2002.

212. Y. C. Martin, M. G. Bures, and P. Willett in K. Lipkowitz and D. Boyd, Eds., *Revisions in Computational Chemistry*, VCH, New York, 1990, pp. 213–263.

213. Y. C. Martin, *J. Med. Chem.*, **35**, 2145–2154 (1992).

214. A. I. Rusinko, R. P. Sheridan, R. Nilakantan, K. S. Haraki, N. Bauman, and R. Venkataraghavan, *J. Chem. Inf. Comput. Sci.*, **29**, 251–255 (1989).

215. A. C. Good, S. J. Peterson, and W. G. Richards, *J. Med. Chem.*, **36**, 2929–2937 (1993).

216. R. S. Pearlman in H. Kubinyi, Ed., *3D QSAR in Drug Design: Theory, Methods and Applications*, ESCOM Scientific, Leiden, Netherlands, 1993, pp. 41–79.

217. J. Sadowski and J. Gasteiger, *Chem. Rev.*, **93**, 2567–2581 (1993).

218. R. L. DesJarlais, R. P. Sheridan, G. L. Seibel, J. S. Dixon, I. D. Kuntz, and R. Venkataraghavan, *J. Med. Chem.*, **31**, 722–729 (1988).

219. R. L. DesJarlais, G. L. Seibel, I. D. Kuntz, P. S. Furth, J. C. Alvarez, P. R. Ortiz de Montellano, D. L. DeCamp, L. M. Babe, and C. S. Craik, *Proc. Natl. Acad. Sci. USA*, **87**, 6644–6648 (1990).

220. R. L. DesJarlais, R. P. Sheridan, J. S. Dixon, I. D. Kuntz, and R. Venkataraghavan, *J. Med. Chem.*, **29**, 2149–2153 (1986).

221. H.-J. Bohm, *J. Comput.-Aided Mol. Des.*, **6**, 61–78 (1992).

222. H.-J. Bohm, *J. Comput.-Aided Mol. Des.*, **6**, 593–606 (1992).

223. P. L. Chau and P. M. Dean, *J. Comput-Aided. Mol. Des.*, **6**, 385–396 (1992).

224. P. L. Chau and P. M. Dean, *J. Comput-Aided. Mol. Des.*, **6**, 397–406 (1992).

225. P. L. Chau and P. M. Dean, *J. Comput-Aided. Mol. Des.*, **6**, 407–426 (1992).

226. C. M. W. Ho and G. R. Marshall, *J. Comput.-Aided Mol. Des.*, **7**, 3–22 (1993).

227. G. Klebe, *J. Mol. Med.*, **78**, 269–281 (2000).

228. H. J. Bohm and M. Stahl, *Med. Chem. Res.*, **9**, 445–462 (1999).

229. H. J. Bohm and M. Stahl, *Curr. Opin. Chem. Biol.*, **4**, 283–286 (2000).

230. R. A. Lewis and P. M. Dean, *Proc. R. Soc. Lond. B*, **236**, 141–162 (1989).

231. R. A. Lewis and P. M. Dean, *Proc. R. Soc. Lond. B*, **236**, 125–140 (1989).

232. Y. Nishibata and A. Itai, *Tetrahedron*, **47**, 8985–8990 (1991).

233. Y. Nishibata and A. Itai, *J. Med. Chem.*, **36**, 2921–2928 (1993).

234. D. A. Pearlman and M. A. Murko, *J. Comput. Chem.*, **14**, 1184–1193 (1993).

235. M. G. Bures, C. Black-Schaefer, and G. Gardner, *J. Comput.-Aided Mol. Des.*, **5**, 323–334 (1991).

236. J. M. Blaney and J. S. Dixon, *Perspect. Drug. Discov. Des.*, **1**, 301–319 (1993).

237. D. L. Bodian, R. B. Yamasaki, R. L. Buswell, J. F. Stearns, J. M. White, and I. D. Kuntz, *Biochemistry*, **32**, 2967–2978 (1993).

238. C. S. Ring, E. Sun, J. H. McKerrow, G. K. Lee, P. J. Rosenthal, I. D. Kuntz, and F. E. Cohen, *Proc. Natl. Acad. Sci. USA*, **90**, 3583–3587 (1993).

239. B. K. Shoichet, R. M. Stroud, D. V. Santi, I. D. Kuntz, and K. M. Perry, *Science*, **259**, 1445–1450 (1993).

240. I. Halperin, B. Ma, H. Wolfson, and R. Nussinov, *Proteins*, **47**, 409–443 (2002).

241. E. Katchalski-Katzir, I. Shariv, M. Eisenstein, A. A. Friesem, C. Aflalo, and I. A. Vakser, *Proc. Natl. Acad. Sci. USA*, **89**, 2195–2199 (1992).

242. I. A. Vakser and C. Aflalo, *Proteins*, **20**, 320–329 (1994).

243. I. A. Vakser, *Protein Eng.*, **9**, 37–41 (1996).

244. I. A. Vakser, *Biopolymers*, **39**, 455–464 (1996).

245. H. A. Gabb, R. M. Jackson, and M. J. E. Sternberg, *J. Mol. Biol.*, **272**, 106–120 (1997).

246. M. Rarey, B. Kramer, T. Lengauer, and G. Klebe, *J. Mol. Biol.*, **261**, 470–489 (1996).

247. S. R. Krystek, Jr., R. E. Bruccoleri, and J. Novotny, *Int. J. Pept. Protein Res.*, **38**, 229–236 (1991).

248. J. Aqvist, *J. Comput. Chem.*, **17**, 1587–1597 (1996).

249. A. N. Jain, *J. Comput.-Aided Mol. Des.*, **10**, 427–440 (1996).

250. A. Alex and P. Finn, *THEOCHEM*, **398**, 551–554 (1997).

251. R. C. Wade, A. R. Ortiz, and F. Gago, *Perspect. Drug Discov. Des.*, **9–11**, 19–34 (1998).

252. R. M. A. Knegtel and P. D. J. Grootenhuis, *Perspect. Drug Discov. Des.*, **9–11**, 99–114 (1998).

253. M. D. Eldridge, C. W. Murray, T. R. Auton, G. V. Paolini, and R. P. Mee, *J. Comput.-Aided Mol. Des.*, **11**, 425–445 (1997).

254. M. K. Gilson, J. A. Given, and M. S. Head, *Chem. Biol.*, **4**, 87–92 (1997).

255. M. K. Gilson, J. A. Given, B. L. Bush, and J. A. McCammon, *Biophys. J.*, **72**, 1047–1069 (1997).

256. T. I. Oprea and G. R. Marshall, *Perspect. Drug Discov. Des.*, **9–11**, 35–61 (1998).

257. M. K. Holloway, *Perspect. Drug Discov. Des.*, **9–11**, 63–84 (1998).

258. H. J. Bohm, *J. Comput.-Aided Mol. Des.*, **12**, 309–323 (1998).

259. T. Hansson, J. Marelius, and J. Aqvist, *J. Comput.-Aided Mol. Des.*, **12**, 27–35 (1998).

260. G. R. Marshall, R. H. Head, and R. Ragno in E. Di Cera, Ed., *Thermodynamics in Biology*, Oxford University Press, Oxford, UK, 2000, pp. 87–111.

261. S. Krystek, T. Stouch, and J. Novotny, *J. Mol. Biol.*, **234**, 661–679 (1993).

262. J. Novotny, R. E. Bruccoleri, and F. A. Saul, *Biochemistry*, **28**, 4735–4749 (1989).

263. R. D. Head, M. L. Smythe, T. I. Oprea, C. L. Waller, S. M. Green, and G. R. Marshall, *J. Am. Chem. Soc.*, **118**, 3959–3969 (1996).

264. C. L. Waller, T. I. Oprea, A. Giolitti, and G. R. Marshall, *J. Med. Chem.*, **36**, 4152–4160 (1993).

265. N. Prattibiraman, M. Levitt, T. E. Ferrin, and R. Langridge, *J. Comput. Chem.*, **6**, 432 (1985).

266. E. C. Meng, B. K. Shoichet, and I. D. Kuntz, *J. Comput. Chem.*, **13**, 505–524 (1992).

267. S. Makino and I. D. Kuntz, *J. Comput. Chem.*, **18**, 1812–1825 (1997).

268. T. J. A. Ewing and I. D. Kuntz, *J. Comput. Chem.*, **18**, 1175–1189 (1997).

269. B. Sandak, R. Nussinov, and H. J. Wolfson, *J. Comp. Biol.*, **5**, 631–654 (1998).

270. C. A. Baxter, C. W. Murray, D. E. Clark, D. R. Westhead, and M. D. Eldridge, *Proteins*, **33**, 367–382 (1998).

271. D. M. Lorber and B. K. Shoichet, *Protein Sci.*, **7**, 938–950 (1998).

272. Y. Sun, T. J. A. Ewing, A. G. Skillman, and I. D. Kuntz, *J. Comput.-Aided Mol. Des.*, **12**, 597–604 (1998).

273. R. Mangoni, D. Roccatano, and A. Di Nola, *Proteins*, **35**, 153–162 (1999).

274. G. M. Morris, D. S. Goodsell, R. S. Halliday, R. Huey, W. E. Hart, R. K. Belew, and A. J. Olson, *J. Comput. Chem.*, **19**, 1639–1662 (1998).

275. M. Liu and S. M. Wang, *J. Comput.-Aided Mol. Des.*, **13**, 435–451 (1999).

276. S. Makino, T. J. A. Ewing, and I. D. Kuntz, *J. Comput.-Aided Mol. Des.*, **13**, 513–532 (1999).

277. M. Rarey, B. Kramer, and T. Lengauer, *Bioinformatics*, **15**, 243–250 (1999).

278. R. M. A. Knegtel and M. Wagener, *Proteins*, **37**, 334–345 (1999).

279. M. L. Lamb, K. W. Burdick, S. Toba, M. M. Young, K. G. Skillman, X. Q. Zou, J. R. Arnold, and I. D. Kuntz, *Proteins*, **42**, 296–318 (2001).

280. R. Abagyan and M. Totrov, *Curr. Opin. Chem. Biol.*, **5**, 375–382 (2001).

281. H. Claussen, C. Buning, M. Rarey, and T. Lengauer, *J. Mol. Biol.*, **308**, 377–395 (2001).

282. C. A. Baxter, C. W. Murray, B. Waszkowycz, J. Li, R. A. Sykes, R. G. A. Bone, T. D. J. Perkins, and W. Wylie, *J. Chem. Inf. Comput. Sci.*, **40**, 254–262 (2000).

283. N. Ota and D. A. Agard, *J. Mol. Biol.*, **314**, 607–617 (2001).

284. S. Naruto, I. Motoc, G. R. Marshall, S. B. Daniels, M. J. Sofia, and J. A. Katzenellenbogen, *J. Am. Chem. Soc.*, **107**, 5262–5270 (1985).

285. K. P. Clark and Ajay, *J. Comput. Chem.*, **16**, 1210–1226 (1995).

286. G. M. Verkhivker, P. A. Rejto, D. K. Gehlhaar, and S. T. Freer, *Proteins*, **25**, 342–353 (1996).

287. D. Q. McDonald and W. C. Still, *J. Am. Chem. Soc.*, **116**, 11550–11553 (1994).

288. F. Guarnieri and W. C. Still, *J. Comput. Chem.*, **15**, 1302–1310 (1994).

289. D. Q. McDonald and W. C. Still, *J. Am. Chem. Soc.*, **118**, 2073–2077 (1996).

290. Z. R. Wasserman and C. N. Hodge, *Proteins*, **24**, 227–237 (1996).

291. J. Desmet, I. A. Wilson, M. Joniau, M. Demaeyer, and I. Lasters, *FASEB J.*, **11**, 164–172 (1997).

292. B. L. King, S. Vajda, and C. Delisi, *FEBS Lett.*, **384**, 87–91 (1996).

293. D. S. Goodsell, H. Lauble, C. D. Stout, and A. J. Olson, *Proteins*, **17**, 1–10 (1993).

294. R. X. Wang and S. M. Wang, *J. Chem. Inf. Comput. Sci.*, **41**, 1422–1426 (2001).

295. P. S. Charifson, J. J. Corkery, M. A. Murcko, and W. P. Walters, *J. Med. Chem.*, **42**, 5100–5109 (1999).

296. N. L. Allinger, Z.-q. S. Zhu, and K. Chen, *J. Am. Chem. Soc.*, **114**, 6120–6133 (1992).

297. B. R. Brooks, R. E. Bruccoleri, B. D. Olafson, D. J. States, S. Swaminathan, and M. Karplus, *J. Comput. Chem.*, **4**, 187–217 (1983).

298. F. A. Momany and R. Rone, *J. Comput. Chem.*, **13**, 888–900 (1992).

299. G. Nemethy, M. S. Pottle, and H. A. Scheraga, *J. Phys. Chem.*, **87**, 1883–1887 (1983).

300. T. A. Halgren, *J. Am. Chem. Soc.*, **114**, 7827–7843 (1992).

301. P. S. Charifson, R. G. Hiskey, L. G. Pedersen, and L. F. Kuyper, *J. Comput. Chem.*, **12**, 899–908 (1991).

302. S. C. Hoops, K. W. Anderson, and K. M. Merz, Jr., *J. Am. Chem. Soc.*, **113**, 8262–8270 (1991).

303. C. J. Casewit, K. S. Colwell, and A. K. Rappe, *J. Am. Chem. Soc.*, **114**, 10035–10046 (1992).

304. C. J. Casewit, K. S. Colwell, and A. K. Rappe, *J. Am. Chem. Soc.*, **114**, 10046–10053 (1992).

305. A. K. Rappe, C. J. Casewit, K. S. Colwell, W. A. Goddard III, and W. M. Skiff, *J. Am. Chem. Soc.*, **114**, 10024–10035 (1992).

306. Y.-D. Wu and K. N. Houk, *J. Am. Chem. Soc.*, **114**, 1656–1661 (1992).

307. K. Houk, J. A. Tucker, and A. Dorigo, *Acc. Chem. Res.*, **23**, 107–113 (1990).

308. F. M. Menger and M. J. Sherrod, *J. Am. Chem. Soc.*, **112**, 8071–8075 (1990).

309. D. P. Riley, P. J. Lennon, W. L. Neumann, and R. H. Weiss, *J. Am. Chem. Soc.*, **119**, 6522–6528 (1997).

310. K. Aston, N. Rath, A. Naik, U. Slomczynska, O. F. Schall, and D. P. Riley, *Inorg. Chem.*, **40**, 1779–1789 (2001).

311. D. H. Williams, *Aldrichimica Acta*, **24**, 71–80 (1991).

312. A. J. Doig and D. H. Williams, *J. Am. Chem. Soc.*, **114**, 338–343 (1992).

313. M. S. Searle and D. H. Williams, *J. Am. Chem. Soc.*, **114**, 10690–10697 (1992).

314. M. S. Searle, D. H. Williams, and U. Gerhard, *J. Am. Chem. Soc.*, **114**, 10697–10704 (1992).

315. D. H. Williams and B. Bardsley, *Perspect. Drug Discov. Des.*, **17**, 43–59 (1999).

316. M. Graffner-Nordberg, K. Kolmodin, J. Aqvist, S. F. Queener, and A. Hallberg, *J. Med. Chem.*, **44**, 2391–2402 (2001).

317. J. Aqvist, V. B. Luzhkov, and B. O. Brandsdal, *Acc. Chem. Rev.*, **35**, 358–365 (2002).

318. G. M. Verkhivker, D. Bouzida, D. K. Gehlhaar, P. A. Rejto, L. Schaffer, S. Arthurs, A. B. Colson, S. T. Freer, V. Larson, B. A. Luty, T. Marrone, and P. W. Rose, *J. Med. Chem.*, **45**, 72–89 (2002).

319. R. Lumry and S. Rajender, *Biopolymers*, **9**, 1125–1227 (1970).

320. I. Muegge and Y. C. Martin, *J. Med. Chem.*, **42**, 791–804 (1999).

321. B. A. Grzybowski, A. V. Ishcheno, J. Shimada, and E. I. Shakhnovich, *Acc. Chem. Res.*, **35**, 261–269 (2002).

322. B. A. Grzybowski, A. V. Ishcheno, C.-Y. Kim, G. Topolov, R. Chapman, D. W. Christianson, G. M. Whitesides, and E. I. Shakhnovich, *Proc. Natl. Acad. Sci. USA*, **99**, 1270–1273 (2002).

323. K. M. Merz, Jr., M. A. Murcko, and P. A. Kollman, *J. Am. Chem. Soc.*, **113**, 4484–4490 (1991).

324. C. F. Wong and J. A. McCammon, *J. Am. Chem. Soc.*, **108**, 3830–3832 (1986).

325. L. M. Hansen and P. A. Kollman, *J. Comput. Chem.*, **11**, 994–1002 (1990).

326. B. G. Rao, R. F. Tilton, and U. C. Singh, *J. Am. Chem. Soc.*, **114**, 4447–4452 (1992).

327. W. E. Harte, Jr. and D. L. Beveridge, *J. Am. Chem. Soc.*, **115**, 3883–3886 (1993).

328. D. M. Ferguson, R. J. Radmer, and P. A. Kollman, *J. Med. Chem.*, **34**, 2654–2659 (1991).

329. J. J. McDonald and C. L. Brooks III, *J. Am. Chem. Soc.*, **114**, 2062–2072 (1992).

330. T. P. Lybrand and J. A. McCammon, *J. Comput.-Aided Mol. Des.*, **2**, 259–266 (1988).

331. W. R. Cannon, J. D. Madura, R. P. Thummel, and J. A. McCammon, *J. Am. Chem. Soc.*, **115**, 879–884 (1993).

332. S. Yun-yu, A. E. Mark, W. Cun-Xin, H. Fuhua, J. C. Berendsen, and W. F. van Gunsteren, *Protein Eng.*, **6**, 289–295 (1993).

333. D. H. Rich, C.-Q. Sun, J. V. N. Vara Prasad, M. V. Toth, G. R. Marshall, P. Ahammadunny, M. D. Clare, R. D. Mueller, and K. Houseman, *J. Med. Chem.*, **34**, 1222–1225 (1991).

334. P. De La Paz, J. M. Burridge, S. J. Oatley, and C. C. F. Blake in C. R. Beddell, Ed., *The Design of Drugs to Macromolecular Targets*, John Wiley & Sons, New York, 1992, pp. 119–172.

335. A. B. Edmundson, J. N. Herron, K. R. Ely, X.-M. He, D. L. Harris, and E. W. Voss, Jr., *Philos. Trans. R. Soc. Lond. Biol.*, **323**, 495–509 (1989).

336. C. Bihoreau, C. Monnot, E. Davies, B. Teutsch, K. E. Bernstein, P. Corvol, and E. Clauser, *Proc. Natl. Acad. Sci. USA*, **90**, 5133–5137 (1993).

337. T. M. Fong, R. R. C. Huang, and C. D. Strader, *J. Biol. Chem.*, **267**, 25664–25667 (1992).

338. U. Gether, T. E. Johansen, R. M. Snider, I. Lowe, J. A., S. Nakanishi, and T. W. Schwartz, *Nature*, **362**, 345–348 (1993).

339. M. F. Hibert, S. Trumpp-Kallmeyer, A. Bruinvels, and J. Hoflack, *Mol. Pharmacol.*, **40**, 8–15 (1991).

340. M. F. Hibert, S. Trumpp-Kallmeyer, J. Hoflack, and A. Bruinvels, *Trends Pharmacol. Sci.*, **14**, 7–12 (1993).

341. G. Nordvall and U. Hacksell, *J. Med. Chem.*, **36**, 967–976 (1993).

342. T. L. Blundell, B. L. Sibanda, M. J. E. Sternberg, and J. M. Thornton, *Nature*, **326**, 347–352 (1987).

343. L. H. Pearl and W. R. Taylor, *Nature*, **329**, 351–354 (1987).

344. I. T. Weber, *Proteins*, **7**, 172–184 (1990).

345. L. M. Balbes and F. I. Carroll, *Med. Chem. Res.*, **1**, 283–288 (1991).

346. R. J. Siezen, W. M. de Vos, J. A. M. Leunissen, and B. W. Dijkstra, *Protein Eng.*, **4**, 719–737 (1991).

347. J. H. Brown, T. Jardetzky, M. A. Saper, B. Samraoui, P. J. Bjorkman, and D. C. Wiley, *Nature*, **332**, 845–850 (1988).

348. L. M. H. Koymans, N. P. E. Vermeulen, A. Baarslag, and G. M. Donne-op den Kelder, *J. Comput.-Aided Mol. Des.*, **7**, 281–289 (1993).

349. R. E. Bruccoleri and M. Karplus, *Biopolymers*, **26**, 137–168 (1987).

350. D. Jones and J. Thornton, *J. Comput.-Aided Mol. Des.*, **7**, 439–456 (1993).

351. J. U. Bowie and D. Eisenberg, *Curr. Opin. Struct. Biol.*, **3**, 437–444 (1993).

352. S. J. Wodak and M. J. Rooman, *Curr. Opin. Struct. Biol.*, **3**, 247–259 (1993).

353. M. J. Sippl, *J. Comput.-Aided Mol. Des.*, **7**, 473–501 (1993).

354. S. Miyazawa and R. C. Jernigan, *Macromolecules*, **18**, 534–552 (1985).

355. S. H. Bryant and C. E. Lawrence, *Proteins*, **16**, 92–112 (1993).

356. D. Frishman and H. W. Mewes, *Nat. Struct. Biol.*, **4**, 626–628 (1997).

357. C. Chothia, *Nature*, **357**, 543–544 (1992).

358. P.-A. Lindgard and H. Bohr in H. Bohr and S. Bunak, Eds., *Protein Folds*, CRC Press, Boca Raton, FL, 1996, pp. 98–102.

359. P. E. Boscott, G. J. Barton, and W. G. Richards, *Protein Eng.*, **6**, 261–266 (1993).

360. D. Frishman and P. Argos, *Proteins*, **27**, 329–335 (1997).

361. R. Srinivasan and G. D. Rose, *Proteins*, **22**, 81–99 (1995).

362. K. A. Dill, H. S. Chan, and K. Yue, *Macromol. Symp.*, **98**, 615–617 (1995).

363. K. Yue and K. A. Dill, *Protein Sci.*, **5**, 254–261 (1996).

364. S. M. Le Grand and K. L. Merz, Jr. in S. M. Le Grand and K. L. Merz, Jr., Eds., *The Protein Folding Problem and Tertiary Structure Prediction*, Birkhauser, Boston, 1994, pp. 109–124.

365. S. Sun, *Protein Sci.*, **2**, 762–785 (1993).

366. J. U. Bowie and D. Eisenberg, *Proc. Natl. Acad. Sci. USA*, **91**, 4436–4440 (1994).

367. J. U. Bowie, K. Zhang, M. Wilmanns, and D. Eisenberg, *Methods Enzymol.*, **266**, 598–616 (1996).

368. S. G. Galaktionov and G. Marshall, *Molecular Graphics and Drug Design: 27th Hawaii International Conference on System Sciences*, IEEE Computer Society Press, Washington, DC, 1994.

369. S. Saitoh, T. Nakai, and K. Nishikawa, *Proteins*, **15**, 191–204 (1993).

370. M. Vendruscolo, E. Kussell, and E. Domany, *Fold. Des.*, **2**, 295–306 (1997).

371. I. D. Kuntz, G. M. Crippen, P. A. Kolman, and D. Kimelman, *J. Mol. Biol.*, **106**, 983–994 (1976).

372. S. Galaktionov, G. V. Nikiforovich, and G. R. Marshall, *Biopolymers*, **60**, 153–168 (2001).

373. A. Aszodi and W. R. Taylor, *Fold. Des.*, **1**, 325–334 (1996).

374. A. Aszodi, R. E. J. Munro, and W. R. Taylor, *Fold. Des.*, **2**, S3–S6 (1997).

375. A. Aszodi and W. R. Taylor, *Comput. Chem.*, **21**, 13–23 (1997).

376. S. R. Holbrook, I. Dubchak, and S.-H. Kim, *Biotechniques*, **14**, 984–989 (1993).

377. B. H. Park, E. S. Huang, and M. Levitt, *J. Mol. Biol.*, **266**, 831–846 (1997).

378. G. M. Crippen and V. N. Maiorov in H. Bohr and S. Bunak, Eds., *Protein Folds*, CRC Press, Boca Raton, FL, 1996, pp. 189–201.

379. P. D. Thomas and K. A. Dill, *J. Mol. Biol.*, **257**, 457–469 (1996).

380. S. Miyazawa and R. L. Jernigan, *J. Mol. Biol.*, **256**, 623–644 (1996).

381. V. J. Hruby, W. Qui, T. Okayama, and V. A. Soloshonok, *Methods Enzymol.*, **343**, 91–123 (2002).

382. M. G. Bursavich and D. H. Rich, *J. Med. Chem.*, **45**, 541–558 (2002).

383. R. Hirschmann, K. C. Nicolaou, S. Pietranico, J. Salvino, E. M. Leahy, P. A. Sprengeler, G. Furst, and A. B. Smith III, *J. Am. Chem. Soc.*, **114**, 9217–9218 (1992).

384. R. Hirschmann, P. A. Sprengeler, T. Kawasaki, J. W. Leahy, W. C. Shakespeare, and A. B. Smith III, *J. Am. Chem. Soc.*, **114**, 9699–9701 (1992).

385. T. W. Ku, F. E. Ali, L. S. Barton, J. W. Bean, W. E. Bondinell, J. L. Burgess, J. F. Callahan, R. R. Calvo, L. Chen, D. S. Eggelston, J. S. Gleason, W. F. Huffman, S. M. Hwang, D. R. Jakas, C. B. Karash, R. M. Keenan, K. D. Kopple, W. H. Miller, K. A. Newlander, A. Nichols, M. F. Parker, C. E. Peishoff, J. M. Samanen, I. Uzinskas, and J. W. Venslavsky, *J. Am. Chem. Soc.*, **115**, 8861–8862 (1993).

386. G. L. Olson, H.-C. Cheung, M. E. Voss, D. E. Hill, M. Kahn, V. S. Madison, C. M. Cook, J. Sepinwall, and G. Vincent, *Concepts and Progress in the Design of Peptide Mimetics: Beta Turns and Thyrotropin Releasing Hormone (Biotechnology USA 1989)*, Conference Management Corporation, Norwald, CT, 1989, pp. 348–360.

387. G. L. Olson, D. R. Bolin, M. P. Bonner, M. Bos, C. M. Cook, D. C. Fry, B. J. Graves, M. Hatada, D. E. Hill, M. Kahn, V. S. Madison, V. K. Rusiecki, R. Sarabu, J. Sepinwall, G. P. Vincent, and M. E. Voss, *J. Med. Chem.*, **36**, 3039–3049 (1993).

388. P. C. Belanger and C. Dufresne, *Can. J. Chem.*, **64**, 1514–1520 (1986).

389. R. Hirschmann, K. C. Nicolaou, S. Pietranico, E. M. Leahy, J. Salvino, B. Arison, M. A. Cichy, P. G. Spoors, W. C. Shakespeare, P. A. Sprengeler, P. Hamley, A. B. Smith III, T. Reisine, K. Raynor, L. Maechler, C. Donaldson, W. Vale, R. M. Friedinger, M. R. Cascieri, and C. D. Strader, *J. Am. Chem. Soc.*, **115**, 12550–12568 (1993).

390. J. H. Arevalo, E. A. Stura, M. J. Taussig, and I. A. Wilson, *J. Mol. Biol.*, **231**, 103–118 (1993).

391. J. H. Arevalo, M. J. Taussig, and I. A. Wilson, *Nature*, **365**, 859–863 (1993).

392. P. Traxler, J. Green, H. Mett, U. Sequin, and P. Furet, *J. Med. Chem.*, **42**, 1018–1026 (1999).

393. P. Traxler, G. Bold, E. Buchdunger, G. Caravatti, P. Furet, P. Manley, T. O'Reilly, J. Wood, and J. Zimmermann, *Med. Res. Rev.*, **21**, 499–512 (2001).

394. R. Bureau, C. Daveu, J. C. Lancelot, and S. Rault, *J. Chem. Inf. Comput. Sci.*, **42**, 429–436 (2002).

395. Y. Kato, A. Itai, and Y. Iitaka, *Tetrahedron Lett.*, **43**, 5229–5236 (1987).

396. W. H. Moos, C. C. Humblet, I. Sircar, C. Rithner, R. E. Weishaar, J. A. Bristol, and A. T. McPhail, *J. Med. Chem.*, **30**, 1963–1972 (1987).

397. D. Mayer, C. B. Naylor, I. Motoc, and G. R. Marshall, *J. Comput.-Aided Mol. Des.*, **1**, 3–16 (1987).

398. G. R. Marshall, C. D. Barry, H. E. Bosshard, R. A. Dammkoehler, and D. A. Dunn in E. C. Olsen and R. E. Christoffersen, Eds., *Computer-Assisted Drug Design*, American Chemical Society, Washington, DC, 1979, pp. 205–226.

399. R. A. Dammkoehler, S. F. Karasek, E. F. B. Shands, and G. R. Marshall, Constrained Search of Conformational Hyperspace: Segmentation and Parallelism, *Abstr. 204th ACS National Meeting*, American Chemical Society, Washington, DC, 1992.

400. G. R. Marshall and R. D. Cramer III, *Trends Pharmacol. Sci.*, **9**, 285–289 (1988).

401. R. D. Cramer III and S. B. Wold, *Comp. Mol. Field Anal. (CoMFA)*, **5**, 388 (editorial) (1991).

402. J. R. Sufrin, D. A. Dunn, and G. R. Marshall, *Mol. Pharmacol.*, **19**, 307–313 (1981).

403. P. R. Andrews, E. J. Lloyd, J. L. Martin, S. L. Munro, M. Sadek, and M. G. Wong in A. S. V. Burgen, G. C. K. Roberts, and M. S. Tute, Eds., *Molecular Graphics and Drug Design*, Elsevier Science, Amsterdam, 1986, pp. 216–255.

404. G. Klopman and S. Srivastava, *Mol. Pharmacol.*, **37**, 958–965 (1989).

405. G. Klopman and M. L. Dimayuga, *J. Comput.-Aided Mol. Des.*, **4**, 117–130 (1990).

406. G. Rum and W. C. Herndon, *J. Am. Chem. Soc.*, **113**, 9055–9060 (1991).

407. C. Silipo and A. Vittoria in C. A. Ramsden, Ed., *Quantitative Drug Design*, Pergamon Press, Oxford, UK, 1990, pp. 153–204.

408. G. M. Crippen in D. Bawden, Ed., *Distance Geometry and Conformational Calculations (Chemometrics Research Studies)*, Vol. **1**, John Wiley & Sons, Chichester, UK, 1981.

409. D. E. Clark, P. Willett, and P. W. Kenny, *J. Mol. Graphics*, **10**, 194–204 (1992).

410. C. A. Pepperrell and P. Willett, *J. Comput.-Aided Mol. Des.*, **5**, 455–474 (1991).

411. A. R. Poirette, P. Willett, and F. H. Allen, *J. Mol. Graphics*, **11**, 2–14 (1993).

412. G. R. Marshall and C. B. Naylor in C. A. Ramsden, Ed., *Quantitative Drug Design*, Pergamon Press, Oxford, UK, 1990, pp. 431–458.

413. A. Davis, B. H. Warrington, and J. G. Vinter, *J. Comput.-Aided Mol. Des.*, **1**, 97–120 (1987).

414. H. Weinstein, R. Osman, S. Topiol, and J. P. Green, *Ann. N. Y. Acad. Sci.*, **367**, 434–448 (1981).

415. N. C. Cohen in B. Testa, Ed., *Advances in Drug Research*, Academic Press, New York, 1985, pp. 40–144.

416. R. C. Wade, K. J. Clark, and P. J. Goodford, *J. Med. Chem.*, **36**, 140–147 (1993).

417. N. Marchand-Geneste, K. A. Watson, B. K. Alsberg, and R. D. King, *J. Med. Chem.*, **45**, 399–409 (2002).

418. C. Hansch, J. Mcclarin, T. Klein, and R. Langridge, *Mol. Pharmacol.*, **27**, 493–498 (1995).

419. C. Hansch, T. Klein, J. McClarin, R. Langridge, and N. W. Cornell, *J. Med. Chem.*, **29**, 615–620 (1986).

420. G. E. Kellogg, S. F. Semus, and D. J. Abraham, *J. Comput.-Aided Mol. Des.*, **5**, 545–552 (1991).

421. G. E. Kellogg and D. J. Abraham, *J. Mol. Graphics*, **10**, 212–217 (1992).

422. D. J. Danziger and P. M. Dean, *J. Theor. Biol.*, **116**, 215–224 (1985).

423. S. K. Kearsley, *J. Comput. Chem.*, **11**, 1187–1192 (1990).

424. G. R. Marshall and C. D. Barry, *Functional Representation of Molecular Volume for Computer-Aided Drug Design*, Abstr. Amer. Cryst. Assoc., Honolulu, HI, 1979.

425. A. J. Hopfinger, *J. Med. Chem.*, **2**, 7196–7206 (1980).

426. Z. Simon, A. Chiriac, S. Holban, D. Ciubotariu, and G. I. Mihalas, *Minimum Steric Difference*, Research Studies Press, Letchworth, UK, 1984.

427. D. Ciubotariu, E. Deretey, T. I. Oprea, T. I. Sulea, Z. Simon, L. Kurunczi, and A. Chiriac, *Quant. Struct.-Act. Relat.*, **12**, 367–372 (1993).

428. H.-D. Holtje and S. Marrer, *J. Comput.-Aided Mol. Des.*, **1**, 23–30 (1987).

429. A. J. Hopfinger, *J. Med. Chem.*, **26**, 990–996 (1983).

430. S. Namasivayam and P. M. Dean, *J. Mol. Graphics*, **4**, 46 (1986).

431. P. L. Chau and P. M. Dean, *J. Mol. Graphics*, **5**, 97 (1987).

432. C. Burt and W. G. Richards, *J. Comput.-Aided Mol. Des.*, **4**, 231–238 (1990).

433. J. Zabrocki, G. D. Smith, J. B. Dunbar, Jr., H. Iijima, and G. R. Marshall, *J. Am. Chem. Soc.*, **110**, 5875–5880 (1988).

434. J. B. Ball, R. A. Hughes, P. F. Alewood, and P. R. Andrews, *Tetrahedron*, **49**, 3467–3478 (1993).

435. J. B. Ball and P. F. Alewood, *J. Mol. Recognit.*, **3**, 55–64 (1990).

436. G. V. Nikiforovich, K. E. Kover, W. J. Zhang, and G. R. Marshall, *J. Am. Chem. Soc.*, **122**, 3262–3273 (2000).

437. J. Labanowski, I. Motoc, C. B. Naylor, D. Mayer, and R. A. Dammkoehler, *Quant. Struct.-Act. Relat.*, **5**, 138–152 (1986).

438. S. Naruto, I. Motoc, and G. R. Marshall, *Eur. J. Med. Chem.*, **20**, 529–532 (1985).

439. R. P. Sheridan and R. Venkataraghavan, *J. Comput.-Aided Mol. Des.*, **1**, 243–256 (1987).

440. E. E. Hodgkin, A. Miller, and M. Whittaker, *J. Comput.-Aided Mol. Des.*, **7**, 515–534 (1993).

441. M. T. Barakat and P. M. Dean, *J. Comput.-Aided Mol. Des.*, **4**, 295–316 (1990).

442. M. T. Barakat and P. M. Dean, *J. Comput.-Aided Mol. Des.*, **4**, 317–330 (1990).

443. T. D. J. Perkins and P. M. Dean, *J. Comput.-Aided Mol. Des.*, **7**, 173–182 (1993).

444. I. Motoc, J. Labanowski, C. B. Naylor, D. Mayer, and R. A. Dammkoehler, *Quant. Struct.-Act. Relat.*, **5**, 99–105 (1986).

445. R. D. Nelson, D. I. Gottlieb, T. M. Balasubramanian, and G. R. Marshall in R. S. Rapaka, G. Barnett, and R. L. Hawks, Eds., *Opioid Peptides: Medicinal Chemistry*, NIDA Office of Science, Rockville, MD, 1986, pp. 204–230.

446. A. K. Ghose and G. M. Crippen, *J. Med. Chem.*, **27**, 901–914 (1984).

447. A. K. Ghose and G. M. Crippen, *J. Med. Chem.*, **28**, 333–346 (1985).

448. A. K. Ghose and G. M. Crippen in C. A. Ramsden, Ed., *Quantitative Drug Design*, Pergamon Press, Oxford, UK, 1990, pp. 716–733.

449. A. K. Ghose and G. M. Chippen, *Mol. Pharmacol.*, **37**, 725–734 (1990).

450. M. R. Linschoten, T. Bultsma, A. P. IJzerman, and H. Timmerman, *J. Med. Chem.*, **29**, 278–286 (1986).

451. G. M. Donne-op den Kelder, *J. Comput.-Aided Mol. Des.*, **1**, 257–264 (1987).

452. T. I. Oprea, D. Ciubotariu, T. I. Sulea, and Z. Simon, *Quant. Struct.-Act. Relat.*, **12**, 21–26 (1993).

453. J. P. Snyder, S. N. Rao, K. F. Koehler, A. Vedani, and R. Pellicciari in C. G. Wermuth, Ed., *Trends in QSAR and Molecular Modelling 92*, ESCOM Scientific, Leiden, Netherlands, 1993, pp. 44–51.

454. G. Klopman, *Quant. Struct.-Act. Relat.*, **11**, 176–185 (1992).

455. I. B. Bersuker and A. S. Dimogo in K. B. Lipkowitz and D. B. Boyd, Eds., *Revisions in Computational Chemistry*, VCH, New York, 1991, pp. 423–460.

456. Y. C. Martin, M. G. Bures, E. A. Danaher, J. DeLazzer, I. Lico, and P. Pavlik, A., *J. Comput.-Aided Mol. Des.*, **7**, 83–102 (1993).

457. G. Jones, P. Willett, and R. C. Glen, *J. Mol. Biol.*, **245**, 43–53 (1995).

458. T. I. Oprea and L. Kurunczi in N. Voiculetz, I. Motoc, and Z. Simon, Eds., *Specific Interactions and Biological Recognition Processes*, CRC Press, Boca Raton, FL, 1993, pp. 295–326.

459. W. E. Klunk, B. L. Kalman, J. A. Ferrendelli, and D. F. Covey, *Mol. Pharmacol.*, **23**, 511–518 (1982).

460. J. A. Calder, J. A. Wyatt, D. A. Frenkel, and J. E. Casida, *J. Comput.-Aided Mol. Des.*, **7**, 45–60 (1993).

461. M. F. Hibert, R. Hoffmann, R. C. Miller, and A. A. Carr, *J. Med. Chem.*, **33**, 1594–1600 (1990).

462. M. F. Hibert, M. W. Gittos, D. N. Middlemiss, A. K. Mir, and J. R. Fozard, *J. Med. Chem.*, **31**, 1087–1093 (1988).

463. A. W. Schmidt and S. J. Peroutka, *Mol. Pharmacol.*, **36**, 505–511 (1989).

464. M. L. Connolly, *Science*, **221**, 709–713 (1983).

465. M. L. Connolly, *J. Appl. Crystallogr.*, **16**, 548–558 (1983).

466. C. E. Kundrot, J. W. Ponder, and F. M. Richards, *J. Comput. Chem.*, **12**, 402–409 (1991).

467. S. M. Le Grand and K. M. Merz, Jr., *J. Comput. Chem.*, **14**, 349–352 (1993).

468. A. H. Beckett and A. F. Casey, *J. Pharm. Pharmacol.*, **6**, 986–999 (1954).

469. L. B. Kier and H. S. Aldrich, *J. Theor. Biol.*, **46**, 529–541 (1974).

470. L. G. Humber, F. T. Bruderlin, A. H. Philipp, M. Gotz, and K. Voith, *J. Med. Chem.*, **22**, 761–767 (1979).

471. G. L. Olson, H. C. Cheung, K. D. Morgan, J. F. Blount, L. Todaro, L. Berger, A. B. Davidson, and E. Boff, *J. Med. Chem.*, **24**, 1026–1034 (1981).

472. H.-D. Holtje and M. Tintelnot, *Quant. Struct.-Act. Relat.*, **3**, 6–9 (1984).

473. W. C. Probst, L. A. Snyder, D. J. Schuster, J. Brosius, and S. C. Sealfon, *DNA Cell Biol.*, **II**, 1–20 (1992).

474. S. Trumpp-Kallmeyer, J. Hoflack, A. Bruinvels, and M. Hibert, *J. Med. Chem.*, **35**, 3448–3462 (1992).

475. D. Timms, A. J. Wilkinson, D. R. Kelly, K. J. Broadley, and R. H. Davies, *Int. J. Quantum Chem. Quantum Biol. Symp.*, **19**, 197–215 (1992).

476. D. Zhang and H. Weinstein, *J. Med. Chem.*, **36**, 934–938 (1993).

477. B. L. Bush and R. B. Nachbar, Jr., *J. Comput.-Aided Mol. Des.*, **7**, 587–619 (1993).

478. J. N. Weinstein, K. W. Kohn, M. R. Grever, V. N. Viswanadhan, L. V. Rubeinstein, A. P. Monks, D. A. Scudiero, L. Welch, A. D. Koutsoukos, A. J. Chiausa, and K. D. Paull, *Science*, **258**, 447–451 (1992).

479. T. A. Andrea and H. Kalayeh, *J. Med. Chem.*, **34**, 2824–2836 (1991).

480. S.-S. So and W. G. Richards, *J. Med. Chem.*, **35**, 3201–3207 (1992).

481. I. V. Tetko, A. I. Luik, and G. I. Poda, *J. Med. Chem.*, **36**, 811–814 (1993).

482. R. D. King, S. Muggleton, R. A. Lewis, and M. J. E. Sternberg, *Proc. Natl. Acad. Sci. USA*, **89**, 11322–11326 (1992).

483. S. A. DePriest, E. F. B. Shands, R. A. Dammkoehler, and G. R. Marshall in C. Silipo and A. Vittoria, Eds., *QSAR: Rational Approaches to*

the *Design of Bioactive Compounds*, Elsevier Science, Amsterdam, 1991, pp. 405–414.

484. S. A. DePriest, D. Mayer, C. B. Naylor, and G. R. Marshall, *J. Am. Chem. Soc.*, **115**, 5372–5384 (1993).

485. C. Hansch, *Acc. Chem. Res.*, **26**, 147–153 (1993).

486. G. Klebe and U. Abraham, *J. Med. Chem.*, **36**, 70–80 (1993).

487. D. P. Getman, G. A. DeCrescenzo, R. M. Heintz, K. L. Reed, J. J. Talley, M. L. Bryant, M. Clare, K. A. Houseman, J. J. Marr, R. A. Mueller, M. L. Vazquez, H.-S. Shieh, W. C. Stallings, and R. A. Stegeman, *J. Med. Chem.*, **36**, 288–291 (1993).

488. G. M. Crippen, *J. Comput. Chem.*, **8**, 943–955 (1987).

489. M. P. Bradley and G. M. Crippen, *J. Med. Chem.*, **36**, 3171–3177 (1993).

490. F. Major, M. Turcotte, D. Gautheret, G. Lapalme, E. Fillion, and R. Cedergren, *Science*, **253**, 1255–1260 (1991).

491. D. Gautheret and R. Cedergren, *FASEB J.*, **7**, 97–105 (1993).

492. P. A. Greenidge, T. C. Jenkins, and S. Neidle, *Mol. Pharmacol.*, **43**, 982–988 (1993).

493. M. G. Cardozo and A. J. Hopfinger, *Mol. Pharmacol.*, **40**, 1023–1028 (1991).

494. M. J. J. Blommers, C. B. Lucasius, G. Kateman, and R. Kaptein, *Biopolymers*, **22**, 45–52 (1992).

495. A. G. Palmer III and D. A. Case, *J. Am. Chem. Soc.*, **114**, 9059–9067 (1992).

496. K. Boehncke, M. Nonella, K. Schulten, and A. H.-J. Wang, *Biochemistry*, **30**, 5465–5475 (1991).

497. J. Xing and H. L. Scott, *Biochem. Biophys. Res. Commun.*, **165**, 1–6 (1989).

498. T. R. Stouch, K. B. Ward, A. Altieri, and A. T. Hagler, *J. Comput. Chem.*, **12**, 1033–1046 (1991).

499. H. L. Scott and S. Kalaskar, *Biochemistry*, **28**, 3687–3691 (1989).

500. P. S. O'Shea and R. Matela, *Biochem. Soc. Trans.*, **14**, 1119–1120 (1986).

501. D. M. Kroll and G. Gompper, *Science*, **255**, 968–971 (1992).

502. L. I. Krishtakik, V. V. Topolev, and Y. I. Kharkats, *Biophysics*, **36**, 257–262 (1991).

503. E. Egberts and H. J. C. Berendsen, *J. Chem. Phys.*, **89**, 3718–3732 (1988).

504. R. P. Mason, D. G. Rhodes, and L. G. Herbette, *J. Med. Chem.*, **34**, 869–877 (1991).

505. L. G. Herbette in C. G. Wermuth, Ed., *Trends in QSAR and Molecular Modelling 92*, ESCOM Scientific, Leiden, Netherlands, 1993, pp. 76–85.

506. H. Heller, M. Schaeffer, and K. Schulten, *J. Phys. Chem.*, **97**, 8343–8360 (1993).

507. W. Im and B. Roux, *J. Mol. Biol.*, **319**, 1177–1197 (2002).

508. T. Kataoka, D. D. Beusen, J. D. Clark, M. Yodo, and G. R. Marshall, *Biopolymers*, **32**, 1519–1533 (1992).

509. G. R. Marshall, *Tetrahedron*, **49**, 3547–3558 (1993).

510. G. V. Nikiforovich and G. R. Marshall, *Biochem. Biophys. Res. Commun.*, **195**, 222–228 (1993).

511. G. V. Nikiforovich and V. J. Hruby, *Biochem. Biophys. Res. Commun.*, **194**, 9–16 (1993).

512. G. Nikiforovich and G. R. Marshall, *Int. J. Pept. Protein Res.*, **42**, 171–180 (1993).

513. G. V. Nikiforovich and G. R. Marshall, *Int. J. Pept. Protein Res.*, **42**, 181–193 (1993).

514. P. Poulin and F. P. Theil, *J. Pharm. Sci.*, **91**, 1358–1370 (2002).

515. G. M. Keseruu and L. Molnar, *J. Chem. Inf. Comput. Sci.*, **42**, 437–444 (2002).

516. H. van de Waterbeemd, *Curr. Opin. Drug Discov. Dev.*, **5**, 33–43 (2002).

517. J. Langowski and A. Long, *Adv. Drug Deliv. Rev.*, **54**, 407–415 (2002).

518. S. Ekins and J. Rose, *J. Mol. Graph. Model.*, **20**, 305–309 (2002).

519. T. I. Oprea, I. Zamora, and A. L. Ungell, *J. Comb. Chem.*, **4**, 258–266 (2002).

520. H. E. Selick, A. P. Beresford, and M. H. Tarbit, *Drug Discov. Today*, **7**, 109–116 (2002).

521. A. P. Li and M. Segall, *Drug Discov. Today*, **7**, 25–27 (2002).

522. A. Kulkarni, Y. Han, and A. J. Hopfinger, *J. Chem. Inf. Comput. Sci.*, **42**, 331–342 (2002).

523. R. D. Brown, M. Hassan, and M. Waldman, *J. Mol. Graph. Model.*, **18**, 427–437, 537 (2000).

524. O. Roche, P. Schneider, J. Zuegge, W. Guba, M. Kansy, A. Alanine, K. Bleicher, F. Danel, E. M. Gutknecht, M. Rogers-Evans, W. Neidhart, H. Stalder, M. Dillon, E. Sjogren, N. Fotouhi, P. Gillespie, R. Goodnow, W. Harris, P. Jones, M. Taniguchi, S. Tsujii, W. von der Saal, G. Zim-

mermann, and G. Schneider, *J. Med. Chem.*, **45**, 137–142 (2002).

525. O. Llorens, J. J. Perez, and H. O. Villar, *J. Med. Chem.*, **44**, 2793–2804 (2001).

526. A. Cheng, D. J. Diller, S. L. Dixon, W. J. Egan, G. Lauri, and K. M. Merz, Jr., *J. Comput. Chem.*, **23**, 172–183 (2002).

527. T. I. Oprea, *J. Comput.-Aided Mol. Des.*, **14**, 251–264 (2000).

528. T. Olsson and T. I. Oprea, *Curr. Opin. Drug Discov. Dev.*, **4**, 308–313 (2001).

529. D. Gorse and R. Lahana, *Curr. Opin. Chem. Biol.*, **4**, 287–294 (2000).

530. M. J. Valler and D. Green, *Drug Discov. Today*, **5**, 286–293 (2000).

531. Y. C. Martin, *Farmaco*, **56**, 137–139 (2001).

532. J. Xu and J. Stevenson, *J. Chem. Inf. Comput. Sci.*, **40**, 1177–1187 (2000).

533. J. S. Mason and B. R. Beno, *J. Mol. Graph. Model.*, **18**, 438–451, 538 (2000).

534. T. I. Oprea and J. Gottfries, *J. Mol. Graph. Model.*, **17**, 261–274, 329 (1999).

535. A. K. Mandagere, T. N. Thompson, and K. K. Hwang, *J. Med. Chem.*, **45**, 304–311 (2002).

CHAPTER FOUR

Drug-Target Binding Forces: Advances in Force Field Approaches

PETER A. KOLLMAN
University of California
School of Pharmacy
Department of Pharmaceutical Chemistry
San Francisco, California

DAVID A. CASE
The Scripps Research Institute
Department of Molecular Biology
La Jolla, California

Burger's Medicinal Chemistry and Drug Discovery
Sixth Edition, Volume 1: Drug Discovery
Edited by Donald J. Abraham
ISBN 0-471-27090-3   © 2003 John Wiley & Sons, Inc.

Contents

1   INTRODUCTION

This chapter describes the forces that hold together complexes between large and small molecules, particularly where the large molecule is a protein or nucleic acid and the small molecule is an inhibitor or substrate. Forces between atoms are conventionally divided into the two categories of covalent and noncovalent "bonds." A covalent bond is an attractive interaction between two atoms in which each contributes a valence electron. For example, such a bond is formed between two hydrogen atoms to make the H_2 molecule: H + H → H-H. It also includes what most chemists might consider "ionic" bonds such as Na + Cl → Na-Cl, even though the valence electron pair in this case is much closer to the chlorine atom than to the sodium atom. The conventional study of chemical reactions is devoted to describing the strengths of covalent bonds and to understanding the ways in which they are formed and broken (1).

Drug–receptor interactions, on the other hand, are generally influenced most by weaker, noncovalent "bonds," where electron pairs are "conserved" in reactants and products. Examples of such interactions are "dative bonds," e.g., H_3N: + BH_3 → H_3N:BH_3 and hydrogen bonds, e.g., H_2O + H_2O → H_2O·· ·HOH. It is these noncovalent bonds that provide the "force" to make drugs interact strongly with their targets.

Some sample potential energy curves for covalent and noncovalent interactions between two atoms are given in Fig. 4.1. The left side shows an interaction curve for the two oxygen atoms in the O_2 molecule. This has a large dissociation energy D_o (about 117 kcal/mol in this case), so that at room temperature where RT approximates 0.6 kcal/mol (R is the universal gas constant and T is the absolute temperature), the fraction of "broken" bonds at equilibrium $e^{-D_o/RT}$ is very small. By contrast, noncovalent bonds are much weaker, typically 1-10 kcal/mol, and thus much easier to break. The right side of Fig. 4.1 shows interaction curves for the two sodium atoms in

the Na_2 dimer; this interaction is somewhat stronger than a typical hydrogen bond but has about the same shape. Also shown is the purely nonbonded interaction between two oxygen atoms in different water molecules. Here the D_o value is so small (about 0.15 kcal/mol) that it really cannot be seen on the scale of this figure. Hence, a significant fraction of nonbonded interactions can be broken at equilibrium at room temperature. It is this weakness of noncovalent bonds that makes them so useful in biological processes, because a small change in the chemical environment (such as temperature, concentrations, or ionic strength) can form or break such a bond. Probably the best known important noncovalent bonds are those between the strands of DNA, where hydrogen bonds hold the double helix together. When the cells begin to replicate, chemical signals (e.g., proteins binding to the DNA) shift the equilibrium to the single-stranded DNA, breaking these hydrogen bonds. Other important examples of noncovalent complexes include those between enzyme and substrate, "receptor" protein and hormone, antibody and antigen, and intercalator and DNA.

Much of our concern in this chapter is with the interaction:

$$\text{drug} + \text{receptor} \underset{k_r}{\overset{k_f}{\rightleftarrows}} \text{complex}$$

The rate constant for association of the complex is k_f; the rate constant for dissociation of the complex is k_r; and the affinity, or association constant $K_{as} = k_f/k_r$. It is usually assumed that the biological activity of a drug is related to its affinity K_{as} for the receptor, although there are processes such as actinomycin D-DNA interactions in which the rate of dissociation k_r is more relevant to the biological activity (2, 3).

The thermodynamic parameters of interest for the reactions above are the standard free energy ($\Delta G°$), enthalpy ($\Delta H°$), and entropy

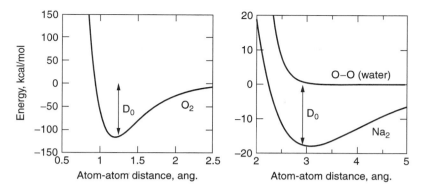

Figure 4.1. Potential energy curves for atom–atom interactions in O_2, Na_2, and the O—O interaction in a water dimer. Note the different energy scales on the left and right.

($\Delta S°$) of association. These are related by the equations

$$\Delta G° = -RT \ln K_{as}$$

$$\Delta G° = \Delta H° - T\Delta S°$$

This measurement of K_{as} allows one to calculate $\Delta G°$, the free energy of association of the complex. To find $\Delta H°$ and $\Delta S°$ separately requires a determination of K_{as} as a function of temperature (if $\Delta H°$ and $\Delta S°$ are relatively temperature independent, a plot of $\ln K_{as}$ vs. $1/T$ can yield $\Delta H°$ and $\Delta S°$) or a calorimetric measurement of $\Delta H°$ directly. Because $\Delta H°$ and $\Delta S°$ themselves are often quite temperature dependent, the latter experiment is more definitive.

This chapter provides some background about the forces that hold molecules together, with emphasis on the noncovalent interactions of interest in biology, and attempts to relate experimental determinations of the thermodynamics of association to the forces involved in the association. The discussion in the remainder of the chapter is divided into two parts. First, we discuss the forces that hold molecules together in the gas phase and solution and describe how these forces can be mathematically modeled by fairly simple functions; second, we discuss biological examples of noncovalent interactions and analyze the binding forces in particular cases.

2 ENERGY COMPONENTS FOR INTERMOLECULAR NONCOVALENT INTERACTIONS

Quantum mechanical calculations on small molecule association suggest that there are five major contributions to the energy of intermolecular interactions in the gas phase (3, 4). The sum of these is the dissociation energy of the intramolecular complex represented in Fig. 4.1. Table 4.1 contains some examples of magnitudes of the different energy components for different interactions. This section provides a qualitative introduction to these forces. Section gives and overview of mathematical models suitable for computer calculations.

2.1 Electrostatic Energy

Given information on the charge distribution of two molecules A and B, we can evaluate the electrostatic interaction energy between them. Although nuclei can be treated as point positive charges, the negative charge of electrons is smeared out over space. Thus, a rigorous evaluation of the electrostatic energy involves an integration over the electron clouds of the two molecules. In most practical calculations, however, the electrons as well as the nuclei are represented by point charges, whose position and magnitude are usually chosen to reproduce known molecular properties. The strength and the directionality of A. . .B electrostatic interactions are usually

Table 4.1   Some Examples of Interaction Energies of Noncovalent Complexes (kcal/mol)

Interaction	Interaction Energies					
	$-\Delta E \equiv D_o$	ΔE_{es}	ΔE_{dis}	ΔE_{ex}	ΔE_{pol}	ΔE_{ct}
He . . . He	0.02^a	0	-0.028	$+0.008$	0	0
Xe . . . Xe	0.64^a	0	-0.86	0.40	0	0
C_6H_6 . . . C_6H_6	$\approx 2^b$	$\neq 0$	$\neq 0$	$\neq 0$	$\neq 0$	$\neq 0$
H_2O . . . H_2O	8.9^c	-9.2	(-1)	4.0	-0.5	-2.2
TCNE . . . OH_2	4.1^d	-3.9	(-1)	2.0	-0.2	-1.0
Li^+ . . . OH_2	48.9^e	-51.1	(-1)	12.7	-7.8	-1.7
F^- . . . OH_2	41.1^e	-37.8	(-1)	20.5	-4.9	-17.9
NH_4^+ . . . F^-	164.7	-181.4	(-1)	61.6	-8.3	-35.6

$-\Delta E$, calculated (or experimental) total interaction energy equal to D_o in Fig. 1, kcal/mol; ΔE_{es}, electrostatic energy; ΔE_{dis}, dispersion energy; ΔE_{ex}, exchange repulsion energy; ΔE_{pol}, polarization energy; ΔE_{ct}, charge transfer energy (values in parentheses are estimated; TCNE, tetracyanoethylene).

[a]See Karplus and Porter (12).
[b]See Janda et al. (13).
[c]See Umeyama and Morokuma (7); this value for ΔE is certainly too large; see better values in Table 3.
[d]See Morokuma et al. (13).
[e]See Kollman (14).

dominated by the first nonvanishing multipole moment $\mathbf{M}_n$ of the charge distribution,

$$\mathbf{M}_n = \sum_{i=1}^{\text{no. charges}} q_i \mathbf{r}_i^n$$

where q_i are the individual charges and $\mathbf{r}_i$ is the vector from the origin of the coordinate system to the ith charge (5, 6). Molecules that are charged have a nonzero zeroth moment $\mathbf{M}_0$. Ionic crystals such as Na^+Cl^- are held together predominantly by electrostatic attraction between oppositely charged ions. Crystals of ice I are mainly held together by *dipolar* electrostatic forces where $\mathbf{M}_0 = 0$ and $\mathbf{M}_1 \neq 0$, because there are virtually no ions in these crystals. It should be noted here that "hydrogen bonding" is not a separate energy component; typically hydrogen bonds contain important energy contributions from all five energy components, although the electrostatic component is usually the largest contributor to this interaction (7).

Of the intermolecular energy components, the electrostatic is the longest range (i.e., it dies off most slowly with distance as the two molecules separate). Ion–ion interactions die off as $1/R$; ion–dipole as $1/R^2$; dipole–dipole as $1/R^3$, etc. In general, if two molecules have as their first nonvanishing multipole moments $\mathbf{M}_n$ and $\mathbf{M}_m$, the electrostatic interaction en-

ergy between them dies off as $1/R^{n+m+1}$. The electrostatic interaction energy between water a dipolar molecule ($n = 1$) and benzene, whose first nonvanishing moment is a quadrupole ($m = 2$), dies off as $1/R^4$.

2.2   Exchange Repulsion Energy

The Pauli principle keeps electrons with the same spin spatially apart. This principle applies whether one is dealing with electrons on the same molecule or on different molecules and is the predominant repulsive force (6) that keeps electrons of different molecules from interpenetrating when noncovalent complexes are formed. This repulsive term is often represented by an analytical function of the form

$$\frac{A}{R^n} \quad n = 9 \text{ or } 12$$

where R is the distance between molecules or nonbonded atoms and A is a constant that depends on the atom types. However, the best available quantum mechanical calculations suggest that this repulsion should diminish with an exponential dependence on the distance between the atoms (6). This difference is only important for very precise calculations: the key point is that the repulsive energy rises very quickly once the electrons from two different atoms overlap significantly. Roughly speaking, this happens with the distance be-

Table 4.2 Selected Atomic van der Waals Radii (in Å)

Element	r_{VDW}
Hydrogen	1.20
Carbon	1.70
Nitrogen	1.55
Oxygen	1.50
Fluorine	1.50
Phosphorus	1.85
Sulfur	1.80
Chlorine	1.70
Bromine	1.80

Values from A. Bondi, *J. Phys. Chem.* **68**, 441 (1964).

tween two atoms is less than the sum of their van der Waals radii. Table 4.2 gives some typical radii for atoms commonly found in organic molecules.

2.3 Polarization Energy

When two molecules approach each other, there is charge redistribution within each molecule, leading to an additional attraction between the molecules. The energy associated with this charge redistribution is invariably attractive and is called the polarization energy. For example, if a molecule with polarizability α is placed in an electric field, $\mathbf{E}$, the polarization energy is

$$E_{pol} = -\frac{1}{2}\,\alpha\mathbf{E}^2$$

If the electric field is caused by an ion, then $\mathbf{E} = q\mathbf{i}/R^2$, where q is the ionic change, $\mathbf{i}$ is the unit vector along the ion–molecule direction, and R the ion–molecule distance, which is the $E_{pol} = -1/2\alpha q^2/R^4$ for this *ion-induced dipole* interaction. The corresponding formula for *dipole-induced dipole* interaction between two dipolar molecules is

$$E_{pol} = -\frac{1}{2}\,\frac{\alpha_1\mu_2^2 + \alpha_2\mu_1^2}{R^6}$$

where the μ's are the dipole moments of the molecules, the α's are their polarizabilities, and R is the distance between molecules. The polarizability of a molecule can be broken down into atomic contributions [atomic polar-

izabilities are additive to a good approximation (8)], and it is roughly proportional to the number of valence electrons, as well as on how tightly these valence electrons are bound to the nuclei. Umeyama and Morokuma (9) have calculated the ion-induced dipole contribution to the proton affinities of the simple alkyl amines. They attributed the order of *gas phase* proton affinities in the alkyl amines [$NH_3 < CH_3NH_2 < (CH_3)_2NH < (CH_3)_3N$] to the greater polarizability of a methyl group than a hydrogen. A simple estimate using the above empirical equation for an ion-induced dipole interaction with $q = +1$, which is the difference in polarizabilities of a methyl and a hydrogen ($\Delta\alpha$) ≈ 4 cm^3, a proton–methyl distance of 2.0 Å, and a proton–proton distance of 1.6 Å, leads to an expected increase of ~ 20 kcal/mol of proton affinity for every methyl group added to NH_3. This very qualitative estimate is of the right magnitude but about two to three times too large (see below).

2.4 Charge Transfer Energy

When two molecules interact, there is often a small amount of electron flow from one to the other. For example, in the equilibrium geometry of the linear water dimer HO—H. . .OH$_2$, the water molecule that is the proton acceptor has transferred about 0.05e$^-$ to the proton donor water (9, 10). The attractive energy associated with this charge transfer is the charge transfer energy and can be thought of as a mixing of an ionic resonance structure H—O$^{(-)}$. . .H——OH$_2$$^{(+)}$ into the overall wave function. Although the charge transfer energy is an important contributor to the interaction energy of most noncovalent complexes, the presence of a "charge transfer" electronic transition in the visible spectrum does not mean that the charge transfer energy is the predominant force holding the complex together in its ground state. For example, the complex between benzene and I$_2$, earlier thought to be a prototype "charge transfer" complex, seems to be held together predominantly by electrostatic, polarization, and dispersion energies in its ground electronic state (11).

2.5 Dispersion Attraction

There are attractive forces existing between all pairs of atoms, even between rare gas atoms (He, Ar, Ne, Kr, Xe), which cause them to condense at a sufficiently low temperature. None of the other attractive forces (electrostatic, polarization, charge transfer) can explain the attraction between rare gas atoms; it is called the dispersion attraction (12). Even though the rare gas atoms have no permanent dipole moments, they are polarizable, and one has instantaneous dipole–dipole attractions in which the presence of a locally asymmetric charge distribution on one molecule induces an asymmetric charge distribution on the other molecule, e.g., $^{\delta-}He^{\delta+}\ldots^{\delta-}He^{\delta+}$.

The net attraction is called dispersion attraction (often known as London or van der Waals attraction) and is dependent on the polarizability and the number of valence electrons of the interacting molecules. It dies off as $1/R^6$, where R is the atom–atom separation. The difference between this attraction and the polarization energy is that the latter involves the interaction of a molecule that is already polar with another polar or nonpolar molecule.

2.6 Summary

Having described the components of the interaction energies, let us consider a number of specific examples in detail (Table 4.1). Unlike the total interaction energy, which can be measured experimentally, the individual energy components cannot. The theoretical estimate of these quantities is often dependent on the method of calculation, but their qualitative features are usually independent of methodology.

Rare gas–rare gas interactions (He. . .He and Xe. . .Xe) have only *dispersion* attraction. The difference between the potential well depth of He. . .He and Xe. . .Xe (Fig. 4.1; D_o) at the equilibrium distance is caused by the greater polarizability of the xenon atoms, and thus to the greater dispersion attraction between them. A simple manifestation of this is the much higher boiling point of xenon than helium, caused by the greater attractive forces in xenon liquid. Although these energies are individually fairly small, they can add in a mo-

lecular environment to significant energies; for example, the single largest attractive free energy contribution to binding in the strongest known small molecule–macromolecule interaction (biotin-avidin) is the dispersion attraction (13).

One might intuitively expect that benzene dimer would pack together like two flat plates, but this is not the case in the gas phase (14); the crystal structure also does not have parallel alignments of benzene molecules (15). Benzene, although having no dipole moment, does have a *quadrupole* moment ($\mathbf{M}_2 \neq 0$). A simple way to think about this quadrupole moment is to realize that a benzene C—H is somewhat electropositive and its electron cloud is rather electronegative. A second benzene molecule would like to approach the first one so that its "electropositive" side approaches the other molecule's "electronegative side." Hence the main component of binding is expected to be electrostatic in nature. The water dimer $(H_2O)_2$ and the ether. . .TCNE interactions are examples of prototypal H bonds and "charge transfer" complexes, but both are also held together mainly by electrostatic forces, although the other attractive energy components contribute significantly to the total ΔE. The electrostatic component is predominant in determining all the structural parameters except the distance between molecules. Similarly, the geometry and net attraction between Li^+ and OH_2, F^- and H_2O, and NH_4^+ and F^- are dominated by the electrostatic energy component.

3 MOLECULAR MECHANICS FORCE FIELDS

We move now from qualitative considerations to a more quantitative approach. It has become clear that a simple molecular mechanical energy expression can represent noncovalent interactions surprisingly well (16). Such energy expressions contain only the first three terms mentioned above: electrostatic, exchange repulsion, and dispersion. By a suitable choice of parameters, change transfer and polarization effects are implicitly included in such an expression, which is simple and easy to evaluate, along with its derivatives, for molecules with thousands of atoms. Over the past quarter century, many interesting applications of such molecular mechanical methods to complex molecules have been carried out (17).

The ideas that are outlined in a qualitative way above can also be cast into a useful mathematical form for computer calculation. The basic idea is to write down a (fairly simple and approximate) function that gives the energy of the system as a function of the positions (or coordinates) of its atoms. Because the derivative (or gradient) of this function yields the forces for Newton's equations, such a function is often called a "force field"; and because molecules are viewed as being made up of balls and springs (so that quantum effects are ignored), the term "molecular mechanics" is used to represent a concrete, mechanical picture of molecular motions and energies.

3.1 Biochemical Force Fields

Equation 4.1 represents about the simplest functional form of a force field that preserves the essential nature of molecules in condensed phases.

On the other hand, biochemists, guided by an interest in proteins and nucleic acids, have more generally followed a "bottom up" approach (16, 19, 20). This approach focuses first on the atomic charges q_i. The most general method to derive the atomic charges is to fit them to quantum mechanically calculated electrostatic potentials on appropriately chosen molecules or fragments. In early attempt to do this, computational limitations in quantum mechanical calculations led to the use of a minimal basis set STO-3G to derive the q_i (16). More recent efforts have used a 6-31G* or larger basis set (19). The 6-31G* basis set has the fortunate property in that it leads to charges (dipole moments) that are enhanced over accurate gas phase experimental values, and thus, implicitly builds in "polarization" effects characteristic of polar molecules in condensed phases. The fact that this basis set enhances the polarity just about the same amount as the popular water models TIP3P (21) and SPC (22), (where the charges are empirically adjusted to reproduce the water enthalpy of vaporization) is a fortunate fact and

$$
\begin{aligned}
U(\mathbf{R}) \quad =& \sum_{\text{bonds}} K_r (r - r_{\text{eq}})^2 && \text{bond} \\
+& \sum_{\text{angles}} K_\theta (\theta - \theta_{\text{eq}})^2 && \text{angle} \\
+& \sum_{\text{dihedrals}} \frac{V_n}{2} \left(1 + \cos[n\phi - \gamma]\right) && \text{dihedral} \\
+& \sum_{i<j}^{\text{atoms}} \frac{A_{ij}}{R_{ij}^{12}} - \frac{B_{ij}}{R_{ij}^6} && \text{van der Waals} \\
+& \sum_{i<j}^{\text{atoms}} \frac{q_i q_j}{\varepsilon R_{ij}} && \text{electrostatic}
\end{aligned}
$$

(4.1)

The earliest force fields, which attempted to describe the structure and strain of small organic molecules, focused considerable attention on more elaborate functions of the first two terms, as well as cross terms (18), representing a "top down" philosophy.

is key in leading to balanced solvent–solvent and solvent–solute interactions.

van der Waals parameters are generally dominated by the inner closed shell of electrons and thus are fortunately far more transferable than atomic charges. Therefore, generally only one set of van der Waals parameters (radius and well depth) per atom type need be employed, with the important exception of hydrogen (23). Unfortunately, it is harder to derive van der Waals parameters than charges using *ab initio* quantum mechanics (6, 24). The alternative that has emerged as a general model is to empirically calibrate results to fit experimental liquid structures and enthalpies (25).

Continuing with the "bottom up" development of a force field, we come to the torsion energy term, where the V_n and γ either come from experiment or quantum mechanical calculations on small molecule models. Whereas "top down" force fields often use many terms in the Fourier series for rotation around a given bond type and attempt to reproduce the conformational energy for a collection of molecules, most "biochemical" force fields take a minimalist approach (16, 19, 20). For example, we would have only a single V_3 torsional term around an X-C-C-Y bond except when X or Y are electronegative, where another term can be rationalized from electronic effects and can be derived directly using quantum mechanical calculations. This helps our model to be more easily generalized to new molecules, albeit in some cases probably at the cost of some accuracy. Exceptions to this minimalist approach are the ψ, ϕ of peptides and χ of nucleic acids, where more terms were added to ensure as accurate as possible a reproduction of the conformational energies around these key bonds.

Finally, to ensure reasonable representation of bond and angle terms, we use empirical data (structures and vibrational frequencies). The use of this simple harmonic model precludes high accuracy, but in our opinion, one would compromise the simplicity and generality of the model with more complex functional forms.

3.2 Force Field Models for Simple Liquids

A key test of this approach is the ability to accurately reproduce liquid structures and energies and free energies of solvation; these have traditionally been considered as key elements in the development of successful force fields for liquids (25). The aqueous solvation free energies of a large number of molecules, including substituted benzenes, methanol, hydrocarbons, N-methyl acetamide, and dimethyl sulfide, as well as the liquid structure and energy of methanol and N-methyl acetamide, show good agreement with experiment, with little or no adjustment of parameters. For example, Fox and Kollman (25) have shown that this approach leads to a density and enthalpy of vaporization of liquid dimethyl sulfoxide (DMSO) within 2% of experiment, using restrained electrostatic potential charges (RESP) and van der Waals parameters taken without modification from the corresponding values in proteins. Similar results have been obtained for other organic liquids.

3.3 Nonadditive and More Complex Models

What are the most important weaknesses in the above-described parameterizational approach and the use of Equation (4.1)? In our opinion, the main ones are the use of an effective two-body potential and the use of only atom-centered charges.

$$E_{pol} = -\frac{1}{2} \sum_i^{\text{atom}} \mu_i \cdot \mathbf{E}_i^{(o)} \quad \text{polarization} \quad (4.2)$$

where μ_i is atomic polarizability. Substantial progress has been made in laying the foundation for the development of a complete force field including explicit nonadditive effects (adding Equation 4.2 to Equation 4.1). First, we have shown that such models, in contrast to additive models, lead to good agreement with experimental solvation free energies of representative organic ions $CH_3NH_3^+$ and $CH_3CO_2^-$ without any adjustment of van der Waals parameters (26). Second, we have shown that such nonadditive terms are essential in accurately describing cation–π interactions (27). Third, we have shown that one can equally well describe liquid CH_3OH and N-methyl-acetamide (NMA) with additive models or a nonadditive model in which the charges are uniformly reduced (by 0.88) (28). Finally, the interaction free energy of Li^+ with hexaanisole spherand is more accurately described by nonadditive than additive molecular mechanical models (29). In addition, considering off-center charges in electrostatic potential fit models of atoms with "lone pairs" shows that they can often be important in leading to very accurate description of H bond directionality (30).

3.4 Long Range Electrostatic Effects

To accurately describe the energy and structure of complex systems, not only are the functional form and parameters of molecular models described by Equations 4.1 and 4.2 important, but also the manner in which the long range electrostatic effects are represented. The standard approach is to use a nonbonded cutoff for both electrostatic and van der Waals interactions, which seems to be a reasonable method for proteins but seems to be a poor method to describe highly charged molecules such as nucleic acids. For periodic systems, Ewald methods (which are too complex to be described here) have been known for a long time to remove most of the artefacts arising from cutoffs, and impressive efficiency and accuracy of a variant called particle-mesh Ewald (PME) has been demonstrated for protein crystals (31) [0.3 Å rms deviation from the observed crystal structure for bovine pancreatic trypsin inhibitor (BPTI) in a 1-ns simulation with an increase in computer time of only ~50% over standard cutoff methods]; the PME method also leads to accurate simulations of proteins, DNA, and RNA in solution (32).

4 THERMODYNAMICS OF ASSOCIATION

We have focused mainly on the energy of association between molecules; in any drug- receptor interaction, we typically want to know the equilibrium constant for association K_{as} and the free energy of association $\Delta G°$. The difference between the free energy ($\Delta G°$) and energy ($\Delta E°$) of association is given by $\Delta G° = \Delta H° - T\Delta S°$, and $\Delta H° = \Delta E° + (\Delta PV)$. For gas phase associations, (ΔPV) is $\sim -RT$, which is -0.6 kcal/mol at room temperature. Thus, this term, when added to ΔE, favors association (the more negative ΔG, the greater tendency for association). However, ΔS, the entropy of association, is typically large and negative. The reason is that one is reducing the "floppy" degrees of freedom, which have large translational and rotational entropies,

by six (six translations and six rotations in the free molecules, three of each in the complex) during complex formation, and replacing these with vibrations, which have lower entropies (33).

4.1 Gas Phase Association

For example, at 300 K, two CH_4 molecules have a translational entropy of 69 eu (entropy unit, or cal/K) and a rotational entropy of 31 eu, whereas $(CH_4)_2$ has a translational entropy of 37 eu and a rotational entropy (assuming a C. . .C distance of 4 Å) of 22 eu. Thus, one can see that the translational and rotational entropy contributions to the reaction $2CH_4 \rightarrow (CH_4)_2$ is -41 eu. These six degrees of freedom become vibrations in the complex $(CH_4)_2$, and as such, might contribute a vibrational entropy of about 20–30 eu. Thus, for the dimerization of CH_4 in the gas phase, we expect $T\Delta S°$ of about -3 to -6 kcal/mol at 300 K.

As stressed in the second law of thermodynamics, the tendency for a chemical process to occur is governed both by the energy released (exothermicity) in the process and the entropy gained (the tendency of the reaction to go to a more random, disordered state). In the case of gas phase association, the energy term is invariably exothermic if the reactants approach each other in an appropriate orientation, and the entropy term is always negative, opposing association. Table 4.3 gives an example of the thermodynamics of association of water molecules in the gas phase. As one can see, the entropy ($\Delta S°$) contribution to association of water molecules in the gas phase is substantial and negative; thus, there is little tendency for water molecules to associate in the gas phase at room temperature and 1 atm pressure, even though the hydrogen bond energy is about 5 kcal/mol.

4.2 Solvation Effects

The thermodynamic cycle (Fig. 4.2) illustrates the problems we face in transferring our knowledge of gas phase intermolecular interactions to solution phase phenomena.

Our real interest is in ΔG_4, the solution phase free energy of association. Until now, our discussion has focused on the energy (ΔE_1), enthalpy (ΔH_1), and free energy (ΔG_1)

Table 4.3 Thermodynamic Functions for Gas Phase Association of Water Molecules: $2H_2O \rightarrow (H_2O)_2$

Thermodynamic Function	Value for H_2O Dimerization (kcal/mol)
ΔE° (0 K)[a]	-6.2
ΔE° (300 K)[a]	-4.2
ΔH° (300 K)[a]	-5.2
ΔS° (300 K)[b]	-9.0
ΔG° (300 K)	$+3.8$

[a]See Joesten and Schaad (13).
[b]Estimated using the vibration frequencies employed by Joesten and Schaad (14).

of association in the gas phase. To be able to calculate ΔG_4, we need to know ΔG_3, the solvation free energy of the drug-receptor complex; ΔG_{2D}, the solvation free energy of the drug; and ΔG_{2R}, the solvation free energy of the receptor. These solvation free energies are the free energies gained (or lost) by taking the molecule from a standard concentration in the gas phase to a corresponding concentration in solution. Using the thermodynamic cycle in Fig. 4.2, it follows that

$$\Delta G_4 = \Delta G_1 - \Delta G_{2D} - \Delta G_{2R} + \Delta G_3 \quad (4.3)$$

Similar relationships hold for ΔH_4 and ΔS_4. There is no reason to expect ΔG_4 and ΔG_1 to be similar, so we face the problem of estimating ΔG_{2D}, ΔG_{2R}, and ΔG_3. We cannot measure ΔG_{2R} or ΔG_3, because this would require us to vaporize a measurable amount of a receptor or drug-receptor complex. For most polar and ionic drugs, ΔG_{2D} is not measurable either. Therefore, one resorts to measuring the free energy of transfer from octanol to water $\Delta G_{2D}(oct)$ rather than the free energy of transfer from the gas phase to water, ΔG_{2D}. This situation underlies the postulate of the Hansch approach (34), which suggests that the differences in $\Delta G_{2D}(oct)$ [$\Delta\Delta G_{2D}(oct)$] may be related to the biological activity of drugs, and in many cases this desolvation (water $\rightarrow$ octanol) does indeed seem to be related to drug binding and/or biological activity.

Because the individual free energies in Equation 4.3 are so hard to measure, one is led to smaller model systems to analyze the major driving force for drug- receptor association, a

step taken by Kauzmann (35) in his classic paper on the forces that affect protein stability and structure. He examined the thermodynamics of association and solution of small nonpolar molecules in aqueous solution. The associations were characterized by a large *positive* entropy term and the solution by a large negative *entropy*, with the enthalpy terms less important. Thus, the well-known lack of solubility of hydrocarbons in water was not caused by a net loss of hydrogen bonds; the hydrocarbons cause the water molecules to become more ordered (thus to lose entropy) so that they can still find a good hydrogen bond partner (ΔH of solution of these hydrocarbons is often negative, but much smaller in magnitude than the $T\Delta S$ of solution). By coming together in aqueous solution, these hydrocarbons "release" some H_2O's, and this favorable $T\Delta S$ association is the driving force for this association. It is generally agreed that this "hydrophobic" effect of hydrocarbon groups is a key feature in many drug-receptor associations. A lucid description of hydrophobic forces is given by Jencks (36) and Dill (37).

Computer simulation approaches have proven very useful in enabling calculation of the association of molecules. For example, the association of two methane molecules in the gas phase would lead to a ΔE° (0 K) of ~ -1 kcal/mol, and by analog with water dimer (Table 4.3), a very positive ΔG° (300 K) and thus no tendency for association. In aqueous solution, one can calculate, using modern statistical mechanical simulation methods, the potential of mean force for association of two molecules, which is the free energy as a function of molecular separation in solution. Although there is some controversy about

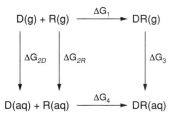

Figure 4.2. A schematic representation of the thermodynamic cycle for molecular association in the gas phase and in solution.

whether there are both "solvent-separated" and "contact" minima for two methane molecules in aqueous solution, there is no question that methane association is quite attractive in aqueous solution compared with the gas phase (38).

One can also apply such approaches to study association of ionic and polar molecules. For example, the association of Na^+ and Cl^- has a free energy of association that is very small in magnitude, in contrast to the gas phase (39). The association of two amides through a $C\!\!=\!\!O \ldots H\!\!-\!\!N$ hydrogen bond is very favorable *in vacuo* and progressively less favorable in non-polar and aqueous solution (40). Thus, water has a significant "leveling" effect on association, making nonpolar associations more favorable and ionic and polar associations less favorable than their gas phase counterparts.

Let us now summarize the foregoing discussion. Unlike the gas phase association, where ΔH_1 and ΔS_1 are invariably negative, for the corresponding thermodynamics in solution, ΔH_4 and ΔS_4 can be of either sign. The enthalpy of association ΔH_4 of two molecules in solution will be *positive* if the interactions of the solvent with the uncomplexed drug and receptor are sufficiently stronger and more exothermic ($\Delta H_3 - \Delta H_{2D} - \Delta H_{2R}$ is more positive than ΔH_1 is negative) than are the interactions of the solvent with the drug-receptor complex. Similarly, the entropy of association in solution ΔS_4 can be positive if $\Delta S_3 - \Delta S_{2D} - \Delta S_{2R}$ is more positive than ΔS_1 is negative. This can come about if the entropy gain from release of solvent from its interaction with the isolated drug and receptor is sufficiently larger than the entropy gain from release of solvent from the drug receptor complex.

An additional important point to keep in mind is that the solution phase thermodynamics may be dominated (as in the case of the hydrophobic effect, the association of nonpolar solutes in water) by changes in *solvent–solvent* interactions in the *presence* of solute.

It is also important to stress that even an analysis of the relative contributions of ΔH and ΔS to ΔG may not give definitive insight into the "nature" of the drug-receptor bond. For example, a large positive ΔS (and small negative ΔH) for association might come from

either a hydrophobic or an ionic association (35). In either case, the driving force for association is likely "release" of H_2O from "tight" binding to the solute.

One final consideration in determining either gas phase or solution phase association constants of drug-receptor complexes is conformational flexibility. Medicinal chemists have often attempted to synthesize *rigid* drug of different stereochemistries in the hopes of finding one that fits "perfectly" into the receptor site. If, for example, the drug has three equal energy conformations and only *one* can fit the receptor site, a price must be paid of $\Delta G = +RT \ln 3$ in binding free energy relative to the drug that is "locked" in the right conformation. If the receptor has to be locked in a conformation to "accept" the drug, one must pay a similar free energy price. A nice example of the latter situation is the difference in binding free energies between "locked" and "unlocked" macrocyclic crown ethers (41) that bind $t\text{-}BuNH_3^+$ cation.

4.3 An Illustrative Example: Protonation of Amines

Before we turn to some examples of drug-receptor interactions, let us present a specific example of the difference between gas phase and solution interactions. We choose the protonation of amines, because of the large literature that attempted to explain the irregular order of pKa's of the alkyl amines [NH_3 = 9.25; CH_3NH_2 = 10.66; $(CH_3)_2NH$ = 10.73; and $(CH_3)_3N$ = 9.81]. This reaction can be represented as

$$\Delta G_1$$
$$R_3N + H^+ \rightarrow R_3N^+H$$

in the gas phase and

$$\Delta G_4$$
$$R_3N(aq) + H^+(aq) \rightarrow R_3NH^+(aq)$$

in aqueous solution. As we noted in connection with Fig. 4.2, the difference between the free energies of protonation in solution and the gas phase is given by

Table 4.4   Free Energies in Cycle (Fig. 4.2) for Protonation of Alkyl Amines (kcal/mol)^a

R_3N	ΔG_1	$\Delta G_2(H^+)$	$\Delta G_2(R_3N)$	$\Delta G_3(R_3NH^+)$	ΔG_4
NH_3	-198.0	269.8	-2.41	-78.0	-3.79
CH_3NH_2	-210.0	269.8	-2.68	-67.7	-5.22
$(CH_3)_2NH$	-216.6	269.8	-2.41	-61.0	-5.39
$(CH_3)_3N$	-220.8	269.8	-1.34	-54.4	-4.06

^{a}See Aue et al. (42).

$$\Delta G_4 - \Delta G_1 = -\Delta G_2(R_3N) - \Delta G_2(H^+)$$
$$+ \Delta G_3(R_3NH^+)$$

Recall also that the solution $pK_a = -\log K_{as}$ $= \Delta G_4{}^0/2.3\,RT$. When the gas phase basicities were measured and showed a regular order, it was clear that the irregular order in solution was caused by a solvation effect. In the gas phase, NH_3 is a weaker base than $(CH_3)_3$ by about 23 kcal/mol; in solution this difference is only about 1 kcal/mol.

Table 4.4 lists the free energies appropriate to the thermodynamic cycle (Fig. 4.2) for the protonation of the amines. Two points deserve strong emphasis.

1. The magnitude of ΔG_4 is much smaller than that of ΔG_1 for protonation, because in aqueous solution, the amines must compete with H_2O for the proton; in the gas phase there is no competition.

2. As clearly analyzed by Aue et al. (42), the smaller the protonated amine, the more effectively solvated it is, and the better base it becomes compared with its relative rank in the gas phase.

5   CALCULATING FREE ENERGIES

Free energy is certainly one of the most important concepts in physical chemistry. The groundwork on calculating free energies was laid by Kirkwood (43) and Zwanzig (43), and the first key "modern" developments and applications came from the work of Postma et al. (44), Jorgensen and Ravimohan (45), Tembe and McCammon (46), and Warshel (47). The fundamentals of computational approaches to calculating free energies are reviewed by Beveridge and Mezei (48), and we attempted to exhaustively review applications up to 1993 (49).

To calculate the relative solvation free energies of molecules A and B in solvent S, we can use a thermodynamic cycle such as in Fig. 4.3. The relative solvation free energy of A and B, determined experimentally, is $\Delta\Delta G_{solv} = \Delta G_{solv}(B) - \Delta G_{solv}(A)$, and because the free energy is a state function, $\Delta\Delta G$ is also $= \Delta G_{mut}(S) - \Delta G_{mut}(g)$, which are the free energies determined by computational means by "mutating" the molecular mechanical model of A into B in solvent S and in the gas phase (g). Of course, if B consists of all "dummy" (non-interacting) atoms, this approach leads to the calculation of the absolute solvation free energies of A.

Being able to accurately calculate free energies of solvation suggests a reasonable balance in solute–solvent and solvent–solvent interactions. The next key challenge is to calculate $\Delta\Delta G_{bind}$ of guests G and G′ to a host H, all in aqueous (or other) solution.

A typical cycle for free energy calculations (45) where H is a host, G is a guest, and HG is the host–guest complex is given in Fig. 4.4.

Now one requires a correct balance of solute (host)–solute (guest), solute (host or guest) –solvent, and solvent–solvent interactions to correctly calculate $\Delta\Delta G_{bind}$, although there clearly can be compensating errors in the calculation of ΔG_{solv} and ΔG_{bind}.

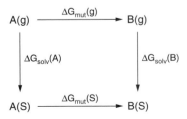

Figure 4.3. Basic thermodynamic cycle for solvation free energy.

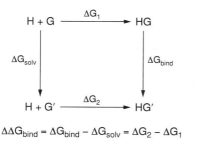

$$\Delta\Delta G_{bind} = \Delta G_{bind} - \Delta G_{solv} = \Delta G_2 - \Delta G_1$$

Figure 4.4. Thermodynamic cycle for host–guest interactions.

6 EXAMPLES OF DRUG-RECEPTOR INTERACTIONS

We discuss three examples of "drug target" interactions: (*1*) biotin-avidin (*2*) dihydrofolate reductase-trimethoprim, and (*3*) DNA-intercalator. The first is the strongest characterized protein–ligand association, the second a prototype enzyme–inhibitor interaction, and the third describes drugs interacting with nucleic acids.

6.1 Biotin-Avidin

Biotin (Fig. 4.5) is involved in the strongest known non-covalent macromolecule–ligand interaction. In fact, given the small size of biotin, it is surprising to many that this association is so strong ($Ka \sim 10^{15}$ corresponding to a $-\Delta G$ of ~ 20 kcal/mol) (50). The X-ray structure of streptavidin (a related protein to avidin with nearly as large a biotin affinity) biotin complex has been solved (51). The ureido group of biotin was thought to be the reason for the uniquely strong binding of this ligand to (strept)avidin.

We have carried out free energy calculations (13) on the relative binding of biotin, aminobiotin, and thiobiotin to streptavidin, as well as absolute free energy calculations of biotin binding. The results of these simulations are instructive in the insight they give us into this association. These free energy calculations can best be understood by considering the thermodynamic cycle in Fig 4.6. The free energy calculations enable one to determine the free energies of the vertical processes by mutating one ligand into another in solution (ΔG_{solv}) and when bound in the active site

(ΔG_{bind}), using molecular dynamics to create an ensemble average of the system. The difference between these calculated free energies $\Delta\Delta G_{bind}$ is equal to the difference in the observed relative free energies of ligand binding.

The biotin–streptavidin system provides a "textbook case" of the relative free energies of the binding of biotin, aminobiotin, and thiobiotin, as illustrated in Table 4.5. First, the calculated relative free energies are in reasonable agreement with experiment; thiobiotin is calculated and observed to bind $\sim 10^3$ or ~ 4 kcal/mol more weakly to streptavidin than biotin, and iminobiotin is calculated and observed to bind $\sim 10^5$ or ~ 7 kcal/mol more weakly than biotin. What is more interesting are the energy components. Thiobiotin is easier to desolvate than biotin by ~ 9 kcal/mol (ΔG_{solv}) but interacts more weakly with the protein by ~ 13 kcal/mol, leading to the observed ≈ 4 kcal/mol preference for biotin binding. On the other hand, iminobiotin is ~ 5 kcal/mol harder to desolvate than biotin, but interacts only ~ 2 kcal/mol more weakly with streptavidin, thus leading $\Delta\Delta G_{bind} = \Delta G_{bind} - \Delta G_{solv} = 2 - (-5)$ to

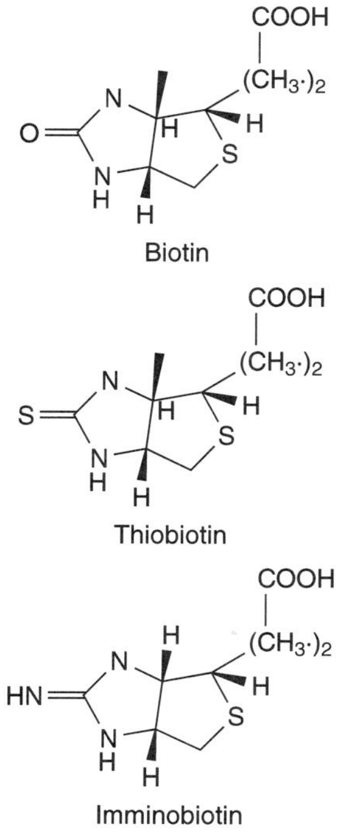

Figure 4.5. Structures of biotin and two analogs.

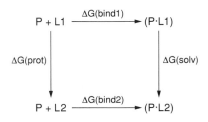

$$\Delta\Delta G_{bind} = \Delta G_{bind2} - \Delta G_{bind1} = \Delta G_{prot} - \Delta G_{solv}$$

Figure 4.6. Thermodynamic cycle for protein-ligand interactions. The experimentally measurable free energies are ΔG_{bind1} and ΔG_{bind1} (horizontal), and the calculated values (ΔG_{solv} and ΔG_{prot}) are the vertical processes.

its ≈ 7 kcal/mol weaker binding to streptavidin than biotin. The above examples illustrate the interesting tradeoff in binding and solvation effects in analysis of ligand–macromolecule interactions.

The fact that one loses only 4–7 kcal/mol out of the ~20 kcal/mol in free energy of binding when mutating the ureido group to its thio and imino analog is strongly suggestive that the "ureido resonance," suggested by the crystallographers (50) who solved the structure as the reason for the unusually high K_{as}, cannot be the main reason. Calculations on the absolute free energy of biotin–streptavidin binding suggest that electrostatic effects, which might include ureido resonance (although perhaps not all of it), contribute ~6 kcal/mol to $\Delta\Delta G_{bind}$, whereas van der Waals effects contribute ~14 kcal/mol.

The large contribution of van der Waals interactions (dispersion plus exchange repulsion) is surprising to many, because an individual van der Waals atom–atom dispersion attraction is very small. But there are many of

them in the streptavidin active site, which, not coincidentally, contains four tryptophan residues.

But why don't the van der Waals interactions with water lost when one moves biotin from water to the streptavidin active site cancel with those gained in the active site? This can be understood by noting, as Sun et al. (52) and Rao and Singh (53) have, that a unique aspect of water as a solvent is its large exchange repulsion contribution to ΔG_{solv}. This exchange repulsion contribution represents the "hydrophobic effect," the fact that methane is less stable by 2 kcal/mol at a 1 M standard state in water than in the gas phase. This exchange repulsion cancels (and sometimes outweighs) the dispersion attraction that occurs for any solute when transferred from the gas phase to a condensed phase. On the other hand, in the streptavidin binding site, preorganized during protein synthesis, one gains dispersion attraction when biotin binds without the compensation from exchange repulsion. The magnitude of this effect is heightened by the large "atom density" both in biotin, with its bicyclic structure and in streptavidin, with its four tryptophan residues. Thus, the key aspects in biotin's tight binding with (strept)avidin is the preorganization and high atom density of the protein active site (54).

Recently, Dixit and Chipot (55) have re-examined this problem, using the improved power of modern computers to expand the sampling of configurations. The results continue to be in good accord with experiments and offer a modern paradigm for how the convergence of these simulations can be monitored.

Table 4.5   Results of Relative Free Energy Calculations^a (kcal/mol)

| | | | $\Delta\Delta G_{bind}$ | |
| | | | Calc. | Exp^b |
Perturbation	ΔG_{solv}	ΔG_{prot}	$\Delta G_{prot} - \Delta G_{solv}$	$\Delta G_{bind2} - \Delta G_{bind1}$
Biotin → thiobiotin	8.8 ± 0.1	12.0 ± 0.3	3.2 ± 0.3	3.6
Biotin → iminobiotin	-5.3 ± 0.1	1.2 ± 0.7	6.5 ± 0.8	6.2

^{a}Errors, where listed, correspond to half the hysteresis between forward and reverse runs.
^{b}Experimental data, Ref. 51.

6.2 Dihydrofolate Reductase–Trimethoprim

A classic example of a drug that works by specificies-specific protein inhibition is trimethoprim (TMP). Because this drug binds to bacterial dihydrofolate reductase (DHFR) $\sim 10^4$ more tightly than to the mammalian enzyme, there is a therapeutic concentration in which the drug can be used as an antibacterial with little deleterious consequences for a mammalian host.

DHFR was the first example where one has solved the X-ray crystal structure of the enzyme protein complexes for both bacteria and mammalian enzymes. Matthews et al. (56) have suggested that it is a key hydrogen bond involving the pyrimidine ring of TMP, which is present in the bacterial but not mammalian enzyme complex, that is responsible for the selectivity. This has not been definitively established with carboxyclic analogs, but analogs have clearly shown an important role of the three methoxy groups in TMP in causing species selectivity. For example, the TMP analog without the three OCH_3 groups have a binding preference for the bacterial enzyme of only ~ 10.

Kuyper (57) has analyzed the structure of the bacterial and mammalian complexes and suggested that the oxygens of the —OCH_3 group plays a key role in species selectivity. The methoxy oxygens are significantly more solvent exposed in the bacterial complex that the mammalian. Thus, because these oxygens do not form hydrogen bonds to enzyme groups in either complex, the desolvation penalty for the oxygen is smaller in the bacterial enzyme and does not as extensively cancel the favorable hydrophobic/dispersion effects on binding of the methoxy methyl groups. This interpretation is supported by the fact that replacing the —OCH_3 with CH_2CH_3 makes the molecules less species selective; such analogs bind only a little better to bacterial DHFR but significantly better to mammalian DHFR (58, 59).

Free energy calculations/molecular dynamics have and will continue to give interesting insight into the DHFR–TMP species selectivity (39–41).

6.3 Nucleotide Intercalator

Because our first two examples have emphasized protein–small molecule interactions, we turn to a nucleic acid–small molecule interaction for our last example. There have been many experimental studies of the "intercalation" of flat, planar dyes into double-stranded DNA and other polynucleotides.

The flexibility of the sugar-phosphate backbone allows the intercalator to be sandwiched between the nucleotides with relatively little "strain." The interaction with polynucleotides by a wide variety of intercalators has been studied by physicochemical techniques. The driving force for association can be primarily hydrophobic, as in actinomycin D, where the driving force for association is $\Delta S°$ (57), or it can contain a large contribution from electrostatic effects as in ethidium bromide and adriamycin analogs, where the driving force for association is $\Delta H°$ (60) (Table 4.6). Both molecules have binding association constants K_{as} to DNA of about 10^6. The role of dispersion binding is not clear at this point, but it is likely to be very important as well (13). As noted above, the ability of these drugs to interfere with DNA replication is apparently related to their rate of dissociation k_r from DNA rather than to their association constant K_{as}. Muller and Crothers (2) showed that both actinomycin and actinomine had values of K_{as} similar to that of DNA, but the former had a much smaller k_r and a much greater effect on the rate of DNA replication.

7 SUMMARY

The foregoing examples illustrate the likely nature of drug-receptor binding. It seems that hydrophobic and dispersion binding do contribute a substantial amount to the net binding affinity. However we have noted some cases (e.g., the ureido group in biotin and the intercalation of

Table 4.6 Thermodynamics of Binding of Drugs to DNA

Drug[a]	$\Delta H°$ (kcal/mol)	$\Delta S°$ (eu)	$\Delta G°$ (kcal/mol)
Proflavin	−6.7	+4.7	−8.1
Ethidium bromide	−6.2	+9.4	−9.0
Actinomycin D	+2.0	+39.0	−9.6
Daunomycin	−6.5	+7.7	−8.8

[a]Conditions in all cases as follows: T = 25°, 0.01 M buffer, pH = 7, 1 = 0.015 [see Quadrifoglio and Crescenzi (60)].

positively charged groups into DNA) in which there might be an important polar or electrostatic driving force for binding. Again, it is difficult to ascertain whether these polar contributions come from "freeing up" water or from direct interactions, but they seem to contribute in a significant fashion to the driving force for association as well as being important in determining biological specificity. The lessons for the medicinal chemist attempting to design a drug to maximize the drug receptor association include the following:

1. Conformational flexibility can decrease the association constants in a straightforwardly predictable way.
2. Hydrophobic effects usually contribute significantly to drug-receptor association, but one must also consider possible specific polar and ionic interactions.
3. Preorganization of the receptor or ligand is a key to obtaining optimal electrostatic or van der Waals interactions.

We have tried to provide examples in this chapter both of the qualitative arguments that are important for understanding ligand–protein or ligand–DNA interactions and of some typical numerical results arising from computer experiments. Understanding these interactions is key to the rational design of inhibitors, and a computer-aided approach is increasingly being used to screen libraries of potential inhibitors and to suggest improvements to lead compounds (61). As force fields and sampling methods improve and as computers become ever-more powerful, the practical use of methods like these should improve as well.

AUTHOR'S NOTE:

Peter Kollman died unexpectedly in May, 2001. He had authored an article on "Drug-Target Binding Forces" for the Fifth Edition of this series. This revision and extension for the Sixth Edition is based primarily on Peter's writings, and is dedicated to his memory.

REFERENCES

1. P. Atkins, *Physical Chemistry*, 4th ed., W. H. Freeman, New York, 1990.

2. W. Muller and D. Crothers, *J. Mol. Biol.*, **35**, 251 (1968).

3. K. Kitaura and K. Morokuma, *Int. J. Quant. Chem.*, **10**, 325 (1976).

4. J. C. G. M. van Duijnevelt-van der Rijdt and F. B. van Duijneveldt, *J. Am. Chem. Soc.*, **93**, 5644 (1971).

5. J. Hirschfelder, C. Curtiss, and R. Bird, *Molecular Theory of Gases and Liquids*, Wiley, New York, 1954.

6. R. H. Margenau and N. Kestner, *Theory of Intermolecular Forces*, 2nd ed., Pergamon Press, Oxford, 1971.

7. H. Umeyama and K. Morokuma, *J. Am. Chem. Soc.*, **99**, 1316 (1977).

8. R. Lefevre, *Adv. Phys. Org. Chem.*, **3**, 1 (1965).

9. H. Umeyama and K. Morokuma, *J. Am. Chem. Soc.*, **98**, 4400 (1976).

10. P. Kollman and L. C. Allen, *Chem. Rev.*, **72**, 283 (1972).

11. M. Hanna, *J. Am. Chem. Soc.*, **90**, 285 (1968); R. Lefevre, D. V. Radford, and P. Stiles, *J. Chem. Soc. B*, **31**, 1297 (1968).

12. M. Karplus and R. Porter, *Atoms and Molecules*, Benjamin, Menlo Park, CA, 1971.

13. K. C. Janda, J. C. Hemminger, J. W. Winna, S. E. Novick, S. J. Harris, and W. Klemperer, *J. Chem. Phys.*, **63**, 1419 (1975); M. Joesten and L. Schaad, *Hydrogen Bonding*, Dekker, New York, 1974; K. Morokuma, S. Iwata, and W. Lathan in R. Daubel and B. Pullman, Eds., *The World of Quantum Chemistry*, D. Reidel, Dordrecht, Holland, 1974, p. 277.

14. P. Kollman. *J. Am. Chem. Soc.*, **99**, 4875 (1977).

15. G. E. Bacon, N. A. Curry, and S. A. Wilson, *Proc. R. Soc. Ser. A*, **279**, 98 (1964).

16. S. J. Weiner, P. A. Kollman, D. A. Case, U. C. Singh, C. Ghio, G. Alagona, S. Profeta, and P. Weiner, *J. Amer. Chem. Soc.*, **106**, 765 (1984).

17. A. McCammon and S. Harvey, *Molecular Dynamics of Proteins and Nucleic Acids*, Cambridge University Press, Cambridge, UK, 1987.

18. U. Bukert and N. L. Allinger, *Molecular Mechanic*, American Chemical Society, Washington, DC, 1982.

19. W. D. Cornell, P. Cieplak, C. I. Bayly, I. R. Gould, K. M. Merz Jr., D. M. Ferguson, D. C. Spellmeyer, T. Fox, J. W. Caldwell, and P. A. Kollman, *J. Am. Chem. Soc.*, **117**, 5179 (1995).

20. A. D. MacKerell Jr., D. Bashford, M. Bellott, R. L. Dunback Jr., J. D. Evanseck, M. J. Field, S. Fischer, J. Gao, H. Guo, S. Ha, D. Joseph-McCarthy, L. Kuchnir, K. Kuczera, F. T. K. Lau, C.

Mattos, S. Michnick, T. Ngo, D. T. Nguyen, B. Prodhom, W. E. Reiher III, B. Roux, M. Schlenkrich, J. C. Smith, R. Stote, J. Straub, M. Watanabe, J. Wirkiewicz-Kuczera, D. Yin, and M. Karplus. *J. Phys. Chem. B*, **102**, 3586 (1998).

21. W. L. Jorgensen, J. Chandrasekhar, J. Madura, R. W. Impey, and M. L. Klein, *J. Chem. Phys.*, **79**, 926 (1983).

22. H. J. C. Berendsen, J. R. Giegera, and T. Straatsma, *J. Phys. Chem.*, **91**, 6269 (1987).

23. D. L. Veenstra, D. M. Ferguson, and P. A. Kollman, *J. Comput. Chem.*, **8**, 971 (1992).

24. J. Pirssette and E. Kochanski, *J. Am. Chem. Soc.*, **100**, 6609 (1978).

25. W. L. Jorgensen and J. Tirado-Rives, *J. Am. Chem. Soc.*, **110**, 1657 (1988); W. L. Jorgensen, D. S. Maxwell, and J. Tirado-Rives, *J. Am. Chem. Soc.*, **118**, 11225 (1996); G. Kaminski and W. L. Jorgensen, *J. Phys. Chem.*, **100**, 18010 (1996); T. Fox and P. A. Kollman, *J. Phys. Chem. B*, **102**, 8070 (1998).

26. E. C. Meng, P. Cieplak, J. W. Caldwell, and P. A. Kollman, *J. Am. Chem. Soc.*, **116**, 12061 (1994).

27. J. W. Caldwell and P. A. Kollman, *J. Am. Chem. Soc.*, **117**, 4177 (1995).

28. J. W. Caldwell and P. A. Kollman, *J. Phys. Chem.*, **99**, 6208 (1995).

29. Y. Sun, J. W. Caldwell, and P. A. Kollman, *J. Phys. Chem.*, **99**, 10081 (1995).

30. R. W. Dixon and P. A. Kollman, *J. Comput. Chem.*, **18**, 1632 (1997).

31. D. M. York, A. Wlodawer, L. Petersen, and T. A. Darden, *Proc. Natl. Acad. Sci. USA*, **91**, 8715 (1994).

32. T. E. CheathamIII, J. L. Miller, T. Fox, T. A. Darden, and P. A. Kollman, *J. Am. Chem. Soc.*, **117**, 4193 (1995).

33. N. Davidson, *Statistical Mechanics*, McGraw-Hill, New York, 1962; M. I. Page and W. P. Jencks, *Proc. Natl. Acad. Sci. USA*, **68**, 1678 (1971).

34. C. Hansch, *Biological Correlations—The Hansch Approach*, ACS, Washington, DC,1973.

35. W. Kauzmann, *Adv. Protein Chem.*, **14**, 1 (1975); C. Tanford, *The Hydrophobic Effect*, Wiley, New York, 1973.

36. W. Jencks, *Catalysis in Chemistry and Enzymology*, McGraw-Hill, New York, 1969.

37. K. Dill, *Biochemistry*, **29**, 7133 (1990).

38. W. Jorgensen, J. K. Buckner, S. Boudon, and J. Tirado-Rives, *J. Chem. Phys.*, **89**, 3742 (1988).

39. L. X. Dang, J. Rice, and P. Kollman, *J. Chem. Phys.*, **93**, 7528 (1990).

40. W. Jorgensen, *J. Amer. Chem. Soc.*, **111**, 3770 (1989).

41. J. Timko, S. Moore, D. Walba, P. Hiberty, and D. Cram, *J. Am. Chem. Soc.*, **99**, 4207 (1977).

42. D. Aue, H. Webb, and M. Bowers, *J. Am. Chem. Soc.*, **31**, 318 (1976).

43. J. Kirkwood, *J. Chem. Phys.*, **3**, 300 (1935); R. Zwanzig, *J. Chem. Phys.*, **22**, 1420 (1954).

44. J. P. M. Postma, H. J. C. Berendsen, and J. R. Houk, *Faraday Symp. Chem. Soc.*, **17**, 55 (1982).

45. W. Jorgensen and C. Ravimohan, *J. Chem. Phys.*, **83**, 3050 (1985).

46. B. L. Tembe and J. A. McCammon, *J. Comput. Chem.*, **8**, 281 (1984).

47. A. Warshel, *J. Phys. Chem.*, **86**, 2218 (1982).

48. D. L. Beveridge and M. Mezei, *Annu. Rev. Biophys. Chem.*, **18**, 431 (1989).

49. P. A. Kollman, *Chem. Rev.*, **93**, 2395 (1993).

50. N. Green, *Biochem. J.*, **101**, 774 (1966).

51. P. C. Weber, J. J. Ohlendorf, and F. R. Salemne, *Science*, **243**, 85 (1989).

52. Y. Sun, D. Spellmeyer, D. Pearlman, and P. Kollman, *J. Amer. Chem. Soc.*, **114**, 6798 (1992).

53. B. C. Rao and U. C. Singh, *J. Amer. Chem. Soc.*, **111**, 3125 (1989); B. C. Rao and U. C. Singh, *J. Amer. Chem. Soc.*, **112**, 3803 (1990).

54. S. Miyamoto and P. Kollman, *Proc. Natl. Acad. Sci. USA*, 8402 (1993); S. Miyamoto and P. Kollman, *Proteins*, **16**, 226 (1993).

55. S. B. Dixit and C. Chipot, *J. Phys. Chem. A*, **105**, 9795 (2001); B. Kuhn and P. A. Kollman, *J. Am. Chem. Soc.*, **122**, 3909 (2000).

56. D. Matthews, J. Bolin, J. Burridge, D. Filman, K. Volz, B. Kaufman, C. Beddell, J. Champness, D. Stammers, and J. Kraut, *J. Biol. Chem.*, **260**, 381 (1985).

57. L. Kuyper in C. Bugg and S. Ealick, Eds., *Crystallographic and Molecular Modeling in Drug Design*, Springer-Verlag, NY, 1989, pp. 56-79.

58. S. Fleischman and C. L. Brooks, *Proteins*, **7**, 52 (1990); C. L. Brooks and S. Fleischman, *J. Amer. Chem. Soc.*, **112**, 3307 (1990).

59. J. J. McDonald and C. L. Brooks, *J. Amer. Chem. Soc.*, **113**, 2295 (1991); J. J. McDonald and C. L. Brooks, *J. Amer. Chem. Soc.*, **114**, 2062 (1992).

60. F. Quadrifoglio and V. Crescenzi, *Biophys. Chem.*, **1**, 319 (1974); F. Quadrifoglio and V. Crescenzi, *Biophys. Chem.*, **2**, 64 (1974).

61. T. J. A. Ewing, S. Makino, A. G. Skillman, and I. D. Kuntz, *J. Comput. Aided Mol. Des.*, **15**, 411 (2001).

CHAPTER FIVE

Combinatorial Library Design, Molecular Similarity, and Diversity Applications

JONATHAN S. MASON
Pfizer Global Research & Development
Sandwich, United Kingdom

STEPHEN D. PICKETT
GlaxoSmithKline Research
Stevenage, United Kingdom

Contents

Burger's Medicinal Chemistry and Drug Discovery
Sixth Edition, Volume 1: Drug Discovery
Edited by Donald J. Abraham
ISBN 0-471-27090-3 © 2003 John Wiley & Sons, Inc.

1 INTRODUCTION

1.1 Scope

This chapter discusses molecular similarity and diversity methods and their main applications to combinatorial library design, the selection of compound subsets, and ligand-based virtual screening. Protein structure-based virtual screening is discussed in chapters 6 and 7. Medicinal chemistry-relevant applications discussed include the design of "diverse," "representative," and "thematic/focused/biased" libraries and subsets. The last application is of particular relevance, in that there is a recent trend to approach drug discovery by a "target class" or "gene family" approach; for example, 7-transmembrane G-protein-coupled receptors (7-TM GPCRs); nuclear hormone receptors (NHRs); ion channels; proteases; kinases; phosphodiesterases.

1.2 Molecular Similarity/Diversity

Molecular similarity and diversity methods have been developed based on the principle that similar molecules exhibit similar activities/properties (1). Molecular similarity is a key concept in the identification of new molecules that have similar biological activity to one or more molecules of known activity. Molecular diversity concepts are used to explore "chemical space," with the scope of application ranging from a particular structure/reaction to a large database of different molecules. The process of evaluating similarity and diver-

sity involves the calculation of descriptors for each structure and the determination of the proximity of compounds within the descriptor (or chemical) space. Virtual screening is the name given to the process by which these computational methods are used to identify a subset of compounds from a database for a specific purpose. The source database may, for example, be compounds in a corporate registry where the goal may be to identify compounds for a biological assay. Alternatively, the source database may be compounds that the chemist believes are synthesizable and the goal of virtual screening is to prioritize compounds for synthesis. Depending on the amount of information available to guide the computational screening, and the method used, different levels of enrichment (number of actives selected in a set relative to a random selection) are obtained. It should be noted also that virtual screening applies not only to the selection of compounds for biological screening but also to the prioritization of compounds based on general properties of biological relevance, for example, selecting compounds more likely to be well absorbed.

Molecules are typically represented by a vector of real-valued properties (molecular weight, $\log P$, etc.) or binary values (e.g., 0 for absence, 1 for presence of a substructure feature) in a bit-string or binary fingerprint. The term *fingerprint* or *key* or *signature* thus refers to an encoding of features/characteristics a molecule exhibits (e.g., substructures present, all possible combinations of 2–4 pharmaco-

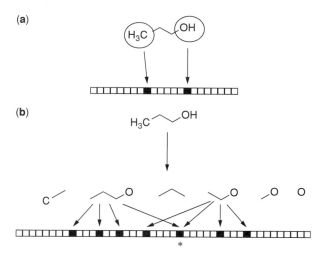

Figure 5.1. A simple illustration of bit-string encoding of chemical structure (7). (a) A fragment dictionary-based approach. (b) Illustration of a hashing scheme using a path-based decomposition of the structure. The asterisk denotes an element in the bit string where a collision has resulted from the hashing procedure.

phoric features) as a string of bits (indicating either the presence or absence of a particular characteristic; see section 2.1.1 and Fig. 5.1), optionally including a count of the number of times the characteristic is exhibited. A wide variety of descriptors is available to evaluate the potential similarity or diversity between structures (2). These range from one-dimensional (1D) descriptors based on molecular properties such as molecular weight, which can be derived from the molecular formula; two-dimensional (2D) substructural fingerprints, topological methods, and atomic/molecular properties [e.g., physicochemical properties such as calculated log P (c log P)] that require knowledge of the "flat" or 2D structure, which represents the bonds between the atoms; to three-dimensional (3D) properties (e.g., pharmacophoric fingerprints), requiring knowledge of the full 3D conformational space available to a molecule. A 3D pharmacophoric fingerprint marks the presence or absence of potential pharmacophores [combinations of different features and distances between them, often for three- or four-point pharmacophore fingerprints (i.e., triplets/triangles or quartets/tetrahedra)] within a molecule.

Three-dimensional properties such as the pharmacophore fingerprints can also be calculated for the target protein binding site, being derived from site points complementary to the functional groups in the protein backbone and side chains, thus bridging the ligand-based

and protein structure-based universes. The pharmacophore fingerprints also represent a simplified approach to the goal of providing molecular descriptors with 3D shape and property content, while obviating the need for molecular superposition or refined pharmacophore hypothesis generation.

Partitioning methods are widely used. The compounds are grouped using either a cell-based approach, in which each dimension of the chemical space is subdivided or "binned," or by a clustering approach, in which islands of similar compounds are formed. Alternatively, the distance between pairs of molecules can be calculated, and this distance minimized (for similarity) or maximized (for diversity). For diversity the goal is normally not to identify a diverse compound in isolation, but to explore a range of diversity through selection of a diverse subset of compounds. Cell-based methods provide the advantage of a common frame of reference in terms of the multidimensional cell positions. It is possible with a cell-based method to evaluate both what is there and what is missing (in terms of empty cells); clustering, by contrast, is based on exploring what is there. The same method/descriptor may thus be used to evaluate both similarity and "diversity." In practice, "dissimilarity" approaches often provide a more acceptable approach to diversity, ensuring that compounds are not too similar, but avoiding a potential pitfall of exploring too frequently the extremes of chemical space. Methods and descriptors are discussed for each of these categories.

1.3 Combinatorial Library Design

Combinatorial library design is an important application of molecular similarity and diversity principles and methods. Combinatorial chemistry approaches can exploit automation and robotics to enable the rapid production of large numbers of compounds. Libraries are synthesized for both lead identification and lead optimization purposes. The resultant libraries consist of products formed by combining "reactants" (reagents, monomers) with each other or with a "scaffold" (template, core). The most efficient use of reactants and automation/robotics would use a strictly combinatorial combination of reactants/scaffold, but other constraints, including the issue of generating products that have suitable properties for biological screening and as potential drugs, often lead to sparse arrays. Parallel synthesis, in which multiple analogs are synthesized at a time, is now a standard part of the drug discovery process.

Many molecular diversity and similarity approaches are brought together in the combinatorial library design process. Either the properties of the reactants/scaffolds are used (reactant-based design) or the properties of the resultant enumerated products are used in selecting appropriate reactants (product-based reactant selection). The latter approach requires much greater computational resources, and a preselection of potential reactants may need to be made to control the total size of the "virtual" (potentially synthesizable) library to be analyzed. Regardless of the method, the required deliverable is sets of reactants/scaffolds to be combined. When working with the properties of the products, the constraint that reactants are to be used as efficiently as possible presents a major optimization problem.

Virtual screening, with experimental verification by biological screening, has provided a validation of many of the molecular similarity/diversity methods used for combinatorial library design, and some ligand-based approaches and examples are discussed in Section 3.

1.4 Subset Selection and Screening Set Enrichment

A related task to combinatorial library design that uses molecular diversity/similarity methods is subset selection of compound screening sets. Initial efforts were focused on small "diverse" or "representative" sets of large corporate compound collections. The increased capabilities of high throughput screening have changed the demand for such sets, and there is a renewed demand for "focused" and "representative" screening subsets of varying sizes; this includes target class (gene family) focus and the identification of "interesting" (e.g., novel) compounds in a large set. Newer biophysical screening methods [e.g., NMR-based screening (3)] still have capacity issues and a need for smaller representative and focused sets. Diverse subset selection can be used to generate sets of compounds to probe a biological assay or to select a subset of reactants to probe the scope of a chemical reaction scheme or screen. However, such methods have a tendency to select compounds at the extremes of chemical space; that is, the selected compounds tend to be less suitable as drug candidates, and hence the approach is less favored for general screening sets. Rather, diversity methods are used to ensure that a random subset of a screening set contains compounds that are representative of the whole, or, in conjunction with a focused method, to ensure a representative sampling of biologically relevant chemical space.

Compound subsets focused/biased to particular target classes (gene families) have become of greater importance, with application to both lead identification and de-orphaning of new targets from genomics studies. Properties important for the target class of interest are identified, using descriptors used for molecular similarity/diversity. A focused subset can then be selected using a combination of all the possible hypotheses for activity for that target class, including the use of one or more molecular similarity approaches to select compounds similar to any known active compounds. For targets that have structural information available, docking methods (one widely used method for virtual screening) can be used to select compounds that are complementary to the binding site(s). Applications encompass both high throughput screening (HTS) and therapeutic area screening where only smaller numbers of compounds can be screened. For HTS, smaller thematic studies using these enriched focused sets enable the rapid prosecution of a set of related targets, and make the use of duplicate runs for all compounds feasible. This enables selectivity to be addressed up front, and the duplicate runs provide potentially higher quality information, with the potential for the identification of hits that might otherwise be missed.

General enrichment of the available screening compound set for lead identification is a major application for both combinatorial library design/synthesis and compound acquisition. The goal of *in silico* (i.e., computer-based) studies in compound acquisition is to evaluate the interest of compounds that could be purchased to add to the screening file, and to select a subset that meets the same type of physicochemical/"druglikeness" criteria discussed for combinatorial libraries. The "interest" of a compound or compound set is evaluated as in combinatorial library design: diversity relative to existing compound, target, target-class focus, and so forth.

2 MOLECULAR SIMILARITY/DIVERSITY

The field of medicinal chemistry is based on the hypothesis that similar compounds will display similar, but probably not identical, activities in some biological screen, and that potency, selectivity, and properties can thus be modulated by analog synthesis. The challenge facing the computational chemist is how to represent compounds in a computer in such a way that "similar" compounds in the *in silico* world are "similar" in the biological world. It is evident that the biological process that is being modeled will influence the nature of the chosen representation. For example, c log P is a useful descriptor for modeling processes involving cell penetration, whereas a pharmacophoric representation would be more appropriate for selecting compounds for screening against a particular protein active site. In this section we review the wide range of representations that have been developed and describe

the various methods for applying these representations to real-world problems. The reader is referred to a number of reviews on various aspects covered by this chapter (2, 4–9). A diverse set of perspectives/reminiscences on computational aspects of molecular diversity has been assembled by Martin (10).

2.1 Descriptors

The problem lies in finding a representation of chemical structure that allows a mapping between the chemical structure and its response in a biological or physical process. The representation must be general enough to be applicable to a range of chemical structures but specific enough to capture the differences between structures that account for differences in response. Once found, this representation or set of descriptors can be said to define a *chemistry space* (11) for the population of compounds of interest. The similarity between two compounds is their distance within this space. Unfortunately, this simple statement hides a number of difficulties. Many descriptors of choice are correlated and it can be difficult to combine categorical (e.g., acid, base, neutral) and real-valued (charge, dipole, c log P) variables. The issue of how to analyze compounds within the chemistry space is covered in Section 2.2.

Methods for describing chemical structures fall into two broad classes. Two-dimensional (2D) methods can be calculated from the 2D graph in which atoms are nodes in the graph and the bonds are the connections between the nodes. Three-dimensional (3D) methods require the generation of a 3D structure (x, y, z coordinates) for a structure. Because a molecule does not exist in a single low energy conformer, the issue of conformer generation also requires addressing with this latter method. Combining the various descriptors, particularly 2D and 3D, is an area of active research.

The potential advantage of 3D descriptors (ligand-protein binding is a 3D spatial/electronic property that can be described only in part using 2D descriptors) (5c) has led many groups to identify 3D descriptors that can handle large numbers of compounds and multiple potential models, and do not require a superimposition in 3D coordinate space (e.g., for review, see Ref. 12). The pharmacophore fingerprints described in Section 2.1.3 are an example of this.

2.1.1 2D Substructural and Topological Descriptors.

The principle behind substructural keys or fingerprints is shown in Figure 5.1. A molecule is encoded by the presence or absence of a set of predefined atoms, atom types, and fragments (e.g., S, aromatic nitrogen, CO_2H). The most widely used set of keys is the publicly available ISIS (MACCS) key set provided by MDL (13a). An alternative to the use of predefined fragments is provided by software packages such as Daylight (13b) and UNITY (13c). In this approach, all possible bond paths in a molecule from zero (the atoms) to a specified number of bonds (usually 7) are identified. A hashing procedure is used to store the paths in a bit string of fixed length. Each path will set several bits in the bit string (giving them the value of 1) and there is the possibility of different paths setting some of the same bits. As a result, individual bits lose any meaning.

The origin of the 2D substructural representation lies in the first chemical registration systems where some means was required to enhance the speed of compound retrieval. Thus, if the query molecule contains a particular combination of features, the whole database can be screened very rapidly using the keys to identify compounds that are likely to contain those features before a more exhaustive graph matching is performed to ensure an exact match with the query. The features represented in the keys (ISIS) or the fingerprint length and density (Daylight) were selected to optimize the process of compound retrieval. Nevertheless, they have proved very useful for a variety of similarity-based tasks (14). Despite these successes, issues surrounding their use in diversity-based approaches have been highlighted (15).

Molecular connectivity indices were first proposed by Randić in 1975 (16) as a means of estimating physical properties of alkanes. This formalism was quickly extended to other types of molecules (17) and, since then, a wide range of indices has been proposed, as reviewed by Hall and Kier (18) and Randić (19). The indices are derived from a graph theoretical representation of the structure where bonds are represented by the edges between nodes (atoms). They provide a direct representation of the topological structure of a molecule encoding information such as the degree of branching ($^1\chi$) and the adjacency of the branch points ($^3\chi$), flexibility, and shape (20a). The superscript describes the number of bonds in the path between atoms used to calculate the index. The software package MOL-CONN-Z (20b) was developed specifically for generating these descriptors. A number of authors have included topological indices or variants thereof in their description of molecules for describing compound collections (21) or large combinatorial libraries, often allied to a dimensionality-reduction algorithm such as principal components analysis (PCA) (6a, 23).

Cahart et al. (24) introduced the concept of atom-pairs, where the topological distance (number of bonds) between atoms of specified element type are encoded in a bit string. This was extended to the topological torsion (25), where elements on all paths of length four are encoded. Kearsley et al. (26) extended this approach to use more generic atom-type properties in place of element type. They termed these types *binding property classes* because they represent key features of intermolecular interactions (positive and negative charge; hydrogen bond donor, hydrogen bond acceptor, and groups that are both of these, such as hydroxyl; hydrophobic atoms; and all others). These descriptors have been used widely for similarity- and diversity-related tasks. The CATS descriptors of Schneider et al. (27a) are a variant on this approach. All topological distances (number of bonds) between a pair of binding property classes (e.g., acid-base) in a molecule are recorded with count information in a correlation vector; that is, how often that topological distance occurs between a specified pair of features in the molecule of interest. Similarity is calculated as the Euclidian distance between the correlation vectors. These CATS descriptors were shown to be useful in scaffold-hopping, identifying actives with a structural type distinct from that of the initial lead structure, and have also been used as the basis for a *de novo* design program, TOPAS (27b).

Functional diversity requirements of compound libraries have been reviewed (28), for which molecular descriptors that relate to both structure and properties are needed, as well as their evaluation in terms of biological relevance.

2.1.2 Atomic/Molecular Properties and 2D/3D Structural Descriptors

2.1.2.1 Physicochemical. The descriptors in the previous section focus largely on the structure of the compound. The binding property classes generalize this to some extent by replacing relationships between elements or atom types with a broader definition, still within the framework of an atoms-and-bonds description of the molecule. An alternative approach would be to describe compounds by whole molecule properties, such as molecular weight and log P. Indeed such properties have been related to important pharmacological and physical properties such as absorption across cell membranes, distribution, and solubility. These properties are represented, in part, by the well-known Lipinski *Rule-of-5* based on molecular weight, calculated log P, and hydrogen bond donor and acceptor counts (29). Thus, such properties have an important role in drug design, and in general assessments of "druggability." However, their use as descriptors for tasks related to similarity or diversity in the context of receptor affinity is less clear and has been questioned (11b). A primary concern is that such properties do not reflect sufficient information regarding chemical structure to enable their use for lead follow-up or similar purposes. For example, a steroid and a benzodiazepine can have identical log P values but are clearly dissimilar from a medicinal chemistry perspective. Another major problem is that many properties (e.g., log P, molecular weight, surface area, volume, molar refractivity, molecular polarizability) are correlated, making it difficult to find a reasonable set of orthogonal descriptors for the calculation of meaningful distances or for cell-based partitioning (see Section 2.2.1). Such whole molecule properties are best used as constraints on a design, to define boundaries of a pharmacologically relevant chemical space or to define a distribution to match. The challenge is then how to combine the measures of diversity while simultaneously maintaining suitable physicochemical properties. This is addressed in later sections. Such properties can also be used to identify particular combinations that are preferred for different gene families, and these are used to focus a design.

2.1.2.2 2/3D Structural. The issues with whole molecule descriptors mentioned above led Pearlman (11) and colleagues to look at an alternative representation ("BCUT" descriptors/metrics) based on atomic properties and on how atoms are connected. The approach stems from original work of Burden (30) to derive a unique signature for a molecule. Pearlman extended the concept to develop the BCUT descriptors suitable for diversity- and similarity-related tasks. Each molecule is described by a series of square matrices with atom labels defining the rows and columns. In a given matrix, the diagonal represents an atomic property such as charge, hydrogen bonding ability (donor/acceptor), or polarizability, with optional weighting by accessible surface area; the off-diagonal terms represent topological or Cartesian interatomic distance or other such property. Molecular descriptors are generated from the lowest and highest eigenvalues of these matrices, and describe the molecular surface distributions of positive or negative charge, H-bond donors, H-bond acceptors, and high or low polarizability.

A number of such matrices can be calculated based on the nature of the diagonal and off-diagonal properties and the scaling between them. An "auto-choose" algorithm [see the DiverseSolutions (DVS) program below] typically finds a 5D or 6D orthogonal chemistry space that best represents the diversity of a given population. This ability to identify relevant (to drug-receptor interactions and reflecting molecular substructure) and orthogonal (noncorrelated) descriptors is critical for the effective use of both distance-based and cell-based methods. Three-dimensional properties may be included by the use of a single conformer to represent atom-atom distances or the inclusion of quantum mechanical properties (bond order or overlap-squared) from semiempirical molecular orbital (MO) calculation. However, the inclusion of 3D/MO information significantly slows down descriptor calculation and does not appear to offer any practical advantage. DVS, a suite of programs, has been written to calculate and manipulate the BCUT (and other) descriptors for a variety of library design, similarity- and diversity-related tasks (31) (see Section 2.2.1.1). DVS uses the power of a cell-based method as a rapid means to derive a chemistry space relevant to the representation of the diversity of large populations of compounds and methods to pick diverse subsets and compare large data sets rapidly. The BCUTs provide an excellent diversity metric based on electronic properties directly related to ligand-receptor interaction that should also relate to biological activity. Indeed, BCUT metrics appear to reflect pharmacophorically important information, albeit in a relatively crude (low dimensional) fashion. They have proven useful for quantitative structure-activity relationship (QSAR) and quantitative structure-property relationship (QSPR) analyses (32a,b), classification of pharmacologically active compounds (32j), diverse and focused combinatorial library design (11d, 32c-e), rational compound acquisition strategies (11c), and various other diversity-related tasks.

2.1.3 3D Properties. The properties and descriptors above are essentially 2D in nature, in that they can be generated from the compound connectivity table, that is, from a knowledge of the bonding pattern within a molecule. There are many advantages to this, not the least of which is the speed of descriptor calculation. Nevertheless, compound interactions with most biological targets are largely 3D in nature. That is, it is the disposition of key functional groups in the molecule in relation to complementary groups within the enzyme or receptor that is important. Thus, there has been much active research into how best to represent the spatial properties of molecules. A particular issue that needs to be handled is conformational flexibility, given that most compounds have rotatable bonds that will change the 3D properties and there is no means *a priori* of identifying which particular conformation is the bioactive conformation (i.e., the conformation of the ligand bound to the biomolecular receptor).

The methods presented below tackle this in one of three ways:

1. A single fixed conformer is used.
2. A relatively small number of representative conformers is generated.
3. An exhaustive enumeration of conformers is used.

3D BCUTs are an example of case (1). They reflect conformational differences, but only to a limited extent because they are inherently low dimensional. This is actually somewhat advantageous because the single low energy conformation from which they are computed may or may not be similar to the bound conformation for a particular receptor. Pearlman (11) has noted that 3D BCUTs appear to be advantageous when the population of interest is a single combinatorial library but that, on average, 2D and 3D BCUTs appear to be equally advantageous when the population of interest is much more diverse. In cases (2) and (3), the descriptors need to be accumulated over all conformers. In the case of bit strings this means "ORing" them over all conformations (combine using logical OR). Herein lies a potential issue with such techniques, in that data from multiple conformations could obscure the signal from the particular bound conformation relevant to a particular target.

2.1.3.1 3D Pharmacophores. The representation of a set of active compounds by a single or small set of pharmacophores that is necessary for that activity was first proposed many years ago and is an excellent model for lead optimization. The development of database systems capable of handling three-dimensional structures in the late 1980s enabled the further exploitation of such methods through giving the ability to search a corporate collection for molecules containing a particular pharmacophore. This approach to lead generation has proved highly successful (e.g., for reviews, see Ref. 33). In particular it is possible to identify active compounds that contain a different core structure from that of the compounds used to generate the model (lead-hopping). This success and the importance of the pharmacophore hypothesis in understanding the interaction of a ligand with a protein target prompted groups to look for ways to use pharmacophores to generate a molecular descriptor for similarity- and diversity-related

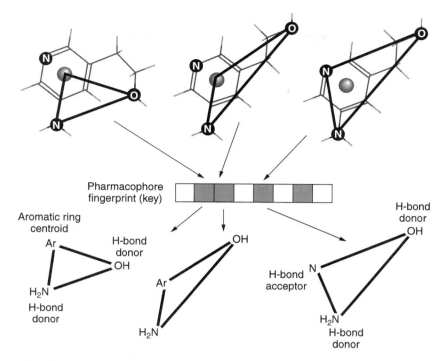

Figure 5.2. Illustration of the creation of a pharmacophore key. As the conformation of a molecule changes, so do the distances between the pharmacophoric groups, shown as spheres. The two different three-point pharmacophores shown each set their own particular bit in the pharmacophore key.

tasks. The diversity-related use was based on the hypothesis that sampling over all potential pharmacophores leads to diversity in a biologically relevant space, in contrast to some other methods that focus on chemical diversity. The descriptor thus generated identifies in a systematic way all the potential pharmacophores that a molecule could exhibit. Triplet (three-point) and quartet (four-point) pharmacophore representations have been extensively used (in addition to two-point/2D approaches), with a variety of features sampled at each point and interfeature distances considered in a discrete set of ranges ("bins") (see Fig. 5.2). The ability of pharmacophores to divorce the three-dimensional structural requirements for biological activity from the two-dimensional chemical makeup of a ligand has been highlighted in a recent review (34).

In an initial implementation from the authors (35), a set of 5916 three-point pharmacophore queries was generated and used to search a database. Compounds were charac-

terized by the pharmacophores that they matched. This method was powerful because it gave precise control over the queries that were generated and ensured that the compounds matched the query, as opposed to satisfying a set of distance constraints; however, it was slow in execution. The Chem-X/Chem-Diverse implementation (36) generates a pharmacophore fingerprint during the course of a single systematic conformational search, with a bump-check and/or rules to eliminate high energy conformers. The details of the conformational search and the definitions of the pharmacophoric features are key components of the system and this methodology has been used extensively for a range of library design and both diversity-and similarity-based tasks (e.g., see Ref. 37). The use of 3D pharmacophores in drug design applications has recently been reviewed (12, 34).

To perform the necessary analyses to generate the pharmacophore fingerprint, relevant features in a molecule need to be identified.

Either substructural definitions to find pharmacophoric features are applied at search time or atom types [and, optionally, additional centroid "dummy" atoms (35)] are used. These can be preassigned (e.g., on database registration) or assigned at search time; a variety of approaches is used (including the use of substructures and connectivity, and of more sophisticated computational approaches). Six properties (features) have commonly been used to describe the potential pharmacophoric features of a structure:

1. Hydrogen bond donor (e.g., amide NH, aromatic amine, and hydroxyl)

2. Hydrogen bond acceptor (e.g., carbonyl, ether, and hydroxyl)

3. Basic ionizable center (positively charged at physiological pH of about 7) (e.g., aliphatic amines, amidines/guanidines, and 4-amino pyridine)

4. Acidic ionizable center (negatively charged at physiological pH of about 7) (e.g., carboxylic acid, unsubstituted tetrazole, acyl sulfonamide)

5. Aromatic rings (ring centroids often used)

6. Hydrophobic regions (e.g., isopropyl, butyl, cyclopentyl, and certain aromatic rings)

It has also been useful to define a seventh feature type in some situations. For example, it may be beneficial to classify separately the groups that can be both hydrogen bond donor and acceptor such as hydroxyl groups or imidazole nitrogens. Alternatively, the seventh feature provides a mechanism to identify an anchor point to substructures of particular interest (see Section 2.2.5).

All combinations of three or four pharmacophoric points (forming triangles or tetrahedra), for all accessible conformations of a given molecule, can be analyzed, with the resultant descriptor bit-string fingerprint (key) containing the pharmacophores from the whole conformational ensemble of the molecule (see Fig. 5.2). Each bit represents a particular combination of pharmacophore points (Donor-Aromatic-Acceptor, Donor-Aromatic-Basic, etc.) and distances between them (defined using discrete ranges, or bins).

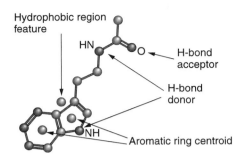

Figure 5.3. Example of how a 3D molecular structure can be broken down into its constituent pharmacophoric elements.

Figure 5.3 illustrates how a molecule can be broken down into pharmacophoric elements. The atom types can be assigned using substructural fragments, taking into account the environment (e.g., a NH group attached to a conjugating group such as C=O is not basic or a H-bond acceptor). Atom types can be automatically assigned when reading a molecule, such as through a customizable substructural fragment database and parameterization file (e.g., Chem-X/ChemDiverse software, 37a). The fragments identify the environment of an atom or group, enabling the correct assignment of a designed feature type. Different options can be set (e.g., a hydroxyl group can be assigned to be both a hydrogen bond donor and acceptor and/or can be assigned to a special feature type for atoms that have both characteristics), and reassignment is possible at search time for structures stored in a database. The identification and representation of hydrophobic regions is one of the most difficult yet critical tasks. Dummy atoms can be used to represent the hydrophobic regions, as a centroid of a group of relevant atoms. This limits the number of hydrophobic features to comparable numbers to other features. An automatic method to add them that uses bond polarities (hydrophobic regions defined for groups of three or more atoms that are not bonded to atoms with a large electronegativity difference) has been implemented in Chem-X/ChemDiverse. Other pharmacophore atom types have also been developed (26).

The extension to four-point pharmacophores enables chirality to be handled and enables some elements of volume/shape linked to electronic properties to be included. This can give a much better performance in similarity searching. It also increases enormously the number of potential pharmacophores that need to be considered. To analyze pharmacophoric patterns in molecules, the distances between pharmacophoric features are divided into a finite number of ranges using a predefined binning scheme (e.g., 0–2, 2–3, 3–5, 5–8 Å, etc.), up to a maximum distance normally between 15 and 20 Å [a nonuniform binning is often used because this mirrors the tolerances (e.g., ±20%) used in 3D database searching that can be more appropriate than fixed increments, given the limited conformational sampling that is possible]. The additional pharmacophoric combinations created in moving from a three- to four-point description provides additional shape information, thus increasing molecular separation in similarity and diversity studies.

Separation has a central role in determining the final result of such calculations, with too little separation resulting in a noisy descriptor and too many molecules being defined as similar, whereas when too large a separation exists, trivial differences can have a disproportionately negative effect on the similarity value. Conformational sampling is necessary, and the granularity of this affects the useful resolution that can be used, as defined by the number and size of the distance bins. The sampling is generally performed by torsional sampling of rotatable bonds.

Thus fewer ranges are generally considered with four-point pharmacophores while concomitantly maintaining or improving on the performance of three-point pharmacophore methods. For example, by the use of 32 distances for three-point pharmacophores with seven different features possible for each of the points, there are about one million possibilities (35). Expanding to four-point pharmacophores, just 15 distance bins generate about 350 million geometrically valid possibilities. Therefore for pragmatic reasons of both memory/disk space, and the limited resolution of the conformational sampling that is normally applied, seven or 10 distance ranges for four-point pharmacophore fingerprints have been used by Mason et al. (37) and recommended for combinatorial library design and virtual

screening applications. Around 2–10 million different potential pharmacophores are resolved in such a fingerprint. A limited sampling of conformations has generally been used to achieve reasonable times (in seconds) for descriptor calculation. For example, Mason et al. (37) use two (conjugated), three (single bonds), or four (sp^2-sp^3 and some conjugated) increments with large data sets, using a systematic analysis for less flexible molecules and random sampling for flexible molecules. See Fig. 5.4 for a comparison of three- and four-point fingerprints. Software companies such as Accelerys, Tripos, the Chemical Computing Group (MOE, http://www.chemcomp.com), and Treweren Consultants (THINK) are developing their versions of pharmacophore fingerprinting methods, with three-point pharmacophore fingerprints already implemented. The automatic assignment of pharmacophore features such as hydrophobes, acids and bases, conformational sampling, and other key options discussed above for the Chem-X software (now no longer supported; owned by Accelerys) such as nonuniform binning are challenges that have variable levels of current implementation; other options and extensions such as overlapping bins are becoming available.

Others have developed similar approaches for library design (38, 39). Horvath (40) generates an autocorrelogram of feature-feature distances for conformers and calculates a dissimilarity score that takes into account separate weightings for each feature and allows fuzziness between the distance bins. These 3D pharmacophoric descriptors were termed *fuzzy bipolar pharmacophore autocorrelograms* (FBPAs), and the use of fuzzy logic to build up and compare the fingerprints avoids the "all-or-nothing" bitwise match of bitstring representations in which sampling artifacts can cause significant differences. The method has been shown useful in library design and for analyzing selectivity profiles in terms of pharmacophore similarity (41).

It is possible to represent not only a ligand by the potential pharmacophores it possesses but also a protein target. In this case the pharmacophore points are identified by the positions where a ligand atom of a particular type (donor, acceptor, acid, base, hydrophobic, aro-

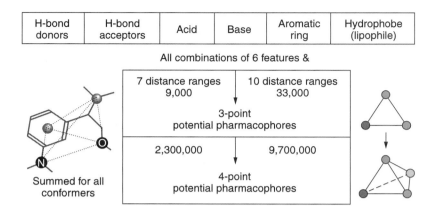

H-bond donors	H-bond acceptors	Acid	Base	Aromatic ring	Hydrophobe (lipophile)

All combinations of 6 features &

7 distance ranges 9,000	10 distance ranges 33,000
3-point potential pharmacophores	
2,300,000	9,700,000
4-point potential pharmacophores	

Summed for all conformers

Figure 5.4. Three- and four-point (triplet/quartet) pharmacophore fingerprint creation. Assignment is often binary (on or off), although a count can be kept, and has been used in more recent studies. The large difference in bin numbers between three- and four-point pharmacophores provides additional shape information, thus increasing molecular separation in similarity and diversity studies.

matic centroid) is likely to bind and so provide a complementary interaction with the adjacent protein residue side chain. The pharmacophore fingerprints are thus generated from these complementary site points. The site points can be positioned in the active site using methods such as GRID (42), in which an energetic survey of the site is made using a variety of functional groups. Figure 5.5 illustrates the favorable energy contours for a variety of pharmacophoric probes for the Factor Xa serine protease active site. Atoms (with associated pharmacophore features) are then added in the positions for the most favorable interaction (also shown in Fig. 5.5).

The resultant ensemble of atoms represents a hypothetical molecule that interacts at all favorable positions in the binding site, and

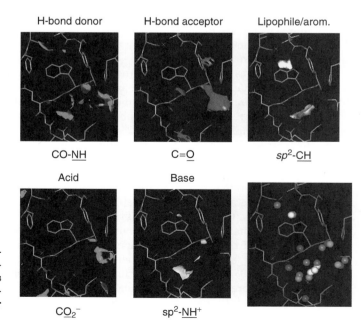

Figure 5.5. GRID probes on Factor Xa site and the combined resultant complementary site points that can be used for pharmacophore fingerprint calculations (lower right). See color insert.

H-bond donor

CO-NH

H-bond acceptor

C=O

Lipophile/arom.

sp^2-CH

Acid

CO_2^-

Base

sp^2-NH$^+$

a pharmacophore fingerprint is calculated from this. This fingerprint represents a form of "protein structure-based diversity," quantifying the range of different pharmacophoric shapes complementary to a target protein binding site. For example, for the Factor Xa serine protease active site, 13 complementary site points generated a fingerprint of 2103 four-point pharmacophore shapes, of which 354 were the same as the 2062 found for the serine protease thrombin, generated from 13 site points. Only 11 significant complementary site points were found for the serine protease trypsin, which has a less defined S4 pocket. Of the 1233 total pharmacophore shapes, 363 were in common with Factor Xa, with 120 in common for all three serine proteases. It is thus possible to identify ensembles of pharmacophores that can be used to both differentiate the sites (selectivity) and identify common features. Comparison of these protein-derived pharmacophore fingerprints with known ligands, using four-point fingerprints, shows that they can be used for searching for novel ligands within a database and that they are specific enough to capture ligand selectivity between similar proteins such as the serine proteases thrombin, Factor Xa, and trypsin (37). With three-point fingerprints, the comparison of ligand- and site-derived fingerprints could identify common binding motifs, although selectivity was not captured (37b).

Pharmacophore fingerprints are relatively slow to calculate, however. Thus, their application to very large virtual libraries requires a great deal of computer power. Researchers at Chiron (12, 43) have developed a pharmacophore-based methodology applicable to reactants, OSPPREYS (Oriented-Substituent Pharmacophore PRopErtY Space). In this approach, reactant pharmacophores are calculated with respect to the reactant attachment atom and combinations of up to nine pharmacophore centers are considered (see Section 4.8). In the Gridding and Partitioning (GaP) approach, developed at GlaxoWellcome (44), reactants are aligned such that the bond between the attachment atom and the first nonhydrogen atom is along the x-axis with the attachment atom at the origin. A conformational analysis is then performed and the pharmacophore features are mapped to a

1-Å grid (Fig. 5.6). Cells occupied by a particular feature are recorded in a bit string. This descriptor is ideally suited to monomer acquisition and reactant diversity.

Topomer shape similarity, developed by Cramer (45) at Tripos, has been used for similarity searching and targeted library design (using Tripos' proprietary software, "ChemSpace"), building on earlier work on steric fields of single "topomeric" conformers, clustering reactants by their 3D steric fields into "bioisosteric" clusters. The descriptor was considered to be useful in describing variations about a fixed molecular core, defining a single, unambiguous, aligned conformation for any nonchiral molecule.

Approaches such as the GaP program that exploit 3D descriptors for monomer selection address a need for an easily accessible set of in-house monomers available for library generation. Such monomers need to be diverse in nature and able to probe regions of space through attachment to known leads, while producing compounds with druglike properties. More detailed conformational searching paradigms can be used for the smaller monomer compounds, and approaches such as GaP and OSPPREYS exploit this opportunity.

For the selection of diverse compound subsets, studies (46a) have compared three-point pharmacophore descriptors and 2D fingerprints. These have highlighted benefits of the different approaches, and the improved performance of some combined descriptors. The use of clustering for the rational selection of compounds for acquisition and for in-house compound collections used for screening has also been investigated (46b), with comparable results obtained with 3D pharmacophore-derived fingerprints to the typically used 2D fingerprints.

2.1.3.2 Shape. Pharmacophores capture the key features of intermolecular interactions. However, they do not explicitly capture the shape and volume of the ligand, even if this is crudely implied by the largest four-point pharmacophore exhibited, and the totality of potential pharmacophores exhibited across a range of conformations encodes shape fragments. Hahn (47) has described a method for three-dimensional shape-based searching implemented in the Catalyst program. Seven

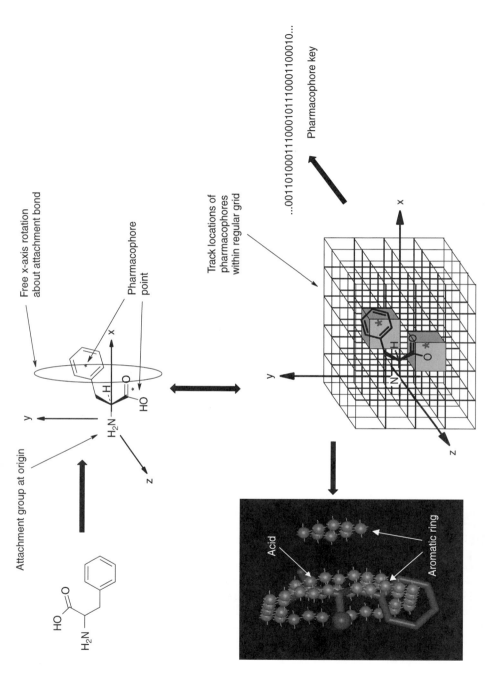

Figure 5.6. Overview of the Gridding and Partitioning (GaP) procedure as applied to monomers, exemplified using phenylalanine as a potential primary amine. This molecule thus contains two pharmacophoric groups (the aromatic ring and the carboxylic acid). During the conformational analysis the locations of these pharmacophoric groups are tracked within a regular grid. See color insert. [Reproduced from A. R. Leach and M. M. Hann, *Drug Discovery Today*, **5**, 326–336 (2000), with permission of Elsevier Science.]

shape indices, positive and negative extents along the three principal axes from the molecular centroid, and the volume of that conformer are computed and stored in a database. These indices can then be used for rapid comparison with a query shape derived from active structures. Conformers passing this filter are then aligned with the query and the similarity is assessed from the volume overlap. Shape-based searching can be used independently, in which case it will complement a 2D similarity search. The method can also be employed in conjunction with a 3D pharmacophore search; however, it is not clear that results are improved in this case (48).

2.1.3.3 Field-Based. A receptor site recognizes the surface properties of a molecule. These can be represented by different types of molecular fields, electrostatic, steric, and hydrophobic, that can be calculated from the atomic composition of the molecule and compared using a measure such as the Carbo index (49). A gaussian representation of the field allows for a more rapid alignment of the molecules (50). Willett's group has developed a program FBSS (51), which uses a genetic algorithm for the alignment of the molecular fields. They have compared the performance of this method with a 2D structural fingerprint (UNITY software, (13c), in searching the WDI, a collection of drug molecules and compounds in development, and the BIOSTER database, a database of functional groups that have been used to replace other groups and retain biological function (e.g., a carboxylic acid and a tetrazole). Although the 2D measure will tend to find more bioactive molecules, the 3D measure gives a greater structural diversity in the hits (52). This seems to be the case for most 3D methods. In these examples conformational flexibility can be considered during the alignment stage but will slow the search down considerably and may also lead to the algorithm becoming stuck in local minima.

An alternative to using the molecule composition in calculating the fields is to use molecular fragments as probes to represent protein side chains. The interaction energy between the probe and the molecule is calculated on a grid surrounding the molecule. These grid fields can then be used in conjunc-

tion with PLS as in the CoMFA (comparative molecular field analysis) 3D-QSAR methodology (53). More recently, these fields have been further transformed to generate 3D molecular descriptors. The VolSurf program (54) calculates a wide range of descriptors from the grid energies [calculated with the program GRID (42)]. These have been shown to correlate to a range of properties such as membrane penetration and solubility (55). The Almond program (56) uses a transform known as the Maximum Auto-Cross Correlation (MACC) between pairs of grid nodes, to give a type of two-point pharmacophoric representation of the fields. Such descriptors have been useful in QSAR studies because they are alignment free; that is, they are independent of the position within the defining grid, and have also been used in reactant selection (Pickett, unpublished results, 1999). However, the limitations of the lack of conformational flexibility have so far precluded their use in more general database searching and diversity applications.

2.1.4 Analysis

2.1.4.1 Descriptor Transformations. A large number of potential descriptors are available and this presents a number of issues. Many descriptors will tend to be correlated with one another to a greater or lesser degree. There is the question of the scale of the descriptors and also the difficulty of combining, say, a fingerprint with a calculated property. Thus the descriptors must first be transformed in some way. A key study in this regard was the work of the Chiron group (57). Groups of similar descriptors were combined using principal components analysis (PCA) and multidimensional scaling (MDS), to give a total of 16 composite descriptors. D-optimal design was then used to further analyze a data set. Also of interest was the use of a "flower plot" to visualize the results. In the DPD (diverse-property derived) methodology (21a), the search was for six noncorrelated descriptors. The selection of relevant BCUT descriptors using a χ^2 test is mentioned below.

2.1.4.2 Similarity and Distance Measures. A variety of measures exist for assessing the similarity or distance between molecules in a given descriptor space (2a), as described above. Similarity measures give a direct measure of similarity between molecules in some property space and give values in the range of 0 to 1, with 1 being identical. Typical examples are the Tanimoto coefficient and the Cosine coefficient. For real-valued properties the Tanimoto is defined as

$$\text{Tanimoto} = \frac{\sum\limits_{i=1}^{i=N} x_{iA}x_{iB}}{\sum\limits_{i=1}^{i=N} (x_{iA})^2 + \sum\limits_{i=1}^{i=N} (x_{iB})^2 - \sum\limits_{i=1}^{i=N} x_{iA}x_{iB}}$$

where x_{iA} is the value of property i of molecule A. When i can take values of only 0 or 1 as in a bit string, then this reduces to

$$\text{Tanimoto} = ab/(a + b - c)$$

where a is the number of on-bits in A and c is the number of bits in common between A and B. The Cosine coefficient can be defined as

$$\text{Cosine} = \frac{\sum\limits_{i=1}^{i=N} x_{iA}x_{iB}}{\sqrt{\sum\limits_{i=1}^{i=N} (x_{iA})^2 \sum\limits_{i=1}^{i=N} (x_{iB})^2}} \Rightarrow \frac{c}{\sqrt{ab}}$$

For field-based measures and overlap of electron density functions then the Carbo index can be used (49), which is equivalent to the Cosine coefficient.

Distance measures give 0 for identical structures and have an upper bound defined by the property space. The Euclidean and Hamming distances are the most common:

$$\text{Euclidean distance} = \sqrt{\sum\limits_{i=1}^{i=N} (x_{iA} - x_{iB})^2} \Rightarrow$$
$$\sqrt{a + b - 2c}$$

$$\text{Hamming distance} = \sum\limits_{i=1}^{i=N} |x_{iA} - x_{iB}| \Rightarrow$$
$$a + b - 2c$$

The fundamental difference between similarity and distance measures is that the latter

expressly include the absence of a feature (or low values for real-valued properties) in the measure of similarity. This has led to the suggestion (58) that, in the chemical domain at least, such measures are best for relative similarity; that is, ranking the similarity of two molecules to a target, as opposed to measuring the absolute similarity of molecules for which similarity measures, are preferred.

Similarity and distance measures form the basis for most of the analysis and selection methods described in the next section and the reader is referred to the reviews by Willett et al. (2, and references therein) for a fuller discussion of the characteristics and specific properties of these measures.

2.2 Analysis and Selection Methods

In this section we describe some general methods for analyzing and partitioning large data sets, with particular reference to selecting representative or diverse subsets. Library design also employs many of the strategies described here and is discussed in more detail in Section 4. The methods fall into two broad categories: cell-based or partitioning methods and distance-based methods. Partitioning methods use the population to define the limits for cells into which the compounds are divided. Adding or comparing to other compound sets requires identifying the cells into which the new compounds would fall based on their descriptors. This is very rapid and the partitioning process provides a frame of reference for many design tasks; for example, compounds can be readily identified to fill empty or poorly represented cells. Potential issues are where to place the cell boundaries and the handling of compounds that fall near to a cell boundary. Also, new compounds may fall outside the range of properties of the initial population. Distance-based methods, such as clustering and dissimilarity-based methods, require the calculation of similarity between members of the population and are thus population dependent. Adding new members to the population requires recalculating similarities and could change the distribution of compounds between the clusters. Identifying poorly represented or empty areas of property space is not possible. Each of these methods is further described below with examples of their application.

2.2.1 Cell-Based Partitioning Methods. Partitioning methods divide chemistry space into hyperdimensional "cells" by "binning" the axes (descriptors) that define the chemistry vector space, just as the eight divisions on the x- and y-axes of a two-dimensional checker board divide the board into 64 squares. A chemical compound occupies a position in chemistry space determined by the descriptors (coordinates) computed based on its structure. Once the compounds have been partitioned, selecting diverse or representative sets of compounds involves selecting a small number of compounds from each occupied cell, either in proportion to the number of compounds in the cell or a specified number from each occupied cell. For focused sets, compounds are sampled from cells neighboring the population of actives. The real advantage of partitioning methods, however, lies in their ability to readily identify underpopulated regions of property space. Selections can then be made from a second population of molecules—a virtual library for instance—to increase the occupancy of underpopulated cells. Usually, such methods require a low dimensional representation of the space, although the pharmacophore methods are a notable exception to this. The low dimensional space may be the result of a dimensionality-reduction algorithm, as described earlier. Alternatively, a small number of descriptors may be judiciously selected. This latter approach was taken by Lewis et al. in their DPD methodology (21a), which is a good example of partition-based selection. The aim was to select a representative set of compounds based on molecular and physicochemical properties for screening. Six properties were chosen from among 49, based on their low pairwise correlation: number of H-bond acceptors, number of H-bond donors, molecular flexibility, an electrotopological state index, c log P, and a measure of aromatic density. Each descriptor (axis) was divided into two to four partitions, to give a total of 576 bins. A major issue was in identifying six relevant and reasonably non-correlated (orthogonal) descriptors, leading to the definition of a new descriptor. The chosen ranges covered more than 85% of a 150,000 subset of the corporate collection and approximately three compounds were taken from each bin. Follow-up of initial hits involves the screening of additional compounds from the cells containing hit molecules. Several leads were identified using this approach (7).

2.2.1.1 Diverse Solutions. DiverseSolutions (DVS) is software developed by Pearlman et al. (11, 31) to generate and use the BCUT descriptors in addition to other DVS-computed or user-provided low dimensional descriptors. (DiverseSolutions is also designed to work with high dimensional metrics such as 2D fingerprints, and includes some novel algorithms for such distance-based work.) DVS uses a χ^2-based "auto-choose" algorithm (11c) to identify the combination of low-D descriptors, which are mutually orthogonal and which most uniformly distribute a given large population of compounds among the cells of the resulting chemistry space. Originally, the binning was performed in a uniform manner along each axis, with a given percentage of outliers to avoid sampling the extremes of space. This could be useful for large sets of diverse compounds where the extremes tend to be undesirable compounds. However, for large (virtual) libraries initial filtering can remove these before the analysis, and thus a nonuniform binning scheme was suggested (59), so that acceptable compounds are not lost as outliers, and is now the preferred option. Often, the large population of compounds used as the basis for defining a chemistry space is the entire compound collection available to a pharmaceutical company for its drug discovery efforts, together optionally with structures from commercial databases of biologically active compounds. The resulting chemistry space can be regarded as the "corporate standard chemistry space" and provides an ideal basis for comparing large sets of compounds such as alternative commercially available compound collections or alternative combinatorial libraries. It is also a good basis for comparing small sets of compounds such as compounds with reasonable affinity for various bioreceptors.

The axes of a corporate standard chemistry space are intended to represent all aspects of molecular structure. Thus, all axes of the corporate chemistry space must be considered for purposes such as general diverse subset selection or rational compound acquisition. However, not all aspects of molecular structure may be important for understanding structure-activity relationships (SARs) for a particular receptor. This led Pearlman and Smith (11d) to introduce the concept of a receptor-relevant subspace (RRSS) of a full chemistry space. For example, starting with a chemistry space of six dimensions, defined to best represent the diversity of all druglike compounds in the MDDR (MDL Drug Data Report) database (13a), they showed how to perceive the three-dimensional subspace that conveys information that is particularly relevant for affinity to the ACE (angiotension converting enzyme) receptor. ACE inhibitors of diverse structure were tightly clustered with respect to the receptor-relevant metrics, thereby providing an obvious near-neighbor strategy for lead follow-up. They (11d) also emphasized the importance of not considering metrics that are not "receptor-relevant" when computing distances for such near-neighbor-based discovery efforts. This also enables diversity in these other dimensions to be explored (e.g., with combinatorial libraries), to obtain compounds with a modified profile for other properties such as bioavailability.

Work on the design and diversity analysis of large combinatorial libraries at Pharmacopeia using BCUT metrics and DiverseSolutions was reported by Schnur (32). A cell-based analysis of synthon-derived libraries was performed, using full product libraries, including library comparisons. Active molecules in these libraries, which involved multiple scaffolds, were found to cluster in various three-dimensional subspaces of the diversity spaces. The utility of a simple property-based reactant/synthon selection tool was also described, targeted at the synthetic chemists, with reactants binned according to patterns based on the ranges of a set of user-selected properties that form a diversity hypothesis.

Chemistry space metrics have been used at Rhône-Poulenc Rorer for diversity analysis, library design, and compound selection (59, 80) using DiverseSolutions to generate a "universal" chemistry space for use as a standard for profiling structural sets of interest. The complementarity of three different diversity measures for comparing and profiling compound collections (a corporate database, com-

binatorial libraries, and the MDDR drugs database) was also shown. The methods used were a 2D structural characterization (Daylight fingerprints), DiverseSolutions, and 3D pharmacophore fingerprints. A combinatorial library of 100,000 structures appeared structurally different from the other databases by the Daylight fingerprint clustering, yet the bulk of its compounds overlapped with druglike compounds (MDDR) in DiverseSolutions BCUT chemistry space and 3D pharmacophore space ("cells" in fingerprints). It was shown and "quantified" that new diversity relative to the company database was explored, with much of this new diversity in desirable areas occupied by MDDR compounds. The nonuniform binning scheme was developed to enable the use of chemistry spaces scaled to include all structures within a set, while maintaining a reasonable distribution of compounds within cells. The method was used to select a subset for initial screening of a large set of combinatorial libraries designed for 7-TM GPCR targets.

2.2.1.2 Pharmacophore Fingerprints. Pharmacophore fingerprints can also be considered as a high dimensional partitioning of the compound space (35). Underrepresented pharmacophores within a population can be identified and act as a possible focus for library design or compound acquisition. Using six feature types (hydrogen bond acceptor, donor, acid, base, hydrophobe, and aromatic ring centroid) with four-point pharmacophores and 7–10 binned distance ranges, it is possible to resolve about 2–10 million different pharmacophoric shapes. Different databases can be compared using this fingerprint, and differences identified. For example, by comparing a corporate screening file (100,000 structures) with the MDDR database (62,000 structures) of biologically active compounds (as discussed above for DiverseSolutions, Refs. 62, 80) "holes" could be identified, in terms of about 1 million 3D pharmacophores exhibited only by MDDR compounds (about 2.7 million were in common and 0.2 million unique to the corporate set). This provides a design space for which combinatorial libraries were designed and synthesized. A total of 100,000 combinatorial library compounds were able to match about 40% (0.4 million) of the pharmacophore "holes" (i.e.,

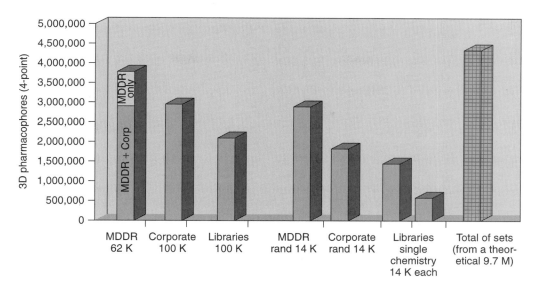

Figure 5.7. Comparisons of the 3D four-point pharmacophore fingerprints exhibited by several sets [MDDR database of 62,000 biologically active compounds, a corporate registry database of 100,000 compounds used for screening, 100,000 compounds from combinatorial libraries (from a four-component Ugi condensation reaction), and 14,000 compound random subsets (MDDR, corporate) or individual libraries]. The four-point potential pharmacophores were calculated using 10 distance range bins and the standard six pharmacophore features.

MDDR pharmacophores not in corporate set), and additionally explore about 0.3 million new pharmacophores. Figure 5.7 illustrates the number of pharmacophores found in these sets, together with those for the ACD (Available Chemicals Directory), random 14,000 subsets of the database sets and some of the combinatorial libraries ($\sim$14,000 each, from a four-component Ugi condensation reaction, $12 \times 12 \times 12 \times 8$ reactants). The relative richness and diversity of the MDDR database, which includes structures from a large number of companies, is clear from the comparisons. The contributions, and eventual diminishing return, of successive libraries using the same chemistry is discussed in Section 5.1.2 (see Fig. 5.24 below).

An example of the use of 3D pharmacophore fingerprints for the design of GPCR libraries (37a) using "relative" fingerprints focused around privileged substructures is described in Section 5.1.2. An approach that combines an optimization of a four-point pharmacophore fingerprint and BCUT chemistry space diversity, using simulated anneal-

ing, has been described (37d; see Section 4.7). Simulated annealing is a widely used optimization methodology whereby the "temperature" of the system is used to control the degree of sampling of solution space. The "temperature" is cooled or annealed as the run progresses so that the system moves into a minimum for the function at low "temperature." In the classical sense, temperature controls the kinetic energy of the system; in a more general sense, the "temperature" has no physical meaning and is a parameter to control the sampling of solution space. Diversity was the goal (function to be optimized) of the studies reported, but the approach can equally be applied to optimize to a desired distribution of properties (e.g., from sets of biologically active compounds). The power of this pharmacophoric approach has been exemplified by Leach et al. in their GaP protocol for monomer acquisition (44).

Pharmacophore fingerprints derived from complementary site points to a target binding site have been used as a quantification of "biological diversity"/structure-based diversity

(37), defining a measure of the intersection between chemical and biological space. They can be compared to the pharmacophore fingerprints calculated from ligands, and the pharmacophore fingerprints of different target binding sites can also be compared to identify similarities (e.g., common binding motifs) and differences (e.g., for selectivity). The four-point pharmacophore fingerprint of a serine protease binding site was used to quantify all the possible binding modes. An example was given of how a combinatorial library could be designed to match as many as possible of these site pharmacophores, with the idea that the biological screening of the resultant library would provide information as to which hypotheses lead to (the best) binding. The site points can be generated by both geometric methods (as implemented in Chem-X/Chem-Protein; see Ref. 133) or through energetic surveys of the site [e.g., by using a variety of probe atoms (as implemented and used for pharmacophore fingerprint generation) (37); see Section 2.1.3.1].

The pharmacophore fingerprinting method thus provides a novel method to measure similarity when comparing ligands to their binding site targets, with applications such as virtual screening and structure-based combinatorial library design, as well as to compare binding sites themselves. Flexibility of the binding site can also be explicitly accounted for by using a composite fingerprint generated from several different binding site conformations.

2.2.2 Cluster-Based Methods.

Clustering methods have a long history of application in chemical information (60). Any set of descriptors can be used in the clustering, but most typically some form of structural fingerprint is used in conjunction with a similarity measure such as the Tanimoto coefficient (see Section 2.1.4.1). The methods fall into two broad classes, hierarchical and nonhierarchical.

Nonhierarchical methods such as that described by Jarvis and Patrick (61) have been widely used for compound selection from large databases (62). The principle behind the Jarvis-Patrick method is to group together compounds that have a large number of nearest neighbors in common. However, the method requires the user to specify the number of clusters desired, and tends to be prone to singletons (clusters of one) and/or a small number of very large clusters. The cascade clustering methodology (59b) was developed to address some of these issues. Parameters were selected to produce an acceptable size distribution for the largest clusters and the small clusters were then reclustered. Doman et al. (63) have developed a fuzzy clustering technique, also based around the Jarvis-Patrick algorithm but which has no user-defined parameters and allows a compound to belong to more than one cluster.

Hierarchical methods can be further subdivided into agglomerative and divisive methods. Agglomerative methods start with each compound in a separate cluster and iteratively join the closest clusters together. Divisive methods start with a single cluster and iteratively subdivide until each compound is a singleton. Hierarchical clustering methods generate a dendrogram showing the relationship between the compounds, the issue being the level at which to cut the hierarchy (i.e., how many clusters to generate). Although heuristics exist, there is no automated method. Such algorithms, however, at best scale to order (N^2) in time, where N is the number of compounds, and so are limited in application to a few hundred thousand compounds at most (64). Nevertheless, they have been shown to be superior to nonhierarchical methods for clustering of chemical compounds (65). Ward's method was shown (5) to be the most effective at separating active from inactive compounds by clustering bit strings that describe the presence or absence of 153 small generic and specific fragments (ISIS structural key descriptors). Even better performance was obtained with the inclusion of pharmacophore distances between site points complementary to hydrogen bonding and charged groups combined with distances between centers of aromatic rings and attachment points for hydrophobic groups.

2.2.3 Dissimilarity-Based Methods.

The methods for compound selection described above essentially group compounds either by partitioning into cells or by clustering. Dissimilarity-based methods (66) avoid this step.

The subset selection can be performed itera- tively. The first compound is chosen at ran- dom and the next compound is selected to be maximally dissimilar to the first; the third is then selected to be maximally dissimilar to the first two, and so on. The selection stops when a prespecified number of compounds have been selected or no more compounds can be chosen that are below a given similarity or above a certain distance to another compound in the selected set. Pearlman (11b,c) refers to such methods as "addition" algorithms because they add compounds to a diverse set of increas- ing size. He notes that such algorithms are quite satisfactory when the size of the desired subset is relatively modest but, given that the time required for such algorithms is propor- tional to the size of the total population and the square of the size of the desired diverse subset, they are far less satisfactory when, for example, selecting a subset of 10,000 from a population of 1,000,000.

Alternatively, the number of desired com- pounds can be predefined and a stochastic al- gorithm used to maximize the diversity of the selected set, although these methods are even slower than addition methods. Sphere-exclu- sion methods, which Pearlman calls "elimina- tion" algorithms because the diverse subset is created by eliminating compounds from the superset, have been implemented in Diverse- Solutions (31) (see Section 2.2.1.1), providing a rapid distance-based diverse subset selection method. The minimum distance between nearest neighbors within the diverse subset is first defined (D_{min}), a compound is chosen at random, all compounds within D_{min} are re- moved, a second compound is chosen at ran- dom, all compounds within D_{min} are removed, and this is repeated until no more compounds can be chosen. The algorithm controls the size of the resulting diverse subset by automati- cally repeating the process with a larger or smaller value of D_{min} as necessary. Because the time required for each elimination sub- set is proportional to the size of the superset and size of the subset (not size of subset squared), the elimination method, despite the need for (typically) four or five automatic repetitions, is far faster than the addition method and yields subsets of essentially equal diversity.

Maximum dissimilarity-based methods tend to give diverse selections, including many out- liers of less potential interest to a medicinal chemist. By contrast, methods such as sphere exclusion (minimum dissimilarity selection) tend to give representative selections that mimic the underlying distribution of com- pounds. The OptiSim method developed by Clark (66c) attempts to achieve a compromise between these two extremes. Three parame- ters are required: the first two, radius or sim- ilarity cutoff and maximum number of selec- tions are common to the other algorithms. A third parameter, K, is required to define a sub- sample size. Up to K selections are added to the subsample at each iteration and the best compound from the subsample added to the selected set. At the limit of $K = 1$, this is equiv- alent to minimum dissimilarity selection, whereas at the limit of $K = N$ (total number of compounds) the algorithm is equivalent to maximum dissimilarity selection. By altering the value of K the user can thus achieve a compromise between the diversity and repre- sentativeness of the selected set. Tests of the algorithm suggest that it is possible to achieve selections similar to those achieved from hier- archical clustering methods at a greatly re- duced computational cost. Maximum dissimi- larity methods were shown (66d) to lead to more stable QSAR models with higher predic- tive power, based on a comparative mean field analysis of angiotensin-converting enzyme inhibitors.

Hudson et al. (67) describe two parameter- based methods for compound selection. The most descriptive compound (MDC) method is aimed at selecting compounds that represent the population as a whole. An information vec- tor is accumulated from the ranked Euclidean distance of each compound in the data set to all others. The most descriptive compound is that with the largest information, which equates to the compound with the smallest overall distance to all other compounds. The next compound is chosen to give the greatest additional information and so on. The sphere- exclusion method used attempts to select com- pounds that most effectively cover the prop- erty space. A compound is selected, say the MDC, and all compounds are removed that are closer to it than a user-defined radius. The next selection is that compound in the remaining set that is closest to the one already selected. This process is repeated until no compounds are left for selection. The methods were applied to the selection of standard sets for biological screening at Wellcome.

The maxmin approach (23c) uses the shortest nearest-neighbor distance as a measure of diversity in the sample:

$$D_{\max \min} = 1 - \frac{1}{N} \sum_{i=1}^{N} \max_{i} [\min_{j \neq i} (d_{ij})]$$

This measure is particularly useful in selecting diverse compound sets from a corporate collection, as exemplified by Higgs et al. (68). They also introduce the concept of a coverage design for lead follow-up, in which compounds are selected to be maximally similar to a set of leads.

An alternative approach is to use the sum of pairwise similarities in the maxsum approach:

$$D_{\max \text{sum}} = 1 - \frac{\displaystyle\sum_{i=1}^{N} \sum_{j=1}^{N} \text{sim}(i,j)}{N^2}$$

This approach is particularly efficient when combined with the Cosine coefficient (69) and was used by Pickett et al. in combination with pharmacophore descriptors (70). In lower dimensional spaces the maxsum measure tends to force selection from the corners of diversity space (6b, 71) and hence maxmin is the preferred function in these cases. A similar conclusion was drawn from a comparison of algorithms for dissimilarity-based compound selection (72).

An excellent discussion of different diversity functions has been given by Waldman et al. (73). A set of ideal behaviors for a diversity function was defined. These are particularly relevant to library design tasks. Thus, although maxmin is suitable for selecting highly diverse molecules, it is less well suited to library design optimization processes. An alternative function was defined based on minimum spanning trees, which had previously been used by Mount et al. (71) and a gaussian error function, erf( ):

$$D = \sum \text{erf}(\alpha d_i)$$

The reader is referred to reviews (e.g., see Refs. 2, 6, 7) for further detail and discussion of other related measures and methods for diversity-based selection.

2.2.4 Biasing to Desired/Desirable Properties.

Any of the methods above can be used to bias the compound selection toward a particular region of property space, for example, by restricting the selection to cells or clusters containing known actives (in the latter case it may mean reclustering if the active is from an external source). However, it is a common experience when applying diversity-based selection to large databases to see a number of compounds that are undesirable for a number of reasons. They may be too large, too flexible, too lipophilic, contain too many acid groups, and so forth. Thus, it is general practice to apply filters during the selection process. These can include limits on property values such as calculated $\log P$ and molecular weight (68) and the application of substructure filters to remove undesirable or reactive compounds (21a, 66a, 74). Lipinski (29) formalized some of these ideas in the *Rule-of-5*, derived from analysis of orally absorbed drugs, with problems more likely if two or more of MW > 500, $c \log P > 5$, sum NH, OH ($\sim$ H-bond donors) > 5, sum N,O ($\sim$ H-bond acceptors) > 10. Variants of this, often stricter (e.g., only one violation and/or lower values for hits/leads), are now widely used in conjunction with other methods for classification of compounds as likely to be orally absorbed or to penetrate the CNS. The reader is directed to several reviews on this topic (75). The usefulness of such approaches has been shown by the work of Pickett et al. (76), where a library was designed using simple descriptors such as polar surface area for oral absorption (77). The designed library showed improved absorption in a Caco-2 system over a previous related library where the products had not been formally designed to these criteria.

In a more general sense, compounds can be selected to reproduce a given set of property profiles for calculated $\log P$, molecular weight, and so forth derived from, say, a set of known

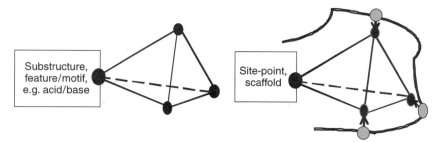

Figure 5.8. Example of privileged four-point pharmacophores, either created from a ligand using a particular feature (e.g., the centroid of a "privileged" substructure) or complementary to a protein site using a site point or attachment point of a docked scaffold. Only pharmacophores that include this special feature are included in the fingerprint, thus providing a relative measure of diversity/similarity with respect to the privileged feature.

drugs. Such an approach is most widely used as an additional constraint in library design algorithms (78) and is further discussed below.

An interesting example of biasing in compound selection is provided by Grassy et al. (79). Lead compounds were used to derive a range of acceptable values for topological indices and other molecular descriptors. These were used to filter a large virtual library and led to an active compound being synthesized.

2.2.5 Relative Diversity/Similarity. This describes an approach that measures "relative" similarity and diversity between chemical objects, in contrast to the use of the concept of a total or "absolute" reference space (80). The ability of 3D pharmacophoric fingerprint descriptors to separate ligand-binding properties from chemical structure has enabled a useful modification to the way the descriptor is evaluated (37). It is possible to identify one of the points of a pharmacophoric description such as a triplet or quartet with a special feature, such as a "privileged" substructure deemed important for binding or a pharmacophore group. A fingerprint can be generated that describes the possible pharmacophoric shapes from the viewpoint of that special point/substructure (see Fig. 5.8). This creates a "relative" or "internally referenced" measure of diversity, enabling new design and analysis methods. The technique has been extensively used to design combinatorial libraries that contain "privileged" substructures focused on GPCRs (37a), and this is described

further in Section 5.1.2. The use of "receptor-relevant" BCUT chemistry spaces from DiverseSolutions provides a different approach to a focused similarity/diversity measure (11d, 32e-h).

3 VIRTUAL SCREENING BY MOLECULAR SIMILARITY

The use of molecular similarity to analyze large databases of structures using information derived from one or several ligands provides a powerful ligand-based virtual screening method (protein structure-based virtual screening methods are by comparison based on docking structures into a binding site). Virtual screening requires that a set of structures is ranked, with the goal of identifying new structures that have similar biological activity, with top-scoring compounds sent for evaluation in a biological assay. Usually, the requirement is to provide a small subset of compounds (10–1000) from a large set (100,000–1,000,000) of possible compounds for screening that is enriched in actives (i.e., contains a greater proportion of actives than that of the full compound set). In this context, enrichment involves identifying the highest number of new chemotypes as opposed to analogs of the query structure(s). Pharmacophoric methods have been found to be particularly effective for this, building on the successful use of 3D database searching for lead generation. Other similarity methods such as the use of 2D descriptors (Section

2.1.1) are also commonly used to identify structures for screening based on the structure of a known ligand. The use of similarity searching in chemical databases has been reviewed by Willett et al. (2a), comparing newer types of similarity measure with existing approaches. In this section the focus is on the use of the 3D pharmacophoric methods, which have been shown to provide a ligand-based virtual screening method that yields new chemotypes.

3.1 Use of Geometric Atom-Pair Descriptors

The topological atom-pair descriptors (24) have been extended by Sheridan and coworkers to geometric atom pairs (26), and shown to be effective at generating hit lists enriched in active molecules of different chemotypes. A set of precalculated conformations ($\sim$10–25) is used for each molecule, and each atom is assigned two different atom types: (1) a binding property (donor, acceptor, acid, base, hydrophobic, polar, and other); (2) a combination of element type, number of neighbors and π-electron count. All combinations of atom pairs are analyzed, for each conformation, and resultant histograms of each probe and database molecule conformation are compared. The technique was compared with its topological equivalent (counting bond connections between atoms to estimate interatomic "distance"). This demonstrated that, although both methods were able to significantly enrich the highest ranking structures with other active molecules for the same target ($\sim$20- to 30-fold enhancement over random in the top 300 compounds), the 3D structure-derived descriptors were able to show their advantage by picking out active chemotypes with greater structural variation relative to those from the 2D searches. The analysis used about 30,000 structures from the Derwent Standard Drug File (SDF; version 6, developed and distributed by Derwent Information Ltd., London, England, 1991, now known as the World Drug Index) using probe molecules with known activity against a particular target to rank the database. Sheridan et al. (26c) have also shown how a single combined atom-pair descriptor from a set of molecules can be used in a single fast search to provide results similar to those from the slower process of individual molecule-by-molecule searches. This provides the ability to search mixtures, which some companies use for high throughput screening, in that both the search query and/or the database being searched can be mixtures of structures.

3.2 Use of 3D Pharmacophore Fingerprints (Three- and Four-Point)

Some research groups have extended the atom-pair descriptors to three-point (triplets) and four-point (quartets) pharmacophore descriptors (35, 37, 76, 81) as described in section 2. These descriptors have a potentially superior descriptive power, and a perceived advantage over atom pairs is the increased "shape" information (intrapharmacophore distance relationships) content of the individual descriptors (37a). The quartet (tetrahedral) four-point descriptors offer further potential 3D content by including information on volume and chirality (37a, 82), compared with the triplets that are components of the quartets and represent planes or "slices" through the 3D shapes.

The fingerprints can be precalculated for database compounds, with conformational sampling, and stored in an efficient format (e.g., four-point pharmacophore fingerprints, where one line of encoded information uses about 11 kilobytes of space for 1000 pharmacophores). Probe fingerprints from one or more structures can be rapidly compared against such databases at speeds of >100,000 compounds/min, even for large four-point pharmacophore fingerprints, representing about 10 million different pharmacophoric shapes. Similarity is measured using potential pharmacophore overlap and similarity indices such as the modified Tanimoto index (37a).

The relative merits of two-, three- and four-point pharmacophore descriptors for different applications is an area of ongoing study (37, 83). Figure 5.9 shows some structurally diverse endothelin antagonists that exhibit low 2D similarity, but maintain significant overlap of their four-point pharmacophore fingerprints (37a).

3.3 Validation Studies

The validation issue for 1D, 2D, and 3D descriptors for similarity searching and virtual

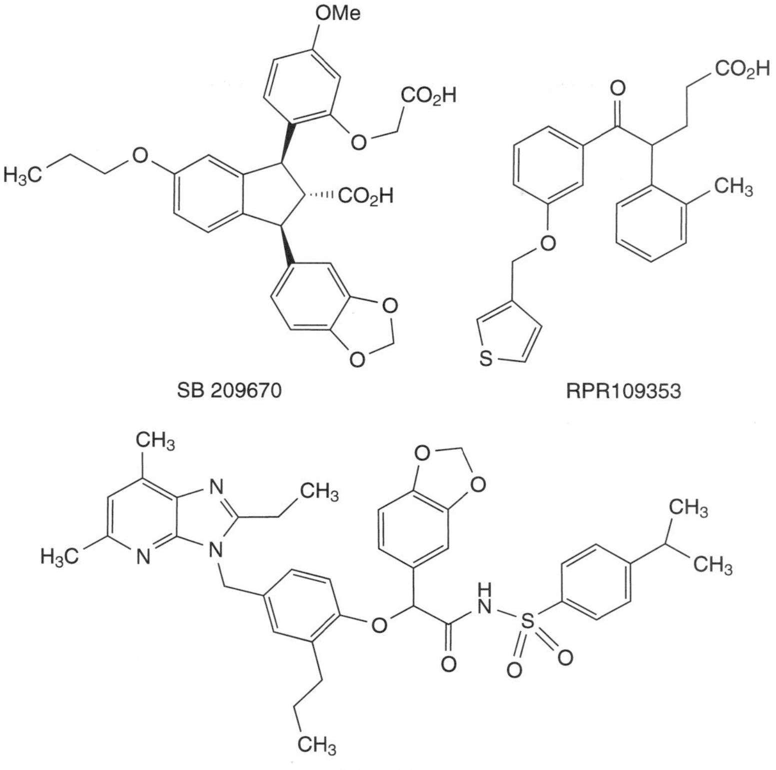

Figure 5.9. Structurally diverse endothelin antagonists exhibiting low 2D similarity while maintaining common pharmacophoric elements crucial to activity.

screening has been addressed in several publications (5d, 14a, 45, 72, 84, 85). Conflicting results have been reported, probably because of the way the different descriptors were used and biases in the test sets. Two primary concepts have been applied to the analysis of biological data. The concept of "neighborhood" behavior (84) as a measure of descriptor utility has been promoted, based on the idea that if a descriptor is able to cluster molecules with a particular biological activity, the descriptor encodes information regarding the requirements for that activity, and by extension is a useful measure for molecular similarity/diversity. Comparisons using 2D fingerprints with pharmacophore fingerprints with this approach led to the conclusion that 2D descriptors performed better than their 1D and 3D counterparts (14a, 45). However, issues with the studies undertaken have been raised (85).

These relate to bias in the data sets arising from the presence of closely related analogs, which by their nature have high 2D substructural similarities, and the way the 3D pharmacophoric descriptors were generated (single conformation only) and used (bin setting, Tanimoto index).

Some comparative studies of ligand-based virtual screening methods have been undertaken within Bristol-Myers Squibb (85) using more optimum settings for pharmacophore fingerprint generation [four-point pharmacophores, 7 distance bins, and full conformational analysis (37a)], which gave quite different results. An example using melatonin as a probe molecule to search against a database of about 150,000 compounds containing about 250 known melatonin antagonists is shown in Fig. 5.10. The graph shows the hit rates obtained by similarity ranking in terms of the

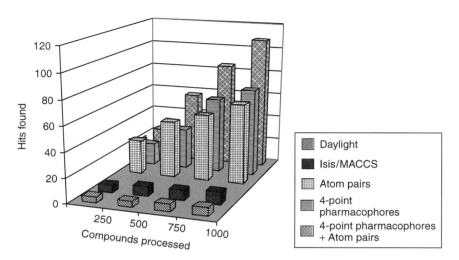

Figure 5.10. Ligand-based virtual screening/similarity analysis of a 150,000-compound database containing about 250 known melatonin antagonists, showing the strong performance of the pharmacophore descriptors, and the complementarity of atom pairs and pharmacophore descriptors when combined.

number of active compounds located across the top-ranking 1000 compounds. In this case, for the 2D descriptors shown, only the atom-pair (26a) descriptor (which has elements of a two-point pharmacophore fingerprint with intercenter distance replaced by bond count) produces comparable results. A 2D similarity search using the UNITY 2D fingerprint (13c), with a very low 50% similarity cutoff, produced a hit list of 1669 compounds containing only 10 melatonin actives (an enrichment of 3.6 relative to random screening of 1669 compounds, for which 2.8 actives would be expected to be found). In contrast, the pharmacophore fingerprint similarity search finds 93 actives in the first 1669 compounds, a total enrichment of >33 relative to random, and a further >ninefold relative to the 2D similarity search. Preliminary studies on systems with a much wider structural diversity in active ligand chemotypes suggested hit rates even more favorable to pharmacophore fingerprints, with cases where no 2D methods were able to improve on random hit rates. It is of interest that averaging the four-point pharmacophore/atom-pair rankings leads to even better results in the melatonin investigation, highlighting (45) the potential advantages of combined descriptors.

The advantage of 3D pharmacophore-based geometric descriptors over topological descriptors in being able to pick up new chemotypes with major 2D structural variations is of particular importance when exploiting a peptide lead. In such cases, the goal of screening is normally a nonpeptidic molecule. Using the pharmacophore fingerprint from relatively large and flexible peptidic molecules (e.g., tetrapeptides), it is possible to identify structures that match just part of the pharmacophoric information; a modified Tanimoto coefficient can be used to reduce the penalty for only a partial match. It is possible to identify a set of reasonable structures that, as an ensemble sample most of the potential pharmacophores exhibited by the peptide. With 2D methods the high ranking molecules will tend to be similarly peptidic, rather than more druglike molecules exhibiting 3D properties of the peptide.

An example of the use of peptidic information comes from the work of Pickett et al. (76) using the known tripeptide RGD (Arg-Gly-Asp) (see Fig. 5.11) motif fibrinogen uses for receptor binding (86). A database of 100,000 compounds, which had been seeded with fibrinogen receptor antagonists covering a wide range of structural classes (see Fig. 5.12), was

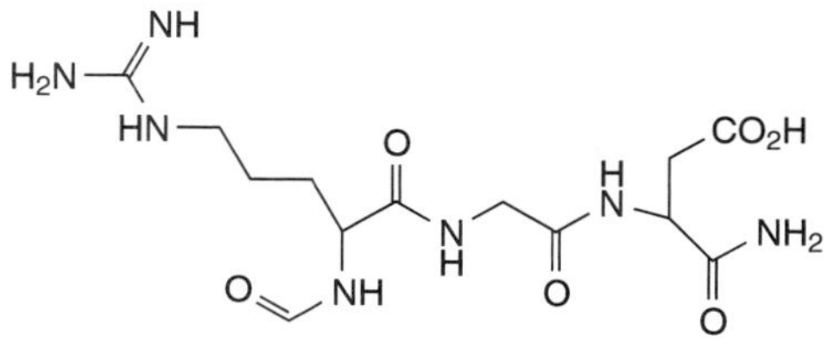

Figure 5.11. Structure of the RGD motif.

screened using a customized (35, 37, 70) version of Chem-X/ChemDiverse with four-point pharmacophore fingerprints. A degree of flexibility of the RGD motif loop region was suggested by structural biology studies, highlighting an advantage of the descriptor encoding a conformational ensemble and not requiring a bioactive conformation to be known. Virtual screening can thus be undertaken even when only limited structural information is available. The four-point pharmacophore fingerprint of the RGD tripeptide probe, N- and C-capped with amide groups, was generated with a full conformational analysis and used for the search. All the actives appeared in the top 3% of the data set, with a reasonable diversity of hits and little 2D resemblance to the probe peptide.

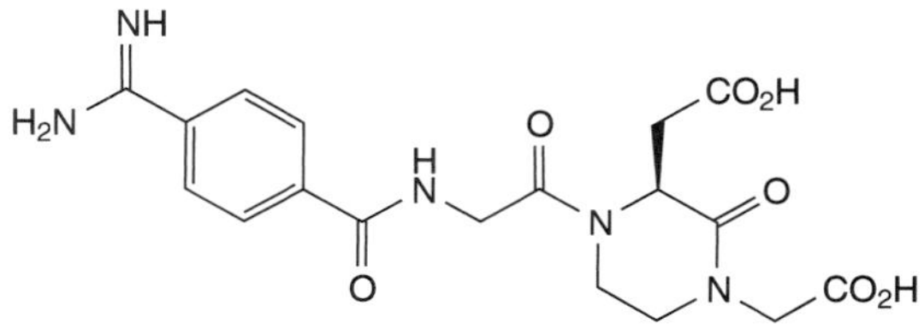

TAK029

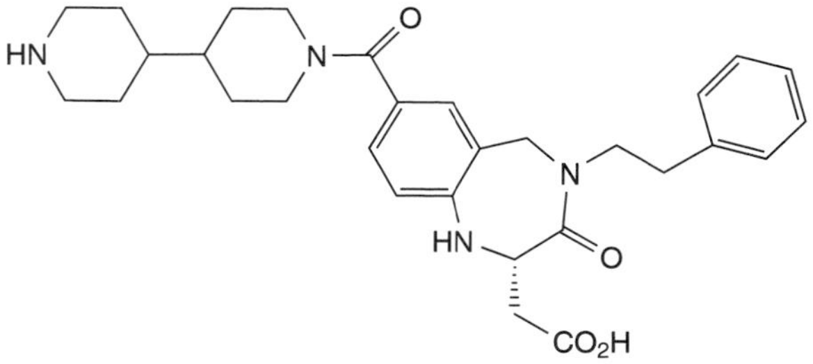

SB214857

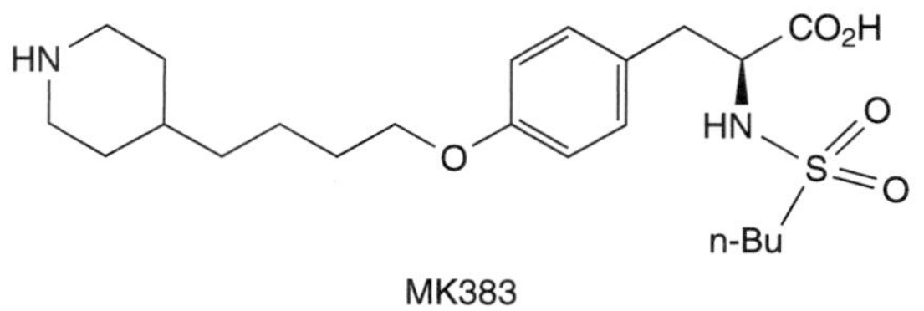

MK383

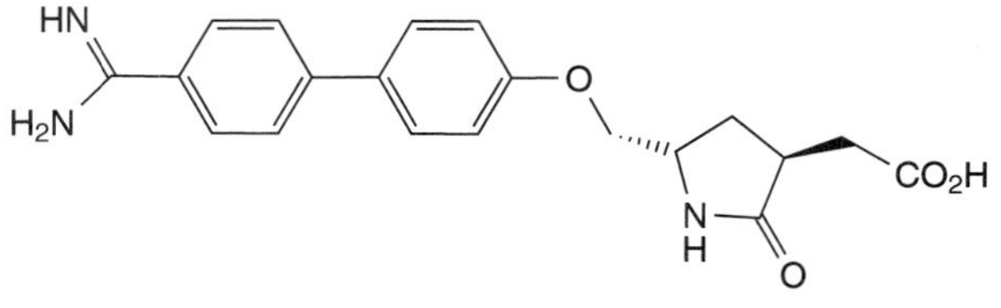

BIBU52

Figure 5.12. Some structurally diverse RGD antagonists matched through virtual screening using the RGD motif and 3D pharmacophore fingerprints.

4 COMBINATORIAL LIBRARY DESIGN

4.1 Combinatorial Libraries

Normally, the term *combinatorial library* implies a library of a few hundred to many thousands of products produced using high throughput robotics in a facility dedicated for such purposes. In contrast, the term *parallel library* implies a library of less than 10 to a few hundred products produced using more or less traditional medicinal chemical synthetic procedures or increasingly common low throughput robotics. In either case, lists of reactants are combined in a combinatorial fashion to yield an array of products. The methods described in this section are applicable equally to both cases. A strictly combinatorial combination of reactants (or reactants and scaffold) produces the most efficient use of reactants and automation/robotics for a library synthesis. However, the issue of generating products that have suitable properties for biological screening and as hit/lead material is key, and constraints discussed below are often used, resulting in not all combinations of reactants being used.

4.2 Combinatorial Library Design

The process of combinatorial library design brings together many molecular diversity and similarity approaches with the aim of identifying a set of reactants that are to be combined (reacted) to form products. Combinatorial library design is, inevitably, an iterative process: software is used to suggest lists of reactants; chemists accept some suggestions but reject others (for various reasons ranging from cost or availability to poor synthetic yield). If software is to be used to suggest replacements for rejected reactants, it must be designed to accommodate this iterative process.

Although the objective is always to identify which reactants should be used to make the products, there are two fundamental approaches to library design: reactant-based methods and product-based methods. Purely product-based methods, which select (or cherry pick) desired products without regard for the number of reactants required to form those products (as in standard similarity searching or diverse set selection), lead to non-combinatorial synthetic schemes and are clearly at odds with the efficiency objectives of a combinatorial synthesis. Reactant-based methods suggest lists of reactants based solely on comparisons of reactant properties without regard for the properties of the resultant virtual products. Thus, reactant-based methods avoid the need for enumerating what could be very large numbers of virtual products and making even greater numbers of comparisons of product properties. Compromise solutions, discussed later, that approximate certain product properties from the reactants without enumeration have been developed. By directly selecting a desired number of each type of reactant, the chemist can ensure an efficient, full-combinatorial array design, and the expediencies of reactant-based methods led to their widespread use for the design of both large and small combinatorial libraries. However, the growing awareness of a need to maintain druglike properties, if only for practical issues such as solubility, has led to the use of reactant-biased methods that consider the properties of the products from enumerated structures, for which the full array of ligand- and structure-based design methods can be applied, and the resultant synthesis of sparse arrays (see later). In addition, the assumption that optimal product diversity can be approximated by using diversity-optimized reactants has been questioned (87), and several product-based diversity methods for selecting reactants have been developed that consider the need for full or sparse combinatorial arrays in the design process. The rationale behind this observation can be understood when one considers that "constraints" based on whole molecule properties and some form of molecular similarity/diversity are usually required.

Pearlman (11, 87c) has made this argument quite convincingly by comparing the results of alternative diverse library design methods in a low dimensional chemistry space, as illustrated in Fig. 5.13. Figure 5.13a depicts a virtual library of 634,721 allowed combinatorial AB products (remaining after optional filtering of the full virtual library based on Lipinski's Rule-of-5 "druglikeness" criteria) in a chemistry space specifically cho-

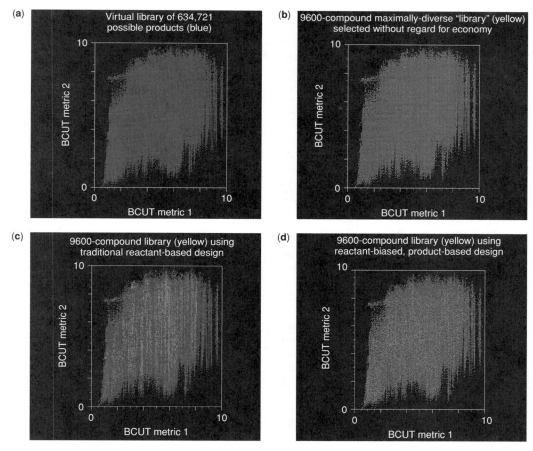

Figure 5.13. (a) A virtual library of 634,721 allowed combinatorial AB products (after filtering out products that failed Lipinski's Rule of 5 "druglike" criteria) shown in a BCUT chemistry space specifically chosen to best represent the diversity of the virtual library. (b) The maximally diverse 9600-compound subset of the virtual library, illustrating the results of purely product-based "library design." Although providing the maximal diversity, synthesis of these 9600 AB products would require the use of 347 A's and 1024 B's-clearly unacceptable from the perspective of synthetic economy (numbers of reactants and robotic control). (c) The 9600-compound library resulting from the traditional, purely reactant-based library design strategy of selecting the 80 most diverse A's and the 120 most diverse B's. Although providing user-selected synthetic economy, the diversity of these 9600 AB products is clearly quite poor. (d) The 9600-compound library resulting from the reactant-biased, product-based (RBPB) algorithm developed by Pearlman and Smith (see Refs. 31, 87c and text). The algorithm selected a different set of 80 A's and a different set of 120 B's, thus providing the same level of user-selected synthetic economy, while also providing substantially greater diversity than could be achieved using a purely reactant-based library design strategy. See color insert.

sen to best represent the diversity of that virtual library. Figure 5.13b illustrates an optimally diverse "library" of 9600 products selected by using cell-based diverse subset selection to cherry pick the 9600 most diverse products without regard for synthetic econ-

omy. Although the diversity of these products is clearly optimal, the fact that 347 A's and 1024 B's would be required to make the 9600 AB products provides an equally clear indication of why purely product-based methods are unsatisfactory from an economical perspective. Figure 5.13c shows the $80 \times 120 = 9600$ products resulting from the typical, purely reactant-based design method in which the 80 most diverse A's (out of 536 commercially available) and the 120 most diverse B's (out of 2061 commercially available) were selected.

The full-array library design is clearly advantageous from an economical perspective, although it is equally clear that the diversity of the resulting products is far from optimal. Clearly, a compromise between the optimal diversity of purely product-based design and the optimal economy of purely reactant-based design would be ideal. Pearlman and Smith have developed a powerful "reactant-biased, product-based" (RBPB) approach to library design in the DVS software (31, 87c) (see Section 2.2.1.1) to enable compromise solutions between reactant economy and design criteria, that can be used for diverse-, property-, and target-focused designs. Figure 5.13d shows the results of using the RBPB algorithm to select a different set of 80 A's and a different set of 120 B's to yield a far more diverse library of 9600 products, while providing exactly the same level of synthetic economy as the traditional, purely reactant-based method used to design the library illustrated in Figure 5.13c. Although these results provide clear indications of the utility of the RBPB algorithm for designing simple diverse libraries, Pearlman notes that the algorithm is of far greater utility for designing focused libraries (and diverse "fill-in" libraries), for which consideration of product properties is clearly essential and purely reactant-based approaches would be inappropriate.

A combinatorial library design generally combines elements of diversity and optionally target or target class focus (from information on ligands and/or the target binding site) together with one or more constraints aimed at restricting the generation of products with unfavorable properties (e.g., poor solubility, absorption). These range from the simple "Lipinski" (i.e., rule of 5) style bounds (molecular weight, log P, hydrogen bonding count) for orally absorbed drugs (29) to more sophisticated "druglikeness" (75a, 87a, 88) to optimizing complementarity to a target protein binding site. Without consideration of some kind of druglike constraint, a combinatorial library

risks generating products that are not only unsuitable as drug leads or candidates, but have such unfavorable physicochemical properties such as solubility that biological screening is not possible. The diversity element ensures that the synthesized library covers the range of accessible compounds within the constraints of similarity to known leads or fit to a protein structure. Design strategies for building druglike chemical libraries have been reviewed (89).

Synthetic constraints clearly have a central role. These include availability and stability of reactants and performance in the desired reaction (yield and product purity). The desired combinatorial efficiency of the final design is an additional constraint: from the most efficient of A + B reactants giving A $\times$ B products (e.g., $10 \times 10 = 100$ products from 20 reactants), through sparse arrays in which only certain combinations of reactants are used [e.g., $(5 \times 5) + (5 \times 5) + (5 \times 5) + (5 \times 5) = 100$ products from 40 reactants] to cherry picking of individual products in which a reactant might be used only once (potentially 100 products from 200 reactants).

The first key step in the design process is thus to identify which reactants are potentially suitable for the desired reaction. This list will often be later modified, requiring a dynamic design process, after trial reactions on a (designed) set of chemically diverse reactants. Pragmatic solutions are often applied, such as to choose some extra reactants initially during the design process, although some compromise on the final design generally results.

The experience of many groups has shown that large, purely combinatorial libraries produce many products of limited use for biological screening, with problems in terms of both yields/purity and physicochemical/druglike properties. The use of sparse arrays and cherry picking from products is thus now common. A desirable and flexible approach is to have full control over which products are synthesized, and a recent publication discusses how this has been achieved at Pfizer UK with the application of noncombinatorial approaches to library production for drug discovery (90), whereby "cherry picking" was enabled in the library production process, allowing the synthesis of a library of compounds with a high degree of control over associated properties.

Thus, the combinatorial library design process brings together many of the methods already described for molecular similarity and molecular diversity coupled to synthetic feasibility considerations. Diversity-based and structure-based approaches to the design of virtual libraries have been reviewed (7, 91a). Both ligand-based and protein structure-based virtual screening methods can be used, with the combinatorial nature of the virtual compounds being exploited to increase the speed of the analysis. Some properties of the products can be estimated rapidly on the fly from the reactants, and products can be generated in the active site. The CombiDOCK approach that can rapidly analyze very large virtual databases in a binding site by connecting reactants to scaffolds docked in multiple orientations is discussed in Section 4.10. A genetic algorithm-based method for the combinatorial docking of reactants has been described by Jones et al. (92), with the application of a ligand-docking genetic algorithm to screening combinatorial libraries.

A challenge in the design of small- and medium-sized focused combinatorial libraries is to harness for use in library design the experience and knowledge gained in generating structure-activity relationships (91b). Screening libraries biased for pharmaceutical discovery are often designed to augment the structural diversity of a chemical library. The approach used in the LASSOO algorithm (93) is based on the identification of compounds from a virtual library that are most different from those already present in a screening set and to a reference set of undesirable compounds, while being simultaneously most similar to a set of compounds with desirable characteristics. An illustration of the method using bit-string structure descriptors is given.

Combinatorial library design approaches have been discussed (94), with the design of library subsets that simultaneously optimize the diversity or similarity of a library to a target, properties (such as druglikeness) of the library members, properties (such as cost or availability) of the reactants required to make them, and the efficiency for array synthesis. They showed that libraries can be designed to

contain molecules constrained to certain drug-like properties with only a small trade-off in terms of the maximum possible diversity.

The design of leadlike combinatorial libraries is an approach of more recent interest. A lower molecular weight starting point is advantageous, in that bulk can be added for potency/selectivity/properties without exceeding "rule of 5" parameters for orally absorbed drugs; otherwise a more labor-intensive step may be needed to identify a smaller active part of the hit. The properties required of library compounds intended to provide leads suitable for further optimization, that may be rather different from final optimized leads, has been reviewed (95).

Thus, library design is a complex optimization problem with often competing constraints, including requirements to have combinatorial efficiency and/or several specified product properties (both desired and nondesired). Methods such as genetic algorithms, simulated annealing, and Monte Carlo optimization have been used, and iterative cyclic approaches applied. The next section describes the application of these methods within the context of library design but the reader should note that some of these methods are applicable only for the design of diverse libraries.

4.3 Optimization Approaches

The most basic product-based selection process used in library design is an order-dependent analysis of products, selecting a compound if it exhibits sufficient "diversity" to products already selected. This approach was used in the Chem-X/ChemDiverse software with three- and four-point pharmacophore fingerprints. A compound was selected if the overlap with the ensemble fingerprint of already selected compounds was less than a user-defined amount; that is, the molecule contains a significant number of pharmacophores not already exhibited in selected compounds. This cherry-picking process is an efficient method for ensuring a high diversity library, but can be a combinatorially inefficient selection for synthesis, with no explicit reference to the constituent reactants being made (see Section 4.2 above for further examples). A preferred selection for combinatorial efficiency is arrays of reactants, in which all reactants from one component of a combinatorial library are reacted with all the reactants in the other components, or sparse arrays, in which subsets of reactants are combined. Additional constraints such as physicochemical properties and flexibility are addressed implicitly by assigning upper and lower bounds for given properties, or controlling the order in which molecules are processed.

To address the issue of using pharmacophore fingerprints in a way that enabled a combinatorially efficient selection of reactants to be selected, and the explicit inclusion of additional molecular properties such as a balance of druglike physicochemical properties and shape descriptors, the HARPick program (78a,b) was created. A stochastic optimization technique [Monte Carlo simulated annealing (96)] was used to enable selections in reactant space, whereas diversity is still calculated in product space. User-defined flexibility for the reactant array sizes was possible, and additional descriptors could be used (e.g., to address the selection of non-drug-like compounds). The pharmacophore fingerprint (three-point, triplets) was used in a nonbinary mode (the frequency of occurrence of each pharmacophore was calculated), and the HARPick diversity measure was tuned to include a term (Conscore) to force molecules to occupy relative rather than absolute voids in pharmacophore space. This avoids the problem of saturation of the fingerprint with large databases in a binary mode, particularly a problem with the three-point pharmacophore descriptors. It was thus possible to design combinatorial libraries that exhibited pharmacophores that were poorly represented in a reference set of compounds. The Conscore constraint score sums the product of the number of times pharmacophore i has been hit for molecules selected from the current data set with the score associated with pharmacophore i for the constraining library. The Conscore term can be inverted, enabling focused designs, in which the selection of products that occupy the more highly occupied bins (e.g., from a set of active compounds) is desired. The flexibility and success of this kind of stochastic optimization methodology has led to its use by many other researchers for library design (5c, 6b, 23d, 78c, 97c,d). Simulated annealing has

been used to perform reactant selection for combinatorial libraries based on three-point pharmacophores (78a,b), as described above, and other metrics (6b, 23d, 97c,d).

Genetic algorithms (GA) are another class of optimization techniques widely used within chemistry (98) that have been explored for library design. A GA is an attempt to utilize the Darwinian process of evolution in an optimization procedure. A solution is represented by a string of fixed length, the chromosome, and is evaluated according to some criterion to give the fitness score, for example, the pharmacophore coverage of the solution (78b). The GA maintains a number of chromosomes (potential solutions) that are ranked on their fitness and are then modified according to operators including mutation, where one element of the string is changed, and crossover, where the string is cut at some position and swapped with equivalent portions of another solution. These new solutions are evaluated and the process is repeated for a defined number of iterations or until all (or most) solutions converge on one result. For library design, the string represents the selected monomers at each variable position of the library. Evaluation involves enumerating the sublibrary defined by the solution and calculating the score associated with the products. The stochastic nature of the process means that the GA is run several times to ensure good convergence.

A GA was used by Sheridan and Kearsley (99) to design peptoid libraries focused to cholecystokinin by scoring on similarity to two peptide leads. Biological activity, rather than a computed fitness, has been used as the score in a directed combinatorial synthesis program (100). Brown and Martin developed GA-LOPED (101) as a way to design combinatorial mixtures. The SELECT program (78c) combines measures of diversity and the physical properties of the designed library. The library can be designed to be both internally diverse and diverse with respect to a reference population. Physical properties are optimized by comparing to a user-defined profile for the property of interest, c log P for example. As for the HARPick approach (78a,b), however, it is necessary to define a weighting scheme between the different elements of the score, which leads to a number of difficulties. Selec-

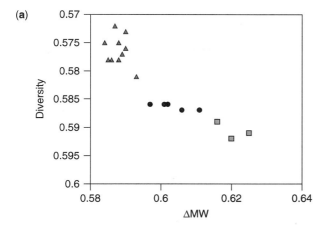

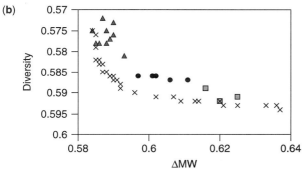

Figure 5.14. (a) Results from multiple SELECT runs with alternative weightings for molecular weight vs. diversity. Filled triangles, 1.0×Div and 1.0×MW; filled circles, 1.0×Div and 0.5×MW; filled squares, 10.0×Div and 1.0×MW. (b) As in a, with results of a single MOGA run shown as crosses. [Reproduced from V. Gillet, et al., *J. Chem. Inf. Comput. Sci.*, **42**, 375–385 (2002) with permission of the American Chemical Society.]

tion of the weights is nonintuitive when comparing different properties or concepts (e.g., diversity and c log P) and the use of weights constrains the search space. Figure 5.14a illustrates how changing the function weight in a SELECT run alters the final solution; note also that the two objectives, molecular weight and diversity, are competing in this example. Given these limitations, a novel modification has been made to the SELECT methodology. The GA in SELECT is replaced by a multiobjective GA (MOGA) (102) that eliminates the need for a weighting scheme. Instead, each element of the scoring function is optimized independently and solutions scored according to the idea of dominance (see Fig. 5.15). The solutions of rank 0, the nondominated solutions, are those solutions for which there is no superior solution when considering all objectives; solutions of rank 1 would be dominated in one objective and so on. It is these ranks that are used to describe the fitness of the solution. Solutions of rank 0 are said to define the pa-

reto surface, as displayed in Fig. 5.14b, which overlays the MOGA results onto the SELECT results. Thus, the MOGA has several advantages over a traditional GA. In one run it generates multiple solutions that are equally valid, more fully explore solution space, and gives the designer an understanding of the relationships between the different objective functions.

The RBPB algorithm of Pearlman and Smith, described above (Section 4.2), considers all possible candidate libraries, which satisfy the user's constraints regarding economy. These include min/max range constraints regarding library size (e.g., number of AB products) and the number of each type of reactant (e.g., number of A's and number of B's). These also include specification of the minimal unit dimensions (MUDs), which define the smallest combinatorial array that the user is willing to address on the robotic table. Each candidate library corresponds to a different way of, at least conceptually, arranging the required

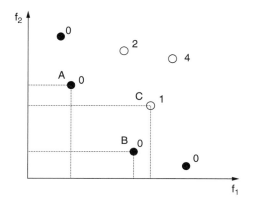

Figure 5.15. Pareto optimality. The filled circles represent rank zero or nondominated solutions for functions $f1$ and $f2$. Point C is rank 1 because it is dominated by point B. (Permission as in Fig. 5.14.)

number of MUDs to construct a library within the user's specified range limits. Each candidate library is scored based on an appropriate function of the scores of the individual products it contains divided by its size; hence, an average product score. The reactants used to make each candidate library are determined by reactant scores, which are functions of product scores and which are updated at each step of the design of that particular candidate library. For example, at a given step in the design process, the reactant score for reactant A_i depends on the scores of the products actually accessible, given the current choices of B-type reactants. The score also depends on the scores of the products that could be made using B-type reactants, which may be selected at a subsequent step in the process. The candidate libraries with the highest scores are output for the user's final decision. In addition to be being remarkably thorough yet fast, the RBPB also makes it very easy to address the iterative nature of library design and to suggest replacements for previously suggested reactants that had to be rejected for one reason or another.

A rapid computational method for lead evolution has been described by the CombiChem (now DeltaGen) group (39). Their 3D computational approach for lead evolution is based on a pharmacophore fingerprint approach using multiple pharmacophore hypotheses. A set or *ensemble* of hypotheses is generated that is most able to discriminate between active and inactive molecules. The ensemble comes from an analysis of a large number of pharmacophore hypotheses, with full conformational sampling for both active and inactive compounds. The ensemble hypothesis is used to rapidly search virtual chemical libraries to identify compounds for synthesis. Large virtual libraries (e.g., a million structures) can be analyzed efficiently. The method was applied to α_1-adrenergic receptor ligands, where heterocyclic α_1-adrenergic receptor ligand leads were evolved to highly dissimilar active N-substituted glycine structures.

LiBrain (103) is a collection of software modules for automated combinatorial library design, including the incorporation of desirable pharmacophoric features and the optimization of the diversity of designed libraries. A Chemistry Simulation Engine module is trained by chemists to determine the suitability of reactants for a specified reaction, to recognize the risk of undesirable side reactions, and to predict the structures of the most likely reaction products, so as to circumvent major bottlenecks associated with automating the process.

Legion and Selector (66c,d, 104) are software from Tripos (13c) for characterization, comparison, and sampling of sets of compounds, including a combinatorial builder (104), with available descriptors including fingerprints and atom-pair distances. Clustering tools (Hierarchical, Jarvis-Patrick, and Reciprocal Nearest Neighbor) and compound selection and diversity comparison methods available include Tanimoto Dissimilarity, the Reciprocal Nearest Neighbor approach, and the OptiSim algorithm (see Section 2.2.3).

4.4 Handling Large Virtual Libraries

The rate-limiting step in a product-based library design process is often the calculation of molecular descriptors. This becomes particularly acute as one moves into the 3D arena, of course, but even the simplest 2D descriptors take a finite time to calculate. In addition, there are the logistics associated with storing virtual libraries of potentially tens of millions of compounds. The ability to search within the possible chemical space of a particular chemistry, as opposed to the limited space of synthesized compounds, is an important component of lead identification because this allows a weak hit from primary screening to be rapidly expanded into a more potent lead. Existing chemical database systems can be used or readily modified to benefit from the combinatorial nature of libraries (64a) but they do not overcome the fundamental issues.

Downs and Barnard (105a) have proposed an elegant solution to these problems using the Markush representation commonly used in chemical patents. The key component of their approach is that descriptor calculation and diversity analysis can be performed without the need for full enumeration of the products. In other words, both storage and calculation will tend to scale as the sum of the number of building blocks in the library rather than the product as in techniques requiring enumeration. The method has been developed into a software suite and released commercially as the LibEngine module of the Cerius^2 suite for combinatorial library analysis and design (105d).

The background and theory behind the approach have been published (105b). In summary, the algorithm relies on identifying a *core* and associated R-groups that define the library. This may or may not be directly related to the manner of synthesis. For example, imagine a tripeptide library synthesized from $20 \times 20 \times 20$ amino acids. The algorithm detects the tripeptide backbone as the core and the amino acid side chains as the R-groups. Partial substructure fingerprints are calculated on a fragment basis representing the core and R-groups taking full account of the attachment to the core and the possibility that a particular path may extend between two R-groups. The partial fingerprints are then combined into the full fingerprint, a relatively fast exercise. The approach is a couple of orders of magnitudes faster than calculating fingerprints from fully enumerated products. Additive molecular properties such as molecular weight, hydrogen-bond donor and acceptor counts, and log P can be calculated in a similar manner as well as topological indices. Fingerprints or property data can also be calculated on demand for use with clustering algorithms, thus avoiding the overhead of storing and retrieving them. There is also interest in extending the approach to 3D property calculation (105c).

An alternative approach has been taken by Agrafiotis and colleagues. In a conference presentation (106) they show how a neural network can be trained on a small sample of enumerated combinatorial products to reproduce 2D molecular descriptors and properties for all library members without the need to construct their connection tables.

A method for rapid similarity searching in large combinatorial spaces using a new algorithm Ftrees-FS was published by Rarey and Stahl (135). The similarity search is based on the feature tree similarity measure representing molecules by tree structures. Combinatorial chemistry spaces are handled as a whole rather than looking at subsets of enumerated compounds. A set of 17,000 fragments of known drugs was used, which could be combined to 10^{18} compounds of reasonable size. A novel ChemSpace approach (45a) for searching large virtual libraries that does not require enumeration has also been developed by Tripos, using shape descriptors (topomeric fingerprints) on the monomers, and has been used for targeted library design (45b).

4.5 Library Comparisons

In the previous sections we described the design of libraries based on a number of user-defined criteria, whether they were focused or whether they were of a more general nature. So far, these designs have been undertaken, treating the library in isolation, with the inclusion of property profiles in methods such as HARPick and SELECT to ensure that the synthesized compounds are of a suitable physical nature. In this sense, the designed library can be said to be internally diverse; that is, the selected compounds are diverse within the limited chemistry space of all virtual products. Even for very large virtual libraries, the chemistry space is still small with respect to the possible chemistry universe. It is very difficult, *a priori*, to address how "diverse" a designed library is compared to a library generated with another set of reactions without having to go through the computationally expensive process of computing all pairwise similarities between members of the libraries.

Nevertheless, questions such as "How diverse is the library compared to the screening collection?" or "Which of the following chemistries should I choose for a library?" are often posed and methods are required to answer them.

Distance-based methods such as clustering can be and have been used but suffer from a number of drawbacks both in terms of speed and the fact that the exercise needs to be repeated for every additional library (i.e., there is no common frame of reference). In addition, all pairwise comparisons would need to be performed. Thus, Shemetulskis et al. (107) used clustering methods to compare the Parke-Davis corporate collection (117,000 compounds) with external compounds from Chemical Abstracts Service (380,000) and Maybridge (42,000). Even today, clustering half a million compounds is a daunting task and interpreting the results is not straightforward. The Jarvis-Patrick method employed by Shemetulskis et al. has several input parameters, including the need to predefine the number of clusters. Voigt et al. (108) compared the National Cancer Institute (NCI) database, a publicly available database of compounds used in the NCI screening program, to a number of compound databases. The diversity of each collection was estimated by the number of compounds selected by use of a diversity-selection algorithm as a function of database size. The similarity overlap between two databases has been determined by calculating the percentage of compounds of the first database for which a compound exists in the second database with a similarity greater or equal to a specified cutoff (109). Such an approach necessitates the calculation of the Tanimoto similarity coefficient of all compounds in a database with all compounds in the other databases. As indicated before, the largest drawback of distance-based methods is that they give no indication of where the voids are within the chemistry space, and searching an additional compound source for interesting compounds would require reclustering.

Therefore, partition/cell-based methods are preferred for such library comparison tasks. They provide a common frame of reference in which it is possible to identify voids within the chemistry space of a population. It must be emphasized that the chemistry space is still defined with respect to a reference population. By comparing the libraries with reference to a population (REFDB), such as a corporate database or a combination of known drug databases, one can make statements such as, library A shows the greatest overlap with REFDB, whereas library B fills the greatest number of empty or low occupancy cells. Cummins et al. (22) used a cell-based approach to compare five databases, including the Wellcome Registry, to select screening sets of diverse compounds. Topological indices and a measure of free energy of solvation were taken as the descriptors and factor analysis was used to combine them and define a four-dimensional chemistry space that was then partitioned. Outliers were removed to allow the partitioning to focus on the most densely populated region. The use of pharmacophore descriptors in such a task was illustrated by Mason and Pickett (4), where the pharmacophore overlap between three libraries was calculated. It was possible to identify the library covering regions of pharmacophore space not covered by the other two. Alternatively, given that library A is synthesized and gives hits in screening, then presumably the library that overlaps best with A should be made. Pearlman and Smith (11d) have adapted their DVS software to identify what they term a *receptor-relevant subspace*, where the BCUT metrics are selected to best group the active compounds within a population (in fact, it is possible to have several groupings of actives within the space) (see Section 2.2.1.1).

Comparing two populations by pharmacophore coverage, although straightforward, does ignore the contribution from individual compounds. This is important, in that two libraries could cover similar regions of pharmacophore space but individual compounds in the two libraries could be displaying different subsets of the total pharmacophores covered. This prompted Pickett et al. (70) to explore an alternative approach. In this case, a number of potential scaffolds were available and the aim was to find which of these would best complement previously synthesized libraries. Virtual libraries were generated using a predefined set of reactants and pharmacophore fingerprints were calculated for these and the previously synthesized libraries. By use of measures proposed by Turner et al. (69b), the virtual libraries were compared to the synthesized libraries at both a whole library level and an individual molecule level. From this analysis it was possible to select the scaffold that best complemented the previously synthesized libraries.

An alternative methodology based on the ring content of a database, using precalculated structure-based *hashcodes* has been proposed (110). The comparison of the hashcode tables can be used to compare two databases and the number of distinct ring-system combinations can be used as an indicator of database diversity. A method for diversity assessment called the *saturation diversity* approach, based on picking as many mutually dissimilar compounds as possible from a database was also proposed. The methods were used to compare a number of public databases and gave similar results.

4.6   Pharmacophore-Based Fingerprints

The examples of GPCR library design (described in Section 5.1.2) and protein-site design for Factor Xa (described in Section 5.3) illustrate the use and relevance of pharmacophore-based fingerprints in library design. A pharmacophoric bias has been a major component of many library designs (111), used in the context of focused or biased libraries. Their broad applicability is important, with the same descriptors being used for diverse library design, screening set selection, and focused library design. This provides a consistent approach that extends to protein-site based pharmacophores as discussed above. Their ability to determine the similarities and differences between structurally diverse molecules and sites is very powerful. An ensemble pharmacophore data set measure is often used, which attempts to condense the individual molecule pharmacophore fingerprints into a single measure that describes the important features of the data set as a whole (36, 37, 78a,b).

McGregor et al. (112) have recently published a version of pharmacophore fingerprinting (the PharmPrint method) applied to QSAR and focused library design that uses a limited basis set of 10,549 three-point pharmacophores. They included the usual six pharmacophoric features, plus an additional definition of *other* for all remaining unassigned atoms. A subset of the MDDR database (13a) was used to define a reference set of bioactive molecules, separated into target classes (gene families). The discriminating power of several molecular descriptors was measured using the target class assignments for this set, and it was found that the pharmacophore fingerprint outperformed other descriptors.

4.7   Combined Pharmacophore Fingerprints and BCUTs

Library design using a simultaneous optimization of BCUT chemistry-space descriptors (11) and four-point pharmacophore fingerprints has been reported (32d, 37d). The authors investigated the feasibility and results in terms of complementarity of simultaneously optimizing two product-based descriptors for reactant selection from large virtual libraries. Diversity around a chosen chemistry was the goal of the studies reported, but the approach could equally be applied to optimize to a desired distribution of properties, say, from sets of biologically active compounds. A simulated annealing algorithm (97) was used to combine both components in a single optimization procedure. The choice was based on the ease of implementation and the ability to include multiple components in the objective (23d), an important goal in many recent designs, if only to modulate physicochemical properties to druglike ranges. In this example a small, fully enumerated virtual library of 86,140 amide compounds was constructed from carboxylic acids and primary amines present in the ACD (Available Chemicals Directory). The products of the optimized and random starting reactant sets were compared using average nearest-neighbor distances, and the Hopkins' statistic (113), which evaluates the degree of clustering in a data set, together with the four-point pharmacophore fingerprint diversity. The potential utility for very large virtual libraries, where precalculation of all the pharmacophore fingerprints would not be feasible, was illustrated by calculating four-point pharmacophore fingerprints for virtual library compounds on the fly. The fingerprints were calculated during the optimization procedure and stored in a compact encoded form, with previously calculated fingerprints reused as needed (calculation times ~1–5 s with conformational sampling per structure on an SGI R10000 machine). Diversity was evaluated for the BCUT chemistry space using the ratio of filled to possible filled cells for the virtual library. Four-point pharmacophore diversity was evaluated by the number of unique pharmacophores and the total number of pharmacophores in the product subset, with the goal to optimize both the pharmacophoric uniqueness of each compound selected and the total number of pharmacophores exhibited. Encouraging results were obtained, with additional work necessary to develop a more general function.

4.8 Oriented-Substituent Pharmacophores

OSPPREYS is a pharmacophore diversity descriptor developed specifically for combinatorial library design by Martin and Hoeffel (43). Advantage is made of the common scaffold, so calculations are performed on the sets of substituents. This enables a more detailed pharmacophoric description of the library products than through calculations that could be practically performed on the enumerated products. By avoiding the problems of having to analyze many products with many conformations per product, and an explicit dependency on the scaffold, a higher spatial resolution could be obtained. The analysis of enumerated combinatorial libraries by pharmacophoric methods is generally limited to smaller virtual libraries, with three- or four-point pharmacophores, and limited conformational sampling, requiring new calculations for every library. The oriented-substituent pharmacophores (OSPs) were developed as a compromise approach between reactant and product-based methods to rectify these limitations. To recapture most of the orienting information that is lost in fragmenting the enumerated products into substituents, two additional points are added to each ordinary one-, two-, and three-point substituent pharmacophore, necessitating approximations through the *combinatorial conformer* and the *template alignment* assumptions. The OSPPREYS analysis does, however, account for up to nine-point pharmacophore similarity in the products of a library with three diversity sites. In addition, the consideration of relatively rigid substituents reduces the number of structures to analyze by up to 10^{10} compared to that of a full product-based analysis. This permits a thorough conformational sampling of very large virtual libraries that would be too slow on enumerated structures. A Euclidean property space for diversity analysis is possible because of the small number of pairwise substituent similarities, enabling options not possible by counting set bits in a library union fingerprint. The database of oriented substituent fingerprints is transferable between libraries, within the restrictions of the noted approximations. A major limitation in using OSPPREYS is that it can be applied only within a combinatorial library, and not between libraries. OSPPREYS is well suited to maximizing the diversity of scaffolds independently, and can be used to build a screening file based on such diversity.

4.9 Integration

The previous sections have outlined the basic methodology that has been developed in the areas of molecular diversity, similarity analysis, and library design. Traditionally, use of these methods was limited to a small number of exponents within a computational chemistry group because it involved bringing together a diverse set of software tools and data sources. Combinatorial and high throughput chemistry is now well integrated into the research process and there is a need for bench chemists to have access to such tools. The Cousin system developed at Upjohn has been in use since 1981 (114) and has recently evolved to the ChemList system. The system includes tools for the browsing of dissimilar compounds from substructure searching, useful in reactant selection, for example. Gobbi et al. (115) have described the development of the CICLOPS system in use at Novartis. The system provides functionality for designing and registering libraries and associated tasks such as accessing reactant availability. Tools are provided for filtering reactant lists and selecting a diverse subset of reactants if required. The system is PC-based and is built around the Daylight chemical information system (13b) and associated tool kits with custom Windows clients to control the process.

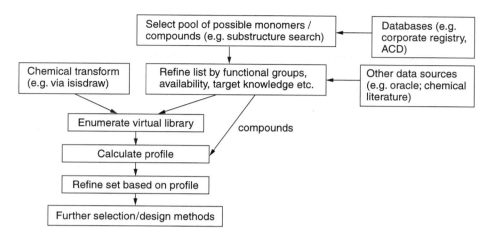

Figure 5.16. The workflow used within ADEPT (A Daylight Enumeration and Profiling Tool; GlaxoWellcome, UK) for compound selection and library design. [Reproduced from A. R. Leach and M. M. Hann, *Drug Discovery Today*, **5**, 326–336 (2000), with permission of Elsevier Science.]

The ADEPT (*A Daylight Enumeration and Profiling Tool*) suite of programs developed at GlaxoWellcome (116) is a Web-based system providing access to a wide range of library design functionality, again based around the Daylight tool kit. Figure 5.16 provides an outline of the process workflow. Reactant lists are generated from searches in databases of in-house and commercially available monomers. A variety of filters can be applied to reduce the size of the lists. These include filters on molecular weight, rotatable bond count, and substructure filters to remove unwanted functionality. After library enumeration, various property histograms are calculated. This allows the user to further refine the reactant choice.

A product-based library design algorithm, PLUMS (117), has been developed to ensure that combinatorial constraints are satisfied in the design. The algorithm successively removes the monomer that adds least value to the library as governed by two terms, the effectiveness (number of molecules meeting user-defined criteria such as property ranges, fit to pharmacophore or dock to protein site) and efficiency (ratio of effectiveness to library size). The algorithm is sufficiently fast to work within the Web-based environment of ADEPT. Figure 5.17 shows screen shots from ADEPT, illustrating how a library can be specified and the resulting product histograms. A

similar system has been implemented at Vertex (118a). A key component of this system is the REOS filtering tool (118b), which applies filters on molecular weight, lipophilicity, unwanted substructures, rotatable bond counts, and so forth to remove "obviously bad" compounds.

4.10 Structure-Based Library Design

Structure-based library design uses 3D structures of the biological targets to direct the design and selection of templates/scaffolds and of reactants that will produce compounds that can fit into the target and thus are likely to bind and have biological activity. The experimental structural information can be derived by a structural biology approach, using X-ray crystallography or NMR spectroscopy. Computational models can be built and used (e.g., homology modeling techniques for closely related proteins), but an experimental structure is always preferred. A structural biology approach can also be used to identify molecules or fragments thereof that bind to a target. For example, NMR screening (3) can be used to identify potential scaffolds or reactants for a combinatorial library that bind to a target site and is able to detect very low affinity binding (in the millimolar range, compared to the low micromolar range from biological screening); this can be done without the need to determine the 3D structure of the target.

(a)

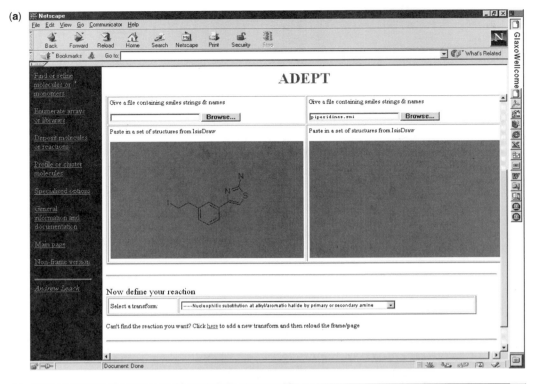

(b)

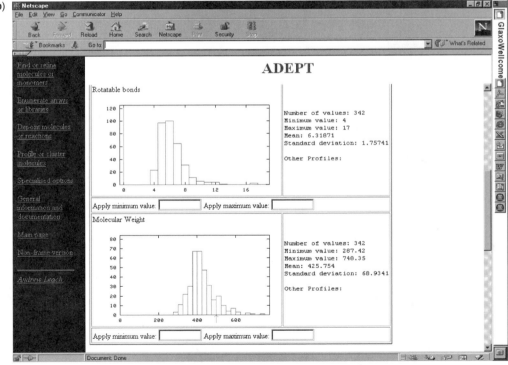

Figure 5.17. Screen-shots from ADEPT. (a) A simple two-component library composed of an aminothiazole template and a series of piperidines specified with ADEPT. (b) Histograms of rotatable bonds and molecular weight for the enumerated virtual library, aiding the medicinal chemist in the design of the library. [Reproduced from A. R. Leach and M. M. Hann, *Drug Discovery Today*, **5**, 326–336 (2000), with permission of Elsevier Science.]

Structure-based drug design (SBDD) is the topic of another chapter, and key issues such as the scoring functions for the ligand-receptor interaction are not discussed further here. The ability to combine SBDD with combinatorial chemistry enables a focused design approach that can explore a range of ideas, reducing the dependency on SBDD limitations (structural information, scoring, conformational sampling, etc.). The ability to obtain the X-ray or NMR structure of new potent molecules complexed with their targets can also be critical for the next iteration, in that computational structure-based design methods may be unable to predict alternative and new binding modes, especially because the protein site is normally kept rigid and unpredicted conformational changes can take place during the binding process. A review by Stahl (119) discusses the technology that directly uses receptor three-dimensional structures, discussing relevant topics such as scoring functions, receptor-ligand docking, and practical applications. Bohm and Stahl (120) have reviewed structure-based library design in terms of molecular modeling merging with combinatorial chemistry.

The synergy between combinatorial chemistry and *de novo* design has been discussed by Leach et al. (121). They present an approach wherein a template (corresponding to the *central core* of a combinatorial library) is positioned within an acyclic carbon chain whose length and bond orders are systematically varied. The conformational space of each resulting structure (core plus chain) is explored, to determine whether it is able to link together two or more strongly interacting functional groups or pharmacophores located within a protein binding site. In a second phase, 2D queries are derived from the molecular skeletons and used to identify possible reactants from a database that would enable the all-carbon linking chains to be replaced by more synthetically feasible groups.

Sheridan et al. (122) have published on designing targeted libraries with genetic algorithms, extending earlier work, to use the GA with 3D scoring methods and showing that the approach of assembling libraries from fragments in high scoring molecules is a reasonable one. Example applications to two situations are described: (*1*) where the 2D structure of some actives (diverse angiotensin II antagonists) is known, with the goal to design a library that best resembles the actives; and (*2*) to simulate the situation where an active site (stromelysin-1 in this case) is available and the requirement is to design a library of structures likely to bind to it.

Tondi (123) discusses several examples in which structure-based drug design and combinatorial library synthesis have worked successfully together in a complementary way. These include the discovery of:

- Potent nonpeptide inhibitors of cathepsin D (124), which uses CombiBUILD (125), a derivative of the DOCK (126a,b) approach, with this structure-based selection approach yielding seven times as many hits as a diversity-based procedure.
- Thrombin inhibitors (127), where Böhm et al. used LUDI to dock and score computationally available primary amines and then score the virtual library generated from benzaldehydes with the top-scoring hit.
- Novel inhibitors of matrix metalloproteinases (128): Rockwell et al. (128a) used a combinatorial library at the beginning of the work to suggest leads suitable for further optimization that required a conformational change at the binding site, and a structure of the complex to enable iterative optimization; Szardenings et al. (128b) used SBDD to design the starting scaffold, with synthesis guiding the introduction of diversity.
- Thymidylate synthase inhibitors (129), using DOCK to identify the starting lead.

The CombiDOCK program (126c), based on DOCK, enables the evaluation of very large virtual libraries by using structure-based combinatorial docking. Multiple docked orientations of the scaffold are used to evaluate reactants separately at each of the substitution positions. The total docking score for each product is rapidly estimated by summing the contributions from reactants at each position (which are attached as in the final product to the docked scaffold, which may be a computationally convenient anchor fragment formed during the reaction rather than a synthetically used chemical). Further checks are made for the highest scoring structures (e.g., for steric interactions between reactants at the different substitution positions). This approximation produces an enormous speed-up over docking all the individual compounds, which, from a time perspective, rapidly becomes prohibitive for large combinatorial libraries. From the scores it is possible to select combinations of reactants that produce compounds complementary to the protein binding site. Combinatorial restraints can be applied as required to obtain the most efficient use of reactants and robotics, with an evaluation of any reduction in the inclusion of higher scoring compounds.

Different strategies for combining diversity and structure-based design in site-focused libraries and the DOCK-based CombiBUILD algorithm are discussed in a review (125), as an example of how lead compounds can be rapidly identified by combining diversity with structure-based design in site-focused libraries.

Lamb et al. (130) have published on the design, docking, and evaluation of multiple libraries against a family of targets, using a similar divide-and-conquer algorithm for side chain selection that enables the exploration of large lists of reactant substituents with linear rather than combinatorial time dependency. The method consists of three main stages: (*1*) docking the scaffold, (*2*) selecting the best substituents at each site of diversity, and (*3*) comparing the resultant structures within and between the libraries. The scaffold docking procedure, in conjunction with a novel vector-based orientation filter, was shown to be effective for several protease targets, reproducing experimental binding modes.

The application of the powerful combination of SBDD and combinatorial chemistry is not limited to lead discovery or the optimization of potency, but also to the optimization of the selectivity (using knowledge of the structures of related targets) and pharmacokinetic/druglike properties of a molecule. For example, the structure of a ligand-receptor complex can clearly indicate areas where chemical modifications could be made to modulate these other properties, without directly affecting binding/potency. Models/structures of ligands

with the cytochrome P450 metabolizing enzymes are also now becoming available.

5 EXAMPLE APPROACHES

5.1 General Target Class-Focused Approaches

5.1.1 Defining the Chemical/Biological Space.
The design of target class (gene family) libraries or compound subsets requires the definition of a biologically relevant chemical space. This "biological" space can then be used for the design and selection of biased/focused libraries and compound subsets. Many approaches can be taken, adapting the use of a wide variety of similarity/diversity descriptors (discussed in Section 2.1) to the identification of *properties* associated with a particular target class or subset thereof. The goal is to identify a feature or set of features that, ideally, is specific, but more generally "enriched" for the target(s) of interest. A common approach is to identify chemical substructures that are characteristic for the target class, and use these for the design. The simplest approach is to include such substructures in the library, but the co-occurrence of other features is often needed, and the quantification of this provides an enhanced design. An example of this combined approach is discussed in the next section, using the pharmacophore fingerprints expressed relative to "privileged" substructures. This provides a convenient cell-based partitioning approach. Alternatively, it is possible to identify properties that are enriched for a particular target class, without reference to any particular substructures: 1D (e.g., physicochemical), 2D (e.g., ISIS keys, BCUTs), and 3D (e.g., pharmacophore fingerprints) properties can all be used. BCUTs have been used within a target (to identify a receptor-relevant subspace, in which actives cluster), to differentiate within a target class (e.g., ion channel openers vs. blockers) and for general target class analysis. BCUT chemical space provides a way to quantify the "diversity" of certain properties within actives for a target class, as well as to identify any particular combination of properties that actives share. BCUTs have

been used to select representative subsets from libraries biased to a target class (59a).

5.1.2 7-Transmembrane G-Protein-Coupled Receptors.

Examples of a product-based combinatorial library design that use four-point pharmacophore fingerprints in a "relative" diversity mode have been described for the design of combinatorial libraries that contain "privileged" substructures focused on 7-TM GPCRs. These are a large family of very important biological targets lacking high resolution experimental 3D structures of the human targets; therefore most design has focused around the ligands. The occurrence of common "privileged" substructures for 7-TM GPCRs, often spanning several targets, provides a useful focused design method. Some example structures are shown in Fig. 5.18.

A useful modification was made to the standard pharmacophore descriptor evaluation (37, 80) by forcing one of the points in the pharmacophoric description to be a *privileged* substructure. This provides a novel quantification of all the 3D pharmacophoric shapes, and thus important 3D information relevant to the biological activity of the ligands, relative to the substructure. This builds on the ability of 3D pharmacophoric descriptors to separate chemical structure from ligand binding properties, and enables a fingerprint to be generated that describes the possible pharmacophoric shapes from the viewpoint of that special point/substructure (see Fig. 5.19). A *relative* or *internally referenced* measure of diversity is thus created, enabling new design and analysis methods (see Section 2.2.5). The goal of the published method was to design novel structures, accessible through combinatorial chemistry, that have one or more privileged substructure reactants/cores, and are enriched in the relative 3D pharmacophoric shapes of known ligands. The method identifies patterns with other key features that need to be present with the privileged substructure, such as acids and bases. The optimization can also include an enrichment in pharmacophoric shapes containing the privileged substructure that are not in existing structures, enabling the exploration of new 3D pharmacophoric diversity focused around a feature known to be important for biological activity.

The Ugi reaction (131), a four-component condensation reaction, was chosen and more than 100,000 compounds were synthesized. Privileged substructures such as biphenyl tetrazole were used, for example, at the amine position (see Fig. 5.20). Other GPCR privileged substructures such as diphenyl methane, biphenyl tetrazole, and indole were used to focus the pharmacophore descriptors (see Figs. 5.18 and 5.21). GPCR ligands reported to be active at receptors with peptidic endogenous ligands were identified from the MDDR (13a). These compounds were used to provide the reference data for the design by calculating the union pharmacophore fingerprint of compounds containing the privileged substructure (see Fig. 5.22 for an example structure). A virtual combinatorial library was then created, and for a particular reactant position, the privileged pharmacophore fingerprints were calculated for each candidate reactant over all the products that would be generated if it were used in the library. Either previously selected or a representative set of reactants were used for the other three components to generate the virtual Ugi products.

The combinatorial library was then designed by comparing for each reactant the fingerprint generated from the resultant products with the fingerprint for the known drug ligands (MDDR-fingerprint). Reactants were selected by identifying, on a position-by-position basis, reactants that gave products that matched the greatest number of these MDDR-exhibited privileged pharmacophores. The design goal was recalculated after the selection of each reactant, by removing the pharmacophores matched by the products generated by that reactant from the target list. Subsequent reactants were thus picked based on their ability to match the remaining pharmacophores. The approach used was to select the first reactant as the one that would give library compounds with the most number of privileged pharmacophores in common with the drug set. The process was continued until no more reactants could be found that contributed a nontrivial number of new privileged pharmacophores. Optimization methods such as the HARPick approach (described in Section 4.3) could be used to enable other properties, such as flexibility and physicochemical

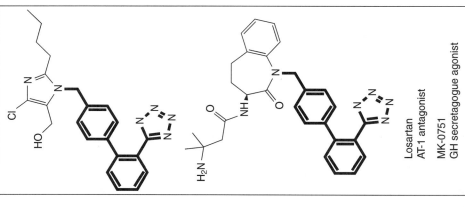

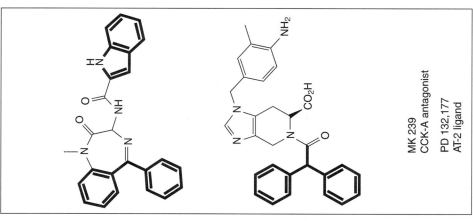

Figure 5.18. Examples of 7-TM GPCR "privileged" substructures highlighted on known active compounds. Note that these often occur together with another characteristic feature (e.g., basic or acidic group).

230

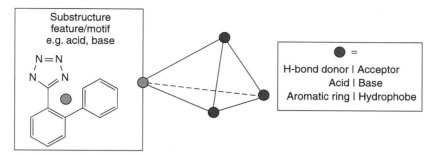

Figure 5.19. Example of a "privileged" four-point pharmacophore. Here biphenyl tetrazole, a substructure seen in a number of GPCR inhibitors, is specifically defined as a pharmacophore feature, using a centroid dummy atom. Only pharmacophores that include this type are included in the fingerprint, thus providing a relative measure of diversity/similarity with respect to the privileged feature.

properties, to be optimized also. The total number of pharmacophores (this time without reference to the privileged substructures) can also be monitored and optimized. Example results from one of the Ugi library optimizations are shown in Fig. 5.23.

This design illustrates an advantage of a partitioning (cell-based) approach. The pharmacophore fingerprint can be used to monitor progress, to quantify how much of the desired goal has been accomplished, and to evaluate whether a given chemistry can yield further compounds that match the design criteria and/or explore new pharmacophoric space.

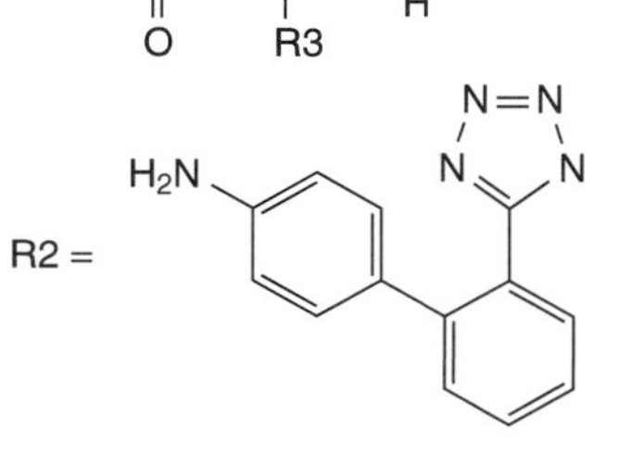

The example here used only a binary fingerprint, but even more powerful results can be obtained when a count for each potential pharmacophore is included. The authors showed that for these designed Ugi libraries the same Ugi chemistry could indeed yield significant new diversity for multiple 14,000 compound libraries, but that after three libraries diminishing returns were obtained. They used the understandable nature of the pharmacophore descriptor by analyzing the remaining MDDR-pharmacophore fingerprint to show that most of the remaining pharmacophores to be matched contained acids and/or bases. A modified chemistry approach was therefore developed using protected acids (t-butyl esters) and bases (BOC protected) in the Ugi reaction. The unmatched cells in the MDDR-fingerprint can be related back to the compounds that

R1COOH + R2NH2 + R3CHO + R4NC

↓ MeOH

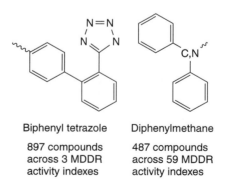

Figure 5.20. Example of the Ugi chemistry with biphenyl tetrazole incorporated as a "privileged" group at the amine position.

Biphenyl tetrazole

897 compounds
across 3 MDDR
activity indexes

Diphenylmethane

487 compounds
across 59 MDDR
activity indexes

Figure 5.21. Examples of 7-TM GPCR "privileged" motifs found in the MDDR database.

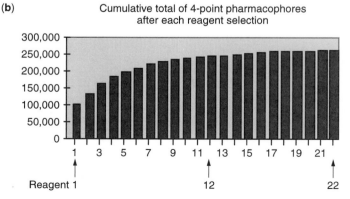

● "Privileged" feature   ● Hydrophobic feature
● Aromatic ring centroid

Total 4-point pharmacophores: 3601
-with "privileged" feature: 1569
(using 10 distance ranges)

Figure 5.22. Example of pharmacophore feature assignments involving the biphenyl tetrazole "privileged" substructure and the total four-point potential pharmacophores calculated for a GPCR antagonist. Note that just the subset (40%) of the total pharmacophores that contained the "privileged" substructure was used for the library design.

generated them, enabling a truly iterative design of further libraries. In Fig. 5.24 the increasing size of the pharmacophore fingerprint from four consecutive Ugi libraries is illustrated, together with the distribution of pharmacophoric features in MDDR pharmacophores that had not been matched.

Another example for GPCR library design is the use of BCUT metrics as the basis for target class-focused approaches to accelerated drug discovery. A particularly interesting example is work done by Wang and Saunders (32e,i) at Neurocrine Bioscience in their effort to discover novel nonpeptidic ligands for a particular member (GPCR-1) of the GPCR-PA+ family of receptors activated by peptides carrying an obligatory positive charge. They and their colleagues performed a thorough search of the literature and identified a few hundred ligands of the various members of the GPCR-PA+ family. Knowing that it is usually not useful to follow up hits or leads showing very poor affinity, they eliminated ligands with less

(a)

Figure 5.23. (a and b) Contributions per acid reactant of pharmacophores for optimization in the Ugi reaction (with biphenyl tetrazole as the "privileged" motif at the amine position). The order shown is the final selected order of reactants, based on obtaining the maximum number of new privileged pharmacophores per additional reactant. Histogram a shows the number of new pharmacophores added by each new selected reactant in the "privileged" pharmacophoric space defined by known GPCR compounds containing the biphenyl tetrazole; shown in histogram b is the matching increase in the total number of pharmacophores for the library for each new selected reactant.

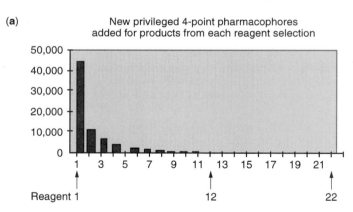

(b)

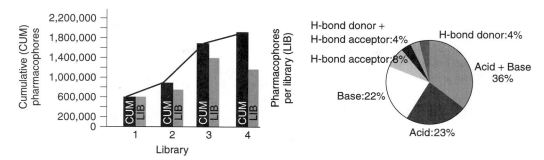

Figure 5.24. On the left is shown the cumulative (black) total number of four-point pharmacophores from consecutive 14,000 sets of Ugi libraries designed for 7-TM GPCR targets, together with the total number of pharmacophores in each library (in gray). Note the diminishing yield of new pharmacophores with later libraries, indicating that a change in strategy is needed. On the right are shown the features present in the resultant unrepresented pharmacophores (i.e., found in 7-TM GPCR biphenyl tetrazole-containing compounds in MDDR but not in synthesized libraries), indicating a strategy change to include more acids and bases together with the biphenyl tetrazole.

than 100 μM affinity for the corresponding receptor. Significantly, they also eliminated ligands with better than 1 μM affinity for the corresponding receptor. This very unusual step was taken in an effort to convince their colleagues that the method they intended to use was not reliant on knowing the answer ahead of time. This left 187 ligands with affinities mostly between 10 and 70 μM for various members of the GPCR-PA+ family of receptors. Using these compounds, they perceived a three-dimensional BCUT subspace within their corporate chemistry space that clusters the ligands of individual members of the GPCR-PA+ family and appears to be appropriate for this target class. The positions of all 187 ligands in the 3D chemistry space shown in Fig. 5.25 were originally indicated by open cyan circles. All ligands of some but not all receptors were then color-coded as indicated. Many red GPCR-2 and yellow GPCR-4 ligands are hidden under the green GPCR-1 ligands. The gray oval provides a crude indication of the region of chemistry space of interest for GPCR-PA+ receptors.

Figure 5.26 indicates the positions of roughly 2000 Neurocrine compounds selected from 14 different combinatorial libraries based on 14 different and proprietary scaffolds. Rather than selecting compounds only near the known ligands of GPCR-1, their receptor of interest, Wang and Saunders also selected compounds spanning the entire GPCR-PA+ receptors. This was done to further convince their colleagues, as explained below. All 2000 compounds were screened for activity against the GPCR-1 receptor. Those testing positive were retested in a secondary, functional assay. All but two compounds having better than 100 nM affinity for the GPCR-1 receptor are colored blue and/or are located within the blue oval. All but one compound having better than 10 nM affinity for the GPCR-1 receptor are colored red and/or are located within the red oval. All compounds with better than 2 nM affinity are colored green and are located within the two small green ovals within the larger green oval, consistent with the two crude clusters of GPCR-1 ligands seen in Fig. 5.25. The fact that these two small ovals each contain products from several different libraries (scaffolds) suggests the possible existence of two binding modes for this receptor. It is also significant to note that, although the authors intentionally synthesized compounds within the entire region of interest for GPCR-PA+ receptors, the only compounds showing significant affinity for the GPCR-1 receptor were located close to the known GPCR-1 ligands (compare with Fig. 5.25), thus supporting the use of BCUT coordinates (on receptor-relevant axes) as a valid approach to virtual high throughput screening. The tight clustering of GPCR-PA+ ligands in both figures clearly suggests that BCUT metrics represent, albeit in a relatively

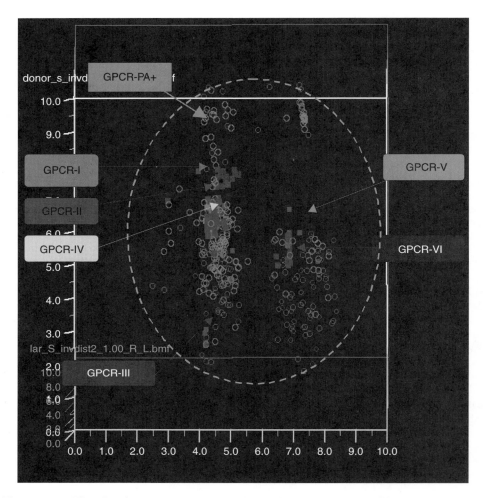

Figure 5.25. The 3D subspace most receptor relevant for members of the GPCR-PA+ family of receptors. Points indicate coordinates of 187 published ligands of various GPCR-PA+ receptors. Some have been color-coded by receptor for illustrative purposes. See Refs. 32e,i and text for further details. See color insert.

crude fashion, the same sort of information as would be represented in a description of the pharmacophore for the receptor of interest.

5.2 Property-Biased Design

The use of pharmacophoric descriptors in enhancing the hit-to-lead properties of lead optimization libraries has been described (76). Pharmacophore fingerprints, based on the Chem-X/ChemDiverse multiple pharmacophore descriptors, were used and several issues in the design of lead optimization libraries were addressed. The applicability of

pharmacophoric methods to the design of focused libraries was demonstrated in this case, where the aim was to design the library toward a known lead or leads. The authors also investigated the design of libraries with improved pharmacokinetic properties. Simple and rapidly computable descriptors applicable to the prediction of drug transport properties were used, and the results illustrate a common problem: to obtain the best results it may be necessary to synthesize libraries in a noncombinatorial manner. A Monte Carlo search procedure was devised to enable the selection of a

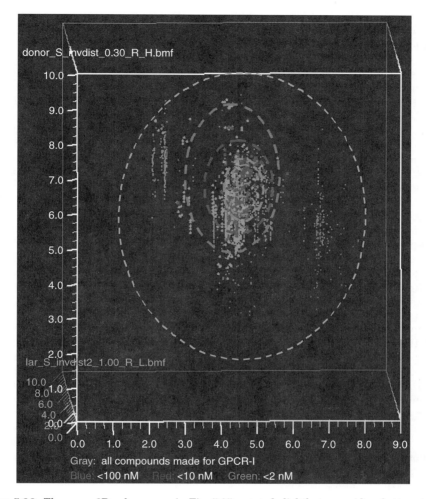

Figure 5.26. The same 3D subspace as in Fig. 5.25, rotated slightly to provide a better viewing perspective. Points indicate coordinates of about 2000 combinatorial products selected from 14 different libraries. Color-coding indicates affinity for the GPCR-1 receptor. See Refs. 32e,i and text for further details. See color insert.

near-combinatorial subset in which all library members satisfy the design criteria. By including calculated log P, molecular weight, and polar surface area in the design of a combinatorial library, it was shown that the compounds with improved absorption characteristics (as determined by experimental Caco-2 measurements) could be obtained.

The use of computational methods such as reactant clustering and library profiling to maximize reactant diversity and optimize pharmacokinetic parameters has been described (132), with four-point pharmacophore

fingerprint analysis used to quantify the added diversity gained by using two independent synthetic routes.

5.3 Site-Based Pharmacophores

Pharmacophore fingerprints generated from complementary site points can be used to direct combinatorial library design and to investigate selectivity. An example of the pharmacophore fingerprinting method for selectivity studies has been validated (37a,b) in studies of three closely related serine proteases: thrombin, Factor Xa, and trypsin. Site points were positioned in the active site of each protein using the results of GRID (42) analyses (see Fig. 5.5), and receptor-based four-point pharmacophore fingerprints were generated. Fingerprints were also generated using full conformational flexibility for some highly selective and potent thrombin and Factor Xa inhibitors. Receptor-based similarity was investigated as a function of common potential three- and four-point pharmacophores for each ligand/receptor pair. The results indicated that the use of just the common potential four-point pharmacophores could give information pertaining to relative enzyme selectivity; when three-point pharmacophores were used, however, poor resolution of enzyme selectivity was observed. The thrombin inhibitor thus exhibited greater similarity with the complementary four-point pharmacophore fingerprints of the thrombin active sites than with the potential pharmacophore keys generated from the other enzymes; a similar result was found for the Factor Xa inhibitors with the Factor Xa site.

Clearly, the inclusion of the shape of the binding site should improve the resolution, and the DiR (Design in Receptor) approach (133) refines the process, requiring that the pharmacophoric match fits the shape of the target site (i.e., is sterically compatible with the site). This clearly provides much additional information at the expense of greatly increased calculation time. Within the DiR approach, two-, three-, and four-point potential site pharmacophores can be used. This provides interesting new library design possibilities, in that it is possible to evaluate which ligands are able to fit in the site by matching at least one set of pharmacophoric features, and to quantify which pharmacophore hypotheses are matched. A subset of ligands can then be designed that match as many different pharmacophoric hypotheses as possible, and the biological screening of the resultant compounds can determine which bind best. Alternatively, pharmacophore constraints can be applied to a shape-driven searching approach, and Good et al. (34) have shown the effectiveness of this with the DOCK virtual screening/docking approach, in which the addition of pharmacophore constraints improved both the enrichment and speed of the process.

The active site of the Factor Xa serine protease (134) has been used for combinatorial library design (37c,d) using the DiR approach. GRID analyses using probes for hydrogen bond donors, acceptors, bases, acids, and hydrophobes resulted in 23 complementary site points being added (see Fig. 5.5). The shape of the active site was defined using 162 protein atoms. To ensure that a relevant area of the binding site was being explored (based on knowledge of X-ray protein-ligand complexes), site pharmacophores were forced to contain a hydrophobe or aromatic ring centroid point from both the S1 and S4 regions of the binding pocket. By using this focused approach, a "diversity" of matched site pharmacophores was obtained, representing a sampling of "reasonable" binding modes related to those experimentally observed and, thus, presumably having a higher probability of giving rise to biological activity. This focused approach reduced the total number of site pharmacophores from 5393 to 775 [using the seven distance ranges setting (37a) and considering all distances in the 1–15 Å range]. The approach was validated by the identification of feasible binding models (37c), similar to that experimentally observed for a known Factor Xa inhibitor. The Ugi four-component condensation reaction (131) (see Fig. 5.20) was used for the study and is capable of producing very large numbers of different structures from commercially available reactants. An example of the power of the method was given, whereby products were selected semimanually from a small virtual library of 432 products (37c,d). Products were constructed from the four reactant sets: carboxylic acids ($R_1 \times 3$), amines ($R_2 \times 2$), aldehydes ($R_3 \times 3$), and isonitriles ($R_4 \times 24$). The pharmacophore-based site analysis showed the optimum positions of substitution and chain length for benzamidine-containing fragments (targeted to the aspartate-containing S1 pocket) and the optimum lengths of other hydrophobic reactants (targeted to the S4 pocket) to produce compounds that would sample the maximum number of binding modes. In this case the groups were always forced to be in the S1 and S4 pockets to maintain "reasonable" binding modes, although this restriction could be excluded to probe even further potential binding modes. Thus, as the identity of the matched site pharmacophore(s) was known for each compound, target site-based diversity of binding modes could be explored in the design process. An optimized selection of reactants was possible, and the value to the design of reactants with different chain lengths could be evaluated.

6 CONCLUSIONS AND FUTURE DIRECTIONS

Similarity and diversity metrics have been successfully used for a variety of tasks, including virtual screening, subset selection, and combinatorial library design. Databases of virtual compounds (e.g., from validated combinatorial chemistry protocols and reactants) can be used for both virtual screening and library design (virtual screening on virtual libraries with additional combinatorial constraints). The ability to exploit rapidly large virtual libraries of compounds that could be made by validated combinatorial chemistry protocols provides very powerful virtual screening and library design approaches. Future directions for library design will involve the application of such approaches in a fully integrated fashion (e.g., the ADEPT tool described in Section 4.10) and further enhancements to the constraints necessary to achieve druglike compounds (e.g., 80% compliance to the Rule of 5, predictive models for metabolism- and toxicity-related issues). Where the goal is lead generation (e.g., to enrich the compound screening file for high throughput screening), a focus will be on target classes (gene families) of interest, and the generation of compounds with leadlike properties, such as a lower molecular weight. The move away from combinatorial libraries to sparse arrays and noncombinatorial (cherry-picked) libraries (90) will continue, enabling more effective designs with control of associated properties. However, as more property constraints are applied to the library designs for leadlike/druglike properties, the need to include positive design elements to ensure good biological activity is emphasized. The goal for drug discovery is thus to identify targets and to generate compounds that are at the intersection of chemical, biological, and druglike property (e.g., absorption,

toxicophores) space. Different targets and different expected routes of administration will require different constraints, and an element of diversity (with constraints toward a drug-occupied chemical space) will remain important, to enable the most effective use of combinatorial library chemistry and to discover new leads for both established and new targets.

REFERENCES

1. M. Johnson and G. Maggiora, *Concepts and Applications of Molecular Diversity*, John Wiley & Sons, New York, 1990.

2. (a) P. Willett, J. M. Barnard, and G. M. Downs, *J. Chem. Inf. Comput. Sci.*, **38**, 983 (1998); (b) P. Willett, *Curr. Opin. Biotechnol.*, **11**, 85 (2000); (c) P. Willett, Ed., Perspectives in Drug Discovery Design, Vols. **7/8**, Kluwer Academic, Dordrecht/Norwell, MA, 1997; (d) J. S. Mason and M. A. Hermsmeier, *Curr. Opin. Chem. Biol.*, **3**, 342 (1999).

3. (a) J. M. Moore, *Curr. Opin. Biotechnol.*, **10**, 54 (1999); (b) P. J. Hajduk, T. Gerfin, J.-M. Boehlen, M. Haeberli, D. Marek, and S. W. Fesik, *J. Med. Chem.*, **42**, 2315 (1999); (c) C. A. Lepre, *Drug Discovery Today*, **6**, 133 (2001); (d) J. Fejzo, C. A. Lepre, J. W. Peng, G. W. Bemis, Ajay, M. A. Murcko, and J. M. Moore, *Chem. Biol.*, **6**, 755 (1999).

4. J. S. Mason and S. D. Pickett, *Perspect. Drug Discov. Des.*, **7/8**, 85 (1997).

5. (a) R. D. Brown, *Perspect. Drug Discov. Des.*, **7/8**, 31 (1997); (b) Y. C. Martin, R. D. Brown, and M. G. Bures in E. M. Gordon and J. F. Kerwin, Jr., Eds., *Combinatorial Chemistry Molecular Diversity Drug Discovery*, Wiley-Liss, New York, 1998, pp. 369–385; (c) M. G. Bures and Y. C. Martin, *Curr. Opin. Chem. Biol.*, **2**, 376 (1998); (d) Y. C. Martin, M. G. Bures, and R. D. Brown, *Pharm. Pharmacol. Commun.*, **4**, 147 (1998).

6. (a) D. K. Agrafiotis in P. v. R. Schleyer, N. L. Allinger, T. Clark, J. Gasteiger, P. A. Kollman, H. F. Schaefer III, and P. R. Schreiner, Eds., *The Encyclopedia of Computational Chemistry*, Vol. **1**, John Wiley & Sons, Chichester, UK, 1998, pp. 742–761; (b) D. K. Agrafiotis and V. S. Lobanov, *J. Chem. Inf. Comput. Sci.*, **39**, 51 (1999); (c) D. K. Agrafiotis, J. C. Myslik, and F. R. Salemme, *Annu. Rep. Comb. Chem. Mol. Diversity*, **2**, 71 (1999); (d) D. K. Agrafiotis, V. S. Lobanov, D. N. Rassokhin, and S. Izrailev,

Methods Princ. Med. Chem., **10** (Virtual Screening for Bioactive Molecules), 265 (2000).

7. R. A. Lewis, S. D. Pickett, and D. E. Clark in K. B. Lipkowitz and D. B. Boyd, Eds., *Reviews in Computational Chemistry*, Vol. **16**, Wiley-VCH, John Wiley & Sons, New York, 2000, pp. 1–51.

8. D. C. Spellmeyer and P. D. J. Grootenhuis, *Annu. Rep. Med. Chem.*, **34**, 287 (1999).

9. H. Matter and M. Rarey in G. Jung, Ed., *Combinatorial Chemistry*, Wiley-VCH Verlag GmbH, Weinheim, Germany, 1999, pp. 409–439.

10. Y. C. Martin, *J. Comb. Chem.*, **3**, 231 (2001).

11. (a) R. S. Pearlman, *Network Sci.*, **2**, (6/7) (1996), available at: http://www.netsci.org/Science/Combichem/feature08.html; (b) R. S. Pearlman and K. M. Smith, *Perspect. Drug Discov. Des.*, **9**, 339/355 (1998); (c) R. S. Pearlman and K. M. Smith, *Drugs Future*, **23**, 885 (1998); (d) R. S. Pearlman and K. M. Smith, *J. Chem. Inf. Comput. Sci.*, **39**, 28 (1999).

12. J. S. Mason, A. C. Good, and E. J. Martin, *Curr. Pharm. Des.*, **7**, 567 (2001).

13. (a) MDL Information Systems Inc., San Leandro, CA, URL: http://www.mdli.com; (b) C. A. James, D. Weininger, and J. Delaney, *Daylight Theory Manual*, version 4.72, Daylight Chemical Information Systems, Inc., URL: http://www.daylight.com/dayhtml/doc/theory/theory.toc.html; (c) UNITY/SLN manual available from Tripos, Inc., 1699 South Hanley Road, Suite 303, St. Louis, MO 63144, URL: http://www.tripos.com.

14. (a) R. D. Brown and Y. C. Martin, *J. Chem. Inf. Comput. Sci.*, **37**, 1 (1997); (b) R. D. Brown and Y. C. Martin, *J. Chem. Inf. Comput. Sci.*, **36**, 572 (1996).

15. D. R. Flower, *J. Chem. Inf. Comput. Sci.*, **38**, 379 (1998).

16. M. Randić, *J. Am. Chem. Soc.*, **97**, 6609 (1975).

17. (a) L. B. Kier, L. H. Hall, W. J. Murray, and M. Randić, *J. Pharm. Sci.*, **64**, 1971 (1975); (b) L. B. Kier, L. H. Hall, and W. J. Murray, *J. Pharm. Sci.*, **64**, 1974 (1975).

18. L. H. Hall and L. B. Kier, *J. Mol. Graph. Modell.*, **20**, 4 (2001).

19. M. Randić, *J. Mol. Graph. Modell.*, **20**, 19 (2001).

20. (a) L. H. Hall and L. B. Kier in K. B. Lipkowitz and D. B. Boyd, Eds., *Reviews in Computational Chemistry*, Vol. **2**, VCH Publishers, New York, 1991, pp. 367–422; (b) MOLCONN-Z, Hall Associates Consulting, 2 Davis Street,

Quincy, MA, available from eduSoft L.C. http://www.eslc.vaviotech.com.

21. (a) R. A. Lewis, J. S. Mason, and I. M. McLay, *J. Chem. Inf. Comput. Sci.*, **37**, 599 (1997).

22. D. J. Cummins, C. W. Andrews, J. A. Bentley, and M. J. Cory, *J. Chem. Inf. Comput. Sci.*, **36**, 750 (1996).

23. (a) W. G. Glen, W. J. Dunn III, and D. R. Scott, *Tetrahedron Comput. Methodol.*, **2**, 349 (1989); (b) C. Cheng, G. Maggiora, M. Lajiness, and M. J. Johnson, *J. Chem. Inf. Comput. Sci.*, **36**, 909 (1996); (c) M. Hassan, J. P. Bielawski, J. C. Hempel, and M. Waldman, *Mol. Div.*, **2**, 64 (1996); (d) D. K. Agrafiotis, *J. Chem. Inf. Comput. Sci.*, **37**, 841 (1997).

24. R. E. Cahart, D. H. Smith, and R. Venkataraghavan, *J. Chem. Inf. Comput. Sci.*, **25**, 64 (1985).

25. R. Nilakantan, N. Bauman, J. S. Dixon, and R. Venkataraghavan, *J. Chem. Inf. Comput. Sci.*, **27**, 82 (1987).

26. (a) S. K. Kearsley, S. Sallamack, E. M. Fluder, J. D. Andose, R. T. Mosley, and R. P. Sheridan, *J. Chem. Inf. Comput. Sci.*, **36**, 118 (1996); (b) R. P. Sheridan, M. D. Miller, D. J. Underwood, and S. K. Kearsley, *J. Chem. Inf. Comput. Sci.*, **36**, 128 (1996); (c) R. P. Sheridan, *J. Chem. Inf. Comput. Sci.*, **40**, 1456 (2000).

27. (a) G. Schneider, W. Neidhart, T. Giller, and G. Schmid, *Angew. Chem. Int. Ed. Engl.*, **38**, 2894 (1999); (b) G. Schneider, O. Clement-Chomienne, L. Hilfiger, P. Schneider, S. Kirsch, H.-J. Bohm, and W. Neidhart, *Angew. Chem. Int. Ed. Engl.*, **39**, 4130 (2000).

28. D. Gorse and R. Lahana, *Curr. Opin. Chem. Biol.*, **4**, 287 (2000).

29. C. A. Lipinski, F. Lombardo, B. W. Dominy, and P. J. Feeney, *Adv. Drug Deliv. Rev.*, **23**, 3 (1997).

30. F. R. Burden, *J. Chem. Inf. Comput. Sci.*, **29**, 225 (1989).

31. DiverseSolutions was developed by R. S. Pearlman and K. M. Smith at the University of Texas, Austin, and is distributed by Tripos, Inc., St. Louis, MO.

32. (a) H. Gao, *J. Chem. Inf. Comput. Sci.*, **41**, 402 (2001); (b) D. Stanton, *J. Chem. Inf. Comput. Sci.*, **39**, 11 (1999); (c) D. Schnur, *J. Chem. Inf. Comput. Sci.*, **39**, 36 (1999); (c) D. Schnur and P. Venkatarangan in A. K. Ghose and V. N. Viswanadhan, Eds., *Combinatorial Library Design and Evaluation*, Marcel Dekker, New York, 2001, pp. 473–501; (d) B. R. Beno and J. S. Mason, *Drug Discovery Today*, **6**, 251

(2001); (e) X.-C. Wang and J. Saunders, Abstracts of Papers, 222nd ACS National Meeting, Chicago, IL, August 26–30, 2001 (2001), MEDI-012; (f) E. L. Stewart, P. J. Brown, J. A. Bentley, and T. M. Willson, Abstracts of Papers, 222nd ACS National Meeting, Chicago, IL, August 26–30, 2001 (2001), COMP-182; (g) Y. Gao and V. Goodfellow, Abstracts of Papers, 221st ACS National Meeting, San Diego, CA, 2001 (2001) MEDI-235; (h) X.-C. Wang and J. Saunders, Abstracts of Papers, 221st ACS National Meeting, San Diego, CA, 2001 (2001), MEDI-207; (i) J. Saunders, Proceedings of the IBC Conference on Drug Discovery by Design, Boston, MA, November 5–8, 2001; (j) B. Pirard and S. D. Pickett, *J. Chem. Inf. Comput. Sci.*, **40**, 1431 (2000) .

33. (a) A. C. Good and J. S. Mason, *Reviews in Computational Chemistry*, Vol. **7**, VCH, New York, 1995, pp. 67–127; (b) G. W. A. Milne, M. C. Nicklaus, and S. Wang, *SAR QSAR Environ. Res.*, **9**, 23 (1998); (c) W. A. Warr and P. Willett, *Design of Bioactive Molecules*, American Chemical Society, Washington, DC, 1998, pp. 73–95.

34. A. C. Good, J. S. Mason, and S. D. Pickett, *Methods Princ. Med. Chem.*, **10** (Virtual Screening for Bioactive Molecules), 131 (2000).

35. S. D. Pickett, J. S. Mason, and I. M. McLay, *J. Chem. Inf. Comput. Sci.*, **36**, 1214 (1996).

36. E. K. Davies in I. M. Chaiken and K. D. Janda, Eds., *Molecular Diversity and Combinatorial Chemistry: Libraries and Drug Discovery*, American Chemical Society, Washington, DC, 1996, pp. 309–316.

37. (a) J. S. Mason, I. Morize, P. R. Menard, D. L. Cheney, C. R. Hulme, and R. F. Labaudiniere, *J. Med. Chem.*, **42**, 3251 (1999); (b) J. S. Mason and D. L. Cheney, *Proc. Pac. Symp. Biocomput.*, **4**, 456 (1999); (c) J. S. Mason, and D. L. Cheney, *Proc. Pac. Symp. Biocomput.*, **5**, 576 (2000); (d) J. S. Mason and B. R. Beno, *J. Mol. Graph. Modell.*, **18**, 438 (2000).

38. (a) M. J. McGregor and S. M. Muskal, *J. Chem. Inf. Comput. Sci.*, **39**, 569 (1999); (b) M. J. McGregor, and S. M. Muskal, *J. Chem. Inf. Comput. Sci.*, **40**, 117 (2000).

39. E. K. Bradley, P. Beroza, J. E. Penzotti, P. D. J. Grootenhuis, D. Spellmeyer, and J. L. Miller, *J. Med. Chem.*, **43**, 2770 (2000).

40. D. Horvath in A. K. Ghose and V. N. Viswanadhan, Eds., *Combinatorial Library Design and Evaluation*, Marcel Dekker, New York, 2001, pp. 429–472.

41. (a) R. Poulain, D. Horvath, B. Bonnet, C. Eckhoff, B. Chapelain, M.-C. Bodinier, and B. Déprez, *J. Med. Chem.*, **44**, 3378 (2001); (b) R. Poulain, D. Horvath, B. Bonnet, C. Eckhoff, B. Chapelain, M.-C. Bodinier, and B. Déprez, *J. Med. Chem.*, **44**, 3391 (2001).

42. (a) P. J. Goodford, *J. Med. Chem.*, **28**, 849 (1985); (b) D. N. A. Bobbyer, P. J. Goodford, and P. M. McWhinnie, *J. Med. Chem.*, **32**, 1083 (1989); (c) The GRID program is developed and distributed by Molecular Discovery Ltd.

43. E. J. Martin and T. J. Hoeffel, *J. Mol. Graph. Modell.*, **18**, 383 (2000).

44. A. R. Leach, D. V. S. Green, M. M. Hann, D. B. Judd, and A. C. Good, *J. Chem. Inf. Comput. Sci.*, **40**, 1262 (2000).

45. (a) K. A. Andrews and R. D. Cramer, *J. Med. Chem.*, **43**, 1723 (2000); (b) R. D. Cramer, M. A. Poss, M. A. Hermsmeier, T. J. Caulfield, M. C. Kowala, and M. T. Valentine, *J. Med. Chem.*, **42**, 3919 (1999); (c) R. D. Cramer, R. D. Clark, D. E. Patterson, and A. M. Ferguson, *J. Med. Chem.*, **39**, 3060 (1996).

46. (a) H. Matter and T. Pötter, *J. Chem. Inf. Comput. Sci.*, **39**, 1211 (1999); (b) V. J. Van Geerestein, H. Hamersma, and S. P. Van Helden in H. Van de Waterbeemd, B. Testa, and G. Folkers, Eds., *Computer-Assisted Lead Finding and Optimization*, Verlag Helvetica Chimica Acta, Basel, Switzerland, 1997, pp. 159–178.

47. M. Hahn, *J. Chem. Inf. Comput. Sci.*, **37**, 80 (1997).

48. O. F. Güner, M. Waldman, R. Hoffmann, and J.-H. Kim in O. F. Güner, Ed., *Pharmacophore Perception, Development and Use in Drug Design*, International University Line, La Jolla, CA, 2000, p. 213.

49. R. Carbo, L. Leyda, and M. Arnau, *Int. J. Quantum Chem.*, **17**, 1185 (1980).

50. A. C. Good, E. E. Hodgkin, and W. G. Richards, *J. Chem. Inf. Comput. Sci.*, **32**, 188 (1992).

51. (a) D. J. Wild and P. Willett, *J. Chem. Inf. Comput. Sci.*, **36**, 159 (1996); (b) D. A. Thorner, D. J. Wild, P. Willett, and P. M. Wright, *J. Chem. Inf. Comput. Sci.*, **36**, 900 (1996).

52. A. Schuffenhauer, V. Gillet, and P. Willett, *J. Chem. Inf. Comput. Sci.*, **40**, 295 (2000).

53. R. D. Cramer III, D. E. Patterson, and J. D. Bunce, *J. Am. Chem. Soc.*, **110**, 5959 (1988).

54. G. Cruciani, P. Crivori, P.-A. Carrupt, and B. Testa, *THEOCHEM*, **503**, 17 (2000).

55. (a) W. Guba and G. Cruciani in K. Guberrtofte and F. S. Jorgensen, Eds., *Molecular Modeling and Prediction of Bioreactivity*, Plenum, New York, 2000, pp. 89–95; (b) P. Crivori, G. Cru-

ciani, P.-A. Carrupt, and B. Testa, *J. Med. Chem.*, **43**, 2204 (2000).

56. M. Pastor, G. Cruciani, I. McLay, S. Pickett, and S. Clementi, *J. Med. Chem.*, **43**, 3233 (2000).

57. E. J. Martin, J. M. Blaney, M. A. Saini, D. C. Spellmeyer, A. K. Wong, and W. H. Moos, *J. Med. Chem.*, **38**, 1431 (1995).

58. C. A. James, D. Weininger, and J. Delaney, Fingerprints-Screening and Similarity, Daylight Theory Manual v4.72, Daylight Chemical Information Systems, Inc., URL: http://www.daylight.com/dayhtml/doc/theory/theory.toc.html.

59. (a) P. R. Menard, J. S. Mason, I. Morize, and S. Bauerschmidt, *J. Chem. Inf. Comput. Sci.*, **38**, 1204 (1998); (b) P. R. Menard, R. A. Lewis, and J. S. Mason, *J. Chem. Inf. Comput. Sci.*, **38**, 497 (1998).

60. P. Willett, *Similarity and Clustering in Chemical Information Systems*, Research Studies Press, Letchworth, UK, 1987.

61. R. A. Jarvis and E. A. Patrick, *IEEE Trans. Comput.*, **C-22**, 1025 (1973).

62. (a) P. Willett, V. Winterman, and D. Bawden, *J. Chem. Inf. Comput. Sci.*, **26**, 109 (1986); (b) J. B. Dunbar, *Perspect. Drug Discov. Des.*, **7/8**, 51 (1997).

63. T. N. Doman, J. M. Cibulskis, M. J. Cibulskis, P. D. McCray, and D. P. Spangler, *J. Chem. Inf. Comput. Sci.*, **36**, 1195 (1996).

64. (a) J. M. Barnard and G. M. Downs, *J. Chem. Inf. Comput. Sci.*, **37**, 141 (1997); (b) J. M. Barnard, and G. M. Downs, *Perspect. Drug Discov. Des.*, **7/8**, 13 (1997).

65. G. M. Downs, P. Willett, and W. Fisanick, *J. Chem. Inf. Comput. Sci.*, **34**, 1094 (1994).

66. (a) M. Lajiness, *Perspect. Drug Discov. Des.*, **7/8**, 55 (1997); (b) V. J. Gillet and P. Willett in A. K. Ghose and V. N. Viswanadhan, Eds., *Combinatorial Library Design and Evaluation*, Marcel Dekker, New York, 2001, pp. 379–398; (c) R. D. Clark, *J. Chem. Inf. Comput. Sci.*, **37**, 1181 (1997); (d) T. Potter and H. Matter, *J. Med. Chem.*, **41**, 478 (1998).

67. B. D. Hudson, R. M. Hyde, E. Rahr, and J. Wood, *Quant. Struct.-Act. Relat.*, **15**, 285 (1996).

68. R. E. Higgs, K. G. Bemis, I. A. Watson, and J. H. Wikel, *J. Chem. Inf. Comput. Sci.*, **37**, 861 (1997).

69. (a) J. D. Holliday, S. S. Ranade, and P. Willett, *Quant. Struct.-Act. Relat.*, **14**, 501 (1995); (b)

D. B. Turner, S. M. Tyrrell, and P. Willett, *J. Chem. Inf. Comput. Sci.*, **37**, 18 (1997).

70. S. D. Pickett, C. Luttmann, V. Guerin, A. Laoui, and E. James, *J. Chem. Inf. Comput. Sci.*, **38**, 144 (1998).

71. J. Mount, J. Ruppert, W. Welch, and A. Jain, *J. Med. Chem.*, **42**, 60 (1999).

72. M. Snarey, N. K. Terrett, P. Willett, and D. J. Wilton, *J. Mol. Graph.*, **15**, 372 (1997).

73. M. Waldman, H. Li, and M. Hasan, *J. Mol. Graph. Modell.*, **18**, 412 (2000).

74. M. Hann, B. Hudson, X. Lewell, R. Lifely, L. Miller, and N. Ramsden, *J. Chem. Inf. Comput. Sci.*, **39**, 897 (1999).

75. (a) D. E. Clark and S. D. Pickett, *Drug Discovery Today*, **5**, 49 (2000); (b) P. J. Eddershaw, A. P. Beresford, and M. K. Bayliss, *Drug Discovery Today*, **5**, 409–414 (2000); (c) H. van de Waterbeemd, D. A. Smith, K. Beaumont, and D. K. Walker, *J. Med. Chem.*, **44**, 1313 (2001).

76. S. D. Pickett, D. E. Clark, and I. M. McLay, *J. Chem. Inf. Comput. Sci.*, **40**, 263 (2000).

77. D. E. Clark, *J. Pharm. Sci.*, **88**, 807 (1999).

78. (a) A. C. Good and R. A. Lewis, *J. Med. Chem.*, **40**, 3926 (1997); (b) R. A. Lewis, A. C. Good, and S. D. Pickett in H. Van de Waterbeemd, B. Testa, and G. Folkers, Eds., *Computer-Assisted Lead Finding and Optimization*, Verlag Helvetica Chimica Acta, Basel, Switzerland, 1997, pp. 135–156; (c) V. J. Gillet, P. Willet, J. Bradshaw, and D. V. S. Green, *J. Chem. Inf. Comput. Sci.*, **39**, 169 (1999).

79. G. Grassy, A. Yasri, R. Lahana, J. Woo, S. Iyer, M. Kaczorek, R. Folc'h, and R. Buelow, *Nat. Biotechnol.*, **16**, 748 (1998).

80. J. S. Mason in P. M. Dean and R. A. Lewis, Eds., *Molecular Diversity Drug Design*, Kluwer Academic, Dordrecht, Netherlands, 1999, pp. 67–91.

81. (a) A. C. Good and I. D. Kuntz, *J. Comput.-Aided Mol. Des.*, **9**, 373 (1995); (b) M. J. Ashton, M. Jaye, and J. S. Mason, *Drug Discovery Today*, **1**, 71 (1996).

82. J. H. Van Drie and R. A. Nugent, *SAR QSAR Environ. Res.*, **9**, 1 (1998).

83. A. C. Good, *Internet J. Chem.*, **3** (2000), http://www.ijc.com/article/2000v3/9/.

84. D. E. Patterson, R. D. Cramer, A. M. Ferguson, R. D. Clark, and L. E. Weinberger, *J. Med. Chem.*, **39**, 3049 (1996).

85. A. C. Good, J. S. Mason, D. V. S. Green, and A. R. Leach in A. K. Ghose and V. N. Viswanadhan, Eds., *Combinatorial Library Design and Evaluation*, Marcel Dekker, New York, 2001, pp. 399–428.

86. C. D. Eldred and B. D. Judkins, *Prog. Med. Chem.*, **36**, 29 (1999).

87. (a) V. J. Gillet, P. Willett, and J. Bradshaw, *J. Chem. Inf. Comput. Sci.*, **38**, 165 (1998); (b) E. A. Jamois, M. Hassan, and M. Waldman, *J. Chem. Inf. Comput. Sci.*, **40**, 63 (2000); (c) R. S. Pearlman and K. M. Smith, Symposium on Combinatorial Chemistry, Abstracts of Papers, 217th ACS National Meeting, Anaheim, CA, March 1999, MEDI-012.

88. (a) J. Sadowski in A. K. Ghose and V. N. Viswanadhan, Eds., *Combinatorial Library Design and Evaluation*, Marcel Dekker, New York, 2001, pp. 291–300; (b) A. K. Ghose, V. N. Viswanadhan, and J. J. Wendoloski, *J. Comb. Chem.*, **1**, 55 (1999); (c) Ajay, W. P. Walters, and M. A. Murcko, *J. Med. Chem.*, **41**, 3314 (1998); (d) J. Sadowski and H. Kubinyi, *J. Med. Chem.*, **41**, 3325 (1998).

89. T. Mitchell and G. A. Showell, *Curr. Opin. Drug Discov. Dev.*, **4**, 314 (2001).

90. J. R. Everett, M. Gardner, F. Pullen, G. F. Smith, M. Snarey, and N. Terrett, *Drug Discovery Today*, **6**, 779 (2001).

91. (a) J. H. Van Drie and M. S. Lajiness, *Drug Discovery Today*, **3**, 274 (1998); (b) R. A. Lewis in P. M. Dean and R. A. Lewis, Eds., *Molecular Diversity Drug Design*, Kluwer Academic, Dordrecht, Netherlands, 1999, pp. 221–248 and other chapters therein.

92. G. Jones, P. Willett, R. Glen, A. Leach, and R. Taylor in A. Parill and M. Reddy, Eds., *Rational Drug Design*, ACS Symposium Series 719, American Chemical Society, Washington, DC, 1999.

93. R. T. Koehler and H. O. Villar, *J. Comput. Chem.*, **21**, 1145 (2000).

94. R. D. Brown, M. Hassan, and M. Waldman, *J. Mol. Graph. Modell.*, **18**, 427 (2000).

95. (a) S. J. Teague, A. M. Davis, P. D. Leeson, and T. Oprea, *Angew. Chem. Int. Ed. Engl.*, **38**, 3743 (1999); (b) T. Oprea, et al., *J. Chem. Inf. Comput. Sci.*, **41**, 1308 (2001).

96. S. Kirkpatrick, C. D. Gelatt, and M. P. Vecchi, *Science*, **220**, 671 (1983).

97. (a) L. Weber, *Drug Discovery Today*, **3**, 379 (1998); (b) W. H. Press, S. A. Teukolsky, W. T. Vetterling, and B. P. Flannery, *The Art of Scientific Computing*, Cambridge University Press, New York, 1984, pp. 444–455; (c) D. K. Agrafiotis, *J. Chem. Inf. Comput. Sci.*, **37**, 576 (1997); (d) W. Zheng, S. J. Cho, C. L. Waller, and A. Tropsha, *J. Chem. Inf. Comput. Sci.*, **39**, 738 (1999).

98. D. E. Clark, Ed., *Evolutionary Algorithms in Molecular Design*, Wiley-VCH, New York, 2000.

99. R. P. Sheridan and S. K. Kearsley, *J. Chem. Inf. Comput. Sci.*, **35**, 310 (1995).

100. J. Singh, M. A. Ator, E. P. Taeger, M. P. Allen, D. A. Whipple, J. E. Soloweji, S. Chowdhray, and A. M. Treasurywala, *J. Am. Chem. Soc.*, **118**, 1669 (1996).

101. R. D. Brown and Y. C. Martin, *J. Med. Chem.*, **40**, 2304 (1997).

102. (a) C. M. Fonseca and P. J. Fleming in S. Forrest, Ed., Genetic Algorithms: Proceedings of the Fifth International Conference, Morgan Kaufmann, San Mateo, CA, 1993, pp. 416–423; (b) C. M. Fonseca and P. J. Fleming in K. De Jong, Ed., Evolutionary Computation, Vol. 3, The Massachusetts Institute of Technology, Cambridge, MA, 1995, pp. 1–16; (c) V. J. Gillet, W. Khatib, P. Willett, P. Fleming, and D. V. S. Green, *J. Chem. Inf. Comput. Sci.*, **42**, 375–385 (2002).

103. A. Polinsky, R. D. Feinstein, S. Shi, and A. Kuki in I. M. Chaiken and K. D. Janda, Eds., *Molecular Diversity and Combinatorial Chemistry: Libraries and Drug Discovery*, American Chemical Society, Washington, DC, 1996, pp. 219–232.

104. D. S. Thorpe, A. E. Chan, L. Binnie, L. C. Chen, A. Robinson, J. Spoonamore, D. Rodwell, S. Wade, S. Wilson, M. Ackerman-Berrier, H. Yeoman, S. Walle, Q. Wu, and K. F. Wertman, *Biochem. Biophys. Res. Commun.*, **266**, 62 (1999).

105. (a) G. M. Downs and J. M. Barnard, *J. Chem. Inf. Comput. Sci.*, **37**, 59 (1997); (b) J. M. Barnard, G. M. Downs, A. von Scholley-Pfab, and R. D. Brown, *J. Mol. Graph. Modell.*, **18**, 452 (2000); (c) G. M. Downs and J. M. Barnard, *2nd Joint Sheffield Conference on Chemoinformatics: Computational Tools for Lead Discovery*, April 9–11, 2001. Presentation available at: http://cisrg.shef.ac.uk/shef2001/talks/downs.ppt; (d) Accelerys, San Diego, CA (previously MSI, Molecular Simulations Inc.), http://www.accelrys.com.

106. D. K. Agrafiotis and V. S. Lobanov, *2nd Joint Sheffield Conference on Chemoinformatics: Computational Tools for Lead Discovery*, April 9–11, 2001. Abstract available at: http://cisrg.shef.ac.uk/shef2001/abstracts.htm.

107. N. E. Shemetulskis, J. B. Dunbar, Jr., B. W. Dunbar, D. W. Moreland, and C. Humblet, *J. Comput.-Aided Mol. Des.*, **9**, 407 (1995).

108. J. H. Voigt, B. Bienfait, S. Wang, and M. C. Nicklaus, *J. Chem. Inf. Comput. Sci.*, **41**, 702 (2001).

109. M. J. McGregor and P. V. Pallai, *J. Chem. Inf. Comput. Sci.*, **37**, 443 (1997).

110. (a) R. B. Nilakantan, N. Bauman, K. S. Haraki, and R. J. Venkataraghavan, *J. Chem. Inf. Comput. Sci.*, **30**, 65 (1990); (b) R. B. Nilakantan, N. Bauman, and K. S. Haraki, *J. Comput.-Aided Mol. Des.*, **11**, 447 (1997).

111. E. J. Martin and R. E. Critchlow, *J. Comb. Chem.*, **1**, 32 (1999).

112. (a) M. J. McGregor and S. M. Muskal, *J. Chem. Inf. Comput. Sci.*, **39**, 569 (1999); (b) M. J. McGregor and S. M. Muskal, *J. Chem. Inf. Comput. Sci.*, **40**, 117 (2000).

113. B. A. Hopkins, *Ann. Bot.*, **18**, 213 (1954).

114. (a) T. R. Hagadone and M. S. Lajiness, *Tetrahedron Comput. Methodol.*, **1**, 219 (1988); (b) T. R. Hagadone, *J. Chem. Inf. Comput. Sci.*, **32**, 515 (1992).

115. A. Gobbi, D. Poppinger, and B. Rohde, *Perspect. Drug Discov. Des.*, **7/8**, 131 (1997).

116. A. R. Leach, J. Bradshaw, D. V. S. Green, and M. M. Hann, *J. Chem. Inf. Comput. Sci.*, **39**, 1161 (1999).

117. G. Bravi, D. V. S. Green, M. M. Hann, and A. R. Leach, *J. Chem. Inf. Comput. Sci.*, **40**, 1441 (2000).

118. (a) W. P. Walters, Presented at the Daylight User Group Meeting, MUG'99. 1999. Available online at http://www.daylight.com/meetings/mug99/Walters/index.html; (b) W. P. Walters, M. T. Stahl, and M. A. Murcko, *Drug Discovery Today*, **3**, 160 (1998).

119. M. Stahl, *Methods Princ. Med. Chem.*, **10** (Virtual Screening for Bioactive Molecules), 229 (2000).

120. H. J. Bohm and M. Stahl, *Curr. Opin. Chem. Biol.*, **4**, 283 (2000).

121. A. R. Leach, R. A. Bryce, and A. J. Robinson, *J. Mol. Graph. Modell.*, **18**, 358 (2000).

122. R. P. Sheridan, S. G. SanFeliciano, and S. K. Kearsley, *J. Mol. Graph. Modell.*, **18**, 320 (2000).

123. D. Tondi and M. P. Costi in A. K. Ghose and V. N. Viswanadhan, Eds., *Combinatorial Library Design and Evaluation*, Marcel Dekker, New York, 2001, pp. 563–603.

124. E. K. Kick, D. C. Roe, A. G. Skillman, G. Lin, T. J. A. Ewing, Y. Sun, I. Kuntz, and J. A. Ellman, *Chem. Biol.*, **4**, 297 (1997).

125. D. C. Roe in P. M. Dean and R. A. Lewis, Eds., *Molecular Diversity Drug Design*, Kluwer Academic, Dordrecht, Netherlands, 1999, pp. 141–173.

126. (a) I. D. Kuntz, J. M. Blaney, S. J. Oatley, R. Langridge, and T. E. Ferrin, *J. Mol. Biol.*, **161**, 269 (1982); DOCK is developed and distributed by the Kuntz Group, Dept. of Pharmaceutical Chemistry, 512 Parnassus, University of California, San Francisco, CA 94143–0446, URL: http://www.cmpharm.ucsf.edu/kuntz; (b) T. J. A. Ewing and I. D. Kuntz, *J. Comput. Chem.*, **18**, 1175 (1997); (c) Y. Sun, T. J. A. Ewing, A. G. Skillman, and I. D. Kuntz, *J. Comput.-Aided Mol. Des.*, **12**, 597 (1998).

127. (a) S. F. Brady, et al., *J. Med. Chem.*, **41**, 401 (1998); (b) H. J. Bohm, D. W. Banner, and L. Weber, *J. Comput.-Aided Mol. Des.*, **13**, 51 (1999).

128. (a) A. Rockwell, M. Melden, R. A. Copeland, K. Hardman, C. P. Decicco, and W. F. DeGrado, *J. Am. Chem. Soc.*, **118**, 10337 (1996); (b) A. K. Szardenings, D. Harris, S. Lam, L. Shi, D. Tien, Y. Wang, D. V. Patel, M. Navre, and D. A. Campbell, *J. Med. Chem.*, **41**, 2194 (1998).

129. D. Tondi, U. Slomiczynska, M. P. Costi, D. M. Watterson, S. Ghelli, and B. K. Shoichet, *Chem. Biol.*, **6**, 319 (1999).

130. M. L. Lamb, K. W. Burdick, S. Toba, M. M. Young, A. G. Skillman, X. Zou, J. R. Arnold, and I. D. Kuntz, *Proteins Struct. Funct. Genet.*, **42**, 296 (2001).

131. I. Ugi and C. Steinbruckner, *Chem. Ber.*, **94**, 734 (1961).

132. T. F. Herpin, G. C. Morton, A. K. Dunn, C. Fillon, P. R. Menard, S. Y. Tang, J. M. Salvino, and R. F. Labaudiniere, *Mol. Diversity*, **4**, 221 (2000).

133. C. M. Murray and S. J. Cato, *J. Chem. Inf. Comput. Sci.*, **39**, 46 (1999).

134. A. Tulinsky, K. Padmanbhan, K. P. Padmanbhan, C. H. Park, W. Bode, R. Huber, D. T. Blankenship, A. D. Cardin, and W. Kisiel, *J. Mol. Biol.*, **232**, 947 (1993).

135. M. Rarey and M. Stahl, *J. Comput.-Aided Mol. Des.*, **15**, 497–520 (2001).

CHAPTER SIX

Virtual Screening

INGO MUEGGE
ISTVAN ENYEDY
Bayer Research Center
West Haven, Connecticut

Contents

Burger's Medicinal Chemistry and Drug Discovery
Sixth Edition, Volume 1: Drug Discovery
Edited by Donald J. Abraham
ISBN 0-471-27090-3   © 2003 John Wiley & Sons, Inc.

1 INTRODUCTION

Virtual screening, sometimes also called *in silico* screening, is a new branch of medicinal chemistry that represents a fast and cost-effective tool for computationally screening compound databases in search for novel drug leads. The roots for virtual screening go back to structure-based drug design and molecular modeling. In the 1970s researchers hoped to find novel drugs designed rationally using a fast growing number of diverse protein structures being solved by X-ray crystallography (1, 2) or nuclear magnetic resonance (NMR) spectroscopy (3). However, only very few drugs have resulted from those early efforts. Examples include captopril as angiotensin-converting enzyme inhibitor (4) and methotrexate as dihydrofolate reductase inhibitor (5). The reasons for this somewhat disappointing drug yield lie in the low resolution of the protein structures as well as limitations in computer power and methods. Researchers have often tried *de novo* to design the final drug candidate on the computer screen. The compounds suggested have often been difficult to synthesize; initial failure in exhibiting potency has often resulted in the termination of structure-based projects. At the end of the 1980s rational drug design techniques became somewhat discredited because of the high failure rate in drug discovery projects.

In the 1990s drastic changes occurred in the way drugs are discovered in the pharmaceutical industry. High throughput synthesis (6, 7) and screening techniques (8) changed the lead identification process that is now governed not only by large numbers of compounds processed but also by fast prosecution of many putative drug targets in parallel. The characterization of the human genome has re-

sulted in a large number of novel putative drug targets. Improved screening techniques also make it possible to look at entire gene families, at orphan targets, or at otherwise uncharacterized putative drug targets. In this environment of data explosion, rational design techniques have experienced a comeback (9). Although the exponentially growing number of solved protein structures at high resolution makes it possible to embark on structure-based design for many drug targets, virtual screening—the computational counterpart to high throughput screening—has become a particularly successful computational tool for lead finding in drug discovery. Whereas proprietary screening libraries typically hold about 10^6 compounds, this is only a tiny fraction of the conceivable chemical space for which estimates range between 10^{60} and 10^{100} compounds (10, 11). The question is, of course, which subset of this enormous space should be synthesized and screened? Virtual screening attempts to answer this question by evaluating large virtual libraries of up to 10^{12} compounds through the use of a cascade of various screening tools to reduce the chemical space. This chapter describes the different concepts and tools used today for virtual screening. They reach from the assessment of the overall "druglikeness" of a small organic molecule to its ability to specifically bind to a given drug target. The interested reader is also referred to a selection of recent books and reviews on the subject of virtual screening (10, 12–18).

2 CONCEPTS OF VIRTUAL SCREENING

The basic goal of virtual screening is the reduction of the enormous virtual chemical space of small organic molecules, to synthesize and/or screen against a specific target protein, to a manageable number of compounds that exhibit the highest chance to lead to a drug candidate (10, 19). The major sources of information to guide virtual screening for a particular target are derived from the following questions:

1. What does a drug look like in general?
2. What is known about compounds that interact with the receptor?
3. What is known about the structure of the target protein and the protein-ligand interactions?

In the following subsections we address these three points, outlining concepts of assessing the overall druglikenss of molecules, the concentration of subsets of molecules in focused libraries, and the identification of specific leads through structure-based virtual screening techniques.

2.1 Druglikeness Screening

Many drug candidates fail in clinical trials because of reasons unrelated to potency against the intended drug target. Pharmacokinetics and toxicity issues are blamed for more than half of all failures in clinical trials. Therefore, the first part of virtual screening evaluates the *druglikeness* of small molecules, mostly independent of their intended drug target (there are specific drug classes such as those acting in the central nervous system that require specific drug profiles). Druglike molecules exhibit favorable absorption, distribution, metabolism, excretion, and toxicological (ADMET) parameters (20–24). They are synthetically feasible and possess pharmacophore features that offer the chance of specific interactions with the intended protein target. Druglikeness is currently assessed using the following types of methods: simple counting methods, functional group filters, topological filters, and pharmacophore filters. Computational techniques used to identify druglikeness include neural networks (25–27), recursive partitioning approaches (25, 28), and genetic algorithms (29). These methods are further discussed below.

Table 6.1 Typical Ranges for Parameters Related to Druglikeness[a]

Parameter	Minimum	Maximum
Log P	−2	5
Molecular weight	200	500
Hydrogen bond acceptors	0	10
Hydrogen bond donors	0	5
Molar refractivity	40	130
Rotatable bonds	0	8
Heavy atoms	20	70
Polar surface area [Å^2]	0	120
Net charge	−2	+2

[a]Data taken from ref. 21.

2.1.1 Counting Schemes. Database collections of known drugs [e.g., CMC (30), WDI (31), or MDDR (32)] are typically used to extract knowledge about structure and properties of potential drug molecules. Key physicochemical properties such as molecular weight, charge, and lipophilicity (33, 34) of drug collections are profiled to extract simple counting rules for relevant descriptors of ADMET-related parameters. Examples include Lipinski's "rule-of-five" (33), which limits the range for molecular weight (MW ≤ 500), computed octanol-water partition coefficient (Clog P ≤ 5), and hydrogen-bond donors and acceptors (OHs + NHs ≤ 5; Ns + Os ≤ 10). Other authors limit the number of rotatable bonds (RB ≤ 8) or rings in a molecule (number of rings ≤ 4) (34). Table 6.1 shows a list of typical boundaries of counting parameters. Figure 6.1 illustrates the profiling procedure for these counting parameters using polar surface area (PSA) (35) as a descriptor. Collections of 776 orally administered CNS drugs and 1590 orally administered non-CNS drugs that reached phase II efficacy studies were analyzed for their PSA. It was found that 90% of the non-CNS compounds have a PSA below 120 Å^2; 90% of CNS drugs have a PSA below 80 Å^2. Although it is possible that drugs have higher PSA values and are still orally bioavailable or penetrate the blood-brain barrier (as the result of active transport or other reasons), the profile suggests that it is much less likely. It is therefore a reasonable assumption in a virtual screening approach to discriminate against compounds outside the most populated descriptor space (in this case, PSA

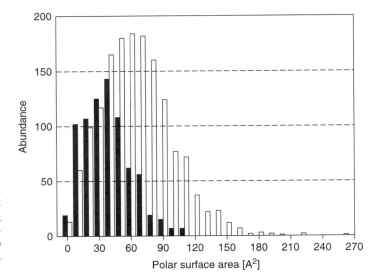

Figure 6.1. Distribution of polar surface area for 776 orally administered CNS drugs (black bars) and for 1590 orally administered non-CNS drugs (white bars) that have reached clinical phase II efficacy studies (35).

< 120 Å^2), especially if the compound lies outside the optimal region for several descriptors (e.g., MW > 500 and Clog $P > 5$).

Simple descriptors as described above are quickly calculated and counted. Therefore, after typically removing compounds with atoms other than C, N, O, S, H, P, Si, Cl, Br, F, and I, counting schemes present the first filter in virtual screening approaches.

2.1.2 Functional Group Filters. Reactive, toxic, or otherwise unsuitable compounds, such as natural product derivatives, are removed using specific substructure filters. Figure 6.2 shows a subset of substructures that lead to the dismissal of compounds in virtual screening. Typical reactive functional groups include, for example, reactive alkyl halides, peroxides, and carbazides. Unsuitable leads may include crown ethers, disulfides, and aliphatic methylene chains seven or more long. Unsuitable natural products may include quinones, polyenes, or cycloheximide derivatives. A list of such fragments coded in Daylight SMARTS is given, for example, by Hann and coworkers (36). It should be noted, however, that natural product derivatives are not always unsuitable leads.

Screening out compounds that contain certain atom groups associated with toxicity provides a practical and fast way to reduce large databases; however, it is only a crude approx-

imation for eliminating potentially toxic compounds. Better descriptions of toxicity may be provided by structure-based methods to assess toxicity of compounds. They draw primarily from mutagenicity, carcinogenicity, and acute toxicity databases assembled, for instance, by the National Toxicology Program (37) and the Toxic Effect of Chemical Substances database, RTECS (38). CASETox (39), TOPKAT (40), and DEREK (41) are commercial software

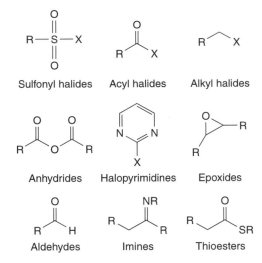

Figure 6.2. Selection of reactive functional groups that should be removed from a virtual screen (examples taken from Ref. 212).

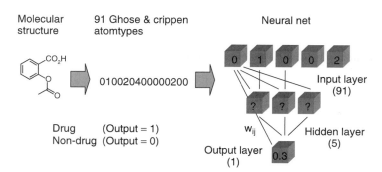

Molecular structure

91 Ghose & crippen atomtypes

Neural net

010020400000200

Drug   (Output = 1)
Non-drug (Output = 0)

Input layer (91)

w_{ij}

Hidden layer (5)

Output layer (1)

Figure 6.3. Neural network architecture for prediction of druglikeness.

products that can be used to evaluate virtual compounds for potential toxicity.

2.1.3 Topological Drug Classification. It is generally assumed that compounds with structural similarity to known drugs may exhibit druglike properties themselves, such as oral bioavailability, low toxicity, membrane permeability, and metabolic stability. Following this idea, drug databases and reagent databases such as the ACD (42) as negative control (assuming they do not contain many drugs) have been analyzed to find structural features of drugs and nondrugs. Neural network approaches have been devised (25, 27) that can discriminate between drugs and nondrugs with about 80% certainty. Recursive partitioning approaches classify drugs and nondrugs with similar accuracy.

2.1.3.1 Artificial Neural Networks and Decision Trees. Figure 6.3 shows an example of a simple neural network that uses Ghose and Crippen atom types (43) to code the molecular

structure. Ninety-one statistically significant atom types correspond to 91 input neurons of the neural net. Typically, five neurons in the hidden layers are used in the net design (25, 27). The single neuron output layer can vary between 0 (nondrugs) or 1 (drugs). Trained on 5000 drugs taken from the WDI and 5000 compounds labeled nondrugs taken from the ACD, the resulting neural net was shown to correctly classify about 80% of other drugs/nondrugs (27).

Recursive partitioning, also known as the decision tree approach, is another powerful method to extract knowledge from a database. Wagener and Geerestein have explored the WDI and ACD databases to train a decision tree for the discrimination of drugs and nondrugs (28). Figure 6.4 shows a partial decision tree derived by the authors. One rule derived from this partial tree is, for example, if a compound possesses no alcohol and a tertiary aliphatic amine but no methylene linker between a heteroatom and a carbon atom, it is not

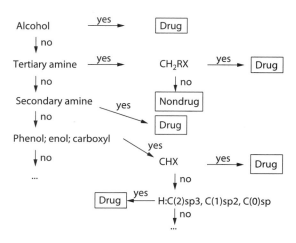

Figure 6.4. Partial decision tree from Wagener and Geerestein (28). $C(n)sp^x$ describes a carbon with hybridization sp^x and formal oxidation number n. X refers to a heteroatom; R refers to any group linked through a carbon. The tree starts at the top left corner. Here is an example of how to read the tree: If a compound contains an alcohol, it is classified as a drug. If it does not contain an alcohol, the presence of a tertiary amine is checked. If it contains a tertiary amine and also contains (does not contain) a CH_2 group with attached heteroatom as well as another R group, it is classified as drug (nondrug).

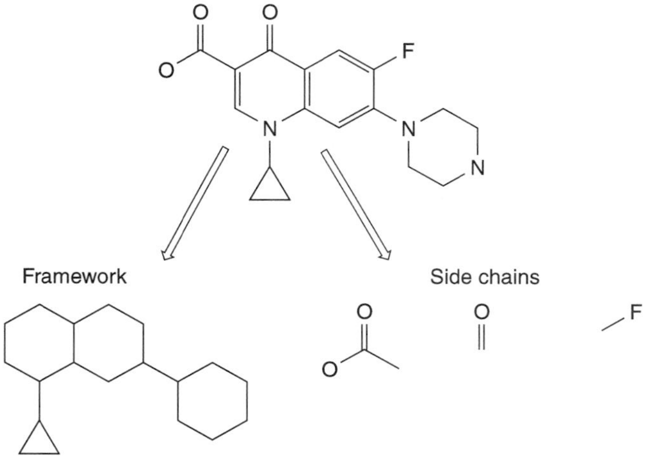

Figure 6.5. Dissection of a drug molecule into framework and side chains.

druglike. It is interesting to note that just by testing the presence or absence of hydroxyl, tertiary and secondary amines, carboxyl, phenol, or enol groups, 75% of all druglike structures in the MDDR and CMC can be recognized.

Once they are trained, neural networks and decision trees are very fast filter tools in virtual screening approaches. They are therefore applied early in the virtual screening filter cascade.

2.1.3.2 Structural Frameworks and Side Chains of Known Drugs. Databases have been mined to find structural motifs and pharmacophore features of small molecules that characterize drugs. Bemis and Murcko (44)

dissected drug molecules from the Comprehensive Medicinal Chemistry (CMC) (30) database into side chains and frameworks (containing ring systems and linkers). They found that only 32 frameworks described the shapes of half the 5120 drugs in the CMC containing 1170 scaffolds. Figures 6.5 and 6.6 show the process of reducing a drug molecule to its framework and a list of the most frequently occurring frameworks in the CMC. Side chains most frequently occurring in drug molecules have also been analyzed (45). It has been found that of the 15,000 side chains contained in the CMC, about 11,000 belong to one of only 20 side chains, including (starting with the most frequent): carbonyl, methyl, hy-

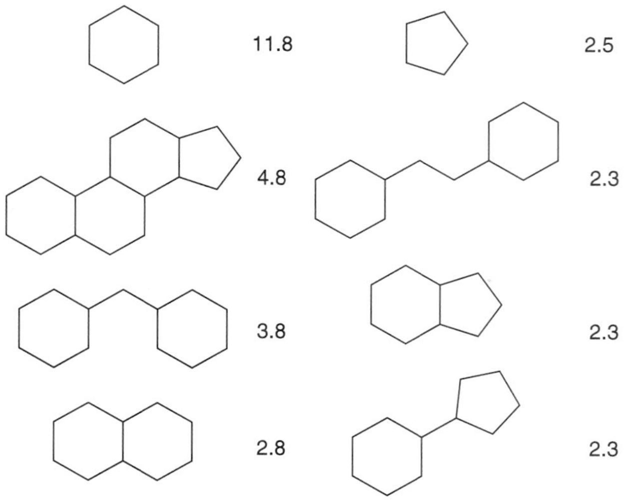

Figure 6.6. Most frequently occurring frameworks in drugs (numbers indicate percentages of occurrence in CMC database). Data are taken from Bemis and Murcko (44).

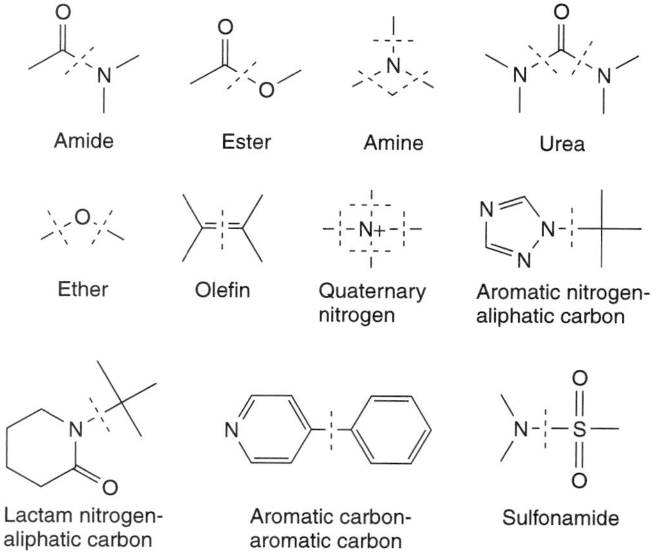

| Amide | Ester | Amine | Urea |

| Ether | Olefin | Quaternary nitrogen | Aromatic nitrogen-aliphatic carbon |

| Lactam nitrogen-aliphatic carbon | Aromatic carbon-aromatic carbon | Sulfonamide |

Figure 6.7. Retrosynthetic reactions in RECAP.

droxyl, methoxy, chloro, methylamine, primary amine, carboxylic acid, fluoro, and sulfone. Most molecules possess between one and five side-chains; more than 20% of the drugs stored in the CMC have two side chains per molecule.

For the analysis of virtual libraries according to the presence or absence of druglike frameworks, side-chains or structural motifs can be used for virtual screening. This idea has been extended in RECAP (retrosynthetic combinatorial analysis procedure), a technique that identifies common motifs in drugs based on fragmenting molecules around bonds formed by common reaction (46) (Fig. 6.7). Extracting rules from RECAP for virtual screening represents a possible way of addressing the questions of ease of synthesis of compounds. A similar approach to assess the occurrence of structural motifs in drug molecules was presented by Wang and Ramnarayan, who developed the concept of multilevel chemical compatibility (MLCC) between a drug database and a test molecule as a measure for druglikeness. In the MLCC method a compound is recognized as druglike if all of its topological motifs occur in other known drugs.

2.1.4 Pharmacophore Point Filter. The topological drug fragmentation approaches discussed above suggest that the occurrence

of a relatively small number of frameworks (ring structures and linkers), an even smaller number of side chains, and a small number of polar groups characterize drugs very well. Although drugs and nondrugs are not completely distinguishable, it has been observed that drugs differ somewhat from nondrugs in their possession of hydrophobic moieties that are well functionalized. Nondrugs often contain underfunctionalized hydrophobic groups (Fig. 6.8). Recent work to characterize the druglikeness of molecules focuses more on the presence of key functional groups in molecules.

A simple pharmacophore point filter has been introduced recently (47). It is based on the assumption that druglike molecules should contain at least two distinct pharmacophore groups (47). Four functional motifs have been identified that guarantee hydrogen-bonding capabilities that are essential for the specific interaction of a drug molecule with its biological target (Fig. 6.9). These motifs can be combined to functional groups that are also referred to here as *pharmacophore points*; they include: amine, amide, alcohol, ketone, sulfone, sulfonamide, carboxylic acid, carbamate, guanidine, amidine, urea, and ester. The following main rules apply to the pharmacophore point filter (PF1):

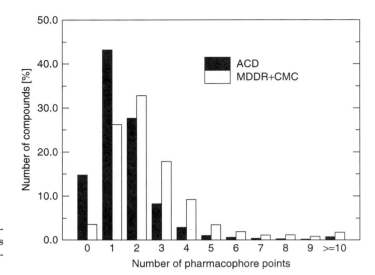

Figure 6.8. Number of pharmacophore points in drug databases (MDDR + CMC) and reagent databases (ACD).

- Pharmacophore points are fused and counted as one if they are separated by less than two carbon atoms.
- Molecules with less than two and more than seven pharmacophore points fail the filter.
- Amines are considered pharmacophore points but not azoles or diazines.
- Compounds with more than one carboxylic acid are dismissed.
- Compounds without a ring structure are dismissed.
- Intracyclic amines in the same ring are fused to one pharmacophore point.

The requirement of two distinct pharmacophore points neglects at least one very important class of drugs: biogenic amine-containing CNS drugs. Therefore, a second pharmacophore filter has been designed that requires only one pharmacophore point in small molecules of the type amine, amidine, guanidine, or carboxylic acid (PF2).

An analysis of drug databases and reagent-type databases reveal that about two thirds of drugs and nondrugs can be classified correctly by PF1. This performance is not as impressive as that of neural networks. However, as a filter for virtual screening, pharmacophore point filters offer some advantages. First, the occurrence and count of pharmacophore points can be evaluated on the building-block level of a virtual combinatorial library. No enumeration is necessary as for druglike neural nets. Second, the results of the pharmacophore point filter can be easily interpreted. Third, the settings of the filter can be easily adjusted (e.g., PF1 for non-CNS drugs, PF2 for CNS drugs).

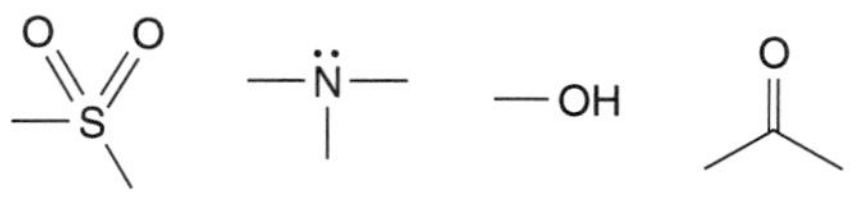

2.2 Focused Screening Libraries for Lead Identification

Without the knowledge about specific drug targets it is sometimes useful to apply virtual screening for the design of focused libraries of a few thousand compounds rather than to find a small number of hits to be tested against a specific target. To save resources it may sometimes be more prudent not to run the entire HTS file against a target protein; instead, a focused library with higher chances of containing hits may be scrutinized. Those focused libraries may be designed to target specific protein families such as GPCRs, kinases, or nuclear hormone receptors. They can also be enriched with privileged structures that occur

Figure 6.9. Functional motifs of drugs used to build pharmacophore points.

more often in drug molecules and/or were found to inhibit members of the protein family.

2.2.1 Targeting Protein Families. Target class-directed libraries can be built from available compounds or be synthesized in combinatorial fashion. The design of target class-directed libraries relies on the identification of structural motifs in small molecules or in building blocks for combinatorial libraries that can be linked to increased activity for the target class. Functional groups that show the propensity to hit a certain target class can be found by examining ligands from the literature. Recurring motifs for GPCRs include, for example, piperazines, morpholines, and piperidines; for kinases they include, for example, heterobicyclic compounds or pyrimidines. Compounds bearing those structural motifs are thought to have a generally higher chance to be active against the respective target classes. A more rigorous approach to identify the "GPCR-likeness" of compounds or building blocks can be provided by a statistical analysis of druglike databases. Neural networks have been shown to be particularly useful in classifying chemical matter, such as CNS-active compounds (26, 48).

A neural network approach similar to that of Sadowski and Kubinyi (27) has been described recently to address the "GPCR-likeness" of small molecules as well as building blocks for combinatorial libraries (49). A feed-forward neural net was trained using 5000 compounds from the MDDR that target GPCRs and 5000 compounds that target other protein classes. Using the "activity-class" field of the database, about 20,000 GPCR-like and 55,000 non-GPCR-like have been identified by entries such as 5HT, leukotriene, and PAF. The resulting neural net classifies GPCR-like compounds correctly with 80% certainty. An independent test of compounds in our proprietary database that were found to hit GPCRs or other targets showed a correct prediction of GPCR-like compounds in 70% of the cases. When several virtual combinatorial libraries were analyzed, it turned out that the property of being GPCR-like could be attributed to the GPCR-likeness of the building blocks alone; that is, the GPCR-likeness of the enumerated

compound correlated very well with the GPCR-likeness of the most GPCR-like building block it contained. This offers an important advantage for the design of combinatorial libraries because, for large virtual libraries, the computer costs for enumeration go with the power of the number of R groups and thus very quickly becomes impractical. For instance, for a 3-R-group library with 1000 building blocks each, the enumerated library would contain 1 billion compounds to be analyzed, whereas the building block-level analysis needs to examine only 3000 compounds. Figure 6.10 shows a list of amine building blocks extracted from the ACD that were found to be most GPCR-like by the neural net.

Not every portion of a GPCR-like molecule has to be GPCR-like. The presence of one GPCR-like moiety (building block or core structure) is sufficient to make a compound GPCR-like. Therefore, the neural network offers two different strategies for the design of GPCR-like libraries: (*1*) GPCR-like core + druglike building blocks (need not be GPCR-like); (*2*) non-GPCR-like core + GPCR-like building blocks. Virtual screening of a database of existing compounds using the described neural net can be applied to assemble a focused screening library. Alternatively, combinatorial libraries can be designed.

2.2.2 Privileged Structures. Privileged structures are structural types of small molecules that are able to bind with high affinity to multiple classes of receptors (50). An enrichment of libraries with privileged structures may increase the chance of finding active compounds. Examples of privileged structures include benzazepine analogs found to be effective ligands for an enzyme that cleaves the peptide angiotensin I, whereas others are effective CCK-A receptor ligands. Cyproheptadine derivatives were found to have peripheral anticholinergic, antiserotonin, antihistaminic, and orexigenic activity. Hydroxamate and benzamidine derivatives have been shown to be privileged structures for metalloproteases and serine proteases, respectively. For the class of 7-transmembrane G-protein-coupled receptors a large number of privileged structures has been found including, for example, diphenylmethane, diazepine, benzazepine, biphenyltetrazole, spiropiperidine, in-

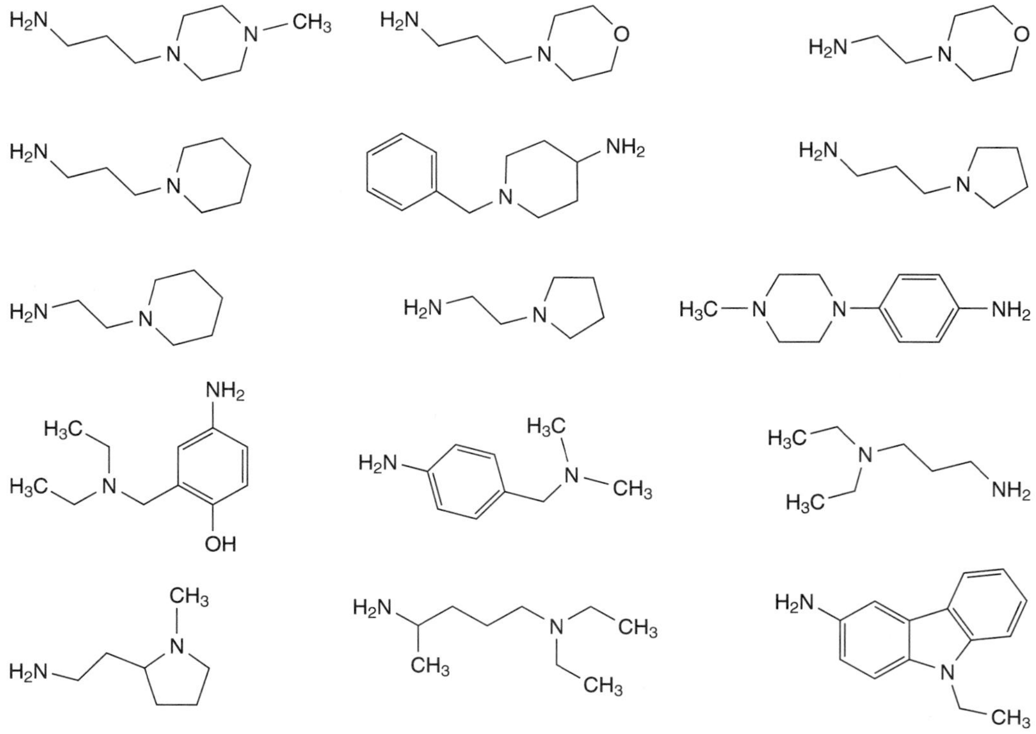

Figure 6.10. Selection of GPCR-like amines from the ACD.

dole, and benzylpiperidine (51). Some ubiquitously privileged structures have recently been identified (52). They include carboxylic acids, biphenyls, diphenylmethane, and, to a lesser extent, naphthyl, phenyl, cyclohexyl, dibenzyl, benzimidazole, and quinoline.

2.3 Pharmacophore Screening

In cases where no structural information about the target protein is given, pharmacophore models can provide powerful filter tools for virtual screening (53). Even in cases where the protein structure is available, pharmacophore filters should be applied early because they are generally much faster than docking approaches (discussed below) and can, therefore, greatly reduce the number of compounds subjected to the more expensive docking applications. For example, a pharmacophore model consisting of three pharmacophore points can be tested against about 10^6 compounds in a few minutes of computer time [disregarding the time it takes to generate three-dimensional (3D) conformations of each molecule] (10). Another interesting aspect of pharmacophores in virtual screening is 3D-pharmacophore diversity. Although the diversity concept for virtual compounds in general is not applicable because of the enormity of the chemical space, diversity in pharmacophore space is a feasible concept. Virtual libraries can therefore be optimized for covering a wide pharmacophore space.

2.3.1 Introduction to Pharmacophores. In 1894 Emil Fischer proposed the "lock-and-key" hypothesis to characterize the binding of compounds to proteins (54). This can be considered the first attempt to explain binding of small molecules to a biological target. Proteins recognize substrates through specific interactions. It is a challenge for the medicinal chemist to synthesize compounds that can capture the 3D arrangement of functional groups in a small molecule that forms the pharmacophore and that is responsible for substrate binding

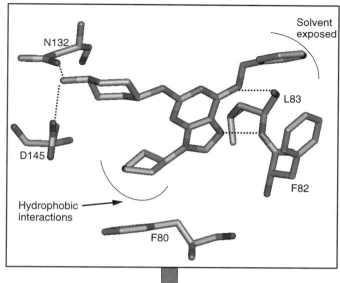

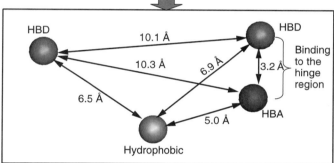

Figure 6.11. Pharmacophore derived based on the interactions between human cyclin-dependent kinase 2 and the adenine-derived inhibitor H717 as observed in the X-ray structure of the complex (PDB entry 1G5S). Dashed lines highlight hydrogen-bonding interactions. HBD, hydrogen-bond donor; HBA, hydrogen-bond acceptor. The hinge region is linking the N- and C-terminal domains of a kinase.

to the protein. The first definition of the pharmacophore formulated by Paul Ehrlich was "a molecular framework that carries (phoros) the essential features responsible for a drug's (pharmacon) biological activity" (55). This definition was slightly modified by Peter Gund to "a set of structural features in a molecule that is recognized at a receptor site and is responsible for that molecule's biological activity" (56). An example is shown in Fig. 6.11. An X-ray structure of CDK2 complexed with the adenine-derived inhibitor H717 (57–59) has been solved. Interactions that are essential to substrate and inhibitor binding to the enzyme will form the pharmacophore that should be captured by inhibitors binding the same way H717 does. As shown in Fig. 6.11, the inhibitor binds to the hinge region (Phe82 and Leu83) through two hydrogen bonds, to a hydropho-

bic region through the cyclopentyl group, and to Asp145 and Asn132 through hydrogen bonds. The pharmacophore that reflects these interactions has a hydrogen-bond donor and a hydrogen-bond acceptor pair that ensures binding to the hinge region, a hydrophobic group that corresponds to the cyclopentyl binding site, and a hydrogen-bond donor that ensures binding to Asp145 and/or Asn132. Note that in addition to distances that describe the 3D relationship among pharmacophore points, angles, dihedrals, and exclusion volumes are also used. Each additional restraint can reduce the number of hits, thus making the compound selection easier for testing. Pharmacophore hypotheses for searching can be generated using structural information from active inhibitors, ligands, or from the protein active site itself (60, 61).

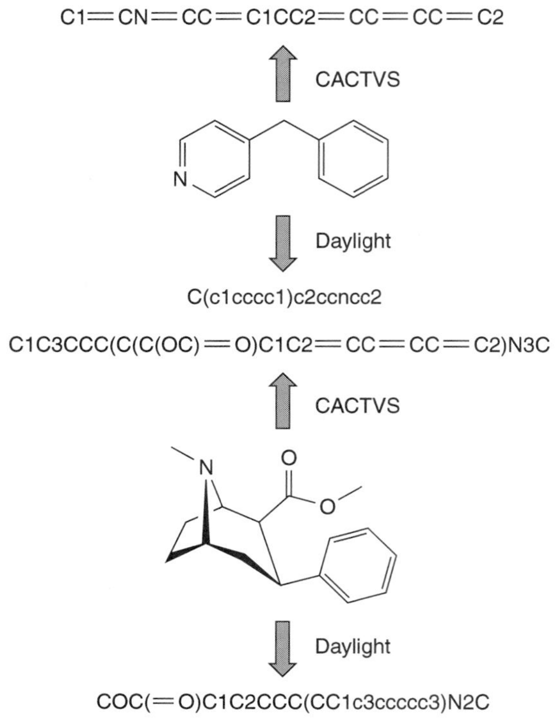

C1═CN═CC═C1CC2═CC═CC═C2

↑ CACTVS

⬇ Daylight

C(c1cccc1)c2ccncc2

C1C3CCC(C(C(OC)═O)C1C2═CC═CC═C2)N3C

↑ CACTVS

⬇ Daylight

COC(═O)C1C2CCC(CC1c3ccccc3)N2C

Figure 6.12. Examples of SMILES notations for two compounds obtained using CACTVS and Daylight.

2.3.2 Databases of Organic Compounds.

Virtual screening is used in general for selecting potentially active compounds from databases of compounds available either in-house or from a vendor. Because virtual screening is not accurate enough to identify only active compounds as hits, it is less risky to screen databases with existing compounds rather than synthesize a new library. Nevertheless, virtual libraries that can be synthesized through combinatorial chemistry and/or rapid analoging can easily be generated using *in silico* methods. These libraries are more often generated for lead optimization and synthesis prioritization (62, 63).

There is a wealth of databases that code available compounds typically in the two-dimensional standard data (2D-SD) format including connectivity from MACCS (32, 64). The most common databases are the Available Chemicals Directory (ACD) (42), Spresi (65), Chemical Abstracts Database (66), and the National Cancer Institute Database (67, 68). Many vendors of chemicals also provide searchable databases with 2D-structure and

property information of their compounds. Sometimes compounds are coded in linear representations such as the SMILES (69, 70) notation. The SMILES codes obtained using CACTVS and Daylight programs for 4-benzyl pyridine and R-cocaine are shown in Fig. 6.12.

The primary source of 3D experimental structures of organic molecules is the Cambridge Structural Database (71). Alternatively, 2D databases of organic compounds can be converted into 3D databases using several software programs (72). Each program starts with generating a crude structure that is subsequently optimized using a force field. CONCORD (73) applies rules derived from experimental structures and a univariate strain function for building an initial structure. CORINA (74) generates an initial structure by use of a standard set of bond lengths, angles and dihedrals, and rules for cyclic systems. RUBICON (75) invokes distance geometry techniques to generate 3D structures based on connectivity tables. This program also uses bond lengths and angle tables to build a matrix containing the upper and lower bounds for distances between all atoms in the molecule. OMEGA (76) uses a torsion-driven approach for building conformers. It generates low energy conformers for each molecule by assembling it from fragments and searching through possible orientations of the subunit added. WIZARD (77) and COBRA (78), AIMB (79) and MIMUMBA (80) employ artificial intelligence techniques for generating a set of user-specified low energy conformations for a compound. MOLGEO (81) uses a depth-first approach for generating 3D structures based on connectivity using bond length and bond angle tables. IDEALIZE (82) is a molecular mechanics program that minimizes 2D structures to generate the corresponding 3D structure.

2.3.3 2D Pharmacophore Searching. Searching 2D databases is of great importance for accelerating drug discovery. Chemical suppliers provide databases of purchasable compounds that medicinal chemists search for starting material for synthesis or analogs of a lead compound. Different strategies are pursued to search a 2D database to identify compounds of interest. Exact structure search is applied to find out whether a compound is present in the database. Substructure searches identify larger molecules that contain the user-defined query, irrespective of the environment in which the query substructure occurs (83) (Fig. 6.13). Furthermore, substructure searching can identify all compounds in a database that share the same core structure. Biochemical data obtained from testing these compounds can be used for generating structure-activity relationships (SARs), even before synthetic plans are made for lead optimization (84). In contrast, superstructure searches are used to find smaller molecules that are embedded in the query (Fig. 6.14). One problem that arises from substructure searches is that the number of compounds identified can reach into the thousands. A solution to this problem is ranking the compounds based on similarity to a reference compound. Similarity searches use one or more structural descriptors for quantifying the similarity between compounds in the database and in the query (85, 86) (Fig. 6.15). A review of descriptors used in similarity searches is provided by Willett et al. (86). Beyond structural similarity, activity similarity

has also been the subject of several studies. Xue et al. showed that compounds with similar activity could be identified using mini-fingerprints (87–89), physicochemical property descriptors (90), or latent semantic structure indexing (91, 92). In addition, similarity searches can be combined with superstructure searches for limiting the number of compounds selected. Flexible match searches are used for identifying compounds that differ from the query structure in user-specified ways. In addition, isomer, tautomer, and parent molecule searches may be done to find in a database isomers, tautomers, or parent molecules of the query.

2.3.4 3D Pharmacophores
2.3.4.1 Ligand-Based Pharmacophore Generation. Ligand-based pharmacophores are typically used when the crystallographic, solution structure, or modeled structure of a protein cannot be obtained. When a set of active compounds is known and it is hypothesized that all compounds bind in a similar way to the protein, then common groups should interact with the same protein residues. Thus, a pharmacophore capturing these common features should be able to identify from a database novel compounds that bind to the same site of the protein as the known compounds do. The process of deriving a pharmacophore, called *pharmacophore mapping*, consists of three steps: (*1*) identifying common binding elements that are responsible for biological activity; (*2*) generating potential conformations that active compounds may adopt; and (*3*) determining the 3D relationship between pharmacophore elements in each conformation generated. To build a pharmacophore based on a set of active compounds, two methods are usually applied. One method is to generate a set of minimum energy conformations for each ligand and search for common structural features. Another method is to consider all possible conformations of each ligand to evaluate shared orientations of common functional groups. Analyzing many low energy conformers of active compounds can suggest a range of the distance between key groups that will take in account the flexibility of the ligands and of the protein. This task can be performed either manually or automatically.

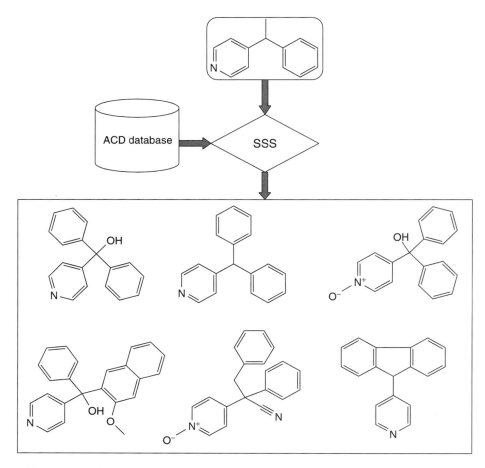

Figure 6.13. Compounds identified from the ACD database through substructure search.

2.3.4.2 Manual Pharmacophore Genera-tion. Manual pharmacophore generation is used when there is an easy way to identify the common features in a set of active compounds and/or there is experimental evidence that some functional groups should be present in the ligand for good activity. An example is the development of a pharmacophore model for dopamine-transporter (DAT) inhibitors (Fig. 6.16). In the first step common structural fea-tures were identified in the selected five DAT inhibitors (93–95) (Fig. 6.16, circles). Four out of five compounds were structurally rigid, whereas the 4-hydroxy piperidinol was flexi-ble. A systematic conformational search for 4-hydroxy piperidinol identified 10 possible conformations. Measuring distances among pharmacophore elements in every inhibitor

and every conformation considered led to the distance ranges among pharmacophore points shown in Fig. 6.16. Because proteins are flex-ible, pharmacophores should also have some flexibility built in, thus justifying the use of distance ranges.

2.3.4.3 Automatic Pharmacophore Genera-tion. Pharmacophore generation through conformational analysis and manual align-ment is a very time-consuming task, especially when the list of active ligands is large and the elements of the pharmacophore model are not obvious. There are several programs, HipHop (96), HypoGen (97), Disco (98), Gasp (99), Flo (100), APEX (101), and ROCS (102), that can automatically generate potential pharma-cophores from a list of known inhibitors. The performance of these programs in automated

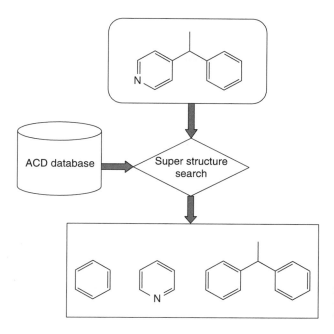

Figure 6.14. Compounds identified from the ACD database through superstructure search.

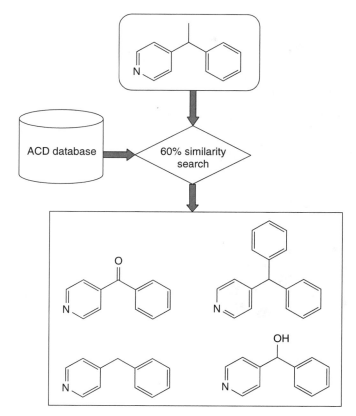

Figure 6.15. Compounds identified from the ACD database through similarity search using 60% similarity as threshold. The lower the specified similarity the higher the number of hits identified.

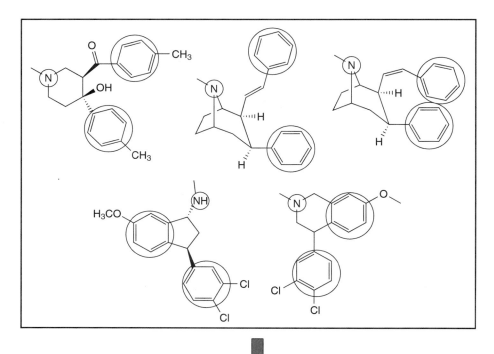

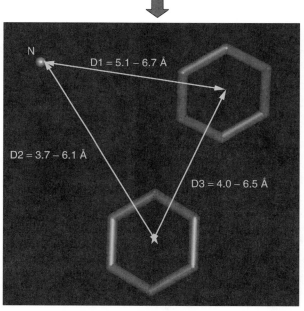

Figure 6.16. Manual pharmacophore mapping by measuring distances between pharmacophore points in every compound and conformation considered. Pharmacophore elements are highlighted with circles. All structures were built and minimized using QUANTA. Conformers of 4-hydroxy piperidinol were generated using the Grid Scan method from QUANTA, followed by clustering, to identify unique conformers.

pharmacophore generation varies depending on the training set. The use of these programs for pharmacophore generations was recently reviewed in detail (103). Here we focus on common features of these programs. All programs use algorithms that identify common pharmacophore features in the training set molecules; they use scoring functions to rank the identified pharmacophores. The following features are identified in each molecule: hydrogen-bond donors, hydrogen-bond acceptors, negative and positive charge centers, and surface accessible hydrophobic regions that can be aliphatic, aromatic, or nonspecific. Most of the programs consider ligand flexibility when generating pharmacophores because compounds might not bind to the protein in the minimum energy conformation.

2.3.4.4 Receptor-Based Pharmacophore Generation. If the 3D structure of a receptor is known, a pharmacophore model can be derived based on the receptor active site. Biochemical data can be used for identifying key residues that are important for substrate and/or inhibitor binding. This information can be used for building pharmacophores targeting the region defined by key residues or for choosing among pharmacophores generated by an automated program. This can greatly improve the chance of finding small molecules that inhibit the protein because the search is focused on a region of the binding site that is crucial for binding substrates and inhibitors. Many ligands bind to proteins through nonbonded interactions such as hydrogen bonds and hydrophobic interactions. Programs such as LUDI (104–106) or POCKET (107) can use the structure of the protein to generate interaction sites or grids to characterize favorable positions that ligand atoms should occupy. Four types of interaction sites are characterized: hydrogen-bond donors, hydrogen-bond acceptors, and hydrophobic groups that can be lipophilic-aliphatic or lipophilic-aromatic. LUDI-generated interaction maps for Cerius^2 Structure-Based Focusing (108) do not differentiate between aliphatic and aromatic interaction sites. This is based on the observation by Burley and Petsko (109) that, besides aromatic side chains, aliphatic and aromatic side chains also pack closely to form the hydrophobic core of proteins. Because proteins are not

rigid, Carlson et al. (110) proposed using molecular dynamics simulation for generating a set of diverse protein conformations to include protein flexibility in the pharmacophore development. In this case distance ranges between pharmacophores are obtained by examining several conformations of the protein. This technique is similar to the one used for the generation of flexible pharmacophores (Fig. 6.16), based on active compounds, when several conformations of the compound and/or many compounds are considered for pharmacophore mapping.

2.3.5 Pharmacophore-Based Virtual Screening. Pharmacophore-based virtual screening is the process of matching atoms and/or functional groups and the geometric relations between them to the pharmacophore in the query. Examples of programs that perform pharmacophore-based searches are 3Dsearch (111), Aladdin (53), UNITY (112), MACCS-3D (113), Catalyst (114), and ROCS (102). There are also web-based applications (115, 116) that can perform pharmacophore searches. Usually pharmacophore-based searches are done in two steps. First, the software checks whether the compound has the atom types and/or functional groups required by the pharmacophore; then it checks whether the spatial arrangement of these elements matches the query. The fastest approach used in the matching step is considering rigid compounds. Because molecules that are not rigid might have a conformation that matches the pharmacophore, flexibility of the ligands should be considered. Flexible 3D searches identify a higher number of hits than rigid searches do (117). However, flexible searches are more time consuming than rigid ones. There are two main approaches for including conformational flexibility into the search: one is to generate a user-defined number of representative conformations for each molecule when the database is created; the other is to generate conformations during the search. By use of the first approach, any rigid search program can be used for doing a flexible search; however, generating the database takes more time and disk space. The second approach gives more flexibility to the user, given that a larger number of conformations can be generated for each molecule during the search. In this case the database search requires more computer resources; however, this approach will not miss conformations that fit the query but were not stored in the database. Pharmacophore queries that define distance ranges between pharmacophore elements compensate for possible conformational changes in the receptor site upon ligand binding. Also, these flexible pharmacophore queries compensate for the difference between using multiconformer databases and generating conformers during the search.

ROCS is using a shape-based superposition for identifying compounds that have similar shape. Grant and Pickup (118) showed that using atomic-centered Gaussians instead of a spherical function can dramatically reduce the time required for a shape alignment of two molecules. This improved routine allows the program to perform shape-based database searches at an acceptable speed (300–400 conformers/s).

There are several methods for generating conformers during *in silico* screening. Torsion optimization (119) is used for minimizing the root-mean-square (rms) deviation between the constraints from the pharmacophore and the corresponding distances in the compound. The "directed tweak" (120) algorithm also uses torsion optimization for minimizing the sum of the squared deviations between distances in the pharmacophore and the corresponding ones in the compound. Chem-DBS-3D (121) generates low energy conformations that can match the pharmacophore using rules similar to those in WIZARD (77). The distance geometry algorithm (122) uses bond length and bond angle information for building a matrix containing upper and lower limits of distances between atoms in the organic compound. These distances can be used for building the conformation that fits the pharmacophore query. The systematic search method (123) is feasible for molecules with few rotatable bonds and thus has limited applicability.

2.4 Structure-Based Virtual Screening

In direct analogy to high throughput screening, docking and scoring techniques can be applied to computationally screen a database of hundreds of thousands of compounds against a specific target protein. Computational methods that predict the 3D structure of a protein-ligand complex are often referred to as *molecular docking* approaches (Fig. 6.17) (124). Protein structures can be employed to dock ligands into the binding site of the protein and to study their interactions (125). For virtual screening, the crucial task at hand is the fast and reliable ranking of a database of putative protein-ligand complexes according to their binding affinities. Depending on ligand and protein flexibility, sampling depth, and optimizing schemes, docking programs used today (Table 6.2) can facilitate this task within a few minutes or sometimes seconds per processor and molecule. Virtual screening as a computation task can be trivially run using parallel computing because the protein-ligand docking events are completely independent of each other. Although docking has initially been developed as a specialist modeling tool run on computer workstations, nowadays inexpensive Linux clusters or distributed computing over networked PCs can be used for virtual screening. This increases the *in silico* throughput into the realm of 100,000 compounds per day on a Linux cluster, thereby reaching the speed of today's high throughput screens. Energy functions that evaluate the

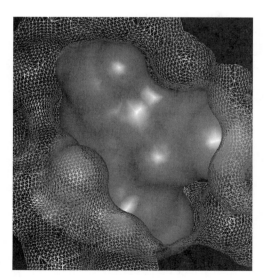

Figure 6.17. Crystal structure (PDB entry 1a4q) of the neuraminidase inhibitor zanamivir bound in the active site (213).

Table 6.2 Selection of Available Protein–Ligand Docking Software for Structure-Based Virtual Screening

Docking Program	Docking/Sampling Method	Scoring Method
GLIDE (www.schrodinger.com)	Rigid protein; multiple conformation rigid docking; grid-based energy evaluation	Empirical scoring, including penalty term for unformed hydrogen bonds; force-field scoring
DOCK (www.cmpharm.ucsf.edu/kuntz/dock.html)	Rigid protein; flexible ligand docking (incremental construction)	Force-field scoring; chemical scoring, contact scoring
FlexX (cartan.gmd.de/FlexX)	Rigid protein; flexible ligand docking (incremental construction)	Empirical scoring intertwined with sampling
DockVision (www.dockvision.com)	Monte Carlo, genetic algorithm	Various force fields
DockIT (www.daylight.com/meetings/emug00/ Dixon)	Ligand conformations generated inside binding-site spheres using distance geometry	PLP, PMF
FRED (www.eyesopen.com/fred.html)	Exhaustive sampling; rigid protein, multiple conformation rigid docking	Chemscore, PLP, ScreenScore, and Gaussian shape scoring
LigandFit (www.accelrys.com)	Monte Carlo	LIGSCORE, PLP, PMF, LUDI
Gold (www.ccdc.cam.ac.uk/prods/gold/)	Genetic Algorithm	Soft core vdW potential and hydrogen bond potentials

binding free energy between protein and ligand sometimes employ rather heuristic terms. Therefore, those functions are more broadly referred to as *scoring functions*.

2.4.1 Protein Structures. A 3D-protein structure of the receptor at atomic resolution is necessary to start a protein-ligand docking experiment. The exponential growth of solved crystal and solution structures in recent years provides a reliable source of protein structures. The protein database (PDB) currently holds more than 18,000 protein structures. It should be noted, however, that the chances of a successful virtual screen very much depend on the quality of the available structure. The crystal structure should be well refined; typically a resolution of at least 2.5 Å is considered to be necessary (126). Small changes in structure can drastically alter the outcome of a computational docking experiment (127). Moreover, many receptor sites are flexible; they often undergo conformational changes upon ligand binding. A good example is the Tyr248 movement of carboxypeptidases upon substrate or ligand binding, which has provided the first structural perspective of Koshland's induced-fit hypothesis (128, 129). Proteins have to be studied carefully in every individual case to decide how promising a virtual screen may be.

For many protein drug targets crystal or solution structures are not available. In such cases homology models (130, 131) and pseudo-receptor models (132) are often used. However, unless there is a very high conservation of receptor site residues the use of homology models for virtual screening is much riskier than using solved structures. On the other hand, the PDB contains a wealth of protein

Table 6.3   Occurrence of Selected Protein Classes Currently Identified in the Human Genome Database (GDB) and the Protein Database (PDB)

Class	GDB[a]	PDB[b]
Nuclear receptors	49	42
G-protein–coupled receptors	408	1
Kinase	945	625
Protease	190	330
Peptidase	108	128
Esterase	106	87
Reductase	210	417
Synthase	191	335
Lyase	38	70
Hydrolase	131	110
Transferase	500	467
Anhydrase	27	156
Sulfatase	26	6
Dehydrogenase	347	338
Desaturase	10	1
Phosphatase	315	184
Phosphodiesterase	63	1
Deacetylase	18	2
Transporter	238	1
Channel	271	24

[a]www.gdb.org

[b]www.rcsb.org/pdb; note that the number of structures available from the PDB often include several structures of the same protein.

structures of a wide variety of enzymes and receptors that can be used for homology modeling. Homology models can be built for a large number of protein classes coded in the human genome (Table 6.3). Because virtual screening is so inexpensive and the possible rewards, if successful, are so high it is generally warranted to run a virtual screening experiment, even if the chances of success are very small, as is often the case when homology models are employed.

2.4.2 Computational Protein-Ligand Docking Techniques. Docking ligands into a receptor site is a geometric search problem. The search has to take protein and ligand conformations as well as their relative orientations into account. The receptor conformation is typically reasonably well known. However, the bioactive conformation of the ligand is usually unknown. Nicklaus and coworkers showed that force-field energies of bioactive conformations of ligands, as represented in

crystal structures of protein-ligand complexes, are typically about 25 kcal/mol^2 higher than minimum conformations in vacuum (133). Therefore, the bioactive conformation of a ligand is hard to guess and a large number of possible ligand conformations have to be considered in docking. Most docking approaches keep the receptor rigid and the ligand flexible during the docking. Although protein flexibility is sometimes included (134–136), we will not discuss protein flexibility here, given that it is currently rarely used for virtual screening because of speed limitations. Some relevant concepts of docking approaches are shortly discussed below (for a broader review we refer the reader to Ref. 125). Scoring protein-ligand complexes will be discussed separately.

2.4.2.1 Rigid Docking. Although ligand and often also protein flexibility are crucial for protein-ligand docking, the simpler rigid ligand docking is sometimes useful. Ligand flexibility can, for example, be simulated by rigidly docking an ensemble of preassigned ligand conformations that represent the relevant conformational space of the molecule. Algorithms such as clique search techniques (137) and geometric hashing (138) are often used to search for distance-compatible matches of protein and ligand features (139). Possible features include complementary hydrogen-bonding interactions, distances, or volume segments of the receptor site of the protein or the ligand.

The program DOCK uses an algorithm for rigid-body docking based on the idea of searching for distance-compatible matches. Starting with the molecular surface of the protein (140–142), a set of spheres is created inside the receptor site. The spheres represent the volume that could be occupied by a ligand molecule (Fig. 6.18). Spheres can represent the ligand also; a direct atom representation is also possible. Early versions of DOCK relied solely on rigid ligand docking. Sets of up to four distance-compatible matches were evaluated. Each set was used for an initial fit of the ligand into the receptor site. Additional compatibility matches were used to improve the fit. The position of the ligand was then optimized and scored.

Since its first introduction in 1982, the DOCK software has been extended in several directions. The matching spheres can be la-

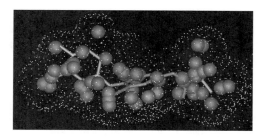

Figure 6.18. Receptor site of thyroid receptor beta filled with spheres (for sake of clarity, sphere centers are depicted; actual size of spheres is larger, so that spheres overlap) and thyronine. Crystal structure taken from the PDB (ID 1bsx).

beled with chemical properties (143) and distance bins are used to speed up the search process (144, 145). Recently, the search algorithm for distance-compatible matches was changed to the clique-detection algorithm introduced by Kuhl (139, 146). Furthermore, several scoring functions are now applied in combination with the DOCK algorithm (147–151).

2.4.2.2 Flexible Ligands. Druglike molecules are typically flexible, with usually up to eight rotatable bonds (34). Energetic differences between alternative ligand conformations are often small compared to the total binding affinity between ligand and target protein. Also, for flexible ligands it is quite common that the bioactive conformations are different from the minimum energy conformations in solution (133). Ligand flexibility is typically handled in docking approaches by combinatorial optimization protocols such as fragmentation, ensembles, genetic algorithms, or simulation techniques.

In fragmentation approaches, the ligand is dissected into pieces that are either rigid or that can be represented by small conformational ensembles. In docking approaches, typically a strategy called *incremental construction* is used to assemble fragments to whole molecules directly in the receptor site. Usually, the largest rigid moiety of the ligand (sometimes called anchor) is docked first in the receptor site. The remaining fragments are subsequently added in a buildup protocol. After each incremental buildup step, torsion angles are sampled and the growing molecule is minimized.

Ligand flexibility can be artificially included into docking by rigidly docking ensembles of pregenerated conformations of the ligand into the receptor site. Rigid docking is faster than flexible docking by use of a fragmentation approach. However, because computing time increases linearly with the number of conformations, computing time and coverage of conformational space have to be balanced. An example of rigid docking of conformation ensembles is given in Flexibase/ FLOG (152). Distance geometry methods (153) are used to generate a small set of diverse conformations for each ligand in the database. A subset of up to 25 conformations per molecule is selected using rms dissimilarity criteria and then docked using a rigid-body-docking algorithm.

Different from the combinatorial approaches for docking mentioned above, simulation methods start with a given configuration of a ligand in the receptor site. Simulation techniques such as simulated annealing (154) are then applied to find energetically more favorable conformations of the ligand. To speed up the docking process, docking programs such as AutoDock (155) precalculate molecular affinity potentials of the protein on a grid. Molecular dynamics (MD) methods (see, e.g., refs. 156 and 157) and Monte Carlo simulation techniques (see, e.g., Refs. 158–162) are also frequently used in protein-ligand docking applications.

A variety of other sampling methods are applied in docking programs, including genetic algorithms, distance geometry methods, random searching, hybrid methods, and generalized effective potential methods. Genetic algorithms have been employed in programs such as Gambler (163), AutoDock (155), and GOLD (126). PRO-LEADS uses an alternative search technique called "tabu search" (164). Starting from a random structure, new structures are created by random moves. A tabu list is maintained during the optimization phase and contains the best and the most recently found binding configurations. Configurations that resemble those stored in the tabu list are rejected, except they are better than the one scoring best. The sampling performance is improved because previously sampled configurations are avoided. Finally, it should be mentioned that multistep hybrid docking procedures have been developed that combine rapid fragment-based searching with sophisticated MC or MD simulations (165, 166).

2.4.3 Scoring of Protein-Ligand Interactions.

The problem of sampling the correct binding geometry (binding mode) of a protein-ligand complex is considered to be solved in many docking programs (167). However, to identify this correct binding mode by its lowest energy or score is a different matter; this is indeed the bottleneck of docking-scoring approaches today. The most important aspect of scoring functions for virtual screening is speed. Therefore, accuracy requirements are low; most functions used do not conceptually describe binding free energies. Therefore, these functions are typically not called energy functions but scoring functions. Three main scoring strategies are typically used in docking applications for virtual screening: force field scoring, empirical scoring, and knowledge-based scoring.

2.4.3.1 Force Field (FF) Scoring.

Nonbonded interaction energy terms of standard force fields are typically used in FF scoring (e.g., *in vacuo* electrostatic terms; sometimes modified by scaling constants that assume the protein to be an electrostatic continuum) and van der Waals (vdW) terms (168–171). DOCK and GREEN (172) use the intermolecular terms of the AMBER energy function (173, 174), with the exception of an explicit hydrogen bonding term (147):

$$E_{\text{nonbond}} = \sum_i^{\text{lig}} \sum_j^{\text{prot}} \left[\frac{A_{ij}}{r_{ij}^{12}} - \frac{B_{ij}}{r_{ij}^{6}} + 332 \frac{q_i q_j}{D r_{ij}} \right]$$

(6.1)

where each term is summed up over ligand atoms i and protein atoms j. A_{ij} and B_{ij} are the vdW repulsion and attraction parameters of the 6–12 potential, r_{ij} is the distance between atoms i and j, q is a point charge at each of the atoms, and D is the dielectric constant. Intraligand interactions are added to the score. Up to a 100-fold gain in docking time can be achieved by precomputing these terms on a 3D grid that represents the protein during docking (155, 175). More recently, solvation terms

have been added to FF scores. Examples include generalized Born/surface area approaches (176) or atomic solvation parameters (177–179).

2.4.3.2 Empirical Scoring.

Empirical scoring functions are multivariate regression methods. They fit coefficients of physically motivated contributions to binding free energy in reproduction of measured binding affinities of a training set of protein-ligand complexes with known 3D structure. As an example, the docking program FlexX (180) uses a scoring function similar to that of Böhm (181, 182). It calculates the sum of free-energy contributions from the number of rotatable bonds in the ligand, hydrogen bonds, ion-pair interactions, hydrophobic and pi-stacking interactions of aromatic groups, and lipophilic interactions:

$$\begin{aligned}
\Delta G = {} & \Delta G_0 + \Delta G_{\text{rot}} N_{\text{rot}} \\
& + \Delta G_{\text{hb}} \sum_{\text{neutral\_Hbonds}} f(\Delta R, \Delta \alpha) \\
& + \Delta G_{\text{io}} \sum_{\text{ionic\_int}} f(\Delta R, \Delta \alpha) \\
& + \Delta G_{\text{aro}} \sum_{\text{aro\_int}} f(\Delta R, \Delta \alpha) \\
& + \Delta G_{\text{lipo}} \sum_{\text{lipo.cont}} f^*(\Delta R)
\end{aligned}$$

(6.2)

where ΔG_0, ΔG_{rot}, ΔG_{hb}, ΔG_{io}, ΔG_{aro}, and ΔG_{lipo} are adjustable parameters that are fitted; $f(\Delta R, \Delta \alpha)$ is a scaling function penalizing deviations from the ideal geometry; and N_{rot} is the number of freely rotatable bonds. The interaction of aromatic groups is an addition to Böhm's original force-field design (181, 182). The lipophilic contributions are calculated as a sum of atom-pair contacts in contrast to evaluating a surface grid as in Böhm's scoring function. Böhm's scoring function and its FlexX implementation are being improved and additional terms are being tested (see, e.g., Refs. 182 and 183).

2.4.3.3 Knowledge-Based Scoring.

Because the forces that govern protein-ligand interactions are so complex, an implicit approach to capture all relevant terms of protein-ligand

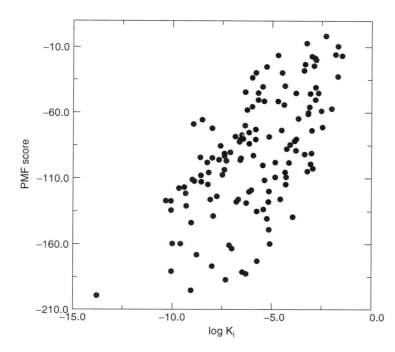

Figure 6.19. PMF score calculated for 132 protein-ligand complexes taken from the PDB without overlap to the set of 697 complexes. The PMF score was derived from Ref. 186.

binding seems very attractive. Borrowing from statistical thermodynamics of liquids, mean-field approaches derived solely from structural information have been applied to protein-ligand binding. Protein-ligand atom-pair potentials can be calculated from structural data (e.g., PDB), assuming that observed crystallographic protein-ligand complexes exhibit optimal placement. As an example, a knowledge-based scoring function was derived recently using 697 protein-ligand complexes from the PDB as knowledge base. Using 16 protein and 34 ligand atom types, a total of 282 statistically significant interaction potentials of atom pairs was derived. The final score is calculated as the sum over all protein-ligand atom-pair interactions.

$$\text{PMF\_score} = \sum_{kl \to r < r_{\text{cutoff}}{}^{ij}} A_{ij}(r); \quad (6.3)$$

$$A_{ij}(r) = -k_B T \ln\left[f_{\text{Vol\_corr}}{}^{j}(r) \; \frac{\rho_{\text{seg}}{}^{ij}(r)}{\rho_{\text{bulk}}{}^{ij}} \right]$$

where kl is a ligand-protein atom pair of type ij; $r_{\text{cutoff}}{}^{ij}$ designates the distance at which atom-pair interactions are truncated (6 Å for carbon-carbon interactions and 9 Å otherwise); all $A_{ij}(r)$ are derived with a reference

sphere radius of 12 Å (184); k_B is the Boltzmann factor; T is the absolute temperature; and $f_{\text{Vol\_corr}}{}^{j}(r)$ is a ligand volume correction factor that is introduced because intraligand interactions are not accounted for (185, 186). $\rho_{\text{seg}}{}^{ij}(r)$ designates the number density of atom pairs of type ij at a certain atom-pair distance r. $\rho_{\text{bulk}}{}^{ij}$ is the number density of a ligand-protein atom pair of type ij in a reference sphere with radius R (184). For use in docking studies, the PMF score is combined with a vdW term to account for short-range interactions (187, 188). The PMF scoring function was implemented into the DOCK4.0 program. For faster scoring it was also implemented on a grid similar to the force-field score in DOCK. Flexible docking experiments on FK506 binding protein (187), neuraminidase (127), and stromelysin (189) showed high predictive power and robustness of the PMF score. Figure 6.19 shows the predictive power of the scoring function applied to 132 protein-ligand complexes taken from the PDB.

2.4.3.4 Consensus Scoring. Consensus scoring is an approach that combines several scoring functions to find common hits. Such an approach seems desirable because of the missing robustness of current scoring functions.

Charifson et al. (163) provided a comprehensive consensus scoring study using DOCK and Gambler, in combination with 13 scoring functions: LUDI (104), ChemScore (190, 191), Score (192), PLP (193), Merck force field (194), DOCK energy score (146, 147), DOCK chemical score, Flog (152), strain energy, Poisson Boltzmann (195), buried lipophilic surface area (196), DOCK contact score (144), and volume overlap (197). Three enzymes were used as test proteins: p38 MAP kinase, inosine monophosphate dehydrogenase, and HIV protease. By comparing the performance of single-scoring functions with consensus scoring schemes involving two or three scoring functions, the authors found that false positives (inactive compounds that have high predicted scores) were significantly reduced in the latter case. The authors estimated that a consensus scoring approach would consistently provide hit rates between 5 and 10% (5–10 out of 100 compounds tested to show low μM activity) for enzymes with reasonably buried binding sites. A comparison of the different scoring functions revealed that ChemScore, PLP, and DOCK energy score performed best as single-

scoring functions and also in consensus combination. Consensus scoring experiments reported by Bissantz et al. found that docking/consensus scoring performances varied widely among targets (198). Stahl and Rarey suggested that the combinations of FlexX and PLP scores are ideal for consensus scoring for a variety of targets including COX-2, ER, p38 MAP kinase, gyrase, thrombin, gelatinase A, and neuraminidase (199).

2.4.4 Docking as Virtual Screening Tool. A virtual screening protocol is schematically shown in Fig. 6.20. The necessary steps include: protein structure preparation, ligand database preparation, docking calculation, and postprocessing.

The protein has to be prepared only once for a virtual screening experiment unless different protein conformations are considered. The receptor site needs to be determined and charges have to be assigned. The protein structure and the receptor site have to be modeled as accurately as possible. Determining protein surface atoms and site points as well as the assignment of interaction data, such as

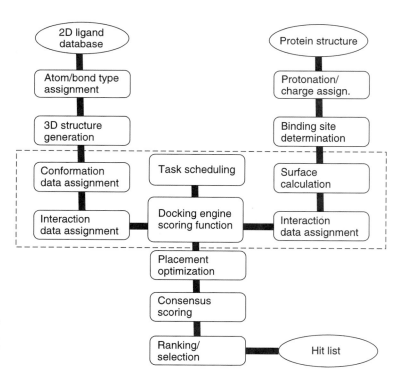

Figure 6.20. Flowchart of docking as virtual screening tool in the example of FlexX.

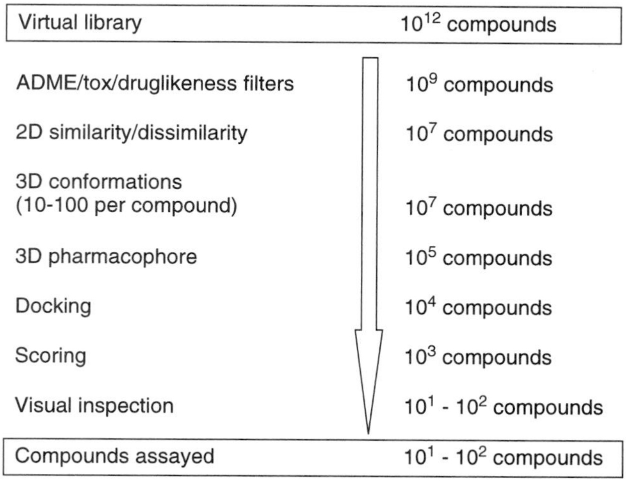

Virtual library	10^{12} compounds

ADME/tox/druglikeness filters	10^9 compounds
2D similarity/dissimilarity	10^7 compounds
3D conformations (10-100 per compound)	10^7 compounds
3D pharmacophore	10^5 compounds
Docking	10^4 compounds
Scoring	10^3 compounds
Visual inspection	10^1 - 10^2 compounds

Compounds assayed	10^1 - 10^2 compounds

Figure 6.21. Virtual screening filter cascade.

marking hydrogen-bond donors/acceptors, and so forth, are sometimes internally included in the docking software (e.g., in FlexX) and sometimes done separately (e.g., DOCK).

Because of the large number of molecules, manual steps in the preparation of ligand databases obviously have to be avoided. Starting typically from 2D structures, bond types have to be checked, protonation states must be determined, charges must be assigned, and solvent molecules removed. 3D coordinates can be generated using a program such as CONCORD or CORINA (74) (see Section 2.3.2). Next, site points for hydrogen-bonding interactions have to be assigned and rotational barriers must be calculated. These tasks are sometimes included in the docking program (e.g., FlexX).

The docking calculation is typically done for one ligand at a time. Depending on optimization and sampling parameters as well as on the flexibility of the compound, typically between a few seconds and a few minutes of CPU time are needed to dock a ligand. Because the individual docking events are independent of each other, they can run on parallel hardware. Task schedulers that distribute ligand docking on available CPUs are used in many docking programs.

Postprocessing steps of hits may include refinement of placement using MD techniques, specific pharmacophore-based filters that penalize certain features, such as unformed hy-

drogen bonds or other constraints that were not met in the primary scoring function. Because of the limitations of scoring functions, a postscoring protocol can be used to reach consensus about hits (discussed above). The recognition of known active ligands mixed within the database can be used to find an appropriate threshold for separating the top-ranking compounds from the rest of the database.

2.5 Filter Cascade

Virtual screening is the process of reducing a given database as quickly and efficiently as possible to a small number of putative lead compounds for a given drug discovery project. The techniques described above form a cascade of different filter functions that are ordered by their speed. Fast ADMET filters are followed by 2D and 3D pharmacophore filters and finally by docking and scoring methods. Figure 6.21 shows a scheme of a possible virtual screening filter cascade.

3 APPLICATIONS

3.1 Identification of Novel DAT Inhibitors through 3D Pharmacophore-Based Database Search

The dopamine transporter (DAT) is a 12-transmembrane helix protein that plays a critical role in terminating dopamine neurotransmission by taking up dopamine released into the synapse. There is no experimental structure available for DAT. However, an extensive SAR of DAT inhibitors (mostly cocaine analogs) is available. DAT is involved in several diseases such as drug addiction and attention deficit disorder (200). For example, ritalin [(±)-threo-methylphenidate], a DAT inhibitor, is marketed for treating attention deficit disorders in children (200, 201). Until recently, all efforts in synthesizing DAT inhibitors were focused on creating analogs around the tropane, piperazine, methylphenidate, and 2,3-dihydro-5-hydroxy-5H-imidazo[2,1-a]isoindole cores. It was shown that, despite structural differences, DAT inhibitors share one or more common 3D pharmacophore models (95, 202, 203). In an effort to identify new chemical cores for developing DAT inhibitors with new pharmacological profiles, a pharmacophore-based 3D database search was proposed (95). For this purpose a pharmacophore

model was derived based on two known potent DAT inhibitors R-cocaine and WIN-35065-2 (Fig. 6.22) (95). The common binding elements of these compounds are a ring N that may be substituted, a carbonyl oxygen, and an aromatic ring that can be defined by the position of its center (Fig. 6.22). Because both compounds have some flexibility, a systematic conformational search was performed to obtain all possible conformations these compounds can have when bound to DAT. To identify structurally diverse conformers, clustering of the generated conformers was done. Measuring distances among chosen pharmacophore elements in the generated conformers led to distances shown in Fig. 6.22.

Recently, analysis of several large chemical databases showed that the NCI database has by far the highest number of unique compounds (204). Thus this database provides a large number of unique synthetic compounds and natural products and is an excellent re-

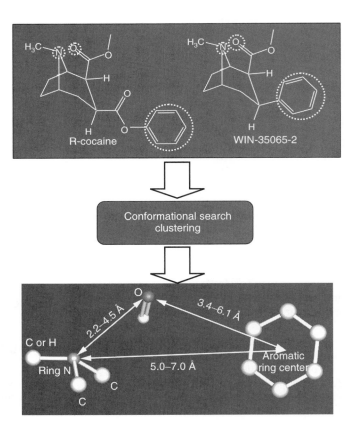

Figure 6.22. Pharmacophore proposed for identifying DAT inhibitors. The pharmacophore was obtained based on two known DAT inhibitors, R-cocaine and WIN-35065-2. Distance ranges between pharmacophore points were obtained through systematic search of all possible conformations that the two compounds may adopt when bound to DAT.

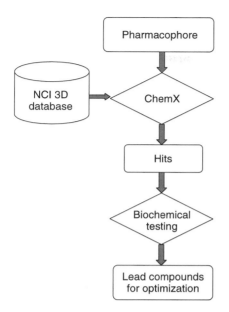

Figure 6.23. Flowchart showing steps used in lead identification using pharmacophore-based 3D database searching.

source for drug lead discovery. Using the 3D pharmacophore from Fig. 6.22, the NCI 3D-database (67) of 206,876 "open compounds" was searched using the program Chem-X (205). The strategy used for identifying leads through virtual screening is shown in Fig. 6.23. During the search each compound was first checked as to whether it had the pharmacophore elements and second as to whether it had any acceptable conformation matching the distance requirements. Up to 3 million conformations were examined for each compound. A total of 4094 compounds, 2% of the database, were identified as "hits." This number was further reduced using filters such as molecular weight, structural novelty, simplicity, diversity, and hydrogen-bond acceptor nitrogen. Seventy compounds were selected for testing in biochemical assays. Forty-four compounds displayed more than 20% inhibition at 10 μM in the [^{3}H]mazindol binding assay, from which three compounds were chosen for deriving an SAR (Fig. 6.24). These results suggested that the 3D pharmacophore-based database search is an efficient tool for identifying novel DAT inhibitors.

3.2 Discovery of Novel Matriptase Inhibitors through Structure-Based 3D Database Screening

Matriptase is a trypsinlike serine protease that was proposed to be involved in tissue remodeling, cancer invasion, and metastasis (206). Potent and selective matriptase inhibitors not only would be useful for further elucidation of the role matriptase has in biological systems but also may be used for the treatment and/or prevention of cancers. Hepatocyte growth factor activator inhibitor 1 (HAI-1) is a natural inhibitor of matriptase. Thus, by analyzing interactions in the complex of matriptase with HAI-1, crucial interactions that an inhibitor should capture can be identified. In consequence, the strategy for identifying inhibitors was to first build the matriptase-HAI-1 Kunitz domain 1 complex, identify binding regions on matriptase, screen the NCI 3D database for hits that capture binding groups of HAI-1 to matriptase, and in the end, biochemical testing (Fig. 6.25). The structure of matriptase was obtained from PDB entry 1EAW (207). Homology modeling, as implemented in MODELLER (208, 209), was chosen to build the 3D structure of the Kunitz domain 1 from KSPI. The complex of matriptase with HAI-1 Kunitz domain 1 was built using a combination of manual docking and molecular dynamics refinement with the program CHARMM (210). The obtained binding mode of HAI-1 Kunitz domain 1 to matriptase (Fig. 6.26) suggests that three regions might be important for inhibitor binding. The S1 binding site Asp185, which is characteristic of trypsinlike serine proteases, is the specificity pocket used to recognize substrates with Arg or Lys as P1 residue. The anionic site, defined by Asp96, Asp60.A, and Asp60.B, is the site at which Arg258 from HAI-1 binds. A hydrophobic region defined by Ile41 and Tyr60.G might also be important for specificity of future matriptase inhibitors.

Thus, the active site used for *in silico* screening with the program DOCK constitutes all three binding regions. Energy scoring was used for ranking docked compounds. The top 2000 compounds were considered for selecting potential inhibitors. Given that

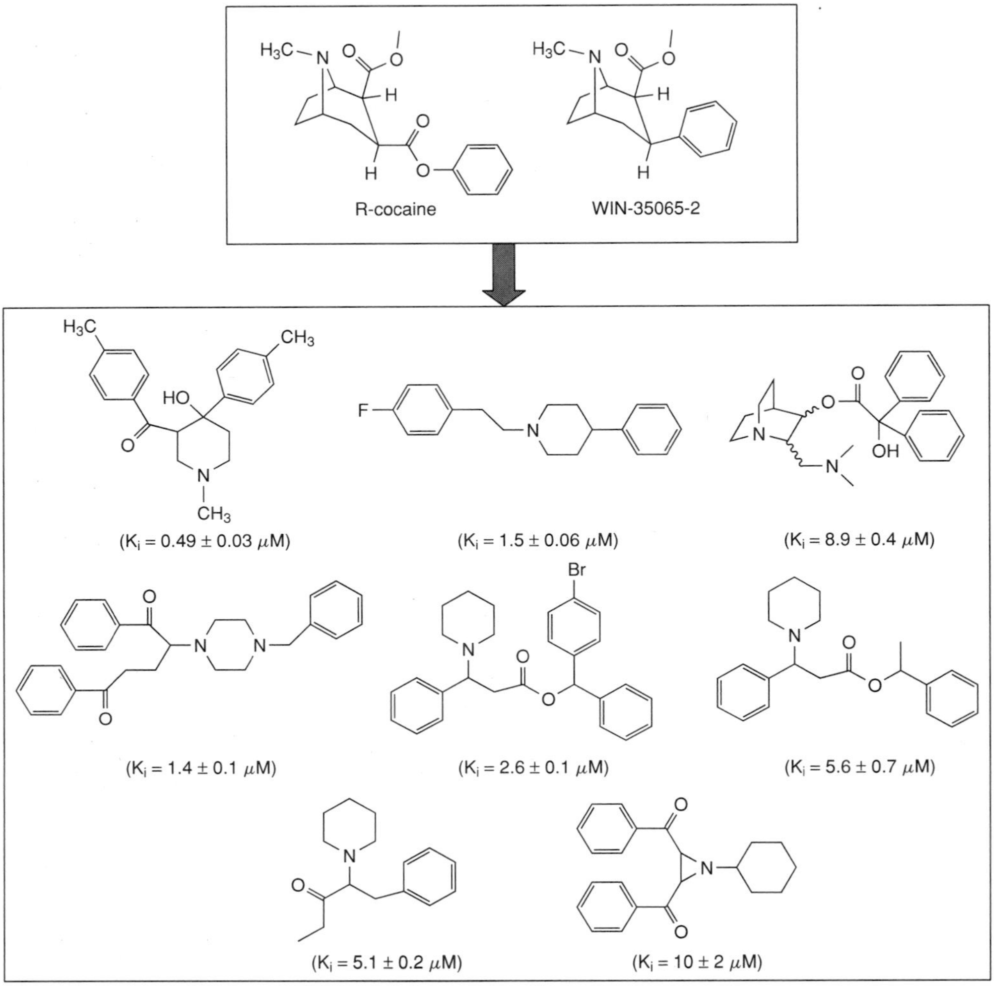

Figure 6.24. Selected DAT inhibitors identified from the NCI database.

matriptase prefers positively charged residues in the P1 position, inhibitors should also have positively charged groups to bind efficiently to Asp185 from the S1 site of matriptase (Fig. 6.27). Note that a more efficient way of doing the virtual screening presented above is to do a pharmacophore search first followed by docking. Thus, 69 compounds were selected for biochemical testing at 75 μM inhibitor and matriptase concentration. Initial screening showed that 50% of compounds tested produced more than 70% inhibition of enzymatic activity when the ratio was one inhibitor mol-

ecule for one protein molecule (Table 6.4). It should be noted that screening results at single dose and IC_{50} depend on the protein concentration, whereas K_i is concentration independent. From the hits in the screening step bis-benzamidines were chosen for K_i determination (Table 6.5) because this class of compounds could bind to both the S1 site and the anionic site. These results show that combining a pharmacophore hypothesis with a structure-based database search can provide an efficient way of identifying leads for a drug design project.

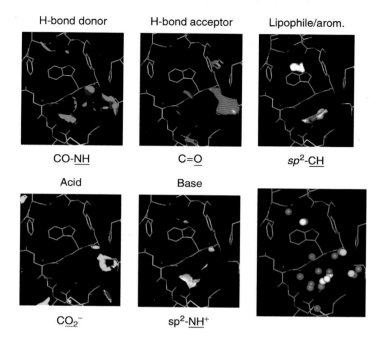

H-bond donor   H-bond acceptor   Lipophile/arom.

CO-$\underline{NH}$ C=$\underline{O}$ sp^2-$\underline{CH}$

Acid   Base

$\underline{CO_2^-}$ sp^2-$\underline{NH}^+$

Figure 5.5. GRID probes on Factor Xa site and the combined resultant complementary site points that can be used for pharmacophore fingerprint calculations (lower right).

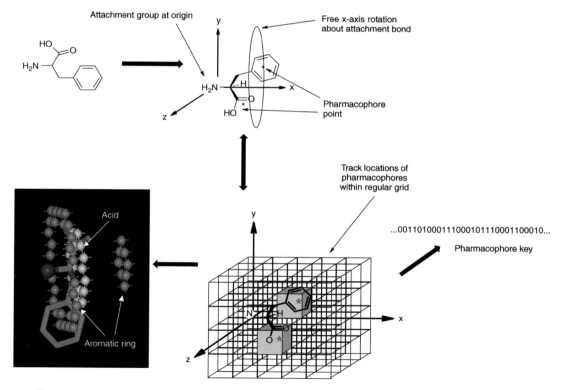

Figure 5.6. Overview of the Gridding and Partitioning (GaP) procedure as applied to monomers, exemplified using phenylalanine as a potential primary amine. This molecule thus contains two pharmacophoric groups (the aromatic ring and the carboxylic acid). During the conformational analysis the locations of these pharmacophoric groups are tracked within a regular grid. [Reproduced from A. R. Leach and M. M. Hann, *Drug Discovery Today,* **5,** 326–336 (2000), with permission of Elsevier Science.]

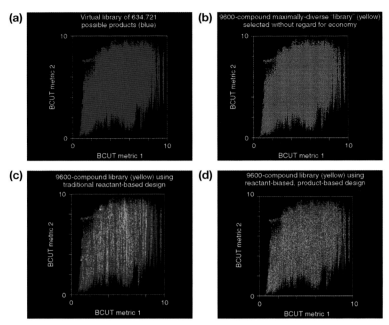

Figure 5.13. (a) A virtual library of 634,721 allowed combinatorial AB products (after filtering out products that failed Lipinski's Rule of 5 "druglike" criteria) shown in a BCUT chemistry space specifically chosen to best represent the diversity of the virtual library. (b) The maximally diverse 9600-compound subset of the virtual library, illustrating the results of purely product-based "library design." Although providing the maximal diversity, synthesis of these 9600 AB products would require the use of 347 A's and 1024 B's—clearly unacceptable from the perspective of synthetic economy (number of reactants and robotic control). (c) The 9600-compound library resulting from the traditional, purely reactant-based library design strategy of selecting the 80 most diverse A's and the 120 most diverse B's. Although providing user-selected synthetic economy, the diversity of these 9600 AB products is clearly quite poor. (d) The 9600-compound library resulting from the reactant-biased, product-based (RBPB) algorithm developed by Pearlman and Smith (see Refs. 31, 87c and text). The algorithm selected a different set of 80 A's and a different set of 120 B's, thus providing the same level of user-selected synthetic economy, while also providing substantially greater diversity than could be achieved using a purely reactant-based library design strategy.

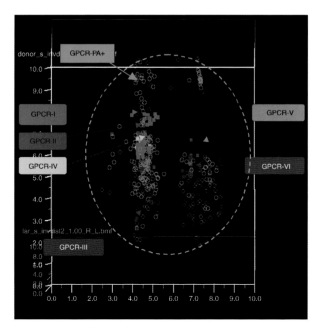

Figure 5.25. The 3D subspace most receptor relevant for members of the GPCR-PA+ family of receptors. Points indicate coordinates of 187 published ligands of various GPCR-PA+ receptors. Some have been color-coded by receptor for illustrative purposes. See Refs. 32e,i and text for further details.

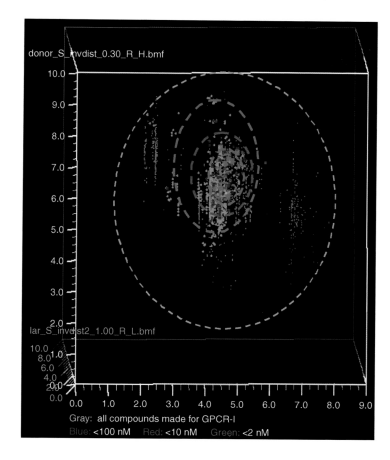

Gray: all compounds made for GPCR-I
Blue: <100 nM   Red: <10 nM   Green: <2 nM

Figure 5.26. The same 3D subspace as in Fig. 5.25, rotated slightly to provide a better viewing perspective. Points indicate coordinates of about 2000 combinatorial products selected from 14 different libraries. Color-coding indicates affinity for the GPCR-1 receptor. See Refs. 32e,i and text for further details.

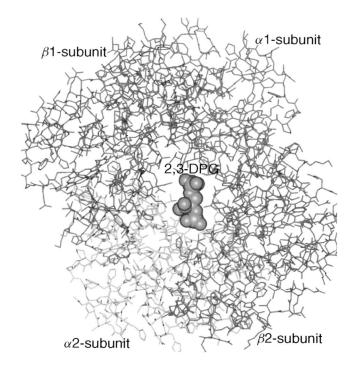

Figure 10.2. View of (**4**) (2,3-DPG) binding site at the mouth of the β-cleft of deoxy hemoglobin.

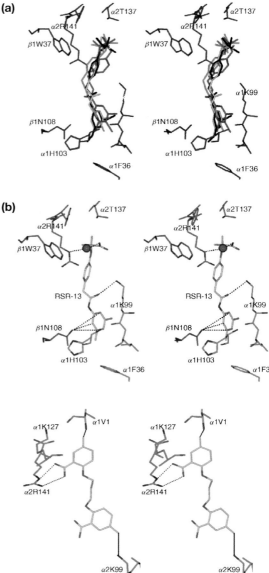

Figure 10.3. Stereoview of allosteric binding site in deoxy hemoglobin. A similar compound environment is observed at the symmetry-related site, not shown here. (a) Overlap of four right-shifting allosteric effectors of hemoglobin: (**6a**) (RSR13, yellow), (**6b**) (RSR56, black), (**7a**) (MM30, red), and (**7b**) (MM25, cyan). The four effectors bind at the same site in deoxy henoglobin. The stronger acting RSR compounds differ from the much weaker MM compounds by reversal of the amide bond located between the two phenyl rings. As a result, in both RSR13 and RSR56, the carbonyl oxygen faces and makes a key hydrogen bonding interaction with the amine of αLys99. In contrast, the carbonyl oxygen of the MM compounds is oriented away from αLys99 amine. The αLys99 interaction with the RSR compounds appear to be critical in the allosteric differences. (b) Detailed interactions between RSR13 (**6a**) and hemoglobin, showing key hydrogen bonding interactions that help constrain the T-state and explain the allosteric nature of the compound and those of other related compounds.

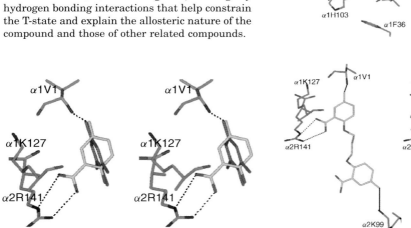

Figure 10.4. Stereoview of superimposed binding sites for (**8b**) (5-FSA, yellow) and (**8a**) (DMHB, magenta) in deoxy hemoglobin. A similar compound enviroment is observed at the symmetry-related site and therefore not shown here. Both compounds form a Schiff base adduct with the α1Val1 N-terminal nitrogen. Whereas the m-carboxylate of 5-FSA forms a salt bridge with the α2Arg141 (opposite subunit), this intersubunit bond is missing in DMHB. The added constraint to the T-state by 5-FSA that ties two subunits together shifts the allosteric equilibrium to the right. On the other hand, the binding of DMHB does not add to the T-state constraint. Instead, it disrupts any T-state salt- or water-bridge interactions between the opposite α-subunits. The result is a left shift of the oxygen equilibrium curve by DMHB.

Figure 10.5. Stereoview of the binding site for (**9**) (n = 3, TB36, yellow) in deoxy Hb. A similar compound environment is observed at the symmetry-related site, not shown here. One aldehyde is covalently attached to the N-terminal α1Val1, whereas the second aldehyde is bound to the opposite subunit, α2Lys99 ammonium ion. The carboxylate on the first aromatic ring forms a bidentate hydrogen bond and salt bridge with the guanidinium ion of α2Arg141 of the opposite subunit. The effector thus ties two subunits together and adds additional constraints to the T-state, resulting in a shift in the Hb allosteric equilbrium to the right. The magnitude of constraint placed on the T-state by the crosslinked αLys99 varies with the flexibility of the linker. Shorter bridging chains form tighter crosslinks and yeild larger shifts in the allosteric equilibrium.

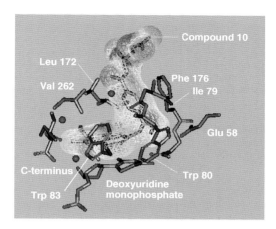

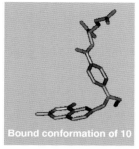

Bound conformation of 10

Figure 10.6. Binding site for (**10**) (*N10-propynyl-5,8-dideazafolate*), within the active site of thymidylate synthase from *Escherichia coli*. The surface of the inhibitor is shown in the left view. The red spheres in the left view are tightly bound water molecules.

(b)

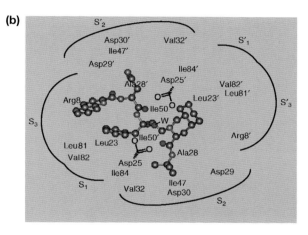

Figure 10.9. (b) Active site with bound (**31**) [saquinavir (PDB code 1HXB)]. Note the asymmetry of inhibitor binding. The flap water that is shown very close to saquinavir is labeled W.

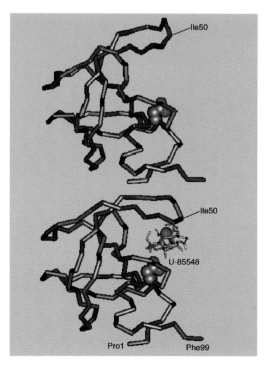

Figure 10.10. Comparison of the structures of HIV-P apoenzyme monomer (top, PDB code 3PHV) and the complex between HIV-P and (**32**) (U-85548; bottom, PDB code 8HVP). The inhibitor is shown as a ball and stick structure. Note the rearrangement of the flap residues; Ile50 is indicated for reference. The van der Waals surface of Asp25 is shown in both structures. The flap water (red ball) is also shown between Ile50 and U-85548. In the bottom structure, the locations of the *N* and C termini of HIV-P are noted.

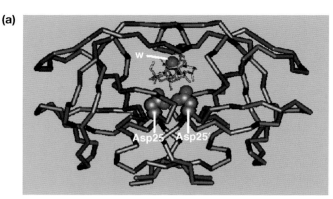

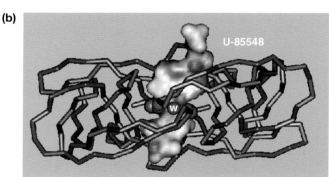

Figure 10.11. Orthogonal views of the complex between HIV-P and (**32**) (U-85548). The view in panel a is rotated approximately 90° (around the long axis of the protein) from the view in panel b. Van der Waals surfaces of Asp25, Asp25′, and the flap water (W) are shown. In panel b, the solvent-accessible surface of the inhibitor is shown.

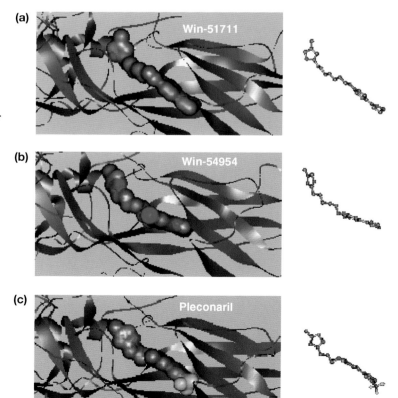

Figure 10.17. Structure of rhinovirus capsid protein VP1 showing the bound conformation of antiviral isoxazole compounds (**78**) [disoxaril, WIN-51711: panel a, top], (**79**) [WIN-54954: panel b, middle], and (**80**) [pleconaril, WIN-63843: panel c, bottom]. The PDB codes for the X-ray structural model coordinates used to create these views are: 1PIV (for **78**), 2HWE (for **79**), and 1C8M (for **80**). On the left side of each panel, the inhibitors are shown as van der Waals surfaces, and the protein as a ribbon diagram. On the right side, the structures of the inhibitor alone are shown, from the same view, as ball and stick representations.

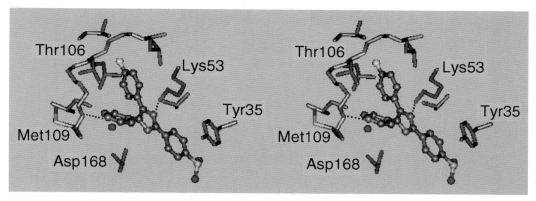

Figure 10.18. Binding of SB203580 (shown as a ball and stick structure) in the active site of MAPK p38α. In addition to the side chains of the labeled residues, the protein backbone between Leu104 and Met109 is shown, as well as several aliphatic side chains and a water molecule (red sphere). Hydrogen bonds (dotted lines) are shown between the backbone amide of Met109 and the inhibitor's pyrimidinyl nitrogen, and between the ε-amino of Lys53 and the inhibitor's imidazole N3. This figure is based on the PDB coordinate set 1A9U (187).

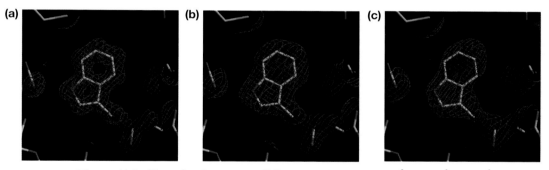

Figure 11.6. Three density maps at differing resolutions: a, 1.3 Å; b, 2.1 Å; c, 3.0 Å.

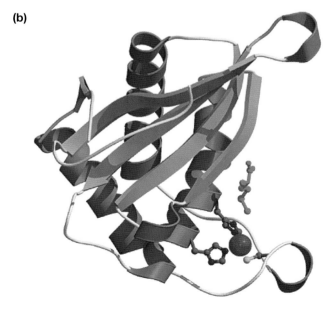

Figure 11.7. (b) Structure of the LuxS monomer highlighting the bound zinc ion (magenta) and methionine (green).

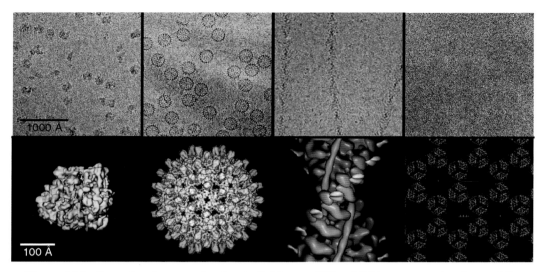

Figure 14.6. Examples of macromolecules studied by cryo-EM and 3D image reconstruction and the resulting 3D structures (bottom row) after cryo-EM analysis. All micrographs (top row) are displayed at above 170,000× magnification and all models at about 1,200,000× magnification. (a) A single particle without symmetry: The micrograph shows 70S *E. coli* ribosomes complexed with mRNA and f Met-tRNA. The surface-shaded density map, made by averaging 73,000 ribosome images from 287 micrographs has a resolution (FSC) of 11.5 Å. The 50S and 30S subunits and the tRNA are colored blue, yellow, and green, respectively. The identity of many of the subunits is known as some RNA double helices are clearly recognizable by their major and minor grooves (e.g., helix 44 is shown in red). [Courtesy of J. Frank (SUNY, Albany), using data from Gabashvili et al. (86).] (b) A single particle with symmetry: The micrograph shows hepatitis B virus cores. The 3D reconstruction, at a resolution of 7.4 Å (DPR), was computed from 6384 particle images taken from 34 micrographs. [From Böttcher et al. (44).] (c) A helical filament: The micrograph shows actin filaments decorated with myosin S1 heads containing the essential light chain. The 3D reconstruction, at a resolution of 30–35 Å, is a composite in which the differently colored parts are derived from a series of difference maps that were superimposed on f-actin. The components include: f-actin (blue), myosin heavy chain motor domain (orange), essential light chain (purple), regulatory light chain (white), tropomyosin (green), and myosin motor domain *N*-terminal beta-barrel (red). [Courtesy of A. Lin, M. Whittaker, and R. Milligan (Scripps Research Institute, LaJolla, CA).] (d) A 2D crystal, light-harvesting complex LHCII at 3.4-Å resolution. The model shows the protein backbone and the arrangement of chromophores in a number of trimeric subunits in the crystal lattice. In this example, image contrast is too low to see any hint of the structure without image processing (see also Fig. 14.3). [Courtesy of W. Kühlbrandt (Max-Planck-Institute for Biophysics, Frankfurt, Germany).]

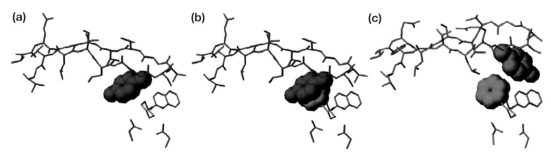

Figure 15.35. GRAB peptidomimetics in action.

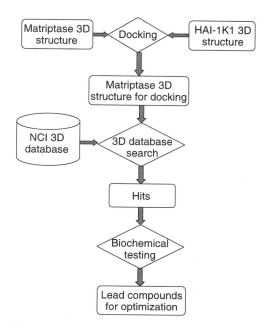

Figure 6.25. Flowchart showing steps to identify matriptase inhibitors through structure-based database searching of the NCI 3D database.

4 CONCLUSIONS

The future development of virtual screening as lead identification technique goes into two main directions. The first direction is to increase the throughput by the use of massively parallel computers, Beowulf clusters, or PC networking. The second direction is to increase the predictive power of scoring functions. There has to be a balance between these developments. To illustrate this point let us remember that the biggest problem of virtual screening is the large number of false positives, that is, inactive compounds with high scores. Typically, the number of diverse compounds available for testing either from inhouse repositories or from vendor libraries is limited to about 10^6–10^7. To increase libraries beyond this point, compound libraries are often generated *in silico* in combinatorial fashion. These virtual libraries are typically biased toward certain chemical classes or types of reactions to generate them. For these libraries it is likely that the ratio of actives versus inactives decreases because it is unlikely that the rate of active compounds remains constant as the number of compounds in the library increases. The number of virtual screening hits that can be tested in a low throughput biological assay, however, is usually constant. If this assumption is true, the number of false positives will increase with the number of compounds subjected to virtual screening, thereby limiting the chances to retrieve active compounds.

To escape this dilemma, virtual screening tools have to become smarter. Protein flexibil-

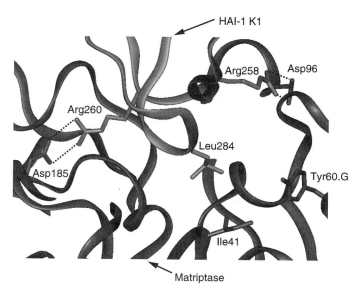

Figure 6.26. Proposed binding mode of HAI-1 Kunitz domain 1 to matriptase as obtained from docking. Highlighted interactions are salt bridges between Arg260 from HAI-1 and Asp185 in matriptase, Arg258 from HAI-1 and Asp96 from matriptase, and hydrophobic interactions between Leu284 from HAI-1 and a hydrophobic pocket defined by Ile41 and Tyr60.G in matriptase.

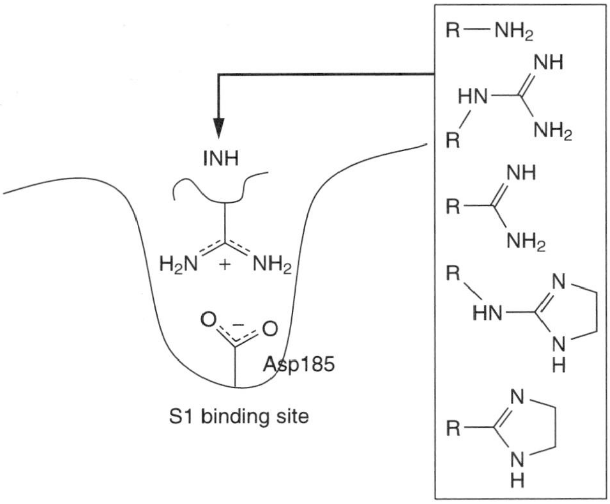

Figure 6.27. Potential functional groups in inhibitors necessary to block the specificity pocket (S1 binding site) of matriptase. The choice reflects the observation that matriptase prefers substrates with Lys or Arg as P1 residue.

ity needs to be included in high-throughput docking. Scoring functions have to improve to make consistently correct predictions of putative protein-ligand binding affinities. Scoring functions, calibrated to reproduce experimental data, have unreliable performance outside their training set. Thus, *de novo* methods using terms describing the thermodynamics of binding should replace the first generation of scoring functions. In consequence, some of the speed gained from low cost parallel computing should be invested into higher accuracy scoring rather than higher throughput. One way of increasing throughput is to keep the number of compounds docked as small as possible by using every bit of knowledge one has to prefilter the database, mainly based on pharmacophore information. In some cases this is

easy. For instance the necessity of having certain features like salt bridges formed on ligand binding [e.g., in influenza virus neuraminidase (211)] or other prevalent information (e.g., hinge region binding for many ATP competitive kinase inhibitors) greatly helps to reduce the number of compounds subjected to docking experiments.

The missing robustness of many structure-based docking/scoring techniques opens the questions of when should one apply it and when should one retreat to pharmacophore-based virtual screening. In many cases it makes sense to prescreen virtual libraries using pharmacophore techniques, particularly if one uses shape representations of the receptor site, such as volume-exclusion spheres, a pharmacophore search can be a very effective prefilter. Also, in cases where receptor-site flexibility is problematic, pharmacophore searching may be less restrictive (unless one tries to deal with protein flexibility in the docking routine—a task that is not easy, usually not applied today, and another future direction of development in virtual screening).

The above tools and pathways show a simple and inexpensive way of discovering novel lead chemical matter for drug discovery programs. However, there are many hurdles to overcome to make virtual screening successful. The properties of druglikeness may not be understood sufficiently enough, resulting in poor pharmacokinetics of the compounds: ex-

Table 6.4 Results Obtained from the Initial Screening of Compounds against Matriptase (206)[a]

% Inhibition	Number of Compounds
Over 95%	15
90–94%	4
70–89%	15
40–69%	13
Below 39%	17
High absorbency	3
Increased activity	3

[a]Testing was done at 75 μM compound and protein concentration. The ratio between compound and protein molar concentration was 1:1.

Table 6.5 K_i Values Obtained for Tested bis-Benzamidines against Matriptase

Compound	Structure	K_i (nM)
1		924
2		191
3		1,160
4		4,500
5		535
6		>10,000
7		208

isting SARs that lead to the generation of pharmacophore models may bias the pharmacophore toward a narrow segment of compounds; structural information of the target protein is often not available; and homology models may not be precise enough. Current scoring functions are often not robust enough to separate actives from inactives. Compounds identified may not be easy to synthesize. Hits may not be selective or patentable.

On one hand, there are obviously many risks involved in virtual screening, many assumptions made, and a positive outcome not at all guaranteed in each and every case. On the other hand, however, the overall process is extremely cost effective and fast. Even if successful in only a few cases, virtual screening can produce leads that may otherwise not have surfaced and so add immense value to a drug discovery program. Especially in cases where high throughput screening cannot identify a viable lead chemical matter, virtual screening applied to vendor databases or combinatorial libraries to be synthesized presents a cost-effective alternative. Mainly because of its speed, cost effectiveness, ease of setup, and increasing robustness, we expect virtual screening to become a mainstream approach throughout the pharmaceutical industry.

5 ACKNOWLEDGMENTS

The authors thank Dr. Matthias Rarey for valuable discussions.

REFERENCES

1. D. J. Abraham, *Intra-Science Chem. Rept.*, **8**, 1 (1974).

2. M. Perutz, *Protein Structure. New Approaches to Disease and Therapy*, Freeman, New York, 1992.

3. S. W. Fesik, *J. Med. Chem.*, **34**, 2937 (1991).

4. D. W. Cushman, H. S. Cheung, E. F. Sabo, and M. A. Ondetti, *Biochemistry*, **16**, 5484 (1977).

5. L. F. Kuyper, B. Roth, D. P. Baccanari, R. Ferone, C. R. Bedell, J. N. Champness, D. K. Stammers, J. G. Dann, F. E. Norrington, D. J. Blaker, and P. J. Goodford, *J. Med. Chem.*, **25**, 1120 (1982).

6. M. A. Gallop, R. W. Barrett, W. J. Dower, S. P. A. Fodor, and E. M. Gordon, *J. Med. Chem.*, **37**, 1233 (1994).

7. E. M. Gordon, R. W. Barrett, W. J. Dower, S. P. A. Fodor, and M. A. Gallop, *J. Med. Chem.*, **37**, 1385 (1994).

8. M. W. Lutz, J. A. Menius, T. D. Choi, R. G. Laskody, P. L. Domanico, A. S. Goetz, and D. L. Saussy, *Drug Discovery Today*, **1**, 277 (1996).

9. H. J. Böhm and G. Klebe, *Angew. Chem. Int. Ed. Engl.*, **35**, 2589 (1996).

10. W. P. Walters, M. T. Stahl, and M. A. Murcko, *Drug Discovery Today*, **3**, 160 (1998).

11. Y. C. Martin, *Perspect. Drug Discovery Des.*, **7/8**, 159 (1997).

12. A. C. Good and J. S. Mason in K. B. Lipkowitz and D. B. Boyd, Eds., *Reviews in Computational Chemistry*, Vol. **7**, VCH, New York, 1995, p. 67.

13. G. Klebe, *Virtual Screening: An Alternative or Complement to High Throughput Screening?*, Kluwer/Escom, Leiden, 2000.

14. H. J. Bohm and G. Schneider, *Virtual Screening for Bioactive Molecules*, Wiley-VCH, Weinheim, 2000.

15. A. C. Good, S. R. Krystek, and J. S. Mason, *Drug Discovery Today*, **5** (Suppl.), S61 (2000).

16. T. Langer and R. D. Hoffmann, *Curr. Pharm. Design*, **7**, 509 (2001).

17. B. Waszkowycz, T. D. J. Perkins, R. A. Sykes, and J. Li, *IBM Systems J.*, **40**, 360 (2001).

18. A. Good, *Curr. Opin. Drug Discovery Dev.*, **4**, 301 (2001).

19. R. F. Burns, R. M. A. Simmons, J. J. Howbert, D. C. Waters, P. G. Threlkeld, and B. D. Gitter, *Exploiting Molecular Diversity, Symposium Proceedings*, Vol. **2**, Cambridge Healthtech Institute, San Diego, 1995, p. 2.

20. W. P. Walters, Ajay, and M. A. Murcko, *Curr. Opin. Chem. Biol.*, **3**, 384 (1999).

21. W. P. Walters and M. A. Murcko, *Methods Principles Med. Chem.*, **10**, 15 (2000).

22. D. E. Clark and S. D. Pickett, *Drug Discovery Today*, **5**, 49 (2000).

23. B. L. Podlogar, I. Muegge, and L. J. Brice, *Curr. Opin. Drug Discovery Dev.*, **4**, 102 (2001).

24. I. Muegge, *Chem. Eur. J.*, **8**, 1976 (2002).

25. Ajay, W. P. Walters, and M. A. Murcko, *J. Med. Chem.*, **41**, 3314 (1998).

26. Ajay, G. W. Bemis, and M. A. Murcko, *J. Med. Chem.*, **42**, 4942 (1999).

27. J. Sadowski and H. Kubinyi, *J. Med. Chem.*, **41**, 3325 (1998).

28. M. Wagener and V. J. vanGeerestein, *J. Chem. Inf. Comput. Sci.*, **40**, 280 (2000).

29. V. J. Gillet, P. Willett, and J. Bradshaw, *J. Chem. Inf. Comput. Sci.*, **38**, 165 (1998).

30. Comprehensive Medicinal Chemistry is available from MDL Information Systems Inc., San Leandro, CA 94577 and contains drugs already on the market.

31. World Drug Index is available from Derwent Information, London, UK. Website: www.derwent.com.

32. MACCS-II Drug Data Report is available from MDL Information Systems Inc., San Leandro, CA 94577 and contains biologically active compounds in the early stages of drug development.

33. C. A. Lipinski, F. Lombardo, B. W. Dominy, and P. J. Feeney, *Adv. Drug Delivery Rev.*, **23**, 3 (1997).

34. T. I. Oprea, *J. Comput.-Aided Mol. Des.*, **14**, 251 (2000).

35. J. Kelder, P. D. J. Grootenhuis, D. M. Bayada, L. P. C. Delbressine, and J. P. Ploemen, *Pharm. Res.*, **16**, 1514 (1999).

36. M. Hann, B. Hudson, X. Lewell, R. Lifely, L. Miller, and N. Ramsden, *J. Chem. Inf. Comput. Sci.*, **39**, 897 (1999).

37. National Toxicology Program. http://ntp-server.niehs.nih.gov.

38. RTECS C2(96–4); National Institute for Occupational Safety and Health (NIOSH), U.S. Department of Health and Human Services: Washington, DC, 1996. URL: http://www.ccohs.ca.

39. G. Klopman and H. S. Rosenkranz, *Mutat. Res.*, **305**, 33 (1994).

40. K. Enslein, V. K. Gombar, and B. W. Blake, *Mutat. Res.*, **305**, 47 (1994).

41. Lhasa Ltd., School of Chemistry, University of Leeds, Leeds, UK. URL: http://www.chem.leeds.ac.uk/LUK/derek/index.html.

42. Available Chemicals Directory is available from MDL Information Systems Inc., San Leandro, CA 94577 and contains specialty bulk chemicals from commercial sources. Website: http://www.mdli.com.

43. V. N. Viswanadhan, A. K. Ghose, G. R. Revankar, and R. K. Robins, *J. Chem. Inf. Comp. Sci.*, **29**, 163 (1989).

44. G. W. Bemis and M. A. Murcko, *J. Med. Chem.*, **39**, 2887 (1996).

45. G. W. Bemis and M. A. Murcko, *J. Med. Chem.*, **42**, 5095 (1999).

46. X. Q. Lewell, D. B. Judd, S. P. Watson, and M. M. Hann, *J. Chem. Inf. Comput. Sci.*, **38**, 511 (1998).

47. I. Muegge, D. Brittelli, and S. L. Heald, *J. Med. Chem.*, **44**, 1841 (2001).

48. B. L. Podlogar and I. Muegge, *Curr. Top. Med. Chem.*, **1**, 257 (2001).

49. R. M. Brunne, G. Hessler, and I. Muegge in K. C. Nicolaou, R. Hanko, and W. Hartwig, Eds., *Handbook of Combinatorial Chemistry*, Vol. 2, Wiley-VCH, Weinheim, 2002, p. 761.

50. B. E. Evans, K. E. Rittle, M. G. Bock, R. M. DiPardo, R. M. Freidinger, W. L. Whitter, G. F. Lundell, D. F. Veber, P. S. Anderson, R. S. L. Chang, V. J. Lotti, D. J. Cerino, T. B. Chen, P. J. Kling, K. A. Kunkel, J. P. Springer, and J. Hirshfield, *J. Med. Chem.*, **31**, 2235 (1988).

51. J. S. Mason, I. Morize, P. R. Menard, D. L. Cheney, C. Hulme, and R. F. Labaudiniere, *J. Med. Chem.*, **42**, 3251 (1999).

52. P. J. Hajduk, M. Bures, J. Praestgaard, and S. W. Fesik, *J. Med. Chem.*, **43**, 3443 (2000).

53. J. H. Van Drie, D. Weininger, and Y. C. Martin, *J. Comput.-Aided Mol. Des.*, **3**, 225 (1989).

54. E. Fischer, *Ber. Dtsch. Chem. Ges.*, **27**, 2985 (1894).

55. P. Ehrlich, *Ber. Dtsch. Chem. Ges.*, **42**, 17 (1909).

56. P. Gund, *Prog. Mol. Subcell. Biol.*, **5**, 117 (1977).

57. F. Bernstein, T. F. Koetzle, G. J. B. Williams, E. F. Meyer, Jr., M. D. Brice, J. R. Rodgers, O. Kennard, T. Schimanouchi, and M. J. Tasumi, *J. Mol. Biol.*, **112**, 535 (1977).

58. H. M. Berman, J. Westbrook, Z. Feng, G. Gilliland, T. N. Baht, H. Weissig, I. N. Shindyalov, and P. E. Bourne, *Nucleic Acids Res.*, **28**, 235 (2000).

59. M. K. Dreyer, D. R. Borcherding, J. A. Dumont, N. P. Peet, J. T. Tsay, P. S. Wright, A. J. Bitonti, J. Shen, and S.-H. Kim, *J. Med. Chem.*, **44**, 524 (2001).

60. G. R. Marshall, C. D. Barry, H. E. Bosshard, R. A. Dammkoehler, and D. A. Dunn in E. C. Olson and R. E. Christoffersen, Eds., *Computer Assisted Drug Design*, American Chemical Society, Washington, 1979, p. 205.

61. J. R. Sufrin, D. A. Dunn, and G. R. Marshall, *Mol. Pharmacol.*, **19**, 307 (1981).

62. R. D. Cramer, D. E. Patterson, R. D. Clark, F. Soltanshahi, and M. S. Lawless, *J. Chem. Inf. Comput. Sci.*, **38**, 1010 (1998).

63. D. Horvath in A. K. Ghose and V. N. Viswanadhan, Eds., *Combinatorial Library Design and Evaluation. Principles, Software Tools, and Applications in Drug Discovery*, Marcel Dekker, New York, 2001, p. 429.

64. A. Dalby, J. G. Nourse, W. D. Hounshell, A. K. I. Gushurst, D. L. Grier, B. A. Leland, and J. Laufer, *J. Chem. Inf. Comput. Sci.*, **32**, 244 (1992).

65. Spresi Chemical Database, InfoChem GMBH, Grobenzell, Germany and Daylight Chemical Information Systems, Irvine, CA (2002).

66. The Chemical Abstracts Database, Chemical Abstracts Service, 2540 Olentangy River Road, PO Box 3012, Columbus, OH (2002).

67. G. W. A. Milne, M. C. Nicklaus, J. S. Driscoll, S. Wang, and D. W. Zaharevitz, *J. Chem. Inf. Comput. Sci.*, **34**, 1219 (1994).

68. G. W. A. Milne and J. A. Miller, *J. Chem. Inf. Comput. Sci.*, **26**, 154 (1986).

69. D. Weininger, *J. Chem. Inf. Comput. Sci.*, **28**, 31 (1988).

70. D. Weininger, A. Weininger, and J. L. Weininger, *J. Chem. Inf. Comput. Sci.*, **29**, 97 (1989).

71. F. H. Allen, S. Bellard, M. D. Brice, B. A. Cartwright, A. Doubleday, H. Higgs, T. Hummelink-Peters, O. Kenard, W. D. S. Motherwell, J. R. Rodgers, and D. G. Watson, *Acta Crystallogr. B*, **35**, 2331 (1979).

72. R. L. DesJarlais in P. S. Charifson, Ed., *Practical Application of Computer-Aided Drug Design*, Vol. **1**, Marcel Dekker, New York, 1997, p. 73.

73. A. RusinkoIII, J. M. Skell, R. Balducci, C. M. McGarity, and R. S. Pearlman, CONCORD, University of Texas, Austin, TX and Tripos Associates, St. Louis, MO (1988).

74. J. Gasteiger, C. Rudolph, and J. Sadowski, *Tetrahedron Comput. Methodol.*, **3**, 537 (1990).

75. D. Weininger, Rubicon, Daylight Chemical Information Systems, Irvine, CA (1995).

76. Omega, Open Eye, Santa Fe, NM (2002).

77. D. P. Dolata, A. R. Leach, and K. Prout, *J. Comput.-Aided Mol. Des.*, **1**, 73 (1987).

78. A. R. Leach and K. Prout, *J. Comput. Chem.*, **11**, 1193 (1990).

79. W. T. Wipke and M. A. Hahn, *Artif. Intell. Appl. Chem.*, **306**, 136 (1986).

80. G. Klebe and T. Mietzner, *J. Comput.-Aided Mol. Des.*, **8**, 583 (1994).

81. E. V. Gordeeva, A. R. Katritzky, V. V. Shcherbukhin, and N. S. Zifirov, *J. Chem. Inf. Comput. Sci.*, **33**, 102 (1993).

82. S. K. Kearsley, D. J. Underwood, R. P. Sheridan, and M. D. Miller, *J. Comput.-Aided Mol. Des.*, **8**, 153 (1994).

83. J. M. Barnard, *J. Chem. Inf. Comput. Sci.*, **33**, 532 (1993).

84. I. J. Enyedy, J. Wang, W. A. Zaman, K. M. Johnson, and S. Wang, *Bioorg. Med. Chem. Lett.*, **12**, 1775 (2002).

85. G. M. Downs and P. Willett in K. B. Lipkowitz and D. B. Boyd, Eds., *Reviews in Computational Chemistry*, VCH, New York, 1996, p. 1.

86. P. Willett, J. M. Barnard, and G. M. Downs, *J. Chem. Inf. Comput. Sci.*, **38**, 983 (1998).

87. L. Xue, J. W. Godden, and J. Bajorath, *J. Chem. Inf. Comput. Sci.*, **39**, 881 (1999).

88. L. Xue, J. W. Godden, and J. Bajorath, *J. Chem. Inf. Comput. Sci.*, **40**, 1227 (2000).

89. L. Xue, F. L. Stahura, J. W. Godden, and J. Bajorath, *J. Chem. Inf. Comput. Sci.*, **41**, 394 (2001).

90. S. K. Kearsley, S. Sallamack, E. M. Fluder, J. D. Andose, R. T. Mosley, and R. P. Sheridan, *J. Chem. Inf. Comput. Sci.*, **36**, 118 (1996).

91. R. D. Hull, E. M. Fluder, S. B. Singh, R. B. Nachbar, S. K. Kearsley, and R. P. Sheridan, *J. Med. Chem.*, **44**, 1185 (2001).

92. R. D. Hull, S. B. Singh, R. B. Nachbar, R. P. Sheridan, S. K. Kearsley, and E. M. Fluder, *J. Med. Chem.*, **44**, 1177 (2001).

93. B. T. Hoffman, T. Kopajtic, J. L. Katz, and A. H. Newman, *J. Med. Chem.*, **43**, 4151 (2000).

94. A. P. Kozikowski, M. K. E. Saiah, K. M. Johnson, and J. S. Bergmann, *J. Med. Chem.*, **38**, 3086 (1995).

95. S. Wang, S. Sakamuri, I. J. Enyedy, A. P. Kozikowski, O. Deschaux, B. C. Bandyopadhyay, S. R. Tella, W. A. Zaman, and K. M. Johnson, *J. Med. Chem.*, **43**, 351 (2000).

96. D. Barnum, J. Greene, A. Smellie, and P. Sprague, *J. Chem. Inf. Comput. Sci.*, **36**, 563 (1996).

97. P. W. Sprague in K. Muller, Ed., *Perspectives in Drug Discovery and Design*, Vol. **1**, ESCOM Science Publishers B.V., Leiden, 1995, p. 1.

98. Y. C. Martin, M. G. Bures, E. A. Danaher, J. DeLazzer, I. Lico, and P. A. Pavlik, *J. Comput.-Aided Mol. Des.*, **7**, 83 (1993).

99. D. Beusen, Alignment of angiotensin II receptor antagonists using GASP, Tripos Technical Notes 1(4), November 1996.

100. Flo96, Thistlesoft, Colebrook, CT (1996).

101. V. Golender, B. Vesterman, O. Eliyahu, A. Kardash, M. Kletzin, M. Shokhen, and E. Vorpagel, Conference Proceedings Barcelona 10th European Symposium on Structure-Activity Relationships, Vol. 10, Prous Science, Barcelona, Spain, 1994, p. 246.

102. ROCS, Open Eye, Santa Fe, NM (2002).

103. O. F. Guner, *Pharmacophore Perception, Development, and Use in Drug Design*, International University Line, La Jolla, CA, 2000.

104. H.-J. Böhm, *J. Comput.-Aided Mol. Des.*, **6**, 593 (1992).

105. H. Böhm, *J. Comput.-Aided Mol. Des.*, **6**, 61 (1992).

106. H. Böhm, *J. Comput.-Aided Mol. Des.*, **8**, 623 (1994).

107. R. Wang, Y. Gao, and L. Lai, *J. Mol. Model.*, **6**, 498 (2000).

108. Cerius2, Accelrys, San Diego, CA (2002).

109. S. K. Burley and G. A. Petsko, *J. Am. Chem. Soc.*, **108**, 7995 (1986).

110. H. A. Carlson, K. M. Masukawa, K. Rubins, F. D. Bushman, W. L. Jorgensen, R. D. Lins,

J. M. Briggs, and J. A. McCammon, *J. Med. Chem.*, **43**, 2100 (2000).

111. R. P. Sheridan, R. Nilakantan, A. Rusinko III, N. Bauman, K. S. Haraki, and R. Venkataraghavan, *J. Chem. Inf. Comput. Sci.*, **29**, 255 (1989).

112. UNITY version 6.6; Tripos Associates, St. Louis, MO. Website: http://www.tripos.com.

113. MDL Information Systems, Inc., San Leandro, CA.

114. J. Greene, S. Kahn, H. Savoj, P. Sprague, and S. Teig, *J. Chem. Inf. Comput. Sci.*, **34**, 1297 (1994).

115. W.-D. Ihlenfeldt, J. H. Voigt, B. Bienfait, F. Oellien, and M. C. Nicklaus, *J. Chem. Inf. Comput. Sci.*, **42**, 46 (2002).

116. X. Fang and S. Wang, *J. Chem. Inf. Comput. Sci.*, **42**, 192 (2002).

117. K. S. Haraki, R. P. Sheridan, R. Venkataraghavan, D. A. Dunn, and R. McCulloch, *Tetrahedron Comput. Methodol.*, **3**, 565 (1990).

118. J. A. Grant, M. A. Gallardo, and B. T. Pickup, *J. Comput. Chem.*, **17**, 1653 (1996).

119. T. E. Mook, D. R. Henry, A. G. Ozkabak, and M. Alamgir, *J. Chem. Inf. Comput. Sci.*, **34**, 184 (1994).

120. T. Hurst, *J. Chem. Inf. Comput. Sci.*, **34**, 190 (1994).

121. N. W. Murrall and E. K. Davies, *J. Chem. Inf. Comput. Sci.*, **30**, 312 (1990).

122. D. E. Clark, P. Willett, and P. W. Kenny, *J. Mol. Graph.*, **11**, 146 (1993).

123. D. E. Clark, G. Jones, and P. Willett, *J. Chem. Inf. Comput. Sci.*, **34**, 197 (1994).

124. J. M. Blaney and J. S. Dixon, *Perspect. Drug Discovery Des.*, **15**, 301 (1993).

125. I. Muegge and M. Rarey in D. B. Boyd and K. B. Lipkowitz, Eds., *Reviews in Computational Chemistry*, Vol. **17**, Wiley-VCH, New York, 2001, p. 1.

126. G. Jones, P. Willett, R. C. Glen, and A. R. Leach, *J. Mol. Biol.*, **267**, 727 (1997).

127. I. Muegge, *Med. Chem. Res.*, **9**, 490 (1999).

128. D. E. Koshland, *Proc. Natl. Acad. Sci. USA*, **44**, 98 (1958).

129. D. W. Christianson and W. N. Lipscomb, *Acc. Chem. Res.*, **22**, 62 (1989).

130. C. Sander and R. Schneider, *Proteins: Struct., Funct., Genet.*, **9**, 56 (1991).

131. T. L. Blundell, B. L. Sibanda, M. J. E. Sternberg, and J. M. Thornton, *Nature*, **326**, 347 (1987).

132. A. Vedani, P. Zbinden, J. P. Snyder, and P. A. Greenidge, *J. Am. Chem. Soc.*, **117**, 4987 (1995).

133. M. C. Nicklaus, S. Wang, J. S. Driscoll, and G. W. A. Milne, *Bioorg. Med. Chem.*, **3**, 411 (1995).

134. H. A. Carlson and J. A. McCammon, *Mol. Pharmacol.*, **57**, 213 (2000).

135. R. M. A. Knegtel, I. D. Kuntz, and C. M. Oshiro, *J. Mol. Biol.*, **266**, 424 (1997).

136. E. A. Sudbeck, C. Mao, R. Vig, T. K. Venkatachalam, L. Tuel-Ahlgren, and F. M. Uckun, *Antimicrob. Agents Chemother.*, **42**, 3225 (1998).

137. C. Bron and J. Kerbosch, *Commun. ACM*, **16**, 575 (1973).

138. Y. Lamdan and H. J. Wolfson in Proceedings of the IEEE International Conference on Computer Vision, 1988, p. 238.

139. F. S. Kuhl, G. M. Crippen, and D. K. Friesen, *J. Comput. Chem.*, **5**, 24 (1984).

140. F. M. Richards, *Annu. Rev. Biophys. Bioeng.*, **6**, 151 (1977).

141. M. L. Connolly, *J. Appl. Crystallogr.*, **16**, 548 (1983).

142. M. L. Connolly, *J. Appl. Crystallogr.*, **18**, 499 (1985).

143. B. K. Shoichet, R. M. Stroud, D. V. Santi, I. D. Kuntz, and K. M. Perry, *Science*, **259**, 1445 (1993).

144. B. K. Shoichet, D. L. Bodian, and I. D. Kuntz, *J. Comput. Chem.*, **13**, 380 (1992).

145. E. C. Meng, D. A. Gschwend, J. M. Blaney, and I. D. Kuntz, *Proteins: Struct., Funct., Genet.*, **17**, 266 (1993).

146. T. J. A. Ewing and I. D. Kuntz, *J. Comput. Chem.*, **18**, 1175 (1997).

147. E. C. Meng, B. K. Shoichet, and I. D. Kuntz, *J. Comput. Chem.*, **13**, 505 (1992).

148. E. C. Meng, I. D. Kuntz, D. J. Abraham, and G. E. Kellogg, *J. Comput.-Aided Mol. Des.*, **8**, 299 (1994).

149. D. A. Gschwend and I. D. Kuntz, *J. Comput.-Aided Mol. Des.*, **10**, 123 (1996).

150. B. K. Shoichet, A. R. Leach, and I. D. Kuntz, *Proteins: Struct., Funct., Genet.*, **34**, 4 (1999).

151. X. Q. Zou, Y. X. Sun, and I. D. Kuntz, *J. Am. Chem. Soc.*, **121**, 8033 (1999).

152. M. D. Miller, S. K. Kearsley, D. J. Underwood, and R. P. Sheridan, *J. Comput.-Aided Mol. Des.*, **8**, 153 (1994).

153. T. F. Havel, I. D. Kuntz, and G. M. Crippen, *Bull. Math. Biol.*, **45**, 665 (1983).

154. S. Kirkpatrik, C. D. J. Gelatt, and M. P. Vecchi, *Science*, **220**, 671 (1983).

155. D. S. Goodsell and A. J. Olson, *Proteins: Struct., Funct., Genet.*, **8**, 195 (1990).

156. T. P. Lybrand in D. B. Boyd and K. B. Lipkowitz, Eds., *Reviews in Comptational Chemistry*, Vol. **1**, VCH, New York, 1990, p. 295.

157. J. A. Given and M. K. Gilson, *Proteins: Struct., Funct., Genet.*, **33**, 475 (1998).

158. T. N. Hart and R. J. Read, *Proteins: Struct., Funct., Genet.*, **13**, 206 (1992).

159. C. McMartin and R. S. Bohacek, *J. Comput.-Aided Mol. Des.*, **11**, 333 (1997).

160. A. Wallqvist and D. G. Covell, *Proteins: Struct., Funct., Genet.*, **25**, 403 (1996).

161. R. Abagyan, M. Totrov, and D. Kuznetsov, *J. Comput. Chem.*, **15**, 488 (1994).

162. J. Apostolakis, A. Pluckthun, and A. Caflisch, *J. Comput. Chem.*, **19**, 21 (1998).

163. P. S. Charifson, J. J. Corkery, M. A. Murcko, and W. P. Walters, *J. Med. Chem.*, **42**, 5100 (1999).

164. C. A. Baxter, C. W. Murray, D. E. Clark, D. R. Westhead, and M. D. Eldridge, *Proteins: Struct., Funct., Genet.*, **33**, 367 (1998).

165. J. Wang, P. A. Kollman, and I. D. Kuntz, *Proteins: Struct., Funct., Genet.*, **36**, 1 (1999).

166. D. Hoffmann, B. Kramer, T. Washio, T. Steinmetzer, M. Rarey, and T. Lengauer, *J. Med. Chem.*, **42**, 4422 (1999).

167. J. S. Dixon, *Proteins: Struct., Funct., Genet.*, Suppl., **1**, 198 (1997).

168. P. D. J. Grootenhuis and P. J. M. vanGalen, *Acta Crystallogr. D*, **51**, 560 (1995).

169. M. K. Holloway, J. M. Wai, T. A. Halgren, P. M. D. Fitzgerald, J. P. Vacca, B. D. Dorsey, R. B. Levin, W. J. Thompson, L. J. Chen, S. J. deSolms, N. Gaffin, A. K. Ghosh, E. A. Giuliani, S. L. Graham, J. P. Guare, R. W. Hungate, T. A. Lyle, W. M. Sanders, T. J. Tucker, M. Wiggins, C. M. Wiscount, O. W. Woltersdorf, S. D. Young, P. L. Darke, and J. A. Zugay, *J. Med. Chem.*, **38**, 305 (1995).

170. M. K. Holloway, *Perspect. Drug Discovery Des.*, **9/10/11**, 63 (1998).

171. N. S. Blom and J. Sygusch, *Proteins: Struct., Funct., Genet.*, **27**, 493 (1997).

172. N. Tomioka and A. Itai, *J. Comput.-Aided Mol. Des.*, **8**, 347 (1994).

173. S. J. Weiner, P. A. Kollman, D. A. Case, U. C. Singh, C. Ghio, G. Alagona, S. Profeta, Jr., and P. Weiner, *J. Am. Chem. Soc.*, **106**, 765 (1984).

174. S. J. Weiner, P. A. Kollman, D. T. Nguyen, and D. A. Case, *J. Comput. Chem.*, **7**, 230 (1986).

175. P. J. Goodford, *J. Med. Chem.*, **28**, 849 (1985).

176. D. Qui, P. S. Shenkin, E. P. Hollinger, and W. C. Still, *J. Phys. Chem.*, **101**, 3005 (1997).

177. D. Eisenberg and A. D. McLachlan, *Nature*, **319**, 199 (1986).

178. P. F. W. Stouten, C. Frommel, H. Nakamura, and C. Sander, *Mol. Simul.*, **10**, 97 (1993).

179. S. Vajda, Z. Weng, R. Rosenfeld, and C. DeLisi, *Biochemistry*, **33**, 13977 (1994).

180. M. Rarey, B. Kramer, T. Lengauer, and G. Klebe, *J. Mol. Biol.*, **261**, 470 (1996).

181. H.-J. Böhm, *J. Comput.-Aided Mol. Des.*, **8**, 243 (1994).

182. H. J. Böhm, *J. Comput.-Aided Mol. Des.*, **12**, 309 (1998).

183. B. Kramer, G. Metz, M. Rarey, and T. Lengauer, *Med. Chem. Res.*, **9**, 463 (1999).

184. I. Muegge, *Perspect. Drug Discovery Des.*, **20**, 99 (2000).

185. I. Muegge and Y. C. Martin, *J. Med. Chem.*, **42**, 791 (1999).

186. I. Muegge, *J. Comput. Chem.*, **22**, 418 (2001).

187. I. Muegge, Y. C. Martin, P. J. Hajduk, and S. W. Fesik, *J. Med. Chem.*, **42**, 2498 (1999).

188. I. Muegge and B. Podlogar, *Quant. Struct.-Act. Relat.*, **20**, 215 (2001).

189. S. Ha, R. Andreani, A. Robbins, and I. Muegge, *J. Comput.-Aided Mol. Des.*, **14**, 435 (2000).

190. M. D. Eldridge, C. W. Murray, T. R. Auton, G. V. Paolini, and R. P. Mee, *J. Comput.-Aided Mol. Des.*, **11**, 425 (1997).

191. C. W. Murray, T. R. Auton, and M. D. Eldridge, *J. Comput.-Aided Mol. Des.*, **12**, 503 (1998).

192. R. X. Wang, L. Liu, L. H. Lai, and Y. Q. Tang, *J. Mol. Model.*, **4**, 379 (1998).

193. D. K. Gehlhaar, G. M. Verkhivker, P. A. Rejto, C. J. Sherman, D. B. Fogel, L. J. Fogel, and S. T. Freer, *Chem. Biol.*, **2**, 317 (1995).

194. T. A. Halgren, *J. Comput. Chem.*, **17**, 520 (1996).

195. B. Honig and A. Nicholls, *Science*, **268**, 1144 (1995).

196. D. R. Flower, *J. Mol. Graphics Modell.*, **15**, 238 (1998).

197. T. R. Stouch and P. C. Jurs, *J. Chem. Inf. Comput. Sci.*, **26**, 4 (1986).

198. C. Bissantz, G. Folkers, and D. Rognan, *J. Med. Chem.*, **43**, 4759 (2000).

199. M. Stahl and M. Rarey, *J. Med. Chem.*, **44**, 1035 (2001).

200. J. G. Millichap, *Ann. N. Y. Acad. Sci.*, **205**, 321 (1973).

201. J. M. Swanson and M. Kinsbourne in G. H. Hale and M. Lewis, Eds., *Attention and Cognitive Development*, Vol. **1**, Plenum, New York, 1979, p. 249.

202. M. Froimowitz, K. S. Patrick, and V. Cody, *Pharm. Res.*, **12**, 1430 (1995).

203. I. J. Enyedy, W. A. Zaman, S. Sakamuri, A. P. Kozikowski, K. M. Johnson, and S. Wang, *Bioorg. Med. Chem. Lett.*, **11**, 1113 (2001).

204. J. H. Voigt, B. Bienfait, S. Wang, and M. C. Nicklaus, *J. Chem. Inf. Comput. Sci.*, **41**, 702 (2001).

205. Chem-X (version 96), Oxford Molecular Group, Inc., Hunt Valley, MD (company no longer exists) (2001).

206. I. J. Enyedy, S.-L. Lee, A. H. Kuo, R. B. Dickson, C.-Y. Lin, and S. Wang, *J. Med. Chem.*, **44**, 1349 (2001).

207. R. Friedrich, P. Fuentes-Prior, E. Ong, G. Coombs, M. Hunter, R. Oehler, D. Pierson, R. Gonzalez, R. Huber, W. Bode, and E. L. Madison, *J. Biol. Chem.*, **277**, 2160 (2002).

208. A. Sali, L. Potterton, F. Yuan, H. van Vlijmen, and M. Karplus, *Proteins: Struct., Funct., Genet.*, **23**, 318 (1995).

209. A. Sali, *Proteins: Struct., Funct., Genet.*, **6**, 437 (1995).

210. B. R. Brooks, R. E. Bruccoleri, B. D. Olafson, D. J. States, S. Swaminathan, and M. Karplus, *J. Comput. Chem.*, **4**, 187 (1983).

211. M. N. Janakiraman, C. L. White, W. G. Laver, G. M. Air, and M. Luo, *Biochemistry*, **33**, 8172 (1994).

212. G. M. Rishton, *Drug Discovery Today*, **2**, 382 (1997).

213. N. R. Taylor, A. Cleasby, O. Singh, T. Skarzynski, A. J. Wonacott, P. W. Smith, S. L. Sollis, P. D. Howes, P. C. Cherry, R. Bethell, P. Colman, and J. Varghese, *J. Med. Chem.*, **41**, 798 (1998).

CHAPTER SEVEN

Docking and Scoring Functions/ Virtual Screening

CHRISTOPH SOTRIFFER
GERHARD KLEBE
University of Marburg
Department of Pharmaceutical Chemistry
Marburg, Germany

MARTIN STAHL
HANS-JOACHIM BÖHM
Discovery Technologies
F. Hoffmann-La Roche AG
Basel, Switzerland

Burger's Medicinal Chemistry and Drug Discovery
Sixth Edition, Volume 1: Drug Discovery
Edited by Donald J. Abraham
ISBN 0-471-27090-3   © 2003 John Wiley & Sons, Inc.

Contents

1 INTRODUCTION

The action of drug molecules and the function
of protein targets are governed by principles of
molecular recognition. Binding events between ligands and their receptors in biological
systems form the basis of physiological activity and pharmacological effects of chemical
compounds. Accordingly, the rational development of new drugs requires an understanding
of molecular recognition in terms of both
structure and energetics (1). With respect to
practical applications, it requires tools that
are based on such knowledge of mutual recognition between molecular structures. Docking
and virtual screening are computational tools
to investigate the binding between macromolecular targets and potential ligands. They
constitute an essential part of structure-based
drug design, the area of medicinal chemistry
that harnesses structural information for the
purpose of drug discovery.

Structure-based design has become an integral part of medicinal chemistry. Although
the knowledge about molecular recognition
and its foundation on structural principles is
still far from being complete, it has already
fueled significant advances and contributed to
many success stories in drug discovery (2).
Convincing evidence has been accumulated
for a large number of targets that the protein
three-dimensional (3D) structure can be used
to design small molecule ligands binding
tightly to the protein. Several marketed compounds can indeed be attributed to successful

structure-based design (3–6), as summarized
in a number of recent reviews (2, 7–9).

Structure-based drug design is an iterative
process (2, 10). It requires as the starting point
the crystal structure or a reliable homology
model of the target protein, preferentially
complexed with a ligand. The first step of the
process is a detailed analysis of the binding
site and a compilation of all aspects possibly
responsible for binding affinity and selectivity.
These data are then used to generate new
ideas how to improve existing ligands or to
develop alternative molecular frameworks.
Computational methods and molecular modeling play an essential role in this phase of
hypothesis generation. They help to exploit information about the binding site geometry by
constructing new molecules *de novo*, by analyzing known molecules with respect to their
affinity and binding geometry, or by searching
compound libraries for potential hits to suggest new leads. Discovered hits that are commercially available or synthetically accessible
are then experimentally tested and their binding properties examined by biochemical, crystallographic, and spectroscopic methods. The
3D structure of new complexes together with
the acquired activity data are subsequently
used to start a new cycle of ligand design to
improve the hypotheses stated in the previous
round.

Since the introduction of computational
structure-based design techniques into the
drug discovery process in the early 1980s, the
impact of these methods has significantly
changed. Initially, computational tools were

often applied *a posteriori* to rationalize and understand the binding and structure-activity relationships in a series of inhibitors and to assist in the manual design of individual compounds. Guided by the creativity of the designer, a novel putative ligand was constructed using computer graphics. Molecular mechanics calculations were performed on the produced protein–ligand complex to assess the properties of the generated ligand in terms of a geometric and energetics analysis. A ligand was assumed to bind with high affinity if satisfactory complementarity in shape and surface properties between the protein and the ligand could be detected.

It has been realized, however, that the design of a single, synthetically accessible, active compound is a larger challenge than anticipated. Many phenomena of molecular recognition are not yet fully understood, nor are current modeling tools able to reflect and accordingly predict them with sufficient reliability. Most important, a fast and accurate computational prediction of binding affinities for new inhibitor candidates is still difficult to obtain. Although the existing tools do certainly not allow the medicinal chemist to design the one perfect ligand, they can help to *enrich* sets of molecules with more active ones, even though the known deficiencies of the methods can still lead to significant rates of both false positives and false negatives. A more moderate goal of current molecular design is thus to improve the hit rates of molecules suggested for biological assaying compared to a mere random compound selection or testing of nontargeted compound libraries. This implies that structure-based design approaches now focus on the processing of large numbers of molecules, arranged in so-called virtual libraries. These can be composed of either existing chemical substances (such as, for example, compound collections of a pharmaceutical company) or hypothetical new molecules that could be synthesized by combinatorial chemistry. The task is then to filter these large libraries by eliminating the majority of molecules that is rather unlikely to bind and by prioritizing the remaining ones. As recent experience shows, this strategy can be success-

ful: several publications have reported quite impressive enrichments of active compounds (11–16).

The change of focus from single molecule to compound library design in modern structure-based drug discovery is also a consequence of major technological advances that have dramatically enhanced the data throughput in a variety of fields:

1. Progress in gene technology, protein chemistry, and structure determination techniques have resulted in a tremendous increase in **protein structure information**. The number of publicly available 3D protein structures continues to grow exponentially, with further acceleration expected from the current initiatives of structural genomics (17). As a consequence, more and more design projects are based on structural information, and structure-based ligand design has become routine at all major pharmaceutical companies. On the other hand, the growing amount of structural knowledge also calls for automated methods that make this new wealth of data accessible and available.

2. Automation and miniaturization have led to the development of **high-throughput screening (HTS)**, which is now a well-established process for large-scale biological testing. Libraries of several hundred thousand compounds are routinely screened against new targets, frequently on a time scale of less than 1 month.

3. The characteristics of synthetic chemistry have significantly changed with the introduction of **combinatorial and parallel chemistry** techniques. The trend continues to move away from the synthesis of individual compounds toward the generation of compound libraries, whose members are accessible through the same type of chemical reaction but different building reagents.

4. Massive data processing and computational tasks formerly requiring expensive supercomputers have become generally feasible by advances in **PC cluster com-**

puting, offering supercomputing power at unprecedented performance-to-price ratios.

To provide competitive advantage, structure-based design tools must be fast enough to process thousands of compounds per day using affordable computing resources. Several algorithms have been developed that allow virtual ligand screening based on high-throughput flexible docking (18, 19). More sophisticated methods may then be applied at a later stage of refinement, where a smaller set of compounds, containing the most promising hits, is subjected to a more detailed analysis. Essential elements of all these docking tools are scoring functions that translate computationally generated protein-ligand binding geometries into estimates of affinity.

Docking as a computational tool of structure-based drug design to predict protein-ligand interaction geometries and binding affinities is the subject of this chapter. Fundamental aspects of the docking process, the scoring of protein–ligand interactions, and the application of docking to virtual ligand screening are discussed. At first, a discussion of the underlying physical principles determining protein-ligand recognition is given (Section 2.1), followed by a description of the general concepts of docking, scoring, and virtual screening (Section 2.2). Subsequently, the current approaches to the docking problem are presented (Section 3.1), focusing on the search methods (Section 3.1.3) and the approaches used to represent protein and ligand structures in an efficient way (Sections 3.1.1–3.1.2). In addition, a number of special aspects is discussed (Section 3.2), including, for example, the issues of protein flexibility (Section 3.2.1) or the consideration of water molecules in the context of docking (Section 3.2.2). This is followed by a section on scoring functions used for docking. Three major classes of scoring functions are presented (Section 4.1) and subjected to critical assessment (Section 4.2). A final section is dedicated to virtual screening, illustrating general strategies (Section 5), special problems (Sections 5.1–5.5), and representative applications (Section 5.6).

Although the goal of this chapter is to highlight the most important aspects of docking in the context of structure-based drug design, a comprehensive discussion of all aspects of current docking methodologies would be beyond its scope. Protein-protein or protein-DNA docking (20–24), as well as the docking of small ligands to DNA or RNA as targets (25), are not explicitly covered. The focus will be on the docking of small molecules to protein binding sites, with emphasis on automated procedures. Interactive docking tools, as provided by some modeling software packages, or fragment-based *de novo* design methods are not discussed. Information about the latter is available through other reviews (26, 27) (see also Ref. 28 for a list of *de novo* design algorithms). Virtual screening is discussed only in the context of docking; database screening techniques based on molecular similarity or pharmacophore models are not considered (29, 30). As a final introductory remark, it should be emphasized that docking methods are actually tools for "ligand" (rather than "drug") design. The identification of a tight-binding ligand is a necessary but not sufficient criterion toward a promising novel lead structure and its development into a drug. Aspects of synthetic accessibility, bioavailability, or toxicity are not the primary subject of docking, but because it is important to consider these factors at early stages of a design project, filters may be applied to compound libraries before docking or to hits obtained from virtual screening on an early stage. This aspect of pre- or postprocessing is discussed only briefly.

2 GENERAL CONCEPTS AND PHYSICAL BACKGROUND

2.1 Protein–Ligand Interactions and the Physical Basis of Biomolecular Recognition

The selective binding of a small-molecule ligand to a specific protein is determined by structural and energetic factors. For ligands of pharmaceutical interest, protein-ligand binding usually occurs through noncovalent interactions. The physical basis of noncovalent interactions is generally well established through the theories of electromagnetic forces or, on a more fundamental level, of quantum mechanics. For macromolecules, liquid systems, or solutions, however, direct application

of these first principles is significantly complicated by the size and complexity of the systems, in which a large number of fluctuating
particles simultaneously interact and influence each other. Principles from classical mechanics and heuristic models are therefore frequently used as an approximation to describe
protein–ligand interactions in aqueous solution.

The primary forces acting between a protein and a ligand are all of electrostatic nature.
It is the interaction between explicit charges,
dipoles, induced dipoles, and higher electric
multipoles that leads to phenomena that are
commonly referred to as salt bridges, hydrogen bonds, or van der Waals interactions. In
simplified classifications, it is only the charge-
charge interaction that is called electrostatic.
This interaction between two charges is of
long range and considerable strength. In vacuum or uniform media it can be described by
Coulomb's law. In aqueous solution of biomolecules, however, its application is complicated
because of the presence of a large number of
water molecules. Unless a sufficiently large
number of water molecules is explicitly included in the calculations [as usually only
tractable in computationally expensive molecular dynamics simulations (31)], the correct
treatment of electrostatic interactions in solution requires solving the Poisson-Boltzmann
equation, where the solvent is considered as a
continuous medium of high dielectric constant
surrounding a low-dielectric solute (32).

Electrostatic interactions, however, do not
only occur between charge monopoles. In a
comprehensive treatment of electrostatics one
has to consider a full power series, and therefore interactions between higher electric moments, such as dipoles and quadrupoles, also
play an essential role. Their interaction energies are orientation dependent and become
shorter in range with increasing electric moment. For example, in contrast to the $1/r$ dependency in Coulomb's law, the energy of the
interaction between a charge and a dipole de-
cays with $1/r^2$, the interaction between two
dipoles with $1/r^3$. This, however, is valid only
for a fixed orientation of the dipoles. If they
are mobile, as in isotropic media (liquids), the
dipole-dipole interaction is thermally averaged and an average interaction proportional

to $1/r^6$ results. This $1/r^6$ dependency is also
encountered in interactions that arise between induced electric moments, such as the
dispersion interaction based on London
forces. The attractive interactions between
(induced) electric multipoles are generally
summarized in the term van der Waals interactions. Accordingly, van der Waals forces are
weak, attractive, short-range forces that decay
with $1/r^6$. These are normally described by intermolecular interaction potentials such as
the Lennard-Jones potential:

$$E \propto A/r^{12} - B/r^6$$

where A and B are parameters depending on
the type of the interacting atoms. The r^{12}
term reflects the short-range repulsive
forces attributed to unfavorable spatial
overlap of electron clouds at short distances.

An interaction deserving special attention
is that of hydrogen bonds (33, 34). In principle,
their origin is of the same nature as the interactions mentioned above. A hydrogen bond is
defined as the interaction of an electronegative atom (the hydrogen-bond acceptor) with a
hydrogen atom covalently bonded to an electronegative atom (the hydrogen-bond donor).
The major component of a hydrogen bond is
the electrostatic interaction of the donor-hy-
drogen dipole with the negative partial charge
of the acceptor. The special characteristics
originate from the fact that the hydrogen
atom is very small and can bear a considerable
positive partial charge, such that the acceptor
can contact the hydrogen atom at a shorter
distance than expected from the van der Waals
radii. Hydrogen bonds are directed interactions showing a high angular dependency.
This directionality arises from the anisotropic
charge distribution around the acceptor atom
(lone pairs) and the fact that the electron
shells of donor and acceptor atom start to
overlap at these short distances unless the
ideal geometry is maintained. Hydrogen
bonds are attributed an important role with
respect to specificity of the protein–ligand interaction. This is based on their directionality
and the fact that they require a well-defined
complementarity in the complex (mutual arrangement of hydrogen-bond donors and ac-

ceptors). However, the importance of hydrogen bonds should not be overemphasized because it is the balance between hydrogen bonds and other forces in protein–ligand complexes that must be appropriately considered (35).

Weakly polar interactions in proteins and protein–ligand complexes are frequently phenomenologically analyzed and classified in terms of the interacting partners (36). This especially includes interactions with π-systems, such as the NH-π, OH-π, or CH-π interaction (37, 38), aromatic-aromatic interactions (parallel π-π stacking versus edge-to-face interaction), and the cation-π interaction (39). All of these can mostly be rationalized in terms of electrostatic interactions outlined above; that is, they involve interactions between monopoles, dipoles, and quadrupoles (permanent and induced). A more distinct character can be attributed to metal complexation, which can play a significant role in individual cases of protein–ligand interactions, as for example in metalloenzymes (2, 40, 41).

Finally, so-called hydrophobic or lipophilic interactions are often mentioned as additional contribution to protein–ligand interactions. These terms are used to describe the preferential association of nonpolar groups in aqueous solution. It should be emphasized, however, that in contrast to what the name suggests, there is no special hydrophobic force. Instead, one should speak of a hydrophobic effect. As further mentioned below, according to the generally accepted view, it arises primarily from the entropically favorable replacement and release of water molecules (42, 43). The association between the nonpolar surfaces itself is simply based on weak London forces (36).

Thermodynamically, the strength of the interaction between a protein and a ligand is described by the binding affinity or (Gibbs) free energy of binding. Assuming a simple equilibrium reaction of the form

$$P + L \rightleftharpoons PL$$

between a protein P and ligand L to give the complex PL, the dissociation constant K_d (or

binding constant K_i) is generally used to describe the stability of complex formation:

$$K_d = [P][L]/[PL].$$

From the experimentally measured equilibrium constant the binding affinity can be calculated as

$$\Delta G^0 = RT \ln K_d$$

where R is the gas constant (8.314 J/molK) and T is the temperature [the equilibrium constant would actually have to be related to a standard concentration to become a dimensionless quantity, but in general this is not explicitly considered (44, 45)]. Experimentally determined binding constants K_i (K_d) are typically in the range of 10^{-2} to 10^{-12} M, corresponding to a Gibbs free energy of binding of roughly -10 to -70 kJ/mol (1, 2).

According to the Gibbs-Helmholtz equation, the free energy of binding consists of an enthalpic and an entropic contribution:

$$\Delta G = \Delta H - T \Delta S.$$

The enthalpy and entropy of binding can be determined experimentally, as, for example, by isothermal titration calorimetry (46, 47). These data, however, are still sparse and not always easy to interpret (48, 49). Substantial compensation between enthalpic and entropic contributions is observed (50–52); this phenomenon and its interpretations have recently been critically reexamined (53). Interestingly, the data also show that binding can be both enthalpy-driven (e.g., streptavidin-biotin, $\Delta G = -76.5$ kJ/mol, $\Delta H = -134$ kJ/mol) or entropy-driven (e.g., streptavidin-HABA, $\Delta G = -22.0$ kJ/mol, $\Delta H = +7.1$ kJ/mol) (54). However, because of strong temperature dependencies, even this partitioning is a question of the temperature used for measuring.

What are the major contributions to the enthalpy and entropy of binding? Direct interactions between the protein and the ligand are obviously very important for the enthalpy of binding. Besides that, an essential factor is that protein–ligand interactions occur in

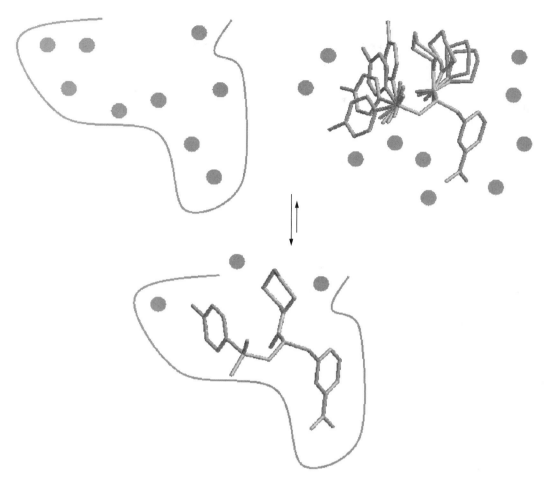

Figure 7.1. Overview of the receptor-ligand binding process. All species involved are solvated by water (symbolized by gray spheres). The binding free energy difference between the bound and unbound state is a sum of enthalpic components (breaking and formation of hydrogen bonds, formation of specific hydrophobic contacts) and entropic components (release of water from hydrophobic surfaces to solvent, loss of conformational mobility of receptor and ligand).

aqueous solution (cf. Fig. 7.1). The unbound reaction partners are solvated and partial desolvation is required before complex formation can occur. This is important for the enthalpy balance because the net energy gain upon complexation can only be the difference between the direct protein–ligand interaction enthalpy and the desolvation enthalpies of the two molecules. In this context, the hydrophobic effect has to be considered again. Upon the formation of lipophilic contacts between apolar parts of the protein and the ligand, unfavorably ordered water molecules are replaced

and released. This leads to an entropy gain that is attributed to the fact that the water molecules are no longer positionally confined. In addition, there is an enthalpic contribution: water molecules occupying lipophilic binding sites are unable to form hydrogen bonds with the protein, but after release they can form strong hydrogen bonds with bulk water. Because the removal of hydrophobic surfaces from contact with water leads to negative changes in the heat capacity (ΔC_p), the buried hydrophobic surface area has frequently been correlated with ΔC_p values measured upon ligand binding. This, however, may be an over-simplification, neglecting other potential contributions to ΔC_p (55). As further noted by Tame, enthalpy-entropy compensation and the temperature dependency of ΔH and $T \Delta S$ (which are both directly related to ΔC_p), make it ultimately impossible to consider polar or apolar contributions as purely enthalpic or entropic, respectively (56).

Entropically unfavorable contributions arise from the loss of translational and rotational degrees of freedom upon complexation, whereas a small gain in entropy can result from low-frequency concerted vibrations in the complex. A more important factor to consider in an actual design process is conformational flexibility. Upon binding, internal degrees of freedom are frozen, the ligand loses a considerable amount of its flexibility, and usually binds in one single orientation. This is also the explanation why rigid analogs of flexible ligands show higher affinity, as, for example, observed for cyclic derivatives of ligands that adopt the same binding mode as the open-chain derivative (57, 58). Accordingly, higher affinity also results if the protein-bound ligand conformation is already preorganized in solution.

From a variety of experiments, quantitative estimates for some of the mentioned energetic contributions to protein-ligand binding could be derived. Based on data from protein mutants, the contribution of individual hydrogen bonds to the binding affinity has been estimated to be 5 ± 2.5 kJ/mol (59–62). This is similar to what has been obtained for the contribution of an intramolecular hydrogen bond to protein stability (63, 64). The consistency of values derived from different proteins suggests some degree of additivity in the hydrogen-bonding interactions. The accurate description of the interplay with water molecules remains, however, a most challenging task. The contribution of hydrogen bonds to the overall affinity strongly depends on local solvation and desolvation effects and can sometimes be very small or even adverse to binding, as illustrated by the comparison of ligand pairs differing by just one hydrogen bond (65). Charge-assisted hydrogen bonds are stronger than neutral ones, but also associated with a higher desolvation penalty.

Thus, the electrostatic interaction of an exposed salt bridge contributes as much as a neutral hydrogen bond (5 ± 1 kJ/mol according to Ref. 66), but the same interaction in the interior of a protein can be significantly stronger (67). Because of the complicated interplay with water, a detailed analysis of the thermodynamics of hydrogen bond formation can sometimes yield surprising results. For a particular hydrogen bond in complexes of the FK506-binding protein, it has been found that its formation is enthalpically unfavorable but entropically favorable (60). The entropy gain appears to be attributable mainly to the replacement of two water molecules (68).

Contributions from hydrophobic interactions have frequently been found to be proportional to the lipophilic surface area buried from solvent, with values in the range of 80–200 J/(mol Å^2) (69–71). The entropic penalty for freezing a single rotatable bond has been estimated to be 1.6–3.6 kJ/mol at 300 K (72, 73); recent estimates derived from NMR shift titrations are much lower (0.5 kJ/mol) (74), but in the systems studied the conformational restriction may not have been as high as in a protein binding site. Finally, the unfavorable entropy contribution from the loss of translational and orientational degrees of freedom has been estimated to be around 10 kJ/mol (75, 76).

Despite many inconsistencies and difficulties in interpretation, most of the experimental data suggest that simple additive models of protein–ligand interactions might be a reasonable starting point for the development of methods to predict binding affinities, that is, for the derivation of empirical scoring functions. Still, it has to be kept in mind that the assumption of additivity in biochemical phenomena is not strictly valid (77). On the other hand, the large body of experimental data on 3D structures of protein–ligand complexes and binding affinities allows one to derive some general characteristics about protein–ligand interactions. Several features are commonly found in complexes of tightly binding ligands:

1. A high steric complementarity between the protein and the ligand, an observation often described as the lock-and-key paradigm (78,

79). This complementarity, however, is frequently not the result of a match between rigid bodies, but rather achieved through significant conformational changes of both binding partners, a phenomenon generally referred to as induced fit. Additionally, electrostatic complementarity can also be induced, for example, by strong pK_a shifts upon ligand binding that result in the release or uptake of protons of different functional groups either of the protein or the ligand.

2. A high complementarity of the surface properties. Lipophilic parts of the ligands are generally in contact with lipophilic parts of the protein, whereas polar groups are usually paired with suitable polar protein groups to form hydrogen bonds or ionic interactions.

3. An energetically favorable conformation of the bound ligand. Significant conformational strain is usually not observed in ligands binding with high affinity.

In addition to insights taken from high-affinity complexes, experimental information about weakly bound complexes could be equally instructive. Such information has indeed been recognized to be vital for the development of scoring functions (80). Structural data on unfavorable protein–ligand interactions, however, are sparse, partly because structures of weakly binding ligands are more difficult to obtain and are usually considered less interesting by many structural biologists. What can be concluded from the available data is that an imperfect steric fit at the lipophilic part of the protein-ligand interface leads to reduced binding affinity and that unpaired buried polar groups at the protein–ligand interface are strongly adverse to binding. Few buried CO and NH groups in folded proteins fail to form hydrogen bonds (81). Therefore, in the ligand design process an important prerequisite to be regarded is that polar functional groups, either of the protein or the ligand, will find suitable counterparts if they become buried on ligand binding.

2.2 Docking, Scoring, and Virtual Screening: The Basic Concepts

The subject of docking is the formation of noncovalent protein–ligand complexes. Given the structures of a ligand and a protein, the task is to predict the structure of the resulting complex. This is the so-called docking problem. Because the native geometry of the complex can generally be assumed to reflect the global minimum of the binding free energy, docking is actually an energy-optimization problem (82), concerned with the search of the lowest free energy binding mode of a ligand within a protein binding site. The macromolecular nature of the protein and the fact that binding occurs in aqueous solution complicate the problem significantly because of the high dimensionality of the configuration space and considerable complexity of the energetics governing the interaction. Accordingly, heuristic approximations are frequently required to render the problem tractable within a reasonable time frame. The development of docking methods is therefore also concerned with making the right assumptions and finding acceptable simplifications that still provide a sufficiently accurate and predictive model for protein–ligand interactions.

Regardless of the nature of the interacting partners, computational docking always requires two components, which may briefly be characterized as "searching" and "scoring" (83). "Searching" refers to the fact that any docking method has to explore the configuration space accessible for the interaction between the two molecules. The goal of this exploration is to find the orientation and conformation of the interacting molecules corresponding to the global minimum of the free energy of binding. Unless the degrees of freedom are restricted to translation and rotation by treating both molecules as rigid bodies, a full systematic search of all "dockings" is normally not feasible because of the huge number of potential solutions and the large amount of computational resources needed to evaluate them. Different strategies are therefore required, which should be accurate and efficient: accurate in the sense that the optimization procedure should not miss any valuable solution (near-global minima), and efficient in terms of computing time and with respect to the fact that the algorithm should not spend unnecessary time by exploring irrelevant regions or by rediscovering previously detected local minima. As will be elaborated in the next section, there are two opposing approaches to simplify the docking problem either by reformulating it to a discrete problem that can be solved with combinatorial algorithms or by using stochastic search algorithms.

"Scoring" refers to the fact that any docking procedure must evaluate and rank the configurations generated by the search process. The scoring scheme most closely related to experiment, the *ab initio* calculation of the free energy of binding, is not easily accessible to computation. Hence, approximate scoring functions must be used that model the binding free energy with sufficient accuracy and correlate well with experimental binding affinities. In particular, the scoring function should be able to discriminate between native and nonnative binding modes.

Scoring is actually composed of three different aspects relevant to docking and design:

1. Ranking of the configurations generated by the docking search for one ligand interacting with a given protein; this aspect is essential to detect the binding mode best approximating the experimentally observed situation.
2. Ranking different ligands with respect to the binding to one protein, that is, prioritizing ligands according to their affinity; this aspect is essential in virtual screening.
3. Ranking one or different ligands with respect to their binding affinity to different proteins; this aspect is essential for the consideration of selectivity and specificity.

If one were able to accurately calculate the free energy of binding, all three aspects would be satisfied simultaneously. Current scoring functions used in docking programs, however, can usually resolve satisfactorily only the first aspect. They provide only a rough estimate with respect to the comparison across different ligand or protein systems. This is the case whenever the scoring scheme neglects certain factors that are virtually constant for different binding modes with respect to one protein, but that matter for comparisons with other proteins.

Following the general paradigm shift in structure-based design from single com-

pounds to compound libraries, state-of-the-art docking and scoring methods have to be sufficiently fast to be applied for virtual screening. The general strategy of a virtual screening process based on the 3D structure of a target typically involves the following steps:

1. Analysis of the 3D protein structure.
2. Selection of key interactions that need to be satisfied by all candidate molecules.
3. Computational search in chemical databases for compounds that potentially satisfy the key interactions, fit into the binding site, and form additional interactions with the protein; this is done by means of docking and/or structure-based pharmacophore searches.
4. Postprocessing by analyzing the retrieved hits and removing undesirable compounds.
5. Synthesis or ordering of the selected compounds.
6. Biological testing, eventually crystallographic confirmation.

All these steps will be discussed in some more detail in section below. Of primary interest in the context of this chapter is step 3. It requires high-throughput docking with efficient search algorithms, and scoring functions that are able to provide a good separation between potentially "binding" and "nonbinding" ligands. The database or library that is screened should consist of a sufficiently large and diverse set of relevant compounds. Thus, library design is increasingly applied to ensure that only reasonably preselected compounds are docked (29, 84, 85).

3 DOCKING

In this section, approaches to the docking problem are presented with respect to the docking algorithm and the search aspect. Scoring is discussed separately in Section 4. It should be noted in this context, that although a specific docking method is frequently associated with a certain scoring procedure, many docking methods could in principle be combined with a variety of different scoring functions, either for postprocessing of the results or as objective function during the optimization. Actually, such strategies are followed by considering multiple scoring schemes to achieve "consensus scoring" (86) or "multidimensional scoring" (87). The emphasis in this section is on general characteristics and principles, rather than individual methods, although occasionally specific docking programs have been selected as representative examples for a more detailed illustration of a general concept. The interested reader is referred to Table 7.1 for an overview of currently used docking programs described in the literature. In addition, a valuable source of information is the corpus of regularly published reviews in the field of docking (18, 19, 26, 27, 83, 88–95).

3.1 General Concepts to Address the Docking Problem

Essential for any docking method is a search algorithm that samples the configuration space of two interacting molecules. These molecules need to be represented in a way that is suitable for efficient handling by the search algorithm. Docking methods may therefore roughly be classified by the way the macromolecular receptor is represented (Section 3.1.1), by the handling of the ligand (Section 3.1.2), and—most important—by the search algorithm itself (Section 3.1.3).

3.1.1 Representation of the Macromolecular Receptor.
The most straightforward approach for representing the macromolecular structure in a docking application would be by atomic coordinates of the entire protein. A full atomic representation, however, is generally impractical because of the size and complexity of protein structures. The structural information therefore needs to be reduced to a manageable yet representative size and form.

A first step into this direction is to limit the search area to the region surrounding the putative binding site. This is general practice in protein-ligand docking (whereas in protein-protein docking often the entire surfaces are searched for appropriate matches). Scanning of the entire surface for potential binding regions of a small-molecule ligand would hardly be feasible with most docking methods. Furthermore, it would be rather unreasonable to

ignore information already available from biochemical experiments or structural data of related complexes. If no such information is available [a situation that we may increasingly be facing as a consequence of the effects of the structural genomics initiatives (17, 96)], methods to identify binding sites are required before the actual docking process can start. Examples are programs for geometric cavity detection, such as LIGSITE (97) or PASS (98), tools to infer protein function from structural homologies (99, 100), or more sophisticated approaches based on a physicochemical and geometrical characterization of binding sites (101). Some docking programs incorporate routines for binding site identification as preprocessing steps (102).

Despite a reduction to only a specified part of the protein surface, a simple representation in terms of atomic coordinates is not practical for most docking procedures. Instead, the space available for ligand binding is frequently characterized by other means that permit more efficient searches. A first alternative is given by geometric shape descriptors, sometimes combined with a physicochemical description. Approaches of this class include molecular surface cubes (103), surface normals at sparse critical points (104), and modified Lee-Richard's dotted surfaces, with each dot coded by chemical property and accessibility (105). A further prominent example is the sphere images of the binding site used in DOCK (106, 107). These spheres are complementary to the molecular surface and represent a space-filling negative image of the binding site. Another important concept that goes beyond a pure geometric description and represents interaction properties of physicochemical relevance is the usage of interaction sites or points, as introduced by the program LUDI (108, 109). These interaction sites are discrete positions and vectors in space serving as dummy representations for atoms capable of forming hydrogen bonds or filling hydrophobic pockets. The docking tool FlexX is based on this concept (110). Also, the program SLIDE (111) and the new approach by Diller and Merz (112) use interaction points for fast docking.

A popular alternative to geometric or physicochemical descriptors is the grid representa-

Table 7.1 Overview of Currently Used Programs for Protein–Ligand Docking

Class of Docking Method[a]	Name of Program	Year Published	Original References	Selected References to Further Developments and Applications
Geometric/combinatorial				
Shape/descriptor matching	DOCK	1982	(106)	(11, 127, 143, 371)
	FLOG	1994	(121)	(366)
	ADAM	1994	(135)	(386)
	LIGIN	1996	(387)	(234)
	SANDOCK	1998	(105)	(13)
	QSDOCK	2000	(136)	
	SLIDE	2000	(111)	
	FRED	2001	(123)	
	(Diller & Merz)	2001	(112)	
Incremental construction	FlexX	1996	(110)	(130, 138, 139, 216, 233)
	Hammerhead	1996	(388)	
	DOCK4.0	1998	(328)	(131)
Systematic search (transl. + rot.)	EUDOC	2001	(125)	(389)
Energy-driven/stochastic				
Monte Carlo simulated annealing	AutoDock	1990	(113)	(95, 115, 390)
	RESEARCH	1992	(146)	(145)
	MCDOCK	1999	(147)	
Monte Carlo minimization	ICM	1994	(116)	(82, 117, 201)
	(Caflisch et al.)	1997	(150)	(151)
	QXP	1997	(152)	
	PRODOCK	1998	(119)	(118)
Molecular dynamics (MD)	MDD	1994	(164)	(165)
	(Luty et al.)	1995	(169)	
	(Vieth et al.)	1998	(166)	
	q-jumping MD	2000	(167)	(168)
Genetic algorithm	GOLD	1995	(176)	(177)
	AutoDock3.0	1998	(115)	(208, 228, 391)
	GAMBLER	1999	(86)	
	DARWIN	2000	(178)	
Tabu search	PRO_LEADS	1998	(188)	(189, 360)
Tabu search + genetic algorithm	SFDock	1999	(392)	
Eigenvector following	Low Mode Search	1999	(211)	
Mining Minima algorithm	Mining Minima	2001	(190)	

[a]The classification provided in the first column can only be approximate for programs that offer a variety of different functionalities or follow multistep strategies.

tion of protein structures. The general principle of this approach is that the protein is represented by a set of affinity grids or maps that cover the entire search region. These reg-ularly spaced, orthogonal grids are calculated before the actual docking process. At every grid point, some sort of scoring value or inter-action energy of a probe atom with the entire protein is calculated, providing a map of pseudo-affinities for each atom type or interaction type possibly present in the ligands to be docked. These maps then serve as look-up tables for the calculation of the interaction energy or scoring value during the docking process. Examples of docking programs using this approach are AutoDock (113–115), ICM (82, 116, 117), or ProDock (118, 119).

It should be noted that most of the mentioned representations of protein structure imply that the protein remains rigid during the docking process. As a matter of fact, docking under the assumption of a rigid protein is still common practice in standard applications. Although an acceptable simplification under certain circumstances, it can represent a serious limitation if only unbound protein structures are available. As a consequence, the inclusion of protein flexibility in the docking process is an active area of research, and a separate section is dedicated to this issue (cf. Section 3.2.1).

3.1.2 Ligand Handling. For the ligand, a complete representation in atomic coordinates is perfectly feasible. Ligand atoms may be used directly for matching with binding site descriptors or in the calculation of interaction energies in the case of energy-driven procedures. The central problem is conformational flexibility. Predicting the binding conformation of a ligand is in fact a major component of the docking problem, given that this conformation can significantly differ from that adopted in other environments.

Two general strategies for ligand handling may be distinguished: whole-molecule approaches and fragment-based methods. In the first case, the ligand is docked as an entire molecule. This is rather straightforward if the ligand is treated as a rigid body and only translational and rotational degrees of freedom are considered. Such rigid docking was common practice in early docking algorithms (106, 120). A straightforward extension to account for flexibility is to separately dock precalculated conformers of a given molecule (variant 1 in Fig. 7.2). Explicit docking of multiple conformers has, for example, been obtained with the FLOG program (121). FLOG deals with conformational flexibility by generating dif-

ferent conformers using distance geometry and docking each conformer in a rigid-body fashion. A similar approach has also been obtained with the DOCK program (122). To avoid redundancy in the docking, a common rigid fragment is identified, which is docked only once for the entire set of pregenerated conformers. The flexible portions of the molecule that determine the different conformations are subsequently scored based on the preplacement of the rigid fragment. Yet other examples for rigid docking of multiple conformers are provided by the programs FRED from OpenEye Scientific Software, which performs a fast exhaustive search over all possible orientations (123), and SYSDOC (124) or EU-DOC (125), which use fast affine transformation to perform systematic searches over the translational and rotational degrees of freedom of the ligand.

Although this multi-conformer docking can be efficient and accurate for molecules with a limited number of discrete, low-energy conformations, it is less suited for larger and highly flexible molecules, simply because the number of possible conformations increases dramatically. Another way of partially accounting for conformational flexibility in whole-molecule rigid-body docking is to subject the initial matches to some kind of optimization that allows for conformational relaxation. This could be done with some standard energy minimization technique (126, 127) or other procedures that resolve clashes of the initial placement by rotation about single bonds, as done, for example, in the docking program SLIDE (111).

A more rigorous treatment of ligand flexibility in whole-molecule docking is performed by sampling ligand conformation space during docking (variant 2 in Fig. 7.2). It normally requires ligand conformational energies to be evaluated besides intermolecular interaction energy. Molecular mechanics force fields are frequently applied for this purpose. Although a more exhaustive sampling of accessible conformations within the binding site is definitely achieved, an obvious disadvantage is the higher computational demand and possibly a reduced efficiency of the algorithm because of lengthy exploration of local minima.

An interesting variant of whole-molecule representations is the use of internal coordi-

Alternative strategies for flexible ligand docking

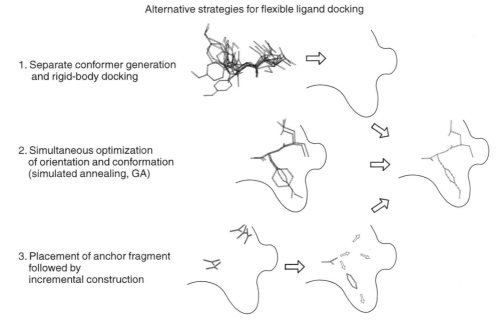

1. Separate conformer generation
   and rigid-body docking

2. Simultaneous optimization
   of orientation and conformation
   (simulated annealing, GA)

3. Placement of anchor fragment
   followed by
   incremental construction

Figure 7.2. Strategies for flexible ligand docking.

nates instead of Cartesian coordinates (82). Internal coordinates help to reduce the number of variables defining the conformation of the molecular system. In Cartesian space, three functionally equivalent variables per atom are required. Internal coordinates, instead, consist of bond lengths, bond angles, and torsion angles. Because bond lengths and angles can be considered rigid to a good approximation, only the torsion angles matter as variables to map conformation space. An efficient implementation of docking algorithms operating on internal coordinates has been obtained, for example, with the ICM method (82, 116, 117).

Fragment-based techniques are an alternative to whole-molecule docking (variant 3 in Fig. 7.2). Here, the molecule is dissected into fragments that can be docked individually in a rigid fashion. The fragments can either be docked separately and then reconnected, or the ligand is built up incrementally following a certain fragmentation scheme. The first variant is very common to programs dedicated to *de novo* design rather than pure docking.

Methods of this class have been reviewed extensively (26, 27). However, the approach has also been applied for docking (128) and compared to the whole-molecule docking approach (129).

The other variant of fragment-based ligand docking is used in incremental construction algorithms (110, 130), sometimes also referred to as "anchor and grow" (131). These search strategies are further described below. They dissect the ligand into modular portions and rebuild it incrementally within the binding site starting from the docking position of a suitable base fragment. The advantage is that many potential combinations are eliminated early in the construction, but success critically depends on the selection and placement of the base fragment.

3.1.3 Strategies for Searching the Configuration and Conformation Space. Search strategies of automated docking procedures may roughly be classified as geometric or combinatorial on the one hand and energy driven or stochastic on the other, although ultimately all methods try to optimize a function that models to some extent the free energy of binding.

3.1.3.1 Geometric/Combinatorial Search Strategies.

Most of the early docking methods were entirely based on the concept of shape complementarity. Until today this is the fundamental idea in most protein-protein docking programs. The observation that protein–ligand complexes frequently show a remarkable shape fit of both binding partners has stimulated the conception of surface or descriptor matching as docking search technique. The molecules are represented by geometric and/or physicochemical descriptors and various alignment procedures are applied to match complementary parts of ligand and protein. An example is the original DOCK method, where the ligand is superimposed onto a negative sphere image of the binding pocket, using a distance matching algorithm followed by least-squares fitting (106, 132). Other examples are the least-squares fitting procedure described by Bacon and Moult to achieve matches between complementary surface patterns (133), or the hierarchical search of geometrically compatible triplets of surface normals on the molecules to be docked, as proposed by Wallqvist and Covell (134). The program ADAM performs a complete combinatorial search over all possible matches between hydrogen bond patterns (135). Recently, a new matching algorithm based on so-called quadratic shape descriptors has been described (QSDock); along with the presentation of their method, the authors also provide an extensive discussion of shape-based docking algorithms (136).

Another recent example of descriptor matching is SLIDE, developed as a tool for ligand database screening by docking (111). The binding site is represented by a template of favorable interaction points onto which ligand atoms are matched during the search. Instead of serving as a purely geometric description, these points address four different types of interactions (hydrogen-bond donor, acceptor, donor/acceptor, or hydrophobic interaction center). The search is then performed such that all triangles of appropriate atoms in the ligand are exhaustively mapped onto triangles of template points with compat-

ible geometry and chemistry. This mapping is used to generate initial placements of molecules in the binding site and followed by a series of steps that refine the initial position, resolve collisions, and consider flexibility of both the ligand and the protein side chains (cf. note on hybrid approaches below). Similarly, the rapid docking approach for library prioritization developed by Diller and Merz (112) is based on rigid-body triplet matching of ligand atoms onto precalculated hot spots; subsequently, pruning is performed to remove any positions with significant steric clash, and the remaining matches are subjected to energy minimization.

Pure descriptor matching is efficient for rigid-body docking only. Flexible docking, in fact, is always faced with the additional problem of a combinatorial explosion of possible conformers depending on the number of rotatable bonds. Systematic searches or explicit consideration of each possible conformation would therefore require enormous computing resources. A popular way to address this problem within the class of geometric/combinatorial docking methods is incremental construction (110, 130, 131, 137). The ligand is dissected into fragments and incrementally reconstructed in the binding site starting from a suitably docked base fragment. To avoid dead-end solutions during construction, multiple placements of the base fragment have to be considered. In addition, it can be useful to perform different fragmentations and hence to use different base fragments as starting points, especially for long and highly flexible molecules. The docking itself, that is, the placement of the base fragment and the attachment of remaining portions, is guided by some descriptor matching procedure.

An example of an incremental construction method is the program FlexX (110, 130, 138, 139). Conformational flexibility is considered using a discrete set of preferred torsion angles about acyclic single bonds, together with multiple conformations for ring systems. These torsion angle preferences are taken from a library compiled from torsional fragments extracted from the Cambridge Structural Database (140). The model of molecular interactions is based on similar rules as implemented in LUDI, originating from a composite crystal-field analysis (141). For each group capable of forming an interaction, a special contact geometry is defined: the group is placed to a center about which an interaction surface is defined, usually as part of a sphere. Two groups form an interaction if the interaction center of one group coincides with the interaction surface of a counter group. To start with the actual docking process, the ligand is fragmented into components by dissecting at all single bonds that are not part of a cycle. Out of these components suitable base fragments are selected. The base fragment is the first portion of the ligand to be placed into the binding site. This is done by superimposing either triples or pairs of interaction centers constructed around the base fragment with triples or pairs of compatible interaction points generated in the binding region. Normally, a large number of initial placements is generated, which is then reduced either by clustering similar solutions or because of clashes with the protein. Next, the incremental construction of the entire ligand is initiated. Starting with the different base placements, the ligand is built up by stepwise linking of the components in compliance with the torsional database. After hooking up additional fragments, new interactions are searched and a scoring function is used to select the best partial solutions, which are expanded in the following step. This is done until the last fragment has been added and placed to result in the complete ligand. The generated ligand positions are finally stored and ranked according to the predicted binding affinity.

An anchor-and-grow algorithm has recently also been incorporated into DOCK (131). Here, after identification of rotatable bonds, the ligand is fragmented into rigid segments, the largest segment is identified as the anchor, and the remaining segments are organized as layers surrounding the anchor. Then the anchor is docked using geometrical matching. Based on the obtained anchor positions, the conformational search is initiated by adding segments from the innermost layer and proceeding outward. This addition is done according to the accessible torsion angle values along the newly added bond. The default is to use two alternative settings for bonds between two sp^2 hybridized atoms, three between two

sp^3 atoms and six between sp^2 and sp^3 atoms. The partial constructs are then locally optimized to minimize the sum of intra- and intermolecular energies and pruned back to an approximately constant size of configurations. Pruning is necessary to cope with combinatorial explosion. It is performed on the basis of the score and the orientation, such that both the best scoring and most deviating orientations are retained from each expansion cycle. Finally, after complete reconstruction of the ligand, the pruned set of binding configurations is again subjected to local energy minimization.

This anchor-and-grow implementation in DOCK represents a combination of a geometric and energy-based approach to docking, due to the intermediate steps of energy minimization. As already encountered for SLIDE (111), such multistep or hybrid approaches are commonly found in current docking protocols. DOCK in general is a prototype of such a program, originally based solely on rigid geometric descriptor matching, later enhanced with a variety of additional features. For example, some degree of flexibility has been introduced into the rigid docking procedure by dissecting the ligand into a small set of rigid fragments that are docked separately and then reconnected (128). The concept of geometric shape complementarity has been extended to consider physicochemical complementarity by assigning properties to binding-site spheres and allowing them to match only those ligand atoms that are of complementary character, an approach referred to as "sphere coloring" (142, 143). Rigid-body minimization has been introduced as refinement after the initial descriptor-matching step (126) or in the variant of on-the-fly optimization using force-field energies precomputed on a grid (127). In summary, the combination of different approaches and algorithms to overcome the limitations of every single approach has provided us with steadily improving solutions to the docking problem.

3.1.3.2 Energy Driven/Stochastic Procedures. As mentioned above, docking is essentially an energy optimization problem because the native binding mode of a ligand can in general be expected to correspond to the global minimum of the binding free energy (82). Accordingly, finding this binding mode by docking corresponds to the identification of the global minimum of the free-energy function. Because the actual free energy of binding is not accessible to computation, approximate energy evaluations or scoring functions are used to guide the search. These functions are required to model the free-energy surface in an appropriate way: although the absolute values are not of relevance for the structural aspect of docking, it is essential that the global minimum of a relative free-energy function models accurately enough the position of the global minimum on the real free energy surface. (It is worth mentioning in this context that in purely geometrical or descriptor-based docking procedures, the central assumption is that the degree of surface complementarity or matching between descriptors is proportional to the interaction energy.)

With a suitable energy function available, docking can be performed by global minimization of the energy with respect to the position, orientation, and conformation of the ligand. However, this apparently straightforward approach bears two fundamental problems, inherently related to characteristics of the energy landscape of protein–ligand interactions: the high dimensionality, which precludes a systematic, exhaustive search; and the ruggedness of the surface, reflected by a large number of local minima. Because of this last aspect, standard energy minimization techniques alone are not useful for docking applications because they can guide the search only to the next local minimum. They *are* used, however, in combination with other techniques and play a valuable role at certain stages of the docking process, primarily to refine docked positions and conformations by exploring the local energy landscape in the vicinity of this position.

To address the docking problem, techniques for a more global exploration of the energy landscape are required. A variety of methods is available, frequently used in the context of other modeling applications and optimization problems as well. Three major classes may be distinguished: Monte Carlo techniques, molecular dynamics simulations, and genetic algorithms. Many different variants exist for all of them and frequently, in

docking procedures, they are applied in combination with other techniques.

Monte Carlo methods consist of two essential components that are repetitively applied: a random walk of the ligand through the receptor-near space (i.e., the random displacement along translational, rotational, and/or torsional degrees of freedom), and the evaluation of the new configuration based on the Metropolis criterion (144). This criterion decides whether a new position is accepted and hence on the configuration from where the search will proceed. If the energy of the new docked position (E_{new}) is more favorable (lower) than the energy of the previous position (E_{old}), the new position is accepted. If it is less favorable, the probability P for its acceptance is given by

$$P = \exp[-(E_{new} - E_{old})/kT]$$

where k is the Boltzmann constant and T is the effective temperature. To turn this sampling technique into an efficient optimization method applicable to docking, it has to be combined either with a temperature lowering protocol or with some local minimization steps. The former approach is known as Monte Carlo simulated annealing, the latter as Monte Carlo minimization.

In simulated annealing, the effective temperature T is initially set to a high value and gradually lowered, after a predefined number of Monte Carlo steps has been performed at a given temperature. At high temperatures, a broad region of configuration space is sampled: energy barriers can be surmounted because of the high acceptance probability for less favorable placements. As the temperature is lowered, this becomes less probable and the configuration is optimized more locally. Given the stochastic nature of the process, multiple independent runs are required to assess convergence (this equally applies to many of the methods further described below). Examples of docking programs using Monte Carlo simulated annealing as a search strategy are AutoDock (113–115), RESEARCH (145, 146), and MCDOCK (147).

In Monte Carlo minimization, an additional step is inserted after the random walk before Metropolis evaluation. This step is a local energy minimization, using techniques such as steepest descent or conjugate gradient. Full local minimization after each random-walk step has been reported to improve the efficiency of the procedure (148, 149). A docking procedure that uses global Monte Carlo minimization is the ICM program of Totrov and Abagyan (82, 116, 117). ICM describes both the relative positions of two molecules and their conformations by a uniform set of internal variables and uses precalculated grids of the interaction energies to speed up calculations. Trosset and Scheraga use Monte Carlo minimization in their ProDock program; computational efficiency is enhanced by a grid-based energy evaluation using Bezier splines, which enables one to evaluate gradients and hence to perform minimization on a 3D grid (118, 119). Further Monte Carlo minimization docking procedures have been reported by Caflisch et al. (150, 151). Also, the QXP program of McMartin and Bohacek relies on Monte Carlo techniques combined with energy-minimization procedures (152).

Molecular dynamics (MD) simulations represent another technique to sample configuration space (31, 153–157). Based on Newton's equation of motion and principles of statistical thermodynamics, the standard application of this technique is to analyze flexibility and dynamic properties of molecular systems and to calculate free energies in a theoretically rigorous manner (158–163). With respect to protein-ligand docking, MD simulations could in principle be used to simulate the actual binding process, thus providing a "realistic" view of how the docking process proceeds, although this is computationally still out of reach. In fact, standard MD requires massive computational resources, which limits its application to a small number of selected systems. In the context of docking, the problem is that standard MD is slow in exploring global features (crossing of large barriers and exploration of multiple binding sites); accordingly, MD is essentially limited to the simulation and refinement of already bound complexes. Di Nola et al. have addressed this problem in their MDD (MD docking) algorithm (164, 165). This method separates the ligand's center of mass motion from its internal motions. A separate coupling to different thermal baths for both types of motion of the ligand and the receptor is performed. Because the temperature and the time constants of coupling to the baths can be varied arbitrarily, it is possible to increase the kinetic energy of the center of mass of the ligand without increasing the temperature of the internal motions of receptor and ligand. This allows for complete control of the search rate. The technique was applied to the docking of phosphocholine to antibody McPC603, starting from distinct positions well separated from the actual binding site. After appropriate sampling, the average structure of the complex in the binding region was found to closely resemble the crystal structure. Still, the method remains computationally expensive, and thus it is not yet suited for a large-scale application to practical drug design docking problems.

Other docking applications of MD have been reported as well. In a comparison of a CHARMM-based MD docking algorithm with a Monte Carlo and a genetic algorithm, Vieth et al. have observed a comparatively good performance of the MD search for the five analyzed test cases (166). Pak et al. have recently presented a docking approach based on so-called q-jumping MD (167, 168); its basic idea is to apply a smoothed generalized effective potential to enhance conformational sampling by MD. Luty et al. have combined a grid representation for the bulk portion of the receptor with MD simulations of the ligand in the flexible binding site (169). Multiple-copy simultaneous search methods (MCSS) can help to speed up energy-based searches. They use numerous ligand copies that are transparent to each other, but subject to the full force of the protein (170, 171). Finally, short MD simulations are occasionally used at some stage of a docking procedure, primarily with the purpose of local refinement, as for example in the multistep docking strategy of Wang et al., where the last step is an MD-based simulated annealing (129).

The third major class of search methods are genetic algorithms (GAs), which are widely used for docking purposes. GAs are stochastic optimization methods inspired by the concepts of evolution (172–174). The optimization problem is generally formulated in the language of genetics. Initially, a random population is generated in which each member corresponds to a potential solution of the problem. A member of the population is represented by its chromosome, in which the variables to be optimized are encoded. This means that each chromosome contains a number of genes, where the genes correspond to the value of a certain variable or set of variables. In the case of docking, the variables for translation and rotation, as well as the torsion angles of the ligand, are encoded in the chromosome. Genetic operators are then applied to the initial population to generate a new population. In general, these operators are "crossover," by which genes from two distinct chromosomes are interchanged to generate two new individuals, and "mutation," by which a given gene is randomly modified. For each newly generated individual the chromosome is decoded (genotype → phenotype) and the fitness of the individual is evaluated. In the context of docking, this fitness is the interaction energy or docking score. Individuals with better scores receive a higher chance for being selected as members of the new population, and thus a higher chance of survival and reproduction into the next generation. Accordingly, the average fitness increases from generation to generation, until, at some point, the process is terminated (by reaching either a fixed number of generations or a constant fitness of the population). The best individual of this final population represents the solution.

Many different variants and implementations of GAs for docking exist, but the general features are always similar. The application of GAs in drug design and docking has been reviewed by Clark et al. (175). A prominent example of a docking program based on a GA is GOLD (176, 177). A special characteristic of GOLD is the direct encoding of hydrogen bonding motifs in the chromosome representation. Upon chromosome decoding, a least-squares fit is used to optimize the overlap of complementary pairs of hydrogen-bonding sites present in the ligand and the receptor. The newest version of AutoDock contains an interesting variant of a GA, a so-called Lamarckian GA (115). This is the combination of a traditional GA with a local search method to perform energy minimization. At each gen-eration, a user-defined fraction of the population is subjected to such a local minimization. This hybrid algorithm was found to be more efficient than a traditional GA, also implemented in AutoDock. A conceptually similar strategy has recently been implemented into the docking program DARWIN (178). Here, a standard GA is combined with a gradient minimization search strategy through an interface to the CHARMM molecular mechanics program (179). Further GA-based docking methods can be found in the literature (86, 180–183).

Another class of evolutionary algorithms that has occasionally found application in the context of docking is known as evolutionary programming (184). Its main difference with respect to GAs is that there is no recombination (crossover) operator, such that evolution is wholly dependent on mutation. Gehlhaar et al. (185) and Westhead et al. (186) have demonstrated the applicability of evolutionary programming to the docking problem, although in a comparative study other algorithms were found to be more effective (186). A new variant called "family competition evolutionary algorithm" has recently been proposed for docking (187).

Besides the three major classes of energy-driven searches (MD, MC, GA), some further heuristic algorithms and search strategies have been developed or adapted for the docking problem. "Tabu search" was found to perform well in comparison with other algorithms (186) and has thus become the main search strategy of the PRO_LEADS docking program (188, 189). Briefly, the tabu search operates on randomly generated positions that are examined on the basis of a tabu list. This list contains a number of previously generated solutions and serves to impose restrictions on the search process: a random move of the ligand is considered "tabu" if it generates a solution that is not sufficiently different from the stored solutions, unless its energy is more favorable than the energy of the best solution so far. Using these restrictions, the search is prevented from revisiting regions of the search space and the exploration of new areas is encouraged. Ideas from tabu search are also used in the recently described adaptation of the Mining Minima algorithm for protein–

ligand docking (190). Here, an exclusion zone is placed around each energy minimum as it is discovered, to avoid rediscovering it in future docking iterations. Mining Minima itself is based on a variety of optimization techniques to gradually focus a large region of random search to areas around the lowest energy minima.

3.2 Special Aspects of Docking

Besides the general characteristics outlined above, there are a number of special issues associated with the docking methodology that deserve explicit consideration: protein flexibility, water molecules, and objective assessment. In addition, the interplay of docking with QSAR methods and homology modeling is of further interest to highlight the possibilities opened by combined application of standard methods in structure-based drug design.

3.2.1 Protein Flexibility. Proteins are inherently dynamic systems (153, 191). A single, fixed conformation, even the average provided by a crystal structure, may not be an adequate representation of the protein, unless the system is very rigid (192). Instead, even under standard equilibrium conditions, the native folded state of a protein is best characterized by a collection or ensemble of energetically nearly equivalent conformations. If the conditions are changed, the local minima and the population of these states may shift, eventually resulting in an observable change of the average structure. Also, the introduction of a ligand corresponds to a change of the environment that may lead to similar effects. Accordingly, the binding conformation of the receptor may already be present in the ensemble of protein conformations (193, 194) and the ligand does not actively deform a fixed state of the protein, as generally inferred from the "induced fit" model.

Whatever the actual mechanism might be, the comparison of experimental protein structures in the ligand-free and in the complexed state frequently shows protein conformational changes induced by or associated with ligand binding (195). The spectrum of phenomena ranges from side-chain rotations to loop rearrangements and the movement of entire domains. Accordingly, in the context of docking it is frequently not justified to neglect protein flexibility (35). If no alternative for docking into the rigid protein is available, at least a protein conformation (possibly from a complex structure) should be used that is compatible with suitable binding modes. Obviously, a preferable docking tool would consider full protein flexibility, but appropriate realization of this goal remains a challenge because of the high dimensionality of protein conformation space. Consideration of protein flexibility also complicates the problem of scoring and selecting the best ligand placement, given the difficulty in accurately evaluating protein conformational free energies in addition to ligand-binding free energies.

Current approaches to the problem of flexible protein docking have recently been reviewed by Carlson and McCammon (196), and more briefly by Abagyan and Totrov (18) and Claussen et al. (197). The methods differ by the degree of flexibility they can cover. The least complex methods are those that model small adjustments of contact residues and side chains in an implicit way using soft docking. The protein itself remains fixed, but either through an adapted geometric representation or using a tolerant scoring function a certain amount of overlap between the protein and the ligand is allowed, emulating some "plasticity" of the receptor. The docking program by Jiang and Kim based on the matching of molecular surface cubes is explicitly based on this soft docking idea (103). Other more recent docking approaches have implemented a soft scoring function (198). The advantage of these simple approaches is that they do not increase the demands on computing time.

The next level is represented by methods that allow for explicit side-chain flexibility. GOLD's genetic algorithm can handle the rotation of a few terminal hydrogen-bond donor and acceptor groups to optimize the hydrogen-bonding network (176, 177). A technique to handle larger side-chain movements is the use of side-chain rotamer libraries, as first demonstrated by Leach. In this approach, heuristic algorithms such as dead-end elimination are used to search the large combinatorial space (199). Schaffer and Verkhivker instead use a rotamer library to first generate likely side-chain conformations, which are then subjected to energy minimization together with the docked ligand (200). Another approach making use of minimization has been described by Apostolakis et al.: after "seeding" the receptor with randomly generated ligand positions that may overlap with the protein, the complex is subjected to minimization, during which nonbonded interactions are gradually switched on, to gently relieve steric overlap by minor conformational changes of the ligand and receptor. The best-ranked solutions are then subjected to further refinement by Monte Carlo minimization (151). Furthermore, the Monte Carlo minimization technique in internal coordinates of the ICM program can sample and optimize side-chain torsions during ligand docking (117, 201). Finally, the docking tool SLIDE allows for some side-chain flexibility at the optimization stage of initial placements. In SLIDE, collisions are resolved by rotations about single bonds in the ligand and the protein side-chains to reduce a maximal number of collisions by minimal conformational changes of both binding partners (111, 202).

An alternative to account, in principle, for an arbitrary degree of protein flexibility is the use of protein structure ensembles. The ensembles could be assembled from multiple crystal structures of a given protein, from NMR structure determination, or from trajectories of molecular dynamics simulations. In addition, a rotamer library can be used to create a minimal set of new conformations (203). Whatever the origin of the individual members of the ensemble, each represents a distinct conformational state of the protein, and may eventually correspond to the preferred ligand-binding state. Three different ways to use protein ensembles for docking can be distinguished: in its most straightforward form, docking is carried out sequentially with each member of the ensemble using rigid-receptor docking (124, 204–206). Another way is to use a weighted-average representation of the ensemble. Knegtel et al. followed this approach by generating composite grids that were used for scoring within the DOCK program (207). Recently, it has also been tested with AutoDock (208). Broughton has developed another method by combining statistical analysis of a conformational ensemble from short

MD simulations with grid-based docking protocols (209). The third and most sophisticated approach to handle protein ensembles is implemented into FlexE, a variant of the FlexX program (197). FlexE is based on a united protein description generated from the superimposed structures of the ensemble. For the parts that differ among the protein structures, discrete alternative conformations are explicitly taken into account on the fly during the incremental construction of the ligand in the binding site. As an important feature, these geometric alternatives are optimally joined to create new valid protein structures in a combinatorial fashion. Thus, conformations of the protein are not limited to those explicitly present in the ensemble, nor are the interactions blurred by averaging over distinct alternative instances, which may correspond to unrealistic protein conformations.

The so-called Low Mode Search (LMOD), originally established as a method for conformational analysis (210), has recently been demonstrated to be applicable also to the problem of docking flexible ligands into flexible protein binding sites (211). To explore the potential energy surface of molecules, LMOD is based on eigenvector following, where eigenvectors correspond to the (low-frequency) "normal modes" of vibration. For the purpose of docking, LMOD has been combined with a limited torsional Monte Carlo movement, as well as random translation and rotation of the ligand.

Generally, however, full consideration of flexibility, either of the binding site or the entire protein, remains the domain of MD simulations. The disadvantage is their high computational demand required to achieve significant sampling. Simplified MD restricted to the binding site has been used by Luty et al., where the bulk of the protein receptor is represented as a grid, whereas a full atomic description is used only for the proximity of the binding site to include flexibility in the docking process (169). The approach of Mangoni et al. mentioned above provides a method for enhanced sampling. It has been used to dock a ligand into a receptor that is treated fully flexible and solvated with explicit water molecules (165). Alternatively, shorter MD runs may be used at intermediate or final stages of a docking procedure to refine complexes generated by rigid-body docking methods. In this case, however, flexibility is not considered simultaneously to the docking process. It thus only refines solutions from rigid receptor docking and does not enhance the scope of the search for possible binding modes.

3.2.2 Water Molecules. Water plays a crucial role in molecular interactions (212, 213). At the interface of a protein–ligand complex, water molecules can have a significant impact on complex formation, either by mediating or improving specificity and affinity of the interaction. They promote adaptability, thus allowing for promiscuous binding (214). Individual conserved ("structural") water molecules can therefore be crucial for the successful design of new inhibitors. A prominent example is the structural water molecule observed in nearly all HIV protease complexes with substrate-like inhibitors. Attempts at replacing it have guided the design of new tight-binding inhibitors [e.g., (215)]. Instead of the usual implicit modeling of solvation effects, explicit consideration of structural water molecules and water-mediated interactions would therefore be a highly desirable feature in docking methods. Ideally, simultaneously to the ligand placement the docking program should be able to predict whether at a particular site water molecules mediating protein–ligand interactions may preferably reside or whether the displacement of these water molecules by appropriate ligand functional groups would be more favorable. No docking tool is yet available to accomplish this task. Obviously, not only the placement of water molecules is demanding, but especially their energy scoring, resulting from the complicated thermodynamics associated with water interactions.

In principle, MD simulations provide the most natural route to the explicit consideration of water molecules. In the MD docking approach described by Mangoni et al., explicit water molecules are indeed used (165). It was found, though, that the presence of explicit water molecules shields the interactions between the ligand and the receptor. Consequently, different weights were applied to the ligand–receptor and ligand–solvent interactions, respectively, to cope with this complica-

tion. Because of the high computational costs, the approach seems affordable only in special cases where the presence of explicit solvent appears important.

An approach to explicitly place water molecules during fast docking has been introduced into FlexX (216). In a preprocessing phase, possible favorable water sites in the binding pocket are calculated and stored. During the incremental construction phase of FlexX, water molecules are switched on at these sites if they provide additional hydrogen bonds to the ligand. Steric constraints produced by these water molecules and the quality of the achieved hydrogen bond geometry are then used to optimize the ligand orientation during the construction process. In several cases, water molecules between protein and ligand could be correctly predicted; however, the overall improvement on the FlexX docking results for a test set of 200 complexes was nearly negligible.

The program SLIDE can consider tightly bound waters while docking potential ligands (111). To select which water molecules to retain and which to remove from the binding pocket before docking, the knowledge-based approach Consolv (217) is applied to determine those waters that are likely to be conserved upon ligand binding and to adjust a penalty for their displacement. Once these waters have been selected to be initially retained upon docking, SLIDE either translates or discards a water molecule to remove overlap with ligand atoms after the ligand has been docked to the binding site. Displacement of a water molecule is performed only if collisions cannot be resolved by iterative translations. Any displacements by nonpolar ligand atoms are penalized upon scoring. In database screening runs on three different target proteins, this procedure was found to produce reasonable results with respect to water-mediated interactions, but no systematic test has been reported so far.

As long as a simultaneous docking of water molecules and ligands is an unsolved problem, it remains common practice to consider essential water molecules as a fixed part of the binding site. Preplaced water molecules may either correspond to recurrently observed waters found in multiple crystal structures of the target protein, or to predicted positions based on estimated water affinity potentials suggested by programs such as GRID (218–220). The latter strategy has been applied by Minke et al. using AutoDock (221), showing that successful docking of carbohydrate derivatives to the heat-labile enterotoxin critically depends on the inclusion of water molecules. Examples for the consideration of experimentally observed water molecules as part of the target during docking are the studies of Rao et al. (docking to factor Xa using AutoDock) (222) and Pospisil (docking to thymidine kinase using AutoDock and FlexX) (223). The influence of explicit water molecules in docking was also investigated in the validation study of the new program DARWIN (178). Inclusion of explicit water molecules was essential in some cases, unless interaction energies were calculated with a Poisson-Boltzmann-based implicit solvent model. Yet another example is a search for metallo-β-lactamase inhibitors (14) with the docking program FLOG. Docking was performed with three different configurations of bound water in the active site. The top-scoring compounds showed an enrichment in biphenyl tetrazoles. A crystal structure of one tetrazole not only confirmed the predicted binding mode but also displayed the water configuration that had, retrospectively, been the most predictive one of the three models. Further examples from virtual screening studies are available that show that the inclusion of conserved water molecules in the docking process can dramatically improve the hit rate (15, 16).

3.2.3 Assessment of Docking Methods. Docking methods are usually assessed by their ability to reproduce the binding mode of experimentally resolved protein–ligand complexes: the ligand is removed from the complex, a search area is defined around the actual binding site, the ligand is redocked into the protein, and the achieved binding mode is compared with the experimental position, usually in terms of a root-mean-square deviation (rmsd). If the rmsd is below 2 Å, it is generally considered a successful prediction. The obvious goal is that such a "near-native" solution is ranked best among the set of ligand poses generated. Virtually any introduction of a new docking method has been accompanied

by such a test. The number of complexes used has varied as much as the reported success rates, which are between 10% (224) and 100% (152). Clearly, success rates of 100% are rather a consequence of the limited test set size than a reflection of the mere quality of the docking method.

Numerous critical issues have to be addressed in this context. Validations carried out on very few complexes (≤20) do not adequately assess the scope of the method, particularly if no attempt was made to select a representative set of structures that appropriately covers a broad range of binding features important to protein–ligand complexes. Up to now, only a few docking methods have been assessed on a broad range of complexes [e.g., FlexX (200 complexes) (139), ScoreDock and DOCK (200 complexes) (225), EUDOC (154 complexes) (125), DOCK, FlexX, and DrugScore (100–150 complexes) (226), GOLD (100 complexes) (177), the method of Diller and Merz (using the GOLD test set) (112), and PRO_LEADS (70 complexes) (189)]. In the case of GOLD it has been explicitly mentioned that the test set was selected by a researcher not involved in the development of the algorithm (177). The definition of an objective and relevant reference test set that could serve as standard benchmark for every new docking method would be highly desirable for both user and developer (18). First efforts in developing a database that could be of use in this context have been reported (227). Suitable test sets should cover a sufficient number of highly diverse protein–ligand complexes, including cases that provide some challenge to docking methods (e.g., water-mediated interactions, interactions with metal ions). To test performance with respect to potential induced fit, the structure of the unligated protein or alternative complexes with different bound ligands should be available as well. The test set should comprise fully resolved crystal structures with a resolution of ≤2.5 Å. Complexes with ligands significantly involved in crystal packing contacts should be avoided. Such cases will likely fail in reproducing the experimental binding mode because of missing contacts present only in the packing environment (228). Finally, the importance to study low-affinity or "non-binding" ligands must be addressed; accordingly, experimental information about the binding geometry and affinity of some weak-binding ligands should also be available.

In addition to the tests usually reported by the authors of a program, comparative studies have been reported on the assessment of different docking and scoring approaches. In part they also address some of the aspects raised above. Westhead et al. have presented a comparison of four heuristic search algorithms (simulated annealing, genetic algorithm, evolutionary programming, and tabu search) (186). In an attempt to provide an unbiased comparison, all algorithms were implemented into the PRO_LEADS program and a single scoring function was used. Other recent examples are the studies of Ha et al., who compared DOCK (using two different scoring functions) and FlexX (229), and, in the context of virtual screening, the work of Bissantz et al., who compared DOCK, FlexX, and GOLD together with seven different scoring functions (230) (cf. also Section 5.2 below).

An unbiased test scenario is guaranteed if researchers are provided with a set of protein–ligand complexes of experimentally resolved, but yet unpublished structure. Two such blind trial competitions have been carried out so far (231, 232). A series of interesting issues regarding docking tests and problems with true predictions have been amply discussed by Dixon (231) and participants in the CASP2 docking competition (117, 145, 233, 234). Unfortunately, the number of targets subjected to such blind tests has so far been rather scarce. A major limitation to such blind comparisons is the availability of experimental data before publication.

3.2.4 Docking and QSAR. As long as the problem of accurate binding free energy prediction on the basis of a given complex geometry has not been resolved (cf. section on scoring functions), computational methods establishing quantitative structure-activity relationships (QSARs) to estimate relative binding affinity differences within a set of ligands remain a pragmatic alternative. Both classical and 3D QSAR methods have been developed as ligand-based approaches (235–237). They rely exclusively on ligand information and try to correlate experimental binding data with features described by a set of relevant descriptors. In 3D QSAR, such as CoMFA (Comparative Molecular Field Analysis), these descriptors are essentially virtual interaction energies (van der Waals and coulombic), calculated using an appropriate probe atom placed at the intersections of a regularly spaced grid surrounding the molecules. The model derived from differences in the various interaction fields provides a quantitative spatial description of those molecular properties that matter for binding. They can be interpreted as a surrogate representation of the binding site. Essential for the success of all 3D QSAR approaches is an appropriate alignment of the ligands: their relative spatial superposition must reflect the differences in binding geometry also experienced at the binding site of the structurally unknown protein. Various strategies have been developed to achieve this goal (235, 236). Increasingly, however, these methods are also applied if the receptor structure is known. This results in "receptor-based 3D QSAR," a combination of a ligand-based QSAR approach with information extracted from receptor structures (238). This additional information is used to generate a ligand alignment based on the experimental or predicted binding mode of the ligands in the binding site. The standard 3D QSAR techniques are subsequently used to derive a correlation model and to ultimately predict the binding affinity of new, appropriately aligned ligands (239). As a practical advantage, receptor-based 3D QSAR provides important information as to which of the protein–ligand interactions are responsible for the variance in biological activity among the given set of ligands.

Obviously, in the case of known receptor structure, the ligand alignment can be obtained by docking. This strategy has indeed been followed in a variety of studies: it has been used to set up CoMFA models [e.g., (240)] or extended to the Comparative Binding Energy (COMBINE) analysis (241–244), that explicitly exploits receptor information to generate the QSAR descriptors. Furthermore, in a GRID/GOLPE (245) analysis, the model generated with the docking alignment has been compared to the traditional CoMFA model based on ligand alignment (238, 246); the alignment generated by docking could be shown to exhibit higher relevance.

Another concept to combine docking with QSAR has recently been proposed by Vieth and Cummins in their DoMCoSAR approach (247). DoMCoSAR is used to statistically determine the docking mode that is consistent with a structure-activity relationship, based on the explicit assumption that all molecules exhibit the same binding mode. In a first step, all molecules of a chemical series with common substructure are docked in an unbiased way to the protein binding site and the results are clustered to establish the most favorable docking modes for the common substructure. Subsequently, constrained docking is performed by forcing all molecules to align with the common substructure in the major docking modes. In a final stage, interaction-energy–based descriptors are calculated for all major docking modes. QSAR models are then derived to determine the statistically significant and most predictive set of descriptors and thus the docking mode that is most consistent with a given structure-activity relationship. As noted by the authors, the appeal of this method is that an objective statistical justification for the selection of a binding mode is obtained. This may especially prove useful in cases where the primary docking scores yield nearly degenerate multiple binding modes and a selection of the most representative result is difficult. However, because one alignment is rendered prominent among others for the sake of best agreement with the derived QSAR model, the danger exists that unconsidered or ill-defined descriptors in the QSAR could possibly distort the final or accepted alignment.

3.2.5 Docking and Homology Modeling. In the absence of an experimental protein structure, a homology model may be used for docking and structure-based design. Such a model can be generated by comparative modeling based on homologous proteins of known structure. Obviously, it is most reliable in the regions of highest homology between the templates and the target protein. Although an overall skeleton of the target protein can frequently be obtained with sufficient accuracy, the structural details of the binding site are often beyond the scope of the method. In fact, members of a homologous protein family may show considerable differences in the binding region. Accordingly, homology models may not be sufficiently accurate to apply standard docking tools, and special methods addressing the docking of ligands to low-resolution structures have been presented (248).

Clearly, flexible-receptor docking could help to alleviate the problem. A frequently followed alternative is to refine the initial complex between the protein model and the ligand, most commonly by relaxation with MD simulations (249–251). This may also be combined with free energy calculations to determine the binding mode most consistent with experimental affinity data (252). However, refinement does not overcome the problem that the initial conformation of the model may preclude the binding of certain ligands. This has, for example, been demonstrated by Schapira et al. in a virtual screening for retinoic acid receptor (RAR) antagonists based on an RAR homology model (201). The automatic selection procedure based on flexible ligand docking was followed by optimization of the selected candidates with flexible protein side chains using the ICM program (82, 116, 117). Nevertheless, some known ligands were repeatedly missed by the screening algorithm because of incompatible binding site conformations. Consideration of side-chain flexibility already in the initial docking simulation was required to accommodate these ligands.

An approach developed especially for the purpose of docking ligands into approximate protein models generated by homology modeling is the DragHome method (253). The binding site is analyzed in terms of putative ligand interaction sites and translated using Gaussian functions into a functional binding-site description represented by physicochemical properties. Similarly, ligands are translated into a description based on Gaussian functions and the docking is computed by optimizing the overlap between the two functional descriptions. The use of "soft" Gaussian functions to describe protein–ligand interactions is one possibility to take into account the limited accuracy of modeled structures for the purpose of docking. The method for generating and optimizing ligand orientations relative to the binding-site representation was adapted from the ligand alignment program SEAL (254–256). For a set of different ligands, the generated solutions are analyzed with respect to the mutual ligand alignment. This alignment is then used to generate 3D QSAR models, which in turn can be interpreted with respect to the surrounding protein model. This can highlight inconsistencies and deficiencies present in the model, and thus information which in future developments of the methods is planned to be fed back into a subsequent modeling step to improve the protein model. The idea behind this is that the cycle of docking and alignment, ligand data analysis (3D QSAR), and protein structure modeling should be repeated until self-consistency is achieved. This would provide a protein homology model optimized with respect to the binding site and suitable to obtain consistent docking results.

4   SCORING FUNCTIONS

This section is dedicated to the scoring aspect of the docking problem. Various approaches are discussed that try to capture the essential elements of protein–ligand interactions in computationally efficient scoring functions. The discussion focuses on general approaches rather than individual functions. The reader is referred to Table 7.2 for original references to the most important scoring functions.

4.1   Description of Scoring Functions for Protein–Ligand Interactions

Reversible protein-ligand binding is an equilibrium between the bound state and the unbound state of the binding partners. The rigorous theoretical description requires full consideration of all species involved: the separate solvated protein, the separate solvated ligand, and the solvated complex, in which the binding partners are partially desolvated and form interactions with each other. The quantity of interest to characterize this equilibrium is the free energy of binding. Its most accurate calculations are based on the evaluation of ensemble averages according to principles of statistical mechanics (45). To obtain reasonably

accurate values of binding free energies, extensive Monte Carlo or MD simulations are necessary, which require large computational resources. Clearly, this is impractical for standard docking applications. Furthermore, even the most advanced techniques are reliable only for calculating binding free energy differences between closely related ligands (162, 163, 257, 258). However, some less rigorous, but faster and, as experience shows, often not less accurate methods have been developed, that are suitable to handle larger numbers of ligands. For example, continuum solvation models are used to replace explicit solvent molecules at least in the final energy evaluation of the simulation trajectory (259), or linear response theory is applied (260–262), sometimes augmented by a surface term (263).

Scoring functions that can be evaluated fast enough to be applied in docking and virtual screening can only estimate the free energy of binding. They usually take into account only one possible configuration of the receptor-ligand complex and disregard ensemble averaging and explicit properties of the unbound states of the binding partners. Furthermore, all methods share the assumption that the free energy can be decomposed into a sum of terms (additivity). In a strict physical sense, this is not allowed, given that the free energy of binding is a state function, although its components are not (77, 264). In addition, simple additive models cannot describe subtle cooperativity effects (265). Nevertheless, it is often useful to interpret receptor-ligand binding in an additive fashion (266–268), and estimates of binding free energy based on the additivity assumption are often accessible at very low computational cost.

Three main classes of fast scoring functions can be distinguished: force field–based methods, empirical scoring functions, and knowledge-based methods. The following sections are dedicated to a separate discussion of each method.

4.1.1   Force Field–Based Methods. An obvious idea to circumvent parameterization efforts for scoring is to use nonbonded energies of existing, well-established molecular mechanics force fields for the estimation of bind-

Table 7.2 Overview of Currently Used Scoring Functions

Type of Function	Name of Function	Year Published	Original References	Selected References to Applications
Force field	Charmm	1998	(274)	
Force field + desolvation	(Schapira, Abagyan et al.)	1999	(280)	
	AMBER + desolvation	1999	(276)	
	Charmm + PB	1999	(393)	
	AMBER + desolvation	1999	(278)	
	MM PB/SA	1999	(343)	(344, 346)
Linear response	LIE	1994	(260)	(261, 263, 394)
Simplified free-energy perturbation	OWFEG Grid	2001	(284)	(395)
Empirical	(Wade, Goodford et al.)	1989, 1993	(220, 298)	GRID (218)
	SCORE1	1994	(294)	LUDI (108, 109); (300, 396)
	(Miller, Sheridan et al.)	1994	(121)	FLOG (121)
	GOLD score	1995	(176, 177)	GOLD (176, 177)
	PLP	1995, 2000	(185, 367)	
	FlexX score	1996	(110)	FlexX (110, 130, 138, 397)
	VALIDATE	1996	(307)	
	(Jain)	1996	(297)	Hammerhead (388)
	ChemScore	1997	(80)	(295)
	SCORE2	1998	(296)	
	(Takamatsu, Itai)	1998	(398)	
	SCORE	1998	(293)	(225)
	AutoDock3	1998	(115)	(391)
	Fresno	1999	(399)	(230)
	ScreenScore	2001	(287)	
Desolvation terms	HINT	1991	(400)	(308)
	(Zhang, DeLisi et al.)	1997	(305)	
Knowledge-based	SMoG	1996	(313)	SMoG (314)
	BLEEP	1999	(315, 316)	(340)
	PMF	1999	(317)	(299, 319, 320, 339, 369)
	DrugScore	2000	(226)	(15, 318)

ing affinity. In doing so, one substitutes estimates of the free energy of binding in solution by an estimate of the gas phase enthalpy of binding. Even this crude approximation can lead to satisfying results. A good correlation was obtained between nonbonded interaction energies calculated with a modified MM2 force field and IC_{50} values of 33 HIV-1 protease inhibitors (269). Similar results were reported in a study of 32 thrombin-inhibitor complexes with the CHARMM force field (270). In both

studies, however, experimental data represented rather narrow activity ranges and covered little structural variation.

The AMBER (271, 272) and CHARMM (179) nonbonded terms are used as scoring function in several docking programs. As mentioned above (Section 3.1.1), protein terms are usually precalculated on a rectangular grid to speed up the energy calculation compared to traditional atom-by-atom evaluations (273). Distance-dependent dielectric constants are usually employed to approximate the long-range shielding of electrostatic interactions by water (274). However, compounds with high formal charges still obtain unreasonably high scores as a result of overestimated ionic interactions. For this reason, a common practice in virtual screening is to separate databases of compounds into subgroups according to their total charges and rank these groups separately. When electrostatic interactions are complemented by a solvation term calculated by the Poisson-Boltzmann equation (32) or faster continuum solvation models (e.g., Ref. 275), effects of high formal charges are usually leveled out. In a validation study on three protein targets, Shoichet and coworkers observed significantly improved ranking of known inhibitors upon correction for ligand solvation (276). The current version of the docking program DOCK calculates solvation corrections based on the generalized Born (277) solvation model (278). The method has been tested in a study where several peptide libraries were docked into various serine protease active sites (279).

In the context of scoring, the van der Waals term of force fields is mainly responsible for penalizing docking solutions with respect to overlap between receptor and ligand atoms. It is often omitted when only the binding of experimentally determined complex structures is analyzed (280–282).

Very recently, a new contribution to the list of force-field-based scoring methods has been developed by Charifson and Pearlman. This so-called OWFEG (one window free energy grid) method (283) is an approximation to the expensive first-principles method of free energy perturbation (FEP). For the purpose of scoring, an MD simulation is carried out with the ligand-free, solvated receptor site. During the simulation, the energetic effects of probe atoms on a regular grid are collected and averaged. Three simulations are run with three different probes: a neutral methyl-type atom, a negatively charged atom, and a positively charged atom. The resulting three grids contain information on the score contributions of neutral, positively, and negatively charged probe atoms located in various positions of the receptor site. They are used for scoring a ligand position by linear interpolation based on the partial charges of the ligand atoms. This approach seems to be successful for K_i prediction as well as virtual screening applications (284). Its conceptual advantage is the implicit consideration of entropic and solvent effects and some protein flexibility.

The calculation of ligand strain energy traditionally also lies in the realm of molecular mechanics force fields. Although effects of strain energy have rarely been determined experimentally (3), it is generally accepted that high-affinity ligands bind in low-energy conformations (285, 286). If a compound must adopt a strained conformation to fit into a receptor pocket, this should lead to a less negative binding free energy. Strain energy can be estimated by calculating the difference between the global energy minimum of the unbound ligand and the current conformation of the ligand in the complex. However, force field estimates of energy differences between individual conformations are not reliable for all systems. In practice, better correlation with experimental binding data is observed when strain energy is used as a filter to weed out unlikely binding geometries rather than including it in the final score. Estimation of ligand strain energy based on force fields can be time-consuming and therefore alternatives are often employed, such as empirical rules derived from small-molecule crystal data (140). Conformations generated by such programs are, however, often not strain-free because only one torsional angle is regarded at a time. Some strained conformations can be excluded when two consecutive dihedral angles are taken into account simultaneously (287).

4.1.2 Empirical Scoring Functions. The underlying idea of empirical scoring functions is that the binding free energy of a noncovalent receptor-ligand complex can be factorized into a sum of localized, chemically intuitive interactions. Such decompositions can be a useful tool to gain some insight into binding phenomena, even without analyzing 3D structures of receptor-ligand complexes. Andrews and colleagues derived average functional group contributions to the binding free energy by analyzing a set of 200 compounds for which the affinity to a receptor had been experimentally determined (266). Such average functional group contributions can then be used to estimate the mean overall binding affinity of a compound independent of a particular binding site. This value can be compared to the experimental binding free energy: if the experimental affinity is similar to or even more favorable than the computed one, the ligand obviously shows a good fit with the receptor and its functional groups are supposedly all involved in interactions with the protein; on the other hand, if it is significantly less favorable, the compound apparently does not fully exploit its potential to form optimal interactions. Similarly, experimental binding affinities have been analyzed on a per-atom basis in quest of the maximal binding affinity of noncovalent ligands (288). It was concluded that in the strongest binding ligands each non-hydrogen atom on average contributes 6.3 kJ/mol to the binding energy.

The analysis of binding phenomena can be performed with much more detail if the 3D structures of receptor-ligand complexes are available. Based on the assumption of additivity, the binding affinity ΔG_{bind} can be estimated as a sum of interactions multiplied by weighting factors:

$$\Delta G_{\text{bind}} \approx \sum \Delta G_i f_i.$$

Here, each f_i corresponds to an interaction term that depends on structural features of the complex and each ΔG_i represents a weighting coefficient, which is determined on the basis of a training set of experimental affinities for crystallographically known protein–ligand complexes. Scoring schemes that use this concept are called empirical scoring functions. Several reviews summarize details of individual parameterizations (26, 44, 56, 289–292). The individual terms in empirical scoring functions are usually chosen such that they intuitively cover important contributions of the total binding free energy. Most empirical scoring functions are derived by evaluating the functions f_i on a set of protein–ligand complexes and fitting the coefficients ΔG_i to experimental binding affinities of these complexes by multiple linear regression or supervised learning. The relative weight of the individual contributions depends on the training set.

Usually, between 50 and 100 complexes are used to derive the weighting factors. In a recent study it has been shown that many more than 100 complexes were necessary to achieve convergence (293). The reason for this finding is probably the fact that the publicly available protein–ligand complexes fall in a few rather strongly populated classes.

Empirical scoring functions usually contain individual terms for hydrogen bonds, ionic interactions, hydrophobic interactions, and binding entropy. Hydrogen bonds are often scored by simply counting the number of donor–acceptor pairs that fall into a given distance and angle range favorable for hydrogen bonding, weighted by penalty functions for deviations from ideal standard values (80, 294–296). The amount of error tolerance in these penalty functions is critical. If large deviations from the ideal are tolerated, the scoring function cannot discriminate sufficiently between different placements of a ligand, whereas too stringent tolerances artificially score similar complexes rather differently. Attempts have been described to reduce the strong distance dependency of such interactions by assigning soft modulating functions on an atom-pair basis (297). Other concepts try to avoid penalty functions and introduce distinct regression coefficients for strong, medium, and weak hydrogen bonds (293). The Agouron group has used a simple four-parameter potential that is a piecewise linear approximation of a potential neglecting angular terms ("PLP scoring function") (185). Most functions consider all types of hydrogen bonds equivalently. Some attempts have been made to distinguish between different donor–acceptor functional group pairs. Hydrogen bond scoring in GOLD (176, 177) is based on a list of hydrogen bond energies, derived from *ab initio* calculations, for any combination of 12 donor and 6 acceptor atom types. A similar differentiation of donor and acceptor groups is attempted in the program GRID (218) for the characterization of binding sites (219, 220, 298). The consideration of such lookup tables in scoring functions might help to avoid false predictions originating from an oversimplification of some individual interactions.

Reducing the weight of hydrogen bonds localized at the solvent-exposed rim of a binding site is a useful concept to avoid false positives in virtual screening. This is achieved by reducing charges of surface-exposed residues in cases where explicit electrostatic terms are used (274) or by multiplying the hydrogen bond contribution with a factor that depends on the accessibility of the involved protein counter group (299).

Ionic interactions are handled in a way similar to hydrogen bonds. Long-distance charge-charge interactions are usually neglected, and it is thus more appropriate to refer to salt bridges or charge-assisted hydrogen bonds. The scoring function by Boehm implemented in LUDI (294) assigns a stronger weight to salt bridges than to neutral hydrogen bonds. This differentiation generally proved successful in scoring series of thrombin inhibitors (295, 300). However, comparable to force field scoring, the danger exists that highly charged molecules receive overestimated scores. Experience with FlexX containing a variant of Boehm's scoring function has shown that more reliable predictions are obtained if charged and uncharged hydrogen bonds are handled equally in a virtual screening application. Similar experience has also been collected using the ChemScore function (80).

Hydrophobic interactions are usually calibrated to the size of the contact surface buried upon receptor-ligand complex formation. Often, a reasonable correlation between experimental binding energies can be achieved considering only a surface term [see, for example, (1, 301, 302) and the discussion in Section 2.1.]. Various approximations for such surface terms have been described, for example, as grid-based (294) or volume-based approaches (cf. the discussion in Ref. 115). Many functions are based on a distance-dependent summation over neighboring receptor-ligand atom pairs. Distance-dependent cutoffs have been introduced in various ways, either short (110) or longer to include atom pairs that are not involved in direct van der Waals contacts (80, 185). The weighting factor ΔG_i of the hydrophobic term depends strongly on the training set. Supposedly, this fact has been underestimated in the development of many empirical scoring functions (35) because in most training sets ligands composed of numerous donor

and acceptor groups are overrepresented (many peptide and carbohydrate fragments).

In most empirical scoring functions, a hydrophobic character is attributed to several atom types, with equivalent weight for all hydrophobic contributions. In a more sophisticated approach, the propensity of particular atom types to be solvent-exposed or embedded in the interior of a protein can be assessed by so-called atomic solvation parameters. These have been derived, for example, from experimental octanol/water partition coefficients (303, 304) or from protein crystal structures (305, 306). Atomic solvation parameters are used in the VALIDATE scoring function (307) and have been tested in DOCK (308).

Entropy terms account for the restriction of conformational degrees of freedom of the ligand upon complex formation. A crude but useful estimate of this entropy contribution is the number of rotatable bonds of a ligand (294, 296). This measure has the advantage of being a function of the ligand only. More sophisticated estimates try to take into account the nature of the ligand portion on either side of a flexible bond, particularly with respect to the interactions formed with the receptor (80, 307). This concept is based on the assumption that purely hydrophobic contacts allow for more residual motion in the ligand fragments.

4.1.3 Knowledge-Based Methods. Empirical scoring functions regard only those interactions that are explicitly part of the model. Less frequent interactions are usually neglected, even though they can be strong and specific, for example, NH-π hydrogen bonds. To generate a comprehensive and consistent description of all these interactions in the framework of empirical scoring functions would be a difficult task. However, the exponentially growing body of structural data on receptor-ligand complexes can be exploited to discover favorable binding geometries. "Knowledge-based" scoring functions try to capture the knowledge about protein-ligand binding that is implicitly stored in the protein data bank by means of statistical analysis of structural data, without referring to often inconsistent experimentally determined binding affinities (309). They are based on the concept of the inverse formulation of the Boltzmann law,

$$E_{ij} = -kT \ln(p_{ijk}) + kT \ln(Z),$$

where the energy function E_{ij} is called a potential of mean force for a state defined by the variables i, j, and k; p_{ijk} is the corresponding probability density; and Z is the partition function. The second term of the sum is constant at constant temperature T and does not have to be regarded, given that $Z = 1$ can be selected by defining a suitable reference state, which leads to normalized probability densities p_{ijk}. The inverse Boltzmann approach has been applied to assemble potentials from databases of protein structures to score protein models in the context of protein structure prediction (310). To establish a function to score protein–ligand complexes, the variables i, j, and k are assigned to address protein and ligand atom types, and their interatomic distances. The occurrence frequency of individual contacts is a measure of their energetic contribution to binding. If a specific contact occurs more frequently than expected by random or seen in an average distribution, it is assumed to be favorable. On the other hand, if it occurs less frequently, repulsive or unfavorable interaction between two atom types is anticipated. The frequencies are thus converted into sets of atom-pair potentials ready for further evaluation.

First applications in drug research (134, 311, 312) were restricted to small data sets of HIV protease-inhibitor complexes and did not result in generally applicable scoring functions. Recent publications (226, 313–318), however, have shown the usefulness of these approaches. The first general-purpose function using such potentials was implemented in the *de novo* design program SMoG (313, 314).

The PMF function by Muegge and Martin (317), consists of a set of distance-dependent atom-pair potentials $E_{ij}(r)$ that are expressed as

$$E_{ij}(r) = -kT \ln[f_j(r)\rho^{ij}(r)/\rho^{ij}].$$

Here, r is the atom pair distance, and $\rho^{ij}(r)$ is the number density of pairs ij in a certain radius interval about r. This density is calculated by the following procedure. First, a maximum search radius is defined. This radius describes a reference sphere around each ligand atom j. Receptor atoms of type i are searched within this sphere. Subsequently, the sphere is subdivided into shells of a predefined thickness. The number of receptor atoms i matching each spherical shell is divided by the volume of this shell and averaged over all occurrences of ligand atoms j in the evaluated data set of protein–ligand complexes. The term ρ^{ij} in the denominator is the average density of receptor atoms i falling into the whole reference volume. It is argued that the spherical reference volume around each ligand atom needs to be corrected by eliminating the occupied volume of the ligand itself, given that ligand-ligand interactions are not regarded in this area. This is done by a volume correction factor $f_j(r)$ that is a function of the ligand atom only and gives a rough estimate of the preference of an atom of type j to be exposed rather than buried in the ligand. Muegge could show that the volume-correction factor contributes significantly to the predictive power of the PMF function (319). Also, reference radii between 7 and 12 Å are applied to implicitly include solvation effects, especially the propensity of individual atom types to be located inside a protein cavity or in contact with solvent (320). To rank docking solutions, the PMF function is evaluated in a grid-based fashion and combined with a repulsive van der Waals potential at short distances.

The DrugScore function by Gohlke et al. (226) is based on roughly the same formalism, albeit with several differences in the derivation that lead to different potential forms. Most notably, the statistical distance distributions $\rho^{ij}(r)/\rho^{ij}$ for the individual atom pairs ij are divided by a common reference state that is taken as the average over the distance distributions of all atom pairs $\rho(r) = \Sigma\Sigma\, \rho^{ij}(r)/ij$. To consider only direct ligand-protein contacts, the upper sample radius has been set to 6 Å. At this distance, no further atoms (e.g., of a water molecule) can mediate a protein–ligand interaction. The individual potentials have the form

$$E_{ij}(r) = -kT(\ln[\rho^{ij}(r)/\rho^{ij}] - \ln[\rho(r)]).$$

These pair potentials are used in combination with potentials depending on single (protein or ligand) atom types that express the propensity of an atom type to be buried within a particular protein environment on complex formation. Contributions of these surface potentials and the pair potentials are weighted equally in the final scoring function. This scoring function has initially been developed with the primary goal to differentiate between correctly docked (near native) ligand poses versus decoy binding modes for the same protein-ligand pair. However, through appropriate scaling also quantitative estimates across different protein–ligand complexes are possible (318).

Mitchell and coworkers choose a different type of reference state for their BLEEP potential (315). The pair interaction energy is written as

$$E_{ij}(r) = kT \ln[1 + m^{ij}\sigma]$$
$$- kT \ln[1 + m^{ij}\sigma\rho^{ij}(r)/\rho(r)].$$

Here, the number density $\rho^{ij}(r)$ is defined as above, but it is normalized by the occurrence frequency of all atom pairs at this same distance instead of by the number of pairs ij in the whole reference volume. The variable m^{ij} is the number of pairs ij found in the evaluated data set, and σ is an empirical factor that defines the weight of each observation. This potential is combined with a van der Waals potential as a reference state to compensate for the lack of sampling at short distances and for certain underrepresented atom pairs.

Besides differences in the functional form and reference state, from a more practical point of view, the knowledge-based potentials differ also with respect to scope of atom type definitions and the amount of structural data used for their derivation. The number of different atom types ranges from 17 in Drug-Score to 40 nonmetal atom types in BLEEP. In all cases, the Protein Data Bank (321) was the source of the solved crystal structures. For BLEEP 351 selected complexes were used, whereas the PMF function was extracted from 697 complexes, and DrugScore was derived using 1376 complexes. In the latter case, the data have been extracted from Relibase (322, 323).

4.2 Critical Assessment of Current Scoring Functions

4.2.1 Influence of the Training Data. All fast scoring functions share a number of defi-

ciencies that one should be aware of in any application. First, most scoring functions are in some way fitted to or derived from experimental data. The functions necessarily reflect the accuracy of the data that were used for their derivation. For instance, a general problem with empirical scoring functions is the fact that the experimental binding energies usually originate from many different sources and therefore consist of a rather heterogeneous data set affected by all kinds of experimental errors. Furthermore, scoring functions mirror not only the quality but also the scope of experimental data used for their development. Virtually all scoring functions are still derived from data mostly based on high-affinity receptor-ligand complexes. Many of these are still of peptidic nature, whereas interesting leads in pharmaceutical research are non-peptidic. This is reflected in the relatively high contributions of hydrogen bonds in the total score. The balance between hydrogen bonding and hydrophobic interactions is a very critical issue in scoring, and its consequences are especially obvious in virtual screening applications, as illustrated in Section 5.3.

4.2.2 Molecular Size. The simple additive nature of most fast scoring functions often leads to gradually increasing scores for molecules of larger size. Although it is true that small molecules with a molecular weight below 200–250 rarely show very high affinity, there is no physical reason why larger compounds should automatically possess higher activity. Comparing the scores of two compounds of significant size difference therefore calls for a term that compensates the size dependency. In some applications, a constant "penalty" term has been added to the score for each heavy atom (324) or a term proportional to the molecular weight has been considered (325). The empirical scoring function implemented in the docking program FLOG has been normalized to remove the linear dependency of the crude score on the number of ligand atoms (121). Originally introduced to improve the correlation between experimental and calculated affinities, entropy terms reflecting the change in conformational mobility upon ligand binding also help to reduce an excessive score for overly large and flexible mol-

ecules (80, 294). The size of the solvent-accessible surface of the ligand in its bound state can also be used as penalty term to discard large ligands not fully buried in the binding site. It should be noted, however, that all these approaches are very pragmatic in nature and do not solve the problem of size dependency, which is closely related to a proper understanding of cooperativity effects (265).

4.2.3 Penalty Terms. In general, scoring functions reward favorable interactions such as hydrogen bonds, but rarely penalize unfavorable ones. They are derived from experimentally determined crystal structures, and thus nonnative and energetically unfavorable orientations of a ligand within the binding site are not observed and can hardly be accounted for in a regression-based scoring function. Knowledge-based scoring functions try to capture such effects by referring to a reference state that corresponds to a mean situation. At first glance, the neglect of angular terms in the compilation of knowledge-based scoring functions results in averaged pair potentials that may not discriminate sufficiently between different binding geometries. However, some degree of angular dependency is considered, given that pair potentials for different atom types are always evaluated in combination with each other (226). Obvious deficiencies in regression-based scoring functions, such as electrostatic repulsions and steric clashes, can be avoided by defining reasonable penalty terms or by importing them from molecular mechanics force fields. This has been realized in the "chemical scoring" function implemented in the docking program DOCK (106, 126, 137, 273, 326, 327), which is a modified van der Waals potential being attractive or repulsive between individual groups of donor, acceptor, and lipophilic receptor and ligand atoms (224, 328). Other situations cannot be avoided by simple "clash" terms, but require a more sophisticated analysis of binding geometry. Among these are incomplete steric filling of the binding cavity by a ligand within the cavity, an unreasonably large ligand surface area remaining solvent exposed in the complex or the formation of voids at the receptor-ligand interface. Possible approaches to resolve these shortcomings are empirical filters that detect such unsatisfactory solutions and

remove them according to user-specified thresholds (329). A promising approach to properly reflect such cases is the inclusion of artificially generated, erroneous, decoy solutions in the optimization of scoring functions as reported for the scoring function of a flexible ligand superposition algorithm (330, 331).

4.2.4 Specific Attractive Interactions. Another general deficiency of scoring functions is the simplified description of attractive interactions. Molecular recognition is not entirely based on hydrogen bonding and hydrophobic contacts. Especially in host-guest chemistry, other specific types of interactions are frequently used to characterize the observed phenomena. For example, hydrogen bonds are formed between acidic protons and π-systems (332). These bonds can substitute for conventional hydrogen bonds in strength and specificity, as has been noted in protein-DNA recognition (333). Another type of less frequently observed interactions is the cation-π interaction, which is especially important at the surface of proteins (39, 334). Current empirical scoring functions usually neglect these interactions. Similarly, the directionality of interactions between aromatic rings is hardly considered (335, 336). Because of the regression-type adjustment, some energy contributions originating from these interactions are already implicitly incorporated into the conventional interaction terms. This might be one explanation why hydrogen bond contributions are traditionally overestimated in regression-based scoring functions. Knowledge-based approaches automatically incorporate these interactions in a scoring function, provided they occur with reasonable frequency in the data set used to develop the potentials.

4.2.5 Water Structure and Protonation State. Uncertainties about protonation states and the involvement of water in ligand binding further complicate scoring. These considerations are relevant for the development as well as the application of scoring functions. As mentioned above, the entropic and enthalpic contributions involving the reorganization of water molecules upon ligand binding are very difficult to predict (see, for example, Ref. 337). Currently, the most pragmatic approach to handle the water problem is the elucidation of "conserved water molecules" and to consider them as part of the receptor. A knowledge-based tool to estimate the "conservation" of water molecules upon ligand binding has been developed (217) and incorporated into a docking procedure (111) (cf. Section 3.2.2). It is based on crystallographic information and tries to extract rules about water sites by analyzing whether they are recurrently occupied by water molecules in series of related protein–ligand complexes.

Scoring functions require predefined atom types for each protein and ligand atom. This also implies the fixed assignment of a protonation state to each acidic and basic group. Knowledge-based functions, which do not consider hydrogen atoms, are equally affected by the problem because the atom type definitions normally imply a certain protonation state. Presently, such estimates might be reliable enough for the situation in aqueous solution; however, significant pK_a shifts are possible upon ligand binding (338) as a result of strong changes of the local dielectric conditions. They give rise to protonation reactions in parallel to the binding process. With respect to scoring, switching from a donor to an acceptor functionality because of altered protonation states has important consequences (279). Accordingly, improved docking and scoring algorithms must incorporate a more detailed and flexible description of protonation states.

4.2.6 Performance in Structure Prediction and Rank Ordering of Related Ligands. Similar to the broad range of available docking tools (cf. Section 3.2.3), the multitude of different scoring schemes calls for an objective assessment to evaluate their scope and limitations. This depends in part on the anticipated application; that is, whether protein–ligand complexes should be predicted (using the scoring scheme as objective function in docking), whether a set of ligands should be ranked with respect to one target protein (K_i prediction), or whether the scoring function is used to select possible hits out of a large database of candidate molecules (virtual screening).

An objective assessment of the available scoring functions is difficult because only very few functions have been tested on the same data sets or with the same docking tool. For structure prediction, several studies have shown that knowledge-based scoring functions are at least equivalent to regression-based functions. The PMF function has been successfully applied to structure prediction of inhibitors of neuraminidase (339) and MMP3 (229) in combination with the program DOCK, yielding superior results to the DOCK force field and chemical scoring. The Drug-Score function was tested on a large set of PDB complexes and gave significantly better results than those of the original FlexX scoring function using solutions generated by FlexX as the docking engine. DrugScore performed similarly to the force field score in DOCK, but outperformed the chemical scoring (226). Moreover, with respect to the correlation between experimental and calculated binding energies, very promising results have been obtained with DrugScore (318) and PMF (229, 317, 319, 339). BLEEP has recently been tested for scoring docked protein–ligand complexes (340). It was found to be slightly better than the DOCK energy function in discriminating decoy situations from near-native binding modes.

Although in many docking programs the same function is applied as an objective function for structure generation and for energy evaluation, better results can sometimes be obtained if different functions are applied. In particular, the docking objective function can be adapted to the docking algorithm used. In a parameter study, Vieth et al. found that using a soft-core van der Waals potential made their MD-based docking algorithm more efficient (274). Using FlexX as the docking engine, we observed that the original FlexX scoring function emphasizes directional interactions (mostly hydrogen bonds) in the docking phase. Subsequently, the ranking of individual library entries can be done successfully with a simple PLP potential that lacks directional terms, but considers general steric fit of receptor and ligand. Results are significantly worse if PLP is used already in the incremental built-up procedure of the docked ligand.

It is even more difficult to draw valid conclusions about the relative performance of scoring functions to rank sets of inhibitors with respect to their binding affinity for the same target. First, there is hardly any published study in which different functions have been applied to the same data sets. Second, experimental data are often not measured under the same conditions but collected from various literature references. This retrieval from various sources usually implies larger uncertainties within the experimental data set.

The task of ranking sets of 10–100 related ligands with respect to one target can also be handled by computationally more demanding methods. The most general approaches are probably force field scores complemented by electrostatic desolvation and surface area terms. An example is the MM-PBSA method that combines Poisson-Boltzmann electrostatics with AMBER molecular mechanics calculations and MD simulations (341, 342). This method has recently been applied to an increasing number of examples, showing quite promising results (343–346). Poisson-Boltzmann calculations have been performed on a variety of targets with many related computational protocols (280–282, 347–350). Alternatively, extended linear response protocols (263) can be used. The OWFEG grid method by Pearlman has also shown very promising results (283).

5   VIRTUAL SCREENING

As outlined in section, virtual screening is a multistep process. Although, in principle, the whole process can be fully automated, it is highly advisable to allow for manual interventions, in that visual inspection and selection still play a major role.

The process usually starts with a detailed analysis of the available 3D protein structures. If possible, highly homologous structures will also be analyzed, either to generate additional ideas about possible ligand structural motifs or to gain some insight on how to achieve selectivity against other proteins of the same class. A superposition of different protein–ligand complexes provides some ideas about key interactions repeatedly found in tight-binding protein–ligand complexes. Such an overlay will also highlight flexible parts of the protein or recurring water molecules in

the binding site that could be included in the docking process. Tools such as Relibase (322, 323) may be used to perform these comparative analyses of protein–ligand complexes in an efficient way. Subsequently, programs like GRID (218), LUDI (108, 109), SuperStar (351, 352), or DrugScore (318) are used to visualize potential binding sites ("hot spots") in the active site; in principle, any scoring function could be used for this purpose.

An important result of the 3D structure analysis is usually the identification of one or more key interactions that all ligands should satisfy. In aspartic proteases, for example, inhibitors should form at least one hydrogen bond to the catalytic Asp side-chains, whereas in metalloproteinases a coordination to the metal seems mandatory. Sometimes, a known ligand portion is used as initial scaffold based on which virtual screening techniques search for optimal side-chains. In principle, this step is not required, and instead one could fully rely on the docking and scoring step. However, following a pragmatic approach, it is important to use any well-founded information that is available about the system under consideration because more valuable results can usually be expected this way.

Once a reasonable hypothesis about the binding-site requirements has been generated, the next level of virtual screening is approached. Whether databases of commercially available compounds or "virtual" libraries of designed compounds are screened, it is advisable not to dock every possible compound, but only those that pass a series of hierarchical filters (cf. also Fig. 7.3). Simple preliminary filters remove

- compounds with reactive groups such as $-SO_2Cl$ or $-CHO$ because they are expected to cause problems in some biological assays as a result of unspecific covalent binding to the protein.

- compounds with molecular weights below 150 or above 500. Small molecules such as benzene are known to bind to proteins rather unspecifically at several sites. Large molecules such as polypeptides are difficult to optimize subsequently, given that good

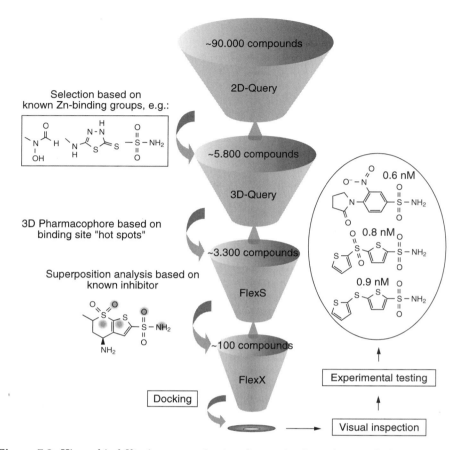

Figure 7.3. Hierarchical filtering process in virtual screening for carbonic anhydrase inhibitors.

bioavailability is usually limited to compounds with molecular weights below 500.

- compounds termed as "non-drug-like" according to criteria extracted from known drugs (353, 354).

After this general preselection, it can be advantageous to apply further steps of hierarchical filtering. As mentioned above, this could involve the selection of functional groups inevitably required to anchor a ligand to the most prominent interaction sites. Subsequently, the information of the "hot spot" analysis—translated into a pharmacophore hypothesis—can be used as matching criterion for a fast database screen. Such tools either involve fast tweak searching (355) or scan over precalculated conformers of the candidate molecules (356). The list of prospective

hits can then be submitted to a similarity search using information about already known active compounds, which could either be ligands already structurally characterized by crystallography or hits from a complementary HTS study. This optional step of the analysis tries to incorporate all available information about known hits and produces a reranking of the candidate molecules to be submitted to docking. As tools for the spatial similarity analysis, Feature Trees (357, 358), FlexS (330), and SEAL (254–256) have all been successfully applied.

All remaining ligands are docked into the target protein and a list of some hundred to several thousand small molecules, each with a computed score, is produced. These have to be further analyzed to discard undesirable structures. Selection criteria could be any of the following:

- *Lipophilicity* (if not addressed before). Highly lipophilic molecules are difficult to test because of their low solubility.
- *Structural class*. If 50% of the docked structures belong to one single chemical class, it is probably not necessary to test all of them (359).
- *Unreasonable docked binding mode*. Fast docking tools cannot produce reliable solutions for all compounds; often, even some well-scoring compounds are simply docked to the outer surface of the protein or adopt rather strained conformations to achieve good surface complementarity within the binding pocket. Computational filters help to detect such situations (329).

Finally, the selected compounds are ordered or synthesized and then tested. If the goal is to identify even weakly binding ligands as first leads, sufficient sensitivity of the biological assay has to be ensured [cf., for example, Ref. 16]. In this context it has also to be considered that limited solubility of the hits in water or water/DMSO mixtures often hampers affinity determinations at high concentrations.

Successful virtual screening has to produce a set of compounds significantly enriched with active compounds compared to random selection. A key parameter to assess the performance of docking and scoring in virtual screening is therefore, at least in theoretical case studies, the so-called enrichment factor. It is simply the ratio of active compounds in the subset selected by docking divided by the number of active compounds found in a randomly selected subset of equal size. To record such enrichment factors also for controlling performance at the various filter steps, a set of known active compounds is mixed with the set of candidate molecules. This strategy, however, requires a set of reasonable size (e.g., 30–50 ligands), which is not always given in a real-life virtual screening study. Furthermore, enrichment factors are far from being ideal indicators, particularly at later filter steps where a (hopefully) increasing amount of active compounds detected among the entries of the database competes with the set of known active ligands and artificially lowers the enrichment factor.

5.1 Combinatorial Docking

Docking of large compound collections requires fast algorithms. If the collection is an unstructured library of more or less unrelated compounds, each individual molecule must be docked independently ("sequential docking"), and only the fastest docking methods are applicable in this context, unless massive computer resources are used, as in the Dock-Crunch project based on the PRO_LEADS program (360). Examples for such fast docking tools are SLIDE (111) or the docking method by Diller and Merz (112). Both have been developed for database screening and library prioritization. Before docking, it is generally advisable to eliminate compounds that would provide only redundant information (similarity filters) or are very unlikely to yield high scores. Clearly, the filter routines need to be faster than the docking and scoring procedure, but this is normally the case.

Complementary to initial filtering, a preorganization of compounds into families exhibiting some kind of similarity has been demonstrated to improve the results of database screening. In the strategy shown by Su et al. (359), all molecules of any family are docked and scored, but only the best-scoring member of a high-ranking family is allowed to remain in the final hit list, whereas the scores of related molecules are recorded as annotations to this representative family member. This increases the diversity of the hit list and helps to identify a higher number of different classes of potential ligands.

An alternative to sequential docking can be followed if combinatorial libraries are evaluated. Quite a few programs have been specifically designed for speed-up by so-called combinatorial docking. They profit from the structured, incremental nature of combinatorial libraries and the fact that molecules of a combinatorial library consist of a common core. This core is assumed to form common specific interactions with the receptor (possibly supported by experimental evidence) and can thus be prepositioned in the binding pocket in one or a few similar orientations. It then serves as skeleton for the addition of substituents. Obviously, this step is ideally suited for incremental construction algorithms (361) and significantly reduces the complexity of the docking problem, limiting the required computation time per ligand. Earlier examples of this combinatorial docking approach are PRO_SELECT (362) and CombiDOCK (324). The latter is based on the DOCK program and has recently been enhanced by a vector-based orientation filter, to ensure productive scaffold poses, and by a free-energy-based scoring procedure (279). Another recent combinatorial docking procedure has been implemented as FlexX^c extension in FlexX (363). It follows a recursive scheme to traverse the combinatorial library space efficiently. The algorithm is based on a tree data structure that allows the efficient reuse of previously calculated docking results. FlexX^c follows the library search tree in a depth-first manner, whereas Combi-DOCK uses a breadth-first approach to evaluate fragments attached to a scaffold. A general advantage of breadth-first searches is that they allow for an efficient pruning of the search tree based on the scoring values.

De novo design tools have also been adapted to the problem of combinatorial docking and combinatorial library design. The program LUDI, for example, has been enhanced by the ability to connect building blocks in a chemically and structurally adequate manner; it can thus be used for combinatorial docking by fitting building blocks onto the interaction sites and simultaneous linking to previously docked core fragments (300). It has been successfully applied in the design of new thrombin inhibitors accessible through a single reaction. Another example is a variant of the Builder program (364) that was used to select substituents for a library of cathepsin D inhibitors (12). Yet another approach is DREAM++, a suite of programs for the design of virtual combinatorial libraries (365). Here, the DOCK algorithm is used for the molecular placement. Variable fragments are joined consecutively in compliance with predefined types of well-characterized organic reactions. Speed-up is achieved by preserving ("inheriting") information about common partial structures across different reactions, such that only the conformations of newly added fragments are searched.

Generally speaking, combinatorial docking approaches work best in cases where a core fragment plays a dominant role in the binding process and can be placed with high confidence in a well-characterized specificity pocket, such as the S1 pocket in thrombin. A further issue to consider is mutual fragment dependencies, that is, when multiple fragments are hooked up to a scaffold in a sequential manner; the results can depend on the sequence by which they are added (see, for example, Ref. 363). Thus, in unfavorable cases, different orders of attachment have to be followed to circumvent this possible limitation.

5.2 Seeding Experiments to Assess Docking and Scoring in Virtual Screening

True enrichment factors can be calculated only if experimental data are available for the full library, although such situations are unusual. Accordingly, studies using enrichment factors as a figure-of-merit to assess the performance of a virtual screening can serve for theoretical validation purposes only. Several authors have tested the predictive ability of docking and scoring tools by compiling an arbitrary set of diverse, drug-like compounds complemented by a number of known active compounds. This "seeded" library is then subjected to the virtual screening, and for the purpose of assessment it is assumed that the added active compounds are the only true actives in the library. Clearly, this is a rather questionable assumption.

Several seeding experiments have been published. An example has been performed at Merck using FLOG (121). A library consisting of 10,000 compounds including inhibitors of various types of proteases and HIV protease was docked into the active site of HIV protease. This resulted in excellent enrichment of the HIV protease inhibitors: all inhibitors but one were among the top 500 library members. However, inhibitors of other proteases were also considerably enriched (366).

Seeding experiments also allow for comparisons of different docking and scoring procedures, as shown, for example, by Charifson et al. (86), Bissantz et al. (230), and Stahl and Rarey (287). Charifson et al. compiled sets of several hundred active molecules for three different targets, p38 MAP kinase, inosine monophosphate dehydrogenase, and HIV protease. These were docked into the corresponding active sites together with 10,000 randomly selected, but drug-like, commercially available compounds using DOCK (327) and the Vertex in-house tool Gambler. ChemScore (80, 188), the DOCK AMBER force field score, and PLP (185) performed consistently well in enriching active compounds. This result was partially attributed to the fact that a rigid-body optimization could be carried out with these functions because they include repulsive terms in contrast to many other tested functions. Stahl and Rarey compared DrugScore (226), PMF (317), PLP (185), and the original FlexX score using FlexX for docking (110, 130, 138). Interestingly, the two knowledge-based scoring functions performed differently. DrugScore achieved better ranking for the tight-binding ligands in narrow lipophilic cavities of COX-2 and the thrombin S1 pocket. In contrast, PMF obtained better enrichment for the case of the very polar binding site of neuraminidase. Obviously, a general strength of PMF is the description of complexes showing multiple hydrogen bonds. This has also been noted in the study by Bissantz et al., in which PMF was found to perform well for the polar target thymidine kinase and less well for the estrogen receptor (230).

5.3 Hydrogen Bonding versus Hydrophobic Interactions

A balanced description of the contribution of hydrogen bonding and hydrophobic interactions to the total score is of general importance, to avoid a bias toward either highly polar or completely hydrophobic molecules. The actual parameterization of a scoring function depends on the compilation of the data set used to develop the function. Empirical scoring functions are more likely affected by the data set composition used for parameterization, but can be quickly reparameterized. In the case of knowledge-based functions such a readjustment is more difficult to perform; however, because of the much larger databases used for their development, they are supposed to be less dependent on special data set compilations.

The PLP function, for example, addresses general steric complementarity and hydrophobic interactions based on rather long-range pair potentials, whereas the FlexX score concentrates on hydrogen-bond complemen-

tarity. This is clearly reflected in results of database-ranking experiments. To combine the virtues of both scoring functions and to construct a more robust general function, a combination of PLP and FlexX called ScreenScore has recently been published (287). It was derived by a systematic optimization of library ranking results over seven targets and covers a wide range of active sites with respect to form, size, and polarity. ScreenScore obtains good enrichments for COX-2 (highly lipophilic binding site) and neuraminidase (highly polar site), whereas the individual functions fail in one of the two cases. The authors of PLP have recently enhanced their scoring function by including directed hydrogen bonding terms (367). Similar to ScreenScore, this could also lead to a more robust scoring function.

5.4 Finding Weak Inhibitors

Seeding experiments are often carried out with a small number of active compounds that are already optimized for binding to the studied target. Enrichment factors based on the retrieval of these compounds are not very conclusive because the recovery of potent inhibitors from a large set of candidate molecules is significantly easier than the discovery of new, but usually rather weak inhibitors from a large majority of nonbinders. In general, as in HTS, one can only expect hits from virtual screening that bind in the low micromolar range.

Nevertheless, a recent study showed that library screening can also successfully detect very weak ligands. Approximately 4000 commercially available compounds had been screened for FKBP-binding by means of the SAR-by-NMR technique (368) and 31 compounds with activity in the low millimolar range were detected. This set of compounds was flexibly docked into the FKBP binding site using DOCK 4.0 with the PMF scoring function (369). Interestingly, significant enrichment factors of 2 to 3 were achieved, whereas scoring with the standard AMBER score of DOCK did not really provide an enrichment.

5.5 Consensus Scoring

Different scoring schemes focus on different aspects as most important contributions to binding. However, these differences do not necessarily become obvious when calculating binding affinities of known active compounds. In contrast, the scoring of non-active compounds could unravel such differences. Vertex has reported good experience with so-called consensus scoring. Here, docking results are scored by several distinct functions and only those hits are considered that are rendered prominent by several of the functions. A significant decrease in false positives has been described (86), but inevitably a number of true positives is lost (see, for example, Ref. 230).

When consensus scoring is applied, one should thus keep in mind that, although the number of false positives can be reduced, the danger exists to discard some active compounds highlighted by only one of the scoring functions. This would, for example, apply to the above-mentioned PLP and FlexX scoring functions, which emphasize different aspects of ligand binding. Here, consensus scoring could be counterproductive. Therefore, along with consensus scoring, the individual scoring results should be consulted. Generally, however, it appears that one can expect more robust results from consensus scoring.

5.6 Successful Identification of Novel Leads through Virtual Screening

A considerable number of publications have proved that virtual screening can be efficiently used to discover novel leads (11, 13, 142, 370–375). Some of the most recent examples are briefly presented in the following.

The program ICM has been used to identify novel antagonists for a nuclear hormone receptor (201) and, together with DOCK, to find inhibitors for the RNA transactivation response element (TAR) of HIV-1 (25). The virtual screening protocol started with 153,000 compounds from the Available Chemicals Directory (ACD) (376) and involved increasingly elaborate docking and scoring schemes as the screening proceeded toward smaller selections of compounds. In the HIV-1 TAR study, the ACD library was first rigidly docked into the binding site using the DOCK program along with a simple contact scoring scheme. Then, 20% of the best-scoring compounds were subjected to flexible docking with ICM and an empirical scoring function specifically tailored to

RNA targets, providing a selection of approximately 5000 compounds. This was followed by two additional steps involving longer sampling of conformational space to retrieve 350 most promising candidates. Of these, a very small fraction was tested experimentally and two compounds were found to significantly reduce the binding of the Tat protein to HIV-1 TAR ($CD_{50} \sim 1~\mu M$).

Recently, Grueneberg et al. discovered subnanomolar inhibitors of carbonic anhydrase II by virtual screening (15). The study was performed following a protocol of several consecutive steps of hierarchical filtering (Fig. 7.3). Carbonic anhydrase II is a metalloenzyme used as prominent target for the treatment of glaucoma. Its binding site is a rather rigid, funnel-shaped pocket. Known inhibitors such as dorzolamide bind to the catalytic zinc ion by a sulfonamide group. In a recent crystallographic study it could be demonstrated that only the sulfonamide group represents an ideal anchor for zinc coordination (377). An initial data set of 90,000 entries from the Maybridge (378) and LeadQuest (379) libraries was converted to 3D structures with Corina (380). In a first filtering step, compounds were requested to possess a known zinc-binding group. These compounds were then processed through UNITY (355) using a protein-derived pharmacophore query. The pharmacophore hypothesis had been constructed from a "hot spot" analysis of the available X-ray structures of the enzyme. This yielded a set of 3314 compounds. In a subsequent filtering step, the known inhibitor dorzolamide was used as a template onto which all potential candidates were flexibly superimposed by means of the program FlexS (330). The top-ranking compounds from this step were docked into the binding site with FlexX (110, 130, 138), taking into account four conserved water molecules in the active site. After visual inspection, 13 top-ranking hits were selected for experimental testing. Nine of these compounds showed activities below 1 μM, and three had K_i values below 1 nM. Two of the hits were also examined crystallographically. The docking solution predicted as best by DrugScore was found to be closer to the experimental structure than the one predicted by the FlexX score.

This strategy of hierarchical filtering starting with a mapping of candidate molecules onto a binding site-derived pharmacophore, followed by a similarity analysis with known ligands using either FlexS (330), SEAL (254–256), or FeatureTrees (357, 358); and concluded by flexible docking with FlexX, which meanwhile was applied to three other proteins in the same laboratory. For t-RNA guanine transglycosylase, thermolysin, and aldose reductase, novel micromolar to submicromolar lead structures could be discovered. Most challenging in this context is aldose reductase because it performs pronounced induced fit changes upon ligand binding. Crystal structure analysis of a micromolar hit retrieved by virtual screening clearly revealed known and new areas of induced fit adaptation. The crystal structure obtained with this hit provides a good starting point for further lead optimization.

The *de novo* design of inhibitors of the bacterial enzyme DNA gyrase, a well-established antibacterial target (381), is another example for successful structure-based virtual screening, reported by Roche (16). HTS performed on the proprietary compound library provided no suitable lead structures. Therefore, a new rational approach was developed to discover potential lead structures using structural information of the ATP binding site in subunit B of the enzyme. At the onset of the project, the crystal structures of DNA gyrase subunit B complexed with a substrate analog and two inhibitors were available. In the buried part of the pocket they all donate a hydrogen bond to an aspartic acid side-chain and accept one from a conserved water molecule. As a design concept, the formation of these two key hydrogen bonds has been defined as mandatory. As an additional requirement, a lipophilic portion forming hydrophobic interactions with the enzyme was demanded. A new assay was established to allow for the detection of weakly binding inhibitors. A computational search of the ACD (376) and the Roche Compound Inventory identified hits having low molecular weights and matching the above-mentioned criteria. Relying on the results of the *in silico* screening [based on docking with LUDI and a pharmacophore search with CATALYST (356)] 600 compounds were tested initially.

Then, close analogs of the first series of hits were assayed, resulting in a total screen of 3000 compounds. This provided 150 hits, clustered into 14 chemical classes. Seven of these classes could be demonstrated as novel DNA gyrase inhibitors competing for the ATP binding site. Subsequent structure-based optimization resulted in inhibitors with potencies equal to or up to 10 times better than those of known antibiotics.

6 OUTLOOK

The first docking programs were introduced about 20 years ago, and the publication of the first generally applicable scoring functions dates back about 10 years. Since then, much experience has been gained in developing and applying docking algorithms, using scoring functions, and assessing their accuracy. Significant progress has been made over the last few years and it appears as if there are now docking tools available to address a variety of goals with considerable accuracy, from the precise and detailed analysis of binding interactions for a small set of ligands up to a fast screening of large compound collections. Similarly, scoring functions are currently available that can be applied to a wide range of different proteins and consistently yield a considerable retrieval of active compounds. As a consequence, the pharmaceutical industry increasingly uses virtual screening to identify possible leads.

In fact, structure-based design is now established as an important approach to drug discovery complementing HTS (382), although HTS has a number of serious disadvantages. It is expensive (383) and it leads to many false positives and a disappointingly small number of real leads (384, 385), particularly if screening is performed on a member of a new protein class. Also, not all assays are easily amenable to HTS requirements. Finally, despite the library sizes of several million entries available to the pharmaceutical industry, these compound collections do not approach the size and diversity needed to even approximately cover the chemical space of drug-like organic molecules. Accordingly, focused design of novel compounds and compound libraries should only gain importance. In light of current trends in structural genomics and patenting strategies, one may speculate that structure-based *de novo* design will become much more important in the near future.

To meet the increasing demands being placed on virtual screening, the development of more reliable scoring functions is certainly vital for success. In addition, novel or improved docking algorithms are required. We conclude by summarizing our perspective on major challenges in the further development of docking procedures and scoring functions:

1. The fact that protein–ligand interactions occur in aqueous solution is generally appreciated, but not yet adequately accounted for in molecular docking procedures. In particular, the simultaneous placement of explicit water molecules upon docking, accurate estimates of the water versus ligand interaction-energy balance, and the fast prediction of protonation states in binding pockets await a more satisfactory solution.

2. The consideration of a sufficient degree of protein flexibility needs to become part of standard docking approaches. This will require faster algorithms. In addition, with respect to scoring, an often overlooked aspect of this problem is that as soon as receptor flexibility is allowed, protein conformational energy changes need to be accounted for appropriately.

3. Although flexible-ligand docking has already become standard practice, the error rate in predictions of interaction geometries is still significant for more flexible ligands. Again, more efficient algorithms will be required to sample the conformation space more thoroughly.

4. Polar interactions are still not treated adequately. It is striking that, even though the role of hydrogen bonds in biology has been appreciated for a long time and the nature of hydrogen bonds is qualitatively well understood, their quantitative energetic description in protein–ligand interactions is still unsatisfactory (65).

5. All scoring functions are essentially expressed as simple analytical functions fitted to experimental binding data. The presently available crystal data on complex structures are strongly biased toward peptidic ligands. Because these data are used for the development of scoring functions, many overestimate the role of polar interactions. The development of improved scoring functions clearly requires access to better data, especially for nonpeptidic, low molecular weight, drug-like ligands, including weakly binding compounds.

6. Unfavorable interactions and unlikely docking modes are not penalized strongly enough. Methods for taking such undesired features into account are still lacking in presently available scoring functions.

7. So far, fast scoring functions cover only part of the whole receptor-ligand binding process. A more detailed picture could be obtained by taking into account properties of the unbound ligand, that is, solvation effects and energetic differences between the low-energy solution conformations and the bound conformation.

7 ACKNOWLEDGMENTS

The authors have benefited from numerous discussions with many researchers active in the field of docking and scoring, especially Holger Gohlke (University of Marburg/ Scripps Research Institute), Ingo Muegge (Bayer), and Matthias Rarey (GMD St. Augustin).

REFERENCES

1. H. J. Boehm and G. Klebe, *Angew. Chem. Int. Ed. Engl.*, **35**, 2588 (1996).

2. R. E. Babine and S. L. Bender, *Chem. Rev.*, **97**, 1359 (1997).

3. J. Greer, J. W. Erickson, J. J. Baldwin, and M. D. Varney, *J. Med. Chem.*, **37**, 1035 (1994).

4. S. W. Kaldor, V. J. Kalish, J. F. Davies, 2nd, B. V. Shetty, J. E. Fritz, K. Appelt, J. A. Burgess, K. M. Campanale, N. Y. Chirgadze, D. K. Clawson, B. A. Dressman, S. D. Hatch, D. A. Khalil, M. B. Kosa, P. P. Lubbehusen, M. A. Muesing, A. K. Patick, S. H. Reich, K. S. Su, and J. H. Tatlock, *J. Med. Chem.*, **40**, 3979 (1997).

5. W. Lew and U. Choung, *Curr. Med. Chem.*, **7**, 663 (2000).

6. M. von Itzstein, W.-Y. Wu, G. B. Kok, M. S. Pegg, J. C. Dyason, B. Jin, T. V. Phan, M. L. Smythe, H. F. White, S. W. Oliver, P. M. Colmant, J. N. Varghese, D. M. Ryan, J. M. Woods, R. C. Bethell, V. J. Hotham, J. M. Cameron, and C. R. Penn, *Nature*, **363**, 418 (1993).

7. R. S. Bohacek, C. McMartin, and W. C. Guida, *Med. Res. Rev.*, **16**, 3 (1996).

8. R. E. Hubbard, *Curr. Opin. Biotechnol.*, **8**, 696 (1997).

9. M. A. Murcko, P. R. Caron, and P. S. Charifson, *Ann. Rep. Med. Chem.*, **34**, 297 (1999).

10. G. Klebe, *J. Mol. Med.*, **78**, 269 (2000).

11. D. A. Gschwend, W. Sirawaraporn, D. V. Santi, and I. D. Kuntz, *Proteins*, **29**, 59 (1997).

12. E. K. Kick, D. C. Roe, A. G. Skillman, G. Liu, T. J. Ewing, Y. Sun, I. D. Kuntz, and J. A. Ellman, *Chem. Biol.*, **4**, 297 (1997).

13. P. Burkhard, U. Hommel, M. Sanner, and M. D. Walkinshaw, *J. Mol. Biol.*, **287**, 853 (1999).

14. J. H. Toney, P. M. Fitzgerald, N. Grover-Sharma, S. H. Olson, W. J. May, J. G. Sundelof, D. E. Vanderwall, K. A. Cleary, S. K. Grant, J. K. Wu, J. W. Kozarich, D. L. Pompliano, and G. G. Hammond, *Chem. Biol.*, **5**, 185 (1998).

15. S. Grueneberg, B. Wendt, and G. Klebe, *Angew. Chem. Int. Ed. Engl.*, **40**, 389 (2001).

16. H. J. Boehm, M. Boehringer, D. Bur, H. Gmuender, W. Huber, W. Klaus, D. Kostrewa, H. Kuehne, T. Luebbers, N. Meunier-Keller, and F. Mueller, *J. Med. Chem.*, **43**, 2664 (2000).

17. S. K. Burley, *Nat. Struct. Biol.* 7 (Suppl.), 932 (2000).

18. R. Abagyan and M. Totrov, *Curr. Opin. Chem. Biol.*, **5**, 375 (2001).

19. I. Muegge and M. Rarey in K. B. Lipkowitz and D. B. Boyd, Eds., *Reviews in Computational Chemistry*, **Vol. 17**, Wiley-VCH, New York, 2001, p. 1.

20. G. R. Smith and M. J. Sternberg, *Curr. Opin. Struct. Biol.*, **12**, 28 (2002).

21. C. J. Camacho and S. Vajda, *Curr. Opin. Struct. Biol.*, **12**, 36 (2002).

22. A. H. Elcock, D. Sept, and J. A. McCammon, *J. Phys. Chem. B*, **105**, 1504 (2001).

23. B. K. Shoichet and I. D. Kuntz, *Chem. Biol.*, **3**, 151 (1996).

24. M. J. Sternberg, H. A. Gabb, and R. M. Jackson, *Curr. Opin. Struct. Biol.*, **8**, 250 (1998).

25. A. V. Filikov, V. Mohan, T. A. Vickers, R. H. Griffey, P. D. Cook, R. A. Abagyan, and T. L. James, *J. Comput.-Aided Mol. Des.*, **14**, 593 (2000).

26. D. E. Clark, C. W. Murray, and J. Li in K. B. Lipkowitz and D. B. Boyd, Eds., *Reviews in Computational Chemistry*, **Vol. 11**, Wiley-VCH, New York, 1997, p. 67.

27. M. A. Murcko in K. B. Lipkowitz and D. B. Boyd, Eds., *Reviews in Computational Chemistry*, **Vol. 11**, Wiley-VCH, New York, 1997, p. 1.

28. G. Schneider and H. J. Boehm, *Drug Discov. Today*, **7**, 64 (2002).

29. W. P. Walters, M. T. Stahl, and M. A. Murcko, *Drug Discov. Today*, **3**, 160 (1998).

30. T. Langer and R. D. Hoffmann, *Curr. Pharm. Des.*, **7**, 509 (2001).

31. W. F. van Gunsteren and H. J. C. Berendsen, *Angew. Chem. Int. Ed. Engl.*, **29**, 992 (1990).

32. B. Honig and A. Nicholls, *Science*, **268**, 1144 (1995).

33. G. A. Jeffrey and W. Saenger, *Hydrogen Bonding in Biological Structures*, Springer-Verlag, Berlin, 1991.

34. G. A. Jeffrey, *An Introduction to Hydrogen Bonding*, Oxford University Press, New York, 1997.

35. A. M. Davis and S. J. Teague, *Angew. Chem. Int. Ed. Engl.*, **38**, 736 (1999).

36. S. K. Burley and G. A. Petsko, *Adv. Protein Chem.*, **39**, 125 (1988).

37. S. Tsuzuki, K. Honda, T. Uchimaru, M. Mikami, and K. Tanabe, *J. Am. Chem. Soc.*, **122**, 11450 (2000).

38. M. Brandl, M. S. Weiss, A. Jabs, J. Suhnel, and R. Hilgenfeld, *J. Mol. Biol.*, **307**, 357 (2001).

39. J. P. Gallivan and D. A. Dougherty, *Proc. Natl. Acad. Sci. USA*, **96**, 9459 (1999).

40. I. L. Alberts, K. Nadassy, and S. J. Wodak, *Protein Sci.*, **7**, 1700 (1998).

41. N. Borkakoti, *Curr. Opin. Drug Disc. Dev.*, **2**, 449 (1999).

42. A. Ben-Naim, *Hydrophobic Interactions*, Plenum, New York, 1980.

43. C. Tanford, *The Hydrophobic Effect*, John Wiley & Sons, New York, 1980.

44. Ajay and M. A. Murcko, *J. Med. Chem.*, **38**, 4953 (1995).

45. M. K. Gilson, J. A. Given, B. L. Bush, and J. A. McCammon, *Biophys. J.*, **72**, 1047 (1997).

46. J. E. Ladbury and B. Z. Chowdhry, *Chem. Biol.*, **3**, 791 (1996).

47. T. Wiseman, S. Williston, J. F. Brandts, and L. N. Lin, *Anal. Biochem.*, **179**, 131 (1989).

48. S. H. Sleigh, P. R. Seavers, A. J. Wilkinson, J. E. Ladbury, and J. R. Tame, *J. Mol. Biol.*, **291**, 393 (1999).

49. M. H. Parker, D. F. Ortwine, P. M. O'Brien, E. A. Lunney, C. A. Banotai, W. T. Mueller, P. McConnell, and C. G. Brouillette, *Bioorg. Med. Chem. Lett.*, **10**, 2427 (2000).

50. D. H. Williams, D. P. O'Brien, and B. Bardsley, *J. Am. Chem. Soc.*, **123**, 737 (2001).

51. P. Gilli, V. Ferretti, G. Gilli, and P. A. Brea, *J. Phys. Chem.*, **98**, 1515 (1994).

52. J. D. Dunitz, *Chem. Biol.*, **2**, 709 (1995).

53. K. Sharp, *Protein Sci.*, **10**, 661 (2001).

54. P. C. Weber, J. J. Wendoloski, M. W. Pantoliano, and F. R. Salemme, *J. Am. Chem. Soc.*, **114**, 3197 (1992).

55. F. Dullweber, M. T. Stubbs, D. Musil, J. Stuerzebecher, and G. Klebe, *J. Mol. Biol.*, **313**, 593 (2001).

56. J. R. Tame, *J. Comput.-Aided Mol. Des.*, **13**, 99 (1999).

57. A. R. Khan, J. C. Parrish, M. E. Fraser, W. W. Smith, P. A. Bartlett, and M. N. James, *Biochemistry*, **37**, 16839 (1998).

58. H. Mack, T. Pfeiffer, W. Hornberger, H. J. Boehm, and H. W. Hoeffken, *J. Enzyme Inhib.*, **9**, 73 (1995).

59. Y. W. Chen, A. R. Fersht, and K. Henrick, *J. Mol. Biol.*, **234**, 1158 (1993).

60. P. R. Connelly, R. A. Aldape, F. J. Bruzzese, S. P. Chambers, M. J. Fitzgibbon, M. A. Fleming, S. Itoh, D. J. Livingston, M. A. Navia, J. A. Thomson, and K. P. Wilson, *Proc. Natl. Acad. Sci. USA*, **91**, 1964 (1994).

61. A. R. Fersht, J. P. Shi, J. Knill-Jones, D. M. Lowe, A. J. Wilkinson, D. M. Blow, P. Brick, P. Carter, M. M. Waye, and G. Winter, *Nature*, **314**, 235 (1985).

62. U. Obst, D. W. Banner, L. Weber, and F. Diederich, *Chem. Biol.*, **4**, 287 (1997).

63. B. P. Morgan, J. M. Scholtz, M. D. Ballinger, I. D. Zipkin, and P. A. Bartlett, *J. Am. Chem. Soc.*, **113**, 297 (1991).

64. B. A. Shirley, P. Stanssens, U. Hahn, and C. N. Pace, *Biochemistry*, **31**, 725 (1992).

65. H. Kubinyi in B. Testa, H. van de Waterbeemd, G. Folkers, and R. Guy, Eds., *Pharmacokinetic Optimization in Drug Research*, Wiley-VCH, Weinheim, Germany, 2001, p. 513.

66. H.-J. Schneider, T. Schiestel, and P. Zimmermann, *J. Am. Chem. Soc.*, **114**, 7698 (1992).

67. A. C. Tissot, S. Vuilleumier, and A. R. Fersht, *Biochemistry*, **35**, 6786 (1996).

68. J. D. Dunitz, *Science*, **264**, 670 (1994).

69. C. Chothia, *Nature*, **254**, 304 (1975).

70. F. M. Richards, *Annu. Rev. Biophys. Bioeng.*, **6**, 151 (1977).

71. K. A. Sharp, A. Nicholls, R. Friedman, and B. Honig, *Biochemistry*, **30**, 9686 (1991).

72. M. S. Searle and D. H. Williams, *J. Am. Chem. Soc.*, **114**, 10690 (1992).

73. M. S. Searle, D. H. Williams, and U. Gerhard, *J. Am. Chem. Soc.*, **114**, 10697 (1992).

74. M. A. Hossain and H. J. Schneider, *Chem. Eur. J.*, **5**, 1284 (1999).

75. J. Hermans and L. Wang, *J. Am. Chem. Soc.*, **119**, 2702 (1997).

76. K. P. Murphy, D. Xie, K. S. Thompson, L. M. Amzel, and E. Freire, *Proteins*, **18**, 63 (1994).

77. K. A. Dill, *J. Biol. Chem.*, **272**, 701 (1997).

78. G. Folkers, Ed., *Pharm. Acta Helv.*, **69**, 175 (1995).

79. D. E. J. Koshland, *Angew. Chem. Int. Ed. Engl.*, **33**, 2408 (1994).

80. M. D. Eldridge, C. W. Murray, T. R. Auton, G. V. Paolini, and R. P. Mee, *J. Comput.-Aided Mol. Des.*, **11**, 425 (1997).

81. I. K. McDonald and J. M. Thornton, *J. Mol. Biol.*, **238**, 777 (1994).

82. M. Totrov and R. Abagyan in R. B. Raffa, Ed., *Drug-Receptor Thermodynamics: Introduction and Applications*, John Wiley & Sons, Chichester, 2001, p. 603.

83. I. D. Kuntz, E. C. Meng, and B. K. Shiochet, *Acc. Chem. Res.*, **27**, 117 (1994).

84. A. R. Leach and M. M. Hann, *Drug Discov. Today*, **5**, 326 (2000).

85. R. A. Lewis, S. D. Pickett, and D. E. Clark in K. B. Lipkowitz and D. B. Boyd, Eds., *Reviews in Computational Chemistry*, **Vol. 16**, Wiley-VCH, New York, 2000, p. 1.

86. P. S. Charifson, J. J. Corkery, M. A. Murcko, and W. P. Walters, *J. Med. Chem.*, **42**, 5100 (1999).

87. G. E. Terp, B. N. Johansen, I. T. Christensen, and F. S. Jorgensen, *J. Med. Chem.*, **44**, 2333 (2001).

88. T. P. Lybrand, *Curr. Opin. Struct. Biol.*, **5**, 224 (1995).

89. G. Jones and P. Willett, *Curr. Opin. Biotechnol.*, **6**, 652 (1995).

90. T. Lengauer and M. Rarey, *Curr. Opin. Struct. Biol.*, **6**, 402 (1996).

91. R. Rosenfeld, S. Vajda, and C. DeLisi, *Annu. Rev. Biophys. Biomol. Struct.*, **24**, 677 (1995).

92. P. Bamborough and F. E. Cohen, *Curr. Opin. Struct. Biol.*, **6**, 236 (1996).

93. J. S. Dixon and J. M. Blaney in Y. C. Martin and P. Willett, Eds., *Designing Bioactive Molecules: Three-Dimensional Techniques and Applications*, American Chemical Society, Washington, DC, 1997, p. 175.

94. P. J. Gane and P. M. Dean, *Curr. Opin. Struct. Biol.*, **10**, 401 (2000).

95. C. A. Sotriffer, W. Flader, R. H. Winger, B. M. Rode, K. R. Liedl, and J. M. Varga, *Methods*, **20**, 280 (2000).

96. R. B. Russell and D. S. Eggleston, *Nat. Struct. Biol.* **7** (Suppl.), 928 (2000).

97. M. Hendlich, F. Rippmann, and G. Barnickel, *J. Mol. Graph. Model.*, **15**, 359 (1997).

98. G. P. Brady, Jr. and P. F. Stouten, *J. Comput.-Aided Mol. Des.*, **14**, 383 (2000).

99. C. A. Orengo, A. E. Todd, and J. M. Thornton, *Curr. Opin. Struct. Biol.*, **9**, 374 (1999).

100. J. M. Thornton, A. E. Todd, D. Milburn, N. Borkakoti, and C. A. Orengo, *Nat. Struct. Biol.* **7** (Suppl.), 991 (2000).

101. S. Schmitt, M. Hendlich, and G. Klebe, *Angew. Chem. Int. Ed. Engl.*, **40**, 3141 (2001).

102. J. Ruppert, W. Welch, and A. N. Jain, *Protein Sci.*, **6**, 524 (1997).

103. F. Jiang and S. H. Kim, *J. Mol. Biol.*, **219**, 79 (1991).

104. D. Fischer, S. L. Lin, H. L. Wolfson, and R. Nussinov, *J. Mol. Biol.*, **248**, 459 (1995).

105. P. Burkhard, P. Taylor, and M. D. Walkinshaw, *J. Mol. Biol.*, **277**, 449 (1998).

106. I. D. Kuntz, J. M. Blaney, S. J. Oatley, R. Langridge, and T. E. Ferrin, *J. Mol. Biol.*, **161**, 269 (1982).

107. C. M. Oshiro and I. D. Kuntz, *Proteins*, **30**, 321 (1998).

108. H. J. Boehm, *J. Comput.-Aided Mol. Des.*, **6**, 593 (1992).

109. H. J. Boehm, *J. Comput.-Aided Mol. Des.*, **6**, 61 (1992).

110. M. Rarey, B. Kramer, T. Lengauer, and G. Klebe, *J. Mol. Biol.*, **261**, 470 (1996).

111. V. Schnecke and L. A. Kuhn, *Perspect. Drug Discov. Des.*, **20**, 171 (2000).

112. D. J. Diller and K. M. Merz, Jr., *Proteins*, **43**, 113 (2001).

113. D. S. Goodsell and A. J. Olson, *Proteins*, **8**, 195 (1990).

114. G. M. Morris, D. S. Goodsell, R. Huey, and A. J. Olson, *J. Comput.-Aided Mol. Des.*, **10**, 293 (1996).

115. G. M. Morris, D. S. Goodsell, R. S. Halliday, R. Huey, W. E. Hart, R. K. Belew, and A. J. Olson, *J. Comput. Chem.*, **19**, 1639 (1998).

116. R. Abagyan, M. Trotov, and D. Kuznetsov, *J. Comput. Chem.*, **15**, 488 (1994).

117. M. Totrov and R. Abagyan, *Proteins* (Suppl.), 215 (1997).

118. J. Y. Trosset and H. A. Scheraga, *J. Comput. Chem.*, **20**, 412 (1999).

119. J. Y. Trosset and H. A. Scheraga, *Proc. Natl. Acad. Sci. USA*, **95**, 8011 (1998).

120. F. S. Kuhl, G. M. Crippen, and W. G. Richards, *J. Comput. Chem.*, **5**, 24 (1984).

121. M. D. Miller, S. K. Kearsley, D. J. Underwood, and R. P. Sheridan, *J. Comput.-Aided Mol. Des.*, **8**, 153 (1994).

122. D. M. Lorber and B. K. Shoichet, *Protein Sci.*, **7**, 938 (1998).

123. M. McGann (2001). FRED [Online]. OpenEye. http://www.eyesopen.com/fred.html [2001, Sept 25].

124. Y. P. Pang and A. P. Kozikowski, *J Comput.-Aided Mol. Des.*, **8**, 669 (1994).

125. Y. P. Pang, E. Perola, K. Xu, and F. G. Prendergast, *J. Comput. Chem.*, **22**, 1750 (2001).

126. E. C. Meng, D. A. Gschwend, J. M. Blaney, and I. D. Kuntz, *Proteins*, **17**, 266 (1993).

127. D. A. Gschwend and I. D. Kuntz, *J. Comput.-Aided Mol. Des.*, **10**, 123 (1996).

128. R. L. DesJarlais, R. P. Sheridan, J. S. Dixon, I. D. Kuntz, and R. Venkataraghavan, *J. Med. Chem.*, **29**, 2149 (1986).

129. J. Wang, P. A. Kollman, and I. D. Kuntz, *Proteins*, **36**, 1 (1999).

130. M. Rarey, B. Kramer, and T. Lengauer, *J. Comput.-Aided Mol. Des.*, **11**, 369 (1997).

131. T. J. Ewing, S. Makino, A. G. Skillman, and I. D. Kuntz, *J. Comput.-Aided Mol. Des.*, **15**, 411 (2001).

132. R. L. DesJarlais, R. P. Sheridan, G. L. Seibel, J. S. Dixon, I. D. Kuntz, and R. Venkataraghavan, *J. Med. Chem.*, **31**, 722 (1988).

133. D. J. Bacon and J. Moult, *J. Mol. Biol.*, **225**, 849 (1992).

134. A. Wallqvist and D. G. Covell, *Proteins*, **25**, 403 (1996).

135. M. Y. Mizutani, N. Tomioka, and A. Itai, *J. Mol. Biol.*, **243**, 310 (1994).

136. B. B. Goldman and W. T. Wipke, *Proteins*, **38**, 79 (2000).

137. S. Makino and I. D. Kuntz, *J. Comput. Chem.*, **18**, 1812 (1997).

138. M. Rarey, S. Wefing, and T. Lengauer, *J. Comput.-Aided Mol. Des.*, **10**, 41 (1996).

139. B. Kramer, M. Rarey, and T. Lengauer, *Proteins*, **37**, 228 (1999).

140. G. Klebe and T. Mietzner, *J. Comput.-Aided Mol. Des.*, **8**, 583 (1994).

141. G. Klebe, *J. Mol. Biol.*, **237**, 212 (1994).

142. R. L. DesJarlais and J. S. Dixon, *J. Comput.-Aided Mol. Des.*, **8**, 231 (1994).

143. B. K. Shoichet and I. D. Kuntz, *Protein Eng.*, **6**, 723 (1993).

144. N. Metropolis, A. W. Rosenbluth, M. N. Rosenbluth, A. H. Teller, and E. Teller, *J. Chem. Phys.*, **21**, 1087 (1953).

145. T. N. Hart, S. R. Ness, and R. J. Read, *Proteins* (Suppl.), 205 (1997).

146. T. N. Hart and R. J. Read, *Proteins*, **13**, 206 (1992).

147. M. Liu and S. Wang, *J. Comput.-Aided Mol. Des.*, **13**, 435 (1999).

148. Z. Li and H. A. Scheraga, *Proc. Natl. Acad. Sci. USA*, **84**, 6611 (1987).

149. R. Abagyan and P. Argos, *J. Mol. Biol.*, **225**, 519 (1992).

150. A. Caflisch, S. Fischer, and M. Karplus, *J. Comput. Chem.*, **18**, 723 (1997).

151. I. Apostolakis, A. Plueckthun, and A. Caflisch, *J. Comput. Chem.*, **19**, 21 (1998).

152. C. McMartin and R. S. Bohacek, *J. Comput.-Aided Mol. Des.*, **11**, 333 (1997).

153. M. Karplus and J. A. McCammon, *Annu. Rev. Biochem.*, **52**, 263 (1983).

154. W. F. van Gunsteren and A. E. Mark, *Eur. J. Biochem.*, **204**, 947 (1992).

155. W. F. van Gunsteren, P. H. Huenenberger, A. E. Mark, P. E. Smith, and I. G. Tironi, *Comput. Phys. Commun.*, **91**, 305 (1995).

156. D. Rognan, *Perspect. Drug Discov. Des.*, **11**, 181 (1998).

157. M. E. Tuckerman and G. J. Martyna, *J. Phys. Chem. B*, **104**, 159 (2000).

158. D. L. Beveridge and F. M. DiCapua, *Annu. Rev. Biophys. Biophys. Chem.*, **18**, 431 (1989).

159. D. A. Pearlman and P. A. Kollman in W. F. van Gunsteren and P. K. Weiner, Eds., *Computer Simulation of Biomolecular Systems: Theoretical and Experimental Applications*, **Vol. 1**, ESCOM, Leiden, The Netherlands, 1989, p. 101.

160. T. P. Straatsma and J. A. McCammon, *Annu. Rev. Phys. Chem.*, **43**, 407 (1992).

161. P. A. Kollman, *Chem. Rev.*, **93**, 2395 (1993).

162. T. P. Straatsma in K. B. Lipkowitz and D. B. Boyd, Eds., *Reviews in Computational Chemistry*, **Vol. 9**, VCH, New York, 1996, p. 81.

163. M. R. Reddy, M. D. Erion, and A. Agarwal in K. B. Lipkowitz and D. B. Boyd, Eds., *Reviews in Computational Chemistry*, **Vol. 16**, Wiley-VCH, New York, 2000, p. 217.

164. A. Di Nola, D. Roccatano, and H. J. Berendsen, *Proteins*, **19**, 174 (1994).

165. M. Mangoni, D. Roccatano, and A. Di Nola, *Proteins*, **35**, 153 (1999).

166. M. Vieth, J. D. Hirst, B. N. Dominy, H. Daigler, and C. L. Brooks, *J. Comput. Chem.*, **19**, 1623 (1998).

167. Y. Pak and S. Wang, *J. Phys. Chem. B*, **104**, 354 (2000).

168. Y. Pak, I. J. Enyedy, J. Varady, J. W. Kung, P. S. Lorenzo, P. M. Blumberg, and S. Wang, *J. Med. Chem.*, **44**, 1690 (2001).

169. B. A. Luty, Z. R. Wasserman, P. W. F. Stouten, C. N. Hodge, M. Zacharias, and J. A. McCammon, *J. Comput. Chem.*, **16**, 454 (1995).

170. A. Miranker and M. Karplus, *Proteins*, **11**, 29 (1991).

171. A. Caflisch, A. Miranker, and M. Karplus, *J. Med. Chem.*, **36**, 2142 (1993).

172. R. Judson in K. B. Lipkowitz and D. B. Boyd, Eds., *Reviews in Computational Chemistry*, **Vol. 10**, VCH, New York, 1997.

173. D. E. Goldberg, *Genetic Algorithms in Search, Optimization, and Machine Learning*, Addison-Wesley, Reading, MA, 1989.

174. L. Davis, *Handbook of Genetic Algorithms*, Van Nostrand Reinhold, New York, 1991.

175. D. E. Clark and D. R. Westhead, *J. Comput.-Aided Mol. Des.*, **10**, 337 (1996).

176. G. Jones, P. Willett, and R. C. Glen, *J. Mol. Biol.*, **245**, 43 (1995).

177. G. Jones, P. Willett, R. C. Glen, A. R. Leach, and R. Taylor, *J. Mol. Biol.*, **267**, 727 (1997).

178. J. S. Taylor and R. M. Burnett, *Proteins*, **41**, 173 (2000).

179. B. R. Brooks, R. E. Bruccoleri, B. D. Olafson, D. J. States, S. Swaminathan, and M. Karplus, *J. Comput. Chem.*, **4**, 187 (1983).

180. R. S. Judson, E. P. Jaeger, and A. M. Treasurywala, *Theochem. J. Mol. Struct.*, **114**, 191 (1994).

181. K. P. Clark and Ajay, *J. Comput. Chem.*, **16**, 1210 (1995).

182. C. M. Oshiro, I. D. Kuntz, and J. S. Dixon, *J. Comput.-Aided Mol. Des.*, **9**, 113 (1995).

183. M. Thormann and M. Pons, *J. Comput. Chem.*, **22**, 1971 (2001).

184. G. Jones in P. v. R. Schleyer, N. L. Allinger, T. Clark, J. Gasteiger, P. A. Kollman, H. F. Schaefer, 3rd, and R. P. Schreiner, Eds., *Encyclopedia of Computational Chemistry*, John Wiley & Sons, New York, 1998.

185. D. K. Gehlhaar, G. M. Verkhivker, P. A. Rejto, C. J. Sherman, D. B. Fogel, L. J. Fogel, and S. T. Freer, *Chem. Biol.*, **2**, 317 (1995).

186. D. R. Westhead, D. E. Clark, and C. W. Murray, *J. Comput.-Aided Mol. Des.*, **11**, 209 (1997).

187. J. M. Yang and C. Y. Kao, *J. Comput. Chem.*, **21**, 988 (2000).

188. C. A. Baxter, C. W. Murray, D. E. Clark, D. R. Westhead, and M. D. Eldridge, *Proteins*, **33**, 367 (1998).

189. C. A. Baxter, C. W. Murray, B. Waszkowycz, J. Li, R. A. Sykes, R. G. Bone, T. D. Perkins, and W. Wylie, *J. Chem. Inf. Comput. Sci.*, **40**, 254 (2000).

190. L. David, R. Luo, and M. K. Gilson, *J. Comput.-Aided Mol. Des.*, **15**, 157 (2001).

191. J. A. McCammon and S. C. Harvey, *Dynamics of Proteins and Nucleic Acids*, Cambridge University Press, London, 1987.

192. H. Frauenfelder, S. G. Sligar, and P. G. Wolynes, *Science*, **254**, 1598 (1991).

193. E. Freire, *Adv. Protein Chem.*, **51**, 255 (1998).

194. B. Ma, S. Kumar, C. J. Tsai, and R. Nussinov, *Protein Eng.*, **12**, 713 (1999).

195. M. Gerstein and W. Krebs, *Nucleic Acids Res.*, **26**, 4280 (1998).

196. H. A. Carlson and J. A. McCammon, *Mol. Pharmacol.*, **57**, 213 (2000).

197. H. Claussen, C. Buning, M. Rarey, and T. Lengauer, *J. Mol. Biol.*, **308**, 377 (2001).

198. D. A. Gschwend, A. C. Good, and I. D. Kuntz, *J. Mol. Recognit.*, **9**, 175 (1996).

199. A. R. Leach, *J. Mol. Biol.*, **235**, 345 (1994).

200. L. Schaffer and G. M. Verkhivker, *Proteins*, **33**, 295 (1998).

201. M. Schapira, B. M. Raaka, H. H. Samuels, and R. Abagyan, *Proc. Natl. Acad. Sci. USA*, **97**, 1008 (2000).

202. V. Schnecke, C. A. Swanson, E. D. Getzoff, J. A. Tainer, and L. A. Kuhn, *Proteins*, **33**, 74 (1998).

203. A. C. Anderson, R. H. O'Neil, T. S. Surti, and R. M. Stroud, *Chem. Biol.*, **8**, 445 (2001).

204. H. A. Carlson, K. M. Masukawa, K. Rubins, F. D. Bushman, W. L. Jorgensen, R. D. Lins, J. M. Briggs, and J. A. McCammon, *J. Med. Chem.*, **43**, 2100 (2000).

205. H. A. Carlson, K. M. Masukawa, and J. A. McCammon, *J. Phys. Chem. A*, **103**, 10213 (2000).

206. D. Bouzida, P. A. Rejto, S. Arthurs, A. B. Colson, S. T. Freer, D. K. Gehlhaar, V. Larson, B. A. Luty, P. W. Rose, and G. M. Verkhivker, *Int. J. Quantum Chem.*, **72**, 73 (1999).

207. R. M. Knegtel, I. D. Kuntz, and C. M. Oshiro, *J. Mol. Biol.*, **266**, 424 (1997).

208. F. Osterberg, G. M. Morris, M. F. Sanner, A. J. Olson, and D. S. Goodsell, *Proteins*, **46**, 34 (2002).

209. H. B. Broughton, *J. Mol. Graph. Model.*, **18**, 247 (2000).

210. I. Kolossváry and W. C. Guida, *J. Am. Chem. Soc.*, **118**, 5011 (1996).

211. I. Kolossváry and W. C. Guida, *J. Comput. Chem.*, **20**, 1671 (1999).

212. J. Israelachvili and H. Wennerstrom, *Nature*, **379**, 219 (1996).

213. M. Levitt and B. H. Park, *Structure*, **1**, 223 (1993).

214. J. E. Ladbury, *Chem. Biol.*, **3**, 973 (1996).

215. P. Y. Lam, P. K. Jadhav, C. J. Eyermann, C. N. Hodge, Y. Ru, L. T. Bacheler, J. L. Meek, M. J. Otto, M. M. Rayner, Y. N. Wong, C. H. Chang, P. C. Weber, D. A. Jackson, T. R. Sharpe, and S. Ericksonviitanen, *Science*, **263**, 380 (1994).

216. M. Rarey, B. Kramer, and T. Lengauer, *Proteins*, **34**, 17 (1999).

217. M. L. Raymer, P. C. Sanschagrin, W. F. Punch, S. Venkataraman, E. D. Goodman, and L. A. Kuhn, *J. Mol. Biol.*, **265**, 445 (1997).

218. P. J. Goodford, *J. Med. Chem.*, **28**, 849 (1985).

219. R. C. Wade and P. J. Goodford, *J. Med. Chem.*, **36**, 148 (1993).

220. R. C. Wade, K. J. Clark, and P. J. Goodford, *J. Med. Chem.*, **36**, 140 (1993).

221. W. E. Minke, D. J. Diller, W. G. Hol, and C. L. Verlinde, *J. Med. Chem.*, **42**, 1778 (1999).

222. M. S. Rao and A. J. Olson, *Proteins*, **34**, 173 (1999).

223. P. Pospisil, L. Scapozza, and G. Folkers in H. D. Hoeltje and W. Sippl, Eds., *Rational Approaches to Drug Design. Proceedings of the 13th European Symposium on Quantitative Structure-Activity Relationships*, Prous Science, Barcelona/Philadelphia, 2001, p. 92.

224. R. M. Knegtel, D. M. Bayada, R. A. Engh, W. von der Saal, V. J. van Geerestein, and P. D. Grootenhuis, *J. Comput.-Aided Mol. Des.*, **13**, 167 (1999).

225. P. Tao and L. Lai, *J. Comput.-Aided Mol. Des.*, **15**, 429 (2001).

226. H. Gohlke, M. Hendlich, and G. Klebe, *J. Mol. Biol.*, **295**, 337 (2000).

227. O. Roche, R. Kiyama, and C. L. Brooks, 3rd, *J. Med. Chem.*, **44**, 3592 (2001).

228. C. A. Sotriffer, H. H. Ni, and J. A. McCammon, *J. Am. Chem. Soc.*, **122**, 6136 (2000).

229. S. Ha, R. Andreani, A. Robbins, and I. Muegge, *J. Comput.-Aided Mol. Des.*, **14**, 435 (2000).

230. C. Bissantz, G. Folkers, and D. Rognan, *J. Med. Chem.*, **43**, 4759 (2000).

231. J. S. Dixon, *Proteins* (Suppl.), 198 (1997).

232. N. C. Strynadka, M. Eisenstein, E. Katchalski-Katzir, B. K. Shoichet, I. D. Kuntz, R. Abagyan, M. Totrov, J. Janin, J. Cherfils, F. Zimmerman, A. Olson, B. Duncan, M. Rao, R. Jackson, M. Sternberg, and M. N. James, *Nat. Struct. Biol.*, **3**, 233 (1996).

233. B. Kramer, M. Rarey, and T. Lengauer, *Proteins* (Suppl.), 221 (1997).

234. V. Sobolev, T. M. Moallem, R. C. Wade, G. Vriend, and M. Edelman, *Proteins* (Suppl.), 210 (1997).

235. H. Kubinyi, Ed., *3D QSAR in Drug Design. Theory, Methods, and Applications*, ESCOM, Leiden, The Netherlands, 1993.

236. H. Kubinyi, G. Folkers, and Y. C. Martin, Eds., *3D QSAR in Drug Design. Recent Advances*, **Vol. 3**, Kluwer/ESCOM, Dordrecht, The Netherlands, 1998.

237. H. Kubinyi, *QSAR, Hansch Analysis and Related Approaches*, VCH, Weinheim, Germany, 1993.

238. W. Sippl, *J. Comput.-Aided Mol. Des.*, **14**, 559 (2000).

239. C. L. Waller, T. I. Oprea, A. Giolitti, and G. R. Marshall, *J. Med. Chem.*, **36**, 4152 (1993).

240. A. M. Gamper, R. H. Winger, K. R. Liedl, C. A. Sotriffer, J. M. Varga, R. T. Kroemer, and B. M. Rode, *J. Med. Chem.*, **39**, 3882 (1996).

241. R. C. Wade in H. D. Hoeltje and W. Sippl, Eds., *Rational Approaches to Drug Design. Proceedings of the 13th European Symposium on Quantitative Structure-Activity Relationships*, Prous Science, Barcelona/Philadelphia, 2001, p. 23.

242. A. R. Ortiz, M. T. Pisabarro, F. Gago, and R. C. Wade, *J. Med. Chem.*, **38**, 2681 (1995).

243. R. C. Wade, A. R. Ortiz, and F. Gago, *Perspect. Drug Discov. Des.*, **9**, 19 (1998).

244. T. Wang and R. C. Wade in H. D. Hoeltje and W. Sippl, Eds., *Rational Approaches to Drug Design. Proceedings of the 13th European Symposium on Quantitative Structure-Activity Relationships*, Prous Science, Barcelona/Philadelphia, 2001, p. 78.

245. G. Cruciani and K. A. Watson, *J. Med. Chem.*, **37**, 2589 (1994).

246. W. Sippl and H. D. Hoeltje, *J. Mol. Struct.-Theochem.*, **503**, 31 (2000).

247. M. Vieth and D. J. Cummins, *J. Med. Chem.*, **43**, 3020 (2000).

248. M. Wojciechowski and J. Skolnick, *J. Comput. Chem.*, **23**, 189 (2002).

249. S. Moro, A. H. Li, and K. A. Jacobson, *J. Chem. Inf. Comput. Sci.*, **38**, 1239 (1998).

250. S. Moro, D. Guo, E. Camaioni, J. L. Boyer, T. K. Harden, and K. A. Jacobson, *J. Med. Chem.*, **41**, 1456 (1998).

251. R. Kiyama, Y. Tamura, F. Watanabe, H. Tsuzuki, M. Ohtani, and M. Yodo, *J. Med. Chem.*, **42**, 1723 (1999).

252. C. A. Sotriffer, W. Flader, A. Cooper, B. M. Rode, D. S. Linthicum, K. R. Liedl, and J. M. Varga, *Biophys. J.*, **76**, 2966 (1999).

253. A. Schafferhans and G. Klebe, *J. Mol. Biol.*, **307**, 407 (2001).

254. S. Kearsley and G. Smith, *Tetrahedron Comput. Methodol.*, **3**, 615 (1990).

255. G. Klebe, T. Mietzner, and F. Weber, *J. Comput.-Aided Mol. Des.*, **8**, 751 (1994).

256. G. Klebe, T. Mietzner, and F. Weber, *J. Comput.-Aided Mol. Des.*, **13**, 35 (1999).

257. P. A. Kollman, *Acc. Chem. Res.*, **29**, 461 (1996).

258. M. L. Lamb and W. L. Jorgensen, *Curr. Opin. Chem. Biol.*, **1**, 449 (1997).

259. M. K. Gilson, J. A. Given, and M. S. Head, *Chem. Biol.*, **4**, 87 (1997).

260. J. Aqvist, C. Medina, and J. E. Samuelsson, *Protein Eng.*, **7**, 385 (1994).

261. T. Hansson, J. Marelius, and J. Aqvist, *J. Comput.-Aided Mol. Des.*, **12**, 27 (1998).

262. J. Aqvist, V. B. Luzhkov, and B. O. Brandsdal, *Acc. Chem. Res.*, **35**, 358 (2002).

263. R. C. Rizzo, J. Tirado-Rives, and W. L. Jorgensen, *J. Med. Chem.*, **44**, 145 (2001).

264. A. E. Mark and W. F. van Gunsteren, *J. Mol. Biol.*, **240**, 167 (1994).

265. D. Williams and B. Bardsley, *Perspect. Drug Discov. Des.*, **17**, 43 (1999).

266. P. R. Andrews, D. J. Craik, and J. L. Martin, *J. Med. Chem.*, **27**, 1648 (1984).

267. H. J. Schneider, *Chem. Soc. Rev.*, **23**, 227 (1994).

268. T. J. Stout, C. R. Sage, and R. M. Stroud, *Structure*, **6**, 839 (1998).

269. M. K. Holloway, J. M. Wai, T. A. Halgren, P. M. Fitzgerald, J. P. Vacca, B. D. Dorsey, R. B. Levin, W. J. Thompson, L. J. Chen, and S. J. deSolms, *J. Med. Chem.*, **38**, 305 (1995).

270. P. D. J. Grootenhuis and P. J. M. van Galen, *Acta Cryst. D*, **51**, 560 (1995).

271. S. J. Weiner, P. A. Kollman, D. A. Case, U. C. Singh, C. Ghio, G. Alagona, S. Profeta, and P. Weiner, *J. Am. Chem. Soc.*, **106**, 765 (1984).

272. S. J. Weiner, P. A. Kollman, D. T. Nguyen, and D. A. Case, *J. Comput. Chem.*, **7**, 230 (1986).

273. E. C. Meng, B. K. Shoichet, and I. D. Kuntz, *J. Comput. Chem.*, **13**, 505 (1992).

274. M. Vieth, J. D. Hirst, A. Kolinski, and C. L. Brooks, *J. Comput. Chem.*, **19**, 1612 (1998).

275. N. Majeux, M. Scarsi, J. Apostolakis, C. Ehrhardt, and A. Caflisch, *Proteins*, **37**, 88 (1999).

276. B. K. Shoichet, A. R. Leach, and I. D. Kuntz, *Proteins*, **34**, 4 (1999).

277. W. C. Still, A. Tempczyk, R. C. Hawley, and T. Hendrickson, *J. Am. Chem. Soc.*, **112**, 6127 (1990).

278. X. Zou, Y. Sun, and I. D. Kuntz, *J. Am. Chem. Soc.*, **121**, 8033 (1999).

279. M. L. Lamb, K. W. Burdick, S. Toba, M. M. Young, A. G. Skillman, X. Zou, J. R. Arnold, and I. D. Kuntz, *Proteins*, **42**, 296 (2001).

280. M. Schapira, M. Totrov, and R. Abagyan, *J. Mol. Recognit.*, **12**, 177 (1999).

281. T. Zhang and D. E. Koshland, Jr., *Protein Sci.*, **5**, 348 (1996).

282. P. H. Hunenberger, V. Helms, N. Narayana, S. S. Taylor, and J. A. McCammon, *Biochemistry*, **38**, 2358 (1999).

283. D. A. Pearlman and P. S. Charifson, *J. Med. Chem.*, **44**, 502 (2001).

284. D. A. Pearlman, *J. Med. Chem.*, **42**, 4313 (1999).

285. J. Bostrom, P. O. Norrby, and T. Liljefors, *J. Comput.-Aided Mol. Des.*, **12**, 383 (1998).

286. M. Vieth, J. D. Hirst, and C. L. Brooks, 3rd, *J. Comput.-Aided Mol. Des.*, **12**, 563 (1998).

287. M. Stahl and M. Rarey, *J. Med. Chem.*, **44**, 1035 (2001).

288. I. D. Kuntz, K. Chen, K. A. Sharp, and P. A. Kollman, *Proc. Natl. Acad. Sci. USA*, **96**, 9997 (1999).

289. J. D. Hirst, *Curr. Opin. Drug Disc. Dev.*, **1**, 28 (1998).

290. R. M. A. Knegtel and P. D. J. Grootenhuis, *Perspect. Drug Discov. Des.*, **9–11**, 99 (1998).

291. T. I. Oprea and G. R. Marshall, *Perspect. Drug Discov. Des.*, **9–11**, 3 (1998).

292. H. J. Boehm and M. Stahl, *Med. Chem. Res.*, **9**, 445 (1999).

293. R. Wang, L. Liu, L. Lai, and Y. Tang, *J. Mol. Model.*, **4**, 379 (1998).

294. H. J. Boehm, *J. Comput.-Aided Mol. Des.*, **8**, 243 (1994).

295. C. W. Murray, T. R. Auton, and M. D. Eldridge, *J. Comput.-Aided Mol. Des.*, **12**, 503 (1998).

296. H. J. Boehm, *J. Comput.-Aided Mol. Des.*, **12**, 309 (1998).

297. A. N. Jain, *J. Comput.-Aided Mol. Des.*, **10**, 427 (1996).

298. D. N. Boobbyer, P. J. Goodford, P. M. McWhinnie, and R. C. Wade, *J. Med. Chem.*, **32**, 1083 (1989).

299. M. Stahl, *Perspect. Drug Discov. Des.*, **20**, 83 (2000).

300. H. J. Boehm, D. W. Banner, and L. Weber, *J. Comput.-Aided Mol. Des.*, **13**, 51 (1999).

301. M. Matsumara, W. J. Becktel, and B. W. Matthews, *Nature*, **334**, 406 (1988).

302. V. Nauchitel, M. C. Villaverde, and F. Sussman, *Protein Sci.*, **4**, 1356 (1995).

303. L. Wesson and D. Eisenberg, *Protein Sci.*, **1**, 227 (1992).

304. S. Vajda, Z. Weng, R. Rosenfeld, and C. DeLisi, *Biochemistry*, **33**, 13977 (1994).

305. C. Zhang, G. Vasmatzis, J. L. Cornette, and C. DeLisi, *J. Mol. Biol.*, **267**, 707 (1997).

306. S. Miyazawa and R. L. Jernigan, *Macromolecules*, **18**, 534 (1985).

307. R. D. Head, M. L. Smythe, T. I. Oprea, C. L. Waller, S. M. Green, and G. R. Marshall, *J. Am. Chem. Soc.*, **118**, 3959 (1996).

308. E. C. Meng, I. D. Kuntz, D. J. Abraham, and G. E. Kellogg, *J. Comput.-Aided Mol. Des.*, **8**, 299 (1994).

309. H. Gohlke and G. Klebe, *Curr. Opin. Struct. Biol.*, **11**, 231 (2001).

310. M. J. Sippl, *J. Comput.-Aided Mol. Des.*, **7**, 473 (1993).

311. G. Verkhivker, K. Appelt, S. T. Freer, and J. E. Villafranca, *Protein Eng.*, **8**, 677 (1995).

312. A. Wallqvist, R. L. Jernigan, and D. G. Covell, *Protein Sci.*, **4**, 1881 (1995).

313. R. S. DeWitte and E. I. Shakhnovich, *J. Am. Chem. Soc.*, **118**, 11733 (1996).

314. R. S. DeWitte, A. V. Ishchenko, and E. I. Shakhnovich, *J. Am. Chem. Soc.*, **119**, 4608 (1997).

315. J. B. O. Mitchell, R. A. Laskowski, A. Alex, and J. M. Thornton, *J. Comput. Chem.*, **20**, 1165 (1999).

316. J. B. O. Mitchell, R. A. Laskowski, A. Alex, M. J. Forster, and J. M. Thornton, *J. Comput. Chem.*, **20**, 1177 (1999).

317. I. Muegge and Y. C. Martin, *J. Med. Chem.*, **42**, 791 (1999).

318. H. Gohlke, M. Hendlich, and G. Klebe, *Perspect. Drug Discov. Des.*, **20**, 115 (2000).

319. I. Muegge, *J. Comput. Chem.*, **22**, 418 (2001).

320. I. Muegge, *Perspect. Drug Discov. Des.*, **20**, 99 (2000).

321. H. M. Berman, J. Westbrook, Z. Feng, G. Gilliland, T. N. Bhat, H. Weissig, I. N. Shindyalov, and P. E. Bourne, *Nucleic Acids Res.*, **28**, 235 (2000).

322. M. Hendlich, *Acta Crystallogr. D Biol. Crystallogr.*, **54**, 1178 (1998).

323. A. Bergner, J. Guenther, M. Hendlich, G. Klebe, and M. Verdonk, *Biopolymers (Nucl. Acid Sci.)*, **61**, 299 (2002).

324. Y. Sun, T. J. Ewing, A. G. Skillman, and I. D. Kuntz, *J. Comput.-Aided Mol. Des.*, **12**, 597 (1998).

325. E. J. Martin, R. E. Critchlow, D. C. Spellmeyer, S. Rosenberg, K. L. Spear, and J. M. Blaney, *Pharmacochem. Libr.*, **29**, 133 (1998).

326. B. K. Shoichet, D. L. Bodian, and I. D. Kuntz, *J. Comput. Chem.*, **13**, 380 (1992).

327. T. J. A. Ewing and I. D. Kuntz, *J. Comput. Chem.*, **18**, 1175 (1997).

328. T. Ewing, Ed., *DOCK User Manual, Version 4.0*, Regents of the University of California, San Francisco, 1998.

329. M. Stahl and H. J. Boehm, *J. Mol. Graph. Model.*, **16**, 121 (1998).

330. C. Lemmen, T. Lengauer, and G. Klebe, *J. Med. Chem.*, **41**, 4502 (1998).

331. C. Lemmen, A. Zien, R. Zimmer, and T. Lengauer, *Pac. Symp. Biocomput.*, 482 (1999).

332. G. R. Desiraju and T. Steiner, *The Weak Hydrogen Bond in Chemistry and Biology*, Oxford University Press, Oxford, 1999.

333. G. Parkinson, A. Gunasekera, J. Vojtechovsky, X. Zhang, T. A. Kunkel, H. Berman, and R. H. Ebright, *Nat. Struct. Biol.*, **3**, 837 (1996).

334. J. P. Gallivan and D. A. Dougherty, *J. Am. Chem. Soc.*, **122**, 870 (2000).

335. G. B. McGaughey, M. Gagne, and A. K. Rappe, *J. Biol. Chem.*, **273**, 15458 (1998).

336. C. Chipot, R. Jaffe, B. Maigret, D. A. Pearlman, and P. A. Kollman, *J. Am. Chem. Soc.*, **118**, 11217 (1996).

337. T. G. Davies, R. E. Hubbard, and J. R. Tame, *Protein Sci.*, **8**, 1432 (1999).

338. G. Klebe, F. Dullweber, and H. J. Boehm in R. B. Raffa, Ed., *Drug-Receptor Thermodynamics: Introduction and Applications*, John Wiley & Sons, Chichester, 2001, p. 83.

339. I. Muegge, *Med. Chem. Res.*, **9**, 490 (1999).

340. I. Nobeli, J. B. O. Mitchell, A. Alex, and J. M. Thornton, *J. Comput. Chem.*, **22**, 673 (2001).

341. I. Massova and P. A. Kollman, *Perspect. Drug Discov. Des.*, **18**, 113 (2000).

342. P. A. Kollman, I. Massova, C. Reyes, B. Kuhn, S. Huo, L. Chong, M. Lee, T. Lee, Y. Duan, W. Wang, O. Donini, P. Cieplak, J. Srinivasan, D. A. Case, and T. E. Cheatham, 3rd, *Acc. Chem. Res.*, **33**, 889 (2000).

343. I. Massova and P. A. Kollman, *J. Am. Chem. Soc.*, **121**, 8133 (1999).

344. O. A. Donini and P. A. Kollman, *J. Med. Chem.*, **43**, 4180 (2000).

345. B. Kuhn and P. A. Kollman, *J. Med. Chem.*, **43**, 3786 (2000).

346. B. Kuhn and P. A. Kollman, *J. Am. Chem. Soc.*, **122**, 3909 (2000).

347. N. Froloff, A. Windemuth, and B. Honig, *Protein Sci.*, **6**, 1293 (1997).

348. J. Shen, *J. Med. Chem.*, **40**, 2953 (1997).

349. C. J. Woods, M. A. King, and J. W. Essex, *J. Comput.-Aided Mol. Des.*, **15**, 129 (2001).

350. G. Archontis, T. Simonson, and M. Karplus, *J. Mol. Biol.*, **306**, 307 (2001).

351. M. L. Verdonk, J. C. Cole, and R. Taylor, *J. Mol. Biol.*, **289**, 1093 (1999).

352. M. L. Verdonk, J. C. Cole, P. Watson, V. Gillet, and P. Willett, *J. Mol. Biol.*, **307**, 841 (2001).

353. D. E. Clark and S. D. Pickett, *Drug Discov. Today*, **5**, 49 (2000).

354. I. Muegge, S. L. Heald, and D. Brittelli, *J. Med. Chem.*, **44**, 1841 (2001).

355. Unity Chemical Information Software, Tripos Inc., St. Louis, MO.

356. Y. Kurogi and O. F. Guener, *Curr. Med. Chem.*, **8**, 1035 (2001).

357. M. Rarey and J. S. Dixon, *J. Comput.-Aided Mol. Des.*, **12**, 471 (1998).

358. M. Rarey and M. Stahl, *J. Comput.-Aided Mol. Des.*, **15**, 497 (2001).

359. A. I. Su, D. M. Lorber, G. S. Weston, W. A. Baase, B. W. Matthews, and B. K. Shoichet, *Proteins*, **42**, 279 (2001).

360. Protherics (2000). DockCrunch [Online]. Protherics Molecular Design Ltd. http://www.protherics.com/crunch/ [2001, Sept 25].

361. H. J. Boehm and M. Stahl, *Curr. Opin. Chem. Biol.*, **4**, 283 (2000).

362. C. W. Murray, D. E. Clark, T. R. Auton, M. A. Firth, J. Li, R. A. Sykes, B. Waszkowycz, D. R. Westhead, and S. C. Young, *J. Comput.-Aided Mol. Des.*, **11**, 193 (1997).

363. M. Rarey and T. Lengauer, *Perspect. Drug Discov. Des.*, **20**, 63 (2000).

364. D. C. Roe and I. D. Kuntz, *J. Comput.-Aided Mol. Des.*, **9**, 269 (1995).

365. S. Makino, T. J. Ewing, and I. D. Kuntz, *J. Comput.-Aided Mol. Des.*, **13**, 513 (1999).

366. M. D. Miller, R. P. Sheridan, S. K. Kearsley, and D. J. Underwood, *Methods Enzymol.*, **241**, 354 (1994).

367. G. M. Verkhivker, D. Bouzida, D. K. Gehlhaar, P. A. Rejto, S. Arthurs, A. B. Colson, S. T. Freer, V. Larson, B. A. Luty, T. Marrone, and P. W. Rose, *J. Comput.-Aided Mol. Des.*, **14**, 731 (2000).

368. S. B. Shuker, P. J. Hajduk, R. P. Meadows, and S. W. Fesik, *Science*, **274**, 1531 (1996).

369. I. Muegge, Y. C. Martin, P. J. Hajduk, and S. W. Fesik, *J. Med. Chem.*, **42**, 2498 (1999).

370. R. L. DesJarlais, G. L. Seibel, I. D. Kuntz, P. S. Furth, J. C. Alvarez, P. R. Ortiz de Montellano, D. L. DeCamp, L. M. Babe, and C. S. Craik, *Proc. Natl. Acad. Sci. USA*, **87**, 6644 (1990).

371. B. K. Shoichet, R. M. Stroud, D. V. Santi, I. D. Kuntz, and K. M. Perry, *Science*, **259**, 1445 (1993).

372. L. R. Hoffman, I. D. Kuntz, and J. M. White, *J. Virol.*, **71**, 8808 (1997).

373. I. Massova, P. Martin, A. Bulychev, R. Kocz, M. Doyle, B. F. Edwards, and S. Mobashery, *Bioorg. Med. Chem. Lett.*, **8**, 2463 (1998).

374. D. Tondi, U. Slomczynska, M. P. Costi, D. M. Watterson, S. Ghelli, and B. K. Shoichet, *Chem. Biol.*, **6**, 319 (1999).

375. T. Toyoda, R. K. Brobey, G. Sano, T. Horii, N. Tomioka, and A. Itai, *Biochem. Biophys. Res. Commun.*, **235**, 515 (1997).

376. Available Chemicals Directory, MDL Information Systems Inc., San Leandro, CA.

377. F. Abbate, C. T. Supuran, A. Scozzafava, P. Orioli, M. T. Stubbs, and G. Klebe, *J. Med. Chem.*, **45**, in press (2002).

378. Maybridge Database (1999), Maybridge Chemical Co. Ltd., UK.

379. LeadQuest Chemical Compound Libraries (2000), Tripos Inc., St. Louis, MO.

380. J. Sadowski, C. Rudolph, and J. Gasteiger, *Tetrahedron Comput. Methodol.*, **3**, 537 (1990).

381. A. Maxwell, *Biochem. Soc. Trans.*, **27**, 48 (1999).

382. D. Bailey and D. Brown, *Drug Discov. Today*, **6**, 57 (2001).

383. R. M. Eglen, G. Schneider, and H. J. Boehm in G. Schneider and H. J. Boehm, Eds., *Virtual Screening for Bioactive Molecules*, VCH, Weinheim, Germany, 2000, p. 1.

384. R. Lahana, *Drug Discov. Today*, **4**, 447 (1999).

385. C. A. Lipinski, F. Lombardo, B. W. Dominy, and P. J. Feeney, *Adv. Drug Delivery Rev.*, **23**, 3 (1997).

386. Y. Iwata, M. Arisawa, R. Hamada, Y. Kita, M. Y. Mizutani, N. Tomioka, A. Itai, and S. Miyamoto, *J. Med. Chem.*, **44**, 1718 (2001).

387. V. Sobolev, R. C. Wade, G. Vriend, and M. Edelman, *Proteins*, **25**, 120 (1996).

388. W. Welch, J. Ruppert, and A. N. Jain, *Chem. Biol.*, **3**, 449 (1996).

389. E. Perola, K. Xu, T. M. Kollmeyer, S. H. Kaufmann, F. G. Prendergast, and Y. P. Pang, *J Med Chem.*, **43**, 401 (2000).

390. D. S. Goodsell, G. M. Morris, and A. J. Olson, *J. Mol. Recognit.*, **9**, 1 (1996).

391. C. A. Sotriffer, H. Ni, and J. A. McCammon, *J. Med. Chem.*, **43**, 4109 (2000).

392. T. Hou, J. Wang, L. Chen, and X. Xu, *Protein Eng.*, **12**, 639 (1999).

393. B. N. Dominy and C. L. Brooks, 3rd, *Proteins*, **36**, 318 (1999).

394. I. D. Wall, A. R. Leach, D. W. Salt, M. G. Ford, and J. W. Essex, *J. Med. Chem.*, **42**, 5142 (1999).

395. D. A. Pearlman and P. S. Charifson, *J. Med. Chem.*, **44**, 3417 (2001).

396. S. S. So and M. Karplus, *J. Comput.-Aided Mol. Des.*, **13**, 243 (1999).

397. M. Rarey, B. Kramer, and T. Lengauer, *Bioinformatics*, **15**, 243 (1999).

398. Y. Takamatsu and A. Itai, *Proteins*, **33**, 62 (1998).

399. D. Rognan, S. L. Lauemoller, A. Holm, S. Buus, and V. Tschinke, *J. Med. Chem.*, **42**, 4650 (1999).

400. G. E. Kellogg, G. S. Joshi, and D. J. Abraham, *Med. Chem. Res.*, **1**, 444 (1991).

CHAPTER EIGHT

Bioinformatics: Its Role in Drug Discovery

David J. Parry-Smith
ChiBio Informatics
Cambridge, United Kingdom

Contents

Burger's Medicinal Chemistry and Drug Discovery
Sixth Edition, Volume 1: Drug Discovery
Edited by Donald J. Abraham
ISBN 0-471-27090-3 © 2003 John Wiley & Sons, Inc.

1 INTRODUCTION

In January 2001, the biopharmaceutical company Millennium announced that, as part of a multimillion-dollar research collaboration with Bayer, an anticancer agent had proceeded to clinical trials (1, 2). The remarkable achievement was not that the collaboration between a pharmaceutical company and a high technology genomics company had been so successful in terms of a product, but that it had ostensibly lopped 2 years from the discovery lifecycle in the process. As a result of perceived improvements in research efficiency such as this, more effort than ever is being placed in development and implementation of genomics and screening technology automation. The substantial volumes of data thus generated are used to pan for innovative lead compounds for novel therapeutic targets to feed ever more voracious development pipelines (3).

The genomics tools of rapid sequence screening, microarray chips, expression analysis, protein interactions, macromolecular structure determination, sequence comparison, and a host more are all biological techniques that generate different types of raw data. Moreover, the data are produced in far greater quantity than has been seen in biology before and the sheer speed with which the data are produced is unprecedented. The improvements in process throughput, which are so exciting to the financial markets and so critical to the alleviation of pain and suffering in human populations, are readily achievable because of the science of information integration and knowledge transformation that are the hallmarks of bioinformatics. It is not enough simply to produce data, even from the most leading edge of techniques; we must be able to manage it effectively and extract useful information that leads to that critical knowledge on which realistic drug discovery decisions can be made.

The opening example illustrates how genomic technologies, including bioinformatics, are making an impact at the drug discovery level. The purpose of this article is to provide some background to the tools and technologies that are used on a daily basis by bioinformatics scientists in an effort to make the subject more accessible to readers with a non-biological training. This article is not, however, a training manual in running programs nor is it an index to the latest resources on the Internet. Both bioinformatics and the World Wide Web grow at such a pace that any journal article is likely to be out-of-date before it goes to press. Nevertheless, the Internet is a crucial resource for the practicing bioinformatics researcher. All of the core uses for executing projects are available there both in terms of databases and computer programs. Lists and descriptions would fill many volumes. Certain resources are mentioned where necessary to illustrate specific examples. Some useful starting points for an exploration of bioinformatics on the Internet are shown in Table 8.1. An introduction to tools and techniques is available in Ref. 4, while Ref. 5 contains a more technical approach based on understanding of machine learning. For a more mathematical treatment, readers are referred to Ref. 6. Reference may be made to the journal *Briefings in Bioinformatics* for reasonably accessible descriptions of systems and processes. The journal *Bioinfomatics* is aimed at a more technical audience. The annual database issue of *Nucleic Acids Research* is the generally accepted place for publication of bioinformatics databases. Some fundamental papers referred to are relatively old for a young science.

Rather than delve into minutiae here, the aim has been to present an overview of the way bioinformatics is being used in the process of drug discovery in 2002. Many important and exciting aspects of the academic research that is being carried out are passed over in silence or referred to only by a brief comment. Computer technologies change rapidly and bioinformatics has always been at the forefront in applying new computing paradigms to biological problems (e.g., use of the Internet, object-oriented programming technologies, neural networks, parallel computing). Some of the molecular biology or bio-

Table 8.1   A Selected List of Key Websites for Further Exploration of Online Bioinformatics Resources

Internet Site	Brief Description
http://www.ncbi.nlm.nih.gov	The National Center for Biotechnology Information (NCBI). Located at the National Library of Medicine in Bethesda, MD, USA. The home of the GenBank DNA sequence database; PubMed literature search engine; sequence search tools (e.g., PSI-BLAST); genomic sequence navigation tools. A substantial repository of resources in all areas of bioinformatics.
http://www.ebi.ac.uk	The European Bioinformatics Institute (EBI). This site is located at Hinxton Hall, Cambridge, UK. The home of the EMBL Nucleotide Sequence Database; data management tools [including publicly accessible version of SRS—the Sequence Retrieval System (7)]; protein family databases; microarray tools; etc. An extensive repository of resources for bioinformatics.
http://www.expasy.ch	The Expert Protein Analysis System. Dedicated to the analysis of protein sequence and structure as well as two-dimensional PAGE. Home of the SWISS-PROT protein knowledgebase and TrEMBL computer annotated supplement (8).
http://www.man.ac.uk	The University of Manchester Bioinformatics Education and Research site (UMBER). Useful because it is the home of the PRINTS (9,10) resource for protein fingerprint analysis and a valuable teaching site for bioinformatics.
http://www.ensembl.org	Site developed by the Sanger Centre, Hinxton Hall, Cambridge, UK and the EMBL-EBI, presenting tools for browsing and researching the human genome sequence (11). This is a public access server providing data and access at no charge. Commercial sites are also available for working on commercially produced human genome sequence.
http://www.mips.biochem.mpg.de	The Munich Information Center for Protein Sequences (MIPS). Provides a different view of several model organism genomes with tools for analysis.

physical aspects of techniques used to generate data are outlined from the point of view of a bioinformatician and not that of a practicing molecular biologist (although it has been reviewed by one, see the acknowledgments) to give a flavor of the kinds of experiments that are performed.

2   BIOINFORMATICS IN DRUG DISCOVERY

Bioinformatics in drug discovery has traditionally been used as a tool for finding new drug targets ("target selection"). This technique of target discovery is an important contribution that bioinformatics has been able to make to the drug discovery process. However, bioinformatics alone without the background of molecular biological and biophysical experiments is sterile. To understand even a bird's-

eye overview such as this, some basic material has to be covered to enable comprehension both of the data and the manner of its analysis.

The second use to which bioinformatics has been put in the drug discovery process is more fundamental. It is concerned with the use of techniques in molecular sequence analysis to generate relationships between sequences that are themselves used to provide fundamental structures for databases of drug discovery information. Relationships between data elements are important because they help to place individual elements in a context that can be readily assimilated by the user of the system. In many situations, observers approach data from different points of view and bring to bear the richness of differing scientific experiences. Whether we care to admit it or not, "biologists" and "chemists" have different training and background and thus offer a range of opinions on similar pieces of information. Even the word "activity" means different things to a chemist who has synthesized a group of compounds or a biologist who has developed an assay to test the compounds. Both aspects are necessary for the discovery of new drugs, but they are different viewpoints that need to be supported by appropriate relationship mining in the data. If the bioinformatics job is done well, both views can be accommodated in the data structures and user interfaces used by both sets of users.

Throughout the pharmaceutical industry, bioinformatics and chemoinformatics groups are working closer together than has been the case hitherto. This is a consequence of the realization that managing data effectively requires integration of thinking (about definitions of common attributes of molecules both small and large), integration of processes, and integration of implementation. The recent rise in popularity in bioinformatics of the ontology is an example of the application of a computer science paradigm to the issue of redundancy in nomenclature in many areas of biology. Application across the chemistry–biology domain interface could well be beneficial for drug discovery effectiveness. The ontology is simply a means to an end, in this instance, that end is improved communication and understanding of basic concepts within and across the boundaries of major scientific disciplines. There may, of course, be a variety of other means to reach that goal.

3 WHAT IS BIOINFORMATICS?

3.1 Definitions

Concisely, bioinformatics is our ability to organize biological data. From another perspective, bioinformatics is our ability to understand how biological information is organized. From this understanding should spring an enhanced view of the interactions between biological molecules. This should, in turn, inform our search strategies for new small molecules that will modulate the behaviour of biological molecules to give a beneficial therapeutic effect. These definitions arise from observation of the way diverse skills are brought to bear in

attempting to answer biological questions. They also stress the importance of organizing and understanding biological data, rather than linking these aspects strongly to specific hardware or software implementations. Use of computers may be involved in the process but the definitions are not limited by the application of any particular technology.

Bioinformatics has also been defined as the application of computer technology to solving biological problems. This definition, perhaps what some would consider to be the canonical one, is broad but restricts the scope of the definition to problems to which computer technology can be applied.

3.2 Integration of Information

Bioinformatics has become a byword for integration; specifically the integration of data across different data resources to generate integrated information resources. Linking data and information in this way is fundamental to bioinformatics activities and so some discussion of the meaning of data, information, and knowledge in the context of bioinformatics for drug discovery is provided in Section 6. Integration is important because it provides context, or at least a background, against which computational analyses are performed. In the past, for single molecule experiments, this background was achieved through reading the literature. Now that multiple molecule experiments are common, even genome-wide or inter-genome analyses, it is simply not practical any longer to rely on the literature in its raw form, unless it is part of an integrated knowledge-based approach that provides connections between disparate pieces of information, backed up by experimental evidence from which to draw conclusions (12).

3.3 Bioinformatics and Skills

The pursuit of bioinformatics involves a number of different skills. Organizing, storing, retrieving, and querying sets of biological data are techniques that lie at the heart of the subject. An ability to analyze the characteristics of particular sets of biological data is fundamental. The translation of those characteristics into electronic representations that can be organized on a large scale is the domain of the bioinformatics software and database developer. The process of analyzing and understanding biological data using the tools available is the domain of the bioinformatics analyst. When new tools are in the course of development, substantial interaction between the two skill sets is essential.

In the pharmaceutical environment, both developer and analyst skills are necessary. This is so even where commercial software is in use, because there is no single system available commercially that provides the level of integration between the worlds of bio- and chemoinformatics necessary to effectively enhance the drug discovery process. Some interfacing of different systems is required and the warehousing of proprietary data is always an issue.

This broad description of bioinformatics and of the two types of bioinformatics scientist is quite abstract. It does not detail the characteristics of the data with which the bioinformatics scientist has to work. Neither does it define the set of tools that the developer should work with or implement. There was a time, in the late 1980s and early 1990s, when the type of data was well defined. Molecular sequence data, the stream of bases in DNA, and the stream of residues at the protein level were the main types of data. Programmers developed code in FORTRAN or C and scripting languages were immature.

Now, as science moves forward into a new millennium, additional types of data have become important; for example, protein–protein interactions and three-dimensional structure, high density gene expression chips, cell imaging, etc. Developers have a wide range of tools to call on, including high performance C and C++ compilers, rich scripting languages (Perl, Python, etc.), and efficient, easily accessible operating systems (particularly Linux) that make porting software to different hardware platforms less of an issue than it was.

Of equal importance to the medicinal chemist, to whom this review is principally directed, is the impact of bioinformatics on the discovery of new medicines. Rather than explain comprehensively all the popular tools and their underlying algorithms, this review focuses on the points in the discovery research process where bioinformatics is making an impact. Technologies will be described to the extent that such understanding is necessary to grasp the relevance of the data being generated and its significance.

3.4 Standardization

Progress in linking items of relevant data and generating integrated information resources would be very limited were it not for efforts in standardization that have been brought about by international collaboration. There is still a long way to go, however. While it is becoming cheaper to obtain each piece of individual data, the proportion of automated experiments is increasing, at least in the life sciences, because of the ready availability of new technologies. It may seem a simple matter to create resources that store and manage streams of DNA bases—represented by the four alphabetic characters A, C, T, and G. However, when we also wish to integrate information on experimentally or computationally determined annotation and cross-reference to other resources using gene names, there are significant problems. The literature abounds with synonyms for gene names and functions; even the labels given to specific cellular functions are not always clearly defined (13).

To be able to process data automatically, it has to be presented in a form that can be parsed by a computer program and must also include all the elements necessary to an understanding of the biological system under study. Reliable information systems should have source data of a consistently high quality to prevent application errors and enable integration into other biological information systems. Some progress is now being made towards consensus in gene naming through the work of the HUGO Gene Nomenclature Committee (see http://www.gene.ucl.ac.uk/nomenclature/). Many researchers now use this system as a source of unique gene names and descriptions in the published literature (14) and in commercial products (e.g., see http://www.biowisdom.com/). Standardizing vocabulary expressing the relationships between the complex network of gene functions is the work of the Gene Ontology (GO) project (see http://www.geneontology.org/).

4 BIOINFORMATICS AND TARGET DISCOVERY

The desire to find new drug targets is grounded in the need of pharmaceutical companies to address the requirements of different disease markets. The literature is full of papers detailing sequence determinations of newly cloned receptors and enzymes along with research on their functional properties. The realization that the sequences of genes could be acquired relatively cheaply through the use of automated sequencing machines using fluorescent base technology, rather than the previous generation of radioactive sequencing gels, meant that sequence data began to flood the public DNA sequence databases. The growth of these databases is reviewed in Fig. 8.1.

Because the translation of DNA coding sequence to protein sequence is straightforward, given the understanding of the genetic code, it is a trivial task to implement software to provide translations of open reading frames (ORFs[1]) to be housed in the annotation sections of the DNA databases. Consequently, protein sequence databases have become swamped with hypothetical proteins—those proteins assumed to exist because open reading frames have been discovered and from which hypothetical protein sequences had been computed. The function of these sequences has been assigned through a comparison of their amino acid residue similarity with that of known sequences (for example, those that have had their biochemical function demonstrated through heterologous expression). In this way, very large numbers of sequences have been processed into the databases and annotated using such sequence comparison techniques.

When it comes to the practical details of how bioinformatics can speed the process of drug discovery, it is reasonable to ask what sorts of data could be valuable in that process. The stages of the drug discovery process where bioinformatics makes an impact are target identification, assay selectivity panel selection, and integration throughout the assay development and screening process. Target identification makes use of sequence data for functional assignment by inference from similarity with known sequences (Fig. 8.2). It also benefits from assessments of differential levels of expression in different cellular contexts and at various stages in the expression process (transcriptome or proteome, see Fig. 8.3 for definitions). Selectivity panel selection relies on a thorough mining of the related gene family and may benefit from phylogenomic analysis (see Section 5.3). Use of bioinformatics for integration of data capitalizes on the generation of relationship information between known genes and the ability to use hyperlinking to create navigational tools and usable interfaces.

A further development, the production of millions of short expressed sequence tags (ESTs), encouraged the focus on target discovery during the 1990s. The development of EST technology itself spawned the genesis of several new genomics companies, including Human Genome Sciences and Incyte, which have worked in collaboration with the pharmaceutical industry to hunt for new disease related targets.

4.1 Functional Genomics and Target Discovery

The collation of gene sequence data, from whatever source, is in effect simply a matter of transferring data from one place to another; for example, from a sequence chromatogram to a computer database. Learning the sequence of a genome, or any of its constituent parts, is a long way from understanding its biological function. The sequencing of genomes has resulted in a technical genome description (at a particular level of detail) through the process of cataloguing an organism's genes. This level of detail is often called the physical map of the genome. There are other ways of mapping the genome that provide different levels (or we may think of it as resolution) of genomic detail; genetic maps indicating the location of genes for specific traits have been known for some time, while single nucleotide polymorphism (SNP) maps can be used to highlight the positions of differences between populations through study of genetic polymorphisms (15). Indeed, the identification

---

[1] ORFs are contiguous strings of residues, uninterrupted by the genetic code's "stop" signal.

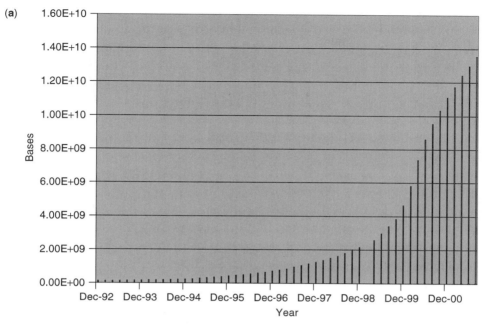

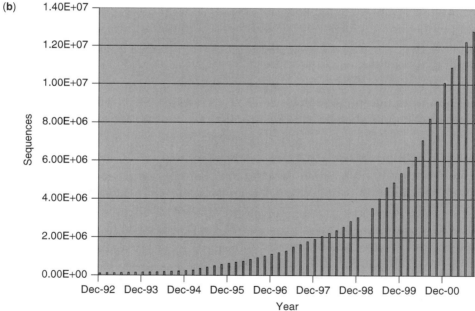

Figure 8.1. Bar charts indicating the growth of GenBank from December 1992 to August 2001 in terms of (a) bases and (b) sequence entries. The release files indicate no release in February 1999. It is evident from the trends in both charts that while there has been explosive growth, particularly from December 1999 until about August 2000, growth is slowing. The base entry curve is showing a distinctly sigmoid shape.

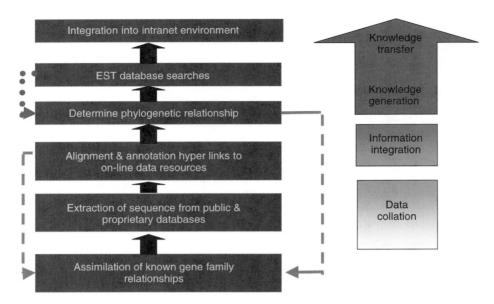

Figure 8.2. A schematic illustrating the bioinformatics process required to create an online gene index by collating data and then integrating related elements to generate value added information through hyperlinking to online resources. Determination of phylogenetic relationships is a relatively late stage in the process. EST analysis is only performed after phylogenetic relationships have been determined because EST data does not cover the whole expressed sequence and may not therefore cover regions that were included in the phylogenetic determination. This is a knowledge generation phase because it is allowing placement of potential new targets within the context of a carefully researched phylogenetic tree. Transfer of knowledge is intimately related to the environment in which the results of analysis are made available, in this case as an online resource.

of genes themselves from genomic sequence is itself a non-trivial matter, especially where those genes are interrupted by non-coding regions (introns) and control regions (expression promoter sites). Functional genomics is the process of creating an understanding of the way genomes function through gene expression. Genes are expressed by a variety of mechanisms, not all of which are fully understood. We can, however, make some measurements of the results of gene expression at the transcript level, mRNA, and at the protein level. Several of the techniques that have been used to assist drug target discovery are presented in the following sections.

4.2   Expression Profiling for Target Discovery

Bioinformatics spans analysis in depth on small quantities of data through to expansive genomic scale analyses, which may be at a lesser level of detail. Historically, the expression of genes at the mRNA level or at the pro-

tein level has been a crucial tool for assessing the significance of specific classes of cells as targets. With the advent of fluorescence-based sequencing techniques and automated sequencing technology, it is now much quicker to generate sequence data on specific molecular targets than ever. Many researchers spend entire careers working on one target type or a restricted part of a target gene family. This approach has yielded many valuable targets for drug discovery. With the new technologies of molecular biology, it is now possible to survey targets in a variety of contexts; perhaps within different types of cells, cells treated with different agents, or even across entire genomes using chip technologies.

There are issues of interpretation of experimental design and results. Does mRNA expression mean anything at a quantitative level? Perhaps even a qualitative view of mRNA expression can be misleading. How is mRNA expression correlated with protein ex-

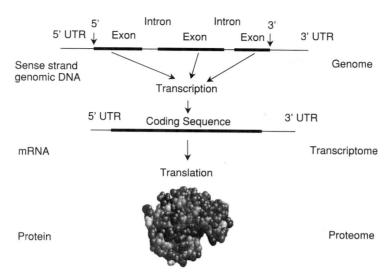

Figure 8.3. A schematic illustration of the relationships between levels of genomic information. Genomic DNA is contained in the nucleus of eukaryotic cells. In many species, including humans, information required to make up the coding sequence of a gene is split into exons (regions that are expressed) interrupted by introns (regions that are not expressed and are edited out of the message at the transcription step). At either end of the gene sequence are untranslated regions (UTRs). 5' and 3' refer to the orientation of the strand of DNA as defined by the sugar-phosphate backbone. The mRNA is the messenger RNA molecule generated by the process of transcription, which is itself mediated by a number of enzymes. The collection of mRNA transcripts that make up the mRNA expression profile of a cell is known as the transcriptome, although the term could also refer to the total possible mRNA transcripts achievable from a genome. Finally, translation of the mRNA occurs on the ribosome and protein sequence is produced, which folds into its final three-dimensional shape—a process that may be assisted by a number of different chaperone proteins. Any post-translational modifications are all part of the proteome, the collection of proteins that represent the expressed genome.

pression? In general, most drugs we discover are likely to interact with proteins and not mRNA, so some understanding of protein expression is an essential adjunct to our genomic knowledge. Hence the proteomics approaches described later. Our exploration of expression profiling begins with a study of mRNA transcript profiling using expressed sequence tags because this technique has led to rapid gene discovery that has, in turn, been able to assist with the annotation of genomic sequences. Then, we consider how whole genome expression profiles can provide a rich new source of data for bioinformatics analysis.

4.2.1 EST Profiling. An EST is a short, single sequence run collecting data over about 200–400 bases from a clone selected from a cDNA library. Typically, cDNA clone libraries contain 1–3 million clones. The library itself is created from mRNA extracted from tissue or refined cell populations. By making a random selection of several thousand clones from a cDNA library it is possible by sequencing ESTs to generate a rapid, if somewhat low resolution, survey of the types of genes represented by the library. The library in turn reflects the composition of genes that are expressed in the tissue or cell line from which it was constructed. Thus, we have a qualitative link between gene expression, at the mRNA level, and the sequence level analysis required for target identification, without the need to go through the full sequencing and validation process across the whole length of each clone. This is a very significant time and cost saving. One of the major issues of EST profiling has been the significance that can be ascribed to expression levels through counting copies of ESTs. This issue is dealt with in some detail in Section 4.2.4.

4.2.2 Sequence Assembly for cDNA Cloning.

To appreciate fully the speed advantage of sequencing tags, rather than fully validating the sequence of an entire clone of a gene, it is useful to step through a brief description of the cloning process from the point of view of a practitioner of bioinformatics.

Sequence assembly is the process of dealing with the bioinformatics of cloning and genomic sequencing (16, 17). When a gene is cloned, it is selected from a set of potential clones in a cDNA library. The gene is present as a piece of cDNA inserted into a cloning vector (a piece of circular DNA) that has been designed for the purpose of cloning. It is necessary to check that the cDNA indeed represents the sequence of the gene that has been cloned. To do this, DNA oligonucleotides are designed that will bind in a complementary fashion (hybridize) to the DNA of the cloning vector and also at 150- to 200-base intervals along the cDNA itself. These oligonucleotides are then extended by adding a base that is complementary to the cDNA insert by using a DNA polymerase. Different polymerases are available commercially that provide high fidelity reproduction of the cDNA insert. In fluorescence-based sequencing, a small proportion of the nucleotides available to the polymerase are fluorescent analogues. Incorporation of one of these into the oligonucleotide terminates the extension, resulting in a population of oligonucleotides of different lengths. These are separated by electrophoresis and the sequence determined.

When the sequence of each oligonucleotide has been determined, the strings of letters that represent the bases are assembled together to generate a full-length sequence of the cDNA that has actually been cloned. Errors in the base sequence can be resolved at this stage, and if necessary, mutagenesis experiments designed to correct any mistakes. The bioinformatics process is intimately linked with the molecular biology techniques of cloning and sequencing. For target discovery, a very high degree of confidence in the sequence of the cDNA clone is required before the clone can be expressed and used in assay development. It can also be seen that this process is much lengthier than taking a single sequence read (a single oligonucleotide string) without correcting errors or considering coverage of the complete gene sequence.

4.2.3 Comparing ESTs with Databases.

Bioinformatics provides the tools necessary to compare each EST with the databases of known genes and a hypothetical functional assignment may be made to a proportion (typically 40–50%) of all the ESTs from a sequencing run.

In this way a rich resource of tags for many clones from many diverse libraries has been built up in the public domain and in commercially available, proprietary databases. One particular approach that generated much interest in the 1990s was that advocated by Incyte. Here, the simple identification of a gene expressed through identification of its EST was not the primary goal. Instead, the approach was based on comparative transcript expression (the so-called "digital Northern"). Here the number of copies of each EST identified was calculated, giving counts for the numbers of each type of EST found in comparing normal with diseased tissue, for example. Subsequent techniques have focussed on more controlled experiments in which specific cell lines are treated with an agent and the expression of genes before administration is compared with the profile afterward. This is the basis of pharmacogenomics (18).

4.2.4 Statistics for Assessing Expression Level Significance.

There are issues with approaches based on counting the number of copies of an EST observed in the output from a sequencing machine. First, the tissue or cell line must be of very high quality and the mRNA harvested in a timely manner because it degrades very quickly. Second, the process of preparing the cDNA library should enable the numbers of clones to be estimated as accurately as possible. Third, the random sampling for the sequencing runs must be controlled carefully so as not to introduce bias into the experiment. The mathematical model for evaluating the meaning of data from such experiments is not well worked out.

Comparison of the differences between the

Table 8.2   Comparative EST Counts for Five Genes Sequenced from Normal Prostate, Stage B2 Cancer, Stage C Cancer, and Benign Prostatic Hyperplasia (BPH) cDNA Libraries

Gene	Normal Prostate Total	Stage B2 Cancer		Stage C Cancer		BPH		All Other Tissue
		Tags	P	Tags	P	Tags	P	
PSA	13	7	0.7–0.8	14	0.6–0.7	22	0.8–0.9	0
PAP	4	1	0.1–0.2	34	>0.999	9	0.7–0.8	1
HGK	1	7	>0.999	6	0.97–0.98	5	0.8–0.9	0
PS1	0	3	0.993–0.994	7	0.997–0.998	1	0.4–0.5	0
PS2	0	2	0.97–0.98	7	0.997–0.998	0	0–<0.1	0
Total clones	4500	1400		3400		4800		732,000

The tag counts are from Ref. 21. The P values are calculated according to Equation 8.1, modified for use with different total EST counts from the source libraries. The web URL http://igs-server.enrs-mrs.fr/~audic/egi-bin/winflat.pl was used to calculate the probability intervals. A P value nearer to 1 indicates that the differential expression is likely to be significant. While prostate specific antigen (PSA) and glandular kallikrein (HGK) have been proposed as prostate cancer markers, both PS1 and PS2 are prostate specific. Thus, the down-regulation of PAP in stage B2 cancer is not significant using this test, whereas, the test shows its up-regulation in the BPH sample to be more significant. So, for lower changes in copy number, where more sensitivity is expected, this test of significance is a valuable tool.

overall profiles obtained from tag counting experiments could be performed using the traditional χ^2 test. However, this is the wrong approach for experiments where the significance of differences between expression levels (i.e., tag counts) of individual genes is to be determined, for example, in diseased and normal tissue states (19). One of the issues in performing tag-sampling experiments is that the experiments themselves are usually not replicated. Thus, the dispersion of results cannot be used to estimate the SEs associated with each expression measurement. This eliminates the possibility of using standard tests of variance. Instead the Poisson distribution, which includes an implicit estimate of standard error, approximates random sampling of tags very well. Audic and Claverie (20) have proposed a significance test (see Equation 8.1) in which the sample size plays no part, so long as it is the same for both experiments, but only depends on the observed tag counts of the same gene from diseased, g_A, and normal, g_B, states:

$$p\langle g_B | g_A \rangle = \frac{(g_A + g_B)!}{g_A! g_B! 2^{(g_A + g_B + 1)}} \qquad (8.1)$$

The equation has also been extended to cover the more practical case of different total numbers of tags. Thus, taking some data from Fannon (21) as an example, we can calculate values for the probability of certain genes expressed at different levels in normal prostate, stage B2 cancer, stage C cancer, and tissue from a benign prostatic hyperplasia (BPH) sample as shown in Table 8.2.

The relationship between gene expression, mRNA level, and protein expression is complex and not one that can be gleaned from collecting copy number information in this type of experiment. Even with careful statistical analysis, such as that described above, the assumption that increases or decreases in copy number reflect real biologically significant events relies on the confidence with which we can compare a library made from one set of cells to a library made from a different set of cells. Thus, most transcript analysis experiments setting out to be quantitative end up simply as target identification exercises. A major goal of proteomics is to generate a factory-type approach to profiling protein level expression that more closely reflects the biological reality. The EST approach has been turned into an industrial scale process but has not been able to impact the drug discovery process significantly because of the biological limitations described and the lack of sound mathematical modeling of the whole process.

Expression experiments are measures of cell population averages, not the contents of individual cells, so it is important to consider to what extent all cells in the candidate popu-

Table 8.3   Brief Descriptions of Three Technologies for Genomic Scale Transcript Profiling

Expression Profiling Technology	Brief Description	Form of Data Generated
cDNA array chip	Tens of thousands of cDNA clones of genes are placed onto a glass slide in a grid formation. Hybridisation of molecular probes (RNA extracts) to the clones is detected using a fluorescence system. By using two sets of probes, labelled with differently coloured fluorescent dyes, it is possible to assess expression differences.	Fluorescence intensities and colours for each spot on the chip. The nature of the clones on the chip is known.
High-density oligonucleotide arrays	Arrays of oligonucleotides are synthesised directly onto the glass chip using special chemistries and light sensitive masking. This generates arrays of known sequences of fixed length. Probes are hybridised to the arrays and computational analysis is necessary to interpret the resulting patterns.	An image of the entire chip is processed using specialised chip scanning software.
Serial analysis of gene expression	A sequence-based approach to the identification of differentially expressed genes through comparative analysis. Allows simultaneous analysis of sequences that derive from different cell populations or tissues. This is not a chip-based method. Identification of sequences relies on completeness of public sequence databases and, therefore, can only be used to analyse known genes.	Sequence data for SAGE tags allows profiling of gene expression.

lation are in the same state (22). Whereas work in single-celled organisms may be more straightforward to control, work in multi-cellular organisms has the added complexity that expression measurements may involve contributions from cells derived from a variety of tissues. Furthermore, when taking into consideration mRNA copy number, it should be understood that absolute transcript abundance measurements do not completely measure mRNA concentration.

Although there was initially some concern that the use of ESTs was a shortcut to discovery of genes for the purposes of patenting and ring-fencing areas of research for profit, in fact, the substantial numbers of quality ESTs in the public domain have helped in the pinpointing of genes in genomic data and have contributed to the speed with which the human genome sequence was completed.

4.2.5 Genome-Wide Expression Analysis. A major step towards understanding how organisms work is the determination of the complete sequence of all genes in the genome. This remarkable goal has been achieved for a num-

ber of organisms, including *Homo sapiens*, the flowering plant *Arabidopsis thaliana*, the single celled yeast *Saccharomyces cerevisiae*, and a large number of bacteria. The analysis of the sequence data then becomes the issue. It is no trivial task even to locate the positions of all the genes in the human genome. Genes for which there are no homologs in the current sequence databases will take some time to elucidate. See Ref. 23 for a detailed analysis of this topic and then Refs. 24 and 25 for detailed studies on the human genomic sequence.

The three basic technologies for generation of genome-wide expression information are cDNA microarrays, high-density oligonucleotide arrays ("GeneChips"), and serial analysis of gene expression (SAGE) (22). These technologies are outlined in Table 8.3.

In terms of quantities of data, a single microarray experiment looking at 40,000 genes from 10 different samples, under 20 different conditions, produces at least 8,000,000 pieces of data (26). Chip technologies, though originally expensive because of the costs of chip fabrication, are now being used to contribute data to public domain databases and are widely used in industrial applications. A recent comparison of array databases available for local installation, public submission of data or public query, listed 13 different systems from sources worldwide (27). One such repository, ArrayExpress, is now being funded by the European Union at the EBI and is in the early stages of development (see http://www.ebi.ac.uk/arrayexpress). It is intended to be compliant with the microarray gene expression database (MGED) standard (see http://www.mged.org/).

The process of gene expression by microarray is shown below, based on (13).

1. Construct array
2. Prepare biological samples for investigation
3. Extract and label sample RNA
4. Hybridize samples to array
5. Image the array
6. Locate spots and evaluate fluorescent intensities
7. Construct gene expression matrix from spot intensities
8. Analyze gene expression matrix

As with any biological experiment, the result of this process should be the accumulation of knowledge concerning the biological processes under study. Interestingly, the first five steps are material handling processes, while the remaining steps only involve information processing.

5 DATABASES, TOOLS, AND APPLICATIONS

Because bioinformatics is about the management of information in the domain of biology, databases play a significant part in acting as repositories for a wide variety of different types of data. The main focus of this section is to give a flavor of the breadth of databases available and to highlight the role of the primary sequence databases and the secondary pattern (or family) databases in assisting with protein functional assignment.

5.1 Databases

Data repositories of DNA sequence, protein sequence, and higher-level resources of integrated information pertaining to the relationships between sequences (for example, pattern and profile databases) are the core tools for performing a wide variety of bioinformatics analyses. Many of these databases are in the public domain and are freely accessible through links available at a range of websites (including those listed in Table 8.1), although copyright is claimed in the annotation sections of some databases. The January 2002 issue of *Nucleic Acids Research* is a special annual database issue. It contains 112 articles describing in some detail different databases in use in the field. These are a subset of the 339 databases (up from 281 in the previous year) listed and briefly described in the Molecular Biology Database Collection, which constitutes an additional article in itself (28, 29). The complete list can also be found at http://www.nar.oupjournals.org. While the list was being prepared for the 2001 edition, 55 new databases were added to the previous total. In 2002, 58 additions were made. This rapid expansion in the number of databases available is indicative of the recognition by the community of the need for accessible, carefully designed databases to meet the needs of a wide diversity of research programs.

Much of the value of databases, assuming the provision of accurate sequence data, arises from the quality of the annotation that is available. This normally includes at least a brief description of the function of the sequence and essential references to the literature. Many databases include a lot more than this. In particular, SWISS-PROT (8) is viewed as the most reliable source for annotation information. SWISS-PROT emerged in the 1980s out of a need to have high quality, robust annotations for the protein sequences that made up its core content. However, the process of annotation is labor intensive and not one that is easily automated. Although the content of SWISS-PROT is well regarded, it lacks the completeness of the source DNA databases because of the necessary delay in incorporating newly annotated sequences. Indeed, a team of annotators is employed at the

EBI solely to perform this task. A computationally annotated supplement, TrEMBL (8), has been made available to make up for this deficiency. Nevertheless, computer annotation still has some way to go before it comes close to the level of competence of skilled human annotators. This is an area of active research (30).

Nevertheless, with the rapid generation of sequence data from genome scale experiments more effective means of characterizing protein sequences and annotation are now required. The database has responded by improving labeling of annotation in both SWISS-PROT and TrEMBL and by adding more advanced and rigorous tagging of evidence for functional statements that have been made (31).

Whereas most patent sequences are available in the public domain for use in research and for commercial exploitation, there is a substantial body that are the subject of patent protection. It is often useful when conducting searches of sequence databases to be aware of the sequences that are patented because this may imply certain restrictions on the use to which these sequences can be put in a commercial context. The commercial repository is maintained by Derwent (Thomson Scientific), which generates the Geneseq database of patented sequences. This is a useful collection because it contains a broad historical collection as well as more recent examples, although the terms for a commercial license to use the database may be off-putting to some potential users. There are also patent sections of GenBank/EMBL DNA databases too, but these are of limited value because they contain only more recent sequence data.

5.2 Sequence Comparison

When dealing with the output of most experiments in target discovery the question "has this gene been seen before?" arises. The answer is, at first sight, straightforward: Compare the sequence obtained from the experimental output with all the known sequences and print the result.

Sequence comparison makes up a major part of the work of the bioinformatics analyst. It demands skill in operating the tools; for example, choosing the appropriate databases to

search and selecting the appropriate search method, followed by insight and experience in assessing the meaning of the results of the search. A search query with a single previously known sequence is likely to return not only the match with itself but also a host of other matches at varying levels of similarity with the query sequence. This extra information can be very valuable in placing the query sequence in the context of many closely related sequences that make up the family of genes to which the query belongs. More distantly related sequence matches can potentially indicate genes with similar function, even if the match is relatively short and of low score.

The experienced analyst should be able to sort the significant matches from the uninteresting ones. Often, this type of experience is difficult, if not impossible with current technologies, to capture in a computer program. Rules that seem to work under some circumstances produce nonsensical results in others. As a result, many of the techniques used for current sequence comparison engines are heuristic rather than strictly algorithmic, that is, the rules that are implemented as part of the process for returning significant hits from the query database tend to produce the correct result but cannot be guaranteed to do so in all circumstances. For a fuller discussion of algorithms and heuristics, albeit outside the context of bioinformatics, see Ref. 32.

One of the key aspects of sequence comparison is the understanding of similarity when applied to molecular sequences. There are essentially two ways of considering this: simple residue identity and residue substitution. In this discussion, we consider the comparison of two protein sequences, but the process is the same for comparison of DNA or RNA sequences. The alphabet used in the comparison is just different because it is 20 for protein sequences and 4 for DNA and RNA. By comparing residues at the same position in each sequence and counting up the number of identities we arrive at, a score that can be expressed as a percentage match for the pair of sequences. The alternative method compares each pair of residues and looks up a score for that pair in a substitution table or scoring matrix. The summed score across the whole sequence length can again be expressed as a percentage match. The two sequences under comparison are, however, likely to be sufficiently different that equivalent residue positions are not in register when the two sequences are laid out, one on top of the other. In this situation, the sequences must be aligned with each other so that equivalent residue positions are in register to make the score meaningful. This may involve insertion of gaps into one or both sequences. The skill here is to create an alignment between the two sequences that reflects some biological reality; it is from this biological reality that we derive the notion of equivalent residue positions. These positions can be deduced from manual manipulation of the alignment on the basis of mutation data or other functional information using a suitable sequence editor (33), or perhaps from understanding the spatial layout of residues if structural data is available. In each of these cases, the resulting sequence alignment will reflect the manner in which equivalent residue positions have been determined—both methods have their place. A variety of methods have been developed for comparing pairs of sequences, including the basic classical methods of Needleman and Wunsch (34) and Smith and Waterman (35).

Extending these pairwise comparison methods to database searching has been carried out, and a plethora of hybrid methods and improvements have been made. The manner in which significant alignments are reported varies from implementation to implementation. Database searching by alignment in this way is computationally intensive and specialized computer hardware is often used to gain speed increases. Because the comparison of pairs of sequence takes place in an exhaustive manner, these types of database searching methods are considered to be the most sensitive. More modern methods of database searching look for shorter matches spread over the lengths of the query and database sequences, and then extend these matches until the score for the match falls below a threshold level. Lists of sequence matches returned are then aligned using a pairwise alignment technique to provide a match and score over the whole length of the comparison sequences. For an example of this type of approach, see FASTA (36). Such methods are readily implemented on standard computer hardware and thus are accessible as Internet resources or as local implementations on UNIX or Linux servers.

The most popular tool currently in use is BLAST (Basic Local Alignment Search Tool) (37) from the NCBI. BLAST is an example of a heuristic that attempts to optimize a specific similarity measure. The most recent revisions to the algorithm are gapped BLAST and PSI-BLAST (38), with improved accuracy for PSI-BLAST using composition-based statistics (39).

5.3 Phylogenomics and Gene Family Databases

Determining protein function from genomic sequences is a central goal of bioinformatics (40), and to achieve this goal, comparing single sequences against databases of DNA or protein sequences is a necessary bioinformatics skill. However, many such searches have already been carried out, and the results are available to analyze at a higher level of abstraction in the protein and gene family databases (9, 10, 14, 41–43). It is the relationships between sequences that form the basis of any gene family database. Many of the current databases did not set out to become gene family databases. However, application of the underlying methodology for defining gene families (whether based on blocks of conserved sequence alignment or on profiles representing entire sequences, or simple regular expressions) has resulted in a number of resources that are particularly valuable in placing drug discovery targets in their biological context.

The processes of evolution by natural selection imply that species are related to each other in a tree-structured hierarchy; but more than this, the history of sequence relationships during evolution is also significant. Organisms are defined by their genes, and their behavior is modified through environmental experience. The relationships between genes within a single organism indicate that genes and their protein products also fall into well-defined families. Protein phylogenetic profiling (40) and phylogenomic analysis (44) are methods that are valuable where functional assignment by sequence similarity alone is

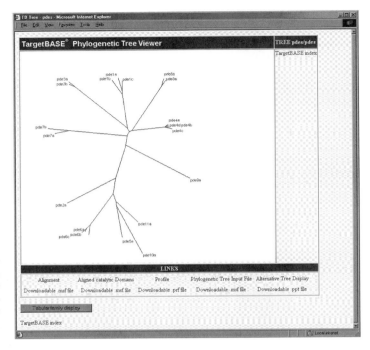

Figure 8.4. An example of a user interface to a phylogenomic-oriented database (48). Relative distances, following black line paths, between nodes on the tree of phosphodiesterases indicate the similarity level between members of the family, based on the regions of the sequences selected for the phylogenetic analysis. Links to aligned domains permit the alignments themselves to be explored. The order in which the genes appear in the tree (the branching order) gives an indication of the homology relationship between members of the family. See Section 5.3.

problematic. This is because phylogenomic analysis is based on understanding the process by which sequences have diverged from common ancestors rather than focusing on the sequence similarity itself, which is an evolutionary endpoint. The approach is to determine the phylogenetic tree of a gene family and then to overlay any known functions of the genes on the tree. The functions of uncharacterized genes are predicted by their phylogenetic positions relative to those of the previously characterized genes. Importantly, depending on the manner of their construction, the trees may indicate similarity distances along connecting branches but it is the order of branching that reflects evolutionary relatedness (otherwise known as homology). For an interesting discussion of the correct use of the terms homology and similarity see Ref. 45.

This approach is illustrated in the phosphodiesterase gene family tree presented in Fig. 8.4. The set of relationships has been determined by comparing not just two sequences with each other, or a database of sequences to one sequence (as in a sequence database search), but by comparing a set of phosphodiesterase sequences to each other in the form of

a multiple sequence alignment (see Fig. 8.5). Conserved regions of un-gapped sequence were chosen from this alignment to use as input to a phylogenetic analysis method (46, 47) and an evolutionary tree was eventually reconstructed. This tree represents a view of the relationships between genes in the phosphodiesterase gene family: more closely related genes are closer together in the diagram (i.e., they are connected by shorter paths); those further away are less closely related to each other. Figure 8.4 is based on an entry in TargetBASE (48), which adopts the phylogenomic paradigm. The relationships between members of gene families are used as the structure for an object-oriented database and associated user interface that provides a navigation tool for a curated gene index.

There are other approaches to family databases that rely more extensively on sequence similarity to define classes of genes or proteins. For example, PROSITE (49) is a resource that uses regular expressions to define patterns of residues that represent biologically significant sequence motifs. Recent versions have incorporated profiles, weight matrices that express the characteristics of a gene

```
                    251                                                    300
pde1a_human    KLHYRWTMAL MEEFFLQGDK EAELGLP.FS PLCDRKSTM. VAQSQIGFID
pde1b_human    LVHSRWTKAL MEEFFRQGDK EAELGLP.FS PLCDRTSTL. VAQSQIGFID
pde1c_human    DLHHRWTMSL LEEFFRQGDR EAELGLP.FS PLCDRKSTM. VAQSQVGFID
pde2a_human    KTTRKIAELI YKEFFSQGDL E.KAMGNRPM EMMDREKA.Y IPELQISFME
pde3a_human    ELHLQWTDGI VNEFYEQGDE EASLGLP.IS PFMDR.SAPQ LANLQESFIS
pde3b_human    DLHLKWTEGI VNEFYEQGDE EANLGLP.IS PFMDR.SSPQ LAKLQESFIT
pde4a_human    ELYRQWTDRI MAEFFQQGDR ERERGME.IS PMCDKHTAS. VEKSQVGFID
pde4b_human    ELYRQWTDRI MEEFFQQGDK ERERGME.IS PMCDKHTAS. VEKSQVGFID
pde4c_human    PLYRQWTDRI MAEFFQQGDR ERESGLD.IS PMCDKHTAS. VEKSQVGFID
pde4d_human    QLYRQWTDRI MEEFFRQGDR ERERGME.IS PMCDKHNAS. VEKSQVGFID
pde5a_human    PIQQRIAELV ATEFFDQGDR ERKELNIEPT DLMNREKKNK IPSMQVGFID
pde6a_human    EVQSQVALLV AAEFWEQGDL ERTVLQQNPI PMMDRNKADE LPKLQVGFID
pde6b_human    EVQSKVALLV AAEFWEQGDL ERTVLDQQPI PMMDRNKAAE LPKLQVGFID
pde6c_human    EVQSQVALMV ANEFWEQGDL ERTVLQQQPI PMMDRNKRDE LPKLQVGFID
pde7a_human    ELSKQWSEKV TEEFFHQGDI EKKYHLG.VS PLCDRHTES. IANIQIGFMT
pde7b_human    EMSKQWSERV CEEFYRQGEL EQKFELE.IS PLCNQQKDS. IPSIQIGFMS
pde8a_human    QYCIEWAARI SEEYFSQTDE EKQQGLPVVM PVFDRNTCS. IPKSQISFID
pde8b_human    DLCIEWAGRI SEEYFAQTDE EKRQGLPVVM PVFDRNTCS. IPKSQISFID
pde9a_human    EVAEPWVDCL LEEYFMQSDR EKSEGLP.VA PFMDRDKVT. KATAQIGFIK
pde10a_human   PVTKLTANDI YAEFWAEGD. EMKKLGIQPI PMMDRDKKDE VPQGQLGFYN
pde11a_human   EISRQVAELV TSEFFEQGDR ERLELKLTPS AIFDRNRKDE LPRLQLEWID
```

Figure 8.5. Part of an alignment of catalytic domains of the human phosphodiesterase gene family. Positions in the alignment where gaps have been introduced into a sequence to bring it into alignment with other sequences are indicated by "." characters.

family using all the sequence information available. The principal value of this resource is that it presents patterns for recognition of gene families that are relatively simple to understand. The downside is that the use of such patterns can produce both true positive hits (members correctly predicted) and false positive hits (members incorrectly predicted). PROSITE lists true and false positives for searches performed in the production of a release of the database, but it is as well to be aware that when the patterns are used in isolation, there is often a false positive hit rate that must be taken into account by reconciling the results of a pattern search with the results of database annotations or other pattern recognition methods.

The PRINTS system (9, 10, 50, 51) is an approach based on an examination of core regions of un-gapped sequence conservation within a set of aligned sequences (multiple sequence alignment). The method rigorously builds up fingerprints for a gene family through use of an iterative database searching technique allied to intelligently applied sequence alignment. The fingerprints themselves can then be used to diagnose new gene family members in novel sequence data or can

be used to identify modules of functional sequence across different gene families.

One of the issues in using different databases of gene family information is that definitions of which genes belong to which gene families can vary depending on the method used. Apweiler et al. have undertaken a useful effort at rationalizing and integrating family database annotation at the EBI in the InterPro resource (52). The databases that make up the membership of the InterPro consortium are PROSITE (49), PRINTS (9), Pfam (53), ProDom (54), and SMART (55). InterProScan is a tool that enables scanning of individual protein sequences against the InterPro member databases (56).

6 THE BIOINFORMATICS KNOWLEDGE MODEL

Up to this point, we have discussed sources of data and means of manipulating and comparing data elements (in terms of sequences, alignments, gene families, etc.), but the end point of all this analytical process must be the acquisition of knowledge. It is through increased understanding that sound decisions can be made in applying the results of bioinformatics analyses to application areas, such as drug discovery. So, in this section we consider the relationships between data, information and knowledge, which are frequently regarded as poor relations to laboratory-based experimental data acquisition. However, as drug discovery organizations, including large pharmaceutical and smaller biotechnical companies, develop a significant history of assays, screens, and leads, it is vital to have strong internal support for managing data flows, integrating related data into information systems, and transforming knowledge thus gleaned into tangible benefits.

6.1 Data, Information, and Knowledge

According to the University of California at Berkeley (57), it has taken 300,000 years for humankind to accumulate 12 exabytes of data.[2] It will take just 2.5 more years to create the next 12 exabytes. (An exabyte is 1,000,000,000,000,000,000 bytes or a billion gigabytes.) This is a truly unimaginable amount of data, equivalent to the data stored on a pile of floppy disks 24 miles high. It is the rate of accumulation of data that is the key point of interest, however, and the fact that it is accelerating.

It is crucial to distinguish between the terms data, information, and knowledge so that we can think clearly about the goal of data accumulation in our own industry sector. There are two views: a tiered hierarchical view and a more formally correct scientific view.

6.2 The Hierarchical View

In the hierarchical view, data is the bottom rung of a ladder leading to the accumulation of information that leads, ultimately, to an increase in knowledge. Apply this hierarchical principal to an everyday example of taking this article to the photocopier: the data represented by the article is the sequence of strokes and dots on the page that make up the page image. The page image is the information represented by the article. The knowledge element only comes later when the observer actually reads and understands the article. Compare this with the act of photocopying a research article, a process that does not in itself add to understanding on the part of either the photocopier or of the researcher. The acquisition of knowledge implies an active relationship between author and recipient of the information. In this, intuitive sense, we know that the hierarchical view works to some extent as a model of the way in which some knowledge is acquired.

6.3 The Scientific View

The second view is the scientific one (58). Here, we start with the piece of information that we are trying to understand, perhaps a gene whose function we plan to determine. Experiments are designed and performed to determine the characteristics of the function of the gene; such experiments yield data that describe aspects of the information. Knowledge comes from understanding and interpreting the results of the experiments. Again, knowledge is accumulated as part of an active relationship between the data describing the information and the investigator reviewing the data and drawing conclusions about the state of the information. Gene function is itself a complicated concept because the functions of gene products can rarely be assessed in isolation, owing to the network of interactions in which most genes are involved. A collection of sequence data, collected at the DNA or protein level, describe the molecular structure of a gene or its product at a primary level—it is not, however, a complete description. There are other biochemical factors to be considered; for example, proteins that assist in the folding process to create an active three-dimensional molecule, post-translational modifications, glycosylation, interactions with other molecules to generate a higher-level function, etc.

6.4 Data is Not Knowledge

Simply increasing the amount of data in the genomic universe does not necessarily increase the speed of knowledge acquisition. In short, data is not knowledge. Knowledge itself requires understanding and demands the active participation of the one acquiring the knowledge.

---

[2] In fact, the referred study uses the word "information." However, within the usage of this article "data" is a more appropriate term.

Most pharmaceutical organizations have in place the means for collating data from a wide variety of sources. Genomic information is available freely in the public domain as well as in proprietary databases. Some successful companies have used the multiple subscription database model to generate revenue to create more data to return to their customers. In this model, data tends to be available on a non-exclusive basis but it is up to the licensee to determine how best to interpret the data in its own research environment. Some information linking is available in such models, more especially those that use Internet portals as a user interface and results delivery mechanism. This has been a valuable means of acquiring data in gene and protein expression. The model is, however, showing signs of age. Pharmaceutical and biotechnology companies have needed to make substantial investments in technology and specialist skills (particularly bioinformatics) just to warehouse the data and make the analytical results available to drug discovery program scientists. Yet, all this effort still remains the heartland of genomics: target discovery groups have sprung up in the pharmaceutical and biotechnology companies to create a process by means of which targets can be gleaned from the genomic morass.

6.5 Drug Targets

When considering drug targets, as opposed to simply gene products, there are a host of characteristics that must be taken into consideration. For all the drugs that are currently available on the commercial market, there are only about 500 drug targets on which they individually act. We now understand the human genome to contain about 30,000 genes, which give rise to a still incalculable number of protein products (59). How many of these products represent tractable drug targets? The gold standard for assessing whether a protein is or is not a target is target validation. In this process, additional biochemical or molecular genetic data are required to determine whether a protein is truly involved in a disease state and thus whether it could be considered to be a suitable target molecule for drug discovery. The prospect of performing such target validation experiments, always assuming the results could be interpreted unambiguously, is a daunting one. Structural determination, by X-ray crystallography or nuclear magnetic resonance, has deepened our understanding of some biological processes immeasurably—particularly in the realm of certain proteases, DNA binding proteins, and some other soluble enzymes. The majority of drug targets are, however, membrane bound (for example, the plethora of ion channels and G-protein–coupled receptors), making structural determination to any degree of critical confidence impossible. Molecular modeling can assist in this process and has been a valuable tool for many years in thinking through possibilities and providing a framework for interpreting other biochemical results. The further away we move from rigorously determined experimental data, however, the less likely is a pharmaceutical company to embark on the commitment of expense to exploratory studies in drug discovery. Finally, in considering the suitability of a gene as a drug discovery target, we must take into consideration the temporal nature of gene expression, an area of research that has not yet been adequately addressed in the analysis of genomic data.

The alternative approach for dealing with the wealth of genomic data, in the context of drug discovery, is to consider the genomic data as a background landscape against which to pick out the currently known and validated drug targets. These targets fall into families whose relationships can be rigorously determined at the sequence level. The related members of these families can then be assessed by analogy to determine their appropriateness as drug targets. These phylogenetic (gene family) relationships then form the basis of the structure for a database and become both a tool for navigating and exploring the relationships themselves as well as a mechanism for integration of other types of drug discovery data—for example, high-throughput screening results and structure activity relationships of bound ligands. Thus, we can see that an integration of information across the realms of genomics, target discovery, screening and lead optimization becomes possible and an achievable goal.

Target analysis belongs in the domain of knowledge for drug discovery. Such knowledge is a representation of our understanding of the function of a gene product in a disease state. This functional understanding is derived from the analysis of experimental data linked to our experience of the functional information, itself defined (if only partially) by the data that characterize that functional information. Setting such knowledge within the context of the genomic universe enhances our ability to select new targets for future validation studies, either through molecular genetic techniques (gene knock-out, anti-sense, etc.) or through mechanistic validation using small molecule tools discovered as part of the assay development and screening process.

There are attempts at capturing this type of knowledge within databases (known as knowledge-bases). Much valuable research is going on in the allied areas of knowledge representation. At present, active human participation is required in accumulating knowledge and deriving ultimate benefit from it in the form of new drugs that fulfill unmet human needs in therapeutic situations.

7   STRUCTURAL GENOMICS

Structural genomics is the process of determining the three-dimensional structures of an organism's proteome (60). Predictions of protein function can be attempted from knowledge of the structure alone, or the additional information gained can be used to inform sequence-based methods of functional prediction (61).

The traditional paradigm of classical structural biology has been to select a protein based on its known biological function, ascertain its molecular structure, and use the data thus gleaned to understand how its biological function is carried out at the molecular level. To this end, more than 12,000 structures have been determined, to varying degrees of resolution and confidence. Much has been learned, as a result, of the complexity of protein structure and the manner of interaction of proteins and their native ligands, or of proteins and small molecule drugs.

The essence of structural genomics is to start from a gene sequence, produce the functional protein, and then determine its three-dimensional structure. The biological function of the protein *in vivo* is then deduced from an understanding of the structure. In this paradigm, there is no limitation on the number of structures that can be determined except the ability to purify sufficient protein for crystallization trials. The usual caveats apply regarding the solution of structures of membrane-bound proteins, for example, G-protein–coupled receptors, ion channels, certain classes of kinases, etc.

Often, the most useful functional information is derived from the structures of protein-ligand complexes because they reveal the nature of the bound ligand and its location in the protein. In the case of enzymes, a catalytic mechanism can often be postulated taking into account the disposition of residues in the active site pocket. While such structures have traditionally been determined by design, the ligand is unknown in the structural genomics approach. Only in rare cases will a ligand be co-crystallized by serendipity from the cloning organism.

From the perspective of bioinformatics, it is important to appreciate that structural determination can only provide data that reflect the biochemical or biophysical properties of the protein. The biological role in the cell or organism is a complex of interactions including spatial and temporal dimensions. Sometimes information can be derived using other techniques—for example, cDNA expression analysis, two-dimensional gel electrophoresis, biochemical assay, etc. Bioinformatics techniques building on other types of data can also assist in providing biological context for the function of a protein—for example, phylogenetic, fingerprint or regular expression analyses, etc. All of these techniques yield data that, when reviewed as a whole, can direct the course of further experiments or influence experimental design.

We have seen that, purely using techniques of sequence comparison, the function of about 40% of genes sequenced from genome projects can be inferred from sequence identity or similarity measures or by motif comparison using a variety of techniques. It is known that proteins exhibiting insignificant sequence similarity often adopt similar tertiary structures, which themselves have similar (or at least related) molecular functions. In fact the variety of types of fold taken up by polypeptides is thought to be quite limited [SCOP (62) and CATH (63)]. Discovering a useful relationship between folding topology and sequence, which can be used to predict folding accurately is, however, not trivial. By comparing the structures newly determined from structural genomics initiatives with structures already deposited in the Protein Data Bank, it may be possible to extend the inference of molecular function further than that achieved from sequence comparisons alone. Once the molecular function has been characterized in this computational way, we may begin to postulate the cellular function of the protein under analysis.

7.1 Predicting Protein Function from Structure

Some consider the "Holy Grail" of computational biology to be the accurate prediction of a protein's function solely from knowledge of its primary sequence. We have already briefly mentioned the role of structural information in guiding and illuminating the process of molecular sequence alignment (Section 5.2.1). Structural data can be a truly effective means of understanding spatial relationships between amino acid side-chains, backbone donor and acceptor groups, and the means of interaction of natural and man-made drugs. For a thorough and insightful overview of these matters see Ref. 64—the entire volume is essential reading.

Many methods have been proposed for capitalizing on our understanding of protein structure by creating algorithms that attempt to predict function from structure, or place proteins in structural categories that may have implications for functional analysis.

7.2 Neural Networks and Protease Function

Stawiski et al. (60) recently performed a study of the unique structural features of proteases in which the authors noted consistent structural similarities among unrelated protease family members. They found that proteases tend to be more tightly packed than other proteins and they tend to have fewer α helical regions and more residues in loop structures.

A neural network was trained to predict protease function with 86% accuracy in a test set. Neural networks are an example of a technique used in bioinformatics for generating a predictive program from a set of weights that can be applied in a learning tool. The tool is trained by using parameters that show discrimination between, in this case, proteases and non-proteases. In this example, 36 proteases were tested. Each protease in turn was used as a test example, the network being trained using the remaining 35 proteases. In 31 of 36 cases (86%), the network was able to identify the remaining protease. By performing the same test on 258 counter-examples, 87% were correctly classified as non-proteases.

7.3 Fold Compatibility Methods

The ability to recognize the way in which a protein sequence is folded in three dimensions should enable us to model the interactions of specific side-chains in a manner that is simply not possible when considering proteins entirely at the sequence level. This notion has resulted in sequence threading algorithms that assess the level of compatibility of a sequence with a database of fold patterns (65, 66). The principal downside to this approach is that novel structural types cannot be predicted, because at least one example of each fold type must be present in the fold pattern database. Structural genomics may be the means whereby fold pattern databases can be populated with sufficient data to make them useful as predictive tools.

7.4 CASP and the State-of-the-Art

Currently, methods of structure prediction from sequence perform poorly. The results of the biannual CASP experiment (Critical Assessment of techniques for protein Structure Prediction see http://predictioncenter.llnl.gov/) are equivocal to say the least. A recent report on the improvements of aligning target sequences to a structural template (67) indicates that over the last four CASP competitions there was no significant improvement in quality in this key step in the prediction process. Alignment remains the major source of error in all models based on less than 30% sequence identity. The subjective impression is that structure prediction is getting better year after year. This analysis, however, seems to suggest there is some way to go before reliable models can be generated for fold types not yet available in the structural databases.

8 THE FUTURE

Bioinformatics is a wide-ranging science that has developed over the last 50 years, since the discovery of the structure of DNA; a period that has resulted in the sequencing of the entire genomic material of major species. Techniques of sequence comparison, database management, design, and curation have resulted in a healthy base on which to build more automated systems. It is this author's view that experienced sequence analysts will always have a place in this process, guiding the design of new algorithms and better knowledge bases. Only in this way will the true synergy between analyst and developer be realized and contribute to the understanding of the fruits of genomic research.

Bioinformatics, allied to drug discovery, is used for discovering new potential drug targets through the use of standard bioinformatics techniques in assigning function to novel gene products at the sequence level and by the informed use of structural, mutational and biochemical data—reflected in sequence level alignment models. Assessment of expression levels of genes and the statistical relevance of differences in levels of expression at the mRNA level has contributed to drug discovery programs in pharmaceutical and biotechnology companies globally. In many respects, it is still early days for seeing the fruits of this work in the products offered on the market by these companies. There should, however, be many clinical candidates on trial in which bioinformatics has contributed, albeit at a level of detail that is frequently far below the level of interest of industry publicists. The fact is that bioinformatics is now engrained in the discovery process for new drugs. The next stage in its development will be integration between chemoinformatics (chemical informatics) and bioinformatics, driven by the need to understand the ways drug interact with

their targets rather than merely exploiting biological assay systems as tools for drug discovery.

9 ACKNOWLEDGMENTS

I thank Jeremy Packer of the BioFocus Plc Bioinformatics Group for casting a critical eye over the manuscript before it went for review and Sue Scott for her proofreading skills.

REFERENCES

1. Millennium and Bayer Announce Genome-derived Oncology Drug Candidate Selected For First Human Studies, available online at http://www.mlnm.com/news/2001/01-10--0.html, accessed on September 12, 2002.
2. J. Owens, A. Hinde, B. Ramster, and R. N. Lawrence, *Drug Discov. Today*, **6**, 229–230 (2001).
3. B. A. Kenny, M. Bushfield, D. J. Parry-Smith, S. Fogarty, and J. M. Treherne, *Prog. Drug Res.*, **51**, 245–269 (1998).
4. T. K. Attwood and D. J. Parry-Smith, *Introduction to Bioinformatics*, Addison Wesley Longman, Harlow, UK, 1999.
5. P. Baldi and S. Brunak, *Bioinformatics: The Machine Learning Approach*, 2nd ed. MIT Press, London, 2000.
6. M. S. Waterman, *Introduction to Computational Biology*. Chapman & Hall, London, 1995.
7. T. Etzold and P. Argos, *Comput. Appl. Biosci.*, **9**, 49–57 (1993).
8. A. Bairoch and R. Apweiler, *Nucleic Acids Res.*, **28**, 45–48 (2000).
9. T. K. Attwood, M. D. Croning, D. R. Flower, A. P. Lewis, J. E. Mabey, P. Scordis, J. N. Selley, and W. Wright, *Nucleic Acids Res.*, **28**, 225–227 (2000).
10. T. K. Attwood, D. R. Flower, A. P. Lewis, J. E. Mabey, S. R. Morgan, P. Scordis, J. N. Selley, and W. Wright, *Nucleic Acids Res.*, **27**, 220–225 (1999).
11. T. Hubbard, D. Barker, E. Birney, G. Cameron, Y. Chen, L. Clark, T. Cox, J. Cuff, V. Curwen, T. Down, R. Durbin, E. Eyras, J. Gilbert, M. Hammond, L. Huminiecki, A. Kasprzyk, H. Lehvaslaiho, P. Lijnzaad, C. Melsopp, E. Mongin, R. Pettett, M. Pocock, S. Potter, A. Rust, E. Schmidt, S. Searle, G. Slater, J. Smith, W. Spooner, A. Stabenau, J. Stalker, E. Stupka, A. Ureta-Vidal, I. Vastrik, and M. Clamp, *Nucleic Acids Res.*, **30**, 38–41 (2002).

12. M. Gerstein, *Nat. Struct. Biol.*, **7** Suppl, 960–963 (2000).

13. A. Brazma, *Bioinformatics*, **17**, 113–114 (2001).

14. J. Packer, E. Conley, N. Castle, D. Wray, C. January, and L. Patmore, *Trends Pharmacol. Sci.*, **21**, 327–329 (2000).

15. B. Destenaves and F. Thomas, *Curr. Opin. Chem. Biol.*, **4**, 440–444 (2000).

16. R. Staden, K. F. Beal, and J. K. Bonfield, *Methods Mol. Biol.*, **132**, 115–130 (2000).

17. R. Staden, D. P. Judge, and J. K. Bonfield, *Methods Biochem. Anal.*, **43**, 303–322 (2001).

18. D. S. Bailey, A. Bondar, and L. M. Furness, *Curr. Opin. Biotechnol.*, **9**, 595–601 (1998).

19. J. M. Claverie, *Hum. Mol. Genet.*, **8**, 1821–1832 (1999).

20. S. Audic and J. M. Claverie, *Genome Res.*, **7**, 986–995 (1997).

21. M. R. Fannon, *Trends Biotechnol.*, **14**, 294–298 (1996).

22. M. Gerstein and R. Jansen, *Curr. Opin. Struct. Biol.*, **10**, 574–584 (2000).

23. F. Sterky and J. Lundeberg, *J. Biotechnol.*, **76**, 1–31 (2000).

24. *Science*, **291**, 1145–1434 (2001).

25. *Nature*, **409**, 745–964 (2001).

26. A. Brazma, A. Robinson, G. Cameron, and M. Ashburner, *Nature*, **403**, 699–700 (2000).

27. M. Gardiner-Garden and T. G. Littlejohn, *Brief Bioinform.*, **2**, 143–158 (2001).

28. A. D. Baxevanis, *Nucleic Acids Res.*, **29**, 1–10 (2001).

29. A. D. Baxevanis, *Nucleic Acids Res.*, **30**, 1–12 (2002).

30. A. G. Rust, E. Mongin, and E. Birney, *Drug Discov. Today*, **7**, S70–S76 (2002).

31. R. Apweiler, *Brief Bioinform.*, **2**, 9–18 (2001).

32. W. D. Hillis, *The Pattern on the Stone*, Weidenfeld & Nicolson, London, 1998, pp. 77–90.

33. D. J. Parry-Smith, A. W. Payne, A. D. Michie, and T. K. Attwood, *Gene*, **221**, GC57–GC63 (1998).

34. S. B. Needleman and C. D. Wunsch, *J. Mol. Biol.*, **48**, 443–453 (1970).

35. T. F. Smith and M. S. Waterman, *J. Mol. Biol.*, **147**, 195–197 (1981).

36. W. R. Pearson, *Methods Mol. Biol.*, **132**, 185–219 (2000).

37. S. F. Altschul, W. Gish, W. Miller, E. W. Myers, and D. J. Lipman, *J. Mol. Biol.*, **215**, 403–410 (1990).

38. S. F. Altschul, T. L. Madden, A. A. Schaffer, J. Zhang, Z. Zhang, W. Miller, and D. J. Lipman, *Nucleic Acids Res.*, **25**, 3389–3402 (1997).

39. A. A. Schaffer, L. Aravind, T. L. Madden, S. Shavirin, J. L. Spouge, Y. I. Wolf, E. V. Koonin, and S. F. Altschul, *Nucleic Acids Res.*, **29**, 2994–3005 (2001).

40. M. Pellegrini, E. M. Marcotte, M. J. Thompson, D. Eisenberg, and T. O. Yeates, *Proc. Natl. Acad. Sci. USA*, **96**, 4285–4288 (1999).

41. E. L. Sonnhammer, S. R. Eddy, E. Birney, A. Bateman, and R. Durbin, *Nucleic Acids Res.*, **26**, 320–322 (1998).

42. E. L. Sonnhammer, S. R. Eddy, and R. Durbin, *Proteins*, **28**, 405–420 (1997).

43. J. G. Henikoff, E. A. Greene, S. Pietrokovski, and S. Henikoff, *Nucleic Acids Res.*, **28**, 228–230 (2000).

44. J. A. Eisen, *Genome Res.*, **8**, 163–167 (1998).

45. G. Theissen, *Nature*, **415**, 741 (2002).

46. J. Felsenstein, *Annu. Rev. Genet.*, **22**, 521–565 (1988).

47. J. Felsenstein, *Methods Enzymol.*, **266**, 418–427 (1996).

48. J. Packer and D. J. Parry-Smith, *Curr. Drug Discov.*, **March**, 29–33 (2002).

49. L. Falquet, M. Pagni, P. Bucher, N. Hulo, C. J. Sigrist, K. Hofmann, and A. Bairoch, *Nucleic Acids Res.*, **30**, 235–238 (2002).

50. W. Wright, P. Scordis, and T. K. Attwood, *Bioinformatics*, **15**, 523–524 (1999).

51. T. K. Attwood, M. J. Blythe, D. R. Flower, A. Gaulton, J. E. Mabey, N. Maudling, L. McGregor, A. L. Mitchell, G. Moulton, K. Paine, and P. Scordis, *Nucleic Acids Res.*, **30**, 239–241 (2002).

52. R. Apweiler, T. K. Attwood, A. Bairoch, A. Bateman, E. Birney, M. Biswas, P. Bucher, L. Cerutti, F. Corpet, M. D. Croning, R. Durbin, L. Falquet, W. Fleischmann, J. Gouzy, H. Hermjakob, N. Hulo, I. Jonassen, D. Kahn, A. Kanapin, Y. Karavidopoulou, R. Lopez, B. Marx, N. J. Mulder, T. M. Oinn, M. Pagni, F. Servant, C. J. Sigrist, and E. M. Zdobnov, *Bioinformatics*, **16**, 1145–1150 (2000).

53. A. Bateman, E. Birney, L. Cerruti, R. Durbin, L. Etwiller, S. R. Eddy, S. Griffiths-Jones, K. L. Howe, M. Marshall, and E. L. Sonnhammer, *Nucleic Acids Res.*, **30**, 276–280 (2002).

54. F. Corpet, F. Servant, J. Gouzy, and D. Kahn, *Nucleic Acids Res.*, **28**, 267–269 (2000).

55. J. Schultz, R. R. Copley, T. Doerks, C. P. Ponting, and P. Bork, *Nucleic Acids Res.*, **28**, 231–234 (2000).

56. E. Zdobnov and R. Apweiler, *Bioinformatics*, **17**, 847–848 (2001).

57. P. Lyman and H. R. Varian, How much information?, available online at http://www.sims.berkeley.edu/research/projects/how-much-info, accessed on September 12, 2002.

58. D. E. Knuth, *Selected Papers on Computer Science*, Cambridge University Press, Cambridge, UK, 1996.

59. J. M. Claverie, *Science*, **291**, 1255–1257 (2001).

60. E. W. Stawiski, A. E. Baucom, S. C. Lohr, and L. M. Gregoret, *Proc. Natl. Acad. Sci. USA*, **97**, 3954–3958 (2000).

61. M. Weir, M. Swindells, and J. Overington, *Trends Biotechnol.*, **19**, 61–66 (2001).

62. C. L. Lo, B. Ailey, T. J. Hubbard, S. E. Brenner, A. G. Murzin, and C. Chothia, *Nucleic Acids Res.*, **28**, 257–259 (2000).

63. C. A. Orengo, F. M. Pearl, J. E. Bray, A. E. Todd, A. C. Martin, C. L. Lo, J. M. Thornton, *Nucleic Acids Res.*, **27**, 275–279 (1999).

64. M. Perutz, *Protein Structure: New Approaches to Disease and Therapy*, Freeman, New York, 1992, pp. 119–137.

65. R. T. Miller, D. T. Jones, and J. M. Thornton, *FASEB J.*, **10**, 171–178 (1996).

66. D. T. Jones, M. Tress, K. Bryson, and C. Hadley, *Proteins*, **37**, 104–111 (1999).

67. C. Venclovas, A. Zemla, K. Fidelis, and J. Moult, *Proteins*, **45** (Suppl 5), 163–170 (2001).

CHAPTER NINE

Chemical Information Computing Systems in Drug Discovery

Douglas R. Henry
MDL Information Systems, Inc.
San Leandro, California

Contents

Burger's Medicinal Chemistry and Drug Discovery
Sixth Edition, Volume 1: Drug Discovery
Edited by Donald J. Abraham
ISBN 0-471-27090-3 © 2003 John Wiley & Sons, Inc.

1 INTRODUCTION

The term drug discovery once encompassed only those activities that were traditionally practiced by synthetic chemists—the design, synthesis, analysis, and testing of new chemical entities. Until the 1980s, most drug discovery was conducted in a serial fashion. Thus, a chemist working on a given project would design a series of structures, then synthesize them one after another, in milligram quantities (large by today's standards), and finally send batches of the compounds for analysis and assay. Based on the assay results, the chemist would design new or modified sets of structures and repeat the cycle until a marketable entity was obtained. This serial, iterative procedure was adequate in a time when a few major drug companies were doing drug design, and the number of therapeutic targets was relatively small. One consequence of this approach was a much higher number of "me-too" drugs on the market than we see today, likely because of the intensive time and resource that was devoted to each new chemical entity.

1.1 Motivation for Chemical Information Management

The serial approach to drug discovery is very costly in time and resource. Figure 9.1 shows an idealized view of the drug discovery "funnel" in which a (very productive) hypothetical chemist could produce 10–20 structures a week. In the 1970s, it was estimated that 1 in 7000 compounds synthesized and tested would eventually reach the market. That number has risen over the years to about 1 in 10,000—a figure that holds true to this day, despite advances in combinatorial and high-throughput chemistry, molecular modeling, structure-based drug design, diversity analysis, and quantitative structure-activity relationships (QSAR). In real dollars, it almost certainly costs as much or more to bring a new drug to market than it did in the 1970s. A commonly quoted figure of $500 million per marketable drug has been questioned by a Ralph Nader watchdog group, but the figure is certainly in the hundreds of millions of dollars (1). To balance the many computational advances that have been made in the past 30 years are factors of increased competition in the field, many more therapeutic targets, increased regulation, and very importantly, the flood of information flowing from high-throughput methods. The advent of high-throughput combinatorial chemistry has increased the number of structures a chemist can generate by 100- to 1000-fold, with a corresponding increase in the amount of data that must be gathered, stored, and processed.

To deal with the flood of information—chemical, biological, and clinical—it became essential over the years to develop chemical information computing systems (i.e., chemical and reaction database systems) from which the chemist and biologist could obtain up-to-

Traditional drug design-
new drug development costs

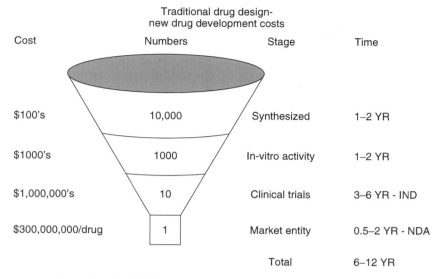

Cost	Numbers	Stage	Time
$100's	10,000	Synthesized	1–2 YR
$1000's	1000	In-vitro activity	1–2 YR
$1,000,000's	10	Clinical trials	3–6 YR - IND
$300,000,000/drug	1	Market entity	0.5–2 YR - NDA
		Total	6–12 YR

1 chemist = 10–20,000 compounds/lifetime = 1–2 drugs

Figure 9.1. Traditional "serial" drug design costs. The drug discovery "funnel" typically shows about a 10-fold reduction at each stage in the process. A chemist who could produce 10-20 structures per week would be lucky to discover a single marketable drug in a 20- to 30-year career.

date information about commercially available and in-house structures, reactions, and data. This chapter briefly describes the history of these systems, the current state of chemical information management as it applies to drug discovery, and a look at future developments in the field. The coverage is primarily aimed at corporate applications of chemical information management, as practiced in the pharmaceutical industry. The expanding use of microcomputers running Microsoft Windows or Linux operating systems means that many of the programs and database systems now used in industry can also be installed and applied in academic settings. Much of the innovation in chemical information management comes from academia, whereas most of the application has been seen in industry. This review is limited to the management and storage of chemical structure information in databases. Other chapters deal with the generation of this information (molecular modeling, property calculation) and with the use of the information in drug discovery (library design, docking and structure-based drug design, and QSAR). By analogy with another rapidly expanding field, bioinformatics, the term chem-informatics (or chemoinformatics) has recently become common to describe the acquisition, management, and use of chemical information.

1.2 Literature, References, Societies, and Research Groups

The literature of chemistry is vast, and chemical information management occupies a small corner of this domain. The chemical information literature overlaps that of computer science, database management, molecular modeling, QSAR, and even mathematics. The primary journals that publish chemical information articles are the American Chemical Society's *Journal of Chemical Information and Computer Sciences,* Kluwer's *Journal of Computer-Aided Molecular Design,* and Elsevier's *Journal of Molecular Graphics and Modeling.* Less frequently, chemical information articles appear in Wiley's *Journal of Computational Chemistry, Quantitative Structure-Activity Relationships,* and *Journal of Chemometrics,* the ACS *Journal of Medicinal Chemistry* and the *Journal of Organic Chemistry,* and Elsevier's *Analytica Chimica Acta, Computers and Chemistry,* and *Chemometrics and Intelligent*

Laboratory Systems. Other journals with articles on chemical information include the University of Bayreuth's *Communications in Mathematical Chemistry (MATCH),* Elsevier's *Drug Discovery Today,* ACS's *Modern Drug Discovery,* and a handful of newer periodicals (2).

The history of chemical information management has recently been catalogued online by the Chemical Heritage Foundation (3). The American Chemical Society has a Division of Chemical Information (CINF), and divisional symposia are held at national meetings of the ACS, often in conjunction with other divisions including Medicinal Chemistry, Computers in Chemistry, and Pesticide Chemistry. The Skolnik Award is given annually by the ACS Division of Chemical Information for "achievement in the areas of computerized information systems, chemical information, chemical indexing and notation systems, nomenclature, structure-activity relationships, and numerical data analysis and correlation." Herman Skolnik, who died in 1994, was the first recipient. He founded the *Journal of Chemical Documentation,* which became the *Journal of Chemical Information and Computer Sciences,* and he made many contributions to the field (4). Besides the ACS, other national and international meetings on chemical information include the Noordwijkerhout Conference on Chemical Structures (5), the Quantitative Structure-Activity Relationship Gordon Conference (6), and the International Conference on Chemical Information (7).

Except for journal articles and some conference proceedings, very recent general books on chemical information management are rather few in number. This is caused in part by the rapid changes in a field so closely tied to computer hardware and software development. Another reason for the paucity of texts is that most chemical information management systems are commercially developed and marketed, not widely used by universities, and in many cases, they use trademarked or even patented technology. Some texts of note in the last decade include several by Collier (8), Martin and Willett (9), and Warr and Suhr (10), one by Wiggins and Emry (11), and ones by Maizell (12) and Ash et al. (13), and a book on chemical searching by Ridley (14).

Most chemical information research and development is conducted by commercial software vendors and in-house at pharmaceutical firms. A small number of academic research groups study chemical information. The Computational Information Systems group at the University of Sheffield, under Peter Willett, has been very active in studying database searching (15). The Computer-Chemie-Centrum at the University of Erlangen under Johann Gasteiger focuses on organic structure representation and reaction classification (16). Numerous other academic groups are active in QSAR and modeling research, described in other chapters in this series.

In addition to the academic groups already mentioned, a number of online resources deal with chemical information management. Examples include the comprehensive CHEMINFO site at Indiana University (17), Cambridge Health Institute's Cheminformatics Glossary (18), the Chemical Structure Association (19), the Computational Chemistry List (CCL) (20), the Molecular Graphics and Modeling Society (21), the Open Molecule Foundation (22), the QSAR and Modeling Society (23), the Royal Society of Chemistry Chemical Information Group (24), and the UK QSAR and Cheminformatics Group (25).

1.3 Brief History of Chemical Data Management

The history of chemical information management parallels the history of computers. It can be roughly viewed in terms of decades of development (Fig. 9.2).

1.3.1 Pre-1980—Flat File Storage of Chemical Structures.

Computers consisted of mainframe machines (e.g., IBM 3090) and small minicomputers (Digital, Prime). Users connected through low speed serial connections, using "dumb" terminals (no graphics capability) or monochrome vector graphics terminals such as Tektronix and Imlac. Chemical structures were mainly stored as either (1) individual structure files, indexed by name, and handled one or a few structures at a time or (2) in a flat-file database accessed by record number (26). A typical corporate database contained up to a few tens of thousands of structures.

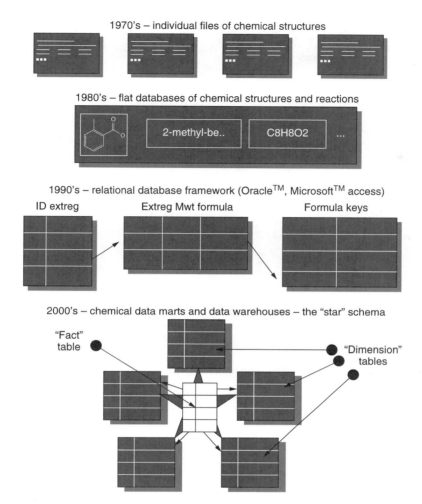

Figure 9.2. Evolution of chemical information storage. The storage of chemical information has typically lagged the development of database management systems, but it is catching up. In the 1970s, structures were stored in individual molecule files or large concatenated files. In the 1980s, proprietary databases of structures and reactions appeared, in which a single record contained all the information for a given structure. In the 1990s, this information was distributed into tables in a relational database. In the 2000s, we see the application of the concepts of data warehousing and data marts that consolidate information from a variety of sources for transactional and/or analytical purposes.

In-house chemical information management systems began to emerge at some of the larger chemical and pharmaceutical firms. These included CONTRAST and SOCRATES at Pfizer, SYNLIB at SmithKline, COUSIN at Upjohn, MSDRL/CSIS at Merck, and CROSS-BOW at ICI (27). The Chemical Abstracts database was made available online in 1967 (28). In 1980 this became CAS ONLINE. A compre-

hensive study of the user acceptance of CAS ONLINE was published in 1988 (29). The first commercial chemical structure database systems appeared in the late 1970s. These offered an in-house solution using a mainframe chemical structure management system with a graphical interface, which could be accessed by interactive graphics terminals. A standard program in widespread use was the MACCS

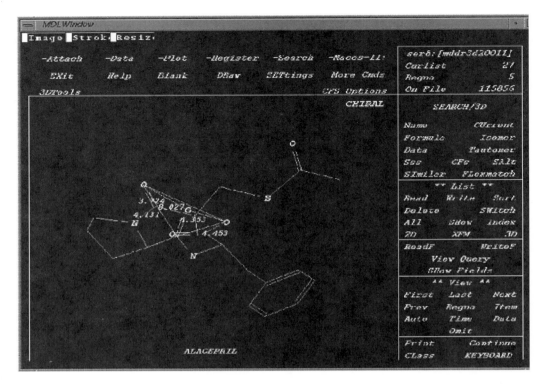

Figure 9.3. MACCS—the Molecular ACCess System—an early structure indexing system. This program originally used fixed menus for searching, registration, and reporting. Later versions allowed users to customize the menus. The figure shows the result of a 3D pharmacophore search for ACE inhibitors. Out of a database of 115,000 structures, 21 fit the 2D and 3D requirements of the search query. The user could typically browse the "hits" from the search, save the list of structures to a list file, and output the structures to a structure-data file (SDFile). The MACCS database was a proprietary flat database system in which data of a given type, say, formula, was stored in a given file, indexed by the compound ID number.

program (Fig. 9.3). Structures could be drawn, registered, searched, and output to files. The systems were only slightly customizable, and the graphics terminals, which used vector displays, were large and expensive.

1.3.2 The 1980s—Flat Database Storage. This was the era of minicomputers (Prime, Vax) and a period of immense growth for chemical information, molecular modeling, and QSAR. In industry, chemical structure databases consisted mainly of custom-designed "flat" databases (where each record in a given table refers to a given structure in the database—much like in a spreadsheet). Client-server architectures appeared, and personal computers replaced graphics terminals

and workstations. Highly successful PC-based "personal" chemical information systems appeared, which included chemical structure drawing and text processing programs (e.g., ChemDraw, ChemText) and personal chemical databases (e.g., ChemBase) (30). Customizable mainframe systems appeared (31), as did reaction indexing and searching systems (32). Additional commercial chemical information vendors appeared including Daylight Chemical Information Systems, Chemical Design Ltd., DARC-Questel, and Cambridge Scientific Corporation. The Beilstein System came online in 1988 (33). In-house and commercial database sizes were typically 100-200K structures in size. The rapid and accurate conversion of two-dimensional (2D) structures to three-dimensional (3D) models became possible using the program CONCORD, introduced by Pearlman in 1987 (34). This enabled the introduction of 3D structural databases with the ability to generate, store, and search 3D molecular models on a large scale. These 3D database systems included ALADDIN by Daylight Chemical Information Systems, UNITY3D by Tripos, CHEMDBS3D by Chemical Design Ltd., and MACCS3D by MDL (35).

1.3.3 The 1990s—Relational Data Storage. This period saw the decline of single-computer mainframe chemical management programs and the rise of server-based systems and distributed computing. By far, the most significant influences on chemical information management were the Internet, the introduction of relational database technology, and the shift to high-throughput combinatorial chemistry. In a relational database, information that formerly was kept in a single large table is stored in numerous smaller tables, indexed by "keys." This is a much more flexible architecture, and combining different fields from several tables into a "view" of the data gives the user the impression of a single large table, as before. At the end of the decade, chemical and pharmaceutical firms could obtain chemical structure, reaction, and 3D model databases from a variety of vendors. These databases were even somewhat integrated with molecular modeling, quantum mechanics, and docking programs, and to literature, spectra, and biological databases. The largest database of known chemical structures, the Chemical Abstracts Registry, grew to about 20 million structures, whereas a typical corporate inventory increased to between 100,000 and 1,000,000 structures. A database of billions of virtual chemical structures was constructed and made available for drug-design purposes by Tripos, Inc. (36).

1.3.4 The 2000s. Like the customization and distributed computing of the 1980s that followed the introduction of mini-mainframe systems, the 2000s are witnessing the customization and further distribution of relational and integrated database systems. Chemical structure-specific and reaction-specific search types can be integrated into rela-

tional databases, to take maximal advantage of the scale and performance of these systems. We see the increasing use of web-based clients, also known as "thin" clients, because they need little software other than a web browser. Former single databases are turning into distributed and replicated database systems, and we see increasing use of data marts and data warehouses, more fully integrated structure, reaction, data, and citation searching, and increasingly "intelligent" database systems.

2 CHEMICAL REPRESENTATION

Chemical structures and reactions can be represented in many ways. At the most fundamental level, the parameters of the time-dependent Schrödinger equation—the atomic and molecular orbitals—do a more or less complete job of characterizing a chemical compound. Storing and representing structures as mathematical wave functions is obviously not suitable for thousands or millions of structures; nor is such a representation useful for drug discovery, except perhaps to a molecular modeler. Synthetic chemists still function in a mostly 2D chemical structure space. Intuition, training, and experience allow a chemist to extrapolate from a flat representation with a few stereochemical hints—dashed and wedged bonds or Z/E double bonds—to a higher-dimensional mental representation of a structure. Chemical representation systems are a compromise of several factors, including the needs of the chemist, the storage and performance characteristics of the chemical database system, and the ultimate 3D reality of chemical structures.

2.1 Types of Chemical Entities

There are several ways to look at chemical representation. One approach is to classify according to the type of chemical data that is stored. The most basic types of chemical structure data are shown in Fig. 9.4, including the following.

2.1.1 Sequences. For linear chemical systems, such as DNA, RNA, and proteins, the sequence of subunits (nucleotide bases or amino acids) provides most of the information

Types of chemical data

- Sequences, names, linear notations - 1-dimensional
  information

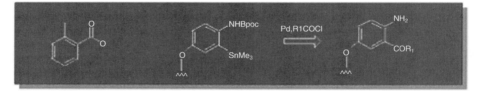

- Structures, reactions - 2-dimensional information

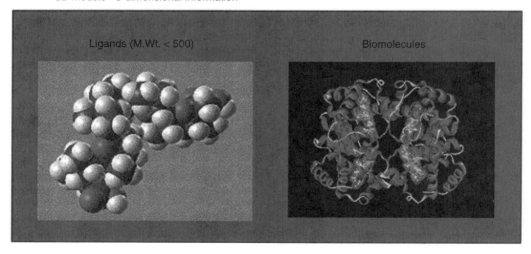

- 3D models - 3-dimensional information

Figure 9.4. Basic types of 2D chemical structure data. The amount of information and the complexity of searching increases with the dimensionality of the data.

about the structure. The deciphering of the human genome and the exploding interest in bioinformatics as a means of identifying new drug targets means there will be an increasing growth in the use of sequence data. The use of a sequence representation depends on a natural "vocabulary" of fixed building blocks. This vocabulary consists of nucleotides in the case of nucleic acids and consists of the amino acids in the case of proteins. If any of the building blocks are unique, or even if the bonding at-

tachment between building blocks differs, a simple sequence notation is not possible or it becomes more complex.

2.1.2 2D Structures. When the building blocks are unique or when dealing with the large variety of ordinary chemical structures, a 2D representation is used. In mathematical terms, this is a "graph" of the structure, which consists of a set of "nodes" (atoms) connected by "edges" (bonds). The important atom information includes atom type (symbol or atomic number), its 2D coordinates, formal charge, valence state, atom stereochemistry, and isotope information. Note that atom stereochemistry can be local (i.e., relative) or it may follow Cahn-Ingold-Prelog (CIP) conventions. Local atom stereochemistry gives the clockwise or counter-clockwise direction of the attachment of neighboring atoms when viewed from some reference attached atom—often a hydrogen atom or the lowest atomic numbered atom (37). The order of atoms in the rotation usually depends on atomic number. CIP stereochemistry is the familiar "*R,S*" nomenclature that relates the stereochemistry of the given atom to the entire structure (38). CIP stereochemistry requires analyzing the entire structure to determine the stereochemistry values. It can occasionally be ambiguous, and if any part of the structure changes, the CIP stereochemistry on distant atoms in the structure may switch. For these reasons, it is common in chemical databases to store local atom stereochemistry, but to perceive CIP stereochemistry "on the fly."

A particular problem with relative stereochemistry is that a given combination of "up" and "down" bonds on a structure implies a mixture of at least two stereoisomers. If all the centers are specified, the structure represents at least the two enantiomers. If some of the stereo centers are not designated, the number of isomers the structure represents is 2^n, where n is the number of undesignated centers. Some database vendors (e.g., MDL) allow a "chiral" designation on the molecule, which indicates that the structure represents only a single stereoisomer, but does not specify which one. One approach to dealing with these problems, which is being adopted in MDL programs, is to allow three kinds of stereo designation at a given tetrahedral center:

1. Absolute—an atom is given a known absolute stereochemistry. If all the stereo centers are so designated, this represents a single stereoisomer of the structure, as drawn.
2. Relative as a single stereoisomer—an up or down bond represents the current relative configuration, with respect to some collection of other chiral centers in the structure.

The structure represents a single stereoisomer among the possible ones. More than one collection of stereo centers may be present in the structure.

3. Relative as a mixture of stereoisomers—an up or down bond represents the current relative configuration, with respect to some collection of other chiral centers in the structure. Now, however, the structure represents a mixture of the possible stereoisomers, considering combinations of the stereo collections that are present.

Examples of these alternatives are shown in Fig. 9.5, which shows the present and the newer stereochemistry options, using a steroid structure as an example.

The bond information usually includes the bonding atoms, the bond type, and bond stereochemistry. Bond types include the common single, double, triple, and aromatic types. They may also include types that are unique to the type of structure, including dative, ionic, hydrogen bonds, etc. The bond stereochemistry for double bonds is usually Z (Zusammen-together), E (Entgegen-opposite), or either (indicating an unknown stereochemistry). For single bonds attached to a chiral or prochiral center it is typically "up" (wedge or thick bond), "down" (dashed or dotted bond), or "either" (often a wiggly bond). Some systems allow the representation of extended stereochemistry, as with the terminal groups of allene systems, which can show a type of tetrahedral stereochemistry if you collapse the allene system to a point. The bonding information—which atoms are attached to which other atoms and the bond types—is collected in the "connection table" of the structure. Table 9.1 shows a simple atom connection table for camphor. The diagonal elements of the table describe the type of atom at a given position in the structure. The off-diagonal elements describe the bonding of that atom with other atoms in the structure. Some information about a structure can be derived implicitly from the connection table. This includes the rings that are present, and the hydrogen atoms that could be attached. When a structure can be represented by more than one isomer, it is common to either (*1*) store multiple

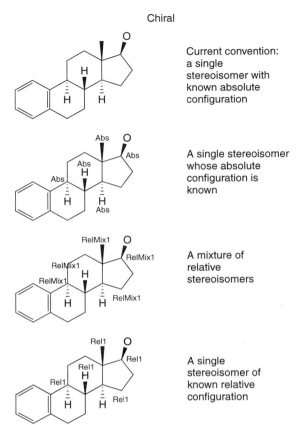

Chiral

Current convention: a single stereoisomer with known absolute configuration

A single stereoisomer whose absolute configuration is known

A mixture of relative stereoisomers

A single stereoisomer of known relative configuration

Figure 9.5. Defining absolute, relative collection, and relative single configuration stereochemistry. The older convention depends on a "chiral" flag on the molecule to specify whether a given structure represents one or several stereoisomers. In the newer convention, collections of stereo centers can be defined, and they can be designated absolute, relative-part-of-a-mixture, or relative-single-configuration.

isomers in the database, or (2) run a structure search using a search query that will hit the desired isomers. This is true for stereoisomers, enantiomers, and tautomers. Because the connection table is often symmetric, it is possible to store only, say, the upper diagonal part of it in the database.

2.1.3 Reactions. Chemical reactions extend the structure representation by adding information about what role the structure plays in the reaction (reactant, catalyst, solvent, product, etc.). Reaction representation may also include information about what bonds are made or broken during the reaction, and which atoms are involved in reacting centers. It is also common to use a hierarchical organization for reaction information (reaction > variation > reactants, catalysts, solvents, products, etc.).

2.1.4 3D Models. These extend the structure representation by adding one or more sets of 3D atomic coordinates for the various conformations that the molecule can adopt. 3D model representation may also include additional atom or bond information such as partial charge or partial bond order. It is common to generate approximate 3D models from 2D structures using fast abbreviated molecular modeling and fragment joining methods such as CONCORD, CORINA, and CONVERTER (39). These programs combine molecular mechanics with rules and heuristics to generate reasonable 3D structures in a fraction of the time required by molecular mechanics or quantum mechanics modeling. Typically, hundreds of structures can be processed per second. Although the resulting models are not the lowest energy models possible, they are quite suitable for 3D pharmacophore searching,

Table 9.1 Connection Table for D-Camphor

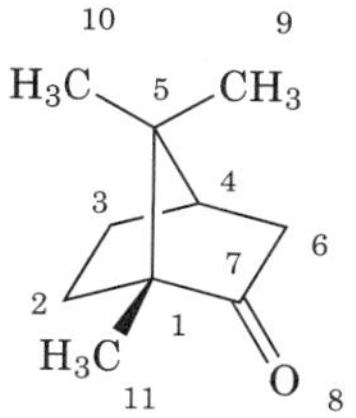
10
9
H_3C
5
CH_3
3
4
6
7
2
1
H_3C
11
O
8

Atom	1	2	3	4	5	6	7	8	9	10	11
1	C	1			1		1				1
2	1	C	1								
3		1	C	1							
4			1	C	1	1					
5	1			1	C				1	1	
6				1		C	1				
7	1					1	C	2			
8							2	O			
9					1				C		
10					1					C	
11	1										C

and they serve as a good starting point for further optimization.

Recently, with the use of combinatorial and high-throughput chemistry, more general types of structure representation, so-called chemical libraries, have become common (Fig. 9.6). These are typically used to represent mixtures and generic structures.

2.1.5 Mixtures. Mixtures are useful to represent isomers, formulations, and the products of reactions. Their representation usually

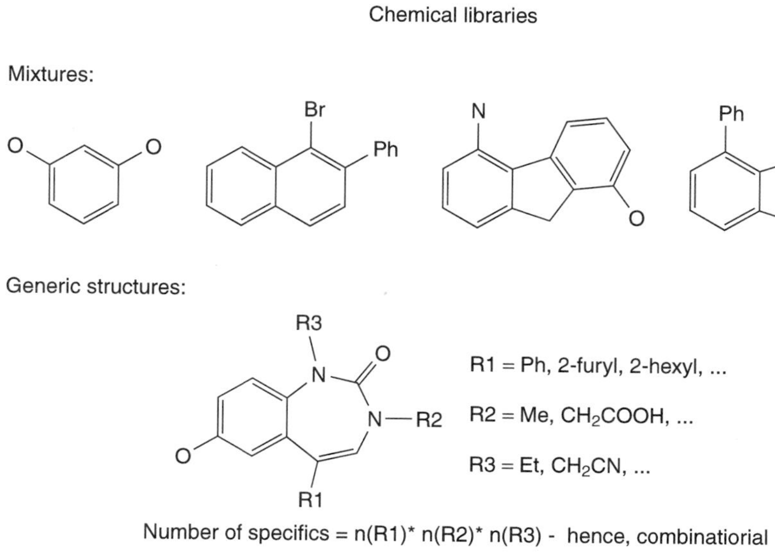
Chemical libraries
Mixtures:
O
O
Br
Ph
N
O
Ph
O
Generic structures:
R3
O
N
N—R2
O
R1
R1 = Ph, 2-furyl, 2-hexyl, ...
R2 = Me, CH_2COOH, ...
R3 = Et, CH_2CN, ...
Number of specifics = n(R1)* n(R2)* n(R3) - hence, combinatiorial

Figure 9.6. Chemical structure data for high-throughput chemistry. The generic structure representation is often referred to as a Markush structure.

requires adding data to specify percent or amount content in the mixture for each component.

2.1.6 Generic Structures.
Generic or Markush structures are commonly used to represent structures for patent purposes. Since the introduction of combinatorial chemistry, generic structures and generic reactions have become a standard means of representing potentially huge numbers of specific compounds in a highly compact representation that is familiar to the chemist (40). The central structure of a generic, which is common to all the structures it represents, is commonly called the "root" or "parent." The variable parts of the structure (R_1, R_2, etc.) are referred to as the "Rgroups." The exact substituents that make up the various Rgroups (e.g., —Cl, —Br, —OH) are referred to as the "members" of the Rgroup. Finally, a specific combination of root and Rgroup members—which constitutes a single, real structure—is referred to as a specific or "enumerated" structure. Some chemical computations, like property and similarity calculations, can be performed on the generic structure without enumerating all the specific structures (41).

The remaining types of chemical data that need representation include substances and search queries.

2.1.7 Substances.
Less common in drug discovery, but very useful for material science and polymer chemistry, is the ability to store "substances." These include unspecified or uncertain chemical structures, polymers, and other chemical entities that cannot be classed with the other chemical representations (42). Polymers pose particular problems, as discussed in the article by Schultz and Wilks (43).

2.1.8 Search Queries.
For all types of chemical representation, there are *query* representations that can be applied to a database to return a list of structures which match or "fit" the query, or that the query "hits" in the database. The same chemical drawing programs that are used to input structures can commonly be used to input chemical structure queries. These drawing programs currently include several programs in the commercial and public domains (44). A comparison of popular drawing programs has recently appeared on the Internet (45). Query structures often contain generalized atom types, bond types, and ring types. They may specify the required presence or absence of certain atom types or functional groups. In the case of 3D models, queries can be devised to represent pharmacophores for certain types of therapeutic activity (46). An important distinction must often be made between the query representation of a pharmacophore used for 3D searching and the conceptual pharmacophore used for drug development.

2.2 Types of Chemical Representation

A second way of looking at chemical representation is to consider the manner in which the chemical structure data is organized and exchanged, either in some file format or in a database. The most common ways of representing structures and reactions include the following.

2.2.1 Linear Notation.
One of the earliest forms of chemical structure representation is Wiswesser line notation (WLN), developed in 1946. This notation used short letter codes to represent functional groups in molecules (47). An alternative early notation is the Beilstein ROSDAL string (48). These two formats are not used much today, having been replaced by the Daylight SMILES notation (49) and its extensions (50). Figure 9.7 shows a drug-like molecule along with WLN, SMILES, and ROSDAL notation. Also shown is a simple chemical reaction represented in SMILES. Note that SMILES and other linear notation schemes do not include 2D coordinates for display of the structure. These are either stored separately or generated on the fly (51). The SMILES notation has become especially popular for property estimation programs, because atom coordinates are not usually needed for connection table-based calculations. It is a very convenient method for web-based input of structures for property calculations (52). Note that the order of atoms in most linear notations is arbitrary, depending on where in the molecule the notation generator (program or chemist) starts. For this reason, some linear notations have a canonical (or "uniquified") form that

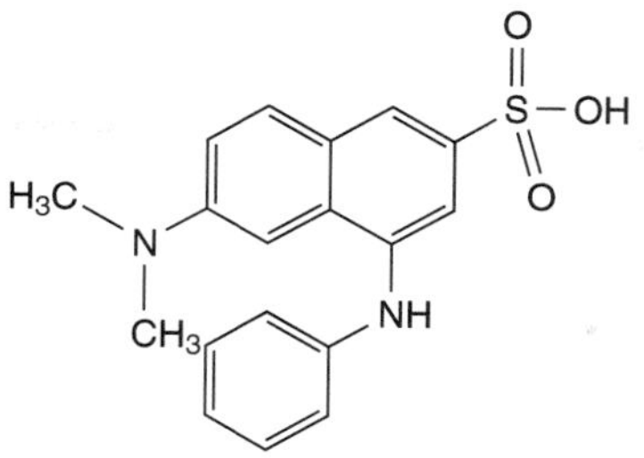

WLN: L66J BMR& DSWQ IN1&1

ROSDAL: 1 = -5- = 10 = 5, 10-1, 1-11N-12- = 17 = 12, 3-18S-19O, 18-20O, 18 = 21O, 8-22N-23, 22-24

SMILES: OS(=O)(=O)c1cc2ccc(cc2c(c1)Nc1ccccc1)N(C)C

SLN: OS(=O)(=O)C[1]:C:C[2]:C:C:C(:C:C(:@2):C(:C(:@1))NC[3]:C:C:C:C:C(:@3)) N(C)C

CHIME: 3aQf713AsUwQDjIMwyMWrSA7AOxHeqiAAPWRmMrZSZIJjTrAEfcsXH1JTUf...

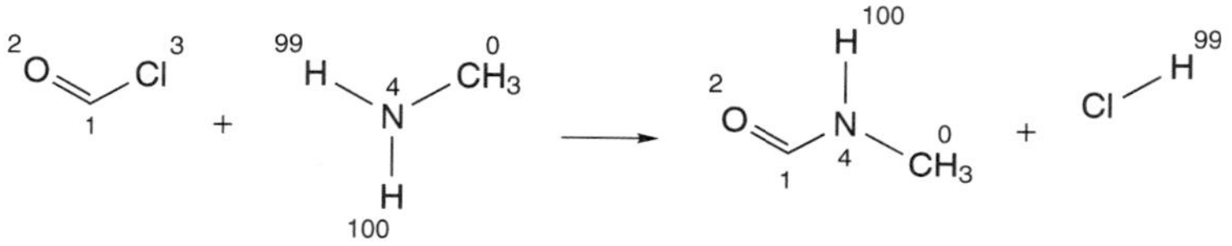

SMILES:

[C:1](=[O:2])[Cl:3].[H:99][N:4]([H:100])[C:0]>> [C:1](=[O:2])[N:4]([H:100])[C:0].[Cl:3][H:99]

Figure 9.7. Various linear notation schemes for chemical representation. Some contain only atom types and connectivity (WLN, ROSDAL, SMILES, SLN) and are chemist-readable. Others are compressed versions of molecule file formats (CHIME) and are meant for computer interpretation.

places the atoms in a topological order, usually reflecting their degree of branching, the types of neighboring atoms and bonds, etc. This canonical ordering of the atoms reduces any user-input ordering to the same string. It can then be used for exact-match lookup of the structure, regardless of how it was drawn or typed. The SMILES notation has also been extended to include reactions as shown in Fig. 9.7 (53). Occasionally, other linear notations are described (54).

2.2.2 Tabular Storage. To preserve more specific information about atoms and bonds, such as coordinates, stereochemistry, charge, and isotope number, it is necessary to store molecule information in a tabular format. Each row of the table typically contains all the information about a single atom or bond. In some formats, the atom and bond information is combined on a single line. Table 9.2 shows three common file formats for a simple struc-

ture. In the MDL molfile format, the atom and bond information is separated into separate blocks. In the Hyperchem HIN file format, the bond information is mixed with the atom information, resulting in fewer records in the file. In the PDB format, the atoms can be assigned to residues. Descriptions of various formats can be found in the reference manuals for chemical management and molecular modeling programs or in the literature (55). The systems that manage reactions typically have their own file formats as well.

Both linear and tabular formats are capable of being transmitted over a network between computers. This allows passing structure information from a server to a workstation for display purposes. It is common to compress and/or encrypt the chemical structure information before it is transmitted, and then have the workstation or display program uncompress or decrypt the resulting structures. This is done for performance and

Table 9.2   Tabular Molecule File Formats

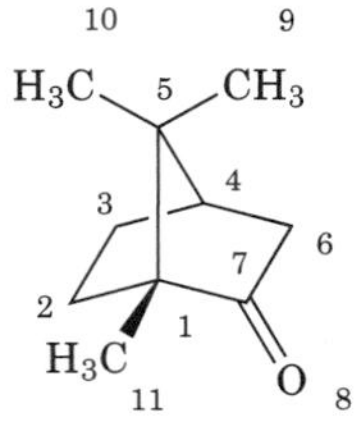

MDL Molfile format:

D-Camphor
 -ISIS- 03130218162D
2D molfile

```
 11  12  0  0  0  0  0  0  0  0 0999 V2000
   -2.0625    -1.1833     0.0000 C   0  0  0  0  0  0  0  0  0  0  0  0
   -1.5583    -0.4167     0.0000 C   0  0  0  0  0  0  0  0  0  0  0  0
   -0.4208     0.1500     0.0000 C   0  0  3  0  0  0  0  0  0  0  0  0
   -0.8292    -0.7542     0.0000 C   0  0  2  0  0  0  0  0  0  0  0  0
   -0.7042     1.1667     0.0000 C   0  0  3  0  0  0  0  0  0  0  0  0
    0.9917    -0.4333     0.0000 C   0  0  0  0  0  0  0  0  0  0  0  0
    0.4667    -1.2167     0.0000 C   0  0  0  0  0  0  0  0  0  0  0  0
    0.5237     1.7332     0.0000 C   0  0  0  0  0  0  0  0  0  0  0  0
   -1.9208     1.6686     0.0000 C   0  0  0  0  0  0  0  0  0  0  0  0
    0.9875    -2.2803     0.0000 O   0  0  0  0  0  0  0  0  0  0  0  0
   -1.6004    -2.0787     0.0000 C   0  0  0  0  0  0  0  0  0  0  0  0
  1  2  1  0  0  0  0
  2  3  1  0  0  0  0
  1  4  1  0  0  0  0
  4  5  1  0  0  0  0
  5  3  1  0  0  0  0
  3  6  1  0  0  0  0
  6  7  1  0  0  0  0
  7  4  1  0  0  0  0
  5  8  1  0  0  0  0
  5  9  1  0  0  0  0
  7 10  2  0  0  0  0
  4 11  1  1  0  0  0
M  END
```

Hyperchem HIN file format:

```
mol 1 C:\TEMP\DCAMPHOR.HIN
atom     1  -  C  **  -  0  0.0000  0.8445  0.0000  2  2  s  4  s
atom     2  -  C  **  -  0  0.3881  1.4347  0.0000  2  1  s  3  s
atom     3  -  C  **  -  0  1.2639  1.8710  0.0000  3  2  s  5  s   6  s
atom     4  -  C  **  -  0  0.9494  1.1749  0.0000  4  1  s  5  s   7  s  11  s
atom     5  -  C  **  -  0  1.0457  2.6537  0.0000  4  4  s  3  s   8  s   9  s
atom     6  -  C  **  -  0  2.3513  1.4219  0.0000  2  3  s  7  s
atom     7  -  C  **  -  0  1.9471  0.8189  0.0000  3  6  s  4  s  10  d
atom     8  -  C  **  -  0  1.9910  3.0899  0.0000  1  5  s
atom     9  -  C  **  -  0  0.1090  3.0401  0.0000  1  5  s
atom    10  -  O  **  -  0  2.3481  0.0000  0.0000  1  7  d
atom    11  -  C  **  -  0  0.3557  0.1552  1.0000  1  4  s
endmol   1
```

Table 9.2 *(Continued)*

Protein Data Bank format:

HEADER	PROTEIN								
COMPND	c:\temp\dcamphor.pdb								
AUTHOR	GENERATED BY BABEL 1.6								
ATOM	1	C	UNK	1	−2.063	−1.183	0.000	1.00	0.00
ATOM	2	C	UNK	1	−1.558	−0.417	0.000	1.00	0.00
ATOM	3	C	UNK	1	−0.421	0.150	0.000	1.00	0.00
ATOM	4	C	UNK	1	−0.829	−0.754	0.000	1.00	0.00
ATOM	5	C	UNK	1	−0.704	1.167	0.000	1.00	0.00
ATOM	6	C	UNK	1	0.992	−0.433	0.000	1.00	0.00
ATOM	7	C	UNK	1	0.467	−1.217	0.000	1.00	0.00
ATOM	8	C	UNK	1	0.524	1.733	0.000	1.00	0.00
ATOM	9	C	UNK	1	−1.921	1.669	0.000	1.00	0.00
ATOM	10	O	UNK	1	0.988	−2.280	0.000	1.00	0.00
ATOM	11	C	UNK	1	−1.600	−2.079	0.000	1.00	0.00
CONECT	1	2	4						
CONECT	2	1	3						
CONECT	3	2	5	6					
CONECT	4	1	5	7	11				
CONECT	5	4	3	8	9				
CONECT	6	3	7						
CONECT	7	6	4	10					
CONECT	8	5							
CONECT	9	5							
CONECT	10	7							
CONECT	11	4							
MASTER	0	0	0	0	0	0	0	0	11	0	11	0
END									

for security. In MDL systems, the Chime linear format is used to transmit structures and reactions, whereas Daylight systems simply use the SMILES representation and depict the structure on the fly (Fig. 9.7).

2.2.3 Graphical Representation. Occasionally it is desirable to store chemical structures as "pictures", usually for document purposes. For example, some chemical drawing packages and many molecular modeling packages can store structures as the following:

- WordPerfect or Microsoft Word document (.doc files)
- Extended postscript (.eps files)
- Windows metafile (.wmf files)
- A proprietary sketch (MDL .skc files)
- A variety of compressed graphics formats including JPEG (.jpg files), bitmap (.bmp files), GIF (.gif files), and TIFF (.tif files)

Often, the graphical format allows the connection table to be stored and transferred transparently with the image—through the computer's clipboard, for instance. This allows the receiving program to "interpret" the image as a chemical structure and manipulate it accordingly.

2.2.4 Markup Languages. The Internet has spawned a host of new "languages" that facilitate the exchange of information. The most common of these are HTML (hypertext markup language) and XML (extensible markup language). A variation of XML that is designed for chemical information exchange is the Chemical Markup Language CML (56). Although it is not widely used as of this writing, it bears watching as more web-based chemical information platforms become available. Problems with markup languages are that they are verbose compared with structure

Table 9.3   Chemical Markup Representation of Acetic Acid

```
⟨molecule convention="MDLMol" id="acetate" title="ACETATE"⟩
⟨date day="23" month="11" year="1995" /⟩
⟨atomArray⟩
⟨atom id="al"⟩
⟨string builtin="elementType"⟩C⟨/string⟩
⟨float builtin="x2"⟩0.27⟨/float⟩
⟨float builtin="y2"⟩0.1217⟨/float⟩
⟨/atom⟩
⟨atom id="a2"⟩
⟨string builtin="elementType"⟩C⟨/string⟩
⟨float builtin="x2"⟩-1.27⟨/float⟩
⟨float builtin="y2"⟩0.1246⟨/float⟩
⟨/atom⟩
⟨atom id="a3"⟩
⟨string builtin="elementType"⟩O⟨/string⟩
⟨float builtin="x2"⟩1.0623⟨/float⟩
⟨float builtin="y2"⟩-1.2937⟨/float⟩
⟨/atom⟩
⟨atom id="a4"⟩
⟨string builtin="elementType"⟩O⟨/string⟩
⟨float builtin="x2"⟩1.1008⟨/float⟩
⟨float builtin="y2"⟩1.4332⟨/float⟩
⟨/atom⟩
⟨/atomArray⟩
⟨bondArray⟩
⟨bond id="b1"⟩
⟨string builtin="atomRef"⟩a1⟨/string⟩
⟨string builtin="atomRef"⟩a2⟨/string⟩
⟨string builtin="order"⟩1⟨/string⟩
⟨/bond⟩
⟨bond id="b2"⟩
⟨string builtin="atomRef"⟩a1⟨/string⟩
⟨string builtin="atomRef"⟩a3⟨/string⟩
⟨string builtin="order"⟩1⟨/string⟩
⟨/bond⟩
⟨bond id="b3"⟩
⟨string builtin="atomRef"⟩a1⟨/string⟩
⟨string builtin="atomRef"⟩a4⟨/string⟩
⟨string builtin="order"⟩2⟨/string⟩
⟨/bond⟩
⟨/bondArray⟩
⟨/molecule⟩
```

files, and they are difficult for chemists to read (although they are not usually meant for chemist interpretation). This is evident in Table 9.3, which shows the CML for acetic acid. By comparison, the SMILES for acetic acid is simply "CC(=O)O".

2.3   Chemical Structure File Conversion

Many chemical information management systems, especially modeling programs, permit a

chemist to import and export structures using a variety of file formats. Commercial programs designed specifically for file conversion are available (57). A widely used public domain program, Babel, is available in source code and in a Windows version (58). It is being extended by the "OpenBabel" programming project (59). It is possible, with a fair amount of accuracy, to convert a chemical structure from a connection table format to an acceptable

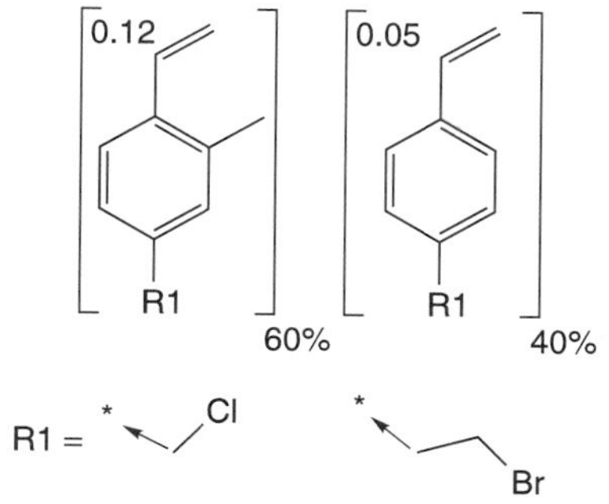

Figure 9.8. Structure representation with additional information, including atom (partial charge) and fragment (percent composition) data and Markush structure features.

IUPAC name (60). The reverse conversion of names to 2D structures is also possible (61).

2.4 Representing Nonstructural Chemical Data

Nonstructural chemical data includes any textual, numeric, or binary data that is not directly a part of the chemical structure. It includes the following:

- Whole-molecule data such as physicochemical properties, spectral data, literature citations, availability, biological or therapeutic activity, etc. In either the molecule file or in a database, data are typically maintained in fields that are separate from the structure field, but linked by some identifier.

- Atom, bond, or fragment-based data such as partial charge, component fraction, secondary structure, various fragment-based physicochemical and QSAR properties, etc. Because these data are linked to particular atoms, bonds, fragments, or components in the structure, they are typically stored along with some indication of the substructural features with which they are associated.

Figure 9.8 shows a structure that contains atom (partial charge) and fragment (component percent) data of various types, as well as Markush structure features (R_1). Table 9.4 shows the corresponding structure-data file (MDL sdfile). Each Rgroup member appears in its own "submolfile" in this file representa-tion. So-called Sgroup data appears as part of the molfile, along with the name of the data field, its value, location on the display, and the atoms and bonds that bear the data.

3 STORING AND SEARCHING CHEMICAL STRUCTURES AND REACTIONS

In the simplest sense, searching chemical information consists of (1) finding structures or reactions that meet the chemist's search criteria and/or (2) finding data that meets the search criteria. Data searching (numbers and text) is a well-established informatics activity, supported by spreadsheets, word processors, and relational database systems. Chemical structures and reactions are a unique form of data. Searching for full or partial matches to structures, models, and reactions requires highly specialized databases and search techniques.

3.1 Storing Chemical Information in Databases

When they were first developed, chemical structure databases consisted of record-oriented flat files, much like a spreadsheet whose columns have each been cut out and placed in a separate file. This organization has limitations in searching, access, and efficiency of storage. Also, it is not the most appropriate form for storing generic structures and reactions, which are more hierarchical in nature. As a result, since the 1990s, chemical information has become increasingly stored in commercial relational database systems, chiefly Oracle and Microsoft Access. Relational storage has the added advantage of combining chemical structure storage with biological data and inventory data (location, cost, units on hand, etc.) that are often stored in the corporation's relational databases. One of the first reports of the relational storage of structures was by Hagadone and Lajiness, who modified the Upjohn COUSIN system (62).

An example of a current commercial relational chemical database is seen in Fig. 9.9, which shows the organization of a basic ISIS chemical database. Each labeled item in the figure is a table in an Oracle relational database. The tables in the database consist of the following.

Table 9.4   Example Molfile Showing Markush Features, Atom, and Fragment Data

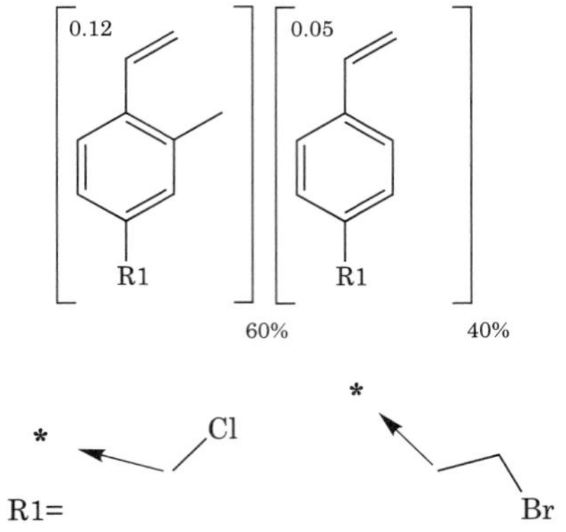

0.12   0.05

60%   40%

R1=   Cl   Br

$MDL REV 1 29AUG01 17:47
$MOL
$HDR
Figure 8 molfile

MACCS-II08290117472D	1	0.00487	0.00000	0

$END HDR
$CTAB

 19 19 0 0 0 0   25 V2000

-5.3301	1.0237	0.0000	C	0	0	0	0	0	0	
-5.3323	-0.5232	0.0000	C	0	0	0	0	0	0	
-3.9956	-1.2952	0.0000	C	0	0	0	0	0	0	
-2.6560	-0.5224	0.0000	C	0	0	0	0	0	0	
-2.6615	1.0307	0.0000	C	0	0	0	0	0	0	
-3.9990	1.7956	0.0000	C	0	0	0	0	0	0	
-3.9881	3.2993	0.0000	C	0	0	0	0	0	0	
-3.9960	-2.8378	0.0000	R#	0	0	0	0	0	0	
-2.6701	4.1134	0.0000	C	0	0	0	0	0	0	
-1.3283	1.8070	0.0000	C	0	0	0	0	0	0	
0.8559	0.9069	0.0000	C	0	0	0	0	0	0	
0.8538	-0.6400	0.0000	C	0	0	0	0	0	0	
2.1904	-1.4121	0.0000	C	0	0	0	0	0	0	
3.5299	-0.6393	0.0000	C	0	0	0	0	0	0	
3.5247	0.9138	0.0000	C	0	0	0	0	0	0	
2.1870	1.6787	0.0000	C	0	0	0	0	0	0	
2.1823	3.2214	0.0000	C	0	0	0	0	0	0	
3.5161	3.9966	0.0000	C	0	0	0	0	0	0	
2.1900	-2.9547	0.0000	R#	0	0	0	0	0	0	

```
  8   3  1  0  0  0
  2   3  1  0  0  0
  7   9  2  0  0  0
  5  10  1  0  0  0
  3   4  2  0  0  0
 11  12  2  0  0  0
  4   5  1  0  0  0
 12  13  1  0  0  0
 13  14  2  0  0  0
  5   6  2  0  0  0
 14  15  1  0  0  0
  6   1  1  0  0  0
 15  16  2  0  0  0
 16  11  1  0  0  0
  1   2  2  0  0  0
 16  17  1  0  0  0
  6   7  1  0  0  0
 17  18  2  0  0  0
 13  19  1  0  0  0
```

Table 9.4 (*Continued*)

```
M    RGP   2    8    1      19    1
M    STY   6    1   GEN    2   GEN    3   DAT    4   DAT    5   DAT    6   DAT
M    SLB   6    1    1      2    2      3    3      4    4      5    5      6    6
M    SAL   1    9    11     12   13     14   15     16   17    18   19
M    SDI   1    4          0.0515      -3.7303        0.0515        4.7961
M    SDI   1    4          4.8009       4.7961        4.8009       -3.7303
M    SAL   2   10    1      2    3      4    5      6    7      8    9      10
M    SDI   2    4         -6.1377      -3.6181       -6.1377        4.9083
M    SDI   2    4         -0.5469       4.9083       -0.5469       -3.6181
M    SAL   3    1    7
M    SDT   3  PCHARGE                              F                            MQ
M    SDD   3         -5.4645          3.7303   DA   ALL        1          5
M    SED   3   0.12
M    SDT   4  PCT                                  F                            MQ
M    SDD   4         -0.9021         -4.9083   DA   ALL        1          5
M    SED   4   60%
M    SDT   5  PCT                                  F                            MQ
M    SDD   5          5.2122         -4.7961   DA   ALL        1          5
M    SED   5   40%
M    SAL   6    1    17
M    SDT   6  PCHARGE                              F                            MQ
M    SDD   6          0.5002          3.9173   DA   ALL        1          5
M    SED   6   0.05
M    SPL   2    4    2      5    1
M    END
$END CTAB
$RGP
    1
$CTAB
  2  1  0  0  0  0              2  V2000
   -3.7453    0.0472    0.0000 C    0    0    0    0    0    0
   -2.5668    1.0385    0.0000 C1   0    0    0    0    0    0
  1  2  1  0  0  0
M  APO  1  1  1
M  END
$END CTAB
$CTAB
  3  2  0  0  0  0              2  V2000
   -5.0122    0.6013    0.0000 C    0    0    0    0    0    0
   -3.5311    1.0229    0.0000 C    0    0    0    0    0    0
   -2.6113   -0.2666    0.0000 Br   0    0    0    0    0    0
  1  2  1  0  0  0
  3  2  1  0  0  0
M  APO  1  1  1
M  END
$END CTAB
$END RGP
$END MOL
```

- The master "data dictionary" table, which describes all the objects in the database, as well as some parameters that are specific to the database (exact match criteria, version of the database, etc.). This is sometimes referred to as "metadata" or "data about data."

- A handful of tables that contain database parameters. These include substructure search key definitions, the periodic table used with the database, and a list of salt moieties that can be considered during searches.

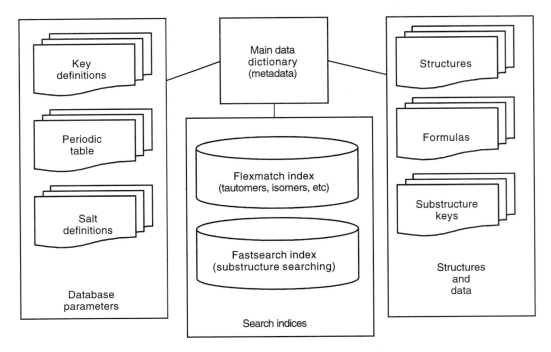

Figure 9.9. ISIS, a relational chemical structure database.

- Structure and data storage is shown on the right. A structure table contains the structures, their internal identifiers, and their external identifiers, if any. The structures are stored in a compact binary representation that includes the connection table, the coordinates, the ring information, and any stereochemical, valence, isomer, isotope, or bond information. Certain types of structure-specific information such as polymer or component designations are stored here, whereas other types of structure-specific information (atom- or bond-specific data, and more verbose text data) are stored in their own tables, referenced by the internal identifier, and the atom or bond numbers to which the data correspond. A formula table contains the molecular formula and various atom and atom-type indexes to enhance formula searching and sorting.

- A table of substructure keys containing a binary or text string of the substructure fingerprint that was identified in the given structure at registration time. These keys represent the presence of either simple functional groups (e.g., phenyl ring, car-

bonyl), or more complex atom/bond combinations (e.g., carbonyl separated from a secondary amine by three bonds). In ISIS, a set of 166 searchable keys can be explicitly used as filters for structure searching. A larger set of 960 keys is used for similarity calculations. For 3D models, it is common to generate 3D pharmacophore keys, which encode all the possible 2- and 3-point pharmacophores represented in the structure, sometimes considering multiple conformations.

- A third kind of information includes indexes to enable structure and substructure searching. A "flexmatch" table contains a numerical hash (see Glossary) of certain features of the molecule, including stereochemistry, charge, and isotopes. This table can be used to retrieve a set of candidate structures quickly for exact match verification (63). It can also be used for "fuzzy" exact-match searches to retrieve tautomers and isomers of the input structure.

- Another index table contains a "fastsearch" index. This contains a single balanced tree (see Glossary) of all the substructural frag-

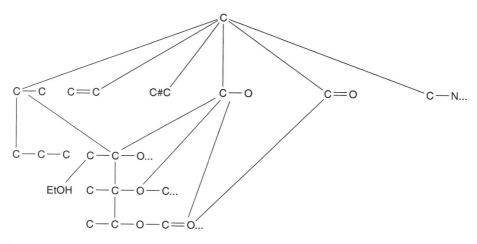

Figure 9.10. Simplified ISIS Fastsearch index—ethanol is a leaf node that can be reached from several substructure nodes.

ments found in structures in the database, up to a fixed pathlength. These are stored in a highly compressed binary format (Fig. 9.10). Similar approaches have appeared in the literature (64). Leaf nodes in the tree contain identifiers of specific structures in the database (simplified in Fig. 9.10). An exact match or substructure search consists of traversing the tree to find structures in the database that have substructural fragments in common with the query structure. Because the fastsearch index is large—often as large as the rest of the structure database, updating it for the addition or removal of structures is time consuming.

This relational chemical database format is extended in ISIS to include 3D models, generic structures, and most recently, reactions. In these cases, additional "trees" in the database hierarchy connect 2D structures with 3D models, connect root structures with corresponding Rgroup members, or connect molecules with reactions.

Other relational structure/reaction database systems are available commercially. These include the Thor system from Daylight (65), Accord and RS³ Discovery from Accelrys (66), and Unity from Tripos (67). Personal database systems that can be implemented on a desktop computer include ISIS/Base (68), Accord for Access (66), and Team Works from Afferent (69).

3.2 Registering Chemical Information

Chemical structure registration is an important activity that is necessary for drug discovery. The structures that have been developed by a pharmaceutical company constitute the "crown jewels" of chemical information, and they must be properly and securely archived. The registration process usually involves the process of extracting, cleaning, transforming, and loading the data—sometimes termed ECTL.

3.2.1 Extract the Data. First, the structures/reactions and corresponding data are extracted, collected, and validated. Increasingly, this is managed automatically, using output from the high-throughput chemistry process. Laboratory information management systems (LIMS) that are "structure smart" can manage chemical structure information starting from the design of a reaction, through the synthesis of the compounds, the chemical analysis of the structures, the *in vitro* biological assay, and finally the storage in the chemical database. Certain steps, such as drawing the initial structures/reactions, still remain an activity for the chemist, although many chemical information systems can take a generic structure, enumerate the many specific combinations, and layout the structures automatically (for example, the Monomer Toolkit by Daylight, the Central Library program by MDL, and CombiLibMaker by Tripos).

3.2.2 Cleaning and Transforming the Data.

Next, the structures/reactions are passed through a filtering program that searches for structure anomalies and corrects the chemical representation. In this step, chemical "business rules" are applied to the structures to insure that representations that can be drawn in different ways, such as nitro groups and tautomers, are represented by a single convention. Specialized chemical manipulation languages such as and Genie Control Language by Daylight, Cheshire by MDL, and Sybyl Programming Language by Tripos are used to implement this step. These languages are versatile and easily programmed, and they can be applied to other steps in the drug discovery process, such as searching, property calculation, and structure manipulation in general.

3.2.3 Loading the Data.

Finally the structures/reactions are handed to a chemical registration system. The chemical registration system will typically "perceive" the structures—identify atoms, bonds, rings, stereochemistry, valence states, isotope values, and other chemical information as needed. In the case of reactions, it notes which structures are reactants, which are products, and which are agents or catalysts. Because there can be many valid ways of drawing a structure depending on which atom you start with, a structure may be given a canonical renumbering of the atoms using a variant of the Morgan algorithm (70). In the case of a linear representation like SMILES, this canonicalization yields a unique string for the structure, which can be generated from any valid SMILES string derived from the structure. In the case of a structure stored in a connection table, the Morgan algorithm results in the atoms being reordered in the connection table to generate a tree, branching outwards from the most highly connected atom in the structure. Because of the efficiency of indexing in modern relational chemical databases, Morgan renumbering is not used as much today as in the past.

The registration system then computes indexes. These include structure-searching indexes, substructure or similarity keys, molecular formula, molecular weight, and other structure-based properties. Substructure keys or "fingerprints" are particularly important. They consist of a number of binary descriptors for the presence of certain functional groups or more generalized atom/bond combinations. These keys can be used to filter structures before searching. They are also used for similarity calculations. Originally, substructure search keys were always used to filter structures before performing a substructure search of the database. If a query structure contains, say, a carbonyl group, then only carbonyl-containing structures should be examined during the substructure search. A key representing the carbonyl group can be used to filter structures that contain the group (the key turned on, or set to 1) from those lacking it (key set to 0). Tree-based substructure searching does not require prior filtering, so today, substructure keys are primarily used for similarity calculations between molecules. If the key values of two structures are compared, the more keys they have in common, the higher their similarity value will be. When registering reactions, the reactants and products may undergo automated or semi-automated perception of reacting bonds and atom centers (71). Generic structures may be analyzed and "clipped" or reverse-transformed to generate root and member structures, which may be stored separately (72).

Before finally storing the structure in the database, the registration program may search the database for some level of match to the input structure or reaction, and skip the registration if it is a duplicate. This is sometimes termed "deduplication" through "exact match" searching. There is usually some redundancy in chemical databases, and to save search time and disk space, most companies do not store duplicate structures or reactions, but rather store pointers to them.

The final step, after registering the structure or reaction, is to assign it a unique registry identifier, which is typically used throughout the company to identify the given structure/reaction and any chemical, biological, or inventory data that is associated with it. Some identifiers, like the Chemical Abstracts Service CAS number and the Beilstein BRN, have wide application, and these may be used in addition to, or instead of, a corporate-assigned external registry identifier.

3.3 Searching Chemical Structures and Reactions

The type of chemical structure and reaction searching that a chemist does usually depends on the current stage of a project. For example, if the chemist is starting a new therapeutic project, a *therapeutic activity* search might be conducted, using a database such as the Derwent World Drug Index, the MDL Drug Data Report, or the MDL Comprehensive Medicinal Chemistry database. Retrieving many search hits, the chemist might organize them by *sorting* on name, molecular weight, ring system, or some topological basis. If the resulting list is too large, the chemist might perform a *cluster analysis* of the structures to see what general classes of compounds have been synthesized in the past. After sampling from the various clusters, and identifying a handful of interesting structures, the chemist might perform a *substructure* search to find structures that contain the features that are felt to be important to activity (i.e., the pharmacophore). If that search returns too many hits, the search query can be refined by making it more specific. If the search returns too few hits, the search query can be relaxed, or a *similarity* search can be used to find structures in the topological neighborhood of the query structure. Eventually, a number of structures will be obtained as candidates for synthesis and/or testing.

The next step is to design a set of reactions to synthesize the compounds. One or more reaction databases can be searched to find whether any reactions give the desired structures as products or give structures that are similar to the desired ones. The chemist may also use reaction similarity searching (73) and searching across reaction schemes (e.g., if A + B → C + D and C + E → F + G; a reaction scheme search will find the query A → F) (74). Once a reaction is found, the chemist needs to decide what reagents to use in the synthesis and where to obtain them. The selection of reagents will usually be based on a combination of physicochemical property considerations (i.e., QSAR and diversity), tempered by

the chemist's experience and preferences, and balanced by synthetic feasibility and economics. The reagents may be located in-house, or they may require ordering from a chemical supplier.

A completely separate approach to reaction discovery is the reaction planning approach implemented in such programs as Logic and Heuristics Applied to Synthetic Analysis (LHASA) (75). This program works by searching a chemical knowledge base that contains information on approximately 2300 retro-reactions or transforms. The chemist draws a target molecule and indicates a strategy for the reverse-synthetic analysis. The program then searches the transform knowledge base for transforms that satisfy the strategy the chemist selected. The program decides which transforms are suitable for the particular target structure and displays the resulting precursors to the chemist. The chemist can then select a precursor for further analysis and choose another strategy option, on which the program returns a second level of precursors in the same way. Processing continues in this manner until the chemist is satisfied that one or more of the precursors correspond to a reasonable starting point for a synthesis. Retrosynthetic methods have not become as widely used in industry as reaction searching, partly because the certainty of the reactions is not guaranteed. Also, searching existing reaction databases generally yields the desired reaction or something close to it. Indeed, a major problem with search results from reaction databases is often an overabundance of hits, which typically need further organization and filtering to be useful. One approach to organizing the results of reaction searching is to apply some clustering or classification to the reactions (76).

To support the workflow just described, a number of structure and reaction search types have come into use (Fig. 9.11). These are briefly described as follows.

3.3.1 Exact Match Searching. Here, the chemist has a particular structure (or reaction) that he wishes to find in the database. The structure/reaction is drawn using a drawing program and then passed to a search program. The program submits the query to the

Searching chemical information

Type of chemical data

	Strings	Structures	Reactions	3D Models	Chemical libraries
Type of search					
Exact match	X	X	X	X	X
Substructure	X	X	X	X	X
Pharmacophore		X		X	
Similarity	X	X	X	X	X
Isomer/tautomer/salt		X			X

Figure 9.11. Search types depend on the nature of the chemical information.

search routine that typically generates index values from the query that are of the same type as those generated for structures/reactions when stored in the database. The index values are then used as filters to retrieve a set of candidate structures/reactions. In ISIS, these filters include the formula, the molecular weight, and the flexmatch index, a numeric hash code based on the presence of isomers, tautomers, isotopes, salts, charges, and stereochemistry (see Glossary). The resulting filtered structures have the minimum set of requirements to fit the search query, but typically only a fraction of these structures will fit the query exactly. Once this set of candidate structures is obtained, the query structure is *mapped* to the candidate structure using a process known as atom-atom mapping, which is known in topology as the "graph isomorphism" problem. This mapping is time-consuming, so the prior filtering step should be as efficient as possible. Each structure that maps exactly to the query is placed in the result set or "hit list." To accommodate various chemists' needs, exact match searching can usually be "relaxed" to permit the finding of isomers, tautomers, salts, charged or uncharged species, etc. In the case of reactions, variations of the reaction can be retrieved by relaxing the constraints on the reaction conditions, solvent, and catalyst (Fig. 9.12).

In a Daylight Thor database, where the

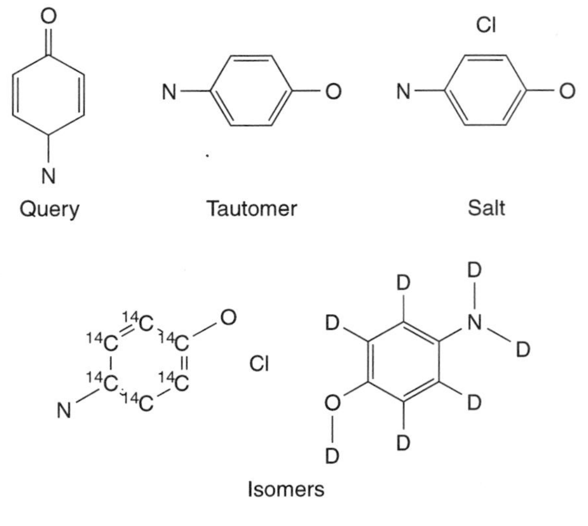

Figure 9.12. Different degrees of exactness can be defined by allowing tautomers, salts, and isomers successively in the search.

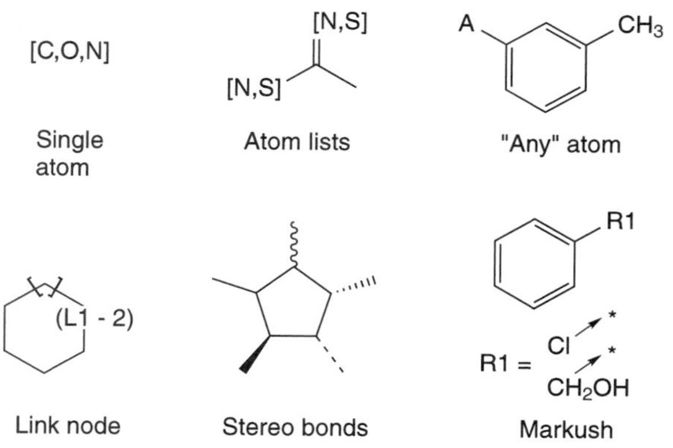

Figure 9.13. Example 2D substructure search queries with various atom and bond query features. The more features that are present, the more flexible the search becomes, but the search may also require more time to complete. There is a trade-off between putting the flexibility into the database (i.e., storing and indexing multiple forms of a structure) and putting the flexibility into the search query and the search software.

structure is stored as unique SMILES, the canonical query SMILES can be compared lexically with strings in the database using fast string comparison and indexing techniques, to find exact match structures and reactions. Because a structure in a Thor database consists of a meaningful, canonical sequence of characters, the computational efficiencies of string searching and comparison can be applied when searching the database. This is in contrast to the highly specialized search techniques used in other structure database formats.

3.3.2 Substructure Searching. A substructure search is performed when a chemist has in mind a pharmacophore consisting of a set of functional groups or a substructure which he knows must be present in the structures to be retrieved. Only part of the molecule is drawn, along with *query features* that generalize atoms, bonds, and rings in the structure. Figure 9.13 shows some typical substructure query features. The features include the following:

- Single atom—specifies a periodic table atom that must be present or a more generalized atom (hetero, metal, etc.) or "superatom" (condensed functional group, such as Ph, Et, Ala, etc.)
- Atom list—a list of atoms, any one of which may be present
- "Any" atom—which simply means some atom must be attached at the given position. As with structures, the hydrogen atoms in substructures are implicit, unless the user

requests a particular hydrogen count or range at a given position

- Link node—which specifies a range of allowed atom or functional group links between atoms
- Stereo bond—including Z/E/either or up/down/either
- Markush feature—used for patent representation, for representation of generic structures for combinatorial chemistry, or to limit the substituents that can be present at a given position. Note that some systems allow logical operations on Markush features (if —OH at R_1, then no —Cl at R_2).

A specialized case of substructure searching is 3D pharmacophore searching, in which a substructure search is combined with the measurement/generation of 3D features, to identify models that could fit a 3D pharmacophore. Figure 9.14 shows an example of a 3D substructure search query that includes various 3D features or constraints. A given conformation of a molecule that is stored in the database may not exactly match a given query, but it could be modified by rotation about single bonds to fit the query. For this reason, conformationally flexible 3D searching is a feature of most 3D database systems (77). When searching conformers, the conformational flexibility can be incorporated into (1) the query, by tethering flexible groups to fixed anchor points in the structure, (2) the database, by storing multiple low energy conformations for each structure, or (3) the search process, by

(a)

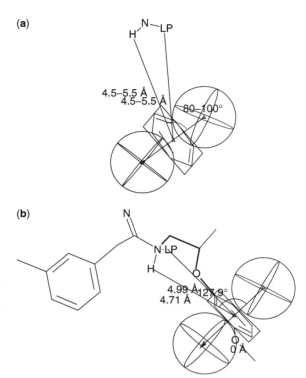

(b)

Figure 9.14. (a) Example 3D pharmacophore search query, showing substructure, distance, angle, and exclusion sphere constraints. (b) Example result of a conformationally flexible 3D search using this query. The molecule was "flexed" in 3D by rotating about the highlighted single bonds to fit the query. The atoms, bonds, and 3D features that match the query are colored. One problem with conformationally flexible 3D searching is that unwanted hits can be conformed to fit the query.

incorporating a rapid conformational analysis into the 3D search algorithm. The last is the most common approach and is a part of database systems from Tripos, Accelrys, and MDL.

Many different approaches to substructure searching have been devised (78). In ISIS, the fastsearch index file is used to retrieve candidate structures. If needed, the query is then mapped onto these structures using a "backtracking" approach. This involves successively matching atoms and bonds in the structure to those in the query in a stepwise manner. When a match fails at any given step, the program backtracks to the last successful step and selects an alternative atom or bond. Once all the atoms and bonds have been matched, the structure is considered a hit. An issue of the *Journal of Chemical Information and Computer Sciences* has been devoted to substructure search methods (79). Hicks and Jochum reported a comparison of several substructure search algorithms in 1990 (80). These authors found the Beilstein-Softron S4 search system to be superior in search speed at that time.

3.3.3 Similarity Searching. The most generalized type of structure/reaction searching is searching for "similar" structures or reactions in the database. Chemical similarity has been a highly debated topic for some time, mostly from the standpoint of what constitutes good descriptors to use in the similarity calculations (81). Nevertheless, there are some general approaches that are widely used, not because of their theoretical soundness, but simply because they work for the chemist. For 2D structures, the most useful and efficient similarity approach is key-based similarity. This involves computing the overlap between a query structure and a candidate structure using substructure or fragment keys. ISIS uses the 960 keys that are generated when the structure is registered. The overlap is typically computed using the Tanimoto metric, which was first used in 2D structure similarity by Willett et al. (82). Depending on the nature and number of the keys, it may be desirable to weight the Tanimoto calculation inversely according to the prevalence of the key in the database. Thus, a cyclopropyl key, which may not be highly prevalent in the database, and would be "swamped" by other, less relevant keys in an unweighted similarity calculation, may have more influence in a weighted calculation. This weighted calculation is used as the default in ISIS chemical databases. It is possible for an ISIS database administrator to regenerate the keys using custom values of the weights to enhance differences in the similarity calculations and select, say, more "drug-like" molecules in the search. In the reaction domain, similarity can be defined in terms of the structures, the reactions, or a combination of the two (83). Other similarity search systems have been described in the literature, including the one used by CAS (84). It is also possible to use 3D pharmacophore keys to compute similarity, although these have typically not performed as well as 2D keys. It is possible that conformational flexibility so vastly expands the "chemical space" of the molecules that a limited number of keys is simply inadequate for 3D similarity calculation. When attempting to predict the type of therapeutic activity a compound has, Briem and Lessel concluded that 2D and 3D keys have complementary information (85).

3.3.4 Reaction Searching.

Reaction searching, sometimes called reaction indexing, has been available for over 20 years. Originally developed as online searching systems, the introduction of in-house systems like REACCS allowed pharmaceutical companies to augment published reaction sources with their own reactions and data (86). As with molecules, reaction storage has moved from proprietary database foundations to storage and access in relational systems. Reaction searching encompasses many of the same types of searches used for molecules. A reaction typically consists of three types of structures: reactants, products, and catalysts or agents, along with textual information about yield, conditions, etc. Reactant and product structures undergo structural changes in the reaction, whereas agents do not. The atom and bond changes that occur in a reaction are isolated in one or more reacting centers of the reactants and products. The atom changes consist of changes in atom valence, charge, number of attached hydrogens, number of bonds at-

tached, retention or inversion of stereochemistry, etc. Bond changes include making and breaking of bonds, and changes in bond order and stereochemistry. When searching reactions, the chemist can search for exact, isomer, or substructure matches in the reactants, in the products, or both. The structure searching can be accompanied by a search of the reaction text information for yield and conditions. Several commercial reaction indexing systems are available from molecule database vendors, and online searching is even possible (87).

In most reactions, the majority of the atoms and bonds are not involved in the reaction, and they remain unchanged between reactants and products. To avoid examining these unchanging atoms and bonds, most reaction indexing systems allow the user to mark, in the reactants and products, those atoms and bonds that are involved in the reaction. These are termed *reacting center* atoms and bonds, and when they are present, they enable much faster reaction searching and they reduce the number of false hits obtained. A simple example is seen in Fig. 9.15. Some systems have semiautomatic perception of reacting centers, which must usually be augmented or checked by a chemist, especially with complex transformations.

As with molecules, it is also possible to do reaction similarity searching. Given a reaction with reactants, products, and agents, one can typically run molecule similarity searches for the reactants, the products, or both. This will retrieve reactions that have similar structures involved in them. This does not guarantee that the molecules undergo the same or even similar transformations. It is possible in some systems to also include the similarity of the transformation as part of the overall similarity search. This is usually carried out using special keys that have been generated for a fixed number of possible transformations. As with molecules, the more keys a query and a reaction have in common, the higher will be the similarity.

3.3.5 Searching Other Data.

Data other than structures and reactions must also be searched in the drug discovery process. Various systems exist for indexing and searching literature and journal contents (88), patents

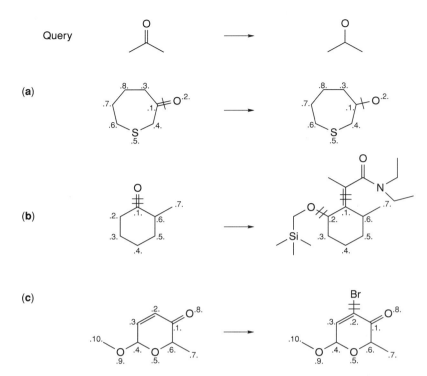

Figure 9.15. Reaction substructure search query and some example hits. If no reacting center or mapping information is used, all three hits are found. If reacting bond information is used, hit c is excluded. If both reacting atom and reacting bond information is included, then false hits b and c are excluded.

(89), material safety data sheets (90), and chemical suppliers (91). Some useful tools include the Accord ChemExplorer program, which allows searching word processor documents and files for particular chemical structures, and the CambridgeSoft ChemFinder for Word (92).

3.4 Chemical Information Management Systems and Databases

A number of software and database vendors provide programs and database systems to implement representation, registration, and searching of chemical information in a corporate environment. Some of these vendors have smaller personal chemical database systems that support registration and searching on a personal computer. A handful of academic and public domain systems are also available. Finally, an increasing number of chemical information systems are being made available on

the Internet. Some representative systems that are being sold or have been discussed recently in the literature are discussed below.

3.5 Commercial Database Systems for Drug-Sized Molecules

Accelrys. A subsidiary of Pharmacopeia, Inc, Accelrys was originally a provider of molecular modeling software. They recently acquired several companies that provide offerings in the chemical information and bioinformatics areas. The company provides unique databases including several for reactions.

- BioCatalysis—biomolecules as catalysts
- BioSter—pairs of biologically similar structures for bioisosterism applications
- Biotransformations—developed in conjunction with the Royal Society of Chemistry

- Failed Reactions—those that did not proceed as expected
- Metabolism—developed in conjunction with the Royal Society of Chemistry
- Methods in Organic Synthesis—33,000 reactions, Protecting Groups—functional group protection with region/stereoselectivity
- Solid Phase Synthesis—with emphasis on small-molecule and combinatorial chemistry

The chemical information programs provided by Accelrys include several database systems.

- Accord for Excel and Access—relational chemical storage for Microsoft programs
- Accord for Oracle—a chemical data cartridge (see Glossary)
- Accord Database Explorer—to access Accelrys reaction databases
- RS^3 Discovery System—with programs for chemical structure, data management, high-throughput screening, and inventory

Accelrys also provides programs for descriptor calculation, QSAR, and data mining (93).

The Beilstein Database. The Beilstein Database, with over 8 million structures, is the oldest in existence, based on the Beilstein Handbook of Organic Chemistry, and contains data that extend back to 1771. The database is produced by the independent Beilstein Institute (94). Access to the database is either through Beilstein Online, available through STN and Dialog, or through the Web using Crossfire Beilstein, which is marketed by MDL GmbH—formerly Beilstein Inc. (95). Data that are stored include the structure, Beilstein and CAS Registry Numbers, names, formula, preparations, reactions, natural product isolations, and chemical derivatives. Physical properties, if available, are also stored, including optical data, mechanical properties, multicomponent system data, spectral and thermodynamic properties, as well as biological function, ecological data, toxicity, and common uses. Citation data, including author, journal,

CODENs, and patent information, are also stored. The data are organized into substance, reaction, and citation contexts, and a user can easily switch from one context to the other. An ACS symposium volume devoted to the Beilstein database has been published (96).

Chemical Abstracts Service. As a division of the American Chemical Society, CAS develops and manages the world's largest databases of chemical structures and reactions.

- CAS Registry—35 million structures—19.5 million distinct structures—13 million biosequences
- CASREACT—4 million reactions
- CHEMCATS—2.5 million commercially available chemicals
- MARPAT—500,000 searchable Markush structures

The CAS databases are maintained online, with searching allowed on a subscription basis. SciFinder is a client/server application to search CAS databases by author, keyword, exact, and substructure. It includes a "keep me posted" update feature, reaction information back to 1974, nucleotide and protein sequence searching, browsing of 1600 journals, and integration of structure, data, and citation information. STN International is a collection of 200 databases covering chemistry, life sciences, engineering, patents, etc. STN Express provides wizard-assisted searching, and STN on the Web serves as a web client for STN. The ChemPort program provides web access to journals (97).

Daylight Chemical Information Systems, Inc. This company provides numerous third-party databases in the Thor format. These include the following:

- Databases of organic structures: Available Chemicals Directory—250,000 structures, Asinex catalog—115,000 structures, Maybridge catalog—62,000 structures, InfoChem SPRESI'95—2.5 million structures
- Drug and biological databases: BioScreen NP and SC—about 52,000 structures including natural products, Pomona College Medchem—36,000 structures with measured Log*P*, National Cancer Institute—

120,000 structures with cancer/HIV screening data, Derwent World Drug Index WDI—60,000 drugs.

- Toxicity: Aquire—5300 EPA structures with aquatic toxicity, TSCA—100,000 EPA substances
- Reactions: InfoChem ChemReact/ChemSynth—390,000 reactions with 470,000 structures, InfoChem SpresiReact—2.5 million reactions and 1.8 million structures

Software and applications from Daylight include the following (98):

- Numerous toolkits: SMILES, Depict, SMARTS, Fingerprint, Monomer, Thor, Merlin, X-Widgets, Program objects, Remote Access, and Reaction Toolkits (see Glossary)
- Daylight chemistry cartridge for Oracle: DayCart (see Glossary)
- Thor database manager—to build and manage thesaurus-oriented databases
- Merlin searching of structures and data
- Clustering package, with Jarvis-Patrick type cluster analysis
- Rubicon, a program for building 3D models using a distance geometry approach
- PCModels for Log$P$ and other physical property calculations
- CombiChem Package to manage high-throughput synthesis
- Reaction Package
- DayCGI—a web development toolkit
- A set of Java tools for chemical information management

Derwent Information. A division of Thomson Scientific, Inc., Derwent is the leading supplier of value-added patent information. The Derwent databases, which are maintained online, include the following:

- Derwent World Patents Index—references to patents, including chemical structure and use patents
- Patents Citations Index—bibliographic and citation data, the Innovations index combined entries from WPI and PCI

- Derwent Selection database—customized subsets of the WPI

The databases are available through several hosting services, including STN, Dialog, and Questel Orbit. User guides for the PCI chemical indexing are available online at Derwent (99). Chemical patents can also be searched using the Merged Markush Service, MicroPatent, and for Japanese patents, the Japanese Patent and Trademark Documents (ISTA) among others (100).

The Gmelin Database. The most comprehensive database of structures, properties, and citations in inorganic and organometallic chemistry is the Gmelin database, based on the Gmelin Handbook of Inorganic and Organometallic Chemistry dating back to 1772. This database includes 1.4 million compounds including coordination compounds, alloys, solid solutions, glasses and ceramics, polymers, and minerals. As such, it is less valuable to drug discovery. The current Gmelin database is owned by the Gesellschaft Deutscher Chemiker and is licensed to MDL GmbH.

MDL Information Systems, Inc. Owned by Elsevier Science Publishing, MDL is a long-time provider of in-house databases and software. Databases include the following:

- Available Chemicals Directory ACD—300,000 structures—reagents and general chemicals, with supplier information
- Bioactivity databases—AIDS database—43,000 structures and data from the National Cancer Institute, Comprehensive Medicinal Chemistry (CMC)—7500 common drug structures, MDL Drug Data Report (MDDR)—120,000 patented drug structures
- Reactions—ChemInform—850,000 reactions and 1.2 million structures, Theilheimer/Chiras/Metalysis—171,000 reactions and 223,000 structures
- Metabolism—Metabolite—53,000 transformations—34,000 structures
- Toxicity—EPA RTECS-based—150,000 structures
- Material safety—OHS Material Safety Data Sheets

Software from MDL includes the ISIS scientific information system (ISIS/Draw, ISIS/Base, and ISIS/Direct), Cheshire for chemical structure manipulation, and Chime and Chemscape for Web access. Combinatorial and high-throughput chemistry programs include Afferent, Central Library, Project Library, Reagent Selector, and Elan. Biological data management programs include Apex and Assay Explorer; literature access through LitLink; reaction access through Reaction Browser/Web; and finally, molecular modeling through Sculpt (101).

Tripos, Inc. Originally the major provider of molecular modeling software, Tripos now offers chemical information content in the form of databases and the tools to manage them. These include the following:

- Several Chapman and Hall databases including ones for organic structures (180,000 structures), inorganic and organometallic structures (40,000 structures), natural products (105,000 structures), and pharmacological agents (22,000 structures)
- The National Cancer Institute structures in a Tripos-compatible format
- The Derwent World Drug Index (60,000 structures)

Chemical information software offered by Tripos now also extends beyond just molecular modeling. Their programs include the following:

- The Unity 3D database system, which features rapid flexible 3D pharmacophore searching
- Concord and Stereoplex—for generating 3D models of database structures including multiple stereochemical isomers
- ChemEnlighten for chemical data mining
- The AUSPYX structure data cartridge for Oracle
- A suite of programs for combinatorial chemistry—Legion to build and store virtual libraries, CombiLibMaker to enumerate structures, Selector to define diversity measures to select diverse subsets of structures, and DiverseSolutions to apply chemi-

cal diversity techniques to chemical populations to characterize and populate chemical space (102)

3.6 Sequence and 3D Structure Databases

Sequence databases of biological macromolecules are useful when defining new therapeutic targets. Databases for DNA, RNA, and proteins are available from such sources as the National Center for Biotechnology Information (NCBI) (103) and the European Bioinformatics Institute (104). Numerous online programs and tools are available to researchers to search and align sequences, generate phylogenetic analyses (chemical evolutionary trees), map genes, and predict secondary structure (105). The Protein Data Bank stores the largest collection of crystallographic, NMR, and molecular-modeling derived protein and nucleic acid 3D models (106). The Cambridge Crystallographic Data Center is the primary source for crystal structure data on small molecules, with more than 250,000 entries. The Cambridge Database can be searched using the programs ConQuest for searching, Mercury for structure visualization, and Vista for numerical display and statistical analysis (107).

3.7 In-House Proprietary and Academic Database Systems

Larger chemical and pharmaceutical firms have, over the years, developed in-house systems with capabilities that are specific to the chemist's needs. Today, the costs of developing from scratch and maintaining an in-house system are prohibitive, especially because commercial chemical information systems are highly efficient and customizable. Personal chemical information software is still being developed and reported in the literature. Examples include a relational database patterned after the Upjohn Cousin system (108), and CheD, which is a SQL-based system with a Web client (109).

Commercial personal database systems are available from several vendors, as described above. These products extend the productivity of an individual chemist or a small workgroup, but are not designed for corporate or enterprise applications. Other personal chemical database programs that are available include ChemFinder from CambridgeSoft, Chem-Folder from ACDLabs, ChemWindow from Softshell, and Aura-Mol from Cybula (110).

4 CHEMICAL PROPERTY ESTIMATION SYSTEMS

The design and screening of drug candidates is increasingly being conducted *in silico*. This is made possible by improvements in programs for property calculation and estimation. Here, the term property *calculation* refers to the generation of some topological (depending only on the 2D structure), topographical (depending on the 3D conformation), or physico-chemical property of a molecule—directly from the structure. The term property *estimation* refers to the generation of some property as a function of other properties—either through a regression equation, a formula, neural network calculation, or some other indirect means.

The distinction between calculation and estimation is important because some properties, like molecular weight, polar surface area, molecular connectivity values, counts of chemical functional groups, partial charges, and other quantum mechanical descriptors, can be calculated precisely and *de novo* from the structure alone. Most of these properties have some fixed definition or algorithm that enables their calculation to be performed unambiguously, with little or no error. What error is present is usually systematic or *deterministic*. A second class of properties, including LogP and other additive-constitutive properties, may be calculated by fragment additivity with various correction terms. These properties differ from *de novo* properties because they are approximations to the true (sometimes measured) values. Often, there are multiple approaches to their calculation. The errors in the calculation of these properties are statistical or *stochastic*. A third class of properties includes those that can only be estimated from other properties, using a regression analysis, neural network, or other linear or nonlinear function of variables. The errors in these properties can be complex and difficult to determine. For all these reasons, it

is important to carefully consider the use of any given property for drug discovery purposes. Too often, properties are calculated simply because they are available, then used in a QSAR analysis, and possibly applied to future predictions—all without proper consideration of their precision, accuracy, and relevance to the chemical problem.

Given this caveat, it must be noted that there are a multitude of programs available for the calculation of properties of structures. Some programs compute only a single property, like LogP. Others calculate a series of values in a given genre of property, like molecular connectivity (111) or BCUT descriptors (112). Still others compute a vast range of properties that include topological, topographical, and physicochemical descriptors alike. It is beyond the scope of this chapter to detail all the programs and vendors that provide property calculation and estimation software. Many of the calculations are provided as part of molecular modeling and QSAR program systems. Some programs and vendors whose products are solely for property calculation are described below.

4.1 Topological Descriptors

Descriptors based on the 2D structure or simply on the connectivity matrix of a structure have long been used for chemical similarity and for property correlations. Because they often lack any relationship to mechanism, these descriptors are best used within a congeneric series or at least a set of similar structures. They may be empirically useful for cluster analysis and chemical library design, because they are effective at representing structure differences and similarities. A few programs and providers of topological descriptors include the following:

- Barnard Chemical Information—provides chemical Fingerprint Generation Pack—to compute fragment-based fingerprints for cluster and diversity analysis (113)
- DRAGON—implementation of about 1400 descriptors of Todeschini and Consonni (114) including constitutional, topological, autocorrelation, geometrical and functional

groups, and including simple molar refractivity, polar surface area, and Moriguchi LogP

- Molconnz-EduSoft LC—provides MOLCONNZ molecular connectivity and electrotopological state descriptors of Kier and Hall (115)

4.2 Physicochemical Descriptors

As a complement to topological descriptors, physicochemical descriptors often have a strong relationship to mechanism, and are widely used in lead optimization and QSAR. The classic triad—steric, electronic, and lipophilic descriptors—are considered the foundation of QSAR, and adequate coverage of the space of these factors is still a major goal in drug discovery. The most common physicochemical descriptor is LogP, the 1-octanol/water partition coefficient. Because it is so important, a number of programs and vendors provide LogP calculations based on a variety of methods. Many of these programs also compute other physicochemical properties, such as pKa and solubility.

- BioByte, Inc.—developers of CLOGP, premier LogP, and molar refractivity calculator (116)
- Syracuse Research Corporation—provide KOWWIN and 11 other structure-based property calculations (117)
- CompuDrug Ltd.—the PALLAS System—including programs for for pK_a, logP, logD predictions, metabolism and toxicity, and high pressure liquid chromatography (HPLC) development (118)
- ACDLabs—physicochemical laboratory program calculates pKa, LogP, logD, aqueous solubility, boiling point and vapor pressure, Hammett electronic constants, and a variety of liquid properties (119)
- XLOGP—The Peking University LogP calculator—a similar version for proteins is available as PLOGP (120)
- EduSoft LC—provider of Hint!-LogP—to accompany the HINT! Hydropathic interaction modeling program (121)
- SciVision—provider of software for chemical property calculation, and to estimate

QSAR, toxicology, oncology, and other biological properties (122)

- Sirius-Analytical—provider of instruments for LogP and pKa determination, and the Absolv program to predict physicochemical properties (123)

Most of the commercial molecular modeling systems also provide some property calculations, which range from simply calculating the polar surface area of a structure to a full range of topological and physicochemical descriptors. These may be based on fragment additivity, like most of the programs mentioned above, or they may involve correlations with quantum mechanical or even molecular dynamics-based calculations.

4.3 Absorption, Distribution, Metabolism, and Excretion Properties

Perhaps the most critical aspect of drug development—the behavior of the drug *in vivo*—is also one of the least predictable. Each year, many drug candidates reach the very expensive stage of clinical trials, only to be discontinued because of problems with absorption, distribution, metabolism, or excretion (ADME). Toxicity is often added to this acronym (ADMET), because we increasingly find critical differences in the way children respond versus adults, males versus females, etc. Among the hopes that accompany the deciphering of the human genome, is that drug selection can someday be tailored to an individual's genotype, to lessen the possibility of untoward drug response. For the present, drug designers are focusing increased attention on the prediction of ADME properties, pharmacokinetics, and *in vivo* behavior. Compared with topological and physicochemical predictions, ADME calculations are still rather crude and approximate. They are usually based on correlations with other properties. And, if the method for obtaining the correlation is a neural network, the predictions may be superior to simpler regression-based approaches, but the interpretability of the model is missing. Some of the programs that are used to predict ADME descriptors include the following:

- LION Bioscience—provider of iDEA, a modular ADME predictive system. The absorption module predicts Caco-2 cell permeation, and performs dose-response modeling of the oral absorption. The metabolism module predicts first-pass effects and models metabolic parameters. Future modules are planned for distribution and elimination (124).
- PASS—prediction of biological activity spectra—compares a test structure with those in a database of about 45,000 structures with known activity/toxicity, using topological descriptors and probability calculations (125).

4.4 Property Calculations Online

Many of the providers of software and databases of chemical properties also provide online calculation services. These include Daylight Chemical Information (126), ACD labs (127), and Syracuse Research (128). In addition, the following sites provide online calculations of a variety of properties:

- Molinspiration—calculates LogP, polar surface area, Lipinski Rule-of-5, and a druglikeness index (129)
- Alogp—VCCLab online LogP calculation (130)
- PETRA—The University of Erlangen property calculation routines (131)
- USEPA Suite—including implementations of the Syracuse Research software (132)

5 DATA WAREHOUSES AND DATA MARTS

Even relational databases have their limitations when dealing with huge amounts of data and high user traffic. The burden of continual data updating and registration—activities known as OnLine Transaction Processing (OLTP), can considerably slow down searching and report generation activity—known as OnLine Analytical Processing (OLAP). For this reason, it is becoming common in the database field to build special large databases designed primarily for searching purposes—so-called data warehouses (133). These warehouses have pre-computed indexes and tables that facilitate repeated searching. A special database architecture, known as the *star schema*, facilitates OLAP activity. In this design, one or more large *fact* tables contain records of frequently searched data for each object (e.g., structure or reaction) in the database. The fact table is joined to smaller *dimension* tables that contain the relational information. The schema is known as a star schema because the architecture resembles a many-pointed star, with the fact table at the center, and dimension tables at the ends of the arms. The design of the fact and dimension tables in the warehouse should reflect the searching habits of the users to get the best performance. Probably the first mention of data warehousing in the pharmaceutical area was that of Axel and Song in 1997 (134).

5.1 Data Warehouses of Chemical Information

A data warehouse is designed to consolidate structures and data from many diverse sources, including relational databases, flat databases, and structure and data files. It is considered to be *multidimensional*. A true chemical data warehouse might contain sequences, 2D structures, 3D models, Markush structures, and reactions—all in the same database. No such commercial database currently exists, but databases presently being developed at MDL and other vendors are examples of chemical data warehouses of structures and their reactions. The MDL data warehouse framework is termed the *concordance*. The fact table for the concordance is the *source* table, which brings together structure and reaction identifiers from all the various data sources and links them to the unique structures in the warehouse (Fig. 9.16). Using the concordance, a substructure search can retrieve a set of unique, unduplicated structures, along with pointers to all the relevant identifiers and reactions in the various data sources. Similar pointers exist to the original citations and stored data.

Physicochemical properties that are based solely on the structure are stored in the data warehouse, but properties that are datasource dependent, such as citation or biological activity, are only referenced. A typical use of a chemical warehouse is to search for a set of

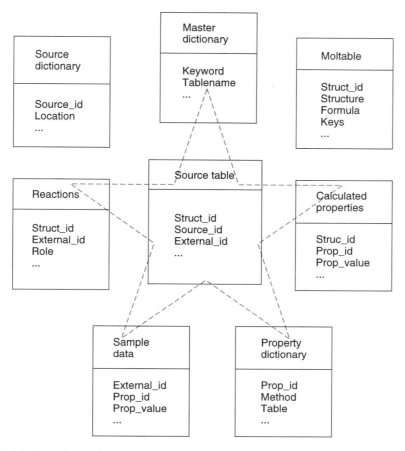

Figure 9.16. Star schema design of a chemical data warehouse. The central source table allows access to the External_ID of every molecule, arranged by source database. These External_ID values can be used to build multidimensional views of the data. For example, to see all the reactions with products that can be found in source database ACD, one would combine data from the source dictionary table (Source_ID for database ACD), the reactions table (Struct_ID, and Role), and moltable (Struct_ID) table, using identifiers (External_ID) from the central source table.

structures that satisfy a search query, then drill-down using a web browser to access the original data sources. In the case of reactions, the user might retrieve and browse a list of reactions that contain the structures that were found in the search. In addition to drill-down, a "hop-into" facility allows passing a set of structures into a search program or web browser that is native to the source database being accessed.

5.2 Data Marts of Chemical Information

For certain purposes, like reagent selection, a data warehouse is too large and comprehen-sive, and a much smaller database is sufficient. Such a data *mart* has the same architecture as a data warehouse, but it has only a single di-mension of structural data—for example, syn-thetic reagents. The MDL Reagent Selector program is one example of a data mart of re-agent structures, with information on their price and availability from various suppliers (Fig. 9.17). It has a fact table that links struc-tures to their identifiers in the various source databases. It stores properties that can be used to filter reagents, such as the molecular weight and LogP, and it has pointers to sup-plier information stored in the MDL Chemical

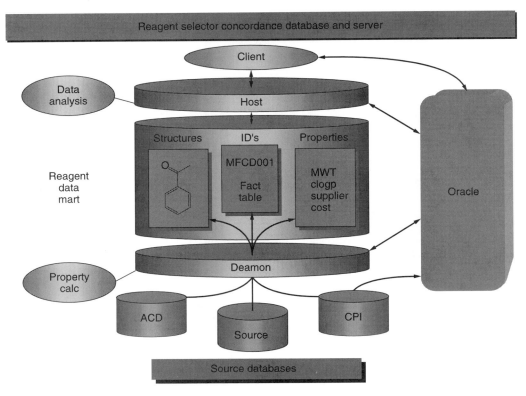

Figure 9.17. Reagent Selector—an example of a chemical data mart. Various components of the system are shown, including the data sources, the daemon program that automatically updates the mart, the concordance database, and the client/server architecture, which is implemented in a three-tier system.

Products Index (CPI) database. To aid in reducing the size of a hit list, a Reagent Selector user can filter reagents and sort on properties, availability, presence or absence of functional groups, etc. (Fig. 9.18). Further list reduction can be achieved by clustering the structures by means of a cluster analysis using substructure keys as descriptors.

An important feature of Reagent Selector is the *daemon* program, which runs in the background. This agent-like program "awakens" on a fixed schedule and checks the various source databases for new or deleted structures or for changes in the structures and data. If any changes or additions are found, the daemon updates the data mart accordingly, so users will see the latest information when they run searches. Another aspect of chemical warehouses and data marts concerns their physical architecture. It is increasingly common to see so-called "multitier" architectures in which the client program (the "application tier") may be a very "thin" Web client that communicates to a more extensive "middle tier" of programs that serve the immediate needs of the client (see Glossary). Requests for searching and registration, which demand database server resources, are passed from the middle tier to a "database tier" that corresponds mostly to the server part of former client-server architectures. There are many advantages to this arrangement. The programs can be distributed onto different computers to optimize performance of the system. The middle tier can be modified independently to accommodate changes in the client and server. From a development point of view, the various tiers in the architecture can be developed and maintained on their own schedule, with minimal dependence on other components.

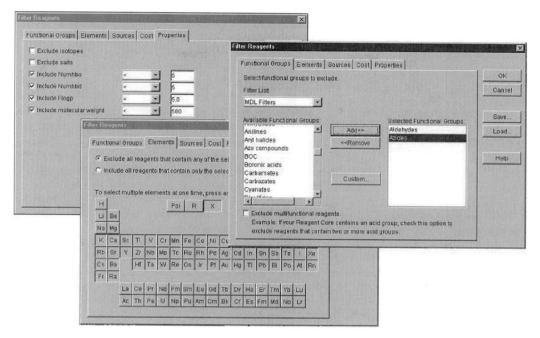

Figure 9.18. Filtering structures as part of the reagent selection process. The filter criteria include criteria for structure complexity, log$P$, Hdonor/acceptor, molecular weight, formula, and substructures.

6 FUTURE PROSPECTS

It is always difficult to predict the direction of advances in information management. Much of the improvement in chemical structure management and searching has been because of advances in hardware and computer systems. Moore's Law, which states that computer power roughly doubles every 18 months, has held since the 1970s, but threatens to break soon (135). As computer manufacturers hasten to avert the leveling off of computer performance gains, new technologies will surely affect the way chemical information is stored and searched. Based on current trends, it is possible to make some short-term predictions about directions in this field.

- Information: Integration of chemical structure data with other types of data. There is a welcome tendency to treat structures, models, and reactions like other relational data. The biggest advantage of this approach is being able to run integrated searches using structures and data together in search que-

ries. An interesting approach pioneered by the Merck group involves generating fingerprints for key words in documents, then searching for combined structure/document similarity (136). This approach and similar ones will be simplified by increasing integration with relational database systems, as described below.

- Knowledge Discovery in Databases: In the past, dating back to the DENDRAL project (137), attempts to apply artificial intelligence and machine learning to problems in chemistry and drug discovery have gained only moderate acceptance. One problem with expert system approaches has been the small base of information to access. In some cases, this has been a single expert chemist or a handful of example structures. This situation is changing as data accumulates in databases. As Fig. 9.19 shows, *data* can be organized, indexed, and stored in databases to produce *information*. This information, depending on how relevant, unique, and complete it is, can be analyzed and modeled

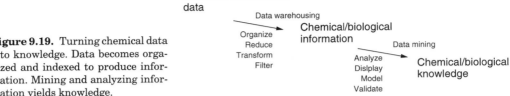

Figure 9.19. Turning chemical data into knowledge. Data becomes organized and indexed to produce information. Mining and analyzing information yields knowledge.

to generate *knowledge* that might not otherwise have been evident. Once this knowledge is materialized, it can be managed, shared, and deployed for future applications. This process is termed Knowledge Discovery in Databases (KDD), and it is becoming more widely practiced (138).

Data mining is the mechanism by which knowledge is derived from databases. It is generally defined as the extraction of predictive models and associations from large volumes of data using statistical and pattern recognition techniques, usually for some competitive advantage. Data mining is already well estab-

lished in the marketing, sales, and telecommunications fields (139). Data mining is being used increasingly by scientists, especially in genomics and proteomics. Example applications include the clustering of DNA array data and using database information for protein secondary structure prediction (140). Examples are starting to appear in the field of drug design. Depending on the stage of a drug discovery project, one can mine chemical structure data for diversity, similarity, or specificity, as shown in Fig. 9.20. This figure shows that the lead discovery, refinement, and optimization phases of drug discovery proceed through mini-cycles, each with their own data

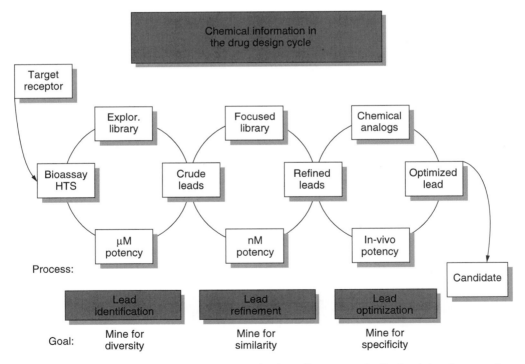

Figure 9.20. Mining chemical information in the drug discovery cycle. Each mini-cycle proceeds until a sufficient number of suitable compounds become available.

mining requirements. So far, data mining has mostly been applied to library design, QSAR, and ADME prediction (141). Whether the techniques become more widely used will depend on the accuracy of predictions made using them, on the availability of convenient software, and most of all, on clean and relevant data.

- Software: Integration with relational systems. A consequence of treating structures as relational data is a tighter integration of once-specialized structure management software techniques with relational database systems. In Oracle, so-called "data cartridges" are being increasingly used to allow a chemist to treat structures like other relational data in a search. Structures, models, and reactions can all be input, registered, and searched using standard SQL to which special operators have been added. SQL stands for Structured Query Language—the standard language for querying relational systems (see Glossary). For example, in the Daylight relational data cartridge, substructure and similarity searches in a reaction database can be conducted directly in SQL as follows:

  Substructure—to find reactions containing benzoic acid as a product:
  SELECT * FROM RXN WHERE CONTAINS (SMILES, '>>O=C(O)c1ccccc1') = 1;
  This statement translates to "Select everything from the table named RXN, where the SMILES field contains the substructure string for benzoic acid as a product". The "=1" clause is an artifact of the data cartridge implementation; it does not necessarily mean that only a single occurrence of the benzoic acid substructure should be found.
  Similarity—to find how many reactions have a solvent that is 80% or more similar to acetic acid:
  SELECT COUNT (*) FROM MEDIUM WHERE SIMILAR (SMILES, 'OC(=O)C', 0.8) = 1;
  This translates to "Tell me the number of rows in the table MEDIUM where the

value of the SMILES column in that table has an 80% similarity to acetic acid." Again, the "=1" parameter is an artifact.

This approach greatly simplifies the development of applications. Also, the searches can take advantage of optimization that is built into the relational database system. Fig. 9.21 shows a Web browser that uses the MDL reaction data cartridge to perform structure and reaction searches. The use of a direct searching approach with an object-relational database for combined retrieval of chemical and biological information was reported by Cargill and MacCuish (142). In the field of data mining, the generation, storage, and deployment of predictive models is fully integrated into SQL Server 2000 and Oracle 9i, and this trend will soon extend to other relational database systems (143).

Another advance in chemical information software that promises to have considerable impact on drug discovery is "meta-layer" searching, as described by Hoctor (144). In this approach, queries entered by the chemist are submitted first to a middle-tier search engine, the meta-layer, which automatically and transparently generalizes and transforms the query into several queries. These are then submitted to various databases to retrieve "more of the same" kinds of information. The results are automatically formatted and presented to the chemist in the context of a Web browser (Fig. 9.22). Thus, a name search might get converted automatically to a structure, for substructure or reaction searching, a literature citation search, or a patent search, etc. The linking of searches across indirectly related literature can also be used to generate new knowledge (145).

- Hardware and Operating Systems: The value of parallel and distributed processing was reported early in the development of structure search systems (146). Since then, some commercial products have adopted parallel processing. These mostly involve CPU-intensive searching like conformationally flexible 3D searching and docking. With the exception of such tasks, the speed of most chemical information searching is determined by data input and output (i.e, "I/O

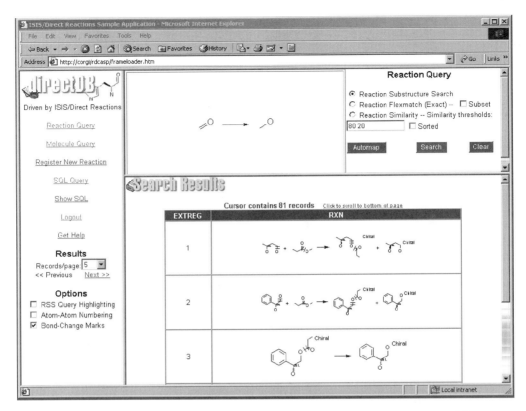

Figure 9.21. Web client for an application that searches a relational reaction database. SQL statements are used to select structures and reactions that satisfy the search query.

bound") because chemical structures are a highly "verbose" type of data. As chemical information systems integrate more with relational systems, they can take advantage of the parallel and distributed processing capability of the relational system. An important development is the "24-7" availability of data in chemical databases (24 h/d, 7 d/wk). This can only be accomplished by distributing and replicating databases across a network.

It would be pure speculation to estimate the impact of changes in hardware and operating systems on chemical information management. Presently, Sun Microsystems is probably the dominant Unix system in chemical database management, largely because of their network presence and their support of Java. Microsoft has released their Windows XP operating system, which merges the Win-

dows 98 and Windows 2000 software streams and will give them continuing dominance in the PC market for a while. Linux is quickly catching on as an inexpensive alternative, and it has a strong foothold in the molecular modeling area, but it requires operating system expertise and lacks a business software base. Small handheld personal data assistants are becoming more capable, and wireless computing is on the rise. The standard desktop computer has at least a 2.0 GHz processor with 512 Mb or more of RAM, about a 30–72 Gb hard drive, and a combination read/write CD with DVD. A relational database of 1 million structures consumes about 3–4 Gb of disk space and can be substructure searched in a few seconds, returning a hit list containing thousands of structures. It takes much longer for a chemist to wade through the results of such a search or to process analytical or bioassay results from a single combinatorial chem-

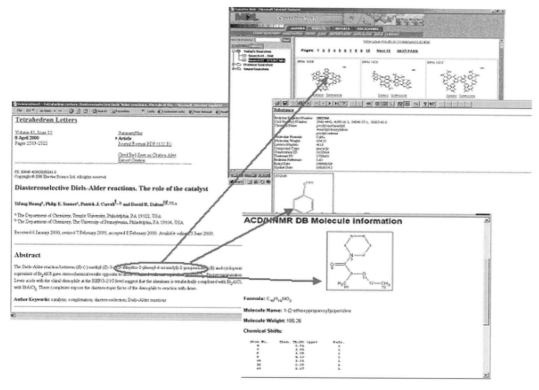

Figure 9.22. Using meta-layer searching to retrieve implicit information. A name search query is converted to a structure, which is then transparently searched to add structure-based search results to the literature citation.

istry experiment than to run most data searches. In light of this, it seems evident that the tools that will succeed are those that will best assist the chemist in extracting relevant, implicit knowledge from the data and deploy that knowledge for future benefit.

7 GLOSSARY OF TERMS

2D Query Feature. A structural feature added to a 2D substructure search query to generalize the query or make it more specific. An example atom query feature would be specifying a list of allowed atoms (Cl, Br, I) or limiting the number of attachments. A bond query feature would be allowing a single or double bond (S/D) or forcing the bond to have a particular stereochemistry. More complex query features can be used to specify which functional groups or substituents are allowed

at a given position (*Rgroups*), specifying a range of chain length size (*link nodes*) and specifying atom, bond, or molecular data queries (*Sgroups*).

2D Structure. In terms of chemical information, a collection of information about atoms and bonds that can be displayed in a manner such that a chemist would recognize it as a chemical structure. The atom and bond types and connections are usually explicit. The layout of the atoms in the display may be explicit (*x,y* coordinates) or implicit—determined at the time of display. Hydrogen atoms may be fully or partially suppressed to save storage space.

3D Model. In terms of chemical information, all the information in a 2D structure plus at least one set of 3D atomic coordinates. This is a single conformation of the structure, which is typically a low energy conformation or even a crystal or spectrometrically determined 3D structure. A 3D model may also contain information about multiple low-energy conformations and atom, bond, and molecular properties such as partial atomic charge, HOMO/LUMO energy, etc.

3D Query Features. Topographical features that relate atoms, bonds, and other 3D features to each other in a pharmacophore or 3D substructure search query. Typical features include (1) *objects* such as atoms, centers of rings, electron lone pairs, and regions of exclusion and (2) *measurements* of distance, angle, dihedral angle, radius of exclusion, etc. Measurements often have a range associated with them (e.g., distance between a carbonyl oxygen and a secondary amino nitrogen is 3.4--5.0 Å).

Agent. A computer program that can run autonomously and on a schedule to perform database searches, maintenance, and reporting activities that a chemist would otherwise have to do manually. An example would be an Internet notification service that sends the chemist an e-mail notification whenever a particular database has been updated.

Application Tier. In a multi-tier architecture, the collection of programs that run on the chemist's client or workstation machine. It is the tier of programs "closest" to the chemist in the architecture. Typically this may be a Web client program or other program with a GUI that allows the chemist to interact with the architecture.

Artificial Intelligence. A branch of information science that attempts to use computer programs to perform or simulate human mental activity. Applications in chemistry include perceiving chemical structures, designing structures to fit topological or topographical criteria, designing 'novel' structures, etc. Many of the activities of AI overlap with, or contain elements of pattern recognition and data mining.

ASCII. American Standard Code for Information Interchange—a widely used system of encoding alphanumeric information into eight-bit bitsets (bytes). The expansion of information to include non-English characters requires the use of larger (16- or 32-bit) character sets such as Unicode.

Atom List. In a substructure search query, a list of allowed (or perhaps disallowed) atom types. Often represented within brackets: [Cl,Br,I].

Atom Stereochemistry. Usually refers to tetrahedral stereochemistry at a given atom, which must be a chiral or prochiral center. The stereochemistry may be *local* (or *relative*) or *global* (based on CIP conventions). If it is local, it usually is termed "parity" or some other nonspecific term, to distinguish it from true global stereochemistry (*R,S*). Local atom stereochemistry is a property of the atom and its nearest attached atoms. Global atom stereochemistry depends on the entire molecule and the stereochemistry at other chiral atoms. In some systems, the atom stereochemistry is perceived from the drawing of the structure, using "up" (wedged) and "down" (dashed) bond marks as cues. In linear notations, characters in the string can be used to specify the counterclockwise (@) or clockwise (@@) orientation of attachments at a given center.

Atom-Atom Mapping. The procedure of assigning each atom in a substructure query to a given atom in a candidate structure. The assigned structure atoms must match the query atom in all characteristics, including atom type, stereochemistry, charge, attachments, etc. In some structures, a query may map onto the structure in many ways (multiple mappings). Additionally, these mappings may *overlap* each other in terms of atoms and bonds, or they may be *non-overlapping*. Some search systems stop after the first mapping, whereas other perform *exhaustive* mapping, until no further mapping can be found.

Automap. A feature implemented in reaction indexing programs like REACCS, which attempts to automatically "discover" which atoms and bonds are involved in a reaction transformation (the *reacting center* atoms and bonds). The chemist draws the reaction or the reaction query as a set of reactants leading to a set of products, then invokes the Automap feature, which causes the reacting atoms and bonds to be marked and identified. When reacting center atoms and bonds are specified in both the query and the reactions in the database, reaction substructure searching is faster and gives fewer false hits.

Backtracking. One process that is used in mapping substructure atoms and bonds to the corresponding atoms and bonds in a candidate structure. Given a certain query—say, an amide group [—C(═O)N—], a backtracking algorithm searches first for a carbon atom, then for an oxygen atom, then checks to see if they are doubly-bonded, and finally checks to see if a nitrogen is singly-attached. At any step in the process, if the check fails (e.g., with an ester, the final check would fail), the program "backtracks" to the last successful step and examines another eligible atom or bond. If no eligible atoms or bonds are found at that step, it backtracks to the next previous step in turn. This procedure is guaranteed to find a mapping, but it can be slow, especially with large or highly symmetric queries or structures, where a multitude of similar paths must be examined. Alternative approaches that use an *indexed tree* can be faster, especially for large databases.

BCUT Descriptors. Descriptors of chemical structure that are derived from an eigen analysis of the connection table of the structure. The class of BCUT descriptor depends on the quantities that are stored in the table (simple connection information versus electronic or steric interaction values). BCUT descriptors have found value in molecular diversity and chemical library design.

Binary Data. Data stored in a file or database that is not chemist-readable, and usually cannot be converted to printable characters. Examples include connection table storage in a database, substructure search keys, and a graphics image of a structure. Note that some other data that is also not chemist-readable, like certain linear notations (e.g., a Chime string), may be made up of printable characters and is not strictly binary data.

Bioinformatics. The application of statistical and mathematical techniques to turn sequence data into useful biological information. The general goal of bioinformatics is to define the structure, location, and function of the proteins and nucleic acids that are the products of the processing of a genome. The application of bioinformatics in drug discovery is primarily the identification of new therapeutic targets.

Biological Data. This includes the results of *in vitro* and *in vivo* assays, toxicology and metabolism studies, DNA and protein array data, etc. It complements the chemical data, and increasingly, both chemical and biological data are being stored in large corporate relational databases. At any given stage in the drug discovery process, obtaining and analyzing the biological data has traditionally been considered the more complex and rate-limiting step in the process. The application of high-throughput methods to screening and pharmacokinetic analysis is yielding considerable benefit in the collection and processing of biological data.

Bitset. A contiguous set of binary digits (bits, 0/1) in computer memory. Bitsets are often used in chemical information to store collections of yes/no, presence/absence, and active/inactive responses in a compact form. Bitsets are used to store substructure search keys for each structure (*fingerprints*), which are used in similarity calculations. Bitset indexes are common features of relational databases, where a collection of bits, one for each structure in the database, can store the presence or absence of a given piece of data for each structure, or, in the case of a substructure search, the a compact representation of the result set from the search. The advantage of bitset representation is that computers can perform very fast logical operations (union and intersection) on bitsets, which enables filtering and subsetting of large lists of structures and data.

BLOB. A Binary Large Object data type. This data type is used in Oracle, for instance, to store large amounts (e.g., up to several Gigabytes) of binary data. Storage of the connection table and all the perceived structural information for a registered structure is one example. Another example is the storage of the entire *fastsearch* index for a database, which can be accessed as a single object by the Oracle data storage and retrieval routines.

Bond Stereochemistry. This complements the atom and molecule stereochemistry of a structure. A given double bond can be assigned Z or E, or *cis* or *trans* stereochemistry based on the attachments. If the stereochemistry is unknown or is a mixture, it can be assigned a value of "either." In a substructure search, the bond stereochemistry can be specified in the query to limit the scope of the search. In some registration systems, the bond stereochemistry of a given structure is perceived from the input drawing of the structure. In the case of linear notations, it can be specified by characters in the string (e.g., Cl\C=C\Cl specifies *trans* dichloroethene).

BRN. The Beilstein Registry Number, which can be used to access structures in the Beilstein database.

Business Rule. An established convention for the representation of data in a given company or laboratory. In the case of chemical structures, an example of a chemical business rule would be "all nitro groups should be drawn as —N(=O)(=O)—and not the charge separated form —N$^+$(—O$^-$)(=O)." In the case of biological data, a business rule might enforce the units in which a given piece of test data is reported (e.g., dosage in mmol/kg). Business rules can be enforced by preprocessing data before it enters the database, or in the case of multiple, diverse data sources feeding into a data warehouse or data mart, the data can be transformed to the correct form before storage in the warehouse.

Canonical Numbering. Reordering the numbering of atoms in a structure to a unique order, based on the extended counting of the number of attachments at each center the atom and bond types, etc.

CAS Number. Chemical Abstracts Registry identification number—very widely used to identify chemical structures.

Chem(o)Informatics. By analogy with bioinformatics, this is the application of statistical and mathematical techniques to turn chemical structure data into useful chemical and biological information. It makes use of techniques from statistics, pattern recognition, artificial intelligence, and data mining to derive useful predictive relationships between structures and their biological or physicochemical properties. Broadly considered, cheminformatics also includes the input, storage, management, and searching of chemical structure information.

Chemical Library. A collection of structures, real or virtual, that is the current starting point for high-throughput screening or analysis. A library may be all the structures in a database, or more commonly, a subset of these. It might consist of diverse structure types or it might represent the enumeration of one or a few generic structures. Libraries can be classified according to the stage of discovery—i.e., diverse libraries for lead discovery, focused libraries for lead development, and optimized libraries for lead optimization.

Chemical Space. A loosely defined concept that all the known or possible chemical structures define some multidimensional space in which the structures are points. Structures that are topologically or topographically similar to each other (i.e., look similar), cluster in chemical space, and by the principle of chemical similarity, should show similar physicochemical and biological properties. This is the basis for diversity analysis of chemical libraries. The challenge is to select or discover properties of the structures that define the chemical space and can be used.

CIP Stereochemistry. Cahn-Ingold-Prelog stereochemistry conventions. An IUPAC approved and widely used set of rules for assigning stereoisomers based on atom and group priorities (see http://www.chem.qmw.ac.uk/iupac/stereo/).

Cleaning and Transforming Data. When importing data from diverse data sources (files, databases, spreadsheets, LIMS systems, etc.) into a database or data warehouse, the data usually needs to be standardized, checked, and sometimes transformed to some common format and content. This allows faster search and retrieval, and serves as a check of data integrity. The rules that define the cleaning/transformation process are often termed "business rules," and in the case of chemical data, they may include checking and modification of chemical structures.

Client-Server Architecture. A computer architecture in which a "server" computer (usually a larger and faster machine at a central location) runs programs that communicate over a network with numerous workstations or "client" machines that reside in offices and laboratories. The server computer performs heavy duty computing tasks such as database searching and molecular and data modeling, in response to commands from the users of the client computers. It then communicates the results back to the client machines. There, depending on whether the client is "thick" (a relatively large and capable application that can display and manipulate data and structures), or "thin" (a small program, possibly running in an Internet browser), the data is displayed, manipulated, and reported. Client-server architecture is two-tier, and is being supplanted by more versatile multi-tier approaches.

Clipping. The computer application of a chemical transformation to a set of structures. One example would be the conversion of a set of *o*-subsituted phenols to a generic representation with the ortho substituents collected into an Rgroup attached to the parent phenol structure. The reverse process, going from the generic structure to all the specific non-generic structures, is termed *enumeration*. Clipping also includes functional group transformations, such as converting a ketone to an alcohol. In the process of cleaning and transforming chemical structure data, clipping may be involved when chemical business rules are "enforced."

CLOB. A Character Large Object data type. This data type is used in Oracle, for instance, to store large amounts (e.g., up to several Gigabytes) of character data. Storage of structures in a relational database in molecule-file format is an example.

Cluster Analysis. The process of discovering "natural" groupings of points in the space of some measurements or descriptors. In chemical information management, one often clusters chemical structures for diversity analysis or to subset the results of a search. Structures are most often clustered using functional group fingerprints as the descriptors. Clustering methods usually consist of either partitioning methods like k-means and Jarvis-Patrick, or hierarchical methods, which may work by successively dividing the points (divisive clustering) or by successively aggregating points (agglomerative clustering). Cluster analysis is an important part of *unsupervised data mining* and pattern recognition.

CML. The Chemical Markup Language. Based on XML and HTML, it provides a standard self-documenting molecule file and information interchange format. Information is described by tags and values. A CML document can be "parsed" by a freely available computer program that can return the structural information on demand.

Combinatorial Chemistry. The application of high-throughput, parallel methods to the synthesis, analysis, screening, and testing of materials. This approach relies on robotics and computer-assisted methods to generate and analyze the results. Synthesis, analysis, and testing of samples occurs in the wells of microtiter plates, which may contain as few as 96 samples or as many as a few thousand. Solid-phase and solution methods are used, and samples may be "one-bead-one-compound" or they may contain mixtures, which require "deconvolution" to determine which component is responsible for observed activity.

CONCORD. Rapid 2D to 3D conversion program introduced by Robert Pearlman's group in 1987. It generates low energy–approximate 3D models from 2D connection tables. It can also do stereo "multiplexing," where multiple configurations of stereochemically ambiguous structures are generated. Marketed by Tripos, Inc.

Concordance. A data warehouse architecture used in MDL relational chemical and reaction databases. The central "fact" table of a concordance has a record for each unique structure in the database, with pointers to the instances of the structure in various "source" databases.

Connection Table. A table or matrix containing topological information about a chemical structure. A structure can be considered a "graph" in 2D space, with atoms as "nodes" and bonds as "edges." The atom connection table has one row and one column for each atom. The diagonal elements of the table are usually the atomic number, and the off-diagonal elements have a zero or null if two atoms are not connected; otherwise they contain the order (1, 2, 3, aromatic, etc.) of the bond connecting the row and column atom. A less common connection table is the *bond* connection table, in which the rows and columns are the bonds in the structure, the diagonal elements are the bond order, and the off-diagonal elements contain information about the atoms at the ends of the bonds.

CONVERTER. A rapid 2D to 3D conversion program marketed by Accelrys. It uses a distance geometry approach to modeling, which covers a wider range of conformations than other methods.

CORINA. A rapid 2D to 3D conversion program developed by the Gasteiger research group at the University of Erlangen. It can handle macrocyclic ring structures, which can be problematic in other conversion programs. Chemists can access CORINA online (http://www2.chemie.uni-erlangen.de/software/corina/free_struct.html).

Daemon. From Unix, a program that runs continually as a background process to perform routine functions on demand or on a schedule. In the context of a chemical data warehouse, an example would be a registration program that periodically checks input databases to see if there are any new structures that need to be added to the warehouse. If there are, the daemon extracts the structures from the source databases, transforms and "cleans" them if needed, and registers them to the warehouse.

Data Cartridge. A popular term for user-customizable search "operators" that can be added to the SQL language of a relational database system. An example in chemical information is the addition of a substructure search (SSS) operator to integrate this type of search directly into a relational database search. One advantage of this approach is that the search "strategy" that the relational search program applies can take the complexity of the custom operator into account (the "cost") when performing the various search operations.

Data Compression. The process of transforming a potentially large amount of data into a smaller dataset, in such a way that reversing the transformation results in no loss of information in the original data. A simple example of a compression operation is to replace a string of blanks with a count plus a number that designates the character to be repeated. Compression programs include personal computer programs like PKZIP and WINZIP, and Unix utilities like gunzip. Compression methods are often used before storing data in a database and before transmitting data over a network. When the data is retrieved from the database or received on the network, it must then be "decompressed" by reversing the steps in the compression process. A chemical information example of compression is the conversion of an MDL molfile to a Chime string, which uses ZIP file compression methods.

Data Mart. Typically, a one-dimensional data warehouse—collecting data from multiple sources, extracting/cleaning/transforming/loading (ECTL) the data, and then indexing it for analytical (OLAP) and data mining purposes, to be used by a given group or department. In chemistry, an example would be an inventory database, with structures, location, and purchasing information from many vendors for use by synthetic chemists. Like a data warehouse, a data mart often has a central *fact* table with each record containing pointers into other *dimension* tables that contain relational data about the items in the fact table. The fact and dimension tables are connected in an organization called the *star schema,* which is a common design for data marts and warehouses. Data marts are often subsets of a data warehouse.

Data Mining. The extraction of previously unknown predictive relationships from a large data set or database. Data mining makes use of descriptive *unsupervised* methods such as association and cluster analysis, as well as predictive *supervised* methods such as decision trees, curve fitting, neural networks, and Bayesian methods. Data mining was once considered "data snooping" and had a poor reputation. The need to analyze huge volumes of data and the success of these methods in marketing and finance have prompted scientists and statisticians to reconsider its use.

Data Warehouse. A large relational database that collects data from multiple diverse sources and organizes it for optimal analytical searching and reporting (OLAP). A data warehouse is a superset of a data mart, containing archival and unchanging data that is important to several groups of researchers (i.e., multidimensional). Data that enters a data warehouse does not usually come from original sources (i.e., chemists, instruments, or assays). It usually comes from intermediate data sources and undergoes cleaning and transformation (ECTL) before registry into the data warehouse. Typically, data is not deleted from a data warehouse, because historical trends are important. For this reason, the warehouse grows very large over a long period of time, and thus its organization and indexing are crucial considerations. An example in chemistry would be a single database containing structures, models, reactions, and data, all cross-referenced, and used by chemists, biologists, and modelers. Typically, each group would extract their own data mart from the warehouse, containing information relevant to their needs. Data warehouses are often used in decision support systems (DSS) to provide data on which to base important corporate decisions.

Database Tier. In a three-tier programming architecture, the database tier resides on a server computer with access to the databases and the programs that manage them.

Deduplication. When registering into a chemical structure database, the process of finding whether the given structure already exists in the database. This usually involves performing an *exact match* search with the given structure as the search query. Note that the definition of exact match may vary with the database, and it may even be configurable. For example, some databases may consider tautomers to be acceptable as exact matches, whereas others may require a more strict definition.

Dimension Tables. In a data mart or warehouse, the dimension tables store non-redundant information about the entries in the *fact table* of the database. For the chemical example of an inventory data mart, the fact table stores the various source database identifiers of each unique structure in the data mart. A dimension table of molecular formulas would store the formula for the unique structure in the mart, rather than storing the same formula for each occurrence of that structure in the various source databases.

Drill-Down. Accessing data with increasing amounts of detail. When examining and browsing the results of a database search, a chemist can often request further information about a structure, even though that information was not included in the search. The process of accessing further information, often stored in a hierarchical manner, is termed

drill-down. The opposite process, which aggregates data, is termed *roll-up*.

ECTL. The process of *Extracting, Cleaning, Transforming,* and *Loading* data into a data mart or data warehouse. The data in a mart or warehouse should be standardized, complete, unambiguous, etc. Raw data from files, instruments, databases, the Internet, etc., must usually be preprocessed before it is "clean" enough to be used in decision making. Structures present special problems because tautomers, isomers, salts, etc. may all represent valid forms. The use of chemical processing languages, which can search for substructures and make modifications of specific atoms and bonds, enables the enforcement of *chemical business rules* during the ECTL process.

Encryption. The conversion of data in a readable or decipherable code into another, possibly undecipherable, code. The most common encryption involves sensitive pieces of data like passwords and identification numbers. In chemistry, it is sometimes necessary to encrypt larger pieces of information, such as chemical structures and the results of assays—at least during passage of such information over networks or the Internet. Decryption of the information typically requires one or more *keys*, which are often built into the encryption and decryption software.

Enumeration. The systematic substitution of all the Rgroup members in a generic structure, giving each possible specific structure the generic structure represents. If some of the Rgroups are not converted in the process, it is termed *partial enumeration*.

Equivalence Class. In the canonicalization of structures that have some element of symmetry, certain atoms that are topologically equivalent may yield the same canonical number. These atoms are considered to be in the same equivalence class. The concept of equivalence class is used, for example, in the Daylight Chemical Information Systems handling of reactions, to examine equivalent atoms when mapping reactant and product atoms.

Exact Match Search. One type of structure searching in which a query molecule is searched for in a database of structures. To exactly match the query, the target structure must be topologically identical and not be a substructure or superstructure of the query.

Extended Stereochemistry. A type of tetrahedral or higher level stereochemistry that appears, for example, in allenes, where the stereochemical center is not a single atom but a system of atoms and bonds that can be conceptually collapsed to a single atom to yield a stereo center.

External Registry Number. A unique "external" identifier assigned to a structure, reaction, 3D model, assay, etc. The external registry number is usually unique across databases, and it can be used as a key to link data from one database or table to another. Any given database may have its own "internal" identifiers that may not be unique across databases.

Fact Table. A central table in a data warehouse whose rows each represent one unit of primary importance in the warehouse. In a chemical warehouse, the rows of the fact table might correspond to unique structures in the database. In a biology data warehouse, each row might correspond to a single experiment. The fields in the fact table are mainly pointers to information stored in other tables, or they contain data that may be repeated in other tables but is stored in the fact table (i.e., denormalized) for rapid access. The fact table connects to other "dimension" tables in the warehouse that contain specific information that is not duplicated.

Fastsearch Index. Term used in MDL databases for a tree-structured index of all the structures in the database. The nodes in the tree represent increasingly complex substructures or properties, where all the structures at or below a given node have in common. The fastsearch index can be very large, but it makes possible very rapid substructure searching.

Field. In database terminology a column of data in a table. Fields are commonly selected in searches of the database, such as "SELECT MOLSTRUCTURE, MOLWEIGHT FROM MOLTABLE WHERE SSS(MOLSTRUCTURE, 'query.mol')=1 AND MOLWEIGHT<500.0". Here, MOLSTRUCTURE and MOLWEIGHT are fields in the MOLTABLE table. SSS is a function that operates on the MOLSTRUCTURE field to find molecules that contain the structure contained in the file query.mol, as a substructure.

Filter. A query or set of criteria designed to select a subset of a given set of data or results. Filters are usually applied to limit the number of hits from a search, or to limit the input to some analysis. Sometimes filters are designed to remove invalid rows of data, to randomly select a subset, or to remove rows based on the values of certain fields. A common application of filtering is in reagent selection, where reactive groups, multi-functional structures, or cost criteria may be applied to the selection of compounds for reactions. A filter can usually be expressed as a SELECT statement in a database search.

Fingerprint. A set of measurements or descriptors, usually binary, that can be used to characterize and identify an object. In the case of a chemical structure, a common fingerprint is a set of substructure keys that represent the presence or absence of specific functional groups. Such keys can be used to compute the topological similarity of the structure to other structures, and can be used as filters in database searching. Other common fingerprints include IR and mass spectral fingerprints, and fingerprints of how a structure behaves in a set of biological assays.

Flat Database or File. Essentially a spreadsheet of data, in which a given row contains all the data about a structure. There are no hierarchical relationships in a flat database. Many older and proprietary structure databases were flat in structure. These are in contrast to relational databases that are more commonly used at present.

Flexmatch Search. Term used in MDL structure searching to allow "relaxed" exact match searching of structures. One can specify, for instance, that everything must match except bond orders, or stereochemistry, or valence at atom centers, etc. By turning on or off various flags, one can for a given structure query, retrieve isomers of various types, salts of a the structure, or instances of the structure that may contain different values of certain types of attached data.

Generic Structure. A structure convention that allows representation of, say, a combinatorial library, as a single, generalized structure. The fixed parts of the structure are represented by the "root" or "parent" structure, and variable parts are represented by

"Rgroups" or "substituent groups" that can each contain multiple substituents or fragments.

Gigabyte. One thousand megabytes, or 10^9 bytes of data. The largest chemical structure databases presently contain a few tens to hundreds of gigabytes of data. A typical structure in a database may require a few thousand bytes of data to store the connection table, coordinates, and other structure-specific data.

Graph and Subgraph Isomorphism. In chemistry, the mapping of a structure or substructure query to a target structure. All the atoms and bonds in the query (the nodes or vertices and edges of the "graph" of the query structure) must be mapped to corresponding atoms and bonds in the target structure to generate a hit.

Hash Code. Converting a set of numeric or character properties into a single, mostly unique, number, for the purpose of rapid lookup and retrieval. For example, in the case of chemical structures, it is common to generate and store a hash of the molecular formula, so that when a user requests a formula search, the search query typed by the user is converted to the same hash number, and a single lookup in the index gives all the structures that correspond to the given formula. A hash code is often generated as a linear combination of the possible values of each of the properties (e.g., $n_1P_1 + n_2P_2 + \cdots$, where the n's are selected such that the products never overlap). If several structures have the same hash code, they are termed "collisions," and typically require further processing—like substructure searching—to differentiate them.

Hierarchical Clustering. One of three main types of clustering applied to chemical structures (hierarchical, partitioning, and fuzzy clustering). In hierarchical clustering, a tree or dendrogram is constructed, with one structure at each of the leaves of the tree. By "trimming" the tree at a given level, one can collect structures into a given number of clusters, such that all the structures in a single cluster have some level of similarity to each other.

High-Throughput Chemistry. Application of parallel processing to the synthesis, analysis, and screening of structures. A subset of high-thoughput chemistry is combinatorial chemistry.

Hit List. Older term for a list of identifiers of structures or other objects obtained from a database search. A more modern term is "result set."

HTML. HyperText Markup Language. The most commonly used specification language of the Internet. Other markup languages of interest in chemistry include XML (eXtensible Markup Language—information in general), CML (Chemical Markup Language—chemical structures), VRML (Virtual Reality Markup Language—3D visualization), and PMML (Predictive Model Markup Language—data mining).

Index. A secondary data field generated from one or more primary data fields, to enhance the searching and retrieval of the primary data. An index in a chemical database may be a characteristic of the database, such as Oracle indexes, or it may be a chemistry-specific index such as a tree index for substructure searching. Indexes require extra space, and they typically must be created and maintained by some administrative process in the database.

Inventory Data. Typically, information about the availability of reagents for chemical synthesis. This includes the suppliers, package sizes, purity, and cost of commercial reagents, and the location, owner, and availability of in-house reagents. Increasingly, this information is being integrated with chemical structure databases and warehouses and with automated ordering and procurement programs.

Inverted Keys. When substructure search keys are generated for a structure, they may be stored in normal order (where each record represents a structure, and the bits or fields for that structure represent the keys). Alternatively, they may be stored in inverted or pivoted order, where each record represents a given substructure key, and the bits represent structures that have that particular key set. This type of storage benefits key searching, where a user wants all the structures that have a particular key set.

Isomer and Tautomer Search. A search types where bond order, hydrogen counts, certain atom valences, and bond or atom stereochemistry may be allowed to vary from those specified in the query. Such searching allows retrieval of keto-enol tautomers, cis-trans isomers, etc. The generalization of the search may be a function of the query of the search process (see Flexmatch search) or both.

Java. Currently the most popular computer language for Internet and middle-tier programming. Java is an object-oriented language developed by Sun Microsystems, that runs on multiple platforms and contains built-in features for networking, database access and security, graphics, etc. Other languages widely used in chemical information programming include C, C++, and Perl. Java and C++ are called "object oriented" languages that focus on the "business objects" of the application—like molecules and reactions. C and Perl are more "procedure oriented" languages that focus on the things the objects do and the process that manages them.

Join. The retrieval of data from more than one table in a relational database, into a single result set. Depending on the structure of the data in the various tables, and the nature of the search query, extra data may need to be added to the result set to fill in certain fields, or the fields may be unpopulated in the result set.

KDD. Knowledge Discovery in Databases— application of analysis and data mining techniques to discover "knowledge" that may be implicit but undiscovered in large amounts of data.

Key Field. Field in a table that uniquely identifies rows in the table (primary key) or contributes to uniquely identifying the rows (secondary or composite key), or that connects the given row to data in another table (foreign key). Key fields are usually indexed for rapid lookup and retrieval.

Kmeans Clustering. Type of partitioning cluster analysis in which an object, such as a chemical structure, is placed into one of K clusters, based on how similar the structure is to the average value (or centroid) of each cluster. The average of the cluster may be an actual structure itself, in which case the technique is referred to as K-medoids clustering.

Linear Notation. Representation of a chemical structure using a linear string of numbers and letters. A linear notation is designed to be interpreted by a chemist. Thus, a SMILES string represents a linear notation, while a Chime string represents a compressed notation.

Linux. Microprocessor version of Unix developed by Linus Torvald. Linux is presently used mostly in network servers and in clusters of microcomputers used for large-scale parallel computation. It is gaining status as an alternative to Microsoft and to Unix, for database applications, because Oracle and other vendors provide Linux versions.

Logic in Query Features. Using AND, OR, and NOT as modifiers on the application of query features. For example, one could run a search to select structures that contain "halogen and not primary or secondary amine, or not halogen and any amine." The logic can be a part of the query substructure, as with Markush queries, or it can be part of the SELECT statement.

Markush Structure. Essentially a generic structure, in which a root or parent structure plus Rgroups and their members can represent an entire combinatorial library. Markush structures were developed for patent purposes, and the specification of substituents are often more general than in the case of generic structures or database queries. Markush structures are also used to represent generic reactions, in which the reactants and products are represented by generic structures.

Member. A single substituent or moiety in an Rgroup collection. If R_1 consists of the substituents (Me, Et, Pr, and Ph), these are termed the members of R_1.

Metadata—"Data About Data". In a database, metadata describes the structure, format, storage, access, and various properties of other stored data, or it describes the characteristics of the database itself. Metadata is sometimes stored in tables termed "dictionaries." As an example, chemical metadata might include the tablename, fieldname, and format of the field containing the chemical structures. This metadata might be stored in a master table in the database and be stored as property_name and property_value pairs, which the application that accesses the database can understand.

Middle Tier. Large-scale database applications often consist of three "tiers" of programs: (1) the application tier, which interacts with the user, (2) the database tier, which interacts with the database management system, and (3) the middle tier, which sits between the user application and the database management. The functions of the middle tier include (1) receiving, transforming, and passing queries and commands from the application tier to the database tier, (2) receiving, consolidating, and formatting data from the database tier and passing it to the user, and (3) performing tasks that may be required by the application but are not available in the database tier, such as managing hit lists and queries. Middle tier programs are often written in Java, a language that runs on many platforms and contains Internet features to communicate with the application tier and database features (JDBC—Java Database Connectivity) to connect to the database tier.

Molecular Connectivity. A class of molecular descriptors derived from the connection table of a structure. For increasing path lengths (1-, 2-, 3-bonds, etc.), the molecular connectivity values are computed as the sum of functions of the connectivity values (number of attachments) of the atoms in the path. Molecular connectivity descriptors can be used to distinguish structures. As such, they can be correlated with physicochemical properties that are functions of structure size, linearity, and degree of branching.

Morgan Algorithm. A procedure for finding a mostly unique (canonical) ordering of atoms in a structure. It involves an iterative process that begins by assigning each atom a score, which is initially computed by counting the number of neighboring atoms attached. In successive iterations, the score at a given atom is computed by summing the previous iteration scores of all the atoms to which it is attached. Eventually, the *order* of the scores becomes invariant (i.e., does not change with further iteration). At that point, atoms that are topologically equivalent have the same score. Atoms in the structure are then renumbered by the order of their Morgan number. One advantage of canonical numbering is that a given structure, drawn two different ways (i.e., different ordering of the atoms) can be reduced to the same Morgan numbering, and thus be matched quickly.

Multidimensional Database. A relational database in which multiple general types of data are stored, indexed, and cross-referenced, for use by several different groups. In chemistry, an example would be a database containing reactions, 2D structures, perhaps generic structures or libraries, and 3D models. Such a database would be used by synthetic, chemical informatic, and molecular modeling scientists. A data warehouse is often a multidimensional database, whereas a data mart is usually single-dimensional.

Multi-tier Architecture. An expansion of a client-server architecture to include a middle layer of software. The middle tier may run on a computer different from either the client or server computers. The middle tier isolates the client and server programs, so that changes in either of them do not require corresponding changes in the other. The middle tier acts as to receive, authenticate, and transform data as it passes between client and server computers. To make middle tier software easy to change and maintain, it is often written in Java, a modern object-oriented computer language that is available free on most of today's computer platforms.

Object-Oriented Language. Procedure-oriented languages like Fortran, Basic, and C operate by calling functions or subroutines with "arguments" that tell the program what to do with certain variables in memory—such as calculate the molecular weight of a collection of atoms. In object-oriented (OO) languages like C++ and Java, an object such as a molecule, has a molecular weight "method" that is specific to that object and is stored with the object in memory. In this way, a slightly different object, like a mixture, could have its own molecular weight method—perhaps a weighted average.

Object Relational Database. A relational database in which data can be collected and combined with methods to fit an object oriented model. Searching in the database is conducted in the context of the objects and their methods, rather than the raw data fields and stored procedures. There is usually considerable overhead in building, maintaining, and using an object relational database, so this type of organization has not so far been widely used in chemical or biological databases.

OLAP. OnLine Analytical Processing. An activity that involves routine searching, analysis, and reporting on data stored in a large database. The database, which may have a data mart or data warehouse organization, is optimized for the kinds of searches and reports that it supports. It may not be optimal in organization for transaction processing (OLTP), which may involve registration of small amounts of data on an irregular schedule, or for data mining, which involves the retrieval and analysis of large volumes of data. In chemistry, an example of OLAP might be an inventory application in which the chemist draws in a structure or substructure, runs one or more filters on the resulting result set, and retrieves and prints a report of structures and inventory data.

OLTP. OnLine Transaction Processing. An activity that involves registration, update, or simple searching in a database of transactions. In chemistry, this might be the routine registration of a new structure and analytical data into a chemical database. Such a database is optimized for registration and may not be suitable either for more analytical types of searching and reporting (OLAP) or for data mining.

Parallel Processing. A technique whereby a given computer task is distributed among several central processing units (CPUs). The CPUs may be part of the same computer (e.g., a multiprocessing computer in which several CPUs share common memory and physical devices), or they may consist of several single-processor computers (a "cluster") that are networked to rapidly share information and disk space. In database management, it is becoming increasingly common to have parallel copies (replications) of a given database at several sites, perhaps worldwide. Special database and networking software provides rapid updates of certain information like data and periodic updates of other information like search indexes.

Pattern Recognition. The application of computers to build descriptive or predictive models (i.e., find patterns) of information from input datasets. The techniques of pattern recognition overlap those used in statistics, chemometrics, and data mining, and include data display, description, and reduction, unsupervised methods such as cluster analysis, and supervised methods such as curve fitting and classification. Engineering applications of pattern recognition include recognizing objects in pictures, and character and voice recognition. Chemical applications are found in the fields of drug discovery, analytical chemistry, and chem/bioinformatics.

Petabyte. One thousand terabytes of data (10^{15} bytes). At present, the largest databases of chemical and biological information are gigabytes (10^9 bytes) in size.

Pharmacophore. The minimum amount of chemical functionality needed in a drug to elicit a given biological response. This functionality is defined in terms of atoms and functional groups and their geometric relationships to each other, including distances, angles, etc. A pharmacophore *query* is the representation of a pharmacophore in a format that can be used to search a chemical database for structures that can satisfy the pharmacohore and may elicit the desired response. Pharmacophore searching is usually conducted on a 3D structural database using search software that combines 2D searching with conformational analysis to find structures that can, by rotating about single bonds, adopt a conformation that satisfies the pharmacophore.

Pharmacophore Keys. Originally designed to speed pharmacophore query searching, pharmacophore keys are bitset fingerprints that indicate the presence or absence of given 3- or 4-point pharmacophores in a structure stored in a database. The 3-point pharmacophore keys represent triangular arrangements of atoms and functional groups separated by given distances or distance ranges. The 4-point pharmacophore keys represent tetrahedral arrangements of atoms and functional groups. When a structure is registered into a 3D database, a rapid conformational analysis is performed involving key single bonds in the structure. From the interatomic distance ranges between given atoms and functional groups in the structure, the various bits in the pharmacophore keys are set. These keys can be used as filters in pharmacophore searching, or increasingly, as filters before docking the structures into a known receptor (virtual screening). Pharmacophore keys have also been used less commonly as descriptors in QSAR and data mining.

Physicochemical Properties. Originally these were just measured properties like melting point, pKa, solubility, and octanol/water Log$P$. Increasingly, they are obtained from programs that can calculate them from the 2D or 3D structure. In QSAR, the classical triad of steric, electronic, and lipophilic properties is still widely used, but it has been enhanced to include enery-based descriptors, measures of binding interaction, 3D-QSAR multivariate descriptors (CoMFA), and others. Quantum mechanical calculations are being used increasingly to estimate physicochemical properties. Once calculated, the properties are used to filter structures which may have undesirable ADMET criteria (e.g., the "rule of five"), or they may be used directly in models to estimate the type or level of biological activity (QSAR).

Pivoting Data. Changing data from row to column values or vice versa. This technique can be a very useful tool for summarizing data. One example of pivoting is to convert substructure keys that are stored by structure (with a bit turned on for each key the structure contains), to storage by key (with one bit turned on for each structure that has the given key). Another example is to convert assay data that is stored by structure, to data that is stored by assay. In the process of pivoting data, it is common to consolidate values, for example, converting raw assay results to ED_{50} values, or taking the average of some physicochemical property.

Proteomics. The conversion of protein sequence data into useful biological information. In general, the goal of proteomics is to characterize a gene product—i.e., protein—as to its structure, subcellular location, and function. Additional information includes how a protein interacts ("networks") with other reactions and cell processes.

QSAR. Quantitative structure-activity relationships—the science of deriving quantitative linear or nonlinear mathematical relationships between physicochemical and topological/topographical properties of chemical structures and their biological activity. Originally, regression analysis was the only tool used to derive QSAR equations. More recently, tools such as partial least squares (PLS), neural networks, and a variety of data mining methods like decision trees and support vector machines have come into use.

Reacting Center. An atom in a reactant which is modified during the course of a reaction. Specifying reacting center information when searching for reactions can speed the search and reduce the number of incorrect hits.

Reaction Scheme. A series of one-step reactions that lead from a given reactant to a given product, by way of intermediate steps. A reaction search system should be able to find reactant/product combinations that span several intermediate reactions.

Refine a Search Query. The process of adding or modifying constraints of a search query to reduce or increase the number of hits. Constraints may be added, removed, relaxed, or tightened to achieve the desired search results.

Registry Number. A unique identifier assigned to a chemical structure or other piece of data when it is registered into a database. The registry number may be internal, primarily for use by the database search system, or external, to be used by chemists and to link the data to other databases and files.

Relational Database. A common database architecture in which related data items are stored in separate tables, accessed by key fields, and indexed for rapid search and retrieval. The dominant relational database systems used in pharmaceutical discovery include Oracle, Microsoft Access and SQL-Server, and IBM DB2.

Result Set. A list of records resulting from a database search. The result set commonly consists of a list of record identifiers (sometimes called a cursor), which can be navigated to select records. In some systems, a result set may also contain related data for each record.

Retrosynthetic Analysis. An approach to computer-assisted synthesis design that starts with the products of a reaction or sequence of reactions and works backwards toward the reactants. An example program that implements retrosynthetic analysis is the LHASA program of E. J. Corey's group.

Rgroup. In a generic or Markush structure, generalized substituents or moieties are given the representation R_1, R_2, etc. These Rgroups represent collections of specific substituents or moieties (members) that can be replaced at the given position.

Roll-up. The agglomeration, summarization, or consolidation of data into a summary presentation. Roll-up often involves summarizing data at a given level in a data hierarchy. Examples would include the average of several ED_{50} values, or a simple yes/no indication that toxicity data for a given structure exists somewhere in the database.

Root Structure. The invariant portion of a Markush or generic structure. The attached Rgroups contain the substituents that vary from one specific structure to the next. Sometimes termed a parent structure.

ROSDAL. Linear notation scheme devised by the Beilstein Institute. It can contain just connection table information, or it may also contain atom coordinates. Several chemical information systems can convert ROSDAL strings to other structure file formats.

Sgroup Data. In MDL structure storage, the attachment of structure-differentiating data directly to the structure. Such data may relate to the structure as a whole, or to atoms, bonds, fragments, or collections of atoms and bonds. Examples would include atomic partial charges on 3D models or percent composition attached to components of a formulation.

Similarity Search. A type of "fuzzy" structure searching in which molecules are compared with respect to the degree of overlap they share in terms of topological and/or physicochemical properties. Topological descriptors usually consist of substructure keys or fingerprints, in which case a similarity coefficient like the Tanimoto coefficient is computed. In the case of calculated properties, a simple correlation coefficient may be used. The similarity coefficient used in a similarity search can also be used in various types of cluster analysis to group similar structures.

SLN—Sybyl Line Notation. Linear notation used in conjunction with Tripos SPL (Sybyl Programming Language) to manipulate chemical structures. It is similar in syntax to SMILES notation.

SMILES. Simplified Molecular Input Line Entry System—linear notation used in Day-light Chemical Information Systems software and widely supported by other systems.

SQL—Structured Query Language. The standard query specification language for searching relational databases. Most database systems support the SQL standard but then add extensions particular to their implementation.

Star Schema. A standard data warehouse architecture, characterized by Ralph Kimball, in which a central "fact" table is connected to various "dimension" tables.

Structural Data Mining. Application of data mining methodology to chemical structure and reaction databases. Currently in its infancy, it remains to be seen whether a "data snooping" approach to information and knowledge discovery can be as useful in drug discovery as it has proven to be in finance, marketing, and merchandising.

Substance. In some structure databases, an entry that lacks a structure completely (a "nostructure"), is only partially characterized, or is an unspecified mixture of known structures. Substances pose obvious problems in database searching.

Substructure Search (SSS) Keys. Originally developed to facilitate substructure searching, these consist of a string of bits that represent a fingerprint of the structure with respect to either (1) a set of known and defined functional groups (e.g., MDL), or (2) a set of discovered atom-bond paths that the structure contains (e.g., Daylight). In MDL systems, the substructure keys are currently either 166 or 960 bits in length. In Daylight systems, the substructure keys are of varying length and can be "folded" to achieve a higher density of bits turned on. Although SSS keys were originally developed to screen candidates for substructure searching, they are currently used more for similarity calculations.

Substructure Search. Application of "subgraph isomorphism" search to chemical structures. This consists of finding a particular arrangement of atoms and bonds as they are embedded in a chemical structure. The arrangement being searched for is termed the query substructure, the structures being searched are termed the candidates, and any particular structure in that set is termed a target structure. If the query substructure is found in the target, the target structure is added to the result set (or hit list). In displaying the results of the search, the atoms and bonds in the substructure as mapped onto the hit may be highlighted (shown darkened or in a different color) in the structure display. In general, more than one occurrence of a substructure may be found in a given structure, and substructure mappings may be overlapping or non-overlapping.

Superstructure Search. Modification of substructure search in which the substructure query becomes the target structure, and the target structure in the database becomes the substructure search query. The search finds structures in the database that are substructures of the query. A similar extension to structure similarity searching yields superstructure similarity searching.

Supervised Data Mining. Searching large volumes of data for hidden predictive relationships. Supervised analysis requires one or more "dependent" or response variables, to be predicted from a set of "independent" or predictor variables. The techniques used include various classification methods (decision tree, support vector, Bayesian) and various estimation methods (regression, neural nets).

Tanimoto Coefficient. Standard coefficient for computing the similarity of chemical structures. If structure A has 20 bits turned on in a fingerprint, and structure B has 30 bits turned on, and the two structures have 10 bits in common, the Tanimoto coefficient is $10/(20 + 30 - 2 \times 10)$ or 0.33. Its value can range from 0 (no similarity) to 1.0 (perfect match). Other similarity coefficients are also used, and in some systems (such as MDL) the various bits are weighted inversely according to their occurrence in the database, so that very common substructures do not contribute much to the similarity.

Terabyte. One thousand gigabytes, or 10^{12} bytes of data. The largest relational databases of any kind are currently a few tens of terabytes in size. At present, the largest databases of chemical and biological information are gigabytes (10^9 bytes) in size.

Thick or Thin Client. A thick client architecture is one in which a significant amount of computing is done on the user's workstation. This is appropriate for user-interface-intensive activity like graphics and calculations on individual molecules. The alternative thin-client architecture either does not require much local computing, or it uses a built-in resource like an Internet browser as a client.

Toolkit. A collection of computer routines that each perform one or a small number of information management tasks. The routines are provided as a library and they can be incorporated into custom user-written application programs to carry out tasks that ordinary application programs may not perform. The interface between the toolkit routines and the user-written program is referred to as the Application Programming Interface, or API.

Topographical. Structure data that is based on the connection table and the 3D structure of a molecule. Examples include surface area and volume and pharmacophore distances between atoms.

Topological. Structure data that is based only on the connection table of the structure, without regard to 2D or 3D coordinates of the atoms. Examples include molecular weight and formula, counts of substructures, and indices like molecular connectivity.

Tree. A data structure that is widely used in chemical information storage. Commonly viewed with the root of the tree at the top (or to the left), successive levels of branching lead to the "nodes" of the tree and ultimately to its "leaves" (terminal nodes). Depending on how they split at a node, trees may be binary or n-nary, and depending on how their nodes are distributed, they may be balanced or unbalanced. A tree is usually traversed from the root to the leaves, and this traversal can be depth-first (following a single path until a leaf node is reached), or breadth-first (looking at all the nodes at a given level). An example of a tree data structure is the fastsearch index used in MDL substructure searching.

Unicode. A 32-bit successor to the ASCII character set. With Unicode, foreign alphabets and special characters can be encoded.

Unix. Widely used operating system for workstations and server computers. Various computer vendors supply their version of Unix, which typically descends from either the Bell Labs or Berkeley versions. A microcomputer version of Unix is Linux, which is rapidly growing in acceptance.

Unsupervised Data Mining. Searching large volumes of data for hidden descriptive relationships. Unlike supervised data mining, no response variables are used. The techniques used include various display and data reduction methods, as well as cluster analysis and association analysis.

VARCHAR, VARCHAR2. SQL data types used to store character data in a relational database system. Storage is limited to about 4000 characters, so larger pieces of data must be stored as CLOB data.

Virtual Screening. Using computer modeling to screen leads for activity. The screening may be through some QSAR or data mining model, which typically requires only 2D structures and data, or it may involve 3D molecular modeling and docking with a known or putative receptor. The speed and increasing accuracy of virtual screening make it a vital step in the drug discovery process.

XML. EXtensible Markup Language—a widely used standard for producing self-documenting text. Documents that subscribe to the XML standard can be freely exchanged over networks and between applications, using standard parsing programs to interpret the document. CML is an extension of XML that can be used to transport structures, reactions, and chemical and biological data. A query language, XMLQuery, is being developed to allow searching of XML documents in a manner similar to the use of SQL to search relational databases.

8 ACKNOWLEDGMENTS

Grateful acknowledgement is made to Guenter Grethe, Stephen Heller, Lingran Chen, and Tim Hoctor of MDL Information Systems, Inc. for useful discussions during the preparation of this chapter.

REFERENCES

1. J. Knight, *Nature*, **412**, 571 (2001).
2. Some examples include Drug Discovery World, New Drugs, Nature Reviews Drug Discovery, Current Drug Discovery and Current Opinion in Drug Discovery and Development.
3. The Chemical Heritage Foundation, available online at http://www.chemheritage.org/HistoricalServices/cheminfo.htm, accessed on September 10, 2002.
4. W. V. Metanomski, *J. Chem. Inf. Comput. Sci.*, **35**, 173–174 (1995).
5. The 5th International Conference on Chemical Structures, June 6-10, 1999, Leeuwenhorst Congress Center, Noordwijkerhout, The Netherlands, available online at http://www.chemweb.com/conference/5iccs/5iccs.html, accessed on September 10, 2002.
6. The 2001 Gordon Research Conference on Quantitative Structure Activity Relationships, August 5–10, 2001, Tilton School, Tilton, NH, available online at http://www.grc.uri.edu/programs/2001/qsar.htm, accessed on August 28, 2001.
7. The 2001 International Chemical Information Conference, October 21–24, 2001, Nimes, France, available online at http://www.infonortics.com/chemical/index.html, accessed on August 26, 2002.
8. H. R. Collier, *Chemical Information*, Springer-Verlag, New York, 1990; H. R. Collier, Ed., *Recent Advances in Chemical Information*, Royal Society of Chemistry, London, 1993; H. R. Collier, Ed., *Further Advances in Chemical Information*, Royal Society of Chemistry, London, 1994.
9. Y. C. Martin and P. Willett, Eds., *Designing Bioactive Molecules: Three-Dimensional Techniques and Applications*, American Chemical Society, Washington, DC, 1998; P. Willett, *Three-Dimensional Chemical Structure Handling*, vol. **1**, Wiley, New York, 1991.
10. W. Warr and C. Suhr, *Chemical Information Management*, Wiley, New York, 1992; W. Warr, Ed., *Chemical Structures. The International Language of Chemistry*, Springer-Verlag, New York, 1988; W. Warr, Ed., *Chemical Structures 2. The International Language of Chemistry*, Springer-Verlag, New York, 1993.
11. G. D. Wiggins and K. Emry, Eds., *Chemical Information Sources*, McGraw Hill, New York, 1991.
12. R. E. Maizell, *How to Find Chemical Information*, Wiley, New York, 1998.
13. J. E. Ash, W. A. Warr, and P. Willett, *Chemical Structure Systems: Computational Techniques for Representation, Searching, and Process of Structural Information*, Ellis Horwood, New York, 1991.
14. D. D. Ridley, *Online Searching: A Scientist's Perspective; A Guide for the Chemical and Life Sciences*, John Wiley & Sons, Chichester, New York, 1996.
15. Computational Informatics Research Group, available online at http://www.shef.ac.uk/~is/research/cirg.html, accessed on June 19, 2002.

16. J. Gasteiger Research Group, available online at http://www2.ccc.uni-erlangen.de, accessed on March 6, 2002.

17. G. D. Wiggins, available online at http://www.indiana.edu/˜cheminfo/, accessed on September 16, 2002.

18. Cambridge Health Institute, Cheminformatics Glossary, available online at http://www.genomicglossaries.com/default.html, accessed on August 21, 2002.

19. The Chemical Structure Association, available online at http://www.chem-structure.org/, accessed on September 10, 2002.

20. The Ohio State University, Computational Chemistry Listserver, available online at http://www.ccl.net/chemistry/, accessed on June 26, 2002.

21. The Molecular Graphics and Modeling Society, available online at http://www.mgms.org/, accessed on September 10, 2002.

22. The Open Molecule Foundation, available online at http://www.xml-cml.org/, accessed on August 14, 2002.

23. The QSAR and Modeling Society, available online at http://www.ndsu.nodak.edu/qsar_soc/, accessed on September 17, 2002.

24. The Royal Society of Chemistry Chemical Information Group, available online at http://www.rsc.org/lap/rsccom/dab/scaf001.htm, accessed on September 10, 2002.

25. The UK QSAR and Cheminformatics Group, available online at http://www.iainm.demon.co.uk/indexnew.htm, accessed on February 1, 2002.

26. M. F. Lynch, J. M. Harrison, W. G. Town, and J. E. Ash, *Computer Handling of Chemical Structure Information*, MacDonald, London, 1971.

27. W. J. Howe, M. M. Milne, and A. F. Pennell, *Retrieval of Medicinal Chemical Information*, ACS Symposium Series **84**, American Chemical Society, Washington, DC, 1978.

28. Chemical Abstracts Service, available online at http://www.cas.org, accessed on September 16, 2002.

29. W. A. Warr and A. R. Haygarth Jackson, *J. Chem. Inf. Comput. Sci.*, **28**, 68–72 (1988).

30. D. E. Meyer, W. A. Warr, and R. Love, Eds, *Chemical Structure Software for Personal Computers*, American Chemical Society, Washington, DC, 1988.

31. E. K. F. Ahrens in W. A. Warr, Ed., *Chemical Structures*, Springer-Verlag, Berlin, 1988, pp. 97–111.

32. P. Willett, Ed., *Modern Approaches to Chemical Reaction Searching*, Gower Press, Brookfield, VT, 1986.

33. S. R. Heller, Ed., *The Beilstein Online Database: Implementation, Content, and Retrieval*, American Chemical Society, Washington, DC, 1990.

34. R. S. Pearlman, *Chem. Design Automation News*, **1**, 5–7 (1987).

35. Y. C. Martin, M. G. Bures, and P. Willett in K. B. Lipkowitz and D. B. Boyd, Eds., *Reviews in Computational Chemistry*, VCH Publishers, New York, 1990, pp. 213–264.

36. R. D. Cramer, D. E. Patterson, R. D. Clark, F. Soltanshahi, and M. S. Lawless, *J. Chem. Inf. Comput. Sci.*, **38**, 1010–1023 (1998).

37. For discussions of relative stereochemistry, see the documentation for various chemical database systems. For example, MDL stereochemistry is described online at http://www.mdli.com/downloads/literature/ctfile.pdf, and Daylight conventions are described at http://www.daylight.com/release/f_manuals.html. CAS stereochemical conventions have been described in L. M. Staggenborg in H. Colier, Ed., *Recent Advances in Chemical Information*, Royal Society of Chemistry, Cambridge, UK, 1993, pp. 89–112.

38. R. S. Cahn, C. K. Ingold, and V. Prelog, *Angew. Chem.*, **78**, 413–447 (1966); V. Prelog and G. Helmchen, *Angew. Chem.*, **94**, 614–631 (1982); D. Seebach and V. Prelog, *Angew. Chem.*, **94**, 696–702 (1982).

39. J. Sadowski and J. Gasteiger, *Chem. Rev.*, **93**, 2567–2581 (1993).

40. M. F. Lynch, J. M. Barnard, S. M. Welford, *J. Chem. Inf. Comput. Sci.*, **21**, 148–150 (1981); J. D. Holliday, G. M. Downs, V. J. Gillet, and M. F. Lynch, *J. Chem. Inf. Comput. Sci.*, **33**, 369–377 (1993).

41. G. M. Downs and J. M. Barnard, *J. Chem. Inf. Comput. Sci.*, **37**, 59–61 (1997).

42. A. J. Gushurst, J. G. Nourse, W. D. Hounshell, B. A. Leland, and D. G. Raich, *J. Chem. Inf. Comput. Sci.*, **31**, 447–454 (1991).

43. J. L. Schultz and E. S. Wilks, *J. Chem. Inf. Comput. Sci.*, **37**, 436–442 (1997).

44. ChemDraw from CambridgeSoft: http://www.cambridgesoft.com/; ChemWindow from Bio-Rad: http://www.bio-rad.com; ChemSketch and Structure Drawing Applet from ACDLabs: http://www.acdlabs.com; Peter Ertls Java molecular editor: http://www.elsevier.com/inca/homepage/saa/eccc3/paper6/; ISIS/Draw from

MDL: http://www/mdl.com; JChemPaint from the JChemPaint Project: http://sourcefor-ge.net/projects/jchempaint; and Marvin from ChemAxon: http://www.chemaxon.com/mar-vin/.

45. A comparison of chemical drawing programs, available online at http://dragon.kite.hu/ ˜gundat/rajzprogramok/dprog.html, accessed on September 13, 2002.

46. O. F. Güner, Ed., *Pharmacophore Perception, Development, and Use in Drug Design*, International University Line, San Diego, CA, 2000.

47. W. J. Wiswesser, *J. Chem. Inf. Comput. Sci.*, **25**, 258–263 (1985).

48. J. M. Barnard in P. V. R. Schleyer, Ed., The Encyclopedia of Computational Chemistry, vol. **4**, Wiley, New York, 1999, pp. 2818–2826.

49. D. Weininger, A. Weininger, and J. L. Wein-inger, *J. Chem. Inf. Comput. Sci.*, **29**, 97–101 (1989); M. A. Siani, D. Weininger, C. A. James, and J. M. Blaney, *J. Chem. Inf. Comput. Sci.*, **35**, 1026–1033 (1995).

50. R. G. A. Bone, M. A. Firth, and R. A. Sykes, *J. Chem. Inf. Comput. Sci.*, **39**, 846–860 (1999).

51. D. Weininger, *J. Chem. Inf. Comput. Sci.*, **30**, 237–243 (1990).

52. For example, the University of Erlangen TORVS research team offers 3D models, phys-icochemical properties, IR and Raman spectral simulation, etc., available online at http://www2.chemie.uni-erlangen.de/services/index. htm, accessed on September 6, 2002.

53. Daylight Chemical Information Systems, Inc., Daylight Reactions, available online at http://www.daylight.com/dayhtml/doc/theory/theory. rxn.html, accessed on September 6, 2002.

54. W. C. Herndon and S. H. Bertz, *J. Comp. Chem.*, **8**, 367–374 (1987); S. Ash, M. A. Cline, R. W. Homer, T. Hurst, and G. B. Smith, *J. Chem. Inf. Comput. Sci.*, **37**, 71–79 (1997).

55. A. Dalby, J. G. Nourse, W. D. Hounshell, A. K. I. Gushurst, D. L. Grier, B. A. Leland, and J. Laufer, *J. Chem. Inf. Comput. Sci.*, **32**, 244–255 (1992).

56. Peter Murray-Rust and Henry S. Rzepor, Chemical Markup Language, available online at http://www.xml-cml.org, accessed on Au-gust 14, 2002.

57. For example, Chemeleon from Exographics Inc., P. O. Box 655, West Milford NJ 07480, and Mol2Mol, http://www.compuchem.com.

58. Babel Chemical File Conversion Program, http://www.smog.com/chem/babel.

59. OpenBabel Project, http://openbabel.source-forge.net.

60. Converting structures to IUPAC names: Nomenclator from Cheminnovation, http://www.cheminnovation.com; Autonom from MDL GmbH, http://www.beilstein.com/products/ autonom.

61. Converting names to structures: NamExpert from Cheminnovation, http://www.cheminno-vation.com; CD/Name from ACD Labs, http://www.acdlabs.com.

62. T. R. Hagadone and M. S. Lajiness, *Tetrahe-dron Comp. Methodol.*, **1**, 219–230 (1988).

63. B. D. Christie, B. A. Leland, and J. G. Nourse, *J. Chem. Inf. Comput. Sci.*, **33**, 545–547 (1993).

64. M. Z. Nagy, S. Kuzics, T. Veszpremi, and P. Bruck in W. A. Warr, Ed., *Chemical Structures*, Springer-Verlag, Berlin, 1988, pp. 127–130.

65. Daylight Thor database system: http://www. daylight.com.

66. Accelrys Accord and RS[3] database systems: http://www.accelrys.com.

67. Tripos Unity database system: http://www. tripos.com.

68. MDL ISIS/Base: http://www.mdli.com

69. Afferent database system for Access: http://www.afferent.com.

70. H. L. Morgan, *J. Chem. Doc.*, **5**, 107–113 (1965); J. Figueras, *J. Chem. Inf. Comput. Sci.*, **33**, 717–718 (1993).

71. T. E. Moock, J. G. Nourse, D. Grier, and W. D. Hounshell in W. E. Warr, Ed., *Chemical Struc-tures 2*, Springer-Verlag, New York, 1993, pp. 303–313; Daylight Theory Manual, Part 7, Re-action Processing, available online at http://www.daylight.com, accessed on September 6, 2002.

72. B. A. Leland, B. D. Christie, J. G. Nourse, D. L. Grier, R. E. Carhart, T. Maffett, S. M. Welford, and D. H. Smith, *J. Chem. Inf. Comput. Sci.*, **37**, 62–70 (1997).

73. T. E. Moock, D. L. Grier, W. D. Hounshell, G. Grethe, K. Cronin, J. G. Nourse, and J. Theo-dosiou, *Tetrahedron Comput. Methodol.*, **1**, 117–128 (1988).

74. B. Christie and T. Moock in W. E. Warr, Ed., *Chemical Structures 2*, Springer-Verlag, Ber-lin, 1993, pp. 469–482.

75. E. J. Corey, *Chem. Soc. Rev.*, **17**, 111–133 (1988).

76. Program CLASSIFY: InfoChem reaction clas-sification, InfoChem GmbH, Munich, http://www.infochem.de/classify.htm; H. Satoh, O.

Sacher, T. Nakata, L. Chen, J. Gasteiger, and K. Funatsu, *J. Chem. Inf. Comput. Sci.*, **38**, 210–219 (1998).

77. D. R. Henry and A. G. Ozkabak in P. V. R. Schleyer, Ed., The Encyclopedia of Computational Chemistry, vol. **3**, Wiley, New York, 1999, pp. 543–552.

78. J. M. Barnard, *J. Chem. Inf. Comput. Sci.*, **33**, 532–538 (1993).

79. G. W. A. Milne, *J. Chem. Inf. Comput. Sci.*, **33**, 531–648 (1993).

80. M. G. Hicks and C. Jochum, *J. Chem. Inf. Comput. Sci.*, **30**, 191–199 (1990).

81. P. Willett, *J. Chem. Inf. Comput. Sci.*, **38**, 983–996 (1998).

82. P. Willett, V. Winterman, and D. Bawden, *J. Chem. Inf. Comput. Sci.*, **26**, 36–41 (1986).

83. G. Grethe and T. E. Moock, *J. Chem. Inf. Comput. Sci.*, **30**, 511–520 (1990).

84. W. Fisanick, A. H. Lipkus, and A. Rusinko III, *J. Chem. Inf. Comput. Sci.*, **34**, 130–140 (1994).

85. H. Briem and U. F. Lessel, *Perspect. Drug Disc. Des.*, **20**, 231–244 (2000).

86. J. E. Mills, C. A. Manryanoff, K. L. Sorgi, R. Stanziione, L. Scott, L. Herring, J. Spink, B. Baughman, and W. Bullock, *J. Chem. Inf. Comput. Sci.*, **28**, 155–159 (1988).

87. J. B. Hendrickson and T. L. Snyder, WebReactions—Organic Reaction Retrieval System, available online at http://webreactions.net, accessed on September 6, 2002.

88. Literature searching—for example: Chemical Abstracts Service: http://www.cas.org; LitLink literature server: http://www.litlink.com; Science Citation Index: http://www.isinet.com; Scirus: http://www.scirus.com; Thomson/ISI Web of Science: http://www.isinet.com/isi/products/citation/wos/.

89. H. Tokuno, *J. Chem. Inf. Comput. Sci.*, **33**, 799–804 (1993).

90. For example, Vermont SIRI MSDS: http://www.hazard.com/msds; Cornell University: http://msds.pdc.cornell.edu/msdssrch.asp.

91. For examples, see the Washington University Libraries Chemical Supplier Catalogs: http://library.wustl.edu/subjects/chemistry/chem/chemical_supplier.html.

92. Accord ChemExplorer: http://www.accelrys.com; CambridgeSoft ChemFinder for Word: http://www.cambridgesoft.com.

93. Accelrys: http://www.accelrys.com.

94. The Beilstein Institute: http://www.beilstein-institut.de.

95. Beilstein Distribution: http://www.beilstein.com.

96. S. R. Heller, Ed., *The Beilstein System: Strategies for Effective Searching*, American Chemical Society, Washington, DC, 1997.

97. Chemical Abstracts Service: http://www.cas.org.

98. Daylight Chemical Information Systems, Inc.: http://www.daylight.com.

99. Derwent: http://www.derwent.com/user-guides/chem.html.

100. Merged Markush Service: http://www.i-mms.com/; MicroPatent: http://www.micropat.com, and ISTA: http://www.paterra.com.

101. MDL Information Systems, Inc: http://www.mdli.com.

102. Tripos, Inc.: http://www.tripos.com.

103. R. M. Woodsmall and D. A. Benson, *Bull. Med. Libr. Assoc.*, **81**, 282–284 (1993).

104. D. B. Emmert, P. J. Stoehr, G. Stoesser, and G. N. Cameron, *Nucleic Acids Res.*, **22**, 3445–3449 (1994).

105. A. D. Baxevanis and B. F. F. Ouellette, *Bioinformatics—A Practical Guide to the Analysis of Genes and Proteins*, Wiley-Interscience, New York, 1998.

106. H. M. Berman, J. Westbrook, Z. Feng, G. Gilliland, T. N. Bhat, H. Weissig, I. N. Shindyalov, and P. E. Bourne, *Nucleic Acids Res.*, **28**, 235–242 (2000).

107. The Cambridge Crystallographic Data Center: http://www.ccdc.cam.ac.uk.

108. T. R. Hagadone and M. W. Schulz, *J. Chem. Inf. Comput. Sci.*, **35**, 879–884 (1995).

109. S. V. Trepalin and A. V. Yarkov, *J. Chem. Inf. Comput. Sci.*, **41**, 100–107 (2001).

110. CambridgeSoft: http://www.camsoft.com; ACDLabs: http://www.acdlabs.com; Softshell: www.softshell.com; Cybula Inc.: http://www.cybula.com.

111. L. H. Hall and L. B. Kier in D. B. Boyd and K. Lipkowitz, Eds., Reviews of Computational Chemistry, vol. **2**., VCH Publishers, New York, 1991, pp. 367–422.

112. R. S. Pearlman and K. M. Smith, *Perspect. Drug Disc. Des.*, **9**, 339–353 (1998).

113. J. M. Barnard and G. M. Downs, *J. Chem. Inf. Comput. Sci.*, **37**, 141–142 (1997); http://www.bci.gb.com/.

114. R. Todeschini and V. Consonni, *Handbook of Molecular Descriptors*, John Wiley & Sons, New York, 2001.

115. L. B. Kier and L. H. Hall, *Molecular Structure Description: The Electrotopological State*, Academic Press, New York, 1999.

116. Biobyte, Inc: http://www.biobyte.com.

117. Syracuse Research Corporation: http://www.syrres.com.

118. CompuDrug Ltd: http://www.compudrug.com.

119. ACD Labs, Inc: http://www.acdlabs.com.

120. Peking University: http://cheminfo.pku.edu.cn/calculator/xlogp/.

121. EduSoft LC: http://www.eslc.vabiotech.com.

122. SciVision, Inc: http://www.scivision.com/.

123. A. M. Zissimos, M. H. Abraham, M. C. Barker, K. J. Box, and K. Y. Tam, *J. Chem. Soc., Perkin Trans.*, **2**, 1–9 (2002).

124. LION Bioscience, Inc: http://www.lionbioscience.com/.

125. V. V. Poroikov, D. A. Filimonov, V. B. Yu, A. A. Lagunin, A. Kos, *J. Chem. Inf. Comput. Sci.*, **40**, 1349–1355 (2000).

126. Daylight Chemical Information online calculator: http://www.daylight.com/cgi-bin/contrib/pcmodels.cgi.

127. ACD Labs online calculator: http://www2.acdlabs.com/ilab/.

128. Syracuse Research online calculator: http://esc.syrres.com/interkow/kowdemo.htm.

129. Molinspiration: http://www.molinspiration.com/.

130. Alogp—VCCLab online LogP calculation: http://vcclab.org/lab.

131. PETRA: http://www2.chemie.uni-erlangen.de/services/petra/.

132. USEPA Suite: http://www.epa.gov/opptintr/exposure/docs/episuite.htm.

133. R. Kimball and M. Ross, *The Data Warehouse Toolkit: The Complete Guide to Dimensional Modeling*, 2nd ed., Wiley, New York, 2002.

134. M. G. Axel and I.-Y. Song, "Data Warehouse Design for Pharmaceutical Drug Discovery Research," in Proceedings of the 8th International Workshop in Database and Expert Systems Applications, IEEE Press, Piscataway, NJ, September 1997.

135. D. Normile, *Science*, **293**, 787 (2001).

136. S. B. Singh, R. P. Sheridan, E. M. Fluder, and R. D. Hull, *J. Med. Chem.*, **44**, 1564–1575 (2001).

137. R. K. Lindsay, B. G. Buchanan, E. A. Feigenbaum, and J. Lederberg, *Applications of Artificial Intelligence for Chemistry: The DENDRAL Project*, McGraw-Hill, New York, 1980.

138. U. Fayyad, G. Piatetsky-Shapiro, P. Smyth, and R. Uythurusamy, Eds., *Advances in Knowledge Discovery and Data Mining*, MIT Press, Cambridge, MA, 1996.

139. M. J. A. Berry and G. Linoff, *Data Mining Techniques: For Marketing, Sales, and Customer Support*, Wiley, New York, 1997.

140. M. J. Zaki, H. T. T. Toivonen, and J. T. L. Wang, Eds., Proceedings of BIOKDD'01: Workshop on Data Mining in Bioinformatics, 7th ACM SIGKDD International Conference on Knowledge Discovery and Data Mining—KDD'01, San Francisco, CA, Association for Computing Machinery, New York, August 26, 2001.

141. R. W. Snyder, "Symposium on Structure-Based Data Mining", *Abstracts, American Chemical Society 221st National Meeting,* San Diego, CA, April 1–5, 2001, American Chemical Society, Washington, DC, 2001.

142. J. F. Cargill and N. E. MacCuish, *Drug Discov. Today*, **3**, 547–551 (1998).

143. B. de Ville, *Microsoft Data Mining*, Digital Press, Boston, MA, 2001.

144. T. Hoctor, "Linking Context-similar Information", *Abstracts, American Chemical Society 222nd National Meeting,* Chicago, IL, August 26–30, 2001, American Chemical Society, Washington, DC, 2001.

145. D. R. Swanson and N. R. Smalheiser, *Artificial Intelligence*, **91**, 183–203 (1997).

146. N. Farmer, J. Amoss, W. Farel, J. Fehribach, and C. Eidner in W. A. Warr, Ed., *Chemical Structures*, Springer-Verlag, Berlin, 1988, pp. 283–295.

CHAPTER TEN

Structure-Based Drug Design

LARRY W. HARDY
Aurigene Discovery Technologies
Lexington, Massachusetts

DONALD J. ABRAHAM
Virginia Commonwealth University
Richmond, Virginia

MARTIN K. SAFO
Virginia Commonwealth University
Richmond, Virginia

Contents

Burger's Medicinal Chemistry and Drug Discovery
Sixth Edition, Volume 1: Drug Discovery
Edited by Donald J. Abraham
ISBN 0-471-27090-3   © 2003 John Wiley & Sons, Inc.

1   INTRODUCTION

Structure-based drug design by use of structural biology remains one of the most logical and aesthetically pleasing approaches in drug discovery paradigms. The first paper on the potential use of crystallography in medicinal chemistry was written in 1974 (1) and was presented at Professor Alfred Burger's retirement symposium in 1972. The excerpted last paragraph in the paper, reproduced below, foresaw the integration of X-ray crystallography into the field of medicinal chemistry.

It is reasonable to assume then that the future of large molecule crystallography in medical chemistry may well be of monumental proportions. The reactivity of the receptor certainly lies in the nature of the environment and position of various amino acid residues. When the structured knowledge of the binding capabilities of the active site residues to specify groups on the agonist or antagonists becomes known, it should lead to proposals for syntheses of very specific agents with a high probability of biological action. Combined with what is known about transport of drugs through a Hansch-type analysis, etc., it is feasible that the drugs of the future will be tailor-made in this fashion. Certainly, and unfortunately, however, this day is not as close as one would like. The X-ray technique for large molecules, crystallization techniques, isolation techniques of biological systems, mechanism studies of active sites and synthetic talents have not been extensively intertwined because of the existing barriers (1).

Since that time there have been numerous successes in advancing new agents into clinical trials by combining crystallography with associated fields in drug discovery. Currently, more structures are solved every year than were in the entire Protein Data Bank in 1972. Although almost every major pharmaceutical company has an X-ray diffraction group, Agouron (now Pfizer) was the first biotechnology startup company to make drug discovery based on X-ray crystallography a central and primary theme (2). Other startup companies (such as BioCryst and Vertex) were soon founded to apply similar approaches. More recent companies, such as Structural Genomix (3) and Astex (4), and the High Throughput Crystallography Consortium, organized by Accelrys (5), have emerged to carry on structure-based drug discovery in a high throughput environment. Third-generation synchrotron sources, such as the Advanced Photon Source (APS) at Argonne National Laboratory outside Chicago, and new detectors, have enormously increased the speed of data collection. It is now possible to collect high resolution data from protein crystals, solve, and refine the structure in days to a few months. This information is covered in an adjacent chapter. Simultaneous advances in computing have added to the speed of obtaining three-dimensional structural information on interesting drug design targets. These developments, coupled with the sequencing of the human genome and the advent of bioinformatics, provide workers in structure-based drug design with a plethora of opportunities for success.

The utility of any drug discovery tool is measured, in the final analysis, by the output of the tool's use. New tools are burdened with unrealistically high expectations. As their application begins, the impact is sometimes more limited than was originally envisaged. Structure-based design methods have had great utility for the design of enzyme inhibitors, tight-binding receptor ligands, and novel proteins. The utility of these methods for the design of *drugs* is somewhat more limited, simply because there are so many factors that must be balanced in the successful design of a drug. Nonetheless, structure-based drug design (SBDD), distinct from the (far easier) structure-based *inhibitor* design, is now a reality and has had significant impact. Aspects of the methods and utility of SBDD have been described in several excellent review articles and monographs (6–12). This chapter focuses on the utility of SBDD in the cases of drugs that have been launched as products, or that have at least entered human clinical trials. In some cases, SBDD has been a remarkable success. In others, it has failed in the sense that, despite its use, the candidate produced did not gain approval to become a marketed drug. In the latter cases, this was usually not truly a failure of SBDD, but rather attributed to the complex criteria that drugs must meet, and to the complex regulatory hurdles that candidates and companies face.

In addition to providing a measure of the impact of SBDD on the creation of actual drugs, these examples will also provide lessons about how to apply SBDD in drug discovery. The chapter is not completely encyclopedic, and some significant instances of SBDD will be missed by the informed reader. However, the discovery programs with drugs and drug candidates that are discussed will provide sufficient diversity that general trends can emerge. In a few cases, relevant results for preclinical compounds that seem likely to enter human trials are described. A growing number of the drugs to which structural design methods are applied are themselves proteins (e.g., cytokines, immunomodulators, monoclonal antibodies). However, this chapter is restricted to small organic molecules that are designed by use of the three-dimensional structure of a target protein.

2 STRUCTURE-BASED DRUG DESIGN

2.1 Theory and Methods

The concept of structure-based drug discovery combines information from several fields: X-ray crystallography and/or NMR, molecular modeling, synthetic organic chemistry, qualitative structure–activity relationships (QSAR), and biological evaluation. Figure 10.1 represents a general road map where a cyclic process refines each stage of discovery. Initial binding site information is gained from the three-dimensional solution of a complex of the target with a lead compound(s). Molecular modeling is usually next applied with the intent of designing a more specific ligand(s) with higher affinity. Synthesis and subsequent *in vitro* biological evaluation of the new agents produces more candidates for crystallographic or NMR analysis, with the hope of correlating

the biological action with precise structural information. It makes good sense at the early stages of design to use lead molecule structural scaffolds that retain low toxicity profiles, given that the latter most often derails successful drug discovery. The most active derivative(s) from this cyclic process can be forwarded for *in vivo* evaluation in animals.

2.2 Hemoglobin, One of the First Drug-Design Targets

2.2.1 History. Perutz and colleagues determined the first three-dimensional structures of proteins. Through use of X-ray crystallography Kendrew determined the structure of myoglobin (13), whereas Perutz determined the structure of hemoglobin (Hb) (14–16). At the present time, new protein and nucleic acid structures and complexes are published weekly. However, for a long period after the first protein structures were solved, progress was slower. Hb was of interest for drug discovery purposes because of the early identification of the mutant 6 Glu → Val, which causes sickle-cell anemia. The crystal structure of sickle Hb (Hbs) was published by Wishner et al. (17) and was later solved at a higher resolution by Harrington et al. (18).

2.2.2 Sickle-Cell Anemia. In 1975, through use of the three-dimensional Hb coordinates, two groups initiated SBDD studies to discover an agent to treat sickle-cell anemia: Goodford et al. in England and Abraham et al. in the United States. Goodford's group was the first to develop an antisickling agent (BW12C), based on structure-based drug design, which reached clinical trials (19, 20). However, Wireko and coworkers were unable to confirm the BW12C binding site proposed by Goodford (21). The second antisickling agent proposed by Abraham et al. to advance to clinical trials was the food additive vanillin (compound **1a**) (22). The crystallographic binding site of BW12C (**1b**) was found to be at the *N*-terminal amino groups of the α-chains (21), whereas that of vanillin shows binding close to αHis103 and also at a minor site between βHis117 and βHis117 (22). A recently redetermined binding site of vanillin at a higher resolution shows weak binding to the *N*-terminal amino group

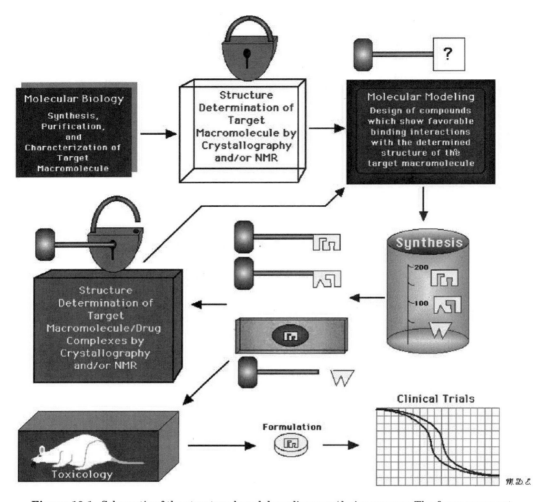

Figure 10.1. Schematic of the structure-based drug discovery/design process. The figure maps out the iterative steps that make use of X-ray crystallography, molecular modeling, organic synthesis, and biological testing to identify and optimize ligand–protein interactions.

(1a) vanillin

(1b) BW12C

of the α-chain (23). A derivative of vanillin has been patented and is a candidate for clinical trials.

Two marketed medicines, ethacrynic acid

(**2**), a diuretic agent, and clofibric acid (**3**), an antilipidemic agent, were reported to have strong antigelling activity (24, 25), and through X-ray analyses of cocrystals, the binding sites of these agents to Hb were elucidated (26). Unfortunately, it was found that high

Cl O

Cl

HO

O

CH₂

O

(2) ethacrynic acid

HO

Cl

O

O

(3) clofibric acid

concentrations of ethacrynic acid were needed to interact with Hb in deformed red cells (27). Clofibric acid, when administered in a 2 gm/day dose (as the ethyl ester clofibrate), appeared to be an ideal potential treatment for sickle-cell anemia, but was subsequently found to be highly bound to serum proteins and not transported in quantities sufficient to interact with sickle Hb. Furthermore, structure-based derivatives were not found to be effective (28, 29).

The major problem with designing a small molecule to treat sickle cell anemia is not so much an issue of specificity, but arises from the treatment of a chronic disease. The potential cumulative toxicity from the amount of drug needed to interact with approximately two pounds of hemoglobin S over a homozygous patient's lifetime is the major concern (22) (for a review, see Vol. 3, Chapter 10. *Sickle Cell Anemia*, by Alan Schecter et al).

2.2.3 Allosteric Effectors. 2,3-Diphosphoglycerate (2,3-DPG, compound **4**), found in most mammalian red cells, is the naturally occurring allosteric effector for Hb. Its physio-

logical role is to right shift the Hb oxygen-binding curve to release more oxygen. The binding site of 2,3-DPG, determined by Arnone (30) lies on the dyad axis at the mouth of the β-cleft (Fig. 10.2) interacting with the N-terminal βVal1, βLys82, and βHis143 of deoxy Hb. A more recent study at a higher resolution, by Richard et al. (31), found DPG to interact with the residues βHis2 and βLys82. Goodford and colleagues were the first to design agents that would bind to the 2,3-DPG site (32–34). An effective allosteric effector that can enter red cells might be used to treat hypoxic diseases such as angina and stroke, to enhance radiation treatment of hypoxic tumors, or to extend the shelf life of stored blood.

Many antigelling agents left shift the oxygen binding curve, producing higher concentrations of oxy-HbS. Given to patients with sickle-cell anemia, this should result in less polymerization, and therefore less red blood cell sickling. It was a surprise therefore when clofibric acid, which blocks sickle-cell Hb polymerization, was found to shift the Hb oxygen binding curve to the right, in a manner similar to that of 2,3-DPG (25). The clofibric acid binding site was found to be far removed from the 2,3-DPG site (25, 35). The determination of the clofibric acid binding site on Hb was the first report of a tense state (deoxy state) allo-

O⁻ CO₂⁻

O⁻—P

O

O

P

O⁻

O

O⁻

O⁻

O

(4) 2,3-DPG

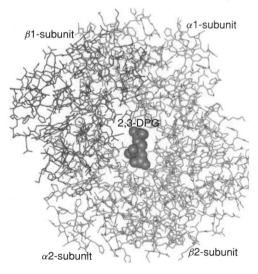

Figure 10.2. View of (**4**) (2,3-DPG) binding site at the mouth of the β-cleft of deoxy hemoglobin. See color insert.

steric binding site different from that of 2,3-DPG (compound **4**). Perutz and Poyart tested another antilipidemic agent, bezafibrate (compound **5**), and found that it was an even more

(5) bezafibrate

potent right-shifting agent than clofibrate (36). Perutz et al. (26) and Abraham (35) determined the binding site of bezafibrate and found it to link a high occupancy clofibrate site with a low occupancy site. Lalezari and Lalezari synthesized urea derivatives of bezafibrate (37), and with Perutz et al. determined the binding site of the most potent derivatives (38). Although these compounds were extremely potent, they were hampered by serum albumin binding (39, 40).

Abraham and coworkers synthesized a series of bezafibrate analogs (39–42). One of these agents, efaproxaril (RSR 13, compound **6a**) is currently in Phase III clinical trials for radiation treatment of metastatic brain tu-

(6a) (RSR-13) R = 3,5-dimethyl
(6b) (RSR-56) R = 3,5-dimethoxy

mors (see, Vol. 4, Chapter 4. *Radiosensitizers and Radioprotective Agents,* by Edward Bump et al). The binding constants and binding sites of a large number of these bezafibrate analogs were measured and agreed with the number of crystallographic binding sites found (42). The degree of right shift in the oxygen-binding curve produced by these compounds was not

solely related to their binding constant, providing a structural basis for E. J. Ariens' theory of intrinsic activity (42).

By use of X-ray crystallographic analyses, the key elements linking allosteric potency with structure were uncovered. In addition, the computational program HINT, which quantitates atom–atom interactions, was used to determine the strongest contacts between various bezafibrate analogs and Hb residues. These analyses revealed that the amide linkage between the two aromatic rings of the compounds must be orientated so that the carbonyl oxygen forms a hydrogen bond with the side-chain amine of αLys99 (41, 43). Three other important interactions were found. The first are the water-mediated hydrogen bonds between the effector molecule and the protein, the most important occurring between the effector's terminal carboxylate and the side-chain guanidinium moiety of residue αArg141. Second, a hydrophobic interaction involves a methyl or halogen substituent on the effector's terminal aromatic ring and a hydrophobic groove created by Hb residues αPhe36, αLys99, αLeu100, αHis103, and βAsn108. Third, a hydrogen bond is formed between the side-chain amide nitrogen of Asn108 and the electron cloud of the effector's terminal aromatic ring (40, 41, 43). Abraham first observed this last interaction while elucidating the Hb binding site of bezafibrate (35). Burley and Petsko had previously pointed out this type of hydrogen bond in a number of proteins, indicating that this contact is involved in a number of other receptor interactions (44, 45). Perutz and Levitte estimated this bond to be about 3 kcal/mol (46). Figure 10.3 shows the overlap of four allosteric effectors (**6a, 6b, 7a** and **7b**) that bind at the same site in deoxy Hb but differ in their allosteric potency.

(7a) (MM-30) R = 3,5-dichloro
(7b) (MM-25) R = 4-chloro

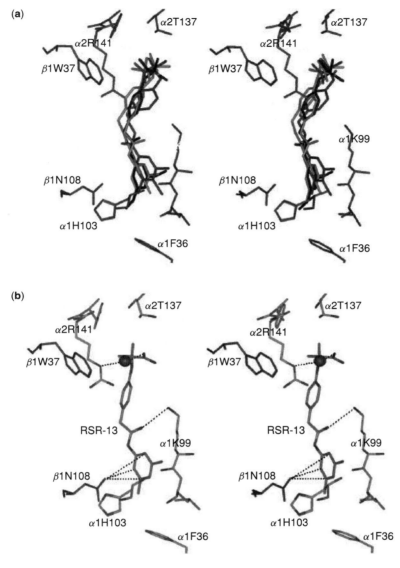

Figure 10.3. Stereoview of allosteric binding site in deoxy hemoglobin. A similar compound environment is observed at the symmetry-related site, not shown here. (a) Overlap of four right-shifting allosteric effectors of hemoglobin: (**6a**) (RSR13, yellow), (**6b**) (RSR56, black), (**7a**) (MM30, red), and (**7b**) (MM25, cyan). The four effectors bind at the same site in deoxy hemoglobin. The stronger acting RSR compounds differ from the much weaker MM compounds by reversal of the amide bond located between the two phenyl rings. As a result, in both RSR13 and RSR56, the carbonyl oxygen faces and makes a key hydrogen bonding interaction with the amine of αLys99. In contrast, the carbonyl oxygen of the MM compounds is oriented away from the αLys99 amine. The αLys99 interaction with the RSR compounds appears to be critical in the allosteric differences. (b) Detailed interactions between RSR13 (**6a**) and hemoglobin, showing key hydrogen bonding interactions that help constrain the T-state and explain the allosteric nature of this compound and those of other related compounds. See color insert.

Over the course of these studies, an interesting anomaly was solved. Allosteric effectors (such as **8a** and **8b**) can bind to a similar site

CHO

H_3C ⬡ CH_3
OH

(8a) DMHB

CHO

⬡ CO_2H
OH

(8b) 5-FSA

and yet effect opposite shifts in the oxygen-binding curve. Agents such as 5-FSA bind to the N-terminal Val and provide groups for hydrogen bonding with the opposite dimer (across the twofold axis) right shift the oxygen-binding curve. In contrast, agents that disrupt the water-mediated linkage between the N-terminal αVal with the C-terminal αArg141 left shift the curve (47) (Fig. 10.4). Structure-based stereospecific allosteric effectors for Hb have also been developed and possess activities and profiles appropriate for clinical efficacy (48, 49).

2.2.4 Crosslinking Agents. The first crosslinking agent that possessed potential as a Hb-based blood substitute was described by Walder et al. (50). Bis(4-formylbenzyl)ethane and bisulfite adducts of similar symmetrical aromatic dialdehydes, previously studied by Goodford and colleagues, provided the starting points that led to these compounds. Chatterjee et al. identified the binding site to deoxy-Hb, and found that the two Lys99 side chains were crosslinked (51). One of the derivatives was proposed as a blood substitute (52), and has been explored commercially (see Vol. 3, Chapter 8. *Oxygen Delivery and Blood Sub-*

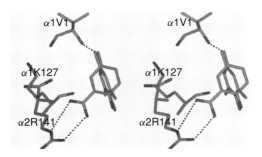

Figure 10.4. Stereoview of superimposed binding sites for (**8b**) (5-FSA, yellow) and (**8a**) (DMHB, magenta) in deoxy hemoglobin. A similar compound environment is observed at the symmetry-related site and therefore not shown here. Both compounds form a Schiff base adduct with the α1Val1 N-terminal nitrogen. Whereas the *m*-carboxylate of 5-FSA forms a salt bridge with the α2Arg141 (opposite subunit), this intersubunit bond is missing in DMHB. The added constraint to the T-state by 5-FSA that ties two subunits together shifts the allosteric equilibrium to the right. On the other hand, the binding of DMHB does not add to the T-state constraint. Instead, it disrupts any T-state salt- or water-bridge interactions between the opposite α-subunits. The result is a left shift of the oxygen equilibrium curve by DMHB. See color insert.

stitutes and Blood Products, by Andeas Mozzarelli et al.). Another crosslinked Hb engineered by Nagai and colleagues, at the MRC-LMB in Cambridge, was developed into a blood substitute that was clinically investigated at Somatogen, now Baxter (53). Boyiri et al. synthesized a number of crosslinking agents (molecular ratchets, such as **9**) whose

OHC ⬡ O-$[CH_2]_n$-O ⬡ CHO
CO_2H CO_2H

(9) $n = 1$–10 TB36, $n = 3$

potency was directly related to the length of the crosslink: the shorter the crosslink (three atoms), the stronger the shift of the oxygen binding curve to the right (54) (Fig. 10.5).

Perutz's hypothesis (55) and the MWC model (56) for allostery, that the more tension is added to the tense (deoxy) state of Hb, the greater the shift to the right of the oxygen-

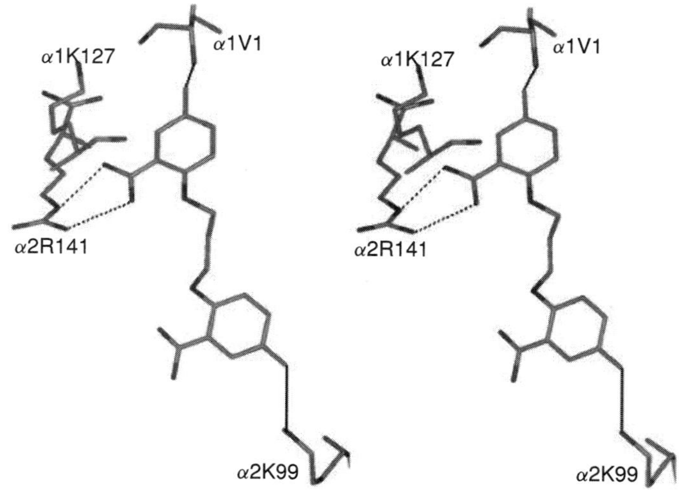

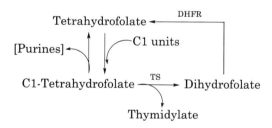

Figure 10.5. Stereoview of the binding site for (**9**) ($n = 3$, TB36, yellow) in deoxy Hb. A similar compound environment is observed at the symmetry-related site, not shown here. One aldehyde is covalently attached to the N-terminal α1Val1, whereas the second aldehyde is bound to the opposite subunit, α2Lys99 ammonium ion. The carboxylate on the first aromatic ring forms a bidentate hydrogen bond and salt bridge with the guanidinium ion of α2Arg141 of the opposite subunit. The effector thus ties two subunits together and adds additional constraints to the T-state, resulting in a shift in the Hb allosteric equilibrium to the right. The magnitude of constraint placed on the T-state by the crosslinked αLys99 varies with the flexibility of the linker. Shorter bridging chains form tighter crosslinks and yield larger shifts in the allosteric equilibrium. See color insert.

binding curve, are generally consistent with the behavior of the allosteric effectors and crosslinking agents.

2.3 Antifolate Targets

2.3.1 Dihydrofolate Reductase. The reduced form of folate (tetrahydrofolate) acts as a one-carbon donor in a wide variety of biosynthetic transformations. This includes essential steps in the synthesis of purine nucleotides and of thymidylate, essential precursors to DNA and RNA. For this reason, folate-dependent enzymes have been useful targets for the development of anticancer and anti-inflammatory drugs (e.g., methotrexate) and anti-infectives (trimethoprim, pyrimethamine). During the reaction catalyzed by thymidylate synthase (TS), tetrahydrofolate also acts as a reductant and is converted stoichiometrically to dihydrofolate. The regeneration of tetrahydrofolate, required for the continuous functioning of this cofactor, is catalyzed by dihydrofolate reductase (DHFR).

Tetrahydrofolate $\xleftarrow{\text{DHFR}}$

[Purines] $\longleftarrow$ $\quad$ C1 units

C1-Tetrahydrofolate $\xrightarrow{\text{TS}}$ Dihydrofolate

Thymidylate

Scheme 10.1.

The first crystal structure of a drug bound to its molecular target was provided by the pioneering X-ray diffraction study of the complex between DHFR and methotrexate (57), albeit in this case the target was a bacterial surrogate for the actual target (the human enzyme). Once X-ray structures of DHFR from eukaryotic sources were also solved, comparisons of the bacterial and eukaryotic DHFR structures revealed the structural basis for the selectivity of the antibacterial drug trimethoprim for the bacterial enzyme. This understanding allowed Goodford and colleagues to rationally design trimethoprim analogs with altered potencies (58). Retrospective studies such as those done by David Matthews and others on DHFR (see, for example, Ref. 59) set the stage for the iterative process of structure-based inhibitor design as it was later developed at Agouron Pharmaceuticals, targeted against another folate-dependent enzyme, TS (60, 61).

2.3.2 Thymidylate Synthase. There are two main modes in which structure-based methods for inhibitor design have been employed. The first mode is structure-guided optimization of the design of a previously known chemical scaffold. The scaffold could be a known drug or inhibitor, substrate analog, or a hit from screening of a random library. The property, which is modified during the optimization, may be, for example, potency, solubility, or target selectivity, or the more challenging aim may be to optimize several properties simultaneously. A second and potentially more powerful mode is for the *de novo* design of inhibitory ligands, sometimes referred to as lead generation. This mode relies strictly on the structure of the target enzyme or receptor as a template. A substrate or an inhibitor may be bound to the crystalline target, and deleted to provide the template. This is advantageous when, as in the case of TS, a substantial conformational change occurs when ligands bind. After a *de novo* design process has provided a new inhibitor that is structurally unique, the properties of the new scaffold can be optimized by continued structural guidance. Both modes of SBDD have been used to generate TS inhibitors that have entered clinical trials.

When the design of inhibitors of human TS at Agouron Pharmaceuticals began, the amounts of the human enzyme required for crystallographic study were unavailable. Because the active site of the enzyme is so highly conserved, it was assumed that an acceptable surrogate for human TS would be the crystal structure of a bacterial TS (60, 62). Figure 10.6 shows the conformation of the quinazoline folate analog **10** (N10-propynyl-5, 8-dideazafolate), bound within the active site of the *Escherichia coli* enzyme with the nucleotide substrate, 2'-deoxyuridine-5'-monophosphate (63, 64). This folate analog, designed by classical medicinal chemistry as an analog of the TS substrate, 5,10-methylenetetrahydrofolate (**11**), is a potent TS inhibitor. Nevertheless, (**10**) failed as an anticancer drug because of its insolubility and resulting nephrotoxicity (65).

2.3.2.1 Structure-Guided Optimization: AG85 and AG337. In the crystalline complex with *E. coli* TS, the quinazoline ring of compound (**10**) binds on top of the pyrimidine of the nucleotide, in a protein crevice surrounded by hydrophobic residues (Fig. 10.6). The bound molecule bends at right angles between the quinazoline and 4-aminobenzoyl rings (at N10), with the D-glutamate portion extending out to the surface of the enzyme. Hydrogen bonds are made with several enzyme sidechains, the terminal carboxylate, and several tightly bound waters. This compound, like folate and most folate analogs, gains entry into cells through a transport system that recognizes its D-glutamate moiety, and intracellular concentrations are elevated because of trap-

(**10**) N10-propynyl-5,8-dideazafolate (also known as PDDF or CB3717)

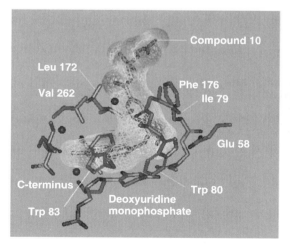

(11) 5,10-methylene, 5,6,7,8-tetrahydrofolate

ping of the compound as highly charged forms after addition of several additional glutamates by a cellular enzyme.

TS inhibitors were designed by Agouron scientists with the aim of providing a drug that could enter cells passively and thus avoid the need for transport or polyglutamylation. The first were designed by structure-guided modification of known antifolates, and others were designed *de novo*. Starting with (**12**), the glutamate moiety was deleted from the structure. [Compound (**12**), the 2-desamino-2-methyl analog of (**10**), had been found to be much more water soluble than (**10**). This eventually led (65) to AstraZeneca's Tomudex,

which is now approved for treatment of colorectal cancer in European markets.] Removal of the glutamate reduced the potency by 2 to 3 orders of magnitude (Table 10.1, **12** versus **13**). The crystal structure solved by use of (**10**) indicated potential interactions that were exploited by substituents such as the m-CF_3 in compound (**14**). The phenyl moiety in (**15**) was added to interact with Phe176 and Ile79 (Fig. 10.6). Combining substituents does not necessarily produce the expected sum of binding free energy (compare **16** with **14** and **15**). Structures of the complexes with several of these compounds revealed that ideal placement of one group does not always accommo-

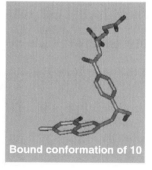

Figure 10.6. Binding site for (**10**) (*N10*-propynyl-5,8-dideazafolate), within the active site of thymidylate synthase from *Escherichia coli*. The surface of the inhibitor is shown in the left view. The red spheres in the left view are tightly bound water molecules. See color insert.

Table 10.1   SAR for 2-Methyl-4-oxo-quinazoline Inhibitors of TS[a]

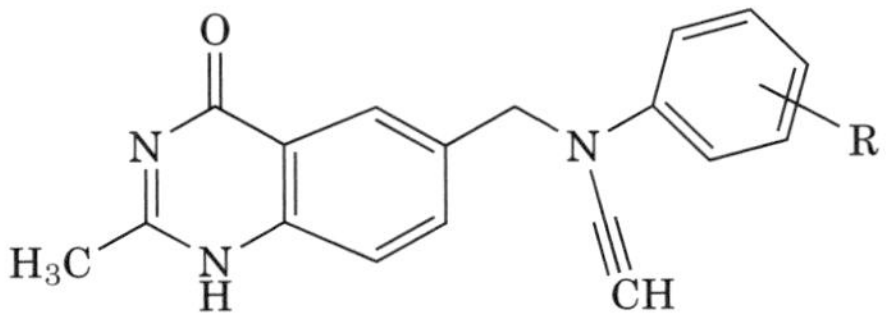

(12)–(17)

Compound	R	K_{is}, μM (*E. coli* TS)	K_{is}, μM (human TS)
(**12**)	*para*-CO-glutamate	0.005	0.009
(**13**)	H	4	2.2
(**14**)	*meta*-trifluoromethyl	0.45	0.4
(**15**)	*para*-SO$_2$-phenyl	0.025	0.013
(**16**)	*meta*-trifluoromethyl, *para*-SO$_2$-phenyl	0.037	0.05
(**17**)	*para*-SO$_2$-(*N*-indolyl)	0.15	0.07

[a]From ref. 60.

date the best interaction for another. (This is a general problem for rigid scaffolds.) Compounds (**15–17**) had significant activity in *in vitro* cell-based assays, which could be reversed by exogenous thymidine. Compound (**17**) (AG85) was tested in human clinical trials for treatment of psoriasis (9).

The structure shown in Fig. 10.6 also suggested another approach to alter the structure of (**12**) to generate a lipophilic inhibitor of TS. The hydrophobic cavity filled by the aromatic ring of the *para*-aminobenzoyl group could be filled instead by a substituent attached to position 5 of the quinazoline nucleus. Four different 5-substituted 2-methyl-4-oxoquinazolines were made to test this idea, and one of these (**18**) was a 1 μM inhibitor of human TS (66).

The X-ray structure of the bacterial enzyme with (**18**) confirmed the hypothetical

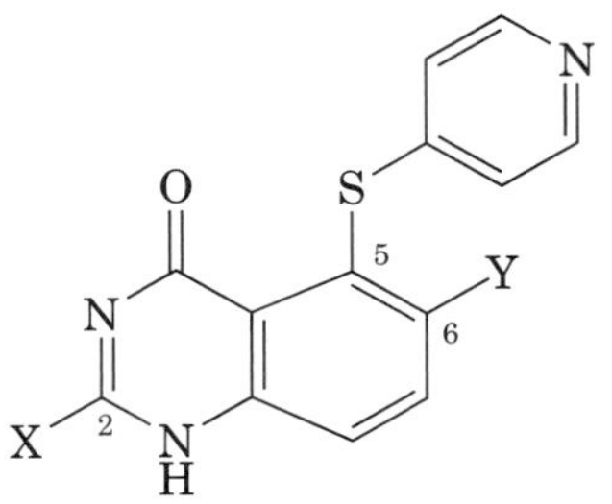

(**18**) X = CH$_3$, Y = H
(**19**) X = NH$_2$, Y = CH$_3$

binding mode. Two dozen 5-substituted quinazolines were made to explore the SAR for this scaffold. However, the eventual clinical candidate (**19**) was only two steps away from (**18**). The methyl group at position 6 was incorporated for favorable interaction with Trp80. This also favorably restricted the torsional flexibility for the 5-substituent, and increased the inhibitory potency against human TS by 10-fold. The 2-methyl was replaced by an amino group, to create a hydrogen bond to a backbone carbonyl in the protein, and increased potency another sixfold. Compound (**19**) (AG337, also known as nolatrexed, and as the hydrochloride, Thymitaq) advanced into human testing and had progressed into later-stage clinical trials as an antitumor agent by 1996 (67).

2.3.2.2 De Novo Lead Generation: AG331.
The *de novo* design effort was initiated through the use of a computational method, Goodford's GRID algorithm (68, 69), to locate a site favorable to the binding of an aromatic system within the TS active site (70). Using computer graphics, naphthalene was visualized and manipulated within this favorable site (Fig. 10.7). This facilitated alterations of the naphthalene scaffold to a benz[*cd*]indole to provide hydrogen-bonding groups to interact with the enzyme and a tightly bound water. Elaboration from the opposite edge of the naphthalene core to extend into the top of the

active site cavity, toward bulk solvent, resulted in (**20**). The use of an amine for the groups attached to position 6 of the benz-

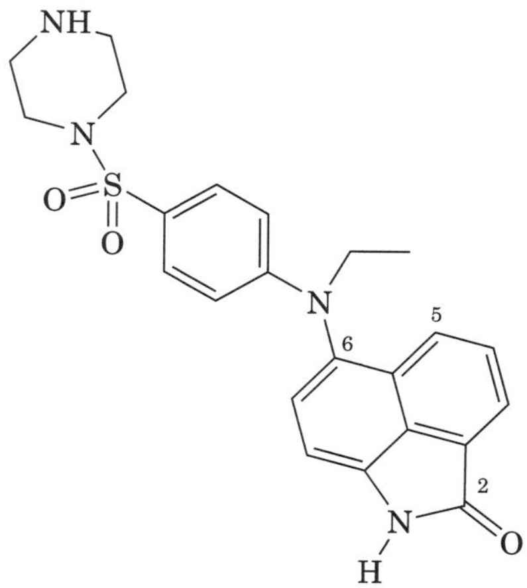

(**20**)

[*cd*]indole improved the synthetic ease for variation of these groups. Compound (**20**) had a K_{is} value of 3 μM for inhibition of human TS and was about 10-fold less potent against the bacterial enzyme.

The X-ray structure of (**20**) bound to *E. coli* TS revealed that the compound actually binds more deeply into the active-site crevice than had been anticipated. Instead of interacting favorably with the enzyme-bound water indicated in Fig. 10.7, the oxygen at position 2 of the benz[*cd*]indole displaces it. This forced the Ala263 carbonyl oxygen to move by about 1 Å. Replacement of the oxygen at position 2 with nitrogen provided a significant increase in inhibitory potency. Structural studies revealed that this also resulted in recovery of the displaced water, and restoration of the original position of the Ala263 carbonyl oxygen. The substituents at position 5, on the tertiary amine nitrogen, and on the sulfonyl group were also varied during the iterative optimization process. The process yielded (**21**) (AG331), which has a K_{is} value of 12 nM for inhibition of human TS. Compound (**21**) entered clinical trials as an antitumor agent (71).

2.3.3 Glycinamide Ribonucleotide Formyltransferase. Glycinamide ribonucleotide formyltransferase (GARFT) catalyzes the N-formyla-

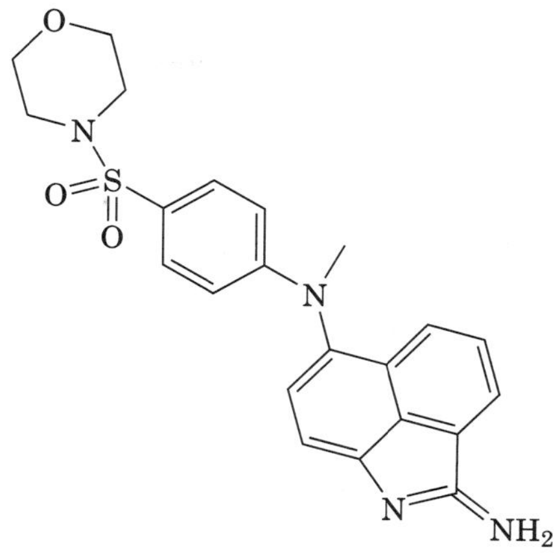

(**21**)

tion of glycinamide ribonucleotide, through use of *N*-10-formyltetrahydrofolate as the one-carbon donor. Because this is an essential step in the synthesis of purine nucleotides, GARFT is a target for blocking the proliferation of malignant cells. Several potent GARFT inhibitors, such as pemetrexed (**22**, ALIMTA,

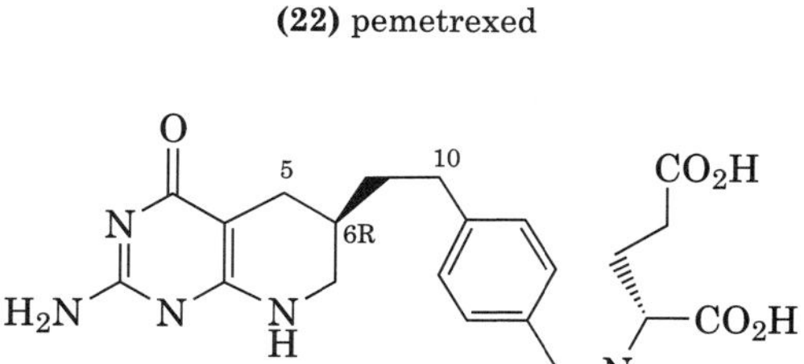

(**22**) pemetrexed

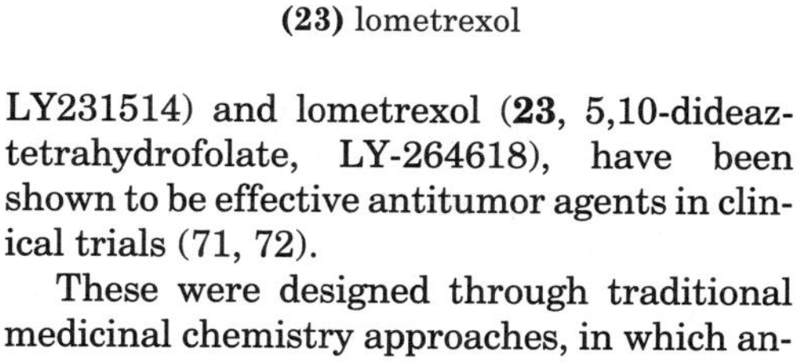

(**23**) lometrexol

LY231514) and lometrexol (**23**, 5,10-dideaztetrahydrofolate, LY-264618), have been shown to be effective antitumor agents in clinical trials (71, 72).

These were designed through traditional medicinal chemistry approaches, in which an-

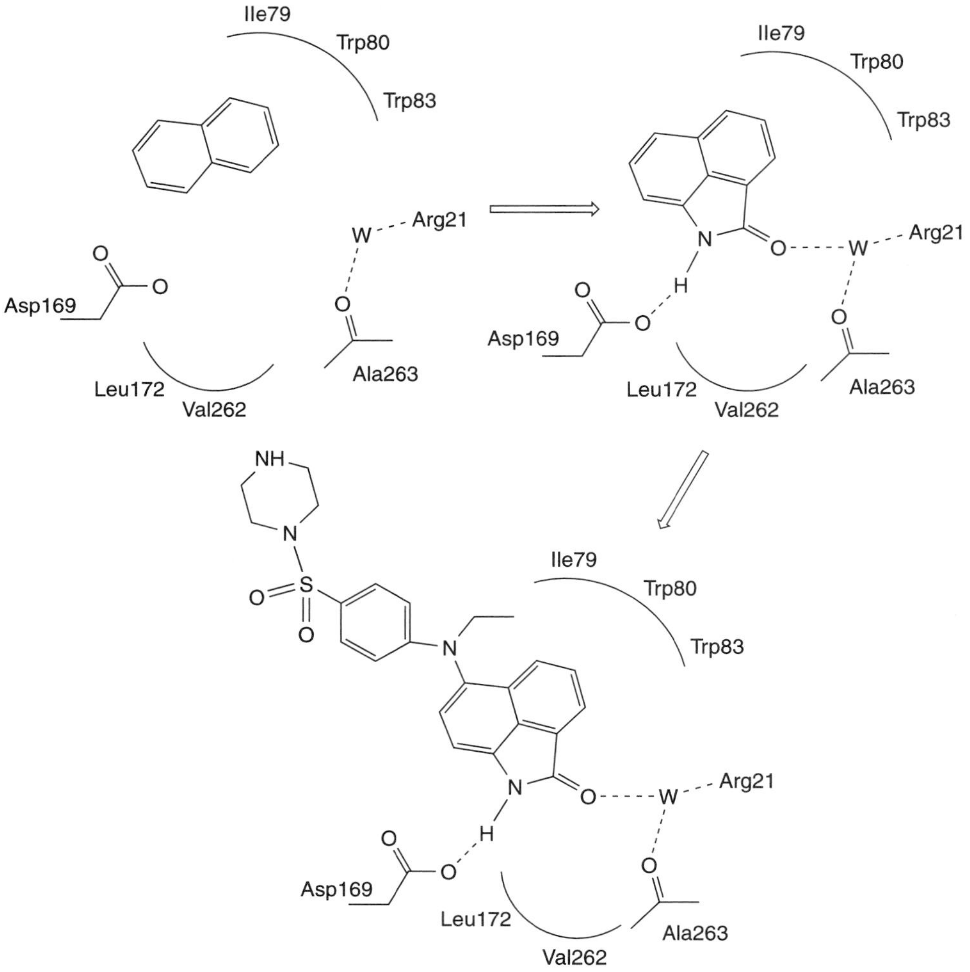

Figure 10.7. Conceptual design of compound (**20**), by use of the active site of *E. coli* TS as a template. W represents a tightly bound water molecule. [Adapted from Babine and Bender (9).]

alogs of folate were synthesized and then tested as inhibitors of tumor cell growth or of the activity of various folate-dependent enzymes (73–75). A recent paper reported the formation *in situ* of a potent bisubstrate analog inhibitor of GARFT, from glycinamide ribonucleotide and a folate analog, apparently catalyzed by the enzyme itself (76). The substrate analog was designed based on consideration of enzyme structure and the GARFT mechanism. This emphasizes the potential to exploit the interplay between binding and catalytic events in the design of new inhibitors.

The development of GARFT inhibitors at

Agouron began with consideration of the structure of the complex between the *E. coli* enzyme and 5-deazatetrahydrofolate (77). An active and soluble fragment of a multifunctional human protein that contained the GARFT activity was provided by recombinant approaches (78), and its structure was also solved (79) in complex with novel inhibitors. Comparison of the two structures subsequently validated the use of the bacterial enzyme as a model for the human GARFT. The design of novel inhibitors also relied on previous studies of the structure–activity relationships (SAR) for substitutents around the core

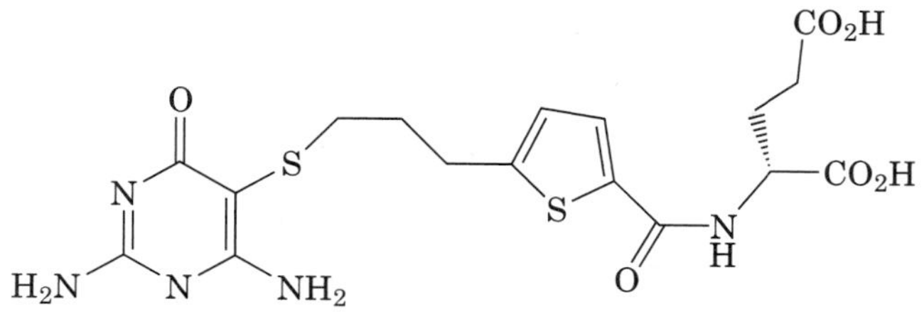

(24)

of (**23**), including some GARFT inhibitors in which the ring containing N5 was opened (80). Inspection of the structure of the bacterial GARFT–inhibitor complex revealed several important features. The pyrimidine portion of the pteridine was fully buried within the GARFT active site, forming many hydrogen bonds with conserved enzymic groups. The D-glutamate moiety was largely solvent exposed, with no immediately obvious potential for building additional interactions. Retention of the D-glutamate unmodified was also desirable for pharmacodynamic reasons. A significant opportunity was presented by the fact that the active site might accommodate a bulkier hydrophobic atom than the methylene group in 5-deazatetrahydrofolate that replaces the naturally occurring N5 in tetrahydrofolate. To test this idea, a series of 5-thiapyrimidinones were synthesized, including compound (**24**). These analogs were more readily prepared than the corresponding cyclic derivatives. This compound had a potency of 30–40 nM in both a cell-based antiproliferation assay and a biochemical assay for human GARFT inhibition. A crystal structure of human GARFT, complexed with (**24**) and glycinamide ribonucleotide, confirmed the structural homology between *E. coli* and human enzymes.

Compounds with one fewer methylene in the linker connecting the thiophenyl moiety to the 5-thia position were much less active. Several other analogs, such as (**25**), were made in attempts to fill the active site more fully, and to restrict the conformational flexibility of the linker. Molecular mechanics calculations failed to correctly predict the conformation on the 5-thiamethylene group of (**25**) bound to GARFT because of unforeseen conformational flexibility of the enzyme revealed by an X-ray structure of this complex. This again emphasizes the importance of interative experimental confirmation of molecular designs. Several functional criteria in addition to GARFT inhibition and cell-based assays were evaluated during the several cycles of optimization. These included the ability of exogenous purine to rescue cells (which indicates selective GARFT inhibition), and the ability of the inhibitors to function as substrates for enzymes involved in the transport and cellular accumulation of antifolate drugs. Balancing these criteria has resulted in the choice of compounds (**26**) and (**27**) (AG2034 and AG2037, respectively) for clinical development at Pfizer. (In 1999, Agouron Pharmaceuticals was acquired by Warner–Lambert, which was subsequently acquired by Pfizer.) It is as yet unclear whether the considerable toxicity of these and other GARFT inhibitors will allow these compounds to be acceptable as anticancer drugs.

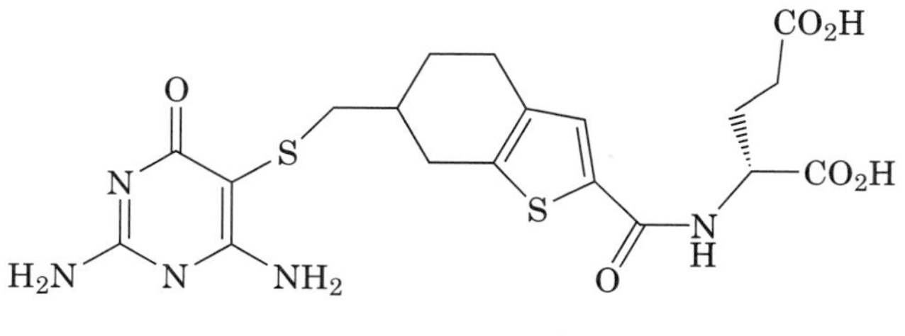

(25)

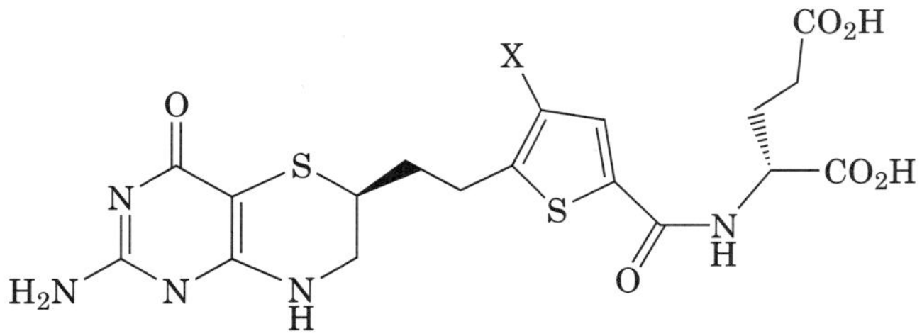

(26) X = H
(27) X = methyl

2.4 Proteases

2.4.1 Angiotensin-Converting Enzyme and the Discovery of Captopril.

The design of captopril was a landmark in the application of structural models for developing enzyme inhibitors (81, 82). This discovery rapidly led to the development of a family of therapeutically useful inhibitors of angiotensin-converting enzyme for the treatment of hypertension (83). The story has been reviewed thoroughly (for a historical perspective, see either Ref. 84 or Ref. 85), and is briefly summarized here. Angiotensin II, a circulating peptide with potent vasoconstriction activity, is generated by the C-terminal hydrolytic cleavage of a dipeptide from angiotensin I, catalyzed by angiotensin-converting enzyme. Therefore, inhibitors of angiotensin-converting enzyme are vasodilators. [*An important aside:* Angiotensin I is generated from a precursor by the action of renin, another exopeptidase that is an aspartyl protease. An orally available renin inhibitor remains an elusive goal, although there are still efforts under way that use SBDD methods (86). Renin inhibitors were early tools in the study of the essential aspartyl protease of human immunodeficiency virus (HIV), which is discussed later.]

10.8). This model was based on the already known X-ray structure of bovine pancreatic carboxypeptidase A. Both enzymes are C-terminal exopeptidases that require zinc ion for activity, but differ in that carboxypeptidase A releases an amino acid, rather than a dipeptide. Hence, the binding site for the angiotensin-converting enzyme was postulated to be longer, and to contain groups to interact with the central peptide linkage. The suggestion had been made (87) that the inhibition of carboxypeptidase A by benzylsuccinate could be explained by viewing benzylsuccinate as a "by-product analog" (Fig. 10.8, top). The hypothesis was that one of the carboxylates bound into a cationic site, whereas the other interacted with the active site zinc. If this were true, then a similar model for angiotensin-converting enzyme predicted that slightly longer diacids, designed with some regard for the sequence preferences of the converting enzyme, should inhibit that enzyme. This hypothesis was quickly confirmed by the inhibitory activity of succinyl-proline (**28a**).

Peptide sequences related to those of snake venom peptides had already been used to define the structural requirements for peptide inhibitors of angiotensin-converting enzyme. Peptides are unstable *in vivo* and poorly ab-

---

Asp-Arg-Val-Tyr-Ile-His-Pro-Phe-His-Leu → Asp-Arg-Val-Tyr-Ile-His-Pro-Phe + His-Leu
Angiotensin I   Angiotensin II

A key tool in the discovery of captopril at Squibb was the use of a model for the active site of angiotensin-converting enzyme (Fig.

sorbed intestinally, and thus are not good drug candidates. However, the best peptide inhibitor was 500-fold more potent than (**28a**). The

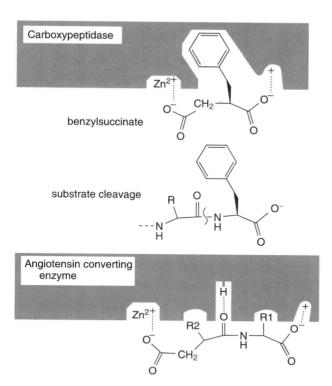

Figure 10.8. Active site models for carboxypeptidase A (top) and angiotensin-converting enzyme (bottom). The design of the dipeptidyl derivative that led to the discovery of captopril is shown bound to the latter enzyme.

information provided by the peptides, the structural model for the active site of angiotensin-converting enzyme, and biochemical and tissue-based pharmacological assays for the enzyme's function were used to guide an iterative design process to improve the potency, selectivity, and stability of small molecules inhibitors. The R1 and R2 substitutents were optimized, and the zinc ligand was changed to a thiol, which significantly increased potency (Table 10.2, compare **28a** with **28c**). This process yielded the orally available and stable small molecule captopril (**28d**) within 18 months of the creation of the model.

The following quotation [from the original research report (81) on the design of captopril] predicted the great promise of SBDD: "The studies described above exemplify the great heuristic value of an active-site model in the design of inhibitors, even when such a model is a hypothetical one."

2.4.2 HIV Protease. The aspartyl endoprotease encoded by human immunodeficiency virus (HIV-P) catalyzes essential events in the maturation of infective virus particles, the cleavage of polyprotein precursors to yield active products. After this was demonstrated in the mid to late 1980s, HIV-P became a target for the development of antiviral drugs to treat acquired immunodeficiency syndrome (AIDS). Several HIV-P inhibitors have been approved for human therapeutic use in the past 10 years, and the speed with which they were developed is attributed in part to the successful use of SBDD methods. There are excellent recent reviews of this area (88, 89). There are numerous reviews of the early work on HIV-P inhibitors (8, 9, 90, 91).

HIV-P is a symmetrical homodimer of identical 99 residue monomers, structurally and mechanistically similar to the pseudosymmetric pepsin family of proteases (92–94), whose members include renin. Because the protease is a minor component of the virion particle, intensive structural studies required overproduction through recombinant DNA methods. One of the first structures was determined with material synthesized nonbiologically (through peptide synthesis). As of June 2002, there were over 100 X-ray structures repre-

Table 10.2 Key Compounds in the Development of Captopril

Compound	Structure	IC_{50} for inhibition of ACE (μM)
(28a) (succinyl-proline)	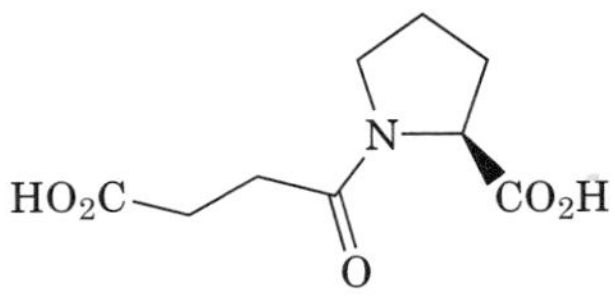	330
(28b)	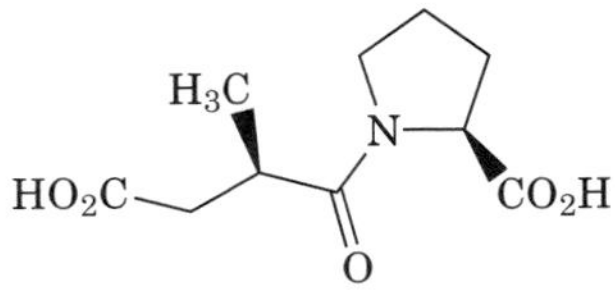	22
(28c)	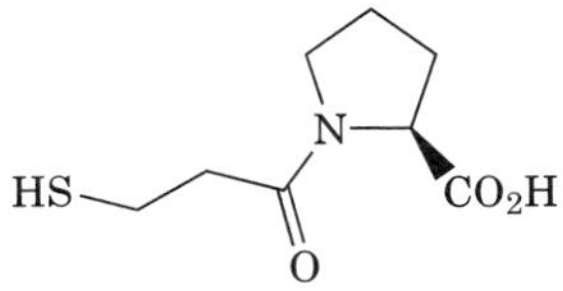	0.2
(28d) (captopril)	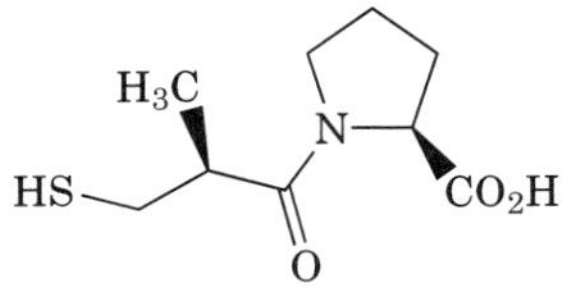	0.023

sented by coordinate sets in the Protein Data Bank, and many hundreds more have been determined in proprietary industrial studies.

The active site of the enzyme is C2 symmetric in the absence of substrates or inhibitors (Fig. 10.9a), and contains two essential aspartic acid residues (Asp25 and Asp25′). The entrance to the active site is partly occluded by "flaps" constructed of two beta strands (residues 43–49 and 52–58) from each monomer, connected by a turn. In the absence of substrate or inhibitor, the flaps seem to be rather flexible. Upon binding of inhibitors and presumably of substrates, the residues within the flaps undergo movements up to several angstroms to interact with the bound ligand (Fig. 10.10). A single tightly bound water is observed in the structures of most HIV-P–inhibitor complexes, accepting hydrogen bonds from the backbone amides of both flap residues Ile50 and Ile50′ and donating to carbonyls of the bound inhibitors. This is referred to as the "flap" water. Despite the presence of this water and several tightly bound water molecules on the floor of the active site, the cavity also contains extensive hydrophobic surface area. The minor differences between the HIV proteases from two major strains of HIV (HIV-1 and HIV-2) are not addressed here. More significant are the HIV-P sequence variants with much reduced sensitivity to existing drugs that have evolved because of selective pressure and the rapid mutation rate of the virus. The reader interested in the differences between the proteases from HIV-1 and HIV-2, or in the issues surrounding drug-resistant variants, is referred to Ref. 91 and Ref. 89, respectively.

The early work on inhibition of HIV-P was much influenced by previous structural and mechanistic work on pepsin and its inhibitors. Both enzymes are thought to catalyze peptide hydrolysis through a tetrahedral transition state, shown below as (29). The previous work

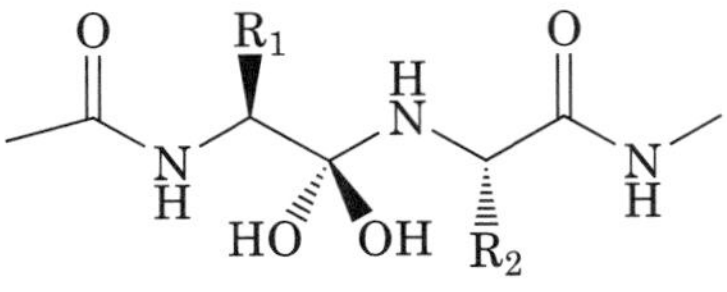

(29)

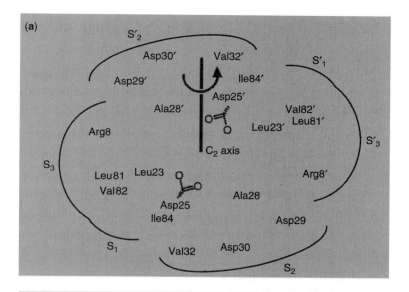

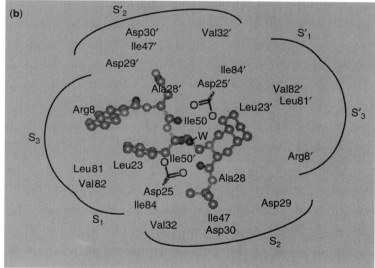

Figure 10.9. (a) Residues in the active site of HIV protease. The C2 axis that relates the residues of the two monomers is indicated. The carboxylates of Asp25 and Asp25′ are the catalytic groups. Not shown in this view are several flap residues (Ile47/Ile47′, Ile50/Ile50′), which move in to interact with inhibitors. (b) Active site with bound (**31**) [saquinavir (PDB code 1HXB)]. Note the asymmetry of inhibitor binding. The flap water that is shown very close to saquinavir is labeled W. See color insert.

on transition state mimics as pepsin inhibitors and the sequence of some cleavage sites for HIV-P led to the discovery at Roche of the R and S versions of (**30**) as submicromolar inhibitors of HIV-P, with the R enantiomer being threefold more potent (95). These inhibitors employ a hydroxyethylamine moiety to replace the P1–P1′ linkage that is normally cleaved (the scissile bond) with a stable group. The lead molecules were optimized without knowledge of the HIV-P crystal structure, to produce (**31**) (Ro 31–8959, saquinavir, Fortovase).

Cbz-Asn—N\
    H

OH CO_2-t-Butyl

(**30**)

Saquinavir (**31**) was the first HIV-P inhibitor approved for human use. Figure 10.9B

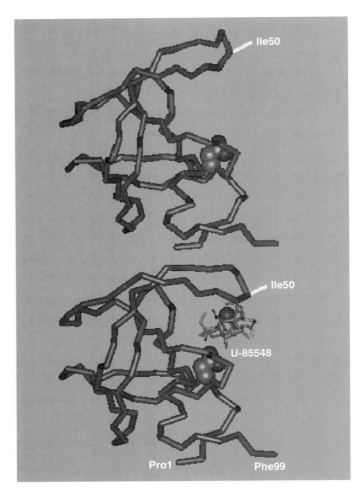

Figure 10.10. Comparison of the structures of HIV-P apoenzyme monomer (top, PDB code 3PHV) and the complex between HIV-P and (**32**) (U-85548; bottom, PDB code 8HVP). The inhibitor is shown as a ball and stick structure. Note the rearrangement of the flap residues; Ile50 is indicated for reference. The van der Waals surface of Asp25 is shown in both structures. The flap water (red ball) is also shown between Ile50 and U-85548. In the bottom structure, the locations of the N and C termini of HIV-P are noted. See color insert.

shows the asymmetrical binding mode of the molecule in the HIV-P active site. Because the metabolic and pharmacokinetic characteristics of this compound and several other early

HIV-P inhibitor drugs are less than ideal, the search for better ones has continued. Many of the deficits arise from the large size and peptidic nature of the inhibitors. Another early

(**31**) saquinavir, Ro 31-8959

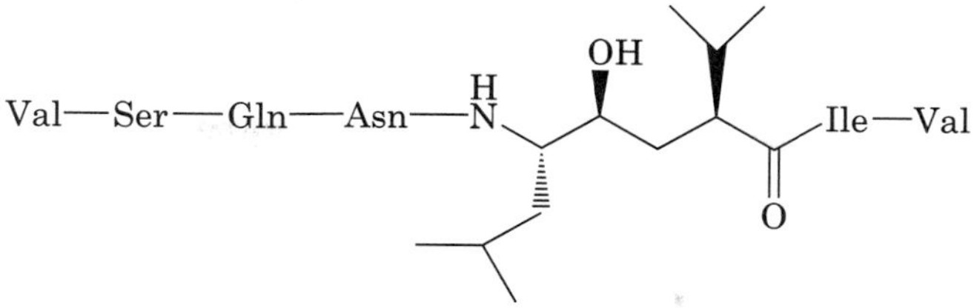

(32)

inhibitor was the modified octapeptide (**32**, U-85548) developed at Upjohn (96).

This subnanomolar inhibitor was used to define the extensive hydrophobic and hydrogen bonding interactions available in the HIV-P active site (97). A common feature in the binding of (**31**) and (**32**) to HIV-P is the interaction of the central hydroxyl group of the inhibitors with the carboxylates of both Asp25 and Asp25'. This hydroxyl group replaces a water molecule that likely binds between these aspartyl side chains during peptide hydrolysis by HIV-P. The inhibitors can therefore be seen as mimics of a "collected substrate." The liberation of this water to bulk solvent probably contributes about 5 kcal mol^{-1} to the free energy of inhibitor binding, based on the studies by Rich and his colleagues on similar inhibitors of pepsin (98, 99). An interesting difference between (**31**) and (**32**) is that (**31**) has R stereochemistry at the hydroxymethyl center, whereas in (**32**) this is an S center. Part of the reason for this is that when (**31**) binds to HIV-P, the decahydroquinoline ring system induces a conformational change in the protein, affecting primar-

ily site S_1'. The optimal stereochemistry at the hydroxymethyl center appears to be whichever one will allow the interaction of the hydroxyl with both catalytic aspartates while accommodating the placement of inhibitor moieties in the S_1, S_2, S_1', and S_2' sites with minimal conformational strain on the inhibitor (9).

Both (**31**) (Fig. 10.9b) and (**32**) (Fig. 10.11) bind to the HIV-P active site asymmetrically. However, after the X-ray studies of crystalline HIV-P apoenzyme revealed it to be a symmetrical dimer, C2 symmetric inhibitors were designed to take advantage of this structural feature (Fig. 10.12). Both alcohol diamines and diol diamines were examined. For example, the C2 symmetric compound (**33**) (A-77003) was synthesized at Abbott and entered clinical trials as an antiviral agent for intravenous treatment of AIDS (100).

The X-ray structures of complexes between HIV-P and diol diamine derivatives like (**33**) showed (101) that, although one of the hydroxyl groups bound between the catalytic aspartyl carboxylates and made contacts with both, the second hydroxyl made only one such

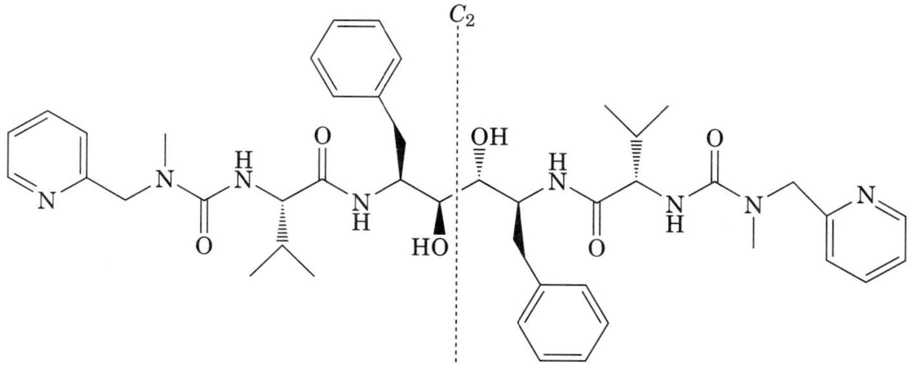

(33) A-77003

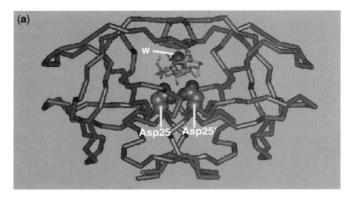

Figure 10.11. Orthogonal views of the complex between HIV-P and (**32**) (U-85548). The view in panel a is rotated approximately 90° (around the long axis of the protein) from the view in panel b. Van der Waals surfaces of Asp25, Asp25′, and the flap water (W) are shown. In panel b, the solvent-accessible surface of the inhibitor is shown. See color insert.

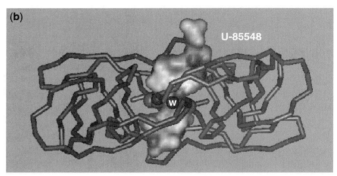

Figure 10.12. Design principle for C2 symmetric inhibitors of HIV-P and the related hydroxyethylene diamine scaffold.

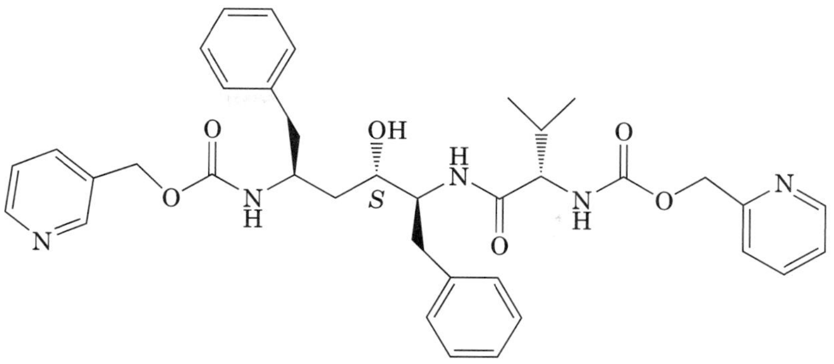

(34) A-80987

contact. Thus the cost of desolvating the second inhibitor hydroxyl upon binding is not compensated by strongly favorable interactions in the complex (8). This led to the deletion of the second hydroxyl, as seen in compound (34), another compound in this program at Abbott. Further structural modifications, to enhance solubility and metabolic stability, were guided by the fact that the "ends" of the protease-bound inhibitors were relatively solvent exposed and made fewer contacts with the enzyme (102). Deletion of a valine residue (33 → 34) gave a smaller compound, presumably aiding solubility and absorption. The eventual product of this program was ritonavir (35, A-84538, ABT-538, or Norvir), which has been successfully launched.

Another C2 symmetric HIV-P inhibitor, discovered at Dupont Merck is compound (36) (DMP-450). This was one of a series of cyclic ureas designed to interact with both the aspartyl carboxylates and the Ile50 and Ile50' backbone amides that hydrogen bond with the flap

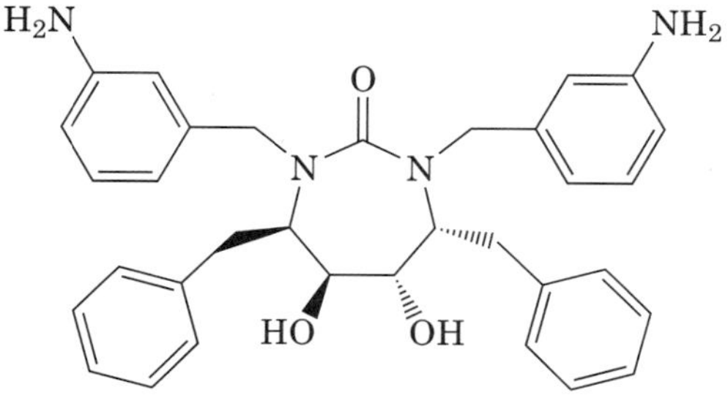

(36)

water (103). The compounds interacted with HIV-P in a highly symmetrical fashion, as they had been designed to do, with the urea oxygen replacing the flap water. Compound (36) was licensed to Triangle Pharmaceuticals, and the mesylate advanced into Phase I clinical trials. Its future is uncertain after the trials were put on hold because of animal toxicity (http://www.tripharm.com/dmp450.html).

One of problems common to many of the HIV-P inhibitors already discussed is their

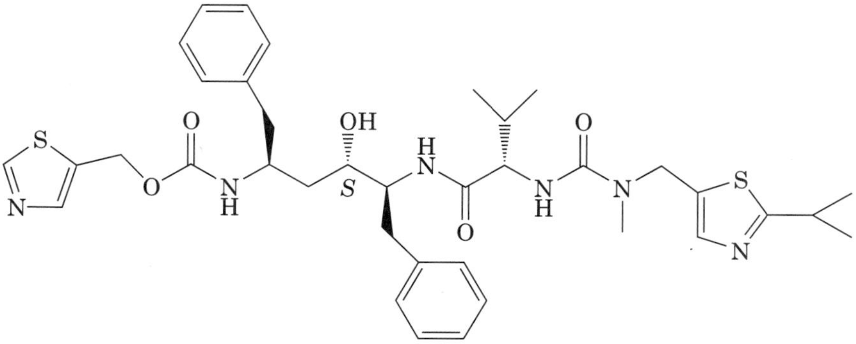

(35) ritonavir

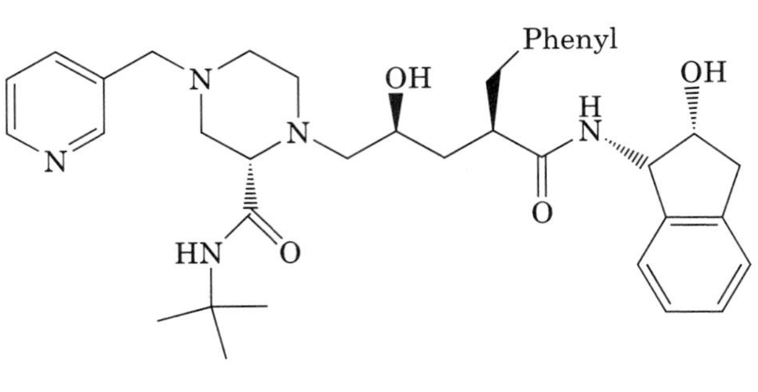

(**37**) indinavir

low solubility, which translates to low bio-availability. The discovery of (**37**) (indinavir, L-735,524) was the result of the successful application of SBDD at Merck to directly address this problem. During an iterative optimization process, the physicochemical properties of HIV-P inhibitors were modified within constraints that were established structurally (104). Crixivan (the sulfate of **37**) was successfully launched for use as an antiviral drug.

The process leading to indinavir (Fig. 10.13) began with (**38**), a hydroxyethylene-containing heptapeptide mimic, originally designed as a renin inhibitor (105). The inhibition of HIV-P by (**38**) was discovered by screening. Classical medicinal chemistry methods allowed a reduction in size, and the discovery of an amino-2-hydroxyindan moiety to replace the terminal dipeptide (corresponding to P_2', thought to bind into the S_2' site). This approach (105, 106) resulted in the generation of (**39**) (L-685,434). Although (**39**) had a subnanomolar IC_{50} for inhibition of HIV-P, it also had very low aqueous solubility, like most peptidomimetics. One way to improve solubility is to insert a charged functional group into the molecule. The tertiary amino group in the HIV-P inhibitor saquinavir (**31**)

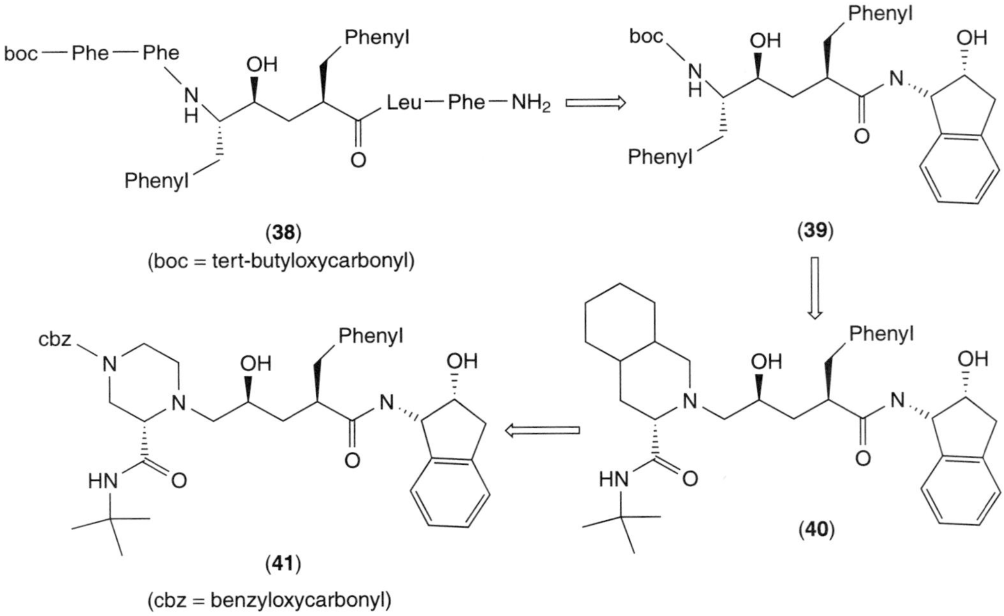

Figure 10.13. Structures of HIV-P protease inhibitors during the optimization process leading to the discovery of (**37**) (indinavir).

was already identified. Piracy of the decahydroisoquinoline *tert*-butylamide from (**31**) provided the idea for the hybrid molecule (**40**). In addition to the charged group, use of this ring system would partly "preorder" the inhibitor's structure, lessening the entropic cost of binding. Molecular modeling was used with known structures of HIV-P–inhibitor complexes to evaluate this idea, and it was judged to be reasonable enough to justify the synthesis of (**40**) (104). This compound was subsequently shown to have much better pharmacokinetic behavior than its antecedents, consistent with improved solubility and dissolution.

A convergent synthetic route was devised to generate (**40**) to improve the accessibility of important analogs. Although (**40**) was an 8 n*M* inhibitor of the isolated enzyme, better potency was needed for acceptable cell-based activity, and still better solubility characteristics were needed. A method for structure-based computational estimation of the interaction energy for HIV protease inhibitors with the enzyme was developed and used to help estimate inhibitor potency before synthesis (107). Variation of the group contributing the tertiary amine led to the discovery of the piperazine derivative (**41**) (L-732,747), which had subnanomolar potency against HIV-P. The X-ray structure of the HIV-P complex with (**41**) confirmed the binding mode predicted by molecular modeling, with the molecule filling the S_1, S_2, S_1', and S_2' pockets, and the S_3 pocket occupied by the terminal benzyloxycarbonyl moiety. Replacement of the benzyloxycarbonyl with more polar heterocycles, chosen to

be accommodated by the S_3 pocket and to further improve aqueous solubility, yielded (**37**).

Several other approved AIDS drugs that act by inhibition of HIV-P have also been developed through use of SBDD methods. Compound (**42**) (amprenavir, Agenerase, also known as VX-478) is the most recent addition to the HIV-P inhibitors approved for human antiviral treatment, and differs significantly from earlier inhibitors. Compound (**42**) was specifically designed by Vertex scientists to minimize molecular weight to increase oral

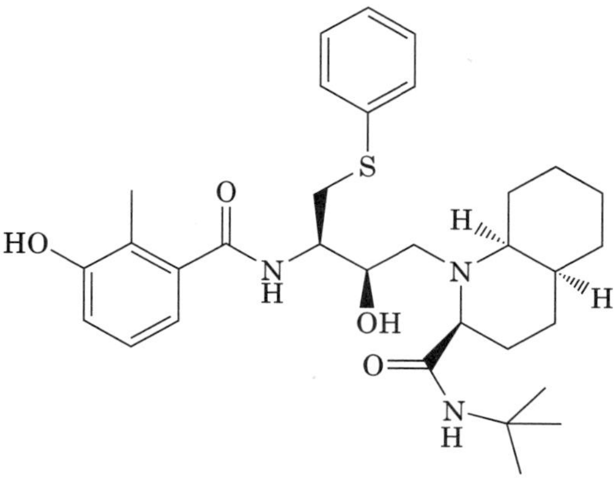

(**43**) nelfinavir

bioavailability (108). Compound (**43**) (nelfinavir, AG-1343, also known as LY312857), like the precursors to the earlier drug (**37**) (indinavir), copied the decahydroisoquinoline *tert*-butylamide group from the first marketed HIV-P inhibitor (**31**) (saquinavir). Compound (**43**) was developed in a collaboration between scientists at Lilly and Agouron (109), and is mar-

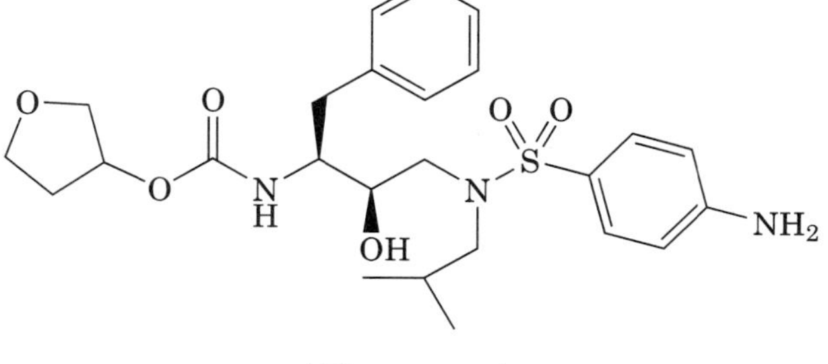

(**42**) amprenavir

keted by Pfizer as Viracept, the mesylate salt of nelfinavir. In both (**42**) and (**43**), the scientists involved used iterative SBDD methods to alter the physicochemical properties of the drug molecule while maintaining potency by optimizing interactions with the active site of the enzyme. An important feature shared by these compounds is the fact that the bound inhibitors appear to be in low energy conformers, so that minimal conformational energy costs must be paid upon binding to the enzyme.

2.4.3 Thrombin. Thromboembolic diseases such as stroke and heart attack are major health problems, especially in many Western countries. This has led to searches for drugs that are effective inhibitors of various serine endoproteases in the blood-clotting cascade, such as factor Xa and thrombin. Existing therapeutic agents such as the coumarins (like warfarin), heparin, and hirudin have problems related to their absorption or unpredictable metabolism and clearance. Recently, new small molecule inhibitors of thrombin have become available for human use in the United States, including (**44**) (argatroban, MD-805, developed by Mitsubishi) and (**45**) (melagatran, developed by AstraZeneca) (110, 111). These nanomolar inhibitors of human thrombin were optimized by classical medicinal chemistry, starting with peptidomimetics similar to the thrombin cleavage site in fibrinogen (see Fig. 10.14a). Poor absorption by an oral route requires that they must be administered intravenously or at best subcutaneously. At present, the only direct inhibitor of thrombin suitable for oral administration is ximelagatran, a prodrug form of melagatran in late development for various cardiovascular indications by AstraZeneca as of mid-2002. The therapeutic need and the availability of high quality crystal structures for human thrombin bound to inhibitors such as (**44**) make this an attractive target for SBDD (112). The significant efforts at Merck to use SBDD approaches to develop orally available inhibitors of thrombin, which have yielded compounds that have entered clinical trials, have been reviewed (113, 114). For a good overview of this area, see the review by Babine and Bender (9). Compound (**46**) [NAPAP, N-alpha-(2-

naphthylsulfonylglycyl)-4-amidinophenylalanine piperidide] is a moderately potent inhibitor of human thrombin, but was found to have an unacceptably short plasma half-life in animals (115). However, (**46**) has been a useful experimental tool and a variety of analogs have been made. The structures of (**44**) and (**46**) bound to human thrombin show that they bind somewhat differently, as shown in Figure 10.14b (112, 116). However, both form hydrogen bonds with the backbone at Gly216 (part of the oxyanion hole), and both fill the S_1 specificity pocket with a permanent cation attached to an extended hydrophobic group. Compound (**46**) was the starting point at Boehringer Ingelheim for the development of the orally bioavailable prodrug (**47**) (BIBR-1048) that generates in vivo a potent inhibitor of human thrombin (117). Compound (**47**) is currently in human clinical trials.

Scientists at Boehringer Ingelheim used the crystal structure of the complex between (**46**) and human thrombin to design a replacement for the central bridging glycine moiety. The hypothesis that a trisubstituted benzimidazole could correctly place groups into the S_1, S_2, and S_4 pockets was confirmed. The first such compound made was (**48**). The IC_{50} for thrombin inhibition by (**48**) was only 1.5 μM, but the compound had an improved serum half-life in rats. Determination of the crystal structure of the thrombin–(**48**) complex showed that (**48**) binds in a similar fashion to (**46**). The N-methyl on the benzimidazole fit into the P_1 pocket, and the phenylsulfonyl group extended into S_4. The low affinity is likely attributable to the fact that (**48**) forms no hydrogen bonds with the backbone of Gly216. An iterative optimization process (Fig. 10.15) was used to regain the lost affinity, eventually surpassing the thrombin affinity of the starting point (**46**) (0.2 μM).

Surprisingly, the N-methyl group could not be replaced with larger alkyl substituents, despite what appeared to be room for them in the P_1 pocket. However, replacing phenyl with larger aryl groups such as naphthyl or quinolinyl on the sulfonamide provided favorable interactions in the P_4 pocket. The crystal structure suggested that the increased lipophilicity of such aryl groups could be balanced by appending charged substituents to

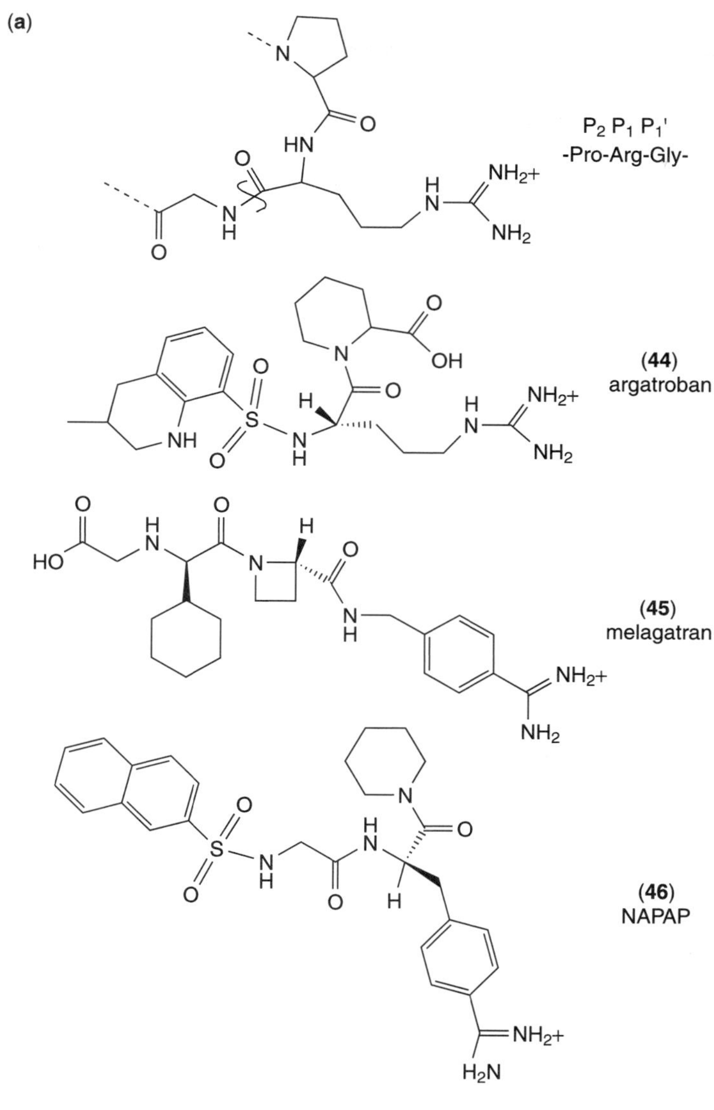

(a)

P$_2$ P$_1$ P$_1$'
-Pro-Arg-Gly-

(44)
argatroban

(45)
melagatran

(46)
NAPAP

Figure 10.14. (a) Sequence in fibrinogen at the thrombin cleavage site (top), and structures of several inhibitors of human thrombin.

the sulfonamide nitrogen. Such substituents appeared likely to extend into solvent and therefore to be tolerated without compromising affinity. This was confirmed (i.e., compound **49**), and this decreased the undesirable affinity for serum-binding proteins. X-ray studies with some of the inhibitors at this point indicated that a longer linker between the central benzimidazole and the benzamidine moiety in the S$_1$ pocket might provide some advantage. This was confirmed with several analogs, with the methylamino linker pro-

viding a 10-fold increase in potency (compound **50**). By this point, the structural basis for interaction of this compound series with thrombin was understood sufficiently to suggest that the amidosulfonyl group could be replaced by a carboxamide. This was confirmed by use of several compounds, such as (**51**). Compound (**51**) (BIBR 953) was quite active as an anticoagulant in animals dosed intravenously, but required conversion to prodrug (compound **47**) to mask its charge and allow oral dosing.

(b)

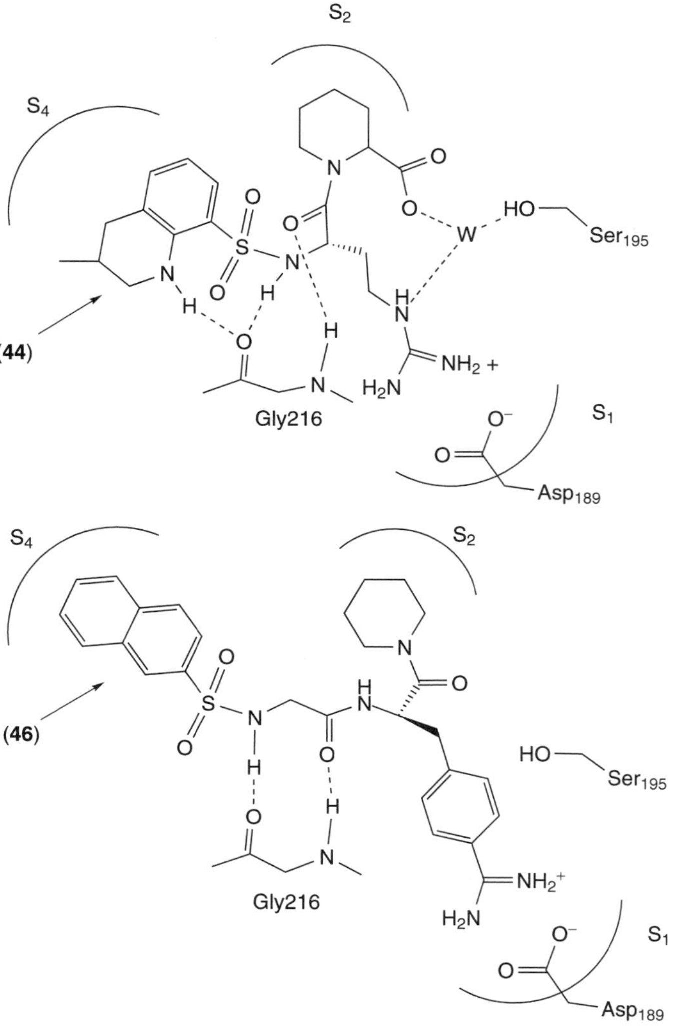

Figure 10.14. (b) Schematic comparison of the binding interactions for (**44**) and (**46**) in X-ray structural models of crystalline thrombin.

2.4.4 Caspase-1. Caspase-1 (interleukin 1-β converting enzyme, or ICE) is a member of a family of cysteine proteases that catalyze the cleavage of key signaling proteins in such processes as inflammatory response and apoptosis. Genetic methods have provided evidence supporting a role for caspase-1 in diseases such as stroke (118) and inflammatory bowel disease (119). The X-ray structure of crystalline human caspase-1 was solved in 1994 by several groups (120, 121), and has been a valuable tool in intensive efforts to design potent and bioavailable inhibitors of the enzyme.

Compound (**52**) (pralnacasan, VX-740) was

developed as a caspase-1 inhibitory therapeutic agent through use of SBDD in a collaboration between Vertex and Aventis. Although the details of the discovery process have not been published, (**52**) probably functions as a prodrug. The cleavage of the lactone of (**52**) would yield a hemiacetal that could hydrolyze to release ethanol and the aldehyde form of the drug, which then can form a covalent thioacetal with the active site thiol of caspase-1, leading to pseudoirreversible inhibition. Clinical trials of compound (**52**) as an anti-inflammatory agent for treatment of rheumatoid arthritis began in 1999 (122). In April 2002, the

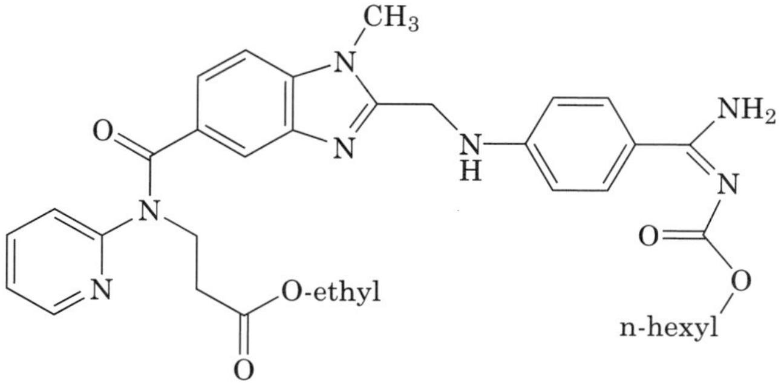

(47) BIBR 1048

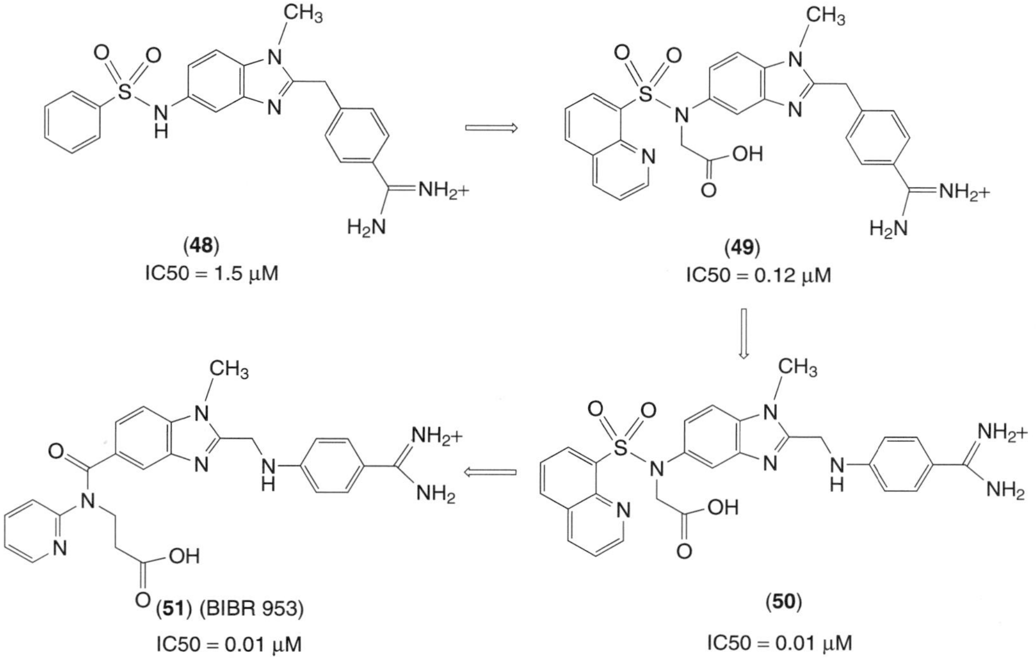

(48)
IC50 = 1.5 μM

(49)
IC50 = 0.12 μM

(51) (BIBR 953)
IC50 = 0.01 μM

(50)
IC50 = 0.01 μM

Figure 10.15. Optimization of structure leading to the discovery of (**51**) (BIBR 953).

companies announced that these trials would continue and would be expanded to include treatment of osteoarthritis.

2.4.5 Matrix Metalloproteases. Matrix metalloproteases (MMPs) are a large and diverse family of zinc endoproteases. Several members of this family (such as the collagenases and the stromelysins) are thought to have important roles in proliferative diseases, including arthritis, retinopathy, and metastatic in-

vasiveness of tumor cells. There are publicly available X-ray structures of enzyme–inhibitor complexes for at least seven different MMPs, as of this writing. Several detailed reviews of the SAR and binding modes for inhibitors of matrix metalloproteases are available (9, 123). All MMP inhibitors contain a moiety that binds to the active site zinc, such as the hydroxamates of (**53**) (prinomastat, AG3340) and (**54**) (CGS-27023) and the carboxylic acid of (**55**) (tanomastat, BAY 12–9566). These

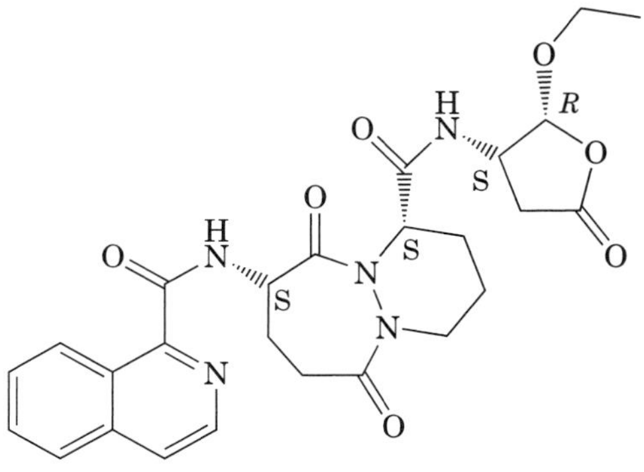

(52) pralnacasan

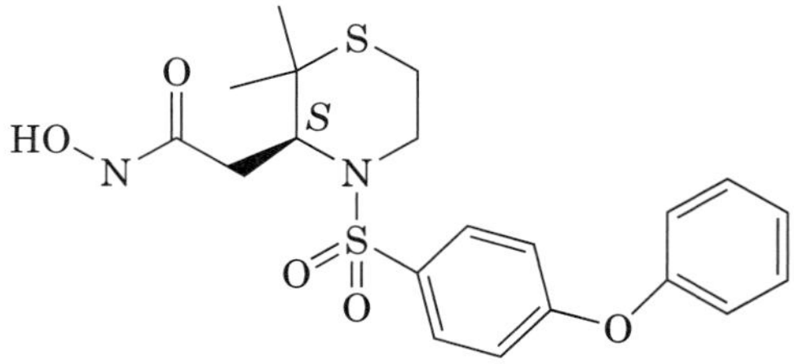

(53) prinomastat, AG3340

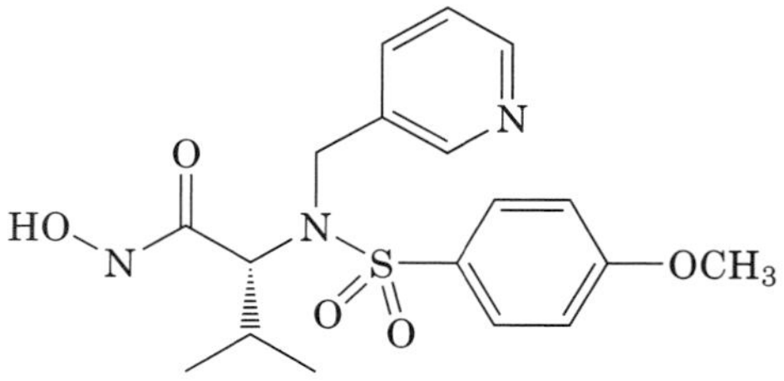

(54) CGS 27023

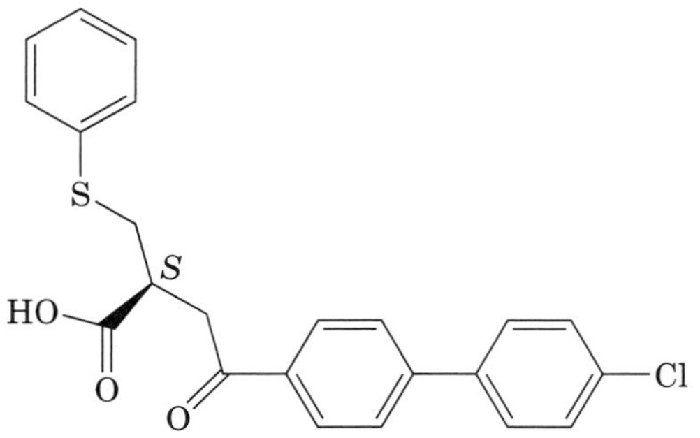

(55) tanomastat, Bay 12-9566

based inhibitor design targeted against the bacterial zinc-protease thermolysin. Compound (**55**), with particularly high affinity for the gelatinases, was also developed with consideration of the structures of other MMP–inhibitor complexes, but not through use of iterative SBDD (127). The clinical trials of compounds (**54**) and (**55**) have been suspended because of their disappointing efficacy (124). It remains somewhat uncertain which MMP is responsible for specific diseases, and the possibility for biological redundancy suggests that inhibition of several MMPs may be required for treatment of some diseases. SBDD clearly could have a major impact on the discovery of selective MMP inhibitors. These could be useful tools in dissecting the disease relevancy of these targets, as well as providing the selectivity and bioavailability required of effective drugs.

2.5 Oxidoreductases

Oxidoreductases catalyze the oxidation or reduction of carbon–carbon, carbon–oxygen, or carbon–nitrogen bonds. Frequently, nicotinamide cofactors are involved, with the oxidized and reduced forms (respectively, NAD^+ or $NADP^+$ and NADH or NADPH) receiving or donating the equivalent of a hydride during this process. Nicotinamide-linked oxidoreductases that have been targeted for the discovery of new therapeutic agents include aromatase, dihydrofolate reductase (mentioned above), aldose reductase, and inosine monophosphate dehydrogenase. SBDD methods have been successfully applied recently to the latter two enzymes to discover agents that are currently

compounds each have affinities in the nanomolar to picomolar range for several MMPs. The inhibitory profiles and ongoing clinical trials of a variety of drug candidates that inhibit MMPs were reviewed in 2000 (124).

Compound (**53**) was developed at Agouron through use of SBDD (125) and is under clinical investigation by Pfizer as an anticancer drug and as a treatment of proliferative retinopathy. Compound (**54**) is a stromelysin inhibitor discovered at Novartis (126), without explicit structural guidance. However, the lead molecule from which (**54**) was developed was originally obtained by X-ray structure-in human testing. The efforts with these two targets are described briefly below.

2.5.1 Inosine Monophosphate Dehydrogenase.

Proliferative cells such as lymphocytes have high demands for the rapid supply of nucleotides to support DNA and RNA synthesis, as do viruses during their proliferative phase. The first dedicated step in the *de novo* biosynthesis of guanine nucleotides is conversion of inosinate to XMP, catalyzed by inosine monophosphate dehydrogenase (IMPDH).

$$IMP + NAD^+ \rightarrow XMP + NADH$$

A prodrug form of (56) (mycophenolic acid), a noncompetitive inhibitor of IMPDH, is approved for human therapeutic use as an im-

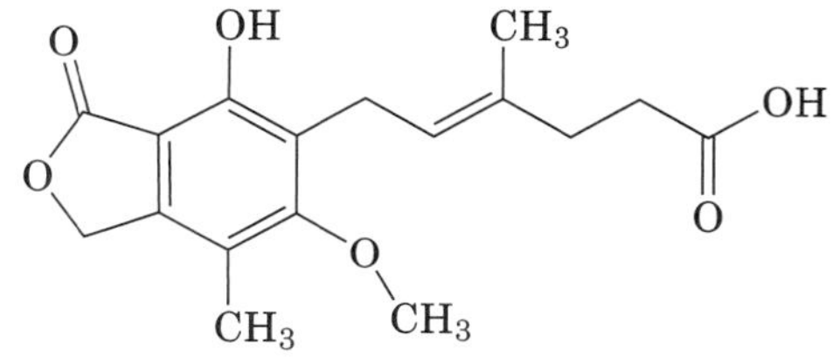

(56) mycophenolic acid

munosuppressant (mycophenolate mofetil, CellCept). The use of this drug is hampered by gastrointestinal side effects probably related to the metabolism of the drug. A second class of IMPDH inhibitors is represented by the nucleoside analog mizoribine (also known as bredinin), a prodrug approved for human use in Japan. Such compounds competitively inhibit IMPDH *in vivo* after phosphorylation (128). These drugs validate the strategy of targeting IMPDH for the discovery of immunosuppres-

sants. Other utilities that have been suggested for IMPDH inhibitors are antiviral and anticancer therapies.

The structure of hamster IMPDH in complex with IMP and (56) was solved at Vertex in the mid-1990s (129). This allowed the visualization of a covalent intermediate, in which a cysteine thiol from the enzyme adds to C2 of the purine ring of the nucleotide substrate. An analogous covalent adduct is postulated to be a key catalytic intermediate during normal turnover (130). The structure was a key tool in the discovery of (57) (VX-497, merimepodip), a novel potent inhibitor of human IMPDH suitable for oral administration (131).

An experimental screen of a diverse library of commercially available compounds for inhibitors of IMPDH identified molecules with the phenyl, phenyloxazole urea scaffold (58) as weak inhibitors. Through use of the compu-

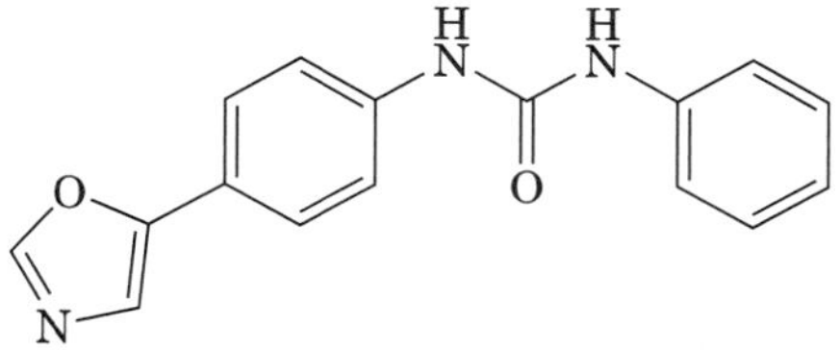

(58)

tational program DOCK (132), the initial inhibitors were built as models into the experimental structure of the crystalline complex of IMPDH, IMP, and (56). Structural analogs were generated to improve potency in an iterative process, guided by the structural modeling and the observed changes in potency for inhibition of human IMPDH.

After this process yielded compound (59), with nanomolar potency, an X-ray structure

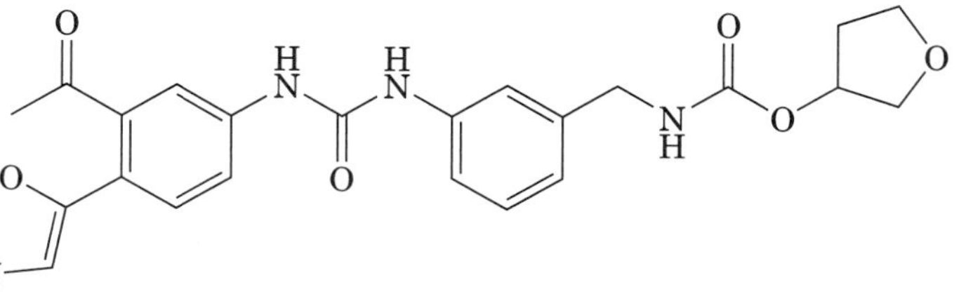

(57) merimepodip

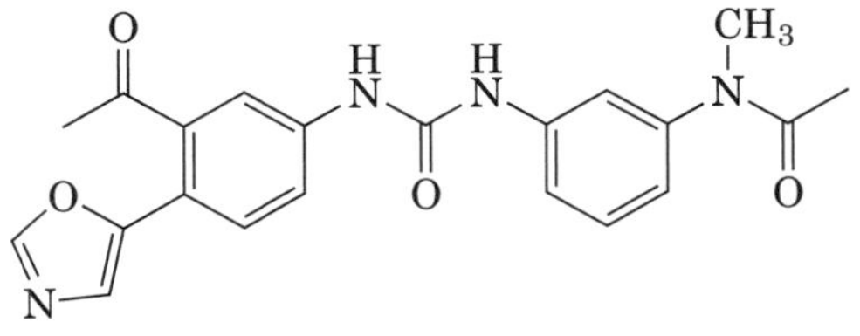

(59)

was determined of (**59**) bound to the hamster enzyme with IMP. This revealed both similarities and differences between the binding modes of (**56**) and (**59**). Aryl groups of both compounds pack against the covalently tethered purine of the nucleotide. Several hydrogen bonding and hydrophobic interactions with the enzyme are also common between the two inhibitors. However, there are several hydrophobic and van der Waals interactions seen in the complex with (**59**) that are not present with (**56**). Importantly, the urea moiety of (**59**) forms a network of hydrogen bonds with an aspartyl carboxylate that is not present in the complex with (**56**). Further modification of the structure was guided by the X-ray study by use of (**59**), to gain potency in a cell-based assay for inhibition of lymphocyte proliferation. This provided compound (**57**), which Vertex has advanced into clinical trials for treatment of hepatitis C infections.

2.5.2 Aldose Reductase. Aldose reductase has been implicated in many of the pathologies resulting from elevated tissue levels of glucose in diabetes mellitus (133, 134). This nicotinamide-dependent enzyme catalyzes the conversion of glucose to sorbitol, accumulation of which ultimately results in damage to the eyes, the nervous system, and the kidneys. Given the enormous damage caused by this disease and the difficulty in regulating blood glucose, selective and potent inhibitors of human aldose reductase offer great potential benefit. However, existing drugs that target aldose reductase have unreliable efficacy (135). For example, compound (**60**) (tolrestat) was withdrawn by Wyeth in 1996 because of poor clinical response. Hence, there is still a need to provide an inhibitor of this enzyme that fulfills the potential in the clinic. To minimize the risk of undesired toxicities, clinical

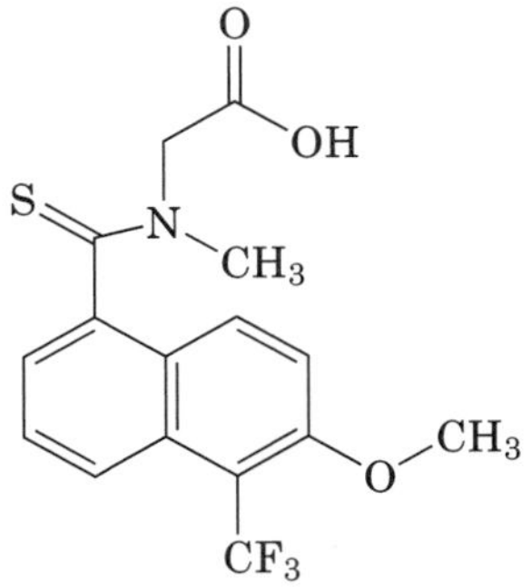

(60) tolrestat

agents that target aldose reductase should not inhibit the closely related aldehyde reductase, an essential hepatic enzyme.

The structure of (**60**) and other inhibitors bound to porcine aldose reductase (136) provided a rich lode of information on the requirements for potent and selective inhibition of aldose reductase. This was mined by scientists at the Institute for Diabetes Discovery, in a project that began in 1996. The Institute for Diabetes Discovery filed an IND application for (**61**) (lidorestat, IDD 676), a potent aldose

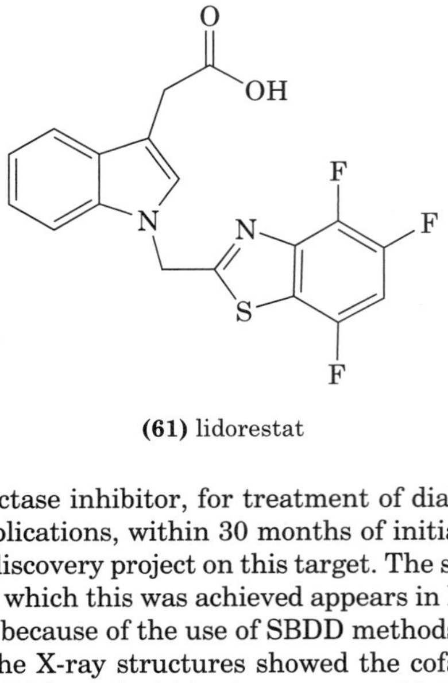

(61) lidorestat

reductase inhibitor, for treatment of diabetic complications, within 30 months of initiating the discovery project on this target. The speed with which this was achieved appears in large part because of the use of SBDD methods.

The X-ray structures showed the cofactor $NADP^+$ buried within the enzyme, with its C4 redox center exposed at the bottom of a deep hydrophobic cleft. An anionic binding site is located near $NADP^+$. Several potent inhibitors bind within the hydrophobic cleft and interact with the anionic site. The binding of potent inhibitors induces a conformational change, opening an adjacent hydrophobic pocket. The conformation induced by (**60**) differs from that caused by other, less selective inhibitors. This "specificity" pocket was thought to offer an opportunity for selective inhibition of aldose reductase while sparing aldehyde reductase. Hence, this structural study provided an initial pharmacophore for both potency and selectivity.

The SAR for this pharmacophore was developed with a series of synthetically accessible salicylic acid derivatives that were scored for potency and selectivity with the purified enzymes, and efficacy in a diabetic rat model (137). One of the most potent and selective of the derivatives was (**62**), containing the benz-

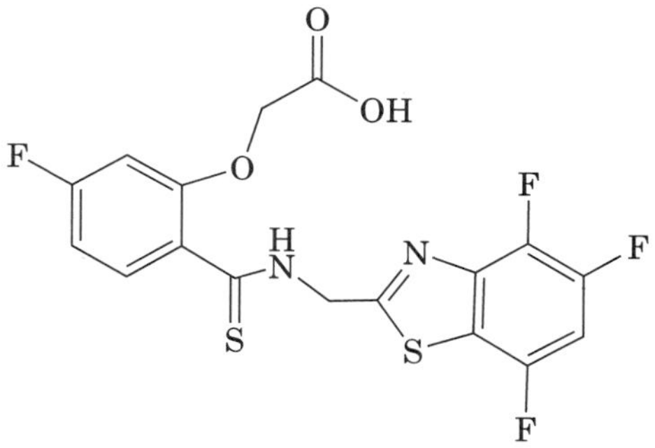

(**62**)

thiazole heterocycle. The SAR was employed, guided by the structures of selected inhibitor complexes, to design a novel indole scaffold to present the pharmacophoric elements (M. Van Zandt, personal communication). The optimization of this series provided the clinical candidate (**61**) (138).

2.6   Hydrolases

Some other hydrolytic enzymes, in addition to proteases, that are important drug targets include protein phophatases, phosphodiesterases, nucleoside hydrolases, acetylhydolases, glycosylases, and phospholipases. Structure-based inhibitor design is currently being applied to a number of these enzymes. The last three mentioned have been successfully tar-

geted in SBDD projects that have produced compounds that are either launched or in clinical trials.

2.6.1 Acetylcholinesterase. A pronounced decrease in the level of the neurotransmitter acetylcholine is one of the most pronounced changes in brain chemistry observed in the sufferers of Alzheimer's disease (139). Several drugs that are approved for the treatment of the dementia thought to result from this neurotransmitter deficit act by inhibiting acetylcholinesterase. These include (**63**) (tacrine, or

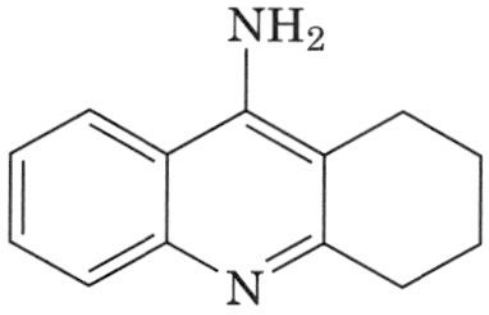

(**63**) tacrine

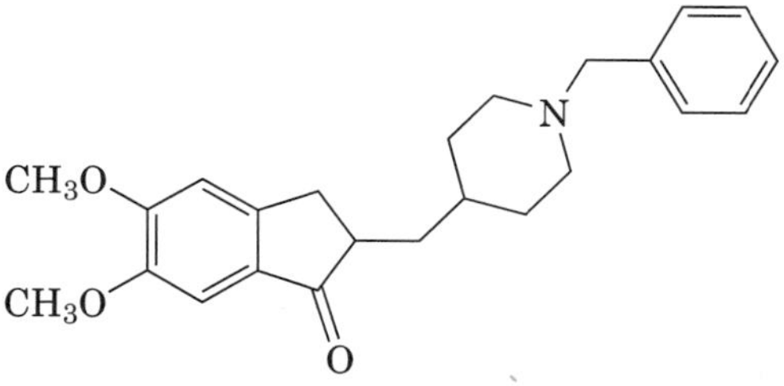

(**64**) donepezil

Cognex, a Pfizer drug that was the first such agent approved for this indication), (**64**) (donezepil), and (**65**) (rivastigmine). Several other agents are in clinical trials. Disappointing ef-

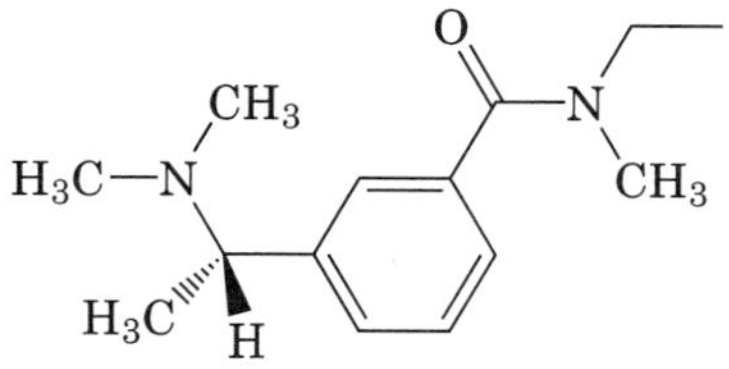

(**65**) rivastigmine

ficacy is observed with the existing drugs, arising from dose limitations that are likely attributable to the inhibition of acetylcholinesterase in peripheral tissues (140). This may be a consequence of the high serum levels required to get these highly cationic molecules to penetrate the blood–brain barrier.

In a discovery project that is reminiscent of the discovery of captopril, scientists at Takeda created a hypothetical structure for the active site of acetylcholinesterase, based on SAR from previous biochemical and medicinal chemical work (141). The model consisted of (in addition to the serine protease-like catalytic machinery) an anionic binding site separating two discrete hydrophobic binding sites. This model was then used to design inhibitors of the enzyme (reviewed in ref. 142). One set of analogs examined were based on the N-(ω-phthalimidylalkyl)-N-(ω-phenylalkyl)-amine (scaffold **66**). An iterative process of testing,

(66)

analysis, design, and synthesis, by use of this and closely related scaffolds, resulted in the production of (**67**) (TAK-147), which is cur-

(67) TAK-147

rently in clinical trials for treatment of the dementia resulting from Alzheimer's disease (142).

The design of (**66**) was partially based on the structures of previously known inhibitors. The two aryl substituents were intended to bind to the hydrophobic binding sites, placing the central amine cation into the anionic bind-

ing site. The length of both alkyl linkers was varied, and the effect of adding a third alkyl substituent was examined. The phthalimide portion of the structure was chosen to improve the synthetic accessibility of the analogs needed for this exercise. The compounds were tested not only for inhibitory potency toward rat cerebral acetylcholinesterase, but also for peripheral response and toxicity in dosed intact rats. After the work was under way, Sussman and coworkers solved the atomic structure of acetylcholinesterase from the electric eel, including complexes with several inhibitors, by X-ray crystallography (143). The availability of this structure made it possible to retrospectively analyze the basis for the SAR in this series of compounds, by use of DOCK (144).

2.6.2 Neuraminidase. Influenza virus infections cause severe human suffering throughout the world and economic damage in the billions of dollars annually, although some years are worse than others. In 1918 a pandemic caused by this disease killed an estimated 40 million people (145). An important protein in the infectious process is the viral neuraminidase, an integral membrane protein whose catalytic domain is exposed on the viral surface. Neuraminidase catalyzes the hydrolytic cleavage of sialic acid (**68**, N-acetylneuraminic acid) from glycoproteins and extracellular mucin on the surface of the host cell. A different viral surface protein tightly binds to terminal sialic acid residues, which promotes the initial infection, but prevents release of viral progeny from the host cells, unless and until the terminal sialic acids are hydrolytically cleaved by viral neuraminidase. Thus, neuraminidase enables the infection to propagate.

The first X-ray structure of influenza neuraminidase was determined in the early 1980s (146). Ten years later, a landmark paper (147) described a highly efficient drug design project at Monash University in Australia. This project yielded antiviral compound (**69**) (zanamivir, Relenza, or Flunet), which was developed into one of the first drugs to be created through use of SBDD. Previous structural work had revealed that the active site of neuraminidase has several rigid pockets and nu-

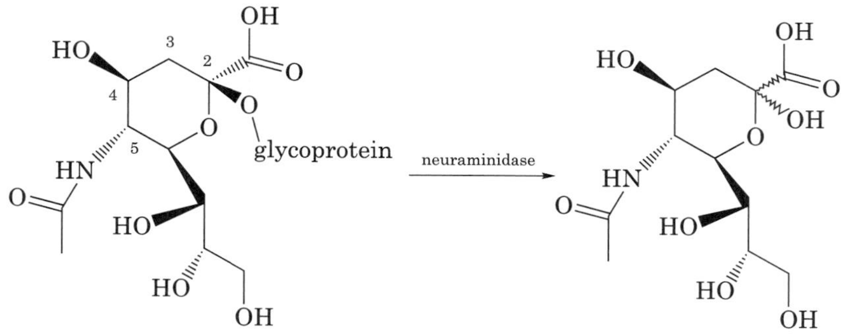

(68) sialic acid

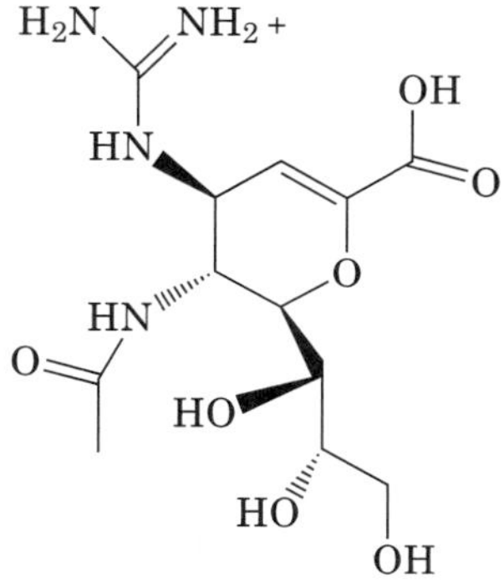

(69) zanamivir

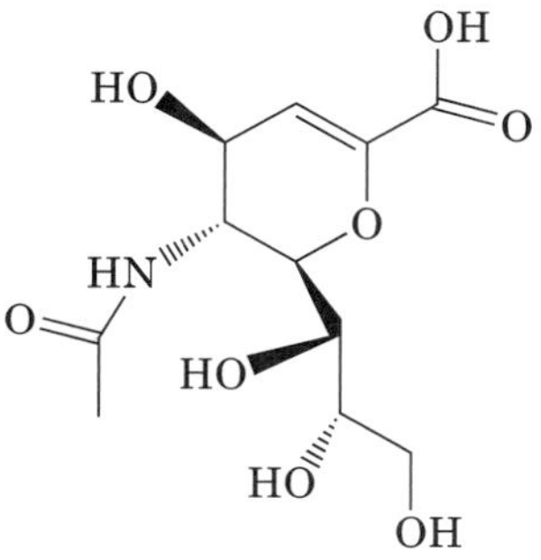

(70) Neu5Ac2en, DANA

merous charged groups. Electrostatic interactions significantly affect the conformation of bound sialic acid, which is deformed into a high energy conformer, attributed in part to the interactions between the 1-carboxylate and arginine side-chains of the protein. This deformation may play a key role in catalysis. Synthesis of a sialic acid analog that is dehydrated across the C2–C6 bond of (68) had provided the putative transition state mimic (70) (sometimes referred to as Neu5Ac2en, or as 2-deoxy-2,3-dehydro-N-acetylneuraminic acid, DANA).

Compound (70) inhibits neuraminidase with micromolar potency (148). Examination of the binding mode of (70) in the active site of neuraminidase (Fig. 10.16) led to the replacement of the 4-hydroxyl by cationic groups, first an amino and then a guanidino group (147). These groups strongly interact with anionic amino acid side chains (corresponding to Glu120 and Glu229 shown in Fig. 10.16) in the

neuraminidase active site. In the case of the guanidine substitution, the binding affinity for neuraminidase was increased about 5000-fold and provided (69), which inhibits viral release in cell cultures and decreases the severity of influenza virus infections in humans. Subsequently, the X-ray structures of neuraminidases from several different influenza subtypes complexed with (69) were analyzed (149). Although the positions of protein residues were well conserved, the water structure seen in these different complexes was quite variable. This may explain the varying potency of (69) against different strains of virus.

One problem with (69) is that it is not well absorbed by an oral route, and so must be administered as an aerosolized powder inhaled into the virus-infected lungs. Two other neuraminidase inhibitors with nanomolar affinities (71 and 72) have been developed through the use of SBDD methods to yield orally bioavailable drugs. The development of these agents was facilitated by the fortuitous discov-

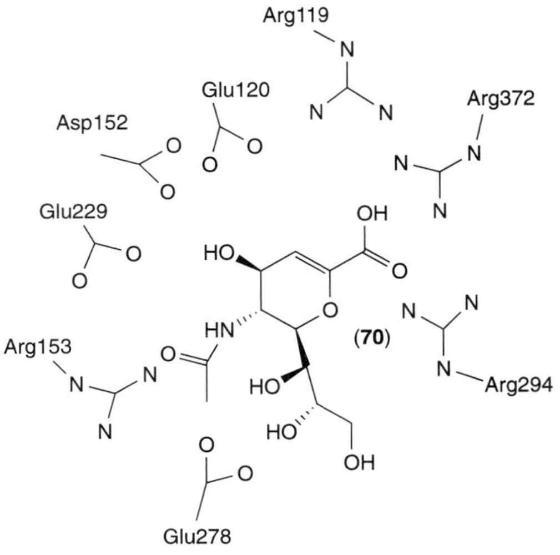

Figure 10.16. View from above: Polar amino acid side-chains surrounding (**70**), bound into the active site of influenza virus neuraminidase (Scheme 10.1 based on PDB code 1NNB, the coordinates of an X-ray structure described in Ref. 148).

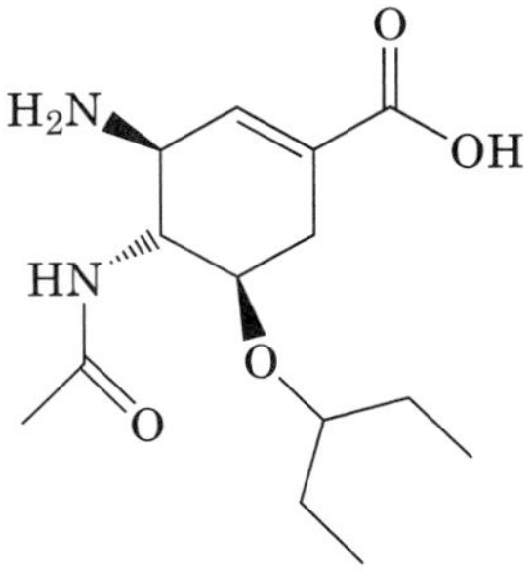

(**71**) GS 4071

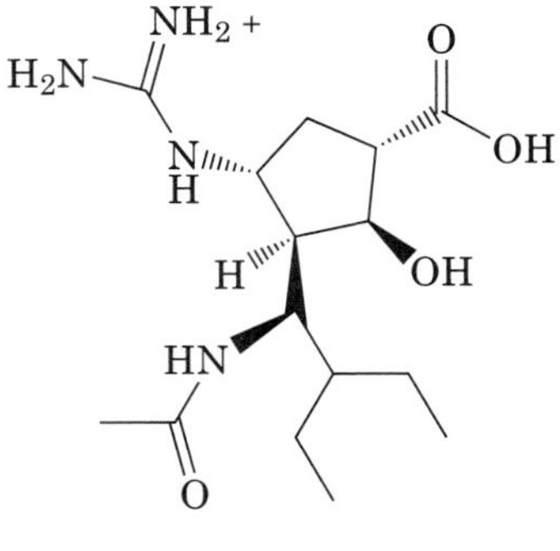

(**72**) BCX 1812

ery by scientists at Biocryst, that analogs of (**69**) in which the cyclic scaffold is a phenyl moiety are much more potent inhibitors if they lack the glycerol side chain! This was subsequently discovered by X-ray structural studies to be attributed to the creation of an unanticipated hydrophobic pocket upon rearrangement of the Glu278 side chain carboxylic acid, which forms several hydrogen bonds with the glycerol portion of (**69**) (Fig. 10.16).

Replacement of the permanently cationic guanidine by an amine (**71**) promoted better intestinal absorption, but also greatly decreased the affinity for neuraminidase. Structure-guided modification of the carbocycle's substituents was used to recover this lost potency. Compound (**71**) (GS 4071) was developed by Gilead Sciences (150). The ethyl ester of (**71**) is a prodrug (oseltamivir or GS 4104) that has been approved for oral dosing to treat influenza infection. Another amphiphilic carbocycle, compound (**72**) (peramivir, RWJ-270201, or BCX 1812) was developed by Bio-Cryst (151) through use of SBDD, and is in clinical trials. The use of clever synthetic routes, biochemical assays for neuraminidase inhibition, a mouse infection model, and X-ray structural information were all valuable tools in the development of both (**71**) and (**72**). Optimization of the affinity required the examination of a variety of alkyl substituents in both cases, to exploit the new hydrophobic pocket created by the conformational change primarily involving Glu278. The ability of the cyclopentyl ring in (**72**) to replace the six-membered ring illustrates that differing central scaffolds can display the essential interacting groups in an effective way.

2.6.3 Phospholipase A2 (Nonpancreatic, Secretory). Phospholipases A2 (PLA2s) are a diverse family of hydrolases that cleave the *sn-2* ester bond of phospholipids. The fatty acid produced is frequently arachidonate, the precursor to the proinflammatory eicosanoids. In several human inflammatory pathologies (e.g., septic shock, rheumatoid arthritis), a nonpancreatic secretory form of PLA2 (hnps-PLA2) is present in extracellular fluids at levels many-fold higher than normal (152). The design of bioavailable inhibitors of this Ca^{2+}-dependent isoform of PLA2 as inflammatory drugs is therefore an attractive goal (153). To be an effective drug, such an inhibitor would also need to be selective for hnps-PLA2 vs. the closely related pancreatic PLA2. Whether selectivity is needed against the quite different cytosolic PLA2 is unclear.

Investigators at an AstraZeneca laboratory (previously Fisons) have used multidimensional NMR and computational techniques to develop an active site model for cytosolic PLA2 (154, 155). Synthesis of compounds based on this model led to (73) (FPL-67047), reported

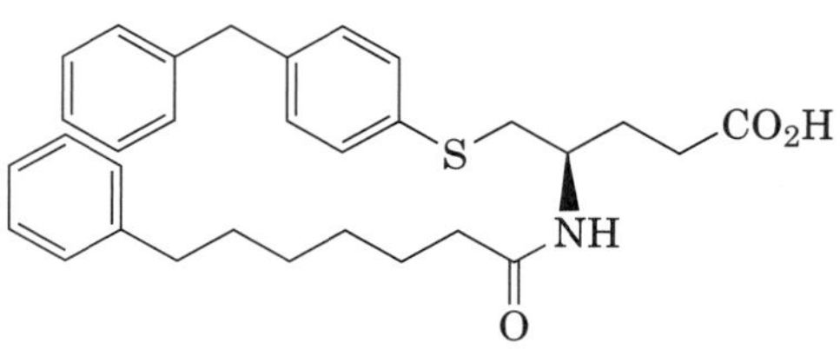

(73) FPL-67047

to be a development candidate for treatment of inflammation (156).

Investigators at Eli Lilly began a project to develop PLA2 inhibitors by investing the effort to clone, overproduce, purify, crystallize, and determine the structure of hnps-PLA2 (157). This also provided the reagent needed for a massive screening campaign to identify hnps-PLA2 inhibitors. They were thus prepared to apply SBDD methods when the screening of Lilly's small molecule collection yielded a weak inhibitor. The hit (74) was sur-

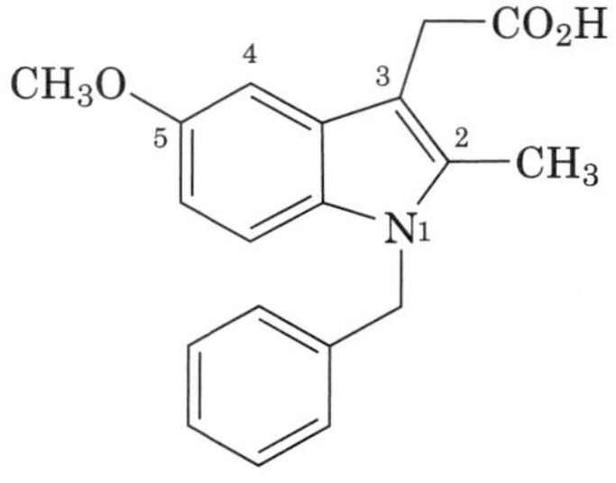

(74)

prisingly similar to indomethacin (75), a nonsteroidal anti-inflammatory drug that acts by inhibiting cyclooxygenase.

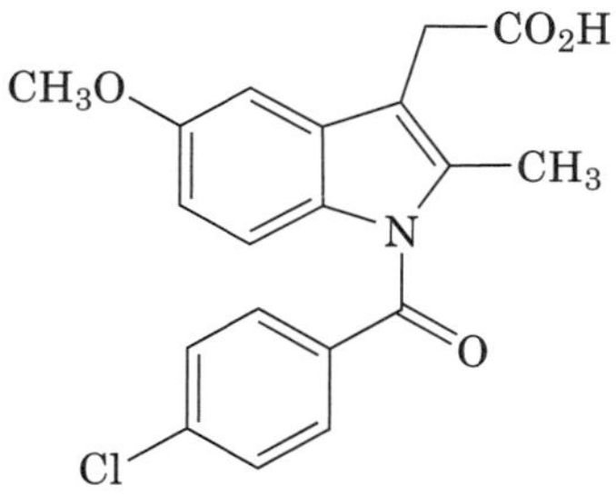

(75) indomethacin

The crystal structures of recombinant hnps-PLA2 bound to (74) and (75) were solved (158), and compared with the previously known structures of PLA2s complexed with substrate mimics (159, 160), including the phosphonate-containing transition state analog (76). The earlier structures revealed sev-

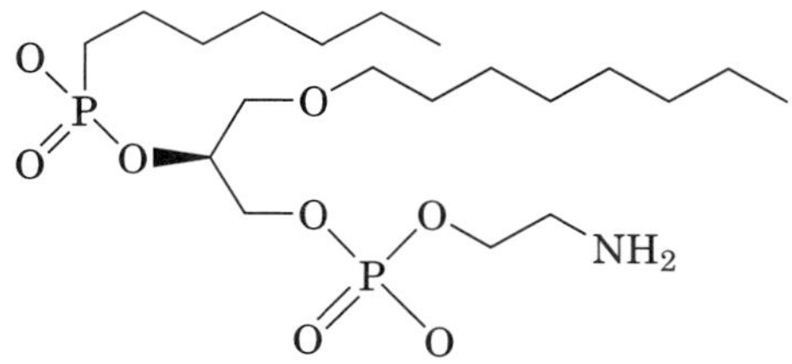

(76) hnps-PLA2 transition-state analog

eral key features. These were: (1) the filling of a significant hydrophobic crevice, (2) the displacement (by the sn-2 alkyl moiety) of the His6 side-chain into a solvent-exposed position to create an adjacent cavity, (3) the coordination of the active site calcium, and (4) formation of hydrogen bonds to His48 and Lys69. The polar contacts were provided by the nonbridging phosphate and phosphonate oxygens in the complex with (76).

The screening hit (74) bound in the hydrophobic crevice, similarly to the substrate mimics, with the 1-benzyl moiety of (74) bound in the adjacent cavity and displacing the enzyme's His6 imidazole. However, there were two surprising findings. First, despite the presence of 10 mM calcium in the crystallization liquor, there was no bound calcium, an essential active-site component, although weakly binding (K_d = 1.5 mM). Second, the carboxylic acid of (74) formed a hydrogen bond with another active-site acid, the side chain of Asp49. The latter finding again emphasizes the importance of experimental structures to guide improvements of inhibitor potency, given that placing two presumed anions so close together would likely never have been predicted by a computational model. Other slight conformational changes were observed to accommodate the 5-methoxy group of (**74**).

The inhibitor's 3-acetate moiety was converted to an acetamide in a successful attempt to restore the active site calcium, form a hydrogen bond to His48, and increase potency. The crystal structure of the complex with the amide version of (**74**) also revealed a significant reorientation of the indole core and 5-methoxy substituent, resulting in an unanticipated 5-Å movement of the terminal methyl. Further changes in inhibitor structure were guided by iterative structural studies and functional assays of potency and selectivity. These changes involved the use of substituents at positions 3 or 4 to optimize coordination of the metal ion, extension of the van der Waals interaction by lengthening the 2-methyl to an ethyl, and conversion of the 3-acetamide to glyoxamide (159, 161). This resulted in the synthesis of (**77**) (compound

(**77**) LY315920

LY315920), which has 6500-fold greater affinity for hnps-PLA2 than did the original hit molecule (**74**). LY315920 effectively inhibits hnps-PLA2 in the serum of transgenic mice dosed with the compound orally or i.v., and is undergoing clinical trials in the United States and Japan (162, 163).

2.7 Picornavirus Uncoating

Picornaviruses, which include the rhinoviruses and enteroviruses, are RNA viruses that cause several infectious human diseases. These diseases include common colds as well as life-threatening infections of the respiratory and central nervous systems. Effective treatments of these diseases would relieve much human suffering, save many lives, and have great economic benefit. There are over 100 serotypes of rhinoviruses alone, making it impossible to generate a vaccine effective against infections by all variants of the virus (164).

The Achilles heel of picornaviruses has been suggested to be that part of the virus structure that interacts with the cell surface receptor because those structural features must be well conserved (165). The virus particle consists of a positive-strand RNA coated by an icosahedral shell, containing 60 copies of four distinct β-barrel proteins (166). These structural proteins contain the binding site for the cellular receptor and undergo significant conformational changes to liberate the viral RNA genome during infection of the cell. A series of isoxazoles that inhibit this picornavirus "uncoating" process were discovered in the early 1980s by scientists at Sterling Winthrop, by use of an *in vitro* cellular assay for antiviral activity (167–170). One of these, compound (**78**) (WIN-51711, disoxaril), gave a 50% suppression of viral plaque formation in this assay at 0.3 μM. Compound (**78**) was also effective in animal models (171) and entered phase I clinical trials, but failed to advance because of its toxicity. Compound (**78**) was shown (172) to bind to viral capsid protein

(**78**) WIN-51711, disoxaril

(a)

(b)

Figure 10.17. Structure of rhinovirus capsid protein VP1 showing the bound conformation of antiviral isoxazole compounds (**78**) [disoxaril, WIN-51711: panel a, top], (**79**) [WIN-54954: panel b, middle], and (**80**) [pleconaril, WIN-63843: panel c, bottom]. The PDB codes for the X-ray structural model coordinates used to create these views are: 1PIV (for **78**), 2HWE (for **79**), and 1C8M (for **80**). On the left side of each panel, the inhibitors are shown as van der Waals surfaces, and the protein as a ribbon diagram. On the right side, the structures of the inhibitor alone are shown, from the same view, as ball and stick representations. See color insert.

(c)

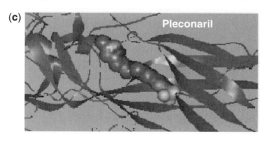

VP1, within a hydrophobic pocket in the floor of the "canyon" that contains the binding site for the cell surface receptor (Fig. 10.17A). Structural changes induced in the canyon floor upon binding of such molecules may also inhibit receptor binding directly (173). X-ray crystallographic studies of (**78**) and analogs bound to the target protein VP1 were an essential part of the iterative optimization process that led to safer and more effective antiviral agents (174–176). The goal of the process was to generate a compound that is potent, chemically and metabolically stable, and effective against as many serotypes of the virus as possible. There was therefore a need to bal-

ance potency and selectivity, and the structural information helped to guide compound design in pursuit of this balance.

A second-generation compound, (**79**) (WIN-54954) also advanced into clinical tests, but had disappointing efficacy in Phase II trials, probably because of extensive metabolism. Modification of the phenylisoxazole, guided by both structural and metabolic considerations (177), allowed the creation of a stable and potent antiviral, the third-generation compound (**80**) (WIN-63843, pleconaril, or Picovir) (178). This compound was evaluated in Phase III clinical trials and showed efficacy in humans. Oral dosing of virally infected patients with

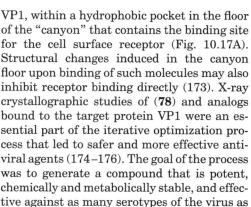

(79) WIN-54954

(80) VP63843, pleconaril

(**80**) three times daily decreased the average time needed to become free of cold symptoms from 10 days to between 8 and 9 days, and also reduced the duration of severe cold symptoms from 4.5 to 3.5 days (179). During the clinical studies to support the new drug application for (**80**), about a quarter of the clinical isolates (of rhinovirus present initially or during the treatment) were resistant to the compound. The majority of these resistant viruses had a single mutation at VP1 residue Ile98, which directly interacts with (**80**) bound to VP1 in wild-type virus. The clinical data also showed the elevation in some patients of hepatic cytochrome P450 levels during treatment with (**80**), raising concerns about potentially hazardous drug–drug interactions. ViroPharma sought and failed in early 2002 to gain the approval of the U.S. Food and Drug Administration for its new drug application for (**80**) for treatment of the common cold.

2.8 Phosphoryl Transferases

Protein kinases and phosphatases play vital roles in intracellular signaling pathways and in the integration and control of major cellular processes. Kinases and other phosphoryl group transferases are essential in the metabolism of lipids, nucleotides, and other small biomolecules. The use of SBDD methods on such targets has expanded as more of their X-ray structures have been solved, and will continue to grow as more targets are validated for their involvement in human diseases.

2.8.1 Mitogen-Activated Protein Kinase p38α. Mitogen-activated protein kinase (MAPK) p38α is a member of a family of Ser/Thr-specific protein kinases that are activated upon exposure of cells to mitogens such as bacterial lipopolysaccharide or environmental stresses such as exposure to UV irradiation or chemical

oxidants. MAPK p38α has a central role in integrating the inputs from a complex signaling network. Activation of MAPK p38α requires the dual phosphorylation of conserved threonine and tyrosine residues on a loop near the enzyme's active site (180). The unactivated (nonphosphorylated) enzyme has a very low affinity for ATP, but can bind to pyridinyl-imidazole inhibitors (181, 182). The activated enzyme in turn phosphorylates numerous substrates, including several transcription factors. This leads to activation of the transcription of many genes and causes the release of proinflammatory cytokines, primarily interleukin-1β (IL-1β) and tumor necrosis factor (TNFα). MAPK p38α was identified as a central player in this inflammatory pathway in a key study by scientists at SmithKline Beecham (183). The study involved the molecular cloning of the genes encoding proteins that bind to anti-inflammatory pyridinyl-imidazole compounds already known to block the biosynthesis of IL-1β and TNFα. The binding proteins turned out to be members of a known kinase family. Since this finding, the enzymes in the MAPK pathway, and especially MAPK p38α, have been attacked by many scientists seeking to discover anti-inflammatory drugs (184).

Compound (**81**) (SB 203580), a specific inhibitor of MAPK p38α, is a prototype for the pyridinyl-imidazole compounds (185). This compound is active in animal models of several inflammatory diseases (186), but was not itself pursued as a clinical candidate because of its inhibition of other enzymes, including hepatic cytochrome P450 reductases. The pyridinyl-imidazole compounds have dissociation constants for MAPK p38α in the nanomolar range, competing with ATP for binding to the enzyme. Because these compounds bind tightly to the unactivated enzyme, which has a

(81) SB203580

low affinity for ATP, they are able to compete effectively even *in vivo*, where the ATP concentration is in the millimolar range. The X-ray structures of (**81**) and several other pyridinyl-imidazole compounds in complexes with human MAPK p38α were solved in a collaborative effort between scientists at SmithKline Beecham and the University of Texas (187). Several X-ray structures of human MAPK p38α with and without bound inhibitors have also been solved by scientists at Vertex (181, 188).

The structures of the inhibited enzyme were useful in understanding what parts of the compounds were responsible for strong binding to MAPK p38α. As shown in Fig. 10.18, both hydrophobic and hydrogen bonding interactions are important components of

the inhibitor binding pocket. This structure suggested that Thr106 is an important structural determinant of the selective inhibition of MAPK p38α by the pyrimidyl-imidazoles, which have low affinity for other closely related kinases. Mutation of Thr106 results in the loss of sensitivity to these inhibitors, whereas the replacement of the corresponding residue in another kinase (ERK2) by threonine caused the mutated variant to become sensitive to these inhibitors (189, 190).

The X-ray structural models were also used at both SmithKline Beecham (later Glaxo-SmithKline) and Vertex to guide the design of new inhibitors. For example, both N1 of the central imidazole and the 2-(*para*-methylsulfonyl)-phenyl substituent in enzyme-bound (**81**) face a channel that opens to bulk solvent. This observation led to the design of (**82**) (VK19911) at Vertex (181) and (**83**) (SB242253) at GlaxoSmithKline (191). Compound (**83**) is fivefold more potent than (**81**) *in vivo*, in a mouse disease model, and was advanced into human clinical trials for treatment of rheumatoid arthritis (192). The piperidine on N1 of (**82**) and (**83**) was designed to form a salt bridge with Asp168. This interaction, and the preservation of other binding interactions, was directly demonstrated (181) for compound (**82**).

Analysis of the structural information from the X-ray models allowed the design at Vertex of a new scaffold for potent inhibition of MAPK p38α, as shown for compound (**84**) (VX-745). This design process, SBDD through

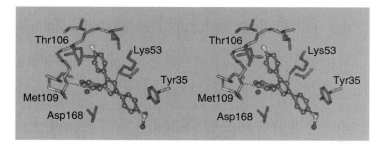

Figure 10.18. Binding of SB203580 (shown as a ball and stick structure) in the active site of MAPK p38α. In addition to the side chains of the labeled residues, the protein backbone between Leu104 and Met109 is shown, as well as several aliphatic side chains and a water molecule (red sphere). Hydrogen bonds (dotted lines) are shown between the backbone amide of Met109 and the inhibitor's pyrimidinyl nitrogen, and between the ε-amino of Lys53 and the inhibitor's imidazole N3. This figure is based on the PDB coordinate set 1A9U (187). See color insert.

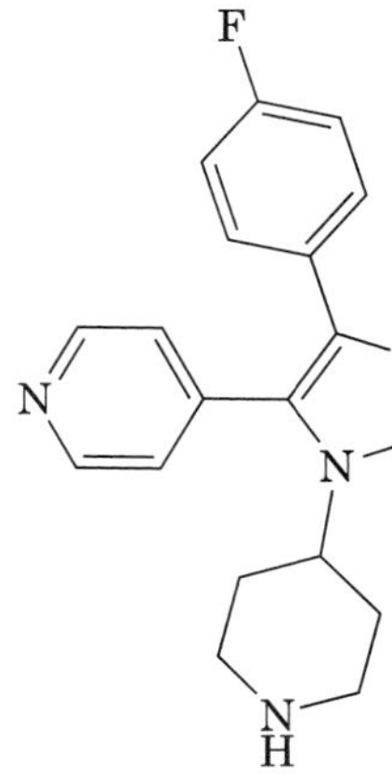

(82) VK19911

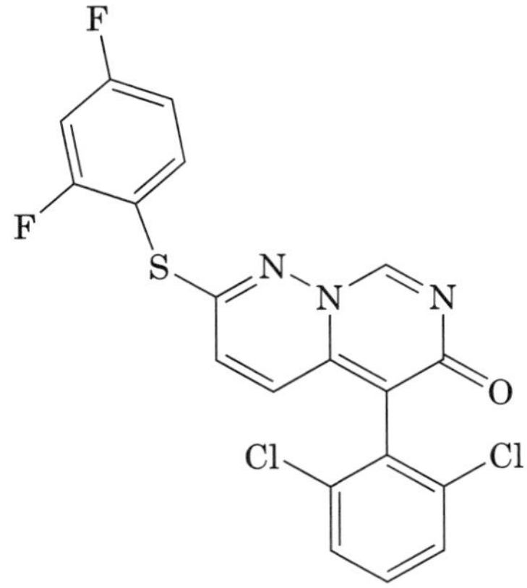

(84) VX-745

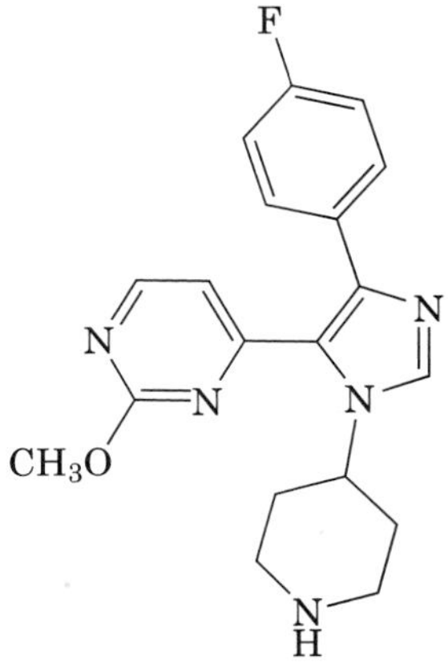

(83) SB242253

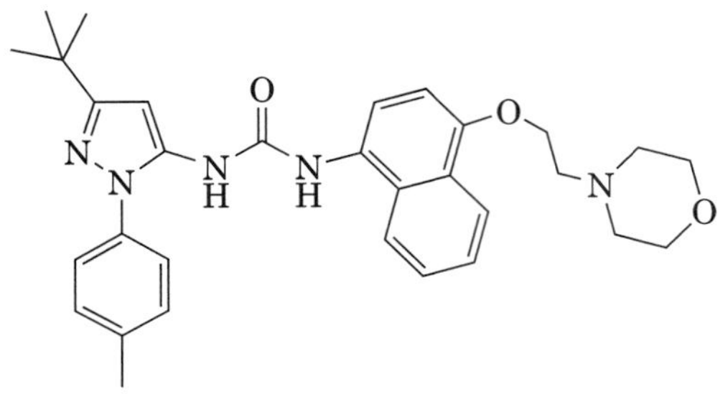

(85) BIRB-796

use of a crystal structure of MAPK p38α to design potent inhibitors with potential utility as human therapeutics, is the subject of an international patent application by Vertex, published in 2000 (193). The binding mode for (84) has not been disclosed, but the compound was advanced into clinical trials (194). Vertex has since discontinued the clinical trials of (84) because of the potential for toxicity, based on animal data, but in mid-2002 Vertex began a phase I clinical study of a new compound targeted against MAPK p38α.

Scientists at Boehringer Ingelheim recently described (195, 196) their discovery of an orally active inhibitor of MAPK p38α, compound (85) (BIRB-796), that is very different

from earlier inhibitors. This compound, whose K_d for MAPK p38α is 100 picomolar, has entered phase II clinical trials for treatment of rheumatoid arthritis. The lead compound that led to compound (85) was a diaryl urea originally identified by high throughput screening. X-ray structural studies revealed novel modes of binding for both the lead compound and (85) in the active site of MAPK p38α. Their binding sites are adjacent to the active site but do not directly overlap with that of ATP; rather, their binding mode changes the conformation of MAPK p38α such that ATP cannot bind. The optimization of the lead compound to clinical candidate (85) was an iterative process using clever synthetic chemical design, biochemical assays for affinity, X-ray crystallographic studies of key complexes, and cell-based and animal models. The development of (85) as a MAPK p38α inhibitor with efficacy *in vivo* makes it evident that there are multiple ways to effectively inhibit this enzyme.

2.8.2 Purine Nucleoside Phosphorylase. Purine nucleoside phosphorylase (PNP) catalyzes the reversible phosphorolysis of purine nucleosides to the purine base and ribosyl or 2-deoxyribosyl-α-1-phosphate.

The vital role of PNP in the proliferation of T-cells is evident from the fact that people with an inherited deficit in this activity have 30- to 100-fold lower numbers of T-lymphocytes than normal (197). The accumulation of dGTP and the resulting inhibition of ribonucleotide reductase in PNP-deficient T-cells causes the suppression of T-cell proliferation. B-lymphocytes are unaffected. Hence, small molecule inhibitors of PNP could be used to treat T-cell lymphomas and other T-cell–mediated diseases such as psoriasis. Adjunct therapy with PNP inhibitors could also block the catabolism of therapeutically useful nucleoside analogs.

Human PNP is a homotrimer of 32-kDa subunits. The X-ray structures of the apoenzyme and some substrate analog complexes were described in 1990. Each of the three identical active sites, located near the subunit interfaces, are composed primarily of residues from one subunit, with Phe159 participating in the active site of the adjacent subunit (198).

Scientists at BioCryst, CIBA–Geigy, Southern Research Institute, and the University of Alabama collaborated to design inhibitors of human PNP (199, 200). The project used an iterative process, in which new compound design was guided by synthetic considerations, computer graphics analysis of X-ray structural models, computational (Monte Carlo and energy minimization) methods, and the inhibitory potency of the compounds against PNP *in vitro*. Evaluation of the most potent inhibitors by use of cell-based assays, followed by pharmacokinetic and pharmacological characterization of several inhibitors in animal models, led to the choice of (**86**) for advancement into clinical trials. Compound (**86**) (BCX-34, peldesine) is being evaluated for treatment of psoriasis and skin cancer (201, 202).

(**86**) peldesine

In the SBDD project that produced (**86**), the design work was initiated through use of the X-ray structure of the PNP apoenzyme, but was more successful when the structures of the PNP–guanine complex and other complexes were available (199). The PNP–guanine crystal structure showed no important interactions with N9, and indicated a potential for hydrophobic interaction in the vicinity of the substrate ribose (Fig. 10.19). To test this, the 9-deaza compound (**87**) was synthe-

(**88**)

(**87**)

sized. This was a weak PNP inhibitor (measured $IC_{50} \sim 1 \mu M$).

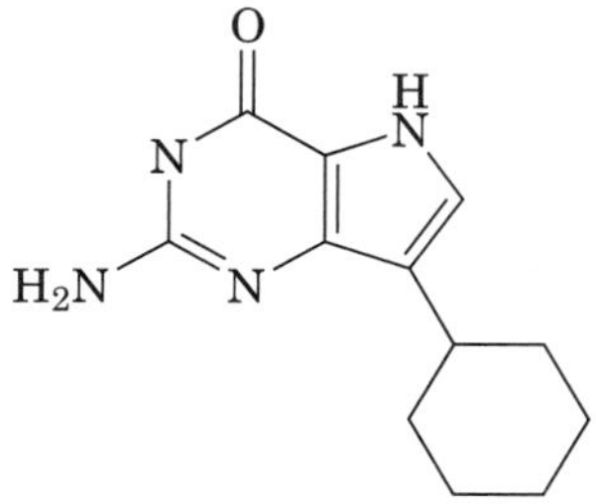

The X-ray structure of the complex between PNP and (**87**) showed that the hydrophobic interaction dominated the binding mode, and resulted in the disruption of the hydrogen bonding interactions seen in the guanine complex (i.e., Fig. 10.19). To increase the spacing between the hydrophobe and the purine mimetic, compounds (**88**) and (**89**)

(**89**)

were made. These had affinities for PNP in the nanomolar range. X-ray crystallographic analysis indicated new hydrogen bonding interactions with these 9-deaza compounds (shown for **89** in Fig. 10.20), made possible because N7 is protonated.

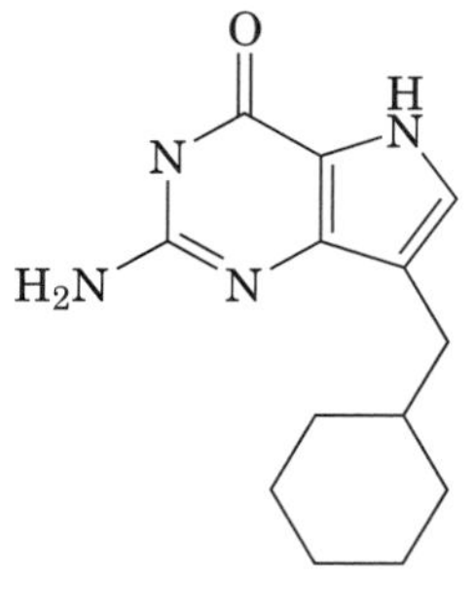

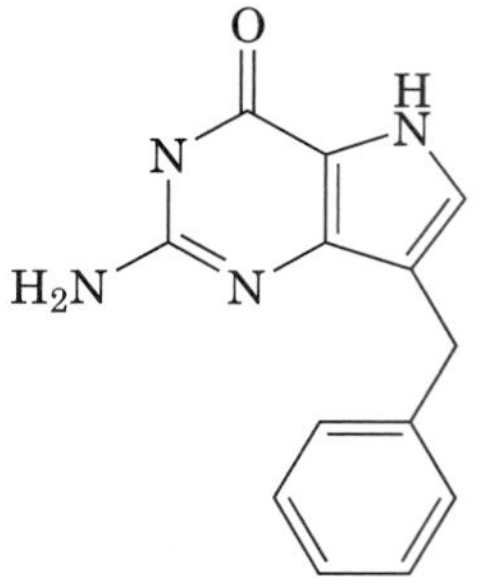

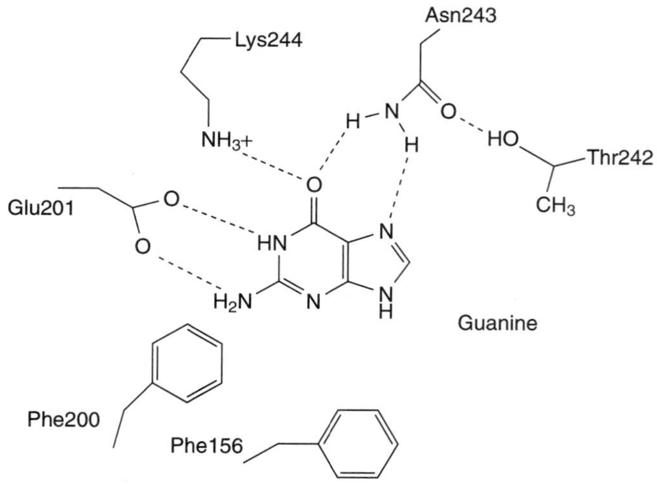

Figure 10.19. Binding interactions in the active site for the complex between guanine and PNP.

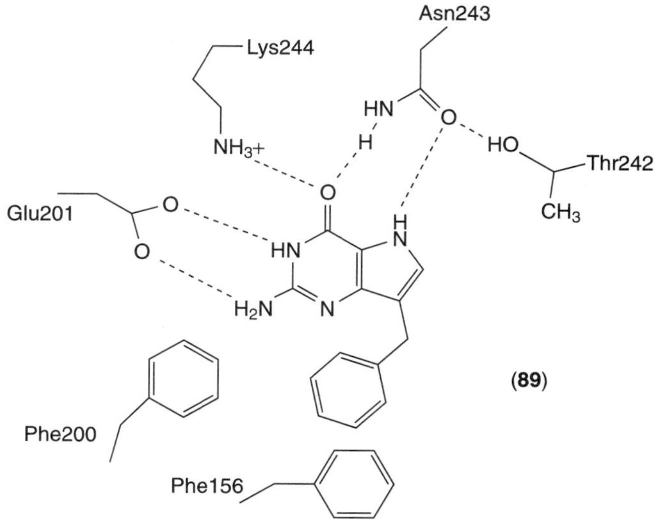

Figure 10.20. Binding interactions in the active site for the complex between PNP and (**89**).

While this work was under way, a Phase I clinical trial was undertaken of PNP inhibitor (**90**) (PD-119229), which was developed by

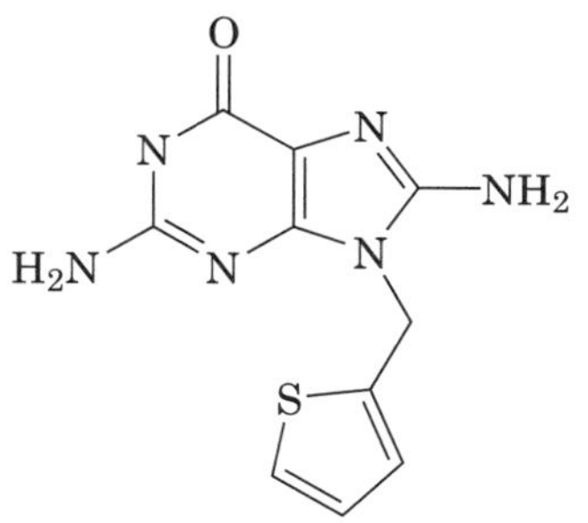

(**90**)

other workers. This led to an exploration of a series of 8-amino, 9-deaza derivatives, although the hydrogen bonding for the simpler 9-deaza compounds turned out to be superior (reviewed in Ref. 202). It may be that compounds such as (**90**) suffer from unfavorable steric interactions between the 8-amino group and the proximal methyl group of Thr242 of the enzyme, or that the energetic cost of dehydrating the 8-amino group cannot be fully repaid by interactions with the enzyme. Other chemical series were also explored, but compound (**86**) had an acceptable safety profile

and superior solubility and pharmacokinetic properties, and so was advanced into human testing.

2.9 Conclusions and Lessons Learned

The projects in which SBDD has been applied to enable the discovery of new drugs and clinical candidates have provided significant lessons for future investigators. Some of these lessons learned are summarized here. Much of the credit for the summary presented here belongs to Michael Varney of Agouron, who provided a copy of a presentation that he made in 1998 to a medicinal chemistry symposium concerning the lessons learned in 10 years of use of SBDD methods.

Experience Matters. In every aspect of SBDD, as in all technical fields, there is no substitute for experience. Given the variety of different techniques that must be incorporated, this means that experience from several different people will be needed for optimal function of a discovery project team. Essential expertise is needed in X-ray crystallographic studies, graphical display of experimental results, initial and iterative design of compounds and synthetic tactics, the creation of databases and database queries, and the analyses of search outputs and of the results of computational simulation experiments.

Combine and Integrate Technologies. Dedicated molecular biology and protein chemistry personnel and equipment are essential for identifying the right constructs for crystallization and to the assurance of a steady supply of protein. Synthetic chemists trained in graphical analysis of protein structures tend to be excellent designers, and will be unlikely to design molecules that they cannot make. Early tactical integration of the synthetic approaches is even more important if combinatorial chemistry is part of the program. The structural information can be used to design combinatorial libraries as effectively as it can to design molecules one at a time. The use of libraries can compensate for the inaccuracies inherent in current computational scoring algorithms. More significantly, the integration of orthogonal technologies will stimulate creative thought and yield much more than the sum of the different technologies applied separately.

Go Big Early and Often. Filling active site space as much as possible will maximize the chance that a compound will be a potent inhibitor. During compound design, it should also be recognized that proteins are flexible, and that accessible conformations are hard to predict. Sometimes, larger functionality can be accommodated than the existing structural model permits. A few compounds should be included to probe this. These may give rise to an unexpected boon, such as access to a significantly altered new protein conformation with novel sites that can be exploited in new rounds of design and synthesis.

Aqueous Solubility is Critical to Success. Both early in SBDD and later on in clinical development, sufficient aqueous solubility is critical. Solubility is important early because the concentrations of compounds must be high during crystallization experiments to saturate the high levels of protein. The ratio of the solubility to the inhibition constant of a compound is also critical to the success of the crystallization experiment. Once some structural information becomes available, both parameters can be manipulated, but usually, soluble inhibitors must be available before the availability of structural information. Solubility matters during animal testing and later in

development because compounds with very low solubility have limited or variable bioavailability.

Binding Sites Can Be Filled Many Ways. More than one small molecule scaffold can provide the necessary and sufficient hydrophobic and polar complementarity to generate potent inhibition. Sometimes, there are many scaffolds that will work. However, the structures of complexes with all the different scaffolds will likely have common features that are distinct from the structure of the apo enzyme, attributed to large-scale conformational changes that occur upon binding any ligand. The most useful X-ray models to use for the design of new compounds will be those that already have some substrate or inhibitor bound. There are several ways to design these compounds: modification of existing inhibitors, *de novo* creation of novel inhibitors, or some combination of these methods.

Not All Inhibitors Are Drugs. Having the X-ray structure of the target protein, or even having used the solved structure to design a potent inhibitor, is only the beginning of solving the difficult problems of drug design. The use of structure to create potent inhibitors can certainly shorten the time to get compounds into human testing, but use of SBDD methods does not guarantee that a potent compound will become a drug. This is an old lesson, actually, but is forgotten at great cost.

Structure of Free Inhibitor Is Important. Desolvation of the free ligand and of the protein's active-site groups upon complex formation are both significant. Both enthalpic and entropic contributions to the binding energy must be considered. Particular attention should be paid to the advantage that can be gained from "preorganization" of the inhibitor before binding, that is, low energy conformers bind with greater apparent avidity.

Bound Water Is Special, But Not All Hydrogen Bonds Are Created Equal. Each of the tightly bound waters present in an X-ray structural model has a unique environment and a unique function. In some cases, liberation of a bound water molecule by displacing it with an inhibitor's functionality can greatly increase inhibitor affinity, although this is not globally applicable. The entropic advantage of releasing a bound water into bulk solvent does not always exceed the enthalpic cost of the displacement. In many situations, the preferred solution will be to retain a water molecule and use it to maximize inhibitor binding. For example, a water molecule that donates two hydrogen bonds and accepts one cannot be isosterically replaced. Electrostatic interactions that are more complex than hydrogen bonds and simple ion pairs are very difficult to model, anticipate, and exploit in inhibitor design.

Retain Potency While Addressing Other Issues. Structural information can be very useful in designing compounds that are not part of a competitor's intellectual property, or that cannot be patented because of information in the public domain. Redesign of a compound that is not itself proprietary, by use of structural information obtained with that compound, can yield valuable new proprietary molecules. Structural information can also guide the modification of physicochemical, metabolic, or pharmacological properties or target selectivity without compromising the potency against the primary therapeutic target.

All Models Are Wrong; Some Are Useful. At present, it is impossible to calculate an accurate value for a binding constant on an absolute scale. However, accurately estimating the relative binding of a series of closely related compounds is possible, and is much more likely to be successful if X-ray structures of target complexes with some of the compounds are available. Thus, although there is much room for improvement, local computational models can sometimes be quite useful. Even in the absence of an experimentally determined X-ray structure of the target, a hypothetical model can be a powerful tool for the design of useful compounds (e.g., captopril and TAK-147).

Iterative SBDD Cycles Are Optimal. Small alterations in ligand structure often cause major changes in binding mode, protein conformation, or both. These changes can go undetected if the structural effects are not analyzed by X-ray analysis iteratively or too infrequently. This can yield confusing or misleading structure-activity relationships, leading to a waste of precious time. Moreover, changes in compound structure seldom affect only one

variable, so multiple orthogonal methods should be used to assess the effects of changes. It is also important during the rational design process to include room for serendipity. Do not reject an idea for a new compound that seems to make intuitive sense based on a single crystal structure or computational calculation.

REFERENCES

1. D. J. Abraham, *Intra-Sci. Chem. Rep.*, **8**, 4 (1974).
2. http://www.agouron.com/
3. http://www.stromix.com/
4. http://www.astex-technology.com
5. http://www.accelrys.com/consortia/htc/
6. P. J. Goodford, *J. Med. Chem.*, **27**, 557 (1984).
7. C. R. Beddell, Ed., *The Design of Drugs to Macromolecular Targets*, John Wiley & Sons, Chichester, UK, 1992.
8. J. Greer, J. W. Erickson, J. J. Baldwin, and M. D. Varney, *J. Med. Chem.*, **37**, 1035–1054 (1994).
9. R. E. Babine and S. L. Bender, *Chem. Rev.*, **97**, 1359 (1997).
10. P. Veerapandian, Ed., *Structure-Based Drug Design*, Marcel Dekker, New York, 1997.
11. R. T. Borchardt, R. M. Freidinger, T. K. Sawyer, and P. L. Smith, Eds., *Integration of Pharmaceutical Discovery and Development. Case Histories* (Pharmaceutical Biotechnology, Band 11), Plenum Press, New York, 1998.
12. K. Gubernator and H.-J. Böhm, Eds., *Structure-Based Ligand Design*, Wiley-VCH, New York/Weinheim, 1998.
13. C. L. Nobbs, H. C. Watson, and J. C. Kendrew, *Nature*, **209**, 339 (1966).
14. M. F. Perutz, *Nature*, **228**, 726 (1970).
15. R. C. Ladner, E. J. Heidner, and M. F. Perutz, *J. Mol. Biol.*, **114**, 385 (1977).
16. G. Fermi, M. F. Perutz, B. Shaanan, and R. Fourme, *J. Mol. Biol.*, **175**, 159 (1984).
17. B. C. Wishner, K. B. Ward, E. E. Lattman, and W. E. Love, *J. Mol. Biol.*, **98**, 179 (1975).
18. D. J. Harrington, K. Adachi, and W. E. Royer Jr., *J. Mol. Biol.*, **272**, 398 (1997).
19. C. R. Beddell, P. J. Goodford, G. Kneen, R. D. White, S. Wilkinson, and R. Wootton, *Br. J. Pharmacol.*, **82**, 397 (1984).
20. M. Merrett, D. K. Stammers, R. D. White, R. Wootton, and G. Kneen, *Biochem. J.*, **239**, 387 (1986).

21. F. C. Wireko and D. J. Abraham, *Proc. Natl. Acad. Sci. USA*, **88**, 2209 (1991).

22. D. J. Abraham, A. S. Mehanna, F. C. Wireko, E. P. Orringer, J. Whitney, and R. P. Thomas, *Blood*, **77**, 1334 (1991).

23. M. K. Safo, S. Nokuri, and D. J. Abraham, Unpublished results.

24. P. E. Kennedy, F. L. Williams, and D. J. Abraham, *J. Med. Chem.*, **27**, 103 (1984).

25. D. J. Abraham, M. F. Perutz, and S. E. V. Phillips, *Proc. Natl. Acad. Sci. USA*, **80**, 324 (1983).

26. M. F. Perutz, G. Fermi, D. J. Abraham, C. Poyart, and E. Bursaux, *J. Am. Chem. Soc.*, **108**, 1064 (1986).

27. E. P. Orringer, D. S. Blythe, J. A. Whitney, S. Brockenbrough, and D. J. Abraham, *Am. J. Hematol.*, **39**, 39 (1992).

28. D. J. Abraham, A. S. Mehanna, F. Williams, E. J. Cragoe Jr., and O. W. Woltersdorf Jr., *J. Med. Chem.*, **32**, 2460 (1989).

29. D. J. Abraham, P. E. Kennedy, A. S. Mehanna, D. Patwa, and F. L. Williams, *J. Med. Chem.*, **27**, 967 (1984).

30. A. Arnone, *Nature*, **237**, 146 (1972).

31. V. Richard, G. G. Dodson, and Y. Mauguen, *J. Mol. Biol.*, **233**, 270 (1993).

32. P. J. Goodford, J. St-Louis, and R. Wootton, *Br. J. Pharmacol.*, **68**, 741 (1980).

33. C. R. Beddell, P. J. Goodford, F. E. Norrington, S. Wilkinson, and R. Wootton, *Br. J. Pharmacol.*, **57**, 201 (1976).

34. F. F. Brown and P. J. Goodford, *Br. J. Pharmacol.*, **60**, 337 (1977).

35. A. S. Mehanna and D. J. Abraham, *Biochemistry*, **29**, 3944 (1990).

36. M. F. Perutz and C. Poyart, *Lancet*, **2**, 881 (1983).

37. I. Lalezari and P. Lalezari, *J. Med. Chem.*, **32**, 2352 (1989).

38. I. Lalezari, P. Lalezari, C. Poyart, M. Marden, J. Kister, B. Bohn, G. Fermi, and M. F. Perutz, *Biochemistry*, **29**, 1515 (1990).

39. D. J. Abraham, R. S. Randad, M. A. Mahran, and A. S. Mehanna, *J. Med. Chem.*, **34**, 752 (1991).

40. D. J. Abraham, F. C. Wireko, R. S. Randad, C. Poyart, J. Kister, B. Bohn, J. F. Leard, and M. P. Kunert, *Biochemistry*, **31**, 9141 (1992).

41. F. C. Wireko, G. E. Kellogg, and D. J. Abraham, *J. Med. Chem.*, **34**, 758 (1991).

42. D. J. Abraham, J. Kister, G. S. Joshi, M. C. Marden, and C. Poyart, *J. Mol. Biol.*, **248**, 845 (1995).

43. M. K. Safo, C. M. Moure, J. C. Burnett, G. S. Joshi, and D. J. Abraham, *Protein Sci.*, **10**, 951 (2001).

44. S. K. Burley and G. A. Petsko, *FEBS Lett.*, **201**, 751 (1986).

45. S. K. Burley and G. A. Petsko, *Science*, **229**, 23 (1985).

46. M. Levitt and M. F. Perutz, *J. Mol. Biol.*, **201**, 751 (1988).

47. D. J. Abraham, M. K. Safo, T. Boyiri, R. E. Danso-Danquah, J. Kister, and C. Poyart, *Biochemistry*, **34**, 15006 (1995).

48. M. P. Grella, R. Danso-Danquah, M. K. Safo, G. S. Joshi, J. Kister, S. J. Hoffman, M. Marden, and D. J. Abraham, *J. Med. Chem.*, **25**, 4726 (2001).

49. A. M. Youssef, M. K. Safo, R. Danso-Danquah, G. S. Joshi, J. Kister, M. Marden, and D. J. Abraham, *J. Med. Chem.*, **45**, 1184 (2002).

50. J. A. Walder, R. H. Zaugg, R. Y. Walder, J. M. Steele, and I. M. Klotz, *Biochemistry*, **18**, 4265 (1979).

51. R. Chatterjee, E. V. Welty, R. Y. Walder, S. L. Pruitt, P. H. Rogers, A. Arnone, and J. A. Walder, *J. Biol. Chem.*, **261**, 9929 (1986).

52. S. R. Snyder, E. V. Welty, R. Y. Walder, L. A. Williams, and J. A. Walder, *Proc. Natl. Acad. Sci. USA*, **84**, 7280 (1987).

53. N. Komiyama, J. Tame, and K. Nagai, *Biol. Chem.*, **377**, 543 (1996).

54. T. Boyiri, M. K. Safo, R. E. Danso-Danquah, J. Kister, C. Poyart, and D. J. Abraham, *Biochemistry*, **34**, 15021 (1995).

55. M. F. Perutz, *Br. Med. Bull.*, **32**, 195 (1976).

56. J. Monod, J. Wyman, and J.-P. Changeux, *J. Mol. Biol.*, **12**, 88 (1965).

57. D. A. Matthews, R. A. Alden, J. T. Bolin, S. T. Freer, R. Hamlin, N. Xuong, J. Kraut, M. Poe, M. Williams, and K. Hoogsteen, *Science*, **197**, 452 (1977).

58. L. F. Kuyper, B. Roth, D. P. Baccanari, R. Ferone, C. R. Beddell, J. N. Champness, D. K. Stammers, J. G. Dann, F. E. Norrington, D. J. Baker, and P. J. Goodford, *J. Med. Chem.*, **25**, 1120 (1982).

59. D. A. Matthews, J. T. Bolin, J. M. Burridge, D. J. Filman, K. W. Volz, and J. Kraut, *J. Biol. Chem.*, **260**, 392 (1985).

60. K. Appelt, R. J. Bacquet, C. A. Bartlett, C. L. J. Booth, S. T. Freer, M. A. Fuhry, M. R. Gehring, S. M. Herrmann, E. F. Howland, C. A. Janson, T. R. Jones, C. C. Kan, V. Kathardekar, K. K. Lewis, G. P. Marzoni, D. A. Matthews, C.

Mohr, E. W. Moomaw, C. A. Morse, S. J. Oatley, R. C. Ogden, M. R. Reddy, S. H. Reich, W. S. Schoettlin, W. W. Smith, M. D. Varney, J. E. Villafranca, R. W. Ward, S. Webber, S. E. Webber, K. M. Welsh, and J. White, *J. Med. Chem.*, **34**, 1925 (1991).

61. S. H. Reich and S. E. Webber, *Perspect. Drug Discov. Des.*, **1**, 371–390 (1993).

62. L. W. Hardy, J. S. Finer-Moore, W. R. Montfort, M. O. Jones, D. V. Santi, and R. M. Stroud, *Science*, **235**, 448–455 (1987).

63. D. A. Matthews, K. Appelt, S. J. Oakley, and N. H. Xuong, *J. Mol. Biol.*, **214**, 923–936 (1990).

64. W. R. Montfort, K. M. Perry, E. B. Fauman, J. S. Finer-Moore, G. F. Maley, L. Hardy, F. Maley, and R. M. Stroud, *Biochemistry*, **29**, 6964–6977 (1990).

65. Y. Takemura and A. L. Jackman, *Anticancer Drugs*, **8**, 3–16 (1997).

66. S. E. Webber, T. M. Bleckman, J. Attard, J. G. Deal, V. Kathardekar, K. M. Welsh, S. Webber, C. A. Janson, D. A. Matthews, W. W. Smith, S. T. Freer, S. R. Jordan, R. J. Bacquet, E. F. Howland, C. L. J. Booth, R. W. Ward, S. M. Herrmann, J. White, C. A. Morse, J. A. Hilliard, and C. A. Bartlett, *J. Med. Chem.*, **36**, 733–746 (1993).

67. I. Niculescu-Duvaz, *Curr. Opin. Invest. Drugs*, **2**, 693–705 (2001).

68. P. J. Goodford, *J. Med. Chem.*, **28**, 849 (1985).

69. P. Goodford, *J. Chemom.*, **10**, 107 (1996).

70. M. D. Varney, G. P. Marzoni, C. L. Palmer, J. G. Deal, S Webber, K. M. Welsh, R. J. Bacquet, C. A. Bartlett, C. A. Morse, C. L. Booth, S. M. Herrmann, E. F. Howland, R. W. Ward, and J. White, *J. Med. Chem.*, **35**, 663–676 (1992).

71. D. R. Newell, *Semin. Oncol.*, **26** (Suppl. 6), 74–81 (1999).

72. P. Norman, *Curr. Opin. Invest. Drugs*, **2**, 1611–1622 (2001).

73. E. C. Taylor, *Adv. Exp. Med. Biol.*, **338**, 387–408 (1993).

74. G. P. Beardsley, B. A. Moroson, E. C. Taylor, and R. G. Moran, *J. Biol. Chem.*, **264**, 328–333 (1989).

75. J. R. Piper, G. S. McCaleb, J. A. Montgomery, R. L. Kisliuk, Y. Gaumont, J. Thorndike, and F. M. Sirotnak, *J. Med. Chem.*, **31**, 2164–2169 (1988).

76. S. E. Greasley, T. H. Marsilje, H. Cai, S. Baker, S. J. Benkovic, D. L. Boger, and I. A. Wilson, *Biochemistry*, **40**, 13538–13547 (2001).

77. R. J. Almassy, C. A. Janson, C. C. Kan, and Z. Hostomska, *Proc. Natl. Acad. Sci. USA*, **89**, 6114–6118 (1992).

78. C. C. Kan, M. R. Gehring, B. R. Nodes, C. A. Janson, R. J. Almassy, and Z. Hostomska, *J. Protein Chem.*, **11**, 467–473 (1992).

79. M. D. Varney, C. L. Palmer, W. H. Romines 3rd, T. Boritzki, S. A. Margosiak, R. Almassy, C. A. Janson, C. Bartlett, E. J. Howland, and R. Ferre, *J. Med. Chem.*, **40**, 2502–2524 (1997).

80. C. Shih, L. S. Gossett, J. F. Worzalla, S. M. Rinzel, G. B. Grindey, P. M. Harrington, and E. C. Taylor, *J. Med. Chem.*, **35**, 1109–1116 (1992).

81. D. W. Cushman, H. S. Cheung, E. F. Sabo, and M. A. Ondetti, *Biochemistry*, **16**, 5484 (1977).

82. M. A. Ondetti, B. Rubin, and D. W. Cushman, *Science*, **196**, 441 (1977).

83. M. J. Wyvratt and A. A. Patchett, *Med. Res. Rev.*, **5**, 483–531 (1985).

84. D. W. Cushman and M. A. Ondetti, *Hypertension*, **17**, 589 (1991).

85. D. W. Cushman and M. A. Ondetti, *Nat. Med.*, **5**, 1110 (1999).

86. J. Rahuel, V. Rasetti, J. Maibaum, H. Rueger, R. Goschke, N. C. Cohen, S. Stutz, F. Cumin, W. Fuhrer, J. M. Wood, and M. G. Grutter, *Chem. Biol.*, **7**, 493–504 (2000).

87. L. D. Byers and R. Wolfenden, *Biochemistry*, **12**, 2070–2078 (1973).

88. J. R. Huff and J. Kahn, *Adv. Protein Chem.*, **56**, 213–251 (2001).

89. A. Wlodawer and J. Vondrasek, *Annu. Rev. Biophys. Biomol. Struct.*, **27**, 249 (1998).

90. T. D. Meek, *J. Enzyme Inhib.*, **6**, 65 (1992).

91. A. Wlodawer and J. W. Erickson, *Annu. Rev. Biochem.*, **62**, 543 (1993).

92. R. Lapatto, T. Blundell, A. Hemmings, J. Overington, A. Wilderspin, S. Wood, J. R. Merson, P. J. Whittle, D. E. Danley, K. F. Geoghegan, et al., *Nature*, **342**, 299–302 (1989).

93. M. A. Navia, P. M. Fitzgerald, B. M. McKeever, C. T. Leu, J. C. Heimbach, W. K. Herber, I. S. Sigal, P. L. Darke, and J. P. Springer, *Nature*, **337**, 615–620 (1989).

94. A. Wlodawer, M. Miller, M. Jaskolski, B. K. Sathyanarayana, E. Baldwin, I. T. Weber, L. M. Selk, L. Clawson, J. Schneider, and S. B. Kent, *Science*, **245**, 616–621 (1989).

95. I. B. Duncan and S. Redshaw, *Infect. Dis. Ther.*, **25**, 27–47 (2002).

96. A. G. Tomasselli, M. K. Olsen, J. O. Hui, D. J. Staples, T. K. Sawyer, R. L. Heinrikson, and C. S. Tomich, *Biochemistry*, **29**, 264–269 (1990).

97. M. Jaskolski, A. G. Tomasselli, T. K. Sawyer, D. G. Staples, R. L. Heinrikson, J. Schneider, S. B. Kent, and A. Wlodawer, *Biochemistry*, **30**, 1600–1609 (1991).

98. M. W. Holladay, F. G. Salituro, and D. H. Rich, *J. Med. Chem.*, **30**, 374–383 (1987).

99. F. G. Salituro, N. Agarwal, T. Hofmann, and D. H. Rich, *J. Med. Chem.*, **30**, 286–295 (1987).

100. D. J. Kempf, K. C. Marsh, D. A. Paul, M. F. Knigge, D. W. Norbeck, W. E. Kohlbrenner, L. Codacovi, S. Vasavanonda, P. Bryant, X. C. Wang, N. E. Wideburg, J. J. Clement, J. J. Plattner, and J. Erickson, *Antimicrob. Agents Chemother.*, **35**, 2209–2214 (1991).

101. M. V. Hosur, N. T. Bhat, D. J. Kempf, E. T. Baldwin, B. Liu, S. Gulnik, N. E. Wideburg, D. W. Norbeck, K. Appelt, and J. W. Erickson, *J. Am. Chem. Soc.*, **116**, 847–855 (1994).

102. D. J. Kempf, H. L. Sham, K. C. Marsh, C. A. Flentge, D. Betebenner, B. E. Green, E. McDonald, S. Vasavanonda, A. Saldivar, N. E. Wideburg, W. M. Kati, L. Ruiz, C. Zhao, L. Fino, J. Patterson, A. Molla, J. J. Plattner, and D. W. Norbeck, *J. Med. Chem.*, **41**, 602–617 (1998).

103. C. N. Hodge, P. E. Aldrich, L. T. Bacheler, C. H. Chang, C. J. Eyermann, S. Garber, M. Grubb, D. A. Jackson, P. K. Jadhav, B. Korant, P. Y. Lam, M. B. Maurin, J. L. Meek, M. J. Otto, M. M. Rayner, C. Reid, T. R. Sharpe, L. Shum, D. L. Winslow, and S. Erickson-Viitanen, *Chem. Biol.*, **3**, 301–314 (1996).

104. B. D. Dorsey, R. B. Levin, S. L. McDaniel, J. P. Vacca, J. P. Guare, P. L. Darke, J. A. Zugay, E. A. Emini, W. A. Schleif, J. C. Quintero, J. H. Lin, I. W. Chen, M. K. Holloway, P. M. D. Fitzgerald, M. G. Axel, D. Ostovic, P. S. Anderson, and J. R. Huff, *J. Med. Chem.*, **37**, 3443–3451 (1994).

105. J. P. Vacca, J. P. Guare, S. J. DeSolms, W. M. Sanders, E. A. Giuliani, S. D. Young, P. L. Darke, I. S. Sigal, W. A. Schleif, J. C. Quintero, E. A. Emini, P. S. Anderson, and J. R. Huff, *J. Med. Chem.*, **34**, 1228–1230 (1991).

106. T. A. Lyle, C. M. Wiscount, J. P. Guare, W. J. Thompson, P. S. Anderson, P. L. Darke, J. A. Zugay, E. A. Emini, W. A. Schleif, J. C. Quintero, R. A. F. Dixon, I. S. Sigal, and J. R. Huff, *J. Med. Chem.*, **34**, 1230–1233 (1991).

107. M. K. Holloway, J. M. Wai, T. A. Halgren, P. M. Fitzgerald, J. P. Vacca, B. D. Dorsey, R. B. Levin, W. J. Thompson, L. J. Chen, S. J. deSolms, N. Gaffin, A. K. Ghosh, E. A. Giuliani, S. L. Graham, J. P. Guare, R. W. Hungate, T. A. Lyle, W. M. Sanders, T. J. Tucker, M. Wiggins, C. M. Wiscount, O. W. Woltersdorf, S. D. Young, P. L. Darke, and J. A. Zuguay, *J. Med. Chem.*, **38**, 305–317 (1995).

108. E. E. Kim, C. T. Baker, M. D. Dwyer, M. A. Murcko, B. G. Rao, R. D. Tung, and M. A. Navia, *J. Am. Chem. Soc.*, **117**, 1181–1182 (1995).

109. S. W. Kaldor, V. J. Kalish, J. F. Davies 2nd, B. V. Shetty, J. E. Fritz, K. Appelt, J. A. Burgess, K. M. Campanale, N. Y. Chirgadze, D. K. Clawson, B. A. Dressman, S. D. Hatch, D. A. Khalil, M. B. Kosa, P. P. Lubbehusen, M. A. Muesing, A. K. Patick, S. H. Reich, K. S. Su, and J. H. Tatlock, *J. Med. Chem.*, **40**, 3979–3985 (1997).

110. M. Moledina, M. Chakir, and P. J. Gandhi, *J. Thromb. Thrombolysis*, **12**, 141–149 (2001).

111. J. Hauptmann, *Eur. J. Clin. Pharmacol.*, **57**, 751–758 (2002).

112. D. W. Banner and P. Hadvary, *J. Biol. Chem.*, **266**, 20085–20093 (1991).

113. P. E. Sanderson and A. M. Naylor-Olsen, *Curr. Med. Chem.*, **5**, 289 (1998).

114. J. P. Vacca, *Curr. Opin. Chem. Biol.*, **4**, 394 (2000).

115. J. Hauptmann, B. Kaiser, M. Paintz, and F. Markwardt, *Biomed. Biochim. Acta*, **46**, 445–453 (1987).

116. H. Brandstetter, D. Turk, H. W. Hoeffken, D. Grosse, J. Sturzebecher, P. D. Martin, B. F. Edwards, and W. Bode, *J. Mol. Biol.*, **226**, 1085–1099 (1992).

117. N. H. Hauel, H. Nar, H. Priepke, U. Reis, J. M. Stassen, and W. Wienen, *J. Med. Chem.*, **45**, 1757–1766 (2002).

118. R. M. Friedlander, V. Gagliardini, H. Hara, K. B. Fink, W. Li, G. MacDonald, M. C. Fishman, A. H. Greenberg, M. A. Moskowitz, and J. Yuan, *J. Exp. Med.*, **185**, 933–940 (1997).

119. B. Siegmund, H. A. Lehr, G. Fantuzzi, and C. A. Dinarello, *Proc. Natl. Acad. Sci. USA*, **98**, 13249–13254 (2001).

120. N. P. Walker, R. V. Talanian, K. D. Brady, L. C. Dang, N. J. Bump, C. R. Ferenz, S. Franklin, T. Ghayur, M. C. Hackett, L. D. Hammill, L. Herzog, M. Hugunin, W. Houy, J. A. Mankovich, L. McGuiness, E. Orlewicz, M. Paskind, C. A. Pratt, P. Reis, A. Summani, M. Terranova, J. P. Welch, L. Xiong, A. Moller, D. E. Tracey, R. Kamen, and W. W. Wong, *Cell*, **78**, 343–352 (1994).

121. K. P. Wilson, J. A. Black, J. A. Thomson, E. E. Kim, J. P. Griffith, M. A. Navia, M. A. Murcko, S. P. Chambers, R. A. Aldape, S. A. Raybuck, and D. Livingstone, *Nature*, **370**, 270–275 (1994).

122. R. Leung-Toung, W. Li, T. F. Tam, and K. Karimian, *Curr. Med. Chem.*, **9**, 979–1002 (2002).

123. M. R. Michaelides and M. L. Curtin, *Curr. Pharm. Des.*, **5**, 787–819 (1999).

124. P. D. Brown, *Expert Opin. Invest. Drugs*, **9**, 2167–2177 (2000).

125. O. Santos, C. D. McDermott, R. G. Daniels, and K. Appelt, *Clin. Exp. Metastasis*, **15**, 499–508 (1997).

126. L. J. MacPherson, E. K. Bayburt, M. P. Capparelli, B. J. Carroll, R. Goldstein, M. R. Justice, L. Zhu, S. Hu, R. A. Melton, L. Fryer, R. L. Goldberg, J. R. Doughty, S. Spirito, V. Blancuzzi, D. Wilson, E. M. O'Byrne, V. Ganu, and D. T. Parker, *J. Med. Chem.*, **40**, 2525–2532 (1997).

127. G. Clemens, B. Hibner, R. Humphrey, H. Kluender, and S. Wilhelm in N. J. Clendeninn and K. Appelt, Eds., *Matrix Metalloproteinase Inhibitors in Cancer Therapy*, Humana Press, Totowa, NJ, 2001, pp. 175–192.

128. T. Kusumi, M. Tsuda, T. Katsunuma, and M. Yamamura, *Cell Biochem. Funct.*, **7**, 201–204 (1989).

129. M. D. Sintchak, M. A. Fleming, O. Futer, S. A. Raybuck, S. P. Chambers, P. R. Caron, M. A. Murcko, and K. P. Wilson, *Cell*, **85**, 921–930 (1996).

130. L. Hedstrom, *Curr. Med. Chem.*, **6**, 545–560 (1999).

131. M. D. Sintchak and E. Nimmesgern, *Immunopharmacology*, **47**, 163–184 (2000).

132. D. A. Gschwend, A. C. Good, and I. D. Kuntz, *J. Mol. Recognit.*, **9**, 175–186 (1996).

133. D. Dvornik, *J. Diabetes Complications*, **6**, 25–34 (1992).

134. D. R. Tomlinson, E. J. Stevens, and L. T. Diemel, *Trends Pharmacol. Sci.*, **15**, 293–297 (1994).

135. C. L. Kaul and P. Ramarao, *Methods Find. Exp. Clin. Pharmacol.*, **23**, 465–475 (2001).

136. A. Urzhumtsev, F. Tete-Favier, A. Mitschler, J. Barbanton, P. Barth, L. Urzhumtseva, J. F. Biellmann, A. Podjarny, and D. Moras, *Structure*, **5**, 601–612 (1997).

137. M. C. Van Zandt, E. O. Sibley, K. J. Combs, E. E. McCann, B. Flam, D. J. Lavoie, D. Sawicki, A. Sabetta, A. Carrington, J. Sredy, V. Calderone, B. Cuevrier, and A. Podjarny, *Poster presented at the 218th National Meeting of the American Chemical Society*, New Orleans, LA, August 22–26, 1999.

138. S. Borman, *Chem. Eng. News*, **80**, 35–39 (2002).

139. E. K. Perry, B. E. Tomlinson, G. Blessed, K. Bergmann, P. H. Gibson, and R. H. Perry, *Br. Med. J.*, **2**, 1457–1459 (1978).

140. B. P. Imbimbo, *CNS Drugs*, **15**, 375–390 (2001).

141. Y. Ishihara, K. Kato, and G. Goto, *Chem. Pharm. Bull. (Tokyo)*, **39**, 3225–3235 (1991).

142. Y. Ishihara, G. Goto, and M. Miyamoto, *Curr. Med. Chem.*, **7**, 341–354 (2000).

143. J. L. Sussman, M. Harel, F. Frolow, C. Oefner, A. Goldman, L. Toker, and I. Silman, *Science*, **253**, 872–879 (1991).

144. Y. Yamamoto, Y. Ishihara, and I. D. Kuntz, *J. Med. Chem.*, **37**, 3141–3153 (1994).

145. A. H. Reid, J. K. Taubenberger, and T. G. Fanning, *Microbes Infect.*, **3**, 81–87 (2001).

146. J. N. Varghese, W. G. Laver, and P. M. Colman, *Nature*, **303**, 35–40 (1983).

147. M. von Itzstein, W.-Y. Wu, G. B. Kok, M. S. Pegg, J. C. Dyason, B. Jin, T. Van Phan, M. L. Smythe, H. F. White, S. W. Oliver, P. M. Colman, J. N. Varghese, D. M. Ryan, J. M. Woods, R. C. Bethell, V. J. Hotham, J. M. Cameron, and C. R. Penn, *Nature*, **363**, 418–423 (1993).

148. P. Bossart-Whitaker, M. Carson, Y. S. Babu, C. D. Smith, W. G. Laver, and G. M. Air, *J. Mol. Biol.*, **232**, 1069–1083 (1993).

149. J. N. Varghese, V. C. Epa, and P. M. Colman, *Protein Sci.*, **4**, 1081–1087 (1995).

150. C. U. Kim, W. Lew, M. Williams, H. Liu, L. Zhang, S. Swaminathan, N. Bischofberger, M. S. Chen, D. Mendel, W. G. Laver, and R. C. Stevens, *J. Am. Chem. Soc.*, **119**, 681 (1997).

151. Y. S. Babu, P. Chand, S. Bantia, P. Kotian, A. Dehghani, Y. El-Kattan, T. H. Lin, T. L. Hutchison, A. J. Elliott, C. D. Parker, S. L. Ananth, L. L. Horn, G. W. Laver, and J. A. Montgomery, *J. Med. Chem.*, **43**, 3482–3486 (2000).

152. J. A. Green, G. M. Smith, R. Buchta, R. Lee, K. Y. Ho, I. A. Rajkovic, and K. F. Scott, *Inflammation*, **15**, 355–367 (1991).

153. P. Vadas, J. Browning, J. Edelson, and W. Pruzanski, *J. Lipid Mediat.*, **8**, 1–30 (1993).

154. C. Bennion, S. Connolly, N. P. Gensmantel, C. Hallam, C. G. Jackson, W. U. Primrose, G. C. Roberts, D. H. Robinson, and P. K. Slaich, *J. Med. Chem.*, **35**, 2939–2951 (1992).

155. S. Connolly, C. Bennion, S. Botterell, P. J. Croshaw, C. Hallam, K. Hardy, P. Hartopp, C. G. Jackson, S. J. King, L. Lawrence, A. Mete, D. Murray, D. H. Robinson, G. M. Smith, L. Stein, I. Walters, E. Wells, and W. J. Withnall, *J. Med. Chem.*, **45**, 1348–1362 (2002).

156. H. G. Beaton, C. Bennion, S. Connolly, A. R. Cook, N. P. Gensmantel, C. Hallam, K. Hardy, B. Hitchin, C. G. Jackson, and D. H. Robinson, *J. Med. Chem.*, **37**, 557–559 (1994).

157. J. P. Wery, R. W. Schevitz, D. K. Clawson, J. L. Bobbitt, E. R. Dow, G. Gamboa, T. Goodson Jr., R. B. Hermann, R. M. Kramer, D. B. Mc-Clure, et al., *Nature*, **352**, 79–82 (1991).

158. R. W. Schevitz, N. J. Bach, D. G. Carlson, N. Y. Chirgadze, D. K. Clawson, R. D. Dillard, S. E. Draheim, L. W. Hartley, N. D. Jones, Mihelich, et al., *Nat. Struct. Biol.*, **2**, 458–465 (1995).

159. D. L. Scott, S. P. White, J. L. Browning, J. J. Rosa, M. H. Gelb, and P. B. Sigler, *Science*, **254**, 1007–1010 (1991).

160. M. M. Thunnissen, E. Ab, K. H. Kalk, J. Drenth, B. W. Dijkstra, O. P. Kuipers, R. Dijkman, G. H. de Haas, and H. M. Verheij, *Nature*, **347**, 689–691 (1990).

161. S. E. Draheim, N. J. Bach, R. D. Dillard, D. R. Berry, D. G. Carlson, N. Y. Chirgadze, D. K. Clawson, L. W. Hartley, L. M. Johnson, N. D. Jones, E. R. McKinney, E. D. Mihelich, J. L. Olkowski, R. W. Schevitz, A. C. Smith, D. W. Snyder, C. D. Sommers, and J. P. Wery, *J. Med. Chem.*, **39**, 5159–5175 (1996).

162. D. W. Snyder, N. J. Bach, R. D. Dillard, S. E. Draheim, D. G. Carlson, N. Fox, N. W. Roehm, C. T. Armstrong, C. H. Chang, L. W. Hartley, L. M. Johnson, C. R. Roman, A. C. Smith, M. Song, and J. H. Fleisch, *J. Pharmacol. Exp. Ther.*, **288**, 1117–1124 (1999).

163. D. M. Springer, *Curr. Pharm. Des.*, **7**, 181–198 (2001).

164. C. Savolainen, S. Blomqvist, M. N. Mulders, and T. Hovi, *J. Gen. Virol.*, **83** (Pt 2), 333–340 (2002).

165. M. G. Rossmann, *Viral Immunol.*, **2**, 143–161 (1989).

166. M. G. Rossmann, E. Arnold, J. W. Erickson, E. A. Frankenberger, J. P. Griffith, H.-J. Hecht, J. E. Johnson, G. Kamer, M. Luo, A. G. Mosser, R. R. Rueckert, B. Sherry, and G. Vriend, *Nature*, **317**, 145–153 (1985).

167. G. D. Diana, M. A. McKinlay, M. J. Otto, V. Akullian, and C. Oglesby, *J. Med. Chem.*, **28**, 1906–1910 (1985).

168. G. D. Diana, M. A. McKinlay, C. J. Brisson, E. S. Zalay, J. V. Miralles, and U. J. Salvador, *J. Med. Chem.*, **28**, 748–752 (1985).

169. M. J. Otto, M. P. Fox, M. J. Fancher, M. F. Kuhrt, G. D. Diana, and M. A. McKinlay, *Antimicrob. Agents Chemother.*, **27**, 883–886 (1985).

170. M. P. Fox, M. J. Otto, and M. A. McKinlay, *Antimicrob. Agents Chemother.*, **30**, 110–116 (1986).

171. B. Jubelt, A. K. Wilson, S. L. Ropka, P. L. Guidinger, and M. A. McKinlay, *J. Infect. Dis.*, **159**, 866–871 (1989).

172. T. J. Smith, M. J. Kremer, M. Luo, G. Vriend, E. Arnold, G. Kamer, M. G. Rossmann, M. A. McKinlay, G. D. Diana, and M. J. Otto, *Science*, **233**, 1286–1293 (1986).

173. D. C. Pevear, M. J. Fancher, P. J. Felock, M. G. Rossmann, M. S. Miller, G. D. Diana, A. M. Treasurywala, M. A. McKinlay, and F. J. Dutko, *J. Virol.*, **63**, 2002–2007 (1989).

174. K. H. Kim, P. Willingmann, Z. X. Gong, M. J. Kremer, M. S. Chapman, I. Minor, M. A. Oliveira, M. G. Rossmann, K. Andries, G. D. Diana, F. J. Dutko, M. A. McKinlay, and D. C. Pevear, *J. Mol. Biol.*, **230**, 206–227 (1993).

175. G. D. Diana, D. Cutcliffe, R. C. Oglesby, M. J. Otto, J. P. Mallamo, V. Akullian, and M. A. McKinlay, *J. Med. Chem.*, **32**, 450–455 (1989).

176. G. D. Diana and D. C. Pevear, *Antiviral Chem. Chemother.*, **8**, 401 (2002).

177. G. D. Diana, P. Rudewicz, D. C. Pevear, T. J. Nitz, S. C. Aldous, D. J. Aldous, D. T. Robinson, T. Draper, F. J. Dutko, C. Aldi, et al., *J. Med. Chem.*, **38**, 1355–1371 (1995).

178. J. M. Rogers, G. D. Diana, and M. A. McKinlay, *Adv. Exp. Med. Biol.*, **458**, 69–76 (1999).

179. F. G. Hayden, T. Coats, K. Kim, H. A. Hassman, M. M. Blatter, B. Zhang, and S. Liu, *Antiviral Ther.*, **7**, 53–65 (2002).

180. B. Dérijard, J. Raingeaud, T. Barrett, I.-H. Wu, J. Han, R. J. Ulevitch, and R. J. Davis, *Science*, **267**, 682–685 (1995).

181. K. P. Wilson, P. G. McCaffrey, K. Hsiao, S. Pazhanisamy, V. Galullo, G. W. Bemis, M. J. Fitzgibbon, P. R. Caron, M. A. Murcko, and M. S. Su, *Chem. Biol.*, **4**, 423–431 (1997).

182. B. Frantz, T. Klatt, M. Pang, J. Parsons, A. Rolando, H. Williams, M. J. Tocci, S. J. O'Keefe, and E. A. O'Neill, *Biochemistry*, **37**, 13846–13853 (1998).

183. J. C. Lee, J. T. Laydon, P. C. McDonnell, T. F. Gallagher, S. Kumar, D. Green, D. McNulty, M. J. Blumenthal, J. R. Heys, S. W. Landvatter, J. E. Strickler, M. M. McLaughlin, I. R. Siemens, S. M. Fisher, G. P. Livi, J. R. White, J. L. Adams, and P. R. Young, *Nature*, **372**, 739–746 (1994).

184. J. C. Lee, S. Kumar, D. E. Griswold, D. C. Underwood, B. J. Votta, and J. L. Adams, *Immunopharmacology*, **47**, 185–201 (2000).

185. A. Cuenda, J. Rouse, Y. N. Doza, R. Meier, P. Cohen, T. F. Gallagher, P. R. Young, and J. C. Lee, *FEBS Lett.*, **364**, 229–233 (1995).

186. A. M. Badger, J. N. Bradbeer, B. Votta, J. C. Lee, J. L. Adams, and D. E. Griswold, *J. Pharmacol. Exp. Ther.*, **279**, 1453–1461 (1996).

187. Z. Wang, B. J. Canagarajah, J. C. Boehm, S. Kassisa, M. H. Cobb, P. R. Young, S. Abdel-Meguid, J. L. Adams, and E. J. Goldsmith, *Structure*, **6**, 1117–1128 (1998).

188. K. P. Wilson, M. J. Fitzgibbon, P. R. Caron, J. P. Griffith, W. Chen, P. G. McCaffrey, S. P. Chambers, and M. S. Su, *J. Biol. Chem.*, **271**, 27696–27700 (1996).

189. T. Fox, J. T. Coll, X. Xie, P. J. Ford, U. A. Germann, M. D. Porter, S. Pazhanisamy, M. A. Fleming, V. Galullo, M. S. Su, and K. P. Wilson, *Protein Sci.*, **7**, 2249 (1998).

190. R. J. Gum, M. M. McLaughlin, S. Kumar, Z. Wang, M. J. Bower, J. C. Lee, J. L. Adams, G. P. Livi, E. J. Goldsmith, and P. R. Young, *J. Biol. Chem.*, **273**, 15605–15610 (1998).

191. J. L. Adams, J. C. Boehm, T. F. Gallagher, S. Kassis, E. F. Webb, R. Hall, M. Sorenson, R. Garigipati, D. E. Griswold, and J. C. Lee, *Bioorg. Med. Chem. Lett.*, **11**, 2867–2870 (2001).

192. T. Fullerton, A. Sharma, U. Prabhakar, M. Tucci, S. Boike, H. Davis, D. Jorkasky, and W. Williams, *Clin. Pharmacol. Ther.*, **67**, 114 (2000).

193. Pat. Appl. Vertex Pharmaceuticals, Inc., assignee, PCT WO 00/36096 (2000).

194. J. J. Haddad, *Curr. Opin. Invest. Drugs*, **2**, 1070 (2001).

195. C. Pargellis, L. Tong, L. Churchill, P. F. Cirillo, T. Gilmore, A. G. Graham, P. M. Grob, E. R. Hickey, N. Moss, S. Pav, and J. Regan, *Nat. Struct. Biol.*, **9**, 268–272 (2002).

196. J. Regan, S. Breitfelder, P. Cirillo, T. Gilmore, A. G. Graham, E. Hickey, B. Klaus, J. Madwed, M. Moriak, N. Moss, C. Pargellis, S. Pav, A. Proto, A. Swinamer, L. Tong, and C. Torcellini, *J. Med. Chem.*, **45**, 2994 (2002).

197. G. R. Boss and J. E. Seegmiller, *Annu. Rev. Genet.*, **16**, 297–328 (1982).

198. S. E. Ealick, S. A. Rule, D. C. Carter, T. J. Greenhough, Y. S. Babu, W. J. Cook, J. Habash, J. R. Helliwell, J. D. Stoeckler, R. E. Parks Jr., S. Chen, and C. E. Bugg, *J. Biol. Chem.*, **265**, 1812 (1990).

199. S. E. Ealick, Y. S. Babu, C. E. Bugg, M. D. Erion, W. C. Guida, J. A. Montgomery, and J. A. Secrist 3rd, *Proc. Natl. Acad. Sci. USA*, **88**, 11540–11544 (1991).

200. J. A. Montgomery, S. Niwas, J. D. Rose, J. A. Secrist 3rd, Y. S. Babu, C. E. Bugg, M. D. Erion, W. C. Guida, and S. E. Ealick, *J. Med. Chem.*, **36**, 55–69 (1993).

201. M. Duvic, E. A. Olsen, G. A. Omura, J. C. Maize, E. C. Vonderheid, C. A. Elmets, J. L. Shupack, M. F. Demierre, T. M. Kuzel, and D. Y. Sanders, *J. Am. Acad. Dermatol.*, **44**, 940–947 (2001).

202. P. E. Morris Jr. and G. A. Omura, *Curr. Pharm. Des.*, **6**, 943–959 (2000).

CHAPTER ELEVEN

X-Ray Crystallography in Drug Discovery

Douglas A. Livingston
Sean G. Buchanan
Kevin L. D'Amico
Michael V. Milburn
Thomas S. Peat
J. Michael Sauder
Structural GenomiX
San Diego, California

Contents

Burger's Medicinal Chemistry and Drug Discovery
Sixth Edition, Volume 1: Drug Discovery
Edited by Donald J. Abraham
ISBN 0-471-27090-3   © 2003 John Wiley & Sons, Inc.

1   INTRODUCTION

The practice of crystallography is undergoing dramatic change because of the advent of new robotics technologies, orders-of-magnitude improvement in X-ray sources and computational power, and the advances in protein production stemming from the recent revolution in molecular biology. This chapter covers these changes in the context of an overview of the techniques of modern crystallography, their application in the identification and characterization of targets and mechanisms for therapeutic intervention, and the nascent field of structural genomics. Structure-based drug design applications are covered elsewhere.

The exponential growth in the rate of determination of new protein structures continues unabated. Technologies developed in the late 1980s (1) have now evolved to the point that they have been implemented in high-throughput (HTS) format, driving the rate even higher. Super-intense, precise, tunable X-rays are now available from undulator beamlines. Three "third-generation" synchrotrons, designed and built for this purpose, are now on line—ESRF in Grenoble, France; SPring-8 in Japan; APS at Argonne National Laboratory in the United States—and others are under construction. In a relative sense, this capability has had minimal impact on medicinal chemistry to date, but that will certainly change. The companies that have successfully built high-throughput protein crystallography systems (SGX and Syrrx in the United States and Astex in the U.K. among others) have all now turned their prodigious capacity to the co-crystallization of small molecules with target proteins for the purpose of drug discovery. The capacity to compare, in parallel, the binding modes of a set of hits from HTS, or a given lead series, will be valuable, but an even greater impact will result from the decrease in turnaround time required to generate co-crystal structures. This has been the most significant hindrance to realizing the full potential of structure-based drug design. A structure is far more useful before the chemist has embarked on the synthesis of the next series, rather than after.

Another important development toward new target identification is the effort in large-scale structural annotation of various genomes, the field of structural genomics. In classifying proteins by function as a step toward validating them as therapeutic targets, structural homology is perhaps the most important tool available. These efforts have been taken up by a number of publicly-funded consortia (2), because the commercial value of genomic databases in general has not been high enough to justify their cost in the private sector. Given that medicinal chemists think and communicate largely in structural terms, this recent growth in the influence of structural biology is very important. It forms the basis of a powerful link between chemistry and biology, and we have only begun to realize its potential.

2   METHODOLOGY

2.1   Theory

X-ray crystallography provides atomic or near atomic resolution of matter. The periodicity of crystals, reflecting the repeating units of molecular structure, diffracts X-rays according to Bragg's law: $n\lambda = 2d\sin\theta$, where n is the order of diffraction, λ the wavelength of the radiation, d the spacing or distance between a family of lattice planes in the crystal, and θ the angle of the diffraction. X-radiation is ideal to analyze atomic structure, because the wavelengths used are in the order of 0.1–2.0 Å with 0.75 Å being about one-half the distance of an aliphatic carbon-carbon bond.

The images of diffracted crystal lattices can be observed with specialized precession photographic equipment, although the modern day image plate detectors used in most laboratories produce a diffraction image that can be analyzed by computer to provide the indices of the lattice diffraction spots (Fig. 11.1, a–c).

The X-ray diffraction from the electron clouds surrounding each nucleus is either reinforced or impeded and gives rise to the dif-

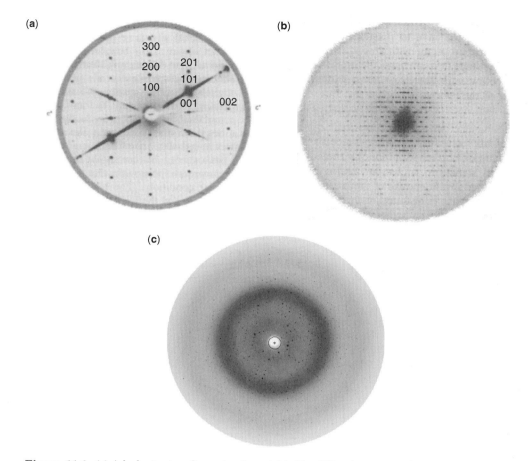

Figure 11.1. (a) A look at a two-dimensional crystal lattice diffraction pattern for a small molecule natural product, MW 222. Each diffraction intensity in the lattice is numbered to give a unique three dimensional address (identification) for that measurement. These numerical addresses are referred to as Miller indices or hkl values. (b) A diffraction pattern from a precession photograph for hemoglobin, MW 65,000. Note the the diffraction lattice spacings are much smaller for the large molecule and reflects the nature of Bragg's law, where the lattice is observed in reciprocal space ($1/d = 2\sin\theta/n\lambda$). (c) An image plate diffraction pattern for a protein. [Adapted with permission from D. J. Abraham, *Computer-Aided Drug Design, Methods, and Applications,* Marcel Dekker, Inc., New York, 1989.]

ference in intensities observed in Fig. 11.1. The steps that one goes through to solve a crystal structure follow, with the intent of providing the non-crystallographer with a simplified and pictorial view of the process.

2.2 Crystallization

Crystallization is the critical first and most important step, because good single crystals usually provide quality diffraction. Linus Pauling once entitled one of his lectures "The Importance of Being Crystalline" (3). Unfor-

tunately, crystallization is still more empirical than scientific. It requires closely monitored matrix changes in growing conditions, i.e., pH, salt concentration, temperature, solvents, and crystallization setups. Most laboratories now use well-known sparse matrix screens pioneered by Jancarik and Kim (4) and further refined and commercially distributed by Hampton Research (5, 6). Screens will typically employ vapor diffusion experiments (hanging drops or sitting drops), and occasionally batch and liquid-liquid diffusion methods.

More recently, batch crystallizations have been rejuvenated by the development of microbatch robots and by the groups of Chayen (7), DeTitta (8), and D'Arcy (9).

Although discovering the crystallization conditions for a new protein or nucleic acid can be tedious, relatively inexperienced individuals can usually succeed at growing crystals once the initial conditions are established. Some of the most successful crystallization methodologies are based on vapor diffusion methods (Fig. 11.2). The general idea behind vapor diffusion crystallization is to dissolve the protein in a buffer, with a non-precipitating amount of the miscible vapor solvent, in a reservoir that is in equilibrium with higher concentration of the vaporizing solvent nearby. Another variation is to set up the crystallization cocktail containing salts, buffers, P (poly) ethylene glycols (PEGS), small molecule solvents, etc., where volume is slowly reduced by the equilibrating mixture, which is placed nearby. McPherson, Carter, and others have developed more quantitative methods for optimizing crystal growth (10).

2.3 Data Collection

Most laboratories have rotating anode sources for production of high intensity X-ray beams. These are coupled with an area detector that has made single crystal diffractometers obsolete. Mirrors and other technology have also been used to provide a more intense and monochromatic radiation source (11). Radiation from rotating anode sources is at a fixed wavelength, usually from high-voltage electrons impinging on either a copper or molybdenum rotating anode, i.e., radiation at 1.54 Å (copper) or 0.71 Å (molybdenum). Radiation from synchrotron sources can often be tuned to a wavelength of interest for multiwavelength anomalous diffraction (MAD) experiments (see below).

X-rays generated by a synchrotron source are typically two orders of magnitude stronger than conventional CuKα radiation generated by a rotating anode. Synchrotron sources have greatly extended the ability to solve new protein structures when only weakly diffracting or small crystals are available. Another advantage in using the stronger synchrotron radiation is that the crystal exposure time is signif-

icantly lower. The typical exposure time for home laboratory CuKα sources ranges from 5 to 60 min for a range of data, whereas the equivalent set of data at an undulator beamline, i.e., the advanced photon source (APS), requires only about 1 s of exposure time. Synchrotron radiation has also allowed the use of MAD, enabling phasing (imaging) of the protein using a derivative with only one heavy element.

A variety of detectors are in common use to record X-ray data and have the advantage of measuring the intensities of large numbers of diffraction spots simultaneously. The most popular detectors are image plates and charge-coupled device (CCD) cameras. Image plates are typically the choice for laboratory rotating anode sources and lower flux synchrotron sources (Fig. 11.3). CCDs have the distinct advantage of speed at the higher flux synchrotron sources, because they simultaneously measure and record diffraction intensities (amplitudes). Current CCD cameras have readout times on the order of a few (typically 2–8) seconds, a speed not dreamed of when the first protein structure data was recorded from photographs (with intensities measured by eye comparison to standard reference spots on a separate film strip). Speed of data collection can be an important advantage at third generation synchrotron sources, with even shorter exposure times. On the other hand, image plates have a greater range of use, being accessible in any X-ray diffraction laboratory, with many of the newer models taking less than 1 min to record the intensity data. Image plate detectors often have more than one image plate, so one can be read while the other is exposed, effectively wasting no time during the collection period. The image plates also offer a larger surface area for data collection than most CCD cameras and are considerably less expensive.

X-ray diffraction data from crystals are either collected at room temperature or under cryogenic conditions at liquid nitrogen temperatures [around 100°K (−170°C)]. For room temperature data collection, crystals are normally mounted in thin-walled glass capillaries, with a small amount of mother liquor about 5 mm from the crystal. The mother liquor in the capillary is critical because protein crystals are 40–80% water—dried protein crystals do not diffract. The nearby mother

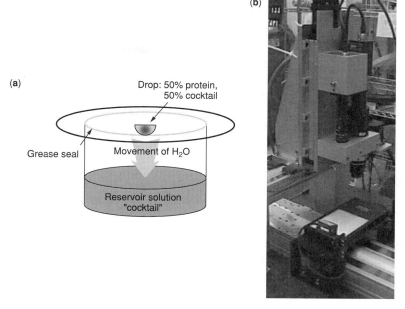

(a)

Drop: 50% protein,
50% cocktail

Movement of H₂O

Grease seal

Reservoir solution
"cocktail"

(b)

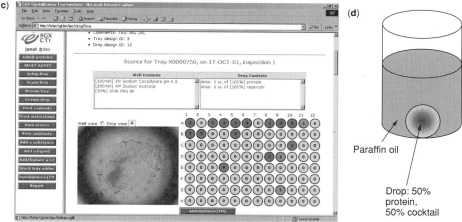

(c)

(d)

Paraffin oil

Drop: 50%
protein,
50% cocktail

Figure 11.2. (a) The drops are typically 1–10 $\mu\lambda$ total volume, with between 100 and 1000 $\mu\lambda$ total volume of cocktail in the well. The smaller the drop size, the faster the equilibrium occurs, in general. There are a variety of plates now available in which to set up these vapor diffusion experiments, the most common being 24-well Limbro plates and 96-well microtiter plates. Several robots have been developed to automatically set up the crystallization experiments; although most are no faster than doing the same procedure by hand (particularly with a multi-channel pipettor), there can be other advantages (e.g., consistency and reducing repetitive stress syndrome). Once plates are set up, they are typically kept at a constant temperature and observed periodically under a microscope. (b and c) Progress in automating this aspect of characterization has occurred, and there are now imagers that will take high resolution, digital pictures of each drop in turn and store these for either manual or automated analysis. (d) Batch experiments are set up such that the protein is mixed with cocktail and there is little concentration or dilution to the sample over time. This can now be done in very high throughput and small scale: 50–200 nL drops under oil in 1536-well plates, for example. This kind of approach has been used to screen hundreds of conditions with small amounts of protein, which may allow for faster optimization later. One caveat is that small crystals don't necessarily lead to larger crystals later, and all structures to date have had crystals of greater than 10 microns in at least one dimension.

(a)

(b)

Figure 11.3. (a) Area detector showing the configuration of the unit. (b) Area detector showing the face.

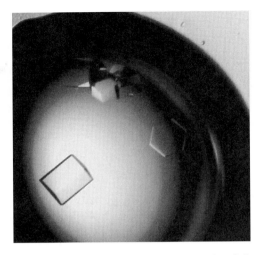

Figure 11.4. The crystals are manipulated by scooping them up with a small loop of nylon that is glued to the end of a pin. Surface tension from the liquid will hold the crystal in the loop, but the crystal can also be held by using a loop that is smaller in size than the crystal of interest. This technique will work particularly well with fragile crystals, thin plates for example, that would normally fall apart in a capillary mount. Once the crystal is frozen, it is placed on an axis in line of both an X-ray source and a stream of nitrogen set to about 100,000 to keep the crystal frozen. The crystal is rotated in increments during the data collection procedure to collect a full data set (typically one or two degrees per frame, depending on the resolution limits, mosaicity of the crystal, unit cell lengths, etc.).

liquor ensures that the crystal is bathed in the vapor of the mother liquor and prevents drying during data collection. The majority of present day data collections in home laboratories and at synchrotrons are done under cryogenic conditions, which allows high intensity X-radiation to be used without the crystal decay observed in room temperature data collection. For cryogenic data collection, crystals are normally mounted in a thin fiber loop with a layer of suitable cryoprotectant solution (Fig. 11.4). The cryoprotectant forms a layer of noncrystalline glass around the crystal to protect it from freeze shock. Simple freezing of the crystal results in the formation of ice in the interior of the crystal and renders it useless. A quick perusal of the literature shows PEG, glycerol, sucrose, and 2-methyl-2,4-pentane diol (MPD) as the most popular cryoprotectants. Oils, such as paraffin oil, have also been used successfully as cryoprotectants (12).

2.4 Phase Problem

X-ray diffraction measurements as described above only provide the amplitudes of the diffracted waves. One must have the phase angles of all measured waves relative to a common origin in order to image the molecule using a Fourier analysis. Figure 11.5 illus-

(a)   Amplitude

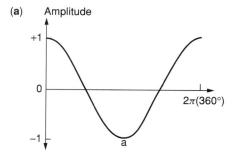

(b)   Phase difference

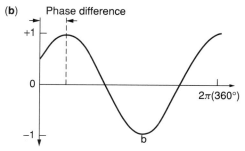

Figure 11.5. Graphs showing the phase relationship of electromagnetic radiation. [Adapted with permission from D. J. Abraham, *Computer-Aided Drug Design, Methods, and Applications,* Marcel Dekker, Inc., New York, 1989.]

trates the differences in the phases and amplitudes for two reflections. The solution of the phase problem that permitted the first image reconstruction of a protein was discovered by Perutz using multiple isomorphous replacement (MIR) (13).

The majority of the earliest structures were solved using MIR to phase the maps. This requires soaking the crystals or co-crystallizing the protein with two or more heavy atoms and hoping that these heavy atoms bind in a specific way to the protein. It also requires that the subsequent crystals are isomorphous with the native protein (i.e., no changes in the unit cell or symmetry of the crystal). Although it is possible to obtain phase information from a single heavy atom derivative using additional information (e.g., anomalous scattering or density modification), one often works diligently to get a second or third derivative to improve the quality of the electron density (Fourier) map.

Two other common methods are used to estimate phases in protein crystallography: molecular replacement (MR), which uses the structural motif of a homologous protein (14), and MAD from a single heavy element (1).

MR methodology requires a structural model that is structurally homologous to the protein that has been crystallized. Phasing is accomplished through a six-dimensional search—a three-dimensional rotation search followed by a three-dimensional translation search, using the model against the crystal data. Molecular replacement is being employed more frequently as the number of known structures has increased, which has made unique structural motifs available for phasing. For highly homologous protein structures, this method is usually straightforward and successful. For marginal cases, the addition of some independent phase information, single isomorphous replacement (SIR) or MIR, in combination with MR can enhance the quality of the Fourier map.

MAD phasing is an alternate methodology for solving the phase problem. MAD requires a single heavy atom with anomalous peak scattering at a wavelength where X-rays both at and near the spectral energy are accessible. Data sets are collected at different wavelengths to optimize the anomalous and dispersive signals from the heavy atom. Certain beamlines have been designed with wavelengths that are tunable "on the fly," and these are often referred to as MAD beamlines. MAD has become the method of choice for rapid structure solution when synchrotron radiation is available. The advantage of MAD phasing is that one often only needs a single crystal to collect all of the data necessary to solve the structure. Although multiple wavelengths are collected (anywhere from two to five sets), data collection is routinely completed in less than a few hours. The peak wavelength choice data set is very important to collect first as it contains the greatest anomalous signal and is often used alone to find the heavy atom sites (Shake'n Bake program, anomalous Patterson maps, etc.). If the crystal degrades quickly in the beam, one can also employ single-wavelength anomalous diffraction phasing (SAD) if a full data set at the peak wavelength was successfully collected. SAD phasing requires additional information, obtained by density modification, to obtain interpretable electron density maps, but has been proven in many instances to result in very high quality maps (15).

Many different heavy atoms have been used for MAD/SAD phasing, the most popular being selenium. Selenium is incorporated into the amino acid sequence of the protein by adding selenomethionine to the growth media when the protein is produced (16). For proteins that bind DNA, 5-bromouracil has been a popular choice for phasing through anomalous scattering. Most heavy elements have good anomalous signals (Hg, Pt, U, Au, etc.). Lanthanides have a particularly good signal and can sometimes substitute for divalent metals found naturally in the protein (e.g., Ca) (17). One of the major advantages of MAD phasing is that the signal does not decay at higher resolution with perfectly isomorphic crystals, so the experimentally phased map can be quite good out to the full resolution of diffraction. This typically has not been the case when using multiple isomorphous replacement, where the experimentally phased map often only extends to around 2.5 Å resolution, because of a lack of isomorphism between the native and heavy metal substituted crystals. Anomalous scattering has been useful in the structure determination of very large structures; the 30S ribosome was recently solved using Os and Lu derivatives (18).

2.5 Computing and Refinement

Raw intensity X-ray crystallographic data is next reduced and scaled to provide structure factors (F) that are used to solve and image the structure. Two of the most popular software packages employed to reduce raw date are Mosflm/CCP4 (19) and the HKL suite (20). Both work very well and are very fast with modern computers. A variety of programs, such as SOLVE (21), Shake and Bake (22), or SHELX (23), can be employed to find the heavy atom positions, including hand searching methods through Patterson maps. Once heavy atom sites are found, they are usually refined with the programs SHARP (24) or MLPHARE (19). The heavy atom positions are next used as phase information input to provide initial phases for electron density maps, which are used to fit the remainder of the protein or nucleic acid. Once a model of the structure is obtained it is refined. In cases where high resolution data is available, pro-

grams such as wARP (24) can automatically provide models of protein structures. When high resolution data is not available, a model is most often built in by hand using such graphics programs like O (25) or XFIT. The models are refined against the data by programs such as REFMAC (19) and CNS (26). All of these programs have become much faster and easier to use because of the incredible increases in speed that new hardware has allowed.

It is worth mentioning that statistical and probabilistic techniques have had a significant impact in how heavy atoms are found and models are refined (e.g., SHARP, SOLVE, REFMAC). Baysian statistics and maximum likelihood methods are now used instead of least-squares methods. One may want to consider how various data collection strategies may affect the later steps in the process by keeping this in mind, i.e., high redundancy in the data makes for better statistics.

The quality of a structure is measured in many ways: how low the R factor or R_{free} is (the fit of observed data to the model), the resolution limit of the data, the ideality of the bonds and angles, etc. How well a structure measures up to other structures of about the same resolution also gives a good idea of how "good" a given structure is (PROCHECK program). SFCHECK is a useful program for assessing the agreement between the atomic model and the experimental X-ray data. The level of confidence one expects from a given model will depend on the resolution of the data. This can be seen clearly in Fig. 11.6, where a residue from a protein structure is shown with three different data cutoffs at different resolution ranges. The model from a 3.0-Å data set may look the same as one from a 1.3-Å data set, but the level of confidence is much higher in the latter. A reasonably well-refined structure will have a crystallographic R factor between 15% and 25% and will have an R_{free} of less than 30% under most circumstances.

2.6 Databases

The Protein Data Bank (PDB) (27, 28) is now coordinated by a consortium of several institutions (Rutgers University, the San Diego Supercomputer Center, and National Institute for Standards and Technology). As of this writing, the PDB has over 18,000 structures,

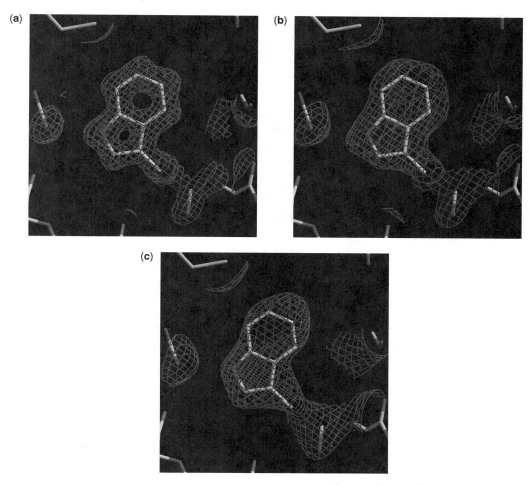

Figure 11.6. Three density maps at differing resolutions: a, 1.3 Å; b, 2.1 Å; c, 3.0 Å. See color insert.

with over 15,000 of these done by X-ray crystallography. Most of the rest were done by NMR. For small molecules, the Cambridge Structural Database (CSD) (29) contains structural information for over 230,000 organic and organometallic compounds. All of these structures have been determined by X-ray or neutron diffraction techniques.

3 APPLICATIONS OF THE USE OF CRYSTALLOGRAPHIC STUDIES IN DRUG DISCOVERY AND DEVELOPMENT

Crystallization of small molecule compounds with a protein or nucleic acid target followed by X-ray crystallographic determination of the combined structure is the basis and hallmark of structure-based drug design. As structural biology moves into the post-genomic age, many companies and academic laboratories are faced with the challenge of co-crystallization of targets and inhibitors or activators on a scale never before attempted. Previously, crystal structure determination of a protein-substrate or inhibitor complex in an academic or industrial environment often yielded the structural information desired to understand the mechanism of action or in the design of a more suitable substrate or inhibitor. However modern day laboratories are now faced with the daunting challenge of crystallizing hundreds of compounds for clues in further ligand design using standard organic synthesis or combinatorial approaches.

A variety of methods have been employed to co-crystallize biological molecules with small molecules. Discovery of crystallization conditions is still an often tedious task, so newer methods for screening crystallization conditions for proteins include the use of semi-automated robots.

Two fundamental methods are available for co-crystallizations. One method is termed "soaking." This employs the addition of the small molecule directly to saturated solutions containing crystals of the biological macromolecule in hopes that the ligand of interest will soak directly into the crystal and bind to its active or binding site, so that the co-crystal structure can be determined. The other method, called co-crystallization, depends on having an ability to add the ligand to the aqueous protein solution in at least stoichiometric amounts, followed by crystallization using either the known crystallization conditions or by setting up a new screen for determining suitable crystallization conditions. Both methods have disadvantages and advantages, and it is primarily up to the investigator to decide which method, or both methods, should be employed for their experiments.

One limitation to the soaking method is that the amount actually dissolved and available to form the complex can often not be easily determined or controlled. In general, an excess of ligand, as a solid, is added to the solution with the crystals of protein with the hope that the ligand dissolves completely and will diffuse into the crystal binding site. One method that has demonstrated success involves equilibration of the crystal with slightly higher concentration of the crystal mother liquor that contains the ligand solubilized by organic cosolvents (i.e., isopropanol, DMSO, ethanol, etc.) as part of the medium for diffusion. However, higher levels of organic solvent often decreases the resolution of diffraction. Lowering the level of solvent after the addition of compound has been found to result in better diffracting crystals. Another major limitation of this method is that it is necessary to collect an entire X-ray diffraction data set to determine if the small molecule is bound to the protein. This trial and error method can be time-consuming or expensive if a high-throughput crystallography approach is the objective.

Co-crystallization permits highly parallel screening for bound ligands through robotic systems. The co-crystallization method is better suited for high-throughput crystallography, because ligand binding can sometimes be determined without the need to solve each structure. Faster spectral analyses, or alternatives such as native gel shifts, gel filtration, and mass spectrometry can provide information on which of the crystals should be taken into X-ray studies. One difficulty in using the co-crystallization method is the problem of determining the concentration of the protein that is most suitable for complexation. For example, if the protein solution is roughly 1 mM in high salt or aqueous buffers, many organic molecules are not as soluble at that level. In these cases a lower concentration of protein is usually employed to attain stoichiometric ratios. As described above, small percentages of organic solvent can be useful for increasing the concentration of the organic compound in solution, but not without affecting the protein stability or crystal quality. In general, lowering the protein concentration sufficiently, followed by addition of the appropriate amount of ligand, and then concentration of the mixture to the desired protein concentration for crystallization is the most successful method.

Once conditions for obtaining the complexed protein have been obtained, the next step is to decide on which crystallization conditions to use. In some cases, those proteins that do not undergo large tertiary structure changes when complexed to ligands can be crystallized under similar conditions as for the uncomplexed ligand. However, in some instances, proteins will change conformation, depending on the type of ligand that they are complexed with, and a large screening of possible new crystallization conditions is required.

In many cases, soaking a compound into a crystal is not possible because of low solubility of the compound in the aqueous mother liquor. Soaking experiments can also be limited when the conformational space of the binding site is hindered, occupied by adjacent molecules in the crystal lattice, or if there are conformational changes in the binding site because of crystal packing effects. On the other hand, co-crystallization of the protein and ligand, by the nature of the process, usually requires more resources in terms of protein and experimental time, leading to greater expense.

Soaking crystals and/or growing crystals in the presence of inhibitors or ligands provides an opportunity to directly observe their binding interactions, along with the often subtle conformational effects that can have a profound effect on the mode of binding. When it is possible to use this in an iterative fashion to guide the design of the next set of compounds to be synthesized in a lead series, this becomes a potent tool. Understanding how and why a compound or series binds to an active site, particularly when the affinity is also known, provides the best understanding, and the highest level at which it is possible to enable drug design. As structural biology becomes a more integrated component of drug discovery, better methods of obtaining crystal structures with and without bound ligands will be developed, with lower costs and faster turnaround times. Many companies and academic laboratories are now focused on solving these challenges. But it is also impressive to see how well we have progressed—Table 11.1 enumerates the structures of known therapeutic targets (with or without ligands) that are available in the public domain. This table is based on the Nature Biotechnology "The Usual Suspects" poster, published in 2001, but is almost identical in content to the 1997 version (30). Reference sequences for 300 of the targets were extracted from NCBI Genbank and the nonredundant database was searched with several iterations of PSI-BLAST. The resulting profile was used to search the PDB + SGX database of known structures. The top hits for each drug target are tabulated below. A great many more reside within pharmaceutical and biotech companies as proprietary structures.

4 STRUCTURAL GENOMICS

4.1 Introduction to Structural Genomics

Until recently, structure determination by protein crystallography was a time-consuming method accessible to a few privileged skilled practitioners. X-ray crystallography was reserved to tackle questions requiring atomic resolution details of a demonstrably important protein, often a drug target. Indeed, to this day, crystallography is almost exclusively used in the pharmaceutical industry to study small molecule interactions with drug targets (see Section 3).

The development of several new methods (described in Section 2) and the availability of the complete genome sequences of both pathogens and hosts provides an unparalleled opportunity to exploit protein structures for drug discovery research in new ways. We can now contemplate using protein structure determination to help annotate genomes, that is, to assess new drug targets as well as provide multiple high-resolution structures that address selectivity issues. This emerging science of high-throughput structural biology has been termed structural genomics.

4.2 Genome Annotation

It is in infectious disease that whole genome information first became available, and it is in this field that structural genomics is having an initial impact (373). A typical approach has been to assess the viability and/or virulence of pathogens by systematic disruption of every predicted gene product. As a consequence, a large number of potential new targets have emerged: genes that are essential for pathogenicity of bacteria in a model system. Often these new genes have been filtered for those that are conserved in a variety of pathogens and that do not have a close human homolog (374). About 30–50% of the genes of a typical pathogen have no reliable functional assignment. A similar fraction of the novel targets shown to be essential also fall into this category, which becomes problematic for assay configuration. Indeed, in target-based approaches, the number of leads emerging has been disappointing. Protein structure can provide the information required to prioritize among these essential genes and to establish assays. Co-complex structures with even low-affinity hits can be used to provide key information for medicinal chemistry.

There are several ways in which structural genomics has promise as a tool for genome annotation and target prioritization. For genes of unknown function, structure can often provide clues to biochemical function. Sequence homology has become a routine method for functional assignment, but even the most

Table 11.1 Known Drug Targets with Published Structures

Target and PDB Reference	Resolution	Source	Homology	Year	Reference
Acetylcholinesterase					
1MAH(A)	3.20 Å	Green mamba	88%	1995	(31)
1B41(A), 1F8U(A)	2.76 Å, 2.90 Å	Green mamba	99%	2000	(32)
1C2B(A), 1C2O(A)	4.50 Å, 4.20 Å	Electric eel	88%	1999	(33)
1MAA(A)	2.90 Å	Mouse	88%	1998	(34)
Adenosine deaminase					
1FKX, 1FKW	2.40 Å	Mouse	82%	1996	(35)
1A4L(A), 1A4M(A)	2.60 Å, 1.95 Å	Mouse	83%	1998	(36)
1UIO, 1UIP	2.40 Å	Mouse	82%	1996	(37)
1ADD	2.40 Å	—	83%	1992	(38)
2ADA	2.40 Å	—	83%	1994	(39)
Alpha-amylase					
1JXJ(A), 1JXK(A)	1.90 Å	Human	99%	2001	(40)
1SMD	1.60 Å	Human	99%	1996	(41)
1C8Q(A)	2.30 Å	Human	99%	2000	(42)
1CPU(A), 2CPU(A)	2.00 Å	Human	97%	1999	(43)
1BSI	2.00 Å	Human	97%	1998	(44)
1HNY	1.80 Å	Human	97%	1995	(45)
3CPU(A)	2.00 Å	Human	96%	1999	(46)
1B2Y(A)	3.20 Å	Human	97%	1998	(47)
1DHK(A)	1.85 Å	Kidney bean	86%	1996	(48)
1JFH	2.03 Å	Pig	86%	1997	(49)
1PIF, 1PIG	2.30 Å, 2.20 Å	Pig	85%	1996	(50)
1OSE	2.30 Å	Pig	86%	1996	(51)
1HX0(A)	1.38 Å	Pig	86%	2001	(52)
1BVN(P)	2.50 Å	S. tendae	85%	1998	(53)
1PPI	2.20 Å	—	86%	1994	(54)
Androgen receptor					
1E3G(A)	2.40 Å	Human	100%	2000	(55)
1I37(A), 1I38(A)	2.00 Å	Rat	99%	2001	(56)
Anticoagulant protein C					
1AUT(C)	2.80 Å	Human	100%	1996	(57)
Aquaporin 1					
1IH5(A)	3.70 Å	Human	100%	2001	(58)
1FQY(A)	3.80 Å	Human	100%	2000	(59)

482

Protein / PDB ID	Resolution	Organism		Year	Ref.
β-Amyloid					
1MWP(A)	1.80 Å	Human	100%	1999	(60)
β-Lactamase[Sa]					
1BTL	1.80 Å	Bacteria	100%	1993	(61)
1FQG(A)	1.70 Å	Bacteria	98%	2000	(62)
1JTD(A)	2.30 Å	Bacteria	99%	2001	(63)
1HTZ(A)	2.40 Å	Bacteria	98%	2001	(64)
1ERM(A), 1ERO(A)	1.70 Å, 2.10 Å	Bacteria	99%	2000	(65)
1ERQ(A)	1.90 Å	Bacteria	99%	2000	(65)
1XPB	1.90 Å	Bacteria	99%	1997	(66)
1ESU(A)	2.00 Å	Bacteria	98%	2000	(67)
1BT5(A)	1.80 Å	Escherichia coli	100%	1998	(68)
1TEM	1.95 Å	Escherichia coli	100%	1996	(69)
1CK3(A)	2.28 Å	Escherichia coli	99%	1999	(70)
1AXB	2.00 Å	Escherichia coli	100%	1997	(71)
β-Tubulin					
1JFF(B)	3.50 Å	Bovine	99%	2001	(72)
1TUB(B)	3.70 Å	Pig	99%	1997	(73)
1FFX(B)	3.95 Å	Rat	99%	2000	(74)
Calcineurin A					
1TCO(A)	2.50 Å	Bovine	99%	1996	(75)
1AUI(A)	2.10 Å	Human	99%	1997	(76)
Carbonic anhydrase 2					
1HEA, 4CAC, 5CAC	2.00 Å, 2.20 Å	Human, HSV-1	100%	1992	(30)
1G6V(A)	3.50 Å	Arabian camel	100%	2000	(77)
1CNW, 1CNX, 1CNY	2.00 Å, 1.90 Å, 2.30 Å	Human	100%	1995	(78)
1IF4(A), 1IF5(A), 1IF6(A)	1.93 Å, 2.00 Å, 2.09 Å	Human	100%	2001	(79)
1IF9(A)	2.00 Å	Human	100%	2001	(80)
1CA3, 1HEB, 1HED	2.30 Å, 2.00 Å	Human	100%	1992	(81)
1DCA, 1DCB	2.20 Å, 2.10 Å	Human	99%	1993	(82)
1CRA	1.90 Å	Human	100%	1992	(83)
1CIL, 1CIM, 1CIN	1.60 Å, 2.10 Å	Human, HSV-1	100%	1993	(84)
1CAY	2.10 Å	Human	100%	1993	(85)
1RZA, 1RZB, 1RZC, 1RZD, 1RZE	1.80 Å–1.90 Å	Human	100%	1993	(86)
2CA2	1.90 Å	Human	100%	1989	(87)
1BN1, 1BN3, 1BN4, 1BNM	2.10 Å, 2.20 Å, 2.60 Å	Human	100%	1998	(88)
1CAH	1.88 Å	Human	100%	1992	(89)

Table 11.1 (Continued)

Target and PDB Reference	Resolution	Source	Homology	Year	Reference
1I8Z(A)	1.93 Å	Human	100%	2001	(90)
1BV3(A)	1.85 Å	Human	100%	1998	(91)
12CA	2.40 Å	Human	99%	1991	(92)
1G53(A)	1.94 Å	Human	100%	2000	(93)
1AM6	2.10 Å	Human	100%	1997	(94)
1CAN, 1CAO	1.90 Å	Human rhinovirus	100%	1992	(95)
1G0E(A), 1G0F(A)	1.60 Å	Human	99%	2000	(96)
1AVN	2.00 Å	Human	100%	1997	(97)
1UGF	2.00 Å	Human	99%	1996	(98)
1HVA	2.30 Å	Human	99%	1992	(99)
5CA2	2.10 Å	Human	99%	1991	(100)
1HCA	2.30 Å	—	100%	1992	(101)
4CA2, 6CA2, 7CA2, 9CA2	2.10 Å–2.80 Å	Human	100%	1991	(102)
1ZNC(A)	2.80 Å	Human	100%	1996	(103)
Catechol methyltransferase					
1VID	2.00 Å	Rat	80%	1996	(104)
Cholecystokinina receptor					
1D6G(A)	NMR	—	95%	1999	(105)
Coagulation factor 10					
1EZQ(A), 1F0R(A), 1F0S(A)	2.20 Å, 2.10 Å	Human	100%	2000	(106)
1C5M(D)	1.95 Å	Human	99%	1999	(107)
1XKA(C), 1XKB(C)	2.30 Å, 2.40 Å	Human	100%	1998	(108)
1FAX(A)	3.00 Å	Human	98%	1996	(109)
1FJS(A)	1.92 Å	Human	100%	2000	(110)
1KIG(H)	3.00 Å	Soft tick	83%	1997	(111)
1HCG(A)	2.20 Å	—	100%	1993	(112)
Coagulation factor 2					
1AI8(H)	1.85 Å	*Hirudo medicinalis*	100%	1997	(30)
1MKW(K), 1MKX(K)	2.30 Å, 2.20 Å	*Bos taurus*	84%	1997	(113)
1BTH(H)	2.30 Å	Bovine	99%	1996	(114)
1HXF(H)	2.10 Å	*Hirudo medicinalis*	100%	1996	(115)
1G30(B)	2.00 Å	*Hirudo medicinalis*	100%	2000	(116)
1A3E(H)	1.85 Å	*Hirudo medicinalis*	100%	1998	(117)
1D3P(B), 1D3Q(B)	2.10 Å, 2.90 Å	*Hirudo medicinalis*	100%	1999	(118)
1HDT(H)	2.60 Å	*Hirudo medicinalis*	100%	1994	(119)
1AD8(H)	2.00 Å	*Hirudo medicinalis*	100%	1997	(120)

1LHC(H), 1LHF(H), 1LHG(H)	1.95 Å, 2.40 Å, 2.25 Å	*Hirudomedicinalis*	100%	1994	(121)
1JOU(B)	1.80 Å	Human	99%	2001	(122)
1DIT(H)	2.30 Å	Human	100%	1995	(123)
1UVS(H)	2.80 Å	Human	100%	1996	(124)
4THN(H)	2.50 Å	Human	100%	1998	(125)
1THP(B)	2.10 Å	Human	99%	1999	(126)
1JOU(B)	1.80 Å	Human	99%	2001	(122)
1DIT(H)	2.30 Å	Human	100%	1995	(123)
1UVS(H)	2.80 Å	Human	100%	1996	(124)
4THN(H)	2.50 Å	Human	100%	1998	(125)
1THP(B)	2.10 Å	Human	99%	1999	(126)
1AY6(H)	1.80 Å	Human	100%	1997	(127)
1C1U(H), 1C1V(H)	1.75 Å, 1.98 Å	Human	100%	1999	(128)
1A4W(H)	1.80 Å	Human	100%	1998	(129)
1G37(A)	2.00 Å	Human	100%	2000	(130)
1EOJ(A), 1EOL(A)	2.10 Å	Human	100%	2000	(131)
1BB0(B)	2.10 Å	Human	100%	1998	(132)
1C4U(2), 1D6W(A), 1D9I(A), 1DOJ(A)	1.70 Å–2.30 Å	Human	100%	1999	(133)
7KME(H)	2.10 Å	Human	100%	1999	(134)
1QBV(H)	1.80 Å	Human	100%	1999	(135)
1DM4(B)	2.50 Å	Human	99%	1999	(136)
1UMA(H)	2.00 Å	Medicinal leech	100%	1996	(137)
1BMM(H), 1BMN(H)	2.60 Å, 2.80 Å	Medicinal leech	100%	1995	(138)
1A2C(H)	2.10 Å	*M. aeruginosa*	100%	1997	(139)
1FPC(H)	2.30 Å	—	100%	1994	(140)
1NRO(H), 1NRR(H)	3.10 Å, 2.40 Å	—	100%	1994	(141)
1HAG(E)	2.00 Å	—	100%	1994	(142)
1HLT(H)	3.00 Å	—	100%	1994	(143)
1TMU(H)	2.50 Å	—	100%	1994	(144)
4HTC(H)	2.30 Å	*Hirudo medicinalis*	100%	1993	(145)
1AIX(H), 1DWB(H), 1DWC(H)	2.10 Å, 3.16 Å, 3.00 Å	—	100%	1992	(146)
2HPP(H)	3.30 Å		100%	1993	(147)
1ABI(H)	2.30 Å		100%	1992	(148)
Coagulation factor 7					
1JBU(H)	2.00 Å	Bacteria	100%	2001	(149)
Coagulation factor 7a					
1QFK(H)	2.80 Å	Human	100%	1999	(150)
1DVA(H)	3.00 Å	Human	100%	2000	(151)

Table 11.1 (*Continued*)

Target and PDB Reference	Source	Homology	Year	Reference	Resolution
1DAN(H)	Human	100%	1997	(152)	2.00 Å
1CVW(H)	Human	100%	1999	(153)	2.28 Å
1FAK(H)	Human	100%	1998	(154)	2.10 Å
Coagulation factor 9					
1RFN(A)	Human	100%	1999	(155)	2.80 Å
1PFX(C)	Pig	88%	1995	(156)	3.00 Å
Cox-1					
1DIY(A)	Sheep	93%	1999	(157)	3.00 Å
1CQE(A), 1PRH(A)	Sheep	92%	1994	(158)	3.10 Å, 3.50 Å
1PTH	Sheep	92%	1995	(159)	3.40 Å
1EBV(A)	Sheep	93%	2000	(160)	3.20 Å
1FE2(A)	Sheep	92%	2000	(161)	3.00 Å
1EQG(A), 1EQH(A), 1HT5(A), 1HTxII\(A)	Sheep	92%	2000	(162)	2.61 Å–2.75 Å
1PGE(A), 1PGF(A), 1PGG(A)	Sheep	92%	1995	(163)	3.50 Å, 4.50 Å
Cox-2					
1CVU(A), 1DDX(A)	Mouse	87%	1999	(164)	2.40 Å, 3.00 Å
1CX2, 3PGH, 4COX, 5COX, 6COX	Mouse	87%	1996	(165)	3.00 Å
Cytochrome P450 reductase					
1B1C(A)	Human	100%	1998	(166)	1.93 Å
1AMO(A)	Rat	93%	1997	(167)	2.60 Å
1J9Z(A), 1JA0(A), 1JA1(A)	Rat	92%	2001	(168)	2.70 Å, 2.60 Å, 1.80 Å
Dihydrofolate reductase					
1BOZ(A)	Human	99%	1998	(169)	2.10 Å
1HFP, 1HFQ, 1HFR	Human	100%	1997	(170)	2.10 Å
1OHJ, 1OHK	Human	100%	1997	(171)	2.50 Å
1DR1, 1DR5, 1DR6, 1DR7	—	75%	1992	(172)	2.20 Å, 2.40 Å
1DR2, 1DR3	—	75%	1992	(173)	2.30 Å
1DR4	—	75%	1992	(174)	2.40 Å
1DHF(A), 2DHF(A)	Human	100%	1989	(175)	2.30 Å
1DLR, 1DLS	Human	99%	1995	(176)	2.30 Å
8DFR	—	75%	1989	(177)	1.70 Å
1DRF	—	100%	1990	(178)	2.00 Å
Dihydroorotate dehydrogenase					
1D3G(A), 1D3H(A)	Human	100%	1999	(179)	1.60 Å, 1.80 Å

Dihydropteroate synthetase[Sa]					
1AD1(A), 1AD4(A)		2.20 Å, 2.40 Å	100%	1997	(180)
DNA helicase pcra[Sa]					
1QHH(A)	S. aureus	2.50 Å	71%	1999	(181)
DNA topoisomerase 1	B. thermophilus				
1EJ9(A)	Human	2.60 Å	99%	2000	(182)
1A36(A)	Human	2.80 Å	99%	1998	(183)
1A31(A), 1A35(A)	Human	2.10 Å, 2.50 Å	90%	1998	(184)
Estrogen receptor 1a					
1QKT(A), 1QKU(A)	Human	2.20 Å, 3.20 Å	98%	1999	(185)
1HCP	Human	NMR	100%	1993	(186)
1A52(A)	Human	2.80 Å	99%	1998	(187)
1ERR(A), 1ERE(A)	Human	2.60 Å, 3.10 Å	99%	1997	(188)
1HCQ(A)	Human	2.40 Å	100%	1993	(189)
3ERT(A), 3ERD(A)	Human	1.90 Å, 2.03 Å	98%	1999	(190)
FK506-binding protein					
1TCO(C)	Bovine	2.50 Å	99%	1996	(75)
1FKD, 2FKE	Human	1.72 Å	100%	1993	(191)
1FKJ, 1FKK, 1FKL	Cow	1.70 Å, 2.20 Å	97%	1995	(193)
1FAP(A)	Human	2.70 Å	100%	1996	(194)
3FAP(A), 4FAP(A)	Human	1.85 Å, 2.80 Å	100%	1999	(195)
1NSG(A)	Human	2.20 Å	100%	1997	(196)
1FKR, 1FKS, 1FKT	Human	NMR	100%	1992	(197)
1EYM(A)	Human	2.00 Å	99%	2000	(198)
1BL4(A)	Human	1.90 Å	99%	1998	(199)
1D6O(A), 1D7H(A), 1D7I(A), 1D7J(A)	Human	1.85 Å–1.90 Å	100%	1999	(200)
1QPF(A), 1QPL(A)	Human	2.50 Å, 2.90 Å	100%	1999	(201)
1F40(A)	Human	NMR	100%	2000	(202)
1B6C(A)	Human	2.60 Å	100%	1999	(203)
1A7X(A)	Human	2.00 Å	100%	1998	(204)
1BKF	Human	1.60 Å	98%	1995	(205)
2FAP(A)	Human	2.20 Å	100%	1998	(206)
1C9H(A)	Human	2.00 Å	83%	1999	(207)
1FKG, 1FKH, 1FKI(A)	—	2.00 Å, 1.95 Å, 2.20 Å	100%	1993	(208)
1FKF	—	1.70 Å	100%	1991	(209)
1FKB	—	1.70 Å	100%	1992	(210)

Table 11.1 (Continued)

Target and PDB Reference	Resolution	Source	Homology	Year	Reference
Follicle stimulating hormone					
1FL7(B)	3.00 Å	Human	99%	2000	(211)
GABA transferase					
1GTX(A)	3.00 Å	Pig	94%	1999	(212)
Glucocorticoid receptor					
1LAT(A)	1.90 Å	Rat	85%	1995	(213)
1GLU(A)	2.90 Å	—	94%	1992	(214)
Glutamate receptor 1					
1EWK(A), 1EWT(A), 1EWV(A)	2.20 Å, 3.70 Å, 4.00 Å	Rat	98%	2000	(215)
Glutathione peroxidase					
1GP1(A)	2.00 Å	—	90%	Jun 1985	(216)
G-CSF 3					
1CD9(A), 1PGR(A)	2.80 Å, 3.50 Å	Mouse	98%	1999	(217)
1BGC, 1BGD, 1BGE(A)	1.70 Å, 2.30 Å, 2.20 Å	—	80%	1993	(218)
1GNC	NMR	—	100%	1994	(219)
1RHG(A)	2.20 Å	—	98%	1993	(220)
Granulocyte-macrophage CSF					
1CSG(A)	2.70 Å	—	100%	1992	(30)
2GMF(A)	2.40 Å	Human	100%	1996	(221)
Growth hormone receptor					
1A22(B)	2.60 Å	Human	100%	1998	(222)
1AXI(B)	2.10 Å	Human	98%	1997	(223)
1HWG(B), 1HWH(B)	2.50 Å, 2.90 Å	Human	100%	1996	(224)
3HHR(B)	2.80 Å	—	100%	1993	(225)
HIV reverse transcriptase					
1DLO	2.70 Å	HIV-1	98%	1996	(226)
1RT3(B)	3.00 Å	HIV-1	99%	1998	(227)
1HPZ, 1HQE, 1HQU	3.00 Å, 2.70 Å	HIV-1	98%	2000	(228)
1BQM, 1BQN	3.10 Å, 3.30 Å	HIV-1	98%	1998	(229)
1TVR(B), 1UWB(B)	3.00 Å, 3.20 Å	HIV-1	99%	1996	(230)
1EET	2.73 Å	HIV-1	98%	2000	(231)
1IKV, 1IKW, 1IKX, 1IKY	2.80 Å – 3.00 Å	HIV-1	98%	2001	(232)
1HVU(B)	4.75 Å	HIV-1	99%	1998	(233)

488

PDB ID	Resolution	Organism	%	Year	Ref.
1C1B(B)	2.50 Å	HIV-1 pol	99%	1999	(234)
2HMI(B)	2.80 Å	Virus	99%	1998	(235)
1HYS	3.00 Å	Virus	98%	2001	(236)
1HMV	3.20 Å	Virus	98%	1994	(237)
1HNI	2.80 Å	Virus	98%	1995	(238)
1HNV	3.00 Å	Virus	98%	1995	(239)
1FKP(B)	2.90 Å	Virus	99%	2000	(240)
1JLA, 1JLB, 1JLC, 1JLE, 1JLF, 1JLG	2.50 Å – 3.00 Å	Virus	99%	2001	(241)
1J5O, 1QE1(B)	3.50 Å, 2.85 Å	Virus	99%	1999	(242)
3HVT(B)	2.90 Å	HIV-1	99%	1994	(243)
Inosine monophosphate dehydrogenase 2					
1JR1(A)	2.60 Å	Chinese hamster	98%	2001	(244)
1B3O(A)	2.90 Å	Human	100%	1998	(245)
Insulin-like growth factor 1					
3LRI(A)	NMR	Human	86%	1999	(246)
1BQT	NMR	Human	100%	1998	(247)
1IMX(A)	1.82 Å	Human	100%	2001	(248)
1B9G(A)	NMR	—	81%	1999	(249)
2GF1, 3GF1	NMR	—	100%	1991	(250)
Insulin-like growth factor 1 receptor					
1IGR(A)	2.60 Å	Human	100%	1998	(251)
1GAG(A)	2.70 Å	Human	78%	2000	(252)
1I44(A)	2.40 Å	Human	79%	2001	(253)
1IR3(A)	1.90 Å	Human	78%	1997	(254)
1IRK	2.10 Å	—	79%	1995	(255)
Insulin-like growth factor 2					
1IGL	NMR	—	100%	1994	(256)
Integrin alpham					
1BHQ(1), 1IDN(1)	2.70 Å	Human	100%	1998	(257)
1JLM	2.00 Å	Human	100%	1996	(258)
1IDO	1.70 Å	Human	96%	1996	(259)
Intercellular adhesion molecule 1					
1IAM	2.10 Å	Human	98%	1998	(260)
1IC1(A)	3.00(?) Å	Human	100%	1998	(261)
1D3E(I), 1D3I(I), 1D3L(A)	2.80 Å, 2.60 Å, 3.25 Å	Human rhinovirus	100%	1999	(262)

Table 11.1 *(Continued)*

Target and PDB Reference	Resolution	Source	Homology	Year	Reference
Interferon α 1					
1ITF	NMR	Human	82%	1997	(263)
1RH2(A)	2.90 Å	Human	83%	1996	(264)
Interferon γ					
1FG9(A)	2.90 Å	Human	100%	2000	(265)
1FYH(A)	2.04 Å	Human	100%	2000	(266)
1EKU(A)	2.90 Å	Human	98%	2000	(267)
1HIG(A)	3.50 Å	—	99%	1991	(268)
Interleukin 1					
2ILA	2.30 Å	—	99%	1991	(269)
Interleukin 1 receptor					
1G0Y(R)	3.00 Å	Human	100%	2000	(270)
1IPA(Y)	2.70 Å	Human	100%	1998	(271)
1ITB(B)	2.50 Å	Human	100%	1997	(272)
Interleukin 10					
1VLK	1.90 Å	Epstein-Barr virus	92%	1997	(273)
2ILK	1.60 Å	Human	100%	1996	(274)
1ILK	1.80 Å	Human	100%	1995	(275)
1J7V(L)	2.90 Å	Human	100%	2001	(276)
1INR	2.00 Å	Human	100%	1995	(277)
Interleukin 12					
1F42(A), 1F45(A)	2.50 Å, 2.80 Å	Human	100%	2000	(278)
Interleukin 13					
1GA3(A)	NMR	Human	100%	2000	(279)
Interleukin 2					
1IRL	NMR	Human	99%	1995	(280)
3INK(C)	2.50 Å	—	99%	1992	(281)
Interleukin 3					
1JLI	NMR	*Escherichia coli*	87%	1995	(282)
Interleukin 4					
1HIJ, 1HIK	3.00 Å, 2.60 Å	Human	100%	1995	(283)
1IAR(A)	2.30 Å	Human	100%	1999	(284)
1HZI(A)	2.05 Å	Human	99%	2001	(285)
1ITM	NMR	—	100%	1994	(286)
1BBN, 1BCN	NMR	—	98%	1992	(287)

Name/ID	Method/Resolution	Species		Identity	Year	Ref.
1CYL	NMR	—		100%	1994	(288)
2CYK	NMR	—		100%	1994	(289)
1TTL	NMR	—		100%	1992	(290)
2INT	2.40 Å	—		100%	1993	(291)
1RCB	2.25 Å	—		100%	1992	(292)
1TTI	NMR	—		98%	1993	(293)
Interleukin 5						
1HUL(A)	2.40 Å	Human		100%	1995	(294)
Interleukin 6						
1IL6, 2IL6	NMR	Human		100%	1997	(295)
1ALU	1.90 Å	Human		100%	1997	(296)
Interleukin 8						
1IKL, 1IKM	NMR	Human		100%	1995	(297)
1ICW(A)	2.01 Å	Human		97%	1996	(298)
1ILP(A), 1ILQ(A)	NMR	Human		100%	1998	(299)
1QE6(A)	2.35 Å	Human		97%	1999	(300)
1ROD(A)	NMR	Human		79%	1995	(301)
3IL8	2.00 Å	Human		100%	1990	(302)
1IL8(A), 2IL8(A)	NMR	—		100%	1990	(303)
Leukotriene A4 hydrolase						
1HS6(A)	1.95 Å	Human		100%	2000	(304)
Lipocortin I						
1AIN	2.50 Å	Human		100%	1992	(305)
1HM6(A)	1.80 Å	Pig		89%	2000	(306)
Luteinizing hormone β						
1QFW(B)	3.50 Å	*Escherichia coli*		83%	1999	(307)
1HCN(B)	2.60 Å	—		83%	1994	(308)
1HRP(B)	3.00 Å	—		83%	1994	(309)
Macrophage CSF 1						
1HMC(A)	2.50 Å	—		100%	1993	(310)
Neuraminidase[int B virus]						
1INF	2.40 Å	Influenza b virus		100%	1995	(311)
1A4G(A), 1A4Q(A)	2.20 Å, 1.90 Å	Influenza b virus		94%	1998	(312)
1IVB	2.40 Å	—		99%	1994	(313)
1NSC(A), 1NSD(A)	1.70 Å, 1.80 Å	—		94%	1993	(314)
1B9S(A), 1B9T(A), 1B9V(A)	2.50 Å, 2.40 Å, 2.35 Å	Influenza b virus		99%	1999	(315)
1INV	2.40 Å	—		99%	1994	(316)
1NSB(A)	2.20 Å	—		94%	1991	(317)

Table 11.1 (*Continued*)

Target and PDB Reference	Resolution	Source	Homology	Year	Reference
Neuropeptide Y					
1RON	NMR	Human	100%	1996	(318)
1F8P(A)	NMR	—	97%	2000	(319)
1FVN(A)	NMR	—	91%	2000	(320)
Parathyroid hormone					
1HTH	NMR	Human	88%	1997	(321)
1FVY(A)	NMR	Human	100%	2000	(322)
1BWX, 1HPY, 1ZWA, 1ZWC	NMR	Human	100%	1998	(323)
1ET1(A)	0.90 Å	Human	100%	2000	(324)
1HPH	NMR	—	100%	1995	(325)
1ZWB, 1ZWD, 1ZWE, 1ZWF, 1ZWG	NMR	—	100%	1996	(326)
PDGF β					
1PDG(A)	3.00 Å	—	100%	1992	
Phospholipase A2					
1BCI	NMR	Human	100%	1998	(327)
1RLW	2.40 Å	Human	98%	1997	(328)
1CJY(A)	2.50 Å	Human	100%	1999	(329)
Potassium channel shaker					
1A68	1.80 Å	Sea hare	87%	1998	(330)
1EOD(A), 1EOE(A), 1EOF(A)	2.45 Å, 1.70 Å, 2.38 Å	Sea hare	87%	2000	(331)
1T1D(A)	1.51 Å	Sea hare	88%	1998	(332)
1EXB(E)	2.10 Å	Rat	100%	2000	(333)
1DSX(A), 1QDV(A), 1QDW(A)	1.60 Å, 2.10 Å	Rat	94%	2000	(334)
PPAR γ					
4PRG(A)	2.90 Å	*Escherichia coli*	97%	1999	(335)
1PRG(A), 2PRG(A)	2.20 Å, 2.30 Å	Human	97%	1998	(336)
1FM6(D), 1FM9(D)	2.10 Å	Human	99%	2000	(337)
3PRG(A)	2.90 Å	Human	99%	1998	(338)
Progesterone receptor					
1E3K(A)	2.80 Å	Human	100%	2000	(55)
1A28(A)	1.80 Å	Human	100%	1998	(339)
Prolactin receptor					
1BP3(B)	2.90 Å	Human	100%	1998	(340)
1F6F(B)	2.30 Å	Rat	72%	2000	(341)
Retinoic acid receptor					
1DSZ(A)	1.70 Å	Human	100%	2000	(342)
1EXA(A), 1EXX(A)	1.59 Å, 1.67 Å	Human	81%	2000	(343)

Name	Resolution	Organism	Identity	Year	Ref.
2LBD	2.00 Å	Human	80%	1997	(344)
3LBD, 4LBD	2.40 Å	Human	80%	1998	(345)
1DKF(B)	2.50 Å	Human	100%	1999	(346)
1FCX(A), 1FCY(A), 1FCZ(A)	1.47 Å, 1.30 Å, 1.38 Å	Human	83%	2000	(347)
1HRA	NMR	—	94%	1993	(348)
Retinoid X receptor					
1FM6(A), 1FM9(A)	2.10 Å	Human	100%	2000	(337)
1DSZ(B)	1.70 Å	Human	100%	2000	(342)
1DKF(A), 1LBD	2.50 Å, 2.70 Å	Human	99%	1999	(349)
2NLL(A)	1.90 Å	Human	100%	1996	(350)
1RXR	NMR	Human	98%	1998	(351)
1G1U(A), 1G5Y(A)	2.50 Å, 2.00 Å	Human	100%	2000	(352)
1FBY(A)	2.25 Å	Human	100%	2000	(353)
1BY4(A)	2.10 Å	Human	100%	1998	(354)
Serotransferrin β					
1JNF(A)	2.60 Å	Rabbit	78%	2001	(355)
Stem cell factor					
1EXZ(A)	2.30 Å	Human	97%	2000	(356)
1SCF(A)	2.20 Å	Human	88%	1998	(357)
Thymidine kinase[HHV]					
1KIM(A)	2.14 Å	HSV-1	98%	1997	(358)
1OHI(A), 2KI5(A)	1.90 Å	HSV-1	98%	1999	(359)
1KI2, 1KI3, 1KI4, 1KI6, 1KI7, 1KI8	2.20 Å – 2.37 Å	HSV-1	99%	1998	(360)
1VTK, 2VTK, 3VTK	2.75 Å, 2.80 Å, 3.00 Å	HSV-1	98%	1997	(361)
1E2H(A), 1E2I(A), 1E2J(A)	1.90 Å, 2.50 Å	—	99%	2000	(362)
1E2M(A), 1E2N(A), 1E2P(A)	2.20 Å, 2.50 Å	HSV-1	99%	2000	(363)
1E2K(A), 1E2L(A)	1.70 Å, 2.40 Å	—	99%	2000	(364)
Tumor necrosis factor receptor 1					
1NCF(A)	2.25 Å	Human	100%	1994	(365)
1EXT(A)	1.85 Å	Human	100%	1996	(366)
1TNR(R)	2.85 Å	—	100%	1994	(367)
Vitamin D receptor					
1IE8(A), 1IE9(A)	1.52 Å, 1.40 Å	Human	83%	2001	(368)
1DB1(A)	1.80 Å	Human	83%	1999	(369)
Xanthine-guanine phosphoribosyltransferase					
1A95(A), 1A97(A), 1A98(A)	2.00 Å, 2.60 Å, 2.25 Å	Escherichia coli	100%	1998	(370)
1NUL(A)	1.80 Å	Escherichia coli	100%	1996	(371)
1A96(A)	2.00 Å	Escherichia coli	100%	1998	(372)

sensitive sequence methods [such as ISS (375)] fail to identify many homologous relationships. Structure is more conserved than sequence, so structural classification schemes (SCOP, CATH) have been a valuable method to assign proteins to functional groups. A now classic example of functional understanding from structural homology was the discovery that the Bcl-2 family of apoptosis proteins are homologous to pore-forming toxins (376). This finding led to the suggestion that Bcl-2 proteins may function by perforating mitochondrial membranes, and has since opened new avenues of fruitful research.

In addition to structural homology as assessed by global similarities, local structural features can give clues to structure even when proteins are not homologous. By identifying surface clusters of polar residues that are well conserved in the sequence family it is possible to identify likely functional sites even when there is no obvious structural homology. These three-dimensional motifs can be compared with a structure database to identify similar motifs with known function. A classic example is found among the serine proteases. Chymotrypsin and subtilisin share a similar catalytic triad (His-Asp-Ser) but are otherwise unrelated structurally. The PLP-dependent enzymes are famed for the diversity of both structure and function, but even among this group, common structural motifs seem to have evolved convergently (377).

Simply searching for large clefts in the protein surface turns out to be an extremely successful method to identify active sites. Nucleic acid binding functions can be particularly obvious from an analysis of surface electrostatics (378, 379). Mice homozygous for tubby loss-of-function mutations show an obese phenotype, and therefore, the tubby protein has attracted considerable interest. However, 3 years after the initial cloning of the *tubby* gene (380, 381), the molecular function of the protein was still a mystery. The structure of the conserved C-terminal domain of tubby was determined by X-ray crystallography, and a large groove of highly positive charge immediately led to the hypothesis that the protein acted as a transcription factor (6). The search is now on for downstream targets of tubby, and further structural work has demonstrated a new role for the tubby protein in G-protein–coupled receptor (GPCR)-mediated signal transduction (382) This on-going story demonstrates the power of structural approaches to determine function.

In tackling structures of proteins of unknown function bound metal ions, natural substrates or even serendipitously bound small molecules arising from the crystal preparation (e.g., buffers) often suggests the location of an active site. If the side-chains contributing to binding are well conserved, then this is good evidence locating an active site and helps assess the "drugability" of the protein. The recent structure of LuxS illustrates the power of this approach (373). A number of genes had been identified that are required for quorum sensing in bacteria by system 2. Quorum sensing by the widely conserved system 2 has emerged as an intriguing mechanism by which bacteria monitor their density and seems to be an important component of the progression to virulence, at least in certain pathogens. LuxS is the product of one of the genes required for system 2, but nothing was known of the molecular function of LuxS in this pathway. Disruption of this pathway has promise in antibacterial drug design, but whether LuxS would be an attractive target for small molecule design was unclear. No information was available to develop a biochemical assay, and besides, it was not clear what kind of library to use in high-throughput screening. The structure of LuxS was solved at Structural GenomiX in less than 2 months, and there are representative X-ray structures from three different bacteria. The structure showed that LuxS forms a dimer in which each monomer has a zinc ion coordinated by a His-His-Cys triad and water molecule. Non-covalently bound methionine molecules were found to have bound in a pocket formed at the dimer interface and close to the zinc ions (see Fig. 11.7, a and b). Methionine was shown to have bound as an artefact of the purification procedure. With this information, it became immediately clear that LuxS is likely a zinc metalloenzyme, and a hypothesis for the likely physiological substrate emerged from molecular modeling studies of the methionine binding site. This example illustrates how structure could rapidly accelerate an early stage project providing the starting point for assay development and selection of an appropriate

(a)

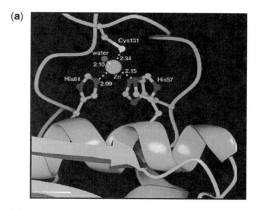

(b)

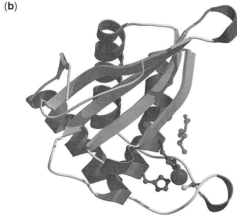

Figure 11.7. (a) The likely active site of LuxS identified by searching for clusters of polar, conserved residues in the structure. (b) Structure of the LuxS monomer highlighting the bound zinc ion (magenta) and methionine (green). See color insert.

screening library (in this case metalloenzyme inhibitor libraries would be desirable). The model of the likely substrate bound to the active site suggests further experiments to test this hypothesis and even provides a starting point for medicinal chemistry exploration.

4.3   Pathways

Increasingly in drug discovery, particular molecular pathways are attracting interest in drug design and often manipulation of any of a number of pathway components would achieve the same end. Pathways controlling apoptosis, the cell cycle, and inflammation all contain multiple biologically validated targets. In microbial disease, several biosynthetic pathways, such as peptidoglycan biosynthesis and translation, are the targets of current drugs and several new pathways are promising targets for the development of novel agents.

Comprehensive, high-resolution structural information of multiple pathway components provides a basis for the rational design of inhibitors targeting the pathway. Interfering with the function of any of a number of enzymes of a pathway may have equally beneficial therapeutic value. Despite this, some enzymes may be more tractable targets for the design of inhibitors than others. Access to high resolution structural information of all the components of a therapeutically relevant pathway enables the rational choice of the best-suited target(s) to pursue for the design of agonists and antagonists. This choice may depend on such pragmatic considerations as the access to libraries targeted to particular enzyme types and available synthetic chemistry expertise. Furthermore, comparison of the binding pockets of consecutive enzymes in the pathway that bind similar (or identical) substrates and products may even enable the design of inhibitors of multiple pathway components. Such a compound may be particularly desirable in the development of novel anti-microbial and cancer agents where compound resistance can rapidly emerge. The evolution of resistance to a drug that inhibits two consecutive enzymes in an essential pathway is theoretically much less probable than evolution of resistance to a single enzyme inhibitor.

The non-mevalonate isopentenyl pyrophosphate biosynthesis pathway has attracted attention in recent years as a novel target for the design of anti-microbial inhibitors (383). At Structural GenomiX, the structures of three consecutive enzymes in this pathway have been solved. There is now a clear understanding of which pathway components may be most tractable to inhibitor discovery, which likely have least structural homology to human proteins, and even how to go about the design of pan-pathway inhibitors.

4.4   Protein Structure Modeling

An aim of structural genomics efforts, to provide high quality three-dimensional structures for every protein sequence, will not be achieved by experimental approaches alone

Many consortiums are selecting targets for X-ray crystallography that would provide the templates for comparative modeling techniques of all other sequences (384, 385). As more structures are determined by NMR and X-ray crystallography, the quality of the models will improve simply because more similar templates will become available but also because and new methods for loop modeling and *ab initio* structure prediction will undoubtedly emerge (386, 387). Efforts are also underway both in industry and academia to assemble databases of homology models for all sequences that can be reasonably well modeled (388).

5 CONCLUSION

Anyone who is involved or interested in drug discovery will recognize the potential of protein crystallography to greatly enhance the process. Whether this promise has been met to date is the subject of considerable debate. What is certain, however, is that in the very near future the advances in crystallography technology will render this question moot. The histograms on the PDB website (27, 28) that show the increasing rate of structures deposited over the last decade are a startling visual indicator of the revolution that is occurring in the field. Clearly, the impact will be felt in drug discovery very soon and perhaps very dramatically, and it serves the audience of this series to be well informed of these advances in technology and their subtle limitations.

It is tempting to draw analogy with the development of other analytical technologies (NMR, FAB-MS) and conclude that protein crystallography will soon leave the incubator of "big machine physics" to become an everyday, routine tool used in the medicinal chemistry laboratory. Hopefully, this chapter has shown some of the subtle complexities of sample preparation and handling, data collection, and refinement, etc. that temper this vision and will likely keep this a specialized field for some time.

REFERENCES

1. W. A. Hendrickson, *Science*, **254**, 51–58 (1991).
2. T. C. Terwilliger, *Nat. Struct. Biol.*, **7**, 935–939 (2000).
3. L. Pauling, Lecture presented at the International Congress of X-ray Crystallography at Stonybrook, NY, August 1973.
4. J. Jancarik and S. H. Kim, *J. Appl. Cryst.*, **24**, 409–411 (1991).
5. Hampton Research, available online at http://www.hamptonresearch.com, accessed on October 9, 2001.
6. G. L. Gilliland, M. Tung, D. M. Blakeslee, and J. Ladner, Biological Macromolecule Crystallization Database (BMCD), available online at http://www.bmcd.nist.gov:8080/bmcd/bmcd.html, accessed on October 9, 2001.
7. N. E. Chayen, *Structure*, **5**, 1269–1274 (1997).
8. I. Jurisica, et al., *IBM Systems J.*, **40**, 394–409 (2001).
9. A. D'Arcy, et al., *J. Cryst. Growth*, **168**, 175–180 (1996).
10. C. W. Carter Jr., *Methods Enzymol.*, **276**, 74–99 (1997).
11. A. C. Bloomer and U. W. Arndt, *Acta Crystallogr. D Biol. Crystallogr.*, **55(Pt 10)**, 1672–1680 (1999).
12. (a) G. A. Petsko, *J. Mol. Biol.*, **96**, 381–392 (1975); (b) R. L. Sutton, *J. Chem. Soc. Faraday Trans.*, **1**, 101–105 (1991); (c) D. W. Rodgers, *Structure*, **2**, 1135–1140 (1994).
13. D. W. Green, V. M. Ingram and M. F. Perutz, *Proc. R. Soc. A*, **225**, 287–307 (1954).
14. M. G. Rossman and D. M. Blow, *Acta Cryst.*, **15**, 24–34 (1962).
15. L. M. Rice, T. N. Earnest, A. T. Brunger, *Acta Cryst. D.*, **56**, 1413–1420 (2000).
16. W. A. Hendrickson, et al., *EMBO J.* **9**, 1665–1672 (1990).
17. W. I. Weis, et al., *Science*, **254**, 1608–1615 (1991).
18. W. M. Clemons, Jr., et al., *J Mol Biol.* **310**, 827–843 (2001).
19. Collaborative Computational Project, Number 4, *Acta Cryst. D*, **50**, 760–763 (1994).
20. Z. Otwinowski and W. Minor, available online at http://www.hkl-xray.com, accessed October 9, 2001.
21. T. Terwilliger, Automated Structure Solution for MIR and MAD, available online at http://www.solve.lanl.gov, accessed October 9, 2001.
22. C. M. Weeks, S. A. Potter, J. Rappleye, R. Miller, available online at http://www.hwi.buffalo.edu/SnB, accessed on October 9, 2001.
23. G. M. Sheldrick, available online at http://shelx.uni-ac.gwdg.de/SHELX/, accessed on October 9, 2001.

24. V. S. Lamzin and A. Perrakis, available online at http://www.embl-hamburg.de/ARP, accessed on October 9, 2001.

25. A. Jones and M. Kjeldgaard, available online at http://www.imsb.au.dk/~mok/o, accessed on October 9, 2001.

26. A. T. Brunger, P. D. Adams, G. M. Clore, W. L. Delano, P. Gros, R. W. Grosse-Kunstleve, J.-S. Jiang, J. Kuszewski, M. Nilges, N. S. Pannu, R. J. Read, L. M. Rice, T. Simonson, and G. L. Warren, Crystallography and NMR System, available online at http://cns.csb.yale.edu/v1.0, accessed on October 9, 2001.

27. H. M. Berman, D. S. Goodsell, and P. E. Bourne, *Am. Scientist*, **90**, 350–359 (2002).

28. H. M. Berman, J. Westbrook, Z. Feng, G. Gilliland, T. N. Bhat, H. Weissig, I. N. Shindyalov, P. E. Bourne, The Protein Data Bank, *Nucleic Acids Res.*, **28**, 235–242 (2000).

29. Information on how to obtain this database is available at: http://www.ccdc.cam.ac.uk/prods/.

30. J. Drews and S. Ryser, *Nature Biotechnol.* **15**, (1997).

31. Y. Bourne, P. Taylor, and P. Marchot, *Cell*, **83**, 503 (1995).

32. G. Kryger, M. Harel, M. Harel, A. Shafferman, I. Silman, and J. L. Sussman, *Acta Crystallogr.*, Sect. D, **56**, 1385 (2000).

33. Y. Bourne, J. Grassi, P. E. Bougis, and P. Marchot, *J. Biol. Chem.*, **274**, 3370 (1999).

34. Y. Bourne, P. Taylor, P. E. Bougis, and P. Marchot, *J. Biol. Chem.*, **274**, 2963 (1999).

35. V. Sideraki, K. A. Mohamedali, D. K. Wilson, Z. Chang, R. E. Kellems, F. A. Quiocho, and F. B. Rudolph, *Biochemistry*, **35**, 7862 (1996).

36. Z. Wang and F. A. Quiocho, *Biochemistry*, **37**, 8314 (1998).

37. V. Sideraki, D. K. Wilson, L. C. Kurz, F. A. Quiocho, and F. B. Rudolph, *Biochemistry*, **35**, 15019 (1996).

38. D. K. Wilson, F. B. Rudolph, and F. A. Quiocho, *Science*, **252**, 1278 (1991).

39. D. K. Wilson and F. A. Quiocho, *Biochemistry*, **32**, 1689 (1993).

40. N. Ramasubbu, C. Ragunath, and Z. Wang, In press.

41. N. Ramasubbu, P. Venugopalan, Y. Luo, G. D. Brayer, and M. J. Levine, In press.

42. N. Ramasubbu, K. Sekar, and D. Velmurugan, *Acta Crystallogr. D Biol. Crystallogr.*, **52**, 435 (1996).

43. G. D. Brayer, G. Sidhu, R. Maurus, E. H. Rydberg, C. Braun, Y. Wang, N. T. Nguyen, C. M. Overall, and S. G. Withers, *Biochemistry*, **39**, 4778 (2000).

44. E. H. Rydberg, G. Sidhu, H. C. Vo, J. Hewitt, H. C. Cote, Y. Wang, S. Numao, R. T. Macgillivray, C. M. Overall, G. D. Brayer, and S. G. Withers, *Protein Sci.*, **8**, 635 (1999).

45. G. D. Brayer, Y. Luo, and S. G. Withers, *Protein Sci.*, **4**, 1730 (1995).

46. G. D. Brayer, G. Sidhu, R. Maurus, E. H. Rydberg, C. Braun, Y. Wang, N. T. Nguyen, C. M. Overall, and S. G. Withers, *Biochemistry*, **39**, 4778 (2000).

47. M. Qian, R. Haser, G. Buisson, E. Duee, and F. Payan, *Biochemistry*, **33**, 6284 (1994).

48. C. Bompard-Gilles, P. Rousseau, P. Rouge, and F. Payan, *Structure (Lond)*, **4**, 1441 (1996).

49. M. Qian, S. Spinelli, H. Driguez, and F. Payan, *Protein Sci.*, **6**, 2285 (1997).

50. M. Machius, L. Vertesy, R. Huber, and G. Wiegand, *J. Mol. Biol.*, **260**, 409 (1996).

51. C. Gilles, J. P. Astier, G. Marchis-Mouren, C. Cambillau, and F. Payan, *Eur. J. Biochem.*, **238**, 561 (1996).

52. M. Qian, V. Nahoum, J. Bonicel, H. Bischoff, B. Henrissat, and F. Payan, *Biochemistry*, **40**, 7700 (2001).

53. G. Wiegand, O. Epp, and R. Huber, *J. Mol. Biol.*, **247**, 99 (1995).

54. M. Qian, R. Haser, G. Buisson, E. Duee, and F. Payan, *Biochemistry*, **33**, 6284 (1994).

55. P. M. Matias, P. Donner, R. Coelho, M. Thomaz, C. Peixoto, S. Macedo, N. Otto, S. Joschko, P. Scholz, A. Wegg, S. Basler, M. Schafer, U. Egner, and M. A. Carrondo, *J. Biol. Chem.*, **275**, 26164 (2000).

56. J. S. Sack, K. F. Kish, C. Wang, R. M. Attar, S. E. Kiefer, Y. Ang. Y. Wu, J. E. Scheffler, M. E. Salvati, S. R. Krystekjr., R. Weinmann, and H. M. Einspahr, *Proc. Nat. Acad. Sci. USA*, **98**, 4904 (2001).

57. T. Mather, V. Oganessyan, P. Hof, R. Huber, S. Foundling, C. Esmon, and W. Bode, *Embo J.*, **15**, 6822 (1996).

58. G. Ren, V. S. Reddy, A. Cheng, P. Melnyk, and A. K. Mitra, *Proc. Nat. Acad. Sci. USA*, **98**, 1398 (2001).

59. K. Murata, K. Mitsuoka, T. Hirai, T. Walz, P. Agre, J. B. Heymann, A. Engel, and Y. Fujiyoshi, *Nature*, **407**, 599 (2000).

60. J. Rossjohn, R. Cappai, S. C. Feil, A. Henry, W. J. Mckinstryd. Galatis, L. Hesse, G. Mul-

thaup, K. Beyreuther, C. L. Masters, and M. W. Parker, *Nat. Struct. Biol.*, **6**, 327 (1999).

61. C. Jelsch, L. Mourey, J. M. Masson, and J. P. Samama, *Proteins*, **16**, 364 (1993).

62. N. C. Strynadka, H. Adachi, S. E. Jensen, K. Johns, A. Sielecki, C. Betzel, K. Sutoh, and M. N. James, *Nature*, **359**, 700 (1992).

63. D. C. Lim, H. U. Park, L. Decastro, S. G. Kang, H. S. Lee, S. Jensen, K. J. Lee, and N. C. J. Strynadka, *Nat. Struct. Biol.*, **8**, 848 (2001).

64. M. C. Orencia, J. S. Yoon, J. E. Ness, W. P. Stemmer, and R. C. Stevens, *Nat. Struct. Biol.*, **8**, 238 (2001).

65. S. Ness, R. Martin, A. M. Kindler, M. Paetzel, M. Gold, J. B. Jones, and N. C. J. Strynadka, *Biochemistry*, **39**, 5312 (2000).

66. E. Fonze, P. Charlier, Y. To'Th, M. Vermeire, X. Raquet, and A. Dubus, *Acta Crystallogr.*, Sect. D, **51**, 682 (1995).

67. E. Fonze, P. Charlier, Y. To'Th, M. Vermeire, X. Raquet, and A. Dubus, *Acta Crystallogr.*, Sect. D, **51**, 682 (1995).

68. L. Maveyraud, L. Mourey, L. P. Kotra, J. D. Pedelacq, V. Guillet, S. Mobashery, and J. P. Samama, *J. Am. Chem. Soc.*, **120**, 9748 (1998).

69. L. Maveyraud, I. Massova, C. Birck, K. Miyashita, J. P. Samama, and S. Mobashery, *J. Am. Chem. Soc.*, **118**, 7435 (1996).

70. P. Swaren, D. Golemi, S. Cabantous, A. Bulychev, L. Maveyraud, S. Mobashery, and J. P. Samama, *Biochemistry*, **38**, 9570 (1999).

71. L. Maveyraud, R. F. Pratt, and J. P. Samama, *Biochemistry*, **37**, 2622 (1998).

72. J. Lowe, H. Li, K. H. Downing, and E. Nogales, *J. Mol. Biol.*, **313**, 1045 (2001).

73. E. Nogales, S. G. Wolf, and K. H. Downing, *Nature*, **391**, 199 (1998).

74. B. Gigant, P. A. Curmi, C. Martin-Barbey, E. Charbaut, S. Lachkar, L. Lebeau, S. Siavoshian, A. Sobel, and M. Knossow, *Cell*, **102**, 809 (2000).

75. J. P. Griffith, J. L. Kim, E. E. Kim, M. D. Sintchak, J. A. Thomson, M. J. Fitzgibbon, M. A. Fleming, P. R. Caron, K. Hsiao, and M. A. Navia, *Cell*, **82**, 507 (1995).

76. C. R. Kissinger, H. E. Parge, D. R. Knighton, C. T. Lewis, L. A. Pelletier, A. Tempczyk, V. J. Kalish, K. D. Tucker, R. E. Showalter, E. W. Moomaw, L. N. Gastinel, N. Habuka, X. Chen, F. Maldonado, J. E. Barker, R. Bacquet, and J. E. Villafranca, *Nature*, **378**, 641 (1995).

77. A. Desmyter, K. Decanniere, S. Muyldermans, and L. Wyns, In press.

78. P. A. Boriack, D. W. Christianson, J. Kingery-Wood, and G. M. Whitesides, *J. Med. Chem.*, **38**, 2286 (1995).

79. C.-Y. Kim and D. W. Christianson, In press.

80. B. A. Grzybowski, A. V. Ishchenko, C.-Y. Kim, G. Topalov, R. Chapman, D. W. Christianson, G. M. Whitesides, and E. I. Shakhnovich, *Proc. Nat. Acad. Sci. USA*, **99**, 1270 (2002).

81. S. K. Nair and D. W. Christianson, *Biochemistry*, **32**, 4506 (1993).

82. J. A. Ippolito and D. W. Christianson, *Biochemistry*, **32**, 9901 (1993).

83. S. Mangani and A. Liljas, *J. Mol. Biol.*, **232**, 9 (1993).

84. G. M. Smith, R. S. Alexander, D. W. Christianson, B. M. McKeever, G. S. Ponticello, J. P. Springer, W. C. Randall, J. J. Baldwin, and C. N. Habecker, *Protein Sci.*, **3**, 118 (1994).

85. K. Hakansson, C. Briand, V. Zaitsev, Y. Xue, and A. Liljas, *Acta Crystallogr. D Biol. Crystallogr.*, **50**, 101 (1994).

86. K. Hakansson, A. Wehnert, and A. Liljas, *Acta Crystallogr. D Biol. Crystallogr.*, **50**, 93 (1994).

87. A. E. Eriksson, P. M. Kylsten, T. A. Jones, and A. Liljas, *Proteins*, **4**, 283 (1988).

88. P. A. Boriack-Sjodin, S. Zeitlin, H. H. Chen, L. Crenshaw, S. Gross, A. Dantanarayana, P. Delgado, J. A. May, T. Dean, and D. W. Christianson, *Protein Sci.*, **7**, 2483 (1998).

89. K. Hakansson and A. Wehnert, *J. Mol. Biol.*, **228**, 1212 (1992).

90. C.-Y. Kim, D. A. Whittington, J. S. Chang, J. Liao, J. A. May, and D. W. Christianson, *J. Med. Chem.*, **45**, 888 (2002).

91. F. Briganti, S. Mangani, A. Scozzafava, G. Vernaglione, and C. T. Supuran, *J. Biol. Inorg. Chem.*, **4**, 528 (1999).

92. S. K. Nair, T. L. Calderone, D. W. Christianson, and C. A. Fierke, *J. Biol. Chem.*, **266**, 17320 (1991).

93. C.-Y. Kim, J. S. Chang, J. B. Doyon, T. T. Bairdjr., C. A. Fierke, A. Jain, and D. W. Christianson, *J. Am. Chem. Soc.*, **122**, 12125 (2000).

94. L. R. Scolnick, A. M. Clements, J. Liao, L. Crenshaw, M. Hellberg, J. May, T. R. Dean, and D. W. Christianson, *J. Am. Chem. Soc.*, **119**, 850 (1997).

95. S. Mangani and K. Hakansson, *Eur. J. Biochem.*, **210**, 867 (1992).

96. D. Duda, C. Tu, M. Qian, P. Laipis, M. Agbandje-Mckenna, D. N. Silverman, and R. Mckenna, *Biochemistry*, **40**, 1741 (2001).

97. F. Briganti, S. Mangani, P. Orioli, A. Scozzafava, G. Vernaglione, and C. T. Supuran, *Biochemistry*, **36**, 10384 (1997).

98. L. R. Scolnick and D. W. Christianson, *Biochemistry*, **35**, 16429 (1996).

99. R. S. Alexander, L. L. Kiefer, C. A. Fierke, and D. W. Christianson, *Biochemistry*, **32**, 1510 (1993).

100. J. F. Krebs, C. A. Fierke, R. S. Alexander, and D. W. Christianso, *Biochemistry*, **30**, 9153 (1991).

101. S. K. Nair and D. W. Christianson, *J. Am. Chem. Soc.*, **113**, 9455 (1991).

102. R. S. Alexander, S. K. Nair, and D. W. Christianson, *Biochemistry*, **30**, 11064 (1991).

103. T. Stams, S. K. Nair, T. Okuyama, A. Waheed, W. S. Sly, and D. W. Christianson, *Proc. Nat. Acad. Sci. USA*, **93**, 13589 (1996).

104. J. Vidgren, L. A. Svensson, and A. Liljas, *Nature*, **368**, 354 (1994).

105. M. Pellegrini and D. F. Mierke, *Biochemistry*, **38**, 14775 (1999).

106. S. Maignan, J. P. Guilloteau, S. Pouzieux, Y. M. Choi-Sledeski, M. R. Becker, S. I. Klein, W. R. Ewing, H. W. Pauls, A. P. Spada, and V. Mikol, *J. Med. Chem.*, **43**, 3226 (2000).

107. B. A. Katz, R. Mackman, C. Luong, K. Radika, A. Martelli, P. A. Sprengeler, J. Wang, H. Chan, and L. Wong, *Chem. Biol.*, **7**, 299 (2000).

108. K. Kamata, H. Kawamoto, T. Honma, T. Iwama, and S. H. Kim, *Proc. Nat. Acad. Sci. USA*, **95**, 6630 (1998).

109. H. Brandstetter, A. Kuhne, W. Bode, R. Huber, W. Vondersaal, K. Wirthensohn, and R. A. Engh, *J. Biol. Chem.*, **271**, 29988 (1996).

110. M. Adler, D. D. Davey, G. B. Phillips, S. H. Kim, J. Jancarik, G. Rumennik, D. L. Light, and M. Whitlow, *Biochemistry*, **39**, 12534 (2000).

111. A. Wei, R. Alexander, J. Duke, H. Ross, S. Rosenfeld, and C.-H. Chang, *J. Mol. Biol.*, **283**, 147 (1998).

112. K. Padmanabhan, K. P. Padmanabhan, A. Tulinsky, C. H. Park, W. Bode, R. Huber, D. T. Blankenship, A. D. Cardin, and W. Kisiel, *J. Mol. Biol.*, **232**, 947 (1993).

113. M. G. Malkowski, P. D. Martin, J. C. Guzik, and B. F. P. Edwards, *Protein Sci.*, **6**, 1438 (1997).

114. A. Vandelocht, W. Bode, R. Huber, B. F. Lebonniec, S. R. Stone, C. T. Esmon, and M. T. Stubbs, *Embo J.*, **16**, 2977 (1997).

115. E. Zhang and A. Tulinsky, *Biophys. Chem.*, **63**, 185 (1997).

116. H. Nar, M. Bauer, A. Schmid, J. Stassen, W. Wienen, H. W. Priepke, I. K. Kauffmann, U. J. Ries, and N. H. Hauel, *Structure*, **9**, 29 (2001).

117. A. Zdanov, S. Wu, J. DiMaio, Y. Konishi, Y. Li, X. Wu, B. F. Edwards, P. D. Martin, and M. Cygler, *Proteins*, **17**, 252 (1993)

118. N. Y. Chirgadze, D. J. Sall, S. L. Briggs, D. K. Clawson, M. Zhang, G. F. Smith, and R. W. Schevitz, *Protein Sci.*, **9**, 29 (2000).

119. L. Tabernero, C. Y. Chang, S. Ohringer, W. F. Lau, E. J. Iwanowicz, W.-C. Han, T. C. Wang, S. M. Seiler, D. G. M. Roberts, and J. S. Sack, *J. Mol. Biol.*, **246**, 14 (1995).

120. J. A. Malikayil, J. P. Burkhart, H. A. Schreuder, R. J. Broersmajunior, C. Tardif, L. W. Kutcheriii, S. Mehdi, G. L. Schatzman, B. Neises, and N. P. Peet, *Biochemistry*, **36**, 1034 (1997).

121. P. C. Weber, S. L. Lee, F. A. Lewandowski, M. C. Schadt, C. W. Chang, and C. A. Kettner, *Biochemistry*, **34**, 3750 (1995).

122. J. A. Huntington and C. T. Esmon, In press.

123. R. Krishnan, A. Tulinsky, G. P. Vlasuk, D. Pearson, P. Vallar, P. Bergum, T. K. Brunck, and W. C. Ripka, *Protein Sci.*, **5**, 422 (1996).

124. R. A. Engh, H. Brandstetter, G. Sucher, A. Eichinger, U. Baumann, W. Bode, R. Huber, T. Poll, R. Rudolph, and W. Vondersaal, *Structure (Lond)*, **4**, 1353 (1996).

125. A. Lombardi, G. Desimone, F. Nastri, S. Galdiero, R. Dellamorte, N. Staiano, C. Pedone, M. Bolognesi, and V. Pavone, *Protein Sci.*, **8**, 91 (1999).

126. E. Guinto, S. Caccia, T. Rose, K. Futterer, G. Waksman, and E. Dicera, *Proc. Nat. Acad. Sci. USA*, **96**, 1852 (1999).

127. B. E. Maryanoff, X. Qiu, K. P. Padmanabhan, A. Tulinsky, H. R. Almondjunior, P. Andrade-Gordon, M. N. Greco, J. A. Kauffman, K. C. Nicolaou, A. Liu, P. H. Brungs, and N. Fusetani, *Proc. Nat. Acad. Sci. USA*, **90**, 8048 (1993).

128. B. A. Katz, J. M. Clark, J. S. Finer-Moore, T. E. Jenkins, C. R. Johnson, M. J. Ross, C. Luong, W. R. Moore, and R. M. Stroud, *Nature*, **391**, 608 (1998).

129. J. H. Matthews, R. Krishnan, M. J. Costanzo, B. E. Maryanoff, and A. Tulinsky, *Biophys. J.*, **71**, 2830 (1996).

130. B. Bachand, M. Tarazi, Y. St-Denis, J. J. Edmunds, P. D. Winocour, L. Leblond, and M. A. Siddiqui, *Bioorg. Med. Chem. Lett.*, **11**, 287 (2001).

131. J. J. Slon-Usakiewicz, J. Sivaraman, Y. Li, M. Cygler, and Y. Konishi, *Biochemistry*, **39**, 2384 (2000).

132. R. Krishnan, E. Zhang, K. Hakansson, R. K. Arni, A. Tulinskym.S. Lim-Wilby, O. E. Levy, J. E. Semple, and T. K. Brunck, *Biochemistry*, **37**, 12094 (1998).

133. R. Krishnan, I. Mochalkin, R. Arni, and A. Tulinsky, *Acta Crystallogr.*, Sect. D, **56**, 294 (2000).

134. I. Mochalkin and A. Tulinsky, *Acta Crystallogr.*, Sect. D, **55**, 785 (1999).

135. R. Bone, T. Lu, C. R. Illig, R. M. Soll, and J. C. Spurlino, *J. Med. Chem.*, **41**, 2068 (1998).

136. R. Krishnan, E. J. Sadler, and A. Tulinsky, *Acta Crystallogr.*, Sect. D, **56**, 406 (2000).

137. M. Nardini, A. Pesce, M. Rizzi, E. Casale, R. Ferraccioli, G. Balliano, P. Milla, P. Ascenzi, and M. Bolognesi, *J. Mol. Biol.*, **258**, 851 (1996).

138. M. F. Malley, L. Tabernero, C. Y. Chang, S. L. Ohringer, D. G. Roberts, J. Das, and J. S. Sack, *Protein Sci.*, **5**, 221 (1996).

139. J. L. R. Steiner, M. Murakami, and A. Tulinsky, *J. Am. Chem. Soc.*, **120**, 597 (1998).

140. I. I. Mathews and A. Tulinsky, *Acta Crystallogr. D Biol. Crystallogr.*, **51**, 550 (1995).

141. I. I. Mathews, K. P. Padmanabhan, V. Ganesh, A. Tulinsky, M. Ishil, J. Chen, C. W. Turck, and S. R. Coughlin, *Biochemistry*, **33**, 3266 (1994).

142. J. Vijayalakshmi, K. P. Padmanabhan, K. G. Mann, and A. Tulinsky, *Protein Sci.*, **3**, 2254 (1994).

143. I. I. Mathews, K. P. Padmanabhan, A. Tulinsky, and J. E. Sadler, *Biochemistry*, **33**, 13547 (1994).

144. J. P. Priestle, J. Rahuel, H. Rink, M. Tones, and M. G. Gruetter, *Protein Sci.*, **2**, 1630 (1993).

145. T. J. Rydel, A. Tulinsky, W. Bode, and R. Huber, *J. Mol. Biol.*, **221**, 583 (1991).

146. D. W. Banner and P. Hadvary, *J. Biol. Chem.*, **266**, 20085 (1991).

147. R. K. Arni, K. Padmanabhan, K. P. Padmanabhan, T-P. Wu, and A. Tulinsky, *Biochemistry*, **32**, 4727 (1993)

148. X. Qiu, K. P. Padmanabhan, V. E. Carperos, A. Tulinsky, T. Kline, J. M. Maraganore, and J. W. Fentonii, *Biochemistry*, **31**, 11689 (1992).

149. C. Eigenbrot, D. Kirchhofer, M. S. Dennis, L. Santell, R. A. Lazarus, J. Stamos, and M. H. Ultsch, *Structure*, **9**, 627 (2001).

150. A. C. W. Pike, A. M. Brzozowski, S. M. Roberts, O. H. Olsen, and E. Persson, *Proc. Nat. Acad. Sci. USA*, **96**, 8925 (1999).

151. M. S. Dennis, C. Eigenbrot, N. J. Skelton, M. H. Ultsch, L. Santell, M. L. Dwyer, M. P. O'Connell, and R. A. Lazarus, *Nature*, **404**, 465 (2000).

152. D. W. Banner, A. D'Arcy, C. Chene, F. K. Winkler, A. Guha, W. H. Konigsberg, Y. Nemerson, and D. Kirchhofer, *Nature*, **380**, 41 (1996).

153. G. Kemball-Cook, D. J. D. Johnson, E. G. D. Tuddenham, and K. Harlos, *J. Struct. Biol.*, **127**, 213 (1999).

154. E. Zhang, R. Stcharles, and A. Tulinsky, *J. Mol. Biol.*, **285**, 2089 (1999).

155. K.-P. Hopfner, A. Lang, A. Karcher, K. Sichler, E. Kopetzkih. Brandstetter, R. Huber, W. Bode, and R. A. Engh, *Structure (Lond)*, **7**, 989 (1999).

156. H. Brandstetter, M. Bauer, R. Huber, P. Lollar, and W. Bode, *Proc. Nat. Acad. Sci. USA*, **92**, 9796 (1995).

157. M. G. Malkowski, S. L. Ginell, W. L. Smith, and R. M. Garavito, *Science*, **289**, 1933 (2000).

158. D. Picot, P. J. Loll, and R. M. Garavito, *Nature*, **367**, 243 (1994).

159. P. J. Loll, D. Picot, and R. M. Garavito, *Nat. Struct. Biol.*, **2**, 637 (1995).

160. P. J. Loll, C. T. Sharkey, S. J. O'Connor, C. M. Dooley, E. O'Brien, M. Devocelle, K. B. Nolan, B. S. Selinsky, and D. J. Fitzgerald, *Mol. Pharmacol.*, **60**, 1407 (2001).

161. E. D. Thuresson, M. G. Malkowski, K. M. Lakkides, C. J. Riekea.M. Mulichak, S. L. Ginell, R. M. Garavito, and W. L. Smith, *J. Biol. Chem.*, **276**, 10358 (2001).

162. B. S. Selinsky, K. Gupta, C. T. Sharkey, and P. J. Loll, *Biochemistry*, **40**, 5172 (2001).

163. P. J. Loll, D. Picot, O. Ekabo, and R. M. Garavito, *Biochemistry*, **35**, 7330 (1996).

164. J. R. Kiefer, J. L. Pawlitz, K. T. Moreland, R. A. Stegeman, J. K. Gierse, W. F. Hood, J. K. Gierse, A. M. Stevens, D. C. Goodwin, S. W. Rowlinson, L. J. Marnett, W. C. Stallings, and R. G. Kurumbail, *Nature*, **405**, 97 (2000).

165. R. G. Kurumbail, A. M. Stevens, J. K. Gierse, J. J. Mcdonald, R. A. Stegeman, J. Y. Pak, D. Gildehaus, J. M. Miyashiro, T. D. Penning, K. Seibert, P. C. Isakson, and W. C. Stallings, *Nature*, **384**, 644 (1996).

166. Q. Zhao, S. Modi, G. Smith, M. Paine, P. D. Mcdonagh, C. R. Wolf, D. Tew, L. Y. Lian, G. C. Roberts, and H. P. Driessen, *Protein Sci.*, **8**, 298 (1999).

167. M. Wang, D. L. Roberts, R. Paschke, T. M. Shea, B. S. Masters, and J. J. Kim, *Proc. Nat. Acad. Sci. USA*, **94**, 8411 (1997).

168. P. A. Hubbard, A. L. Shen, R. Paschke, C. B. Kasper, and J. J. Kim, *J. Biol. Chem.*, **276**, 29163 (2001).

169. A. Gangjee, A. P. Vidwans, A. Vasudevan, S. F. Queener, R. L. Kisliuk, V. Cody, R. Li, N. Gal-

itsky, J. R. Luft, and W. Pangborn, *J. Med. Chem.*, **41**, 3426 (1998).

170. V. Cody, N. Galitsky, J. R. Luft, W. Pangborn, R. L. Blakley, and A. Gangjee, *J. Mol. Biol.*, **221**, 583 (1991).

171. V. Cody, N. Galitsky, J. R. Luft, W. Pangborn, A. Rosowsky, and R. L. Blakley, *Biochemistry*, **36**, 13897 (1997).

172. M. A. McTigue, J. F. Davies II, B. T. Kaufman, and J. Kraut, *Biochemistry*, **31**, 7264 (1992).

173. M. A. McTigue, J. F. Davies II, B. T. Kaufman, and J. Kraut, *Biochemistry*, **32**, 6855 (1993).

174. M. A. McTigue, J. F. Daviesii, B. T. Kaufman, N.-H. Xuong, and J. Kraut, In Press.

175. J. F. Davies, T. J. Delcamp, N. J. Prendergast, V. A. Ashford, J. H. Freisheim, and J. Kraut, *Biochemistry*, **29**, 9467 (1990).

176. W. S. Lewis, V. Cody, N. Galitsky, J. R. Luft, W. Pangborn, S. K. Chunduru, H. T. Spencer, J. R. Appleman, and R. L. Blakley, *J. Biol. Chem.*, **270**, 5057 (1995).

177. J. F. Davies, D. A. Matthews, S. J. Oatley, B. T. Kaufman, N.-H. Xuong, and J. Kraut, In press.

178. C. Oefner, A. D'Arcy, and F. K. Winkler, *Eur. J. Biochem.*, **174**, 377 (1988).

179. S. Liu, E. A. Neidhardt, T. H. Grossman, T. Ocain, and J. Clardy, *Structure (Lond)*, **8**, 25 (2000).

180. I. C. Hampele, A. D'Arcy, G. E. Dale, D. Kostrewa, J. Nielsenc. Oefner, M. G. Page, H. J. Schonfeld, D. Stuber, and R. L. Then, *J. Mol. Biol.*, **268**, 21 (1997).

181. P. Soultanas, M. S. Dillingham, S. S. Velankar, and D. B. Wigley, *J. Mol. Biol.*, **290**, 137 (1999).

182. M. R. Redinbo, J. J. Champoux, and W. G. Hol, *Biochemistry*, **39**, 6832 (2000).

183. L. Stewart, M. R. Redinbo, X. Qiu, W. G. Hol, and J. J. Champoux, *Science*, **279**, 1534 (1998).

184. M. R. Redinbo, L. Stewart, P. Kuhn, J. J. Champoux, and W. G. Hol, *Science*, **279**, 1504 (1998).

185. M. Ruff, M. Gangloff, S. Eiler, S. Duclaud, J. M. Wurtz, and M. Dino, In press.

186. J. W. R. Schwabe, L. Chapman, J. T. Finch, D. Rhodes, and D. Neuhaus, *Structure (Lond)*, **1**, 187 (1993).

187. D. M. Tanenbaum, Y. Wang, S. P. Williams, and P. B. Sigler, *Proc. Nat. Acad. Sci. USA*, **95**, 5998 (1998).

188. A. M. Brzozowski, A. C. W. Pike, Z. Dauter, R. E. Hubbard, T. Bonn, O. Engstrom, L. Ohman, G. L. Greene, J.-A. Gustaffson, and M. Carlquist, *Nature*, **389**, 753 (1997).

189. J. W. R. Schwabe, L. Chapman, J. T. Finch, and D. Rhodes, *Cell*, **75**, 567 (1993).

190. A. K. Shiau, D. Barstad, P. M. Loria, L. Cheng, P. J. Kushner, D. A. Agard, and G. L. Greene, *Cell*, **95**, 927 (1998).

191. J. W. R. Schwabe, L. Chapman, J. T. Finch, and D. Rhodes, *Cell*, **75**, 567 (1993).

192. A. K. Shiau, D. Barstad, P. M. Loria, L. Cheng, P. J. Kushner, D. A. Agard, and G. L. Greene, *Cell*, **95**, 927 (1998).

193. K. P. Wilson, M. M. Yamashita, M. D. Sintchak, S. H. Rotsteinm.A. Murcko, J. Boger, J. A. Thomson, M. J. Fitzgibbon, and M. A. Navia, *Acta Crystallogr.*, Sect. D, **51**, 511 (1995).

194. J. Choi, J. Chen, S. L. Schreiber, and J. Clardy, *Science*, **273**, 239 (1996).

195. J. Liang, J. Clardy, and J. Choi, *Acta Crystallogr.*, Sect. D, **55**, 736 (1999).

196. J. Liang, J. Choi, and J. Clardy, *Acta Crystallogr. D Biol. Crystallogr.*, **55**, 736 (1999).

197. S. W. Michnick, M. K. Rosen, T. J. Wandless, M. Karplus, and S. L. Schreiber, *Science*, **252**, 836 (1991)

198. C. T. Rollins, V. M. Rivera, D. N. Woolfson, T. Keenan, M. Hatada, S. E. Adams, L. J. Andrade, D. Yaeger, M. R. Vanschravendijk, D. A. Holt, M. Gilman, and T. Clackson, *Proc. Nat. Acad. Sci. USA*, **97**, 7096 (2000).

199. T. Clackson, W. Yang, L. Rozamus, J. Amara, M. H. Hatada, C. T. Rollins, L. F. Stevenson, S. R. Magari, S. A. Wood, N. L. Courage, X. Lu, F. Cerasolijunior, M. Gilman, and D. Holt, *Proc. Nat. Acad. Sci. USA*, **95**, 10437 (1998).

200. P. Burkhard, P. Taylor, and M. D. Walkinshaw, *J. Mol. Biol.*, **295**, 953 (2000).

201. J. W. Becker, J. Rotonda, J. G. Cryan, M. Martin, W. H. Parsons, P. J. Sinclair, G. Wiederrecht, and F. Wong, *J. Med. Chem.*, **42**, 2798 (1999).

202. C. Sich, S. Improta, D. J. Cowley, C. Guenet, J. P. Merly, M. Teufel, and V. Saudek, *Eur. J. Biochem.*, **267**, 5342 (2000).

203. M. Huse, Y. G. Chen, J. Massague, and J. Kuriyan, *Cell*, **96**, 425 (1999).

204. L. W. Schultz and J. Clardy, *Bioorg. Med. Chem. Lett.*, **8**, 1 (1998).

205. S. Itoh, M. T. Decenzo, D. J. Livingston, D. A. Pearlman, and M. A. Navia, *Bioorg. Med. Chem. Lett.*, **5**, 1983 (1995).

206. J. Liang, J. Choi, and J. Clardy, *Acta Crystallogr.*, Sect. D, **55**, 736 (1999).

207. C. C. S. Deivanayagam, M. Carson, A. Thotakura, S. V. L. Narayana, and C. S. Chodavarapu, *Acta Crystallogr.*, Sect. D, **56**, 266 (2000).

208. D. A. Holt, J. I. Luengo, D. S. Yamashita, H.-J. Oh, A. L. Konialian, H.-K. Yen, L. W. Rozamus, M. Brandt, M. J. Bossard, M. A. Levy, D. S. Eggleston, T. J. Stout, J. Liang, L. W. Schultz, and J. Clardy, *J. Am. Chem. Soc.*, **115**, 9925 (1993).

209. G. D. VanDuyne, R. F. Standaert, P. A. Karplus, S. L. Schreiber, and J. Clardy, *Science*, **252**, 839 (1991).

210. G. D. Vanduyne, R. F. Standaert, S. L. Schreiber, and J. Clardy, *J. Am. Chem. Soc.*, **113**, 7433 (1991).

211. K. M. Fox, J. A. Dias, and P. Vanroey, *Mol. Endocrinol.*, **15**, 378 (2001).

212. P. Storici, G. Capitani, D. Debiase, M. Moser, R. A. John, J. N. Jansonius, and T. Schirmer, *Biochemistry*, **38**, 8628 (1999).

213. D. T. Gewirth and P. B. Sigler, *Nat. Struct. Biol.*, **2**, 386 (1995).

214. B. F. Luisi, W. X. Xu, Z. Otwinowski, L. P. Freedman, K. R. Yamamoto, and P. B. Sigler, *Nature*, **352**, 497 (1991).

215. N. Kunishima, Y. Shimada, Y. Tsuji, T. Sato, M. Yamamoto, T. Kumasaka, S. Nakanishi, H. Jingami, and K. Morikawa, *Nature*, **407**, 971 (2000).

216. O. Epp, R. Ladenstein, and A. Wendel, *Eur. J. Biochem.*, **133**, 51 (1983)

217. M. Aritomi, N. Kunishima, T. Okamoto, R. Kuroki, Y. Ota, and K. Morikawa, *Nature*, **401**, 713 (1999).

218. B. Lovejoy, D. Cascio, and D. Eisenberg, *J. Mol. Biol.*, **234**, 640 (1993).

219. T. Zink, A. Ross, K. Luers, C. Ciesler, R. Rudolph, and T. A. Hola, *Biochemistry*, **33**, 8453 (1994).

220. C. P. Hill, T. D. Osslund, and D. Eisenberg, *Proc. Natl. Acad. Sci. USA*, **90**, 5167 (1993).

221. D. Rozwarski, K. Diederichs, R. Hecht, T. Boone, and P. A. Karplus, *Proteins*, **26**, 304 (1996).

222. T. Clackson, M. H. Ultsch, J. A. Wells, and A. M. Devos, *J. Mol. Biol.*, **277**, 1111 (1998).

223. S. Atwell, M. Ultsch, A. M. Devos, and J. A. Wells, *Science*, **278**, 1125 (1997).

224. M. Sundstrom, T. Lundqvist, J. Rodin, L. B. Giebel, D. Milligan, and G. Norstedt, *J. Biol. Chem.*, **271**, 32197 (1996).

225. A. M. Devos, M. Ultsch, and A. A. Kossiakoff, *Science*, **255**, 306 (1992).

226. Y. Hsiou, J. Ding, K. Das, A. D. Clarkjunior, S. H. Hughes, and E. Arnold, *Structure*, **4**, 853 (1996).

227. J. Ren, R. M. Esnouf, A. L. Hopkins, E. Y. Jones, I. Kirby, J. Keeling, C. K. Ross, B. A. Larder, D. I. Stuart, and D. K. Stammers, *Proc. Nat. Acad. Sci. USA*, **95**, 9518 (1998).

228. Y. Hsiou, J. Ding, K. Das, A. D. Clark, P. L. Boyer, P. Lewi, P. A. J. Janssen, J.-P. Kleim, M. Roesner, S. H. Hughes, and E. Arnold, *J. Mol. Biol.*, **309**, 437 (2001).

229. Y. Hsiou, K. Das, J. Ding, A. D. Clarkjunior, J. P. Kleim, M. Rosner, I. Winkler, G. Riess, S. H. Hughes, and E. Arnold, *J. Mol. Biol.*, **284**, 313 (1998).

230. K. Das, J. Ding, Y. Hsiou, A. D. Clarkjunior, H. Moereels, L. Koymans, K. Andries, R. Pauwels, P. A. Janssen, P. L. Boyer, P. Clark, R. H. Smithjunior, M. B. Kroegersmith, C. J. Michejda, S. H. Hughes and E. Arnold, *J. Mol. Biol.*, **264**, 1085 (1996).

231. M. Hogberg, C. Sahlberg, P. Engelhardt, R. Noreen, J. Kangasmetsa, N. G. Johansson, B. Oberg, L. Vrang, H. Zhang, B. L. Sahlberg, T. Unge, S. Lovgren, K. Fridborg, and K. Backbro, *J. Med. Chem.*, **43**, 304 (2000).

232. J. Lindberg, S. Sigurdsson, S. Lowgren, H. O. Andersson, C. Sahlberg, R. Noreen, K. Fridborg, H. Zhang, and T. Unge, *Eur. J. Biochem.*, **269**, 1670 (2002).

233. J. Jaeger, T. Restle, and T. A. Steitz, *Embo J.*, **17**, 4535 (1998).

234. A. L. Hopkins, J. Ren, H. Tanaka, B. Baba, M. Okamato, D. I. Stuart, and D. K. Stammers, *J. Med. Chem.*, **42**, 4500 (1999).

235. J. Ding, K. Das, H. Yu, S. G. Sarafianos, A. D. Clarkjunior, A. Jacobo-Molina, C. Tantillo, S. H. Hughes, and E. Arnold, *J. Mol. Biol.*, **284**, 1095 (1998).

236. S. G. Sarafianos, K. Das, C. Tantillo, A. D. Clarkjr., J. Ding, J. Whitcomb, P. L. Boyer, S. H. Hughes, and E. Arnold, *Embo J.*, **20**, 1449 (2001).

237. D. W. Rodgers, S. J. Gamblin, B. A. Harris, S. Ray, J. S. Culp, B. Hellmig, D. J. Woolf, C. Debouck, and S. C. Harrison, *Proc. Natl. Acad. Sci. USA*, **92**, 1222 (1995).

238. J. Ding, K. Das, C. Tantillo, W. Zhang, A. D. Clarkjunior, S. Jessen, X. Lu, Y. Hsiou, A. Jacobo-Molina, K. Andries, R. Pauwels, H. Moereels, L. Koymans, P. A. J. Janssen, R. H. Smithjunior, M. K. Koepke, C. J. Michejda, S. H. Hughes, and E. Arnold, *Structure (Lond)*, **3**, 365 (1995).

239. J. Ding, K. Das, H. Moereels, L. Koymans, K. Andries, P. A. J. Janssen, S. H. Hughes, and E. Arnold, *Nat. Struct. Biol.*, **2**, 407 (1995).

240. J. Ren, J. Milton, K. L. Weaver, S. A. Short, D. I. Stuart, and D. K. Stammers, *Structure*, **8**, 1089 (2000).

241. J. Ren, C. Nichols, L. Bird, P. Chamberlain, K. Weaver, S. Short, D. I. Stuart, and D. K. Stammers, *J. Mol. Biol.*, **312**, 795 (2001).

242. S. G. Sarafianos, K. Das, C. Tantillo, A. D. Clarkjr., J. Ding, J. Whitcomb, P. L. Boyer, S. H. Hughes, and E. Arnold, *Proc. Nat. Acad. Sci. USA*, **96**, 10027 (1999).

243. J. Wang, S. J. Smerdon, J. Jager, L. A. Kohlstaedt, P. A. Rice, J. M. Friedman, T. A. Steitz, *Proc. Natl. Acad. Sci. USA*, **91**, 7242 (1994).

244. M. D. Sintchak, M. A. Fleming, O. Futer, S. A. Raybuck, S. P. Chambers, P. R. Caron, M. A. Murcko, and K. P. Wilson, *Cell*, **85**, 921 (1996).

245. T. D. Colby, K. Vanderveen, M. Strickler, G. D. Markham, and B. M. Goldstein, *Proc. Nat. Acad. Sci. USA*, **96**, 3531 (1999).

246. L. G. Laajoki, G. L. Francis, J. C. Wallace, J. A. Carver, and M. A. Keniry, *J. Biol. Chem.*, **275**, 10009 (2000).

247. A. Sato, S. Nishimura, T. ohkubo, Y. Kyogoku, S. Koyama, M. Kobayashi, T. Vasuda, and Y. Kobayashi, *Int. J. Pept. Protein Res.*, **41**, 433 (1993).

248. F. F. Vajdos, M. Ultsch, M. L. Schaffer, K. D. Deshayes, J. Liu, N. J. Skelton, and A. M. Devos, *Biochemistry*, **40**, 11022 (2001).

249. E. Dewolf, R. Gill, S. Geddes, J. Pitts, A. Wollmer, and J. Grotzinger, *Protein Sci.*, **5**, 2193 (1996).

250. R. M. Cooke, T. S. Harvey, and I. D. Campbell, *Biochemistry*, **30**, 5484 (1991).

251. T. P. J. Garrett, N. M. Mckern, M. Lou, M. J. Frenkel, J. D. Bentley, G. O. Lovrecz, T. C. Elleman, L. J. Cosgrove, and C. W. Ward, *Nature*, **394**, 395 (1998).

252. K. Parang, J. H. Till, A. J. Ablooglu, R. A. Kohanski, S. R. Hubbard, and P. A. Cole, *Nat. Struct. Biol.*, **8**, 37 (2001).

253. J. H. Till, A. J. Ablooglu, M. Frankel, S. M. Bishop, R. A. Kohanski, and S. R. Hubbard, *J. Biol. Chem.*, **276**, 10049 (2001).

254. S. R. Hubbard, *Embo J.*, **16**, 5572 (1997).

255. S. R. Hubbard, L. Wei, L. Ellis, and W. A. Hendrickson, *Nature*, **372**, 746 (1994).

256. A. M. Torres, B. E. Forbes, S. E. Aplin, J. C. Wallace, G. L. Francis, and R. S. Norton, *J. Mol. Biol.*, **248**, 385 (1995).

257. E. T. Baldwin, R. W. Sarver, G. L. Bryantjunior, K. A. Currym.B. Fairbanks, B. C. Finzel, R. L. Garlick, R. L. Heinrikson, N. C. Horton,

L. L. Kelley, A. M. Mildner, J. B. Moon, J. E. Mott, V. T. Mutchler, C. S. Tomich, K. D. Watenpaugh, and V. H. Wiley, *Structure (Lond)*, **6**, 923 (1998).

258. J. O. Lee, L. A. Bankston, M. A. Arnaout, and R. C. Liddington, *Structure (Lond)*, **3**, 1333 (1995).

259. J. O. Lee, P. Rieu, M. A. Arnaout, and R. Liddington, *Cell*, **80**, 631 (1995).

260. J. Bella, P. R. Kolatkar, C. W. Marlor, J. M. Greve, and M. G. Rossmann, *Proc. Nat. Acad. Sci. USA*, **95**, 4140 (1998).

261. J. M. Casasnovas, T. Stehle, J. H. Liu, J. H. Wang, and T. A. Springer, *Proc. Nat. Acad. Sci. USA*, **95**, 4134 (1998).

262. P. R. Kolatkar, J. Bella, N. H. Olson, C. Bator, T. S. Baker, and M. G. Rossmann, *Embo J.*, **18**, 6249 (1999).

263. W. Klaus, B. Gsell, A. M. Labhardt, B. Wipf, and H. Senn, *J. Mol. Biol.*, **274**, 661 (1997).

264. R. Radhakrishnan, L. J. Walter, A. Hruza, P. Reichert, P. P. Trotta, T. L. Nagabhushan, and M. R. Walter, *Structure*, **4**, 1453 (1996)

265. D. J. Thiel, M.-H. Ledu, R. L. Walter, A. D'Arcy, C. Chene, M. Fountoulakis, G. Garotta, F. K. Winkler, and S. E. Ealick, *Structure Fold Des.*, **8**, 927 (2000).

266. M. Randal and A. A. Kossiakoff, *Structure*, **9**, 155 (2001).

267. A. Landar, B. Curry, M. H. Parker, R. Digiacomo, S. R. Indelicato, T. L. Nagabhushan, G. Rizzi, and M. R. Walter, *J. Mol. Biol.*, **299**, 169 (2000).

268. S. E. Ealick, W. J. Cook, S. Vijay-Kumar, M. Carson, T. L. Nagabhushan, P. P. Trotta, and C. E. Bugg, *Science*, **252**, 698 (1991).

269. B. J. Graves, M. H. Hatada, W. A. Hendrickson, J. K. Miller, V. S. Madison, and Y. Satow, *Biochemistry*, **29**, 2679 (1990).

270. G. P. A. Vigers, D. J. Dripps, C. K. Edwards, and B. J. Brandhuber, *Structure*, **275**, 36927 (2000).

271. H. Schreuder, C. Tardif, S. Trump-Kallmeyer, A. Soffientini, E. Sarubbi, A. Akeson, T. Bowlin, S. Yanofsky, and R. W. Barrett, *Nature*, **386**, 194 (1997).

272. G. P. Vigers, L. J. Anderson, P. Caffes, and B. J. Brandhuber, *Nature*, **386**, 190 (1997).

273. A. Zdanov, C. Schalk-Hihi, S. Menon, K. W. Moore and A. Wlodawer, *J. Mol. Biol.*, **268**, 460 (1997).

274. A. Zdanov, C. Schalk-Hihi, and A. Wlodawer, *Protein Sci.*, **5**, 1955 (1996).

275. A. Zdanov, C. Schalk-Hihi, A. Gustchina, M. Tsang, J. Weatherbee, and A. Wlodawer, *Structure*, **3**, 591 (1995).

276. K. Josephson, N. J. Logsdon, and M. R. Walter, *Immunity*, **15**, 35 (2001).

277. M. R. Walter and T. L. Nagabhushan, *Biochemistry*, **34**, 12118 (1995).

278. C. Yoon, S. C. Johnston, J. Tang, M. Stahl, J. F. Tobin, and W. S. Somers, *Embo J.*, **19**, 3530 (2000).

279. E. Z. Eisenmesser, D. A. Horita, A. S. Altieri, and R. A. Byrd, *J. Mol. Biol.*, **310**, 231 (2001).

280. H. R. Mott, B. S. Baines, R. M. Hall, R. M. Cooke, P. C. Driscoll, M. P. Weir, and I. D. Campbell, *J. Mol. Biol.*, **248**, 979 (1995).

281. D. B. McKay, *Science*, **257**, 412 (1992).

282. Y. Feng, B. K. Klein, and C. A. Mcwherter, *J. Mol. Biol.*, **259**, 524 (1996).

283. T. Mueller, F. Oehlenschlaeger, and M. Buehner, *J. Mol. Biol.*, **247**, 360 (1955).

284. T. Hage, W. Sebald, and P. Reinemer, *Cell*, **97**, 271 (1999).

285. M. Hulsmeyer, C. Scheufler, and M. K. Dreyer, *Acta Crystallogr.*, Sect. D, **57**, 1334 (2001).

286. C. Redfield, L. J. Smith, J. Boyd, G. M. P. Lawrence, R. G. Edwards, C. J. Gershater, R. A. G. Smith, and C. M. Dobson, *J. Mol. Biol.*, **238**, 23 (1994).

287. R. Powers, D. S. Garrett, C. J. March, E. A. Frieden, A. M. Gronenborn, and G. M. Clore, *Science*, **256**, 1673 (1992).

288. T. Mueller, T. Dieckmann, W. Sebald, and H. Oschkinat, *J. Mol. Biol.* **237**, 423 (1994)

289. T. Mueller, T. Dieckmann, W. Sebald, and H. Oschkinat, *J. Mol. Biol.*, **237**, 423 (1994).

290. L. J. Smith, C. Redfield, J. Boyd, G. M. P. Lawrence, R. G. Edwards, R. A. G. Smith, and C. M. Dobson, *J. Mol. Biol.*, **224**, 899 (1992).

291. M. R. Walter, W. J. Cook, B. G. Zhao, R. Cameronjunior, S. E. Ealick, R. L. Walterjunior, P. Reichert, T. L. Nagabhushan, P. P. Trotta, and C. E. Bugg, *J. Biol. Chem.*, **267**, 20371 (1992).

292. A. Wlodawer, A. Pavlovsky, and A. Gustchina, *Febs Lett.*, **309**, 59 (1992).

293. R. Powers, D. S. Garrett, C. J. March, E. A. Frieden, A. M. Gronenborn, and G. M. Clore, *Biochemistry*, **32**, 6744 (1993).

294. M. V. Milburn, A. M. Hassell, M. H. Lambert, S. R. Jordan, A. E. I. Proudfoot, P. Graber, and T. N. C. Wells, *Nature*, **363**, 172 (1993).

295. G. Y. Xu, H. A. Yu, J. Hong, M. Stahl, T. Mcdonagh, L. E. Kay, and D. A. Cumming, *J. Mol. Biol.*, **268**, 468 (1997).

296. W. Somers, M. Stahl, and J. S. Seehra, *Embo J.*, **16**, 989 (1997).

297. K. Rajarathnam, I. Clark-Lewis, and B. D. Sykes, *Biochemistry*, **34**, 12983 (1995).

298. C. Eigenbrot, H. B. Lowman, L. Chee, and D. R. Artis, *Proteins*, **27**, 556 (1997).

299. N. J. Skelton, C. Quan, D. Reilly, and H. Lowman, *Structure Fold Des.*, **7**, 157 (1999).

300. N. Gerber, H. Lowman, D. R. Artis, and C. Eigenbrot, *Proteins: Struct., Funct., Genet.*, **38**, 361 (2000).

301. H. Sticht, M. Auer, B. Schmitt, J. Besemer, M. Horcher, T. Kirsch, J. D. Lindley, and P. Roesch, *Eur. J. Biochem.*, **235**, 26 (1996).

302. E. T. Baldwin, I. T. Weber, R. St. Charles, J.-C. Xuan, E. Appella, M. Yamada, K. Matsushima, B. F. P. Edwards, G. M. Clore, A. M. Gronenborn, and A. Wlodawer, *Proc. Nat. Acad. Sci. USA*, **88**, 502 (1991).

303. G. M. Clore, E. Appella, M. Yamada, K. Matsushima, and A. M. Gronenborn, *Biochemistry*, **29**, 1689 (1990).

304. M. M. G. M. Thunnissen, P. N. Nordlund, and J. Z. Haeggstrom, *Nat. Struct. Biol.*, **8**, 131 (2001).

305. X. Weng, H. Luecke, I. S. Song, D. S. Kang, S-H. Kim, and R. Huber, *Protein Sci.*, **2**, 448 (1993).

306. A. Rosengarth, V. Gerke, and H. Luecke, *J. Mol. Biol.*, **306**, 489 (2001).

307. M. Tegoni, S. Spinelli, M. Verhoeyen, P. Davis, and C. Cambilla, *J. Mol. Biol.*, **289**, 1375 (1999).

308. H. Wu, J. W. Lustbader, Y. Liu, R. E. Canfield, and W. A. Hendrickson, *Structure*, **2**, 545 (1994).

309. A. J. Lapthorn, D. C. Harris, A. Littlejohn, J. W. Lustbader, R. E. Canfield, K. J. Machin, F. J. Morgan, and N. W. Isaacs, *Nature*, **369**, 455 (1994).

310. A. Bohm, J. Pandit, J. Jancarik, R. Halenbeck, K. Koths, and S.-H. Kim, *Science*, **258**, 1358 (1992).

311. M. J. Jedrzejas, S. Singh, W. J. Brouillette, W. G. Laver, G. M. Air, and M. Luo, *J. Mol. Biol.*, **267**, 584 (1997).

312. N. R. Taylor, A. Cleasby, O. Singh, T. Skarzynski, A. J. Wonacott, P. W. Smith, S. L. Sollis, P. D. Howes, P. C. Cherry, R. Bethell, P. Colman, and J. Varghese, *J. Med. Chem.*, **41**, 798 (1998).

313. M. J. Jedrzejas, S. Singh, W. J. Brouillette, W. G. Laver, G. M. Air, and M. Luo, *Biochemistry*, **34**, 3144 (1995).

314. W. P. Burmeister, B. Henrissat, C. Bosso, S. Cusack, and R. W. H. Ruigrok, *Structure*, **1**, 19 (1993).

315. J. B. Finley, V. R. Atigadda, F. Duarte, J. J. Zhao, W. J. Brouillette, G. M. Air, and M. Luo, *J. Mol. Biol.*, **293**, 1107 (1999).

316. C. L. White, M. N. Janakiraman, W. G. Laver, C. Philippon, A. Vasella, G. M. Air, and M. Luo, *J. Mol. Biol.*, **245**, 623 (1995).

317. W. P. Burmeister, R. W. H. Ruigrok, and S. Cusack, *Embo J.*, **11**, 49 (1992).

318. S. A. Monks, G. Karagianis, G. J. Howlett, and R. S. Norton, *J. Biomol. NMR*, **8**, 379 (1996).

319. R. Bader, A. Bettio, A. G. Beck-Sickinger, and O. Zerbe, *J. Mol. Biol.*, **305**, 307 (2001).

320. C. Cabrele, M. Langer, R. Bader, H. A. Wieland, H. N. Doods, O. Zerbe, and A. G. Beck-Sickinger, *J. Biol. Chem.*, **275**, 36043 (2000).

321. G. Seidel, W. Schaefer, A. Esswein, E. Hofmann, and P. Roesch, In Press.

322. Z. Chen, P. Xu, J.-R. Barbier, G. Willick, and F. Ni, *Biochemistry*, **39**, 12766 (2000).

323. U. C. Marx, K. Adermann, P. Bayer, W.-G. Forssmann, and P. Roesc, *Biochem. Biophys. Res. Comm.*, **267**, 213 (2000).

324. L. Jin, S. L. Briggs, S. Chandrasekhar, N. Y. Chirgadze, D. K. Clawson, R. W. Schevitz, D. L. Smiley, A. H. Tashjian, and F. Zhang, *J. Biol. Chem.*, **275**, 27238 (2000).

325. U. C. Marx, S. Austermann, P. Bayer, K. Adermann, A. Ejcharth. Sticht, S. Walter, F.-X. Schmid, R. Jaenicke, W.-G. Forssmann and P. Roesch, *J. Biol. Chem.*, **270**, 15194 (1995).

326. U. C. Marx, *Strukturen Verschiedener Parathormonfragmente in Loesung*, University of Bayreuth Thesis, Bayreuth, 1996.

327. G. Y. Xu, T. Mcdonagh, H. A. Yu, E. A. Nalefski, J. D. Clark, and D. A. Cumming, *J. Mol. Biol.*, **280**, 485 (1998).

328. O. Perisic, S. Fong, D. E. Lynch, M. Bycroft, and R. L. Williams, *J. Biol. Chem.*, **273**, 1596 (1998).

329. A. Dessen, J. Tang, H. Schmidt, M. Stahl, J. D. Clark, J. Seehra, and W. S. Somers, *Cell*, **97**, 349 (1999).

330. A. Kreusch, P. J. Pfaffinger, C. F. Stevens, and S. Choe, *Nature*, **392**, 945 (1998).

331. S. J. Cushman, M. H. Nanao, A. W. Jahng, D. Derubeis, S. Choe, and P. J. Pfaffinger, *Nat. Struct. Biol.*, **7**, 403 (2000).

332. K. A. Bixby, M. H. Nanao, N. V. Shen, A. Kreusch, H. Bellamy, P. J. Pfaffinger, and S. Choe, *Nat. Struct. Biol.*, **6**, 38 (1998).

333. J. M. Gulbis, M. Zhou, S. Mann, and R. Mackinnon, *Science*, **289**, 123 (2000).

334. D. L. Minorjr., Y.-F. Lin, B. C. Mobley, A. Avelar, Y. N. Janl.Y. Jan, and J. M. Berger, *Cell*, **102**, 657 (2000).

335. J. L. Oberfield, J. L. Collins, C. P. Holmes, D. M. Goreham, J. P. Cooper, J. E. Cobb, J. M. Lenhard, E. A. Hull-Ryde, C. P. Mohr, S. G. Blanchard, D. J. Parks, L. B. Moore, J. M. Lehmann, K. Plunket, A. B. Miller, M. V. Milburn, S. A. Kliewer, and T. M. Wilson, *Proc. Nat. Acad. Sci. USA*, **96**, 6102 (1999).

336. R. T. Nolte, G. B. Wisely, S. Westin, J. E. Cobb, M. H. Lambert, R. Kurokawa, M. G. Rosenfeld, T. M. Willson, C. K. Glass, and M. V. Milburn, *Nature*, **395**, 137 (1998).

337. R. T. Gampejr., V. G. Montana, M. H. Lambert, A. B. Miller, R. K. Bledsoe, M. V. Milburn, S. A. Kliewer, T. M. Willson, and H. E. Xu, *Mol. Cell*, **5**, 545 (2000).

338. J. Uppenberg, C. Svensson, M. Jaki, G. Bertilsson, L. Jendeberg, and A. Berkenstam, *J. Biol. Chem.*, **273**, 31108 (1998).

339. S. P. Williams and P. B. Sigler, *Nature*, **393**, 392 (1998).

340. W. Somers, M. Ultsch, A. M. Devos, and A. A. Kossiakoff, *Nature*, **372**, 478 (1994).

341. P. A. Elkins, H. W. Christinger, Y. Sandowski, E. Sakal, A. Gertler, A. M. Devos, and A. A. Kossiakoff, *Nat. Struct. Biol.*, **7**, 808 (2000).

342. F. Rastinejad, T. Wagner, Q. Zhao, and S. Khorasanizadeh, *Embo J.*, **19**, 1045 (2000).

343. B. P. Klaholz, A. Mitschler, M. Belema, C. Zusi, and D. Moras, *Proc. Nat. Acad. Sci. USA*, **97**, 6322 (2000).

344. J. P. Renaud, N. Rochel, M. Ruff, V. Vivat, P. Chambon, H. Gronemeyer, and D. Moras, *Nature*, **378**, 681 (1995).

345. B. P. Klaholz, J. P. Renaud, A. Mitschler, C. Zusi, P. Chambon, H. Gronemeyer, and D. Moras, *Nat. Struct. Biol.*, **5**, 199 (1998).

346. W. Bourguet, V. Vivat, J. M. Wurtz, P. Chambon, H. Gronemeyer, and D. Moras, *Mol. Cell*, **5**, 289 (2000).

347. B. P. Klaholz, A. Mitschler, and D. Moras, *J. Mol. Biol.*, **302**, 155 (2000).

348. R. M. A. Knegtel, M. Katahira, J. G. Schilthuis, A. M. J. J. Bonvin, R. Boelens, D. Eib, P. T. Vandersaag, and R. Kaptein, *J. Biomol. NMR*, **3**, 1 (1993).

349. W. Bourguet, M. Ruff, P. Chambon, H. Gronemeyer, and D. Moras, *Nature*, **375**, 377 (1995).

350. F. Rastinejad, T. Perlmann, R. M. Evans, and P. B. Sigler, *Nature*, **375**, 203 (1995).

351. S. M. Holmbeck, M. P. Foster, D. R. Casimiro, D. S. Sem, H. J. Dyson, and P. E. Wright, *J. Mol. Biol.*, **281**, 271 (1998).

352. R. T. Gampejr., V. G. Montana, M. H. Lambert, G. B. Wisely, M. V. Milburn, and H. E. Xu, *Genes Dev.*, **14**, 2229 (2000).

353. P. F. Egea, A. Mitschler, N. Rochel, M. Ruff, P. Chambon, and D. Moras, *Embo J.*, **19**, 2592 (2000).

354. Q. Zhao, S. A. Chasse, S. Devarakonda, M. L. Sierk, and B. A. Rastinejad, *J. Mol. Biol.*, **296**, 509 (2000).

355. D. R. Hall, J. M. Hadden, G. A. Leonard, S. Bailey, M. Neu, M. Winn, and P. F. Lindley, *Acta Crystallogr.*, Sect. D, **58**, 70 (2002).

356. Z. Zhang, R. Zhang, A. Joachimiak, J. Schlessinger, and X. Kong, *Proc. Nat. Acad. Sci. USA*, **97**, 7732 (2000).

357. X. Jiang, O. Gurel, E. A. Mendiaz, G. W. Stearns, C. L. Clogston, H. S. Lu, T. D. Osslund, R. S. Syed, K. E. Langley, and W. A. Hendrickson, *Embo J.*, **19**, 3192 (2000).

358. J. N. Champness, M. S. Bennett, F. Wien, R. Visse, W. C. Summers, P. Herdewijn, E. Declercq, T. Ostrowski, R. L. Jarvest, and M. R. Sanderson, *Proteins*, **32**, 350 (1998).

359. M. S. Bennett, F. Wien, J. N. Champness, T. Batuwangala, T. Rutherford, W. C. Summers, H. Sun, G. Wright, and M. R. Sanderson, *Febs Lett.*, **443**, 121 (1999).

360. J. N. Champness, M. S. Bennett, F. Wien, R. Visse, W. C. Summers, P. Herdewijn, E. Declerq, T. Ostrowski, R. L. Jarvest, and M. R. Sanderson, *Proteins Struct. Funct. Genet.*, **32**, 350 (1998).

361. K. Wild, T. Bohner, G. Folkers, and G. E. Schulz, *Protein Sci.*, **6**, 2097 (1997).

362. J. Vogt, R. Perozzo, A. Pautsch, A. Prota, P. Schelling, B. Pilger, G. Folkers, L. Scapozza, and G. E. Schulz, *Proteins Struct. Funct. Genet.*, **41**, 545 (2000).

363. C. Wurth, U. Kessler, J. Vogt, G. E. Schulz, G. Folkers, and L. Scapozza, *Protein Sci.*, **10**, 63 (2001).

364. A. Prota, J. Vogt, B. Pilger, R. Perozzo, C. Wurth, V. Marquez, P. Russ, G. E. Schulz, G. Folkers, and L. Scapozza, *Biochemistry*, **39**, 9597 (2000).

365. J. H. Naismith, T. Q. Devine, B. Brandhuber, and S. R. Sprang, *J. Biol. Chem.*, **270**, 13303 (1995).

366. J. H. Naismith, T. Q. Devine, H. Khono, and S. R. Sprang, *Structure*, **4**, 1251 (1996).

367. D. W. Banner, A. D'Arcy, W. Janes, R. Gentz, H-J. Schoenfeld, C. Broger, H. Loetscher, and W. Lesslauer, *Cell*, **73**, 431 (1993).

368. G. Tocchini-Valentini, N. Rochel, J. M. Wurtz, A. Mitschler, and D. Moras, *Proc. Nat. Acad. Sci. USA*, **98**, 5491 (2001).

369. N. Rochel, J. M. Wurtz, A. Mitschler, B. Klaholz, and D. Moras, *Mol. Cell*, **5**, 173 (2000).

370. S. Vos, R. J. Parry, M. R. Burns, J. Dejersey, and J. L. Martin, *J. Mol. Biol.*, **282**, 875 (1998).

371. S. Vos and J. De Jersey, *Biochemistry*, **36**, 4125 (1997).

372. S. Vos, R. J. Parry, M. R. Burns, J. Dejersey, and J. L. Martin, *J. Mol. Biol.*, **282**, 875 (1998).

373. H. A. Lewis, *Structure (Camb)*, **9**, 527–537 (2001).

374. C. Freiberg, *Drug Discov. Today*, **6**, S72–S80 (2001).

375. J. M. Sauder, J. W. Arthur, and R. L. Dunbrack, *Proteins*, **40**, 6–22 (2000).

376. S. W. Muchmore, *Nature*, **381**, 335–341 (1996).

377. K. A. Denessiouk, A. I. Denesyuk, J. V. Lehtonen, T. Korpela, and M. S. Johnson, *Proteins*, **35**, 250–261 (1999).

378. M. Teplova, *Protein Sci.*, **9**, 2557–2566 (2000).

379. T. J. Boggon, W. S. Shan, S. Santagata, S. C. Myers, and L. Shapiro, *Science*, **286**, 2119–2125 (1999).

380. P. W. Kleyn, *Cell*, **85**, 281–290 (1996).

381. K. Noben-Trauth, J. K. Naggert, M. A. North, and P. M. Nishina, *Nature*, **380**, 534–538 (1996).

382. S. Santagata, *Science*, **292**, 2041–2050 (2001).

383. W. Eisenreich, *Chem. Biol.*, **5**, R221–R233 (1998).

384. R. Sanchez, *Nat. Struct. Biol.*, **7**, 986–990 (2000).

385. A. Sali, *Nat. Struct. Biol.*, **5**, 1029–1032 (1998).

386. A. Fiser, R. K. Do, and A. Sali, *Protein Sci.*, **9**, 1753–1773 (2000).

387. K. T. Simons, C. Strauss, and D. Baker, *J. Mol. Biol.*, **306**, 1191–1199 (2001).

388. R. Sanchez, *Nucleic Acids Res.*, **28**, 250–253 (2000).

CHAPTER TWELVE

NMR and Drug Discovery

David J. Craik
Richard J. Clark
Institute for Molecular Bioscience
Australian Research Council Special Research
Centre for Functional and Applied Genomics
University of Queensland
Brisbane, Australia

Burger's Medicinal Chemistry and Drug Discovery
Sixth Edition, Volume 1: Drug Discovery
Edited by Donald J. Abraham
ISBN 0-471-27090-3   © 2003 John Wiley & Sons, Inc.

Contents

1   INTRODUCTION

NMR spectroscopy has been widely used as a front-line tool in the pharmaceutical industry for several decades. In the past, the main use of NMR was in the structural characterization of organic molecules synthesized in the course of medicinal chemistry programs. Indeed, medicinal chemists have long regarded NMR as the premier tool to be used in the structure characterization process, to confirm the identity of intermediates or to determine the conformation of lead molecules. Over the last decade major developments in both instrumentation and methods have resulted in this traditional use of NMR in the pharmaceutical industry being augmented by a range of exciting new applications. Two of the most important of these are the use of NMR in structure-based drug design and in screening for drug discovery. Both applications differ from the traditional use of NMR in that now the macromolecular binding partner of the medicinal compound is included in the mixture to be analyzed; that is, contemporary applications of NMR in drug discovery are predominantly focused on the *interaction* between drug molecules and their macromolecular targets.

The aim of this chapter is to describe how NMR spectroscopy is used in modern drug discovery. The term *discovery* is used generically throughout to include processes that involve rational drug design as well as those that involve discovery through NMR screening. The latter is a relatively recent development and refers to the use of NMR as a tool to screen a compound library, to identify a molecule or molecules that bind to a chosen macromolecular target. Of course, the distinction between "design" and "discovery" is often quite blurred. This is nowhere more evident than in the recently developed SAR-by-NMR approach (1), in which the discovery of several weakly bound ligands from a screening program is intimately linked to a design process to chemically join them. SAR-by-NMR represents an exciting new technique for lead generation and is described in more detail later in this chapter.

Drug design/discovery represents only the first stage in the whole drug development process. As is clear from the other chapters in this volume, there are many other steps that need to be made once a lead molecule has been designed or discovered. Although other stages of the process, including lead optimization, tox-

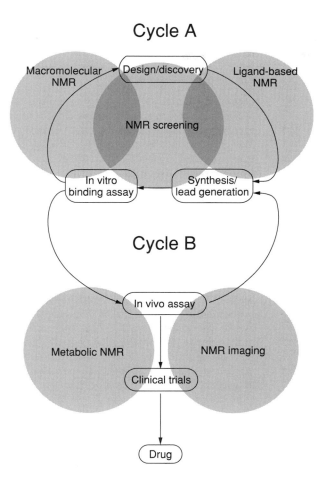

Figure 12.1. Overview of the drug development process and summary of various types of NMR experiments that contribute at different stages.

icity studies, preclinical investigations, and clinical monitoring, do not fall within the scope of this chapter, it is worth mentioning that NMR spectroscopy contributes significantly across the whole spectrum of drug development, right through into the clinical domain. For example, NMR spectroscopy has been applied for the detection of drug metabolites in biological fluids and magnetic resonance imaging, which is based on the fundamental principles of NMR, plays an important role in clinical investigations. It is increasingly being used to monitor the functional outcomes of drug therapy. We briefly address these broader applications of NMR before returning to the main topic of NMR in drug discovery.

1.1 Overview of Drug Development

To give an overview of the breadth of applications of NMR, Fig. 12.1 summarizes the drug development process and indicates the role of NMR at various stages. Drug development is an iterative process and can be simplified by representing it with two interconnected cycles of activity. Cycle A involves the design or discovery of an initial lead followed by its synthesis and bioassay. Based on the initial assay results there may be several loops around this cycle before commencing the *in vivo* studies represented in Cycle B. At this stage consideration of bioavailability, metabolism, and pharmacokinetic profiles must be made and this may involve synthetic modifications of the lead molecules to improve their druglike properties. Again, several loops around Cycle B may be necessary before one or more development candidates are identified. Ultimately one or two of these development candidates are identified for progression through clinical trials.

As indicated in Fig. 12.1, it is convenient to envisage five broad categories of NMR experiments that may contribute to this overall drug development process.

1. *Small molecule, or ligand-based, NMR.* This involves studies of drugs and drug leads, typically organic molecules with a molecular weight <500 Da, but also including small proteins of up to a few kDa. These studies may be used to characterize natural products or synthetic drug leads, or to determine their conformation.

2. *Macromolecular NMR.* This involves studies of the macromolecular targets of drugs, typically to determine their three-dimensional structure and/or the nature of their complexes with ligands.

3. *NMR screening.* This involves the use of NMR to identify lead molecules that bind to a macromolecular target. These studies typically involve both small molecules and macromolecules and seek to detect the presence of binding interactions between them.

4. *Metabolic NMR.* This involves studies of endogenous molecules whose levels may be modified by drug treatment, or studies of the metabolites of drugs themselves.

5. *NMR imaging.* Such studies provide anatomical information in an animal model or human patient. This includes, for example, monitoring the size of plaques or tumors in the brains of Alzheimer's or cancer patients, respectively, during drug therapy.

It is clear from these descriptions that NMR covers a wide range of applications in the pharmaceutical industry, although for the remainder of this chapter we will focus on NMR in the drug design/discovery phase of drug development, that is, on categories 1–3 of the preceding list. Together, the studies in cat-

egories 1 and 2 may be classified as structure-based design, whereas category 3 relates to drug discovery.

1.2 Scope of Chapter

Our aim is to give a broad overview on the use of NMR as a tool in structure-based design and in screening approaches to drug discovery. The chapter also contains a description of the relevant NMR methods, which are highlighted by illustrative examples. We briefly describe the instrumentation required for such studies and emerging trends in the field are discussed. This includes developments in the field of drug discovery in the postgenomic era that are likely to have an impact on the way in which NMR is used, as seen for example by the recent interest in structural genomics programs. NMR instrument developments are also described. For example, recent advances in cryoprobe technology promise to dramatically increase the sensitivity of NMR spectroscopy and increase its application across the pharmaceutical industry. Finally, a section outlining some of the practical considerations in structure-based design and screening is included. Future directions for the field are mentioned throughout the discussion.

There have been a number of reviews that describe applications of NMR in drug discovery or screening and the reader is referred to these for additional information (2–15). Recent books covering aspects of NMR in drug design are also available (16, 17).

It is assumed that most readers will be familiar with the basic principles of NMR. However, for completeness and to define some of the terms that will be used in this chapter it is useful to give a brief overview of the principles. Excellent texts are available to provide more detail (18, 19).

1.3 Principles of NMR Spectroscopy

The underlying basis of NMR is that when nuclei with a nonzero spin quantum number are placed in a magnetic field they take up one of a discrete number of quantized states. The application of radiofrequency (rf) energy produces transitions between these states. The energy changes associated with these transitions are detected as small voltages induced in

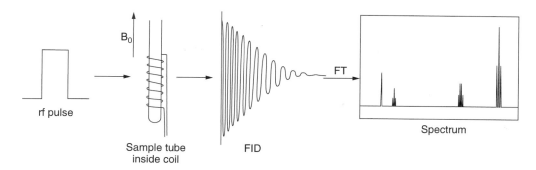

Figure 12.2. Overview of the principles of NMR spectroscopy. Polarization of nuclear spins by a magnetic field is perturbed by application of a radiofrequency (rf) pulse. The resultant signal is Fourier transformed, to yield a spectrum reflecting the number and environments of nuclei in the sample.

a receiver coil that are subsequently amplified, digitized, and processed to yield spectra, as illustrated in Fig. 12.2. The most commonly studied NMR-active nucleus is the proton, ^{1}H, but in modern NMR experiments ^{2}H, ^{13}C, and ^{15}N nuclei are also very important. For these heteronuclei it is common to isotopically enrich the sample because of their low natural abundance. This is particularly important for studies of proteins, as will become apparent later in this chapter. Occasionally, other nuclei find specialist applications. For example, in fluorine-containing drugs it is possible to use sensitive ^{19}F-NMR signals to monitor interaction with target proteins, as described later in this chapter.

In modern spectrometers the rf energy is supplied in the form of short pulses (typically, $\sim$10 μs) that simultaneously excite all nuclei of a given isotope type (e.g., all protons or all ^{13}C nuclei). Nuclei of a given isotope that are in different chemical environments by virtue of their atomic locations in the molecule have slightly different resonance frequencies and lead to different oscillating voltages in the receiver coil. The resultant combined signal, termed a *free induction decay* (FID), is Fourier transformed to give a spectrum that is basically a plot of peak intensity vs. frequency, with one peak for each chemically distinct nucleus. These features are schematically illustrated in Fig. 12.2. The frequency axis is termed the *chemical shift* because it reflects the local chemical environment of each nucleus. The range of chemical environments of

nuclei in a molecule is such that chemical shifts range up to only a few hundred parts per million (ppm) of the base resonance frequency for ^{13}C and ^{15}N. For ^{1}H the range is smaller still, covering only about 10 ppm. Despite this small range, chemical shifts provide valuable diagnostic information on the environment of the nucleus giving rise to the signal.

The chemical shift is an extremely important NMR parameter but there are many other parameters that can be discerned from NMR spectra. Indeed, NMR is unique among many forms of spectroscopy in that there are so many parameters associated with a spectrum other than just peak intensity and frequency. These include coupling constants, which provide information on local conformations and also on molecular connectivities; nuclear Overhauser effects (NOEs), which provide information on internuclear distances; and relaxation parameters, which provide information on molecular dynamics. Table 12.1 summarizes the main NMR parameters that may be measured and highlights their applications in the drug discovery process.

The following sections of this chapter provide specific examples of how these various parameters are useful in the drug discovery process. Before doing this, though, it is useful to consider some of the limitations of one-dimensional (1D) NMR spectroscopy, particularly when the detected nucleus is ^{1}H, as is most commonly the case. With one signal coming from each chemically distinct proton and with those signals spread only over 10 ppm, it is

Table 12.1   NMR Parameters and Their Applications in Drug Design/Discovery

Parameter	Information Provided Relevant to Drug Design
Chemical shift	Reflects local chemical environment; provides a fingerprint marker of structure (particularly in HSQC spectra)
Coupling constants	Conformational analysis, establishing molecular connectivity
Nuclear Overhauser effect	Determining interproton distances, three-dimensional structures
Relaxation times	Molecular dynamics
Line-shape	Detecting and quantifying chemical exchange processes
Peak intensities	Reflect relative number of nuclei, molecular symmetry
Amide exchange rates / temperature coefficients	Hydrogen bonding or solvent exposure of amide protons

clear that spectral overlap can potentially be a major problem for anything but the simplest of molecules. The development of higher field NMR spectrometers, which effectively provide greater dispersion in the frequency dimension, has contributed significantly to overcoming this limitation and increasing the application of NMR for studying pharmaceutically relevant molecules. In addition to such instrumental developments, methodological advances have also played a key role in extending the use of NMR. Multidimensional NMR methods have revolutionized biomolecular NMR spectroscopy by removing the limitations of a single frequency dimension, leading to the development of 2D, 3D, and 4D spectra.

A simple way of illustrating multidimensional NMR is through reference to heteronuclear correlation spectroscopy, in which two or more separate frequency dimensions are correlated with one another. For example, a particularly valuable 2D experiment is ^{1}H-^{15}N heteronuclear single quantum correlation (HSQC) spectroscopy, in which the resultant spectrum has two frequency axes, corresponding to ^{1}H and ^{15}N frequency dimensions, and one intensity axis. Analogous ^{1}H-^{13}C HSQC spectra are also widely used. Such spectra are normally represented with the intensity axis in contour form so that they may be drawn in two dimensions as a set of contour peaks. Spectral peaks occur for pairs of ^{15}N/^{1}H or ^{13}C/^{1}H nuclei that are directly bonded to one another, and with each frequency being characteristic for the local chemical environment they represent a relatively simple, but highly characteristic fingerprint of the sample. Figure 12.3 shows the relationship between 1D and 2D spectra for the immunosuppressive

drug cyclosporin, and includes a region of both the ^{1}H/^{15}N and ^{1}H/^{13}C HSQC spectra. In HSQC spectra-overlap problems are alleviated because, even if two protons have the same chemical shift and would hence be overlapped in a 1D spectrum, chances are that the respective heteronuclear signals will not be overlapped, allowing the signals to be resolved in the 2D spectrum. HSQC spectra are widely used in NMR-based drug screening and we will return to them later.

Multidimensional NMR spectra are not restricted to cases where the separate frequency axes encode signals from different nuclear types. Indeed, much of the early work on the development of 2D NMR was performed on cases where both axes involved ^{1}H chemical shifts. The main value in such spectra comes from the information content in cross peaks between pairs of protons. In COSY-type spectra (COSY = COrrelation SpectroscopY) cross peaks occur only between protons that are scalar coupled (i.e., within 2 or 3 bonds) to each other, whereas in NOESY (NOE SpectroscopY) spectra cross peaks occur for protons that are physically close in space (<5 Å apart). A combination of these two types of 2D spectra may be used to assign the NMR signals of small proteins and provides sufficient information on internuclear distances to calculate three-dimensional structures. Figure 12.3 includes a panel showing the COSY spectrum of cyclosporin and highlights the relationships between 1D ^{1}H-NMR spectra and corresponding 2D homonuclear (COSY) and heteronuclear (HSQC) spectra.

Homonuclear 2D spectra are generally applicable for the study of proteins up to only approximately 80 amino acids in size. For

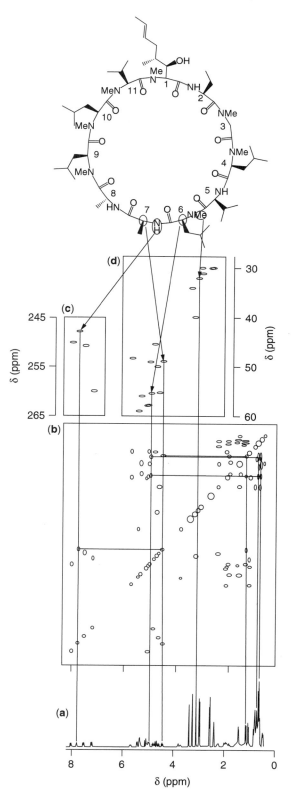

Figure 12.3. A schematic representation of the (a) 1D ^{1}H; (b) 2D DQF-COSY; (c) ^{15}N/^{1}H-HSQC; and (d) ^{13}C/^{1}H-HSQC spectra of the immunosuppressive agent cyclosporin. Example resonances/correlations from residues 6 and 7 have been highlighted to illustrate the assignment process.

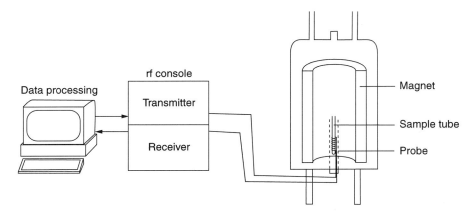

Figure 12.4. Block diagram of a modern NMR spectrometer. These systems use superconducting magnets that are based on a solenoid of a suitable alloy (e.g., niobium/titanium or niobium/tin) immersed in a dewar of liquid helium. The extremely low temperature of the magnet itself (4.2 K) is well insulated from the sample chamber in the center of the magnet bore. The probe in which the sample is housed usually incorporates accurate temperature control over the range typically of 4 to 40°C for biological samples. The rf coil in the probe is connected in turn to a preamplifier, receiver circuitry, analog-to-digital converter (ADC), and a computer for data collection.

larger proteins the increased number of signals leads to overlap problems and, in addition, COSY-type spectra suffer from poor sensitivity when the signal linewidths are of the same order as or larger than ^{1}H, ^{1}H scalar coupling constants. Such limitations are reduced by use of spectra of higher dimensionality (i.e., 3D or 4D spectra) that are based on correlations involving heteronuclear rather than homonuclear coupling constants. Such spectra are important in the structure determination process for larger proteins and are typically recorded for samples that incorporate uniform labeling with ^{15}N, or both ^{13}C and ^{15}N nuclei. Multidimensional spectra that involve irradiation of ^{1}H, ^{13}C, and ^{15}N nuclei are referred to as *triple resonance spectra*.

The details of how multidimensional spectra are obtained is beyond the scope of this chapter, but it suffices to say that, like most other modern NMR experiments, they involve irradiation of the sample with a set of rf pulses of defined length, frequency, and phase, with specific interpulse delays. The pulse programs for such experiments are commonly provided with the spectrometer as part of a standard library of experiments and may easily be run by novice users after input of an appropriate set of parameters to define the relevant spectral widths and type of experiment required.

The above discussion provides a basic overview of some of the methods important in modern NMR spectroscopy. Before examining specific applications in drug discovery it is useful to describe the instrumental requirements for such studies.

1.4 Instrumentation

NMR spectrometers constitute a powerful and homogeneous magnet, a radiofrequency console for generating appropriate rf pulses, a probe for applying this rf energy to the sample and receiving the resultant signals, and a computer console for controlling the experiments and acquiring the resultant data. These features are summarized in Fig. 12.4. Spectrometers are normally specified in terms of the resonant frequency of protons at the given magnetic field (e.g., 500 MHz corresponds to a magnetic field of 11.7 Tesla). Both sensitivity and dispersion of signals increase with increasing magnetic field.

There have been some major breakthroughs in both NMR instrumentation and methodology over the last decade that have greatly increased the utility of NMR for drug discovery applications. These are summarized in Table 12.2, which also includes some of the earlier milestones in the development of NMR. Most notable among recent innovations

Table 12.2 Milestones in the Development of NMR Spectroscopy

Year	Development	Nature
1970	FT NMR	Instrumental
1975	Superconducting magnets	Instrumental
1980	2D NMR	Methodological
1985	Protein structure determination	Methodological
1990	Isotope labeling/multidimensional NMR	Methodological
1990	Pulsed field gradients	Instrumental/methodological
1995	NMR screening	Methodological
1997	TROSY	Methodological
1998	LC-NMR/LCMS-NMR	Instrumental
2000	Cryoprobes	Instrumental

are the use of pulsed-field gradient methods for improving spectral quality and allowing new types of experiments to be performed, transverse relaxation-optimized spectroscopy (TROSY) methods (20) for increasing the size of macromolecules that can be examined, and cryoprobes for enhancing sensitivity. The development of cryoprobes has resulted in the biggest single gain in sensitivity over recent years, effectively giving 500-MHz spectrometers the sensitivity of 800-MHz spectrometers (although without the gain in resolution!). The enhanced sensitivity is obtained by cooling the receiver coil and associated circuitry to near liquid helium temperatures, thereby reducing thermal noise. There were considerable technical barriers to be overcome in developing such probes because of the large difference in temperature between the receiver coils and the sample, which are only a few millimeters apart. These barriers have now been overcome and cryoprobes are being installed in a large number of laboratories. They are also becoming available for higher field systems (800 MHz), thus providing further sensitivity gains.

Although the basic configurations of instruments tailored for structure-based design or for NMR drug screening are similar, there are some minor differences. For structure-based design applications a relatively high field spectrometer is required (>500 MHz), usually equipped with three or four radiofrequency channels for the simultaneous irradiation of ^1H, ^{13}C, ^{15}N, and in some cases ^2H nuclei. The greatest sensitivity and dispersion are obtained with the highest possible magnetic field. Instruments of up to 900 MHz are

currently available but, at the time of writing, only a few have been installed. Numerous 800-MHz systems dedicated to structure-based design have been installed in pharmaceutical laboratories. The high field instruments provide another advantage in that TROSY experiments (20) can be used to produce a marked improvement in spectral quality for larger proteins. Such developments promise to push higher the size of proteins whose structure can be determined by NMR.

For NMR drug screening programs, the basic requirement of a spectrometer of 500 MHz or greater remains, but in addition, an interface that allows the spectrometer to sample a library of compounds of potential binding ligands needs to be present. This may be done either by use of a discrete sample changer or a flow-type system. Flow systems have the potential advantage of increased throughput but have the potential disadvantage of precipitation of protein samples. In practice this appears not to have been a major problem and both types of systems are in use in the pharmaceutical industry. Sample changer systems currently have the advantage that they may be adapted for use with cryoprobe technology (currently unavailable for flow systems). Cryoprobes allow dramatically enhanced sensitivity gains, which bring particular advantages to the study of macromolecule-ligand interactions used in screening programs (21).

Pulsed-field gradients have become integral to most modern NMR spectrometers and are routinely used both for structure determination and screening experiments. Another recent development has been the interface of NMR spectrometers with other instrumenta-

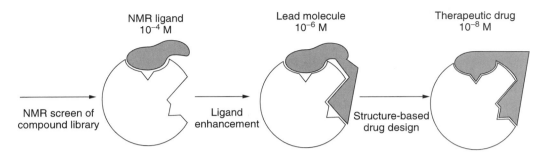

Figure 12.5. A summary of the relationship between NMR screening and structure-based design. (Adapted from Ref. 15.)

tion such as liquid chromatography (LC) and/or mass spectrometry (MS). The applications of these instrumental developments to drug discovery have been recently reviewed (8, 13, 22).

1.5   Applications of NMR in Drug Design and Discovery

Our focus here is on the use of NMR in the discovery and design phase of drug development. The major role of NMR in the *design* process comes about by its exquisite ability to provide structural information, whereas the major role of NMR in *discovery* comes through its use as a screening tool to detect the binding of novel ligands to macromolecular targets. As already noted, the latter application is a relatively recent development but has created much interest in the pharmaceutical industry and promises to significantly enhance applications of NMR in this industry. The impact of the methodology is already becoming evident even at this early stage, with several SAR-by-NMR-derived leads currently in clinical development. As already noted, though, the discovery and design phases are often intimately connected, with lead molecules discovered in screening programs routinely being optimized by use of structure-based design approaches (Fig. 12.5).

In the context of this chapter structure-based design refers to the process of determining the three-dimensional structure of a lead molecule or macromolecular target, or determining the structure of the macromolecule-ligand complex, and using this information to

design new drugs. The questions that may be asked when embarking on structure-based design projects are:

- What are the solution and bound conformations of the ligand?
- What is its charge/tautomeric state?
- Which functional groups bind to the receptor and what charge state are they in?
- What is the structure of the receptor?
- Which parts interact with the ligand?
- What is the geometry of the ligand-receptor complex?
- What are the kinetics of binding and are there dynamic motions of ligand, receptor, or the complex?

Table 12.3 summarizes these and other questions and indicates the type of NMR approaches that can provide answers. Remaining sections of this chapter are organized around the headings identified in Table 12.3.

In considering these questions it is convenient to distinguish between ligand-based design, where the structural focus is on the small lead molecule, and receptor-based design, where the aim is to determine the structure of the macromolecular target. The NMR methods used in ligand-based design have been well established for many years, based on the use of NMR by organic and natural product chemists for more than four decades. However, there have been some important recent advances in NMR methods such as the use of pulsed field gradients, and in the combination of NMR

Table 12.3 Information on Ligands, Macromolecules, and Their Complexes Sought in Structure-Based Design and Relevant NMR Technologies Used to Derive This Information

Target	Information	NMR Technology
Ligand	Solution conformation	1D/2D NMR
	Charge/tautomeric state	Chemical shift/titrations
	Solution dynamics	Line-shape/relaxation analysis
	Pharmacophore models	All of the above, and TrNOE, of multiple ligands
	Bound ligand conformation	TrNOE
Macromolecule	3D structure	2D/3D/4D NMR
	Macromolecular dynamics	Relaxation time measurements
	Structure of articulated macromolecules (e.g., multimeric or membrane-bound receptors)	TROSY
Ligand–macromolecular complex	Stoichiometry of complex	Chemical shift/titration
	Kinetics of binding	Line width, titration analysis
	Location of interacting sites	HSQC, isotope editing
	Orientation of bound ligand	NOE docking
	Bound ligand conformation	TrNOE
	Structure of complex	3D/4D NMR
	Dynamics of complex	Relaxation time measurements

with other technologies such as LC and MS that promise to enhance applications in this field (13). The use of NMR to determine the three-dimensional structures of macromolecules is a newer field, commencing only in around 1985, and is one that is still rapidly evolving. NMR screening is a still newer approach, developed since around 1996. Ligand-based and receptor-based design are examined in Sections 2 and 3, respectively, and screening-based approaches are examined in Section 4.

2 LIGAND-BASED DESIGN

Many naturally occurring molecules have potent bioactivity that renders them useful leads in the drug design process. These may be naturally occurring hormones, neurotransmitters, or other endogenous molecules, or they may be bioactive molecules from plants or microorganisms. Furthermore, screening programs on synthetic compound libraries frequently result in the discovery of bioactive molecules that then become starting points in drug design. The general aim of ligand- or analog-based design is to determine the structure and conformation of a known bioactive molecule and then mimic this conformation in

a designed lead compound, with the aim of improving the activity or druglike properties. The following sections examine various aspects of ligand-based design and illustrate them with examples.

2.1 Structure Elucidation

If the bioactive molecule is a synthetic product, its structure may be rapidly deduced by a simple comparison of NMR parameters (often combined with MS) of the product relative to those of the known precursor, to see whether the desired chemical transformation has taken place. If the bioactive compound is an unknown molecule discovered in an active fraction in bioassay-guided screening, then the first step is to elucidate its structure. Typical molecules that form the basis of such natural products-based drug discovery studies include "organic" natural products as well as small peptides and proteins. The approaches to structure elucidation for natural products and peptides/proteins are a little different from each other and are described in turn.

2.1.1 Structure Elucidation of Natural Products. In the case of nonpeptidic natural products the main structural focus initially is to elucidate the carbon framework. This nor-

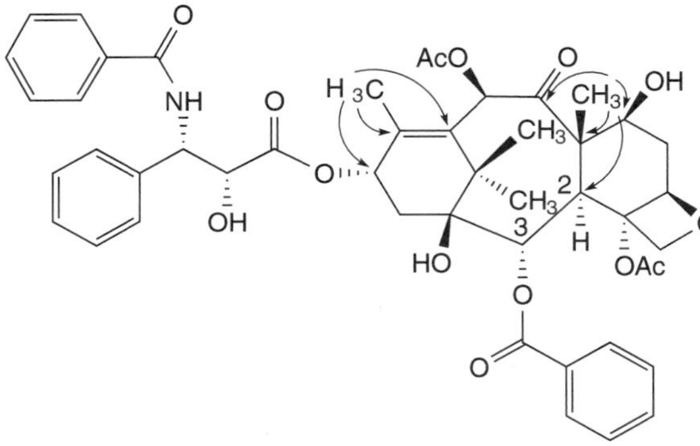

Figure 12.6. Illustration of the HMBC correlations (arrows) used to assign the positions of two of the methyl quaternary methyl groups in taxol.

mally involves a combination of 1D ^{1}H and ^{13}C-NMR, followed by homonuclear (DQF-COSY, TOCSY, ROESY, or NOESY) and heteronuclear (HSQC, HMBC) 2D experiments. Heteronuclear multiple bond correlation (HMBC) spectra are particularly valuable because they assist in tracing the backbone of the molecule. Such spectra display cross peaks between a ^{13}C nucleus and protons connected within two or three bonds and, in doing so, provide valuable information on molecular connectivity. Figure 12.6 shows typical HMBC correlations seen for selected regions of taxol, a plant-derived natural product that is currently a leading treatment for breast and ovarian cancers. Although the structure of taxol itself was originally deduced from a combination of X-ray crystallography on a degradation product and a range of ^{1}H and ^{13}C spectra in the 1970s, before HMBC spectra had been invented, HMBC spectra have been widely used for studies of the many taxol derivatives that have been examined in the last decade.

Elucidation of the carbon framework of natural products often yields substantial information about the three-dimensional structure at the same time, but if there are remaining questions on the stereochemistry of chiral centers or other factors affecting the three-dimensional structure, these can usually be resolved from NOESY spectra and/or an analysis of coupling constants. We will return to the taxol example later in Section 2.2 when describing conformational analysis.

2.1.2 Structure Determination of Bioactive Peptides. In contrast to the process described for organic molecules, the structure elucida-

tion of peptide-based natural products involves two distinct steps: (1) the elucidation of the primary structure (amino acid sequence) followed by (2) a determination of secondary/tertiary structure. The primary structure determination is routinely done through Edman sequencing, or more recently by MS-MS methods. NMR plays a key role in the elucidation of the secondary and tertiary structure of peptides, mainly based on 2D homonuclear NMR spectroscopy. A combination of DQF-COSY and TOCSY (TOtal Correlalation SpectroscopY) spectra are used to assign spin systems to amino acid types and then NOESY spectra are used to sequentially assign the resonances to individual protons in the peptide (23). The three-dimensional structure is then determined by deriving a series of internuclear distance restraints from the NOESY spectrum and using them in a simulated annealing algorithm to calculate a family of structures consistent with them.

Because the structure determination of peptides and proteins represents a very important contribution of NMR to the drug development process, it is informative to describe the process in more detail. To do this we will use the recently developed peptide-based drug MVIIA as an example.

2.1.2.1 NMR Structure of Ziconotide: A Novel Treatment for Pain. MVIIA, now known as Ziconotide, is a 25-amino acid peptide originally discovered from the venom of the marine cone snail, *Conus magus*. Like other ω-conotoxins it is a potent blocker of N-type calcium channels, giving it a wide range of potential therapeutic applications. When delivered intrathecally (i.e., through spinal infusion), it is approximately 1000 times more potent than morphine as an analgesic and has great potential for the treatment of intractable cancer pain (24). Figure 12.7 shows the peptide sequence and illustrates selected regions of the TOCSY and NOESY spectra.

As seen in Fig. 12.7, the TOCSY experiment is useful for classifying spin systems to amino acid type, with typically the most useful region being the "skewers" emanating from individual NH shifts (~7–10 ppm). For each NH proton in the peptide a series of cross peaks to the α, β, and other side-chain protons is observed and these patterns define the spin system as belonging to a particular type of amino acid. Note, however, that there is some degeneracy in the resultant patterns. The NH side-chain pattern is truncated if there is a break of more than three bonds between protons within the spin system. This means, for example, that the skewers for aromatic residues extend only as far as the β-protons and they therefore appear similar to other "AMX" residues such as Cys, Ser, Asp, or Asn. Nevertheless, the ability to assign signals to either individual amino acid types or to the AMX group is a useful starting point in the assignment. However, such spectra provide no information about the sequential location of an amino acid if it is not unique in the sequence. These sequential assignments are obtained from the NOESY spectrum, as illustrated in the sequential walk shown in the middle panel of Fig. 12.7. The aim of the sequential assignment process is to locate adjacent amino acid spin systems, principally through a cross peak between the αH proton of one residue (i) and the NH of the following residue ($i + 1$), often denoted as dαN(i, $i + 1$). Additional support for the assignment is usually also sought in dβN(i, $i + 1$) and dNN(i, $i + 1$) correlations. At the early stages of an assignment it is impossible to be certain whether a particular cross peak is a sequential or longer range cross peak; however, as the assignment procedure progresses, ambiguities become resolved. The assignment process is generally highly convergent, in that once a series of correct assignments is made the number of choices for remaining cross peaks diminishes, in principle making their assignment easier.

Because peptides are polymers of amino ac-

ids units, the repeated NH, Hα, and side-chain protons tend to fall in characteristic chemical shift ranges that can be useful in looking for patterns to identify amino acid types. Table 12.4 shows typical chemical shifts for each of the 20 common amino acids when located in a "random-coil" environment (23, 25, 26). It is important to stress that these shifts can vary quite considerably in structured proteins (by up to several ppm) and are more useful for pattern recognition purposes than for exact identification of a particular residue. In the case of the Hα protons, the differences between the actual shifts in a structured protein and these random-coil values have an additional important use, in that they provide an indication of the local secondary structure. Intuitively, the further a chemical shift is from a random-coil value, the more likely it is attributed to that residue's being in a structured environment.

After the assignment is complete it is possible to derive substantial information about the secondary structure from an analysis of chemical shifts, coupling constants, and NOEs, even before the three-dimensional structure calculations are commenced. Figure 12.8 shows a typical summary of the relevant NMR information, again using the data for MVIIA as an example (27, 28). Trends in these data provide a general indication of major elements of secondary structure. For example, a series of strong dαN(i, $i + 1$), relative to dNN(i, $i + 1$) NOEs often indicates an extended or β-type structure, whereas strong dNN(i, $i + 1$) NOEs indicate local helical structure or turns. Large JαN coupling constants (>8.5 Hz) are associated with extended structure and small ones (<5 Hz) with helical structure. Similarly, deviations of chemical shifts from random-coil values, often represented in terms of "chemical shift indices" (29), indicate extended (positive values) or helical structure (negative values).

An additional useful parameter is the exchange rate of amide protons after dissolution of the sample in D$_2$O. Slowly exchanging amide protons indicate protection from solvent and possible involvement in intramolecular hydrogen bonds associated with elements of secondary structure. All of the NMR and slow exchange data can be consolidated to give

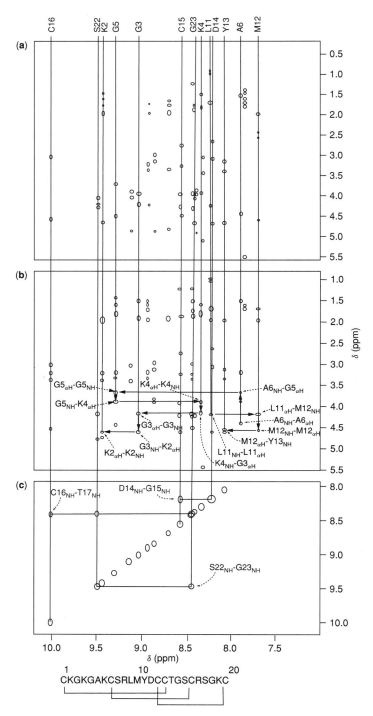

Figure 12.7. Schematic representations of 2D-NMR spectra of the conotoxin MVIIA. (a) The fingerprint region of the TOCSY spectrum with selected spin systems marked. (b) Fingerprint region of the NOESY spectrum showing two (K2-A6 and L11-Y13) sequential walks. (c) NH-NH region of the NOESY spectrum showing correlations between the NH protons of D14 and G15; C16 and T17; and S22 and G23.

Table 12.4 **¹H Chemical Shifts for the 20 Common Amino Acid Residues in Random-Coil Peptides**[a]

Residue	NH	αH	βH	Others
Ala	8.24	4.32	1.39	
Arg	8.23	4.34	1.89, 1.79	γCH_2 1.70, 1.70
				δCH_2 3.32, 3.32
				NH 7.17, 6.62
Asn	8.40	4.74	2.83, 2.75	γNH_2 7.59, 6.91
Asp	8.34	4.64	2.84, 2.75	
Cys	8.43	4.71	3.28, 2.96	
Gln	8.32	4.34	2.13, 2.01	γCH_2 2.38, 2.38
				δNH_2 6.87, 7.59
Glu	8.42	4.35	2.09, 1.97	γCH_2 2.31, 2.28
Gly	8.33	3.96		
His	8.42	4.73	3.26, 3.20	2H 8.12
				4H 7.14
Ile	8.00	4.17	1.90	γCH_2 1.48, 1.19
				γCH_3 0.95
				δCH_3 0.89
Leu	8.16	4.34	1.65, 1.65	γH 1.64
				δCH_3 0.94, 0.90
Lys	8.29	4.32	1.85, 1.76	γCH_2 1.45, 1.45
				δCH_2 1.70, 1.70
				εCH_2 3.02, 3.02
				εNH_3^+ 7.52
Met	8.28	4.48	2.15, 2.01	γCH_2 2.64, 2.64
				εCH_3 2.13
Phe	8.30	4.62	3.22, 2.99	2,6H 7.30
				3,5H 7.39
				4H 7.34
Pro		4.42	2.28, 2.02	γCH_2 2.03, 2.03
				δCH_2 3.68, 3.65
Ser	8.31	4.47	3.88, 3.88	
Thr	8.15	4.35	4.22	$\gamma CH3$ 1.23
Trp	8.25	4.66	3.32, 3.19	2H 7.24
				4H 7.65
				5H 7.17
				6H 7.24
				7H 7.50
				NH 10.22
Tyr	8.12	4.55	3.13, 2.92	2,6H 7.15
				3,5H 6.86
Val	8.03	4.12	2.13	γCH_3 0.97, 0.94

[a]The backbone shifts (αH and NH, ppm) are from Wishart et al. (26). The remaining shifts are from Wüthrich 1986 (23).

an accurate representation of secondary structure, as indicated in the lower panel of Figure 12.8. In the case of MVIIA a triple-stranded β-sheet may be deduced on the basis of the local NOE, coupling, chemical shift, and amide-exchange NMR data.

Once all peaks in the 2D spectra have been assigned, cross peaks in the NOESY spectrum are used to derive a series of interproton distance restraints. These are then used in a simulated annealing algorithm to calculate a family of 3D structures consistent with the input restraints. Fig. 12.9 shows two commonly used methods of representing such NMR-derived structures, either as a stereoview of the superimposed family of structures or as a ribbon diagram, in which elements of secondary structure are highlighted. For the latter rep-

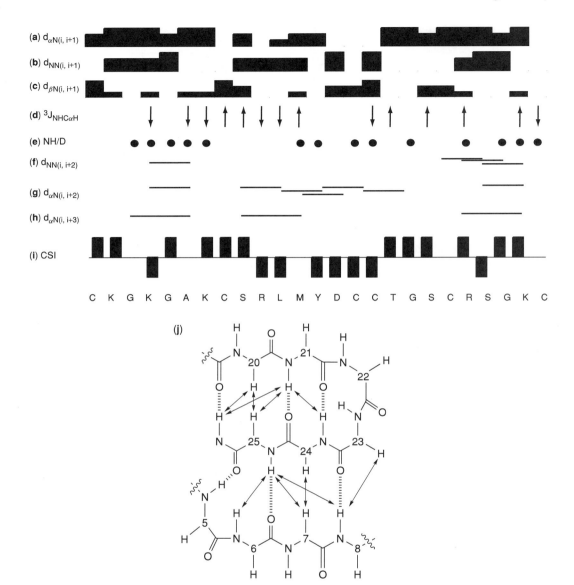

Figure 12.8. A summary of the NMR data observed for MVIIA. (a) Hα-NH sequential NOEs. (b) NH-NH sequential NOEs. (c) Hβ-NH sequential NOEs. (f-h) Other short-range NOEs. The thickness of the bar indicates the strength of the observed NOE (weak, medium, or strong). (d) Three-bond NH-Hα coupling data, where upward-pointing arrows indicate a large coupling (>8 Hz) and downward-pointing arrows indicate a small coupling (<5 Hz). (e) H/D exchange data, where a filled circle represents a slow exchanging NH. (i) Chemical shift index (CSI) data. The CSI uses a scoring system that compares Hα shifts to random-coil chemical shifts. A sequence of consecutive +1 scores is indicative of β-structure, whereas a sequence of consecutive −1 scores suggests helical structure. (j) The β-sheet of MVIIA. Double-headed arrows indicate observed NOEs and broken lines indicate proposed H-bonds.

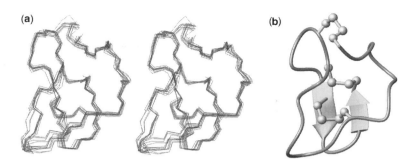

Figure 12.9. (a) A stereoview of the superimposed backbone structures of the 20 lowest energy conformations for MVIIA (27). (b) Ribbon diagram of MVIIA.

resentation the lowest energy or average member of the ensemble is often chosen as representative of the structure. It is important, however, to examine the full ensemble to gain a complete understanding of the structure. Regions of disorder in the ensemble can be indicative of a lack of sufficient distance restraints, perhaps attributable to overlap or assignment errors, or may be related to local flexibility.

In the case of MVIIA the peptide itself is being clinically developed as the active drug for administration through the intrathecal (spinal infusion) route. However, in general, peptides have a range of potential disadvantages as drugs, including poor bioavailability and susceptibility to proteolytic breakdown. Thus, for many cases involving peptide-based leads the structural information of the type described above might be used as a starting point to design smaller constrained peptides or nonpeptidic mimics. This is the case, for example, in the development of endothelin antagonists described below.

2.1.2.2 Endothelin as a Lead in Ligand-Based Design. Endothelin (ET), shown in Fig. 12.10, is a 21-amino acid endothelial-derived constricting factor that has gained prominence as a pharmacological lead molecule. Interest in the peptide arose because of its potent renal, pulmonary, and neuroendocrine activities. Endothelin and its isoforms have been implicated in a wide variety of disease states including ischemia, cerebral vasospasm, stroke, renal failure, hypertension, and heart failure (30). It exerts its pharmacological effect by acting on specific G-protein–

coupled receptors. In mammalian species two receptors, ET_A and ET_B, have been cloned; both are widely distributed in human tissue and are distinguished by different responses to various ET isoforms.

The NMR-derived three-dimensional structure of ET-1 consists of several distinct regions, including a random-coil N-terminus, a β-turnover residues 5–8, followed by a short helical region and a flexible C-terminal tail (as summarized in Ref. 31). The presence of the flexible tail in solution is not surprising, as may be imagined from the primary sequence shown in Fig. 12.10. Although solution structures of ET and its analogs (32–46) determined by NMR have been valuable in defining the gross conformation of these molecules, the flexibility of the tail in solution makes it difficult to extrapolate to the bound state. Indeed, an X-ray structure of ET-1 has quite a different structure for the C-terminal tail than for the random-coil arrangement in solution (46). The bound conformation may be different again.

There is clearly an advantage to having lead molecules with reduced flexibility, given that their solution conformation will intrinsically provide a better reflection of the bound conformation. In addition, the development of a more rigid drug will reduce unfavorable entropic contributions to binding energy. Indeed, a range of small cyclic peptides that are ET_A- or ET_B-selective antagonists have been discovered and provide valuable leads to the development of potential therapeutics. NMR studies have been instrumental in determining their solution conformations. For exam-

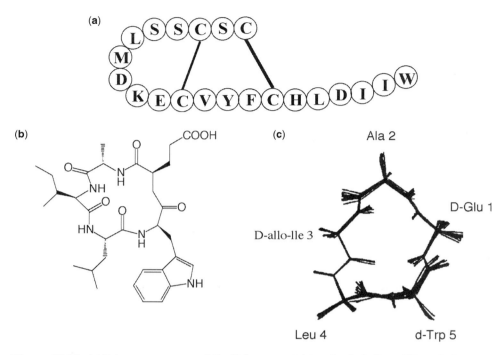

Figure 12.10. (a) Primary sequence and disulfide connectivities of endothelin-1 (ET-1). (b) Primary structure of the cyclic endothelin antagonist BE18257B and (c) a family of 36 NMR structures, which demonstrate the well-defined nature of the cyclic peptide backbone.

ple, the rather well defined solution conformation (47) of the ET_A-selective antagonist BE18257B (shown in Fig. 12.10) contrasts with the flexibility of the tail region of ET that this peptide is thought to mimic. The discovery and development of these molecules illustrate the principle that cyclic peptides are often more suitable than linear peptides as lead ligands in drug design. In addition to their better-defined and less-flexible conformations than those of their linear counterparts, they generally have improved bioavailability and resistance to protease attack.

We shall return to endothelin as a lead in drug design, in relation to a nonpeptidic antagonist. The underlying theme illustrated by the endothelin example is that ligand-based design often proceeds from initial studies of flexible endogenous molecules (particularly peptides) to constrained mimics (e.g., cyclic peptides) and often culminates in the development of nonpeptidic drug leads. NMR assists by defining the structures of the lead and subsequent molecules.

2.1.3 Instrumental Advances and their Impact on Structure Elucidation. Over the last few years there have been several exciting instrumental developments that promise to dramatically expand the role NMR will play in the drug discovery process. These relate to the combination of NMR with other technologies such as LC and/or MS and the use of NMR to directly monitor reactions carried out on solid-phase resins (8, 13, 22). The latter promises to indirectly enhance drug discovery programs by improving the monitoring and hence efficiency of solid-phase combinatorial synthesis. Effectively, resin-based syntheses can be monitored at successive stages without the need to cleave intermediate products from the resin.

As already mentioned, the additional sensitivity brought about by cryoprobe technology promises to enhance a wide range of NMR applications, but will be particularly important in natural products-based drug discovery. In many cases only limited amounts of pure compounds are isolated from natural products extracts and sensitivity has been a major limit-

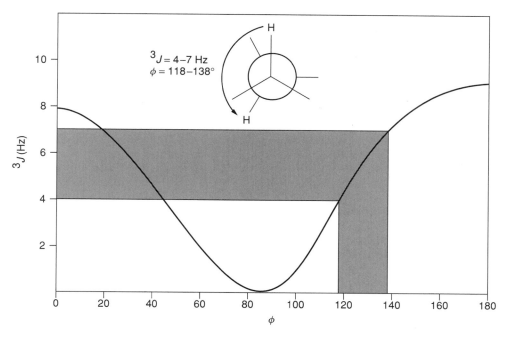

Figure 12.11. Illustration of the Karplus relationship between three-bond scalar coupling constants and the dihedral angle of the intervening bond. The relationship is indicated for the ϕ torsion angle of the H2 and H3 protons within the rigid core of taxol and related derivatives. See Fig. 12.6 for the structure of taxol.

ing factor on structure elucidation. LC/MS/ NMR systems will greatly improve the efficiency of such analyses by minimizing the need for separate sample-handling steps for the different analytical technologies.

2.2 Conformational Analysis

Usually only 1D or 2D NMR methods are required to determine the solution conformation of bioactive ligands. Useful tools include analysis of chemical shifts, coupling constants, and NOEs. An assumption inherent in the application of such studies to drug design is that the solution conformation will be maintained on binding to the receptor. This is justified in the case of relatively rigid ligands. However, for potentially flexible ligands the possibility of changes in conformation on binding must be considered, as noted above for the case of endothelin.

Coupling constants and NOEs are the main NMR parameters used in determining the solution conformations of drug leads. NOEs provide information about through-space proxim-

ity. Three-bond vicinal coupling constants are particularly valuable because their dependency on the intervening dihedral angle through the Karplus relationship allows local geometry to be determined. This is illustrated in Fig. 12.11 for taxol. Although there are several vicinal coupling constants in this molecule (Fig. 12.6), only one $^3J_{H2H3}$ occurs in a region of the molecule that is expected to be conformationally rigid and thus suitable for conformational determination by use of coupling constants. In taxol and a range of analogs this coupling is in the range 4–7 Hz, consistent with partially eclipsed dihedral angles of approximately 120–140° for this ring-constrained structure. This is in good agreement with the X-ray structure of a taxol analog, where the angle is 120°. Note that, in general, such a Karplus analysis does not give a unique solution unless several coupling constants sampling the same dihedral angle are present and is reliant on the assumption that the molecule exists only in a single conformation in solution. Although it is generally believed that

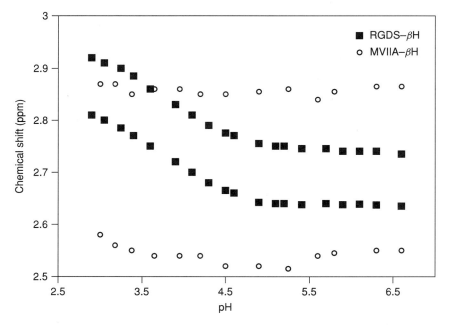

Figure 12.12. Chemical shift changes of the β-protons of Asp14 in MVIIA illustrating the lack of titration of the adjacent carboxyl group, indicating its invovlment in salt bridge. By contrast, the shift of a control random-coil peptide varies with an apprent pK_a value of 3.7, as expected for an uncomplexed carboxyl moiety in peptides.

this is the case for the core of taxol, recent relaxation data (48) described in section 2.4 suggest that this conclusion may need to be reexamined.

In addition to studies of the taxol core, there have been a large number of studies of the conformations of the side chains of taxol and it appears that these are certainly flexible and that the molecule may adopt both extended and folded conformations of the side chains. In a case like this the observed vicinal-coupling constants are a weighted average of those from the participating conformers.

2.3 Charge State

An advantage of NMR over other structural techniques such as X-ray crystallography is that it has the potential to provide information not only on structure but also on the electronic properties of molecules. Many drug leads contain ionizable groups and a determination of their charge state in solution and/or at the bound site is important in the design of analogs. Simple plots of chemical shifts as a

function of pH for nuclei near these ionizable groups provide a convenient way of determining the pK_a value and hence charge state. This is illustrated for ziconotide in Fig. 12.12, where it was suspected that one of the ionizable groups in the molecule, Asp[14], may be involved in a stabilizing salt-bridge interaction (28). This was confirmed by noting that the pK_a value for this residue is lowered considerably relative to the usual value for Asp. The β-proton chemical shifts were essentially independent of pH over the range 3–7 (indicating a $pK_a < 3$), whereas those of a control, random-coil peptide, titrated as expected over this range.

2.4 Tautomeric Equilibria

Tautomerization is a relatively common feature of drug molecules that is amenable to analysis through the use of chemical shifts or coupling constants as probes. This was recently demonstrated in a study of some non-peptide endothelin analogs (49). Starting from the modestly active compound (**1**) (Table

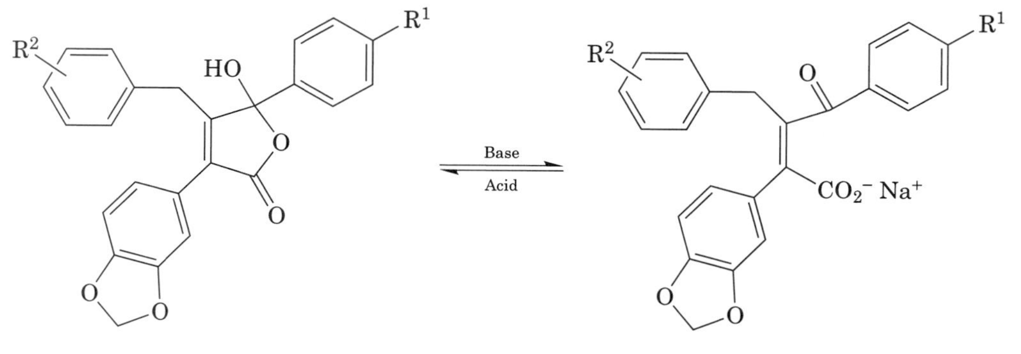

12.5), derived by screening a compound library for ET_A antagonists, the nanomolar inhibitor (**2**) was developed. Further optimization through examination of electronic and structural requirements led to the subnanomolar inhibitor (**3**), which was subsequently put forward for evaluation in a number of preclinical disease models for stroke.

These molecules display keto-enol tautomerization, as illustrated in the following structures. The open form keto-acid salts and the closed form butenolides exist in a pH-dependent equilibrium in solution, and at physiological pH both forms exist. In principle, the biological activity could reside in either or both forms.

The extent of tautomerization was established by evaluation of NMR spectra as a function of pH, from 2.65–9.05 (49). At acidic pH, compound (**2**) exists essentially in the closed butenolide form. Because the pH is slowly raised by addition of NaOD, the spectrum begins to exhibit properties associated with the open form keto-acid, and at basic pH the compound is essentially all in the open form. The coupling pattern shown by the benzylic protons is a particularly characteristic marker of the tautomeric process. At acidic pH the benzylic protons exhibit an AB quartet pattern consistent with the ring-closed structure. As the pH is raised this pattern coalesces to a singlet, broad at neutral pH and sharp at basic pH, as would be expected with the open form keto-acid structure. After the pH was basic, addition of DCl to acidify the solution caused

Table 12.5 Substitution Pattern and Receptor-Binding Affinity of Nonpeptidic Endothelin Antagonists[a]

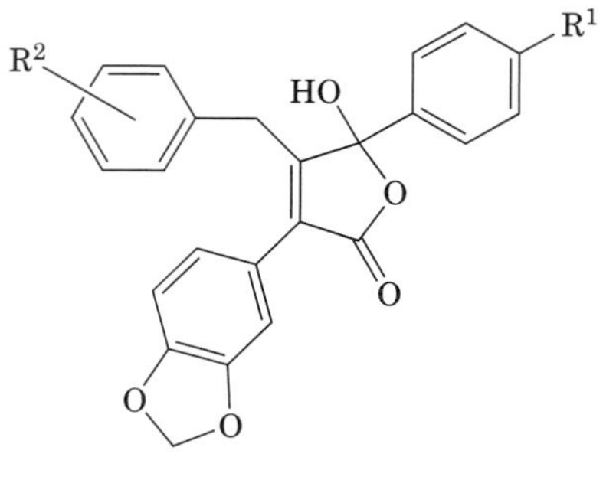

Compound	R^1	R^2	IC$_{50}$(nM)	
			ET$_A$	ET$_B$
(**1**) PD012527	Cl	H	430	27000
(**2**) PD155080	OCH$_3$	H	>0.4	4550
(**3**) PD156707	OCH$_3$	3,4,5-OCH$_3$	0.3	780
(**4**)	OCH$_3$	3,5-OCH$_3$,4-O(CH$_2$)$_3$SO$_3$Na	0.38	1600

[a]From Refs. 49 and 50.

the spectrum to return to its original appearance, consistent with a reversible tautomerization process.

Identical biological results were obtained with the salt and closed butenolide form in all pharmacological assays, reflecting equilibration at physiological pH. This made it difficult to identify the biologically active form from these experiments alone, although methylation of the OH group in compounds 1–3 resulted in a loss of activity. Because these analogs cannot tautomerize to form open ketoacids, it seems likely that the open form is responsible for activity.

In addition to its impact on the biologically active form, the tautomeric process has profound implications for formulation of drug candidates, as illustrated in some recent further development work on compound (3) (50). Although it is easy to synthesize and isolate water-soluble salts of the keto-acids, once they are placed in aqueous solution the tautomeric equilibrium determines how much of each form is present. Indeed, if the closed butenolide tautomer is sufficiently water insoluble, it can precipitate out of solution and the equilibrium can drive the complete precipitation of the compound. Although (3) has good oral activity, its intravenous use is limited by the insolubility of the closed-form butenolide tautomer without the use of a specific and complex buffered formulation. Thus in recent work a series of water-soluble butenolides was developed (50) to overcome this limitation for parenteral uses. This culminated in the development of (4) (Table 12.5), currently in pre-clinical evaluation.

This description of the development of (4) provides a good illustration of the fact that the availability of an active molecule is not the end of the drug development pathway, and that formulation considerations can be critical. In this case NMR played a significant role in understanding tautomeric processes that had a direct bearing on solubility and hence formulation.

2.5 Ligand Dynamics: Line-Shape and Relaxation Data

It is increasingly being recognized that the solution molecular dynamics of drugs may have an important role in modulating biological activity (48, 51–54). For example, dynamics may influence entropic contributions to the free energy of binding. In general the more flexible a ligand is, the more unfavorable will be the loss in entropy on binding, assuming a relatively rigid bound state of the ligand. However, in some cases flexibility of a ligand may be a positive factor. This applies, for example, if a degree of flexibility is required to allow a ligand access to a buried active site, or if activation of a receptor requires a conformational change mediated by ligand binding (9). Therefore, a knowledge of the flexibility of lead molecules is an important supplement to the structural and electronic information available from NMR.

The two major NMR methods for obtaining information on ligand flexibility are line-shape analysis and relaxation measurements (usually ^{13}C or ^{15}N T_1, T_2, or heteronuclear NOE measurements). In general terms, the former is sensitive to motions on the milli- to microsecond timescale and the latter to nanosecond timescales. To some extent, structure calculations on peptide-based lead molecules can also give an indication of regions of flexibility from an examination of local regions of disorder among a family of calculated structures. Caution must be exercised because other factors can contribute to disorder, although in many cases there is a connection between disorder in a structural ensemble and molecular flexibility (55). A recent example concerns the solution structures of three isomers of the α-conotoxin GI (56). Attempts to increase structural diversity through the engineering of nonnative disulfide bonds showed that nonnative isomers were not only different in conformation but were also considerably more flexible than the native isomer and had reduced activity.

In an example that illustrates the application of NMR relaxation measurements for studying ligand flexibility, Kessler and colleagues (57) investigated the role of disulfide bonds in the α-amylase inhibitor tendamistat. This small protein contains two disulfide bonds (C11-C27 and C45-C73) and opening of the latter is known to reduce the melting temperature of the protein (i.e., reduce its stability), but in this case does not affect its α-amylase inhibitor function. The latter observation

Table 12.6 ^{13}C-NMR Chemical Shift and Relaxation Data for Thyroxine

Position	Chemical Shift (ppm)	Experimental[a]		Isotropic Motion		Two-State Internal Motion	
		T_1 (s)	NOE	T_1 (s)	NOE	T_1 (s)	NOE
C2′,6′	127.3	0.63	2.53	0.63	2.96	0.63	2.53
C2,6	142.6	0.63	2.63	0.63	2.96	0.63	2.58
C-α	57.1	0.51	2.37	0.51	2.94	0.51	2.57
C-β	36.4	0.64	2.29	0.64	2.96	0.64	2.51

[a]Measured relaxation data at 75 MHz and 305 K (52).

[b]Theoretical best-fit values based on the indicated models for molecular motion.

suggests that this disulfide bond may not affect either the structure or the dynamics of the molecule. To investigate this possibility heteronuclear NOE measurements were used to determine the effect of the selective removal of the single disulfide bond on the dynamics of the molecule. To assess structural changes, chemical shift differences, intrastrand NOE effects, and protected amide protons were measured for the disulfide-deficient variant C45A/C73A and wild-type tendamistat. Removal of the C45-C73 bond by the C45A/C73A mutation was found to have no influence on the β-barrel structure, apart from very local changes at the mutation sites. ^{13}C and ^{15}N relaxation data showed that the only region of significant internal mobility in either the wild type or variant protein was at the N-terminus. There was little difference in internal flexibility between the two proteins, consistent with their similar α-amylase inhibitory activity. In this case it appears that the role of the C45/C73 disulfide bond is more related to thermodynamic stability and the secretion efficiency of the protein rather than to limiting its flexibility.

The thyroid hormones, exemplified by thyroxine (**5**), provide an example of the use of both line-shape analysis and NMR relaxation time measurements, to give an insight into the internal flexibility, and perhaps the mode of action, of pharmaceutically important molecules (52, 53, 58). The thyroid hormones act by binding to a nuclear receptor and appear to control receptor function by inducing a conformational change that directs the alignment of functionally critical secondary-structure elements of the receptor (59). Synthetic thyrox-

(5)

ine is widely used for the treatment of thyroid disorders and indeed has been the second most widely prescribed drug in the United States over recent years.

Table 12.6 shows experimental ^{13}C T_1 and heteronuclear NOE data for thyroxine together with best-fit theoretical values for these parameters based on two different models of the motion of thyroxine in solution. A simple isotropic model, in which the drug is regarded as a rigid body tumbling in solution, is clearly unable to simultaneously fit the T_1 and NOE data, whereas the two-state jump model, which incorporates a degree of internal mobility based on rapid flipping between conformational states, is better able to account for the experimental data. Although it is difficult to precisely define the nature of the internal motion, it is clear that rapid (nanosecond) internal motion is present, and this appears to involve small-amplitude torsions about the aromatic ring axes. The correlation time for overall tumbling of thyroxine was deduced to be approximately 0.35 ns, with the internal motion approximately 30-fold faster (52).

Studies of the ^{1}H-NMR spectrum of thyroxine as a function of temperature had earlier

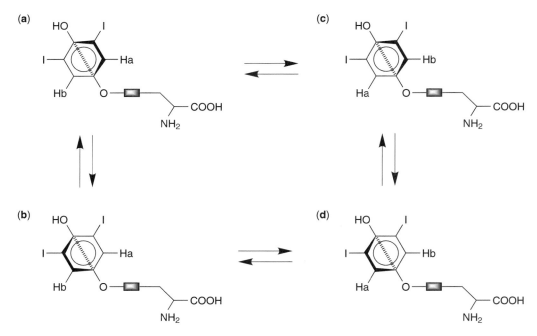

Figure 12.13. Schematic illustrations of motions of the outer ring of thyroxine. The dotted line through the outer ring shows the jump axis about which the ring rotates. (a) Ha is shown in the proximal position and is closer to the viewer than Hb because the torsion angle ϕ' is greater than 0°. This conformation corresponds to one of the two states of the two-state jump model and agrees with the "twist" of the outer ring observed in the crystal structure. (b) Rotation about the dotted line through the center of the outer ring moves Ha away from the viewer and brings Hb toward the viewer. This corresponds to the second state of the outer ring in the two-state jump model. (c) Hb is now in the proximal position and closer to the viewer than Ha. (d) Hb is in the proximal position and is now further from the viewer than Ha. Transition from a to b and from c to d involves small amplitude jumps on the nanosecond timescale and is detected by NMR relaxation measurements. Although not illustrated in the figure, the inner ring also exhibits this type of motion. Transitions a to c and b to d result in 180° flips of the outer ring and exchange of the environments of Ha and Hb. This ring flip occurs on a microsecond timescale and is detected by variable temperature line-shape studies. (Reprinted with permission from Ref. 52. Copyright 1996 American Chemical Society.)

demonstrated the presence of additional larger amplitude, but slower ring flips (60). At low temperature two signals were seen for the H2′ and H6′ protons. These signals broadened with increasing temperature, then coalesced and sharpened as the temperature was further increased. This was attributed to exchange of the environments of the two protons brought about by 180° rotation of the "outer" ring of thyroxine. Substitution of the observed coalescence temperature (T_c) and the chemical shift difference of the two signals at low temperature ($\delta\nu$) allowed the free energy of activation for this slow ring flip process to be established from equation 12.1 (53, 60).

$$\Delta G^{\neq} = 19.14 T_c [9.97 + \log(T_c/\delta\nu)] \quad (12.1)$$

The derived barriers for several thyroid hormones are in the range 36–38 kJ/mol, which corresponds to large-amplitude ring flips on the milli- to microsecond timescale. From a combination of the relaxation data and the dynamic line-shape analysis data it was possible to propose a unified model that accounts for both the fast and slow internal motions, as summarized in Fig. 12.13.

In this model, both aromatic rings of the thyroid hormones jump rapidly between two energetically equivalent conformations on a nanosecond timescale (a ⇔ b and c ⇔ d in Fig. 12.13). The half-angle of the jump varies, depending on the solvent, corresponding to an average displacement of about 90° between the two extreme jump positions. These separate states are not detectable on the chemical-shift timescale but lead to an average proximal environment for Ha and an average distal environment for Hb (attributed to rapid interchange between a and b in Fig. 12.13), which are seen in the low temperature spectra. However, these fast motions are detected by relaxation studies. Although the rate of this motion is rapid, its amplitude is not sufficient to average the environment of proximal and distal protons. Occasionally (about once every 1000 jumps) the outer ring jumps further than the nominal 90° range, exchanging the environments of the proximal and distal protons (a ⇔ c and b ⇔ d in Fig. 12.13). Although the actual rate of an individual ring flip is rapid, the effective rate of the process is on the microsecond timescale, because on average a large number of small amplitude jumps occur for every large amplitude ring flip. It is the exchanging of proximal and distal protons on the microsecond timescale that is detected by the variable-temperature line-shape studies.

The fact that thyroxine is apparently able to so freely move over a moderately large region of conformational space has implications for receptor binding. The crystal structure of the thyroid receptor ligand-binding domain complexed with the thyroid agonist 3,5-dimethyl-3'-isopropylthyronine (59) shows that the thyroid hormones bind at the center of the hydrophobic core of the ligand-binding domain and may play a structural role in the conformational changes that activate the receptor. The structures of the retinoid-X receptor ligand-binding domain (61) and the retinoic acid-retinoic acid receptor ligand-binding domain complex (62) indicate that significant conformational changes accompany ligand binding in those cases. The conformational flexibility exhibited by the thyroid hormones may also be required for binding. It has been suggested that the rapid "wiggling" of the aromatic rings could enable the hormone to work its way to the center of the ligand-binding domain as the protein reorders itself about

the ligand and may in fact trigger receptor conformational changes (52).

As briefly mentioned earlier, taxol provides another example where relaxation time measurements provide an insight into dynamics processes. Although it is generally thought that the taxane core is rigid, ^{13}C relaxation data suggest that a degree of flexibility (on the nanosecond timescale) may be present and may vary for different taxol analogs (48). In particular it appears that the removal of certain side chains may introduce additional flexibility into the core region that would not easily be predicted based on a simple inspection of the structure.

Another example of the application of line-shape analysis to ligand dynamics is described in Section 3.2 for the drug trimetrexate when bound to dihydrofolate reductase (DHFR). From that example and earlier studies on DHFR (63–65), it is clear that the techniques described above can equally be applied to ligands when bound to their receptor. In some cases significant but highly specific mobility appears to be present at the bound site.

2.6 Pharmocophore Modeling: Conformations of a Set of Ligands

Determination of the conformations of a range of ligands that all act at the same receptor site can provide significantly more information than just a single ligand structure. With a sufficiently broad range of ligands, it is often possible to generate a pharmocophore model of the receptor site, deduced based on conserved structural features and the conformations of the ligands. This has been done recently, for example, for the ω-conotoxins, the broad class of conotoxins to which Ziconotide (or MVIIA, mentioned above) belongs. From structural studies of a range of ω-conotoxins and from literature data on various mutants with altered binding affinities, it was determined that only a localized region of the surface of these molecules is involved in receptor binding (66). This allowed a pharmocophore model of putative receptor-binding pockets to be developed.

The advantage of such a pharmocophore model is that smaller, nonpeptide molecules that might have improved stability and bioavailability over their peptidic counterparts can, in principle, be designed. The NMR approach used in such pharmocophore modeling often involves a combination of many of the techniques already described. By determining information about structure and electronic properties for a range of different ligands, all acting at the same receptor site, it is often possible to infer information about the binding site, even if direct structural studies of this site are not possible.

2.7 Limitations of Analog-Based Design

Although a determination of the structure of bioactive molecules is of key importance, there are distinct limitations on the use of solution structures for drug design. In particular, unless the molecule is rigid there is no certainty that the solution conformation is the same as the bioactive bound conformation. For this reason there has been a shift over recent years to approaches in which information about the bound state is obtained. The other approach has been to probe the bound conformation by making a range of constrained analogs of a flexible lead molecule, as illustrated earlier for endothelin.

The most direct way of determining the conformation of a drug lead is to determine the full three-dimensional structure of its receptor complex. This has now been achieved in a significant number of cases but represents a substantial undertaking, as described in later sections of this chapter. A simpler approach that has also been applied is to use transferred NOE methods, as described below. This approach fits at the interface of ligand-based design and receptor-based design. It fits with the former because no knowledge of the receptor structure is required, but it also fits with the latter because it requires the macromolecule of interest to be included in the mixture to be analyzed. It is appropriate therefore to introduce the topic here but also to discuss it further in Section 3.

2.8 Conformation of Bound Ligands: Transferred NOEs

In ligand-based drug design it is not necessary to know the structure of the receptor, or even the location of the binding site, although the conformation of the ligand bound to the recep-

tor is crucial. It is clearly better if this can be measured directly rather then be inferred from the conformation of the free ligand. In certain circumstances this information on the bound conformation can be obtained from the transferred NOE (TrNOE) technique (67, 68). This method takes advantage of the fact that NOEs build up more rapidly in a ligand-macromolecule complex than they do in free ligand, and given appropriate exchange conditions for a mixture of ligand and macromolecule (typically satisfied for $K_D \geq 10^{-7}$ M^{-1}), then signals from a free ligand may be used to determine the bound conformation.

The theory of the technique was reviewed previously (69) and recent developments that minimize potential artifacts from spin diffusion have been described (5). Because it is not necessary to monitor signals from the macromolecule in this technique, it is usually present in substoichiometric amounts, thus requiring only minimal amounts of what is sometimes the more expensive component of ligand-macromolecule complexes. In addition, the molecular weight restrictions inherent in full 3D-structure determinations of complexes are ameliorated and the conformations of ligands bound to very large macromolecules may be determined. For example, the technique was recently used to determine the structure of an antibiotic bound to the ribosome (70). A range of other applications including enzyme-substrate, protein-carbohydrate, and protein-peptide interactions have recently been summarized (5).

In addition to its application as a tool for determining bound conformations of ligands, the TrNOE method has also been used recently as a screening aid for the identification of ligands from mixtures th atbind to a protein of interest. This application is addressed in more detail later in this chapter.

3 RECEPTOR-BASED DESIGN

Receptor-based design refers to the process of determining the three-dimensional structure of a macromolecular target and using this information to design ligands to interact with it. In general there have been few cases where the structure of a macromolecule or receptor alone has been successfully used to design, *de novo*, a ligand to interact with that receptor. However, such an approach is likely to become more common with improved computer-based approaches to molecular design in the future (71). Currently, the most common approach is to study a ligand-macromolecule complex and to initiate the design process based on the interaction of the lead ligand with the macromolecule.

Although the structure of the macromolecule alone is of less interest than that of the complex, in many cases a determination of the structure of the complex follows from earlier studies on the unbound macromolecule. It is thus useful to describe the approaches to structure determination of macromolecular targets. This is followed by a discussion of the dynamic aspects of protein structures in Section 3.2, before addressing the main topic of macromolecule-ligand interactions in Section 3.3.

3.1 Macromolecular Structure Determination

The two major techniques for determining three-dimensional structures of proteins or nucleic acids are X-ray crystallography and NMR spectroscopy. The crystallographic approach to structure determination is described elsewhere of this volume and here the focus is on NMR. NMR has been used to determine the structures of proteins only for about the last 15 years, with the first NMR structure determination being made in 1985. NMR has a number of advantages over X-ray crystallography, including the fact that the requirement that the protein needs to be crystallized is avoided, and that the dynamic information available from NMR studies complements the structural information. A major disadvantage of NMR spectroscopy, though, is that it is currently limited to the determination of structures of <35 kDa. With the development of new NMR techniques, such as TROSY (20), this seems certain to increase significantly over coming years, although the fact remains that among all structures currently deposited in the protein database the average size of NMR structures is about 8 kDa (72), substantially smaller than the average size of protein structures determined by X-ray crystallogra-

phy. Despite this limitation, NMR has made major inroads into the macromolecular structure determination process, and currently approximately one-fifth of all new structures deposited in the protein database have been determined by NMR spectroscopy.

3.1.1 Overview of Approach. The basis for structure determination by NMR is that, by determining a large number of distance restraints between pairs of protons, it is possible to reconstruct a three-dimensional image of the molecule. These distance restraints are derived primarily from nuclear Overhauser effect (NOE) measurements, which detect distances up to about 5 Å. Over recent years such distance restraints have been supplemented by a range of other restraints, including dihedral angle restraints derived from coupling constant measurements and orientation restraints derived from residual dipolar couplings. These restraints are input into a simulated annealing algorithm, which is used to calculate a family of structures consistent with the restraints.

NMR is unique in that it can provide detailed and specific information on molecular dynamics in addition to structural information. The use of relaxation time measurements allows the relative mobility of individual atomic positions within a macromolecule to be determined. The dynamic information obtained includes not only the rates or frequencies of internal motions but also their amplitudes. Such amplitudes are often expressed by order parameters. Not surprisingly, it is observed in many cases that the termini of proteins are more flexible than internal regions. More interestingly, NMR has provided a number of examples where internal loops in proteins have been shown to have dynamics that may be associated with their function. A good example of this is HIV protease, where NMR studies have identified reduced-order parameters in the flap region of the molecule that may reflect flexibility to allow entry of substrates or inhibitors into the active site.

In summary, a major strength of NMR is that a global picture not only of the structure but also of the dynamics of the macromolecular target is obtained. Further, NMR provides information on ionization states of titratable groups and other electronic features within macromolecules that may have an impact on ligand binding and function.

3.1.2 Sample Requirements and Assignment Protocols.
Structure determination by NMR typically requires 500 μL of a 1–2 mM solution of the protein of interest. It is important that the macromolecule does not aggregate because this causes spectral broadening and may preclude assignment. The sample should preferably be stable in solution over the extended period of time required to collect the range of NMR experiments (73–75) needed for assignment and structure determination. Individual experiments may last from a few hours to several days, with several weeks of data acquisition required for studies of larger proteins.

The particular set of NMR experiments required for NMR structure determination depends on the size of the protein. It suffices to say that for smaller proteins ($\leq$7 kDa) it is usually possible to determine the structures, mainly using 2D NMR, without the need for isotopic labeling, by use of procedures described in detail above for Ziconotide. For proteins in the range 7–14 kDa, ^{15}N-labeling and a combination of 2D/3D NMR experiments is usually sufficient, whereas for larger proteins ^{13}C/^{15}N labeling and 3D or 4D NMR is more or less mandatory. For proteins at the top end of the currently accessible range (25–35 kDa), there are additional advantages associated with partial deuteration of the protein.

3.1.3 Recent Developments.
A number of recently developed methods offer the potential for improving the quality of NMR structures and for increasing the size of proteins that can be examined. In particular, the use of residual dipolar couplings and of anisotropic contributions to relaxation provide new kinds of restraints that promise to lead to more accurate NMR structures (74, 76). As already mentioned the TROSY method (20) exploits relaxation phenomena to produce spectra with narrow lines and promises to significantly expand the size of protein targets that can be examined by NMR, from the current limit of about 35 kDa to perhaps >100 kDa.

Another development that is likely to have a significant impact is the increasing number of structural genomics programs being developed. The demands arising from such programs will no doubt stimulate new methods for the large-scale production of labeled proteins (77, 78), and for speeding up the rate of structure determination by both NMR and crystallography.

3.1.4 Dynamics.
Proteins exhibit a range of internal motions, from the millisecond to nanosecond timescale, and a full understanding of how small drugs might interact with such a "moving target" requires more than just the time-averaged macromolecular structure. Thus, over recent years much effort has been directed toward defining motions within proteins.

The most commonly applied approach has been to use ^{13}C or ^{15}N relaxation parameters such as T_1, T_2, and the heteronuclear NOE to derive correlation times for overall motion, together with rates and amplitudes of internal motions (79). Although the precise interpretation of the NMR relaxation data in terms of motional parameters remains dependent on the appropriateness of the motional model chosen, the results from many studies on the dynamics of proteins are sufficiently clear to confirm that nanosecond timescale motions in proteins are common. The functional significance of motions on the nanosecond timescale remains unclear and so far there have been few cases where significant differences in motions on this timescale between ligand-free and ligand-bound forms of proteins have been measured. It will be interesting to assess the functional significance of such motions as more data become available. However, slower motions have been correlated with function in a number of proteins, with a good example being HIV protease, described in more detail in Section 4.2.

Relaxation measurements require a considerable investment of spectrometer time and in some cases it may be possible to derive basic information about molecular dynamics from the structural ensemble alone. Although regions of disorder can reflect factors other than dynamics, a recent analysis (55) suggests that ill-defined regions in structural ensembles often do reflect slow, large-amplitude motions. Even if relaxation measurements are done, it is often not necessary to undertake extensive analysis to derive correlation times, given that trends are often apparent from the raw experimental data. For example, in the case of tendamistat, described above, it is clear directly from the heteronuclear NOE data that significant internal mobility is present at the N-terminus.

3.1.5 Nucleic Acid Structures. Most of the discussion on macromolecular targets so far has focused on proteins. DNA represents another valuable target in drug design. Most studies in which DNA is the target are done using short model oligonucleotides to mimic the binding region of DNA. The regular repeating nature of DNA structures makes this a more successful approach than similar attempts to dissect out binding regions of receptor proteins, where often the whole protein must be present to maintain a viable binding site. Similar comments apply for RNA, where improvements in synthetic methods have led to an increasing number of structure determinations over recent years. The principles involved in structure determination of nucleic acid targets are similar to those of proteins, but in practice nucleic acid structures are somewhat more difficult to solve.

3.1.6 Challenges for the Future: Membrane-Bound Proteins. The majority of targets for currently known drugs are membrane-bound receptors, yet this represents the class of proteins for which least structural information is known. Membrane proteins are notoriously difficult to characterize at a structural level because they are difficult to crystallize, thus inhibiting X-ray crystallographic studies, and are both too large and too difficult to reconstitute in suitable media for NMR studies. Nevertheless, solid-state NMR methods are beginning to show promise that eventually such targets may be structurally characterized (80). Rotational resonance solid-state NMR measurements, for example, allow precise distances to be measured in membrane-bound proteins (81).

3.2 Macromolecule-Ligand Interactions

Macromolecule-ligand interactions are integral to a wide range of biological processes, including hormone, neurotransmitter or drug binding, antigen recognition, and enzyme-substrate interactions. Fundamental to each of these interactions is the recognition by a ligand of a unique binding site on the macromolecule. Through an understanding of the specific interactions involved it may be possible to design or discover analogous ligands with altered binding properties that might inhibit the biochemical function of the macromolecule in a highly specific manner. The study of macromolecule-ligand interactions thus forms the cornerstone of most structure-based drug design applications. The macromolecule of interest may be a protein or a nucleic acid, although the majority of drug design applications have focused on protein-ligand interactions. For this reason we will refer mainly to protein-ligand interactions in the following discussion, but will include some examples of drug-DNA interactions.

3.2.1 Overview. There are several important aspects of macromolecule-ligand interactions that have a bearing on structure-based design. The simplest question that might be asked is "what is the strength of the binding interaction?," whereas the most detailed task would be to precisely define the atomic coordinates of the complete protein-ligand complex. In between these extremes there are many other questions important to the drug design process; these include questions about the binding stoichiometry and kinetics, the conformation of the bound ligand, and about the nature of functional group interactions between the protein and bound ligand. These and other important questions were introduced briefly in Table 12.3 and are examined in more detail later in this section. Before doing this it is first necessary to consider NMR timescales because the ability of NMR methods to answer questions about macromolecule-ligand complexes depends critically on the kinetics of the binding interaction. Section 3.2.2 thus describes how various NMR parameters depend on binding kinetics and in particular how fast- and slow-exchange conditions affect the interpretation of NMR data.

Having identified the exchange regime, the task then becomes to decide which NMR parameters can be used to answer the questions

Table 12.7   NMR Parameters and Their Changes on Binding

Parameter	Difference	Typical Magnitude[a]
Chemical shift	$\nu_L - \nu_{ML}$	0–1000 s^{-1}
	$\nu_M - \nu_{ML}$	0–500 s^{-1}
Coupling constant	$J_L - J_{ML}$	0–12 s^{-1}
	$J_M - J_{ML}$	0–12 s^{-1}
Relaxation rate[b]	$1/T_L - 1/T_{ML}$	0–50 s^{-1} (for T_1, larger for T_2)
	$1/T_M - 1/T_{ML}$	0–10 s^{-1}

[a]Ranges are approximate only and larger effects may be seen in some cases.
[b]$1/T$ refers to either $1/T_1$ or $1/T_2$.

posed above about the complex. Many of the NMR parameters that were described earlier for deriving information about ligands are also applicable to studies of complexes. These include chemical shifts, NOEs, and relaxation parameters. However, the presence of two interacting partners means that there are some differences in the way such parameters are measured and this has led to the development of several techniques that are particularly important for the study of macromolecule-ligand interactions, including chemical-shift mapping, isotope editing, and various NMR titrations. Section 3.2.3 describes these techniques. Finally, illustrative examples of the application of these techniques to specific drug design problems are given in Section 3.2.4.

3.2.2 Influence of Kinetics and NMR Timescales.

Macromolecule-ligand interactions are characterized by an equilibrium reaction that potentially has a wide range of affinities and rates:

$$M + L \leftrightarrow ML$$

The rate constant for the forward reaction is referred to as the on rate (k_{on}), whereas dissociation of the complex is characterized by the reverse rate constant, k_{off}. The equilibrium constant for this interaction, represented in terms of the dissociation constant of the complex K_D, reflects a balance of the on and off rates, as shown in Equation 12.2:

$$K_D = [M][L]/[ML] = k_{off}/k_{on} \quad (12.2)$$

For many protein-ligand interactions k_{on} is of the order of $10^8 \, M^{-1} \, s^{-1}$, and is typically quite similar for different ligands. The observation

that K_D values may vary over a wide range, typically from millimolar to nanomolar (i.e., $K_D = 10^{-3}$–$10^{-9} \, M$) for cases of interest, is a reflection of a variation in k_{off} for different ligands. Consideration of the k_{on} value above and the range of K_D values noted suggests a range in k_{off} from 10^{-1} to $10^5 \, s^{-1}$. The lifetime of the bound complex ($\tau_{ML} = 1/k_{off}$) may thus vary from much less than a millisecond to tens of seconds (10^{-5} to 10 s based on the above off rates). The exchange rate for the second-order binding process is given by (82):

$$
\begin{aligned}
k = 1/\tau &= 1/\tau_{ML} + 1/\tau_L \\
&= k_{off}(1 + p_{ML}/p_L)
\end{aligned}
\quad (12.3)
$$

where p_{ML} and p_L are the mole fractions of bound and free ligand, respectively.

The appearance of an NMR spectrum of a protein-ligand complex is dependent on the rate of chemical interchange between free and bound states. In particular, the effects of exchange on an individual NMR parameter (e.g., chemical shift, coupling constant, or relaxation rate) depend on the relative magnitude of the exchange rate and the difference in the NMR parameter between the two states. The cases where the rate of interchange is greater than, about equal to, or less than, the parameter difference are referred to as fast, intermediate, and slow exchange, respectively, as indicated in Table 12.7.

Table 12.7 shows that the changes in chemical shifts on ligand binding (for signals either from the ligand or from the macromolecule) are in general greater than those for coupling constants or relaxation rates. Given that 100 s^{-1} might represent a typical exchange rate between free and bound states, it is clear that

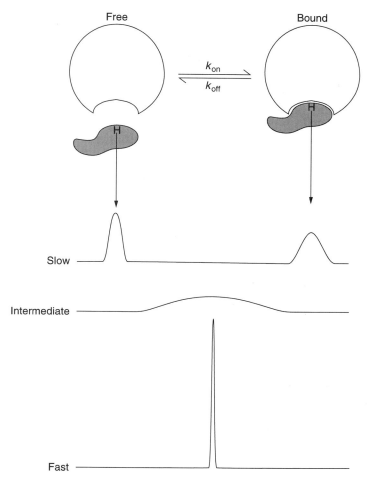

Figure 12.14. Schematic illustration of the effects of slow, intermediate, and fast exchange on the appearance of peaks in NMR spectra of macromolecule-ligand complexes. In the slow exchange case separate peaks are seen for free and bound forms. Note the broader peak for the bound ligand because it now adopts the correlation time of the macromolecule. In the fast exchange case only an averaged peak is observed.

individual NMR signals may be found in either slow, fast, or intermediate exchange on the chemical-shift timescale, but it is more likely that couplings or relaxation parameters will be in fast exchange. Thus, in most cases where the term "NMR timescale" is used in the literature, it refers to the chemical-shift timescale. The table also emphasizes that there are two types of signals that can be monitored, those from the ligand and those from the macromolecule. In general, the typical magnitude of changes to chemical shifts or couplings of either type of signal on binding are similar, although the changes to ligand signals may be larger than those from the macromolecule. However, changes to relaxation parameters for signals from ligands are much more likely to be greater than those for protein signals.

This reflects the sensitivity of relaxation parameters to molecular mobility: a ligand undergoes a greater relative change in mobility on binding than does a protein, given that the relative increase in molecular weight in the complex is much greater for the ligand than for the protein.

The exchange regime (slow, intermediate, or fast) determines how a spectrum of a protein-ligand mixture changes during a titration, or as a function of temperature. Figure 12.14 schematically illustrates the various exchange regimes for macromolecule-ligand binding interactions. Slow exchange, corresponding to tight binding, is potentially the most useful regime, given that much detailed information on the nature of a complex can be deduced in this case. Nevertheless, fast exchange also allows valuable kinetic and thermodynamic parameters to be derived. The analysis is more complex for intermediate exchange and few quantitative studies are attempted for this situation.

3.2.2.1 Slow Exchange. This situation applies when the rate of exchange is much slower than the difference in chemical shifts between the two states (i.e., $k \ll \nu_B - \nu_F$), where we now change to a nomenclature using subscripts to refer to the bound (B) and free (F) states. It should be understood that the signals may derive from either ligand or macromolecule, so although B is always related to the ML complex, F might refer to either free ligand (L) or free macromolecule (M). In this situation separate peaks are potentially observable for both free and bound states at their respective chemical shifts. Whether such signals are *actually* observed depends on the mole ratio of the individual species in a titration, and on the signals not being obscured by overlap or broadening.

Addition of a ligand to a solution of a protein results in the appearance of new signals attributed to bound protein resonances, with a concurrent decrease in the intensity of the free protein resonances, reflecting the decreased proportion of free protein during the titration. Once a stoichiometric mole ratio is achieved (usually 1:1, but sometimes 2:1 or higher if multiple binding sites are present on the protein), peaks from free ligand appear with increasing intensity as the excess of free ligand increases.

From such a titration it is possible to determine the stoichiometry of the complex, together with the chemical shifts of the bound states of the ligand and protein. In 1D NMR spectra, overlap of peaks makes it difficult to monitor more than a few resonances from either species and such studies are most readily done when there is a well-resolved signal on one of the interacting species. Selective isotope labels have been used in the past for such studies but it is now more common to use uniform ^{15}N- or ^{13}C-labeling of the protein and detect the chemical shifts in 2D HSQC spectra. It is often more difficult to label the ligand but in some cases the presence of rare nuclei such as ^{19}F can be used to advantage. A good example is the binding of the inhibitor 4-flu-

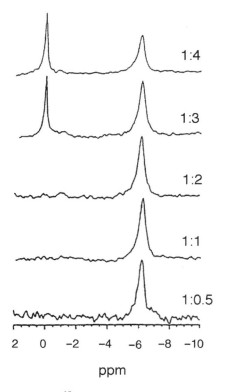

Figure 12.15. ^{19}F NMR spectra at 282 MHz of the 4-fluorobenzenesulfonamide-carbonic-anhydrase-1 system at various ratios of enzyme to inhibitor, as indicated on the traces. The peak at −6 ppm is caused by bound inhibitor. The enzyme concentration was 1 m*M* at pH 7.2 in D_2O at 25°C. (Reprinted with permission from Ref. 83; © 1998, American Chemical Society.)

orobenzenesulfonamide to the enzyme carbonic anhydrase (83). Figure 12.15 shows ^{19}F spectra of the enzyme inhibitor complex at various mole ratios. The broadened peak for the bound ligand has a chemical shift of approximately 6 ppm and is in slow exchange with the peak from free ligand at 0 ppm. The stoichiometry of the complex in this case is 2:1, so that no signal from free ligand is visible until more than 2 moles of inhibitor are present. Addition of increasing amounts of ligand results in an increase in the free ligand signal, but no change in the bound ligand signal.

Determination of the binding constant from slow exchange spectra is not usually attempted. Generally for slow-exchange conditions to exist in the first place, the binding is submicromolar in affinity and non-NMR methods are more suitable for determining affinities in these cases. NMR studies are done at millimolar concentrations, making it difficult to determine K_D with any accuracy for tight binding systems.

In principle, kinetic information on the complex can be obtained from slow-exchange spectra, as seen from the expressions for T_2 for free and bound ligand signals:

$$1/T_{2F,obs} = 1/T_{2F} + k_{off}p_B/p_F \quad (12.4)$$

$$1/T_{2B,obs} = 1/T_{2B} + k_{off} \quad (12.5)$$

Because the linewidth of a peak is related to T_2 by

$$LW = 1/\pi T_2 \quad (12.6)$$

then measurements of linewidth during a titration can be used to derive k_{off}. Equations 12.4 and 12.5 show that, although the signal from bound ligand is independent of concentration, that of the free ligand decreases in linewidth as more ligand is added. A plot of linewidth vs. ligand/macromolecule mole ratio allows k_{off} to be determined, as illustrated for example in a study of the ^{31}P linewidths for the 2'-phosphate of $NADP^+$ binding to dihydrofolate reductase (84).

Although the determination of off rates is of significance in assessing the stability of the complex, the major interest in more recent studies of complexes in the slow-exchange limit has centered on determining the complete geometry of the complex through the use of intra- and intermolecular NOEs. A recent, but already classic, example of this approach is illustrated by the binding of immunosuppressant peptides such as cyclosporin and FK506 to their receptors. These types of examples are discussed in more detail in Section 3.2.4.2.

3.2.2.2 Fast Exchange. When exchange between free and bound states is very fast, observed NMR parameters are a simple weighted average of those from the two contributing states, illustrated by Equation 12.7 for chemical shifts and Equation 12.8 for linewidths.

$$\nu_{obs} = p_F\nu_F + p_B\nu_B \quad (12.7)$$

$$1/T_{2,obs} = p_F/T_{2,F} + p_B/T_{2,B} \quad (12.8)$$

These equations show that, in the fast-exchange limit, addition of a ligand to a protein solution will cause a progressive change in chemical shift. Signals from the protein initially reflect the free state, but as ligand is added the population of bound protein increases and the observed signals move toward those of the bound state. Similarly, when ligand signals are first detected they reflect predominantly the bound state, but with increasing amounts of ligand they move toward the chemical shift of the free state. By regression analysis to Equation 12.7, taking into account the dependency of the mole fractions on K_D by the standard quadratic binding equation, it is possible to obtain estimates of both K_D and the bound shift. The procedure works best for rather weakly binding ligands (e.g., millimolar dissociation constants) (85).

When exchange is somewhat slower, but still within the fast-exchange limit, there is an exchange contribution to linewidth, as shown in Equation 12.9:

$$\begin{aligned} 1/T_{2,obs} = p_F/T_{2,F} + p_B/T_{2,B} \\ + \{[p_Bp_F^2 4\pi^2(\nu_B - \nu_F)^2]/k_{off}\} \end{aligned} \quad (12.9)$$

In this case a maximum in the broadening of ligand or protein peaks occurs during the titration at a mole ratio of approximately 0.3, as illustrated below.

The spectral changes that occur in the fast-exchange regime can conveniently be illustrated by studies on the binding of a series of terephthalamide ligands to an oligonucleotide model of DNA. The ligands, referred to as L(NO_2), L(NH_2), and L(Gly), were synthesized as precursors for potential anticancer agents (86). To establish whether they bind in the minor groove of AT-rich DNA, a series of NMR titration experiments was undertaken.

Figure 12.16 shows an expansion of the aliphatic region of a series of ^1H-NMR spectra of 0.5 mM of the oligonucleotide d(GGTAAT-TACC)$_2$, to which increasing amounts of L(NH_2) were added (86), of which the spectra cover mole ratios of ligand to DNA duplex

(6) L(NH$_2$) R = NH$_2$
(7) L(NO$_2$) R = NO$_2$
(8) L(Gly) R = NHCOCH$_2$NH$_2$

ranging from 0:1 to 2.6:1. Although the spectra are complicated by overlap in some regions, it is clear that addition of the ligand causes significant changes to the DNA peaks. A typical example is seen for the T6 methyl peak, for which addition of ligand causes both an upfield shift and broadening of the peak at certain stages of the titration. The chemical shift moves monotonically with ligand concentration up to a mole ratio of 1:1 and then reaches a plateau, remaining constant as larger amounts of ligand are added. Broadening of the peak reaches a maximum at a ligand: DNA mole ratio of approximately 0.3. Both observations are consistent with there being moderately fast exchange on the chemical-shift timescale between the free and ligand-bound forms of the DNA in solution. In this case, the observed spectral peaks reflect neither the free nor the bound form of DNA, but are averaged signals.

Ligand peaks are also in fast exchange, as seen with the L(NH$_2$) methyl peak, which first appears at a ligand:DNA ratio of 1.36:1 as a shoulder on the overlapped T3 and T7 methyl peaks at approximately 1.27 ppm. This peak is not initially visible in spectra at low ligand: DNA mole ratios because of the small population of bound species and the overlapping DNA peaks. It moves upfield with increasing ligand concentration and, again, represents an averaged peak intermediate in chemical shift between free and bound forms, reflecting fast-exchange kinetics. Eventually, the chemical shift of this signal approaches that of the free ligand at 1.1 ppm, measured in a separate experiment with a solution of ligand alone.

In the fast-exchange cases such as this it is possible to obtain an estimate of the dissociation constant for the complex (K_D) and the bound chemical shift (ν_B) of DNA resonances by fitting the observed chemical shift changes as a function of ligand concentration to equation 12.7 (85). The parameters that best fit the experimental data for the T6 methyl peak were $K_D = 1.2 \times 10^{-6}$ M and ($\nu_B - \nu_F$) = 46 Hz. Limitations on the accuracy of K_D values derived in this way were described previously (85).

To further define the thermodynamic constants associated with binding, the linewidth data were also quantitatively examined by use of Equations 12.6 and 12.9. In the case of moderately fast exchange, a maximum linewidth is predicted at a ligand:DNA mole ratio of 0.33 (82, 85), and this was indeed observed in the current case. Derived binding parameters were $K_D \leq 1.0 \times 10^{-6}$ M, $k_{off} \approx 250$ s^{-1}, ($\nu_B - \nu_F$) = 49 Hz, and LW$_B$ = 12 Hz, consistent with the values derived from the analysis of chemical shifts. Subsequent studies with the related ligand L(NO$_2$) showed similar binding to L(NH$_2$). However, a third ligand, L(Gly), was found to bind somewhat more tightly, with some signals in the intermediate exchange regime.

3.2.2.3 Intermediate Exchange. In this regime the rate of exchange between bound and free states is comparable to the differences in NMR parameters associated with the exchange. In general the spectral peaks often become very broad and analysis is difficult. This is the case, for example, for L(Gly). In the methyl region of the spectra shown in Fig.

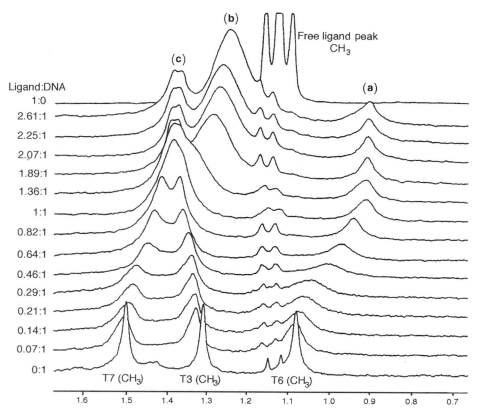

Figure 12.16. Expanded regions from 300-MHz ^{1}H-NMR spectra for complexes between L(NH$_2$) and d(GGTAATTACC)$_2$ recorded at 10°C. The two small peaks at 1.12 and 1.14 ppm arise from an impurity. Increasing ligand concentration causes an upfield shift of the T6 methyl resonance (a), and causes the T7 and T3 resonances to become overlapped at later stages of the titration (c). Peak (b) is an averaged resonance from the ligand methyl groups intermediate in shift between the bound and free forms of the ligand. (Reprinted with permission from Ref. 86.)

12.17, the T7 CH$_3$ signal moves upfield and the T3 CH$_3$ signal moves slightly downfield with increasing ligand concentration, as seen previously for L(NH$_2$) and L(NO$_2$). However, in contrast to the case for the other ligands, the characteristic broadening of peaks at intermediate ratios is non-Lorentzian, suggesting kinetics in the intermediate exchange regime. The T6 CH$_3$ peak does not shift in the characteristic fast-exchange manner but, instead, a new broad resonance appears close to the expected position of the bound T6 CH$_3$ chemical shift on the first addition of ligand, and increases in intensity with increasing ligand concentration. This observation is consistent with the ligand being in slow to inter-

mediate exchange between the free and bound forms, with $k_{off} \approx (\nu_B - \nu_F)$. Based on the magnitude of $\nu_B - \nu_F$ for this resonance, k_{off} for L(Gly) is estimated to be 50 s^{-1}, which is significantly slower than that for L(NO$_2$) and L(NH$_2$).

At a ligand:DNA ratio of approximately 1:1, the ratio of the integrals of the T6 methyl peak and the overlapped T3 and T7 methyl peaks is about 1:6. The expected value is 1:2, which indicates that the bound ligand methyl peak (4 × CH$_3$) is overlapped with the T7 and T3 methyl peaks, as observed with L(NH$_2$) and L(NO$_2$). When the ligand:DNA ratio is increased beyond a 1:1 ratio, a new peak appears at about 1.15 ppm and increases in intensity as

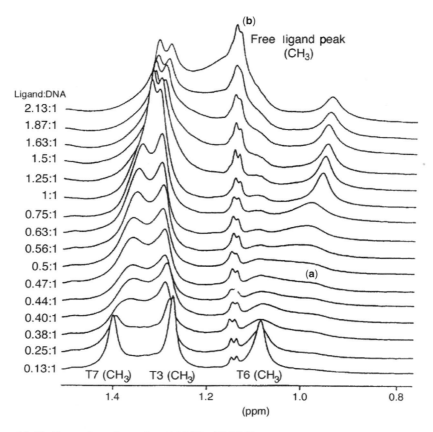

Figure 12.17. Expansions from the 600-MHz ^{1}H-NMR spectra for complexes formed between L(Gly) and d(GGTAATTACC)$_2$ showing the methyl resonances. The two small peaks at 1.12 and 1.14 ppm are attributed to an impurity. The complex nature of the T6 methyl resonance at ligand:DNA ratios less than 1:1 (a), and the manner in which signal intensity increases at about 1.15 ppm at DNA:ligand ratios greater than 1:1 (b), are indicative of intermediate exchange. (Reprinted with permission from Ref. 86.)

the ligand concentration is increased. This new peak corresponds to the methyl peak of the free ligand and its appearance in this manner is consistent with slow exchange on the chemical-shift timescale. To confirm this, spectra of a 2:1 mixture of L(Gly) and d(GG-TAATTACC)$_2$ were acquired at different temperatures (86), as illustrated in Fig. 12.18.

At low temperatures, signals at 1.15 and 1.30 ppm (overlapped with the T7 CH$_3$ and T3 CH$_3$ peaks) attributable to the methyl groups from the free ligand and bound ligand, respectively, are distinguishable. As the temperature is increased, a broad peak appears between these two signals (at $\sim$1.22 ppm). At the lower temperatures $k_{off} \leq (\nu_B - \nu_F)$, so that

methyl resonances of the ligand have complex characteristics reflecting slow-intermediate exchange. At higher temperatures, $k_{off} \geq (\nu_B - \nu_F)$, so the signal appears as a fast-exchanged average between the free and bound resonances. From a qualitative analysis of the spectra, k_{off} for L(Gly) was estimated to be 50–60 s^{-1} at 283 K.

The fact that some peaks (e.g., the oligonucleotide T7 and T3 methyl signals) exhibit fast exchange, whereas others in the same spectrum of the same complex exhibit slow-intermediate characteristics, is a reflection of the different $(\nu_B - \nu_F)$ values for different peaks. This emphasizes the point made earlier that the "exchange regime" is a relative expression

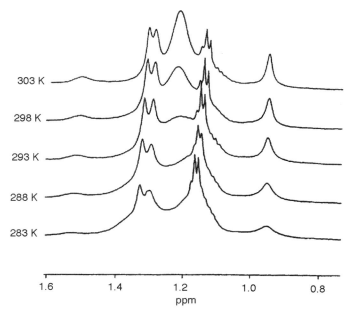

303 K

298 K

293 K

288 K

283 K

1.6   1.4   1.2   1.0   0.8

ppm

Figure 12.18. Expansions of ^{1}H-NMR spectra of a 2:1 mixture of L(Gly) and d(GGTAATTACC)$_2$ acquired at different temperatures. (Reprinted with permission from Ref. 86.)

and depends not only on the rate of exchange, but also the size of the chemical shift differences involved. In summary, the observations suggest that k_{off} for binding of L(Gly) to the oligonucleotide duplex is much slower than that for the other two derivatives. This provides an illustration of the value of NMR as a quick method for comparing the binding of different ligands and for confirming ligand-binding hypotheses.

The change in binding kinetics may be rationalized by considering the different structure of the L(Gly) ligand relative to the other ligands. It was anticipated that, upon binding to the minor groove, the terephthalamides would adopt a conformation in which the substituent on the central ring would form part of the convex edge of the ligands and therefore be directed toward the "mouth" of the groove. Given this binding arrangement, the ligand L(Gly) would have a positively charged alkylamine group positioned to interact with the negatively charged phosphate groups of the DNA backbone. The L(Gly) derivative also has a bulkier substituent than that of the other ligands and this is also consistent with some differences in its binding.

3.2.3 NMR Techniques. NMR is a particularly versatile tool for the analysis of protein-

ligand interactions. As well as being able to observe different nuclei, measurements may be made of a range of different NMR parameters, including chemical shifts, linewidths, coupling constants, and relaxation parameters. In addition, there are several specific NMR techniques that have been applied for the measurement of these parameters. The techniques that are particularly valuable for the study of macromolecule-ligand interactions are described in the following sections.

3.2.3.1 Chemical-Shift Mapping. Chemical shifts are exquisitely sensitive markers of the local charge state and environment. Although it is not possible to construct an accurate model of a binding site from a knowledge of the chemical shifts of a bound ligand, a qualitative interpretation of *changes* in chemical shifts of the macromolecule on binding provides significant insight into the location of the binding site. Traditionally, such studies were done using 1D NMR but are now increasingly done by 2D HSQC spectra. By simultaneously obtaining information on chemical shifts for a large number of sites in a macromolecule and seeing which ones change when a ligand binds, and which ones do not, it is possible to deduce the location of the binding site. This procedure is referred to as chemical-shift mapping. A prerequisite of the approach

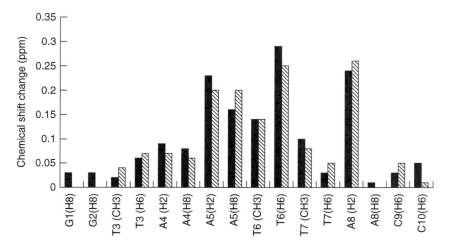

Figure 12.19. Chemical-shift perturbations of DNA protons upon ligand binding. The lighter and darker columns represent shifts attributed to L(NO_2) and L(NH_2) derivatives, respectively.

is that the chemical shifts have been assigned. Chemical-shift mapping by use of HSQC spectra is widely used in NMR screening approaches and we will defer a more detailed discussion on it until Section 4.

The relative simplicity of how chemical shift information localizes binding sites may be illustrated by continuing with the example introduced above of terephthalamide binding to DNA. Figure 12.19 shows that, upon binding of the terephthalamides L(NH_2) and L(NO_2) to d(GGTAATTACC)$_2$, the DNA protons on the four base pairs between A5 and A8 are perturbed to a much larger degree than protons in the rest of the sequence. It is thus likely that these four residues form the binding site.

A more detailed analysis allows the binding site to be further localized to the minor, rather than to the major, groove in the region of these bases. A4, A5, and A8 are the only residues containing easily detectable minor groove protons (H2). These resonances, which originate from the floor of the minor groove, are shifted downfield with ligand binding, whereas most other resonances are shifted upfield. This observation is consistent with the ligands binding in the minor groove and has been reported for other minor groove binders such as Hoechst 33258 (87) and SN-6999 (88, 89), where adenine H2 protons on the floor of the groove experience deshielding ring current ef-

fects. However, significant chemical shift changes were also observed for some major groove protons. This illustrates the general point that sometimes allosteric effects can cause changes at sites not directly involved in binding. In the case of DNA, binding perturbations in the major groove have also been observed for other established minor groove binders such as distamycin (90), netropsin (91), and Hoechst 33258 (92). Based on NOE and crystallographic data, it was concluded that the effects were caused by distortions of the B-DNA duplex, including changes in the "base roll" of residues within the binding site, upon complexation. Electronic effects arising from the close proximity of charged groups on the ligand to neighboring nucleotides were also found to perturb major groove resonances.

In the case of the terephthalamides a comparison of the minor and major groove perturbations for a particular residue shows that the minor groove protons are affected to a much greater extent. This is particularly evident for A8, where the H2 proton shifts by approximately 0.25 ppm and the H8 proton is not affected (Fig. 12.19). It is difficult to conceive of a binding mode in the major groove that would account for such a large effect on the minor groove A8 H2 resonance without a simultaneous effect on the major groove protons of T7 and A8. The observed 1:1 stoichiometry of the complex excludes the possibility that the ligand binds to the major and minor groove at the same time. It is more likely, therefore, that binding in the minor groove causes distortion of the DNA structure so that perturbations are observed for the major groove protons of A5 and T6, but not neighboring nucleotides.

Other examples of the use of chemical-shift mapping to locate binding sites have been made for ligands binding to a range of drug targets, including immunophilins, matrix metalloproteases, and DHFR. Some of these examples are described in more detail in section 3.2.5.

3.2.3.2 NMR Titrations. There are a number of advantages in undertaking a titration of ligand against macromolecule or vice versa rather than just examining the final complex. These include introducing the possibility of distinguishing signals from the individual components on the basis of intensities at intermediate stages of the titration in the slow-exchange case and obtaining kinetic and thermodynamic parameters associated with the interaction in the fast-exchange case. Such titrations may be done using either 1D or 2D spectra and are very useful for establishing the exchange regime of the complex, as described in Section 2.1. A variety of parameters may be monitored in the titration, although the two most common are chemical shifts and linewidths. Examples of such titrations are given in Figs. 12.16 and 12.17.

3.2.3.3 Isotope Editing and Filtering. Isotope editing provides a powerful way of distinguishing between the components in a complex without the need for a titration. It is one of the most useful tools for the study of macromolecule-ligand complexes, and indeed the background NMR technology that underpins isotope editing was developed specifically for the study of complexes. The principle of the approach is illustrated in Fig. 12.20 and is based on the use of isotopes to select for signals from either the ligand or macromolecule, or signals exclusively linking both of them.

The conceptually simplest approach is to uniformly deuterate the macromolecule, thereby removing its signals from ^{1}H-detected NMR spectra, and allowing signals from only the ligand to be observed. This substantially simplifies the spectrum and allows, for example, the bound conformation of a ligand to be determined from NOESY data recorded in D_2O. By rerunning the spectrum in H_2O, additional NOEs to exchangeable amide protons on the protein may be detected, thereby providing information on contacts between ligand and protein. Alternatively, ^{15}N or ^{13}C signals may be introduced selectively into either the ligand or protein and editing techniques used to select only signals attached to these labels and their proximate protons. This was used in the first example of an isotope-edited study, in this case to examine the binding of a ^{15}N-labeled peptide-based inhibitor to pepsin (93).

Potentially, the most useful approach involves uniform labeling of one of the components with either ^{15}N or ^{13}C and leaving the other component unlabeled. It is then possible to edit the spectrum by selecting for interactions (either through bond or through space) that connect protons that are both one-bond coupled to ^{15}N or ^{13}C. Alternatively, the spectrum may be filtered to specifically remove such signals, thereby selecting only signals involving protons coupled to ^{14}N or ^{12}C (i.e., on the unlabeled component). It is generally easier to uniformly label the protein rather than the ligand, and editing methods are highly efficient, thus making it easy to visualize just the protein. However, because ligand signals are often of interest, filtering experiments play a valuable role in visualizing them. Unfortunately, filtering experiments are more susceptible to artifacts than are editing experiments, although there have been recent advances in reducing artifacts (94).

Another possibility is to use half-edited/half-filtered 2D experiments to detect NOEs that specifically involve interactions between protons attached to ^{15}N or ^{13}C and those that are not. This approach is used, for example, to detect intermolecular NOEs between a labeled protein and an unlabeled ligand. Examples of isotope editing/filtering are given in section 3.2.4.

3.2.3.4 NOE Docking. In many cases the study of a complex may follow a previous structure determination of the isolated macromolecule and in that case it may be possible to determine much information about a complex by obtaining a relatively small number of NOEs linking the ligand and macromolecule.

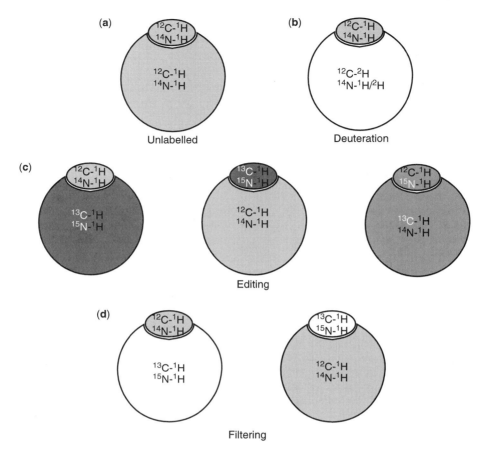

Figure 12.20. Isotope editing and filtering can be used to select signals from either the ligand or the protein. (a) Normal protein and ligand with no filtering or editing. (b) Selection of the ligand signals by 2H labeling of the protein. (c) Selection of protein or ligand signals by ^{13}C and/or ^{15}N labeling/editing. (d) Removal of protein or ligand signals by ^{13}C or ^{15}N filtering.

Gradwell and Feeney (95) recently analyzed factors important in such NOE docking experiments. In their analysis, a high resolution X-ray structure of a protein-ligand complex was used to simulate loose distance restraints of varying degrees of quality that might typically be estimated from experimental NOE intensities. These simulated data were used to examine the effect of the number, distribution, and representation of the experimental constraints on the precision and accuracy of the calculated structures. A standard simulated annealing protocol was used, as well as a more novel method based on rigid-body dynamics. The results showed some parallels with those from similar studies on complete protein NMR

structure determinations, but it was found that more constraints per torsion angle are required to define docked structures of similar quality. This is because the conformation and orientation of the ligand are defined only by NOEs and not by covalent attachment, as is the case for amino acid side chains in a protein structure. The effectiveness of different NOE-constraint averaging methods was explored and the benefits of using "R^{-6} averaging" rather than "center averaging" with small sets of NOE constraints were demonstrated. With these considerations in mind it appears that NOE docking can be a very cost-efficient procedure for defining the environment, orientation, and conformation of ligands.

(9)

3.2.4 Selected Examples. Applications of the various NMR techniques described are now illustrated with selected examples. The examples have been chosen to give a broad perspective on the types of NMR experiments that can be done and the types of information they provide. Specifically, the first example covers the case of drug-nucleic acid binding and focuses on more traditional NMR experiments, involving relatively standard homonuclear methods. The second example covers binding of moderately large ligands to immunophilins and highlights modern isotope editing techniques. The third example, covering ligand binding to a matrix metalloproteinase, also highlights the importance of these techniques and shows how relatively simple spectra involving ^{19}F-containing ligands can be very informative. The fourth example describes ligand binding to DHFR, one of the most extensively studied systems by NMR, and illustrates the derivation of a broad range of kinetic and geometric information on intermolecular complexes. The final example, on HIV protease, describes how NMR complements X-ray studies and provides information on dynamic motions within complexes.

3.2.4.1 DNA-Binding Drugs. The NMR approaches that have been used to examine the interactions of minor groove binding drugs with DNA can be illustrated with studies on the bisbenzimidazole-based compound, Hoechst 33258, (**9**). It has been used widely as a fluorescent cytological DNA stain and is also active as an anthelmintic agent. It has activity against intraperitoneally implanted L1210 and P388 leukemias in mice (96).

Footprinting studies (96) have shown that sequences of four AT base pairs are a prerequisite for strong binding to DNA, consistent with similar observations for other structur-

ally related molecules such as distamycin and netropsin (87, 90, 91, 97–99). The first structural studies of Hoechst 33258 complexed to short sequences of synthetic oligonucleotides were done using X-ray crystallographic methods (100–102). NMR and further X-ray studies followed (92, 103–107). Three of the X-ray studies (100, 101, 103) used the *Eco*RI sequence d(CGCGAATTCGCG)$_2$ and another (102) used the sequence d(CGCGATATCGCG)$_2$. Both sequences fulfil the requirement of at least four consecutive AT base pairs, and the resulting complexes showed similar modes of binding. In all of the X-ray studies, the Hoechst ligand was found to bind to the minor groove.

The NMR studies of complexes between Hoechst 33258 and oligonucleotide sequences provided complementary information to the crystal structure data (92, 103–106). Because the binding is reversible, the NMR data offer the opportunity to derive information about the kinetics of the interaction. As with the crystallographic studies, the oligonucleotide sequences were designed to contain runs of AT base pairs. Some NMR studies were performed with dodecanucleotide sequences used in crystallographic studies, including d(CGCGAATTCGCG)$_2$, which allowed a direct comparison with the crystallographic data. Experiments were also performed with sequences specifically designed to investigate different aspects of the interaction. The sequence d(CTTTTGCAAAAG)$_2$ was designed to offer two binding sites, and it was shown that two Hoechst molecules interacted with the DNA duplex in symmetry-related orientations at the 5'-TTTT-3' and 5'-AAAA-3' sites (92).

3.2.4.1.1 Stoichiometry and Kinetics. The starting point in studies of ligand-DNA complexes is usually a titration experiment to es-

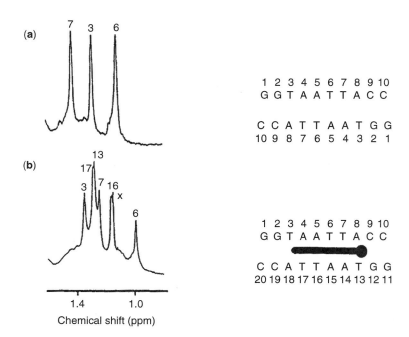

Figure 12.21. 1D ^{1}H-NMR spectra (recorded at 20°C) illustrating the thymine methyl region for the symmetrical ligand-free duplex (a) and for the 1:1 Hoechst:d(GGTAATTACC)$_2$ complex (b), which is no longer symmetrical because of the ligand binding. x corresponds to a small impurity peak. The DNA strands are numbered to the right of the spectra and the approximate location of the ligand is indicated by a black bar. (Reprinted with permission from Ref. 105. Copyright 1993, Blackwell Publishing Science.)

tablish the nature and stoichiometry of the complex. Complexes between the ligand and DNA duplex are obtained by adding small aliquots of ligand solution to a sample of the DNA duplex with one-dimensional ^{1}H NMR spectra acquired after each addition. The effects observed on the NMR spectrum after each addition reveal whether an interaction is taking place and allow the interaction to be characterized as fast or slow exchange on the NMR timescale. The stoichiometry of the interaction can also be determined from the titration.

In general, the addition of Hoechst 33258 to the oligonucleotide duplexes causes a decrease in the intensity of free DNA resonances and a concomitant increase in the intensity of new resonances, which appear in previously unoccupied spectral regions. This is consistent with the free and bound forms of the DNA duplex being in slow exchange with each other. For example, when Hoechst 33258 is

added to d(GGTAATTACC)$_2$, the free DNA signals completely disappear at a DNA:drug ratio of 1:1, and the number of new resonances is twice the number of previously observed free DNA resonances (Fig. 12.21). This is a common feature of complexes with 1:1 stoichiometry and reflects a loss of the dyad symmetry of the duplex attributed to ligand binding.

Upon addition of Hoechst 33258 to d(CTTTTCGAAAAG)$_2$, the free DNA signals completely disappeared at a ratio of 2:1 drug: DNA and there was no doubling of the number of DNA resonances in the spectrum (92). From this, it could be concluded that two molecules were bound per duplex in a manner that retained the dyad symmetry of the DNA duplex. The binding was also determined to be cooperative, in that no intermediate 1:1 complex was detected (92). The formation of a 1:1 complex would have resulted in a very complicated spectrum at intermediate ligand:DNA ratios, given that resonances arising from the free

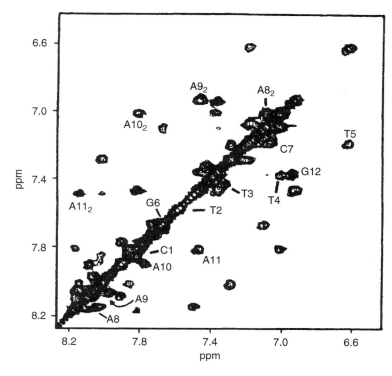

Figure 12.22. Aromatic region of NOESY spectrum of a 1:1 mixture of Hoechst vs. d(CTTTTCGAAAAG)$_2$ recorded with a 200-ms mixing time. Chemical exchange cross peaks between protons of the free DNA and the 2:1 Hoechst:DNA complex are labeled with their identifying base pair. Below the diagonal the H6 and H8 cross peaks are shown, whereas those of the adenine H2 resonances are highlighted in the upper portion of the figure and labeled with a subscript 2. (Reprinted with permission from Ref. 92. Copyright 1990, Oxford University Press.)

DNA, the 1:1 complex and the 2:1 complex, would have produced four times as many observable peaks relative to the free DNA species. At intermediate ligand concentration, however, only two sets of peaks arising from DNA molecules were detected. In the 2:1 complex only four thymine methyl resonances were detected (1.0–1.5 ppm), as expected for a symmetrical DNA duplex. These are all overlapped in the free DNA spectrum. In the 1:1 mixture, only signals from free DNA and from the 2:1 complex were detected.

The reversible nature of the Hoechst:DNA interaction is illustrated by the observation of chemical-exchange cross peaks in NOESY spectra of mixtures of free and complexed oligonucleotides (92, 104). This may be seen in the NOESY spectrum of a mixture of free and complexed d(CTTTTCGAAAAG)$_2$, shown in Fig. 12.22, in which many chemical exchange cross peaks are observed between resonances arising from the free and bound oligonucleotide. In a NOESY spectrum acquired at lower temperature, the intensity of these chemical-exchange cross peaks is significantly reduced,

indicating that the exchange is slowed at lower temperatures. The exchange rate was estimated to be <10 s^{-1} at 10°C (92).

The ability to observe such dynamic exchange phenomena is one of the strengths of NMR relative to X-ray crystallography and several examples of these phenomena are described later in the chapter.

3.2.4.1.2 Binding Site. A combination of chemical shift and NOE information can be used to locate and characterize binding sites. Chemical-shift differences between resonances arising from free and bound forms of DNA are indicative of the nature of the interaction. In all studies of the Hoechst complexes described above (92, 104–107) significant changes to the chemical shifts of thymine H1′ protons and adenine H2 protons were observed, in contrast to the generally small perturbations observed for the base H8/H6 and CH$_3$ resonances located in the major groove. Perturbations of this nature are consistent with binding to the minor groove. In some instances, significant perturbations were observed to major groove protons located well

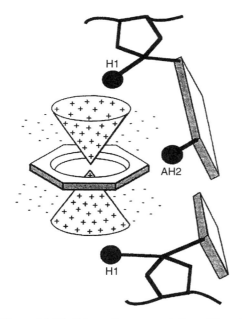

Figure 12.23. Schematic representation of ligand-induced ring-current effects on nucleotide protons that form the walls (deoxyribose H1') and floor (adenine H2) of the minor groove. (+), shielding effects; (−), deshielding effects. (Reprinted with permission from Ref. 105. Copyright 1993. Blackwell Publishing Science.)

within the binding site, reflecting changes in the conformation of the DNA duplex (e.g., base roll, propeller twisting) (92, 106).

Further evidence of minor groove binding is provided by the fact that resonances arising from protons on the floor of the groove, such as the adenine H2 and imino resonances, are shifted downfield, whereas resonances from protons on the minor groove walls, such as the H1' protons, are shifted upfield. This is a consequence of the ligand's being inserted edge-on into the minor groove. The deoxyribose protons that form the walls of the minor groove are positioned above the π-plane of the aromatic rings and consequently receive upfield perturbations to their chemical shifts. Protons positioned on the floor of the groove, however, generally lie in the plane of the aromatic rings and experience downfield perturbations to their chemical shifts, as illustrated in Fig. 12.23.

The magnitude of chemical shift changes is a strong indicator of the location of the bind-

ing site. In the case of the 2:1 complex with d(CTTTTCGAAAAG)$_2$ (92), the largest chemical shift changes occur over the 5'-TTTT-3' and 5'-AAAA-3' regions of the duplex. In the case of 1:1 complexes, where the DNA duplex contains an AT base-pair segment located at the center of the sequence, greater chemical shift perturbations are observed for resonances in that region (104–106), consistent with the binding site's being located there.

Assignment of the bound ligand and DNA resonances enables the identification of intermolecular NOEs, which are required for a precise determination of the binding site. The interaction of Hoechst 33258 with the oligonucleotides produced a large number of intermolecular NOEs (~25–30), placing considerable constraints on the structure of the complex and enabling the orientation of the ligand within the binding site to be determined. The NOE contacts observed for different complexes have a few features in common. The contacts generally involve DNA protons associated with the minor groove, such as ribose H1' and adenine H2, clearly locating Hoechst in the minor groove. Protons of all four spin systems of the ligand show NOEs to protons of the DNA, demonstrating that the interaction occurs along the entire length of the drug. Typically, protons along one edge of the ligand (e.g., NH and H4'/H4") exhibit close contacts to protons on the floor of the minor groove, showing that the bound drug is crescent shaped and isohelical with the DNA (92, 104–106).

Models of the interaction of Hoechst 33258 with the oligonucleotides studied were generated based on the intermolecular NOEs. The models of the 1:1 complexes indicated that the ligand interacted with the four AT base pairs located at the center of the sequence. Interestingly, there was no evidence for interactions with GC base pairs on the periphery of the binding sites. In the 2:1 complex reported by Searle and Embrey (92), the array of contacts observed located the ligand in the minor groove at the center of the 5'-TTTT-3' and 5'-AAAA-3' sites, as illustrated in Fig. 12.24.

As well as defining the *location* of the binding site, intermolecular NOEs can be used to determine the *orientation* of the ligand at that site. In the case of the 2:1 complex, the *N-*

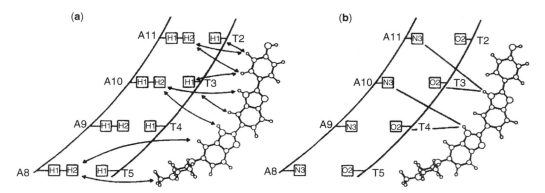

Figure 12.24. Schematic representation of Hoechst 33258 bound to the minor groove of the 5′-TTTT sequence. (a) Highlights of some of the NOEs that determine the position and orientation of the Hoechst molecule within the minor groove. (b) Intermolecular hydrogen-bonding scheme. Molecular-modeling studies with an idealized B-DNA helical structure indicate that the benzimidazole H3′ is capable of forming bifurcated interactions with A11N3 and T302, whereas the benzimidazole H3″ hydrogen bonds in a similar manner, but with A10N3 and T402. In the proposed model of the complex, these distances fall within 3.5 Å and are thus within acceptable hydrogen-bonding limits. (Reprinted with permission from Ref. 92. Copyright 1990 Oxford University Press.)

methylpiperizine moieties were found to point toward the center of the duplex, as indicated by NOEs between the protons from the piperizine ring and the 5′-terminus of the adenine tract (Fig. 12.24). Corresponding NOEs were also observed between the drug phenolic protons and the 5′-terminus of the thymine tract, as well as the 3′-terminus of the adenine tract of the complementary strand. This model did not indicate any interaction with the central GC base pairs (92).

The orientation of the ligand was similarly determined in the 1:1 complexes based on intermolecular NOEs between protons located at the extremities of the Hoechst molecule and protons of the binding site. For example, in the interaction with d(GTGGAATTCCAC)$_2$, Fede et al. (106) reported NOEs between protons from the piperizine moiety and the H2 and H1′ protons of the dinucleotide fragment d(A5T5)·d(A6T6).

3.2.4.1.3 Dynamic Processes. The binding of the Hoechst molecule to the self-complementary oligonucleotide duplexes in a 1:1 ratio lifts the dyad symmetry of the duplexes so that two sets of DNA resonances are observed. This indicates that the drug is in slow exchange between the free and the bound forms. Close examination of the 2D NOE data, however, reveals the presence of chemical-ex-

change cross peaks between symmetry-related protons on opposite sides of the dyad axis of the DNA duplex. The mechanism by which this occurs has been described as dissociation of the Hoechst molecule from the duplex, followed by a 180° reorientation and rebinding (105, 106). The self-complementary nature of the sequences ensures that the same complex is formed for either ligand orientation but with the net effect of interchanging the two strands with respect to the orientation of the Hoechst molecule. The rate at which this process occurs was estimated using cross-peak intensities in the NOESY spectrum (106). When interacting with d(GGTAATTACC)$_2$ and d(GTGGAATTCCAC)$_2$, the lifetime of the complex in each state ($1/k_{ex}$) was reported to be approximately 0.8 and 0.45 s, respectively (105, 106). These values indicate a small but significant difference in the affinity of Hoechst for TAATTA and GAATTC sites.

Intramolecular dynamic processes that are fast on the NMR timescale are also observable in the ^{1}H-NMR spectrum of the bound Hoechst molecule. Resonance averaging is observed for the H2/H6 and H3/H5 protons of the phenol group, which is consistent with the environments on either side of the ring being averaged by rapid ring-flipping motions about the C4-C2′ axis. This occurs despite the apparent tight fit between the phenyl ring and the walls of the minor groove, which, in a static model of the complex, must present a large barrier for rotation. It was estimated (105) that the rate for this process is as high as 1000 s^{-1}. This is much higher than the rate of interconversion between free and bound forms of the duplex; thus, dissociation of the drug from the complex cannot be the rate-limiting factor for phenol ring flipping. Dynamic fluctuations of the DNA conformation are more likely to provide the rate-limiting step.

3.2.4.1.4 Summary of Solution Studies. The data obtained from these NMR studies are consistent with the bound ligand fitting tightly within the minor groove of AT tetramers, with the aromatic rings of the ligand being roughly coplanar. The AT tract provides the key recognition features required for binding, including the narrowness of the minor groove. The importance of van der Waals interactions is evident, given the large number of NOE contacts between the ligand and the walls and floor of the groove. Hydrogen bonding also plays a significant role in stabilizing the interaction, as do electrostatic interactions between the positively charged piperizine ring and the minor groove. Electrostatic interactions are also likely to play a significant role in orienting the ligand within the binding site, as shown in the 2:1 complex, where the piperazine rings point toward the center of the duplex where the positive charge is best stabilized (92). The information derived from these studies, as well as from NMR studies of the interactions of other minor groove binders with DNA, is useful for the design of ligands with altered specificity or increased binding affinity, with the overall goal being the development of novel drugs.

3.2.4.2 Immunophilins: Studies of FK506 Analog Binding to FKBP. Some of the most detailed investigations of the interaction between ligands and their target proteins have been made for the immunophilin class of proteins. The major FK506 binding protein (FKBP) has a molecular mass of about 11.8 kDa, whereas cyclophilin (Cp) has a mass of about 17 kDa. These proteins are unrelated in amino acid sequence but both have peptidyl-prolyl *cis-trans* activities that are inhibited by immunosuppressants that block signal trans-

duction pathways leading to T-lymphocyte activation. FK506 (**10**) and rapamycin (**11**) in-

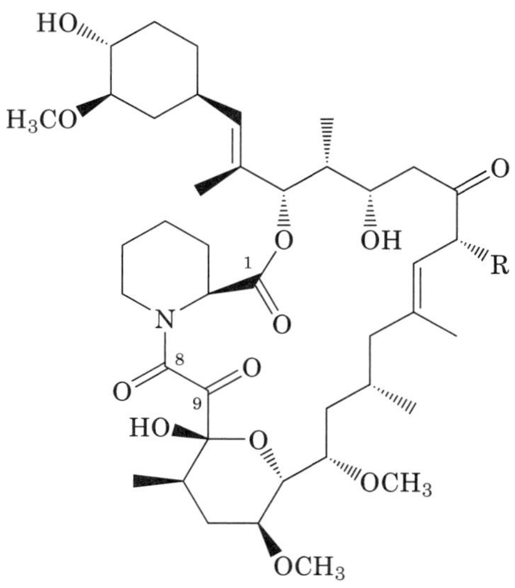

(**10**) FK506 R = CH₂CHCH₂
(**12**) Ascomycin R = CH₂CH₃

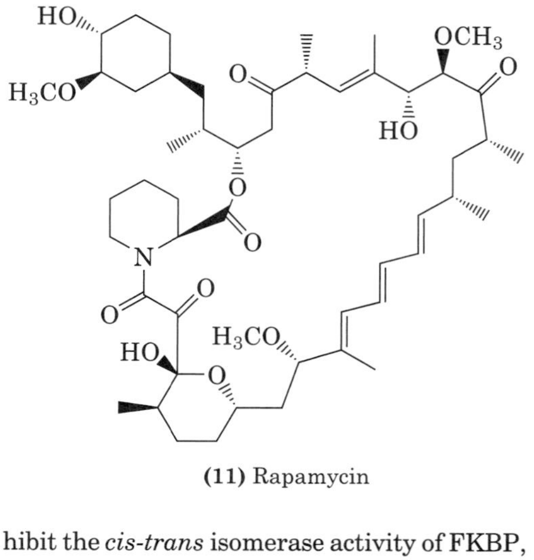

(**11**) Rapamycin

hibit the *cis-trans* isomerase activity of FKBP, whereas cyclosporin A (structure shown in Fig. 12.3) inhibits that of Cp. NMR has contributed significantly to the understanding of binding interactions to both proteins.

Initial studies on FK506 focused on the structure of the free ligand to aid in the design of further analogs (108–110). However, it was established from studies of the cyclosporin A-

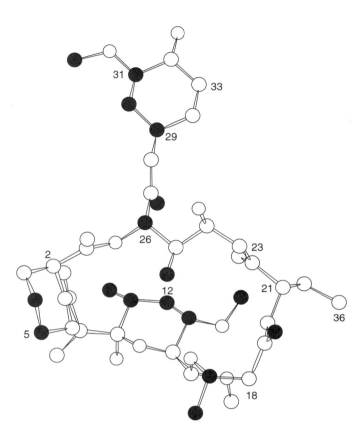

Figure 12.25. Three-dimensional structure of ascomycin bound to FKBP. Protons on the ligand that showed NOEs to the protein are denoted by a black shading of the carbons to which they are attached. Although no NOEs were observed from the protons at position 3 to the protein, the upfield shift of their resonances, -1.09 and 0.25 ppm, suggests that they are in close proximity to an aromatic region of FKBP. (Reprinted with permission from Ref. 116; © 1991, American Chemical Society.)

cyclophilin complex that the conformation of a molecule bound to its target site may be very different from that in the free state (111–113). In addition, analog design is assisted by knowing the location of the binding region of the ligand. Studies were therefore undertaken to determine the bound state of the ligand as well as to identify those portions of the drug interacting with the binding protein.

The first investigations involved the analysis of ^{13}C carbonyl chemical shifts of C8 and C9 and the ^1H chemical shifts of the piperidine ring of FK506 bound to FKBP (114, 115). The upfield shifts of the piperidine ring protons, as well as NOEs observed between these protons and aromatic protons of FKBP, suggested that the bound site on FKBP resided in an aromatic-rich domain, and allowed a putative binding site on FKBP to be proposed. It was also evident that the pipecolinyl functionality of FK506 and analogs was involved in the binding face of the ligand.

In another study (116), a uniformly ^{13}C-labeled ascomycin, (**12**), was prepared, allowing the bound conformation of ascomycin to be determined in the presence of FKBP. The enhanced ^{13}C signals were used to edit the ^1H NOESY spectra used for the structural analysis. Not only were the assignments of side-chain methyls made possible by the ^{13}C enrichment, but ligand resonances could be distinguished readily from those of the protein. The conformation of the ligand was determined from NOEs observed in a 3D HMQC-NOESY spectrum. The resulting ascomycin structure (Fig. 12.25) differed considerably from that of the uncomplexed FK506 obtained by X-ray crystallography, but was similar to that of rapamycin. In particular, the bound ascomycin displayed a *trans* orientation of the 7,8-amide bond, whereas this bond is *cis* in free FK506 and *trans* in rapamycin. The backbone structure of the macrocyclic ring differed from that of uncomplexed FK506, but showed a similarity in the piperidine ring region to that of rapamycin. This study also showed that both the piperidine ring and the pyranoyl moiety of ascomycin are involved in the binding interface in the complex with FKBP. Ligand protons that show NOEs to the protein are in bold in Fig. 12.25. X-ray studies since have confirmed these results for both the FK506-FKBP and rapamycin-FKBP complexes, showing the *trans* orientation of the ligand amide bond in the bound conformation, and verifying the involvement of the piperidine and pyranoyl regions of the ligand in the binding interface (117).

The binding site of the FKBP complex has also been investigated through use of NMR spectroscopy. Michnick et al. (118) and Moore et al. (119) solved the structure of uncomplexed FKBP by use of ^{1}H-NMR methods. Although spectral overlap did not allow every structural constraint present to be identified unambiguously, convergent structures defining the global fold of the 107-residue FKBP protein were obtained. Previous biochemical data allowed the extensive aromatic cluster within the core of the structure to be identified as the ligand-binding pocket. The loop regions of the protein between residues 37–43 and 83–90, situated at the open end of the binding pocket, were also of interest. The loops were the least well defined regions of FKBP and were thought to be flexible, and perhaps involved in the binding interaction. Examination of ^{1}H and ^{15}N chemical-shift changes on addition of ligand supported this notion and suggested that significant structural changes in these loop regions occurred upon ligand binding (118).

In a later study, a high resolution structure of the complete ascomycin-FKBP complex was calculated by heteronuclear 3D and 4D NMR by Meadows et al. (120). Uniformly labeled [^{15}N]FKBP and [^{13}C,^{15}N]FKBP were prepared and incubated with unlabeled ascomycin to form the complexes. Three-dimensional NOESY-HSQC spectra, resolved according to ^{15}N shifts, were used to obtain the NH-NH NOEs within FKBP. CH-NH NOEs were derived from a 4D [^{13}C,^{1}H,^{15}N,^{1}H]-NOESY spectrum of the doubly labeled material in H$_2$O

and CH-CH NOEs from the same experiment repeated in D$_2$O. Hydrogen bond constraints were obtained by the identification of slowly exchanging amide protons from a series of HSQC spectra acquired over several days. Torsional angle constraints were obtained from coupling constants measured in a 2D HMQC-J spectrum of [U-^{15}N]FKBP/ascomycin. In all, 1958 distance constraints were applied to the structure calculation, with the extra resolution afforded by isotopic labeling, as compared with the 590 and 1047 restraints used in earlier homonuclear studies (118, 119). Restraints defining the structure of bound ascomycin were obtained from the previously reported data of Petros et al. (116) and, along with the intermolecular NOE-derived distance constraints also reported in their study, the complete ascomycin/FKBP solution structure was calculated.

The extra detail afforded by the multidimensional NMR approach allowed the ligand-protein contact area to be located unambiguously and even specific intermolecular hydrogen bonds identified. The structure of the complexed FKBP was essentially similar to that of the uncomplexed structure, except that the "ill-defined" loop regions between residues 36–45 and 78–92 were found to adopt well-defined conformations in the complexed proteins, as preempted by previous studies. Although this difference may partially be a result of the differences in resolution achieved in the complexed and uncomplexed FKBP NMR studies, generally it was thought that binding involved some rearrangement of the 36–45 and 78–92 loops. This provides a good example of the dynamic nature of protein binding as revealed by NMR spectroscopy.

The dynamic aspects of the ligand-FKBP complex formation were pursued by Cheng et al. (121) through analysis of ^{15}N-NMR relaxation data. In particular, the increased backbone mobility for several residues within the 36–45 and 78–95 loops compared with that of the rest of the protein was noted. From analysis of the ^{15}N relaxation rates of FKBP complexed with FK506, it was found that flexibility was restricted along the entire polypeptide chain (122). This confirmed the proposition that the binding interaction of FKBP with ligand involves stabilization and structuring of the protein loops adjacent to the binding site.

In summary, it was possible not only to define the free and bound conformations of the ligand but also to identify the two binding interfaces involved in the interaction and demonstrate a reduction in protein mobility in a defined region of the protein upon binding. This level of analysis was possible because of the tight binding of the FKBP-ligand complex, its small size, and the availability of labeled species. The information proved to be complementary to X-ray crystallographic studies and will help to clarify the role of FKBP complex formation in immunoregulation.

3.2.4.3 Matrix Metalloproteinases.

Matrix metalloproteinases (MMPs), including stromelysin, collagenase, and gelatinase, are involved in tissue remodeling associated with embryonic development, growth, and wound healing. Unregulated or overexpressed MMPs have been implicated in several pathological conditions, including arthritis and cancer, and inhibitors of stromelysin and other MMPs have attracted much interest because of their potential for the treatment of these diseases.

Several NMR structural studies of stromelysin (123–127) and collagenase (128, 129) complexes have been reported. The secondary structure and global fold have been found to be quite similar for the catalytic domains of both enzymes and their various complexes with ligands. The active site in each enzyme is a cleft spanning the width of the enzyme, with a catalytic zinc atom coordinated by three histidine residues located in the center. Different dynamic properties of active-site residues in stromelysin/ligand complexes (3) and of collagenase with and without bound inhibitor (128, 129) have been reported. It has been proposed that structural and dynamic differences can be exploited in structure-based drug design to achieve broad inhibitor activity against several MMPs or to obtain more selective inhibition (3).

Of recent interest have been structural data on a novel class of MMP-binding inhibitors, represented by PNU-107859 (**13**) and PNU-142372 (**14**), which contain a thiadiazole moiety that coordinates the catalytic zinc atom through its exocyclic sulfur atom (130).

(13)

(14)

Isotope editing/filtering studies played an important role in defining interactions between the ligands and stromelysin. For example, for the stromelysin/PNU-107859 complex a 3D ^{12}C-filtered, ^{13}C-edited NOESY spectrum recorded on the [^{12}C,^{14}N]PNU-107859/[^{13}C,^{15}N]-stromelysin complex was used to assign protein/ligand NOEs. Of the 11 observed NOEs between the ligand and protein aliphatic protons, nine involved the aromatic ring of (**13**) and one involved the terminal methyl group. NOEs were observed between (**13**) and protons of Tyr^{155}, His^{166}, Try^{168}, and Ala^{169}. All four of these residues are located in the S_1-S_3 binding sites on one side of the active site. Comparison of 2D ^{1}H-^{15}N HSQC spectra showed that differences between the ^{1}H and ^{15}N chemical shifts for the stromelysin/**13** and stromelysin/**14** complexes are concentrated in the active site, indicating that no gross conformational differences in protein structure exist. The aromatic rings of (**13**) and (**14**) bind in the same region of the protein.

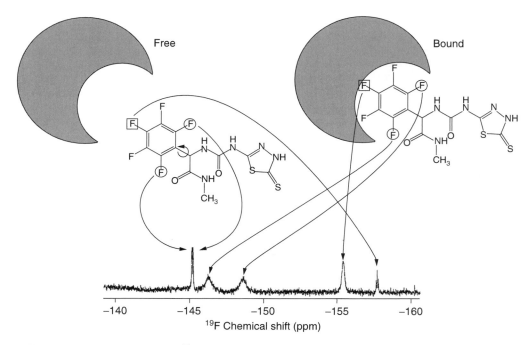

Figure 12.26. Region of the 1D ^{19}F spectrum of the stromelysin/PNU-142372 complex. Signals from free (sharp) and bound (broad) PNU-142372 are observed. (Adapted from Ref. 3 and reprinted with permission from Elsevier Science.)

A region of the 1D ^{19}F spectrum of the stromelysin/**14** complex is shown in Figure 12.26 (3). Two separate resonances were observed for the two *ortho* fluorine atoms of the bound ligand in contrast to the single resonance observed for both *ortho* protons of stromelysin-bound (**13**), indicating that the ring flip rate (rotation about the C$^{\beta}$-C$^{\gamma}$ bond) is reduced for stromelysin-bound (**14**) compared to stromelysin-bound (**13**). A ring flip rate of approximately 100 s^{-1} was estimated from the difference in linewidths for the bound *ortho* and *para* fluorine atom resonances of (**14**), more than two orders of magnitude slower than the ring flip rate for (**13**). The ^{19}F spectrum in Figure 12.26 illustrates several general principles that are useful in NMR studies of ligand macromolecule complexes. First, note that the use of a rare probe nucleus such as ^{19}F produces spectra of elegant simplicity. Because there is no naturally occurring ^{19}F in the macromolecule, it generates no interfering signals. Second, the offset

in chemical shift between bound and free signals reflects the different environment of the bound and free states. Third, signals from the bound ligand are broader than those from the free ligand because of the higher molecular weight of the complex but are still clearly visible for a complex of this size.

NMR studies have also been reported for ligands bound to collagenase. Interest so far has focused on hydroxamate-containing ligands, where it has been shown that binding causes a decrease in mobility of some but not all active-site residues (128, 129). Interestingly, some active-site residues adjacent to residues that interact directly with inhibitor were found to have high mobility both in the presence and the absence of inhibitor (129). This contrasts with what is observed for stromelysin complexed to hydroxamate ligands and a more complete understanding of the dynamics of the respective interactions may provide critical information for drug design (3).

Hydroxamate-containing ligands have also featured in other NMR studies, this time using transferred NOEs to determine their bioactive conformations (131). TrNOE data were used to determine the conformation of the inhibitors when bound to stromelysin. The NOE-derived structures of the bound inhibitors were used as templates to screen a database of 260,000 compounds. Eighteen of the 23 compounds identified for which stromelysin binding data were available had affinities less than 200 nM, demonstrating the value of deriving a conformationally restricted template for structure-based drug design (131). This study also demonstrates the close synergy that exists between structure-based design and screening approaches, either *in silico* or experimental.

3.2.4.4 Dihydrofolate Reductase. Dihydrofolate reductase (DHFR) is an important intracellular enzyme that is the target of several clinically used drugs, including methotrexate (**15**), an anticancer compound, and trimethoprim (**16**), an antibacterial. These act by inhibiting the enzyme in malignant cells and parasites, respectively. The small size of DHFR (18–20 kDa) makes it amenable to structural studies and there have been numerous complexes determined using both X-ray and NMR methods. The focus here will be on a recent illustrative example of the structure of a new complex of DHFR with trimetrexate (**17**). Trimetrexate was initially investigated as an antimalarial agent but has subsequently been found to have antineoplastic activity against breast, neck, and head cancers. It has also been used as an antibacterial for the treatment of *Pneumocystis carinii* pneumonia in AIDS patients. As seen from the following structures, trimetrexate combines some of the features of trimethoprim and methotrexate:

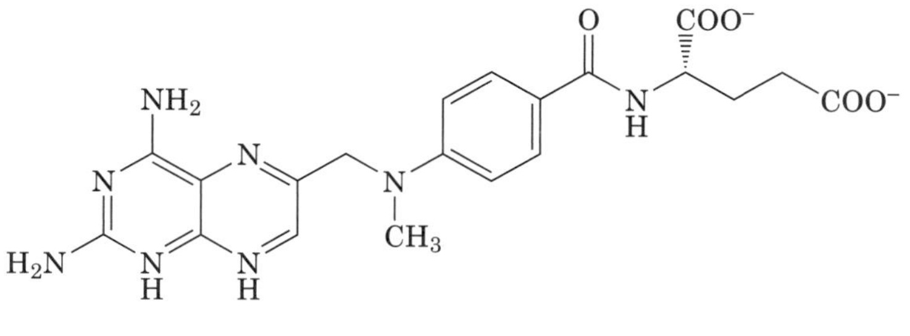

(15)

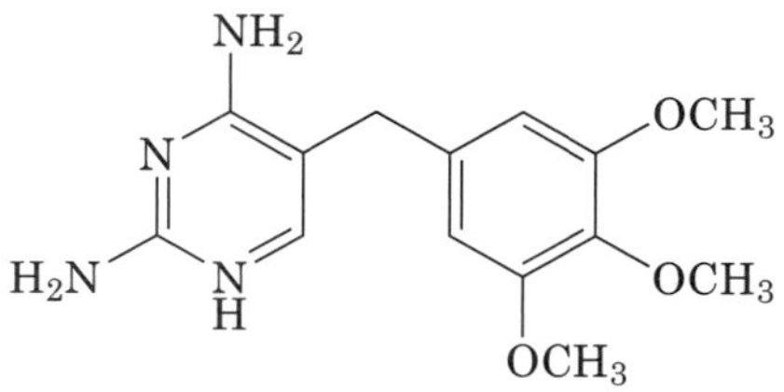

(16)

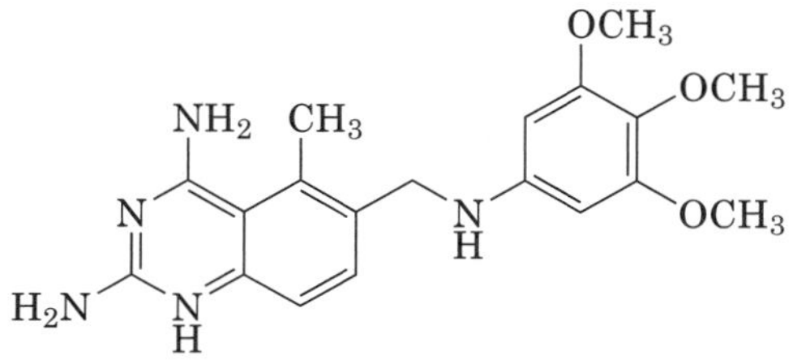

(17)

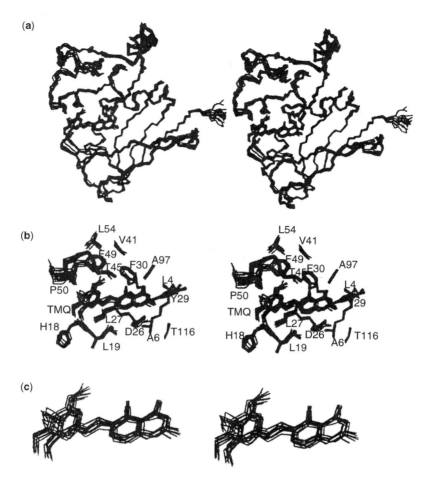

Figure 12.27. Stereoview of a superposition over the backbone atoms (N, Cα, and C) of residues 1–162 of the final 22 structures of the DHFR-trimetrexate complex. (a) View of the protein backbone and the trimetrexate heavy atoms. (b) View of trimetrexate in the binding site of enzyme. (c) Conformation of trimetrexate in the binding site of enzyme. The orientation of trimetrexate is identical for (a)-(c) and only its heavy atoms are shown. (Reprinted with permission from Ref. 132. Copyright 1999 The Protein Society.)

The three-dimensional structure of the complex of DHFR with trimetrexate was determined using about 2000 distance restraints, 300 angle restraints, and 100 hydrogen-bonding restraints (132). Simulated annealing calculations produced a family of 22 structures consistent with the constraints. Several intermolecular protein-ligand NOEs were obtained by using a novel approach that monitored temperature effects of NOE signals resulting from dynamic processes in the bound ligand. At low temperature (5°C) the trimethoxy ring of bound trimetrexate flips

sufficiently slowly to give narrow signals in slow exchange, which give good NOE cross peaks. At higher temperature these broaden and their NOE cross peaks disappear, thus allowing the signals in the lower temperature spectrum to be identified as NOEs involving ligand protons. Figure 12.27 shows the structure of the complex, including the orientation of the ligand in the binding site.

The binding site for trimetrexate is well defined and was compared with the binding sites in related complexes formed with methotrexate and trimethoprim. No major conforma-

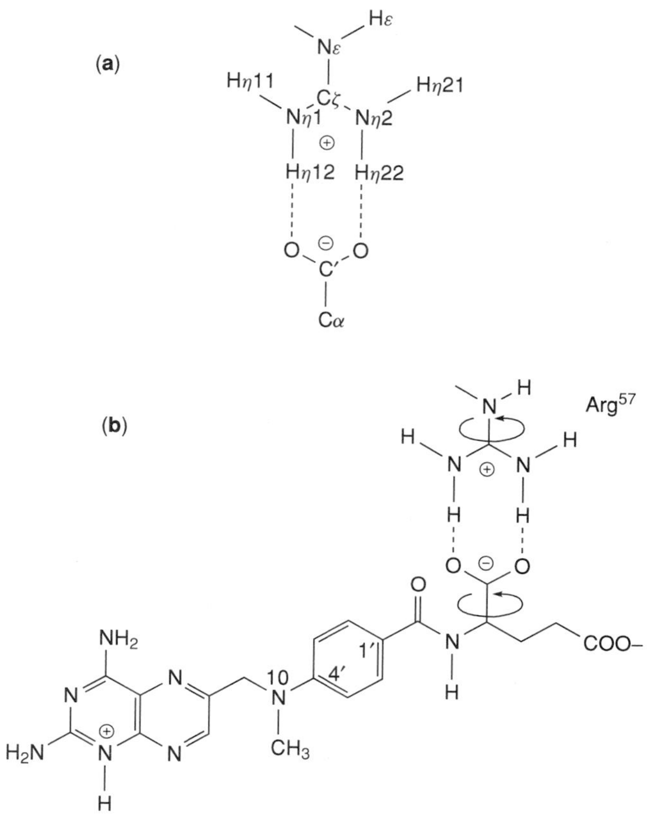

(a)

(b)

Figure 12.28. Correlated motions of a carboxylate group from methotrexate and Arg57 of DHFR detected by NMR. (a) Structure of an arginine-carboxylate complex formed with symmetrical end-on interactions and (b) structure of methotrexate showing its interaction with the guanidine group of Arg57 of DHFR. (Reprinted with permission from Ref. 133.)

tional differences were detected between the different complexes. The 2,4-diaminopyrimidine-containing moieties in the three drugs bind essentially in the same binding pocket and the remaining parts of their molecules adapt their conformations such that they can make effective van der Waals interactions with essentially the same set of hydrophobic amino acids. The side-chain orientations and local conformations are not greatly changed in the different complexes.

The ring flipping of the trimethoxy aromatic ring mentioned above was detected by variable-temperature studies of the spectral line shape. The presence of such dynamics processes involving the ligand appear to be not uncommon in macromolecule-ligand complexes and the ability of NMR methods to detect such phenomena represents one distinct advantage of NMR over X-ray methods of structure determination. Relaxation measure-

ments were also used to probe dynamics of the protein and no large amplitude motions were found, apart from that at the C-terminus (132). The power of NMR methods for studying dynamics of complexes is further illustrated by an earlier study of the complex of DHFR with methotrexate (133). In this case a correlated dynamic rotation of a carboxylate group on the ligand and Arg[57] of the protein was detected, as illustrated in Fig. 12.28.

3.2.4.5 HIV Protease. Because of its essential role in the HIV life cycle, HIV protease is a major target for structure-based design of anti-AIDS drugs. There are now more than 100 structures of HIV protease and protease inhibitor complexes in the HIV-protease structure database (134–136) and the availability of this wealth of high resolution structural information has been the driving force behind numerous structure-based design programs (134, 135, 137). Most of the high reso-

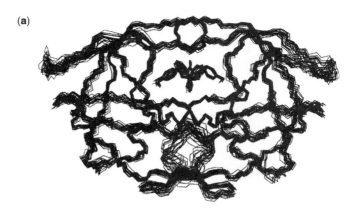

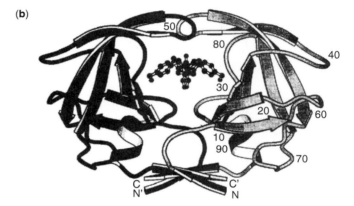

Figure 12.29. (a) View of the superimposed heavy atom (N, Ca, C) of the ensemble of structures of the HIV-1 proteases/DMP323 complex. (b) Ribbon diagram of the minimized average structure of the complex. (Reprinted with permission from Ref. 138. Copyright 1996 The Protein Society.)

lution structural information on HIV protease has been obtained from X-ray crystallography data (136). Although there are relatively few examples of HIV protease/inhibitor complexes that have been determined by use of NMR spectroscopy, the NMR data, taken together with the structural data from X-ray experiments, have contributed to an understanding of protease-inhibitor recognition and dynamics. Indeed, studies of HIV protease/inhibitor complexes are a powerful example of the way in which complementary information obtained from X-ray crystallography and NMR spectroscopy can be used to facilitate structure-based drug design.

HIV protease/inhibitor complexes have a molecular weight of approximately 22 kDa. Although NMR spectroscopy is well suited to determination of the structure of molecules in this size range, efforts to determine the solution structure of the complex were hampered

by the fact that the protease undergoes rapid autocatalysis in solution. It required the development of potent inhibitors before NMR studies of the complex became feasible. The first solution structure (Fig. 12.29) of HIV protease bound to the cyclic urea inhibitor DMP-323 (**18**) was reported in 1996 (138).

(18)

The protease exists as a homodimer. Each 99-residue monomer contains 10 β-strands and the dimer is stabilized by a four-stranded antiparallel β-sheet formed by the N- and C-terminal strands of each monomer. The active site of the enzyme is formed at the interface, where each monomer contributes a catalytic triad (Asp[25]-Thr[26]-Gly[27]) that is responsible for cleavage of the protease substrates. The "flap region" is located above the reactive site and is formed by a hairpin from each monomer of two antiparallel β-strands joined by a β-turn. There is little difference between the solution and crystal structures of protease-inhibitor complexes, except in those regions where the polypeptide chain is disordered. However, experiments in solution have allowed access to parameters that are not directly accessible from crystal data. These parameters, such as the amplitude and frequency of backbone dynamics, the protonation states of the catalytic aspartate residues, and the rate of monomer interchange, are essential in understanding the interaction of HIV protease with potent inhibitors.

The cyclic urea inhibitor DMP-323 was designed by analysis of crystal structures of HIV protease/inhibitor complexes. A feature common to many of the complexes of HIV protease is a buried water molecule that bridges the inhibitor and Ile[50] in the flaps. Interactions with this water molecule are thought to induce the fit of the flaps over the inhibitor (139). In contrast, mammalian aspartic-protease/inhibitor complexes are unable to accommodate an equivalent water molecule (135). This observation led to the design of a series of cyclic urea-based inhibitors that are capable of displacing the buried water molecule (139). As well as improving the specificity of inhibitors to the viral protease, displacement of the water molecule was expected to increase the entropic contribution to inhibitor binding and thus enhance the affinity of complex formation. The cyclic urea inhibitors are highly potent and specific inhibitors of HIV protease (139) and for DMP-323 it has been shown in both the crystal structure (139) and in solution (140) that the urea moiety does indeed replace the buried water molecule.

Although DMP-323 replaces one buried water molecule, several others are observed in the crystal structure of the complex. A more recent NMR study investigated the role of these water molecules to determine whether any had a structural role in the formation of the HIV protease/DMP-323 complex (141). In favorable cases, NMR can be used to estimate the residence times of hydration water molecules (142), thus providing information about the timescale of the interaction of buried water with the bulk solvent. This analysis led to the identification of a symmetry-related pair of water molecules that may have a structural role in formation of the complex. Such information may prove useful in the design of future cyclic urea inhibitors. An interesting finding in this study was the fact that each of the hydroxyl protons of DMP-323 is in rapid exchange with solvent. This is a surprising result, given that two of these hydroxyl protons are completely buried and form a network of hydrogen bonds with the catalytic Asp[25]/Asp[125] side chains (143). Furthermore, the dissociation rate of DMP-323 is less than $1\ \mathrm{s}^{-1}$ under the conditions of the experiment, which is too small to average the chemical shifts of the hydroxyl protons and the bulk water. The observation is ascribed to local fluctuations in the complex that allow solvent molecules to penetrate into the binding site. This conclusion is supported by the observation that the catalytic protons of the Asp[25]/Asp[125] side chains in the protease/DMP-323 complex undergo H-D exchange with solvent, even though they are buried and hydrogen bonded to the inhibitor (143). These studies highlight that even well-ordered structures such as the protease/DMP-323 complex may be flexible on the millisecond to microsecond timescale.

Interestingly, in the DMP-323 complex, both of the catalytic Asp[25]/Asp[125] side chains are protonated over the pH range 2–7 (143). The protonated Asp[25]/Asp[125] residues form a network of hydrogen bonds with the hydroxyl groups of DMP-323. In contrast it has been shown that in the complex with the asymmetric inhibitor KNI-272, the side chain of Asp[25] is protonated, whereas that of Asp[125] is not. A suggested explanation for this is that both oxygens of the Asp[125] side chain are deprotonated to accept two hydrogen bonds, one from a bound water molecule and one from the inhibitor. In contrast the side chain of Asp[25] is protonated so that it can donate a hydrogen bond to the inhibitor (144). Consequently, the protonation state of the enzyme is influenced strongly by interaction with specific inhibitors and this knowledge is essential for a detailed understanding of the protease/drug interactions.

NMR has also been used to study the relationship between flexibility and enzymatic function for HIV protease. For the protease/DMP-323 complex, ^{15}N spin-relaxation studies determined that residues that are flexible correlate well with residues that are disordered in the NMR structure of the complex (145). For example, residues in poorly defined loops were found to undergo large-amplitude internal motions on the nanosecond-picosecond timescale. In contrast, two regions of the molecule were found to exhibit motions on the millisecond-microsecond timescale. The first of these is at the N-terminus of the protein around Thr^4-Leu^5. This is adjacent to the major site of autolysis of the protease and it has been suggested that the rate of cleavage may regulate HIV protease activity *in vivo* (146). Consequently, the observed flexibility may be important for regulation of protein function. The second region found to be undergoing millisecond-microsecond motion was the tips of the flaps around Ile^{50}-Gly^{51}. In crystal structures, this region of the protease is well ordered and not involved in crystal contacts, although its conformation varies from structure to structure. This motion is interpreted as a dynamic conformational exchange process, which is fast relative to the chemical-shift timescale. Thus when the protease is bound to a symmetric inhibitor in solution, this conformational exchange results in the chemical shifts of the flap residues in the two monomers being identical (138, 145). In contrast, when the protease is bound to an asymmetric inhibitor, such as KNI-272, crystal structures show that each monomer interacts with the inhibitor in a different way (144). This is reflected in the fact that the chemical shifts of the monomers are different when asymmetric inhibitors are bound (141, 147). Analysis of spectra from such an asymmetric complex has revealed that the inhibitor is capable of "flipping" its orientation with respect to the two monomers without dissociating from the com-

plex (148). These data again highlight the importance of defining both the structural and dynamic aspects of binding to understand the requirements for potent interactions between HIV protease and its inhibitors.

The development of inhibitors of HIV protease represents a major success for structure-based drug design. When HIV was first identified in the early 1980s there were no known drugs effective for treatment of infection. A combination of X-ray crystallography, NMR spectroscopy, computer modeling, and chemical synthesis has resulted in the development of several effective HIV protease inhibitors. However, in common with other retroviruses, HIV has a high transcription error rate that results in a rapid mutational rate. One of the results of this is the production of a divergent population of viruses in which the sequence of the HIV protease produced may differ substantially (149, 150). As a consequence, drug-resistant strains of the virus emerge. Clearly, knowledge of the structural principles that govern inhibition of the protease and the mechanism by which the virus develops resistance will continue to be important in the development of effective new drugs.

4 NMR SCREENING

In the past, NMR was predominantly used in the design stage of drug discovery rather than the screening stages. Recently, new methods that make use of NMR to screen ligands for binding to a protein target have been developed and are proving to be a powerful tool in the discovery of new drug leads. This section gives an overview of the various experimental methods, summarized in Table 12.8, which can be used to screen mixtures of ligands for binding to a drug target. There will also be a brief discussion on the practical considerations that need to be made when designing an NMR screening program.

4.1 Methods

4.1.1 Chemical-Shift Perturbation. Chemical shift is a function of the chemical (and hence magnetic) environment that individual nuclei experience. Perturbations of chemical

Table 12.8 Summary of the Methods Available for NMR Screening and Their Respective Characteristics

Screening Methodology	Signals Observed	Protein Size Limit	Labeling	Binding Information Obtained	K_D Limit	K_D Determined	Suitable for HTS	Mixture Deconvolution Required?
Chemical shift perturbation (e.g., SAR by NMR)	Protein	<30 kDa	^{15}N protein	Location	10^{-3}–10^{-7} M	√	√	√
STD	Ligand	None	None	Orientation	10^{-3}–10^{-7} M	x	√	x
Diffusion-based (e.g., affinity NMR)	Ligand	None	^{2}H protein for isotope editing	None	10^{-3}–10^{-7} M	√	√	x
Relaxation-based	Ligand	None	None	None	10^{-3}–10^{-7} M	√	√	x
trNOE	Ligand	None	None	Bound conformation	10^{-3}–10^{-7} M	x	√	x
NOE pumping	Ligand or protein (reverse)	None	None	Bound conformation	10^{-3}–10^{-7} M	x	√	x / √[a]
Spin labeling	Ligand or protein	None	Spin label for either ligand or protein	Orientation, simultaneous binding	10^{-3}–10^{-6} M	x	√[b]	√[b]

[a]For reverse NOE pumping.
[b]For primary screening if the protein is spin-labeled or for second-site screening if the first-site ligand is spin-labeled.

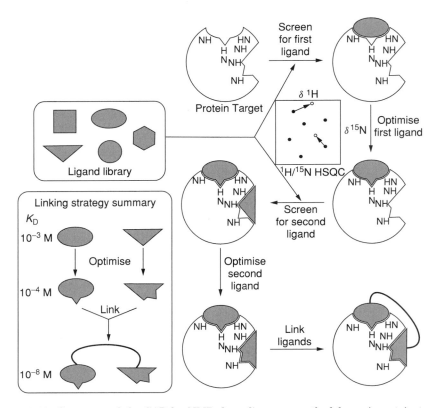

Figure 12.30. Summary of the SAR by NMR drug discovery methodology. A protein target is screened against a library consisting of small organic molecules by use of the $^1H/^{15}N$ HSQC experiment. When two ligands that bind in close proximity are identified, they are linked to form a composite ligand with an increased affinity for the target.

shifts can be used to detect binding of a ligand to a protein target. When a ligand binds to a protein the local chemical environment is changed, and this is reflected by a change in the chemical shifts of nuclei in close proximity to the ligand-binding site. The most common experiment used in this screening methodology is the $^1H/^{15}N$ HSQC that generates a discrete signal for each amide group within the protein. A reference $^1H/^{15}N$ HSQC spectrum, which is acquired in the absence of potential ligands, is compared to a spectrum recorded in the presence of ligands and any changes in the amide chemical shifts are indicative of a ligand binding to a location close to the corresponding amide groups. The major advantage of this technique is that, if the NMR assignment of the amide resonances is known, then the site of binding for each ligand can be determined.

This is a valuable piece of information in the development of more potent second-generation drug leads. Binding affinities can also be determined by measuring the change in chemical shift as a function of ligand concentration.

One technique that utilizes this screening method for drug design is "SAR-by-NMR," developed by Fesik and coworkers (1, 4, 151–155). SAR-by-NMR is a fragment-based drug design approach in which a potent drug candidate is derived by chemically linking two or more small low affinity ligands for a target. In theory, the binding energy of the linked compounds will be the sum of the binding energies of the two individual compounds plus contributions to binding energy attributed to linkage. Therefore, it is possible to generate a drug lead with a nanomolar dissociation constant (K_D) from two milli- to micromolar fragments.

The first step in this process (Fig. 12.30) involves screening a library of ligands (typically with a MW < 400) in mixtures of up to 10 for binding to a protein target by comparing the $^1H/^{15}N$ HSQC spectrum of a ^{15}N-enriched protein in both the presence and the absence of ligands. Any ligand-induced changes in the chemical shift of the nitrogen and amide proton signals indicate binding of one or more ligands in the mixture to the protein target. The mixture containing the binding ligand(s) is deconvoluted and each individual compound screened to identify the individual ligand(s) responsible for the observed chemical-shift perturbations. Once a binding ligand is identified analogs can be screened to optimize binding.

A second ligand, which binds at a proximal site, is then identified either from the original screen or by repeating the library screening with the first ligand site bound to the protein. This ligand is then optimized and the structure of the ternary complex determined by use of either NMR or X-ray crystallography. The ternary complex structure provides information on the conformation and orientation of the bound ligands, which facilitates the synthesis of hybrid molecules where the two ligands are joined by a suitable linking moiety.

There are several examples that illustrate the potential of SAR by NMR. As noted earlier, FK506 binding protein (FKBP) inhibits calcineurin and blocks T-cell activation when complexed to the immunosuppressant FK506. This protein was used as a target for SAR by

NMR screening and subsequently two ligands, (**19**) and (**20**), were identified with K_D values

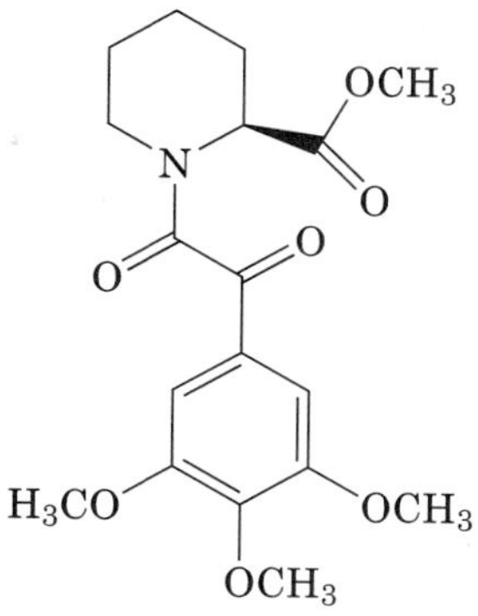

(**19**)

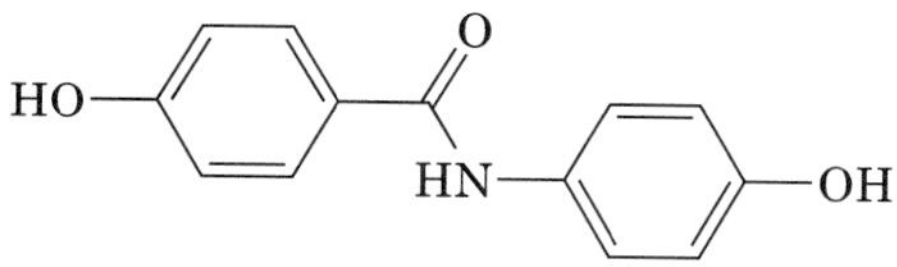

(**20**)

of 2 μM and 0.1 mM, respectively. A model of the ternary complex between the protein and both ligands was produced, which indicated that the methyl ester of (**19**) was close to the benzoyl hydroxyl group in (**20**). These two groups were linked with alkyl chains of various lengths, with the most active compound

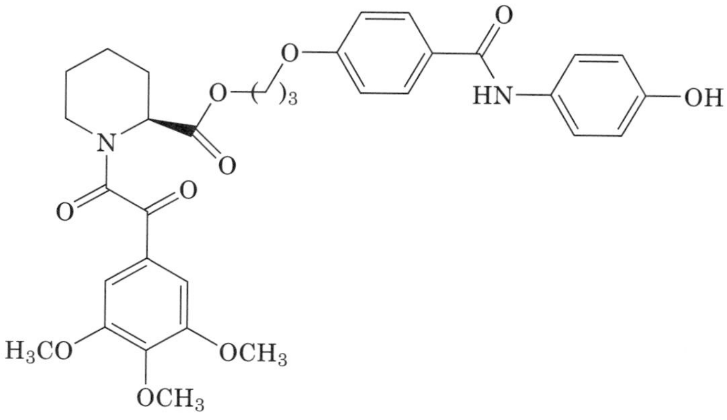

(**21**)

(21) having a three-carbon linker and a K_D value of 19 nM (151).

Inhibitors of the matrix metalloproteinase (MMP) stromelysin have also been designed through the use of the SAR-by-NMR screening methodology. As mentioned in section 3.2.5.3, MMPs are involved in matrix degradation and tissue remodeling, with overexpression of these enzymes being associated with arthritis and tumor metastasis. Acetoxyhydroxamate (22) was used as one ligand be-

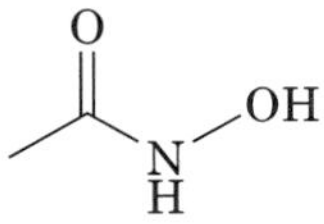

(22)

cause it was known previously that MMP inhibitors contain a hydroxamate moiety. The K_D value of (22) was determined to be 17 mM. To identify a second ligand the protein was screened against a ligand library in the presence of saturating amounts of (22). The library was biased for hydrophobic compounds, given that stromelysin demonstrates a substrate preference for a hydrophobic amino acids and structural studies had identified a hydrophobic binding pocket supporting this observation. From the library screen a series of biphenyl compounds were identified and analogs of these compounds were synthesized. A biphenyl derivative (23) was produced with a

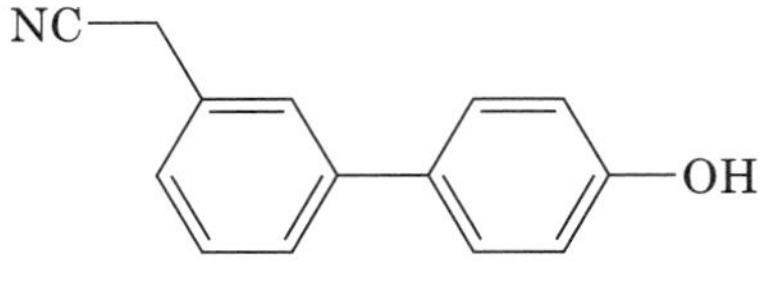

(23)

K_D value of 0.02 mM. The NMR structure of a ternary complex, consisting of stromelysin (22) and the biaryl derivative (24) (chosen for its superior aqueous solubility), was determined and indicated that the methyl group of (22) was in close proximity to the pyrimidine ring of (24). With this information (22) and (23) were subsequently linked by different

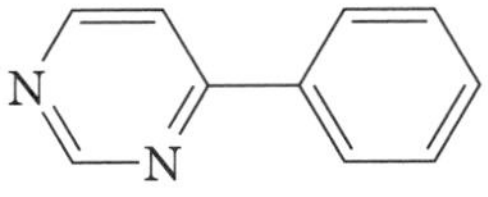

(24)

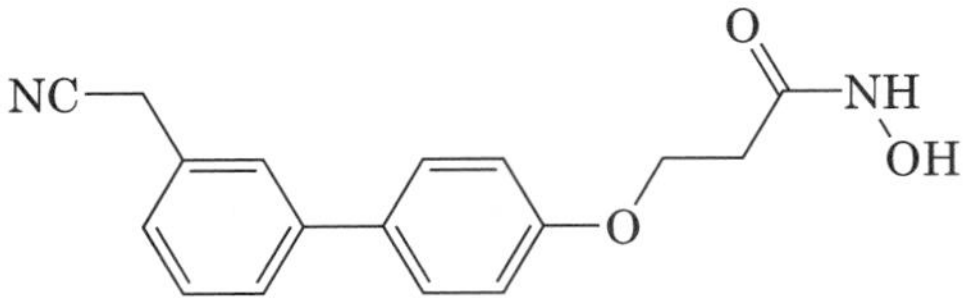

(25)

length linkers and the most active compound produced, (25), had a K_D value of 15 nM (154).

A variation of SAR-by-NMR is to optimize binding or improve the pharmacological properties of known drug leads generated by other methods (e.g., natural products isolation or combinatorial chemistry). A compound can be fragmented into individual subunits and then alternative fragments identified through use of $^1H/^{15}N$ HSQC screening. These fragments can then be incorporated into the molecular structure in the hope of improving the binding and/or pharmacological properties of the parent compound (Fig. 12.31). The alternative fragment must bind in the same location as the corresponding section of the original molecule, making $^1H/^{15}N$ HSQC screening method ideal as it provides information on the binding site of ligands.

In a demonstration of this fragmentation method, an antagonist of the interaction between leukocyte function-associated antigen 1 (LFA-1) and intracellular adhesion molecule 1 (ICAM-1) was used as a starting molecule. This interaction plays a role in the inflammatory response and specific T-cell immune responses, and inhibitors have applications in the treatment of inflammation and organ transplant rejection. The p-arylthio cinnamide antagonist (26) had an IC_{50} value of 44 nM; however, it was envisaged that the molecule's activity and physical properties could be improved by replacing the isopropyl phenyl group with a more hydrophilic moiety. Screening of a 2500-compound library provided several hits, and analogs of (26) were made that

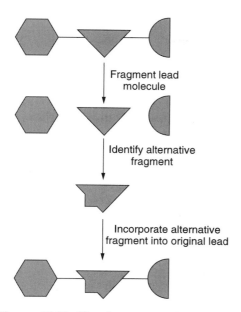

Figure 12.31. The fragment optimization approach developed from SAR by NMR. A known ligand of a protein is broken into fragments and small molecules based on the fragments are screened for binding. Any molecules that are found to bind can then be incorporated into the original lead compound with the hope of improving its binding and/or physicochemical properties.

(26)

incorporated these ligands in place of the isopropyl phenyl group. Compounds (27) and (28) had both improved aqueous solubility and pharmacokinetic profiles, with similar or im-

(27)

(28)

proved activity (IC_{50} values of 20 and 40 nM, respectively) when compared to that of the parent compound (26) (156).

Many compounds bind to human serum albumin (HSA), which significantly reduces their *in vivo* activity and hence their potential as a drug lead. The fragmentation method has recently been used to find analogs of diflusinal (29) that have a reduced affinity toward HSA

(29)

(157). Diflusinal is a nonsteroidal anti-inflammatory that is more potent, longer acting, and is more tolerated *in vivo* than aspirin. However, 99% of diflusinal is bound to albumin in plasma and, as a result, high doses are required for it to be effective. Structural studies of the diflusinal/HSA-III complex indicated that, by introducing polar functionality to the difluorophenyl moiety, binding affinity to HAS may be reduced without affecting activity. A series of organic compounds analogous to the difluorophenyl fragment were screened using the $^1H/^{15}N$ HSQC chemical-shift perturbation method and several alternative fragments were identified. These were incorporated into the diflusinal structure, resulting in a number of analogs (e.g., 30 and 31) with reduced affinity for HSA but still maintaining some activity. It was predicted that the next

(30)

(31)

generation of compounds would incorporate the fluorine atoms present in (**29**) into leads such as (**30**) and (**31**) because the presence of fluorine increased activity without increasing the affinity for HSA.

Overall, SAR-by-NMR is already showing great promise as a major new tool in drug design, but it has a few limitations. First, the screening of large ligand libraries and any subsequent deconvolution steps can be expensive, in that a large amount of ^{15}N-labeled protein is needed. However, as already noted, recent advances in cryoprobe technology have reduced the amount of protein required. The use of smaller, more druglike libraries such as that described in the SHAPES ideology (7) could also be used to reduce the amount of protein required for screening. The second limitation is that the NMR assignments for the protein must be known or determined. This limits the size of the protein target to 30 kDa or less, although this value will presumably increase

as TROSY-based NMR technologies for structure determination advance. The method also requires the comparison of ^{1}H/^{15}N HSQC of the protein in both the absence and the presence of ligands. Changes in solvent conditions such as pH, polarity, salt concentration, and viscosity may cause shifts in amide resonances, leading to false positives (158).

Recently, a protocol was described that screens ligands using the ^{1}H/^{15}N HSQC experiment but includes a mass spectrometry pre-screening step. Ligand mixtures are added to a protein target and then subjected to size-exclusion chromatography, which separates ligand/protein complexes from free ligands. The identity of the bound ligands can then be determined using MS data and, once identified, screened by ^{1}H/^{15}N HSQC to determine the location and specificity of binding (159). This MS/NMR methodology reduces the amount of ^{15}N-labeled protein required because only a fraction of the library is screened by NMR and there is no deconvolution step.

4.1.2 Magnetization Transfer Experiments. Proteins are composed of a large network of dipole-dipole interactions, resulting in the efficient transfer of magnetization throughout the molecule. The saturation transfer difference (STD) experiment (Fig. 12.32) uses this phenomenon to detect the binding of ligands to a protein. It relies on the fact that saturation of a single protein resonance results in saturation of all protein resonances and any ligands that bind to the protein, provided they are not affected directly by the selective saturation pulse (160). The STD experiment is able to detect the binding of ligands with a K_{D} between 10^{-3} and 10^{-8} M.

An STD experiment consists of irradiating an isolated protein resonance (either at low or high field) with a series of pulses that saturate the entire protein and any binding ligands. This results in a spectrum containing reduced signal intensities from both the protein and the ligands that bind to it. A second spectrum of the protein and ligand library is then recorded with the saturation pulses off-resonance. Subtraction of these two spectra results in the STD spectrum that shows only those ligands that bind to the protein (residual protein resonances are removed using a T_2 re-

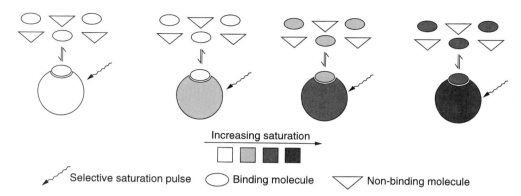

Figure 12.32. A schematic representation of the saturation transfer effect. The protein resonances are saturated (indicated by shading) by a selective pulse by spin diffusion. Resonances of nonbinding ligands (triangles) are not affected by this pulse but ligands that are interacting with the protein (ellipses) will also become saturated. These interacting ligands are transferred to solution through chemical exchange where they are detected. (Reprinted with permission from Ref. 16. Copyright 1999, Wiley-VCH.)

laxation filter). This subtraction occurs internally through phase cycling after every scan, to reduce artifacts attributed to temperature or magnetic field variations (160, 161).

An STD can be added to most forms of NMR experiments including COSY, TOCSY, NOESY, and inversely detected ^{13}C or ^{15}N spectra. A high resolution magic angle spinning (HR-MAS) STD experiment has been developed to study the binding of ligands to a protein immobilized on a solid support. HR-MAS STD NMR provides a way of obtaining ligand-binding information for proteins that are difficult to work with in solution attributed to either poor solubility or conformational changes (161). In addition to screening ligands for binding to a protein (160, 162, 163), the binding epitope of a molecule can also be determined by examining the intensities of ligand resonances (164–166). The proton signals having the strongest signals will correspond to those that are part of the ligand's binding epitope. For example, it was shown that when methyl β-D-galactoside (**32**) bound to *Ricinus communis* agglutin I the H2, H3, and H4 were saturated to the highest degree (values on structure indicate relative signal strengths) and hence were in close proximity to the protein protons. This analysis was subsequently extended to the decasaccharide NA$_2$ (**33**) and demonstrated that the Gal-6' and

GlcNAc-5' residues bind edge-on to the protein, with the binding contribution of the terminal galactose residue being the greater (165).

STD NMR studies have also been performed on membrane-bound receptors by embedding the protein in the phospholipid bilayer of a liposome (166).

There are a number of advantages in using the STD experiment for the detection of ligand binding. The saturation transfer effect is an efficient process, which results in high sensitivity, and hence only small quantities of protein are required (nanomolar concentrations of a protein with MW > 10 kDa) (160, 165). In addition, protein size is noncritical; in fact, as the protein becomes larger, the saturation transfer effect becomes more efficient. The acquisition time for each experiment is also quite short and, because the experiment is ligand observed, no deconvolution of mixtures

(33)

is required, making this a good technique for high throughput screening of large ligand libraries. Unlike the chemical-shift perturbation techniques, STD experiments provide no information on the site of ligand binding.

A second variation of saturation transfer experiment has been devised by Dalvit and coworkers that uses the transfer of magnetization from the water (167). Water is intimately associated with proteins being bound either within or on the surface of the macromolecular structure. Saturation of the water resonance will lead to protein saturation through a variety of mechanisms, including saturation of the αH resonances, saturation of exchanging protein resonances, and NOE interactions between water and the protein. If a compound is bound to the protein it will also become saturated, and this effect can be used as an indication of ligand binding (167).

4.1.3 Molecular Diffusion. Molecules can be distinguished based on their diffusion coefficients, which are related to molecular size. Large macromolecules, such as proteins, diffuse more slowly than small molecules and it is this size difference that can be exploited to screen for ligand binding. If a small molecule binds to a protein target its diffusion coefficient is altered to a value more like that of the protein. Therefore, by utilizing a diffusion fil-

ter, resonances generated by small molecules that do not bind to the protein can be removed from the spectrum.

Diffusion editing is achieved with the use of a pair of gradient pulses. If field homogeneity is ignored, then all spins experience an identical magnetic field despite having different positions throughout the sample. The application of a field gradient has the effect of making field strength dependent on position. Under the influence of the gradient pulse, the phase of individual spins become dependent on their position within the sample and hence the spins are spatially "encoded." If diffusion does not occur, this spatial encoding is fully reversible by a second gradient of inverse polarity and no loss of NMR signal will occur. However, the second gradient pulse will be unable to "decode" the spins that have undergone diffusion and the resulting NMR signal will be reduced. Acquiring spectra of a sample with and without the diffusion filter and then subtracting them allows the ligands binding to the protein to be identified. This filtering method can be used for both 1D and 2D experiments and can be "tuned" by altering the strength and duration of the gradients.

Because the ligand signals are being observed in this screening method, no convolution of the ligand mixture is required, given that any signals can be assigned directly to individual compounds within the mixture. However, signals from the protein are always present, which can pose a problem in interpreting spectra. An isotope-edited version of the diffusion experiment has been designed to avoid this problem, although labeled protein is required (168). Generally, there is no requirement for labeling of the protein target or for the protein resonances to be assigned and thus, in theory, there is no size limit on the proteins that can be screened by use of this method, although no information is obtained on the location of ligand binding. However, if the protein is large, then the transverse relaxation time may be too short to observe the bound ligands in the diffusion-edited spectrum (169). Only one sample, containing protein and ligands, is used to obtain both reference and screening data and therefore differences between the sample and reference spectra caused by addition of the ligands (pH, salt concentration, etc.) are avoided.

Diffusion-filtered NMR screening requires that there is a significant difference in observed translational diffusion between the free and bound states. The ligands are in fast exchange on the diffusion timescale and as a consequence the observed diffusion coefficient for binding ligands is an average between the free and bound diffusion values. Free ligands diffuse at a much faster rate than those in the bound state and thus only a small amount of free ligand has a considerable effect on the observed average diffusion coefficient. This effect may be significant enough to reduce the difference between binding and nonbinding ligands, making it more difficult to interpret results (169). It has also been demonstrated that chemical exchange and NOE can affect the interpretation of diffusion experiments and that these factors need to be taken into consideration (170, 171).

Shapiro and coworkers developed a methodology based on diffusion filtering, named "affinity NMR," that they have used to screen for binding (172–175). Diffusion-edited NMR experiments were able to identify two known binding tetrapeptide ligands of vancomycin from a mixture of 10 peptides (176). Hajduk et al. demonstrated the application of diffusion-editing experiments by differentiating ligands of stromelysin from a mixture containing non-binding compounds (177).

4.1.4 Relaxation. Like diffusion, the transverse relaxation time (T_2) of molecules is also dependent on molecular size. Large molecules, such as proteins, have a short T_2 and hence exhibit broad NMR signals, whereas small molecules have a longer T_2 and hence narrower line widths. Therefore, if a small molecule ligand binds to a protein, its T_2 value will decrease and a line-broadening effect of bound ligand signals can be observed. Alternatively, a relaxation filter can be used to remove signals from molecules with a short T_2 value. Subtraction from a reference spectrum will result in a spectrum containing only those ligands that bind to the protein.

The ability to identify binding ligands using relaxation filters has been demonstrated using FKBP. A mixture of nine compounds consisting of one known ligand of FKBP, 2-phenylimidazole (**34**), and eight nonbinding compounds (e.g., **35–37** were screened and only signals from (**34**) were observed (177).

(34)

(35)

(36)

4.1.5 NOE. NOE experiments can also be used to identify ligands that bind to protein targets (178–180). Small molecules have a fast

(37)

tumbling rate and, as a consequence, generally exhibit small positive NOEs. In contrast, large molecules such as proteins generate strong negative NOEs because of their slow tumbling time. When a small molecule binds to a protein, its tumbling rate is slowed to that of the protein and it exhibits strong negative NOEs. On dissociation, these are transiently retained and are known as transfer NOEs (TrNOEs) (Fig. 12.33). TrNOEs and those arising directly from the free ligand can be distinguished by the rate of signal build up. Transfer NOEs accumulate significantly faster and therefore can be selected for by use of shorter mixing times in the NOE experiment (179).

In practice, a 2D NOESY spectrum of the mixture of potential ligands in the absence of protein is recorded and all molecules exhibit

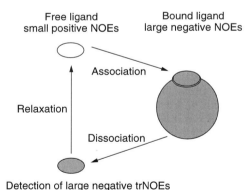

Detection of large negative trNOEs

Figure 12.33. A schematic representation of the TrNOE experiment used to detect ligand binding (180). The free ligand (white ellipse) exhibits only small positive NOEs, although binding to the large protein target results in the generation of large negative TrNOEs. The appearance of these large negative TrNOE signals can be used to identify ligands within a mixture that are binding to the protein and also provide some information on the bound conformation of the ligand.

small positive NOEs. The experiment is then repeated in the presence of protein and molecules that bind display negative TrNOEs. Subtraction of the two spectra provides signals arising from only those compounds that bind. These TrNOEs can be interpreted to provide information regarding the bound conformation of the active ligands. However, when analyzing the conformational data care must be taken to ensure that the ligands are in fast exchange and that the observed TrNOEs are not affected by contributions from spin-diffusion (179). Relative binding affinities between ligands can also be determined by comparison of TrNOE signal strength but, again, the fast-exchange regime and spin-diffusion effects need to be taken into account (178, 179). If all ligands are in fast exchange, the stronger binding ligands occupy more binding sites and thus give larger TrNOE intensities. Because of the need for an averaging effect, brought about by fast chemical exchange, TrNOE experiments are limited to those ligands with a K_D value from 10^{-3} to 10^{-7} M. The spectral properties of excess ligand in solution are evoked by small fractions of bound molecules, greatly enhancing sensitivity (178).

Transfer NOE experiments have been used to identify a bioactive disaccharide from a library of 15 mono- and disaccharides that bound to *Aleuria aurantia* agglutinin (179). Another study has described the identification of a silalyl Lewis mimetic (**38**) that binds to

(38)

E-selectin from a library of 10 compounds (178). As well as being used to detect binding, TrNOEs may also be used to determine bound ligand conformations, as described earlier in this chapter.

A second technique that uses NOEs to detect binding is NOE pumping. This method was designed to alleviate some of the problems associated with the diffusion-edited screening methods (169). Signals from ligand molecules are removed using a diffusion filter and then transfer of signal from the protein to bound ligands by NOE occurs. The inverse of this is possible (known as reverse NOE pumping), which uses a relaxation filter to attenuate the protein resonances, after which the signal is transferred to the protein by NOE. Ligands may lose signal either by relaxation (for a free ligand) or through relaxation and NOE transfer (for a bound ligand). Therefore by subtracting spectra (which is done internally to reduce subtraction artifacts) from experiments with and without NOE pumping to the protein, the binding ligands can be detected (181).

The ability of NOE and reverse NOE pumping to identify ligands has been demonstrated through the use of human serum albumin (HSA) and several known binding and nonbinding compounds (169, 181).

4.1.6 Spin Labels. Spin-spin relaxation rates are proportional to the product of the squares of the gyromagnetic ratios of the involved spins. The gyromagnetic ratio of an unpaired electron is significantly larger than that of a proton and therefore any spins influenced by this electron will have substantially shortened relaxation times. The resonances of protons that are within 15–20 Å from the unpaired electron will experience this effect and be significantly broadened. The introduction of a short spin-lock period will significantly reduce the intensity or quench these signals.

The spin-label method can be used as either a primary screening method or to identify a second ligand-binding site. The primary screening method requires residues around the binding pocket of the target to be spin labeled. Residues suitable for this labeling include lysine, cysteine, histidine, glutamate, aspartate, tyrosine, and methionine. Any ligands that bind to the protein in close proximity to the spin-labeled residues will be able to be identified. To screen for second-site ligand binding, the known first-site binding ligand is spin labeled. A reduced signal will be observed for any ligands that bind simultaneously and in close proximity to the first ligand-binding site. In addition, the degree of reduction in signal intensity gives an indication of the orientation of the second ligand in relation to the first, given that the effect of the spin label is inversely proportional to the distance separating the electron and proton. This information is valuable in the design of linkers to join the two ligands.

There are several advantages to using the spin-label screening method. Currently, it is the only method that can detect ligands that bind to the protein simultaneously, unlike other methods that can produce false positives if the first ligand-binding site is not fully saturated. The concentration of protein required for screening is relatively small ($\sim 10 \mu M$) because of the substantial enhancement of the relaxation rate by the spin label. The protein can also be unlabeled and partially purified and there is no molecular weight limit. The spin labels also quench protein signals, making interpretation of spectra easier. The experiment is easy to set up and analyze, making it amenable to automation. It is also insensitive to small changes in solvent conditions that can generate false positives in other methods. The information obtained on the orientation of ligands is also valuable and makes it an alternative to the chemical-shift perturbation methods when the proteins are large and NMR assignments have not been made.

A disadvantage of the method is the requirement for spin-labeled proteins and ligands. In addition, any ligands with slow dissociation rates will show no averaging of relaxation rates and therefore tightly binding compounds ($K_D < 10^{-6} M$) will produce false negatives. Protein spin labeling must occur adjacent but not within the binding site to minimize alteration of its binding properties.

The antiapoptopic protein Bcl-xL is responsible for the reduced susceptibility of cancer cells to undergo apoptosis and is therefore a target for the development of new anticancer agents. The structure of a previously identified ligand for Bcl-xL (**39**) was modified to incorporate a TEMPO spin label (**40**). By use of spin-labeled (**40**), an eight-compound library was screened for simultaneous binding to Bcl-xL. From this library an aromatic ketoxime

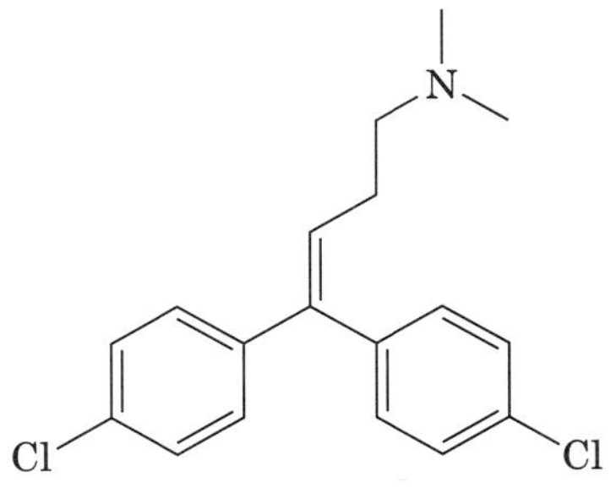

(39)

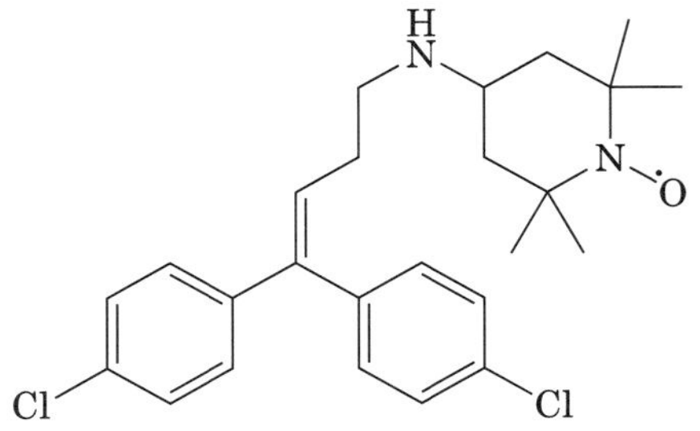

(40)

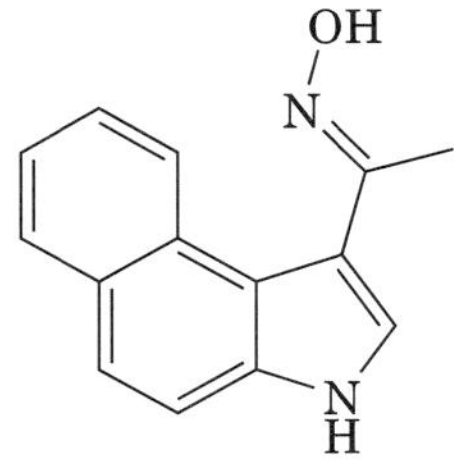

(41)

(**41**) was identified as binding simultaneously with and in the vicinity of (**40**). Analysis of relaxation enhancements revealed that the protons around the indole ring were closest to the spin label.

4.2 Practical Considerations

4.2.1 Screening Approach. The first step in using NMR screening is to select a suitable screening method for the target protein being used. Table 12.8 lists the characteristics of each NMR screening method. The choice of experiment will be determined by the characteristics of the protein target and the informa-

tion that is desired from the screen. For example, the SAR-by-NMR method is suitable only for small, easily expressed proteins because the NMR assignments for the target need to be known so the location of binding can be determined and a large amount of ^{15}N-enriched protein is required. If a simple "yes/no" answer on ligand binding is wanted, then the shorter, less resource intensive ligand-observed experiments (e.g., STD, diffusion-edited, or TrNOE) may suffice.

It is also important to determine the correct NMR solvent conditions for the screening procedure. These should facilitate good solubility, with little precipitation or aggregation, and acquisition of good quality data; maintain protein structure and activity; and provide a sufficient buffering effect to allow for ligands to be added. Two methods that permit the screening of a range of solvent conditions to determine without the need for a large amount of protein are the microdialysis button test (182) and the microdrop-screening method (183). A review on this subject has recently been published (184), to which the reader is referred for a more in-depth discussion on the subject of solvent conditions for NMR.

4.2.2 Library Design. Effective design and management of the ligand library to be used for screening is essential if successful results are to be obtained. The major considerations in library design are only briefly described here. There are a number of reviews that provide a more in-depth discussion of library design (15, 185).

4.2.2.1 Ligand Properties. Diversity of ligands is an important factor to consider in the design of a library for NMR screening and there are a number of factors to take into consideration. Although it would seem logical to maximize diversity, this may not always be the most efficient approach. If the system being studied exhibits neighborhood behavior, then maximizing diversity is a good option. Neighborhoods are regions of multidimensional molecular space defined by a set of molecular descriptors. By the choice of a molecule that is in the center of a neighborhood, it is possible, in theory, to represent all molecules within that molecular space. By spreading out the mole-

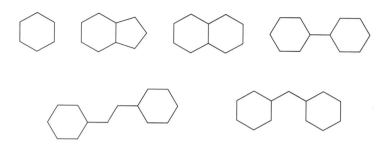

Figure 12.34. Examples of molecular frameworks from the SHAPES library.

cules that are selected for the library so that each neighborhood does not overlap, diversity is maximized.

However, if neighborhoods are only small then compound libraries must be very large so that the neighborhoods overlap and hence all molecular space is covered. In addition, some systems do not exhibit neighborhood behavior and relatively small changes to the structure of a compound may lead to large changes in its binding affinity for the target. Maximizing diversity may also be inefficient because many molecules do not possess physicochemical characteristics that are suitable as the basis for a drug. In practice, the more that is known about the drug target, the less diverse and more focused the library can be. However, if the library is too focused then some outlying "new" ligand type for the target being screened may be missed.

One strategy for library design is to select compounds that have druglike characteristics. A simple set of rules, determined by Lipinski and coworkers, for determining whether a compound is druglike is known as "the rule of 5." According to this set of criteria, the majority of orally available drugs have five or fewer hydrogen bond donors, 10 or fewer hydrogen bond acceptors, a log P of less than 5, and a molecular weight less than 500 (186). Additional factors that can be taken into consideration include the number of heavy atoms, rotatable bonds, and ring systems (187–189). Another study has revealed that there are a number of frameworks and side-chains that commonly occur in many drugs. Drug molecules, from the comprehensive medicinal database, were broken down into systems consisting of frameworks (Fig. 12.34) and side-chains. Analysis of these two structural features revealed that approximately 50% of

all known drugs could be represented by only 32 different frameworks. When atom type and bond order were included in the analysis, 41 frameworks were found to describe 24% of all drugs (190). A similar analysis of side-chain frequency indicated that approximately 70% of all side chains present in the compound database analyzed were from the top 20 occurring side chains (191).

The presence of these common frameworks and side-chains has been exploited in the SHAPES methodology (7) for NMR screening. This strategy employs a small focused library based on these common frameworks and side chains to screen against protein targets through the use of relaxation and NOE experiments. The advantages of this approach are that the library is small and hence only relatively small amounts of protein are required and any hits from the library will possess druglike characteristics. However, a disadvantage of the method is that it is unlikely to yield new drug types, given that the library is based on known drug frameworks.

Diversity of molecular type is not the only factor that must be taken into account when designing a library to be used in an NMR screening program. Because the screening occurs in an aqueous solution, the organic compounds chosen for the library must demonstrate reasonable solubility in the aqueous conditions used. In general, compounds are dissolved in DMSO and then added to the protein solution at the appropriate concentration. Currently, there are no good methods for determining the solubility of a wide range of compounds before screening commences. A simple method is to dilute the DMSO solution in buffer and observe whether any precipitation or aggregation occurs. However, this method will not be suitable for compounds that precipitate or aggregate over several hours, and solutions that appear clear may still contain high MW aggregates, which will cause false positives in experiments such as the diffusion, relaxation, and TrNOE methods (15, 185).

It is also preferable to choose ligands that are synthetically accessible and/or possess suitable moieties to build upon or link to other fragments. This is especially important in the SAR-by-NMR screening methodology because this relies on the ability to link individual fragments to form a more potent drug lead. If the ligands to be linked are not synthetically accessible or do not possess suitable linking functional groups then this process is severely hindered.

4.2.2.2 Mixture Design. The optimal number of compounds per mixture is dependent on the screening method. For ligand-observed experiments the limiting factor for the number of compounds in a mixture is spectral overlap. Ligands need to be chosen so that spectral overlap is minimized, making interpretation of the resulting data far simpler. In theory, protein-observed experiments could have a large number of compounds per mixture that would both minimize screening time and the requirement for large amounts of protein. However, because the experiments are protein observed then deconvolution of the mixtures and rescreening of each individual compound are required to identify any hits. Therefore, the number of compounds per mixture is dependent on the hit rate in the screening procedure, given that the greater the hit rate, the more deconvolution steps required and consequently more protein and spectrometer time are needed. The number of experiments required is at a minimum when the number of compounds is equal to $1/(\text{hit rate})^{1/2}$; thus, with a hit rate of 10% the optimal number of compounds per mixture is three (185). In addition to these factors, if the hit rate is high then it is likely that several compounds within a mixture containing a large number of compounds may compete for the same binding pocket, which may lead to false negatives.

In mixtures of organic compounds the possibility of interactions between compounds, such as reactions or ion pairing, is also possible and should be taken into consideration,

especially when one uses large numbers of compounds per mixture. It has been demonstrated that in random mixtures of 10 compounds in DMSO, the probability of a reaction occurring between two of the mixture's components is 26%. This value can be reduced by careful selection of mixture components (e.g., separating acids from bases) to approximately 9% (192).

4.2.3 Hardware and Automation. Automation is a requirement if libraries containing a large number of compounds are to be screened. Technology has been developed that allows the automation of almost all steps of the NMR screening process from sample preparation through to data analysis (193).

The general setup for NMR screening consists of a robot for just-in-time preparation of each sample, which is then transferred to the magnet either through a flow system or as discrete samples on a rail system. There are several disadvantages in using a flow system, including the possibility of contamination of samples by previously screened compounds, the capillary line can be blocked if the protein or ligands precipitate or form aggregates, recovery of the sample is more laborious because it has been diluted, and cryoprobe technology (discussed later in this section) is not yet available in the flow system. Many of these problems can be overcome by using the discrete samples with the rail system.

Data acquisition is easily automated and there are several software packages that will automate data processing for 2D spectra. The processing of 1D spectra automatically is reported to be less reliable because of the large solvent signal and usually require manual adjustment of phasing (193). One of the most laborious tasks in NMR screening is the analysis and comparison of the resulting spectra. For 1D ligand-observed experiments difference methods (e.g., STD) provide the most reliable method for interpretation of results, in that the presence of signals in the spectra will correspond to the ligands that are binding. In 2D protein-observed experiments (e.g., ^{1}H/^{15}N HSQC) a more statistically rigorous analysis of changes in chemical shift is required and a discussion of this is beyond the scope of this chapter. A more in-depth account of data analysis is provided by Ross and Senn (193).

Currently, approximately 50–100 samples can be screened per day and if mixtures contain 10 compounds each this provides a substantial throughput. This throughput rate will increase as technology improves, as has been demonstrated by the use of cryoprobes. Cryogenic NMR probes, where the preamplifier and radio frequency coils are cooled to low temperatures, can significantly increase the signal-to-noise ratio of an NMR spectrum. By use of these probes NMR data can be obtained in much faster times and by use of lower protein concentrations, which subsequently increases throughput, the total amount of protein needed to screen a library is reduced. Hajduk and coworkers (21) demonstrated the substantial improvements made through the use of a CryoProbe instead of a conventional probe in $^1H/^{15}N$ chemical-shift perturbation screening. Stromelysin (50 μM) was screened against mixtures of 100 compounds (50 μM each), facilitating the screening of more than 10,000 compounds in one day. The use of lower concentrations of both protein and ligands increases the stringency levels for the binding strength of ligands. At a protein/ligand concentration of 0.5 mM, ligands with dissociation constants in the millimolar range can be detected, although at a protein/ligand concentration of 50 μM this dissociation constant limit is reduced to approximately 0.15 mM. Although using higher protein/ligand concentrations can be advantageous when screening libraries containing small low affinity ligands, a higher stringency is required when screening large libraries, to reduce the number of hits obtained to a manageable number (21).

5 CONCLUSIONS

In this chapter we have given an overview of the two major approaches used in NMR and drug discovery, structure-based design and NMR-based screening. Both areas are flourishing and, together with more traditional uses of NMR, they demonstrate the versatility of NMR as a tool in medicinal chemistry. The power of NMR has been dramatically enhanced over the last decade by developments in both instruments and methodology. On the instrumental side, increases in magnetic field strengths and the development of cryoprobes have greatly increased sensitivity. Linkages of NMR to LC and MS have increased versatility. On the methods front there have been a range of new approaches discovered that will enhance the study of larger molecular complexes. Advances in protein expression and labeling have played a major role in stimulating the development of new NMR pulse sequences to extract information from such complexes.

6 ACKNOWLEDGMENTS

Work in our laboratory on NMR in drug design and development is supported by the Australian Research Council. D.J.C. is an ARC Professorial fellow. We thank Norelle Daly for assistance with some of the figures and Robyn Craik, Shaiyena Williams, and David Ireland for help in preparation of the manuscript.

REFERENCES

1. P. J. Hajduk, R. P. Meadows, and S. W. Fesik, *Science*, **278**, 497 (1997).
2. S. W. Fesik, *J. Biomol. NMR*, **3**, 261 (1993).
3. B. Stockman, *Prog. Nucl. Magn. Reson. Spectrosc.*, **33**, 109 (1998).
4. P. J. Hajduk, R. P. Meadows, and S. W. Fesik, *Q. Rev. Biophys.*, **32**, 211 (1999).
5. G. C. Roberts, *Curr. Opin. Biotechnol.*, **10**, 42 (1999).
6. J. M. Moore, *Curr. Opin. Biotechnol.*, **10**, 54 (1999).
7. J. Fejzo, C. A. Lepre, J. W. Peng, G. W. Bemis, Ajay, M. A. Murcko, and J. M. Moore, *Chem. Biol.*, **6**, 755 (1999).
8. P. A. Keifer, *Curr. Opin. Biotechnol.*, **10**, 34 (1999).
9. G. C. Roberts, *Drug Discovery Today*, **5**, 230 (2000).
10. D. C. Fry and S. D. Emerson, *Drug Des. Discov.*, **17**, 13 (2000).
11. D. J. Craik and M. J. Scanlon in G. A. Webb, Ed., Annual Reports on NMR Spectroscopy, Vol. **42**, Academic Press, San Diego, 2000, pp. 115–173.
12. T. Diercks, M. Coles, and H. Kessler, *Curr. Opin. Chem. Biol.*, **5**, 285 (2001).
13. R. P. Hicks, *Curr. Med. Chem.*, **8**, 627 (2001).
14. M. Shapiro, *Farmaco*, **56**, 141 (2001).

15. J. W. Peng, C. A. Lepre, J. Fejzo, N. Abdul-Manan, and J. M. Moore, *Methods Enzymol.*, **338**, 202 (2001).

16. D. J. Craik, Ed., *NMR in DrugDesign*, CRC Press, Boca Raton, FL, 1996, pp. 1–476.

17. U. Holzgrabe, I. Wawer, and B. Diehl, *NMR Spectroscopy in Drug Development and Analysis*, Wiley-VCH, Weinheim, Germany, 1999, pp. 1–299.

18. A. E. Derome, *Modern NMR Techniques for Chemistry Research*, Pergamon, New York, 1987.

19. H. Gunther, *NMR Spectroscopy—An Introduction*, John Wiley & Sons, Chichester, UK, 1980, pp. 1–436.

20. K. Pervushin, R. Riek, G. Wider, and K. Wuthrich, *Proc. Natl. Acad. Sci. USA*, **94**, 12366 (1997).

21. P. Hajduk, T. Gerfin, J. Boehlen, M. Haberli, D. Marek, and S. Fesik, *J. Med. Chem.*, **42**, 2315 (1999).

22. P. A. Keifer, *Prog. Drug Res.*, **55**, 137 (2000).

23. K. Wuthrich, *NMR of Proteins and Nucleic Acids*, John Wiley & Sons, New York, 1986, pp. 1–292.

24. C. E. Heading, *Drugs*, **4**, 339 (2001).

25. D. S. Wishart, B. D. Sykes, and F. M. Richards, *J. Mol. Biol.*, **222**, 311 (1991).

26. D. S. Wishart, C. G. Bigam, A. Holm, R. S. Hodges, and B. D. Sykes, *J. Biomol. NMR*, **5**, 67 (1995).

27. K. J. Nielsen, L. Thomas, R. J. Lewis, P. F. Alewood, and D. J. Craik, *J. Mol. Biol.*, **263**, 297 (1996).

28. L. K. MacLachlan, D. A. Middleton, A. J. Edwards, and D. G. Reid in D. G. Reid, Ed., Protein NMR Techniques, Vol. **60**, Humana Press, Totowa, NJ, 1997, pp. 337–362.

29. D. S. Wishart, B. D. Sykes, and F. M. Richards, *Biochemistry*, **31**, 1647 (1992).

30. F. Dasgupta, A. K. Mukherjee, and N. Gangadhar, *Curr. Med. Chem.*, **9**, 549 (2002).

31. D. J. Craik, K. J. Nielsen, and K. A. Higgins in G. A. Webb, Ed., Annual Reports on NMR Spectroscopy, Vol. **32**, Academic Press, San Diego, 1995, pp. 143–213.

32. S. R. Krystek Jr., D. A. Bassolino, J. Novotny, C. Chen, T. M. Marschner, and N. H. Andersen, *FEBS Lett.*, **281**, 212 (1991).

33. N. H. Andersen, C. P. Chen, T. M. Marschner, S. R. Krystek Jr., and D. A. Bassolino, *Biochemistry*, **31**, 1280 (1992).

34. V. Saudek, J. Hoflack, and J. T. Pelton, *FEBS Lett.*, **257**, 145 (1989).

35. S. Endo, H. Inooka, Y. Ishibashi, C. Kitada, E. Mizuta, and M. Fujino, *FEBS Lett.*, **257**, 149 (1989).

36. R. G. Mills, S. I. O'Donoghue, R. Smith, and G. F. King, *Biochemistry*, **31**, 5640 (1992).

37. S. Munro, D. Craik, C. McConville, J. Hall, M. Searle, W. Bicknell, D. Scanlon, and C. Chandler, *FEBS Lett.*, **278**, 9 (1991).

38. M. D. Reily and J. B. Dunbar Jr., *Biochem. Biophys. Res. Commun.*, **178**, 570 (1991).

39. H. Tamaoki, Y. Kobayashi, S. Nishimura, T. Ohkubo, Y. Kyogoku, K. Nakajima, S. Kumagaye, T. Kimura, and S. Sakakibara, *Protein Eng.*, **4**, 509 (1991).

40. A. Aumelas, L. Chiche, S. Kubo, N. Chino, H. Tamaoki, and Y. Kobayashi, *Biochemistry*, **34**, 4546 (1995).

41. A. Aumelas, L. Chiche, E. Mahe, D. Le-Nguyen, P. Sizun, P. Berthault, and B. Perly, *Int. J. Pept. Protein Res.*, **37**, 315 (1991).

42. D. C. Dalgarno, L. Slater, S. Chackalamannil, and M. M. Senior, *Int. J. Pept. Protein Res.*, **40**, 515 (1992).

43. Y. Boulanger, E. Biron, A. Khiat, and A. Fournier, *J. Pept. Res.*, **53**, 214 (1999).

44. K. Arvidsson, T. Nemoto, Y. Mitsui, S. Ohashi, and H. Nakanishi, *Eur. J. Biochem.*, **257**, 380 (1998).

45. C. M. Hewage, L. Jiang, J. A. Parkinson, R. Ramage, and I. H. Sadler, *J. Pept. Sci.*, **3**, 415 (1997).

46. B. A. Wallace, R. W. Janes, D. A. Bassolino, and S. R. Krystek Jr., *Protein Sci.*, **4**, 75 (1995).

47. M. Coles, V. Sowemimo, D. Scanlon, S. L. Munro, and D. J. Craik, *J. Med. Chem.*, **36**, 2658 (1993).

48. D. J. Detlefsen, S. E. Hill, S. H. Day, and M. S. Lee, *Curr. Med. Chem.*, **6**, 353 (1999).

49. W. C. Patt, J. J. Edmunds, J. T. Repine, K. A. Berryman, B. R. Reisdorph, C. Lee, M. S. Plummer, A. Shahripour, S. J. Haleen, J. A. Keiser, M. A. Flynn, K. M. Welch, E. E. Reynolds, R. Rubin, B. Tobias, H. Hallak, and A. M. Doherty, *J. Med. Chem.*, **40**, 1063 (1997).

50. W. C. Patt, X. M. Cheng, J. T. Repine, C. Lee, B. R. Reisdorph, M. A. Massa, A. M. Doherty, K. M. Welch, J. W. Bryant, M. A. Flynn, D. M. Walker, R. L. Schroeder, S. J. Haleen, and J. A. Keiser, *J. Med. Chem.*, **42**, 2162 (1999).

51. S. L. Munro, P. R. Andrews, D. J. Craik, and D. J. Gale, *J. Pharm. Sci.*, **75**, 133 (1986).

52. B. M. Duggan and D. J. Craik, *J. Med. Chem.*, **39**, 4007 (1996).

53. B. M. Duggan and D. J. Craik, *J. Med. Chem.*, **40**, 2259 (1997).

54. M. G. Casarotto and D. J. Craik, *J. Pharm. Sci.*, **90**, 713 (2001).

55. R. Abseher, L. Horstink, C. W. Hilbers, and M. Nilges, *Proteins: Struct., Funct., Genet.*, **31**, 370 (1998).

56. J. Gehrmann, P. F. Alewood, and D. J. Craik, *J. Mol. Biol.*, **278**, 401 (1998).

57. J. Balbach, S. Seip, H. Kessler, M. Scharf, N. Kashani-Poor, and J. W. Engels, *Proteins*, **33**, 285 (1998).

58. D. J. Craik, B. M. Duggan, and S. L. A. Munro in M. I. Choudary, Ed., Biological Inhibitors, Vol. **2**, Harwood Academic, Amsterdam, 1996, pp. 255–302.

59. R. L. Wagner, J. W. Apriletti, M. E. McGrath, B. L. West, J. D. Baxter, and R. J. Fletterick, *Nature*, **378**, 690 (1995).

60. D. J. Gale, D. J. Craik, and R. T. C. Brownlee, *Magn. Reson. Chem.*, **26**, 275 (1988).

61. W. Bourguet, M. Ruff, P. Chambon, H. Gronemeyer, and D. Moras, *Nature*, **375**, 377 (1995).

62. J. P. Renaud, N. Rochel, M. Ruff, V. Vivat, P. Chambon, H. Gronemeyer, and D. Moras, *Nature*, **378**, 681 (1995).

63. J. Feeney and B. Birdsall in G. C. K. Roberts, Ed., *NMR of Macromolecules: A Practical Approach*, Oxford University Press, Oxford, UK, 1993, pp. 181–215.

64. J. Feeney in I. Bertini, H. Molinari, and N. Niccolai, Eds., *NMR and Biomolecular Structure*, VCH, New York, 1991, pp. 189–205.

65. J. Feeney, *Biochem. Pharmacol.*, **40**, 141 (1990).

66. K. J. Nielsen, D. Adams, L. Thomas, T. Bond, P. F. Alewood, D. J. Craik, and R. J. Lewis, *J. Mol. Biol.*, **289**, 1405 (1999).

67. A. P. Campbell and B. D. Sykes, *Annu. Rev. Biophys. Biomol. Struct.*, **22**, 99 (1993).

68. B. D. Sykes, *Curr. Opin. Biotechnol.*, **4**, 392 (1993).

69. D. J. Craik and K. A. Higgins in G. A. Webb, Ed., Annual Reports on NMR Spectroscopy, Vol. **22**, Academic Press, London, 1990, pp. 61–138.

70. G. Bertho, J. Gharbi-Benarous, M. Delaforge, and J. P. Girault, *Bioorg. Med. Chem.*, **6**, 209 (1998).

71. R. E. Hubbard, *Curr. Opin. Biotechnol.*, **8**, 696 (1997).

72. G. Wider and K. Wuthrich, *Curr. Opin. Struct. Biol.*, **9**, 594 (1999).

73. G. M. Clore and A. M. Gronenborn, *Curr. Opin. Chem. Biol.*, **2**, 564 (1998).

74. N. Tjandra and A. Bax, *Science*, **278**, 1111 (1997).

75. G. F. King and J. P. Mackay in D. J. Craik, Ed., *NMR in Drug Design*, CRC Press, Boca Raton, FL, 1996, pp. 101–200.

76. N. Tjandra, A. M. Garrett, A. M. Gronenborn, A. Bax, and G. M. Clore, *Nat. Struct. Biol.*, **4**, 443 (1997).

77. A. M. Edwards, C. H. Arrowsmith, D. Christendat, A. Dharamsi, J. D. Friesen, J. F. Greenblatt, and M. Vedadi, *Nat. Struct. Biol.*, **7** (Suppl.), 970 (2000).

78. D. Christendat, A. Yee, A. Dharamsi, Y. Kluger, M. Gerstein, C. H. Arrowsmith, and A. M. Edwards, *Prog. Biophys. Mol. Biol.*, **73**, 339 (2000).

79. L. E. Kay, *Nat. Struct. Biol.*, **5**, 513 (1998).

80. F. M. Marassi and S. J. Opella, *Curr. Opin. Struct. Biol.*, **8**, 640 (1998).

81. A. Watts, *Curr. Opin. Biotechnol.*, **10**, 48 (1999).

82. L.-Y. Lian and G. C. K. Roberts in G. C. K. Roberts, Ed., *NMR of Macromolecules: A Practical Approach*, Oxford University Press, Oxford, UK, 1993.

83. L. Dugad and J. T. Gerig, *Biochemistry*, **27**, 4310 (1988).

84. E. I. Hyde, B. Birdsall, G. C. Roberts, J. Feeney, and A. S. V. Burgen, *Biochemistry*, **19**, 3746 (1980).

85. J. Feeney, J. G. Batchelor, J. P. Albrand, and G. C. K. Roberts, *J. Magn. Reson.*, **33**, 519 (1979).

86. S. Pavlopoulos, M. Rose, G. Wickham, and D. J. Craik, *Anticancer Drug Des.*, **10**, 623 (1995).

87. G. J. Pelton and D. E. Wemmer, *Proc. Natl. Acad. Sci. USA*, **86**, 5723 (1990).

88. W. Leupin, W. J. Chazin, S. Hyberts, W. A. Denny, and K. Wuthrich, *Biochemistry*, **25**, 5902 (1986).

89. S. M. Chen, W. Leupin, M. Rance, and W. J. Chazin, *Biochemistry*, **31**, 4406 (1992).

90. R. E. Klevit, D. E. Wemmer, and B. R. Reid, *Biochemistry*, **25**, 3296 (1986).

91. D. J. Patel and L. Shapiro, *J. Biol. Chem.*, **261**, 1230 (1986).

92. M. S. Searle and K. J. Embrey, *Nucleic Acids Res.*, **18**, 3753 (1990).

93. S. W. Fesik, J. R. Luly, J. W. Erickson, and C. Abad-Zapatero, *Biochemistry*, **27**, 8297 (1988).

94. C. Zwahlen, P. Legault, S. J. F. Vincent, J. Greenblatt, R. Konrat, and L. E. Kay, *J. Am. Chem. Soc.*, **119**, 6711 (1997).

95. M. J. Gradwell and J. Feeney, *J. Biomol. NMR*, **7**, 48 (1996).

96. K. D. Harshman and P. B. Dervan, *Nucleic Acids Res.*, **13**, 4825 (1985).

97. J. G. Pelton and D. E. Wemmer, *Biochemistry*, **27**, 8088 (1988).

98. J. G. Pelton and D. E. Wemmer, *Proc. Natl. Acad. Sci. USA*, **86**, 5723 (1989).

99. M. Coll, J. Aymami, G. A. van der Marel, J. H. van Boom, A. Rich, and A. H. Wang, *Biochemistry*, **28**, 310 (1989).

100. P. E. Pjura, K. Grzeskowiak, and R. E. Dickerson, *J. Mol. Biol.*, **197**, 257 (1987).

101. M. K. Teng, N. Usman, C. A. Frederick, and A. H. Wang, *Nucleic Acids Res.*, **16**, 2671 (1988).

102. M. A. Carrondo, M. Coll, J. Aymami, A. H. Wang, G. A. van der Marel, J. H. van Boom, and A. Rich, *Biochemistry*, **28**, 7849 (1989).

103. J. R. Quintana, A. A. Lipanov, and R. E. Dickerson, *Biochemistry*, **30**, 10294 (1991).

104. J. A. Parkinson, J. Barber, K. T. Douglas, J. Rosamond, and D. Sharples, *Biochemistry*, **29**, 10181 (1990).

105. K. J. Embrey, M. S. Searle, and D. J. Craik, *Eur. J. Biochem.*, **211**, 437 (1993).

106. A. Fede, A. Labhardt, W. Bannwarth, and W. Leupin, *Biochemistry*, **30**, 11377 (1991).

107. A. Fede, M. Billeter, W. Leupin, and K. Wuthrich, *Structure*, **1**, 177 (1993).

108. T. Taga, H. Tanaka, T. Goto, and S. Tada, *Acta Crystallogr.*, **C43**, 751 (1987).

109. B. E. Bierer, P. K. Somers, T. J. Wandless, S. J. Burakoff, and S. L. Schreiber, *Science*, **250**, 556 (1990).

110. P. Karuso, H. Kessler, and D. F. Mierke, *J. Am. Chem. Soc.*, **112**, 9434 (1990).

111. S. W. Fesik, R. T. Gampe Jr., T. F. Holzman, D. A. Egan, R. Edalji, J. R. Luly, R. Simmer, R. Helfrich, V. Klahore, and D. H. Rich, *Science*, **250**, 1406 (1990).

112. S. W. Fesik, R. T. Gampe Jr., H. L. Eaton, G. Gemmecker, E. T. Olejniczak, P. Neri, T. F. Holzman, D. A. Egan, R. Edalji, R. Simmer, R. Helfrich, J. Hochlowski, and M. Jackson, *Biochemistry*, **31**, 6574 (1991).

113. C. Weber, G. Wider, K. von Freyberg, R. Truber, W. Braun, H. Widner, and K. Wuetrich, *Biochemistry*, **30**, 6564 (1991).

114. M. K. Rosen, R. F. Standaert, A. Galat, M. Nakatsuka, and S. L. Schreiber, *Science*, **248**, 863 (1990).

115. T. J. Wandless, S. W. Michnick, M. K. Rosen, M. Karplus, and S. L. Schreiber, *J. Am. Chem. Soc.*, **113**, 2339 (1991).

116. A. M. Petros, R. T. Gampe Jr., G. Gemmecker, P. Neri, T. F. Holzman, R. Edalji, J. Hochlowski, M. Jackson, J. McAlpine, J. R. Luly, et al., *J. Med. Chem.*, **34**, 2925 (1991).

117. G. D. van Duyne, R. F. Standaert, M. Karplus, S. L. Schreiber, and J. Clardy, *Science*, **252**, 839 (1991).

118. S. W. Michnick, M. K. Rosen, T. J. Wandless, M. Karplus, and S. L. Schreiber, *Science*, **252**, 836 (1991).

119. J. M. Moore, D. A. Peattie, M. J. Fitzgibbon, and J. A. Thomson, *Nature*, **351**, 248 (1991).

120. R. P. Meadows, D. G. Nettesheim, R. X. Xu, E. T. Olejniczak, A. M. Petros, T. F. Holzman, J. Severin, E. Gubbins, H. Smith, and S. W. Fesik, *Biochemistry*, **32**, 754 (1993).

121. J. W. Cheng, C. A. Lepre, S. P. Chambers, J. R. Fulghum, J. A. Thomson, and J. M. Moore, *Biochemistry*, **32**, 9000 (1993).

122. J. W. Cheng, C. A. Lepre, and J. M. Moore, *Biochemistry*, **33**, 4093 (1994).

123. P. R. Gooley, B. A. Johnson, A. I. Marcy, G. C. Cuca, S. P. Salowe, W. K. Hagmann, C. K. Esser, and J. P. Springer, *Biochemistry*, **32**, 13098 (1993).

124. P. R. Gooley, J. F. O'Connell, A. I. Marcy, G. C. Cuca, S. P. Salowe, B. L. Bush, J. D. Hermes, C. K. Esser, W. K. Hagmann, J. P. Springer, and B. A. Johnson, *Nat. Struct. Biol.*, **1**, 111 (1994).

125. P. R. Gooley, J. F. O'Connell, A. I. Marcy, G. C. Cuca, M. G. Axel, C. G. Caldwell, W. K. Hagmann, and J. W. Becker, *J. Biomol. NMR*, **7**, 8 (1996).

126. S. R. Van Doren, A. V. Kurochkin, Q.-Z. Ye, L. L. Johnson, D. J. Hupe, and E. R. P. Zuiderweg, *Biochemistry*, **32**, 13109 (1993).

127. S. R. Van Doren, A. V. Kurochkin, W. Hu, Q.-Z. Ye, L. L. Johnson, D. J. Hupe, and E. R. Zuiderweg, *Protein Sci.*, **4**, 2487 (1995).

128. M. A. McCoy, M. J. Dellwo, D. M. Schneider, T. M. Banks, J. Falvo, K. J. Vavra, A. M. Mathiowetz, M. W. Qoronfleh, R. Ciccarelli, E. R. Cook, T. A. Pulvino, R. C. Wahl, and H. Wang, *J. Biomol. NMR*, **9**, 11 (1997).

129. F. J. Moy, M. R. Pisano, P. K. Chanda, C. Urbano, L. M. Killar, M. L. Sung, and R. Powers, *J. Biomol. NMR*, **10**, 9 (1997).

130. E. J. Jacobsen, M. A. Mitchell, S. K. Hendges, K. L. Belonga, L. L. Skaletzky, L. S. Stelzer, T. J. Lindberg, E. L. Fritzen, H. J. Schostarez, T. J. O'Sullivan, L. L. Maggiora, C. W. Stuchly, A. L. Laborde, M. F. Kubicek, R. A. Poorman, J. M. Beck, H. R. Miller, G. L. Petzold, P. S. Scott, S. E. Truesdell, T. L. Wallace, J. W. Wilks, C. Fisher, L. V. Goodman, P. S. Kaytes, et al., *J. Med. Chem.*, **42**, 1525 (1999).

131. N. C. Gonnella, R. Bohacek, X. Zhang, I. Kolossvary, C. G. Paris, R. Melton, C. Winter, S. I. Hu, and V. Ganu, *Proc. Natl. Acad. Sci. USA*, **92**, 462 (1995).

132. V. I. Polshakov, B. Birdsall, T. A. Frenkiel, A. R. Gargaro, and J. Feeney, *Protein Sci.*, **8**, 467 (1999).

133. P. M. Nieto, B. Birdsall, W. D. Morgan, T. A. Frenkiel, A. R. Gargaro, and J. Feeney, *FEBS Lett.*, **405**, 16 (1997).

134. A. Wlodawer, M. Miller, M. Jaskolski, B. K. Sathyanarayana, E. Baldwin, I. T. Weber, L. M. Selk, L. Clawson, J. Schneider, and S. B. Kent, *Science*, **245**, 616 (1989).

135. A. Wlodawer and J. W. Erickson, *Annu. Rev. Biochem.*, **62**, 543 (1993).

136. A. Wlodawer and J. Vondrasek, *Annu. Rev. Biophys. Biomol. Struct.*, **27**, 249 (1998).

137. D. J. Kempf and H. L. Sham, *Curr. Pharm. Des.*, **2**, 225 (1996).

138. T. Yamazaki, A. P. Hinck, Y. X. Wang, L. K. Nicholson, D. A. Torchia, P. Wingfield, S. J. Stahl, J. D. Kaufman, C. H. Chang, P. J. Domaille, and P. Y. Lam, *Protein Sci.*, **5**, 495 (1996).

139. P. Y. Lam, P. K. Jadhav, C. J. Eyermann, C. N. Hodge, Y. Ru, L. T. Bacheler, J. L. Meek, M. J. Otto, M. M. Rayner, Y. N. Wong, et al., *Science*, **263**, 380 (1994).

140. S. Grzesiek, A. Bax, L. K. Nicholson, T. Yamazaki, P. T. Wingfield, S. J. Stahl, C. J. Eyermann, D. A. Torchia, C. N. Hodge, P. Y. Lam, P. K. Jadhav, and C. H. Chang, *J. Am. Chem. Soc.*, **116**, 1581 (1994).

141. Y. X. Wang, D. I. Freedberg, S. Grzesiek, D. A. Torchia, P. T. Wingfield, J. D. Kaufman, S. J. Stahl, C. H. Chang, and C. N. Hodge, *Biochemistry*, **35**, 12694 (1996).

142. G. Otting, E. Liepinsh, and K. Wuthrich, *Science*, **254**, 974 (1991).

143. T. Yamazaki, L. K. Nicholson, P. Wingfield, S. J. Stahl, J. D. Kaufman, P. J. Domaille, and D. A. Torchia, *J. Am. Chem. Soc.*, **116**, 10791 (1994).

144. E. T. Baldwin, T. N. Bhat, S. Gulnik, B. Liu, I. A. Topol, Y. Kiso, T. Mimoto, H. Mitsuya, and J. W. Erickson, *Structure*, **3**, 581 (1995).

145. L. K. Nicholson, T. Yamazaki, D. A. Torchia, S. Grzesiek, A. Bax, S. J. Stahl, J. D. Kaufman, P. T. Wingfield, P. Y. Lam, P. K. Jadhav, et al., *Nat. Struct. Biol.*, **2**, 274 (1995).

146. J. R. Rose, R. Salto, and C. S. Craik, *J. Biol. Chem.*, **268**, 11939 (1993).

147. D. I. Freedberg, Y. X. Wang, S. J. Stahl, J. D. Kaufman, P. Wingfield, Y. Kiso, and D. A. Torchia, *J. Am. Chem. Soc.*, **120**, 7916 (1998).

148. E. Katoh, T. Yamazaki, Y. Kiso, P. Wingfield, S. J. Stahl, J. D. Kaufman, and D. A. Torchia, *J. Am. Chem. Soc.*, **121**, 2607 (1999).

149. D. L. Winslow, S. Stack, R. King, H. Scarnati, A. Bincsik, and M. J. Otto, *AIDS Res. Hum. Retroviruses*, **11**, 107 (1995).

150. J. H. Condra, W. A. Schleif, O. M. Blahy, L. J. Gabryelski, D. J. Graham, J. C. Quintero, A. Rhodes, H. L. Robbins, E. Roth, M. Shivaprakash, et al., *Nature*, **374**, 569 (1995).

151. S. B. Shuker, P. J. Hajduk, R. P. Meadows, and S. W. Fesik, *Science*, **274**, 1531 (1996).

152. E. Olejniczak, P. Hajduk, P. Marcotte, D. Nettesheim, R. Meadows, R. Edalji, T. Holzman, and S. Fesik, *J. Am. Chem. Soc.*, **119**, 5828 (1997).

153. P. J. Hajduk, J. Dinges, J. M. Schkeryantz, D. Janowick, M. Kaminski, M. Tufano, D. J. Augeri, A. Petros, V. Nienaber, P. Zhong, R. Hammond, M. Coen, B. Beutel, L. Katz, and S. W. Fesik, *J. Med. Chem.*, **42**, 3852 (1999).

154. P. Hajduk, G. Sheppard, D. Nettesheim, E. Olejniczak, S. Shuker, R. Meadows, D. Steinman, G. Carrera, P. Marcotte, J. Severin, K. Walter, H. Smith, E. Gubbins, R. Simmer, T. Holzman, D. Morgan, S. Davidsen, J. Summers, and S. Fesik, *J. Am. Chem. Soc.*, **119**, 5818 (1997).

155. P. Hajduk, M. Zhou, and S. Fesik, *Bioorg. Med. Chem. Lett.*, **9**, 2403 (1999).

156. G. Liu, J. R. Huth, E. T. Olejniczak, R. Mendoza, P. DeVries, S. Leitza, E. B. Reilly, G. F. Okasinski, S. W. Fesik, and T. W. von Geldern, *J. Med. Chem.*, **44**, 1202 (2001).

157. H. Mao, P. J. Hajduk, R. Craig, R. Bell, T. Borre, and S. W. Fesik, *J. Am. Chem. Soc.*, **123**, 10429 (2001).

158. A. Ross, G. Schlotterbeck, W. Klaus, and H. Senn, *J. Biomol. NMR*, **16**, 139 (2000).

159. F. J. Moy, K. Haraki, D. Mobilio, G. Walker, R. Powers, K. Tabei, H. Tong, and M. M. Siegel, *Anal. Chem.*, **73**, 571 (2001).

160. M. Mayer and B. Meyer, *Angew. Chem. Int. Ed. Engl.*, **38**, 1784 (1999).

161. J. Klein, R. Meinecke, M. Mayer, and B. Meyer, *J. Am. Chem. Soc.*, **121**, 5336 (1999).

162. W. Hellebrandt, T. Haselhorst, T. Köhli, E. Bäuml, and T. Peters, *J. Carbohydr. Chem.*, **19**, 769 (2000).

163. M. Vogtherr and T. Peters, *J. Am. Chem. Soc.*, **122**, 6093 (2000).

164. H. Maaheimo, P. Kosma, L. Brade, H. Brade, and T. Peters, *Biochemistry*, **39**, 12778 (2000).

165. M. Mayer and B. Meyer, *J. Am. Chem. Soc.*, **123**, 6108 (2001).

166. R. Meinecke and B. Meyer, *J. Med. Chem.*, **44**, 3059 (2001).

167. C. Dalvit, P. Pevarello, M. Tato, M. Veronesi, A. Vulpetti, and M. Sundstrom, *J. Biomol. NMR*, **18**, 65 (2000).

168. N. Gonnella, M. Lin, M. J. Shapiro, J. R. Wareing, and X. Zhang, *J. Magn. Reson.*, **131**, 336 (1998).

169. A. Chen and M. J. Shapiro, *J. Am. Chem. Soc.*, **120**, 10258 (1998).

170. A. Chen, C. S. Johnson Jr., M. Lin, and M. J. Shapiro, *J. Am. Chem. Soc.*, **120**, 9094 (1998).

171. A. Chen and M. J. Shapiro, *J. Am. Chem. Soc.*, **121**, 5338 (1999).

172. A. Chen and M. J. Shapiro, *Anal. Chem.*, **71**, 669A (1999).

173. M. Lin and M. J. Shapiro, *J. Org. Chem.*, **61**, 7617 (1996).

174. M. Lin, M. J. Shapiro, and J. R. Wareing, *J. Am. Chem. Soc.*, **119**, 5249 (1997).

175. M. Lin, M. J. Shapiro, and J. R. Wareing, *J. Org. Chem.*, **62**, 8930 (1997).

176. K. Bleicher, M. Lin, M. J. Shapiro, and J. R. Wareing, *J. Org. Chem.*, **63**, 8486 (1998).

177. P. J. Hajduk, E. T. Olejniczak, and S. W. Fesik, *J. Am. Chem. Soc.*, **119**, 12257 (1997).

178. D. Henrichson, B. Ernst, J. L. Magnani, W. Wang, B. Meyer, and T. Peters, *Angew. Chem. Int. Ed. Engl.*, **38**, 98 (1999).

179. B. Meyer, T. Weimar, and T. Peters, *Eur. J. Biochem.*, **246**, 705 (1997).

180. M. Mayer and B. Meyer, *J. Med. Chem.*, **43**, 2093 (2000).

181. A. Chen and M. J. Shapiro, *J. Am. Chem. Soc.*, **122**, 414 (2000).

182. S. Bagby, K. I. Tong, D. Liu, J. R. Alattia, and M. Ikura, *J. Biomol. NMR*, **10**, 279 (1997).

183. C. Lepre and J. Moore, *J. Biomol. NMR*, **12**, 493 (1998).

184. S. Bagby, K. I. Tong, and M. Ikura, *Methods Enzymol.*, **339**, 20 (2001).

185. C. A. Lepre, *Drug Discovery Today*, **6**, 133 (2001).

186. C. A. Lipinski, F. Lombardo, B. W. Dominy, and P. J. Feeney, *Adv. Drug Delivery Rev.*, **46**, 3 (2001).

187. A. K. Ghose, V. N. Viswanadhan, and J. J. Wendoloski, *J. Comb. Chem.*, **1**, 55 (1999).

188. T. I. Oprea, J. Gottfries, V. Sherbukhin, P. Svensson, and T. C. Kuhler, *J. Mol. Graph. Model.*, **18**, 512 (2000).

189. J. Xu and J. Stevenson, *J. Chem. Inf. Comput. Sci.*, **40**, 1177 (2000).

190. G. Bemis and M. Murcko, *J. Med. Chem.*, **39**, 2887 (1996).

191. G. Bemis and M. Murcko, *J. Med. Chem.*, **42**, 5095 (1999).

192. M. Hann, B. Hudson, X. Lewell, R. Lifely, L. Miller, and N. Ramsden, *J. Chem. Inf. Comput. Sci.*, **39**, 897 (1999).

193. A. Ross and H. Senn, *Drug Discovery Today*, **6**, 583 (2001).

CHAPTER THIRTEEN

Mass Spectrometry and Drug Discovery

RICHARD B. VAN BREEMEN
Department of Medicinal Chemistry and Pharmacognosy
University of Illinois at Chicago
Chicago, Illinois

Contents

Burger's Medicinal Chemistry and Drug Discovery
Sixth Edition, Volume 1: Drug Discovery
Edited by Donald J. Abraham
ISBN 0-471-27090-3 © 2003 John Wiley & Sons, Inc.

1   INTRODUCTION

At the beginning of the 20th century, mass spectrometers were invented to help physicists and physical chemists prove the existence of isotopes of the elements. As radioactivity and nuclear physics was explored, specialized mass spectrometers were used to characterize the fission products of radioactive elements as they were created or discovered. In addition, mass spectrometers were used for the measurement of isotopic enrichment of radioactive elements, their inorganic derivatives, and even the isotopic purification of radioactive elements as inorganic compounds. As this era of mass spectrometry reached maturity in the 1940s, some physicists announced that there would no longer be any need for mass spectrometry because virtually all of the elements had been discovered and characterized. Of course, these prognosticators were wrong because the entire field of organic mass spectrometry was about to begin.

While mass spectrometers were being used for the purification of fissionable material for atomic weapons as part of the Manhattan Project of World War II, organic mass spectrometry was being invented for the analysis and quality control of aviation fuel. In 1945, the application of mass spectrometry to organic chemistry emerged as a productive new area of research and discovery. Commercial production of organic mass spectrometers began immediately, and petroleum companies became the first customers for these new analytical instruments. Early commercial mass spectrometers used electron impact (EI) ionization (see Equations 13.1 and 13.2) to generate ions from gas–phase molecules that were separated by acceleration through an electromagnetic field provided by either a fixed magnet or an electromagnet. After separation, the ions were detected using a simple impact detector such as a Faraday cup. This basic design is still in use today for the identification and quantitative analysis of volatile organic compounds.

$$M + e^- \ (70 \ eV) \ \rightarrow \ M^{+\cdot} + 2e^- \ \text{formation}$$

$$\text{of positive molecular ions using} \quad (13.1)$$

$$\text{EI ionization}$$

$$M + e^- \ (2\text{--}10 \ eV) \ \rightarrow \ M^{-\cdot} \qquad (13.2)$$

$$\text{electron capture EI ionization}$$

Toward the late 1950s, organic mass spectrometers began to be used for the analysis of a wider variety of organic molecules and eventually became a fundamental analytical tool for the characterization of synthetic organic compounds. Today, mass spectrometers are used routinely to confirm the molecular weights of organic compounds and to verify their structures based on fragmentation patterns. Fragmentation results from the cleavage of chemical bonds within an ion, resulting in the formation of a product ion of lower mass and one or more neutral products. Of course, only the fragment ions and not the neutral species are detected in a mass spectrometer because this instrument measures the mass-to-charge ratio (m/z) of ions in the gas phase. The energy for fragmentation is the result of excess energy imparted to the molecular ion or during a process known as collision-induced dissociation (CID), which will be discussed along with tandem mass spectrometry (MS-MS) below. Because the fragmentation pattern reflects the relative strengths of chemical bonds in a compound, mass spectra (a plot of ion relative abundance versus m/z) provide structurally significant fragment ions for compound identification. Rules for structure elucidation of chemical structures through the interpretation of mass spectra have been developed. (For a review of EI and ion fragmentation pathways, see McClafferty et al. 1997, Section 4).

In many cases, EI imparts so much excess energy into a molecule that only fragment ions and no molecular ions are produced. Therefore, "softer" ionization techniques were developed to enhance molecular weight information. The first of these ionization methods was chemical ionization (CI). Developed by re-

Table 13.1 Types of Mass Spectrometers and Tandem Mass Spectrometers

Instrument	Resolving Power	m/z Range	Tandem MS
Magnetic sector	100,000	12,000	Low resolution
Quadrupole	< 4,000	4,000	none
Triple quadrupole	< 4,000	4,000	Low resolution
Time-of-flight (TOF)	15,000	> 200,000	none
FTICR	> 200,000	< 10,000	MS^n, high resolution
Ion trap	< 4,000	< 10,000	MS, low resolution
QTOF	12,000	4,000	High resolution
TOF-TOF	15,000	> 10,000	High resolution

searchers in the petroleum industry (1), CI became another standard ionization technique for organic mass spectrometry. During CI, high energy electrons (as in EI) are used to ionize a gas called a reagent gas at a constant pressure (usually ~1 Torr) in the mass spectrometer ionization source. The reagent gas in turn ionizes the sample molecules through ion–molecule reactions that usually involve the exchange of protons. Less frequently, sample molecule ionization might involve a charge exchange. Two of the most common ionization mechanisms in CI are summarized in Equations 13.3 and 13.4.

$$M + RH^+ \rightarrow MH^+ + R \text{ CI through proton}$$

$$\text{transfer, R = reagent gas} \quad (13.3)$$

$$M + R^{+\cdot} \rightarrow M^{+\cdot} + R$$
$$(13.4)$$
$$\text{CI through charge exchange}$$

During the 1960s, high resolution double-focusing magnetic sector instruments became available and are now standard tools for the determination of elemental compositions using a type of analysis called exact mass measurement. In mass spectrometry, resolution is defined as M/ΔM, where M is the m/z value of a singly charged ion, and ΔM is the difference (measured in m/z) between M and the next highest ion. Alternatively, ΔM may be defined in terms of the width of the peak. High resolution is typically regarded as a value of at least 10,000. At this resolution, the molecular ions of most drug-like molecules (that is compounds with molecular weights less than ~500) can be resolved from each other. After resolving a sample ion from others in a mass spectrum, an exact mass measurement may be

carried out by accurately weighing the unknown ion and comparing its m/z value to that of a calibration standard. Since the 1960s, other types of mass spectrometers capable of high resolution exact mass measurements have become available as commercial products, including Fourier transform ion cyclotron resonance (FTICR) mass spectrometers, reflectron TOF instruments, and recently, quadrupole time-of-flight hybrid (QqTOF) mass spectrometers (see Table 13.1 for a listing of types of organic mass spectrometers and a comparison of their performance characteristics). By the early 2000s, FTICR and QqTOF instruments became more popular than magnetic sector mass spectrometers for exact mass measurements, high resolution measurements, and drug discovery applications. As will be discussed below, exact mass measurements are essential to many types of mass spectrometry-based screening and drug discovery today.

Biomedical applications of mass spectrometry began during the 1960s both at academic institutions and pharmaceutical companies. These applications depended on the volatilization (usually by heating) of pharmaceutical compounds and biochemicals before their gas-phase ionization using EI or CI. To increase the thermal stability and volatility of these compounds, a variety of derivatization methods were developed to mask polar functional groups and reduce hydrogen bonding between molecules. These methods were particularly effective for use with gas chromatography–mass spectrometry (GC-MS), which was introduced during the 1960s as a practical and powerful tool for qualitative and quantitative analysis of compounds in mixtures. Both EI and CI were immediately useful for GC-MS, because both of these ionization methods require that the analytes be in the gas phase. When capillary GC was incorporated into GC-MS, this technique reached maturity. GC-MS may be used to select, identify, and quantify organic compounds in complex mixtures at the femtomole level. The speed of GC-MS is determined by the chromatography step, which typically requires several minutes to 1 h per analysis. By the 1970s, some organic chemists were announcing that organic mass spectrometry had reached maturity and that no new applications were possible. Like the physicists and physical chemists who had pronounced the end of mass spectrometry a generation earlier, this group would soon be proved wrong.

Although GC-MS remains important for the analysis of many organic compounds, this technique is limited to volatile and thermally stable compounds that comprise only a small fraction of all organic compounds and even fewer biomedically important molecules. Therefore, thermally unstable compounds, including many pharmaceutical compounds such as nucleic acid analogs and biomolecules such as proteins, carbohydrates, and nucleic acids, cannot be analyzed in their native forms using GC-MS. (For more details regarding GC-MS and its applications, see Watson 1997, Section 4.) Although derivatization facilitates the GC-MS analysis of many of these compounds, alternative ionization techniques were needed for the analysis of the vast majority of polar and non-volatile compounds of interest to drug discovery.

During the 1970s and early 1980s, desorption ionization techniques such as field desorption (FD), desorption EI, desorption CI (DCI), and laser desorption were developed to extend the use of mass spectrometry toward the analysis of more polar and less volatile compounds (see Watson 1997, Section 4, for more information regarding desorption ionization techniques including DCI and FD). Although these techniques helped extend the mass range of mass spectrometry beyond a traditional limit of m/z 1000 and toward ions of m/z 5000, the first breakthrough in the analysis of polar, non-volatile compounds occurred

in 1982 with the invention of fast atom bombardment (FAB) (2). FAB and its counterpart, liquid secondary ion mass spectrometry (LSIMS), facilitated the formation of abundant molecular ions, protonated molecules, and deprotonated molecules of non-volatile and thermally labile compounds such as peptides, chlorophylls, and complex lipids up to approximately m/z 12,000. FAB and LSIMS use energetic particle bombardment (fast atoms or ions from 3 to 30,000 V of energy) to ionize compounds dissolved in non-volatile matrices such as glycerol or 3-nitrobenzyl alcohol and desorb them from this condensed phase into the gas phase for mass spectrometric analysis (see Fig. 13.1). Protonated or deprotonated molecules are usually abundant and fragmentation is minimal.

Introduced in the late 1980s, matrix-assisted laser desorption ionization (MALDI) has helped solve the mass limit barriers of laser desorption mass spectrometry so that singly charged ions may be obtained up to m/z 500,000 and sometimes higher (3). For most commercially available MALDI mass spectrometers, ions up to m/z 200,000 are readily obtained. Like FAB and LSIMS, MALDI samples are mixed with a matrix to form a solution that is loaded onto the sample stage for analysis. Unlike the other matrix-mediated techniques, the solvent is evaporated before MALDI analysis, leaving sample molecules trapped in crystals of solid phase matrix. The MALDI matrix is selected to absorb the pulse of laser light directed at the sample. Most MALDI mass spectrometers are equipped with a pulsed UV laser, although IR lasers are available as an option on some commercial instruments. Therefore, matrices are often substituted benzenes or benzoic acids with strong UV absorption properties. During MALDI, the energy of the short but intense UV laser pulse obliterates the matrix and in the process desorbs and ionizes the sample. Like FAB and LSIMS, MALDI typically produces abundant protonated or deprotonated molecules with little fragmentation.

By the time that GC-MS had become a standard technique in the late 1960s, LC-MS was still in the developmental stages. Producing gas-phase sample ions for analysis in a vacuum system while removing the high perfor-

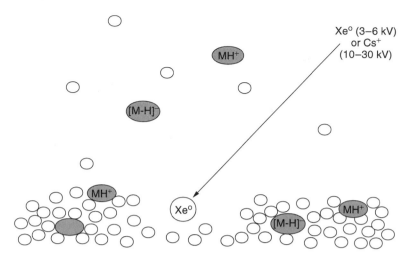

Figure 13.1. Scheme for desorption ionization using FAB or LSIMS from a liquid matrix (○).

mance liquid chromatography (HPLC) mobile phase proved to be a challenging task. Early LC-MS techniques included a moving belt interface to desolvate and transport the HPLC eluate into an CI or EI ion source or a direct inlet system in which the eluate was pumped at a low flow rate (1–3 μL/min) into a CI source. However, neither of these systems was robust enough or suitable for a broad enough range of samples to gain widespread acceptance.

Because FAB (or LSIMS) requires that the analyte be dissolved in a liquid matrix, this ionization technique was easily adapted for infusion of solution-phase samples into the FAB ionization source in an approach known as continuous-flow FAB. Then, continuous-flow FAB was connected to microbore HPLC columns for LC-MS applications (4). Because this method is limited to microbore HPLC applications at flow rates of <10 μL/min and requires considerable operator intervention, it is not ideal for the analysis of large sample sets. Instead, more robust techniques have been developed to fulfill this requirement. However, continuous-flow FAB is still in use in some laboratories.

Like continuous-flow FAB, the popularity of particle beam interfaces is diminishing, but systems are still available from commercial sources. During particle beam LC-MS, the HPLC eluate is sprayed into a heated chamber connected to a vacuum pump. As the droplets evaporate, aggregates of analyte (particles) form and pass through a momentum separator that removes the lower molecular weight solvent molecules. Finally, the particle beam enters the mass spectrometer ion source where the aggregates strike a heated plate from which the analyte molecules evaporate and are ionized using conventional EI or CI. Particle beam LC-MS is limited to the analysis of volatile and thermally stable compounds that are amenable to flash evaporation and EI or CI mass spectrometry. Therefore, this approach is not used for polar biochemicals such as carbohydrates, sugars, peptides, proteins, or nucleic acids.

Because thermospray became the first widely used LC-MS technique (during the late 1970s and early 1980s), this technique should be mentioned here. Thermospray facilitates the interfacing of standard analytical HPLC systems at flow rates up to 1 mL/min with mass spectrometers. Although the interface between the HPLC and mass spectrometer is inefficient and exhibits low sensitivity for most analytes, thermospray has been useful for the LC-MS analysis of many types of small molecules. During thermospray, the HPLC eluate is sprayed through a heated capillary into a heated desolvation chamber at reduced pressure. Gas phase ions remaining after desolvation of the droplets are extracted through a

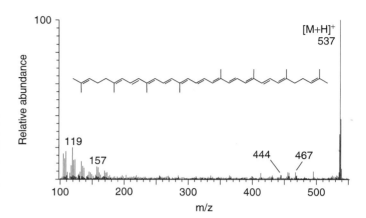

Figure 13.2. Positive ion APCI mass spectrum of the red carotenoid lycopene in a solution of methanol and *tert*-butyl methyl ether (1:1; v/v). In this analysis, lycopene formed a protonated molecule instead of a molecular ion, $M^{+\cdot}$.

skimmer into the mass spectrometer for analysis. The sensitivity of thermospray is poor because there is no mechanism or driving force to enhance the number of sample ions entering the gas phase from the spray during desolvation. Also, thermally labile compounds tend to decompose in the heated source. These problems were solved when thermospray was replaced by electrospray during the late 1980s.

During the 1990s, electrospray and atmospheric pressure chemical ionization (APCI) became the standard interfaces for LC-MS. Today, APCI and electrospray ionization are the most widely used ionization sources and HPLC interfaces for drug discovery using mass spectrometry. Unlike thermospray, particle beam or continuous-flow FAB, electrospray and APCI interfaces operate at atmospheric pressure and do not depend on vacuum pumps to remove solvent vapor. As a result, they are compatible with a wide range of HPLC flow rates. Also, no matrix is required. Both APCI and electrospray are compatible with a wide range of HPLC columns and solvent systems. Like all LC-MS systems, the solvent system should contain only volatile solvents, buffers or ion pair agents to reduce fouling of the mass spectrometer ion source. In general, APCI and electrospray form abundant molecular ion species. When fragment ions are formed, they are usually more abundant in APCI than electrospray mass spectra.

The APCI interface uses a heated nebulizer to form a fine spray of the HPLC eluate, which is much finer than the particle beam system

but similar to that formed during thermospray. A cross-flow of heated nitrogen gas is used to facilitate the evaporation of solvent from the droplets. The resulting gas-phase sample molecules are ionized by collisions with solvents ions, which are formed by a corona discharge in the atmospheric pressure chamber. Molecular ions, $M^{+\cdot}$ or $M^{-\cdot}$, and/or protonated or deprotonated molecules can be formed. The relative abundance of each type of ion depends on the sample itself, the HPLC solvent, and the ion source parameters. Next, ions are drawn into the mass spectrometer analyzer for measurement through a narrow opening or skimmer that helps the vacuum pumps to maintain very low pressure inside the analyzer, while the APCI source remains at atmospheric pressure. For example, the positive ion APCI mass spectrum of lycopene is shown in Fig. 13.2. The carotenoid lycopene is the red pigment of ripe tomatoes and is under clinical investigation for the prevention of prostate cancer (5).

During electrospray, the HPLC eluate is sprayed through a capillary electrode at high potential (usually 2000–7000 V) to form a fine mist of charged droplets at atmospheric pressure. As the charged droplets migrate towards the opening of the mass spectrometer because of electrostatic attraction, they encounter a cross-flow of heated nitrogen that increases solvent evaporation and prevents most of the solvent molecules from entering the mass spectrometer. Molecular ions, protonated or deprotonated molecules, and cationized species such as $[M + Na]^+$ and $[M + K]^+$ can be formed. (For additional information on electrospray ionization, see Cole 1997, Section 4). In addition to singly charged ions, electrospray is unique as an ionization technique in that multiply charged species are common and often constitute the majority of the sample ion abundance. The relative abundance of each of these species depends on the chemistry of the analyte, the pH, the presence of proton donating or accepting species, and the levels of trace amounts of sodium or potassium salts in the mobile phase. In contrast, APCI, MALDI, EI, CI, and FAB/LSIMS usually produce singly charged species. A consequence of forming multiply charged ions is that they are detected at lower m/z values (i.e., $z > 1$) than the corresponding singly charged species. This has the benefit of allowing mass spectrometers with modest m/z ranges to detect and measure ions of molecules with very high masses. For example, electrospray has been used to measure ions with molecular weights of hundreds of thousands or even millions of Daltons on mass spectrometers with m/z ranges of only a few thousand. (For a review of LC-MS techniques, see Niessen 1999, Section 4.)

An example of the C_{18} reversed phase HPLC-negative ion electrospray mass spectrometric (LC-MS) analysis of an extract of the botanical, *Trifolium pratense* L. (red clover), is shown in Fig. 13.3. Extracts of red clover are used as dietary supplements by menopausal and postmenopausal women and are under investigation as alternatives to estrogen replacement therapy (6). The two-dimensional map illustrates the amount of information that may be acquired using hyphenated techniques such as LC-MS. In the time dimension, chromatograms are obtained, and a sample computer-reconstructed mass chromatogram is shown for the signal at m/z 269. An intense chromatographic peak was detected eluting at 12.4 min. In the m/z dimension, the negative ion electrospray mass spectrum recorded at 12.4 min shows a base peak at m/z 269. Based on comparison with authentic standards (data not shown), the ion of m/z 269 was found to correspond to the deprotonated molecule of genistein, which is an estrogenic isoflavone (6). Because almost no fragmentation of the genistein ion was observed, additional charac-

terization would require CID and MS-MS as discussed in the next section.

When analyzing complex mixtures such as the botanical extract shown in Fig. 13.3, the use of chromatographic separation before mass spectrometric ionization and analysis is essential to distinguish between isomeric compounds. Even simple mixtures of synthetic compounds might contain isomers that would require LC-MS for adequate characterization. Another problem overcome by using a chromatography step before mass spectrometric analysis is ion suppression. No matter what ionization technique is used, the presence of multiple compounds in the ion source might enhance the ionization of one compound while suppressing the ionization of another. Usually, only some of the compounds in a complex mixture can be detected by mass spectrometry without chromatographic separation. The presence of salts and buffers in a sample can also suppress sample ionization. Therefore, LC-MS has become a powerful tool for analyzing natural products, synthetic organic compounds, and pharmaceutical agents and their metabolites.

In general, APCI facilitates the ionization of non-polar and low molecular weight species, and electrospray is more useful for the ionization of polar and high molecular weight compounds. In this sense, APCI and electrospray are often complementary ionization techniques. However, during the analysis of large or diverse combinatorial libraries, both polar and non-polar compounds are usually present. As a result, no one set of ionization conditions using APCI or electrospray is adequate to detect all the compounds contained in the library of compounds. Therefore, a UV ionization technique called atmospheric pressure photoionization (APPI) has been developed for use with combinatorial libraries and LC-MS (7). Recently, APPI became a commercially available ionization alternative to APCI and electrospray. During APPI, a liquid solution or HPLC eluate is sprayed at atmospheric pressure, as in APCI. Instead of using a corona discharge as in APCI, ionization occurs during APPI because of irradiation of the analyte molecules by an intense UV light source. Obviously, the carrier solvent must not absorb UV light at the same wavelengths, or interfer-

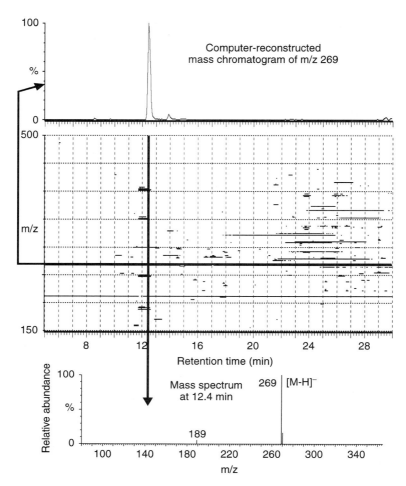

Figure 13.3. Two-dimensional map showing the LC-MS analysis of an extract of red clover under investigation for the management of menopause. Reversed phase separation was carried out using a C_{18} HPLC column in the time dimension and negative ion electrospray mass spectrometry was used for compound detection and molecular weight determination in the second dimension.

ence would prevent sample ionization and detection. The use of APPI as an alternative to APCI and electrospray for drug discovery applications is under investigation.

Desorption ionization techniques like FAB, MALDI, and electrospray facilitate the molecular weight determination of a wide range of polar, non-polar, and low, and high molecular weight compounds including drugs and drug targets such as proteins. However, the "soft" ionization character of these techniques means that most of the ion current is concentrated in molecular ions, and few structurally significant fragment ions are formed. To en-

hance the amount of structural information in these mass spectra, CID may be used to produce more abundant fragment ions from molecular ion precursors formed and isolated during the first stage of mass spectrometry. Then, a second mass spectrometry analysis may be used to characterize the resulting product ions. This process is called tandem mass spectrometry or MS-MS and is illustrated in Fig. 13.4.

Another advantage of the use of tandem mass spectrometry is the ability to isolate a particular ion such as the molecular ion of the analyte of interest during the first mass spec-

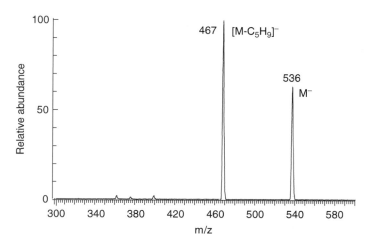

Figure 13.4. Scheme illustrating the selectivity of MS-MS and the process by which CID facilitates fragmentation of preselected ions. Negative ion electrospray tandem mass spectrum of lycopene. CID was used to induce fragmentation of the molecular ion of m/z 536. As a result, the fragment ion of m/z 467 was formed by the loss of a terminal isoprene unit. This fragment ion may be used to distinguish lycopene from isomeric α-carotene and β-carotene, which lack terminal isoprene groups.

trometry stage. This precursor ion is essentially purified in the gas-phase and free of impurities such as solvent ions, matrix ions, or other analytes. Finally, the selected ion is fragmented using CID and analyzed using a second mass spectrometry stage. In this manner, the resulting tandem spectrum contains exclusively analyte ions without impurities that might interfere with the interpretation of the fragmentation patterns. In summary, CID may be used with LC-MS-MS or desorption ionization and MS-MS to obtain structural information such as amino acid sequences of peptides and sites of alkylation of nucleic acids, or to distinguish structural isomers such as β-carotene and lycopene. Beginning in 2001, TOF-TOF tandem mass spectrometers became available from instrument manufacturers. These instruments have the potential to deliver high resolution tandem mass spectra with high speed that should be compatible with the chip-based chromatography systems now under development.

Over the course of the last century, mass spectrometry has become an essential analytical tool for a wide variety of biomedical applications including drug discovery and development. By combining mass spectrometry with chromatography as in LC-MS or by adding another stage of mass spectrometry as in MS-MS, the selectivity of the technique increases considerably. As a result, mass spectrometry offers all of the analytical elements that are essential to modern drug discovery namely speed, sensitivity, and selectivity.

2 CURRENT TRENDS AND RECENT DEVELOPMENTS

Since the early 1990s, pharmaceutical research has focused on combinatorial chemistry (8, 9) and high-throughput screening (10) in an effort to accelerate the pace of drug discovery. The goal has been to produce, in a short time, large numbers of synthetic organic compounds representing a great diversity of chemical structures through a process called combinatorial chemistry and then quickly screen them *in vitro* against pharmacologically significant targets such as enzymes or receptors. The "hits" identified through these high-throughput screens may then be optimized by quickly and efficiently synthesizing and then screening large numbers of analogs called targeted or directed libraries. As a result, lead compounds might emerge from such combinatorial chemistry drug discovery programs in a few weeks instead of several years. Furthermore, a single organic chemist using combinatorial synthetic methods might synthesize thousands of compounds or more in a single week instead of less than five in the same time using conventional techniques, and a single medicinal chemist might identify hundreds of lead compounds per month instead of just one or two in the same period of time.

Accompanying this new drug discovery paradigm, new scientific journals have been established such as *Combinatorial Chemistry & High Throughput Screening, Journal of Combinatorial Chemistry, Journal of Biomolecular Screening*, and *Molecular Diversity* (see list of journal websites in Section 4). The variety of topics published in these journals reflects the multidisciplinary nature of the current drug discovery process and ranges from organic chemistry, medicinal chemistry, molecular modeling, molecular biology, and pharmacology, to analytical chemistry. As described below, the most significant analytical component of drug discovery has become mass spectrometry. Only mass spectrometry has become an essential element at all stages of the drug discovery and development process.

Although a variety of spectroscopic and chromatographic techniques, including infrared spectroscopy, nuclear magnetic resonance spectroscopy, fluorescence spectroscopy, gas chromatography, HPLC, and mass spectrometry, are being used to support drug discovery in various capacities, some of them, such as gas chromatography and fluorescence spectroscopy, are not applicable to most new chemical entities, some are not specific enough for chemical identification (e.g., infrared spectroscopy), and other techniques suffer from low throughput (e.g., nuclear magnetic resonance spectroscopy). Unlike gas chromatography, HPLC is compatible with virtually all drug-like molecules without the need for chemical derivatization to increase thermal stability or volatility. In addition, mass spectrometry provides a universal means to characterize and distinguish drugs based on both molecular weight and structural features while at the same time providing high throughput. With the development of routine LC-MS interfaces and ionization techniques such as electrospray and APCI, mass spectrometry has also become an ideal HPLC detector for the analysis of combinatorial libraries (11), and LC-MS, MS-MS, and LC-MS-MS have become fundamental tools in the analysis of combinatorial libraries and subsequent drug development studies (12–14).

The application of combinatorial chemistry and high-throughput screening to drug discovery has altered the traditional serial process of lead identification and optimization that previously required years of human effort. Consequently, neither the synthesis of new chemical entities nor their screening is limiting the pace of drug discovery. Instead, a new bottleneck is the verification of the structure and purity of each compound in a combinatorial library or of each lead compound obtained from an uncharacterized library using high-throughput screening. Because the number of lead compounds entering the drug development process has increased, in part because compounds are entering development at earlier stages than in the past, the traditional drug development investigations concerning absorption, distribution, metabolism, and excretion (ADME) and even toxicology evaluations of new drug entities have become additional bottlenecks. As a solution to the drug development bottlenecks, high-throughput assays to assess the metabolism, bioavailability, and toxicity of lead compounds are being developed and applied earlier than ever during the drug discovery process, so that only those compounds most likely to become successful drugs enter the more expensive and slower preclinical pharmacology and toxicology studies. In support of these new combinatorial chemistry synthetic programs and new high-throughput assays, mass spectrometry has emerged as the only analytical technique with sufficient throughput, sensitivity, selectivity, and robustness to address all of these bottlenecks.

2.1 LC-MS Purification of Combinatorial Libraries

Although combinatorial libraries were originally synthesized as mixtures, today most libraries are prepared in parallel as discrete compounds and then screened individually in microtiter plates of 96-well, 384-well, or 1536-well formats. To facilitate subsequent structure-activity analyses and to assure the validity of the screening results, many laboratories verify the structure and purity of each compound before high-throughput screening. Semi-preparative HPLC has become the most

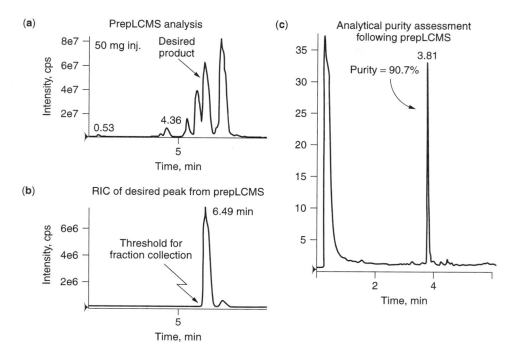

Figure 13.5. Mass-directed purification of a combinatorial library. Chromatographic separation was carried out using gradient elution of 10–90% acetonitrile in water for 7 min after an initial hold at 10% acetonitrile for 1 min. (a) Total ion chromatogram showing desired product and impurities. (b) Computer-reconstructed ion chromatogram (RIC) corresponding to the expected product. (c) Post-purification analysis of the isolated component with a purity >90%. (Reproduced from Ref. 15 by permission of Elsevier Science.)

popular technique for the purification of combinatorial libraries on the milligram scale because of high throughput and the ease of automation. Typically during semi-preparative HPLC, fraction collection is initiated whenever a UV signal is observed above a predetermined threshold. This procedure usually results in the collection of several fractions per analysis and hence creates additional issues such as the need for large fraction collector beds and the need for secondary analysis using flow-injection mass spectrometry, LC-MS, or LC-MS-MS to identify the appropriate fractions. When purification of large numbers of combinatorial libraries is required, this approach can become prohibitively time consuming and expensive.

To enhance the efficiently of this purification procedure, the steps of HPLC purification and mass spectrometric analysis may be combined into automated mass-directed fraction-

ation (15–17). Any size HPLC column may be used, and only a small fraction of the eluant ($\sim \mu L/min$) is diverted to the mass spectrometer equipped for APCI or electrospray ionization. Because all of the components, including autosampler, injector, HPLC, switching valve, mass spectrometer, and fraction collector, are controlled by computer, the procedure may be fully automated. For greatest efficiency, the system may be programmed to collect only those peaks displaying the desired molecular ions, or alternatively, all peaks displaying abundant ions within a specified mass range. An example of the MS-guided purification of a compound synthesized during the parallel synthesis of a combinatorial library of discrete compounds is shown in Fig. 13.5. Although the crude yield of the reaction product was only 30% (Fig. 13.5a), the desired product was detected based on its molecular ion (Fig. 13.5b). After MS-guided fractionation, re-analysis using LC-MS showed that the desired product was >90% pure (Fig. 13.5c).

The use of MS-guided purification of combinatorial libraries provides a means for reducing the number of HPLC fractions collected per sample and eliminates the need for post-purification analysis to further characterize and identify each compound as would be necessary when using UV-based fractionation. The ionization technique (i.e., electrospray, APCI, or APPI), and ionization mode (positive or negative) must be suitable for the combinatorial compound so that molecular ion species are formed. Also, a suitable mobile phase and HPLC column must be selected. As an alternative to HPLC, supercritical fluid chromatography-mass spectrometry (SFC-MS) has been used for the high-throughput analysis of combinatorial libraries (18, 19). The advantages of SFC-MS relative to conventional LC-MS for the purification of combinatorial libraries of compounds are the lower viscosities and higher diffusivities of condensed CO_2 compared with HPLC mobile phases and the ease of solvent removal and disposal after analysis. However, SFC instrumentation remains more expensive and less widely available than conventional HPLC systems.

2.2 Confirmation of Structure and Purity of Combinatorial Compounds

The determination of molecular weights, elemental compositions, and structures of compounds used for high-throughput screening, whether discrete compounds or combinatorial library mixtures, is typically carried out using mass spectrometry, because traditional spectroscopic and gravimetric techniques are too slow to keep pace with combinatorial chemical synthesis. In addition, mass spectrometry may be used to assess the purity of compounds being used for high-throughput screening. The highest-throughput technique for confirming molecular weights and structures of drug candidates is flow injection analysis of sample solutions using electrospray, APCI, or APPI mass spectrometry. Typically, no sample preparation is necessary.

Although any organic mass spectrometer may be used to confirm the molecular weight of a compound, tandem mass spectrometers provide additional structural information through the use of CID to produce fragment ions. As discussed above (see also Table 13.1), tandem mass spectrometers include triple quadrupole instruments, QqTOF mass spectrometers, ion trap mass spectrometers, multiple sector magnetic sector instruments, FTICR instruments, and the new TOF-TOF mass spectrometers. In most applications, APCI or electrospray ionization is used.

In addition to molecular weight and fragmentation patterns, high precision and high resolution mass spectrometers such as QqTOF instruments, reflectron TOF mass spectrometers, double focusing magnetic sector mass spectrometers, and FTICR instruments are necessary for the measurement of exact masses of drugs and drug candidates for the determination of elemental compositions. The combination of high resolution and high precision is especially useful for determining the elemental compositions of compounds in combinatorial library mixtures without having to isolate each compound using chromatography or some other separation technique. Because FTICR instruments and the hybrid QqTOF mass spectrometers are capable of simultaneously measuring exact masses at high resolution of both molecular ions and fragment ions generated during MS-MS, these instruments are becoming extremely popular within drug discovery programs.

As an example of the exact mass measurement of a combinatorial library mixture, the FTICR negative ion electrospray mass spectra of a 36- and a 120-compound peptide library mixture are shown in Fig. 13.6. The resolution achieved in this experiment was 20,000–40,000. Although the exact masses of all components in a small combinatorial library can often be measured during a single infusion experiment, on-line HPLC separation or the analysis of discrete compounds is sometimes required to overcome ion suppression problems. However, LC-MS is a relatively slow process because of the slow chromatographic separation step. Because LC-MS is required in many instances for the analysis of mixtures and to eliminate interfering salts or buffers, two approaches have emerged to increase the throughput of this technique; parallel LC-MS and fast LC-MS. One approach to increasing

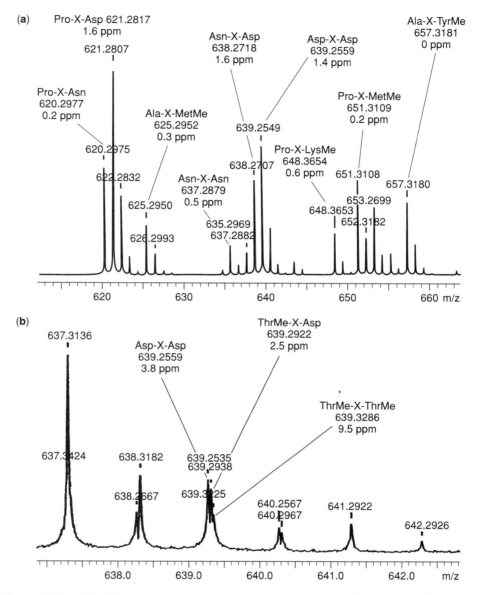

Figure 13.6. (a) Partial negative ion electrospray mass spectrum of a 36-component library mixture. Both the measured mass and the difference between the measured and theoretical values (in ppm) are shown. (b) Negative ion electrospray spectrum of the 120-component library showing the resolution of three nominally isobaric peaks. (Reproduced from Ref. 24 by permission of Bentham Science Publishers).

throughput of the rate-limiting chromatographic separation has been to simultaneously interface multiple HPLC columns to a single mass spectrometer. This approach is called parallel LC-MS. Commercial parallel electro-spray interfaces and HPLC systems are now available that can accommodate up to eight HPLC columns simultaneously (20–22). Although the multiple sprays are introduced to the ion source simultaneously, these streams may be sampled in a time-dependent manner to minimize cross contamination between channels.

Another solution to increasing the throughput of LC-MS has been to minimize the time required for HPLC separation through an approach called fast HPLC. HPLC separations are accelerated by using shorter columns and higher mobile phase flow rates. Because coelution of some species is likely to occur during fast chromatographic separations, the selectivity of the mass spectrometer is essential for the characterization and/or quantitative analysis of the target compound. However, samples of compounds prepared using combinatorial chemistry are usually simple mixtures of reagents, by-products, and products that require only partial chromatographic purification to prevent ion suppression effects during mass spectrometric analysis.

In addition to molecular weight determination using conventional MS or high exact mass measurement and structural confirmation using MS-MS, fast LC-MS is also used to assess the purity and yield of combinatorial products (15, 23). Before high-throughput screening, many researchers analyze combinatorial libraries for both purity and structural identity using mass spectrometry to assure the validity of structure-activity relationships that might be derived from the screening data. Fast LC-MS and LC-MS-MS may be carried out to satisfy this requirement using gradients (usually a step gradient with a reverse phase HPLC column) with a total cycle time of 1–3 min (24) or using an isocratic system requiring less than 1 min per analysis. A variety of HPLC columns are used for fast LC-MS that include narrow bore (2-mm) and analytical bore (4.6-mm) columns with length typically from 0.5–5 cm. The mobile phase flow rate for these fast LC-MS analyses is usually from 1.5–5 mL/min.

2.3 Encoding and Identification of Compounds in Combinatorial Libraries and Natural Product Extracts

The use of mass spectrometric identification in combinatorial chemistry is not limited to the analysis of synthetic products as a means of quality control, but also for the identification of active compounds or "hits" during high-throughput screening. Although the synthesis and screening of discrete compounds (25) enables them to be followed through the entire process by using partial encoding or bar-coding, it is sometimes advantageous to screen libraries prepared as mixtures (26) and use a technique such as mass spectrometry to rapidly identify the hit(s) in the mixture. One approach to the rapid deconvolution of combinatorial library mixtures is to prepare libraries containing compounds of unique molecular weight and then identify them using mass spectrometry. However, such libraries are necessarily small because the molecular weight of most drug-like molecules is between 150–400 Da. Because of the molecular weight degeneracy of larger combinatorial libraries, several encoding strategies have been devised to rapidly identify active compounds in these mixtures (27–29).

Because most combinatorial libraries contain compounds with degenerate molecular weights, various tagging strategies have been devised to uniquely identify library compounds bound to beads. Most of these tagging approaches are based on the synthesis of encoding molecules. For example, peptide (30) or oligonucleotide (31) labels have been synthesized on the beads in parallel to the target molecules and then sequenced for bead decoding. Alternatively, haloarene tags have been incorporated during synthesis and then identified with high sensitivity using electron-capture gas chromatography detection (32). In addition to the increased time and cost for the synthesis of a library containing tagging moieties, the tagging groups themselves might interfere with screening giving false positive or negative results.

For peptide libraries, one solution to this problem uses matrix-assisted laser desorption ionization (MALDI) mass spectrometry to directly desorb and identify peptides from beads that were screened and found to be hits (33). This technique is called the termination synthesis approach. Because the peptide library compounds are analyzed directly, products with amino acid deletions or substitutions, side-reaction products, or incomplete deprotection are readily observed. Also, because there are no extra molecules used for chemical tagging, this source of interference is avoided. However, this approach is specific to peptide libraries and is not necessarily applicable to other types of combinatorial libraries.

Another approach that eliminates possible interference from the chemical tags, "ratio encoding," has been developed for the mass spectrometric identification of bioactive leads using stable isotopes incorporated into the library compounds (29, 34). Within the ligand itself, the code might be a single-labeled atom that is conveniently inserted whenever a common reagent transfers at least one atom to the target compound or ligand. The code consists of an isotopic mixture having one of the many predetermined ratios of stable isotopes and can be incorporated in the linker or added through a reagent used during the synthesis. The mass spectrum of the compound shows a molecular ion with a unique isotope ratio that codes for a particular library compound. For example, Wagner et al. (29) used isotope ratio encoding during the synthesis of a 1000-compound peptoid library and was able to identify uniquely all the components based on their isotopic patterns and molecular weights. Because isotope ratio codes are contained within each combinatorial compound, a chemical tag is not required. The speed of MS-based decoding outperforms most other decoding technologies, which are time consuming and decode a restricted set of active compounds.

Although combinatorial synthesis provides rapid access to large numbers of compounds for screening during drug discovery and lead optimization, these libraries are usually based on a small number of common structures or scaffolds. There is a constant need for increasing the molecular diversity of combinatorial libraries and finding new scaffolds, and natural products have always been a rich source of chemical diversity for drug discovery. The traditional approach to screening natural products for drug leads uses bioassays to test organic solvent extracts for activity. If strong activity is detected, then activity-guided fractionation of the crude extract is used to isolate the active compound(s), which is identified using mass spectrometry (including tandem mass spectrometry and exact mass measurements), IR, UV/VIS spectrometry, and NMR. Recently, a variety of mass spectrometry–

based affinity screening methods have been developed to streamline the tedious process of activity-guided fractionation. These approaches are discussed in Section 2.4.

Whether lead compounds in natural product extracts are isolated using bioassay-guided fractionation or mass spectrometry–based screening, there is a high probability that the structure of the active compound(s) has already been reported in the natural product literature. In such cases, the tedious process of complete structure elucidation using a battery of spectrometric tools should be unnecessary. Instead, mass spectrometry alone may be used to quickly "dereplicate" or identify the known compounds based on molecular weight, fragmentation patterns, and elemental composition in combination with natural product database searching (35–39). Commercially available natural products databases include NAPRALERT (40), Scientific & Technical Information Network (STN) (41), and the Dictionary of Natural Products (42). Because some of these databases also contain UV/VIS absorbance data, it is also advantageous to use a photodiode array detector between the HPLC and mass spectrometer to obtain additional spectrometric data during LC-UV-MS dereplication (36, 37).

2.4 Mass Spectrometry–Based Screening

The earliest approaches to combinatorial synthesis used portioning and mixing (26) and enabled the synthesis of combinatorial libraries containing hundreds of thousands to millions of compounds. Today, this approach remains the most efficient method for preparing enormous libraries of compounds. However, until the mid-1990s, efficient screening techniques did not exist to rapidly identify the "hits" within large combinatorial mixtures. Therefore, chemists were motivated to develop ways to prepare large numbers of discreet compounds using massively parallel synthesis, which could be assayed quickly for pharmacological activity using high throughput screening one compound at a time. Recently, several mass spectrometry–based screening assays have been developed that are suitable for screening combinatorial library mixtures, and some are even useful for screening natural product extracts which have always been a

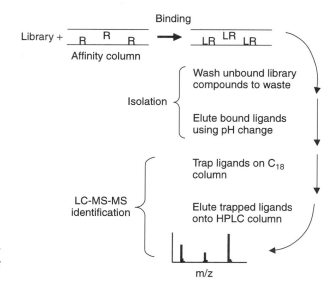

Figure 13.7. Affinity chromatography combined with LC-MS-MS for screening combinatorial library mixtures.

source of molecular diversity for drug discovery. All of the mass spectrometry–based screening methods use receptor binding of ligands as the basis for identification of lead compounds.

2.4.1 Affinity Chromatography–Mass Spectrometry. Since the introduction of affinity chromatography more than 30 years ago, this technique has become a standard biochemical tool for the isolation and identification of new binding partners to specific target molecules. Therefore, the coupling of affinity chromatography to mass spectrometry is a logical extension of this technique, and the application of affinity LC-MS to the screening of combinatorial libraries has been demonstrated by several groups (43, 44). During affinity LC-MS screening, a receptor molecule such as a binding protein or enzyme is immobilized on a solid support within a chromatography column. The library mixture is pumped through the affinity column in a suitable binding buffer so that any ligands in the mixture with affinity for the receptor would be able to bind. Then, unbound material is washed away. Finally, the specifically bound ligands are eluted using a destabilizing mobile phase and identified using mass spectrometry. This affinity-column LC-MS assay is summarized in Fig. 13.7.

In some applications (43), ligands are eluted from the affinity column and then trapped on a second column such as a reverse phase HPLC column. LC-MS or LC-MS-MS identification of the ligands (hits) is then carried out using the trapping column. In other systems, ligands are identified directly from the affinity column using mass spectrometry (44). For example, Kelly et al. (44) prepared an affinity column containing immobilized phosphatidylinositol-3-kinase and used it for direct LC-MS screening of a 361-component peptide library. Electrospray mass spectrometry and tandem mass spectrometry were used to identify the ligands released from the affinity column using pH gradient elution.

Advantages of affinity chromatography–mass spectrometry for screening during drug discovery include versatility and re-use of the column. Both combinatorial libraries and natural product extracts can be screened using this approach, and a wide range of binding buffers may be used. Mass spectrometry–compatible mobile phases are only required during the final LC-MS detection step. Furthermore, a single column may be used multiple times to screen different samples for ligands unless the destabilization solution irreversibly denatures, releases, or inhibits the receptor.

Despite these advantages, affinity chromatography has numerous drawbacks that have

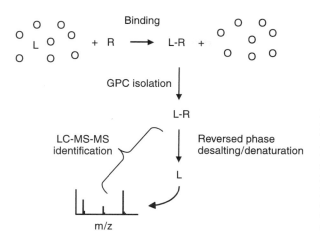

Figure 13.8. GPC followed by LC-MS-MS for screening mixtures of combinatorial libraries. After incubation of a receptor with a library of compounds, the ligand-receptor complexes (L-R) are separated from the low molecular weight unbound library compounds using GPC. Next, the L-R complexes are denatured during reversed phase HPLC to release the ligands for MS-MS identification.

prompted the development of alternative mass spectrometer screening tools. For example, immobilization of the receptor might change its affinity characteristics causing false negative or false positive hits. This is particularly problematic for receptors that are solution-phase in their native state. Also, developing and then implementing an immobilization scheme is often a slow, tedious, and even expensive process, and this process is unique for each new receptor. Finally, false positive hits are often obtained when screening large molecularly diverse libraries, because there are usually compounds in such mixtures that have affinity for the stationary phase or linker molecule instead of the receptor.

2.4.2 Gel Permeation Chromatography–Mass Spectrometry. Another type of chromatography that has been combined with mass spectrometry as a screening system for drug discovery is gel permeation chromatography (GPC) (45, 46). Also called size-exclusion chromatography, GPC separates molecules according to size as they pass through a stationary phase containing particles with a defined pore size. During GPC-based screening, a library mixture is pre-incubated with a macromolecular receptor to allow any ligands in the library to bind, and then GPC is used to separate the large receptor–ligand complexes from the unbound low molecular weight compounds in the mixture. Finally, ligands are released from the receptor during reversed

phase HPLC and identified either on-line or off-line using tandem mass spectrometry. This screening method is illustrated in Fig. 13.8.

During the pre-incubation and GPC steps, any binding buffer may be used, because the binding buffer will be removed during reverse phase LC-MS analysis. However, the GPC separation step must be carried out quickly, because ligands begin to dissociate from the receptor immediately and can become lost into the size exclusion gel. Despite this disadvantage, this approach allows both receptor and ligand to be screened in solution, which avoids some of the problems associated with the use of affinity columns for screening. The GPC LC-MS-MS screening method should also be suitable for screening natural product extracts as well as combinatorial library mixtures.

2.4.3 Affinity Capillary Electrophoresis–Mass Spectrometry. Affinity capillary electrophoresis was originally used for the determination of the binding constants of small molecules to proteins (47–49). This solution-based technique is rapid and requires only small amounts of ligands. Affinity constants are measured based on the mobility change of the ligand on interaction with the receptor present in the electrophoretic buffer (50). By combining affinity capillary electrophoresis with on-line mass spectrometric detection and

(a)

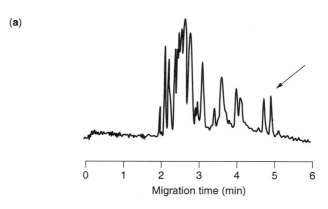

Migration time (min)

Figure 13.9. Affinity capillary electrophoresis–UV-mass spectrometry of a 100-tetrapeptide library screened for binding to vancomycin (104 μM in the electrophoresis buffer). (a) The elution of peptides was monitored with UV absorbance during capillary electrophoresis, and the elution time increased with increasing affinity for vancomycin. (b) Positive ion electrospray mass spectrum with CID of the Tris adduct of the protonated peptide detected at ~5 min in the electropherogram shown in a. (Reproduced from Ref. 52 by permission of the American Chemical Society.)

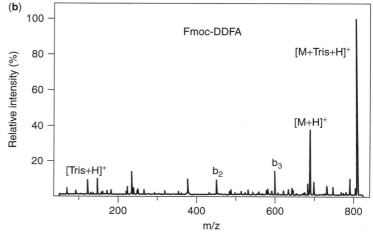

identification, affinity constants for multiple compounds can be measured in a single analysis (51). Recognizing that on-line mass spectrometric detection was helpful for the identification of each ligand, Chu et al. (52) extended this approach to include the screening of combinatorial libraries as a means of drug discovery. The data in Fig. 13.9 show the results of screening a 100-tetrapeptide library for affinity to vancomycin using affinity capillary electrophoresis–mass spectrometry. Without vancomycin in the electrophoresis buffer, all the peptides eluted within 3 min. When vancomycin was present, the peptides eluted in order of affinity, with the highest affinity compounds being detected between 4.5 and 5 min. Positive ion electrospray tandem mass spectrometry was used to identify the highest affinity ligands (see Fig. 13.9b).

Note that some peptide ligands such as Fmoc-DDFA were detected as adducts with Tris, which was used in the electrophoresis buffer. Although the identification of this peptide was not prevented by the formation of this adduct, some buffers used during electrophoresis might interfere with mass spectrometric ionization and detection. Also, the types of libraries that have been screened using this approach have contained modest numbers of synthetic analogs such as peptides. Libraries exceeding 400 members required preliminary purification using affinity chromatography to reduce the number of compounds (52). As a result, this approach is probably not ideal for screening libraries containing molecularly diverse compounds or for screening natural product extracts. However, affinity capillary electrophoresis–mass spectrometry is fast; each analysis requires less than 10 min. Also, it may be used to measure affinity constants for ligand–receptor interactions.

2.4.4 Frontal Affinity Chromatography–Mass Spectrometry.

Like affinity chromatography–mass spectrometric screening (see Section 2.4.1), frontal affinity chromatography uses an affinity column containing immobilized receptor molecules (53). The difference between the two screening methods is that the ligands are continuously infused into the column during frontal affinity chromatography and detected using mass spectrometry. Compounds with no affinity for the immobilized receptor elute immediately in the void volume, but the elution of the ligands is delayed. As compounds compete for binding sites on the affinity column, these sites become saturated until ligands begin to elute from the column at their infusion concentration. In this manner, frontal affinity chromatography may be used to measure affinity constants for ligands, and by using a mass spectrometer for on-line identification of ligands, this technique becomes a screening method (54, 55).

During frontal affinity chromatography–mass spectrometry, signals for all compounds eluting from the affinity column are recorded by the mass spectrometer, and the last compounds to elute at their infusion concentrations represent the highest affinity compounds or "hits." An example of the screening of six oligosaccharides with different binding affinities for an immobilized monoclonal carbohydrate-binding antibody is shown in Fig. 13.10. Compounds 1–3 eluted immediately (no affinity), whereas compounds 4–6 eluted in order of increasing affinity for the antibody. Dissociation constants were determined to be 185, 12.6, and 1.8 μM for compounds 4–6, respectively (54).

Because frontal affinity chromatography uses a conventional affinity column, this technique provides additional applications of this type of column to investigators already using affinity–mass spectrometry (See Section 2.4.1). However, the same limitations and disadvantages of using immobilized receptors still apply, such as non-specific binding to the stationary phase, the development time and cost of preparing the affinity columns, and the possibility that immobilizing the receptor might alter its binding characteristics and specificity. In addition, mass spectrometric detection creates some additional limitations.

Because all library compounds must be monitored simultaneously, the compounds must be selected so that they have unique molecular weights. Also, one compound in the mixture should not suppress the ionization of another. Therefore, this approach is probably restricted to the screening of small combinatorial libraries that are similar in chemical structure and ionization efficiencies. Finally, the binding buffer used for affinity chromatography must be compatible with on-line APCI or electrospray mass spectrometry. This means that the mobile phase must be volatile and usually of low ionic strength (i.e., typically <40 mM for electrospray ionization).

2.4.5 Bioaffinity Screening using Electrospray FTICR Mass Spectrometry.

Although FTICR mass spectrometry may be used to determine the exact masses of combinatorial library compounds and to confirm their structures using CID and high resolution tandem mass spectrometry (see definitions of CID and MS-MS in Section 1), electrospray FTICR mass spectrometry may be used for the direct screening of combinatorial libraries without the need for any pre-purification or chromatography. In this application, a combinatorial library is pre-incubated with a receptor in solution and then analyzed directly using electrospray to identify receptor–ligand complexes in the gas phase (56–60). Once a receptor–ligand complex is ionized and trapped in the FTICR mass spectrometer, the mass difference between the complex and the receptor alone might be measured with sufficient resolution and accuracy to determine the mass(es) and perhaps elemental composition(s) of the ligand(s). If the ligand carries a charge, then CID may be used to dissociate the ligand for subsequent analysis using tandem mass spectrometry. This elegant and simple screening approach is summarized in Fig. 13.11.

An extension of this FTICR mass spectrometry–based screening technique has been to screen a combinatorial library for ligands to two receptors simultaneously (59, 60). In this example, the two receptors consisting of RNA constructs representing the prokaryotic (16s) rRNA and eukaryotic (18s) rRNA A-site were incubated simultaneously with an aminogly-

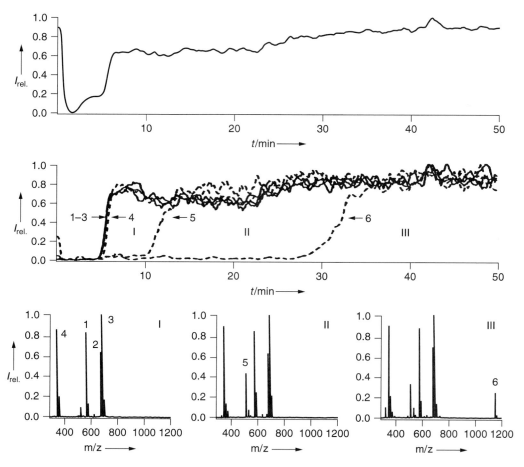

Figure 13.10. Frontal affinity chromatography–mass spectrometry screening of a 6-oligosaccharide mixture for affinity to an immobilized carbohydrate-binding monoclonal antibody. Top: positive ion electrospray total ion chromatogram. Middle: computer-reconstructed mass chromatograms for the molecular ion species of all six compounds. Compounds 1–3 (solid line) eluted in the void volume, indicating no binding to the antibody. Break through signals for compounds 4, 5, and 6 appear at successively later times, indicating increasing affinity for the immobilized antibody. Bottom: positive ion electrospray mass spectra recorded at times I, II, and III as indicated in the middle trace. The protonated molecules of compounds 1–6 are labeled.

coside library to identify potential ligands. By screening a target mixture against the same library, screening efficiency is enhanced and the number of analyses required is reduced.

The advantage of this screening method over other approaches is the elimination of purification steps before mass spectrometric identification. Also, the disadvantages associated with chromatographic separations are eliminated. However, the use FTICR mass spectrometric screening restricts the binding buffer and receptors that may be used. Only low ionic strength and volatile buffers are compatible with this approach (such as 10 mM ammonium acetate). Also, the receptor and ligand must be highly purified to avoid impurities that might interfere with ionization and detection. Therefore, this technique is probably more suitable for the screening of combinatorial libraries than complex natural product mixtures. Finally, the receptor–ligand complex must ionize efficiently during electro-

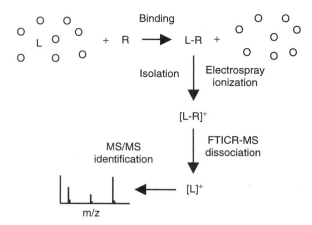

Figure 13.11. Bioaffinity electrospray FTICR mass spectrometry. The isolation and mass spectrometric identification of receptor-specific ligands are carried out entirely in the mass spectrometer without chromatography or other separation steps.

spray under solvent and ion source conditions that do not cause dissociation of the complex.

2.4.6 Pulsed Ultrafiltration–Mass Spectrometry. A versatile approach to screening solution phase combinatorial libraries and natural product extracts is pulsed ultrafiltration–mass spectrometry (61, 62), which uses a standard LC-MS system with an ultrafiltration chamber substituted for the HPLC column.

The principle of pulsed ultrafiltration screening of combinatorial libraries is shown in Fig. 13.12. During pulsed ultrafiltration, ligand–receptor complexes remain in solution in the ultrafiltration chamber while unbound library compounds and buffer are washed away. After unbound compounds are removed, the hits from the library are eluted from the chamber by destabilizing the ligand–receptor complex using an organic solvent, a pH change, or a

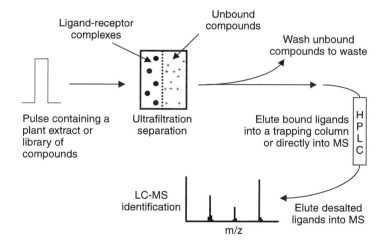

Figure 13.12. Combinatorial library screening using pulsed ultrafiltration mass spectrometry. During the loading step (left), ligands are bound to the receptor either on-line (top) using a flow-through approach or off-line (bottom two incubations). Unbound compounds and binding buffer, cofactors, etc. are washed out of the ultrafiltration chamber to waste during a separation step (middle). Bound ligands are dissociated from the receptor molecules and eluted from the chamber by introducing a destabilizing solution such as methanol, pH change, etc. Finally, released ligands are identified using mass spectrometry, tandem mass spectrometry, or LC-MS (right). (Reproduced from Ref. 64 by permission of John Wiley & Sons.)

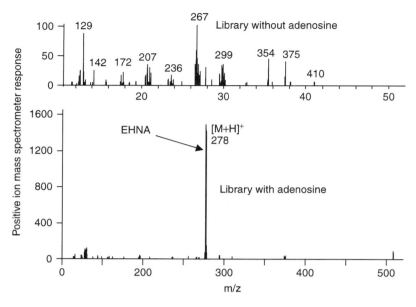

Figure 13.13. Identification of EHNA as the highest affinity ligand for adenosine deaminase in a combinatorial library of 20 adenosine analogs using ultrafiltration electrospray mass spectrometry. (Reproduced from Ref. 61 by permission of the American Chemical Society.)

combination of both. The released ligands are identified on-line using APCI or electrospray mass spectrometry (61) or collected and analyzed off-line using mass spectrometry, LC-MS, or LC-MS-MS (63).

An example of pulsed ultrafiltration mass spectrometry for the screening of a library of 20 adenosine analogs for ligands to adenosine deaminase is shown in Fig. 13.13. After a 15-min preincubation of the library compounds (17.5 μM each except for EHNA, which was present at 1.75 μM) with 2.1 μM adenosine deaminase in 50 mM phosphate buffer, an aliquot containing 420 pmol of the receptor was injected into the ultrafiltration and washed for 8 min at 50 μL/min with water to remove the phosphate buffer and unbound or weakly binding library compounds. Methanol was introduced into the mobile phase to dissociate the enzyme–ligand complex and release bound ligands for identification by electrospray mass spectrometry. During methanol elution, only EHNA [erythro-9-(2-hydroxy-3-nonyl) adenine] was detected as the [M+H]$^+$ ion of m/z 278 (Fig. 13.13). In control experiments using the library without enzyme, no library compounds were detected during methanol elu-

tion (Fig. 13.13, Control). Despite being present at a 10-fold lower concentration than the natural substrate adenosine analogs, EHNA was easily identified because it had the highest affinity among the library compounds ($K_d = 1.9$ nM). This demonstrates the use of ultrafiltration electrospray mass spectrometry for identifying a high affinity ligand among a set of analogs that bind to a specific receptor. In a follow-up lead optimization study using pulsed ultrafiltration mass spectrometry, a synthetic combinatorial library of EHNA analogs was screened for binding to adenosine deaminase, and structure-activity relationships for EHNA binding were identified (65).

As an illustration of the versatility of pulsed ultrafiltration–mass spectrometry, binding assays for a variety of receptors have been reported including dihydrofolate reductase (63), cyclooxygenase-2 (62), serum albumin (66, 67) and estrogen receptors (68). Not only is pulsed ultrafiltration useful for identifying ligands to different receptors, but a wide range of combinatorial libraries and natural product extracts in any suitable binding buffer may be screened. In addition to combinatorial libraries, complex natural product extracts have been screened (68), and neither plant nor fermentation broth matrices were found to interfere with screening (62). As another example of the flexibility of this screening system, a centrifuge tube equipped with an ultrafiltration membrane (69) has been used instead of an on-line ultrafiltration chamber. Other applications of pulsed ultrafiltration–mass spectrometry include screening drugs and drug candidates for metabolic stability (70), metabolic activation to reactive metabolites (71), and the measurement of affinity constants for ligand–receptor interactions (66, 67).

Metabolism and toxicity screening applications of pulsed ultrafiltration use hepatic microsomes in the ultrafiltration chamber. For metabolic screening drugs and the cofactor nicotinamide dinucleotide phosphate (NADPH) are flow-injected through the ultrafiltration chamber (oxygen is dissolved in the mobile phase), and the metabolites formed by microsomal cytochrome P450 and any unreacted compounds flow out of the chamber for mass spectrometric identification and/or quantitative analysis (70). On-line applications require the use of volatile buffers, but LC-MS and LC-MS-MS may be used off-line to analyze the ultrafiltrate no matter what buffer had been used. Screening drugs for metabolic activation using pulsed ultrafiltration–mass spectrometry is carried out in a similar manner, except that glutathione is coinjected along with NADPH and the drug substrate (71). MS-MS may be used on-line or LC-MS-MS may be used off-line to screen for glutathione adducts as an indication that the drug was metabolized to a reactive intermediate(s) that was trapped by reaction with glutathione. Finally, pulsed ultrafiltration may be used with UV or mass spectrometric detection to measure affinity constants of individual compounds (66).

To measure affinity constants and other physico-chemical properties of binding such as the number of binding sites, two pulsed ultrafiltration measurements are carried out. First, an aliquot or pulse of a liquid is injected through the chamber, and the elution profile is recorded. Then, the chamber is loaded with a receptor, and the ligand is reinjected. If binding occurs, the elution profile will be delayed in proportion to the affinity constant. The con-

trol injection is used to control for non-specific binding to the apparatus. Because the concentration of receptor and total amount of liquid are known, and because the concentration of free ligand is measured as it elutes from the chamber over a wide range of concentrations, the affinity constant and other binding parameters may be calculated.

In most of the applications of pulsed ultrafiltration to date, serial analyses were carried out with a throughput of approximately one or two assays per hour. Because the purpose of these assays was to screen complex mixtures or to obtain metabolism data for new drug entities, the throughput of these analyses was acceptable, but was not high throughput. The rate limiting step in these analyses was the ultrafiltration separation and not the mass spectrometric detection. Two solutions have been reported to increase the throughput of pulsed ultrafiltration mass spectrometry. In the first solution, van Breemen et al. (70) used a multiplex ultrafiltration system in which up to 60 ultrafiltration chambers could be arranged in parallel and interfaced to a single mass spectrometer. This scheme is shown in Fig. 13.14. In this system, a continuous flow of the buffer or mobile phase is maintained through the ultrafiltration chambers, but the mass spectrometer samples each ultrafiltrate solution at 1-min intervals. The sampling time would be selected to correspond to the time at which a maximum concentration of metabolites would be expected to elute from the chamber. This approach was demonstrated to increase the throughput of metabolic screening using ultrafiltration mass spectrometry by 60-fold. Although used originally for metabolic screening, this approach would be applicable to toxicity screening and drug discovery screening as well.

The second solution to increasing the throughput of pulsed ultrafiltration mass spectrometry has been to miniaturize the ultrafiltration chamber volume while maintaining the flow rate and chamber pressure. Because the ultrafiltration membrane cannot withstand high pressure without rupturing, the ultrafiltration process cannot be accelerated simply by increasing the flow rate through the chamber. The approach of Beverly et al. (72) was to fabricate a 35-μL ultra-

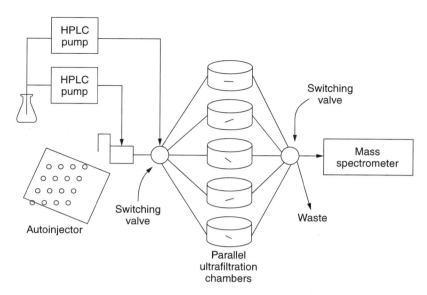

Figure 13.14. High-throughput pulsed ultrafiltration mass spectrometry system for screening drug candidates for metabolic transformation. Multiple ultrafiltration chambers are connected in parallel to a single mass spectrometer detector. After loading each chamber with liver microsomes, a different drug is injected into each chamber at intervals of 1 min (for 60 screens/h using 60 chambers). Constant flow of incubation buffer is maintained through all chambers, but only one chamber at a time is connected on-line to the mass spectrometer. Drug metabolite profiles are recorded using mass spectrometry for up to 1 min per chamber. (Reproduced from Ref. 70 by permission of the American Society for Pharmacology and Experimental Therapeutics.)

filtration chamber that was approximately threefold lower in volume than the smallest reported by van Breemen et al. (61). As a result, ultrafiltration mass spectrometric analyses could be carried out at the rate of at least three per hour, which corresponded to a threefold enhancement of throughput. This study suggests that chip-based ultrafiltration mass spectrometry would have the potential to result in a truly high-throughput system.

The advantages of pulsed ultrafiltration–mass spectrometry include the variety of different applications that may be carried out, the convenience of on-line screening, solution-phase screening, the ability to screen either combinatorial libraries or natural product extracts, the diversity of receptors that may be screened, and the freedom to use either volatile or non-volatile binding buffers. For metabolic and toxicity screening, flow injection analyses have the additional advantages that product feedback inhibition is prevented so that the metabolic profile more closely approximates the *in vivo* system (70). Finally, the

disadvantages of pulsed ultrafiltration screening for drug discovery include the washing step, during which dissociation and loss of weakly bound ligands might occur, and the slow speed of each experiment, which can take up to 1 h.

2.4.7 Solid Phase Mass Spectrometric Screening. Because drugs are usually in a soluble form to be transported to the active sites in cells and tissues, it is logical that most mass spectrometry–based screening methods use solution-phase analysis of these compounds, and it is no surprise that most successful mass spectrometry screening assays use electrospray ionization or APCI. However, solid phase ionization techniques such as matrix-assisted laser desorption ionization (MALDI) might be effective, provided that ligand–receptor interactions are allowed to take place in an environment similar to *in vivo* conditions and that a suitable separation step is carried out before the preparation of the MALDI sample.

To use MALDI mass spectrometry for screening, several research groups have developed immobilized receptors on MALDI targets or on solid supports that can be placed on a MALDI target for use in the affinity purification of potential drugs from test solutions. Following procedures originally developed for affinity chromatography, the preparation of affinity surfaces for MALDI mass spectrometry has been achieved quite easily. However, the use of these affinity MALDI chips for screening mixtures of small molecules during drug discovery has been unproductive. One of the problems has been the high background noise at low m/z values caused by the matrix used for MALDI. This problem may be mitigated by eliminating the matrix or using alternative sample stages such as porous silicon chips (73, 74). However, noise persists because of the affinity support and immobilized receptor molecules. Another problem that has yet to be overcome is the elimination of the high background noise caused by non-specific binding of test compounds to the affinity target. Although this problem is similar to the false positive results and non-specific binding that occurs during affinity chromatography–mass spectrometry (see Section 2.4.1), the signals for non-specific binding are magnified by the fact that the actual affinity surface is being irradiated and sampled by the MALDI laser beam. As a result, affinity-based screening coupled with MALDI mass spectrometry has not been a successful drug discovery approach.

However, progress is being made in the use of affinity probes for the capture of proteins and other macromolecules from biological solutions followed by MALDI mass spectrometric detection and identification (75–77). One affinity MALDI mass spectrometry method has been paired with the affinity probes using in surface plasmon resonance systems (78). These affinity-based MALDI mass spectrometry screening assays are promising approaches for testing blood or other biological fluids for the presence of specific proteins or other macromolecules. As a result, these have the potential to become clinical diagnostic tools or might even lead to the identification of new therapeutic targets. However, they are unlikely to become useful for screening combinatorial libraries or natural product extracts for the purpose drug discovery.

3   THINGS TO COME

Mass spectrometry has become an essential analytical tool at every stage of the drug discovery and development process. In this chapter, the various applications of mass spectrometry to combinatorial chemistry and drug discovery have been highlighted. Although the speed of mass spectrometry matches the demands of combinatorial chemistry, the slow and serial nature of chromatography in the various LC-MS applications remains a bottleneck that limits their throughput. Because mass spectrometry is highly selective, only partial chromatographic separations are needed for most measurements. In fact, the primary function of the chromatography step is usually to separate species that might otherwise interfere with the ionization process. Recognizing this limited function of chromatography during LC-MS–based screening assays, manufacturers of chromatography columns are addressing this need by developing high-throughput columns for fast chromatography for LC-MS. Improvements in this direction should continue to reduce the time required for LC-MS from a few minutes to a few seconds. Meanwhile chip-based technology is beginning to emerge for miniaturized capillary electrophoresis–mass spectrometry (CE-MS) (79). These chips are being developed to enable ultrafast and highly sensitive electrospray mass spectrometric analysis. Because of their microscopic size, CE-MS chips have the potential to hold large arrays of samples that would facilitate high-throughput analysis.

In terms of mass spectrometry instrumentation, the currently available instruments such as time-of-flight (TOF) analyzers and hybrid quadrupole-TOF analyzers are able to acquire complete mass spectra at rates compatible with fast CE separations. As CE or ultrafast chromatography replaces conventional, slow HPLC applications, TOF-based mass spectrometers will be needed to replace the less efficient scanning types of instruments such as quadrupoles and ion traps for most high-throughput applications. FTICR mass spectrometry remains unsurpassed in terms of resolution and mass accuracy for both MS and MS-MS applications. However, the throughput of FTICR mass spectrometric analysis needs to be increased to remain useful for combinatorial chemistry applications. Advances in increasing the throughput of FTICR mass spectrometry are anticipated.

Hyphenated technologies such as LC-NMR-MS are being developed to support structure elucidation of combinatorial libraries (80). Although such technologies are still in a developmental stage, they have great potential for analyses of combinatorial libraries and for natural product drug discovery (81–83). The main impediments of applying LC-NMR-MS to combinatorial chemistry remain poor sensitivity of the NMR, the obligatory use of deuterated solvents for chromatography, and the low throughput of NMR analyses. However, efforts are in progress to improve the throughput of NMR analyses (84–86).

In conclusion, mass spectrometry provides rapid, reliable, sensitive, and selective analysis of combinatorial libraries for structure confirmation, purity analysis, and library deconvolution. In addition, mass spectrometric screening methods have been developed and are beginning to be applied to drug discovery. In the case of natural products, mass spectrometry facilitates the screening of natural product extracts and facilitates the dereplication and characterization of lead compounds. At different times during the last 100 years, first physicists and physical chemists and then organic chemists pronounced that mass spectrometry had run out of new applications and had no future. Fortunately, they were wrong. Today, medicinal chemists recognize that the potential of mass spectrometry to contribute to all facets of drug discovery has only just begun to be explored. Furthermore, applications of mass spectrometry to drug development are even less developed and are waiting to be developed. Mass spectrometry has become a fundamental analytical tool for drug discovery, and this role should continue to grow in the future.

4 WEB SITE ADDRESSES AND RECOMMENDED READING FOR FURTHER INFORMATION

- http://www.asms.org   Homepage of the American Society for Mass Spectrometry. This web site contains additional information about mass spectrometry and links to a variety of reference materials regarding biomedical mass spectrometry.
- http://www.bentham.org/cchts/ *Combinatorial Chemistry & High Throughput Screening*
- http://pubs.acs.org/journals/jcchff/ Journal of Combinatorial Chemistry
- http://www.5z.com/moldiv/ Molecular Diversity
- http://www.liebertpub.com/BSC/default1.asp Journal of Biomolecular Screening
- R. B, Cole, Ed, *Electrospray Ionization Mass Spectrometry*, John Wiley and Sons, New York, 1997.
- F. W. McClafferty, and F. Turecek, *Interpretation of Mass Spectra*, 4th ed, University Science Books, Mill Valley, CA, 1993.
- W. M. Niessen, *J. Chromatogr. A*, **856**, 179–189 (1999).
- J. T. Watson, *Introduction to Mass Spectrometry*, 3rd ed, Lippincott-Raven, Philadelphia, PA, 1997.

5 ACKNOWLEDGMENTS

I thank Young Geun Shin, Benjamin Johnson, and Jennifer Mosel for help in writing and preparing this chapter.

REFERENCES

1. F. Field, *J. Am. Soc. Mass Spectrom*, **1**, 277–283 (1990).
2. M. Barber, R. S. Bordoli, G. J. Elliott, R. D. Sedgwick, and A. N. Tyler, *Anal. Chem.*, **54**, 645A–657A (1982).
3. F. Hillenkamp, M. Karas, R. C. Beavis, and B. T. Chait, *Anal. Chem.*, **63**, 1193A–1203A (1991).
4. Y. Ito, T. Takeuchi, D. Ishii, and M. Goto, *J. Chromatogr*, **346**, 161–166 (1985).
5. L. Chen, M. Stacewicz-Sapuntzakis, C. Duncan, R. Sharifi, L. Ghosh, R. van Breemen, D. Ashton, and P. E. Bowen, *J. Natl. Cancer Inst.*, **93**, 1872–1879 (2001).
6. J. Liu, J. E. Burdette, H. Xu, C. Gu, R. B. van Breemen, K. P. L. Bhat, N. Booth, A. I. Constantinou, J. M. Pezzuto, H. H. S. Fong, N. R. Farnsworth, and J. L. Bolton, *J. Agric. Food Chem.*, **49**, 2472–2479 (2001).
7. J. A. Syage and M. D. Evans, *Spectroscopy*, **16**, 14–21 (2001).

8. E. M. Gordon, M. A. Gallop, and D. V. Patel, *Acc. Chem. Res.*, **29**, 144–154 (1996).

9. L. A. Thompson and J. A. Ellman, *Chem. Rev.*, **29**, 132–143 (1996).

10. J. A. Loo, *Eur. Mass Spectrom*, **3**, 93–104 (1997).

11. J. N. Kyranos and J. C. Hogan, *Anal. Chem.*, **70**, 389A–395A (1998).

12. Y. Dunayevskiy, P. Vouros, T. Carell, E. A. Wintner, and J. Rebek. *Anal Chem.*, **67**, 2906–2915 (1995).

13. Y. Dunayevskiy, Y. V. Lyubarskaya, Y. H. Chu, P. Vouros, and B. L. Karger, *J. Med. Chem.*, **41**, 1201–1204 (1998).

14. P. A. Demirev and R. A. Zubarev, *Anal. Chem.*, **69**, 2893–2900 (1997).

15. L. Zeng, L. Burton, K. Yung, B. Shushan, and D. B. Kassel, *J. Chromatogr. A.*, **794**, 3–13 (1998).

16. L. Zeng and D. B. Kassel, *Anal. Chem.*, **70**, 4380–4388 (1998).

17. J. P. Kiplinger, R. O. Cole, S. Robinson, E. J. Roskamp, R. S. Ware, H. J. O'Connell, A. Brailsford, and J. Batt, *Rapid Commun. Mass Spectrom*, **12**, 658–664 (1998).

18. M. C. Ventura, W. P. Farrell, C. M. Aurigemma, and M. J. Greig, *Anal. Chem.*, **71**, 2410–2416 (1999).

19. M. C. Ventura, W. P. Farrell, C. M. Aurigemma, and M. J. Greig, *Anal. Chem.*, **71**, 4223–4231 (1999).

20. V. de Biasi, N. Haskins, A. Organ, R. Bateman, K. Giles, and S. Jarvis, *Rapid Commun. Mass Spectrom.*, **13**, 1165–1168 (1999).

21. T. Wang, L. Zeng, J. Cohen, and D. B. Kassel, *Comb. Chem. High Throughput Screen.*, **2**, 327–334 (1999).

22. L. Yang, N. Wu, R. P. Clement, and P. J. Rudewicz, *Proc. 48th ASMS Conf. Mass Spectrom. Allied Topics*, 861–862 (2000).

23. H. N. Weller, M. G. Young, S. J. Michalczyk, G. H. Reitnauer, R. S. Cooley, P. C. Rahn, D. J. Loyd, D. Fiore, and S. J. Fischman, *Mol. Diversity*, **3**, 61–70 (1997).

24. A. S. Fang, V. Vouros, and C. C. Stacey, et al., *Comb. Chem. High Throughput Screen.*, **1**, 23–33 (1998).

25. D. G. Powers and D. L. Coffen, *Drug Discovery Today*, **4**, 377–383 (1999).

26. A. Furka and W. D. Bennett, *Comb. Chem. High Throughput Screen.*, **2**, 105–122 (1999).

27. K. D. Janda, *Proc. Natl. Acad. Sci. USA*, **91**, 10779–10785 (1994).

28. A. W. Czarnik, *Proc. Natl. Acad. Sci. USA*, **94**, 12738–12739 (1997).

29. D. S. Wagner, C. J. Markworth, C. D. Wagner, F. J. Schoenen, C. E. Rewerts, B. K. Kay, and H. M. Geysen, *Comb. Chem. High Throughput Screen.*, **1**, 143–153 (1998).

30. J. M. Kerr, S. C. Banville, and R. N. Zuckermann, *J. Am. Chem. Soc.*, **115**, 2529–2531 (1993).

31. S. Brenner and R. A. Lerner, *Proc. Natl. Acad. Sci. USA*, **89**, 5381–5383 (1992).

32. M. H. J. Ohlmeyer, R. N. Swanson, L. W. Dillard, J. C. Reader, G. Asouline, R. Kobayashi, M. Wigler, and W. C. Still, *Proc. Natl. Acad. Sci. USA*, **90**, 10922–10926 (1993).

33. R. S. Youngquist, G. R. Fuentes, M. P. Lacey, and T. Keough, *Rapid Commun. Mass Spectrom.*, **8**, 77–81 (1994).

34. G. Karet, *Drug Discovery Dev.*, **1**, 32–38 (1999).

35. O. Potterat, K. Wagner, and H. Haag, *J. Chromatogr. A*, **872**, 85–90 (2000).

36. G. A. Cordell and Y. G. Shin, *Pure Appl. Chem.*, **71**, 1089–1094 (1999).

37. Y. G. Shin, G. A. Cordell, Y. Dong, J. M. Pezzuto, AVNA Rao, M. Ramesh, B. R. Kumar, and M. Radhakishan, *Phytochem. Anal.*, **10**, 208–212 (1999).

38. C. L. Zani, T. M. A. Alves, R. Queiroz, M. A. L. Chaves, E. S. Fontes, Y. G. Shin, and G. A. Cordell, *Phytochemistry*, **53**, 877–880 (2000).

39. H. L. Constant and C. W. W. Beecher, *Nat. Prod. Lett.*, **6**, 193–196 (1995).

40. D. G. Corley and R. C. Durley, *J. Nat. Prod.*, **57**, 1484–1490 (1994).

41. S. Stinson, *Chem. Eng. News*, **72**, 18–18 (1994).

42. W. E. Running, *J. Chem. Info. Comp. Sci.*, **33**, 934–935 (1993).

43. M. L. Nedved, S. Habibi-Goudarzi, B. Ganem, and J. D. Henion, *Anal. Chem.*, **68**, 4228–4236 (1996).

44. M. A. Kelly, H. B. Liang, I. I. Sytwu, I. Vlattas, N. L. Lyons, B. R. Bowen, and L. P. Wennogle, *Biochemistry*, **35**, 11747–11755 (1996).

45. S. Kaur, L. McGuire, D. Tang, G. Dollinger, and V. Huebner, *J. Protein Chem.*, **16**, 505–511 (1997).

46. M. M. Siegel, K. Tabei, G. A. Bebernitz, and E. Z. Baum, *J. Mass Spectrom.*, **33**, 264–273 (1998).

47. Y. H. Chu, L. Z. Avila, H. A. Biebuyck, and G. M. Whitesides, *J. Med. Chem.*, **35**, 2915–2917 (1992).

48. Y. H. Chu and G. M. Whitesides, *J. Org. Chem.*, **57**, 3524–3525 (1992).

49. K. L. Rundlett and D. W. Amstrong, *J. Chromatogr.*, **721**, 173–186 (1996).

50. L. Z. Avila, Y. H. Chu, E. C. Blossey, and G. M. Whitesides, *J. Med. Chem.*, **36**, 126–133 (1993).

51. B. L. Karger, *J. Med. Chem.*, **41**, 1201–1204 (1998).

52. Y. H. Chu, Y. M. Dunayevskiy, D. P. Kirby, P. Vouros, and B. L. Karger, *J. Am. Chem. Soc.*, **118**, 7827–7835 (1996).

53. K. Kasai and Y. Oda, *J. Chromatogr.*, **376**, 33–47 (1986).

54. D. C. Schriemer, D. R. Bundle, L. Li, and O. Hindsgaul, *Angew Chem. Int. Ed.*, **37**, 3383–3387 (1998).

55. B. Zhang, M. M. Palcic, H. Mo, I. J. Goldstein, and O. Hindsgaul, *Glycobiology*, **11**, 141–147 (2001).

56. X. H. Cheng, R. D. Chen, J. E. Bruce, B. L. Schwartz, G. A. Anderson, S. A. Hofstadler, D. C. Gale, R. D. Smith, J. M. Gao, G. B. Sigal, M. Mammen, and G. M. Whitesides, *J. Am. Chem. Soc.*, **117**, 8859–8860 (1995).

57. J. Gao, X. Cheng, R. Chen, G. B. Sigal, J. E. Bruce, B. L. Schwartz, S. A. Hofstadler, G. A. Anderson, R. D. Smith, and G. M. Whitesides, *J. Med. Chem.*, **39**, 1949–1955 (1996).

58. M. Wigger, J. P. Nawrocki, C. H. Watson, J. R. Eyler, and S. A. Benner, *Rapid Commun. Mass Spectrom.*, **11**, 1749–1752 (1997).

59. S. A. Hofstadler, K. A. Sannes-Lowery, S. T. Crooke, D. J. Ecker, H. Sasmor, S. Manalili, and R. H. Griffey, *Anal. Chem.*, **71**, 3436–3440 (1999).

60. K. A. Sannes-Lowery, J. J. Drader, R. H. Griffey, and S. A. Hofstadler. *Trends Anal. Chem.*, **19**, 481–491 (2000).

61. R. B. van Breemen, C. R. Huang, D. Nikolic, C. P. Woodbury, Y. Z. Zhao, and D. I. Venton, *Anal. Chem.*, **69**, 2159–2164 (1997).

62. D. Nikolic, S. Habibi-Goudarzi, D. G. Corley, S. Gafner, J. M. Pezzuto, and R. B. van Breemen, *Anal. Chem.*, **72**, 3853–3859 (2000).

63. D. Nikolic and R. B. van Breemen, *Comb. Chem. High Throughput Screen.*, **1**, 47–55 (1998).

64. Y. G. Shin and R. B. van Breemen. *Biopharm. Drug Dispos.*, **22**, 353–372 (2001).

65. Y. Z. Zhao, R. B. van Breemen, D. Nikolic, C. R. Huang, C. P. Woodbury, A. Schilling, and D. L. Venton, *J. Med. Chem.*, **40**, 4006–4012 (1997).

66. C. Gu, D. Nikolic, J. Lai, X. Xu, and R. B. van Breemen, *Comb. Chem. High Throughput Screen.*, **2**, 353–359 (1999).

67. R. B. van Breemen, C. P. Woodbury, and D. L. Venton in B. S. Larson and C. N. McEwen, Eds.,

Mass Spectrometry of Biological Materials, Marcel Dekker, New York, pp. 99–113 (1998).

68. J. Liu, J. E. Burdette, H. Xu, C. Gu, R. B. van Breemen, K. P. Bhat, N. Booth, A. I. Constantinou, J. M. Pezzuto, H. H. Fang, N. R. Farnsworth, and J. L. Bolton, *J. Agric. Food Chem.*, **49**, 2472–2479 (2001).

69. R. Wieboldt, J. Zweigenbaum, and J. Henion, *Anal. Chem.*, **69**, 1683–1691 (1997).

70. R. B. van Breemen, D. Nikolic, and J. L. Bolton, *Drug Metabol. Dispos.*, **26**, 85–90 (1998).

71. D. Nikolic, P. W. Fan, J. L. Bolton, and R. B. van Breemen, *Comb. Chem. High Throughput Screen.*, **2**, 165–175 (1999).

72. M. B. Beverly, P. West, and R. K. Julian, *Comb. Chem. High Throughput Screen.*, **5**, 65–73 (2002).

73. Z. Shen, J. J. Thomas, C. Averbuj, K. M. Broo, M. Engelhard, J. E. Crowell, M. G. Finn, and G. Siuzdak, *Anal. Chem.*, **73**, 612–619 (2001).

74. J. Wei, J. M. Buriak, and G. Siuzdak, *Nature*, **399**, 243–246 (1999).

75. T. W. Hutchens and T. T. Yip, *Rapid Commun. Mass Spectrom.*, **7**, 576–580 (1993).

76. R. W. Nelson, *Mass Spectrom. Rev.*, **16**, 353–376 (1997).

77. R. W. Nelson, D. Nedelkov, and K. A. Tubbs, *Anal. Chem.* **72**, 404A–411A (2000).

78. R. W. Nelson and J. R. Krone, *J. Mol. Recognit*, **12**, 77–93 (1999).

79. T. Wachs and J. Henion, *Anal. Chem.* **73**, 632–638 (2000).

80. R. M. Holt, M. J. Newman, F. S. Pullen, D. S. Richard, and A. G. Swanson, *J. Mass Spectrom.*, **32**, 64–70 (1997).

81. J. L. Wolfender, K. Ndjoko, and K. Hostettmann, *Phytochem. Anal.*, **12**, 2–22 (2001).

82. S. C. Bobzin, S. T. Yang, and T. P. Kasten, *J. Chromatogr. B*, **748**, 259–267 (2000).

83. R. T. William, E. L. Chapin, A. W. Carr, J. R. Gilbert, P. R. Graupner, P. Lewer, P. McKamey, J. R. Carney, and W. H. Gerwick, *Org. Lett.*, **2**, 289–292 (2000).

84. J. Chin, J. B. Fell, M. J. Shapiro, J. Tomesch, J. R. Wareing, and A. M. Bray, *J. Org. Chem.*, **62**, 538–539 (1997).

85. J. Chin, J. B. Fell, M. Jarosinski, M. J. Shapiro, and J. R. Wareing, *J. Org. Chem.*, **63**, 386–390 (1998).

86. P. A. Keifer, S. H. Smallcombe, E. H. Williams, K. E. Salomon, G. Mendez, J. L. Belletire, and C. D. Moore, *J. Combin. Chem.*, **2**, 151–171 (2000).

CHAPTER FOURTEEN

Electron Cryomicroscopy of Biological Macromolecules

RICHARD HENDERSON
Medical Research Council Laboratory of Molecular Biology
Cambridge, United Kingdom

TIMOTHY S. BAKER
Purdue University
Department of Biological Sciences
West Lafayette, Indiana

Contents

Burger's Medicinal Chemistry and Drug Discovery
Sixth Edition, Volume 1: Drug Discovery
Edited by Donald J. Abraham
ISBN 0-471-27090-3　© 2003 John Wiley & Sons, Inc.

1 MACROMOLECULAR STRUCTURE DETERMINATION BY USE OF ELECTRON MICROSCOPY

The two principal methods of macromolecular structure determination that use scattering techniques are electron microscopy and X-ray crystallography. The most important difference between the two is that the scattering cross section is about 105 times greater for electrons than it is for X-rays, so significant scattering using electrons is obtained for specimens that are 1 to 10 nm thick, whereas scattering or absorption of a similar fraction of an illuminating X-ray beam requires crystals that are 100 to 500 μm thick. The second main difference is that electrons are much more easily focused than X-rays because they are charged particles that can be deflected by electric or magnetic fields. As a result, electron lenses are greatly superior to X-ray lenses and can be used to produce a magnified image of an object as easily as a diffraction pattern. This then allows the electron microscope to be switched back and forth instantly between imaging and diffraction modes so that the image of a single molecule at any magnification can be obtained as conveniently as the electron diffraction pattern of a thin crystal.

In the early years of electron microscopy of macromolecules, electron micrographs of molecules embedded in a thin film of heavy atom stains (1, 2) were used to produce pictures that were interpreted directly. Beginning with the work of Klug and Berger (3), a more rigorous approach to image analysis led first to the interpretation of the two-dimensional (2D) images as the projected density summed along the direction of view and then to the ability to reconstruct the three-dimensional (3D) object from which the images arose (4, 5), with subsequent more sophisticated treatment of image contrast transfer (6).

Later, macromolecules were examined by electron diffraction and imaging without the use of heavy atom stains by embedding the specimens in either a thin film of glucose (7) or in a thin film of rapidly frozen water (8–10), which required the specimen to be cooled while it was examined in the electron microscope. This use of unstained specimens thus led to the structure determination of the molecules themselves rather than the structure of a "negative stain" excluding volume, and has created the burgeoning field of 3D electron microscopy of macromolecules.

Many medium resolution structures of macromolecular assemblies (e.g., ribosomes), spherical and helical viruses, and larger protein molecules have now been determined by electron cryomicroscopy in ice. Four atomic resolution structures have been obtained by electron cryomicroscopy of thin 2D crystals embedded in glucose, trehalose, or tannic acid (11–14), where specimen cooling reduced the effect of radiation damage. One of these, the structure of bacteriorhodopsin (11) provided the first structure of a seven-helix membrane protein. The medium resolution density distributions can often be interpreted in terms of the chemistry of the structure if a high resolution model of one or more of the component pieces has already been obtained by X-ray, electron microscopy, or NMR methods. As a result, the use of electron microscopy is becoming a powerful technique for which, in some cases, no alternative approach is possible. Useful reviews [e.g., Dubochet et al. (9), Amos et al. (15), Walz and Grigorieff (16), and Baker et al. (17)] and a book [Frank (18)] have been written.

2 ELECTRON SCATTERING AND RADIATION DAMAGE

A schematic overview of scattering and imaging in the electron microscope is depicted in Fig. 14.1. The incident electron beam passes through the specimen and individual electrons are either unscattered or scattered by the atoms of the specimen. This scattering occurs either elastically, with no loss of energy and therefore no energy deposition in the specimen, or inelastically, with consequent energy loss by the scattered electron and accompanying energy deposition in the specimen, resulting in radiation damage. The electrons emerging from the specimen are then

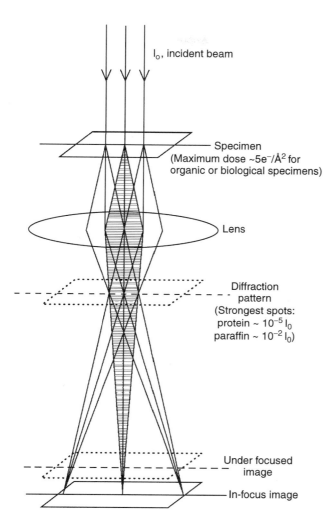

I_0, incident beam

Specimen
(Maximum dose ~5e$^-$/Å^2 for
organic or biological specimens)

Lens

Diffraction
pattern
(Strongest spots:
protein ~ $10^{-5} I_0$
paraffin ~ $10^{-2} I_0$)

Under focused
image

In-focus image

Figure 14.1. Schematic diagram showing the principle of image formation and
diffraction in the transmission electron
microscope. The incident beam I_0 illumi-
nates the specimen. Scattered and unscattered electrons are collected by the
objective lens and focused back to form
first an electron diffraction pattern and
then an image. For a 2D or 3D crystal,
the electron diffraction pattern would
show a lattice of spots, each of whose intensity is a small fraction of that of the
incident beam. In practice, an in-focus
image has no contrast, so images are recorded with the objective lens slightly
defocused to take advantage of the out-
of-focus phase-contrast mechanism.

collected by the imaging optics, shown here for
simplicity as a single lens, but in practice consisting of a complex system of five or six lenses,
with intermediate images being produced at
successively higher magnification at different
positions down the column. Finally, in the
viewing area, either the electron diffraction
pattern or the image can be seen directly by
eye on the phosphor screen, or detected by a
TV or CCD camera, or recorded on photographic film or image plate.

3 ELASTIC AND INELASTIC SCATTERING

The coherent, elastically scattered electrons
contain all the high resolution information

describing the structure of the specimen.
The amplitudes and phases of the scattered
electron beams are directly related to the
amplitudes and phases of the Fourier components of the atomic distribution in the
specimen. When the scattered beams are re-
combined with the unscattered beam in the
image, they create an interference pattern
(the image), which, for thin specimens, is
related approximately linearly to the density
variations in the specimen. The information
about the structure of the specimen can then
be retrieved by digitization and computer-
based image processing, as described later.
The elastic scattering cross sections for electrons are not as simply related to the atomic

composition as happens with X-rays. With X-ray diffraction, the scattering factors are simply proportional to the number of electrons in each atom, normally equal to the atomic number. Given that elastically scattered electrons are in effect diffracted by the electrical potential inside atoms, the scattering factor for electrons depends not only on the nuclear charge but also on the size of the surrounding electron cloud that screens the nuclear charge. As a result, electron scattering factors in the resolution range of interest in macromolecular structure determination (up to $1/3$ Å^{-1}), are sensitive to the effective radius of the outer valency electrons and therefore depend sensitively on the chemistry of bonding. Although this is a fascinating field in itself, with interesting work already carried out by the gas phase electron diffraction community [e.g., Hargittai and Hargittai (19)], it is still an area where much work remains to be done. At present, it is probably adequate to think of the density obtained in macromolecular structure analysis by electron microscopy as roughly equivalent to the electron density obtained by X-ray diffraction but with the contribution from hydrogen atoms being somewhat greater relative to carbon, nitrogen, and oxygen.

Those electrons that are inelastically scattered lose energy to the specimen by a number of mechanisms. The energy loss spectrum for a typical biological specimen is dominated by the large cross section for plasmon scattering in the energy range 20–30 eV, with a continuum in the distribution that decreases up to higher energies. At discrete high energies, specific inner electrons in the K shell of carbon, nitrogen, or oxygen can be ejected with corresponding peaks in the energy loss spectrum appearing at 200–400 eV. Any of these inelastic interactions produces an uncertainty in the position of the scattered electron (by Heisenberg's uncertainty principle) and, as a result, the resolution of any information present in the energy loss electron signal extends only to low resolutions of around 15 Å (20). Consequently, the inelastically scattered electrons are generally considered to contribute little except noise to the images.

4   RADIATION DAMAGE

The most important consequence of inelastic scattering is the deposition of energy into the specimen. This is initially transferred to secondary electrons, which have an average energy (20 eV) that is 5 or 10 times greater than the valency bond energies. These secondary electrons interact with other components of the specimen and produce numerous reactive chemical species, including free radicals. In ice-embedded samples, these would be predominantly highly reactive, hydroxyl free radicals that arise from the frozen water molecules. In turn, these react with the embedded macromolecules and create a great variety of radiation products such as modified side chains, cleaved polypeptide backbones, and a host of molecular fragments. From radiation chemistry studies, it is known that thiol or disulfide groups react more quickly than aliphatic groups and that aromatic groups, including nucleic acid bases, are the most resistant. Nevertheless, the end effect of the inelastic scattering is the degradation of the specimen to produce a cascade of heterogeneous products, some of which resemble the starting structure more closely than others. Some of the secondary electrons also escape from the surface of the specimen, causing it to charge up during the exposure. As a rough rule for 100-kV electrons, the dose that can be used to produce an image in which the starting structure at high resolution is still recognizable is about 1 e$^-$/Å^2 for organic or biological materials at room temperature, 5 e$^-$/Å^2 for a specimen near liquid nitrogen temperature ($-170°$C), and 10 e$^-$/Å^2 for a specimen near liquid helium temperature (4–8 K). However, individual experimenters will often exceed these doses if they wish to enhance the low resolution information in the images that is less sensitive to radiation damage. The effects of radiation damage attributed to electron irradiation are essentially identical to those from X-ray or neutron irradiation for biological macromolecules except for the amount of energy deposition per useful coherent elastically scattered event (21). For electrons scattered by biological structures at all electron energies of interest, the number of inelastic events exceeds the number of elastic events by a factor of 3 to 4, so that 60 to 80 eV of energy is deposited for each elastically scattered electron. This limits the amount of information in an image of a single biological macromolecule. Consequently, the 3D atomic structure cannot be determined from a single molecule but requires the averaging of the information from at least 10,000 molecules in theory, and even more in practice (21). Crystals used for X-ray or neutron diffraction contain many orders of magnitude more molecules.

It is possible to collect both the elastically and the inelastically scattered electrons simultaneously with an energy analyzer and, if a fine electron beam is scanned over the specimen, then a scanning transmission electron micrograph displaying different properties of the specimen can be obtained. Alternatively, conventional transmission electron microscopes to which an energy filter has been added can be used to select out a certain energy band of the electrons from the image. Both types of microscope can contribute in other ways to the knowledge of structure, but in this presentation, we concentrate on high voltage, phase-contrast electron microscopy of unstained macromolecules most often embedded in ice because this is the method of widest impact in structural biology.

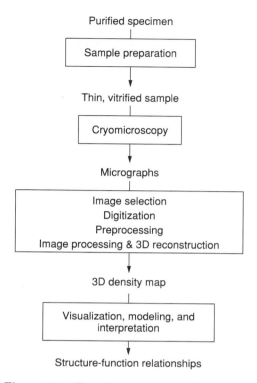

Figure 14.2. Flow diagram showing all the procedures involved in electron cryomicroscopy from sample preparation to map interpretation.

5 REQUIRED PROPERTIES OF ILLUMINATING ELECTRON BEAM

The important properties of the image in terms of defocus, astigmatism, and the presence and effect of amplitude or phase contrast are discussed later. The best quality incident electron beam is produced by a field emission gun (FEG). This is because the electrons from a FEG are emitted from a very small volume at the tip, which is the apparent source size. Once these electrons have been collected by the condenser lens and used to produce the illuminating beam, that beam of electrons is then nearly parallel (divergence of $\sim 10^{-2}$ mrad) and therefore spatially coherent. Similarly, because the emitting tip of a FEG is not heated as much as a conventional thermionic tungsten source, the thermal energy spread of the electrons is relatively small (0.5–1.0 eV) and, as a result, the illuminating beam is

closer to being monochromatic. Electron beams can also be produced by a normal, heated tungsten source, which gives a less parallel beam with a larger energy spread, but is nevertheless adequate for electron cryomicroscopy if the highest resolution images are not required.

6 THREE-DIMENSIONAL ELECTRON CRYOMICROSCOPY OF MACROMOLECULES

The determination of 3D structure by cryo-EM methods follows a common scheme for all macromolecules (Fig. 14.2). A more detailed discussion of the individual steps as applied to different classes of macromolecules appears in subsequent sections. Briefly, each specimen must be prepared in a relatively homogeneous, aqueous form (1D or 2D crystals or a suspension of single particles in a limited number of states) at relatively high concentration, rapidly frozen (vitrified) as a thin film, transferred into the electron microscope, and photographed by means of low dose selection and focusing procedures. The resulting images, if recorded on film, must then be digitized. Digitized images are then processed by the use of computer programs that allow different views of the specimen to be combined into a 3D reconstruction that can be interpreted in terms of other available structural, biochemical, and molecular data.

7   OVERVIEW OF CONCEPTUAL STEPS

Radiation damage by the illuminating electron beam generally allows only one good picture (micrograph) to be obtained from each molecule or macromolecular assembly. In this micrograph, the signal-to-noise ratio of the 2D projection image is normally too small to accurately determine the projected structure. This implies, first, that it is necessary to average many images of different molecules taken from essentially the same viewpoint to increase the signal-to-noise ratio and, second, that many of these averaged projections, taken from different directions, must be combined to build up the information necessary to determine the 3D structure of the molecule. Thus, the two key concepts are: (1) averaging to a greater or lesser extent depending on resolution, particle size and symmetry to increase the signal-to-noise ratio; and (2) the combination of different projections to build a 3D map of the structure.

In addition, there are various technical corrections that must be made to the image data to allow an unbiased model of the structure to be obtained. These include correction for the phase-contrast transfer function (CTF) and, at high resolution, for the effects of beam tilt. For crystals, it is also possible to combine electron diffraction amplitudes with image phases to produce a more accurate structure (7), and in general to correct for loss of high resolution contrast for any reason by "sharpening" the data by application of a negative temperature factor (22).

The idea of increasing the signal-to-noise ratio in electron images of unstained biological macromolecules by averaging was dis-

cussed in 1971 (23) and demonstrated in 1975 (7, 24), although earlier work on stained specimens had shown the value of averaging to increase the signal-to-noise ratio. The improvement obtained, as in all repeated measurements, gives a factor of $\sqrt{N}$ improvement in signal-to-noise ratio, where N is the number of times the measurement is made. The effect of averaging to produce an improvement in signal-to-noise ratio is seen most clearly in the processing of images from 2D crystals. Figure 14.3 shows the results of applying a sequence of corrections, beginning with averaging, to two-dimensional crystals of bacteriorhodopsin in 2D space group p3. The panels show: (a, b) 2D averaging, (c) correction for the microscope contrast transfer function (CTF), and (d) threefold crystallographic symmetry averaging of the phases and combination with electron diffraction amplitudes. At each stage in the procedure the projected picture of the molecules gets clearer. The final stage results in a virtually noise-free projected structure for the molecule at near atomic (3Å) resolution.

The earliest successful application of the idea of combining projections to reconstruct the 3D structure of a biological assembly was made by DeRosier and Klug (4). The idea is that each 2D projection corresponds after Fourier transformation to a central section of the 3D transform of the assembly. If enough independent projections are obtained, then the 3D transform will have been fully sampled and the structure can then be obtained by back transformation of the averaged, interpolated, and smoothed 3D transform. This procedure is shown schematically for a three-dimensional object in the shape of a duck, which represents the molecule whose structure is being determined (Fig. 14.4).

In practice, the implementation of these concepts has been carried out in a variety of ways, given that the experimental strategy and type of computer analysis used depend on the type of specimen, especially the molecular weight of the individual molecule, its symmetry, and whether it assembles into an aggregate with one-dimensional (1D), two-dimensional (2D), or three-dimensional (3D) periodic order.

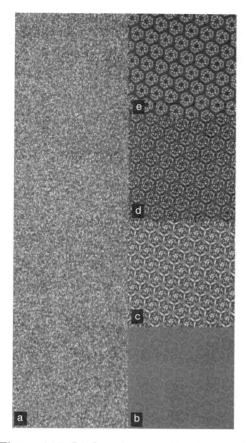

Figure 14.3. Display of the results at different stages of image processing of a digitized micrograph of a 2D crystal of bacteriorhodopsin. The left panel (a) shows an area of the raw digitized micrograph in which only electron noise is visible. The lower right panel (b) shows the results of the averaging of unit cells from the whole picture by unbending in real space and filtering in reciprocal space. The scale of the density in (b) is the same as that in the original micrograph, showing that the signal is very much weaker than the noise. Panel (c) shows the same density as that in (b) but with contrast increased 10-fold to show that the signal in the original picture is approximately $10\times$ below the noise level. Panel (d) shows the density after correction for contrast transfer function (CTF) attributed in this case to a defocus of 6000 Å. Panel (e) shows the density after further threefold crystallographic averaging (the space group is p3) and replacement of image amplitudes by electron diffraction amplitudes. Panel (e) therefore shows an almost perfect atomic resolution image of the projected structure of bacteriorhodopsin. The trimeric rings of molecules are centered on the crystallographic threefold axis and the internal structure shows α-helical segments in the protein.

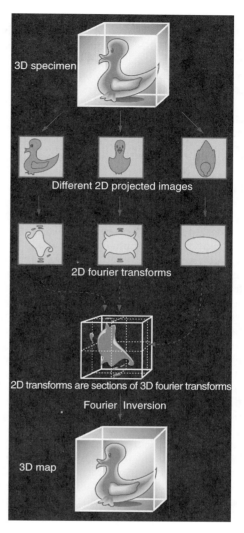

Figure 14.4. Schematic diagram to illustrate the principle of 3D reconstruction. Each 2D projected image, as recorded on the micrograph and after CTF correction, represents a section through the 3D Fourier transform. This is called the projection theorem. After accumulation of enough information from enough different views, a 3D map of the structure can be calculated by Fourier inversion.

8 CLASSIFICATION OF MACROMOLECULES

The symmetry of a macromolecule or supramolecular complex is the primary determinant of how specimen preparation, microscopy, and 3D image reconstruction are performed. The classification of molecules according to their level of periodic order and symmetry (Table 14.1) provides a logical and convenient way to consider the means by which specimens are studied in 3D by microscopy.

Each type of specimen offers a unique set of challenges in obtaining 3D structural information at the highest possible resolution. The best resolutions achieved by 3D EM methods to date, at about 3–4 Å, have been obtained with several thin, 2D crystals, in large part because of their excellent order.

With the exception of true 3D crystals, which must be sectioned to make them thin enough to study by transmission electron microscopy, the resolutions obtained with biological specimens are generally dictated by the preservation of periodic order, and the symmetry and complexity of the object. Hence, studies of the helical acetylcholine receptor tubes (36), the icosahedral hepatitis B virus capsid (44), the 50S ribosome (45), and the centriole (26) have yielded 3D density maps at resolutions of 4.6, 7.4, 15, and 280 Å, respectively.

If high resolution were the sole objective of EM, it would be necessary, given the capabilities of existing technology, to try to form well-ordered 2D crystals or helical assemblies of each macromolecule of interest. Indeed, a number of different crystallization techniques have been devised [e.g., Horne and Pasquali-Ronchetti (46); Yoshimura et al. (47); Kornberg and Darst (48); Jap et al. (49); Kubalek et al. (50); Rigaud et al. (51); Hasler et al. (52); Reviakine et al. (53); Wilson-Kubalek et al. (54)], and some of these have yielded new structural information about otherwise recalcitrant molecules like RNA polymerase (55). However, despite the obvious technological advantages of having a molecule present in a highly ordered form, most macromolecules function not as highly ordered crystals or helices but instead as single particles (e.g., many enzymes) or, more likely, in concert with other macromolecules as occurs in supramolecular assemblies. Also, crystallization tends to constrain the number of conformational states a molecule can adapt and the crystal conformation might not be functionally relevant. Hence, although resolution may be restricted

to much below that realized in the bulk of current X-ray crystallographic studies, cryo-EM methods provide a powerful means to study molecules that resist crystallization in 1D, 2D, or 3D. These methods allow one to explore the dynamic events, different conformational states (as induced, for example, by altering the microenvironment of the specimen), and macromolecular interactions that are the key to understanding how each macromolecule functions.

9 SPECIMEN PREPARATION

The goal in preparing specimens for cryomicroscopy is to keep the biological sample as close as possible to its native state to preserve the structure to atomic or near-atomic resolution in the microscope and during microscopy. The methods by which numerous types of macromolecules and macromolecular complexes have been prepared for cryo-EM studies are now well established (9, 56, 57). Most such methods involve cooling samples at a rate fast enough to permit vitrification (solid, glasslike state) rather than crystallization of the bulk water. Noncrystalline biological macromolecules are typically vitrified by applying a small (often <10 μL) aliquot of a purified, approximately 0.2–5 mg/mL suspension of sample to an EM grid coated with a carbon or holey carbon support film. The grid, secured with a pair of forceps and suspended over a container of ethane or propane cryogen slush (maintained near its freezing point by a reservoir of liquid nitrogen), is blotted nearly dry with a piece of filter paper. The grid is then plunged into the cryogen, and the sample, if thin enough (~0.2 μm or less), is vitrified in millisecond or shorter time periods (58–60).

The ability to freeze samples with a time resolution of milliseconds affords cryo-EM one of its unique and, as yet, perhaps most underutilized advantages: capturing and visualizing dynamic structural events that occur over time periods of a few milliseconds or longer. Several devices that allow samples to be perturbed in a variety of ways as they are plunged into cryogen have been described [e.g., Subramaniam et al. (61); Berriman and Unwin (59); Siegel et al. (62); Trachtenberg (63); White et

Table 14.1 Classification of Macromolecules According to Periodic Order and Symmetry

Periodic Order	Type Symmetry	Example Macromolecule/Complex	Representative Reference
0D	Point group	Ribosome	25
	C_1	Centriole	26
	C_1	BacteriophageΦ29 head	27
	C_5	Ribonucleoprotein vault	28
	C_8	TMV disk	29
	C_{17}	β-galactosidase	30
	D_2	Clathrin coats	31
	D_8	*Lumbricus terrestris* hemoglobin	32
	D_6	Dps protein	33
	T	*Azotobacter* pyruvate dehydrogenase core	34
	O	Icosahedral viruses	17
	I		
1D	Screw axis (helical)[a]	Acto-myosin filament	35
		Acetylcholine receptor tubes	36
		Microtubule	37
		Bacterial flagella	38
		Tobacco mosaic virus	39
2D	2D space group (2D crystal)	Bacterial rhodopsin membrane	11
	p3	Aquaporin membrane	40
	$p42_12$	Gap junction membrane	41
	p6	Light harvesting complex II	12
	p321	Tubulin sheet	13
	$p12_1$		
3D	3D space group (3D crystal)	Myosin S1 protein crystal	42
	$P2_12_12_1$	Insect flight muscle	43
	$P6_5$ or $P6_4$		

[a]The symmetry of a helical structure is defined by an n_m space axis, which combines a rotation of $2\pi/n$ radius about an axis followed by a translation of m/n of the repeat distance.

619

al. (60)]. Examples of the use of such devices include spraying acetylcholine onto its receptor to cause the receptor channel to open (64), or lowering the pH of an enveloped virus sample to initiate early events of viral fusion (65), or inducing a temperature jump with a flash-tube system to study phase transitions in liposomes (66), or mixing myosin S1 fragments with F-actin to examine the geometry of the crossbridge powerstroke in muscle (67).

Crystalline (2D) samples fortunately can often be prepared for cryo-EM by means of simpler procedures, and vitrification of the bulk water is not always essential to achieve success (68). Such specimens may be applied to the carbon film on an EM grid by normal adhesion methods, washed with 1–2% solutions of solutes like glucose, trehalose, or tannic acid, blotted gently with filter paper to remove excess solution, air dried, loaded into a cold holder, inserted into the microscope, and, finally, cooled to liquid nitrogen temperature.

10   MICROSCOPY

Once the vitrified specimen is inserted into the microscope and sufficient time is allowed (~15 min) for the specimen stage to stabilize to minimize drift and vibration, microscopy is performed to generate a set of images that, with suitable processing procedures, can later be used to produce a reliable 3D reconstruction of the specimen at the highest possible resolution. To achieve this goal, imaging must be performed at an electron dose that minimizes beam-induced radiation damage to the specimen, with the objective lens of the microscope defocused to enhance phase contrast from the weakly scattering, unstained biological specimen, and under conditions that keep the specimen below the devitrification temperature and minimize its contamination.

The microscopist locates specimen areas suitable for photography by searching the EM grid at very low magnification ($\leq 3000\times$) to keep the irradiation level very low (<0.05 e$^-$/Å^2) while assessing sample quality. In microscopes operated at 200 keV or higher, where image contrast is very weak, it is helpful to perform the search procedure with the assistance of a CCD camera or a video-rate TV-intensified camera system. For some specimens, like thin 2D crystals, searching is conveniently performed by viewing the low magnification, high contrast image produced by slightly defocusing the electron diffraction pattern by use of the diffraction lens.

After a desired specimen area is identified, the microscope is switched to high magnification mode for focusing and astigmatism correction. These adjustments are typically performed in a region about 2–10 μm away from the chosen area at the same or higher magnification than that used for photography. The choice of magnification, defocus level, accelerating voltage, beam coherence, electron dose, and other operating conditions is dictated by several factors. The most significant ones are the size of the particle or crystal unit cell being studied, the anticipated resolution of the images, and the requirements of the image processing needed to compute a 3D reconstruction to the desired resolution. For most specimens at required resolutions from 3 to 30 Å, images are typically recorded at 25,000–50,000$\times$ magnification, with an electron dose of between 5 and 20 e$^-$/Å^2. These conditions yield micrographs of sufficient optical density (OD 0.2–1.5) and image resolution for subsequent image-processing steps. Most modern EMs provide some mode of low dose operation for imaging beam-sensitive, vitrified biological specimens.

The intrinsic low contrast of unstained specimens makes it impossible to observe and focus on specimen details directly, as is routine with stained or metal-shadowed specimens. Focusing, aimed to enhance phase contrast in the recorded images but minimize beam damage to the desired area, is achieved by judicious defocusing on a region that is adjacent to the region to be photographed and preferably situated on the microscope tilt axis. The appropriate focus level is set by adjusting the appearance of either the Fresnel fringes that occur at the edges of holes in the carbon film or the "phase granularity" from the carbon support film.

Unfortunately, electron images do not give a direct rendering of the specimen density distribution. The relationship between image and specimen is described by the contrast transfer function (CTF), which is characteristic of the particular microscope used, the specimen, and the conditions of imaging. The microscope CTF arises from the objective lens focal setting and from the spherical aberration present in all electromagnetic lenses and varies with the defocus and accelerating voltage according to a formula (see below) that includes both phase- and amplitude-contrast components. First, however, it might be useful to describe briefly the essentials of amplitude contrast and phase contrast, two concepts carried over from optical microscopy. Amplitude contrast refers to the nature of the contrast in an image of an object that absorbs the incident illumination or scatters it in any other way, so that a proportion of it is lost. As a result, the image appears darker where greater absorption occurs. Phase contrast is required if an object is transparent (i.e., it is a pure phase object) and does not absorb but only scatters the incident illumination. Biological specimens for cryo-EM are almost pure phase objects and the scattering is relatively weak, so that the simple theory of image formation by a weak phase object applies (69, 70). An exactly in-focus image of a phase object has no contrast variation because all the scattered illumination is focused back to equivalent points in the image of the object from which it was scattered. In optical microscopy, the use of a quarter wave plate can retard the phase of the direct unscattered beam, so that an in-focus image of a phase object has very high "Zernicke" phase contrast. However, there is as yet no simple quarter wave plate for electrons, so instead, phase contrast is created by introducing phase shifts into the diffracted beams by adjustment of the excitation of the objective lens so that the image is slightly defocused. In addition, because all matter is composed of atoms and the electric potential inside each atom is very high near the nucleus, even the electron scattering behavior of the light atoms found in biological molecules deviates from that of a weak phase object; however, for a deeper discussion of this the reader should refer to Reimer (70) or Spence (69). In practice, the proportion of "amplitude" contrast is about 7% at 100 kV, 5% at 200 kV, and 4% at 300 kV for low dose images of protein molecules embedded in ice.

The overall dependency of CTF on resolution, wavelength, defocus, and spherical aberration is given by

$$CTF(\nu) = -\{(1 - F_{amp}^2)^{1/2}\sin[\chi(\nu)] + F_{amp}\cos[\chi(\nu)]\}$$

where $\chi(\nu) = \pi\lambda\nu^2(\Delta f - 0.5C_s\lambda^2\nu^2)$; ν is the spatial frequency (in $\mathring{A}^{-1}$); F_{amp} is the fraction of amplitude contrast; λ is the electron wavelength (in $\mathring{A}$), where

$$\lambda = 12.3/\sqrt{V + 0.000000978 \cdot V^2}$$

(=0.037, 0.025, and 0.020 $\mathring{A}$ for 100, 200, and 300 keV electrons, respectively); V is the voltage (in volts); Δf is the underfocus (in $\mathring{A}$); and C_s is the spherical aberration of the objective lens of the microscope (in $\mathring{A}$).

In addition, this CTF is attenuated by an envelope or damping function, which depends on the coherence of the beam, specimen drift, and other factors (6, 71, 72). Figure 14.5 shows a few representative CTFs for different amounts of defocus on a normal and a FEG microscope. Thus, for a particular defocus setting of the objective lens, phase contrast in the electron image is positive and maximal only at a few specific spatial frequencies. Contrast is either lower than maximal, completely absent, or it is opposite (inverted or reversed) from that at other frequencies. Hence, as the objective lens is focused, the electron microscopist selectively accentuates image details of a particular size.

Images are typically recorded 0.8–3.0 μm underfocus to enhance specimen features in the 20–40 $\mathring{A}$ size range and thereby facilitate phase origin and specimen orientation search procedures carried out in the image-processing steps. However, this level of underfocus also enhances the contrast envelope in lower resolution maps, which may help in interpretation. To obtain results at better than 10–15 $\mathring{A}$ resolution, it is essential to record, process, and combine data from several micrographs that span a range of defocus levels [e.g., Unwin and Henderson (7); Böttcher et al. (44)]. This strategy ensures good information transfer at all spatial frequencies up to the limiting reso-

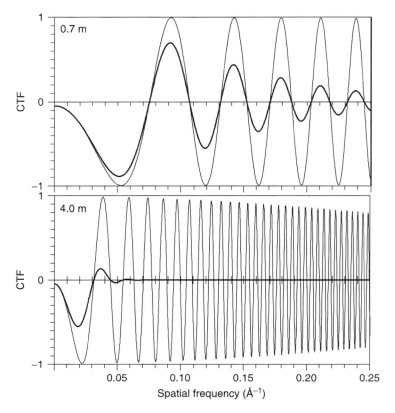

Figure 14.5. Representative plots of the contrast transfer function (CTF) as a function of spatial frequency, for two different defocus settings (0.7 and 4.0 μm underfocus) and for a field emission (light curve) or tungsten (dark curve) electron source. All plots correspond to electron images formed in an electron microscope operated at 200 kV and with objective lens aberration coefficients, $C_s = C_c$ = 2.0 mm, and assuming amplitude contrast of 4.8% (73). The spatial coherence, which is related to the electron source size and expressed as β, the half-angle of illumination, for tungsten and FEG electron sources was fixed at 0.3 and 0.015 milliradians, respectively. Likewise, the temporal coherence (expressed as ΔE, the energy spread) was fixed at 1.6 and 0.5 eV for tungsten and FEG sources. The combined effects of the poorer spatial and temporal coherence of the tungsten source leads to a significant dampening, and hence loss of contrast, of the CTF at progressively higher resolutions compared to that observed in FEG-equipped microscopes. The greater number of contrast reversals with higher defocus arises because of the greater out-of-focus phase shifts.

lution but requires careful compensation for the effects of the microscope CTF during image processing. Also, the recording of image focal pairs or focal series from a given specimen area can be beneficial in determining origin and orientation parameters for processing of images of single particles [e.g., Cheng et al. (74); Trus et al. (75)].

Many high resolution cryo-EM studies are now performed with microscopes operated at 200 keV or higher and with field emission gun (FEG) electron sources [e.g., Zemlin (76, 78); Zhou and Chiu (77); Mancini et al. (79)]. The high coherence of a FEG source ensures that phase contrast in the images remains strong out to high spatial frequencies ($>1/3.5$ Å^{-1}), even for highly defocused images. The use of higher voltages provides potentially higher resolution [greater depth of field (i.e., less curvature of the Ewald sphere) attributed to smaller electron beam wavelength], better beam penetration (less multiple scattering), reduced problems with specimen charging that plague microscopy of unstained or uncoated vitrified specimens (80), and reduced phase shifts associated with beam tilt.

Images are recorded on photographic film or on a CCD camera with either flood beam or spot-scan procedures. Film, with its advantages of low cost, large field of view, and high resolution ($\sim$10 μm), has remained the primary image recording medium for most cryo-EM applications, despite disadvantages of high background fog and need for chemical development and digitization. CCD cameras provide image data directly in digital form and with very low background noise, but suffer from higher cost, limited field of view, limited spatial resolution caused by poor point spread characteristics, and a fixed pixel size (typically between 14 and 24 μm). They are useful, for example, for precise focusing and adjustment of astigmatism [e.g., Krivanek and Mooney (81); Sherman et al. (82)].

For studies in which specimens must be tilted to collect 3D data, such as with 2D crystals, or single particles that adopt preferred orientations on the EM grid, or specimens requiring tomography, microscopy is performed in essentially the same way as described above. However, the limited tilt range ($\pm$60–70°) of most microscope goniometers can lead to nonisotropic resolution in the 3D reconstructions (the "missing cone" problem), and tilting generates a constantly varying defocus across the field of view in a direction normal to the tilt axis. The effects caused by this varying defocus level must be corrected in high resolution applications.

11 SELECTION AND PREPROCESSING OF DIGITIZED IMAGES

Before any image analysis or classification of the molecular images can be done, a certain amount of preliminary checking and normalization is required to ensure there is a reasonable chance that a homogeneous population of molecular images has been obtained. First, good quality micrographs are selected in which the electron exposure is correct, there is no image drift or blurring, and there is minimal astigmatism and a reasonable amount of

defocus to produce good phase contrast. This is usually done by visual examination and optical diffraction.

Once the best pictures have been chosen, the micrographs must be scanned and digitized on a suitable densitometer. The sizes of the steps between digitization of optical density and the size of the sample aperture over which the optical density is averaged by the densitometer must be sufficiently small to sample the detail present in the image at fine enough intervals (83). Normally, a circular (or square) sample aperture of diameter (or length of side) equal to the step between digitizations is used. This avoids digitizing overlapping points, without missing any of the information recorded in the image. The size of the sample aperture and digitization step depends on the magnification selected and the resolution required. A value of 1/4 to 1/3 of the required limit of resolution (measured in μm on the emulsion) is normally ideal because it avoids having too many numbers (and therefore wasting computer resources), without losing anything during the measurement procedure. For a 40,000$\times$ image, on which a resolution of 10 Å at the specimen is required, a step size of 10 μm $\{= \frac{1}{4} \times [(10 \text{ Å} \times 40{,}000)/(10{,}000 \text{ Å}/\mu\text{m})]\}$ would be suitable.

The best area of an image of a helical or 2D crystal specimen can then be boxed off using a soft-edge mask. For images of single particles, a stack of individual particles can be created by selecting out many small areas surrounding each particle. Because, in the later steps of image processing, the orientation and position of each particle are refined by comparing the amplitudes and phases of their Fourier components, it is important to remove spurious features around the edge of each particle and to make sure the different particle images are on the same scale. This is normally done by masking off a circular area centered on each particle and floating the density so that the average around the perimeter becomes zero (83). The edge of the mask is apodized by applying a soft cosine bell shape to the original densities so they taper toward the background level. Finally, to compensate for variations in the exposure attributed to ice thickness or electron dose, most microscopists normalize the stack of individual particle images so that the mean density and mean density variation over the field of view are set to the same values for all particles (84).

Once some good particles or crystalline areas for 1D or 2D crystals have been selected, digitized, masked, and their intensity values normalized, true image processing can begin.

12 IMAGE PROCESSING AND 3D RECONSTRUCTION

Although the general concepts of signal averaging, together with combining different views to reconstruct the 3D structure, are common to the different computer-based procedures that have been implemented, it is important to emphasize one or two preliminary points. First, a homogeneous set of particles must be selected for inclusion in the 3D reconstruction. This selection may be made by eye, to eliminate obviously damaged particles or impurities, or by the use of multivariate statistical analysis (85) or some other classification scheme. This allows a subset of the particle images to be used to determine the structure of a better defined entity. All image-processing procedures require the determination of the same parameters that are needed to specify unambiguously how to combine the information from each micrograph or particle. These parameters are: the magnification, defocus, astigmatism, and, at high resolution, the beam tilt for each micrograph; the electron wavelength used (i.e., accelerating voltage of the microscope); the spherical aberration coefficient (C_s) of the objective lens; and the orientation and phase origin for each particle or unit cell of the 1D, 2D, or 3D crystal. There are 13 parameters for each particle, of which eight may be common to each micrograph and two or three (C_s, kV, magnification) to each microscope. The different general approaches that have been used in practice to determine the 3D structure of different classes of macromolecular assemblies from one or more electron micrographs are listed in Table 14.2.

The precise way in which each general approach codes and determines the particle or unit cell parameters varies greatly and is not described in detail. Much of the computer software used in image reconstruction studies is relatively specialized compared to that used in the more mature field of macromolecular X-ray crystallography. In part, this may be attributed to the large diversity of specimen types amenable to cryo-EM and reconstruction methods. As a consequence, image-reconstruction software is evolving quite rapidly, and references to software packages cited in Table 14.2 are likely to become quickly outdated. Extensive discussion of algorithms and software packages in use at this time may be found in a number of recent special issues of the *Journal of Structural Biology* [volumes 116(1), 120(3), 121(2), and 125(2/3)].

In practice, attempts to determine or refine some parameters may be affected by the inability to determine accurately one of the other parameters. The solution of the structure is therefore an iterative procedure in which reliable knowledge of the parameters that describe each image is gradually built up to produce an increasingly accurate structure, until no more information can be squeezed out of the micrographs. At this point, if any of the origins or orientations is wrongly assigned, there will be a loss of detail and signal-to-noise ratio in the map. If a better determined or higher resolution structure is required, it would then be necessary to record images on a better microscope or to prepare new specimens and record better pictures.

The reliability and resolution of the final reconstruction can be measured by use of a variety of indices. For example, the differential phase residual (DPR) (133), the Fourier shell correlation (FSC) (134), and the Q-factor (135) are three such measures. DPR is the mean phase difference, as a function of resolution, between the structure factors from two independent reconstructions, often calculated by splitting the image data into two halves. FSC is a similar calculation of the mean correlation coefficient between the complex structure factors of the two halves of the data as a function of resolution. The Q-factor is the mean ratio of the vector sum of the individual structure factors from each image divided by the sum of their moduli, again calculated as a function of resolution. Perfectly accurate measurements would have values of DPR, FSC, and Q-factor of 0°, 1.0, and 1.0 respectively, whereas random data containing no informa-

Table 14.2 Methods of Three-Dimensional Image Reconstruction

Type Structure (symmetry)	Method	Reference(s) to Technical/ Theoretical Details
Asymmetric (Point group C_1)	Random conical tilt	18, 87, 88
	•Software package	89
	Angular reconstitution	32, 90
	•Software package	91
	Weighted back projection	92, 93
	Radon transform alignment	94
	Reference-based alignment	95
	Reference free alignment	96, 97
	Fourier reconstruction and alignment	98
	Tomographic tilt series and remote control of microscope[a]	99–102
Symmetric (Point groups C_n, D_n; $n > 1$)	Angular reconstitution	32, 90
	•Software packages	91, 103
	Fourier–Bessel synthesis	27
	Reference-based alignment and weighted back projection	104
Icosahedral (Point group I)	Fourier–Bessel synthesis (common-lines)	79, 105–107
	•Reference-based alignment	108–111
	•Software packages	112–116
	Angular reconstitution	90, 117
	Tomographic tilt series	
Helical	Fourier–Bessel synthesis	4, 36, 83, 119–123
	•Software packages and filament straightening routines	112, 123–126
2D Crystal	Random azimuthal tilt	11, 15, 24, 127, 128
	•Software packages	112, 129
3D Crystal	Oblique section reconstruction	43, 130
	•Software package	131
	Sectioned 3D crystal	42

[a]Note: Electron tomography is the subject of an entire issue of *J. Struct. Biol.* [120, 207–395 (1997)] and a book edited by Frank (132).

tion would have values of 90°, 0.0, and 0.0. The spectral signal-to-noise ratio (SSNR) criterion has been advocated as the best of all (136): it effectively measures, as a function of resolution, the overall signal-to-noise ratio (squared) of the whole of the image data. It is calculated by taking into consideration how well all the contributing image data agree internally.

An example of a typical strategy for determination of the 3D structure of a new and unknown molecule without any symmetry and that does not crystallize might be as follows:

1. Record a single axis tilt series of particles embedded in negative stain, with a tilt range from $-60°$ to $+60°$.

2. Calculate 3D structures for each particle by use of an R-weighted back-projection algorithm (93).

3. Average 3D data for several particles in real or reciprocal space to get a reasonably good 3D model of the stain excluding the region of the particle.

4. Record a number of micrographs of the particles embedded in vitreous ice.

5. Use the 3D negative stain model obtained in (3) with inverted contrast to determine the rough alignment parameters of the particle in the ice images.

6. Calculate a preliminary 3D model of the average, ice-embedded structure.

7. Use the preliminary 3D model to determine more accurate alignment parameters for the particles in the ice images.

8. Calculate a better 3D model.

9. Determine defocus and astigmatism to allow CTF calculation and correct 3D model so that it represents the structure at high resolution.

10. Keep adding pictures at different defocus levels to get an accurate structure at as high a resolution as possible.

For large single particles with no symmetry or for particles with higher symmetry or for crystalline arrays, it should be possible to miss out the negative staining steps and go straight to alignment of particle images from ice-embedding because the particle or crystal tilt angles can be determined internally from comparison of phases along common lines in reciprocal space or from the lattice or helix parameters from a 2D or 1D crystal.

The following discussion briefly outlines for a few specific classes of macromolecule the general strategy for carrying out image processing and 3D reconstruction (see Fig. 14.6).

12.1　2D Crystals

For 2D crystals, the general 3D reconstruction approach consists of the following steps: First, a series of micrographs of single 2D crystals are recorded at different tilt angles, with random azimuthal orientations. Each crystal is then unbent using cross-correlation techniques, to identify the precise position of each unit cell (127), and amplitudes and phases of the Fourier components of the average of that particular view of the structure are obtained for the transform of the unbent crystal. The reference image used in the cross-correlation calculation can either be a part of the whole image masked off after a preliminary round of averaging by reciprocal space filtering of the regions surrounding the diffraction spots in the transform, or it can be a reference image calculated from a previously determined 3D model. The amplitudes and phases from each image are then corrected for the CTF and beam tilt (11, 22, 127) and merged with data from many other crystals by scaling and origin refinement, taking into account the proper symmetry of the 2D space group of the crystal. Finally, the whole data set is fitted by least squares to constrained amplitudes and phases along the lattice lines (137) before calculating a map of the structure. The initial determination of the 2D space group can be carried out by a statistical test of the phase relationships in one or two images of untilted specimens (138). The absolute hand of the structure is automatically correct, given that the 3D structure is calculated from images whose tilt axis and tilt angle are known. Nevertheless, care must be taken not to make any of a number of trivial mistakes that would invert the hand.

12.2　Helical Particles

The basic steps involved in processing and 3D reconstruction of helical specimens include: Recording a series of micrographs of vitrified particles suspended over holes in a perforated carbon support film. The micrographs are digitized and Fourier-transformed to determine image quality (astigmatism, drift, defocus, presence, and quality of layer lines, etc.). Individual particle images are boxed, floated, and apodized within a rectangular mask. The parameters of helical symmetry (number of subunits per turn and pitch) must be determined by indexing the computed diffraction patterns. If necessary, simple spline-fitting procedures may be employed to "straighten" images of curved particles (124), and the image data may be reinterpolated (126) to provide more precise sampling of the layer line data in the computed transform. Once a preliminary 3D structure is available, a much more sophisticated refinement of all the helical parameters can be used to unbend the helices onto a predetermined average helix so that the contributions of all parts of the image are correctly treated (123). The layer line data are extracted from each particle transform and two phase origin corrections are made, one to shift the phase origin to the helix axis (at the center of the particle image) and the other to correct for effects caused by having the helix axis tilted out of the plane normal to the electron beam in the electron microscope. The layer line data are separated out into near- and far-side data, corresponding to contributions from the near and far sides of each particle imaged. The relative rotations and

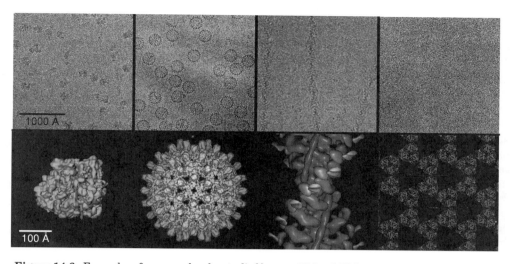

Figure 14.6. Examples of macromolecules studied by cryo-EM and 3D image reconstruction and the resulting 3D structures (bottom row) after cryo-EM analysis. All micrographs (top row) are displayed at about 170,000× magnification and all models at about 1,200,000× magnification. (a) A single particle without symmetry: The micrograph shows 70S *E. coli* ribosomes complexed with mRNA and fMet-tRNA. The surface-shaded density map, made by averaging 73,000 ribosome images from 287 micrographs has a resolution (FSC) of 11.5 Å. The 50S and 30S subunits and the tRNA are colored blue, yellow, and green, respectively. The identity of many of the subunits is known and some RNA double helices are clearly recognizable by their major and minor grooves (e.g., helix 44 is shown in red). [Courtesy of J. Frank (SUNY, Albany), using data from Gabashvili et al. (86).] (b) A single particle with symmetry: The micrograph shows hepatitis B virus cores. The 3D reconstruction, at a resolution of 7.4 Å (DPR), was computed from 6384 particle images taken from 34 micrographs. [From Böttcher et al. (44).] (c) A helical filament: The micrograph shows actin filaments decorated with myosin S1 heads containing the essential light chain. The 3D reconstruction, at a resolution of 30–35 Å, is a composite in which the differently colored parts are derived from a series of difference maps that were superimposed on f-actin. The components include: f-actin (blue), myosin heavy chain motor domain (orange), essential light chain (purple), regulatory light chain (white), tropomyosin (green), and myosin motor domain *N*-terminal beta-barrel (red). [Courtesy of A. Lin, M. Whittaker, and R. Milligan (Scripps Research Institute, La Jolla, CA).] (d) A 2D crystal, light-harvesting complex LHCII at 3.4-Å resolution. The model shows the protein backbone and the arrangement of chromophores in a number of trimeric subunits in the crystal lattice. In this example, image contrast is too low to see any hint of the structure without image processing (see also Fig. 14.3). See color insert. [Courtesy of W. Kühlbrandt (Max-Planck-Institute for Biophysics, Frankfurt, Germany).]

translations needed to align the different transforms are determined so the data may be merged and a 3D reconstruction computed by Fourier-Bessel inversion procedures (83). Determination of the absolute hand requires comparison of a pair of images recorded with a small tilt of the specimen between the views (139).

12.3   Icosahedral Particles

The typical strategy for processing and 3D reconstruction of icosahedral particles consists of the following steps: First, a series of micro-graphs of a monodisperse distribution of particles, normally suspended over holes in a perforated carbon support film, is recorded. After digitization of the micrographs, individual particle images are boxed and floated with a circular mask. The astigmatism and defocus of each micrograph is measured from the sum of intensities of the Fourier transforms of all particle images (140). Autocorrelation techniques are then used to estimate the particle phase origins, which coincide with the center of each particle, where all rotational symmetry axes intersect (141). The view orientation of each particle, defined by three Eulerian angles, is determined either by means of common and cross-common lines techniques or with the aid of model-based procedures [e.g., Crowther (106); Fuller et al. (107); Baker et al. (17)]. Once a set of self-consistent particle images is available, an initial, low resolution 3D reconstruction is computed by merging these data with Fourier-Bessel methods (106). This reconstruction then serves as a reference for refining the orientation, origin, and CTF parameters of each of the included particle images, for rejecting "bad" images, and for increasing the size of the data set by including new particle images from additional micrographs taken at different defocus levels. A new reconstruction, computed from the latest set of images, serves as a new reference and the above refinement procedure is repeated until no further improvements, as measured by the reliability criteria mentioned above, are made. Determination of the absolute hand of the structure requires the recording and processing of a pair of images taken with a known, small relative tilt of the specimen between the two views (142).

13 VISUALIZATION, MODELING, AND INTERPRETATION OF RESULTS

Once a reliable 3D map is obtained, computer graphics and other visualization tools may be used as aids in interpreting morphological details and understanding biological function in the context of biochemical and molecular studies and complementary X-ray crystallographic and other biophysical measurements.

14 TRENDS

The new generation of intermediate voltage (~300 kV) FEG microscopes is now making it much easier to obtain higher resolution images that, by use of larger defocus values, have good image contrast at both very low and very high resolution. The greater contrast at low resolution greatly facilitates particle-alignment procedures, and the increased contrast resulting from the high coherence illumina-

tion helps to increase the signal-to-noise ratio for the structure at high resolution. Cold stages are constantly being improved, with several liquid helium stages now in operation (143, 144). Two of these are commercially available from JEOL and FEI/Philips.

Finally, three additional likely trends include: (1) increased automation, including the recording of micrographs, the use of spotscan procedures in remote microscope operation (145, 146), and in every aspect of image processing; (2) production of better electronic cameras (e.g., CCD or pixel detectors); and (3) increased use of dose-fractionated, tomographic tilt series, to extend EM studies to the domain of larger supramolecular and cellular structures (102, 147).

15 ACKNOWLEDGMENTS

We are greatly indebted to all our colleagues at Purdue and Cambridge for their insightful comments and suggestions; to B. Böttcher, R. Crowther, J. Frank, W. Kühlbrandt, and R. Milligan for supplying images used in Figure 14.6; and J. Brightwell for editorial assistance. T.S.B. was supported in part by Grant GM33050 from the National Institutes of Health.

16 ABBREVIATIONS

0D	zero-dimensional (single particles)
1D	one-dimensional (helical)
2D	two-dimensional
3D	three-dimensional
CCD	charge coupled device (slow scan TV detector)
cryo-EM	electron cryomicroscopy
CTF	contrast transfer function
EM	electron microscope/microscopy
FEG	field emission gun

REFERENCES

1. S. Brenner and R. W. Horne, *Biochem. Biophys. Acta—Prot. Struct.*, **34**, 103–110 (1959).

2. H. E. Huxley and G. Zubay, *J. Mol. Biol.*, **2**, 10–18 (1960).

3. A. Klug and J. E. Berger, *J. Mol. Biol.*, **10**, 565–569 (1964).

4. D. J. DeRosier and A. Klug, *Nature*, **217**, 130–134 (1968).

5. W. Hoppe, R. Langer, G. Knesch, and C. Poppe, *Naturwissenschaften*, **55**, 333–336 (1968).

6. H. P. Erickson and A. Klug, *Philos. Trans. R. Soc. Lond. B*, **261**, 105–118 (1971).

7. P. N. T. Unwin and R. Henderson, *J. Mol. Biol.*, **94**, 425–440 (1975).

8. J. Dubochet, J. Lepault, R. Freeman, J. A. Berriman, and J.-C. Homo, *J. Microsc.*, **128**, 219–237 (1982c).

9. J. Dubochet, M. Adrian, J.-J. Chang, J.-C. Homo, J. Lepault, A. W. McDowall, and P. Schultz, *Q. Rev. Biophys.*, **21**, 129–228 (1988).

10. K. A. Taylor and R. M. Glaeser, *Science*, **186**, 1036–1037 (1974).

11. R. Henderson, J. M. Baldwin, T. A. Ceska, F. Zemlin, E. Beckmann, and K. H. Downing, *J. Mol. Biol.*, **213**, 899–929 (1990).

12. W. Kühlbrandt, D. N. Wang, and Y. Fujiyoshi, *Nature*, **367**, 614–621 (1994).

13. E. Nogales, S. G. Wolf, and K. H. Downing, *Nature*, **391**, 199–203 (1998).

14. K. Murata, K. Mitsuoka, T. Hirai, T. Waltz, P. Agre, J. B. Heymann, A. Engel, and Y. Fujiyoshi, *Nature*, **407**, 599–605 (2000).

15. L. A. Amos, R. Henderson, and P. N. T. Unwin, *Prog. Biophys. Mol. Biol.*, **39**, 183–231 (1982).

16. T. Walz & N. Grigorieff, *J. Struct. Biol.*, **121**, 142–161 (1998).

17. T. S. Baker, N. H. Olson, and S. D. Fuller, *Microbiol. Mol. Biol. Rev.*, **63**, 862–922 (1999).

18. J. Frank, *Three-Dimensional Electron Microscopy of MacromolecularAssemblies*, Academic Press, San Diego, CA, 1996, 342 pp.

19. I. Hargittai and M. Hargittai, Eds., *Stereochemical Applications of Gas-Phase Electron Diffraction*, VCH, New York, 1988.

20. M. Isaacson, J. Langmore, and H. Rose, *Optik*, **41**, 92–96 (1974).

21. R. Henderson, *Q. Rev. Biophys.*, **28**, 171–193 (1995).

22. W. A. Havelka, R. Henderson, and D. Oesterhelt, *J. Mol. Biol.*, **247**, 726–738 (1995).

23. R. M. Glaeser, *J. Ultrastruct. Res.*, **36**, 466–482 (1971).

24. R. Henderson and P. N. T. Unwin, *Nature*, **257**, 28–32 (1975).

25. J. Frank, *Curr. Opin. Struct. Biol.*, **7**, 266–272 (1997).

26. J. Kenney, E. Karsenti, B. Gowen, and S. D. Fuller, *J. Struct. Biol.*, **120**, 320–328 (1997).

27. Y. Tao, N. H. Olson, W. Xu, D. L. Anderson, M. G. Rossmann, and T. S. Baker, *Cell*, **95**, 431–437 (1998).

28. L. B. Kong, A. C. Siva, L. H. Rome, and P. L. Stewart, *Structure*, **7**, 371–379 (1999).

29. A. C. Bloomer, J. Graham, S. Hovmoller, P. J. G. Butler, and A. Klug, *Nature*, **276**, 362–368 (1978).

30. R. H. Jacobson, X.-J. Zhang, R. F. DuBose, and B. W. Matthews, *Nature*, **369**, 761–766 (1994).

31. G. P. A. Vigers, R. A. Crowther, and B. M. F. Pearse, *EMBO J.*, **5**, 529–534 (1986).

32. M. Schatz, E. V. Orlova, P. Dube, J. Jager, and M. van Heel, *J. Struct. Biol.*, **114**, 28–40 (1995).

33. R. A. Grant, D. J. Filman, S. E. Finkel, R. Kolter, and J. M. Hogle, *Nat. Struct. Biol.*, **5**, 294–303 (1998).

34. A. Mattevi, G. Obmolova, E. Schulze, K. H. Kalk, A. H. Westphal, A. D. Kok, and W. G. J. Hol, *Science*, **255**, 1544–1550 (1992).

35. R. A. Milligan, *Proc. Natl. Acad. Sci. USA*, **93**, 21–26 (1996).

36. A. Miyazawa, Y. Fujiyoshi, M. Stowell, and N. Unwin, *J. Mol. Biol.*, **288**, 765–786 (1999).

37. K. Hirose, W. B. Amos, A. Lockhart, R. A. Cross, and L. A. Amos, *J. Struct. Biol.*, **118**, 140–148 (1997).

38. K. Namba and F. Vonderviszt, *Q. Rev. Biophys.*, **30**, 1–65 (1997).

39. T.-W. Jeng, R. A. Crowther, G. Stubbs, and W. Chui, *J. Mol. Biol.*, **205**, 251–257 (1989).

40. A. Cheng, A. N. van Hoek, M. Yeager, A. S. Verkman, and A. K. Mitra, *Nature*, **387**, 627–630 (1997).

41. V. M. Unger, N. M. Kumar, N. B. Gilula, and M. Yeager, *Science*, **283**, 1176–1180 (1999).

42. D. A. Winkelmann, T. S. Baker, and I. Rayment, *J. Cell Biol.*, **114**, 701–713 (1991).

43. K. A. Taylor, J. Tang, Y. Cheng, and H. Winkler, *J. Struct. Biol.*, **120**, 372–386 (1997).

44. B. Böttcher, S. A. Wynne, and R. A. Crowther, *Nature*, **386**, 88–91 (1997).

45. A. Malhotra, P. Penczek, R. K. Agrawal, I. S. Gabashvili, R. A. Grassucci, R. Jünemann, N. Burkhardt, K. H. Nierhaus, and J. Frank, *J. Mol. Biol.*, **280**, 103–116 (1998).

46. R. W. Horne and I. Pasquali-Ronchetti, *J. Ultrastruct. Res.*, **47**, 361–383 (1974).

47. H. Yoshimura, M. Matsumoto, S. Endo, and K. Nagayama, *Ultramicroscopy*, **32**, 265–274 (1990).

48. R. Kornberg and S. A. Darst, *Curr. Opin. Struct. Biol.*, **1**, 642–646 (1991).

49. B. Jap, M. Zulauf, T. Scheybani, A. Hefti, W. Baumeister, and U. Aebi, *Ultramicroscopy*, **46**, 45–84 (1992).

50. E. W. Kubalek, S. F. J. LeGrice, and P. O. Brown, *J. Struct. Biol.*, **113**, 117–123 (1994).

51. J.-L. Rigaud, G. Mosser, J.-J. Lacapere, A. Olofsson, D. Levy, and J.-L. Ranck, *J. Struct. Biol.*, **118**, 226–235 (1997).

52. L. Hasler, J. B. Heymann, A. Engel, J. Kistler, and T. Walz, *J. Struct. Biol.*, **121**, 162–171 (1998).

53. I. Reviakine, W. Bergsma-Schutter, and A. Brisson, *J. Struct. Biol.*, **121**, 356–361 (1998).

54. E. M. Wilson-Kubalek, R. E. Brown, H. Celia, and R. A. Milligan, *Proc. Natl. Acad. Sci. USA*, **95**, 8040–8045 (1998).

55. A. Polyakov, C. Richter, A. Malhotra, D. Koulich, S. Borukhov, and S. A. Darst, *J. Mol. Biol.*, **281**, 465–473 (1998).

56. M. Adrian, J. Dubochet, J. Lepault, and A. W. McDowall, *Nature*, **308**, 32–36 (1984).

57. J. R. Bellare, H. T. Davis, L. E. Scriven, and Y. Talmon, *J. Electron Microsc. Technol.*, **10**, 87–111 (1988).

58. E. Mayer and G. Astl, *Ultramicroscopy*, **45**, 185–197 (1992).

59. J. Berriman and N. Unwin, *Ultramicroscopy*, **56**, 241–252 (1994).

60. H. D. White, M. L. Walker, and J. Trinick, *J. Struct. Biol.*, **121**, 306–313 (1998).

61. S. Subramaniam, M. Gerstein, D. Oesterhelt, and R. Henderson, *EMBO J.*, **12**, 1–18 (1993).

62. D. P. Siegel, W. J. Green, and Y. Talmon, *Biophys. J.*, **66**, 402–414 (1994).

63. S. Trachtenberg, *J. Struct. Biol.*, **123**, 45–55 (1998).

64. N. Unwin, *Nature*, **373**, 37–43 (1995).

65. S. D. Fuller, J. A. Berriman, S. J. Butcher, and B. E. Gowen, *Cell*, **81**, 715–725 (1995).

66. D. P. Siegel and R. M. Epand, *Biophys. J.*, **73**, 3089–3111 (1997).

67. M. Walker, X.-Z. Zhang, W. Jiang, J. Trinick, and H. D. White, *Proc. Natl. Acad. Sci. USA*, **96**, 465–470 (1999).

68. M. Cyrklaff and W. Kühlbrandt, *Ultramicroscopy*, **55**, 141–153 (1994).

69. J. C. H. Spence, *Experimental High-Resolution Electron Microscopy*, Oxford University Press, Oxford, UK, 1988.

70. L. Reimer, *Transmission Electron Microscopy*, Springer-Verlag, Berlin, 1989.

71. R. H. Wade and J. Frank, *Optik*, **49**, 81–92 (1977).

72. R. H. Wade, *Ultramicroscopy*, **46**, 145–156 (1992).

73. C. Toyoshima, K. Yonekura, and H. Sasabe, *Ultramicroscopy*, **48**, 165–176 (1993).

74. R. H. Cheng, N. H. Olson, and T. S. Baker, *Virology*, **186**, 655–668 (1992).

75. B. L. Trus, R. B. S. Roden, H. L. Greenstone, M. Vrhel, J. T. Schiller, and F. P. Booy, *Nat. Struct. Biol.*, **4**, 413–420 (1997).

76. F. Zemlin, *Ultramicroscopy*, **46**, 25–32 (1992).

77. Z. H. Zhou and W. Chiu, *Ultramicroscopy*, **49**, 407–416 (1993).

78. F. Zemlin, *Micron*, **25**, 223–226 (1994).

79. E. J. Mancini, F. D. Haas, and S. D. Fuller, *Structure*, **5**, 741–750 (1997).

80. J. Brink, M. B. Sherman, J. Berriman, and W. Chiu, *Ultramicroscopy*, **72**, 41–52 (1998).

81. O. L. Krivanek and P. E. Mooney, *Ultramicroscopy*, **49**, 95–108 (1993).

82. M. B. Sherman, J. Brink, and W. Chiu, *Micron*, **27**, 129–139 (1996).

83. D. J. DeRosier and P. B. Moore, *J. Mol. Biol.*, **52**, 355–369 (1970).

84. J. L. Carrascosa and A. C. Steven, *Micron*, **9**, 199–206 (1978).

85. M. van Heel and J. Frank, *Ultramicroscopy*, **6**, 187–194 (1981).

86. I. S. Gabashvili, R. K. Agrawal, C. M. T. Spahn, R. A. Grassucci, D. I. Svergun, J. Frank, and P. Penczek, *Cell*, **100**, 537–549 (2000).

87. M. Radermacher, T. Wagenknecht, A. Verschoor, and J. Frank, *J. Microsc.*, **146**, 113–136 (1987).

88. M. Radermacher, *J. Electron Microsc. Technol.*, **9**, 359–394 (1988).

89. J. Frank, M. Radermacher, P. Penczek, J. Zhu, Y. Li, M. Ladjadj, and A. Leith, *J. Struct. Biol.*, **116**, 190–199 (1996).

90. M. van Heel, *Ultramicroscopy*, **21**, 111–124 (1987a).

91. M. van Heel, G. Harauz, and E. V. Orlova, *J. Struct. Biol.*, **116**, 17–24 (1996).

92. M. Radermacher in D.-P. Hader, Ed., *Image Analysis in Biology*, CRC Press, Boca Raton, FL, 1991, pp. 219–246.

93. M. Radermacher in J. Frank, Ed., *Electron Tomography*, Plenum Press, New York, 1992, pp. 91–115.

94. M. Radermacher, *Ultramicroscopy*, **53**, 121–136 (1994).

95. P. A. Penczek, R. A. Grassucci, and J. Frank, *Ultramicroscopy*, **53**, 251–270 (1994).

96. M. Schatz and M. van Heel, *Ultramicroscopy*, **32**, 255–264 (1990).

97. P. Penczek, M. Radermacher, and J. Frank, *Ultramicroscopy*, **40**, 33–53 (1992).

98. N. Grigorieff, *J. Mol. Biol.*, **277**, 1033–1046 (1998).

99. D. E. Olins, A. L. Olins, H. A. Levy, R. C. Durfee, S. M. Margle, E. P. Tinnel, and S. D. Dover, *Science*, **220**, 498–500 (1983).

100. U. Skoglund and B. Daneholt, *Trends Biochem. Sci.*, **11**, 499–503 (1986).

101. J. C. Fung, W. Liu, W. J. DeRuijter, H. Chen, C. K. Abbey, J. W. Sedat, and D. A. Agard, *J. Struct. Biol.*, **116**, 181–189 (1996).

102. W. Baumeister, R. Grimm, and J. Walz, *Trends Cell Biol.*, **9**, 81–85 (1999).

103. A. K. Shah and P. L. Stewart, *J. Struct. Biol.*, **123**, 17–21 (1998).

104. F. Beuron, M. R. Maurizi, D. M. Belnap, E. Kocsis, F. P. Booy, M. Kessel, and A. C. Steven, *J. Struct. Biol.*, **123**, 248–259 (1998).

105. R. A. Crowther, L. A. Amos, J. T. Finch, D. J. DeRosier, and A. Klug, *Nature*, **226**, 421–425 (1970).

106. R. A. Crowther, *Philos. Trans. R. Soc. Lond.*, **261**, 221–230 (1971).

107. S. D. Fuller, S. J. Butcher, R. H. Cheng, and T. S. Baker, *J. Struct. Biol.*, **116**, 48–55 (1996).

108. R. H. Cheng, V. S. Reddy, N. H. Olson, A. J. Fisher, T. S. Baker, and J. E. Johnson, *Structure*, **2**, 271–282 (1994).

109. R. A. Crowther, N. A. Kiselev, B. Böttcher, J. A. Berriman, G. P. Borisova, V. Ose, and P. Pumpens, *Cell*, **77**, 943–950 (1994).

110. T. S. Baker and R. H. Cheng, *J. Struct. Biol.*, **116**, 120–130 (1996).

111. J. R. Castón, D. M. Belnap, A. C. Steven, and B. L. Trus, *J. Struct. Biol.*, **125**, 209–215 (1999).

112. R. A. Crowther, R. Henderson, and J. M. Smith, *J. Struct. Biol.*, **116**, 9–16 (1996).

113. J. A. Lawton and B. V. V. Prasad, *J. Struct. Biol.*, **116**, 209–215 (1996).

114. P. A. Thuman-Commike and W. Chiu, *J. Struct. Biol.*, **116**, 41–47 (1996).

115. I. M. Boier Martin, D. C. Marinescu, R. E. Lynch, and T. S. Baker, *J. Struct. Biol.*, **120**, 146–157 (1997).

116. Z. H. Zhou, W. Chiu, K. Haskell, H. J. Spears, J. Jakana, F. J. Rixon, and L. R. Scott, *Biophys. J.*, **74**, 576–588 (1998).

117. P. L. Stewart, C. Y. Chiu, S. Huang, T. Muir, Y. Zhao, B. Chait, P. Mathias, and G. R. Nemerow, *EMBO J.*, **16**, 1189–1198 (1997).

118. J. Walz, T. Tamura, N. Tamura, R. Grimm, W. Baumeister, and A. J. Koster, *Mol. Cell*, **1**, 59–65 (1997).

119. M. Stewart, *J. Electron Microsc. Technol.*, **9**, 325–358 (1988).

120. C. Toyoshima and N. Unwin, *J. Cell Biol.*, **111**, 2623–2635 (1990).

121. D. G. Morgan and D. DeRosier, *Ultramicroscopy*, **46**, 263–285 (1992).

122. N. Unwin, *J. Mol. Biol.*, **229**, 1101–1124 (1993).

123. R. Beroukhim and N. Unwin, *Ultramicroscopy*, **70**, 57–81 (1997).

124. E. H. Egelman, *Ultramicroscopy*, **19**, 367–374 (1986).

125. B. Carragher, M. Whittaker, and R. A. Milligan, *J. Struct. Biol.*, **116**, 107–112 (1996).

126. C. H. Owen, D. G. Morgan, and D. J. DeRosier, *J. Struct. Biol.*, **116**, 167–175 (1996).

127. R. Henderson, J. M. Baldwin, K. H. Downing, J. Lepault, and F. Zemlin, *Ultramicroscopy*, **19**, 147–178 (1986).

128. J. M. Baldwin, R. Henderson, E. Beckman, and F. Zemlin, *J. Mol. Biol.*, **202**, 585–591 (1988).

129. S. Hardt, B. Wang, and M. F. Schmid, *J. Struct. Biol.*, **116**, 68–70 (1996).

130. R. A. Crowther and P. K. Luther, *Nature*, **307**, 569–570 (1984).

131. H. Winkler and K. A. Taylor, *J. Struct. Biol.*, **116**, 241–247 (1996).

132. J. Frank in J. Frank, Ed., *Electron Tomography: Three-Dimensional Imaging with the Transmission Electron Microscope*, Plenum Press, New York, 1992, 399 pp.

133. J. Frank, A. Verschoor, and M. Boublik, *Science*, **214**, 1353–1355 (1981).

134. M. van Heel, *Ultramicroscopy*, **21**, 95–100 (1987b).

135. M. van Heel and J. Hollenberg in W. Baumeister and W. Vogell, Eds., *Electron Microscopy at Molecular Dimensions*, Springer-Verlag, Berlin, 1980, pp. 256–260.

136. M. Unser, B. L. Trus, J. Frank, and A. C. Steven, *Ultramicroscopy*, **30**, 429–434 (1989).

137. D. A. Agard, *J. Mol. Biol.*, **167**, 849–852 (1983).

138. J. M. Valpuesta, J. L. Carrascosa, and R. Henderson, *J. Mol. Biol.*, **240**, 281–287 (1994).

139. J. T. Finch, *J. Mol. Biol.*, **66**, 291–294 (1972).

140. Z. H. Zhou, S. Hardt, B. Wang, M. B. Sherman, J. Jakana, and W. Chiu, *J. Struct. Biol.*, **116**, 216–222 (1996).

141. N. H. Olson and T. S. Baker, *Ultramicroscopy*, **30**, 281–298 (1989).

142. D. M. Belnap, N. H. Olson, and T. S. Baker, *J. Struct. Biol.*, **120**, 44–51 (1997).

143. Y. Fujiyoshi, T. Mizusaki, K. Morikawa, H. Yamagishi, Y. Aoki, H. Kihara, and Y. Harada, *Ultramicroscopy*, **38**, 241–251 (1991).

144. F. Zemlin, E. Beckmann, and K. D. vanderMast, *Ultramicroscopy*, **63**, 227–238 (1996).

145. N. Kisseberth, M. Whittaker, D. Weber, C. S. Potter, and B. Carragher, *J. Struct. Biol.*, **120**, 309–319 (1997).

146. M. Hadida-Hassan, S. J. Young, S. T. Peltier, M. Wong, S. Lamont, and M. H. Ellisman, *J. Struct. Biol.*, **125**, 235–245 (1999).

147. B. F. McEwen, K. H. Downing, and R. M. Glaeser, *Ultramicroscopy*, **60**, 357–373 (1995).

CHAPTER FIFTEEN

Peptidomimetics for Drug Design

M. Angels Estiarte
Daniel H. Rich
School of Pharmacy—Department of Chemistry
University of Wisconsin-Madison
Madison, Wisconsin

Contents

Burger's Medicinal Chemistry and Drug Discovery
Sixth Edition, Volume 1: Drug Discovery
Edited by Donald J. Abraham
ISBN 0-471-27090-3 © 2003 John Wiley & Sons, Inc.

1 INTRODUCTION

Protein–protein interactions are central to biology and provide one mechanism to convert genomic information into regulated biological responses. Important examples of protein-peptide interactions include the binding of peptide ligands to proteases, the binding of peptide hormones to peptide receptors, the recruitment of proteins to effect signal transduction, and apoptosis. Peptides also act as neurotransmitters, neuromodulators, hormones, and autocrine and paracrine factors. Unfortunately, their use as pharmaceutical drugs is made difficult by their poor pharmacokinetic profiles; they are easily proteolyzed, poorly transported, and rapidly excreted. Although modern formulation techniques have improved delivery of peptides (e.g., inhalation of insulin), there remains a need for small potent molecules that can be administered orally.

For these reasons, much effort has been expended to find ways to replace portions of peptides with non-peptide structures, called peptidomimetics, in the hope of obtaining orally bioavailable entities. Several types of peptidomimetics have been developed, and the field has emerged as one of the important approaches to drug design and discovery. This review will describe the various methods developed to design peptidomimetics. Due to space limitations, the biological rationale for each peptidomimetic and its chemical synthesis can not be covered. Selected examples of the strategies employed to obtain peptidomimetics are provided to illustrate the breadth of research in this field.

2 CLASSIFICATION OF PEPTIDOMIMETICS

The term peptidomimetic is often used in the literature to indicate a multitude of structural types that differ in fundamental ways. Comparisons between peptidomimetics suffer from the lack of accepted definitions of what a peptidomimetic is (1). The term is often applied to highly modified analogs of peptides without distinguishing how these differ from classical analogs of peptides. For example, peptide (**2**) is derived from the decapeptide LH-RH (**1**); (**2**) contains only five amino acids, none of which is present in the parent compound, yet it is a powerful antagonist of the LH-RH receptor (Fig. 15.1) (2). Is (**2**) a peptide analog or a peptidomimetic?

In the 1970s, Hughes et al. were the first to show that two very different chemical structures have similar agonist properties (3). The opioid natural product, morphine (**3**), was found to resemble the *N*-terminal structure of the endogenous opioid peptides, enkephalins, (**4a**) and (**4b**), and β-endorphin (**5**) (Fig. 15.2). The remarkable similarity between the morphine phenol system and the *N*-terminal tyrosine residue in the peptide opioids implied that these units reacted with opioid receptors in a similar fashion to elicit comparable responses (4–6).

The realization that a non-peptide natural product was mimicking the action of a natural peptide effector led Farmer to postulate that other non-peptide structures might be found that would mimic other peptide effectors (7). His concept of "peptide mimetic," which later was called "peptidomimetic," proposed that

*p*Glu-His-Trp-Ser-Tyr-Gly-Leu-Arg-Pro-Gly-NH$_2$

LH-RH

(**1**)

$\downarrow$

(4-fluorophenyl)propionyl-*D*1Nal-*N*MeTyr-*D*Lys(Nic)-Lys(Isp)-*D*Ala-NH$_2$

A-76154   ED$_{50}$ = 10.3 µg/ml

(**2**)

Figure 15.1. Reduced-size antagonist of LH-RH.

Met-enkephalin   Tyr-Gly-Gly-Phe-Met
(**4a**)

Leu-Enkephalin   Tyr-Gly-Gly-Phe-Leu
(**4b**)

β-Endorphin   Tyr-Gly-Gly-Phe-Met-Thr-Ser-Glu-Lys-Ser-
Gln-Thr-Pro-Leu-Val-Thr-Phe-Lys-Asn-Ala-
Ile-Ile-Lys-Asn-Ala-Tyr-Lys-Lys-Gly-Glu
(**5**)

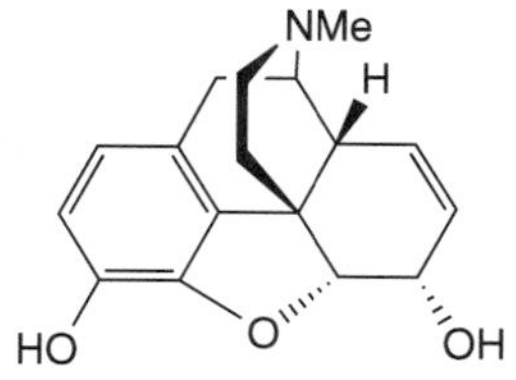

Morphine
(**3**)

Figure 15.2. Examples of peptidic and non-peptidic opioid receptor ligands.

novel scaffolds could be designed to replace the entire peptide backbone while retaining isosteric topography of the enzyme-bound peptide (or assumed receptor-bound) conformation. Farmer's definition went beyond simple replacement of amide bonds and the concept of stringing together conformationally restricted amino acid derivatives to mimic the native peptide structure. In the intervening years, many non-peptide and partially peptide structures have been found that mimic (or antagonize) the action of the peptide ligand at its receptor; this is particularly true with substances active at G-protein–coupled receptors.

The pyrrolinone unit (**6**) designed by Smith and Hirschmann illustrates a modern use of these two concepts (Fig. 15.3) (8). Pyrrolinones constrain the peptide-like side-chains into an extended β-structure topography that fits the active sites of most peptidases; pyrrolinones are resistant to normal proteolysis because no α-amino acid units remain, and the units impart sophisticated partitioning properties to the final inhibitor. Pyrrolinones, like many peptide-derived peptidomimetics, retain an atom-to-atom correspondence to the parent peptide, especially with respect to the peptide backbone structure. Most of these structures contain elements that accomplish one of two objectives: they replace amide bonds with metabolically stable units, and they affect a conformational constraint on peptides or on the peptide replacement. In contrast, heterocyclic natural products or screening leads that bind to peptide receptors also have been called peptidomimetics by virtue of their mimicking (or antagonizing) the function of the natural peptide. Although structural data confirming mimicry of the designed mimetics are rarely available for receptor bound ligands, ample ev-

idence is available from X-ray crystallography that heterocyclic inhibitors are mimicking the extended β-strand of enzyme-bound substrate-derived inhibitors (*vide infra*).

Based on these considerations, four distinct types of peptidomimetics have been identified to date (9, 10). The first invented were structures that contain one or more mimics of the local topography about an amide bond (amide bond isosteres). Strictly speaking, these are properly classified as *pseudopeptides* (11), but in recent years, they have been called peptidomimetics on occasion. For historical reasons, we classify the peptide backbone mimetics as *type I* mimetics (Table 15.1). These

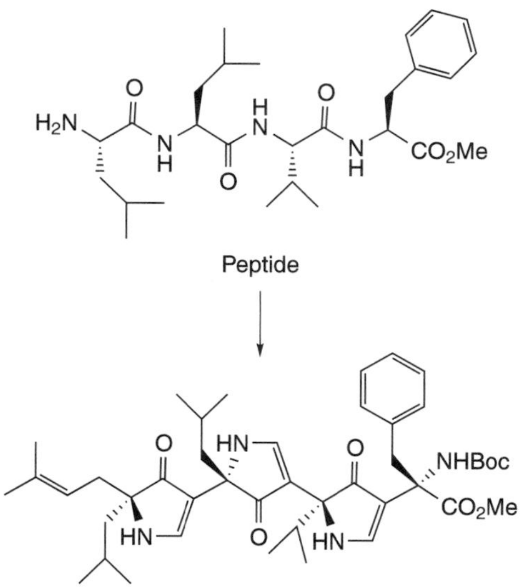

Peptide

↓

Pyrrolinone analog
(**6**)

Figure 15.3. Correlation of pyrrolinone-based peptidomimetics and the parent peptide.

Table 15.1 Peptidomimetic Types

Peptidomimetic			Examples
Type I	Peptide backbone mimetics	Substrate-based design	Pseudopeptides
Type II	Functional mimetics	Molecular modeling, HTS	GPCR antagonists
Type III	Topographical mimetics	Structure-based design	Non-peptide protease inhibitors
Type IV	Non-peptide peptidomimetics	Group Replacement Assisted Binding	Piperidine inhibitors

mimetics often match the peptide backbone atom-for-atom while retaining functionality that makes important contacts with binding sites. Some units mimic short portions of secondary structure (e.g., β-turns) and have been used to generate lead compounds. Many early protease inhibitors were designed from transition state analog mimetics or from collected substrate/product mimetics, each designed to mimic reaction pathway intermediates of the enzyme-catalyzed reaction. These are mimics of the peptide bond in a putative transition state or product state and will be classified here as peptidomimetics.

The second type of mimetic to emerge was the *functional mimetic*, or *type II* mimetic, which is a small non-peptide molecule that binds to a peptide receptor. Morphine was the first well-characterized example of this type of peptidomimetic. Initially, type II mimetics were presumed to be direct structural analogs of the natural peptide, but characterization of both the endogenous peptide and antagonist's binding sites by site-directed mutagenesis has raised doubts about this interpretation (12). The mutagenesis data indicate that antagonists for a large number of receptors seem to bind to receptor subsites different than those used by the parent peptide. Consequently, functional mimetics may not mimic the structure of the parent hormone; this remains to be determined. Despite this uncertainty, the approach has been quite successful and produced a number of potential drug lead structures.

Type III mimetics represent the *Farmer definition of peptidomimetics* in that they *possess novel templates, which appear unrelated to the original peptides but contain the essential groups, positioned on a novel non-peptide scaffold to serve as topographical mimetics*. Several type III peptidomimetic protease inhibitors have been characterized where direct

X-ray structural determination of both the peptide-derived inhibitor and the heterocyclic non-peptide inhibitor complexes have been compared. These examples demonstrate that alternate scaffolds can display side-chains so that they interact with proteins in fashion closely related to that of the parent peptide.

Recently, a fourth type of peptidomimetic called a *GRAB-peptidomimetic* (group replacement–assisted binding) has been identified (10). These structures might share structural-functional features of type I peptidomimetics, but they bind to an enzyme form not accessible with type I peptidomimetics.

Previous reviews on peptidomimetics have addressed pseudopeptides (11), macrocyclic mimetics (13), natural product mimetics (14), cyclic protease inhibitors (15), mimetics for receptor ligands (16–22), and earlier general overviews (23–29). This review will focus on the design process itself. Novel peptidomimetics in which the structural relationship between parent peptide and the peptidomimetic has been established by biophysical methods are used to clarify the principles. Successful approaches are highlighted to illustrate how these concepts are currently used.

3 DESIGN OF CONFORMATIONALLY RESTRICTED PEPTIDES

Peptide derivatives that contain conformationally restricting amino acid units or other conformational constraints were first called conformationally constrained (or restricted) peptide analogs. The use of steric hindrance or cyclization to limit rotational degrees of freedom in biologically active molecules has a long history and was originally applied to non-peptide neurotransmitters (30). Subsequently, it was applied to amino acid substituents and to

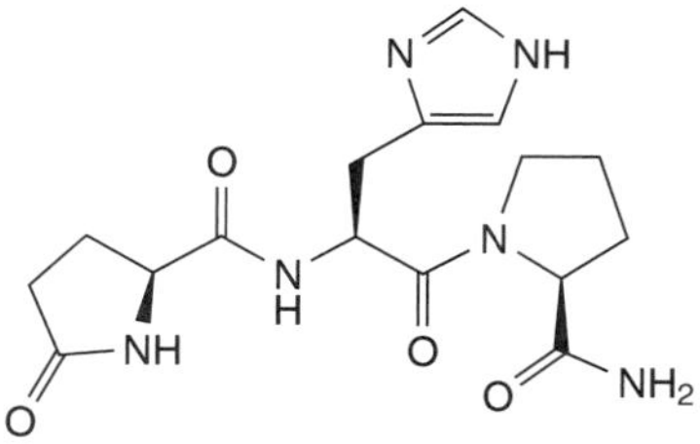

Figure 15.4. Structure of TRH tripeptide.

cyclic peptides (31, 32) and to control secondary structure in model proteins.

Conformational restriction is a very powerful method for probing the bioactive conformations of peptides. Small peptides have many flexible torsion angles so that enormous numbers of conformations are possible in solution. For example, a simple tripeptide such as thyrotropin-releasing hormone (TRH; **7**) (Fig. 15.4) with six flexible bonds could have over 65,000 possible conformations. The number of potential conformers for larger peptides is enormous, and some method is needed to exclude potential conformers. Modern biophysical methods, e.g., X-ray crystallography or isotope edited nuclear magnetic resonance (NMR), (33) can be used to characterize peptide–protein interactions for soluble proteins, but most biophysical methods cannot yet determine the conformation of a ligand bound to constitutive receptors, e.g., G-protein–coupled receptors (34, 35).

Cyclization is one of the earliest techniques applied to design peptidomimetics. Cyclic peptides are more stable to amide bond hydrolysis and allow less conformational flexibility; consequently, the resulting analogs are anticipated to be more selective and less toxic. Methods for restricting conformations include peptide backbone cyclization, disulfide bond formation, side-chain cyclization, and metal ion chelation.

The first successful application of conformational restriction to peptide chemistry was carried out by Veber et al. at Merck, (36), who were trying to simplify the structure of somatostatin (**8**) (Fig. 15.5) to produce an orally active derivative. Their approach was to introduce conformational restraints into the mac-

rocyclic peptide ring system to reduce the number of conformations available to the analog. Not all substitutions were expected to produce biologically active products, but those that retained activity were assumed to be able to adopt conformations close to the normal bioactive conformation. This work began from the earlier discovery by Rivier et al. (37) that replacement of L-tryptophan in the position-8 of somatostatin by D-tryptophan produced an analog that retained biological activity. This unusual biological result is possible when a D,L-sequence (D-Trp-Lys) replaces an L,L-sequence (Trp-Lys) in a peptide at a type II β-turn, because the topography of the amino acid side-chains at these positions is essentially identical in these turns (38). These results led Veber et al. to postulate that the amino acid sequence Phe-Trp-Lys-Thr might be part of a type II β-turn, and that this tetrapeptide sequence might comprise the active pharmacophore. Although this hypothesis was highly speculative for its time, it was shown to be essentially correct by applying the principle of conformational restriction (Fig. 15.5). Deletion of the N-terminal dipeptide, followed by insertion of the D-Trp at position-8, and replacement of the disulfide sulfurs with carbons produced analog (**9**). NMR and other data suggested that the two Phe side-chains were clustered, thus they were replaced by a transannular disulfide bond limiting the available conformation, as in compound (**10**). After several iterations of this process, a biologically active cyclic hexapeptide (**11**) was discovered that retained only 6 of the original 14 amino acids in somatostatin yet produced a fully active derivative (31).

The work of Veber et al. established that valuable information about the bioactive conformation of a flexible peptide could be obtained by applying the principles of conformational restriction, and several additional examples soon were reported that followed this strategy. Conformationally restricted enkephalin analogs, e.g., (**12–13**), were formed by cyclizing between positions 2 and 5 of enkephalins (**4a-b**) (39). Cyclization of α-melanotropin (**14**) gave the unusually active analog (**15**) (40). Small cyclic analogs of endothelin (**16**) (41) have been discovered by applying these methods, as illustrated by (**17**) (Fig.

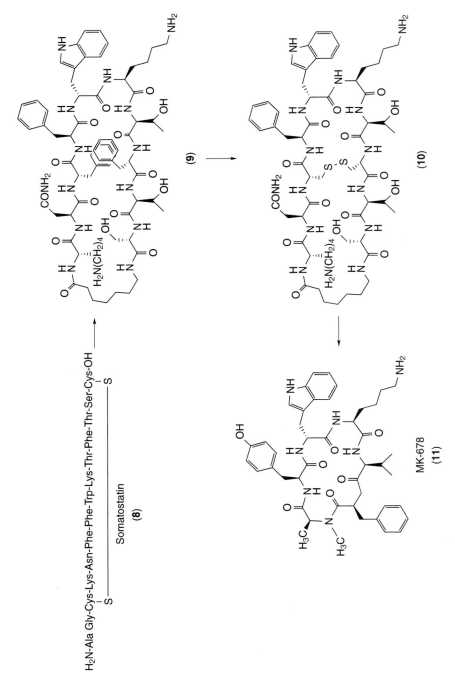

Figure 15.5. Conformationally restricted somatostatin analogs.

638

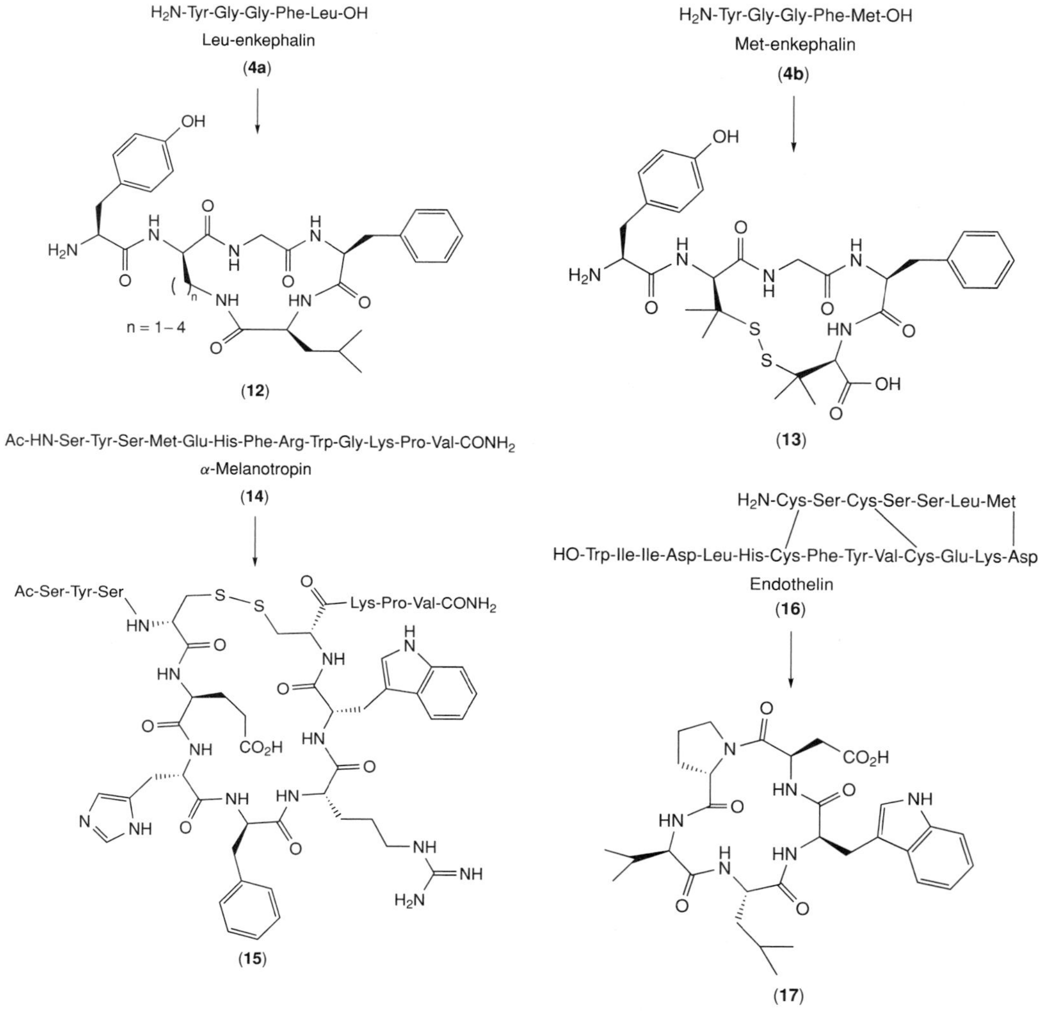

H2N-Tyr-Gly-Gly-Phe-Leu-OH
Leu-enkephalin
(4a)

H2N-Tyr-Gly-Gly-Phe-Met-OH
Met-enkephalin
(4b)

(12)

(13)

Ac-HN-Ser-Tyr-Ser-Met-Glu-His-Phe-Arg-Trp-Gly-Lys-Pro-Val-CONH2
α-Melanotropin
(14)

H2N-Cys-Ser-Cys-Ser-Ser-Leu-Met

HO-Trp-Ile-Ile-Asp-Leu-His-Cys-Phe-Tyr-Val-Cys-Glu-Lys-Asp
Endothelin
(16)

(15)

(17)

Figure 15.6. Cyclic hormone peptide analogs.

15.6). Peptide chemists routinely apply conformational restriction in their attempts to determine possible bioactive conformations.

Flexible peptides can be conformationally restricted by a variety of methods other than macrocyclization of the peptide. For example, Marshall et al. introduced α-methyl amino acid substituents into peptides as a way to decrease the conformational space available to the resulting peptide (42); these types of approaches led to his "Active Analog" approach for determining bioactive conformations of flexible molecules (43). Some other traditional

modifications of the peptide substrate are the replacement of the amino acids of the P_1-P_1' cleavage site by D-amino acids or the employment of α-C or α-N alkylated amino acids and cyclic or β-amino acids (Fig. 15.7).

Mimicking the secondary structure of peptides has become one of the most important tools for rational drug design (44–47). These methods induce the synthetic analog to adopt a set of target conformations, which are designed to mimic the bioactive conformation predicted in the native substrate from biophysical techniques. Molecular surrogates

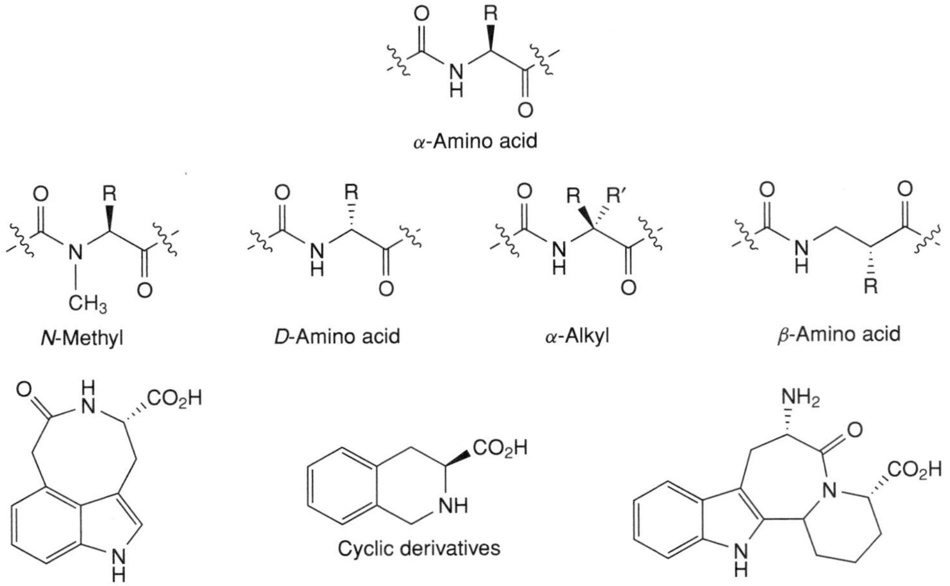

α-Amino acid

| *N*-Methyl | *D*-Amino acid | α-Alkyl | β-Amino acid |

Cyclic derivatives

Figure 15.7. Representative amino acid mimetics.

have been found that efficiently mimic turns, strands, sheets, and helices. By far, the major efforts have focused on the design of β-turn mimetics. Some of the templates used to constrain the conformational torsion angles of the peptide chain are summarized in Figs. 15.8–15.14.

In a very early example, Freidinger et al. developed a series of cyclic lactams that stabilized β- and γ-turn structures in linear peptides (Fig. 15.8). This strategy was applied to determine conformations of LH-RH that are consistent with the turn structure permitted by the constraint. For example, the 3-aminolactam (**18**) was used to mimic a β-turn conformation. When inserted in LH-RH, com-

pound (**19**) retained good biological activity so that the bioactive conformation of LH-RH probably contains a β-turn around residues 6 and 7 (48).

Conformational restriction has also been used to determine the bioactive conformation of enzyme-inhibitor systems for which no X-ray crystal structure is available. Thorsett et al. (49) synthesized conformationally restricted bicyclic lactam derivatives of the angiotensin converting enzyme (ACE) inhibitors enalapril (**20**) and enalaprilat (**21**) (Fig. 15.9) to characterize torsion angles in the bioactive conformation. Analog (**22**) was used to constrain the torsion angle psi (Ψ). Flynn et al.

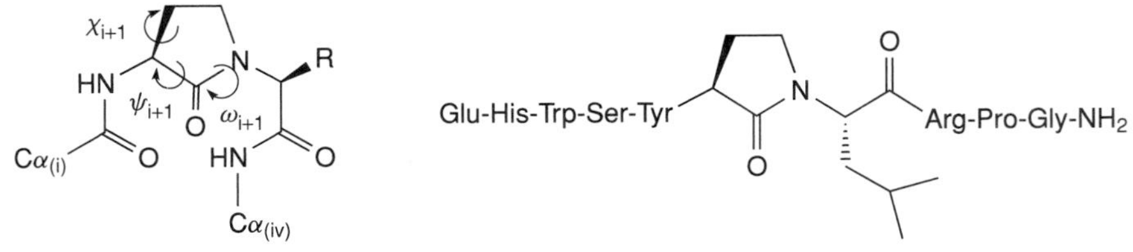

| β-Turn | LH-RH β-turn mimetic |
| (**18**) | (**19**) |

Figure 15.8. γ-Lactam analog of LH-RH. A β-turn mimetic.

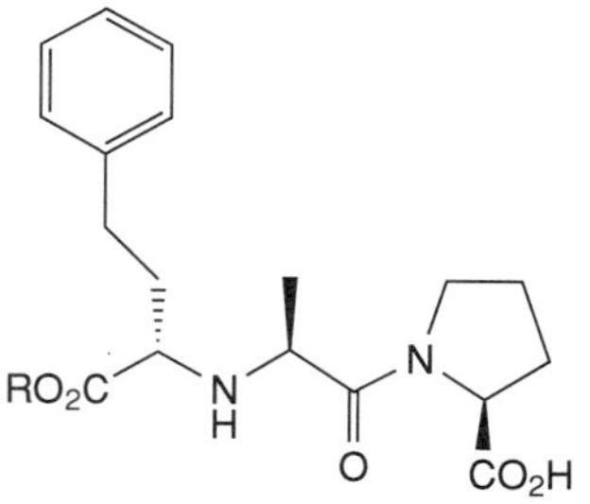

Enalapril, R = Et, (**20**)
Enalaprilat, R = H, (**21**)

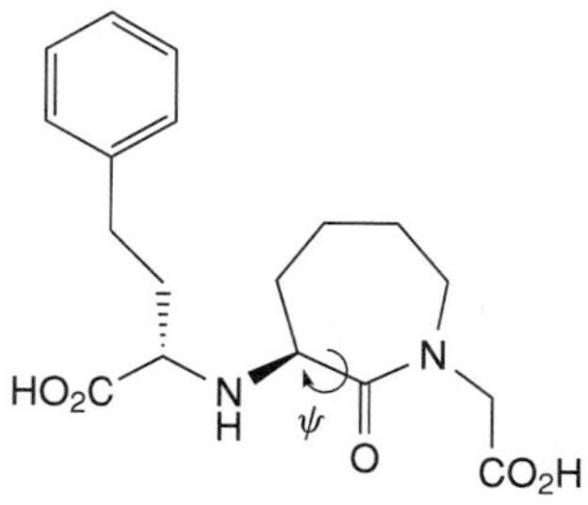

(**22**)

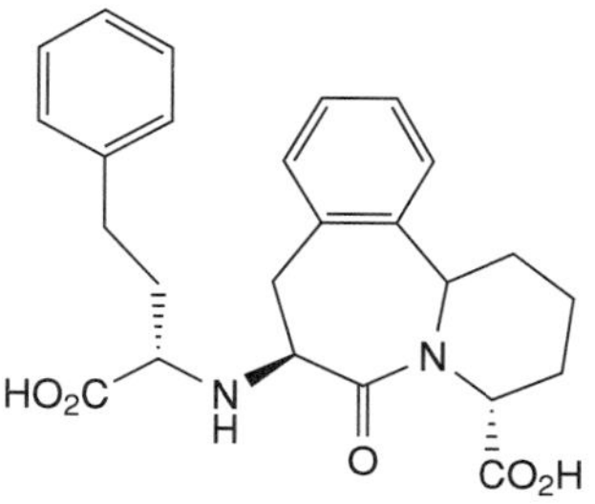

(**23**)

Figure 15.9. Conformationally restricted ACE inhibitors.

(50) extended this principle to prepare the very tight-binding tricyclic ACE inhibitor (**23**) (Fig. 15.9).

Several other γ-, δ-, and ε-lactam derivatives have been prepared and inserted into receptor antagonists or agonists. For instance, the thiazolidine lactam (**24**) (Fig. 15.10) has

been shown to induce the desired secondary structure in a gramicidine S analog. Later, it was used to prepare a conformationally restricted cyclosporin A analog (51). Several β-turn and γ-turn mimetics are shown in Figs. 15.10–15.12, and many other examples are available in the recent literature (52–54).

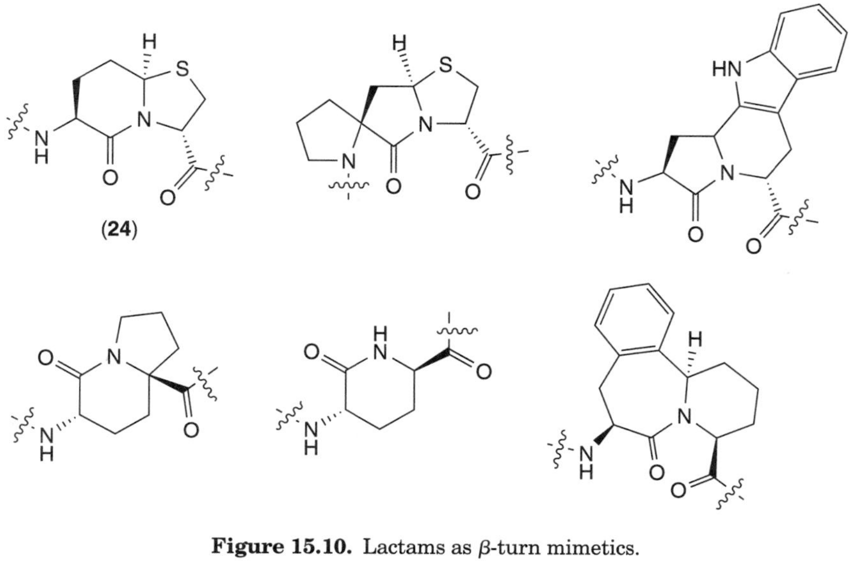

(**24**)

Figure 15.10. Lactams as β-turn mimetics.

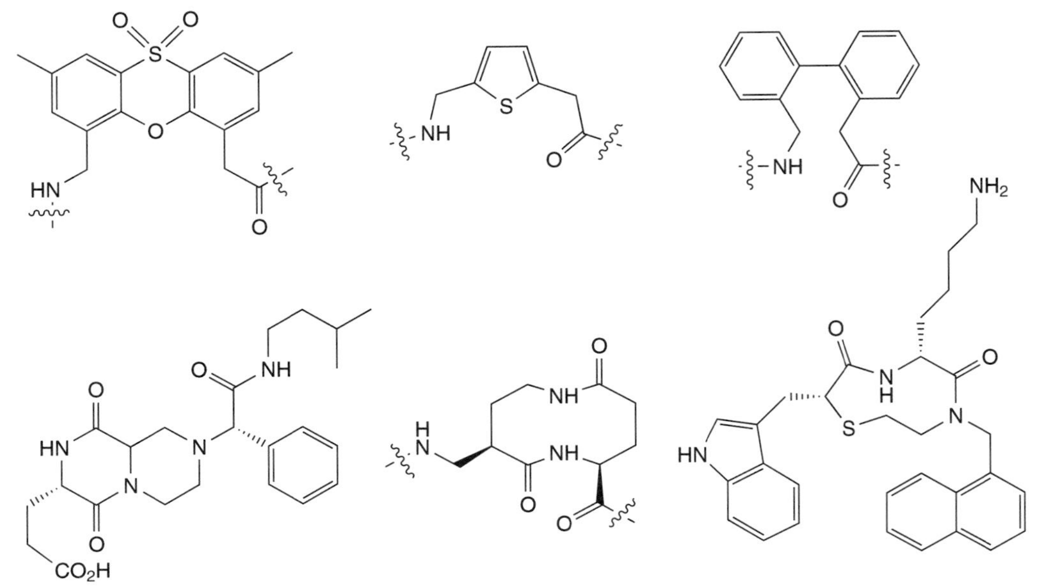

Figure 15.11. Other β-turn mimetic scaffolds.

The β-sheet is an important, biologically relevant secondary structure. As noted earlier, the pyrrolinones invented by Smith et al. (Fig. 15.3) adopt a β-strand conformation, which was corroborated by computer modeling and by X-ray crystallography (55). The diacylaminoeindolidinone (**25**), the dibenzofuran (**26**), and the *N,N*-linked oligourea (**27**) illustrate other molecular templates designed to stabilize peptides in a β-strand conformation (Fig. 15.13) (56, 57).

Peptidomimetic structures that support α-helixes (**28–30**) (Fig. 15.14) and loops have been reported less frequently because of the difficulty in presenting the side-chains correctly. However, newer approaches have pro-

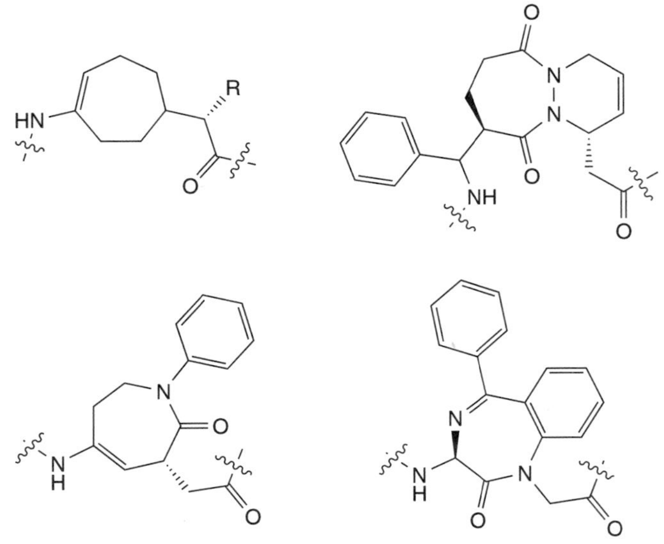

Figure 15.12. Some γ-turn mimetic scaffolds.

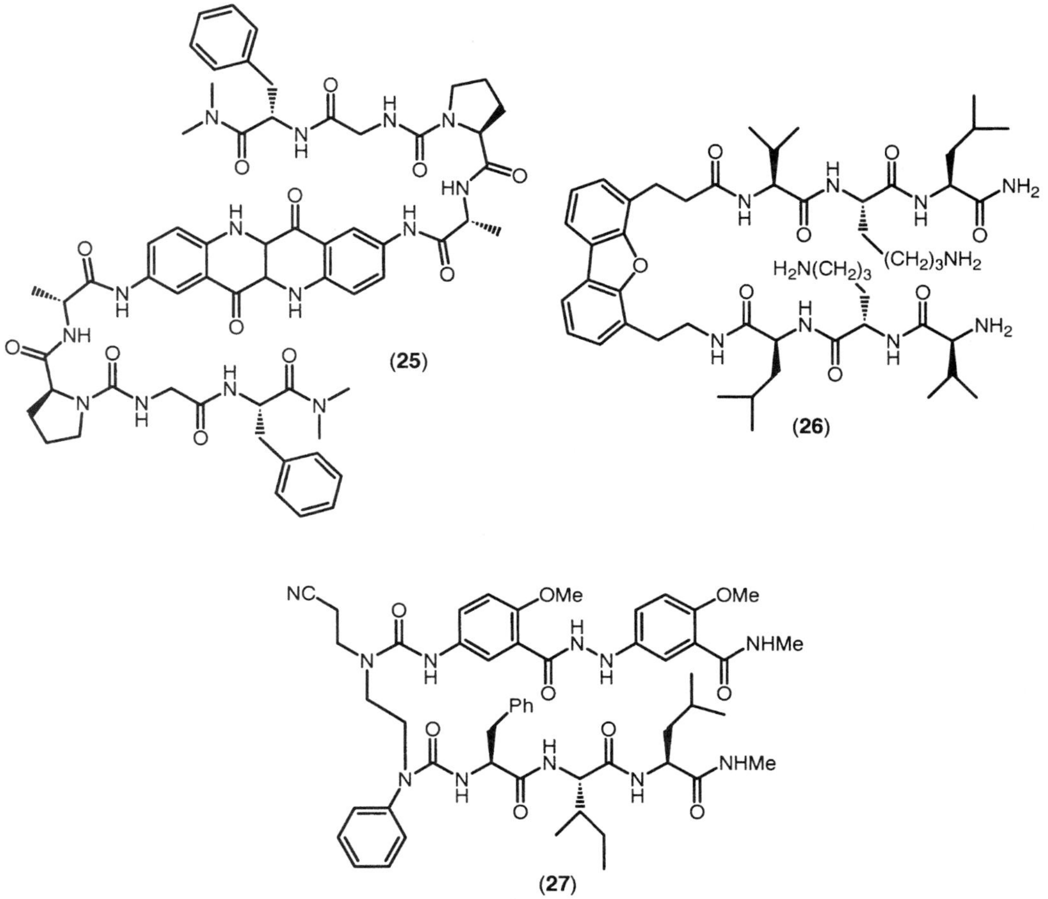

Figure 15.13. Structures of β-sheet mimetics.

vided mimetics of multiple discontinuous protein surfaces (56). Over the last few years, the Gellman, Seebach, and Hanessian research groups have invented novel helical structures (e.g., **31, 32**) by use of β-, γ-, and δ-peptides (58).

It is important to stress that even a small change in the structure or in a single torsional angle can be sufficient to dramatically modify the conformation of the resulting peptide. Numerous additional conformational constraints have been developed, and the reader is encouraged to consult these reviews for additional examples (32, 59–63).

4 TEMPLATE MIMETICS

Highly functionalized molecular scaffolds have proven to be very successful in mimick-

ing specific protein–protein interactions. Insertion of the key pharmacophoric groups into a nonpeptidic framework has provided good inhibitors of a variety of biological targets.

This technique has been successfully applied in those biological targets where the key structural amino acids of the native peptide for peptide recognition are known. Miscellaneous examples are found in glycoprotein GbIIb/IIIa inhibitors (**33**) that mimic the RGD sequence (64) or in Ras-farnesyltransferase inhibitors (**34**) that mimic the CAAX sequence (Fig. 15.15) (65).

An early example of this concept was developed by Hirschmann et al. in the design of a somatostatin analog (Fig. 15.15) (55). Three of the four crucial amino acid side-chains of the parent peptide (Tyr, Trp, and Lys) were posi-

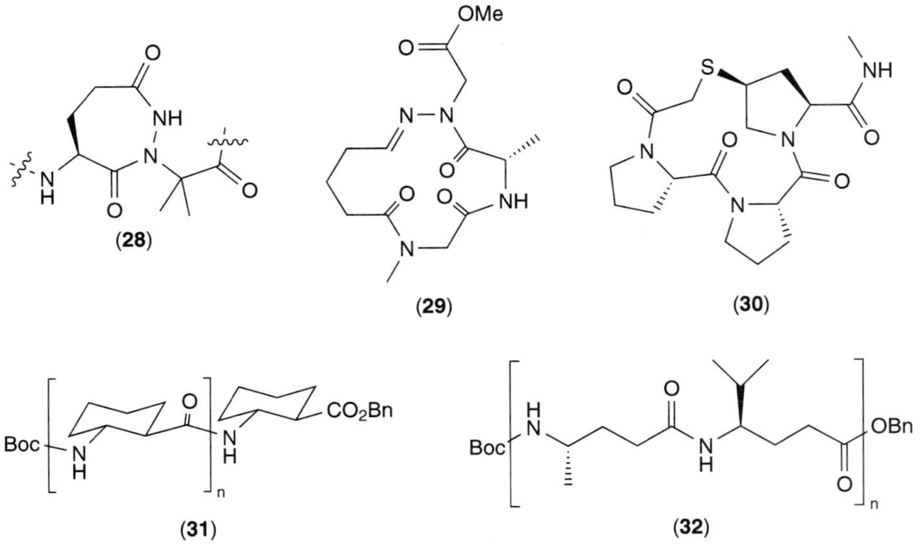

Figure 15.14. Newer templates found in helical or loop structures.

tioned on a sugar template (**35**). Although originally designed as a somatotropin release inhibitory factor (SRIF) antagonist, compound (**35**) also proved to be a good Substance P antagonist. These sugar derivatives, as well as the benzodiazepine, diphenylmethane, and spiropiperidine scaffolds, are elements found in a variety of inhibitors of receptors, and have been designated as "privileged structures" (66). Thus, these common scaffolds can often provide a template for further optimization of a desired activity. Evans et al. have noted that the essential surface area of biologically active peptides is similar to the surface area of benzodiazepines, one type of non-peptide scaffold known to bind to G-protein–coupled receptors (67).

The quest for functionalized lead structures that effectively mimic the "hot spots" within the biological ligand is not easy (68). Molecular modeling and high-throughput screening (HTS) are techniques that are currently used for this purpose and have been summarized elsewhere.

The design and synthesis of antifungal analogs of the cyclic peptide rhodopeptin (**36**) (Fig. 15.16) illustrate a recent application of peptidomimetic scaffolding, where the structure of the biological target is not known. After structure-activity relationship (SAR) studies, the important side-chains of the peptide

ligand were identified; then, NMR and molecular modeling techniques were used to model these side-chains onto known scaffolds and to compare with the original three-dimensional (3D) structure of the native peptide. Compound (**37**) (Fig. 15.16) is a potent peptidomimetic derivative with improved solubility in water that functions the same as the cyclic tetrapeptide (69, 70).

5 PEPTIDE BOND ISOSTERES

The replacement of amide bonds by retro-inverso amide replacements (71, 72) and other amide bond isosteres generates pseudopeptides (11). This process was first used to stabilize peptide hormones *in vivo*, and later to prepare transition state analog (TSA) inhibitors. Systematic efforts to convert good *in vitro* inhibitors into good *in vivo* inhibitors became the driving force for further development of peptidomimetics. Figure 15.17 illustrates some of the peptide backbone modifications that have been made in an effort to increase bioavailability. Replacement of scissile amide (CONH) bonds with groups insensitive to hydrolysis (e.g., CH_2NH) has been extensively practiced. Reviews of this work have appeared (11, 73). Removal of the proton donors and

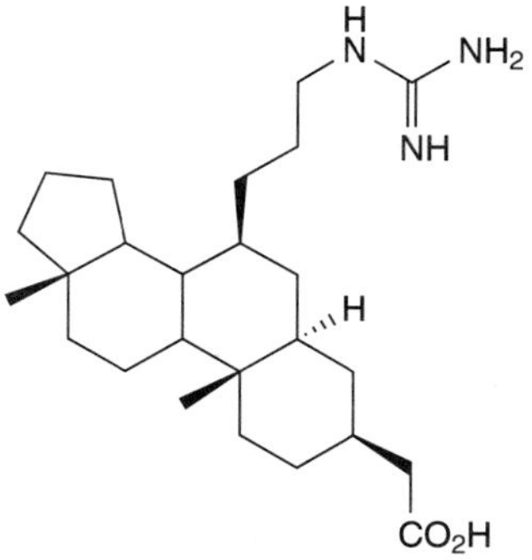

GpIIb/IIIa
(**33**)

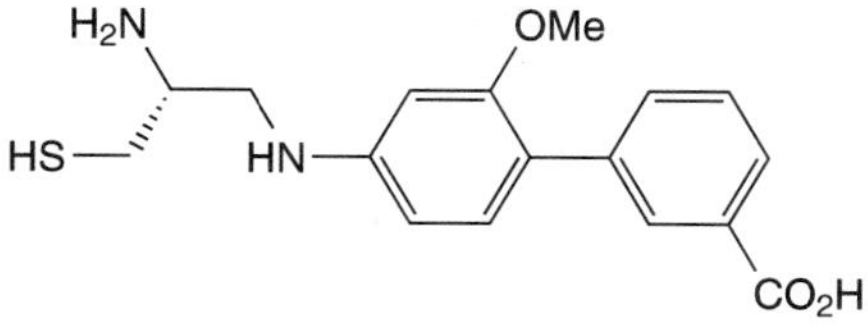

Ras-farnesyltransferase
(**34**)

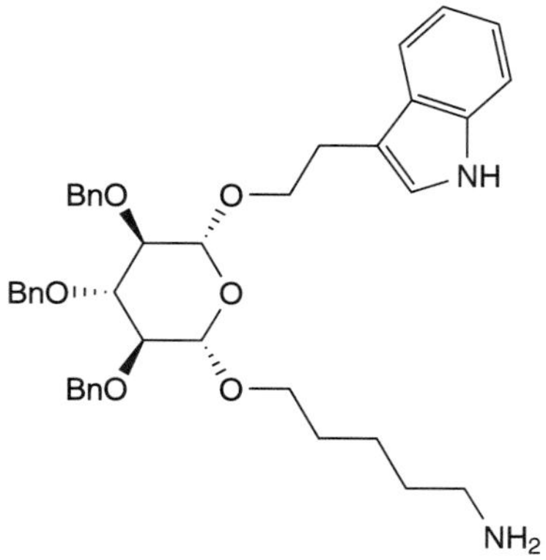

Somatostatin
(**35**)

Figure 15.15. Biologically active template mimetics.

acceptors in an amide bond also reduces hydration, which improves the ability of the compounds to penetrate lipid membranes (74). These approaches represent important first steps in development of peptidomimetics. However, removal of an amide bond also affects the geometry and increases the flexibility of the molecule at this position, which decreases ligand binding. Effective analogs have been obtained when conformational restriction, transition-state analog design, and amide bond replacements have been applied to

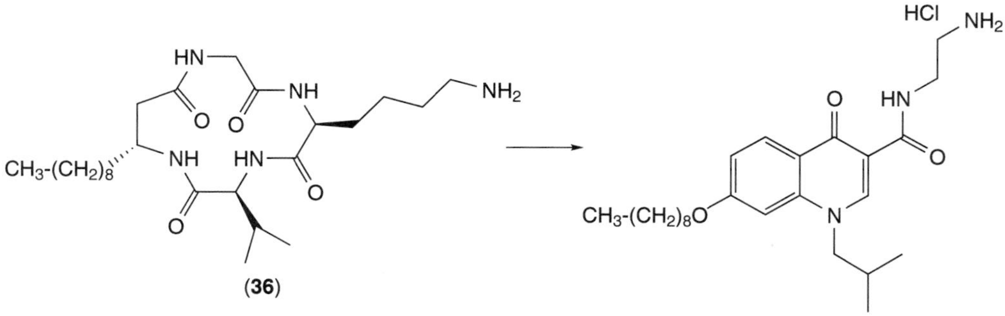

(**37**)

Figure 15.16. Rhodopeptin analogs. Representative example of scaffolding methodology.

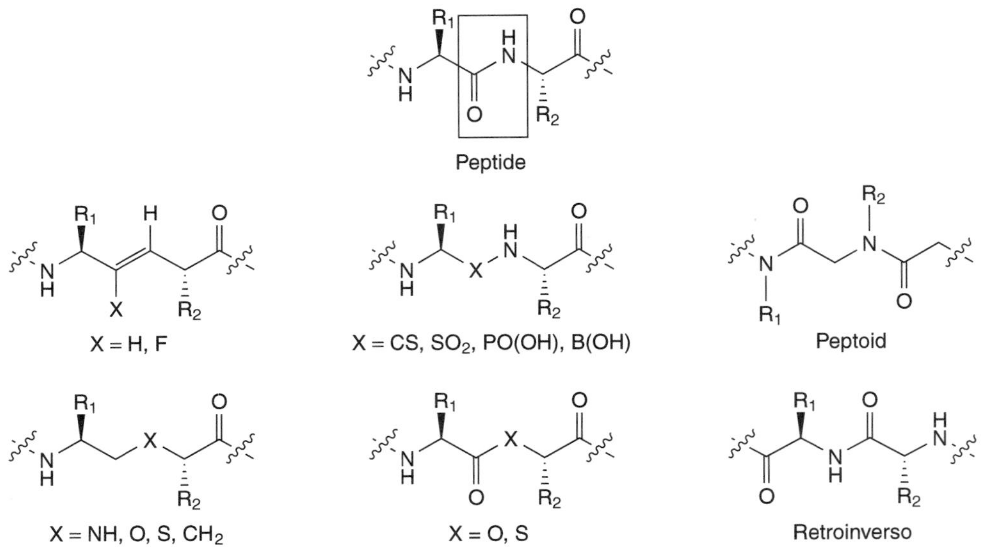

Figure 15.17. Isosteres that replace peptide backbone amide groups to generate pseudopeptides.

scaffolds with molecular weights below 500–600 (75, 76), but at present this process is very labor intensive.

6   FROM TRANSITION-STATE ANALOG INHIBITORS TO NON-PEPTIDE INHIBITORS: EXAMPLES IN PROTEASE INHIBITORS

Many peptidomimetics derived from the design of TSA inhibitors, molecules designed according to the hypothesis provided by Pauling (77) and implemented by Wolfenden (78, 79). TSA protease inhibitors are stable analogs of the tetrahedral intermediate for peptide bond hydrolysis that inhibit the enzyme (Fig. 15.18). The first successful commercial application was the development of captopril (**38**) by Ondetti et al. (80), and many applications have been reported over the past quarter century.

Figures 15.19–15.32 list examples of analogs of peptidyl transition states that have been employed to develop inhibitors of four classes of peptidases (81, 82). These units are used to replace the scissile amide bond in a substrate sequence with either an amino acid or dipeptide isostere, or with a chelating moiety in the case of metallo peptidases. The

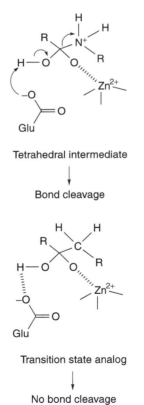

Figure 15.18. TSA inhibit peptide bond hydrolisis.

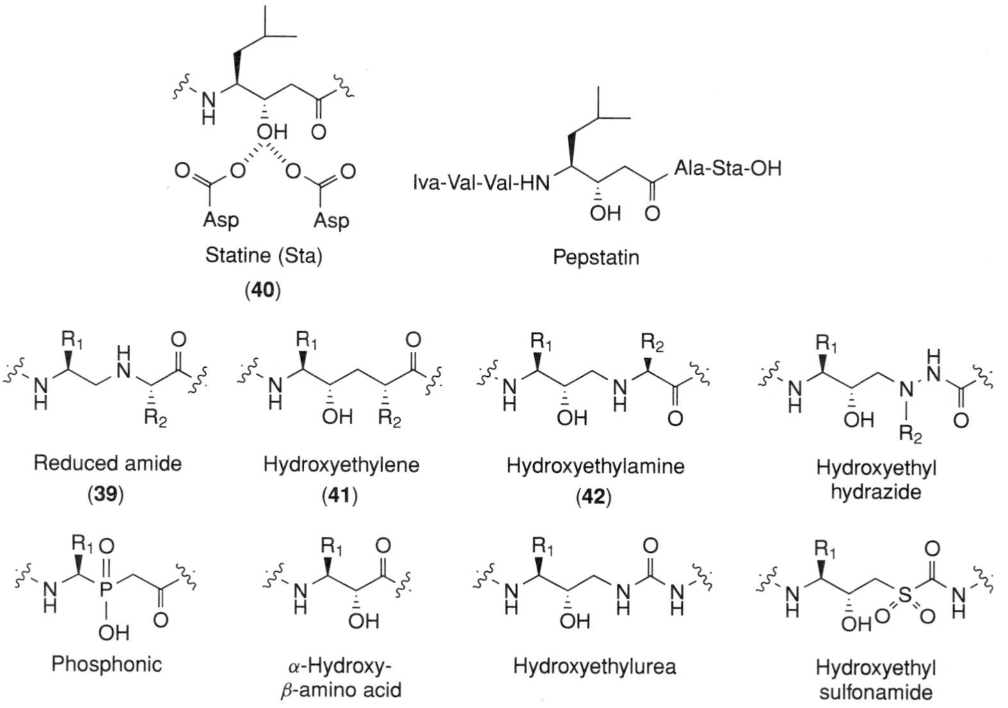

Figure 15.19. TSA used to inhibit aspartic peptidases.

dipeptide TSA provides the functionality that interacts tightly with the enzyme catalytic groups while the amino acid sequence up- and downstream on the peptide chain provides interactions that lead to selective inhibition of the target enzyme. The enzyme active site typically is buried in a cleft capable of accommodating up to three to nine amino acid residues of the substrate/inhibitor depending on the minimum amino acid sequence necessary for hydrolysis. The inhibitor's exquisite selectivity derives from the interactions of the ligand's P_6–P_3' residues with the enzyme binding sites (S_6–S_3') (83). Recently, some aspartic and serine peptidase inhibitors have been found that access an additional binding site sub-pocket (S_3^{sp}) to increase both inhibitor potency and selectivity (84–86).

6.1 TSA in Aspartic Peptidase Inhibitors

The reduced amide isostere (**39**), developed by Szelke, and the statine (hydroxylmethylene) isostere (**40**) were early transition-state analogs used to design inhibitors of various aspar-

tic proteases, (87–89), and their success led to other tetrahedral intermediate mimics such as the hydroxylethylene (**41**) and hydroxyethylamine (**42**) isosteres (Fig. 15.19) (90–92). The statine subunit, which mimics the tetrahedral intermediate, represents one of the earlier examples of TSA, although statine is one atom short in backbone length to be a true dipeptide or two atoms too long to be an isosteric replacement for a single amino acid.

Early work focused on developing inhibitors of renin as potential antihypertensive agents, but these compounds failed to become drugs primarily because of difficulties in obtaining orally active drugs. As a result, the first pharmaceutical attempts to develop renin inhibitors for treatment of hypertension through TSA-based inhibitors failed (93). It was eventually realized after extensive modifications to the ancillary peptide functionality that developing bioavailable peptide-derived inhibitors critically depended on the molecular weight of the inhibitor. Developing inhibitors for HIV protease was substantially easier

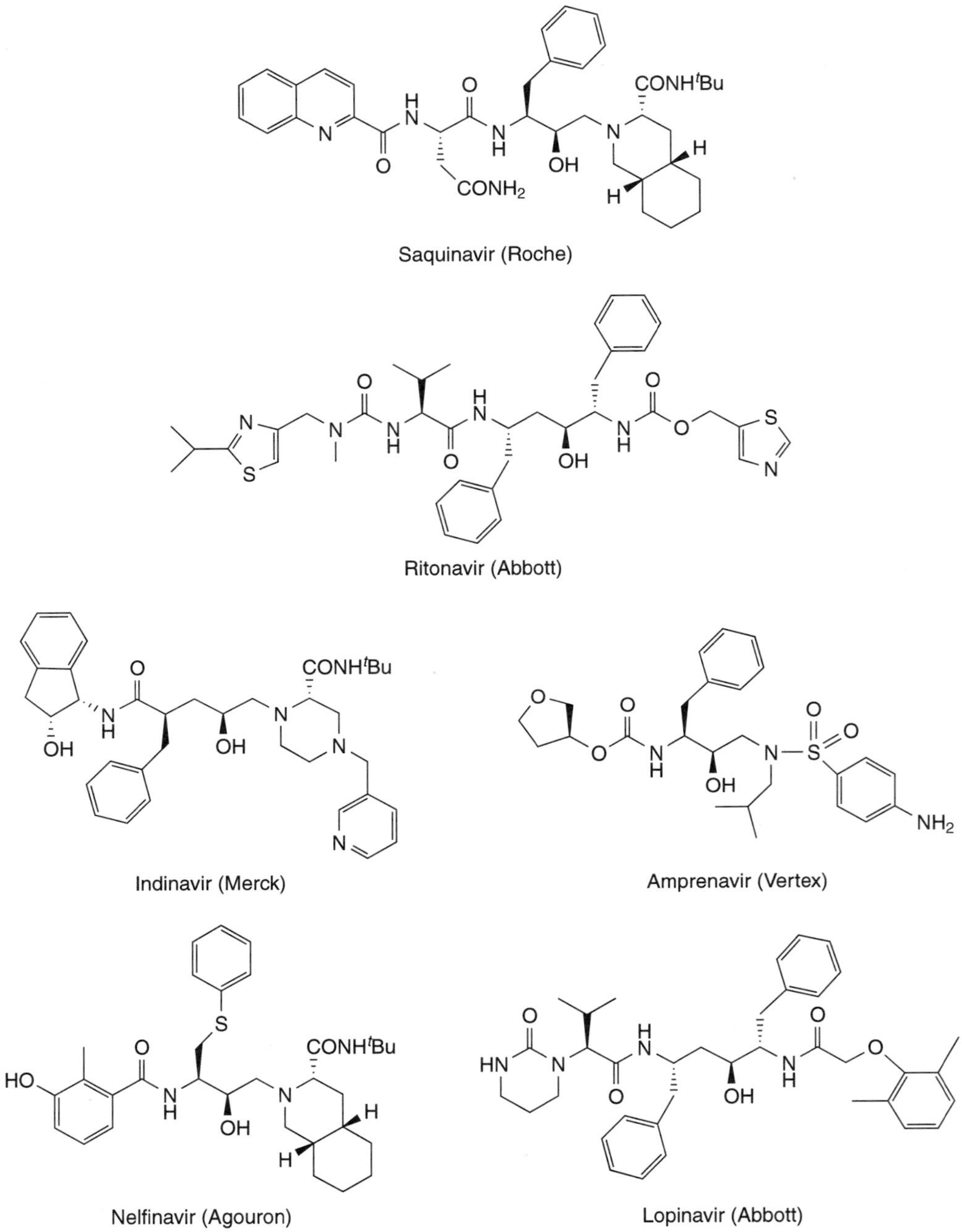

Saquinavir (Roche)

Ritonavir (Abbott)

Indinavir (Merck)

Amprenavir (Vertex)

Nelfinavir (Agouron)

Lopinavir (Abbott)

Figure 15.20. Peptide-derived TSA inhibitors used clinically in HIV protease-based AIDS therapies.

than for renin because HIV protease requires a significantly smaller minimum substrate sequence (94). In addition, the principles elucidated to develop renin inhibitors were known and could be applied to the development of HIV protease inhibitors. Variations on the hydroxyethyl amine moiety proved to be very successful. Some of the highly modified HIV

β-Secretase cleaves APP at:

-Ser-Glu-Val-Lys-Met-⦙-Asp-Ala-Glu-Phe-Arg-

-Ser-Glu-Val-Asn-Leu⦙--Asp-Ala-Glu-Phe-Arg-

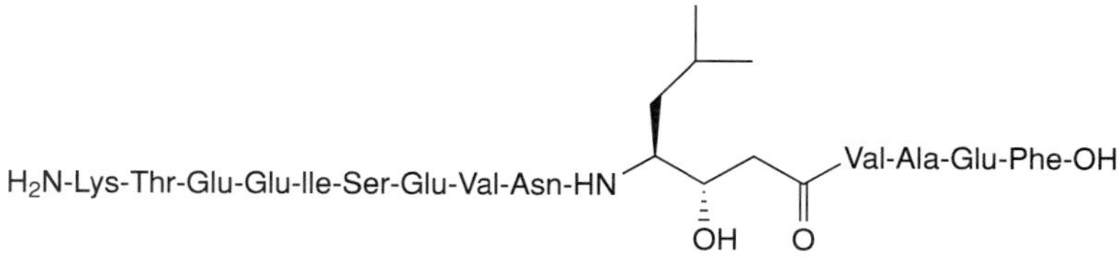

H$_2$N-Lys-Thr-Glu-Glu-Ile-Ser-Glu-Val-Asn-HN

Val-Ala-Glu-Phe-OH

OH O

IC$_{50}$ = 30 nM

(**43**)

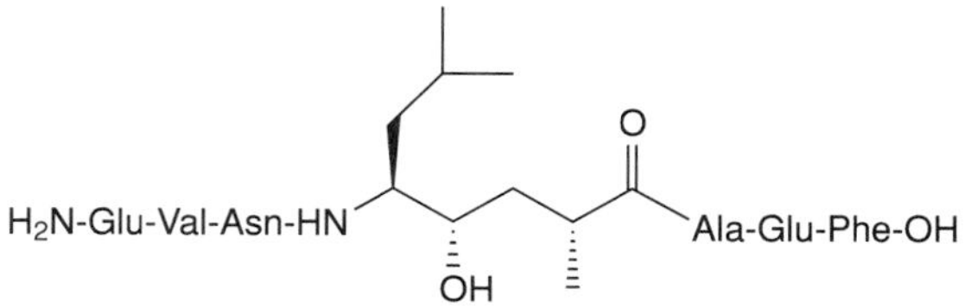

H$_2$N-Glu-Val-Asn-HN

Ala-Glu-Phe-OH

OH

K$_i$ = 1.6 nM

(**44**)

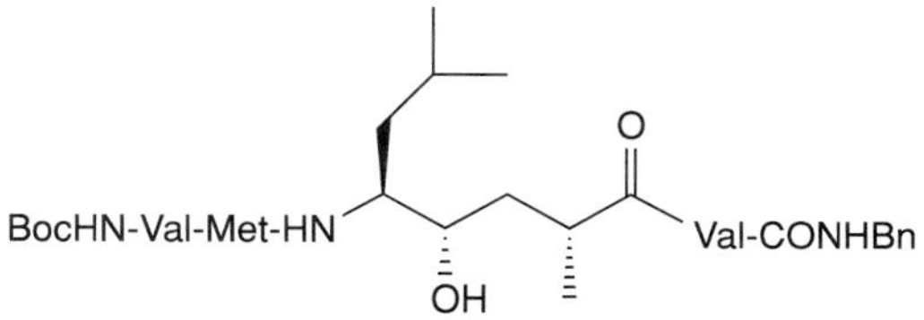

BocHN-Val-Met-HN

Val-CONHBn

OH

K$_i$ = 2.5 nM

(**45**)

Figure 15.21. Peptide-derived TSA inhibitors as β-secretase inhibitors.

protease inhibitors now in clinical use (Fig. 15.20) have excellent oral bioavailability and establish that application of the transition state analog design process can be very successful in favorable cases.

More recently, the principles for designing inhibitors of aspartic proteases have been applied to the design of inhibitors of β-secretase (BACE or Memapsin-2) as potential agents for treating or preventing Alzheimer's disease (95, 96). Both statine-derived inhibitors (**43**) and hydroxyethylene-derived BACE inhibitors have been reported (Fig. 15.21) (97, 98). A crystal structure of (**44**) bound to β-secretase has been reported (99). As expected, the hy-droxyl group is hydrogen bonded to Asp32 and Asp228, like in other hydroxyethylene derivatives, and the inhibitor binds in an extended conformation. Because the target β-secretase is within the CNS, successful inhibitors have to penetrate the brain blood barrier readily, a property not yet achieved with any of the peptidomimetic inhibitors currently available.

With the crystal structure in hand, structure-based modification of the parent lead compound has just started to provide new peptidomimetic structures with lower molecular weight and fewer hydrogen bonds (e.g., **45**) (Fig. 15.21), opening further avenues to pharmacologically useful compounds (100).

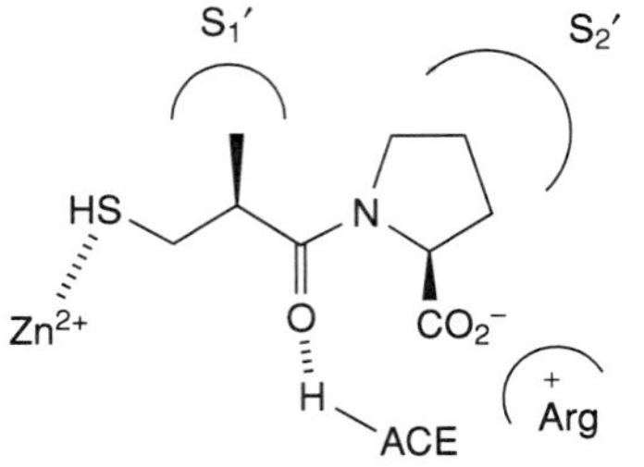

Captopril
(**38**)

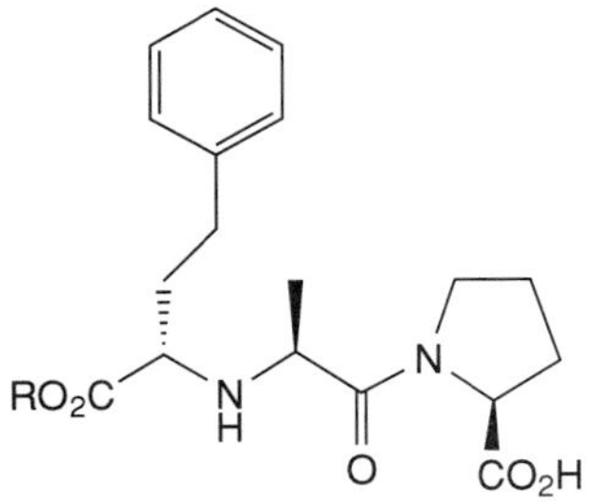

Enalapril, R = Et
Enalaprilat, R = H, (**46**)

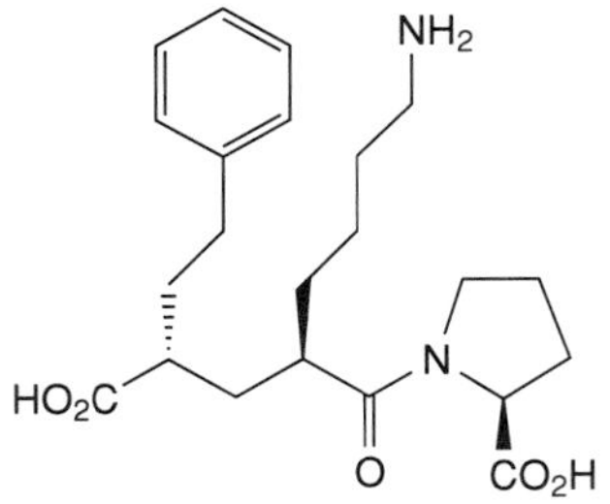

Lisinopril
(**47**)

Figure 15.22. Examples of TSA as ACE inhibitors.

6.2 TSA in Metallo Peptidase Inhibitors

The discovery of the angiotensin converting enzyme inhibitors in the middle 1970s constitutes one of the major advances in the rational design of drugs, the consequences of which are still being realized. The discovery of these metallo peptidase inhibitors was carried out by Ondetti et al. as part of a long-term study to develop antihypertensive drugs (80); in 1999 they received the Lasker Prize in Clinical Medicine for their work.

Angiotensin converting enzyme (ACE) is a carboxy zinc metallo dipeptidase that cleaves His-Leu from the C-terminus of angiotensin-I. Ondetti et al. reasoned that the product of normal reaction, the carboxyl group, could bind to the active site zinc ion, and that the carboxyl group of a collected-product inhibitor also could bind weakly. To improve the interaction between inhibitor and enzyme zinc ion, they replaced the carboxyl group with a sulfhydryl group, which binds zinc about 1000 times more tightly. This led to captopril (Capoten) (**38**) (Fig. 15.22) (80). Later developments by other companies led to many ACE inhibitors. Some are illustrated by enalaprilat (**46**) and lisinopril (**47**) (Fig. 15.22) (101, 102).

Most metallopeptidase inhibitors append a zinc chelating functionality to a peptide or peptidomimetic that is recognized by the S1′–S3′ subsites in the target enzyme. Successful clinical candidates invariably contain groups that replace the initial di- and tri-peptide moieties to achieve selectivity and orally activity. For example, neutral endopeptidase (NEP), another endopeptidase involved in degrading the larger opioid peptides dynorphan and/or endorphan, is inhibited by thiorphan (**48**) (103) and a variety of NEP inhibitors: retrothiorphan (**49**) (104) and kelatorphan (**50**) (Fig. 15.23) (105). The hydroxamic acid moiety is used in many inhibitors of metallopeptidases.

Inhibition of NEP also prevents the degradation of atrial natriuretic factor (ANF), a natural hypotensive peptide. Dual inhibitors of NEP and ACE have been designed successfully because both enzymes share significant structural homology, particularly in their active sites. Simultaneous inhibition of both peptidases produces a more powerful hypoten-

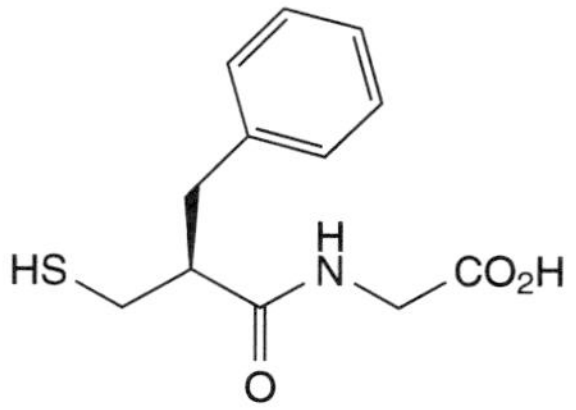

Thiorphan

(48)

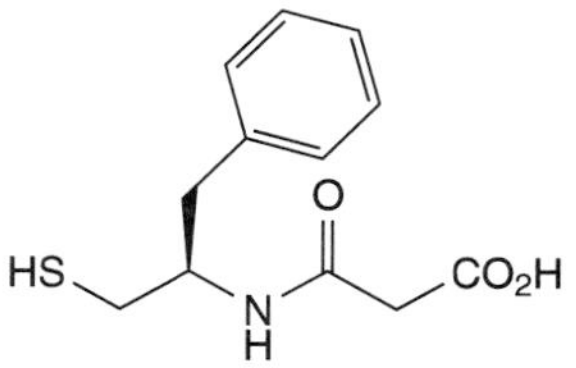

Retrothiorphan

(49)

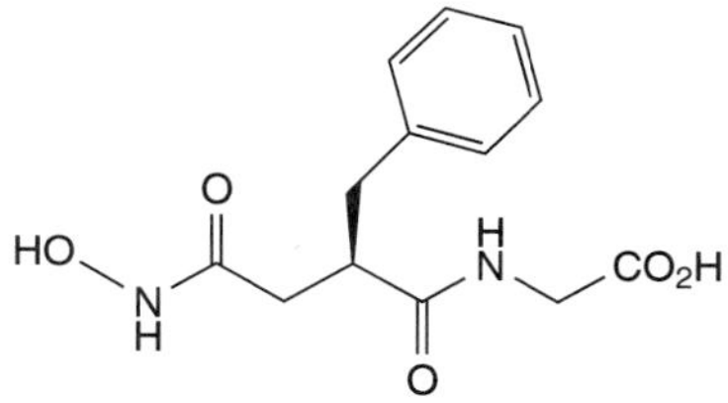

Kelatorphan

(50)

Figure 15.23. Examples of TSA as NEP inhibitors.

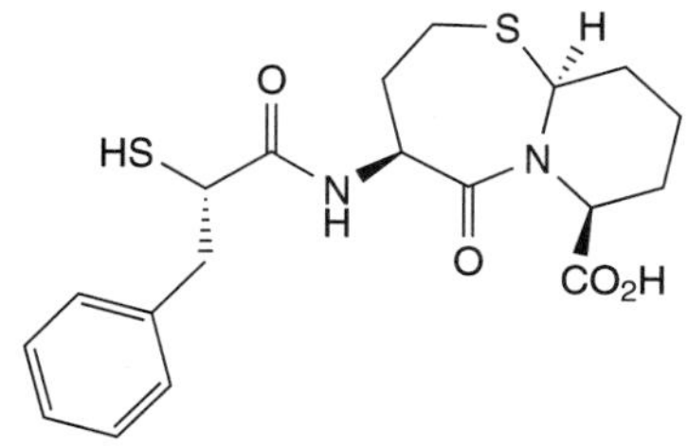

Omapatrilat
ACE IC_{50} = 6 nM
NEP IC_{50} = 9 nM

(51)

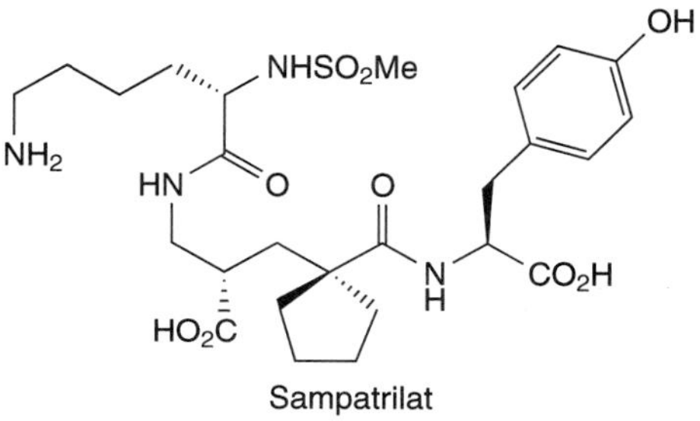

Sampatrilat
ACE IC_{50} = 7 nM
NEP IC_{50} = 20 nM

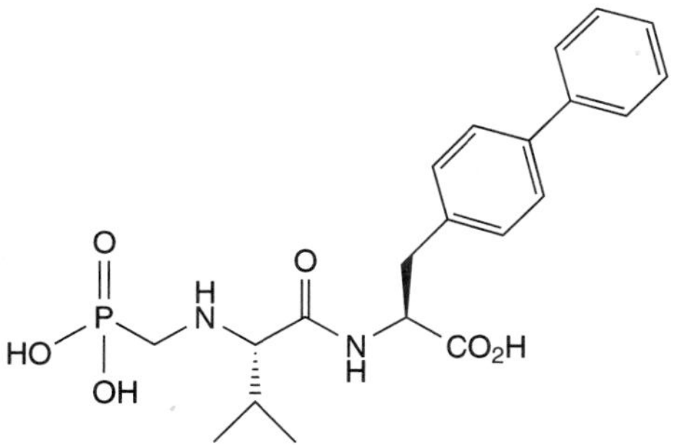

ACE IC_{50} = 25 nM
NEP IC_{50} = 3 nM

Figure 15.24. Examples of TSA as dual ACE/NEP inhibitors.

sive response (106, 107). Several dual inhibitors are in phase III clinical trial for treating hypertension (Fig. 15.24). Omapatrilat (**51**, BMS-189921) is the furthest along as of late 2001 (105).

Matrix metalloproteases (MMP) are also inhibited by hydroxamic acids and/or thiols. Over 25 variants of these enzymes are known, and some are involved in diseases ranging from inflammation to metastatic cancer (108). MMPs contain a zinc ion in the active site and function through the metallopeptidases catalytic mechanism already discussed. However, subtle differences between enzymes enable selective inhibitors to be developed (109). Fig. 15.25 lists some of the reported MMP inhibitors that use carboxylic acid (**52–53**), a hydroxamic acid (**54–55**), or thiol groups (**56**) as metal chelators.

Other reported zinc binding chelators used in matrix metalloproteinase inhibitors are summarized in Fig. 15.26. For instance, one of the oxygens in the phosphonamide (**57**) binds strongly to the zinc ion, whereas the other one coordinates weakly with the metal (110). More recently, "suicide substrate" MMP inhibitors have appeared (**58**) (Fig. 15.26) (111). The se-

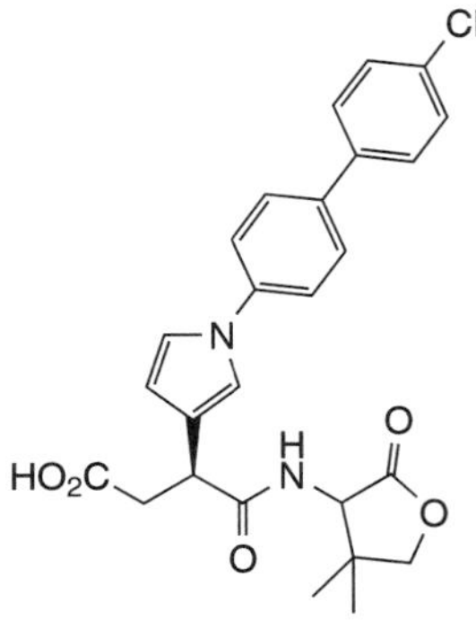

Figure 15.25. Traditional TSA used to inhibit metallopeptidases.

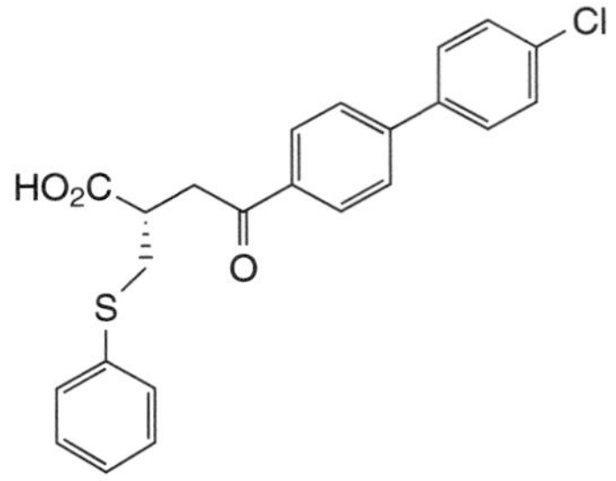

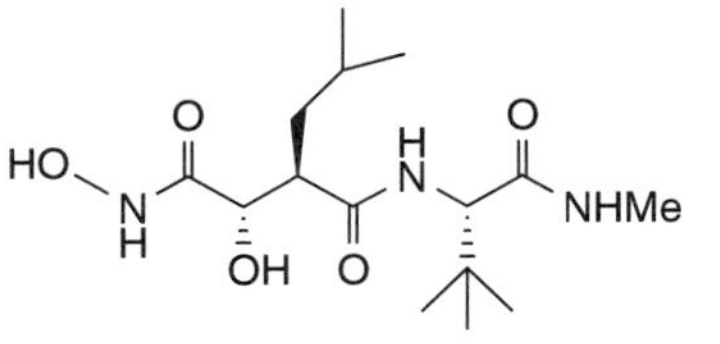

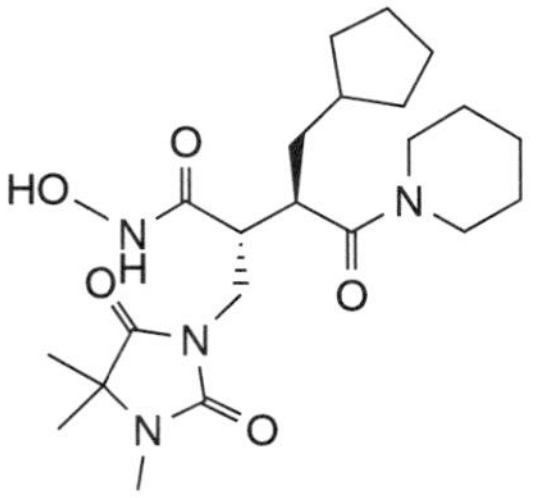

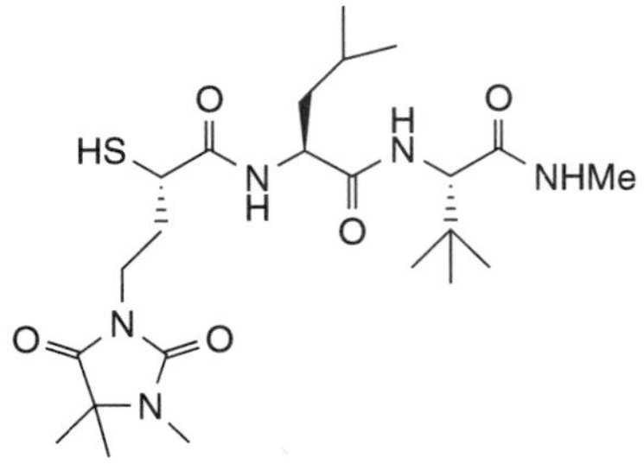

lectivity of this type of compound arises from the specific coordination of the thiirane with the active-site zinc ion, which facilitates thiirane ring opening by nucleophilic attack by neighboring Glu-404. This novel mode of binding was assessed by X-ray absorption studies because of the difficulty to obtain a suitable crystal structure (111, 112).

ADAMs are membrane proteins that contain *a d*isintegrin and *a m*etalloprotease domain. Disintegrins are RGD-containing proteins that inhibit cell/matrix interactions (adhesion) and cell/cell interactions (aggregation) through the integrin receptors. In addition, ADAMs have two other domains that are involved in signaling and transport (113).

There are more than 25 ADAMs proteases identified so far. ADAM 17 has been shown to be TNF-α converting enzyme (TACE) (114). Inhibition of TACE slows the production of TNF-α, a potent cytokine involved in inflammatory responses to infection. Normally TNF-α produces a useful response, but in some cases, too much TNF-α is released and inhibition of TNF-α production would be ther-

apeutically useful. Synthetic analogs have been synthesized that inhibit this enzyme. Clinical candidates like Ro-32,7315 (**59**) (Fig. 15.27) are starting to emerge, and more are expected in the near future (115, 116).

Aminopeptidases, enzymes that cleave off the *N*-terminal amino acid from a peptide chain, are bismetallo peptidases, a class of metallopeptidase that contain two metals ions in the catalytic site (117, 118). These can be inhibited by compounds related to bestatin (**60**) (Fig. 15.28), which contains the *N*-terminal α-hydroxy-β-amino acid residue, sometimes referred to as norstatine. In leucine amino peptidase, chelation occurs between both the amide carbonyl group and the adjacent hydroxyl and the hydroxyl and the *N*-terminal amino group (119, 120).

6.3 TSA-Derived Cysteine and Serine Peptidase Inhibitors

Classical TSA inhibitors of cysteine and serine proteases differ from the metallo- and aspartic protease inhibitors in that they mimic the tet-

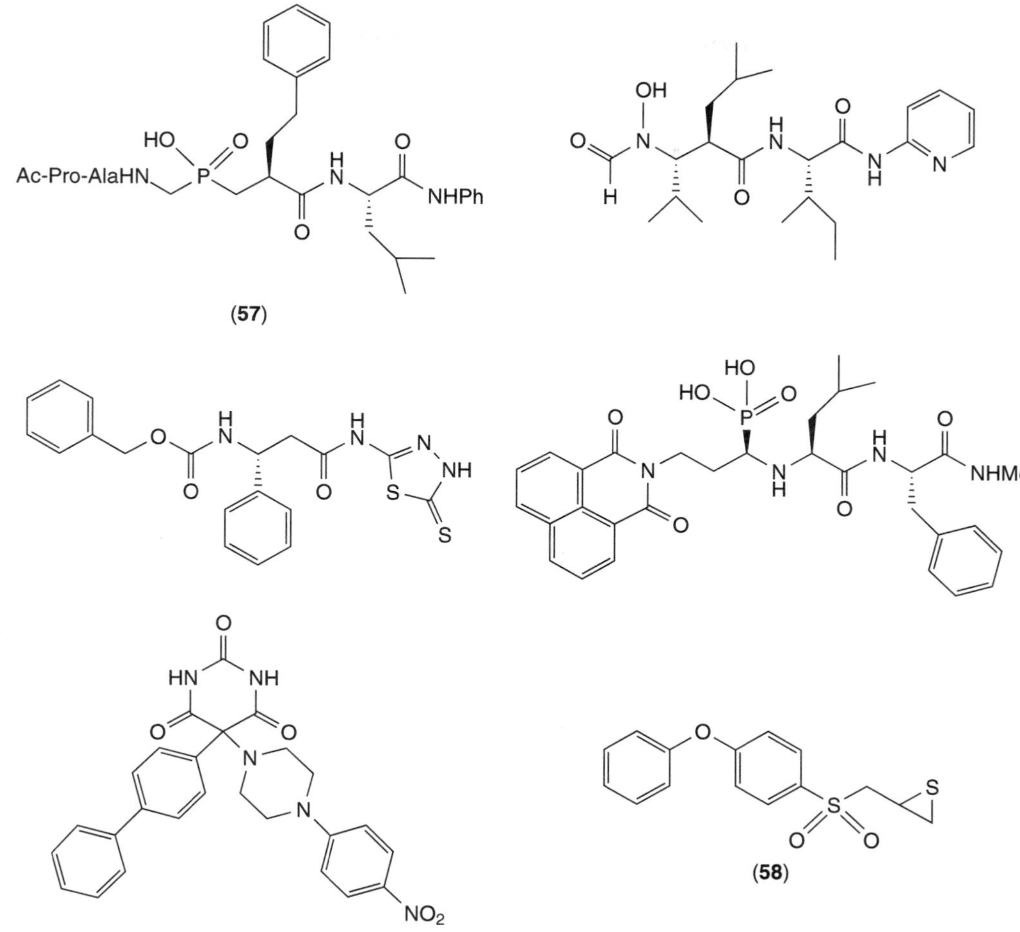

(57)

Figure 15.26. Novel TSA used to inhibit metallopeptidases.

(58)

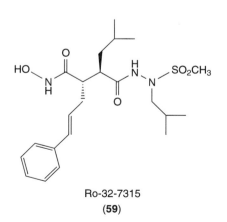

Ro-32-7315

(59)

Figure 15.27. Example of TSA as an TNF-α inhibitor.

rahedral intermediates for enzyme-catalyzed amide bond hydrolysis only after a *reversible* chemical reaction between enzyme and inhibitor takes place. Usually this involves the addition of the enzyme catalytic nucleophile (the serine protease hydroxyl group or the cysteine protease thiol group) to an electrophilic group in the inhibitor to generate ketal-like species (121).

Some of the serine and cysteine TSA moieties are shown in Fig. 15.29. Selective inhibition between these two classes of protease can be achieved easily. For example, trifluoromethylketones (**61**) and peptidyl boronic acids (**62**) do not efficiently inhibit cysteine proteases. However, selective inhibition of enzymes within each class can be very difficult.

Bestatin
(**60**)

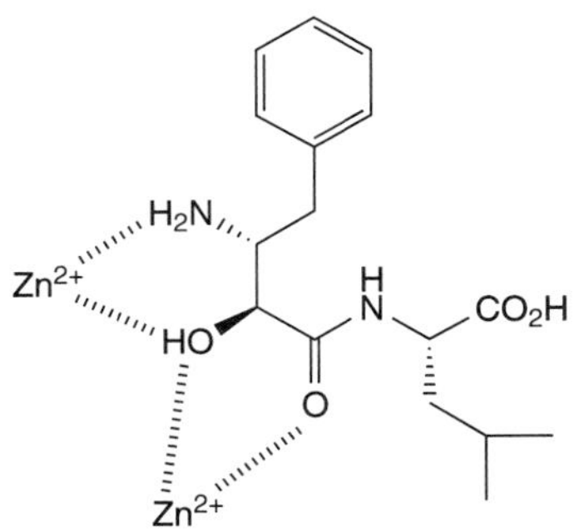

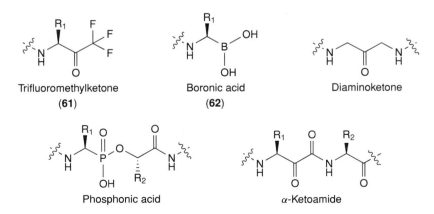

Figure 15.28. Proposed binding mode of bestatin.

Many cysteine peptidases are involved in the biosynthesis and degradation of biologically important peptides. Most early work was done with papain, a cysteine peptidase isolated from the papaya fruit and used in meat tenderizer many years ago. The readily available source of this enzyme led to one of the very first X-ray crystal structures of any peptidase (122, 123), despite the fact that no cysteine peptidase was then known to be important in human pathology. Since then, cathepsins B, H, L, and S were discovered to be involved in biosynthetic steps in human immune response, inflammation, and other biologies. For example, cathepsin B is clearly involved in the metastatic process and must act at some stage to permit transformed tumor cells to migrate to other parts of the body; for 20 years, people have sought inhibitors of cathepsin B as potential anti-metastatic drugs (124). Cathepsin K was recently discovered and shown to be involved in osteoporosis and bone regulation (125).

Inhibitors of cathepsin K illustrate the principles developed to inhibit this class of enzyme. This enzyme sequence was detected in 1994 by sequencing of human DNA for the human genome project (126). Cathepsin K was found to be inhibited by leupeptin (**63**) and by compound (**64**), which surprisingly binds "backwards" to the active site (Fig. 15.30). A hypothesis to develop symmetrical inhibitors of cathepsin K derived from the superposition of both aldehydes on the carbonyl carbon; this led to the diamino ketone TSA (**65**). The diamino ketone moiety seems to work in several classes of cysteine proteases (127).

Based on these results, Marquis et al. have recently described the design and synthesis of conformationally constrained cyclic ketones as highly potent and selective cathepsin K inhibitors (**66–67**) (Fig. 15.31) (128). The labile stereogenic group in position α of the ketone was shown to be important for the binding mode and pharmacokinetic profile of these type of inhibitors. The crystal structure of the two epimers showed two alternate directions of binding to the enzyme active site. In both structures, the primed region of the enzyme was occupied by these inhibitors. Further investigation, resulted in the azepanone derivative (**68**) as a configurationally stable template for the selective inhibition of this cysteine protease (K_i = 4.8 pM) (129).

Trifluoromethylketone
(**61**)

Boronic acid
(**62**)

Diaminoketone

Phosphonic acid

α-Ketoamide

Figure 15.29. TSA used to inhibit serine or cysteine peptidases.

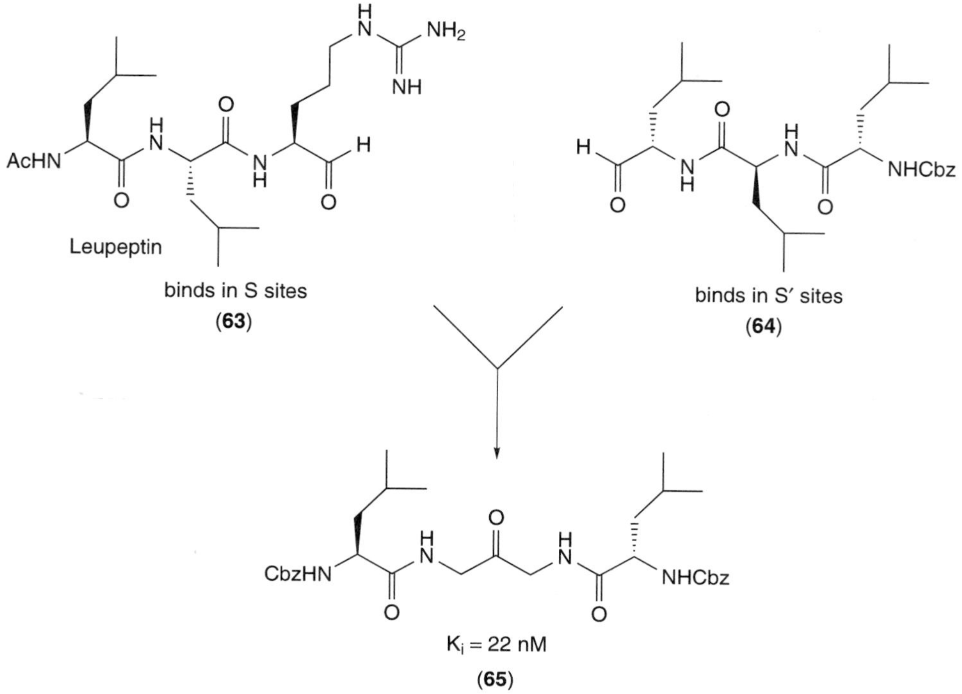

Figure 15.30. Structure-based design of cathepsin K inhibitors.

Caspases are involved in a variety of cell functions, especially in programmed cell death (apoptosis). These enzymes recognize tetrapeptide sequences ending in an aspartic acid recognition point: X-Y-Z-Asp-NHR. Much effort has been expended in trying to obtain selective inhibitors of the 14 different types identified to date. In this context, selective inhibitors of caspase 1 or of caspase 3/7 have recently been reported (130).

Peptidomimetic modifications of the tetrapeptide sequence have led to the conformationally constrained compound (**69**) as a selective inhibitor of caspase-1 or interleukin-1β converting enzyme (ICE) as potential anti-inflammatory compounds (131). Recently, new non-peptide peptidomimetic diphenyl ether sulfonamides have been described as novel lead structures (**70**) (Fig. 15.32) (132).

Finally, researchers from SmithKline Glaxo have identified potent and selective inhibitors of caspases 3 and 7 that lack the required carboxyl group in P_1 (**71**) (Fig. 15.32). The X-ray co-crystal structure reveals the for-

mation of the typical tetrahedral intermediate of the isatin type structures, which may compromise its selective inhibition of proteases (133, 134).

These reversible caspase inhibitors differ from inhibitors that form irreversible covalent bonds, the so-called "dead-end" or "suicide" inhibitors of these enzymes, For example, the α-acetoxy ketone (**72**) in Fig. 15.32 is an alkylating irreversible inhibitor; the enzyme cysteinyl group displaces the α-acetoxy group to form an irreversible covalent bond (135).

7 SPEEDING UP PEPTIDOMIMETIC RESEARCH

As mentioned before, combinatorial chemistry, high-throughput screening, and analogous techniques have become powerful tools to promote drug discovery in peptidomimetic research. It is not the intention of this chapter to summarize all these methods, and excellent

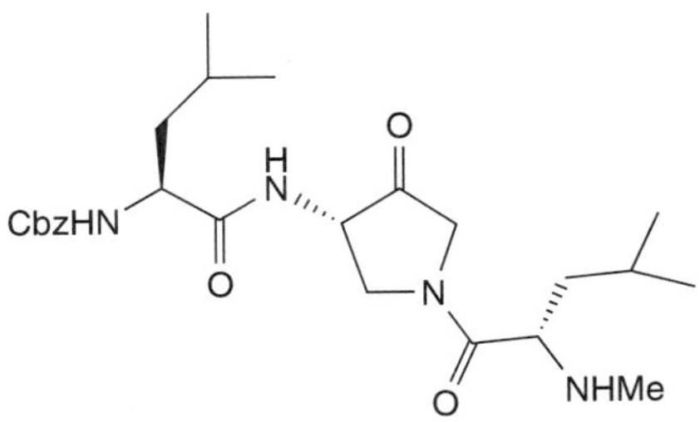

(66)

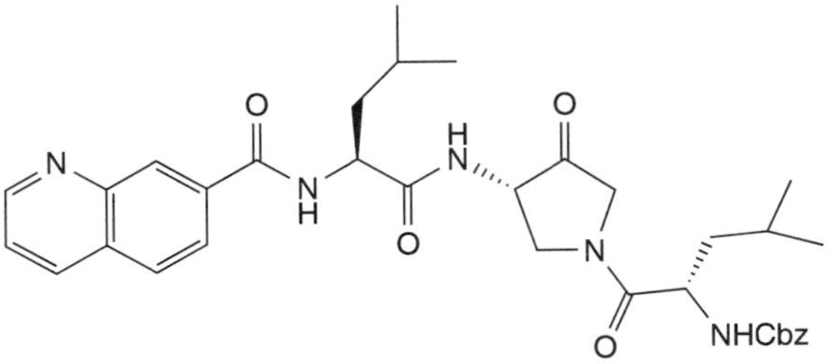

(67)

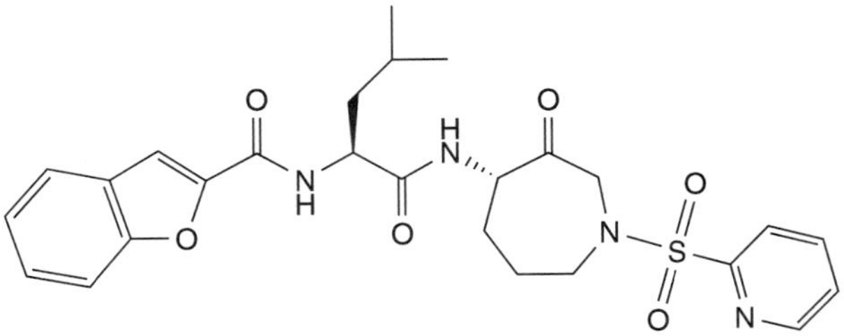

(68)

Figure 15.31. Cyclic ketones in novel cathepsin K inhibitors.

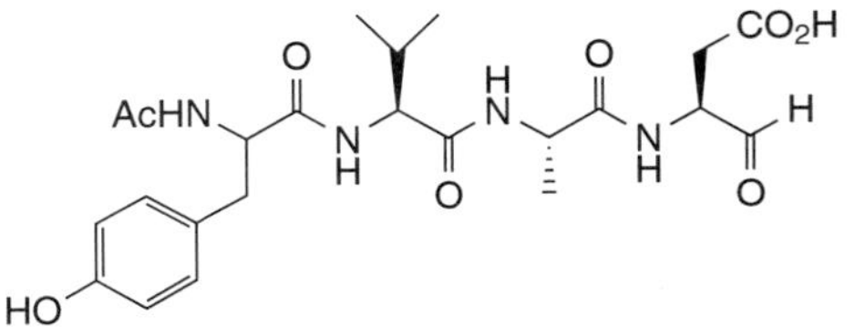

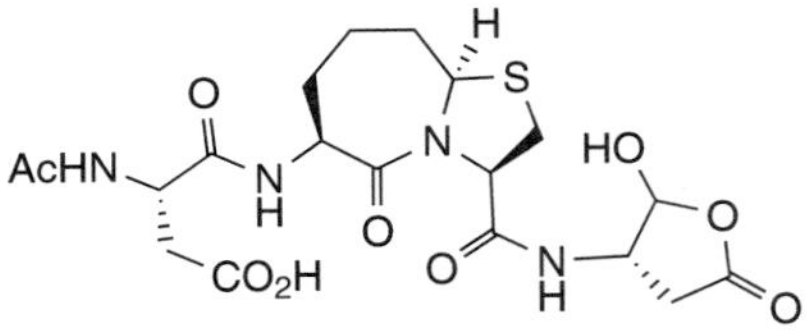

(69)

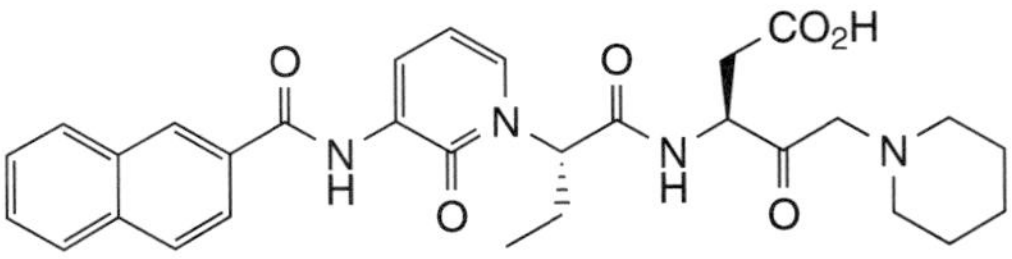

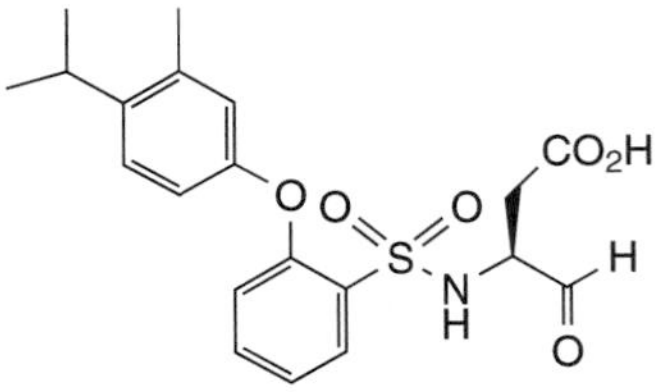

(70)

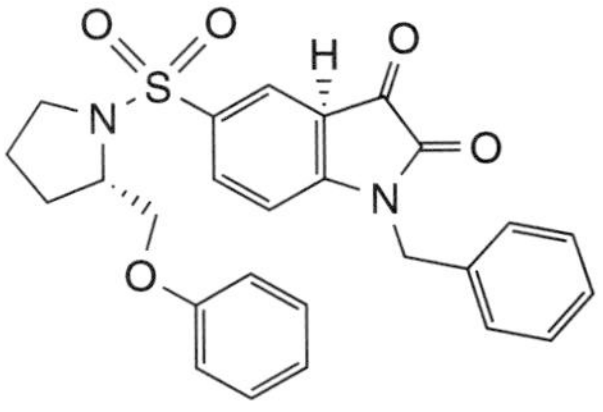

(71)

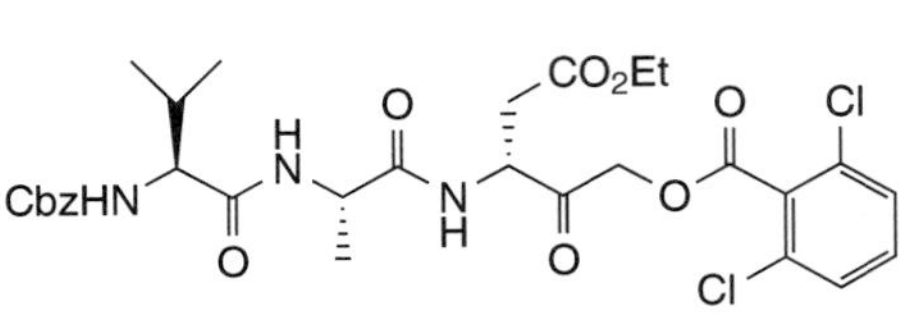

(72)

Figure 15.32. Examples of TSA as caspase inhibitors.

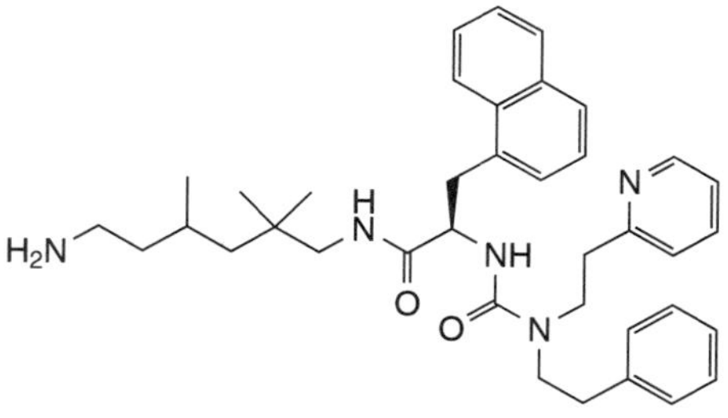

L-797, 591

(**73**)

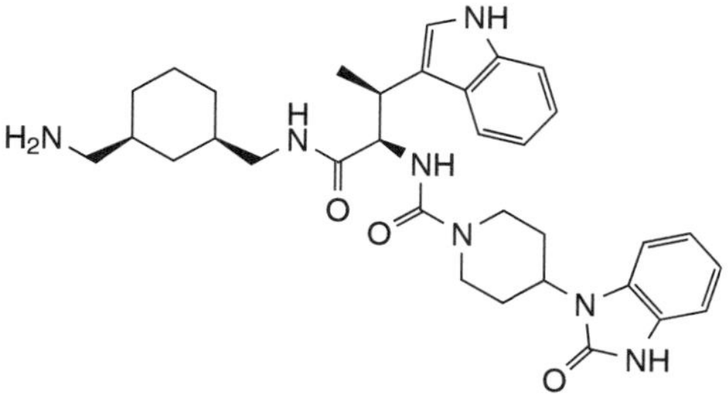

L-779, 976

(**74**)

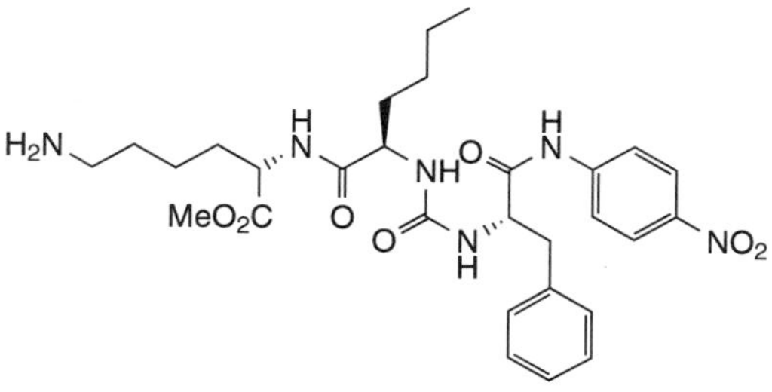

L-796, 778

(**75**)

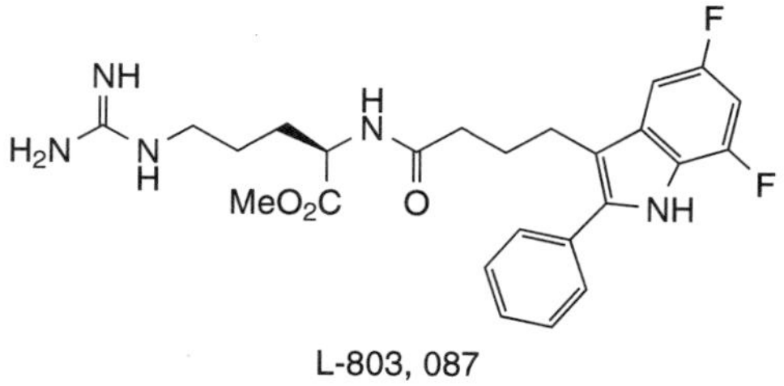

L-803, 087

(**76**)

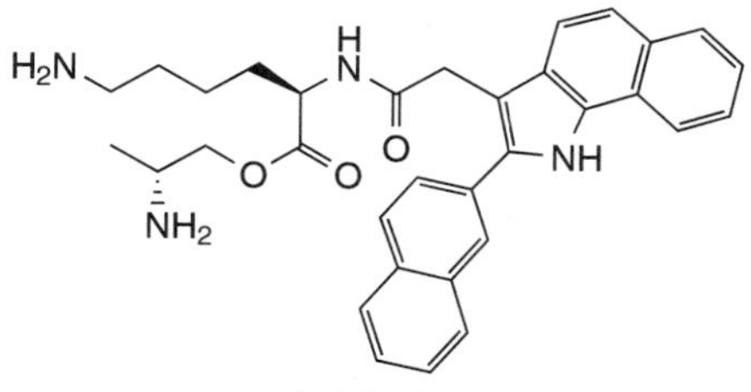

L-817, 818

(**77**)

Figure 15.33. Somatostatin receptor agonists found through combinatorial chemistry.

reviews are available in the literature (136–140). However, one successful approach developed at Merck for the rapid identification of selective agonists of the somatostatin receptor through combinatorial chemistry should be highlighted, because it illustrates the evolution of a constrained peptide into a non-peptide peptidomimetic structure (141).

A series of combinatorial libraries were constructed on the basis of molecular modeling of known peptide agonists like MK-678 and ocreotide. A chemical collection of 200,000 compounds was screened, giving priority to the residues Tyr-Trp-Lys, important pharmacophores in somatostatin determined first by Veber et al. (31) This approach yielded five compounds (**73–77**) (Fig. 15.33), each being selective for one of the somatostatin receptor subtypes: sst1 (**73**), sst2 (**74**), sst3 (**75**), sst4 (**76**), and sst5 (**77**).

8 TOWARD RATIONAL DRUG DESIGN: DISCOVERY OF NOVEL NON-PEPTIDE PEPTIDOMIMETICS

Current pharmaceutical research has taken advantage of newer computational methods, the so-called computer-aided drug design, and other physicochemical techniques such as X-

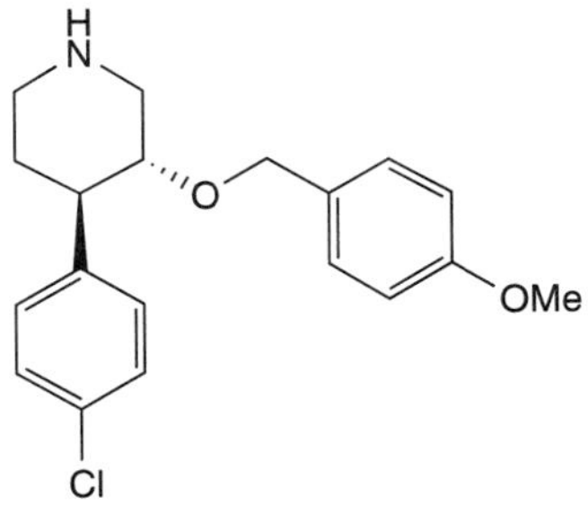

Roche
(**78**)

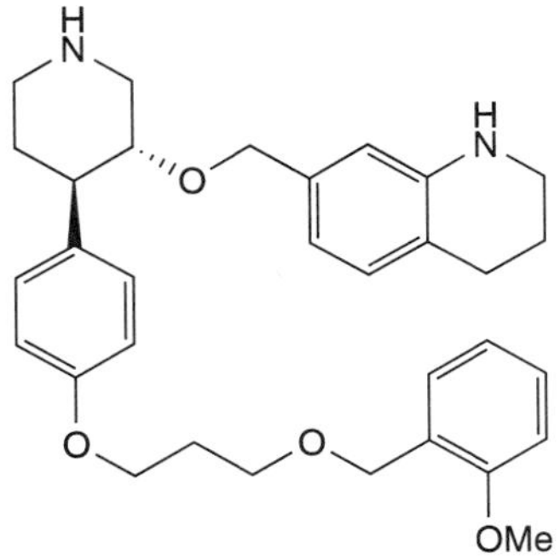

(**79**)

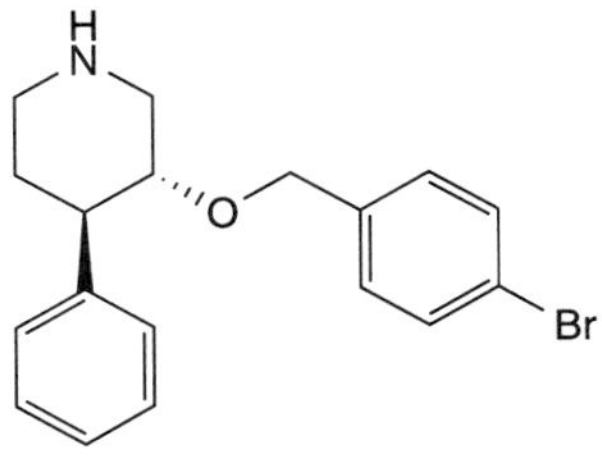

Rich et al.
(**80**)

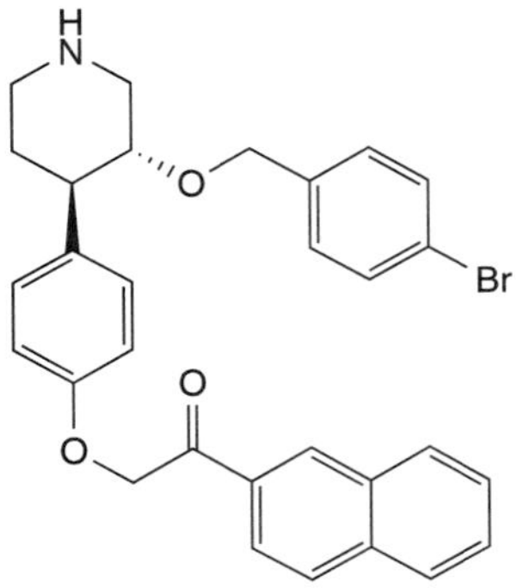

(**81**)

Figure 15.34. Examples of GRAB peptidomimetics.

ray crystallography and NMR (142). The main goal in rational drug design is to translate the structural information in the native peptide into low molecular weight non-peptidic molecules. Over the past years, many 3D structures of biological targets have been solved and have been successfully used to design new, pharmacologically useful compounds (*vide infra*). Different computer-aided design methods, e.g., 3D pharmacophore model, 3D quantitative SAR (QSAR), docking, and *de novo* design, have been extensively reviewed elsewhere (75, 143–146).

Recently, the importance of generating inhibitors that target receptor conformational ensembles has been pointed out (10). This method goes beyond the current docking of known structures to known active site conformers and can lead to type III and GRAB peptidomimetics.

The concept of Group Replacement Assisted Binding (GRAB) peptidomimetics derives from the discovery at Roche of the piperidine class of renin inhibitors. The non-

peptide inhibitors of renin (**78**) ($K_i = 26\ \mu M$) and (**79**) ($K_i = 31\ nM$) (Fig. 15.34)(84, 147–149) stabilize an enzyme active site conformation different than the β-strand binding enzyme conformation typical for other peptidase inhibitors. A close analysis of the X-ray crystal structure of the enzyme inhibitor complex shows that the piperidine C4-phenyl group binds to the enzyme to replace Tyr[75] that has rotated to another position. Interestingly, Leu[73] also rotates to fill some of the vacated Tyr[75] pocket, and this in turn allows Trp[39] to occupy a new site formed in part by the vacated Leu[73] (Fig. 15.35). This cascade of conformational transitions in the renin active site allows the optimized inhibitor to stabilize an enzyme conformation not observed when the classic peptide-derived peptidomimetics bind. This stabilization process is defined as group replacement process, and the piperidine inhibitors constitute a new type of peptidomimetic: GRAB peptidomimetics.

Comparison of (**78**) and (**79**) with the structures of other peptide-derived inhibitors re-

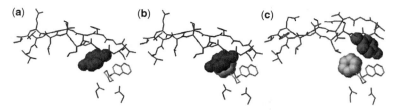

Figure 15.35. GRAB peptidomimetics in action. See color insert.

vealed how the different enzyme active site conformation were found. Bursavich et al. have successfully extended the initial renin modeling to the design of inhibitors of two other aspartic peptidases: pepsin and *R. chinensis* pepsin (**80**) (K_i = 2 μM) and (**81**) (K_i = 0.2 μM) (Fig. 15.34) (150).

The extended β-strand binding conformation could be changed into the piperidine binding conformation by a series of low-energy, mechanistically related conformational changes in active site side-chains. The discovery of the Roche inhibitors and the correlation of these structures with peptide-derived inhibitors are analogous to a peptidomimetic "Rosetta Stone." This design strategy has the potential for designing novel types of peptidomimetic structures.

9 HISTORICAL DEVELOPMENT OF IMPORTANT NON-PEPTIDE PEPTIDOMIMETICS

9.1 HIV Protease

Type-I HIV-1 protease inhibitors, Saquinavir, Ritonavir, Indinavir, Amprenavir, Viracept (neflinavir mesilate), and Lopinavir (Fig. 15.20) are established drugs for the treatment of AIDS. All these inhibitors employ the central hydroxyl transition state mimetic as a scaffold on which varying functionality was systematically added until the required balance between potency, *in vivo* activity and oral absorption was achieved. In general, the binding interactions were optimized through iterative synthesis and co-crystallization of inhibitor with enzyme, molecular modeling, and re-designing the inhibitor side-chains. Pharmacokinetic properties were addressed only after the initial inhibitor was identified and optimized. Compounds (**82–83**) (Fig. 15.36)

are highly modified peptidic structures that stabilize the enzyme-bound extended β-conformation (151, 152).

Another approach to achieve greater *in vivo* activity is to start with a molecular template with proven useful pharmacokinetics and oral bioavailability and to selectively modify it to achieve the desired activity. Identification of the orally active anticoagulant warfarin (**84**) (Fig. 15.37) as a weak inhibitor (IC_{50} = 18 μM) of HIV protease was followed by two reports of 4-hydroxycoumarins as possible type III HIV inhibitors. Subsequent SAR studies led to the more potent 5,6-dihydro-4-hydroxy-3-pyrone inhibitor (**85**) (IC_{50} = 2.7 nM), which has good anti-viral activity (EC_{50} = 0.5 μM) and is orally bioavailable (153). Upjohn researchers also used a structure-based design approach based on warfarin to obtain (**86**), their clinical candidate PNU-140690 (154). It should be noted that both inhibitors bind to the extended β-strand binding active site conformation.

Workers at DuPont used a pharmacophore model and database search to develop the first type III mimetic inhibitor of HIV protease, DuP 450 (**87**) (Fig. 15.38). This evolved from a 3D phamacophore that retained two key interactions: replacement of the flap-bound water and a hydroxyl transition-state isostere (155). Molecular modeling led to a cyclohexanone as a better spacer between these groups, and finally the seven-membered cyclic urea (**87**) was created (Fig. 15.38). The development of these inhibitors illustrates the importance of conformational analysis in the design of constrained analogs.

Surprisingly, the symmetric cyclic sulfonyl-urea derivative analog (**88**) (Fig. 15.38, K_i = 3 nM) binds differently in the active site and adopts a flipped conformation (156).

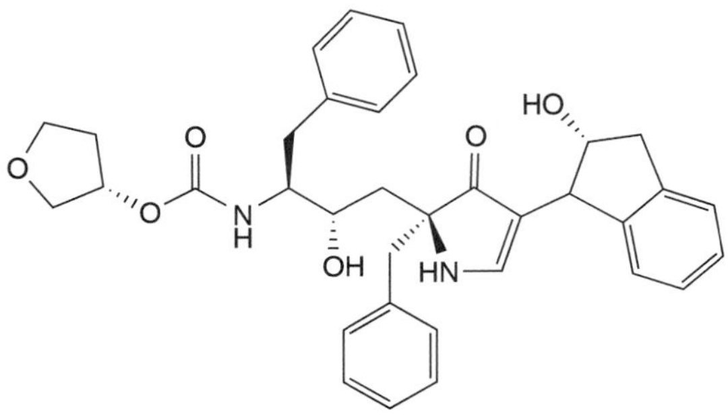

(82)

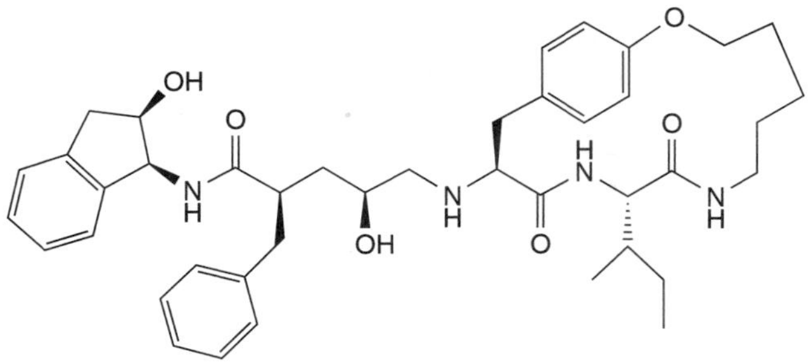

Figure 15.36. β-Strand HIV protease
inhibitors.   (83)

Moreover, SAR of the cyclic urea and cyclic
sulfamide inhibitors do not follow a straightforward pattern. These contradictory results
clearly illustrate the structural diversity created by a subtle structural modification in two
otherwise related peptidomimetic protease inhibitors.

The peptidase inhibitors, (**82**) and (**83**), are
actually amino acid and transition-state mimics pieced together to emulate the typical
ligand-bound extended β-strand inhibitor conformation. The structurally distinct heterocyclic aspartic protease inhibitors (**85–86**) and
(**87–88**) are non-peptide peptidomimetics because of their remote structural relationship
to native peptide substrates. Yet these two distinct peptidomimetic classes bind to the same
active site topography. These structurally distinct peptidomimetics selectively stabilize
closely related enzyme conformations.

9.2   Thrombin

Thrombin and Factor Xa are both serine proteases involved in the blood coagulation cascade. Inhibition of these two enzymes is providing novel anticoagulants (157–159).

The development of thrombin inhibitors
that lack the functionalized TSA highlights a
major new approach to type I peptidomimetics. In 1995, a Lilly group found that D-Phe-
Pro-Agmatine analogs showed increased selectivity for thrombin over other fibrinolytic
enzymes despite a 100-fold loss in potency
caused by the removal of the aldehyde group
(160). These studies paved the way for
Merck's development of picomolar thrombin
inhibitors (161, 162), which use a similar motif. Removal of an α-ketoamide transition
state mimic from L-370,518 (**89**) (Fig. 15.39,
$K_i = 0.09$ nM) led to an expected 100-fold drop
in potency for (**90**) ($K_i = 5$ nM). However, systematic modification of the P_3 position re-
stored potency and led to an inhibitor (**91**)
with a $K_i = 2.5$ pM. Interestingly, potency
seems to be enhanced by a fortuitous hydrophobic collapse into a favorable binding conformation.

Thrombin inhibitors (**92**) and (**93**) illustrate a novel type III peptidomimetic. Most
protease inhibitors bind in an extended
β-strand conformation that is stabilized by
multiple enzyme ligand hydrogen bonds.

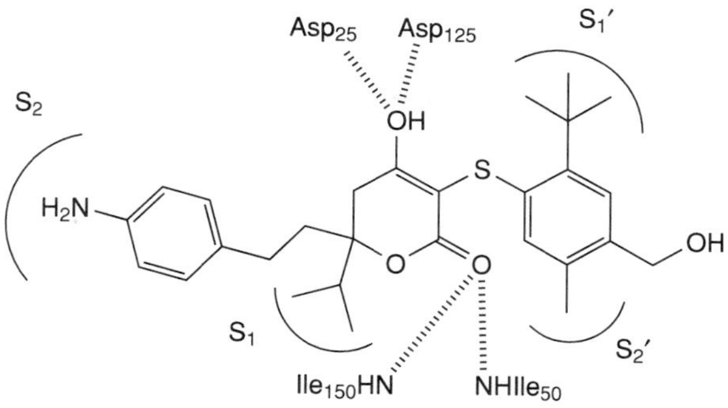

Parke-Davis X-ray structure

(85)

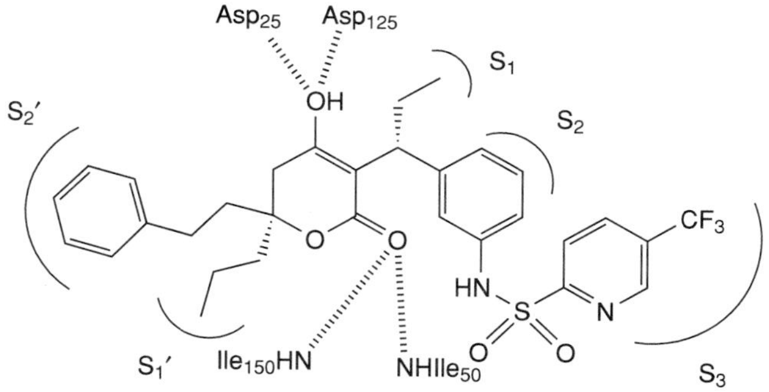

Pharmacia-Upjohn X-ray structure

(86)

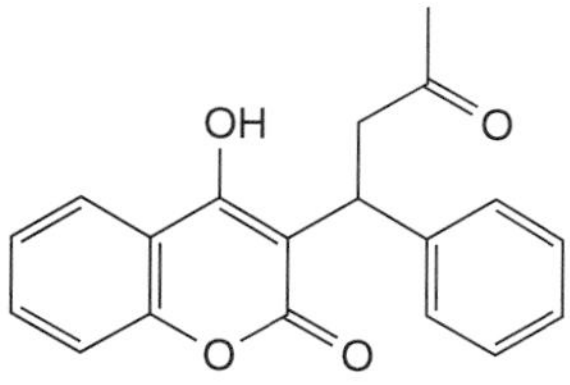

Warfarin

(84)

Figure 15.37. Warfarin analogs as non-peptide HIV protease inhibitors.

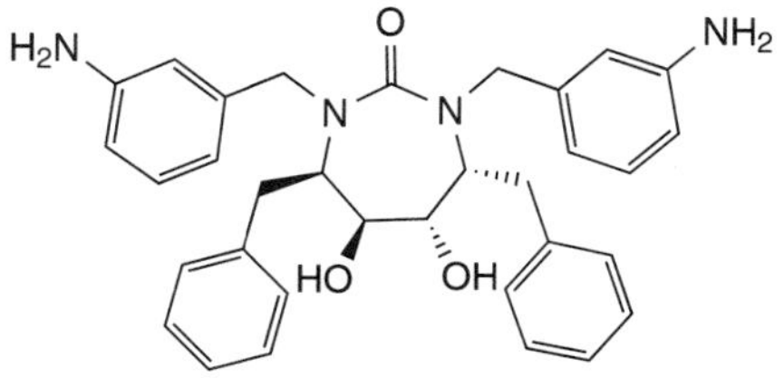

(87)

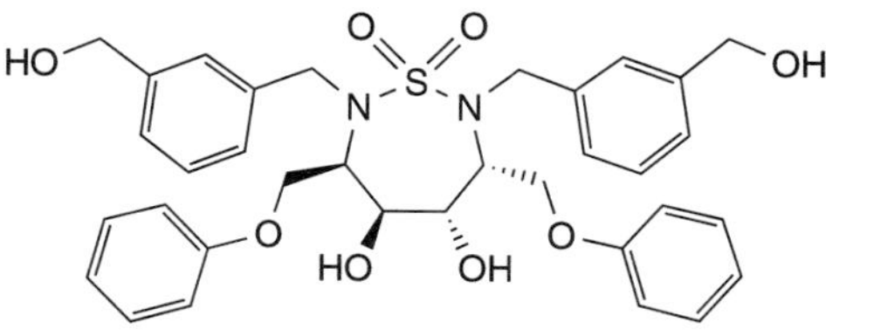

(88)

Figure 15.38. Cyclic ureas as non-peptide HIV protease inhibitors.

However, Boehringer Mannheim developed thrombin inhibitors (**92**) (Fig. 15.40) that lack these H-bonds (163). This idea was exploited by researchers at 3D Pharmaceuticals, who were able to crystallize (**93**) in the active site (164, 165). In this example, the benzene ring acts as a scaffold to display the three different substituents to fill the three principal binding pockets.

Other type III peptidomimetic inhibitors of thrombin have been developed from screening leads (166, 167) such as inhibitor (**94**) (Fig. 15.40). SAR led to the design of (**95**). Inhibitor (**96**) was derived from docking studies with the 5-amidino indole nucleus, followed by addition of a lipophilic side-chain to interact with the important S_3 subsite of thrombin. The crystal structures of both (**95**) and (**96**) in the active site of thrombin shows that the aromatic core, binds in the S_1 site as expected, but

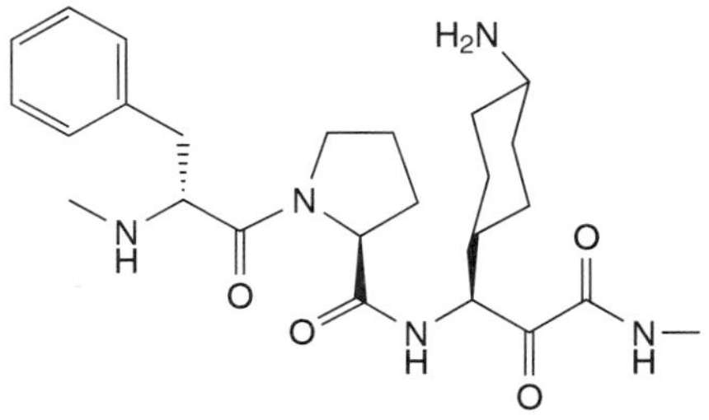

(89)

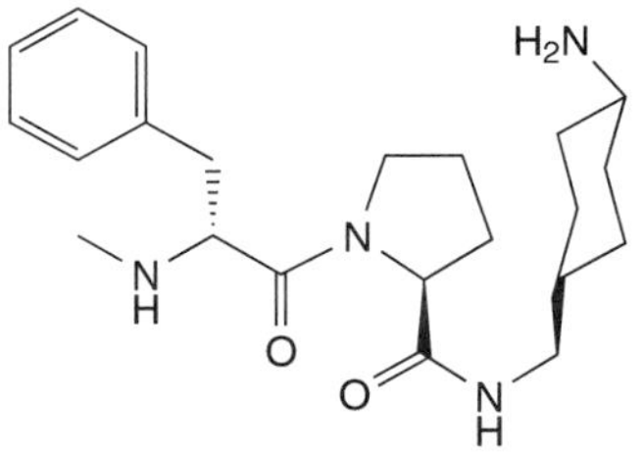

(90)

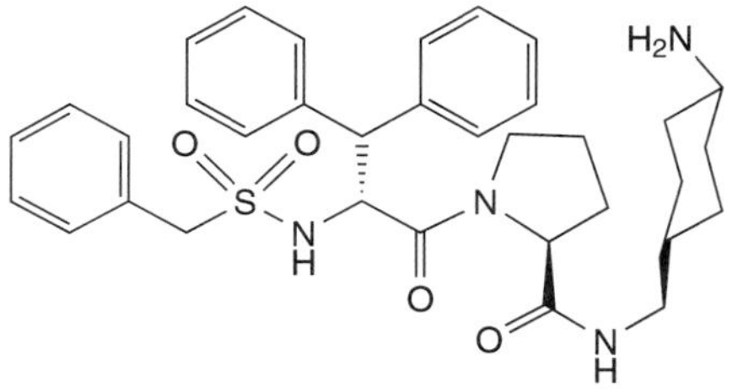

(91)

Figure 15.39. Non-TSA thrombin inhibitors.

does not pick up hydrogen bonding from the important active site sequence Ser214–Gly216 (168). Both crystal structures showed a similar binding mode; where interaction of the C-2 side-chain with Trp60D might explain the high thrombin selectivity observed for this series (169).

Another type III peptidomimetic inhibitor was derived from the crystal structure of a bicyclic [3.1.3] inhibitor (170) complexed to thrombin (**97**) (Fig. 15.41). The X-ray structure revealed that one of the carbonyls was oriented towards the hydrophobic P-pocket (S$_2$). The desolvation necessary to place a carbonyl in a hydrophobic pocket is unfavorable and various alkyl groups were used as possible replacements. This led to the potent ($K_i = 13$ nM) and selective (>760 for thrombin over trypsin) inhibitor (**98**).

One of the major drawbacks in thrombin inhibitor design was the requirement for a basic side-chain in P$_1$ needed to form a salt bridge to enzyme Asp189. However, the other amino acid side-chains in S$_1$ are largely lipophilic and neutral. This feature suggested that the strongly basic group in P$_1$ could be replaced by a weaker base or even with hydrogen-bonding groups. Compounds (**99–100**) are representative of this strategy (Fig. 15.40) (171). An X-ray crystal structure of (**99**) shows a new binding mode in which the formamide group points out of the S$_1$ pocket and forms new hydrogen bonds with Gly219 (172). The ability to obtain crystal structures of thrombin inhibitors complexes for many of the inhibitors shown in Figs. 15.40–15.41 establishes that most are type III peptidomimetics.

9.3 Factor Xa

New approaches to design inhibitors of Factor Xa as potential anticoagulants have been reviewed (173), and important type III mimetics have been described (Fig. 15.42). All inhibitors contain amidine or basic groups that bind in the enzyme's S$_1$ site; none of the inhibitors contains a classical electrophilic center of the type employed in TSA inhibitors (174–180).

Compound (**101**) (Fig. 15.42) was designed from a strategy involving connection of a three-point pharmacophore derived from molecular modeling. Beginning with the X-ray structure of the Factor Xa dimer, Gong et al. (176) envisioned three important enzymatic contact points: a phenylamidine in the S$_1$ subsite, a phenylamidine in the S$_4$ site, and a carboxylate moiety postulated by a group at Daiichi to confer selectivity over thrombin through an interaction with Gln192 of Factor Xa. Systematic iterative modifications led to the potent inhibitor (**101**) ($K_i = 9$ nM), which has 350-fold selectivity over thrombin. This approach highlights a truly *de novo* method where fragments were docked into the active site and an appropriate spacer was chosen to connect them. Further SAR data led to modifications that improved both potency and selectivity (176).

9.4 Glycoprotein IIb/IIIa (GP IIb/IIIa)

Some outstanding examples of the use of conformational restriction to characterize the

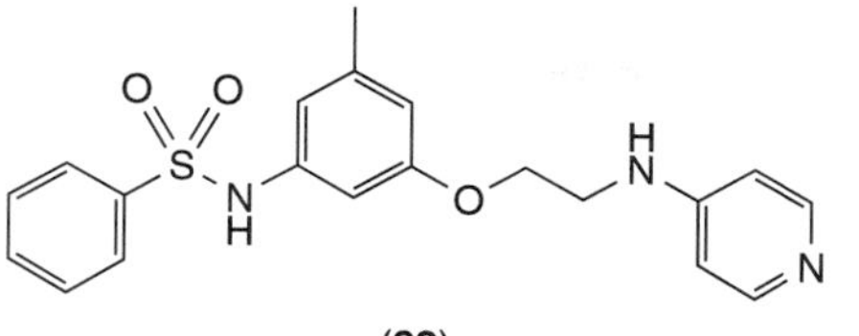

Figure 15.40. Non–H-bonding–based thrombin inhibitors.

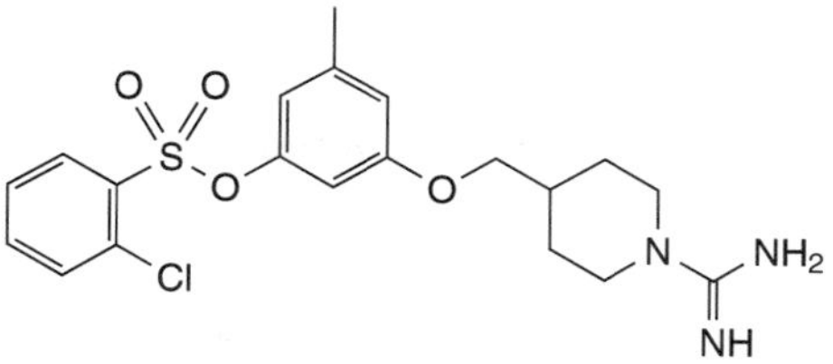

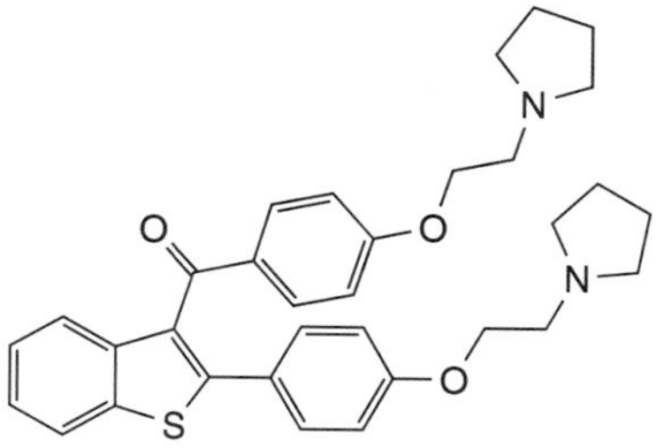

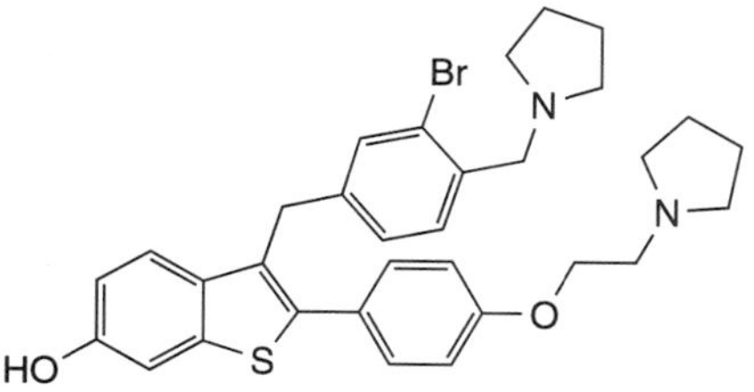

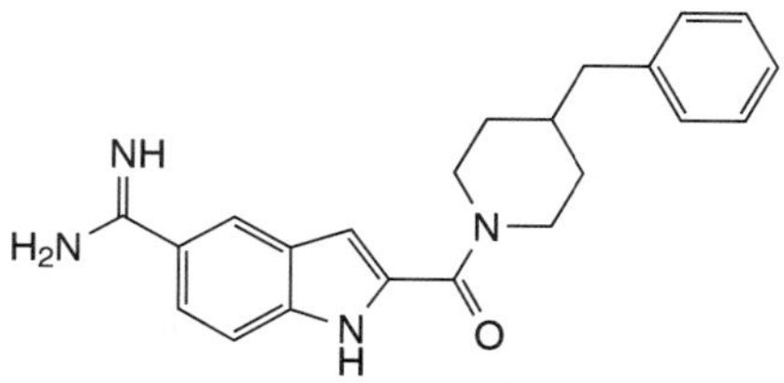

bioactive conformations of Arg-Gly-Asp peptidometic antagonists illustrate the present state-of-the-art. Members of the integrin family of receptors recognize and bind the peptide sequence, Arg-Gly-Asp, as an important step in platelet aggregation and other physiological processes (181), and competitive antagonists for this process could serve as potential drug candidates. Much effort has been directed toward identifying small ligands that might mimic the RGD peptide sequence (182). This drug design concept was supported by the fact that protein antagonists of integrin receptors are known that contain the RGD sequence (183) and that small peptide sequences containing the RGD moiety weakly antagonize the endogenous ligand (184). Consequently, several groups synthesized conformationally restricted derivatives of small peptides as starting points for developing metabolically stable peptides or peptidemimetics. Ali et al. (185) synthesized a series of disulfide derivatives of the RGD sequence, which were designed by analogy with the somatostatin work (*vide supra*). Excellent antagonists related to (**102**) were obtained. Further constraint of the peptide system by use of the o-thiol benzene derivatives led to the novel antagonist SKF 107260 (**103**) (Fig. 15.43), a good inhibitor of both platelet aggregation and binding to GPIIb/IIIa. Barker et al. (186) followed a similar strategy but used cyclic sulfides as an additional conformationally restricting element. These derivatives had the advantage of being rapidly synthesized by solid phase methods. Systematic structure-activity studies with respect to the amino acid preceding the RGD sequence and the chirality of sulfoxide derivatives led to the discovery of G-4120 (**104**), a potent, biologically active derivative.

The conformation of both (**103**) and (**104**) in water was found to be highly constrained, and a single predominant conformation could be characterized in aqueous solution by use of

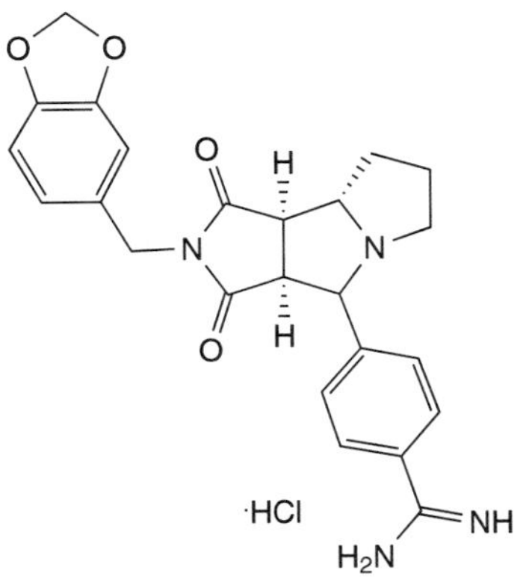

(97)

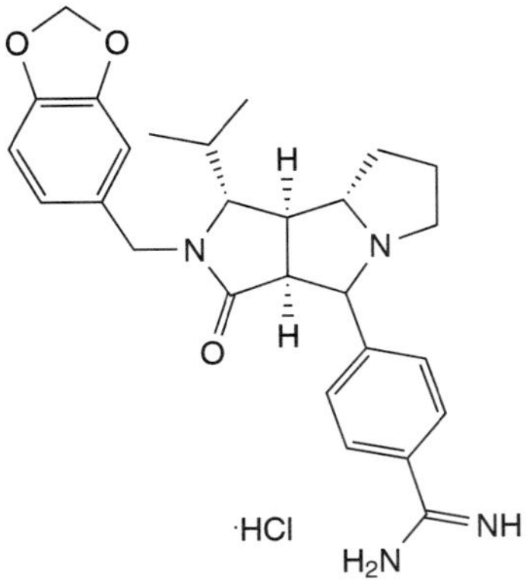

(98)

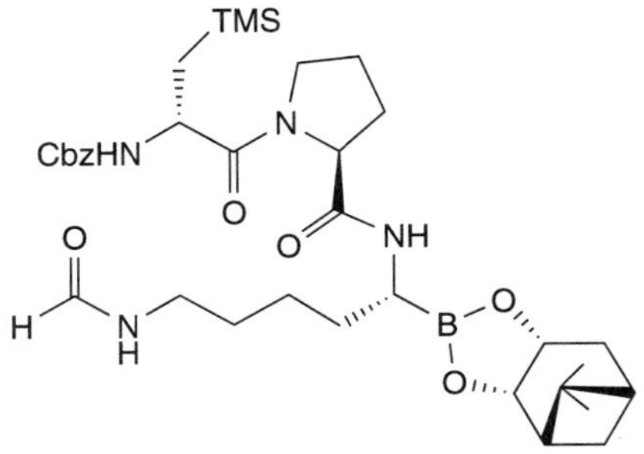

(99)

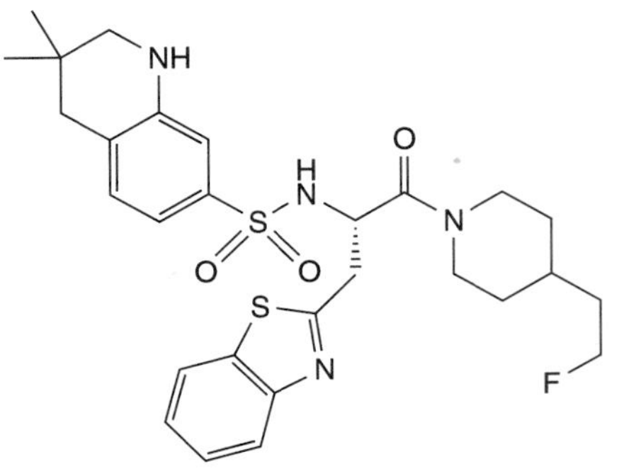

(100)

Figure 15.41. Optimized P_1 and P_2 thrombin inhibitors.

NMR methods and computational chemistry (185, 187). This bioactive conformation defined the topographical placement of the arginine guanidine group and the aspartic carboxyl group, and was superimposed onto a conformationally restricted template of a class of compounds with generally suitable pharmacodynamic properties. In this case, the benzodiazepine ring system was used, and the strategy generated the low molecular non-peptide RGD receptor antagonist (**105–107**) (Fig. 15.44), which contain at least two conformational restrictions, the bicyclic heterocycle and the acetylene linker. The compounds shown in Fig. 15.44 represent what can be achieved by applying the principles of conformational restriction to peptides when no X-ray or NMR structural information are available for the complex between ligand and receptor. Benzodiazepines (**105–107**) represent the first type III peptidomimetics designed *de novo* by sys-

tematically modifying a natural receptor-binding peptide (187–189).

A variety of other scaffolds have been developed by exploiting the idea that glycine represents a spacer between the two important recognition residues Arg and Asp. This template-based approach positions the key side functionality, a basic function and an acid one within a distance of 11–17 Å, required for presentation to the receptor. Several examples of these scaffolds are shown in Fig. 15.45 (190–195).

Recent results suggest that the RGD tripeptide can adopt multiple conformations that allow tight binding to the receptor. This theory is supported by the fact that nonpeptide RGD peptidomimetics can adopt a range of different topographies such as found in cupped, turn-extended-turn, or β-turn conformations (196).

RGD type I peptidomimetics are usually poorly bioavailable compounds because of the presence of multiple hydrogen-bonding sites

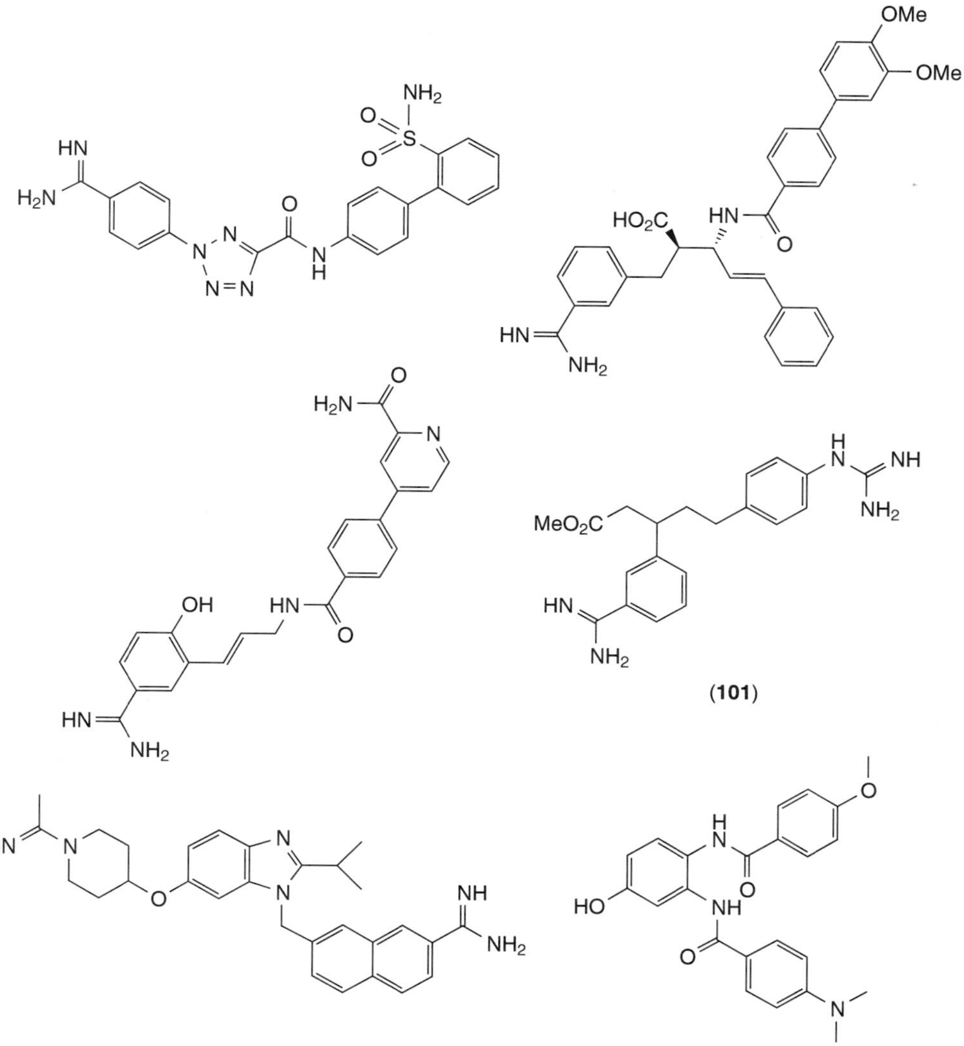

Figure 15.42. Examples of FXa protease inhibitors.

plus the charged polar functional groups at both ends. Esters or coumarin (197) linkers have been used to provide orally available pro-drugs, and bioisosteric replacements of the guanidinium group by a pyridine, (198) tetra-hydronaphtyridine, (199), or aminobenzimidazole (200) moieties provided more bioavailable analogs.

9.5 Ras-Farnesyltransferase

Inhibitors of Ras-farnesyltransferase have been developed by mimicking the C-terminal CAAX motif (where C is a cysteine residue, A is any aliphatic amino acid, and X is usually Met, Ser, or Ala). This tetrapeptide is the signal for farnesylation of Ras proteins. Ras-far-nesyltransferase is one of the most promising targets for novel anti-cancer drugs, because at least 30% of the human cancers contain mutated Ras (201, 202).

Two types of peptidomimetic structures have been used to develop inhibitors (203). Some typical type I inhibitors were generated by replacing the amide backbone with differ-

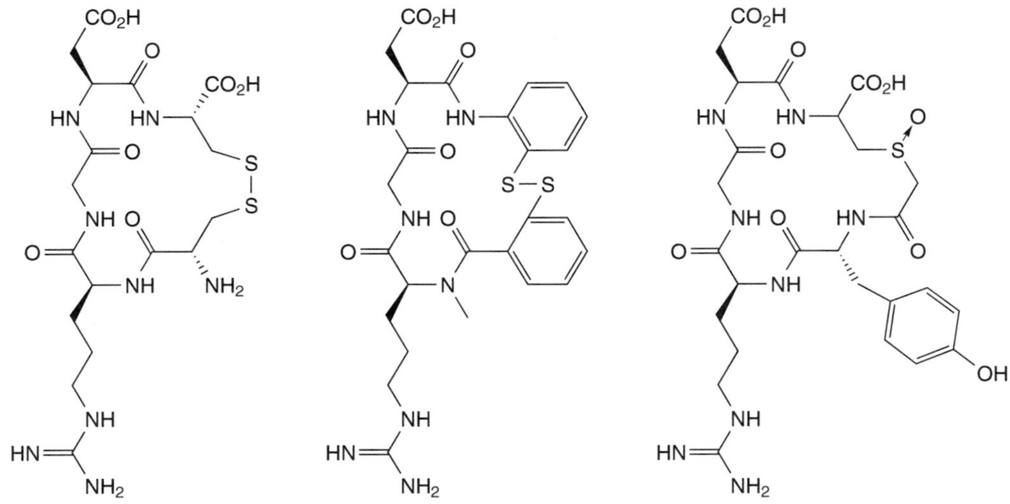

Figure 15.43. Conformationally restricted RGD cyclic peptides.

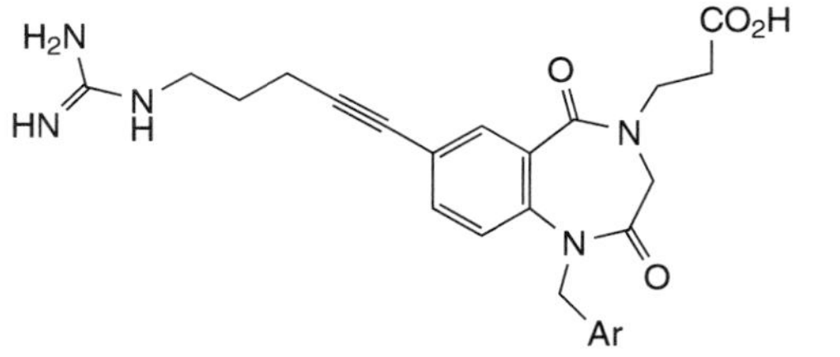

Figure 15.44. Benzodiazepines RGD analogs.

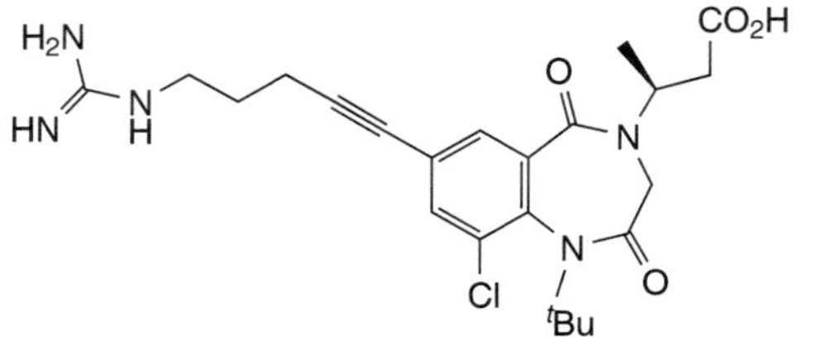

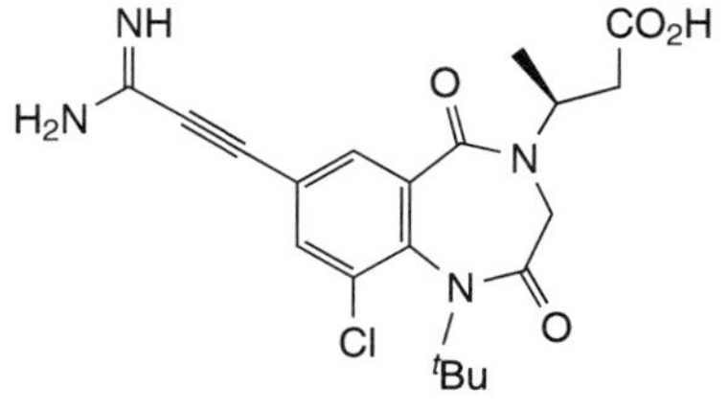

ent isosteres like the oxymethylene amide bond in (**108**) (Fig. 15.46, $IC_{50} = 60$ nM) (204). The central dipeptide segment of CA_1A_2X has been replaced with rigid linkers like the 3-aminomethylbenzoic acid (AMBA) in (**109**) (205). This novel inhibitor was not farnesylated, showing that the two amino acids in the middle of the CAAX tetrapeptide are required for farnesylation. An imidazole group has been used to replace the thiol group of the CAAX motif to produce compound (**110**) (206).

An outstanding example in peptidomimetic design evolved from these studies. Truncation and conformational restriction of a reduced isostere of the parent peptide substrate, followed by systematic replacement of the peptide-like side-chains provided the potent non-peptidic inhibitor (**111**) (Fig. 15.46) (207). This approach highlights the transition from a peptide-derived structure to a compound with no apparent resemblance to the original peptide.

Recently, crystal structures of farnesyltransferase complexed with a farnesyl group donor and the native substrate or a type I peptidomimetic show the structural basis for inhibition of this enzyme. The X-ray data show that the CAAX motif adopts an extended conformation rather than a β-turn, which is the conformation observed by transferred nuclear

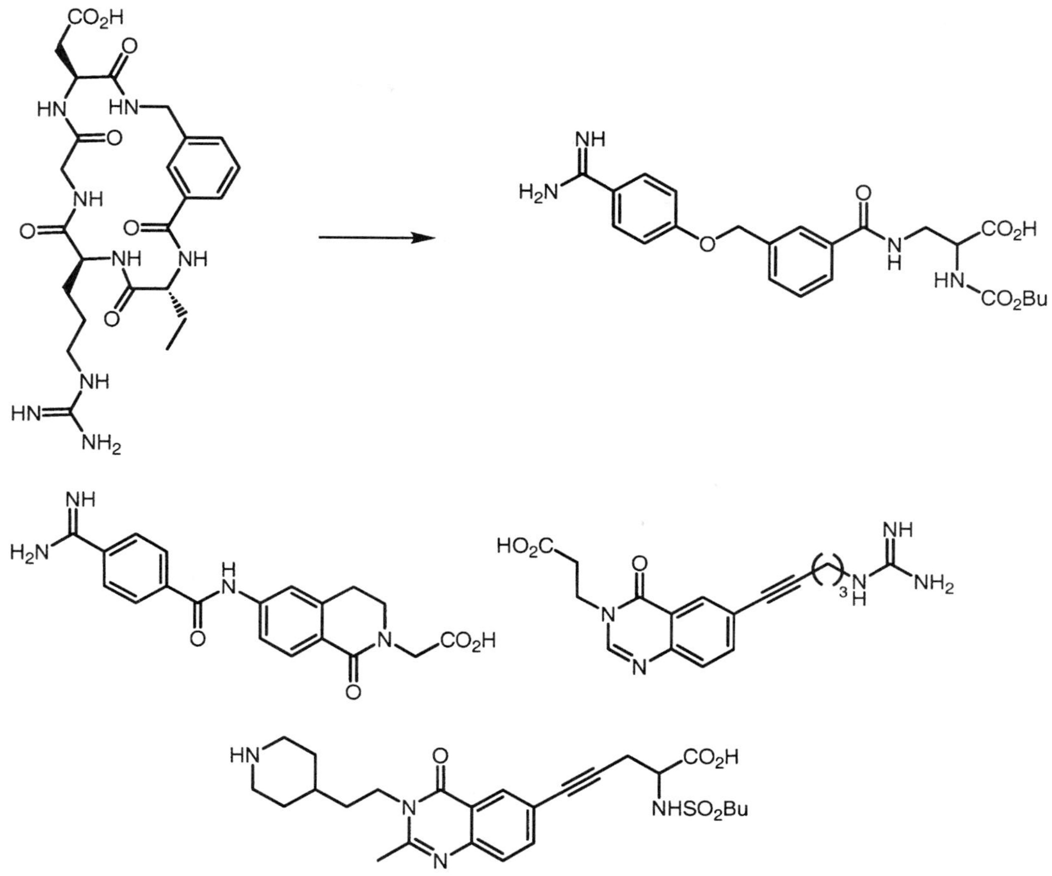

Figure 15.45. Different scaffolds used in RGD mimetics.

Overhauser effect experiments; coordination of the cysteine side-chain to the Zn ion promotes the conformational change in the peptide backbone. Moreover, differences in the conformation binding mode of peptides and peptidomimetics is one of the bases for selective farnesylation (208).

Other type III peptidomimetic inhibitors of this enzyme have also been reported. Inhibitor (**112**) (Fig. 15.47) was developed by replacing the A_1A_2 dipeptidyl sequence with a benzodiazepine scaffold (209). Later, SAR modifications of the benzodiazepine nucleus that included a hydrophilic 7-cyano group and a 4-sulfonyl group provided the potent, orally available and *in vivo* active (**113**) (210).

HTS also produced several non-peptide leads typified by inhibitor SCH 47307 (**114**)

(Fig. 15.47) (211). Subsequent SAR work led to the potent inhibitor SCH 66701 (**115**) ($K_i = 1.7 \ \mu M$), which was crystallized within the enzyme active site (212). This series of compounds is completely non-peptidic and also lacks the free sulfhydryl or imidazole seen in the other inhibitors discussed here. This is a breakthrough that shows that potency can be achieved even without the "essential" cysteine or sulfhydryl mimic.

9.6 Non-Peptidic Ligands for Peptide Receptors

This section illustrates the successful development of non-peptide peptidomimetics from a screening lead by assuming the inhibitor binds to the receptor in the same way as does the native peptide hormone. These assump-

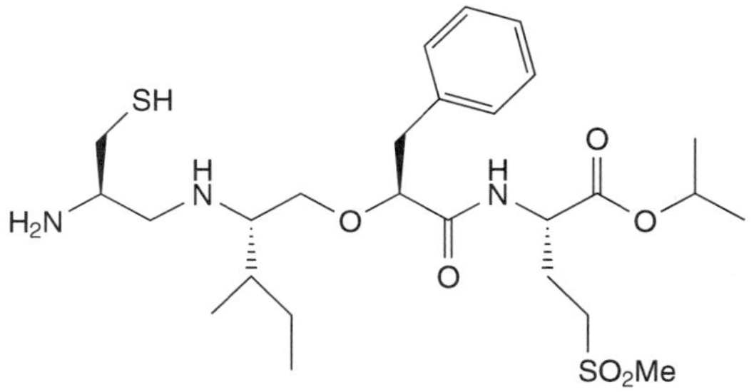

(108)

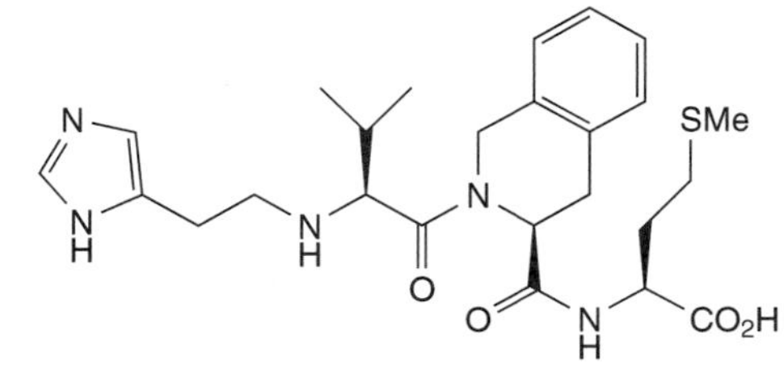

(110)

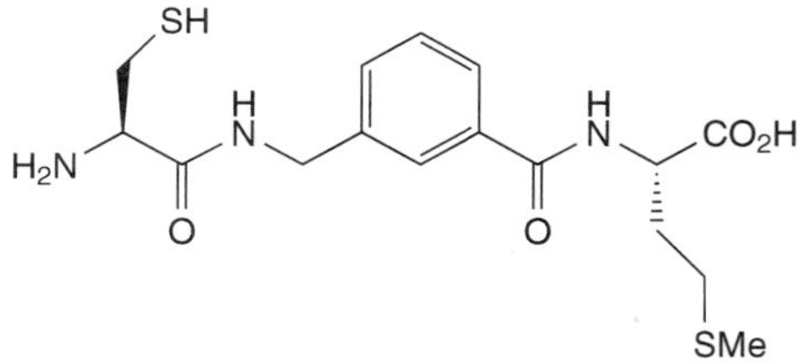

(109)

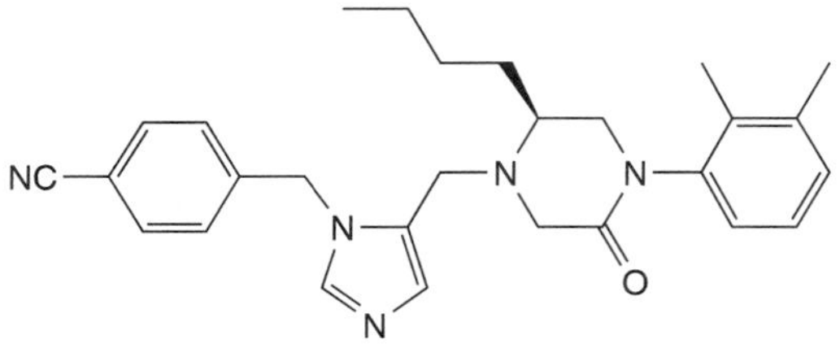

(111)

Figure 15.46. Peptide-like Ras-farnesyltransferase inhibitors.

tions actually led to effective inhibitors of the receptor. Later, site-directed mutagenesis of target receptors suggested that for many of these compounds, the mimetic was binding to the receptor at ancillary, perhaps overlapping, sites on the receptor. Later still, pharmacological studies indicated that peptide receptors adopted multiple states, suggesting that different antagonists might bind to different receptor forms. Of course, if compounds do not bind to the same receptor site as the endogenous hormone, SAR data collected on the natural peptide substrate is not applicable to these antagonists. Most of these peptidomimetics are probably type II or functional mimetics. Yet the success of this approach suggests that at least for some non-peptide antagonists, there may be some congruent structure that interacts with the receptor. These issues will only be determined unambiguously when high-resolution structures of the G-protein–coupled receptors (213) and other constitutive receptor systems are determined.

9.6.1 Angiotensin II. The first non-peptide antagonists of the AT1 receptor were found by HTS. The imidazole (**116**) (Fig. 15.48) is a

weak (IC_{50} = 43 μM) but quite selective A-II receptor antagonist (214). Using this as a lead compound, DuPont and SmithKline Glaxo researchers independently developed potent small molecule A-II receptor antagonists. The DuPont group used the conformation suggested by Smeby and Fernandjian to guide the design (215). It was speculated that the carboxyl group and the imidazol group of (**116**) were bound to the A-II terminal carboxyl group and to the imidazole group, respectively. This rationalization culminated in the synthesis of nanomolar inhibitors, with compound (**117**) as a clear representative (216).

Although workers at SmithKline Glaxo used the same conformation as starting point, they postulated other binding modes to the receptor. One of their alternative hypothesis considered compound (**116**) as a constrained analog in which the benzyl and the carboxyl groups corresponded to the Tyr side-chain and the C-terminal carboxyl group of A-II. Following this hypothesis, modification of lead compound (**116**) eventually led to compound (**118**) (Fig. 15.48) with an IC_{50} = 1.45 nM and oral activity of 30% (217).

Site-directed mutagenesis studies on the AT1 receptor revealed differences in the bind-

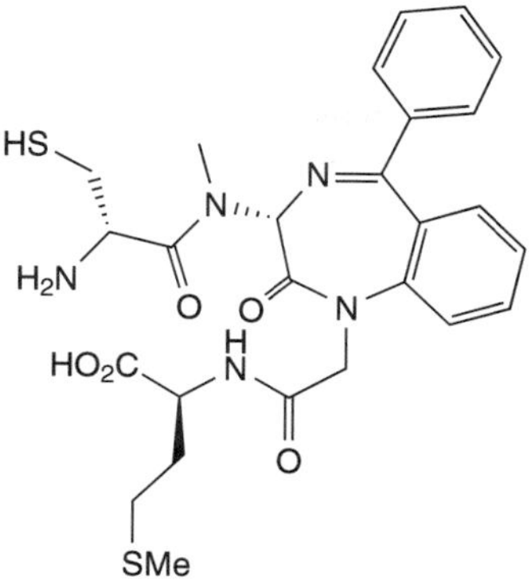

(112) 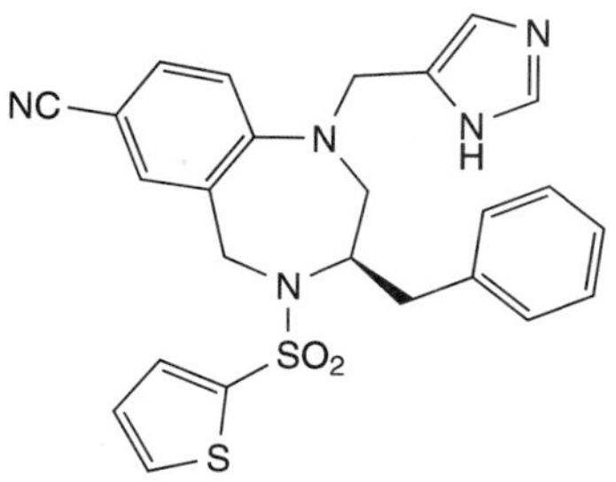   (113)

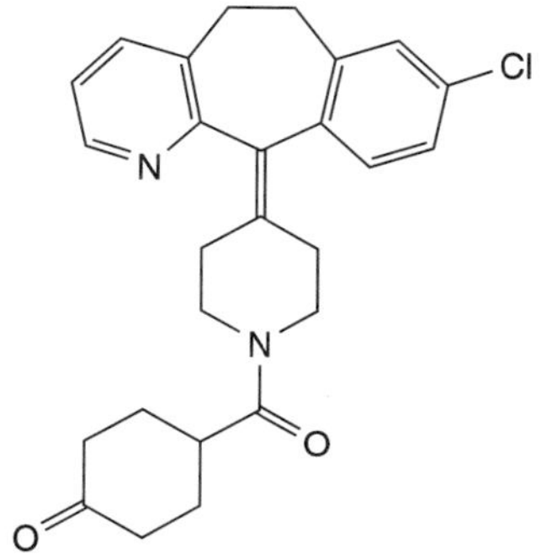

(114) 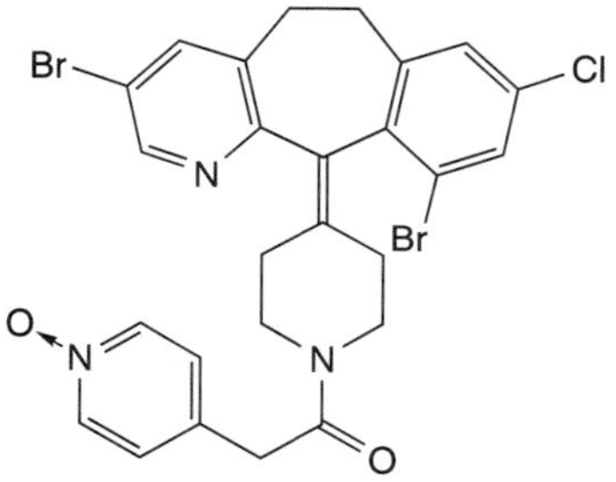   (115)

Figure 15.47. Non-peptidic Ras-farnesyltransferase inhibitors.

ing site of angiotensin and the small molecule non-peptide compounds (**119–120**) (Fig. 15.49). There is no evidence that the single residues involved in inhibitor binding overlap with endogenous peptide binding.

Some other non-peptide agonists have also appeared in the literature. Surprisingly, their binding mode differs from the binding mode of the peptide agonist (**121**), as well as that of the structurally similar non-peptide antagonist (**122**) (Fig. 15.49) (218). However, angiotensin and L-162,313 (**122**) require common critical residues for angiotensin AT1 receptor activation (219).

9.6.2 Substance P. The tachykinin receptors (NK-1, NK-2, and NK-3) and their endogenous ligands, the tachykinins, and neurokinins are important neurotransmitters (220–

222). Antagonists of tachykinin receptors produce beneficial effects in several CNS disease states such as pain, asthma, emesis, and depression.

A general approach for converting a variety of peptide structures into small, type II peptidomimetic antagonists was devised by Horwell and colleagues and is illustrated here for antagonists to Substance P. An alanine scan of the parent undecapeptide revealed that the Phe^4–Phe^5 sequence was required for binding. Replacement of one these residues by Trp, followed by introduction of conformational constraints by α-alkylation, provided the subnanomolar inhibitor (**123**) (Fig. 15.50) (223). Improved brain penetration was achieved by amine (**124**) (224).

Chemical screening of corporate compound libraries resulted in the discovery of another

Asp-Arg-Val-Tyr-Ile-His-Pro-Phe-OH

Angiotensin II

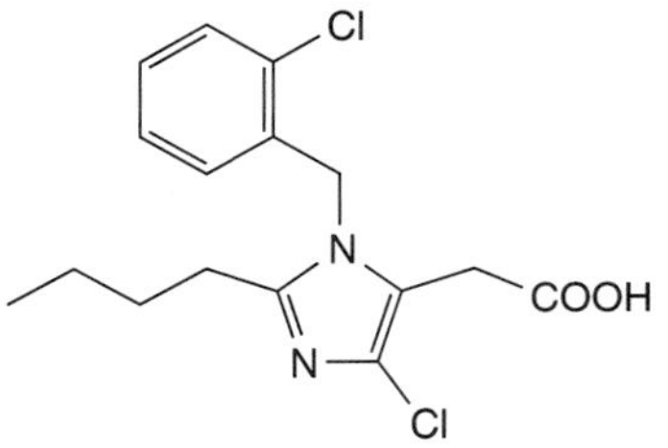

IC$_{50}$ = 43 μM

(116)

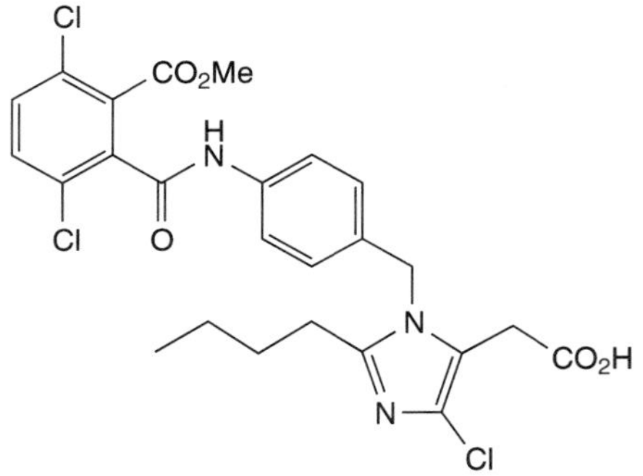

IC$_{50}$ = 18 nM

(117)

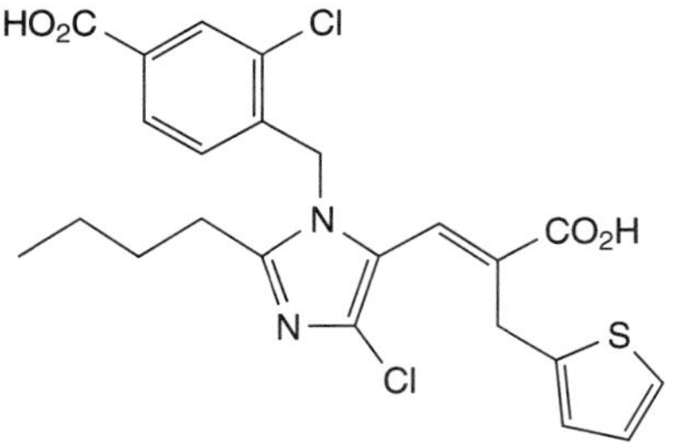

IC$_{50}$ = 1.45 nM

(118)

Figure 15.48. Angiotensin II inhibitors derived from a HTS lead.

type of non-peptidic NK-1 antagonist, CP-96,345 (**125**) (K_i = 0.66 nM, Fig. 15.51). This compound heralded a breakthrough in the design of these potential drugs (225, 226). Replacement of the basic quinuclidine ring with a morpholine core improves duration of action and insertion of an amino triazole unit confers excellent solubility and CNS penetration (**126**) (K_i = 0.19 nM, Fig. 15.51) (227).

Dual NK-1/NK-2 inhibitors, e.g., (**127**) (Fig. 15.52), have recently been designed by determining the important sites for maintaining NK-2 selectivity of the lead compound SR-48968 and introducing NK-1 pharmacophore groups (228–230).

Fewer NK-3 selective receptor antagonists have been described, but a quinoline scaffold previously reported to be a selective NK-3 receptor antagonist, has been converted to a dual NK-2/NK-3 inhibitor (**128**) (K_i = 0.8 nM NK-2 and 0.8 nM NK-3). The lead optimization was carried out by docking potential structures into a novel receptor model. The theoretical model compares closely with the recently published crystal structure of rhodopsin (231).

9.6.3 Neuropeptide Y. Neuropeptide Y (NPY) is a 36 amino acid polypeptide that is involved in hormonal, sexual, and cardiac effects (232, 233). In 1994, two "first generation" type I NY-1 selective antagonists, BIBP 3226 (**129**) (234) and SR120107A (**130**) (235) were reported (Fig. 15.53). BIBP 3236 (**129**) corresponds to a truncated and modified peptide in which the D-Arg is assumed to correspond with Arg[33] in Neuropeptide Y.

More recently, a series of indole Y1 antagonists discovered by screening (236) led to the benzimidazole (**131**) (K_i = 0.052 nM). In this type of compound, the diamino moieties are postulated to mimic the two C-terminal arginines of NPY (237).

Selective NPY-Y5 inhibitors have been shown to inhibit food intake activity *in vivo*. Most inhibitors found by HTS and lead optimization gave nanomolar and selective antagonists. It is not known whether these are functional or topological mimetics (Fig. 15.54) (238–240).

9.6.4 Growth Hormone Secretagogues. Growth hormone (GH) releasing peptide mimetics have become attractive alternatives to

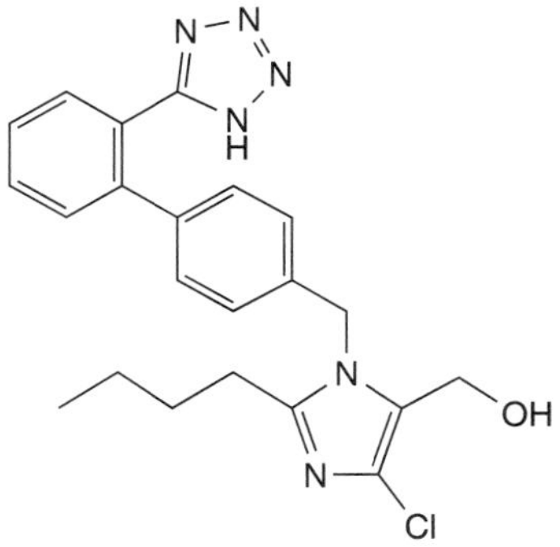

DuP753
(119)

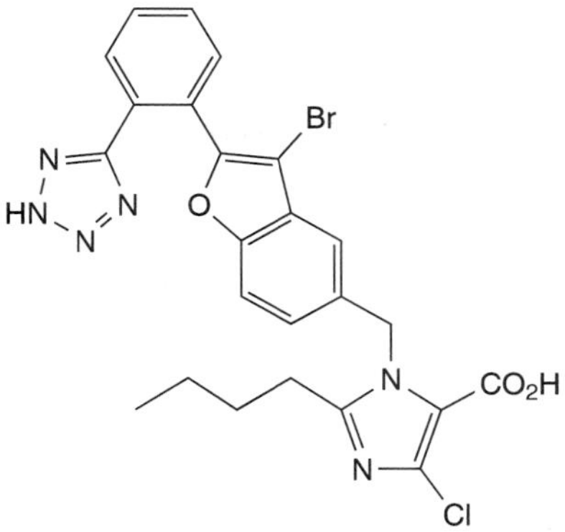

GR 117289
(120)

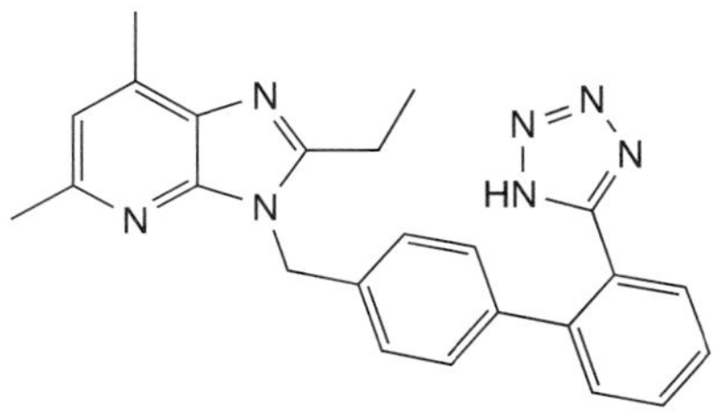

L 158809
(121)

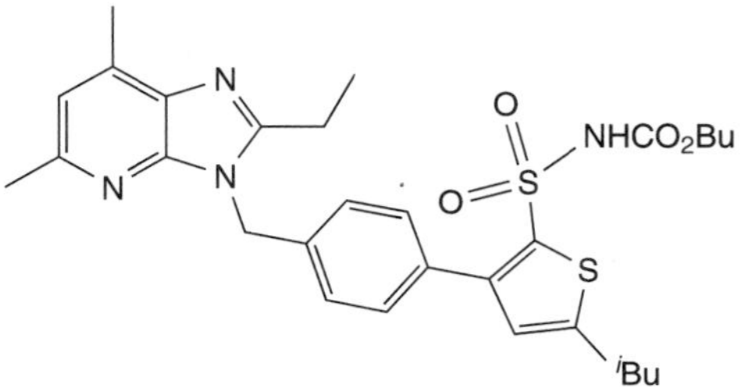

L 162313
(122)

Figure 15.49. Examples of angiotensin II inhibitors.

GH replacement therapy (241). The peptidyl GH secretagogue GHRP-6 (242) was used to develop the clinical candidate MK-0677 (**132**) (Fig. 15.55, $EC_{50} = 1.3$ nM) (243, 244). After

Arg-Pro-Lys-Pro-Gln-Gln-Phe-Phe-Gly-Leu-Met-NH$_2$
Substance P

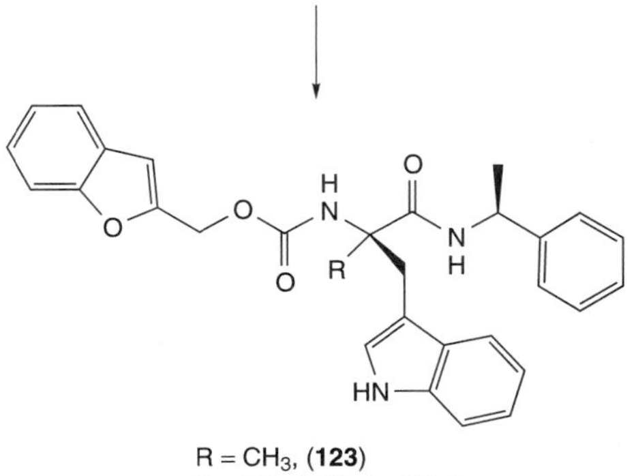

R = CH$_3$, (**123**)
R = CH$_2$N(CH$_3$)$_2$, (**124**)

Figure 15.50. Development of a substance P inhibitor.

identifying the important residues for bioactivity in GHPR-6, the Merck group began searching other receptor libraries for known "privileged structures" in a combinatorial synthetic fashion (see Section 4) (66). The more active derivative contained a spiropiperidine moiety attached to an indoline ring.

More recently, ghrelin has been isolated and identified as an endogenous ligand of the GHS receptor and some new peptidomimetic structures [e.g., **133** (Fig. 15.55)] have started to appear (245).

In another approach, SAR studies and systematic simplification of GHPR-6 at Novo Nordisk produced the orally bioavailable derivative NN-703 (**134**). Molecular modeling overlapping of NN703 (**134**) (Fig. 15.56) and MK-0677 (**132**) (Fig. 15.55) showed structural similarities between both compounds. Highly potent hybrids of Ipamorelin and NN-703 (e.g., **135**) (Fig. 15.56) have also been described (246, 247).

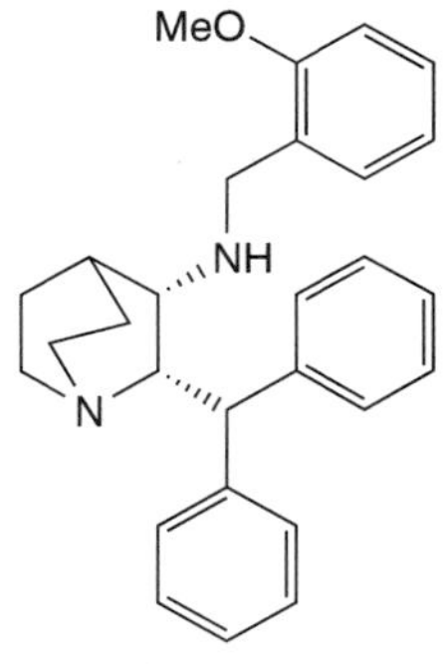

CP-96345
(**125**)

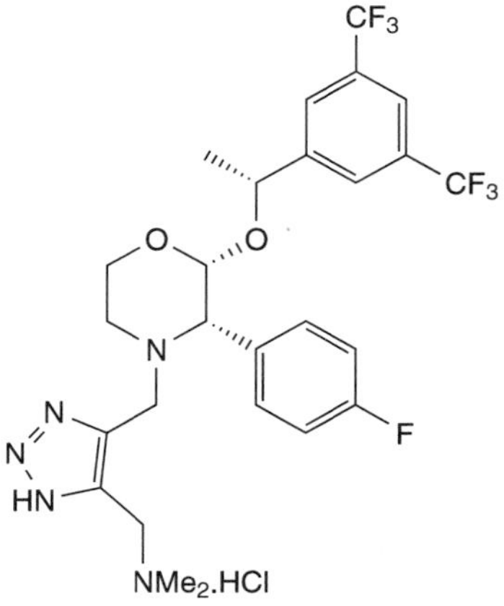

(**126**)

Figure 15.51. Examples of NK-1 antagonists.

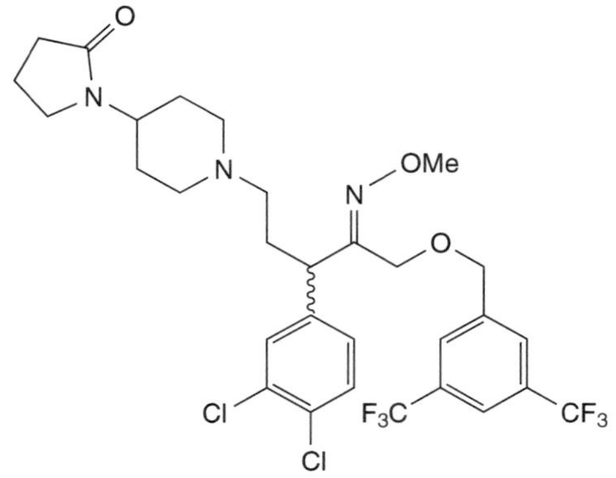

(**127**)

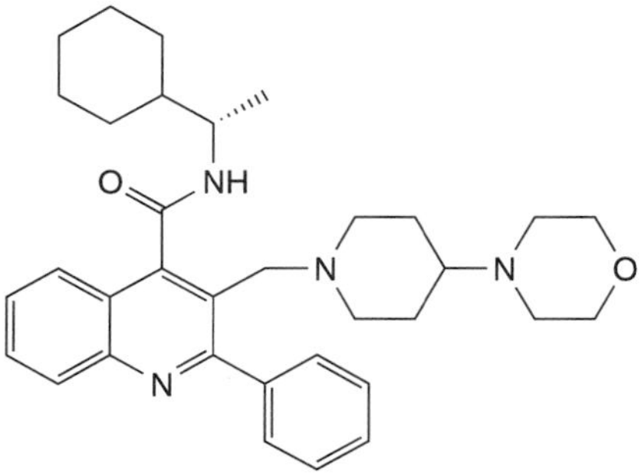

(**128**)

Figure 15.52. Dual NK-1/NK-2 and NK-1/NK-2 inhibitors.

A common 3D pharmacophore was recently described for peptidic and non-peptidic GH secretagogues by means of computational chemistry. After QSAR analysis, four pharmacophoric sites were found: two aromatic rings, a proton acceptor, and a protonated amine. Using these strategies, some nanomolar antagonists [e.g., **136** (Fig. 15.56)] were discovered (248).

9.6.5 Endothelin. The first report of endothelin in 1988 stimulated a huge effort to develop selective and non-selective endothelin receptor (ET_A and ET_B) antagonists (249, 250). One successful approach derived from the postulate that the phenyl groups of the screening lead might mimic two of the aromatic side-chains (Tyr[13], Phe[14], or Trp[21]) of

ET-1 (251, 252). Knowing that the carboxylic acid was also necessary for good activity, researchers at SmithKline overlaid their inhibitor with the aromatic groups Tyr[13], Phe[14], and Asp[18] in ET-1. After using a conformationally constrained analog of ET-1 to further define their NMR-derived structure of ET-1, the final overlay suggested that a carboxylic acid attached with a linker of two to three atoms on the 2-position of the phenyl ring would provide further binding interaction by mimicking the C-terminal carboxylic acid. This led to compound (**137**) (Fig. 15.57), a potent antagonist of both the ET_A and ET_B receptors with $K_i = 0.43$ and 15.7 nM, respectively. Analogs based on a pyrrolidine scaffold are also effective (e.g., **138**) (Fig. 15.57) (253).

The Kohonen neural network has been used to develop bioisosteres of the methylendioxyphenyl group found in a variety of antag-

Tyr-Pro-Ser-Lys-Pro-Asp-Asn-Pro-Gly-Glu-Asp-Ala-Pro-Ala-Glu-Asp-Leu-Ala
Arg-Tyr-Tyr-Ser-Ala-Leu-Arg-His-Tyr-Ile-Asn-Leu-Ile-Thr-Arg-Gln-Arg-Tyr-NH$_2$

Neuropeptide Y

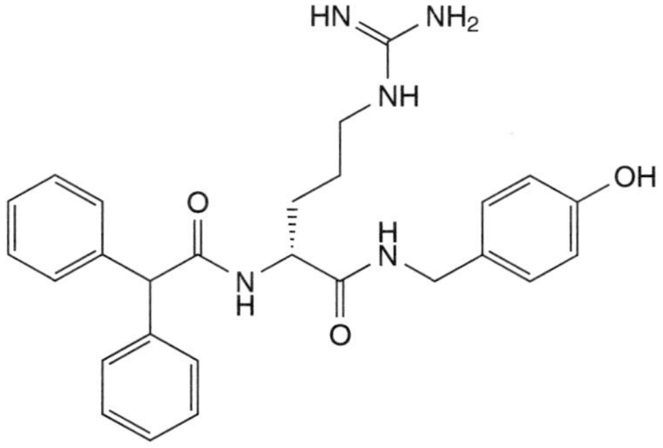

BIBP3226
(**129**)

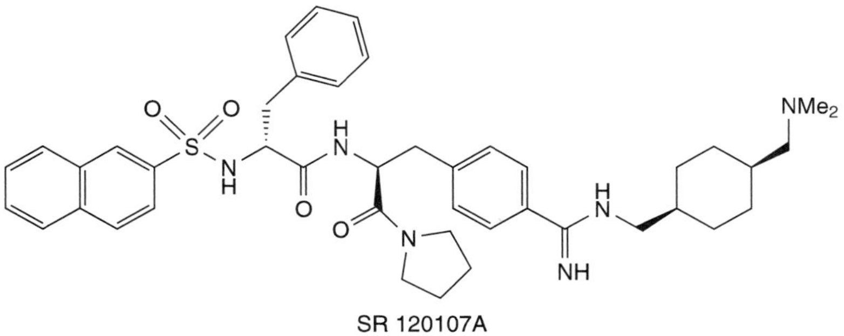

SR 120107A
(**130**)

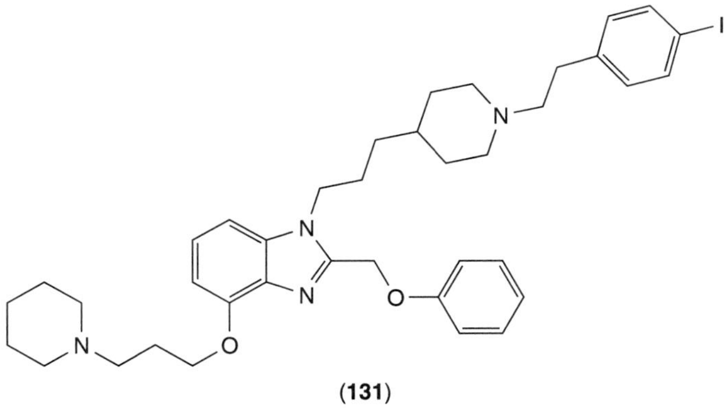

(**131**)

Figure 15.53. Examples of neuropeptide Y1 inhibitors.

onists [e.g., **139** (Fig. 15.58)]. The benzothia-diazole (**140**) functions as a bioisostere that retains and sometimes improves binding to the ET$_A$ receptor (254–256).

Since the discovery of Ro46–2005 (**141**) (Fig. 15.58), the first orally active ET inhibitor, major efforts have been made to modify arylsulfonamide derivatives. An isoxazole as

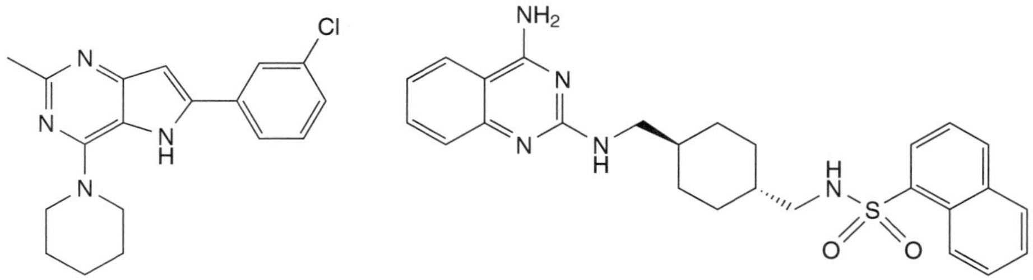

Figure 15.54. Examples of neuropeptide Y5 inhibitors.

the heterocycle attached to the amino function-ality provided selectivity against ET_A receptor (257) and led to BMS 193884 (**142**) ($K_i = 1.4$ nM) (258) and others, e.g., TBC 3214 (**143**) ($K_i = 0.04$ nM) (259), which are potent, selective, and orally available ET_A antagonists.

Different binding modes have been pro-posed for ET antagonists. The acid or sulfon-amido groups are needed to interact with a cationic site in the receptor, and an aromatic interaction with Tyr^{129} is postulated to be re-sponsible for ET_A selectivity. However, be-cause all these receptors are members of the GPCR, there is no assurance that any bind as

modeled. Thus, they must be classified as type II peptidomimetics until structural data can resolve the issue.

10   SUMMARY AND FUTURE DIRECTIONS

The "Holy Grail" of peptidomimetic research in drug discovery has been to find ways to transform the structural information con-tained in peptides into non-peptide structures that have drug-like pharmacodynamic proper-ties. Many different strategies have been

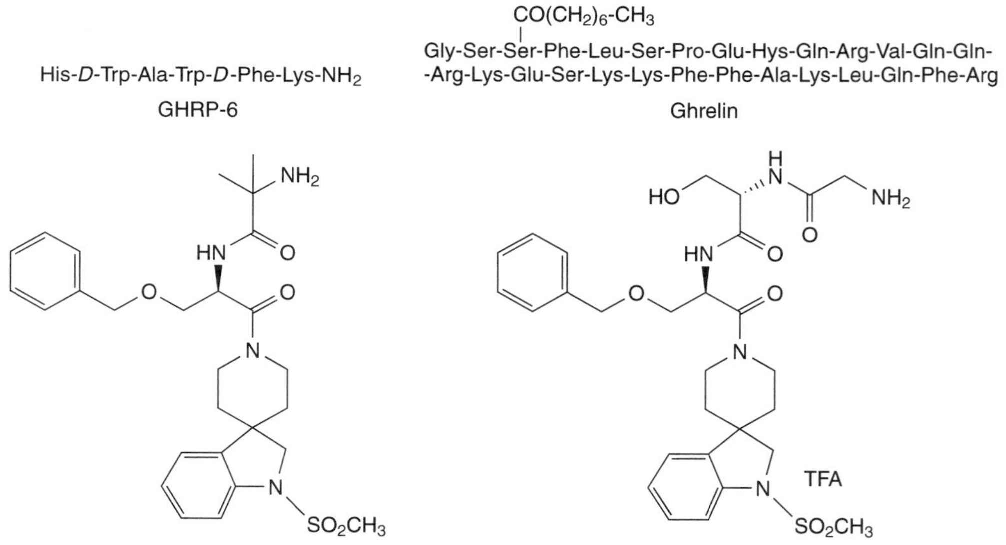

Figure 15.55. GHRP-6 and ghrelin non-peptide derivatives as growth hormone secretagogues inhibitors.

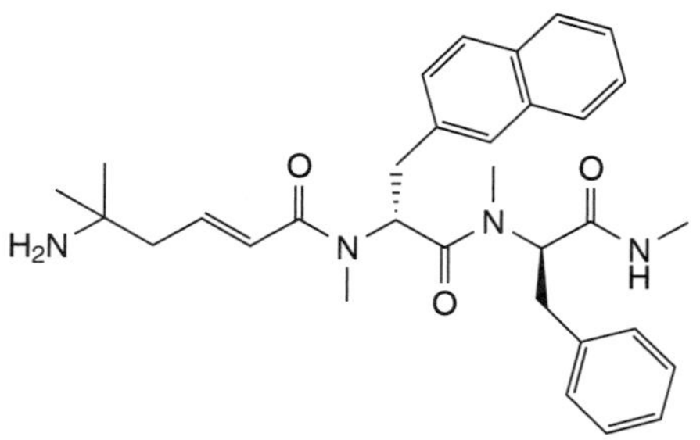

NN-703
(134)

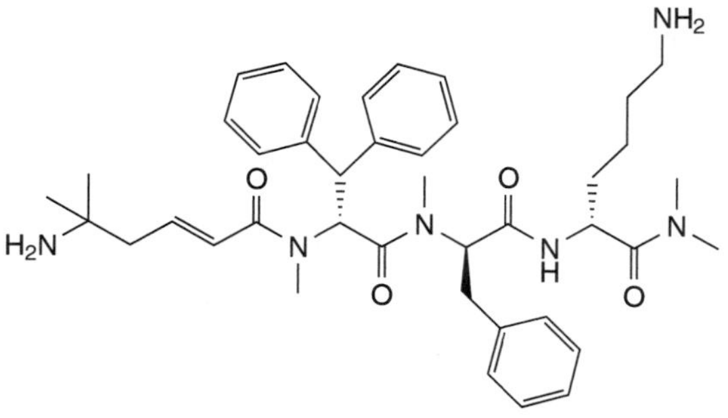

(135)

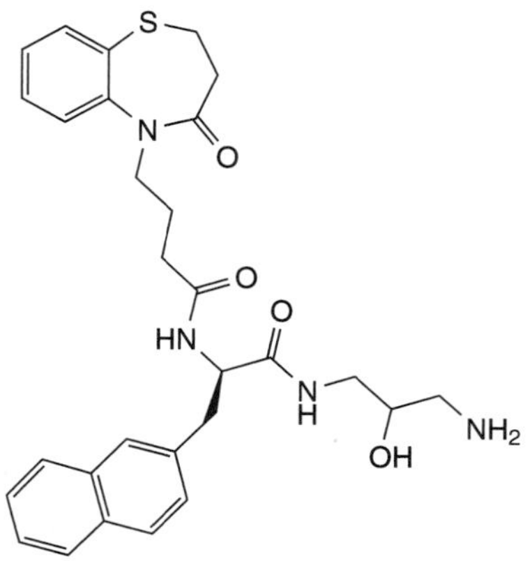

S-37435
(136)

Figure 15.56. Newer approaches to growth hormone secretagogues inhibitors.

Cys-Ser-Cys-Ser-Ser-Leu-Met

Trp-Ile-Ile-Asp-Leu-His-Cys-Phe-Tyr-Val-Cys-Glu-Lys-Asp

Endothelin-1

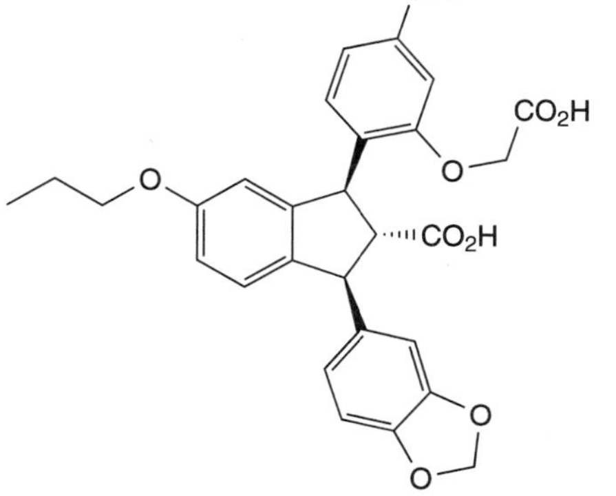

SB 209670
(137)

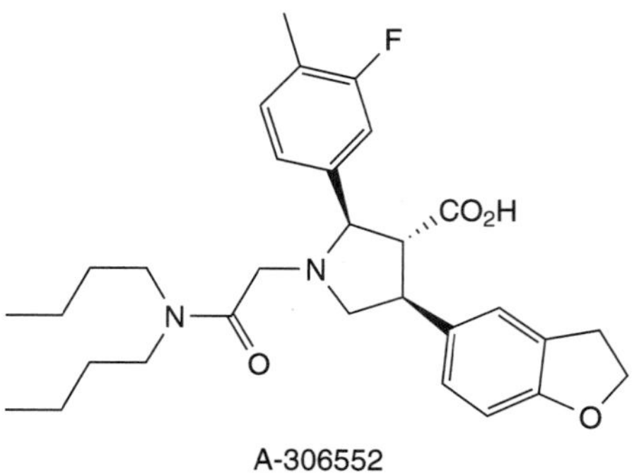

A-306552
(138)

Figure 15.57. Non-peptide endothelin analogs.

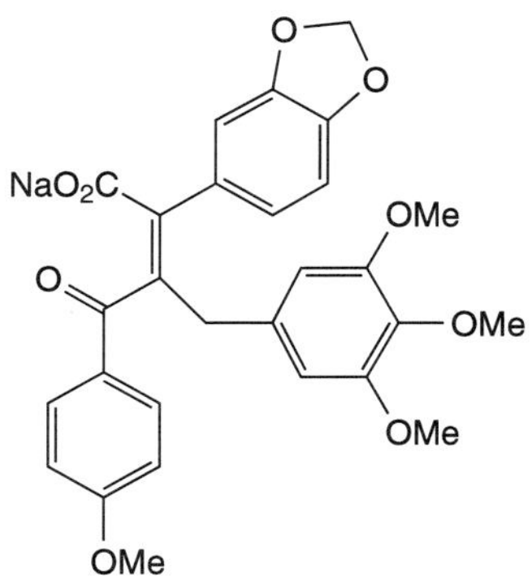

PD156707
(**139**)

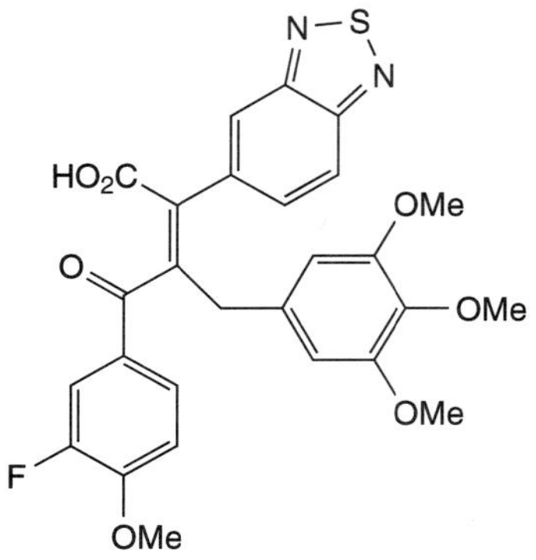

EMD 122946
(**140**)

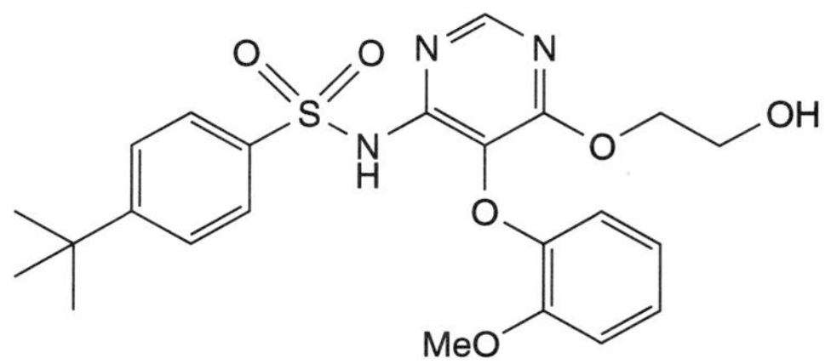

Ro 46-2005
(**141**)

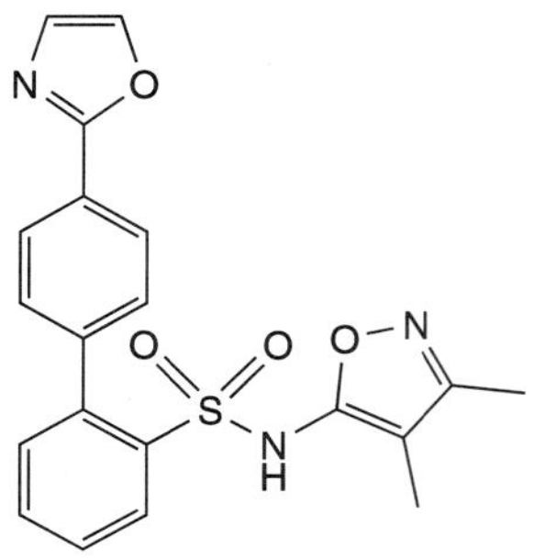

BMS 193884
(**142**)

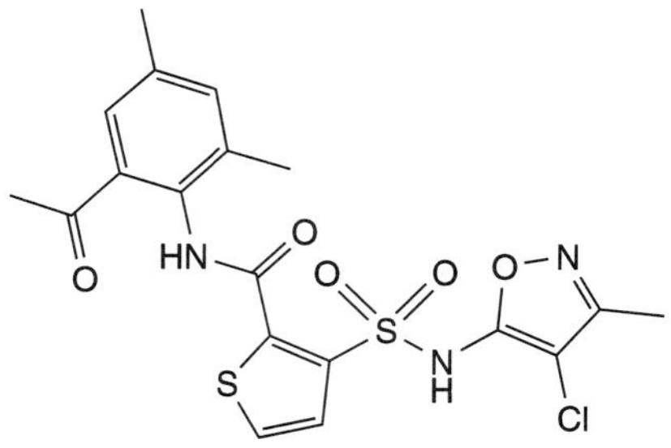

TBC3214
(**143**)

Figure 15.58. Examples of ET_A inhibitors.

employed in the search for useful peptidomimetics—rational design of amide bond replacements, mimics of turn structures, and the like, as well as both designed and discovered scaffolds that replace the amide bond core of peptides. The field has a long way to go before rational design of type III peptidomimetics can be achieved routinely. However, the progress made to date suggests that this goal will be achieved. We know that some nonpeptide scaffolds are topographical mimetics of the extended β-strand of enzyme-bound protease inhibitors because we have the biophysical methods for characterizing both types of enzyme-inhibitor complexes. Type III peptidomimetic inhibitors of peptidases have been designed from the substrate sequences and they have been revealed by HTS processes and optimized by application of structural biology. At this point, we have learned more about the design of inhibitors by studying how screening leads inhibit enzymes than from the design of inhibitors from our current, limited knowledge of enzyme catalysis. Probably the most important recent discovery is that some screening leads inhibit proteases by binding to a different enzyme active site conformation that is related mechanistically to the well-characterized extended β-strand of enzyme-bound protease inhibitors. This result emphasizes the importance of considering the entire ensemble of protein conformations when designing inhibitors of peptide-protein interactions.

Our understanding of peptide mimicry for ligands of constitutive receptors, such as G-protein–coupled receptors (GPCR), is much more primitive because high resolution structural data for agonist- and/or antagonist-receptor complexes are not yet available. For this reason, all attempts to rationalize the interactions between ligand and receptor contain a considerable element of speculation. It is too early to know whether small non-peptide structures that bind to GPCR are functional or topographical mimetics. However, based on the results obtained by studying peptidase inhibitors, it seem likely that at least some of the known functional peptidomimetics receptors ligands will be shown to be topographical mimetics. Others may be found to act more like GRAB-peptidomimetics in that they bind to receptor conformations closely related in energy and mechanism to native conformations. Still others will no doubt be found that inhibit or stimulate the receptor system by allosteric mechanisms or by interfering with some multi-step binding process preceding the formation of the active ligand-receptor complex. In any case, it is clear that successful design of functional mimetics by assuming some structural relationship between a screening lead and the parent peptide can work (see Section 9.6), as can the systematic modification of the parent peptide. The application of the principles of peptidomimetic research has become very important to drug discovery. Although our present knowledge about protein–protein interactions is still quite limited, the rapid growth of structural information and methods will eventually allow us to design rationally peptidomimetic compounds suitable for use in human therapy.

REFERENCES

1. M. D. Fletcher and M. M. Campbell, *Chem. Rev.*, **98**, 763 (1998).
2. F. Haviv, T. D. Fitzpatrick, C. J. Nichols, E. N. Bush, G. Diaz, G. Bammert, A. T. Nguyen, E. S. Johnson, J. Knittle, and J. Greer, *J. Med. Chem.*, **37**, 701 (1994).
3. J. Hughes, T. W. Smith, H. W. Kosterlitz, L. A. Fothergill, B. A. Morgan, and H. R. Morris, *Nature*, **258**, 577 (1975).
4. G. D. Smith and J. F. Griffin, *Science*, **199**, 1214 (1978).
5. A. Aubry, N. Birlirakis, M. Sakarellos-Daitsiotis, C. Sakarellos, and M. Marraud, *Biopolymers*, **28**, 27 (1989).
6. A. F. Bradbury, D. G. Smyth, and C. R. Snell, *Nature*, **260**, 165 (1976).
7. P. S. Farmer in E. J. Ariens, Ed., *Drug Design*, Academic Press, New York, 1980.
8. A. B. Smith III, T. P. Keenan, R. C. Holcomb, P. A. Sprengeler, M. C. Guzman, J. L. Wood, P. J. Carroll, and R. Hirschmann, *J. Am. Chem. Soc.*, **114**, 10672 (1992).
9. A. S. Ripka and D. H. Rich, *Curr. Opin. Chem. Biol.*, **2**, 441 (1998).
10. M. G. Bursavich and D. H. Rich, *J. Med. Chem.*, **45**, 541 (2002).
11. A. F. Spatola in B. Weistein, Ed., *Chem. Biochem. Amino Acids, Pept., Proteins*, Marcel Dekker, New York, 1983.
12. U. Gether, *Endocr. Rev.*, **21**, 90 (2000).
13. D. P. Fairlie, G. Abbenante, and D. R. March, *Curr. Med. Chem.*, **2**, 654 (1995).
14. R. A. Wiley and D. H. Rich, *Med. Res. Rev.*, **13**, 327 (1993).
15. J. D. A. Tyndall and D. P. Fairlie, *Curr. Med. Chem.*, **8**, 893 (2001).
16. N. R. A. Beeley, *Drug Discov. Today.*, **5**, 354 (2000).
17. R. M. Freidinger, *Trends Pharmacol. Sci.*, **10**, 270 (1989).
18. R. M. Freidinger, *Curr. Opin. Chem. Biol.*, **3**, 395 (1999).
19. A. Giannis and T. Kolter, *Angew. Chem. Intl. Ed. Engl.*, **32**, 1244 (1993).

20. G. J. Moore, *Trends Pharmacol. Sci.*, **15**, 124 (1994).

21. D. C. Rees, *Curr. Med. Chem.*, **1**, 145 (1994).

22. E. E. Sugg, *Annu. Rep. Med. Chem.*, **32**, 277 (1997).

23. T. K. Sawyer, *Drugs Pharm. Sci.*, **101**, 81 (2000).

24. B. A. Morgan and J. A. Gainor, *Annu. Rep. Med. Chem.*, **24**, 243 (1989).

25. G. J. Moore, *Proc. West. Pharmacol. Soc.*, **40**, 115 (1997).

26. A. Giannis and F. Rubsam, *Adv. Drug Res.*, **29**, 1 (1997).

27. M. Goodman and S. Ro in M. E. Wolff, Ed., *Burger's Medicinal Chemistry and Drug Discovery*, **vol. 1**, Wiley-Interscience, San Diego, CA, 1995, pp. 803–861.

28. D. Obrecht, M. Altorfer, and J. A. Robinson, *Adv. Med. Chem.*, **4**, 1 (1999).

29. J. Gante, *Angew. Chem. Intl. Ed. Engl.*, **33**, 1699 (1994).

30. P. A. Hart and D. H. Rich, *Pract. Med. Chem.*, 393–412 (1996).

31. D. F. Veber, *Pept.: Chem. Biol.*, in J. A. R. Smith and E. Jean, Eds., *Proc. Am. Pept. Symp., 12th*, ESCOM, Leiden, (1992).

32. G. R. Marshall, *Tetrahedron*, **49**, 3547 (1993).

33. J. W. Erickson and S. W. Fesik, *Annu. Rep. Med. Chem.*, **27**, 271 (1992).

34. G. Muller, *Curr. Med. Chem.*, **7**, 861 (2000).

35. D. F. Mierke and C. Giragossian, *Med. Res. Rev.*, **21**, 450 (2001).

36. D. F. Veber, F. W. Holly, R. F. Nutt, S. J. Bergstrand, S. F. Brady, R. Hirschmann, M. S. Glitzer, and R. Saperstein, *Nature*, **280**, 512 (1979).

37. J. Rivier, M. Brown, and W. Vale, *Biochem. Biophys. Res. Commun.*, **65**, 746 (1975).

38. G. D. Rose, L. M. Gierasch, and J. A. Smith, *Adv. Protein Chem.*, **37**, 1 (1985).

39. P. W. Schiller, The Peptides: Analysis, Synthesis and Biology, Vol. **6**, Academic Press, Orlando, FL, 1984.

40. T. K. Sawyer, V. J. Hruby, P. S. Darman, and M. E. Hadley, *Proc. Natl. Acad. Sci. USA*, **79**, 1751 (1982).

41. K. Ishikawa, T. Fukami, T. Nagase, K. Fujita, T. Hayama, K. Niiyama, T. Mase, M. Ihara, and M. Yano, *J. Med. Chem.*, **35**, 2139 (1992).

42. G. R. Marshall, F. A. Gorin, and M. L. Moore, *Annu. Rep. Med. Chem.*, **13**, 227 (1978).

43. G. R. Marshall, C. D. Barry, H. E. Bosshard, R. A. Dammkoehler, and D. A. Dunn, *ACS Symp. Ser.*, **112**, 205 (1979).

44. R. M. J. Liskamp, *Recl. Trav. Chim. Pays-Bas.*, **113**, 1 (1994).

45. G. Hölzemann, *Kontakte (Darmstadt)*, **1**, 3 (1991).

46. G. Hölzemann, *Kontakte (Darmstadt)*, **2**, 55 (1991).

47. M. Kahn, *Synlett*, 821–826 (1993).

48. R. M. Freidinger, D. F. Veber, D. S. Perlow, J. R. Brooks, and R. Saperstein, *Science*, **210**, 656 (1980).

49. E. D. Thorsett, E. E. Harris, S. D. Aster, E. R. Peterson, J. P. Snyder, J. P. Springer, J. Hirshfield, E. W. Tristram, A. A. Patchett, E. H. Ulm, and T. C. Vassil, *J. Med. Chem.*, **29**, 251 (1986).

50. G. A. Flynn, E. L. Giroux, and R. C. Dage, *J. Am. Chem. Soc.*, **109**, 7914 (1987).

51. U. Nagai, K. Sato, R. Nakamura, and R. Kato, *Tetrahedron*, **49**, 3577 (1993).

52. S. Hanessian, G. McNaughton Smith, H. G. Lombart, and W. D. Lubell, *Tetrahedron*, **53**, 12789 (1997).

53. A. J. Souers and J. A. Ellman, *Tetrahedron*, **57**, 7431 (2001).

54. K. Burgess, *Acc. Chem. Res.*, **34**, 826 (2001).

55. A. B. Smith III, M. C. Guzman, P. A. Sprengeler, T. P. Keenan, R. C. Holcomb, J. L. Wood, P. J. Carroll, and R. Hirschmann, *J. Am. Chem. Soc.*, **116**, 9947 (1994).

56. K. D. Stigers, M. J. Soth, and J. S. Nowick, *Curr. Opin. Chem. Biol.*, **3**, 714 (1999).

57. J. S. Nowick, E. M. Smith, and M. Pairish, *Chem. Soc. Rev.*, **25**, 401 (1996).

58. R. P. Cheng, S. H. Gellman, and W. F. De-Grado, *Chem. Rev.*, **101**, 3219 (2001).

59. J. Venkatraman, S. C. Shankaramma, and P. Balaram, *Chem. Rev.*, **101**, 3131 (2001).

60. D. P. Fairlie, M. L. West, and A. K. Wong, *Curr. Med. Chem.*, **5**, 29 (1998).

61. V. J. Hruby and P. M. Balse, *Curr. Med. Chem.*, **7**, 945 (2000).

62. V. J. Hruby, *Acc. Chem. Res.*, **34**, 389 (2001).

63. H. Nakanishi and M. Kahn, *Bioorg. Chem. Pept. & Protein*, **12**, 395 (1998).

64. R. Hirschmann, P. A. Sprengeler, T. Kawasaki, J. W. Leahy, W. C. Shakespeare, and A. B. Smith III, *Tetrahedron*, **49**, 3665 (1993).

65. Y. Qian, A. Vogt, S. M. Sebti, and A. D. Hamilton, *J. Med. Chem.*, **39**, 217 (1996).

66. B. E. Evans, K. E. Rittle, M. G. Bock, R. M. DiPardo, R. M. Freidinger, W. L. Whitter, G. F. Lundell, D. F. Veber, P. S. Anderson, R. S. L. Chang, V. J. Lotti, D. J. Cerino, T. B. Chen, P. J. Kling, K. A. Kunkel, J. P. Springer, and J. Hirshfield, *J. Med. Chem.*, **31**, 2235 (1988).

67. B. E. Evans, K. E. Rittle, M. G. Bock, R. M. DiPardo, R. M. Freidinger, W. L. Whitter, N. P. Gould, G. F. Lundell, C. F. Homnick, and D. F. Veber, *J. Med. Chem.*, **30**, 1229 (1987).

68. T. Clackson and J. A. Wells, *Science*, **267**, 383 (1995).

69. H. C. Kawato, K. Nakayama, H. Inagaki, and T. Ohta, *Org. Lett.*, **3**, 3451 (2001).

70. K. Nakayama, H. C. Kawato, H. Inagaki, and T. Ohta, *Org. Lett.*, **3**, 3447 (2001).

71. M. D. Fletcher and M. M. Campbell, *Chem. Rev.*, **98**, 763 (1998).

72. M. Chorev and M. Goodman, *Acc. Chem. Res.*, **26**, 266 (1993).

73. R. M. Williams, *Biomed. Appl. Biotechnol.*, **1**, 187 (1993).

74. C. A. Lipinski, *J. Pharmacol. Toxicol. Methods*, **44**, 235 (2001).

75. G. Klebe, *J. Mol. Med.*, **78**, 269 (2000).

76. H.-J. Böhm and G. Klebe, *Angew. Chem. Intl. Ed. Engl.*, **35**, 2589 (1996).

77. L. Pauling, *Chem. Eng. News*, **24**, 1375 (1946).

78. R. Wolfenden, *Annu. Rev. Biophys. Bioeng.*, **5**, 271 (1976).

79. R. Wolfenden, *Acc. Chem. Res.*, **5**, 10 (1972).

80. M. A. Ondetti, B. Rubin, and D. W. Cushman, *Science*, **196**, 441 (1977).

81. D. Leung, G. Abbenante, and D. P. Fairlie, *J. Med. Chem.*, **43**, 305 (2000).

82. R. E. Babine and S. L. Bender, *Chem. Rev.*, **97**, 1359 (1997).

83. I. Schechter and A. Berger, *Biochem. Biophys. Res. Commun.*, **27**, 157 (1967).

84. H. P. Marki, A. Binggeli, B. Bittner, V. Bohner-Lang, V. Breu, D. Bur, P. Coassolo, J. P. Clozel, A. D'Arcy, H. Doebeli, W. Fischli, C. Funk, J. Foricher, T. Giller, F. Gruninger, A. Guenzi, R. Guller, T. Hartung, G. Hirth, C. Jenny, M. Kansy, U. Klinkhammer, T. Lave, B. Lohri, F. C. Luft, E. M. Mervaala, D. N. Muller, M. Muller, F. Montavon, C. Oefner, C. Qiu, A. Reichel, P. Sanwald-Ducray, M. Scalone, M. Schleimer, R. Schmid, H. Stadler, A. Treiber, O. Valdenaire, E. Vieira, P. Waldmeier, R. Wiegand-Chou, M. Wilhelm, W. Wostl, M. Zell, and R. Zell, *Farmaco*, **56**, 21 (2001).

85. D. Banner, J. Ackermann, A. Gast, K. Gubernator, P. Hadvary, K. Hilpert, L. Labler, K. Mueller, G. Schmid, T. Tschopp, H. van de Waterbeemd, and B. Wirz, *Perspect. Med. Chem.*, 27–43 (1993).

86. J. Rahuel, V. Rasetti, J. Maibaum, H. Rueger, R. Goschke, N. C. Cohen, S. Stutz, F. Cumin, W. Fuhrer, J. M. Wood, and M. G. Grutter, *Chem. Biol.*, **7**, 493 (2000).

87. M. J. Parry, A. B. Russell, and M. Szelke in J. Meienhofer, Ed., *Chem. Biol. Pept., Proc. Am. Pept. Symp., 3rd*, Ann Arbor Science, Ann Arbor, MI, 1972, p. 541.

88. M. Szelke, D. M. Jones, B. Atrash, A. Hallett, and B. J. Leckie in V. J. Hruby and D. H. Rich, Eds., *Pept.: Struct. Funct., Proc. Am. Pept. Symp., 8th*, Pierce Chemical Co., Rockford, 1983, p. 579.

89. D. H. Rich and E. T. O. Sun, *Biochem. Pharmacol.*, **29**, 2205 (1980).

90. E. M. Gordon, J. D. Godfrey, J. Pluscec, D. Von Langen, and S. Natarajan, *Biochem. Biophys. Res. Commun.*, **126**, 419 (1985).

91. D. H. Rich, *J. Med. Chem.*, **28**, 263 (1985).

92. D. H. Rich, J. Green, M. V. Toth, G. R. Marshall, and S. B. H. Kent, *J. Med. Chem.*, **33**, 1285 (1990).

93. E. J. Lien, H. Gao, and L. L. Lien, *Prog. Drug Res.*, **43**, 43 (1994).

94. F. Lebon and M. Ledecq, *Curr. Med. Chem.*, **7**, 455 (2000).

95. E. D. Thorsett and L. H. Latimer, *Curr. Opin. Chem. Biol.*, **4**, 377 (2000).

96. M. S. Wolfe, *J. Med. Chem.*, **44**, 2039 (2001).

97. S. Sinha, J. P. Anderson, R. Barbour, G. S. Basi, R. Caccavello, D. Davis, M. Doan, H. F. Dovey, N. Frigon, J. Hong, K. Jacobson-Croak, N. Jewett, P. Keim, J. Knops, I. Lieberburg, M. Power, H. Tan, G. Tatsuno, J. Tung, D. Schenk, P. Seubert, S. M. Suomensaari, S. Wang, D. Walker, J. Zhao, L. McConlogue, and V. John, *Nature*, **402**, 537 (1999).

98. A. K. Ghosh, D. Shin, D. Downs, G. Koelsch, X. Lin, J. Ermolieff, and J. Tang, *J. Am. Chem. Soc.*, **122**, 3522 (2000).

99. L. Hong, G. Koelsch, X. Lin, S. Wu, S. Terzyan, A. K. Ghosh, X. C. Zhang, and J. Tang, *Science*, **290**, 150 (2000).

100. A. K. Ghosh, G. Bilcer, C. Harwood, R. Kawahama, D. Shin, K. A. Hussain, L. Hong, J. A. Loy, C. Nguyen, G. Koelsch, J. Ermolieff, and J. Tang, *J. Med. Chem.*, **44**, 2865 (2001).

101. H. H. Rotmensch, P. H. Vlasses, B. N. Swanson, J. D. Irvin, K. E. Harris, D. G. Merrill, and R. K. Ferguson, *Am. J. Cardiol.*, **53**, 116 (1984).

102. A. A. Patchett, E. Harris, E. W. Tristram, M. J. Wyvratt, M. T. Wu, D. Taub, E. R. Peterson, T. J. Ikeler, J. ten Broeke, L. G. Payne, D. L. Ondeyka, E. D. Thorsett, W. J. Greenlee, N. S. Lohr, R. D. Hoffsommer, H. Joshua, W. V. Ruyle, J. W. Rothrock, S. D. Aster, A. L. Maycock, F. M. Robinson, R. Hirschmann, C. S. Sweet, E. H. Ulm, D. M. Gross, T. C. Vassil, and C. A. Stone, *Nature*, **288**, 280 (1980).

103. B. P. Roques, M. C. Fournie-Zaluski, E. Soroca, J. M. Lecomte, B. Malfroy, C. Llorens, and J. C. Schwartz, *Nature*, **288**, 286 (1980).

104. B. P. Roques, E. Lucas-Soroca, P. Chaillet, J. Costentin, and M. C. Fournie-Zaluski, *Proc. Natl. Acad. Sci. USA*, **80**, 3178 (1983).

105. R. Bouboutou, G. Waksman, J. Devin, M. C. Fournie-Zaluski, and B. P. Roques, *Life Sci.*, **35**, 1023 (1984).

106. J. Bralet and J. C. Schwartz, *Trends Pharm. Sci.*, **22**, 106 (2001).

107. S. DeLombaert, R. Chatelain, C. A. Fink, and A. J. Trapani, *Curr. Pharm. Design*, **2**, 443 (1996).

108. J. W. Skiles, L. G. Monovich, and A. Y. Jeng, *Annu. Rep. Med. Chem.*, **35**, 167 (2000).

109. M. R. Michaelides and M. L. Curtin, *Curr. Pharm. Design*, **5**, 787 (1999).

110. M. Whittaker, C. D. Floyd, P. Brown, and A. J. H. Gearing, *Chem. Rev.*, **99**, 2735 (1999).

111. S. Brown, M. M. Bernardo, Z. H. Li, L. P. Kotra, Y. Tanaka, R. Fridman, and S. Mobashery, *J. Am. Chem. Soc.*, **122**, 6799 (2000).

112. O. Kleifeld, L. P. Kotra, D. C. Gervasi, S. Brown, M. M. Bernardo, R. Fridman, S. Mobashery, and I. Sagi, *J. Biol. Chem.*, **276**, 17125 (2001).

113. M. L. Moss, J. M. White, M. H. Lambert, and R. C. Andrews, *Drug Discov. Today*, **6**, 417 (2001).

114. R. A. Black, C. T. Rauch, C. J. Kozlosky, J. J. Peschon, J. L. Slack, M. F. Wolfson, B. J. Castner, K. L. Stocking, P. Reddy, S. Srinivasan, N. Nelson, N. Boiani, K. A. Schooley, M. Gerhart, R. Davis, J. N. Fitzner, R. S. Johnson, R. J. Paxton, C. J. March, and D. P. Cerretti, *Nature*, **385**, 729 (1997).

115. F. Anon, *Expert Opin. Ther. Pat.*, **10**, 1617 (2000).

116. H. Hilpert, *Tetrahedron*, **57**, 7675 (2001).

117. H. Umezawa, T. Aoyagi, H. Suda, M. Hamada, and T. Takeuchi, *J. Antibiot. (Tokyo)*, **29**, 97 (1976).

118. T. Aoyagi, H. Tobe, F. Kojima, M. Hamada, T. Takeuchi, and H. Umezawa, *J. Antibiot. (Tokyo)*, **31**, 636 (1978).

119. H. Kim and W. N. Lipscomb, *Proc. Natl. Acad. Sci. USA*, **90**, 5006 (1993).

120. W. T. Lowther, A. M. Orville, D. T. Madden, S. J. Lim, D. H. Rich, and B. W. Matthews, *Biochemistry*, **38**, 7678 (1999).

121. For in-depth analysis, see D. F. Veber and S. K. Thompson, *Curr. Opin. Drug Discov. Dev.*, **3**, 362 (2000); A. Krantz, *Bioorg. Med. Chem. Lett.*, **2**, 1327 (1992).

122. J. Drenth, J. N. Jansonius, and B. G. Wolthers, *J. Mol. Biol.*, **24**, 449 (1967).

123. J. Drenth, J. N. Jansonius, R. Koekoek, J. Marrink, J. Munnik, and B. G. Wolthers, *J. Mol. Biol.*, **5**, 398 (1962).

124. S. Michaud and B. J. Gour, *Expert Opin. Ther. Pat.*, **8**, 645 (1998).

125. D. S. Yamashita and R. A. Dodds, *Curr. Pharm. Des.*, **6**, 1 (2000).

126. K. Tezuka, Y. Tezuka, A. Maejima, T. Sato, K. Nemoto, H. Kamioka, Y. Hakeda, and M. Kumegawa, *J. Biol. Chem.*, **269**, 1106 (1994).

127. D. S. Yamashita, W. W. Smith, B. Zhao, C. A. Janson, T. A. Tomaszek, M. J. Bossard, M. A. Levy, H.-J. Oh, T. J. Carr, S. K. Thompson, C. F. Ijames, S. A. Carr, M. McQueney, K. J. D'Alessio, B. Y. Amegadzie, C. R. Hanning, S. Abdel-Meguid, R. L. DesJarlais, J. G. Gleason, and D. F. Veber, *J. Am. Chem. Soc.*, **119**, 11351 (1997).

128. R. W. Marquis, Y. Ru, J. Zeng, R. E. L. Trout, S. M. LoCastro, A. D. Gribble, J. Witherington, A. E. Fenwick, B. Garnier, T. Tomaszek, D. Tew, M. E. Hemling, C. J. Quinn, W. W. Smith, B. Zhao, M. S. McQueney, C. A. Janson, K. D'Alessio, and D. F. Veber, *J. Med. Chem.*, **44**, 725 (2001).

129. R. W. Marquis, Y. Ru, S. M. LoCastro, J. Zeng, D. S. Yamashita, H.-J. Oh, K. F. Erhard, L. D. Davis, T. A. Tomaszek, D. Tew, K. Salyers, J. Proksch, K. Ward, B. Smith, M. Levy, M. D. Cummings, R. C. Haltiwanger, G. Trescher, B. Wang, M. E. Hemling, C. J. Quinn, H. Y. Cheng, F. Lin, W. W. Smith, C. A. Janson, B. Zhao, M. S. McQueney, K. D'Alessio, C.-P. Lee, A. Marzulli, R. A. Dodds, S. Blake, S.-M. Hwang, I. E. James, C. J. Gress, B. R. Bradley, M. W. Lark, M. Gowen, and D. F. Veber, *J. Med. Chem.*, **44**, 1380 (2001).

130. R. V. Talanian, K. D. Brady, and V. L. Cryns, *J. Med. Chem.*, **43**, 3351 (2000).

131. D. S. Karanewsky, X. Bai, S. D. Linton, J. F. Krebs, J. Wu, B. Pham, and K. J. Tomaselli, *Bioorg. Med. Chem. Lett.*, **8**, 2757 (1998).

132. A. B. Shahripour, M. S. Plummer, E. A. Lunney, T. K. Sawyer, C. J. Stankovic, M. K. Connolly, J. R. Rubin, N. P. C. Walker, K. D. Brady, H. J. Allen, R. V. Talanian, W. W. Wong, and C. Humblet, *Bioorg. Med. Chem. Lett.*, **11**, 2779 (2001).

133. D. Lee, S. A. Long, J. L. Adams, G. Chan, K. S. Vaidya, T. A. Francis, K. Kikly, J. D. Winkler, C.-M. Sung, C. Debouck, S. Richardson, M. A. Levy, W. E. DeWolf Jr., P. M. Keller, T. Tomaszek, M. S. Head, M. D. Ryan, R. C. Haltiwanger, P.-H. Liang, C. A. Janson, P. J. McDevitt, K. Johanson, N. O. Concha, W. Chan, S. S. Abdel-Meguid, A. M. Badger, M. W. Lark, D. P. Nadeau, L. J. Suva, M. Gowen, and M. E. Nuttall, *J. Biol. Chem.*, **275**, 16007 (2000).

134. D. Lee, S. A. Long, J. H. Murray, J. L. Adams, M. E. Nuttall, D. P. Nadeau, K. Kikly, J. D. Winkler, C.-M. Sung, M. D. Ryan, M. A. Levy, P. M. Keller, and W. E. DeWolf Jr., *J. Med. Chem.*, **44**, 2015 (2001).

135. R. E. Dolle, J. Singh, J. Rinker, D. Hoyer, C. V. C. Prasad, T. L. Graybill, J. M. Salvino, C. T. Helaszek, R. E. Miller, and M. A. Ator, *J. Med. Chem.*, **37**, 3863 (1994).

136. B. K. Kay, A. V. Kurakin, and R. Hyde-DeRuyscher, *Drug Discov. Today*, **3**, 370 (1998).

137. F. Al-Obeidi, V. J. Hruby, and T. K. Sawyer, *Mol. Biotechnol.*, **9**, 205 (1998).

138. A. E. P. Adang and P. H. H. Hermkens, *Curr. Med. Chem.*, **8**, 985 (2001).

139. K. S. Lam, M. Lebl, and V. Krchnak, *Chem. Rev.*, **97**, 411 (1997).

140. A. C. Good, S. R. Krystek, and J. S. Mason, *Drug Discov. Today*, **5**, 61 (2000).

141. S. P. Rohrer, E. T. Birzin, R. T. Mosley, S. C. Berk, S. M. Hutchins, D.-M. Shen, Y. Xiong, E. C. Hayes, R. M. Parmar, F. Foor, S. W. Mitra, S. J. Degrado, M. Shu, J. M. Klopp, S.-J. Cai, A. Blake, W. W. S. Chan, A. Pasternak, L. Yang, A. A. Patchett, R. G. Smith, K. T. Chapman, and J. M. Schaeffer, *Science*, **282**, 737 (1998).

142. R. S. Bohacek, C. McMartin, and W. C. Guida, *Med. Res. Rev.*, **16**, 3 (1996).

143. Y. Kurogi and O. F. Guner, *Curr. Med. Chem.*, **8**, 1035 (2001).

144. J. S. Mason, A. C. Good, and E. J. Martin, *Curr. Pharm. Des.*, **7**, 567 (2001).

145. F. Ooms, *Curr. Med. Chem.*, **7**, 141 (2000).

146. J. Wouters and F. Ooms, *Curr. Pharm. Des.*, **7**, 529 (2001).

147. E. Vieira, A. Binggeli, V. Breu, D. Bur, W. Fischli, R. Guller, G. Hirth, H. P. Marki, M. Muller, C. Oefner, M. Scalone, H. Stradler, M. Wilhelm, and W. Wostl, *Bioorg. Med. Chem. Lett.*, **9**, 1397 (1999).

148. G. Guller, A. Binggeli, V. Breu, D. Bur, W. Fischli, G. Hirth, C. Jenny, M. Kansay, F. Montavon, M. Muller, C. Oefner, H. Stradler, E. Vieira, M. Wilhelm, W. Wostl, and H. P. Marki, *Bioorg. Med. Chem. Lett.*, **9**, 1403 (1999).

149. C. Oefner, A. Binggeli, V. Breu, D. Bur, J.-P. Clozel, A. D'Arcy, A. Dorn, W. Fischli, F. Gruninger, R. Guller, G. Hirth, H. P. Marki, S. Mathews, M. Muller, R. G. Ridler, H. Stadler, E. Vieira, M. Wilhelm, F. K. Winklier, and W. Wostl, *Chem. Biol.*, **6**, 127 (1999).

150. M. G. Bursavich, C. W. West, and D. H. Rich, *Org. Lett.*, **3**, 2317 (2001).

151. A. B. Smith III, R. Hirschmann, A. Pasternak, W. Yao, P. A. Sprengeler, M. K. Holloway, L. C. Kuo, Z. Chen, P. L. Darke, and W. A. Schleif, *J. Med. Chem.*, **40**, 2440 (1997).

152. J. D. A. Tyndall, R. C. Reid, D. P. Tyssen, D. K. Jardine, B. Todd, M. Passmore, D. R. March, L. K. Pattenden, D. A. Bergman, D. Alewood, S.-H. Hu, P. F. Alewood, C. J. Birch, J. L. Martin, and D. P. Fairlie, *J. Med. Chem.*, **43**, 3495 (2000).

153. S. E. Hagen, J. V. N. V. Prasad, F. E. Boyer, J. M. Domagala, E. L. Ellsworth, C. Gajda, H. W. Hamilton, L. J. Markoski, B. A. Steinbaugh, B. D. Tait, E. A. Lunney, P. J. Tummino, D. Ferguson, D. Hupe, C. Nouhan, S. J. Gracheck, J. M. Saunders, and S. VanderRoest, *J. Med. Chem.*, **40**, 3707 (1997).

154. T. M. Judge, G. Phillips, J. K. Morris, K. D. Lovasz, K. R. Romines, G. P. Luke, J. Tulinsky, J. M. Tustin, R. A. Chrusciel, L. A. Dolak, S. A. Mizsak, W. Watt, J. Morris, S. L. V. Velde, J. W. Strohbach, and R. B. Gammill, *J. Am. Chem. Soc.*, **119**, 3627 (1997).

155. G. V. De Lucca, S. Erickson-Viitanen, and P. Y. S. Lam, *Drug Discov. Today*, **2**, 6 (1997).

156. W. Schaal, A. Karlsson, G. Ahlsen, J. Lindberg, H. O. Andersson, U. H. Danielson, B. Classon, T. Unge, B. Samuelsson, J. Hulten, A. Hallberg, and A. Karlen, *J. Med. Chem.*, **44**, 155 (2001).

157. J. P. Vacca, *Curr. Opin. Chem. Biol.*, **4**, 394 (2000).

158. P. E. J. Sanderson, *Med. Res. Rev.*, **19**, 179 (1999).

159. P. E. J. Sanderson and A. M. Naylor-Olsen, *Curr. Med. Chem.*, **5**, 289 (1998).

160. M. R. Wiley, N. Y. Chirgadze, D. K. Clawson, T. J. Craft, D. S. Gifford-Moore, N. D. Jones, J. L. Olkowaki, A. L. Schacht, L. C. Weir, and G. F. Smith, *Bioorg. Med. Chem. Lett.*, **5**, 2835 (1995).

161. S. F. Brady, K. J. Stauffer, W. C. Lumma, G. M. Smith, H. G. Ramjit, S. D. Lewis, B. J. Lucas, S. J. Gardell, E. A. Lyle, S. D. Appleby, J. J. Cook, M. A. Holahan, M. T. Stranieri, J. J. Lynch Jr., J. H. Lin, I. W. Chen, K. Vastag, A. M. Naylor-Olsen, and J. P. Vacca, *J. Med. Chem.*, **41**, 401 (1998).

162. T. J. Tucker, W. C. Lumma, A. M. Naylor-Olsen, S. D. Lewis, R. Lucas, R. M. Freidinger, A. M. Mulichak, Z. Chen, and L. C. Kuo, *J. Med. Chem.*, **40**, 830 (1997).

163. R. A. Engh, H. Brandstetter, G. Sucher, A. Eichinger, U. Baumann, W. Bode, R. Huber, T. Poll, R. Rudolph, and W. Von der Saal, *Structure*, **4**, 1353 (1996).

164. T. B. Lu, B. Tomczuk, R. Bone, L. Murphy, F. R. Salemme, and R. M. Soll, *Bioorg. Med. Chem. Lett.*, **10**, 83 (2000).

165. T. B. Lu, R. M. Soll, C. R. Illig, R. Bone, L. Murphy, J. Spurlino, F. R. Salemme, and B. E. Tomczuk, *Bioorg. Med. Chem. Lett.*, **10**, 79 (2000).

166. N. Y. Chirgadze, D. J. Sall, V. J. Klimkowski, D. K. Clawson, S. L. Briggs, R. Hermann, G. F. Smith, D. S. Gifford-Moore, and J.-P. Wery, *Protein Sci.*, **6**, 1412 (1997).

167. D. J. Sall, J. A. Bastian, S. L. Briggs, J. A. Buben, N. Y. Chirgadze, D. K. Clawson, M. L. Denney, D. D. Giera, D. S. Gifford-Moore, R. W. Harper, K. L. Hauser, V. J. Klimkowski, T. J. Kohn, H.-S. Lin, J. R. McCowan, A. D. Palkowitz, G. F. Smith, K. Takeuchi, K. J. Thrasher, J. M. Tinsley, B. G. Utterback, S.-C. B. Yan, and M. Zhang, *J. Med. Chem.*, **40**, 3489 (1997).

168. M. F. Malley, L. Tabernero, C. Y. Chang, S. L. Ohringer, D. G. M. Roberts, J. Das, and J. S. Sack, *Protein Sci.*, **5**, 221 (1996).

169. N. Y. Chirgadze, D. J. Sall, S. L. Briggs, D. K. Clawson, M. Zhang, G. F. Smith, and R. W. Schevitz, *Protein Sci.*, **9**, 29 (2000).

170. U. Obst, D. W. Banner, L. Weber, and F. Diederich, *Chem. Biol.*, **4**, 287 (1997).

171. A. von Matt, C. Ehrhardt, P. Burkhard, R. Metternich, M. Walkinshaw, and C. Tapparelli, *Bioorg. Med. Chem.*, **8**, 2291 (2000).

172. U. Baettig, L. Brown, D. Brundish, C. Dell, A. Furzer, S. Garman, D. Janus, P. D. Kane, G.

173. R. Rai, P. A. Sprengeler, K. C. Elrod, and W. B. Young, *Curr. Med. Chem.*, **8**, 101 (2001).

174. J. M. Fevig, D. J. Pinto, Q. Han, M. L. Quan, J. R. Pruitt, I. C. Jacobson, R. A. Galemmo Jr., S. Wang, M. J. Orwat, L. L. Bostrom, R. M. Knabb, P. C. Wong, P. Y. S. Lam, and R. R. Wexler, *Bioorg. Med. Chem. Lett.*, **11**, 641 (2001).

175. S. I. Klein, M. Czekaj, C. J. Gardner, K. R. Guertin, D. L. Cheney, A. P. Spada, S. A. Bolton, K. Brown, D. Colussi, C. L. Heran, S. R. Morgan, R. J. Leadley, C. T. Dunwiddie, M. H. Perrone, and V. Chu, *J. Med. Chem.*, **41**, 437 (1998).

176. Y. Gong, H. W. Pauls, A. P. Spada, M. Czekaj, G. Y. Liang, V. Chu, D. J. Colussi, K. D. Brown, and J. B. Gao, *Bioorg. Med. Chem. Lett.*, **10**, 217 (2000).

177. D. K. Herron, T. Goodson, M. R. Wiley, L. C. Weir, J. A. Kyle, Y. K. Yee, A. L. Tebbe, J. M. Tinsley, D. Mendel, J. J. Masters, J. B. Franciskovich, J. S. Sawyer, D. W. Beight, A. M. Ratz, G. Milot, S. E. Hall, V. J. Klimkowski, J. H. Wikel, B. J. Eastwood, R. D. Towner, D. S. Gifford-Moore, T. J. Craft, and G. F. Smith, *J. Med. Chem.*, **43**, 859 (2000).

178. Y. K. Yee, A. L. Tebbe, J. H. Linebarger, D. W. Beight, T. J. Craft, D. Gifford-Moore, T. Goodson, D. K. Herron, V. J. Klimkowski, J. A. Kyle, J. S. Sawyer, G. F. Smith, J. M. Tinsley, R. D. Towner, L. Weir, and M. R. Wiley, *J. Med. Chem.*, **43**, 873 (2000).

179. M. R. Wiley, L. C. Weir, S. Briggs, N. A. Bryan, J. Buben, C. Campbell, N. Y. Chirgadze, R. C. Conrad, T. J. Craft, J. V. Ficorilli, J. B. Franciskovich, L. L. Froelich, D. S. Gifford-Moore, T. Goodson, D. K. Herron, V. J. Klimkowski, K. D. Kurz, J. A. Kyle, J. J. Masters, A. M. Ratz, G. Milot, R. T. Shuman, T. Smith, G. F. Smith, A. L. Tebbe, J. M. Tinsley, R. D. Towner, A. Wilson, and Y. K. Yee, *J. Med. Chem.*, **43**, 883 (2000).

180. Z. S. Zhao, D. O. Arnaiz, B. Griedel, S. Sakata, J. L. Dallas, M. Whitlow, L. Trinh, J. Post, A. Liang, M. M. Morrissey, and K. J. Shaw, *Bioorg. Med. Chem. Lett.*, **10**, 963 (2000).

181. J. M. Smallheer, R. E. Olson, and R. R. Wexler, *Annu. Rep. Med. Chem.*, **35**, 103 (2000).

182. W. Wang, R. T. Borchardt, and B. Wang, *Curr. Med. Chem.*, **7**, 437 (2000).

183. E. Ruoslahti and M. D. Pierschbacher, *Science*, **238**, 491 (1987).

184. D. M. Haverstick, J. F. Cowan, K. M. Yamada, and S. A. Santoro, *Blood*, **66**, 946 (1985).

185. F. E. Ali, D. B. Bennett, R. R. Calvo, J. D. Elliott, S. M. Hwang, T. W. Ku, M. A. Lago, A. J. Nichols, T. T. Romoff, D. H. Shah, J. A. Vasko, A. S. Wong, T. O. Yellin, C. K. Yuan, and J. M. Samanen, *J. Med. Chem.*, **37**, 769 (1994).

186. P. L. Barker, S. Bullens, S. Bunting, D. J. Burdick, K. S. Chan, T. Deisher, C. Eigenbrot, T. R. Gadek, R. Gantzos, M. T. Lipari, C. D. Muir, M. A. Napier, R. M. Pitti, A. Padua, C. Quan, M. Stanley, M. Struble, J. Y. K. Tom, and J. P. Burnier, *J. Med. Chem.*, **35**, 2040 (1992).

187. R. S. Mcdowell and T. R. Gadek, *J. Am. Chem. Soc.*, **114**, 9245 (1992).

188. W. E. Bondinell, R. M. Keenan, W. H. Miller, F. E. Ali, A. C. Allen, C. W. De Brosse, D. S. Eggleston, K. F. Erhard, R. C. Haltiwangerc, W. F. Huffmana, S.-M. Hwangd, D. R. Jakasa, P. F. Kosterf, T. W. Kua, C. P. Leee, A. J. Nicholsf, S. T. Rossa, J. M. Samanena, R. E. Valocikf, J. A. Vasko-Moserf, J. W. Venslavskya, A. S. Wongd, and C.-K. Yuana, *Bioorg. Med. Chem.*, **2**, 897 (1994).

189. B. K. Blackburn, A. Lee, M. Baier, B. Kohl, A. G. Olivero, R. Matamoros, K. D. Robarge, and R. S. McDowell, *J. Med. Chem.*, **40**, 717 (1997).

190. J. D. Prugh, R. J. Gould, R. J. Lynch, G. X. Zhang, J. J. Cook, M. A. Holahan, M. T. Stranieri, G. R. Sitko, S. L. Gaul, R. A. Bednar, B. Bednar, and G. D. Hartman, *Bioorg. Med. Chem. Lett.*, **7**, 865 (1997).

191. N. J. Liverton, D. J. Armstrong, D. A. Claremon, D. C. Remy, and J. J. Baldwin, *Bioorg. Med. Chem. Lett.*, **8**, 483 (1998).

192. B. C. Askew, C. J. McIntyre, C. A. Hunt, D. A. Claremon, J. J. Baldwin, P. S. Anderson, R. J. Gould, R. J. Lynch, C. C. T. Chang, J. J. Cook, J. J. Lynch, M. A. Holahan, G. R. Sitko, and M. T. Stranieri, *Bioorg. Med. Chem. Lett.*, **7**, 1531 (1997).

193. E. J. Topol, T. V. Byzova, and E. F. Plow, *Lancet*, **353**, 227 (1999).

194. C. B. Xue, J. Roderick, S. Jackson, M. Rafalski, A. Rockwell, S. Mousa, R. E. Olson, and W. F. DeGrado, *Bioorg. Med. Chem.*, **5**, 693 (1997).

195. M. J. Fisher, B. Gunn, C. S. Harms, A. D. Kline, J. T. Mullaney, A. Nunes, R. M. Scarborough, A. E. Arfsten, M. A. Skelton, S. L. Um, B. G. Utterback, and J. A. Jakubowski, *J. Med. Chem.*, **40**, 2085 (1997).

196. G. D. Hartman, M. E. Duggan, W. F. Hoffman, R. J. Meissner, J. J. Perkins, A. E. Zartman, A. M. Naylor-Olsen, J. J. Cook, J. D. Glass, R. J. Lynch, G. Zhang, and R. J. Gould, *Bioorg. Med. Chem. Lett.*, **9**, 863 (1999).

197. B. Wang, W. Wang, G. P. Camenisch, J. Elmo, H. Zhang, and R. T. Borchardt, *Chem. Pharm. Bull. (Tokyo)*, **47**, 90 (1999).

198. M. S. Smyth, J. Rose, M. M. Mehrotra, J. Heath, G. Ruhter, T. Schotten, J. Seroogy, D. Volkots, A. Pandey, and R. M. Scarborough, *Bioorg. Med. Chem. Lett.*, **11**, 1289 (2001).

199. M. E. Duggan, L. T. Duong, J. E. Fisher, T. G. Hamill, W. F. Hoffman, J. R. Huff, N. C. Ihle, C.-T. Leu, R. M. Nagy, J. J. Perkins, S. B. Rodan, G. Wesolowski, D. B. Whitman, A. E. Zartman, G. A. Rodan, and G. D. Hartman, *J. Med. Chem.*, **43**, 3736 (2000).

200. A. Peyman, V. Wehner, J. Knolle, H. U. Stilz, G. Breipohl, K.-H. Scheunemann, D. Carniato, J.-M. Ruxer, J.-F. Gourvest, T. R. Gadek, and S. Bodary, *Bioorg. Med. Chem. Lett.*, **10**, 179 (2000).

201. A. D. Cox, *Drugs*, **61**, 723 (2001).

202. H. Waldmann and M. Thutewohl, *Top. Curr. Chem.*, **211**, 117 (2001).

203. A. Wittinghofer and H. Waldmann, *Angew. Chem. Int. Ed.*, **39**, 4192 (2000).

204. N. E. Kohl, C. A. Omer, M. W. Conner, N. J. Anthony, J. P. Davide, S. J. Desolms, E. A. Giuliani, R. P. Gomez, S. L. Graham, K. Hamilton, L. K. Handt, G. D. Hartman, K. S. Koblan, A. M. Kral, P. J. Miller, S. D. Mosser, T. J. Oneill, E. Rands, M. D. Schaber, J. B. Gibbs, and A. Oliff, *Nat. Med.*, **1**, 792 (1995).

205. M. Nigam, C.-M. Seong, Y. Qian, A. D. Hamilton, and S. M. Sebti, *J. Biol. Chem.*, **268**, 20695 (1993).

206. J. T. Hunt, V. G. Lee, K. Leftheris, B. Seizinger, J. Carboni, J. Mabus, C. Ricca, N. Yan, and V. Manne, *J. Med. Chem.*, **39**, 353 (1996).

207. C. J. Dinsmore, J. M. Bergman, D. D. Wei, C. B. Zartman, J. P. Davide, I. B. Greenberg, D. M. Liu, T. J. O'Neill, J. B. Gibbs, K. S. Koblan, N. E. Kohl, R. B. Lobell, I. W. Chen, D. A. McLoughlin, T. V. Olah, S. L. Graham, G. D. Hartman, and T. M. Williams, *Bioorg. Med. Chem. Lett.*, **11**, 537 (2001).

208. S. B. Long, P. J. Hancock, A. M. Kral, H. W. Hellinga, and L. S. Beese, *Proc. Natl. Acad. Sci. USA*, **98**, 12948 (2001).

209. G. L. James, J. L. Goldstein, M. S. Brown, T. E. Rawson, T. C. Somers, R. S. McDowell, C. W. Crowley, B. K. Lucas, A. D. Levinson, and J. C. Marsters Jr., *Science*, **260**, 1937 (1993).

210. J. T. Hunt, C. Z. Ding, R. Batorsky, M. Bednarz, R. Bhide, Y. Cho, S. Chong, S. Chao, J.

Gullo-Brown, P. Guo, S. H. Kim, F. Y. F. Lee, K. Leftheris, A. Miller, T. Mitt, M. Patel, B. A. Penhallow, C. Ricca, W. C. Rose, R. Schmidt, W. A. Slusarchyk, G. Vite, and V. Manne, *J. Med. Chem.*, **43**, 3587 (2000).

211. F. G. Njoroge, R. J. Doll, B. Vibulbhan, C. S. Alvarez, W. R. Bihop, J. Petrin, P. Kirschmeier, N. I. Carruthers, J. K. Wong, M. M. Albanese, J. J. Piwinski, J. Catino, V. Girijavallabhan, and A. K. Ganguly, *Bioorg. Med. Chem.*, **5**, 101 (1997).

212. C. L. Strickland, P. C. Weber, W. T. Windsor, Z. Wu, H. V. Le, M. M. Albanese, C. S. Alvarez, D. Cesarz, J. del Rosario, J. Deskus, A. K. Mallams, F. G. Njoroge, J. J. Piwinski, S. Remiszewski, R. R. Rossman, A. G. Taveras, B. Vibulbhan, R. J. Doll, V. M. Girijavallabhan, and A. K. Ganguly, *J. Med. Chem.*, **42**, 2125 (1999).

213. M. Gurrath, *Curr. Med. Chem.*, **8**, 1605 (2001).

214. F. Yushiyasu, K. Shoji, and N. Kohei, US Patent 4,355,040, 1982.

215. R. R. Smeby and S. Fermandjian, *Chem. Biochem. Amino Acids, Pept., Proteins*, 1978.

216. J. V. Duncia, A. T. Chiu, D. J. Carini, G. B. Gregory, A. L. Johnson, W. A. Price, G. J. Wells, P. C. Wong, J. C. Calabrese, and P. B. M. W. M. Timmermans, *J. Med. Chem.*, **33**, 1312 (1990).

217. J. Weinstock, R. M. Keenan, J. Samanen, J. Hempel, J. A. Finkelstein, R. G. Franz, D. E. Gaitanopoulos, G. R. Girard, J. G. Gleason, D. T. Hill, T. M. Morgan, C. E. Peishoff, N. Aiyar, D. P. Brooks, T. A. Fredrickson, E. H. Ohlstein, R. R. Ruffolo, E. J. Stack, A. C. Sulpizio, E. F. Weidley, and R. M. Edwards, *J. Med. Chem.*, **34**, 1514 (1991).

218. S. Perlman, H. T. Schambye, R. A. Rivero, W. J. Greenlee, S. A. Hjorth, and T. W. Schwartz, *J. Biol. Chem.*, **270**, 1493 (1995).

219. B. Vianello, E. Clauser, P. Corvol, and C. Monnot, *Eur. J. Pharmacol.*, **347**, 113 (1998).

220. C. Swain and N. M. J. Rupniak, *Annu. Rep. Med. Chem.*, **34**, 51 (1999).

221. J. A. Lowe III, *Med. Res. Rev.*, **16**, 527 (1996).

222. S. McLean, *Med. Res. Rev.*, **16**, 297 (1996).

223. D. C. Horwell, J. A. H. Lainton, J. A. O'Neill, M. C. Pritchard, and J. Raphy, *Spec. Publ. R. Soc. Chem.*, **264**, 95 (2001).

224. V. A. Ashwood, M. J. Field, D. C. Horwell, C. Julien-Larose, R. A. Lewthwaite, S. McCleary, M. C. Pritchard, J. Raphy, and L. Singh, *J. Med. Chem.*, **44**, 2276 (2001).

225. R. M. Snider, J. W. Constantine, J. A. Lowe III, K. P. Longo, W. S. Lebel, H. A. Woody, S. E.

Drozda, M. C. Desai, F. J. Vinick, R. W. Spencer, and H. J. Hess, *Science*, **251**, 435 (1991).

226. J. A. Lowe III, S. E. Drozda, R. M. Snider, K. P. Longo, S. H. Zorn, J. Morrone, E. R. Jackson, S. McLean, D. K. Bryce, J. Bordner, A. Nagahisa, Y. Kanai, O. Suga, and M. Tsuchiya, *J. Med. Chem.*, **35**, 2591 (1992).

227. T. Harrison, A. P. Owens, B. J. Williams, C. J. Swain, A. Williams, E. J. Carlson, W. Rycroft, F. D. Tattersall, M. A. Cascieri, G. G. Chicchi, S. Sadowski, N. M. J. Rupniak, and R. J. Hargreaves, *J. Med. Chem.*, **44**, 4296 (2001).

228. P. C. Ting, J. F. Lee, J. C. Anthes, N. Y. Shih, and J. J. Piwinski, *Bioorg. Med. Chem. Lett.*, **11**, 491 (2001).

229. P. C. Ting, J. F. Lee, J. C. Anthes, N. Y. Shih, and J. J. Piwinski, *Bioorg. Med. Chem. Lett.*, **10**, 2333 (2000).

230. G. A. Reichard, Z. T. Ball, R. Aslanian, J. C. Anthes, N. Y. Shih, and J. J. Piwinski, *Bioorg. Med. Chem. Lett.*, **10**, 2329 (2000).

231. F. E. Blaney, L. F. Raveglia, M. Artico, S. Cavagnera, C. Dartois, C. Farina, M. Grugni, S. Gagliardi, M. A. Luttmann, M. Martinelli, G. M. M. G. Nadler, C. Parini, P. Petrillo, H. M. Sarau, M. A. Scheideler, D. W. P. Hay, and G. A. M. Giardina, *J. Med. Chem.*, **44**, 1675 (2001).

232. A. W. Stamford and E. M. Parker, *Annu. Rep. Med. Chem.*, **34**, 31 (1999).

233. J. Wright, *Drug Discov. Today*, **2**, 19 (1997).

234. K. Rudolf, W. Eberlein, W. Engel, H. A. Wieland, K. D. Willim, M. Entzeroth, W. Wienen, A. G. Beck-Sickinger, and H. N. Doods, *Eur. J. Pharmacol.*, **271**, R11 (1994).

235. R. E. Malmstroem, A. Modin, and J. M. Lundberg, *Eur. J. Pharmacol.*, **305**, 145 (1996).

236. P. A. Hipskind, K. L. Lobb, J. A. Nixon, T. C. Britton, R. F. Bruns, J. Catlow, D. K. Dieckman-McGinty, S. L. Gackenheimer, B. D. Gitter, S. Iyengar, D. A. Schober, R. M. A. Simmons, S. Swanson, H. Zarrinmayeh, D. M. Zimmerman, and D. R. Gehlert, *J. Med. Chem.*, **40**, 3712 (1997).

237. H. Zarrinmayeh, D. M. Zimmerman, B. E. Cantrell, D. A. Schober, R. F. Bruns, S. L. Gackenheimer, P. L. Ornstein, P. A. Hipskind, T. C. Britton, and D. R. Gehlert, *Bioorg. Med. Chem. Lett.*, **9**, 647 (1999).

238. M. H. Norman, N. Chen, Z. D. Chen, C. Fotsch, C. Hale, N. H. Han, R. Hurt, T. Jenkins, J. Kincaid, L. B. Liu, Y. L. Lu, O. Moreno, V. J. Santora, J. D. Sonnenberg, and W. Karbon, *J. Med. Chem.*, **43**, 4288 (2000).

239. O. Dellaz Uana, M. Sadlo, M. Germain, M. Feletou, S. Chamorro, F. Tisserand, C. de Montrion, J. F. Boivin, J. Duhault, J. A. Boutin, and N. Levens, *Int. J. Obes.*, **25**, 84 (2001).

240. H. Rueeger, P. Rigollier, Y. Yamaguchi, T. Schmidlin, W. Schilling, L. Criscione, S. Whitebread, M. Chiesi, M. W. Walker, D. Dhanoa, I. Islam, J. Zhang, and C. Gluchowski, *Bioorg. Med. Chem. Lett.*, **10**, 1175 (2000).

241. R. P. Nargund, A. A. Patchett, M. A. Bach, M. G. Murphy, and R. G. Smith, *J. Med. Chem.*, **41**, 3103 (1998).

242. C. Y. Bowers, F. A. Momany, G. A. Reynolds, and A. Hong, *Endocrinology*, **114**, 1537 (1984).

243. M. H. Chen, M. G. Steiner, A. A. Patchett, K. Cheng, L. T. Wei, W. W. S. Chan, B. Butler, T. M. Jacks, and R. G. Smith, *Bioorg. Med. Chem. Lett.*, **6**, 2163 (1996).

244. A. A. Patchett, R. P. Nargund, J. R. Tata, M. H. Chen, K. J. Barakat, D. B. R. Johnston, K. Cheng, W. W. S. Chan, B. Butler, G. Hickey, T. Jacks, K. Schleim, S. S. Pong, L. Y. P. Chaung, H. Y. Chen, E. Frazier, K. H. Leung, S. H. L. Chiu, and R. G. Smith, *Proc. Natl. Acad. Sci. USA*, **92**, 7001 (1995).

245. B. L. Palucki, S. D. Feighner, S. S. Pong, K. K. McKee, D. L. Hreniuk, C. Tan, A. D. Howard, L. H. Y. Van der Ploeg, A. A. Patchett, and R. P. Nargund, *Bioorg. Med. Chem. Lett.*, **11**, 1955 (2001).

246. T. K. Hansen, M. Ankersen, B. S. Hansen, K. Raun, K. K. Nielsen, J. Lau, B. Peschke, B. F. Lundt, H. Thogersen, N. L. Johansen, K. Madsen, and P. H. Andersen, *J. Med. Chem.*, **41**, 3705 (1998).

247. T. K. Hansen, M. Ankersen, K. Raun, and B. S. Hansen, *Bioorg. Med. Chem. Lett.*, **11**, 1915 (2001).

248. P. Huang, G. H. Loew, H. Funamizu, M. Mimura, N. Ishiyama, M. Hayashida, T. Okuno, O. Shimada, A. Okuyama, S. Ikegami, J. Nakano, and K. Inoguchi, *J. Med. Chem.*, **44**, 4082 (2001).

249. M. L. Webb and T. D. Meek, *Med. Res. Rev.*, **17**, 17 (1997).

250. A. Ray, L. G. Hegde, A. Chugh, and J. B. Gupta, *Drug Discov. Today*, **5**, 455 (2000).

251. E. H. Ohlstein, P. Nambi, S. A. Douglas, R. M. Edwards, M. Gellai, A. Lago, J. D. Leber, R. D. Cousins, A. M. Gao, J. S. Frazee, C. E. Peishoff, J. W. Bean, D. S. Eggleston, N. A. Elshourbagy, C. Kumar, J. A. Lee, T. L. Yue, C. Louden, D. P. Brooks, J. Weinstock, G. Feuerstein, G. Poste, R. R. Ruffolo, J. G. Gleason, and J. D. Elliott, *Proc. Natl. Acad. Sci. USA*, **91**, 8052 (1994).

252. J. D. Elliott, M. A. Lago, R. D. Cousins, A. M. Gao, J. D. Leber, K. F. Erhard, P. Nambi, N. A. Elshourbagy, C. Kumar, J. A. Lee, J. W. Bean, C. W. Debrosse, D. S. Eggleston, D. P. Brooks, G. Feuerstein, R. R. Ruffolo, J. Weinstock, J. G. Gleason, C. E. Peishoff, and E. H. Ohlstein, *J. Med. Chem.*, **37**, 1553 (1994).

253. H.-S. Jae, M. Winn, T. W. von Geldern, B. K. Sorensen, W. J. Chiou, B. Nguyen, K. C. Marsh, and T. J. Opgenorth, *J. Med. Chem.*, **44**, 3978 (2001).

254. S. Anzali, W. W. K. R. Mederski, M. Osswald, and D. Dorsch, *Bioorg. Med. Chem. Lett.*, **8**, 11 (1998).

255. W. W. K. R. Mederski, D. Dorsch, M. Osswald, S. Anzali, M. Christadler, C.-J. Schmitges, P. Schelling, C. Wilm, and M. Fluck, *Bioorg. Med. Chem. Lett.*, **8**, 1771 (1998).

256. W. W. K. R. Mederski, M. Osswald, D. Dorsch, S. Anzali, M. Christadler, C.-J. Schmitges, and C. Wilm, *Bioorg. Med. Chem. Lett.*, **8**, 17 (1998).

257. P. D. Stein, J. T. Hunt, D. M. Floyd, S. Moreland, K. E. J. Dickinson, C. Mitchell, E. C. K. Liu, M. L. Webb, N. Murugesan, J. Dickey, D. Mcmullen, R. G. Zhang, V. G. Lee, R. Serafino, C. Delaney, T. R. Schaeffer, and M. Kozlowski, *J. Med. Chem.*, **37**, 329 (1994).

258. N. Murugesan, Z. X. Gu, P. D. Stein, S. Spergel, A. Mathur, L. Leith, E. C. K. Liu, R. A. Zhang, E. Bird, T. Waldron, A. Marino, R. A. Morrison, M. L. Webb, S. Moreland, and J. C. Barrish, *J. Med. Chem.*, **43**, 3111 (2000).

259. C. D. Wu, E. R. Decker, N. Blok, J. Li, A. R. Bourgoyne, H. Bui, K. M. Keller, V. Knowles, W. Li, F. D. Stavros, G. W. Holland, T. A. Brock, and R. A. F. Dixon, *J. Med. Chem.*, **44**, 1211 (2001).

CHAPTER SIXTEEN

Analog Design

Joseph G. Cannon
The University of Iowa
Iowa City, Iowa

Contents

Burger's Medicinal Chemistry and Drug Discovery
Sixth Edition, Volume 1: Drug Discovery
Edited by Donald J. Abraham
ISBN 0-471-27090-3   © 2003 John Wiley & Sons, Inc.

1 INTRODUCTION

This chapter is limited to nonprotein therapeutic candidates. The subject of peptide analogs and peptidomimetic agents merits separate consideration. Contemporary search for
new drugs makes extensive use of robotic
techniques of combinatorial chemistry and
high throughput synthesis, whereby huge
numbers of compounds can be prepared for
high throughput screening. However, this
nonselective synthetic method based on a random screening philosophy should not replace
the strategy of analog design, but rather it
should be considered as a useful prelude to
analog design.

In any strategy aimed at designing new
drug molecules or analogs of known biologically active compounds, there are no absolute
guidelines or rules for procedure; the knowledge, imagination, and intuition of the medicinal chemist are the most important contributors to success. Analog design is as much an
art as it is a science. The concept of analog
design presupposes that a lead has been discovered; that is, a chemical compound has
been identified that possesses some desirable
pharmacological property. The search for and
identification of leads is a challenge and is a
separate topic. It is sufficient for the present
discussion to note that lead compounds are
frequently identified as endogenous participants (hormones, neurotransmitters, second
messengers, or enzyme cofactors) in the
body's biochemistry and physiology, or a lead
may result from routine, random biological
screening of natural products or of synthetic
molecules that were created for purposes
other than for use as drugs.

Analog design is most fruitful in the study
of pharmacologically active molecules that are
structurally specific: their biological activity
depends on the nature and the details of their
chemical structure (including stereochemistry). Hence, a seemingly minor modification of
the molecule may result in a profound change
in the pharmacological response (increase, diminish, completely destroy, or alter the nature
of the response). In pursuing analog design

and synthesis, it must be recognized that the
newly created analogs are chemical entities
different from the lead compound. It is not
possible to retain all and exactly the same solubility and solvent partition characteristics,
chemical reactivity and stability, acid or base
strength, and/or *in vivo* metabolism properties of the lead compound. Thus, although the
new analog may demonstrate pharmacological
similarities to the lead compound, it is not
likely to be identical to it, either chemically or
biologically, nor will its similarities and differences always be predictable.

The goal of analog design is twofold: (*1*) to
modify the chemical structure of the lead compound to retain or to reinforce the desirable
pharmacologic effect while minimizing unwanted pharmacological (e.g., toxicity, side effects, or undesired routes of and/or unacceptable rates of metabolism) and physical and
chemical properties (e.g., poor solubility and
solvent partition characteristics or chemical
instability), which may result in a superior
therapeutic agent; and (*2*) to use target analogs as pharmacological probes (i.e., tools used
for the study of fundamental pharmacological
and physiological phenomena) to gain better
insight into the pharmacology of the lead molecule and perhaps to reveal new knowledge of
basic biology. Studies of analog structure-activity relationships may increase the medicinal chemist's ability to predict optimum
chemical structural parameters for a given
pharmacological action.

Analog design is greatly facilitated if the
medicinal chemist can initially define the
pharmacophore of the lead compound: that
combination of atoms within the molecule
that is responsible for eliciting the desired
pharmacologic effect. Analog design may be
directed toward maintaining this combination
of atoms intact in a newly designed molecule
or toward a carefully planned, systematic
modification of the pharmacophore. If the medicinal chemist is uncertain about the structural features of the pharmacophoric portion
of the molecule, a prime initial goal of analog
design should be to define the pharmacophore.
The medicinal chemist should address the fol-

lowing questions: What change(s) can be made in the lead molecule that permit(s) retention or reinforcement of pharmacological action? and What change(s) can be made in the molecule that diminish, destroy, or qualitatively change the basic pharmacologic action? The ideal program of analog design should involve a *single* structural change in the lead molecule with each new compound designed and synthesized. An analog in which multiple changes in the structure of the lead molecule have been made simultaneously may occasionally reveal highly desirable pharmacologic effects. However, relatively little useful structure-activity information will be gained from such a molecule. It cannot be readily determined which change (or combination of changes) was responsible for the change in the pharmacological effect. On a practical basis, it is frequently chemically impossible to effect only one discrete change in the lead molecule; one simple molecular structural alteration can influence many structural and chemical parameters. Nonetheless, the medicinal chemist should be cognizant of the disadvantages inherent in "shotgun" (nonsystematic, multiparametric) modification of lead molecules.

In analog design, molecular modification of the lead compound can involve one or more of the following strategies:

1. Bioisosteric replacement.
2. Design of rigid analogs.
3. Homologation of alkyl chain(s) or alteration of chain branching, design of aromatic ring-position isomers, alteration of ring size, and substitution of an aromatic ring for a saturated one, or the converse.
4. Alteration of stereochemistry, or design of geometric isomers or stereoisomers.
5. Design of fragments of the lead molecule that contain the pharmacophoric group (bond disconnection).
6. Alteration of interatomic distances within the pharmacophoric group or in other parts of the molecule.

None of these strategies is inherently preferable to the others; all merit the medicinal chemist's attention and consideration. Application of a combination of these strategies to the lead molecule may be advantageous. Considering the possible permutations and combinations of these changes that are possible within a single lead molecule, it is obvious that the number of analogs that can be designed from a lead molecule is potentially extremely large. Some structural changes that might be proposed are chemically impracticable (e.g., the molecule is incapable of existence) or the proposed analog may represent an overwhelmingly formidable synthetic challenge. These negative factors will diminish the population of possible analogs to be considered for synthesis; nevertheless, the medicinal chemist will always be confronted with a multitude of possible target molecules. Rational decisions must be made concerning which compounds should be synthesized, and synthetic priorities must be established for target compounds. All other factors being equal, the medicinal chemist should synthesize the less-challenging compounds first. Beyond this truism, the medicinal chemist's best resources are intuition and imagination. Selection and application of specific molecular modification strategies depend on the chemical structure of the lead compound and, to a certain extent, on the pharmacological action to be studied.

All of the strategies of analog design as well as subsequent decisions concerning target compounds to be synthesized can be facilitated by the use of computational chemistry (computer-assisted molecular modeling) techniques. These may give the medicinal chemist further insights into structural, stereochemical, and electronic implications of the proposed molecular modification.

2 BIOISOSTERIC REPLACEMENT AND NONISOSTERIC BIOANALOGS (NONCLASSICAL BIOISOSTERES)

The concept of bioisosterism derives from Langmuir's (1) observation that certain physical properties of chemically different substances (e.g., carbon monoxide and nitrogen, ketene and diazomethane) are strikingly similar. These similarities were rationalized on the basis that carbon monoxide and nitrogen both have 14 orbital electrons and, similarly,

Table 16.1 Bioisosteric Atoms and Groups

1. Univalent
   –F –OH –NH$_2$ –CH$_3$ –Cl
   ⠀⠀⠀⠀⠀⠀–SH –PH$_2$
   ⠀⠀⠀⠀⠀⠀–I⠀⠀t-C$_4$H$_9$
   ⠀⠀⠀⠀⠀⠀–Br⠀⠀i-C$_3$H$_7$
2. Bivalent
   –O– –S– –Se– –CH$_2$– –NH–
3. Tervalent
   –N= –CH=
   –P= –As=
4. Quadrivalent
   –C– –Si–
5. Ring equivalents
   –CH=CH– –S–(e.g., benzene, thiophene)
   =CH–⠀⠀⠀⠀=N–(e.g., benzene, pyridine)
   –O– –S– –CH$_2$– –NH–

diazomethane and ketene both have 22 orbital electrons. Medicinal chemists have expanded and adapted the original concept to the analysis of biological activity. The following definition has been proposed: "Bioisosteres are groups or molecules which have chemical and physical properties producing broadly similar biological properties" (2). This definition might be modified to include the concept that bioisosteres may produce *opposite* biological effects, and these effects are frequently a reflection of some action on the same biological process or at the same receptor site. Bioisosteric similarity of molecules is commonly assigned on the basis of the number of valence electrons of an atom or a group of atoms rather than on the total number of orbital electrons, as was originally specified by Langmuir. In a remarkable number of instances, compounds result that have similar (or even diametrically opposite) pharmacological effects compared with those of the parent compound. Categories of classic bioisosteres have been described (2) (Table 16.1).

A more recent comprehensive review of bioisosterism appeared in 1996 (3). In a short communication, Burger (4) discussed and provided valuable insights into isosterism and bioanalogy in drug design.

Many compounds have been identified that comply with the "biology" aspect of the bioisostere concept but that do not fit the strict chemical (steric and electronic) definition of

bioisosteres. Floersheim et al. (5) proposed that such compounds be designated as *nonisosteric bioanalogs*, replacing the older term, "nonclassical bioisosteres." However, most of the contemporary literature retains the nonclassical bioisostere terminology. Table 16.2 lists representative nonclassical bioisosteres.

Dihydromuscimol (**1**) and thiomuscimol (**2**) are cyclic analogs of γ-aminobutyric acid (GABA) (**3**), in which the C=N moiety of the

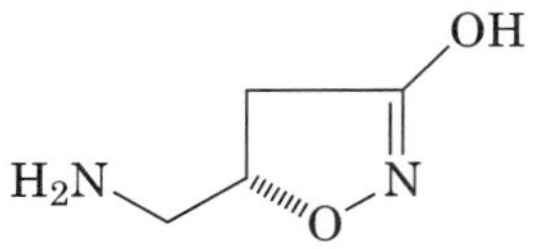

(1)

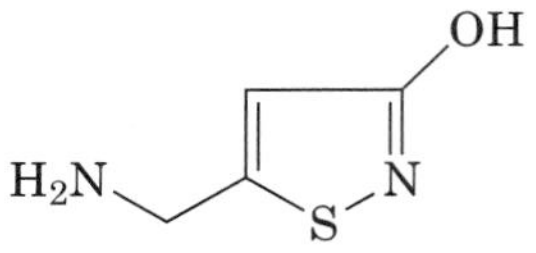

(2)

(3)

heterocyclic ring is considered to be bioisosteric with the C=O of GABA. The –S– moiety of thiomuscimol is bioisosteric with the ring –O– of dihydromuscimol. Both (**1**) and (**2**) are highly potent agonists at GABA$_A$ receptors, as determined in an electrophoresis-based assay (6).

Because of its bioisosteric similarity to the normal physiological substrate L-dopa (**4**), L-mimosine (**5**) inhibits catechol oxidation by the enzyme tyrosinase (7). These compounds exemplify a situation in which bioisosteres display *opposite* pharmacologic effects at the same receptor.

The sulfonium bioisostere (**6**) of *N,N*-dimethyldopamine (**7**) retains the dopaminergic agonist effect displayed by (**7**) (8). The fact that (**6**) bears a permanent unit positive charge was invoked in support of the hypoth-

Table 16.2 Nonclassical Bioisosteres

1. Carbonyl group

2. Carboxylic acid group

$$-\overset{O}{\overset{\|}{C}}-OH \qquad -\overset{O}{\underset{O}{\overset{\uparrow}{\underset{\downarrow}{S}}}}-N-H \atop R \qquad -\overset{O}{\underset{O}{\overset{\uparrow}{\underset{\downarrow}{S}}}}-OH \qquad -\overset{O}{\underset{NH_2}{\overset{\uparrow}{P}}}-OH \qquad -\overset{O}{\underset{OC_2H_5}{\overset{\uparrow}{P}}}-OH$$

3. Hydroxy group

$$-OH \qquad -\overset{}{\underset{H}{N}}-\overset{O}{\overset{\|}{C}}-R \qquad -NHSO_2R \qquad -CH_2OH \qquad -NH-\overset{O}{\overset{\|}{C}}-NH_2$$

$$-NHCN \qquad -CH(CN)2$$

4. Catechol

$$X = O, NR$$

5. Halogen
X CF_3 CN $N(CN)_2$ $C(CN)_3$

6. Thioether

7. Thiourea

$$-NH-\overset{S}{\overset{\|}{C}}-NH_2 \qquad -NH-\overset{\overset{\textstyle CN}{\diagup}}{\overset{N}{\overset{\|}{C}}}-NH_2 \qquad -NH-\overset{H-C-NO_2}{\overset{\|}{C}}-NH_2$$

8. Pyridine

9. Hydrogen
H F

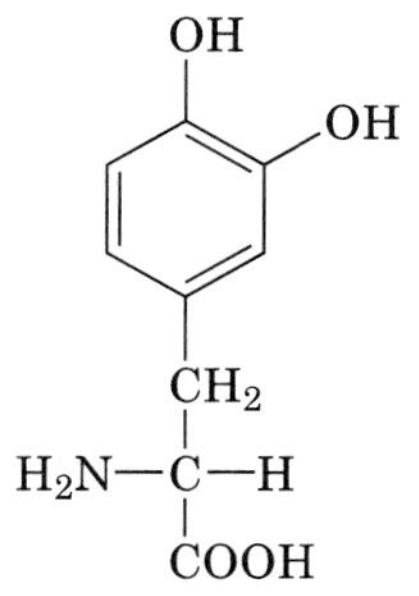

(4)

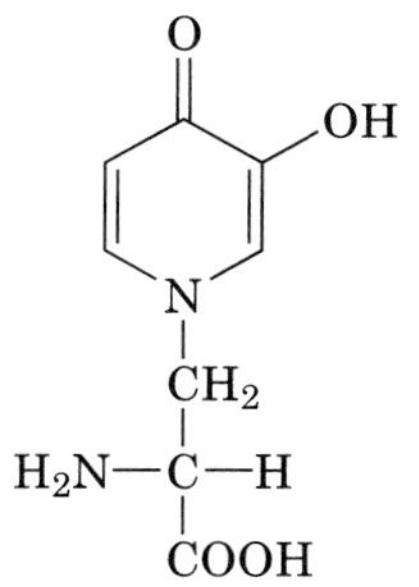

(5)

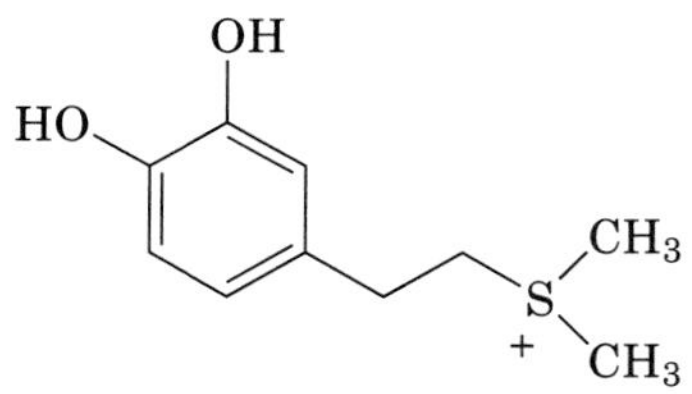

(6)

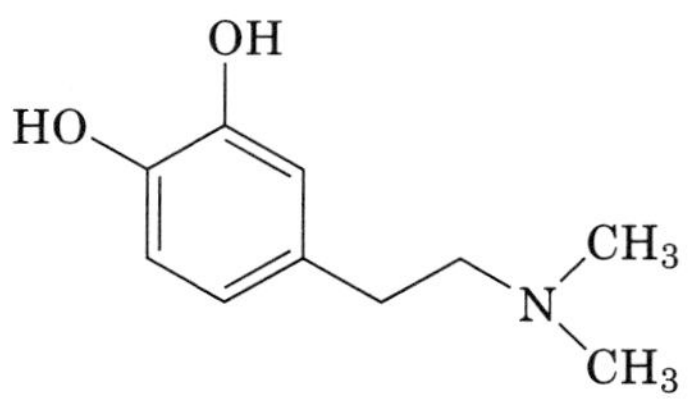

(7)

esis that β-phenethylamines such as (7) interact with dopamine receptors in their protonated (cationic) form.

Bioisosteric replacement strategy has been fruitful in design of psychoactive agents, by use of the antidepressant dibenzazepine derivative imipramine (8) as the lead. The structural similarity between imipramine and the phenothiazine antipsychotics [typified by chlorpromazine (9)] is apparent. Although these two

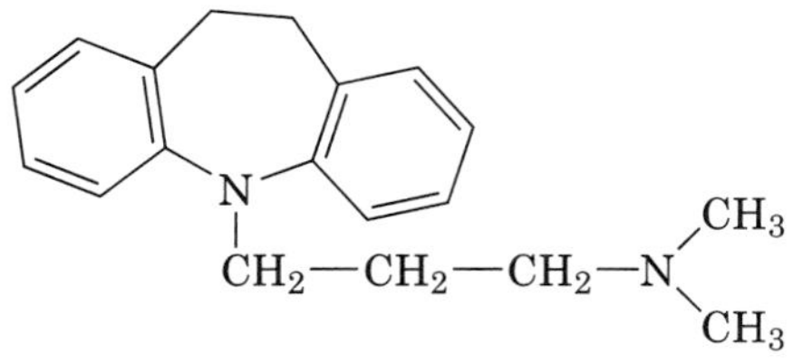

(8)

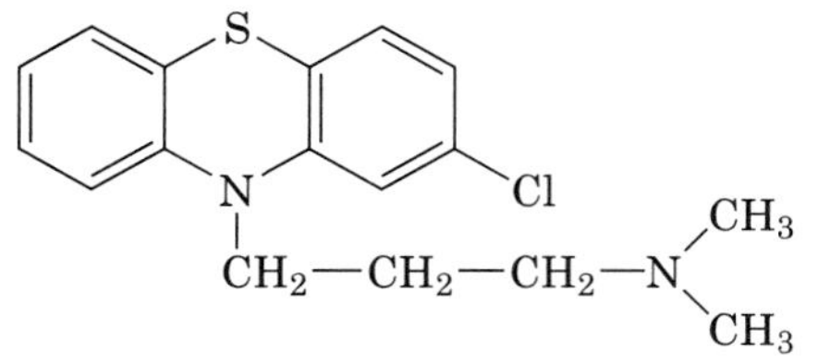

(9)

bioisosteric molecules have different pharmacological properties and therapeutic uses and likely have different mechanisms and sites of action in the central nervous system (9), they share the property of being psychotropic agents. They illustrate the observation that bioisosteric manipulation of a molecule may change its mode of action. In the antidepressant dibenzocycloheptene derivative amitriptyline (10), the ring nitrogen of imipramine is

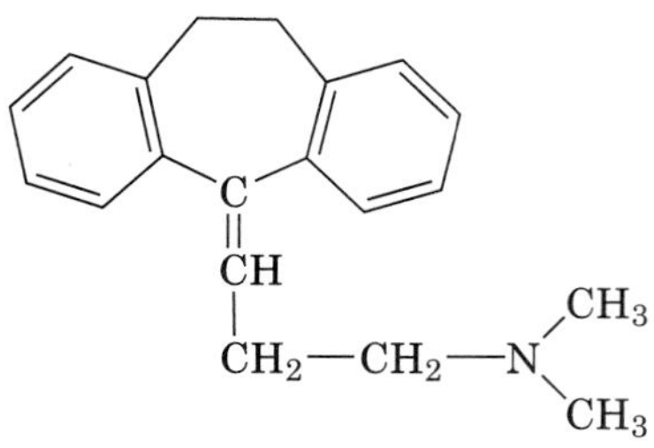

(10)

replaced by an exocyclic olefin moiety. Demexiptiline (11), doxepin (12), and dothiepin (13) represent other bioisosteric modifications of imipramine that possess antidepressant ac-

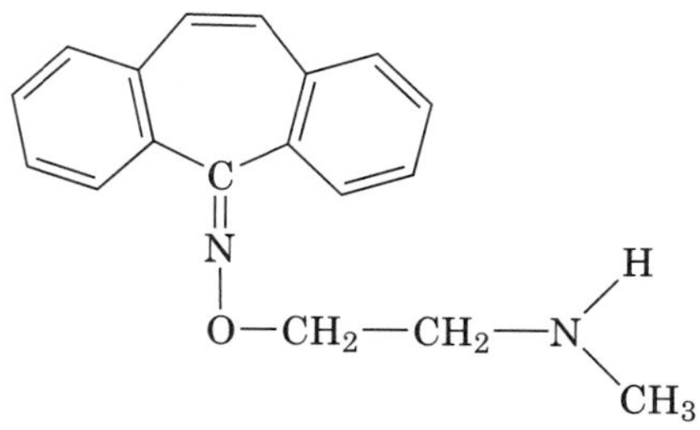

(11)

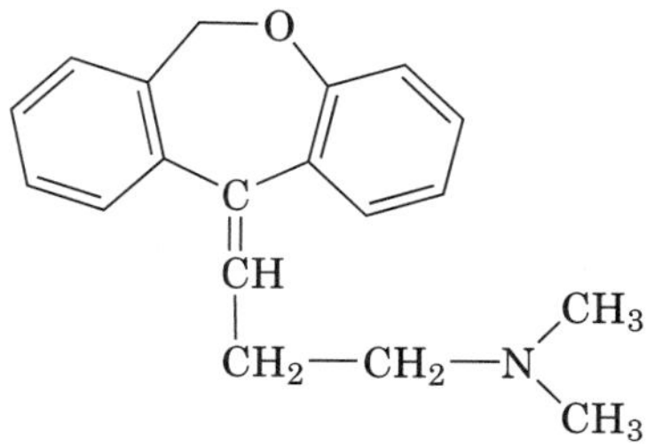

(12)

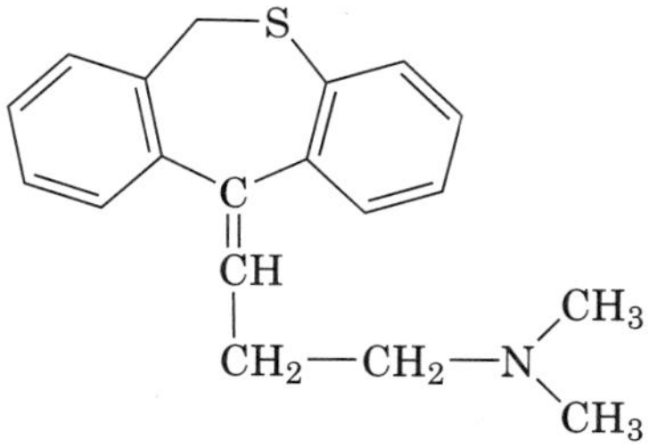

(13)

tivity (10). Variations in the precise nature of psychotropic effects manifested by compounds (**8–13**) may be ascribed to the marked changes in orientation in space of the two benzene ring components of the tricyclic portion of these molecules, imposed on them by the isosteric moieties ($-CH=CH-$, $-CH_2-CH_2-$, $-S-$, $-CH_2O-$, CH_2S-). The Z-isomer of oxepin is a more potent antidepressant than the E-isomer, but the drug is marketed as a mixture of isomers (11). Doxepin is also a potent antagonist at histamine H_1 receptors. The Z-isomer is somewhat more potent than the E-isomer against histamine in the guinea pig ileum (12). The identity of the geometric isomer of dothiepin (**13**) used in pharmacological testing is

uncertain. Apparently, attempts were made (13) to isolate the E- and Z-isomers of all of the compounds prepared in the series studied, but no information was provided about the stereochemistry of the dothiepin material used in the pharmacological studies.

Replacement of the entire indole ring system of melatonin (**14**) by a naphthalene ring

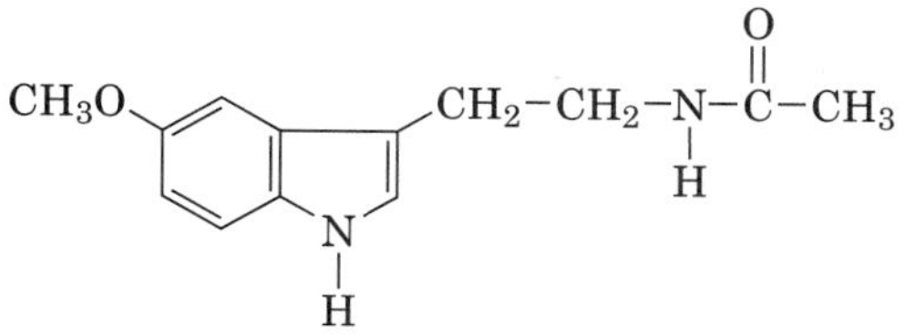

(14)

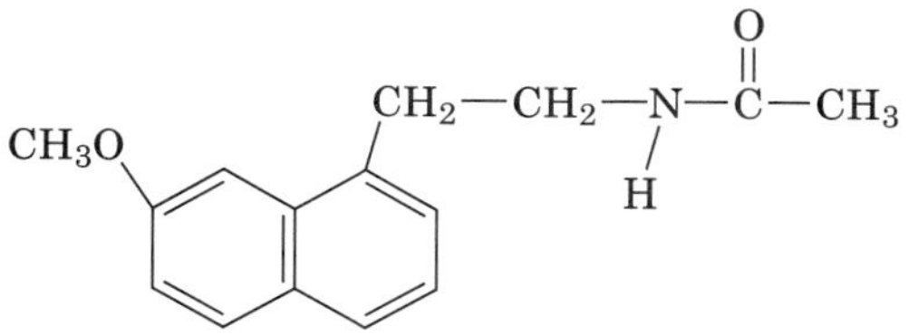

(15)

(**15**) permitted retention of binding affinity in an ovine pars tuberalis membrane assay (14).

From a study (15) of a series of muscarinic M_1 agonists derived from the structure of arecoline (**16**) and typified by (**17**), it was con-

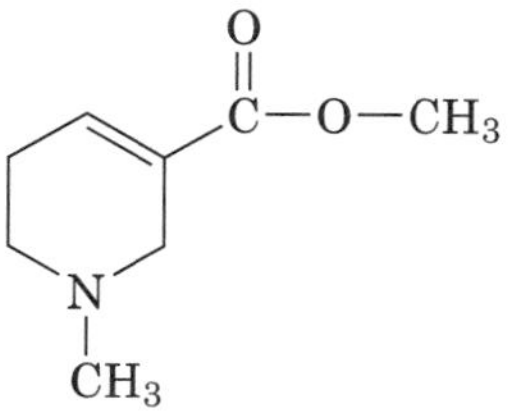

(16)

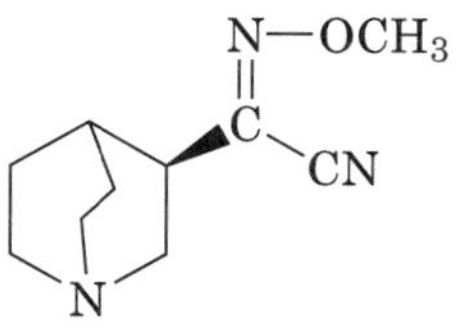

(17)

cluded that the Z-N-methoxyimidoyl nitrile group serves as a stable methyl ester bioisostere. The Z-isomer has an 18-fold higher affinity than its E-isomer for the rat cerebral cortex tissue used in the binding studies.

Replacement of the methyl ester moiety of the muscarinic partial agonist arecoline (16) by the putative nonclassical bioisosteric 1,2,4-oxadiazole ring system (18), where R = un-

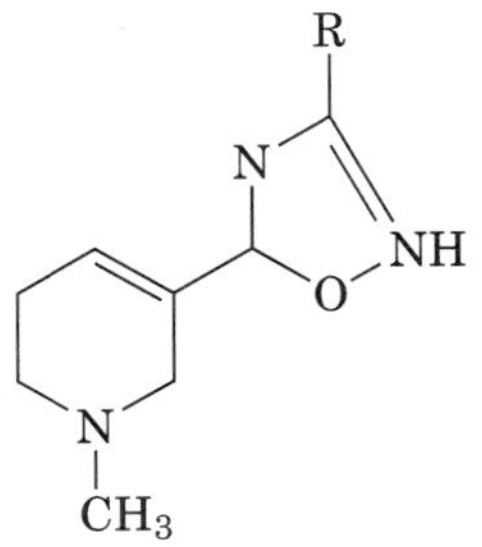

(18)

branched C_1-C_8 alkyl) permits retention of muscarinic agonism (16).

The 1,2,4-oxadiazole ring system of quisqualic acid (19), an agonist at a subpopulation

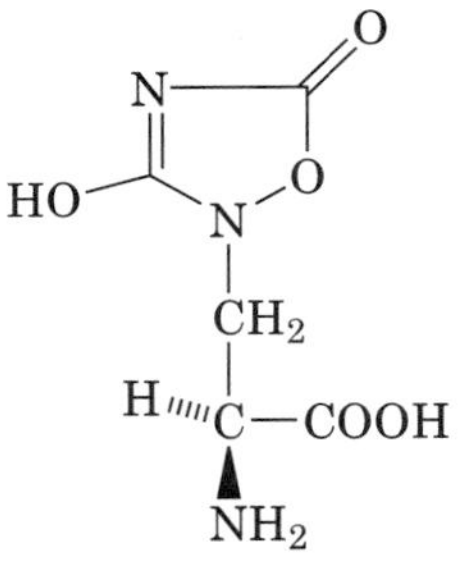

(19)

of glutamate receptors (17), can be considered to be a nonclassical bioisostere of the corresponding carboxyl group of glutamic acid (20).

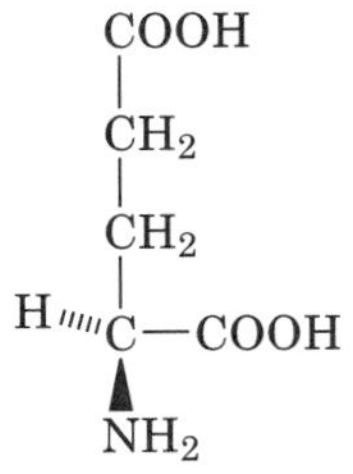

Compounds (21–23) illustrate further examples of nonclassical bioisosteres. Compound (21) was reported to display antitrypanosomal activity (18). The analogs (22) and (23) also displayed antitrypanosomal activity (19). Compound (22) demonstrated the most impressive activity (IC_{50} values of 40 and

(20)

165 nM) with respect to potency and effect on two arsenic-resistant strains of the organism.

Although the strategy of bioisosteric replacement may be a powerful and highly productive tool in analog design, Thornber (2) has emphasized that fundamental chemical and physical chemical changes can be expected to result from these molecular modifications, which may in themselves profoundly affect the pharmacological action of the resulting molecules. Contributing factors include change in the size of the atom or group introduced, which may affect the overall shape and size of the molecule; changes in bond angles; change in partition coefficient; change in the pK_a of the molecule; alteration of chemical reactivity and chemical stability of the molecule, with accompanying qualitative and quantitative alteration of in vivo metabolism of the molecule; and change in hydrogen-bonding capacity. The chemical and biological results and pharmacological significance of many of these factors are unpredictable and must be determined experimentally.

3 RIGID OR SEMIRIGID (CONFORMATIONALLY RESTRICTED) ANALOGS

Imposition of some degree of molecular rigidity on a flexible organic molecule (e.g., by incorporation of elements of the flexible molecule into a rigid ring system or by introduction of a carbon-carbon double or triple bond) may result in potent, biologically active agents that show a higher degree of specificity of pharmacologic effect. There are possible advantages to this technique (20): the key functional groups are held in one steric disposition or, in

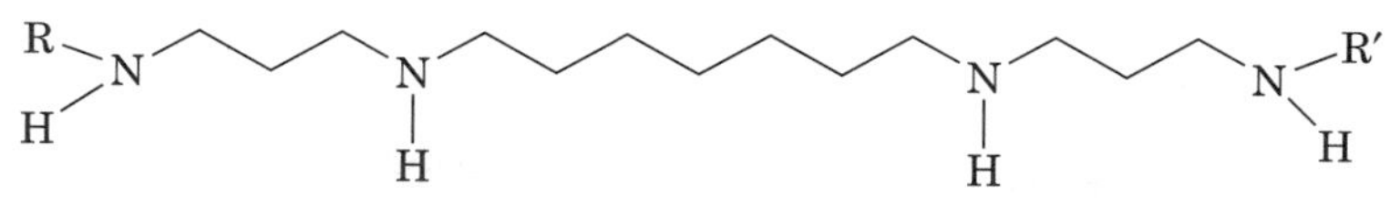

(21) R = R′ = CH$_2$—C$_6$H$_5$

(22) R = 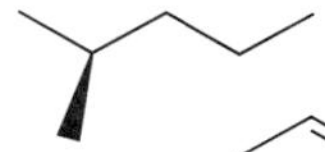   R′ =

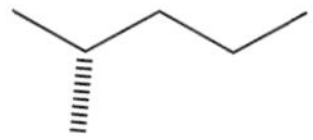

(23) R = R′ = CH$_2$—

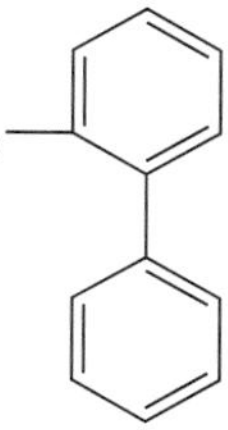

the case of a semirigid structure, the key functional groups are constrained to a limited range of steric dispositions and interatomic distances. By the rigid analog strategy, it is possible to approximate "frozen" conformations of a flexible lead molecule that, if an enhanced pharmacological effect results, may assist in defining and understanding structure-activity parameters, including the three-dimensional geometry of the pharmacophore. These data may be useful in constructing a model of the topography of the receptor. Computational chemistry strategies can be a valuable tool in designing rigid analogs.

The semirigid tetralin congeners (24) and (25) of N,N-dimethyldopamine (7) represent

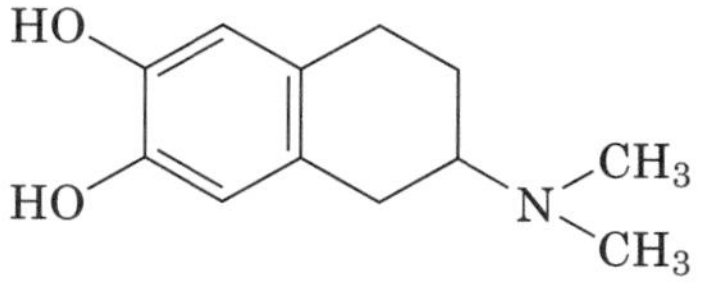

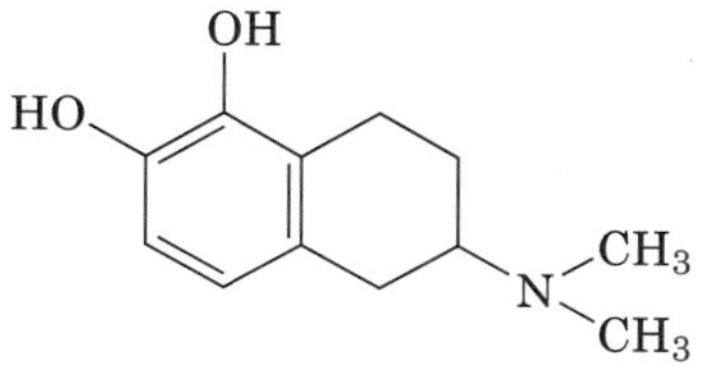

(25)

ferent conformations assumed by the flexible dopamine molecule at its various *in vivo* sites of action.

Restriction of conformational freedom of the acyl moiety in 4-DAMP (26), an antimus-

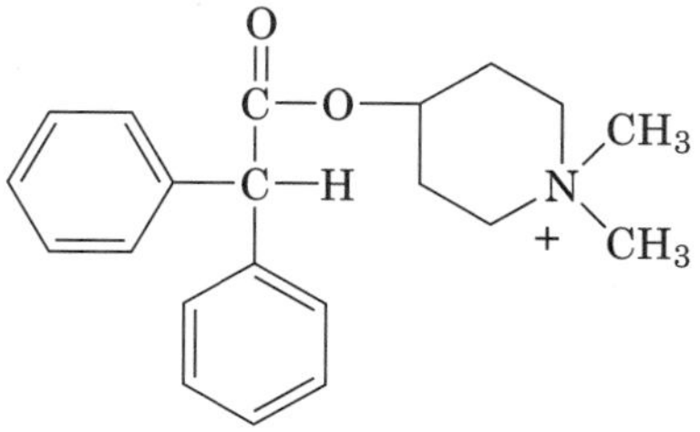

(24)

the two rotameric conformational extremes of the spatial relationship of the aromatic ring of dopamine to the ethylamine side chain when the ring and the side chain are coplanar. Compounds (24) and (25) display effects at different subpopulations of dopamine receptors (21), which have been proposed to reflect dif-

(26)

carinic compound displaying higher affinity at ileal M$_3$ acetylcholine receptors than at atrial M$_2$ receptors) was imposed by the structure of the *spiro*-compound (27) (22).

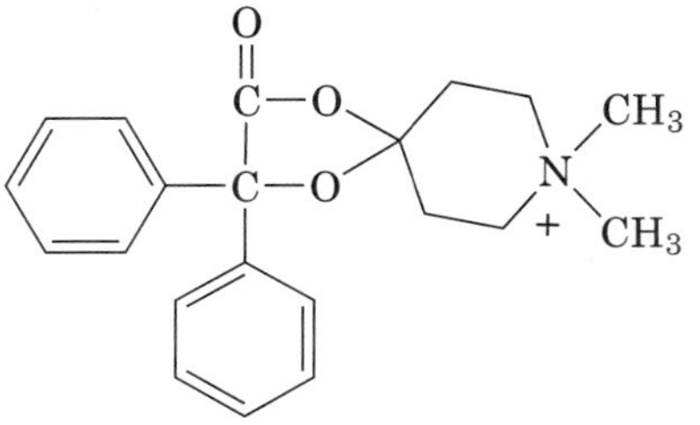

(27)

Spiro-DAMP (**27**) was slightly more potent at M_2 muscarinic receptors than at M_3 receptors. It was proposed that the geometry of the *spiro*-molecule might reflect the receptor-bound conformation of 4-DAMP (**26**); this conformation differs from that observed in the crystal structure of 4-DAMP.

Conformational restriction was introduced into the side chain of a nonclassical serotonin bioisostere (**28**), a selective 5-HT$_{1A}$ and 5-HT$_{1D}$ receptor agonist) by its incorporation into a fused six-membered ring (**29**) (23).

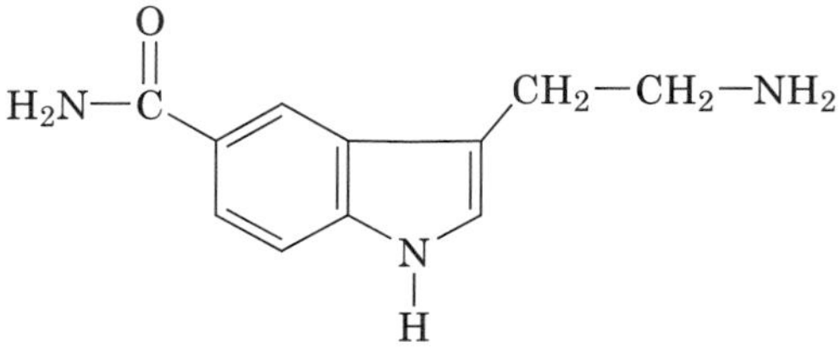

(28)

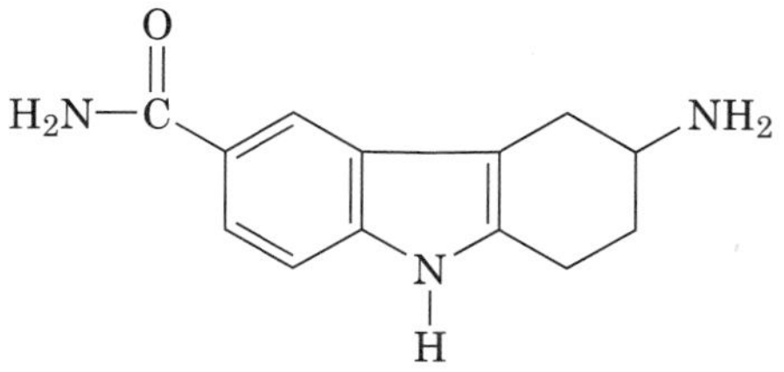

(29)

The conformational restrictions imposed on the indole-3-ethylamine moiety permitted retention of affinity for the 5-HT$_{1D}$ receptor but it diminished affinity for the 5-HT$_{1A}$ receptor by a factor of 1000. In two functional assays, (**29**) exhibited potency equal to or marginally greater than that of serotonin. Com-

pound (**29**) was described as a partial agonist. It was concluded that the conformation of the indole-3-ethylamine portion of the fused system (**29**) reflects the conformation assumed by the flexible system (**28**) when it binds to the 5-HT$_{1D}$ receptor.

Imposition of rigidity into the piperidine ring of the opioid analgesic meperidine (**30**) by introduction of a methylene bridge between carbons 2 and 5 resulted in epimers (**31**) and (**32**), representing "frozen" conformations of meperidine (**24**).

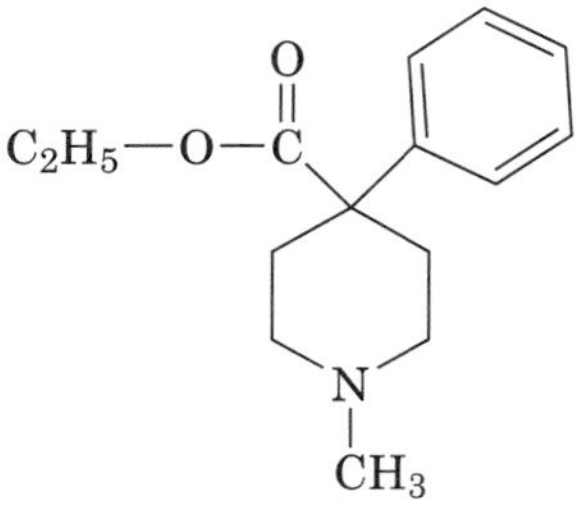

(30)

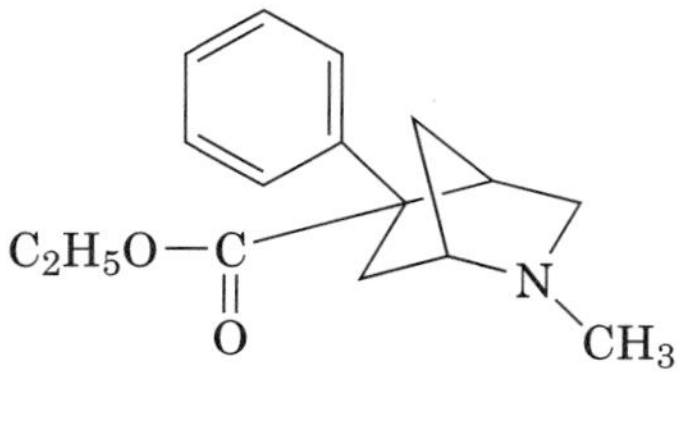

(31)

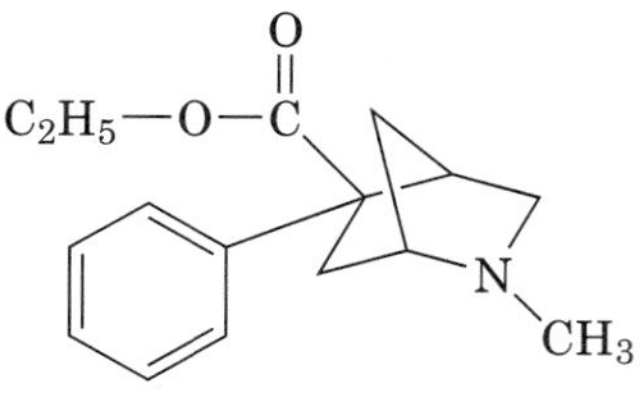

(32)

The *exo*-phenyl isomer (**32**) was six times as potent as the *endo*-phenyl isomer (**31**), and it was twice as potent as meperidine itself in a benzoquinone-induced writhing assay for analgesic effect.

Rigid analogs (**33**), (**34**), and (**35**) of phencyclidine (**36**) possess a rigid carbocyclic struc-

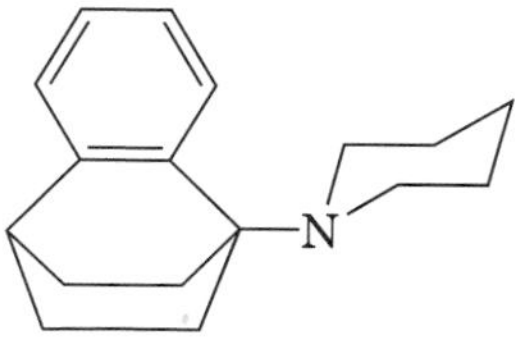

(33)

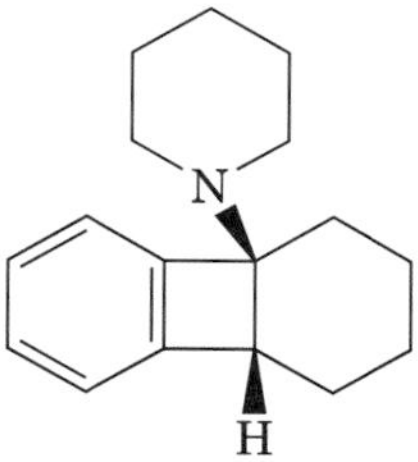

(34)

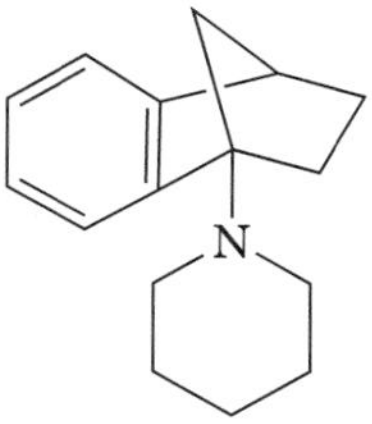

(35)

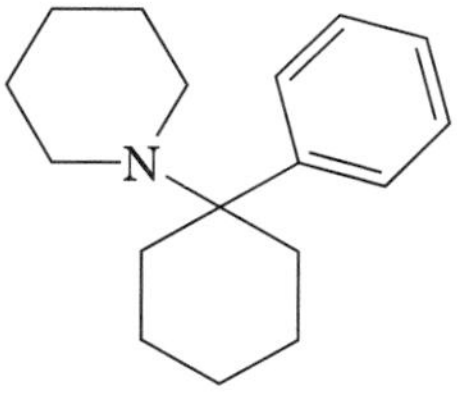

(36)

Incorporation of the choline portion of acetylcholine (37) into a cyclopropane ring system resulted in cis- and trans-1,2-disubstituted molecules, (38) and (39), in which the

(37)

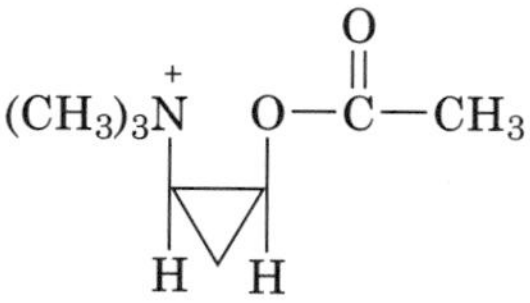

(38)

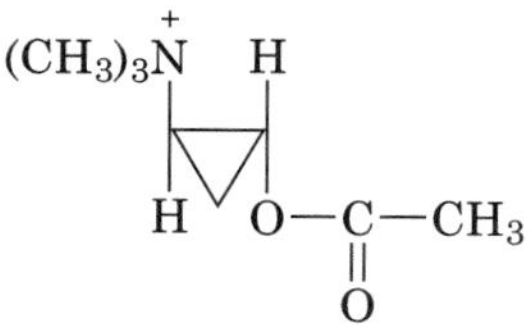

(39)

acetylcholine molecule is locked into folded ("cisoid") and extended ("transoid") conformations.

The (1S),(2S)-(+)-trans-isomer (39) was somewhat more potent than acetylcholine itself in tissue and whole-animal assays for muscarinic agonism (27) and it was an excellent substrate for acetylcholinesterase. The (1R),(2R) enantiomer of (39) was exponentially less potent than its (1S),(2S) enantiomer in the assays cited, but it was a good substrate for acetylcholinesterase. The (±)-cis-isomer (38) was almost inert at nicotinic and muscarinic receptors and it was a poor substrate for acetylcholinesterase. These data were taken as evidence that the flexible acetylcholine molecule interacts with muscarinic receptors in an extended geometry of the chain of atoms (28). When this semirigid analog strategy was applied to a cyclobutane ring system (compound 40), there was a marked loss of pharmacologic effect (29). This result is enigmatic; differences in interatomic distances and bond angles in the pharmacophoric moiety as well

ture and an attached piperidine ring that is free to rotate. All three rigid analogs showed low to no affinity for the PCP receptor, but they had good affinity in a σ-receptor-binding assay (25). These binding data were proposed to be useful in defining a model for the σ-receptor pharmacophore. This study also provided additional evidence that the σ-receptor is independent of the PCP-binding site (cf. Ref. 26 and references therein).

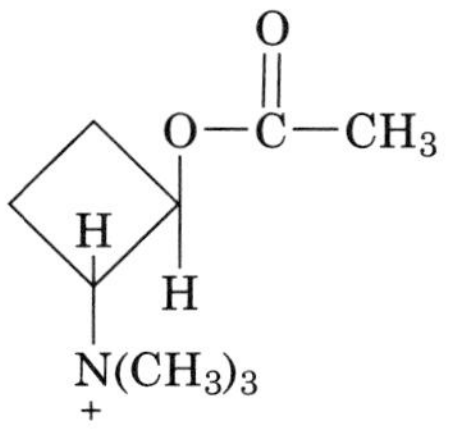

(40)

as differences in extraneous molecular bulk seem insufficient to account for the dramatic difference in pharmacological potencies between the three- and the four-membered ring systems.

The cyclopropane ring was employed to impart a degree of rigidity to the side chain of dopamine (structures **41** and **42**) (30).

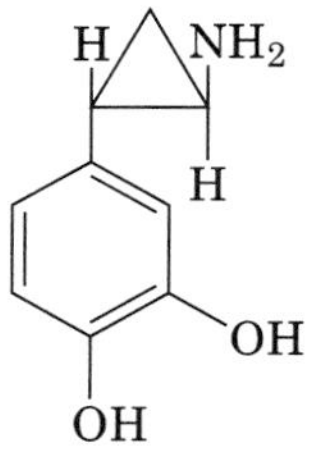

(41)

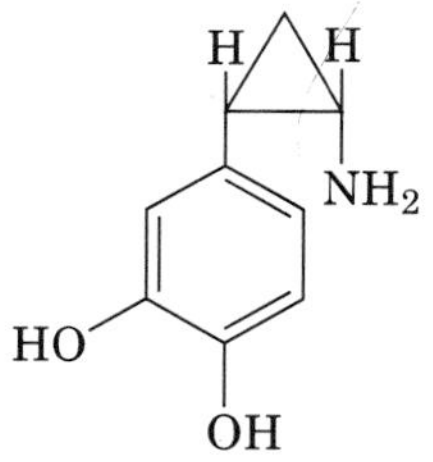

(42)

Neither isomer displayed effects at dopamine receptors, but both were α-adrenoceptor agonists, with the $(\pm)$-*trans*-isomer (**41**) being approximately five times more potent than the $(\pm)$-*cis*-isomer (**42**). It was suggested (31) that these findings may contribute to determining the preferred conformation of β-phenethylamines at the α-adrenoceptor. The racemic *trans*-cyclobutane congeners (**43a**) and (**43b**) are more potent than their racemic *cis*-isomers (**44a**) and (**44b**) in binding studies on rat

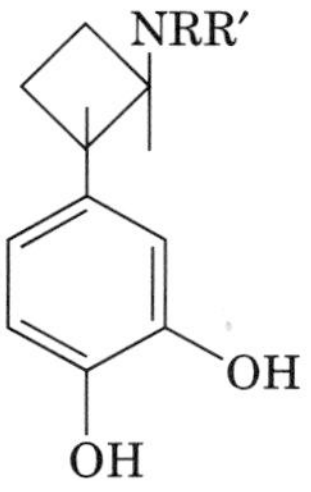

(43a) R = R′ = CH$_3$
(43b) R = R′ = H

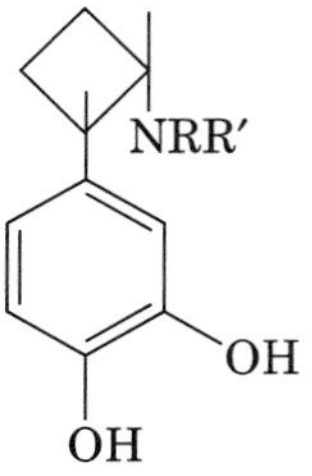

(44a) R = R′ = CH$_3$
(44b) R = R′ = H

corpus striatum tissue, but the binding affinities for (**43a**) and (**43b**) are much less than that of dopamine (32). Racemic *trans*-(**43a**) was more potent than the *trans*-primary amine (**43b**), but it was still much less potent than dopamine. The racemic *cis*-isomer of (**44b**) demonstrated very low affinity for the receptor.

A β-phenethylamine moiety was incorporated into the *trans*-decalin ring system (**45**)

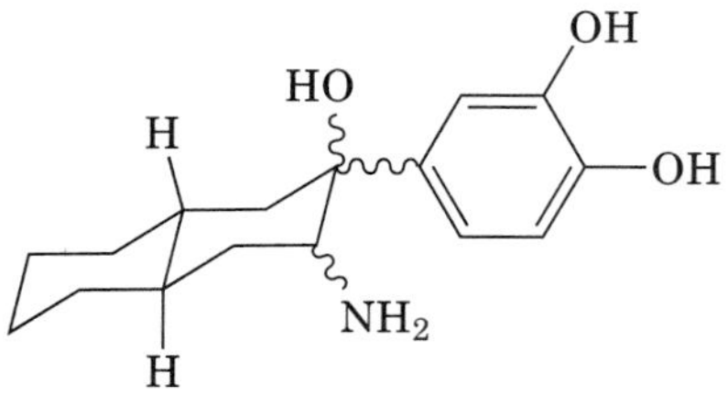

(45)

and the racemic modifications of all four possible isomers were prepared as "frozen" analogs of possibly significant conformations of the flexible norepinephrine molecule (33). All four compounds displayed approximately equal (extremely low) potency. This result illustrates that the achievement of conformational integrity by incorporation of a flexible pharmacophore into a bulky, complex molecule may be at the expense of biological activity.

Rigidity was introduced into the glutamic acid moiety in a series of bioisosteric congeners (**46–48**) (34). These systems showed po-

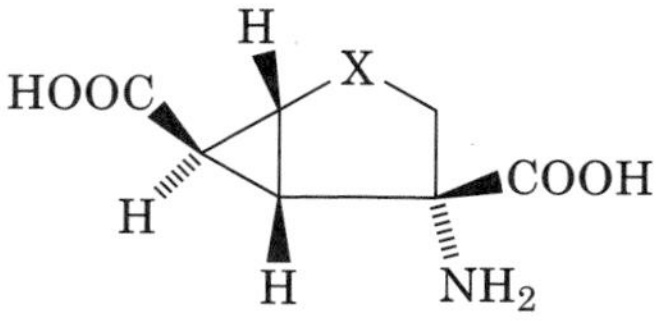

(**46**) X = CH$_2$
(**47**) X = O
(**48**) X = S

tent agonist activity at subpopulations of metabotropic glutamate receptors. The geometry of these congeners led to the conclusion that glutamic acid itself interacts with the metabotropic glutamate receptors in a fully extended conformation.

The rotational orientation of the ester moieties of the myoneural blocking agent succinyldicholine (**49**) was restricted by introduction of a double bond into the succinic acid portion (**50**), (**51**) (35). The E-fumarate ester

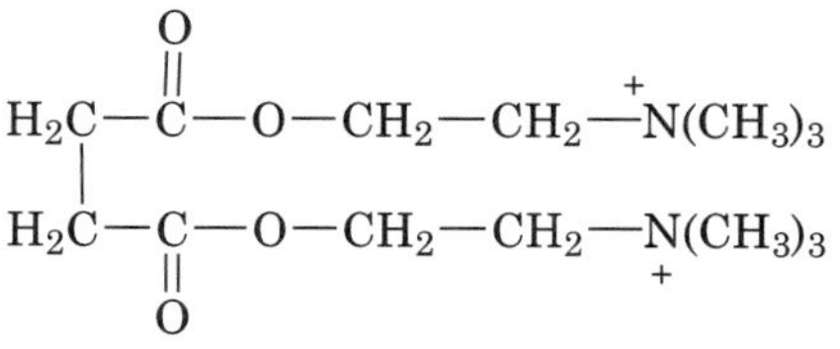

(**49**)

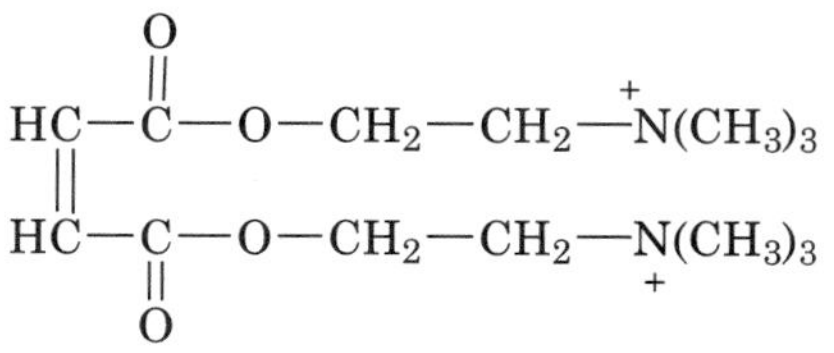

(**50**)

(**51**) was approximately one-half as potent as the flexible succinate ester (**49**), whereas the Z-maleate ester (**50**) was 1/40 as potent as the

succinate. These results led to the conclusion that the molecular shape of the E-ester (**51**) more closely approximates that assumed by succinylcholine when it interacts with myoneural nicotinic receptors.

Restricted rotation was also introduced into the succinic acid moiety of succinyldicholine by preparation of the choline esters of cis- and trans-cyclopropane-1,2-dicarboxylic acids (**52**) and (**53**) (36, 37). Myoneural blocking activity was assessed in dogs (37) and cats (36) and, as indicated above for the E- and Z-olefinic esters (**51**) and (**50**), the extended trans-isomer (**53**) demonstrated much greater potency and a longer duration of action than those of the cis-isomer (**52**). The cyclobutane congeners (**54**) and (**55**) presented unexpected results that are difficult to rationalize: the cis-isomer (**54**) was much less potent than the trans-isomer (**55**) in a cat assay for myoneural blockade, but it presented a decidedly longer duration of action than that of the trans-isomer (36).

4 HOMOLOGATION OF ALKYL CHAIN OR ALTERATION OF CHAIN BRANCHING; CHANGES IN RING SIZE; RING-POSITION ISOMERS; AND SUBSTITUTION OF AN AROMATIC RING FOR A SATURATED ONE, OR THE CONVERSE

Change in size or branching of an alkyl chain on a bioactive molecule may have profound (and sometimes unpredictable) effects on physical and pharmacological properties. Alteration of the size and/or shape of an alkyl substituent can affect the conformational preference of a flexible molecule and may alter the spatial relationships of the components of the pharmacophore, which may be reflected in the ability of the molecule to achieve complementarity with its receptor or with the catalytic surface of a metabolizing enzyme. The alkyl group itself may represent a binding site with the receptor (through hydrophobic interactions), and alteration of the chain may alter its binding capacity. Position isomers of substituents (even alkyl groups) on an aromatic ring may possess different pharmacological properties. In addition to their ability to affect electron distribution over an aromatic ring sys-

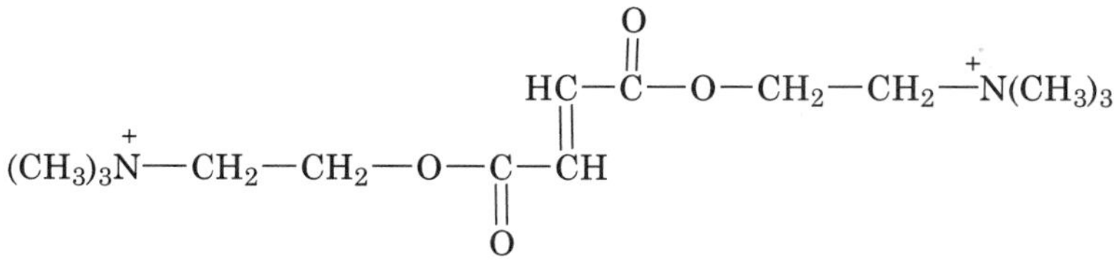

$$HC-\overset{\overset{\displaystyle O}{\|}}{C}-O-CH_2-CH_2-\overset{+}{N}(CH_3)_3$$

$$(CH_3)_3\overset{+}{N}-CH_2-CH_2-O-\underset{\underset{\displaystyle O}{\|}}{C}-CH$$

(51)

tem, position isomers may differ in their comple-mentarity to receptors, and the position of a sub-stituent on a ring may influence the spatial occupancy of the ring system with respect to the remainder of a conformationally variable mole-cule. What has sometimes been trivialized as "methyl group roulette" may indeed be an im-portant parameter in the design of analogs.

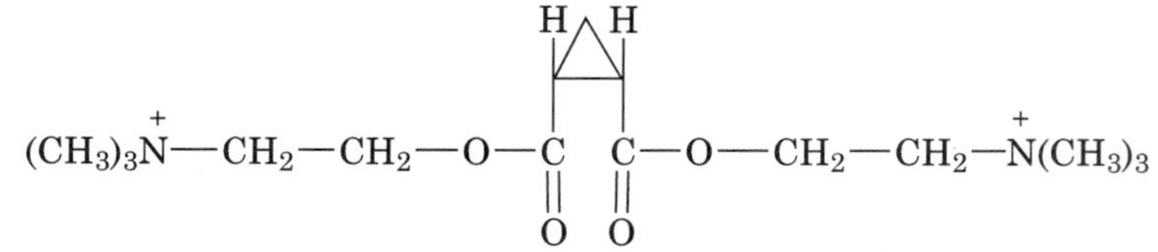

$$(CH_3)_3\overset{+}{N}-CH_2-CH_2-O-\underset{\underset{\displaystyle O}{\|}}{C}\quad\underset{\underset{\displaystyle O}{\|}}{C}-O-CH_2-CH_2-\overset{+}{N}(CH_3)_3$$

(52)

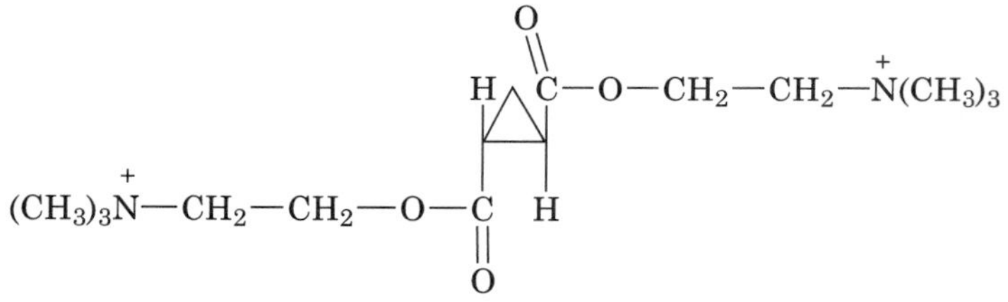

$$(CH_3)_3\overset{+}{N}-CH_2-CH_2-O-\underset{\underset{\displaystyle O}{\|}}{C}\quad H$$

$$H\quad\overset{\overset{\displaystyle O}{\|}}{C}-O-CH_2-CH_2-\overset{+}{N}(CH_3)_3$$

(53)

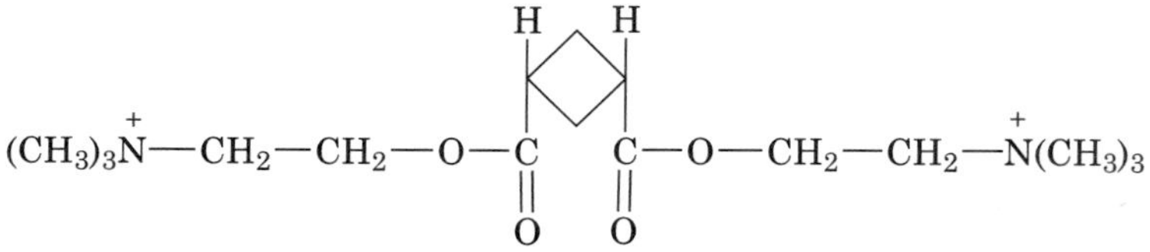

$$(CH_3)_3\overset{+}{N}-CH_2-CH_2-O-\underset{\underset{\displaystyle O}{\|}}{C}\quad\underset{\underset{\displaystyle O}{\|}}{C}-O-CH_2-CH_2-\overset{+}{N}(CH_3)_3$$

(54)

$$(CH_3)_3\overset{+}{N}-CH_2-CH_2-O-\underset{\underset{\displaystyle O}{\|}}{C}\quad H$$

$$H\quad\overset{\overset{\displaystyle O}{\|}}{C}-O-CH_2-CH_2-\overset{+}{N}(CH_3)_3$$

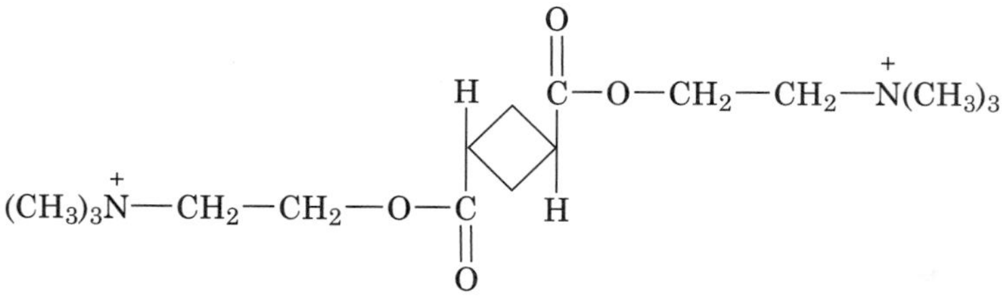

(55)

Homologation of the *N*-alkyl chain in norapomorphine (**56**) from methyl (**57**) to *n*-propyl (**59**) produced incremental increases in emetic response in dogs and in stereotypy responses in rodents (38, 39).

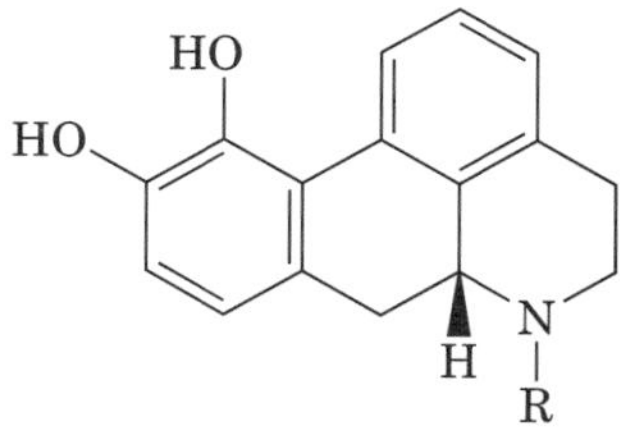

(**56**) R = H
(**57**) R = CH$_3$
(**58**) R = C$_2$H$_5$
(**59**) R = *n*-C$_3$H$_7$
(**60**) R = *n*-C$_4$H$_9$

The next member of the series, the *n*-butyl homolog (**60**), demonstrated a tremendous loss in potency and activity compared to that of the lower homologs (39). Studies of *N,N*-dialkyl dopamines (**61–64**) revealed that some

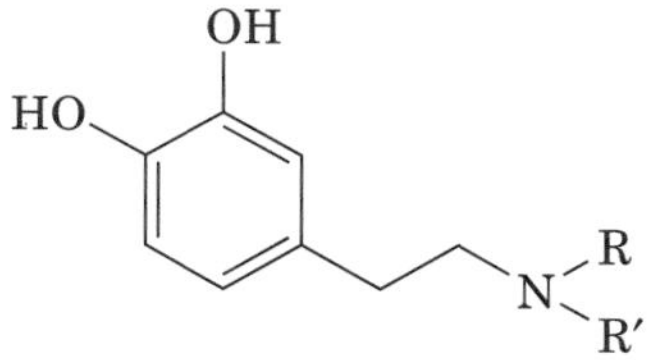

(**61**) R = R′ = CH$_3$
(**62**) R = R′ = *n*-C$_3$H$_7$
(**63**) R = *n*-C$_3$H$_7$; R′ = *n*-C$_4$H$_9$
(**64**) R = R′ = *n*-C$_4$H$_9$

combinations of alkyl groups may impart a high degree of dopamine agonist effects (40).

N,N-dimethyldopamine (**61**) is extremely potent in assays for dopaminergic agonism (pigeon pecking, emesis in dogs, and inhibition of cat cardioaccelerator nerve), as is *N,N*-di-*n*-propyldopamine (**62**) (41). *N-n*-Propyl-*N-n*-butyldopamine (**63**) is potent in behavioral assays in nigra-lesioned rats (42). However, *N,N*-di-*n*-butyldopamine (**64**) is virtually inert in these assays (41, 42). *N,N*-di-*n*-Pentyldopamine was reported (43) to be inert in a caudectomized mouse behavioral assay and in a rotatory behavioral assay in nigra-lesioned

rats. It seems likely that the enhanced dopaminergic agonist effects conferred by *N*-ethyl and *N-n*-propyl groups on aporphine and β-phenethylamine-derived molecules are not related merely to enhanced lipophilic character or to partitioning phenomena, but rather to the likelihood that the two- and three-carbon chains have a positive affinity for subsites on certain dopamine receptors. It may be speculated that these receptor subsites do not accommodate longer alkyl chains (e.g., *n*-butyl or *n*-pentyl). However, different assays for dopaminergic stimulant effects and different animal species were used in refs. 41, 42, and 43, and care must be exercised in drawing firm structure-activity relationship conclusions based on these data.

The alkyl linker between the two heterocyclic ring systems in structure (**65**) was modi-

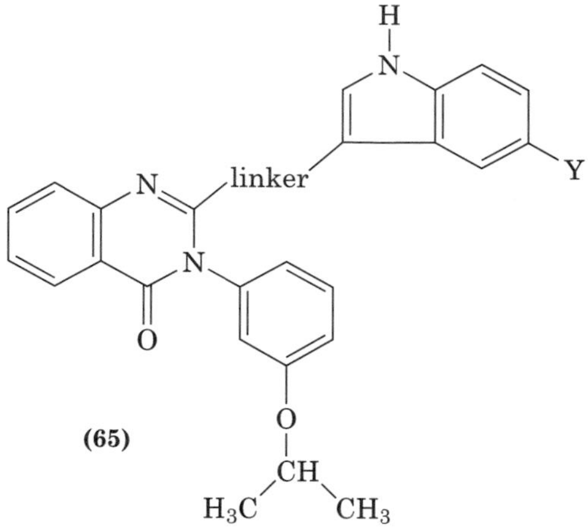

(**65**)

(**66**) linker = $\diagdown\diagup\diagdown$ Y = H

(**67**) linker = $\diagup\!\!\!\!\diagdown\diagup\diagdown$ Y = H

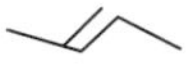

(**68**) linker = $\overset{CH_3}{\diagdown\diagup\diagdown}$ Y = H

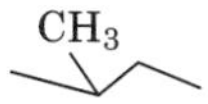

(**69**) linker = $\diagdown\diagup\underset{CH_3}{\diagup\diagdown}$ Y = H

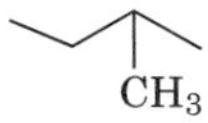

fied in studies of the ability of analogs to bind to the cholecystokinin-B receptor (44). When this linking group was –CH$_2$—CH$_2$–, the compound (structure **66**) was extremely potent in radioligand displacement assays on mouse brain membranes. Introduction of carbon-carbon unsaturation (*E*-olefin) into the linker (structure **67**) resulted in a 16-fold decrease in binding ability; this suggests that conformational restriction and limitation of molecular flexibility have deleterious effects on biological activity. However, no data were reported on the *Z*-isomer of this olefinic molecule, so that caution should be exercised in drawing conclusions. Introduction of a bromine substituent (**65**, Y = Br) into (**66**) produced a threefold increase in potency, whereas the same structural modification of the olefin (**67**) resulted in a threefold decrease in potency.

Branching the linker chain with a methyl group adjacent to the quinazolinone ring (**68**) resulted in a 350-fold decrease in affinity. However, chain branching with a methyl group in the alternate position on the ethylene chain produced compound (**69**), whose receptor affinity was of the same order of magnitude as the extremely potent lead compound (**66**). The exponential difference in receptor-binding ability exhibited by the two isomeric branched-chain linker compounds (**68**) and (**69**) was ascribed to unfavorable steric interactions between the receptor and the linker methyl group of (**68**) (44). This conclusion may be compromised by the fact that both (**68**) and (**69**) were evaluated as their racemates.

A study (45) of 2-(phosphonomethoxy)ethylguanidines (**70–73**) as antiviral (herpes and

ished toxicity 16-fold compared to that of the nonmethylated system (**70**).

In contrast, (*R*)-(**71**), the 2′-methyl congener, exhibited only a fivefold decrease in antiviral potency compared to that of compound (**70**), but it also exhibited a 30-fold lessening of toxicity, to produce a substantial increase in therapeutic index over that of (**70**). The (*S*)-(**71**) enantiomer was somewhat less potent than its (*R*)-enantiomer. The *gem*-dimethyl congener (**73**) was also somewhat less potent than the (*R*)-2′-monomethyl compound (**71**) and it was markedly more toxic. The (*S*)-2′-methyl stereoisomer of (**71**) exhibited a decidedly lower therapeutic index than that of its (*R*)-enantiomer.

Closely related to alteration of chain length and/or chain branching is alteration of ring size. Compound (**74**) showed nano-

(**74**) $n = 1$; $m = 2$
(**75**) $n = m = 1$
(**76**) $n = 1$; $m = 3$
(**77**) $n = 1$; $m = 4$
(**78**) $n = 2$; $m = 2$
(**79**) $n = 2$; $m = 3$

(**70**) R = R′ = H
(**71**) R = H; R′ = CH$_3$
(**72**) R = CH$_3$; R′ = H
(**73**) R = R′ = H; R″ = *gem*-dimethyl

HIV) agents revealed that branching of the ethylene chain by introduction of a methyl group at the 1′-position (as in racemic **72**) diminished antiviral activity 25-fold and dimin-

molar-level activity as an inhibitor of 5-lipoxygenase (46). The size of the oxygen-containing ring as well as the position of the oxygen member with respect to the methoxy and aryl substituents was varied. The (seven-membered) oxepane ring derivative (**79**) and the (six-membered) tetrahydropyran ring derivative (**78**) showed two- to 10-fold enhanced potency over that of the tetrahydrofuran lead compound (**74**). The other analogs shown demonstrated much weaker enzyme inhibitory activity.

In a series of *spiro*-tetraoxacycloalkanes (**80**), with varying heterocyclic ring sizes, it was found that the compound where $n = 1$ demonstrated marked antimalarial activity against *P. bergei* and *P. falciparum*, and showed low toxicity (47). The analog in which

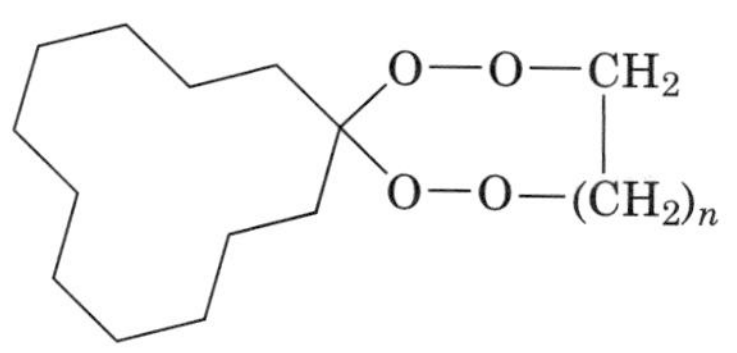

(80) $n = 1, 2, 3, 4, 5, 6$

$n = 4$ showed strong activity against *P. falciparum* but it was unimpressive in the *P. bergei* assay.

In a series of arylsulfonamidophenethanolamines (**81**) (48), derivatives bearing the sul-

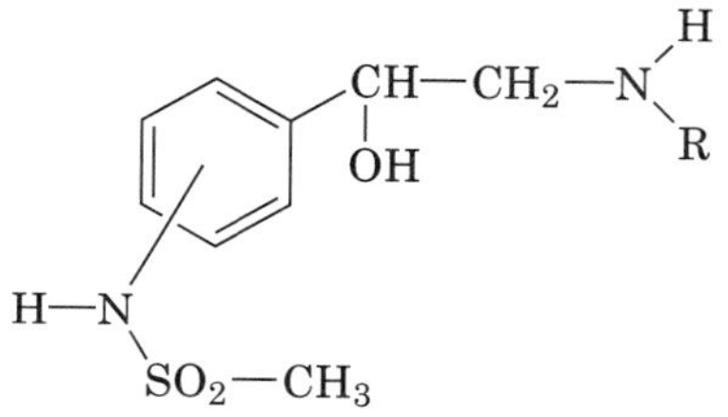

(81)

fonamido group *meta* to the ethanolamine side chain displayed properties of a β-adrenoceptor partial agonist, whereas 19 compounds bearing the sulfonamido group in the *para* position were β-antagonists.

Changing the positions of attachment of the two benzene rings linking the quinolinium moieties of the calcium-activated potassium channel blocker (**82**) reduced activity 10- to

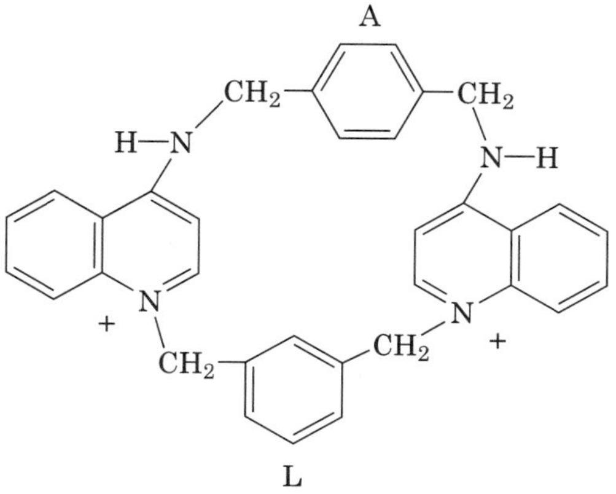

(82)

60-fold in a rat superior cervical ganglion assay (49). Other structural variations studied included benzene ring A *meta*-substituted and benzene ring L *meta*-substituted; benzene ring A *meta*-substituted and benzene ring L *para*-substituted; and benzene ring A *para*-substituted and benzene ring L *para*-substituted. All of these variations were much less potent than those of (**82**).

The phenolic group of serotonin (**83**) was incorporated into a pyran ring (**84**) (50), thus

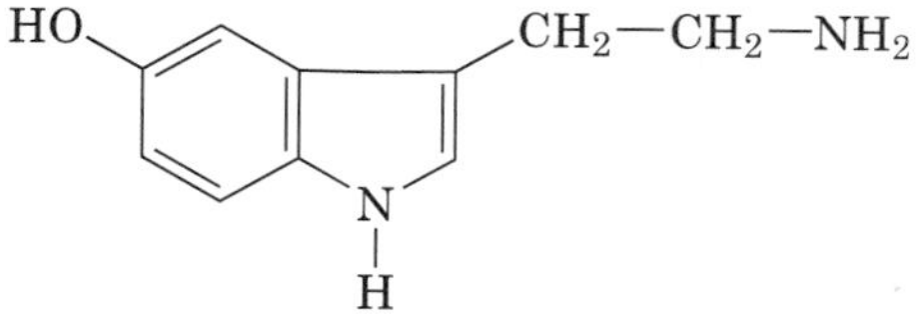

(83)

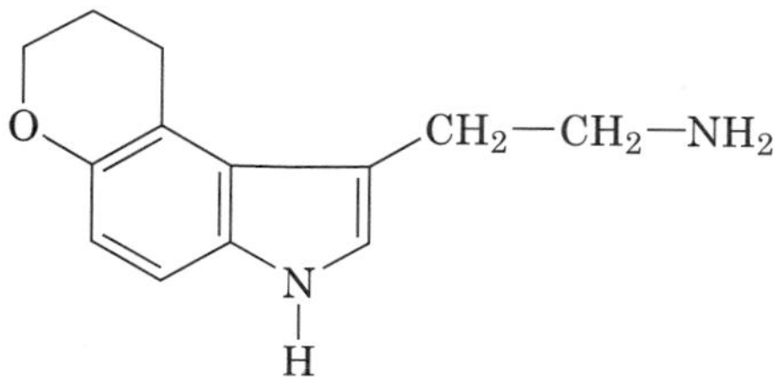

(84)

also introducing an alkyl substituent at position 4 of the indole ring system.

This tricyclic analog (**84**) lacked serotonin-like affinity for 5-HT$_1$ receptors, but it demonstrated high and selective affinity for 5-HT$_2$ receptors. Like serotonin, it stimulated phosphatidyl inositol turnover in rat brain slices. The low affinity for 5-HT$_1$ receptors was rationalized, in part, on the basis of steric interference between the dihydropyran ring and the aminoethyl side chain, which inhibits the tryptamine system from assuming the folded ergotlike conformation, as illustrated in (**85**), which probably approximates the conformation of serotonin at 5-HT$_1$ receptors. The methyl ether of serotonin exhibits approximately the same affinity for 5-HT$_{1A}$ sites as does serotonin (51). The methyl ether also has marked affinity for 5-HT$_{1C}$ and 5-HT$_{1D}$ receptors, but it has diminished affinity (compared with serotonin) at 5-HT$_{1B}$ receptors. It was

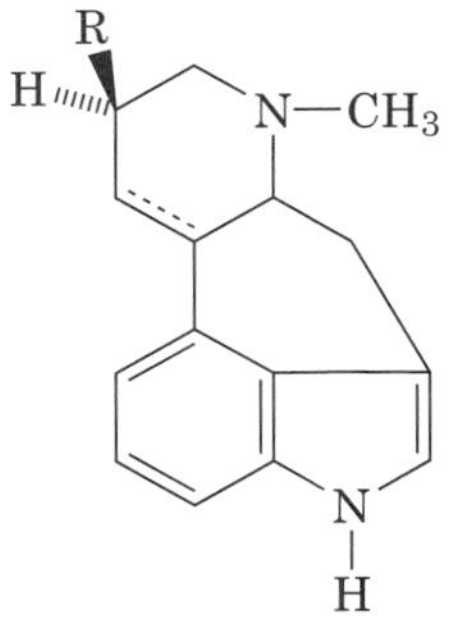

(85)

suggested (50) that the high affinity for the 5-HT$_2$ receptor exhibited by such compounds as (84) demonstrates that the C-5 hydroxyl group of serotonin can function as a hydrogen-bond acceptor at the receptor.

Replacement of the benzene ring of the potent indirect acting central noradrenergic stimulant methamphetamine (86) by a cyclohexane ring (compound 87) results in some

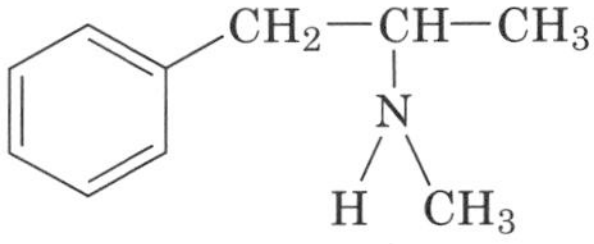

(86)

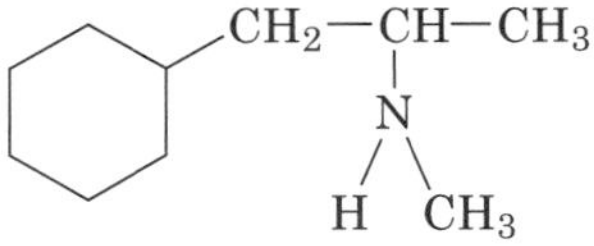

(87)

loss of pressor effect, but the drug, like amphetamine, has been used as a nasal decongestant, and it has CNS-mediated anorexigenic effect (52, 53). It is said to have somewhat less central stimulant action than the corresponding aromatic ring derivatives (54a-d).

The benzene (88) and cyclohexane (89) congeners have almost identical effects in blocking bronchoconstriction produced by histamine, serotonin, or acetylcholine in the guinea pig *in vivo* (55). They also showed identical LD$_{50}$ values in mice. The stereochemistry of these compounds was not addressed.

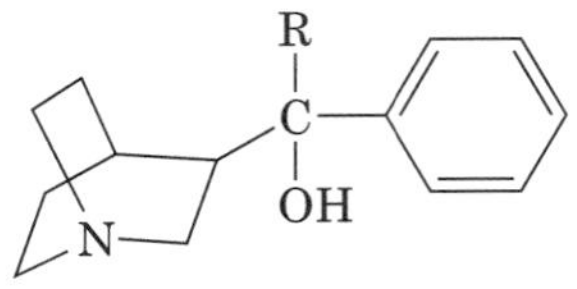

(88) R = C$_6$H$_5$
(89) R = c-C$_6$H$_{11}$

In a study of anticonvulsant agents, the (S)-benzene ring analog (90) was somewhat more potent in a mouse assay than was the (S)-cyclohexane analog (91) (56). There was

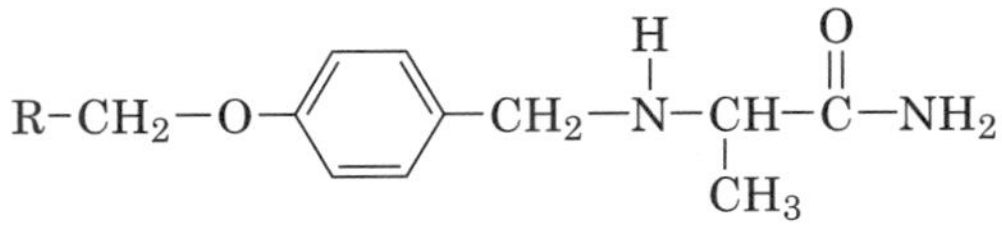

(90) R = C$_6$H$_5$
(91) R = c-C$_6$H$_{11}$

only a slight difference in potency between (R)- and (S)-(90). The (R)-enantiomer of (91) was not reported.

5 ALTERATION OF STEREOCHEMISTRY AND DESIGN OF STEREOISOMERS AND GEOMETRIC ISOMERS

The earlier, almost universally accepted belief that if one enantiomer of a chiral molecule demonstrates pharmacological activity, the other enantiomer will be pharmacologically inert, is not valid. It must be anticipated that all stereoisomers of an organic molecule will exhibit pharmacological effects, frequently widely different and unpredictable. Many examples of qualitative and quantitative differences in metabolism of enantiomers are documented (57).

(±)-3-(3-Hydroxyphenyl)-N-n-propylpiperidine (3-PPP, 92) was originally described (58) as having highly selective activity at dopaminergic autoreceptors.

At high doses (R)-(92) selectively stimulated presynaptic dopaminergic receptor sites, whereas at lower doses it selectively stimulated postsynaptic receptor sites (59). In contrast, the (S)-enantiomer stimulated presynaptic dopamine receptors and at the same dose level, it blocked postsynaptic dopamine recep-

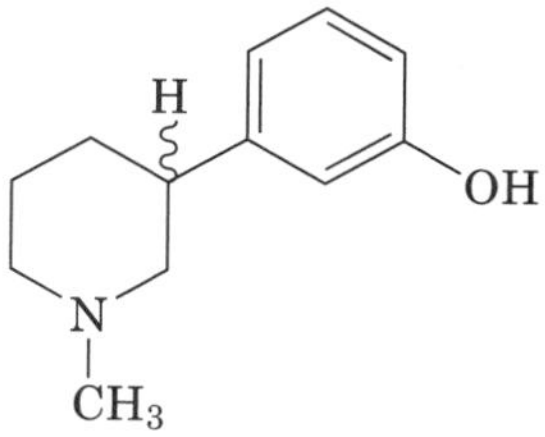

(92)

tors. Thus, this enantiomer exhibits a bifunctional mode of dopaminergic attenuation: that of presynaptic agonism and postsynaptic antagonism. The observed pharmacological effects of the racemic modification are the sum total of the complex activities of the two enantiomers, and the pharmacology of racemic 3-PPP is not an accurate reflection of the pharmacological properties of the individual enantiomers. The contemporary literature strongly reflects the philosophy that pharmacological testing only of a racemic mixture is inadequate and may be misleading.

(R)-$(-)$-11-Hydroxy-10-methylaporphine **(93)** is a highly selective serotonergic 5-HT$_{1A}$ agonist (60).

Remarkably, the (S)-enantiomer **(94)** is a potent antagonist at this same subpopulation of serotonin receptors (guinea pig ileum prep-

(93)

(94)

aration) (61). Both enantiomers bind strongly to 5-HT$_{1A}$ receptors from rat forebrain membrane. The phenomenon of enantiomers that possess opposite effects (agonist-antagonist) at the same receptor, once considered to be extremely rare, has recently been noted more often, probably because of the increasing recognition by medicinal chemists and pharmacologists that each member of an enantiomeric pair may possess its own unique and unpredictable pharmacology.

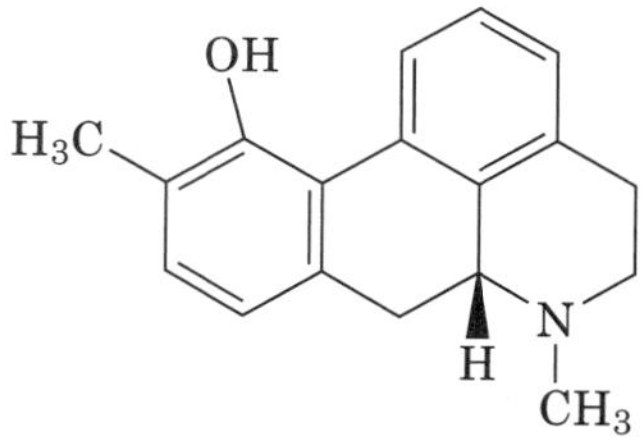

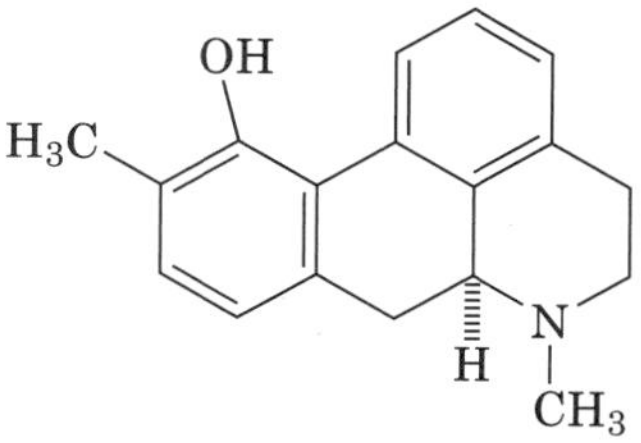

In addition to stereochemistry about a carbon center, other potentially chiral atoms offer possibilities for pharmacological significance. A gastroprokinetic compound **(95)** with

(95)

serotonergic activity bears a chiral sulfoxide moiety (62). The enantiomers are equipotent, but the (S)-enantiomer demonstrates a greater intrinsic activity than that of the (R)-enantiomer.

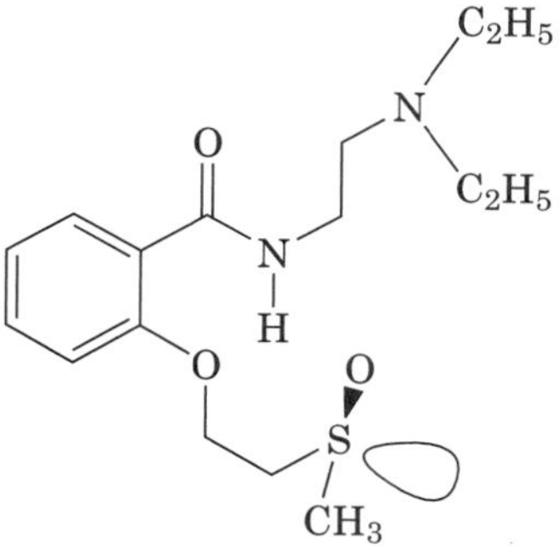

Casy (63) cited pharmacological differences between stereoisomers of chiral sulfoxide moieties in cholinergic oxathiolane congeners **(96–99)** of muscarine.

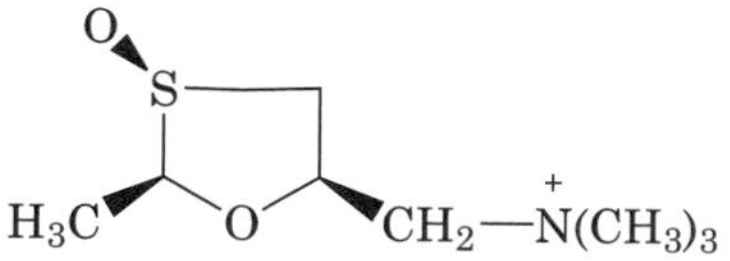

(96)

cis- and trans-4-Aminocrotonic acids **(100)** and **(101)** were prepared (64) as congeners of γ-aminobutyric acid (GABA) (6).

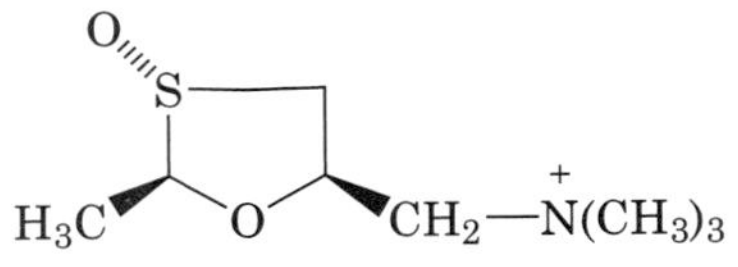

(97)

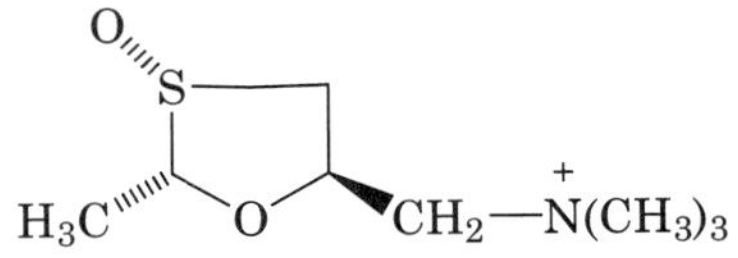

(98)

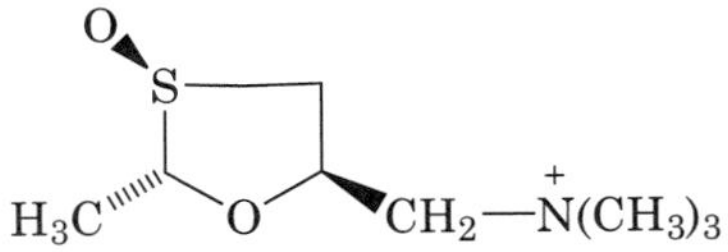

(99)

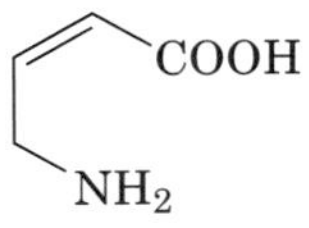

(100)

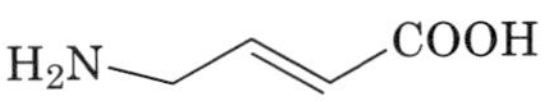

(101)

(102)

(103)

(104)

geometric isomer is an open-chain analog of the natural estrogen estradiol (**105**) (66). In dienestrol (**106**), the geometric isomerism possible with olefinic moieties has been further

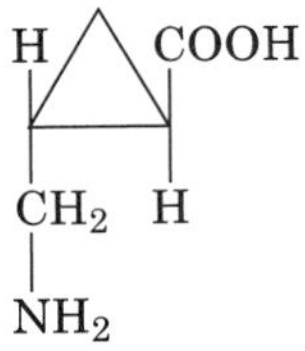

(105)

(106)

The folded *Z*-isomer (**100**) was inactive in assays for GABA agonism, whereas the extended *E*-isomer (**101**) was active. These data demonstrate biological differences of geometric isomers, which in turn involve a parameter discussed previously: imposition of a degree of structural rigidity on the molecule. A strategy analogous to this *E/Z* olefinic GABA congener design addressed *cis*- and *trans*-1,2-disubstituted cyclopropane derivatives (**102**) and (**103**), whose relative effects at GABA receptors paralleled those of the olefinic derivatives (65).

The *E*-isomer of the diethylstilbestrol structure (**104**) has 10 times the estrogenic potency of the *Z*-isomer; this effect has been rationalized from the conclusion that the *E*-

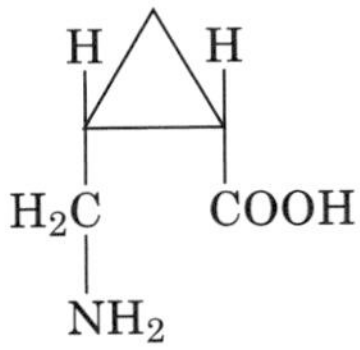

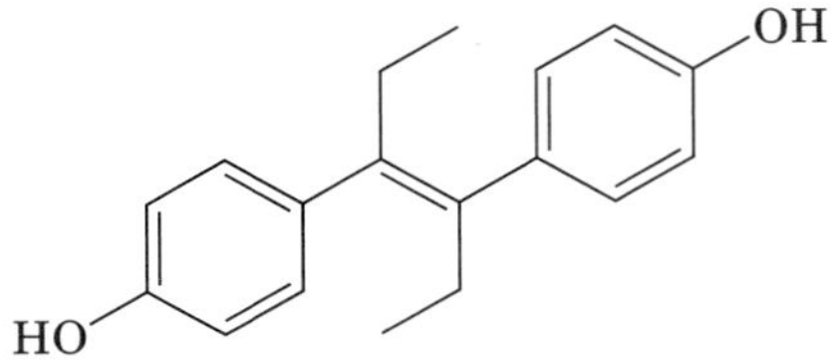

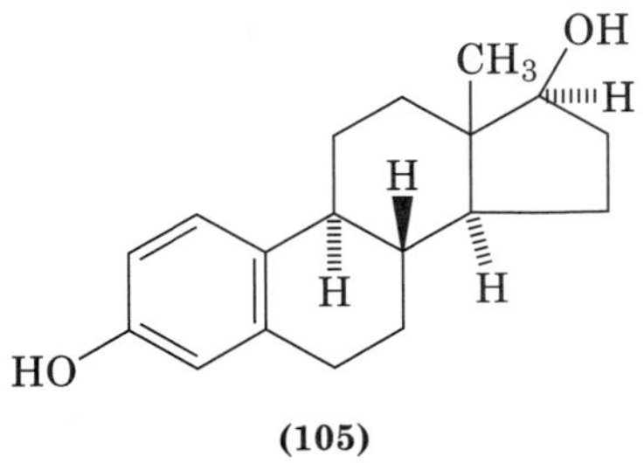

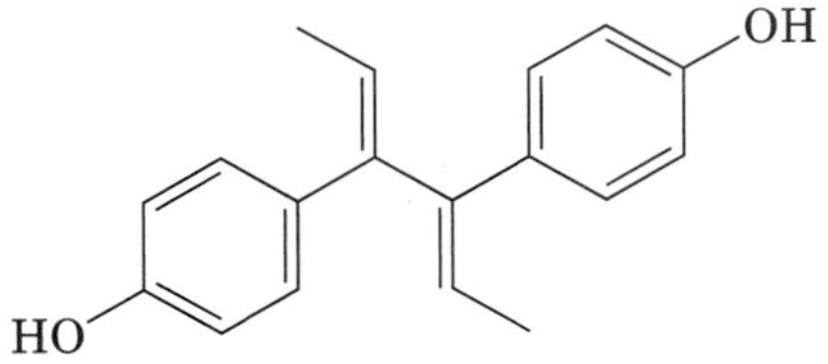

exploited to achieve a similar kind of open-chain analogy to the steroid ring system as in diethylstilbestrol, and a high level of estrogenic activity results.

Hexestrol (**107**), the saturated congener of

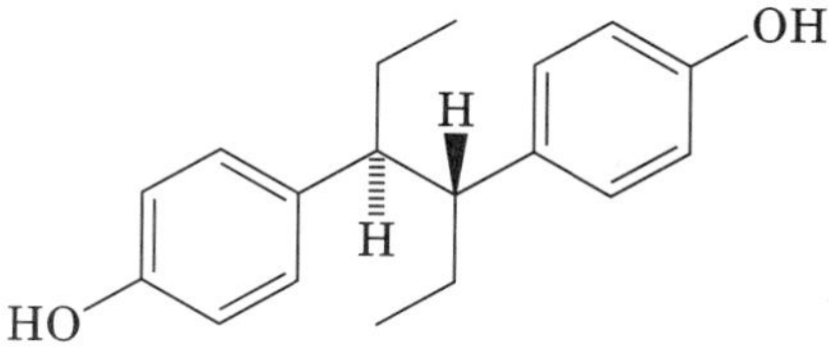

(**107**)

diethylstilbestrol (**104**), is the *meso*-form of the molecule. It has the greatest estrogenic potency of the three possible stereoisomers; however, it is less potent than diethylstilbestrol (**67**).

A partial restriction of side-chain flexibility in retinoic acid (**108**) was achieved by incorporating portions of the side chain into a benzene ring and a cyclopropane ring (**109**) (68).

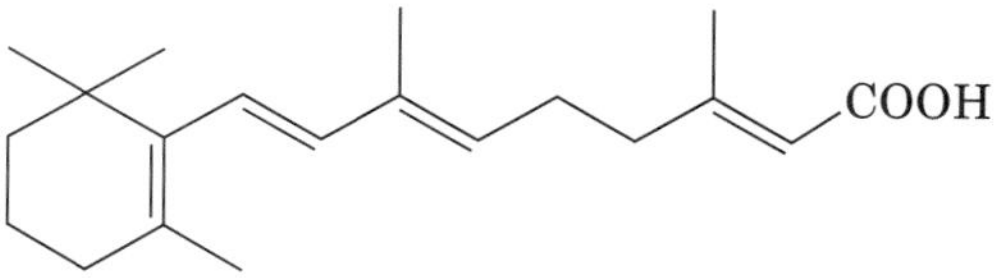

(**108**)

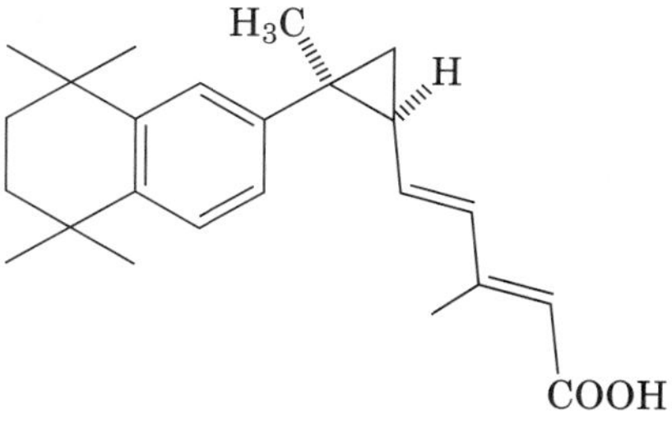

(**109**)

Introduction of the cyclopropane ring changes the corresponding *trans*-olefinic moiety of (**108**) to a cisoid disposition in (**109**), thus changing the overall steric disposition of the side chain. Moreover, the cyclopropane ring introduces chirality into the molecule. The (*S,S*)-enantiomer shown is a potent reti-

noid X-receptor ligand and it is inactive at the retinoic acid receptor, whereas the (*R,R*)-enantiomer is an extremely weak agonist at the retinoid X-receptor, although it has some effect at the retinoic acid receptor. Thus, the molecular modifications shown in (**109**) result in selectivity of action at these two receptors.

6 FRAGMENTS OF THE LEAD MOLECULE

Design of fragments of a lead molecule is based on the premise that some lead molecules, especially polycyclic natural products, may be much more structurally complex than is necessary for optimal pharmacologic effect. It is hypothesized that a pharmacophoric moiety may be buried within the complex structure of the lead compound and, if this pharmacophore can be clearly defined, it may be possible to "dissect" it out chemically. The result may be biologically active, simpler molecules that may themselves be used as leads in further analog design. A bond disconnection strategy may be employed in which bonds in the polycyclic structure are broken or removed to destroy one or more of the rings. The result may be a valuable drug that is more accessible (through chemical synthesis) than the original lead molecule. A possible disadvantage to this strategy of analog design is that the greater flexibility that is introduced into a rigid molecule may compromise or destroy the conformational integrity that may have existed in the pharmacophoric portion, at the expense of activity and/or potency. There may be a similar destruction of chiral centers, which may be undesirable. Morphine (**110**) typifies a lead molecule for which fragment analog design has been used.

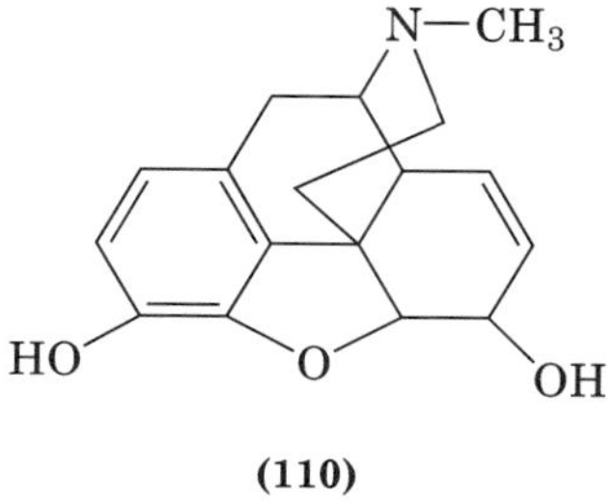

(**110**)

The analgesic μ-receptor pharmacophore of morphine has been defined (69) as comprising the basic nitrogen atom, the aromatic ring located three carbon atoms from the nitrogen, and a quaternary carbon adjacent to the aromatic ring, which provides a region of molecular bulk. A bond disconnection strategy involved disruption of the hydrofuran ring to give rise to morphinan derivatives [e.g., levorphanol (111)], whose pharmacologic effects

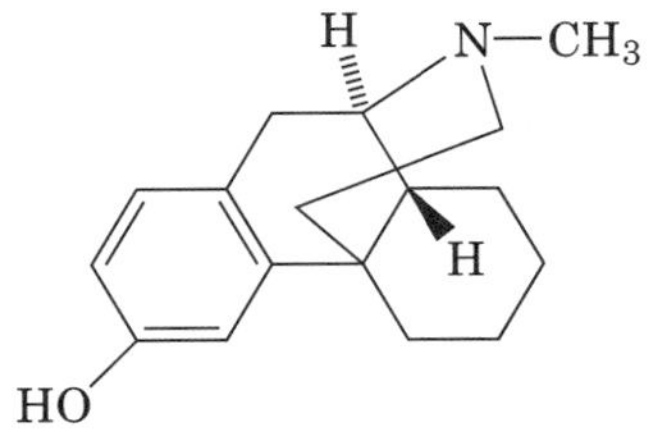

(111)

closely parallel those of morphine (70). Further simplification of the morphine ring system led to benzomorphan derivatives, typified by metazocine (112), in which morphinelike

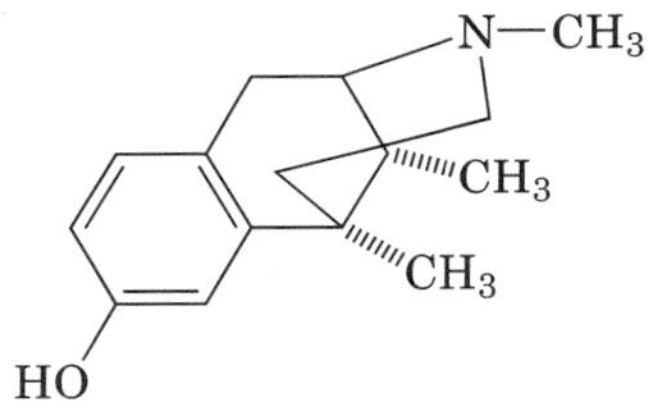

(112)

analgesic activity is retained. Finally, 4-phenylpiperidine derivatives typified by meperidine (113) and the nonheterocyclic system

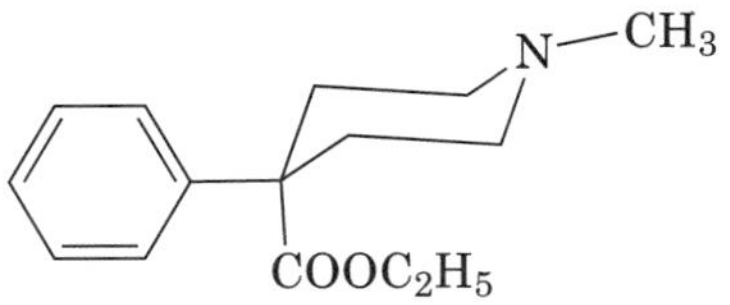

(113)

methadone (114) present the putative analgesic pharmacophore with a seemingly minimal number of extraneous atoms. These simple compounds retain opioid analgesic activity. It

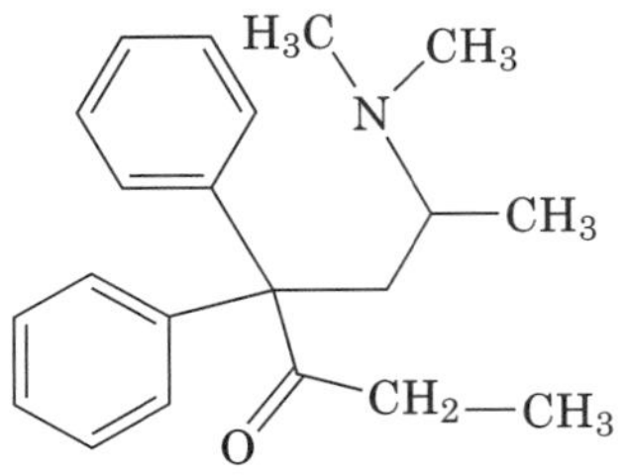

(114)

must be noted, however, that the discovery of analgesic activity in 4-phenylpiperidine derivatives was not a result of a systematic structure-activity study of the morphine molecule, but was serendipitous (71).

Asperlicin (115), a potent cholecystokinin-A antagonist, was subjected to two different bond disconnection strategies, as indicated (72).

Path A leads to tryptophan derivatives (116), some of which are potent cholecystokinin antagonists (73). Some quinazolinone derivatives (117) of disconnection pathway B showed extremely high potency and excellent selectivity as cholecystokinin-B receptor subtype ligands (44). A combination of X-ray crystallography and computational chemistry was used in the decision-making process in the bond disconnection (44) and in the design of the specific target molecules.

The myoneural-blocking pharmacophore in d-tubocurarine (118) was speculated to include the two cationic heads (the quaternary ammonium group and the protonated tertiary amine); the cationic heads are separated by 10 atoms (nine carbons and one oxygen).

Based on these parameters, a simple molecule, decamethonium (119), in which two trimethylammonium heads are separated by 10 methylene groups to approximate the internitrogen distance in d-tubocurarine, was designed independently by two groups of investigators (74, 75). This synthetic fragment/analog of d-tubocurarine exhibits a high degree of potency and activity in production of flaccid paralysis of skeletal muscles, superficially like that of the lead compound. However, the myoneural blockade from d-tubocurarine is of the nondepolarizing type, whereas decamethonium produces a depolarizing skel-

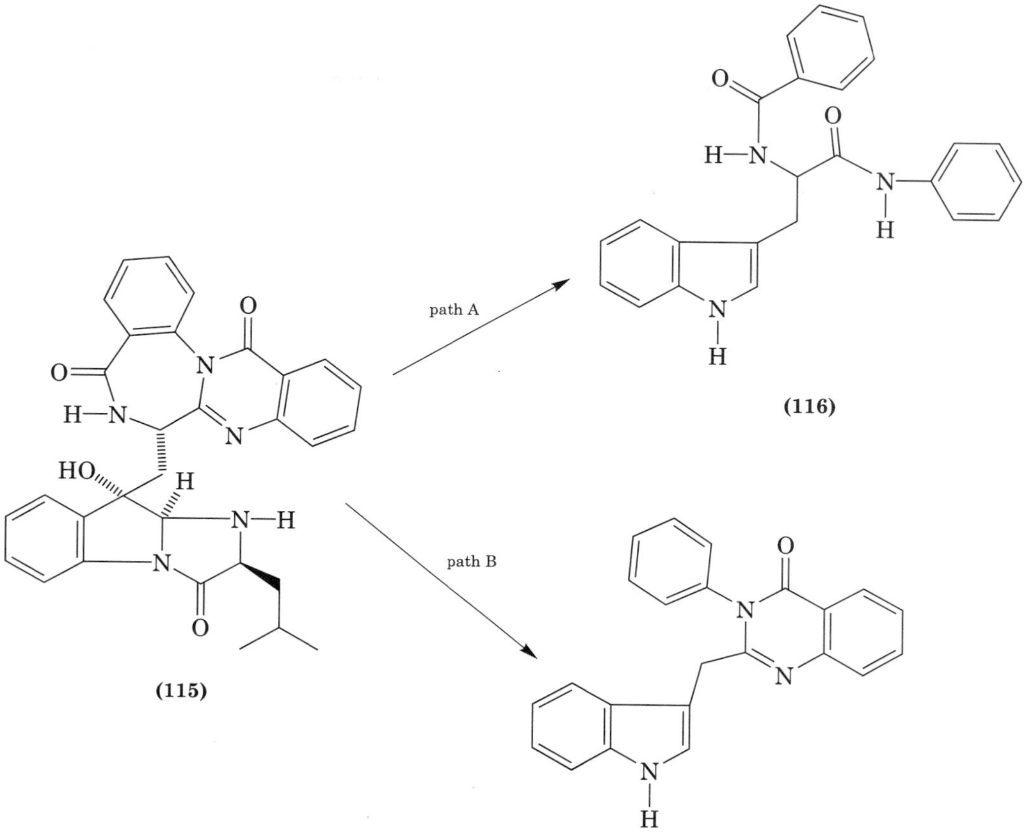

(115)

path A

(116)

path B

(117)

etal muscle blockade. This fundamental mechanistic difference is probably attributed, at least in part, to the flexibility of the decamethonium molecule compared with that of d-tubocurarine. There is a considerable differ-ence in the spectrum and severity of side effects and in the technique of employment of these two drugs in clinical practice. In all types of analog design, changes in chemical structure may result in unanticipated changes in

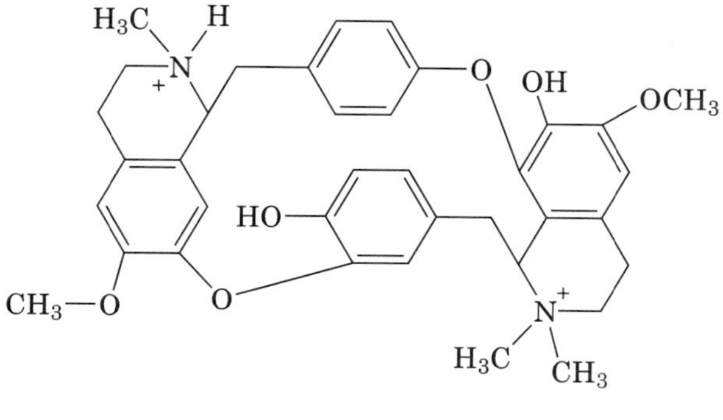

(118)

$$(CH_3)_3\overset{+}{N}-(CH_2)_{10}-\overset{+}{N}(CH_3)_3$$

(119)

mechanism of action, even though the chemical nature of the pharmacophore may not be altered.

7 VARIATION IN INTERATOMIC DISTANCES

Alteration of distances between portions of the pharmacophore of a molecule (or even between other portions of the molecule) may produce profound qualitative and/or quantitative changes in pharmacological actions. In α,ω-bis-trimethylammonium polymethylene compounds (**120–123**), maximal activity for

$$(CH_3)_3\overset{+}{N}-(CH_2)_{\overline{n}}-\overset{+}{N}(CH_3)_3$$

(120) $n = 5$
(121) $n = 6$
(122) $n = 16$
(123) $n = 18$

blockade of autonomic ganglia (nicotinic N_2 receptors) resides in those derivatives where $n = 5$ or 6 (compounds **120** and **121**) (76, 77).

Ganglionic effects drop drastically when $n = 4$ or 7. These observations have been rationalized as being a reflection of attainment of optimal interquaternary distance in the penta- and hexamethylene congeners, for optimal interaction with ganglionic receptor subsites. Remarkably, as the number of methylene groups in (**120**) is greatly increased, a high level of ganglionic-blocking potency returns. The hexadecyl and octadecyl congeners (**122**) and (**123**) are approximately four times as potent at autonomic ganglia as the penta- and hexamethylene compounds. As was mentioned previously, polymethylene bis-quaternary systems, in which the cationic heads are separated by 10 methylene groups, have potent effects at myoneural junctions (nicotinic N_1 receptors) and have little ability to affect nerve activity at autonomic ganglia. Thus, extension of a bis-quaternary polyalkylene molecule from five or six methylenes to 10 pro-

duces a pharmacological change from ganglionic blockade to myoneural blockade, and further extension to 16–18 methylenes results in loss of myoneural effects and a return of ganglionic blocking action.

Hemicholinium (**124**) competitively inhibits the high affinity, sodium-dependent uptake of choline into the nerve terminal (the rate-determining step in acetylcholine synthesis in the nerve terminal), thus depleting stores of acetylcholine and producing slow onset, long-duration myoneural blockade (78, 79). In a series of congeners of hemicholinium, the central biphenyl portion of the molecule was changed to terphenyl (**125**) and to p-phenylene (**126**). Both changes resulted in profound loss of the myoneural blockade characteristic of hemicholinium (68). This result was ascribed to alteration of the proposed opti-

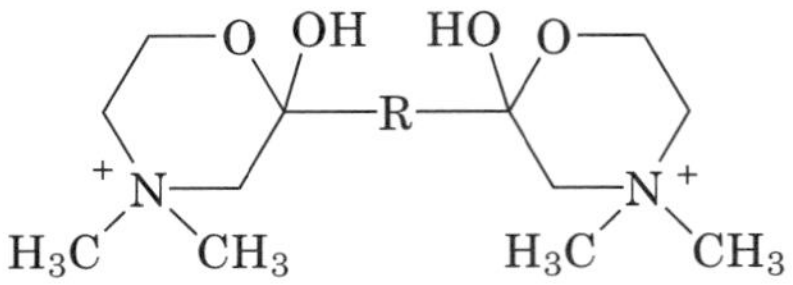

(124) R =

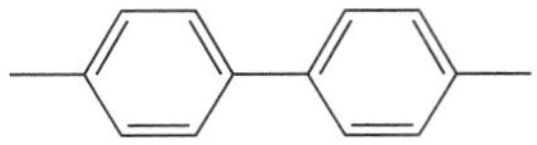

(125) R =

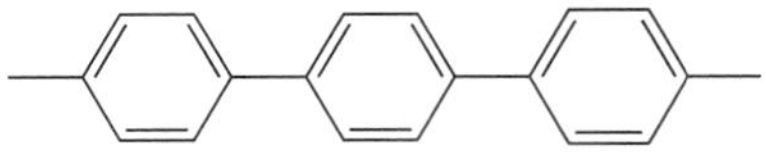

(126) R =

(127) R =

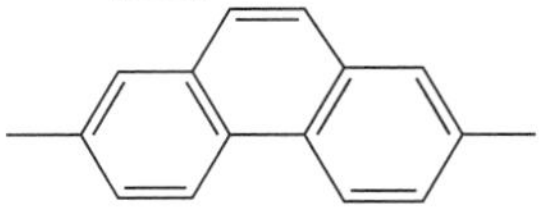

(128) R =

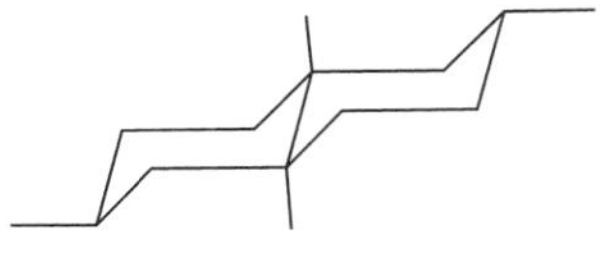

(129) R =

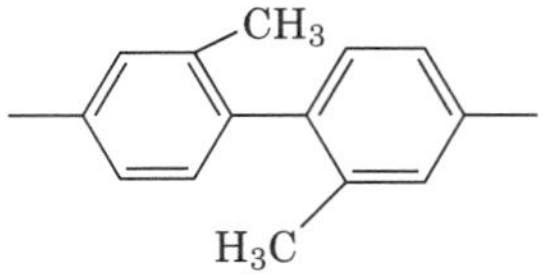

(130) R = $-(CH_2)_6-$

(131) R = $-(CH_2)_7-$

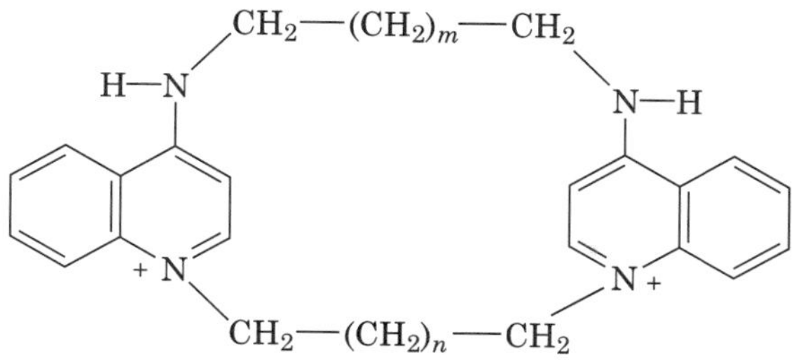

$$CH_2-(CH_2)_m-CH_2$$

(132)

mum interquaternary nitrogen distance of 14.4 Å in hemicholinium (124), to 18.4 Å in the terphenyl analog, and to 10.2 Å in the p-phenylene analog.

The central biphenyl spacer in hemicholinium was changed to a 2,7-disubstituted phenanthrene (127), $trans,trans$-4,4'-bicyclohexyl (128), and 2,2'-dimethylbiphenyl (129). In all three of these systems the 14.4-Å interquaternary distance found in hemicholinium was maintained; all of these congeners were qualitatively and quantitatively similar to hemicholinium in inhibition of neuromuscular transmission. Conformational analysis of the polyalkylene congeners (130) and (131) demonstrated that, when the flexible polyalkylene chain is maximally extended and is in a staggered conformation, the interquaternary distance in the hexamethylene congener (130) is approximately 14 Å, and in the heptamethylene congener (131) it is approximately 15 Å. Both compounds exhibited hemicholinium-like inhibition of neuromuscular transmission, although they were less potent than hemicholinium (80). This diminution of potency might be ascribed to the compromising of another structural parameter in the hemicholinium molecule: the rigidity of the central biphenyl spacer unit that maintains the internitrogen distance.

Replacement of the benzene ring linkers of (82) (see above) by alkyl linkers (structure 132) permitted retention of blocking activity on calcium-activated potassium channels (81). The most potent member of the series studied was that in which $m = n = 3$. In this compound the two respective internitrogen distances closely approximate those in the benzene ring-linked compound (82).

In a series of phenylalkylenetrimethylammonium derivatives (133–136), nicotinic agonism is maximal when $n = 3$ (compound 136).

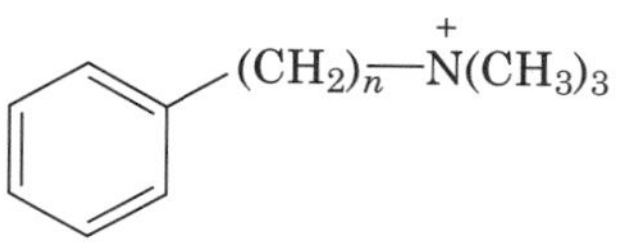

$$(CH_2)_{\overline{n}}\overset{+}{N}(CH_3)_3$$

(133) $n = 0$
(134) $n = 1$
(135) $n = 2$
(136) $n = 3$

It was concluded (82) that a moiety (here, a benzene ring) with high electron density three or four single bond lengths (~6Å) from the cationic center is a requirement for nicotinic agonism in the series. These conclusions may be compromised by the fact that the alkylene series was not extended beyond the three-carbon spacer chain. Therefore, it is not known whether the four-carbon homolog would display greater or lesser potency than that of the three-carbon molecule. Peculiarly, the first two members of the series have only very weak nicotine-like activity in the presence of atropine.

A series of compounds, illustrated by (137), was evaluated for in $vitro$ affinity for α_1 and α_2-adrenoceptors by radioligand-binding assays (83). All compounds showed good affinities for the α_1 adrenoceptor, with K_i values in the low nanomolar range. The polymethylene chain spacer between furoylpiperazinylpyradizinone and aryl piperazine moieties was shown to influence the affinity and selectivity of these compounds. A gradual increase in affinity for the α_1 adrenoceptor was observed, by length-

(137)

ening the polymethylene chain, up to a maximum of seven carbon atoms.

The α_2/α_1 ratio of adrenoceptor-binding affinities for the series of compounds did not parallel the α_1 adrenoceptor-binding affinities for the series, although all of the seven (C_1-C_7) congeners of (137) had somewhat more affinity for the α_2 receptor.

REFERENCES

1. I. Langmuir, *J. Am. Chem. Soc.*, **41**, 868 (1919).

2. C. W. Thornber, *Chem. Soc. Rev.*, **8**, 563 (1979).

3. G. A. Patani and E. G. La Voie, *Chem. Rev.*, **96**, 3147 (1996).

4. A. Burger, *Med. Chem. Res.*, **4**, 89 (1994).

5. P. Floersheim, E. Pombo-Villar, and G. Shapiro, *Chimia*, **46**, 323 (1992).

6. P. Krogsgaard-Larsen, H. Hjeds, E. Falk, F. S. Jørgenson, and L. Nielsen, *Adv. Drug Res.*, **17**, 38 (1988).

7. H. Hashiguchi and H. Takahashi, *Mol. Pharmacol.*, **13**, 362 (1977).

8. K. Anderson, A. Kuruvilla, N. Uretsky, and D. Miller, *J. Med. Chem.*, **24**, 683 (1981).

9. R. J. Baldessarini in A. G. Gilman, L. S. Goodman, T. W. Rall, and F. Murad, Eds., *Goodman and Gilman's The Pharmacological Basis of Therapeutics*, 7th ed., Macmillan, New York, 1985, pp. 393–397, 414.

10. S. I. Ankier in G. P. Ellis and G. B. West, Eds., *Progress in Medicinal Chemistry*, Vol. **23**, Elsevier, Amsterdam, 1986, p. 121.

11. E. I. Isaacson in J. N. Delgado and W. A. Remers, Eds., *Wilson and Gisvold's Textbook of Organic Medicinal and Pharmaceutical Chemistry*, 10th ed., Lippincott-Raven, Philadelphia, 1998, p. 473.

12. L. Otsuki, J. Ishiko, M. Sakai, K. Shiniahara, and T. Momiyama, *Pharmacometrics (Tokyo)*, **6**, 973 (1972).

13. M. Protiva, M. Rajsner, V. Seidlova, E. Adlerova, and Z. Vejdelek, *Experientia*, **18**, 326 (1962).

14. S. Yous, J. Andrieux, H. E. Howell, P. J. Morgan, P. Reynard, B. Pfeiffer, D. Lessieur, and B. Guardiola-Lemaitre, *J. Med. Chem.*, **35**, 1484 (1992).

15. S. M. Bromidge, F. Brown, F. Cassidy, M. S. G. Clark, S. Dabbs, M. S. Hadley, J. Hawkins, J. M. Loudon, C. B. Naylor, B. S. Orlek, and G. J. Riley, *J. Med. Chem.*, **40**, 4265 (1997).

16. P. Sauerberg, J. W. Kindtler, L. Nielsen, M. J. Sheardown, and T. Honoré, *J. Med. Chem.*, **34**, 687 (1991).

17. M. Hollmann, A. O'Shea-Greenfield, S. W. Rohers, and S. Heinemann, *Nature*, **342**, 643 (1989).

18. F. H. Bellevue, M. L. Boahbedason, R. H. Wu, R. A. Casero Jr., D. Rattendi, C. J. Bacchi, and P. M. Woster, *Bioorg. Med. Chem. Lett.*, **6**, 2765 (1996).

19. P. M. Woster, *Annu. Rep. Med. Chem.*, **36**, 99 (2001).

20. E. Mutschler and G. Lambrecht in E. Ariëns, W. Soudjin, and P. B. M. W. M. Timmermans, Eds., *Stereochemistry and Biological Activity of Drugs*, Blackwell, Oxford, UK, 1983, p. 65.

21. J. G. Cannon in E. Jucker, Ed., *Progress in Drug Research*, Vol. **29**, Birkhäuser Verlag, Basel, Switzerland, 1985, pp. 324–334.

22. C. Melchiorre, A. Chiarini, M. Gianella, D. Giardina, W. Quaglia, and V. Tumiatti in V. Claasen, Ed., *Trends in Drug Research*, Vol. **13**, Elsevier, Amsterdam, 1990, pp. 37–48.

23. F. D. King, A. M. Brown, L. M. Gaster, A. J. Kaumann, A. D. Medhurst, S. G. Parker, A. A. Parsons, T. L. Patch, and P. Raval, *J. Med. Chem.*, **36**, 1918 (1993).

24. P. S. Portoghese, A. A. Mikhail, and H. J. Kupferberg, *J. Med. Chem.*, **11**, 219 (1968).

25. R. M. Moriarty, L. A. Enache, L. Zhao, R. Gilardi, M. V. Mattson, and O. Prakash, *J. Med. Chem.*, **41**, 468 (1998).

26. C. M. Bertha, B. J. Vilner, M. V. Mattson, W. D. Bowen, K. Becketts, H. Xu, R. B. Rothman, J. L. Flippen-Anderson, and K. C. Rice, *J. Med. Chem.*, **38**, 4776 (1995).

27. C. Y. Chiou, J. P. Long, J. G. Cannon, and P. D. Armstrong, *J. Pharmacol. Exp. Ther.*, **166**, 243 (1969).

28. J. G. Cannon and P. D. Armstrong, *J. Med. Chem.*, **13**, 1037 (1970).

29. J. G. Cannon, T. Lee, V. Sankaran, and J. P. Long, *J. Med. Chem.*, **18**, 1027 (1975).

30. P. W. Ehrhardt, R. J. Gorczynski, and W. G. Anderson, *J. Med. Chem.*, **22**, 907 (1975).

31. R. R. Ruffolo in G. Kunos, Ed., *Adrenoceptors and Catecholamine Action, Part B*, Wiley-Interscience, New York, 1983, pp. 10–11.

32. H. J. Komiskey, J. F. Bossart, D. D. Miller, and P. N. Patel, *Proc. Natl. Acad. Sci. USA*, **75**, 2641 (1978).

33. E. E. Smissman and W. H. Gastrock, *J. Med. Chem.*, **11**, 860 (1968).

34. J. A. Monn, M. J. Valli, S. M. Massy, M. M. Hansen, T. J. Kress, J. P. Wepsiec, A. R. Harkness, J. L. Grutsch Jr., R. A. Wright, B. G. Johnson, S. L. Andés, A. Kingston, R. Tomlinson, R. Lewis, K. R. Griffey, J. P. Tizzano, and D. D. Schoepp, *J. Med. Chem.*, **42**, 1027 (1999).

35. J. F. McCarthy, J. G. Cannon, and J. P. Buckley, *J. Pharm. Sci.*, **52**, 1168 (1963).

36. J. F. McCarthy, J. G. Cannon, J. P. Buckley, and W. J. Kinnard, *J. Med. Chem.*, **7**, 72 (1964).

37. A. Burger and G. R. Bedford, *J. Med. Chem.*, **6**, 402 (1963).

38. M. V. Koch, J. G. Cannon, and A. M. Burkman, *J. Med. Chem.*, **11**, 977 (1968).

39. E. R. Atkinson, F. J. Bullock, F. E. Granchelli, S. Archer, F. J. Rosenberg, D. G. Teiger, and F. C. Nachod, *J. Med. Chem.*, **18**, 1000 (1975).

40. J. G. Cannon in ref. 21, pp. 309–310.

41. J. G. Cannon, F.-L. Hsu, J. P. Long, J. R. Flynn, B. Costall, and R. J. Naylor, *J. Med. Chem.*, **21**, 248 (1978).

42. J. Z. Ginos and F. C. Brown, *J. Med. Chem.*, **21**, 155 (1978).

43. J. Z. Ginos, G. C. Cotzias, and D. Doroski, *J. Med. Chem.*, **21**, 160 (1978).

44. M. J. Yu, J. R. McCowan, N. R. Mason, J. B. Deeter, and L. G. Mendelsohn, *J. Med. Chem.*, **35**, 2534 (1992).

45. K.-L. Yu, J. J. Bronson, H. Yang, A. Patick, M. Alam, V. Brankovan, R. Datema, M. J. M. Hitchcock, and J. C. Martin, *J. Med. Chem.*, **35**, 2958 (1992).

46. G. C. Crawley, R. I. Dowell, P. N. Edwards, S. J. Foster, R. M. McMillan, E. R. H. Walker, and D. Waterson, *J. Med. Chem.*, **35**, 2600 (1992).

47. H.-S. Kim, Y. Nagai, K. Ono, K. Begum, Y. Wataya, Y. Hamada, K. Tsuchiya, A. Masuyama, M. Nojima, and K. J. McCullough, *J. Med. Chem.*, **44**, 2357 (2001).

48. R. H. Uloth, J. R. Kirk, W. A. Gould, and A. A. Larsen, *J. Med. Chem.*, **9**, 88 (1966).

49. J. Campos-Rosa, D. Galanakis, A. Piergentili, K. Bhandari, C. R. Ganellin, P. M. Dunn, and D. H. Jenkinson, *J. Med. Chem.*, **43**, 420 (2000).

50. J. E. Macor, C. B. Fox, C. Johnson, B. K. Koe, L. A. Label, and S. H. Zorn, *J. Med. Chem.*, **35**, 3625 (1992).

51. R. A. Glennon in J. G. Cannon, Ed., *Advances in CNS Drug-Receptor Interactions*, JAI Press, Greenwich, CT, 1991, pp. 147–148.

52. B. B. Hoffman in J. G. Hardman and L. E. Limbird, Eds., *Goodman and Gilman's The Pharmacological Basis of Therapeutics*, 10th ed., McGraw-Hill, New York, 2001, p. 218.

53. R. L. Johnson in J. N. Delgado and W. A. Remers, Eds., *Wilson and Gisvold's Textbook of Organic Medicinal and Pharmaceutical Chemistry*, 10th ed., Lippincott-Raven, Philadelphia, 1998, p. 493.

54. (a) A. M. Lands, J. R. Lewis, and V. L. Nash, *J. Pharmacol. Exp. Ther.*, **83**, 253 (1945); (b) A. M. Lands, V. L. Nash, and B. L. Dertlinger, *J. Pharmacol. Exp. Ther.*, **89**, 382 (1947); (c) D. F. Marsh and D. A. Herring, *J. Pharmacol. Exp. Ther.*, **97**, 68 (1949); (d) E. J. Fellows, E. Macko, and R. A. McLean, *J. Pharmacol. Exp. Ther.*, **100**, 267 (1950).

55. M. D. Mashkovsky, L. N. Yakhoutov, M. E. Kaminka, and E. E. Mikhlina in E. Jucker, Ed., *Progress in Drug Research*, Vol. **27**, Birkhäuser Verlag, Basel, Switzerland, 1983, pp. 35–38.

56. P. Pevarello, A. Bonsignori, P. Dostert, F. Heidemperger, V. Pinciroli, M. Colombo, R. A. McArthur, P. Salvati, C. Post, R. G. Fariello, and M. Varasi, *J. Med. Chem.*, **41**, 579 (1998).

57. A. F. Casy, *The Steric Factor in Medicinal Chemistry. Dissymmetric Probes of Pharmacological Receptors*, Plenum, New York/London, 1993, pp. 52–61.

58. S. Hjorth, A. Carlsson, H. Wikström, P. Lindberg, D. Sanchez, U. Hacksell, L.-E. Arvidsson, O. Svensson, and J. L. G. Nilsson, *Life Sci.*, **28**, 1225 (1981).

59. H. Wikström, D. Sanchez, P. Lindberg, U. Hacksell, L.-E. Arvidsson, A. M. Johansson, S.-O. Thorberg, J. L. G. Nilsson, K. Svensson, S.

Hjorth, D. Clark, and A. Carlsson, *J. Med. Chem.*, **27**, 1030 (1984).

60. J. G. Cannon, P. Mohan, J. Bojarski, J. P. Long, R. K. Bhatnagar, P. A. Leonard, J. R. Flynn, and T. K. Chatterjee, *J. Med. Chem.*, **31**, 313 (1988).

61. J. G. Cannon, S. T. Moe, and J. P. Long, *Chirality*, **3**, 19 (1991).

62. B. T. Butler, G. Silvey, D. M. Houston, D. R. Borcherding, V. L. Vaughn, A. T. McPhail, D. M. Radzik, H. Wynberg, W. Ten Hoeve, E. Van Echten, N. K. Ahmed, and M. D. Linnik, *Chirality*, **4**, 155 (1992).

63. See ref. 57, pp. 244–247; see also M. Pigini, L. Brasili, M. Gianella, and F. Gualtieri, *Eur. J. Med. Chem.*, **16**, 415 (1981).

64. G. A. Johnston, D. R. Curtis, P. M. Beart, C. J. A. Game, R. M. McColloch, and B. Twitchin, *J. Neurochem.*, **24**, 157 (1975).

65. R. D. Allan, D. R. Curtis, P. M. Headley, G. A. Johnson, D. Lodge, and B. Twitchen, *J. Neurochem.*, **34**, 652 (1980).

66. D. S. Fullerton in R. F. Doerge, Ed., *Wilson and Gisvold's Textbook of Organic Medicinal and Pharmaceutical Chemistry*, 8th ed., Lippincott, Philadelphia, 1982, p. 670.

67. R. W. Brueggemeier, D. D. Miller, and D. T. Witiak in W. O. Foye, T. L. Lemke, and D. A. Williams, Eds., *Principles of Medicinal Chemistry*, 4th ed., Williams & Wilkins, Media, PA, 1995, p. 474.

68. V. Vuligonda, S. M. Thacher, and R. A. S. Chandraratna, *J. Med. Chem.*, **44**, 2298 (2001).

69. T. Nogrady, *Medicinal Chemistry*, 2nd ed., Oxford University Press, New York, 1988, p. 457.

70. J. H. Jaffe and W. R. Martin in ref. 9, p. 513.

71. J. V. Aldrich in M. E. Wolff, Ed., *Burger's Medicinal Chemistry and Drug Discovery*, 5th ed., Vol. **3**, Wiley-Interscience, New York, 1996, p. 357.

72. M. J. Yu, K. J. Thrasher, J. R. McCowan, N. R. Mason, and L. G. Mendelsohn, *J. Med. Chem.*, **34**, 1505 (1991).

73. F. W. Hahne, R. T. Jensen, G. F. Lemp, and J. D. Gardner, *Proc. Natl. Acad. Sci. USA*, **78**, 6304 (1981).

74. R. B. Barlow and H. R. Ing, *Nature*, **161**, 718 (1948).

75. W. D. M. Paton and E. J. Zaimis, *Nature*, **161**, 718 (1948).

76. D. J. Triggle, *Neurotransmitter-Receptor Interactions*, Academic Press, New York, 1971, p. 360.

77. V. Trcka in D. A. Kharkevich, Ed., *Handbook of Experimental Pharmacology*, Vol. **53**, Springer-Verlag, New York, 1980, p. 138.

78. J. G. Cannon, T. M.-L. Lee, A. M. Nyanda, B. Bhattacharyya, and J. P. Long, *Drug Des. Deliv.*, **1**, 209 (1987).

79. J. G. Cannon, *Med. Res. Rev.*, **14**, 505 (1994).

80. J. G. Cannon, T. M.-L. Lee, Y.-a. Chang, A. M. Nyanda, B. Bhattacharyya, J. R. Flynn, T. Chatterjee, R. K. Bhatnagar, and J. P. Long, *Pharm. Res.*, **5**, 359 (1988).

81. J.-Q. Chen, D. Galanakis, C. R. Ganellin, P. M. Dunn, and D. H. Jenkinson, *J. Med. Chem.*, **43**, 3478 (2000).

82. W. C. Holland in E. J. Ariëns, Ed., Proceedings of the Third International Pharmacological Meeting, Vol. 7, Pergamon, Oxford, UK, 1966, pp. 295–303; K. C. Wong and J. P. Long, *J. Pharmacol. Exp. Ther.*, **137**, 70 (1962).

83. R. Barbaro, L. Betti, F. Corelli, G. Giannacinni, L. Maccari, F. Manetti, G. Straoaghetti, and S. Corsano, *J. Med. Chem.*, **44**, 2118 (2001).

CHAPTER SEVENTEEN

Approaches to the Rational Design of Enzyme Inhibitors

MICHAEL J. MCLEISH
GEORGE L. KENYON
Department of Medicinal Chemistry
University of Michigan
Ann Arbor, Michigan

Contents

Burger's Medicinal Chemistry and Drug Discovery
Sixth Edition, Volume 1: Drug Discovery
Edited by Donald J. Abraham
ISBN 0-471-27090-3   © 2003 John Wiley & Sons, Inc.

1 INTRODUCTION

Many of the top 100 drugs sold worldwide are enzyme inhibitors. In recent years, enzyme inhibitors not only have provided an increasing number of potent therapeutic agents for the treatment of diseases, but also have significantly advanced the understanding of enzymatic transformations. The aim of this chapter is to present current approaches to so-called rational inhibitor design, which uses knowledge of enzymic mechanisms and structures in the design process. Rational inhibitor design is intended to complement laborious and resource-consuming screening processes, which consist of testing large numbers of synthetic chemicals or natural products for inhibitory activity against a chosen target enzyme.

1.1 Enzyme Inhibitors in Medicine

A human cell contains thousands of enzymes each of which can, theoretically, be selectively inhibited. These enzymes constitute the various metabolic pathways that, in concert, provide the requirements for the viability of the cell. A selective inhibitor may block either a single enzyme or a group of enzymes, leading to the disruption of a metabolic pathway(s). This will result in either a decrease in the concentration of enzymatic products or an increase in the concentration of enzymatic substrates. The effectiveness of an enzyme inhibitor as a therapeutic agent will depend on (1) the potency of the inhibitor, (2) its specificity toward its target enzyme, (3) the choice of metabolic pathway targeted for disruption, and (4) the inhibitor or a derivative possessing appropriate pharmacokinetic characteristics. Higher potency will mean less drug is required to obtain a physiological response, whereas high specificity means that the inhibitor will react only with its target enzyme and not with other sites in the body. Taken together, low

dosage and high specificity combine to reduce both the toxicity caused by inhibition of other vital enzymes and the problems arising from the formation of toxic decomposition products. Further, high specificity will generally avoid depletion of the inhibitor concentrations in the host by nonspecific pathways. The areas of potency and specificity will both be addressed in this chapter. Clearly, the choice of target enzyme is also of prime importance for chemotherapy, although that is beyond the scope of this review. However, there are a number of texts available that provide a good introduction to this subject (1–4). Good bioavailability of the drug is also crucial for the drug to reach its site of action in the body in effective therapeutic concentrations. For example, highly polar or charged compounds, such as phosphorylated compounds, frequently cannot readily cross cell membranes and are therefore generally less useful as drugs. Physical approaches to facilitate the transport of this class of compounds into the cell include the use of liposomes or nanoparticles (5–7). Chemical approaches may also be employed. These include the use of prodrugs, in which functional groups on the inhibitor are modified in such a manner that they are able to be taken up by the cell and, later, metabolically converted to the active drug. Prodrugs are discussed in more detail in Volume 2, Chapter 14.

As indicated earlier, a wide variety of enzyme inhibitors have found use in the clinic. Tables 17.1–17.3 show a number of these compounds and, although they provide by no means an exhaustive list, they do give an indication of the range of human disease states that can be ameliorated with the use of enzyme inhibitors.

The human body, even though its defenses are constantly on guard, is still susceptible to invasion by foreign pathogens. Since the de-

Table 17.1 Examples of Enzyme Inhibitors Used in the Treatment of Bacterial, Fungal, Viral, and Parasitic Diseases

Clinical Use	Enzyme Inhibited	Inhibitor
Antibacterial	Dihydropteroate synthetase	Sulphonamides
Antibacterial	Dihydrofolate reductase	Trimethoprim, methotrexate
Antibacterial	Alanine racemase	D-Cycloserine
Antibacterial	Transpeptidase	Penicillins, cephalosporins
Antifungal	Fungal sterol 14α-demethylase	Clotrimazole, ketoconazole
Antifungal	Fungal squalene epoxidase	Terbinafine, naftifine
Antiviral	Thymidine kinase and thymidylate kinase	Idoxuridine
Antiviral	DNA, RNA polymerases	Cytosine arabinoside (Ara-C)
Antiviral	Viral DNA polymerase	Acyclovir, vidarabine
Antiviral	HIV reverse transcriptase	Dideoxyinosine, zidovudine
Antiviral	HIV protease	Saquinavir
Antiviral	Influenza virus neuraminidase	Zanamavir, oseltamivir
Antiprotozoal	Pyruvate dehydrogenase	Organoarsenical agents
Antiprotozoal	Ornithine decarboxylase	α-Difluoromethylornithine

velopment of the sulfa drugs (sulfonamides), enzyme inhibitors have played a vital role in controlling these infectious agents. Table 17.1 provides a list of enzyme inhibitors that have been used in the treatment of the various diseases caused by these agents. All these compounds needed to satisfy the usual requirements for specificity and low toxicity.

This can be achieved in a variety of ways. For instance, it is possible to inhibit an essential pathway in the pathogen that does not exist in the host. D-Cycloserine (1) (Fig. 17.1), for example, inhibits alanine racemase, an enzyme involved in bacterial cell wall biosynthesis and not found in humans (8, 9). D-Cycloserine is active against a broad spectrum of both gram-positive and gram-negative bacteria (10), but plays its major role in the treatment of tuberculosis (11). Conversely, even if both host and pathogen contain the same enzymes, it may be possi-

ble to exploit subtle structural differences between the isozymes to obtain a highly specific inhibitor that preferentially binds to the invader's version. Trimethoprim (2) shows this selective toxicity. An inhibitor of dihydrofolate reductase, trimethoprim is a potent antibacterial agent because the bacterial enzyme is inhibited at a concentration several thousand times lower than that required for inhibition of the mammalian isozyme (12). Acyclovir (3a), an antiviral drug used for the treatment of herpes infections (13, 14), also fits into this category. It binds very tightly to the *Herpes simplex* DNA polymerase with an estimated half-life of about 40 days. Acyclovir is a prodrug because it requires transformation by a viral thymine kinase and cellular phosphotransferases to the corresponding triphosphate (3b) to serve *in vivo* as an inhibitor of the viral DNA polymerase (15).

Table 17.2 Examples of Enzyme Inhibitors Used in the Treatment of Cancer

Type of Cancer	Enzyme Inhibited	Inhibitor
Benign prostatic hyperplasia	Steroid 5α-reductase	Finasteride
Estrogen-mediated breast cancer	Aromatase	Aminoglutethimide, fadrozole
Leukemia, osteosarcoma, head, neck, and breast cancer	Dihydrofolate reductase	Methotrexate
Colorectal cancer	Thymidylate synthase	5-Fluorouracil
Leukemia	Glutamine-PRPP amidotransferase	6-Mercaptopurine, azathioprine
Small-cell lung cancer, non-Hodgkin's lymphoma	Topoisomerase II	Etoposide
Hairy-cell leukemia	Adenosine deaminase	Pentostatin

Table 17.3   Examples of Enzyme Inhibitors Used in Various Human Disease States

Clinical Use	Enzyme Inhibited	Inhibitor
Epilepsy	GABA transaminase	γ-Vinyl GABA
Epilepsy	Carbonic anhydrase	Sulthiame
Epilepsy	Succinic semialdehyde dehydrogenase	Sodium valproate
Antidepressant	Monoamine oxidase (MAO)	Tranylcypromine, phenelzine
Antihypertensive	Angiotensin converting enzyme	Captopril, enalaprilat
Cardiac disorders	Na^+,K^+-ATPase	Cardiac glycosides
Gout	Xanthine oxidase	Allopurinol
Ulcer	H^+,K^+-ATPase	Omeprazole
Hyperlipidemia	HMG-CoA reductase	Atorvastatin, simvastatin
Anti-inflammatory	Prostaglandin synthase, Cyclooxygenase (COX) I and II	Aspirin, naproxen, ibuprofen
Arthritis	Cyclooxygenase (COX) II	Celecoxib
Glaucoma	Acetylcholinesterase	Neostigmine
Glaucoma	Carbonic anhydrase II	Acetazolamide, dichlorphenamide

Although their inhibitors are not specifically therapeutic agents in themselves, the β-lactamases are another important target for drug design. These are bacterial enzymes and, as with the alanine racemases, are not found in humans. Inhibitors of β-lactamases include clavulanic acid (**4**) (16–20) and sulbactam (penicillanic acid sulfone) (**5**) (18, 21–24). These two compounds act to prevent the bacterial degradation of penicillins and cephalosporins by β-lactamases, thereby extending their lifetime and effectiveness. Accordingly, both clavulanic acid (**4**) and sulbactam (**5**) have reached the market as drugs that act synergistically with these commonly prescribed antibacterial agents.

Even though it has proved possible to selectively inhibit the enzymes of a number of pathogens, the enzymes of cancer cells have proved to be a far more elusive target. Indeed, the majority of the currently employed antitumor agents can be described as antiproliferative agents. These take advantage of the fact that many, but not all, tumor cells grow and divide more rapidly than normal cells. Lymphomas, for example, proliferate more rapidly than solid tumors, whereas, conversely, acute leukemia cells divide more slowly than the surrounding bone marrow cells. Most of the enzyme inhibitors used as these antiproliferative agents (Table 17.2) can also be described as antimetabolites (i.e., they inhibit a metabolic pathway), often those involved in DNA biosynthesis, which are important for cell survival or replication. 5-Fluorouracil (**6**), the

prodrug form of an inactivator of thymidylate synthase (25), and methotrexate (**7**), an inhibitor of dihydrofolate reductase (26, 27), both fit into this category. Unfortunately, rapidly dividing normal cells, such as hair follicles, the cells lining the gastrointestinal tract, and the bone marrow cells involved in the immune system are also significantly affected. The resultant hair loss, nausea, and susceptibility to infection means that this type of chemotherapy is seldom employed as a first-line defense against cancer.

The inhibition of enzymes involved in metabolic pathways is not restricted to anticancer agents. A variety of diseases have been correlated with either the dysfunction of an enzyme or an imbalance of metabolites. A cross section of the disease states treated with enzyme inhibitors is shown in Table 17.3. Practically, these may be treated by the inhibition of an individual enzyme or by using enzyme inhibitors to regulate the metabolite concentration in the body. For example, an imbalance of the two neurotransmitters, glutamate and γ-aminobutyric acid, is responsible for the convulsions observed in epileptic seizure. The latter is metabolized by γ-aminobutyric acid aminotransferase (GABA-T) and, consequently, inhibitors of this enzyme offered themselves as potential antiepileptic candidates. This led to the development of the GABA-T inhibitor, vigabatrin (**8**) (28), which clinically results in an increase of the brain concentration of γ-aminobutyric acid and cessation of epileptic convulsions. As with the anticancer agents, block-

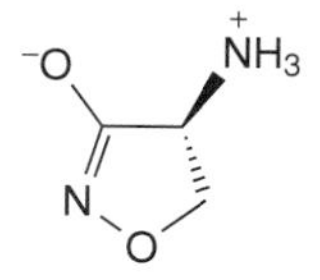

(1)

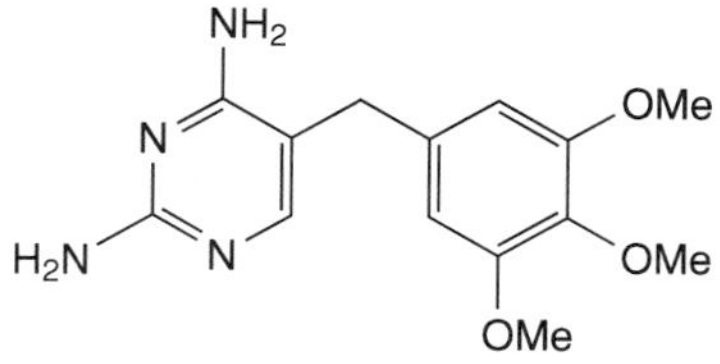

(2)

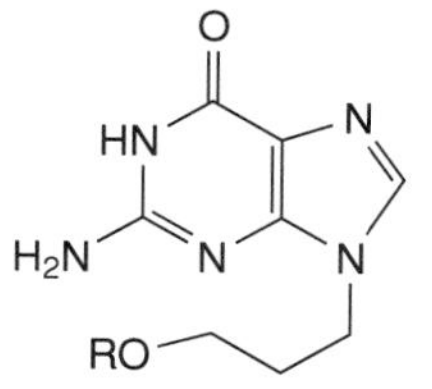

(3a) R = H
(3b) R = PPP

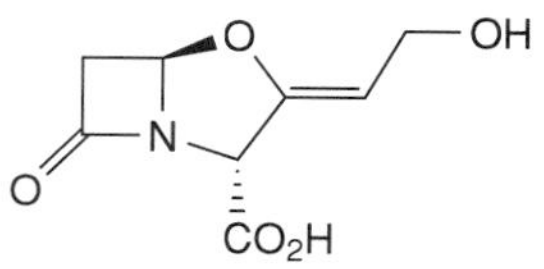

(4)

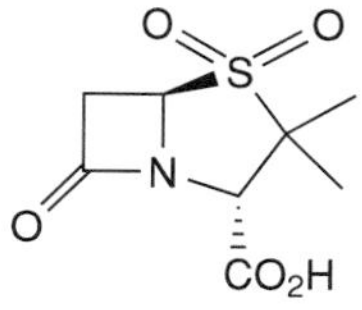

(5)

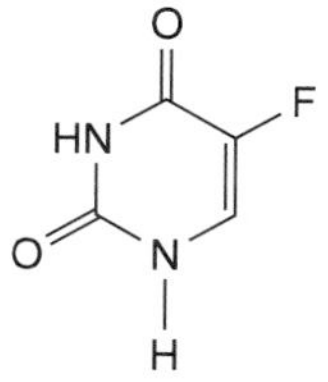

(6)

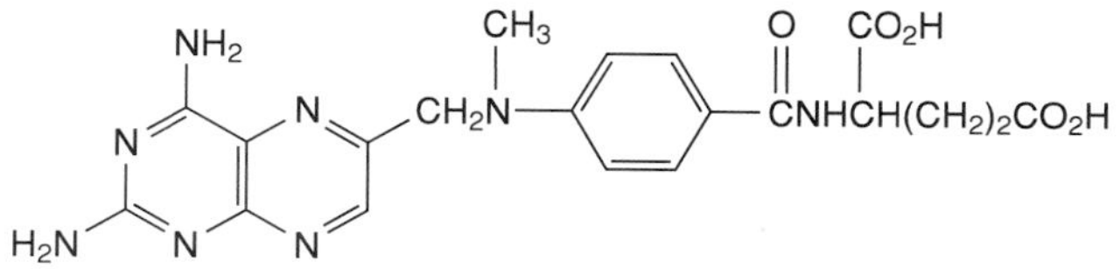

(7)

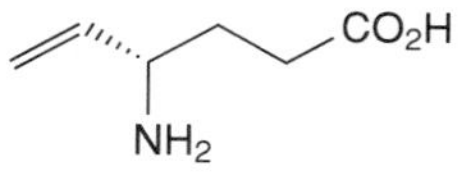

(8)

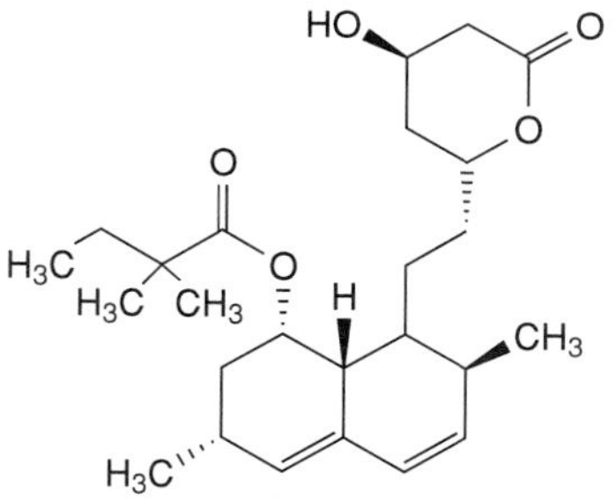

(9)

Figure 17.1. Examples of enzyme inhibitors used clinically.

ade of a metabolic pathway may also have therapeutic benefits. The statins, a group of serum cholesterol-lowering drugs, are inhibitors of hydroxymethylglutaryl-CoA (HMG-CoA) reductase (29). HMG-CoA reductase catalyzes the irreversible conversion of HMG-CoA to mevalonic acid, the rate-determining step in cholesterol biosynthesis (30–32). Inhibitors such as simvastatin (**9**) have been found to be effective in the treatment of hyperlipidemia and familial hypercholesteremia (33, 34) and have become some of the world's best-selling drugs.

Finally, enzyme inhibitors can also be used to induce an animal model of a genetic disease. Inactivation of γ-cystathionase by propargylglycine, for example, produces an experimental model of the disease state known as cysta-

Table 17.4 Classification of Enzyme Inhibitors Employed in This Chapter

Noncovalent Inhibitors	Covalent Inhibitors
Rapid reversible inhibitors (ground-state analogs)	Chemical modifiers
Tight, slow, slow-tight binding inhibitors	Affinity labels
Multisubstrate analogs	Mechanism-based inhibitors
Transition-state analogs	Pseudoirreversible inhibitors

thioninuria (35). Deficiency of this enzyme leads to the accumulation of cystathionine in the urine and has sometimes been associated with mental retardation (36).

1.2 Enzyme Inhibitors in Basic Research

In basic research enzyme inhibitors have found a multitude of uses. They serve as useful tools for the elucidation of structure and function of enzymes, as probes for chemical and kinetic processes, and in the detection of short-lived reaction intermediates (37). Product inhibition patterns provide information about an enzyme's kinetic mechanism and the order of substrate binding (38). Covalently binding enzyme inhibitors have been used to identify active-site amino acid residues that could potentially be involved in substrate binding and catalysis of the enzyme (39, 40). Reversible enzyme inhibitors are routinely used to facilitate enzyme purification by using the inhibitor as a ligand for affinity chromatography (41, 42) or as eluants in affinity-elution chromatography (43). Immobilized enzyme inhibitors can also be used to identify their intracellular targets (44), whereas irreversible inhibitors can be used to localize and quantify enzymes *in vivo* (45).

In Table 17.4 we have provided the classification of the various types of enzyme inhibitors that we employ in this chapter. The classification may appear somewhat arbitrary, in that some inhibitors may fit into more than one category. This can arise because these categories are attempting to bring together some nonrelated properties such as structure, mechanism of action, and kinetic behavior. Thus, what we have classed as a reversible inhibitor may, simply because it has a slow dissociation rate, be described elsewhere in the literature as being irreversible. In each instance we will discuss approaches to the design of that type of inhibitor, as well as indi-

cating how it may be evaluated. The discussion will be accompanied by references to recent, representative examples from the literature. Where appropriate, these examples will be of inhibitors of therapeutic interest.

It should be noted that we will concentrate on inhibitors directed at the active site of the enzyme. While recognizing that there are inhibitors that bind to regions other than the active site, such as allosteric effectors, these are not the focus of this chapter and will not be included. There are many reviews of enzyme inhibitors available in the literature (37, 46–48) and the reader is referred to them for more detailed analysis.

2 RATIONAL DESIGN OF NONCOVALENTLY BINDING ENZYME INHIBITORS

As their name indicates, this class of inhibitors binds to the enzyme's active site without forming a covalent bond. Therefore the affinity and specificity of the inhibitor for the active site will depend on a combination of the electrostatic and dispersive forces, and hydrophobic and hydrogen-bonding interactions. Traditionally, noncovalently binding enzyme inhibitors were analogs of substrates, products, or reaction intermediates. More recently, an explosion in the use of combinatorial chemistry and rapid screening techniques has seen the development of large numbers of enzyme inhibitors that bear little or no resemblance to the substrate or products, yet still bind selectively to their target enzyme. Computer-aided drug design, in the broadest sense, encompasses both structure-based drug design and quantitative structure-activity relationship (QSAR) methods. A complement to the rapid screening techniques, computer-aided methods provide a more focused approach to the design and discovery of both substrate and nonsubstrate analog inhibitors.

In structure-based design, the structure of a drug target interacting with small molecules is used to guide drug discovery. Consequently, either the three-dimensional enzyme structure or, at a minimum, the pharmacophore structure must be known. A pharmacophore represents the nature of the chemical groups of a given ligand and their relative orientation important for inhibitor binding. Today, structure-based design, used in conjunction with docking techniques, combinatorial chemistry, and rapid screening not only leads more quickly to novel enzyme inhibitors but also greatly reduces the number of compounds that must be synthesized. More information on these approaches may be found in Chapter 10 and some recent monographs (49–52).

Traditionally, an increase in inhibitory or biological activity was achieved by synthesizing an analog of the substrate and then making gradual empirical changes in the structure by adding or removing functional groups. QSAR methods provide a means of making this empirical testing more focused. In this technique there is no need to know the structure of the active site. Instead, computer algorithms are employed to correlate the biological activity of a series of inhibitors with their chemical structure, thereby allowing better predictions as to how to change the structure to obtain a more potent inhibitor. This topic is discussed further in Chapter 1, and detailed reviews are also available (53–56).

Table 17.4 shows the classification of noncovalent inhibitors we use in this chapter. Based on their kinetics it is possible to distinguish among rapid reversible, tight-binding, slow-binding, slow-tight-binding, irreversible, and pseudoirreversible inhibitors. Conversely, inhibitors classified on the basis of structure, such as ground-state analogs, multisubstrate inhibitors, and transition-state analogs, which mimic the structures of substrates and products, reaction intermediates, and transition states, may fall into any of the kinetic categories. However, before introducing these categories, it is important to have an understanding of the forces involved in the binding of substrates and inhibitors to an enzyme's active site.

2.1 Forces Involved in Forming the Enzyme-Inhibitor Complex

To understand the design concepts of the various types of noncovalently binding enzyme inhibitors, a basic knowledge of the binding forces between an enzyme's active site and its inhibitors is required. The forces involved in a substrate or an inhibitor binding to an enzyme's active site are, as with a drug binding to a receptor, the same forces that are experienced by all interacting organic molecules. These include ionic (electrostatic) interactions, ion-dipole and dipole-dipole interactions, hydrogen bonding, hydrophobic interactions, and van der Waals interactions. A brief overview of the forces involved follows. More comprehensive treatments can be found in Chapter 4 and elsewhere (57–60).

The binding of an inhibitor is dependent on a variety of interactions, and it is the sum of these interactions that will determine the degree of affinity of an inhibitor for the particular enzyme. The reversible binding of an inhibitor to an enzyme's active site can be described as shown in Equation 17.1.

$$\text{E} + \text{I} \underset{k_{-1}}{\overset{k_1}{\rightleftharpoons}} \text{E} \cdot \text{I} \qquad (17.1)$$

There is an equilibrium between the free enzyme (E), inhibitor (I), and the enzyme-inhibitor complex (E · I). The affinity of an inhibitor for the enzyme is measured by the inhibition constant K_i, which is the dissociation constant of the enzyme-inhibitor complex, at equilibrium (Equation 17.2).

$$K_i = \frac{[\text{E}][\text{I}]}{[\text{E} \cdot \text{I}]} \qquad (17.2)$$

The lower the K_i value, the better the inhibitor, given that the equilibrium lies more in favor of enzyme-inhibitor complex formation. The affinity of an inhibitor for an enzyme may be related to the standard free energy ($\Delta G°$) of a system by Equation 17.3.

$$\Delta G^\circ = RT \ln K_i \qquad (17.3)$$

where R is the universal gas constant and T the temperature in degrees Kelvin. The more negative the value of ΔG°, the more favorable the interaction at equilibrium, and the smaller the K_i value. It should be noted that, from Equation 17.3, at physiological temperature relatively small changes in free energy, only 2–3 kcal/mol, will have a significant effect on K_i.

The standard free energy (ΔG°) can also be expressed in terms of enthalpic (ΔH°) and entropic (ΔS°) components (Equation 17.4).

$$\Delta G^\circ = \Delta H^\circ - T\Delta S^\circ \qquad (17.4)$$

Equation 17.4 states that the free energy of a system is lowered (i.e., the reaction is made more favorable) by either a decrease in enthalpy or an increase in entropy. This is also an important concept because there are both enthalpic and entropic components to the forces that contribute to the strength of the enzyme-inhibitor interaction.

When discussing the forces involved in the noncovalent binding of a substrate/inhibitor to an enzyme, or drug to a receptor, it must be recognized that these interactions will be carried out in an aqueous medium. The physical properties of water mean that noncovalent interactions in aqueous solution will be significantly different from those interactions observed in either an organic medium or in the gas phase. A water molecule has electronic asymmetry; the strongly electronegative oxygen atom withdraws electron density from the hydrogen atoms. This creates partial positive charges on the hydrogens and a partial negative charge on the oxygen. As a result a water molecule possesses a permanent dipole moment, facilitating strong interactions with other water molecules as well as with any charged or polar species.

Water is both a donor and acceptor of hydrogen bonds. Consequently, in bulk solvent, water molecules are extensively hydrogen bonded to each other. These are relatively weak bonds ($\sim$5 kcal/mol) and, at physiological temperature, are rapidly broken and reformed. However, the hydrogen-bonding network affects many of the properties of water.

For example, water has a higher melting point, boiling point, and heat of vaporization than those of comparable hydrides such as H_2S and NH_3. The heat capacity of water indicates that it is highly structured and its surface tension (73 dyne cm^{-1} at 20°C) is considerably higher than that of most liquids (20–40 dyne cm^{-1}). The dielectric constant of water (80) is also considerably higher than that of most liquids, which are generally less than 30. Ethanol, for example, has a dielectric constant of 24, whereas those of benzene and hexane are 2.3 and 1.9, respectively. All told, water is a unique solvent, and one that has a major influence on binding interactions between an enzyme and an inhibitor.

Hydrogen bonds are readily formed between water and biologically important atoms such as the hydrogen bond acceptors N and O and, to a lesser extent, S. The conjugate acids NH and OH may act as hydrogen bond donors. Molecules containing these atoms have the capacity for many hydrogen-bonding interactions with water and, as a result, are usually soluble in water. However, solute-solute hydrogen bonding interactions are less favorable because their formation will require the disruption of favorable solute-water hydrogen bonds. Thus, what may be strong hydrogen bonds in the gas phase, or in organic media, are often considerably weaker in aqueous media.

Water's high dielectric constant makes it extremely effective in solvating, dissociating, and dissolving most salts. Because of its permanent dipole, water is readily able to interact with ionic species, with the result that ionic solute-solute interactions are less favored. The situation is analogous to that observed for hydrogen bonding and again results in a weakening of the normally strong interactions between ions that occur in the gas phase or nonpolar media. This is sometimes described as a "leveling effect."

Small amounts of many nonpolar substances can also dissolve in water. However, these substances do not interact very well with water and prefer to interact with each other. The force driving this interaction, known as the hydrophobic force, is not so much an attraction between hydrophobic molecules as an entropic effect arising from the displacement of water. Indeed, there are no hydrophobic forces in the gas phase or in non-polar solvents. However, collectively, hydrophobic forces are thought to transcend other types of forces, particularly in the folding of proteins, in all biological systems.

2.1.1 Electrostatic Forces. Although we recognize that, in essence, all forces between atoms and molecules are electrostatic, here we use the term to describe ion-ion, ion-dipole, and dipole-dipole interactions. At physiological pH, the side-chains of basic residues such as lysine and arginine and, to a lesser extent, the imidazole ring of histidine will be protonated, whereas the acidic groups on the side chains of aspartic and glutamic acid residues will be deprotonated. In addition, the N-terminal amino groups and C-terminal carboxylates will be ionized. Therefore, in addition to atoms with permanent and induced dipoles, an enzyme potentially will have several charged groups available for binding to charged or polarized groups on a substrate or inhibitor. As described by Equation 17.5, the electrostatic force (F) between the charged atoms (q_1 and q_2) will depend on the distance between the charged groups (r) and the dielectric constant of the surrounding medium (D).

$$F = \frac{q_1 q_2}{r^2 D} \quad (17.5)$$

The strength of an ion-ion interaction is inversely related to the square of distance between the ions, whereas ion-dipole and dipole-dipole interactions have $1/r^4$ and $1/r^6$ relationships, respectively. Because the strength of the interaction decreases more slowly with distance, ion-pair interactions can be thought of as long-range interactions. Conversely, interactions involving dipoles are effective over only a short range, although, because they are much more prevalent, dipole interactions may be more significant to the overall binding process. Clearly, the dependency of the strength of interaction on the distance between atoms is an important consideration when designing potential enzyme inhibitors.

Equation 17.5 also leads to the fact that electrostatic interactions are less favorable in polar solvents. As discussed above, because of its high net permanent dipole moment, water is very polar and has a large dielectric constant. The high polarity of water greatly diminishes the attraction or repulsion forces between any two charged groups giving rise to the leveling effect of water. It is somewhat difficult to predict the exact strength of a charge-charge interaction between an enzyme and an inhibitor. For example, the formation of a salt bridge (charge-charge) interaction between an enzyme (Enz) and an inhibitor (I) may be described by Equation 17.6.

$$\text{Enz}-\overset{\oplus}{\text{N}}\text{H}_3 \cdot (\text{H}_2\text{O})_x + \text{I}-\text{CO}_2^{\ominus} \cdot (\text{H}_2\text{O})_y \rightleftharpoons$$

$$\text{Enz}-\overset{\oplus}{\text{N}}\text{H}_3 \cdot {}^{\ominus}\text{O}_2\text{C}-\text{I} + (\text{H}_2\text{O})_{x+y} \quad (17.6)$$

Both the charged species are initially solvated by water, and to form the salt bridge both ions must be desolvated. This comes at some enthalpic cost, but the freeing of water molecules leads to a concomitant, favorable increase in entropy. The strength of the ion pair will depend on the stability of the salt bridge vs. that of the individual solvated ions. If the salt bridge is buried in a relatively hydrophobic active site, it is less solvated and will be more favored than the same interaction in a solvent-exposed active site.

2.1.2 van der Waals Forces. Also called nonpolar interactions or London dispersion forces, these are the universal attractive interactions that occur between atoms. As two molecules closely approach each other there is an interpenetration of their electron clouds. As a consequence, temporary local fluctuations in the electron density occur, giving rise to a temporary dipole in each molecule, even though the molecules may, in themselves, have no net dipole moment. Thus there will be an attractive force between the two molecules, with the magnitude of the force depending on the polarizability of the particular atoms involved and the distance between each other. Electronegative oxygen has, for example, a much lower polarizability than that of a nonpolar methylene group. Accordingly, dispersion forces are considerably stronger between nonpolar compounds than between nonpolar compounds and water. The optimal distance between the atoms is the sum of each of their van der Waals radii, so these forces come into play only when there is good complementarity between enzyme and inhibitor. Although van der Waals forces are quite weak, usually around 0.5-1.0 kcal/mol for an individual atom-atom interaction, they are additive and can make an important contribution to inhibitor binding.

2.1.3 Hydrophobic Interactions. Hydrophobic interactions may be described as entropy-based forces. When a nonpolar compound is dissolved in water, the strong water-water interactions around the solute lead to an effective "ordering" of the structure of the solvent. This is entropically unfavorable; that is, there is negative entropy of dissolution. When a nonpolar inhibitor binds to a nonpolar region of an enzyme, all the ordered water molecules become less ordered as they associate with bulk solvent, leading to an increase in entropy. According to Equation 17.4 any increase in entropy will lead to a decrease in free energy and, through Equation 17.3, stabilization of the enzyme-inhibitor complex. It has been calculated that a single methylene-methylene interaction releases about 0.7 kcal/mol of free energy. Even though this figure is not high, given that enzymes and inhibitors usually have large regions of hydrophobic surface, this type of bonding may also play a significant role in inhibitor binding.

2.1.4 Hydrogen Bonds. A hydrogen bond occurs when a proton is shared between two electronegative atoms (i.e., $-X-H\cdots Y$). Electron density is pulled from the hydrogen by X, giving the hydrogen a partial positive charge that is strongly attracted to the non-bonded electrons of Y. The bond is usually asymmetric, with one of the heteroatoms, the hydrogen bond donor, having a normal covalent bond distance to the proton. The other heteroatom, the hydrogen bond acceptor, is usually at a distance somewhat shorter than the van der Waals contact distance and, for optimal hydrogen bonding, the atoms should be arranged linearly. A hydrogen bond is a special type of dipole-dipole interaction and, as we have seen, although these forces can be quite

significant in nonpolar solvents, water greatly diminishes their magnitude. The energy of the amide-amide $NH\cdots O$ hydrogen bond is about 5 kcal/mol, and is typical for hydrogen bonds (60).

It should be remembered that, for a hydrogen bond to form between an enzyme and an inhibitor, any hydrogen bonds between the inhibitor and water, as well as those between the enzyme and water, must be broken (Equation 17.7).

$$(17.7)$$

Overall, the total number of hydrogen bonds remains constant and, provided that the hydrogen bonds between the inhibitor and enzyme are not significantly more favorable than those between water and the inhibitor or those between water and the enzyme, the net change in enthalpy is usually insignificant. On the other hand, formation of the enzyme-inhibitor complex usually leads to an overall increase of entropy because the inhibitor remains bound to the enzyme and the formerly bound water molecules are released.

2.1.5 Cation-π Bonding. Recently it has become apparent that there is another important noncovalent binding force that may be exploited when designing enzyme inhibitors. Cations, from simple ions such as Li^+ to more complex organic molecules such as acetylcholine, are strongly attracted to the electron-rich (π) face of benzene and other aromatic compounds (61, 62). Cation-π bonds, as well as other amino-aromatic interactions, are common in structures in the protein data bank (63), and it has been estimated that more than 25% of tryptophan residues are involved in interactions of this type (64). The finding that the cationic group of acetylcholine was bound primarily by aromatic residues, most especially by a tryptophan residue, not by the ex-

pected carboxylate anion, provided evidence that cation-π interactions may play an important role in ligand binding (65, 66). Model systems suggest that, energetically, the cation-π interaction can compete with full aqueous solvation in binding cations (61), and there is now significant effort being expended in studying the contribution of these interactions to molecular recognition (62, 66).

In summary, the K_i provides an indication of the relative stability of the enzyme-inhibitor complex compared to stability of the enzyme and inhibitor free in solution. Moreover, it is clear that entropy, enthalpy, and water will all have a major impact on the binding of an inhibitor to an enzyme.

2.2 Steady-State Enzyme Kinetics

Just as an appreciation of the forces involved is essential to comprehending the binding of an inhibitor to an enzyme, so is an understanding of the kinetic analysis of an enzyme-catalyzed reaction essential to any kinetic evaluation of an inhibitor. In this section we provide a brief introduction to the study of enzyme kinetics, particularly steady-state kinetics. Regardless, the reader is advised to refer to other sources for more in-depth reviews of the kinetic equations and mathematical derivations involved (38, 60, 67–71).

2.2.1 The Michaelis-Menten Equation. In the simplest case, an enzyme-catalyzed reaction involves the conversion of a single substrate to a single product, as shown in Equation 17.8.

$$\text{E} + \text{S} \rightleftharpoons \text{E} \cdot \text{S} \rightleftharpoons \text{E} \cdot \text{P} \rightleftharpoons \text{E} + \text{P} \quad (17.8)$$

The free enzyme (E) binds the substrate (S) to form a noncovalent enzyme-inhibitor complex (E $\cdot$ S). This is assumed to be a rapid, reversible process, not involving any chemical changes, and with the affinity of the substrate for the enzyme's active site being determined by the binding forces discussed above. A chemical transformation of substrate to product (P), initially in complex with enzyme (E $\cdot$ P), then takes place. Finally, the product (P) is released into the medium with concomitant regeneration of free enzyme (E).

As can be seen from the following discussion, it is not difficult to carry out a kinetic analysis of a single-substrate reaction such as that described in Equation 17.8. However, as more substrates are added the task becomes more complex. Fortunately, kinetic analysis of enzymatic reactions involving two or more substrates can be made easier by varying the concentration of only one substrate at a time. By keeping all but one of the substrates at fixed, saturating concentrations, the reaction rate will depend only on the concentration of the varied substrate. This permits the use of the kinetic analysis employed for enzyme-catalyzed, single-substrate reactions even for complex multisubstrate reactions. In a further simplification, the dissociation of the E $\cdot$ P complex is assumed not to be rate limiting, and the reversion of product to substrate is assumed to be negligible. The latter assumption is valid under what are known as initial velocity conditions, that is, when less than about 5% of substrate has been consumed. Under these conditions, the concentration of P is low, and Equation 17.8 simplifies to Equation 17.9.

$$\text{E} + \text{S} \underset{k_{-1}}{\overset{k_1}{\rightleftharpoons}} \text{E} \cdot \text{S} \overset{k_2}{\longrightarrow} \text{E} + \text{P} \quad (17.9)$$

Generally, kinetic analyses are carried out by studying the reaction under steady-state conditions, that is, when the concentration of the enzyme is well below that of the substrate. Under those circumstances, following a brief preequilibrium period, the concentrations of the various enzyme-bound species, E $\cdot$ S and E $\cdot$ P in Equation 17.8, become effectively constant and the rate of conversion of substrate to product will greatly exceed the change in concentration of any enzyme species. This is an approximation but, provided the substrate concentration does not greatly change (e.g., under initial velocity conditions), it is a very useful approximation. Given steady-state conditions, the Michaelis-Menten equation (Equation 17.10) is a quantitative description of the reaction described by Equation 17.9.

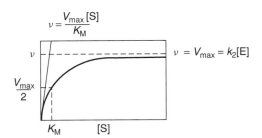

Figure 17.2. Plot showing dependency of the initial velocity (v) on substrate concentration [S] for an enzyme-catalyzed reaction obeying Michaelis-Menten (saturation) kinetics.

$$v = \frac{[S]V_{max}}{K_M + [S]}$$

$$V_{max} = k_2[E] \tag{17.10}$$

This implies that the initial velocity (v) is directly proportional to the enzyme concentration [E], and that v follows saturation kinetics with respect to the substrate concentration [S]. This is shown graphically in Figure 17.2 and explained as follows: at very low substrate concentrations v increases in a linear fashion, so that $v = V_{max}[S]/K_M$. As the substrate concentration increases, the observed increase in v is less than the increase in [S]. This trend continues until, at high (saturating) substrate concentrations, v becomes effectively independent of [S] and tends toward the limiting value V_{max}.

V_{max} is the maximal velocity that can be achieved at a specific enzyme concentration. In the simple Michaelis-Menten mechanism described by Equation 17.9, there is only one E · S complex and all binding steps are rapid. In this instance, V_{max} is the product of the enzyme concentration [E] and k_2 (also known as k_{cat}), which is the first-order rate constant for the chemical conversion of the E · S complex to free enzyme and product. The catalytic constant k_{cat} is often referred to as the turnover number because it represents the maximum number of substrate molecules converted to products per active site per unit time. In a more complicated reaction, k_{cat} is a function of all the first-order rate constants and, effectively, sets a lower limit on all the chemical rate constants.

The Michaelis-Menten constant K_M is a combination of rate constants and is independent of enzyme concentration under steady-state conditions. It is equal to the substrate concentration at which half the maximum velocity of the enzyme-catalyzed reaction is reached; that is, when [S] = K_M, then $v = \frac{1}{2}V_{max}$. For the reaction illustrated in Equation 17.9, K_M is described by Equation 17.11.

$$K_M = \frac{k_2 + k_{-1}}{k_1} \tag{17.11}$$

If, for a given reaction, $k_{-1} \gg k_2$, then Equation 17.11 simplifies to $K_M = K_S$, where K_S is the dissociation constant for the enzyme substrate complex. It is important to remember that the Michaelis-Menten equation holds true not only for the mechanism as stated above, but for many different mechanisms that are not included in this treatment. In summary, K_M can be described as an apparent dissociation constant for all enzyme-bound species and, in all cases, it is the substrate concentration at which the enzyme operates at half-maximal velocity.

Another parameter often referred to when discussing Michaelis-Menten kinetics is k_{cat}/K_M. This is an apparent second-order rate constant that relates the reaction rate to the free (not total) enzyme concentration. As described above, at very low substrate concentrations when the enzyme is predominantly unbound, the velocity (v) is equal to $[E][S]k_{cat}/K_M$. The value of k_{cat}/K_M sets a lower limit on the rate constant for the association of enzyme and substrate. It is sometimes referred to as the specificity constant because it determines the specificity of the enzyme for competing substrates.

Again, for more detailed treatment of this subject the reader should refer to more specialized texts (38, 60, 67–69).

2.2.2 Treatment of Kinetic Data. Analysis of Michaelis-Menten kinetics is greatly facilitated by a linear representation of the data. Converting the Michaelis-Menten Equation 17.10 into Equation 17.12 leads to the popular Lineweaver-Burk plot.

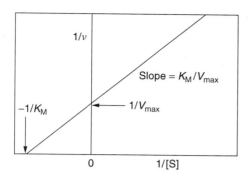

Figure 17.3. The Lineweaver-Burk plot.

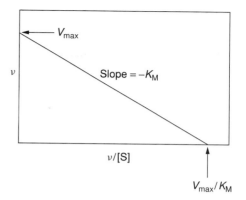

Figure 17.4. The Eadie-Hofstee plot.

$$\frac{1}{v} = \frac{1}{V_{\max}} + \frac{K_M}{V_{\max}[S]} \qquad (17.12)$$

If you plot $1/v$ against $1/[S]$ (Fig. 17.3), the y-intercept gives a value of $1/V_{\max}$ and the x-intercept gives a value of $-1/K_M$. The slope of the line is equal to $K_M/V_{\max}$. Although very popular, the Lineweaver-Burk plot suffers from the disadvantage that it emphasizes points at lower concentrations and compresses data points obtained at high concentrations (67). As a result it is not recommended for obtaining accurate kinetic constants.

A preferable, alternative form of the Michaelis-Menten equation is that of the Eadie-Hofstee plot (Equation 17.13)

$$v = V_{\max} - \frac{K_M v}{[S]} \qquad (17.13)$$

As shown in Fig. 17.4, plotting v against $v/[S]$ results in the y-intercept providing a value of $V_{\max}$, whereas the x-intercept provides $V_{\max}/K_M$, and the slope of the line is equal to $-K_M$.

Another linear representation of the Michaelis-Menten equation is the Hanes-Woolf plot (Equation 17.14).

$$\frac{[S]}{v} = \frac{1}{V_{\max}} [S] + \frac{K_M}{V_{\max}} \qquad (17.14)$$

Thus a plot of $[S]/v$ vs. $[S]$ is linear, with a slope of $1/V_{\max}$ (Fig. 17.5). The y-intercept gives $K_M/V_{\max}$ and the x-intercept gives $-K_M$.

Finally, it is possible to directly plot pairs of v, $[S]$ data in such a way as to directly determine K_M and $V_{\max}$ values. Of the linear graphical methods, the direct linear plot of Eisenthal and Cornish-Bowden (72), shown in Fig. 17.6, is often considered to provide the best estimates of K_M and $V_{\max}$ values. In this method pairs of v and $[S]$ values are obtained in the usual manner. A v value is plotted on the y-axis and a corresponding negative value of $[S]$ is plotted on the x-axis. A straight line is then drawn, passing through the points on the two axes and extending beyond the "point of intersection." This is repeated for each set of v and $[S]$ values. Thus, there are n sets of lines for n pairs of values. A horizontal line drawn from the point of intersection to the y-axis provides the $V_{\max}$ value, whereas a vertical line from the point of intersection to the x-axis provides the K_M value.

Each of these linear plots has its own merits, particularly for plotting inhibition data

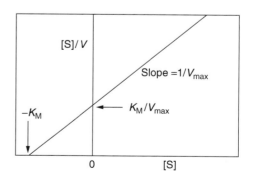

Figure 17.5. The Hanes-Woolf plot.

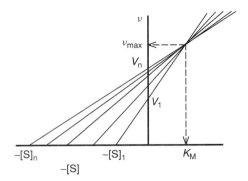

Figure 17.6. Eisenthal-Cornish-Bowden direct linear plot of enzyme kinetic data fitting the Michaelis-Menten equation.

(38), and its own drawbacks (67). However, the rapid advances in personal computing make it relatively easy to fit kinetic data to the Michaelis-Menten equation (or other appropriate hyperbolic functions) by use of a variety of commercial graphical or spreadsheet packages. One simple package, HYPER, which is readily available on the Internet (http://www.ibiblio.org/pub/academic/biology/molbio/ibmpc/hyper102.zip), simultaneously furnishes Michaelis-Menten parameters obtained using hyperbolic regression analysis, as well as those obtained using three of the plots described here. As such, it provides a rapid contrast of these graphical methods but, unfortunately, is not suitable for the study of inhibition kinetics. In addition, the recent monograph by Copeland (71) provides a list of useful computer software and Internet sites for the study of enzymes.

2.3 Rapid, Reversible Inhibitors

This class of inhibitors acts by binding to the target enzyme's active site in a rapid, reversible, and noncovalent fashion. The net result is that the active site is blocked and the substrate is prevented from binding. Accordingly, in designing inhibitors of this type, optimization of the noncovalent binding forces between the inhibitor and the active site of the enzyme is of paramount importance.

2.3.1 Types of Rapid, Reversible Inhibitors.
Binding of these inhibitors follows simple

Michaelis-Menten kinetics and, depending on their preference of binding to the free enzyme and/or the enzyme-substrate complex, competitive, uncompetitive, and noncompetitive inhibition patterns can be distinguished. For the purposes of this discussion it will be assumed that the initial equilibrium of free and bound substrate is established significantly faster than the rate of the chemical transformation of substrate to product, that is, $k_1, k_{-1} \gg k_2$ (Equation 17.9). As discussed in section 2.2.1, this reduces K_M to the dissociation constant K_S of the $E \cdot S$ complex.

2.3.1.1 Competitive Inhibitors. A competitive inhibitor often has structural features similar to those of the substrates whose reactions they inhibit. This means that a competitive inhibitor and enzyme's substrate are in direct competition for the same binding site on the enzyme. Consequently, binding of the substrate and the inhibitor are mutually exclusive. A kinetic scheme for competitive inhibition is shown in Equation 17.15.

$$
\begin{array}{c}
\overset{\text{S}}{E \rightleftharpoons E \cdot S} \rightarrow E + P \\
\Big\Uparrow I \qquad\qquad\qquad\qquad (17.15) \\
E \cdot I
\end{array}
$$

The enzyme-bound inhibitor may either lack an appropriate functional group for further reaction, or may be bound in the wrong position with respect to the catalytic residues or to other substrates. In any event, the enzyme-inhibitor complex $E \cdot I$ is unreactive (it is sometimes referred to as a dead-end complex) and the inhibitor must dissociate and substrate bind before reaction can take place.

Solving this kinetic scheme for simple Michaelis-Menten kinetics leads to Equation 17.16.

$$
v = \frac{[S]V_{max}}{[S] + K_M\left(1 + \dfrac{[I]}{K_i}\right)} \qquad (17.16)
$$

Here, K_i, sometimes called the inhibition constant, is the equilibrium constant for the dissociation of the enzyme-inhibitor complex, and is described by Equation 17.17.

$$K_i = \frac{[\mathrm{E}][\mathrm{I}]}{[\mathrm{E}\cdot\mathrm{I}]} \qquad (17.17)$$

Competitive inhibitors do *not* change the value of V_{max}, which is reached when sufficiently high concentrations of the substrate are present so as to completely displace the inhibitor. However, the affinity of the substrate for the enzyme *appears* to be decreased in the presence of a competitive inhibitor. This happens because the free enzyme E is not only in equilibrium with the enzyme-substrate complex $\mathrm{E}\cdot\mathrm{S}$, but also with the enzyme-inhibitor complex $\mathrm{E}\cdot\mathrm{I}$. Competitive inhibitors increase the *apparent* K_{M} of the substrate by a factor of $(1 + [\mathrm{I}]/K_i)$. The evaluation of the kinetics is again greatly facilitated by the conversion of Equation 17.15 into a linear form using Lineweaver-Burk, Eadie-Hofstee, or Hanes-Woolf plots, as shown in Fig. 17.7.

2.3.1.2 Uncompetitive Inhibitors. Uncompetitive inhibitors do not bind to the free enzyme. They bind only to the enzyme-substrate complex to yield an inactive $\mathrm{E}\cdot\mathrm{S}\cdot\mathrm{I}$ complex (Equation 17.18).

$$E \underset{}{\overset{\mathrm{S}}{\rightleftharpoons}} \mathrm{E}\cdot\mathrm{S} \rightarrow \mathrm{E} + \mathrm{P}$$
$$\big\Updownarrow\, \mathrm{I} \qquad\qquad (17.18)$$
$$\mathrm{E}\cdot\mathrm{S}\cdot\mathrm{I}$$

Uncompetitive inhibition is rarely observed in single-substrate reactions but is frequently observed in multisubstrate reactions. An uncompetitive inhibitor can provide information about the order of binding of the different substrates. In a bisubstrate-catalyzed reaction, for example, a given inhibitor may be competitive with respect to one of the two substrates and uncompetitive with respect to the other. The linear plots for classical uncompetitive inhibition patterns are described by Equation 17.19 and are illustrated in Fig. 17.8.

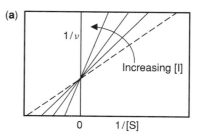

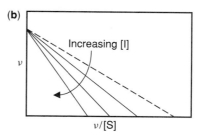

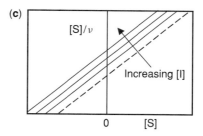

Figure 17.7. (a) Lineweaver-Burk, (b) Eadie-Hofstee, and (c) Hanes-Woolf plots exhibiting competitive inhibition patterns. The dashed line indicates the reaction in the absence of inhibitor, whereas the solid lines represent enzymatic reactions in the presence of increasing concentrations of inhibitor.

$$v = \frac{\dfrac{[\mathrm{S}]V_{\mathrm{max}}}{1 + [\mathrm{I}]/K_i}}{[\mathrm{S}] + \dfrac{K_{\mathrm{M}}}{1 + [\mathrm{I}]/K_i}} \qquad (17.19)$$

As with a competitive inhibitor, the apparent K_{M} for the substrate decreases by a factor of $(1 + [\mathrm{I}]/K_i)$ because the formation of $\mathrm{E}\cdot\mathrm{S}\cdot\mathrm{I}$ will use up some of the $\mathrm{E}\cdot\mathrm{S}$, thereby shifting the equilibrium further in favor of $\mathrm{E}\cdot\mathrm{S}$ formation. However, uncompetitive inhibitors also decrease V_{max} by the same factor because

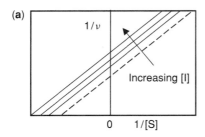

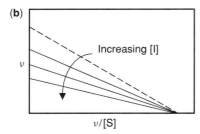

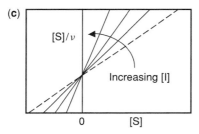

Figure 17.8. (a) Lineweaver-Burk, (b) Eadie-Hofstee, and (c) Hanes-Woolf plots exhibiting uncompetitive inhibition patterns. The dashed line indicates the reaction in the absence of inhibitor, whereas the solid lines represent enzymatic reactions in the presence of increasing concentrations of inhibitor.

some of the enzyme remains in the $E \cdot S \cdot I$ form, even at infinite substrate concentration.

2.3.1.3 Noncompetitive Inhibitors. Classical noncompetitive inhibitors have no effect on substrate binding and vice versa, given that they bind randomly and reversibly to different sites on the enzyme. They also bind with the same affinity to the free enzyme and to the enzyme-substrate complex. Both the enzyme-inhibitor complex $E \cdot I$ and the enzyme-substrate-inhibitor complex $E \cdot S \cdot I$ are catalytically inactive. The equilibria are outlined in Equation 17.20.

$$E \underset{}{\overset{S}{\rightleftharpoons}} E \cdot S \longrightarrow E + P$$

Equation (17.20) reaction scheme:

$$
\begin{array}{ccc}
E & \overset{S}{\rightleftharpoons} & E \cdot S \longrightarrow E + P \\
\Updownarrow I & & \Updownarrow I \\
E \cdot I & \overset{S}{\rightleftharpoons} & E \cdot S \cdot I
\end{array}
\tag{17.20}
$$

Simple Michaelis-Menten kinetics of noncompetitive inhibitors are described in Equation 17.21.

$$v = \frac{\dfrac{[S]V_{\max}}{1 + [I]/K_i}}{[S] + K_M} \tag{17.21}$$

From Equation 17.21 it is clear that noncompetitive inhibitors have an effect only on $V_{\max}$, decreasing it by a factor of $(1 + [I]/K_i)$, consequently giving the impression of reducing the total amount of enzyme present. As with an uncompetitive inhibitor, a portion of the enzyme will always be bound in the nonproductive enzyme-substrate-inhibitor complex $E \cdot S \cdot I$, causing a decrease in maximum velocity, even at infinite substrate concentrations. However, because noncompetitive inhibitors do not affect substrate binding, the K_M value of the substrate remains unchanged. Linear plots for noncompetitive inhibition are shown in Fig. 17.9.

Again, this type of inhibition is rarely seen in single-substrate reactions. It should also be noted that, frequently, the affinity of the noncompetitive inhibitor for the free enzyme, and the enzyme-substrate complex, are different. These nonideally behaving noncompetitive inhibitors are called mixed-type inhibitors, and they alter not only $V_{\max}$ but also K_M for the substrate. Further discussion of inhibitors of this type may be found in Segel (38).

Sometimes steady-state kinetics are insufficient to analyze the mechanism of inactivation for a given inhibitor. For example, irreversible enzyme inhibitors that bind so tightly to the enzyme that their dissociation rate (k_{off}) is effectively zero also exhibit noncompetitive inhibition patterns. They act by destroying a portion of the enzyme through irreversible binding, thereby lowering the overall enzyme concentration and decreasing $V_{\max}$. The apparent K_M remains unaffected because irre-

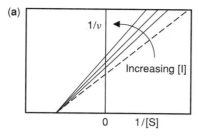

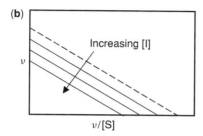

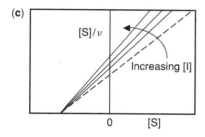

Figure 17.9. (a) Lineweaver-Burk, (b) Eadie-Hofstee, and (c) Hanes-Woolf plots exhibiting noncompetitive inhibition patterns. The dashed line indicates the reaction in the absence of inhibitor, whereas the solid lines represent enzymatic reactions in the presence of increasing concentrations of inhibitor.

versible inhibitors do not influence the dissociation constant of the enzyme-substrate complex. A simple experiment to distinguish between a reversible noncompetitive inhibitor and irreversible inhibitor is shown in Fig. 17.10, and a comprehensive review describing the kinetic evaluation of irreversibly binding enzyme inhibitors is available (73). Allosteric effectors may also show noncompetitive kinetic patterns by rendering the enzyme in the $E \cdot S \cdot I$ complex less active than that in the $E \cdot S$ complex. Again, additional analyses are often required in these less well defined

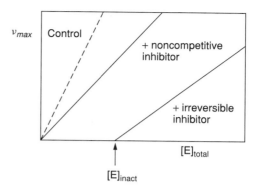

Figure 17.10. Plot showing dependency of V_{max} on the total enzyme concentration, $[E]_{total}$. An irreversible inhibitor will titrate a fraction of the enzyme $[E]_{inact}$.

situations. Such analyses may include more in-depth steady-state kinetics, as well as pre-steady-state kinetics, and testing for irreversible inhibition. Irreversible covalently binding enzyme inhibitors are discussed extensively later in this chapter.

2.3.2 Dixon Plots. Another linear method for plotting inhibition data, the Dixon plot, is shown in Fig. 17.11 (74). In this method the initial velocity is measured as a function of inhibitor concentration at two or more fixed substrate concentrations. By plotting $1/v$ against $[I]$ for each substrate concentration, the different types of inhibition can easily be distinguished. Further, in cases of competitive or noncompetitive inhibition, the value of K_i may be determined from the x-axis value at which the lines intercept. Overall, the Dixon plot is probably the simplest and most rapid graphical method for obtaining a K_i value.

2.3.3 IC$_{50}$ Values. The potencies of enzyme inhibitors evaluated using rapid screening techniques are often reported in terms of IC$_{50}$ values rather than K_i values. An IC$_{50}$ value is the inhibitor concentration that is required to halve the activity of the enzyme, that is, that concentration that leads to 50% enzyme inactivation. It is important to recognize that an IC$_{50}$ value is not a constant, except in the case of noncompetitive inhibition, and is dependent on the substrate concentration used in the experiment. IC$_{50}$ values are com-

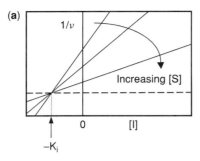

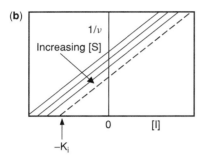

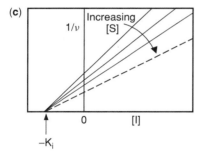

Figure 17.11. Dixon plots for (a) competitive, (b) uncompetitive, and (c) noncompetitive inhibitors. The solid lines represent enzymatic reactions in the presence of increasing concentrations of substrate. The dashed line represents the reaction at infinite substrate concentration.

monly determined by keeping the concentration of the substrate and the enzyme constant and incrementally varying the concentration of the inhibitor. This simple experimental approach makes it relatively easy to screen large numbers of potential inhibitors. Industrial high-throughput screens often employ half-log increments, and the value of IC_{50} provides a ready means of comparing the extent of in-

hibition. It should also be noted that the IC_{50} value can be no less than half the concentration of the enzyme, a factor that becomes important if the inhibitor is very potent or if high concentrations of enzyme are employed.

For a competitive inhibitor, a K_i value may be obtained using the relationship described by Equation 17.22.

$$IC_{50} = K_i\left(1 + \frac{[S]}{K_M}\right) \qquad (17.22)$$

Provided that a reasonable substrate concentration ($\leq 0.1\ K_M$) is employed for the experiment, the IC_{50} value may be a reasonable approximation of the true K_i. Equation 17.22 indicates that substrate concentrations greater than about 0.1-fold of the K_M value will lead to an underestimation of the K_i value, an underestimation that becomes quite significant at high substrate concentrations.

The dependency of the IC_{50} value on the substrate concentration for uncompetitive inhibitors is given in Equation 17.23.

$$IC_{50} = K_i\left(1 + \frac{K_M}{[S]}\right) \qquad (17.23)$$

In this instance it is at high concentrations of the substrate that the K_i value is comparable to the IC_{50} value, and a significant underestimation will occur at lower substrate concentrations.

From these two equations it is clear that, for preliminary screening when the type of inhibition is unknown, substrate concentrations close to the K_M value should be used. This minimizes the deviation of the IC_{50} value from the K_i value to, in the cases of competitive and uncompetitive inhibitors, a factor of 2. If necessary, a Dixon plot can be used to provide a quick indication of the K_i and the type of inhibition (38, 74). It should be noted that the relationship between IC_{50} and K_i requires the initial velocity to be linearly dependent on the concentration of inhibitor. In the cases of mixed-competitive and irreversible inhibitors, the dependency of the inhibitor concentration and the initial velocity is nonlinear. Therefore, in those cases, the use of the IC_{50} value is limited.

2.3.4 Examples of Rapid Reversible Inhibitors. Competitive inhibitors are often similar in structure to one of the substrates of the reaction they are inhibiting. Inhibitors of this type are sometimes called substrate analogs and their binding affinity (K_i) usually approximates that of the substrate. One of the first reactions inhibited by a substrate analog was that catalyzed by succinate dehydrogenase (Equation 17.24).

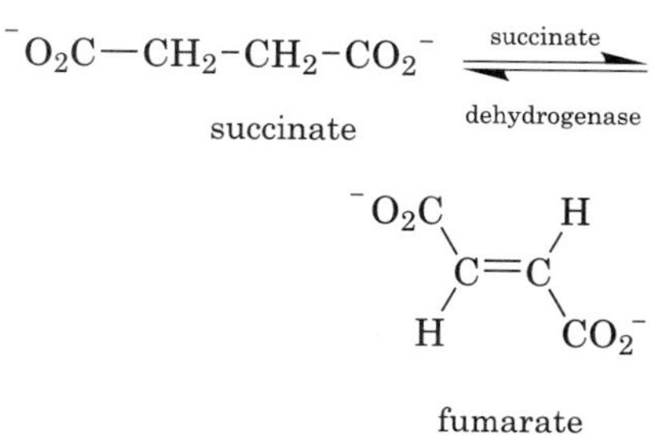

$$^-O_2C-CH_2-CH_2-CO_2^- \xrightleftharpoons[\text{dehydrogenase}]{\text{succinate}}$$

succinate

$$(17.24)$$

fumarate

This reaction is competitively inhibited by malonate ($^-OOCCH_2COO^-$) that has, like succinate, two carboxylate groups. It is therefore able to bind to the enzyme's active site but, with only one carbon atom between the carboxylates, further reaction is impossible.

Substrate analogs are rarely useful as enzyme inhibitors, given that large concentrations are required for inhibition, and their inhibition is readily overcome by any buildup of substrate. However, they are often useful probes for determining enzyme specificity and even mechanism. Phenylethanolamine N-methyltransferase (PNMT) catalyzes the terminal step in epinephrine (adrenaline) biosynthesis, the conversion of norepinephrine to epinephrine (Equation 17.25), with concomitant conversion of S-adenosyl-L-methionine (SAM, AdoMet) to S-adenosyl-L-homocysteine (SAH, AdoHcy).

S-Adenosyl-L-homocysteine (**10**) (Fig. 17.12), the product of the reaction, and 2-(2,5-dichlorophenyl)cyclopropylamine (**11**) are analogs of S-adenosyl-L-methionine and norepinephrine, respectively. Using these inhibitors it was possible to ascertain the binding order of the two substrates (75). Kinetic analyses showed that SAH was a competitive inhibitor of SAM and a noncompetitive inhibitor of norepinephrine, whereas (**11**) was a competitive inhibitor of norepinephrine and an uncompetitive inhibitor of SAM. This indicates that the binding of substrates is ordered, with SAM binding first. If norepinephrine bound first, it would be expected that SAH would be an uncompetitive inhibitor and (**11**) would be noncompetitive with respect to SAM. If a random

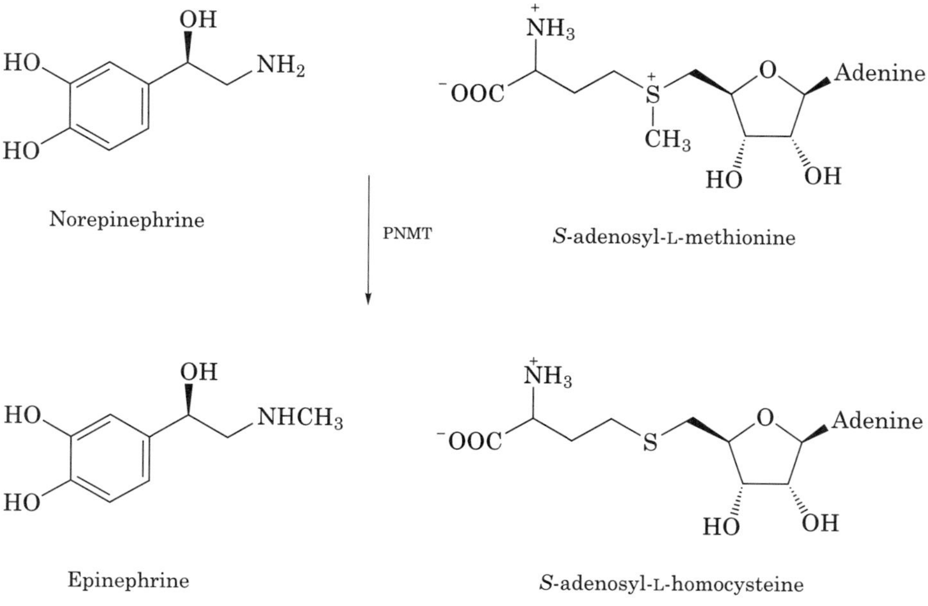

Norepinephrine

S-adenosyl-L-methionine

$$(17.25)$$

Epinephrine

S-adenosyl-L-homocysteine

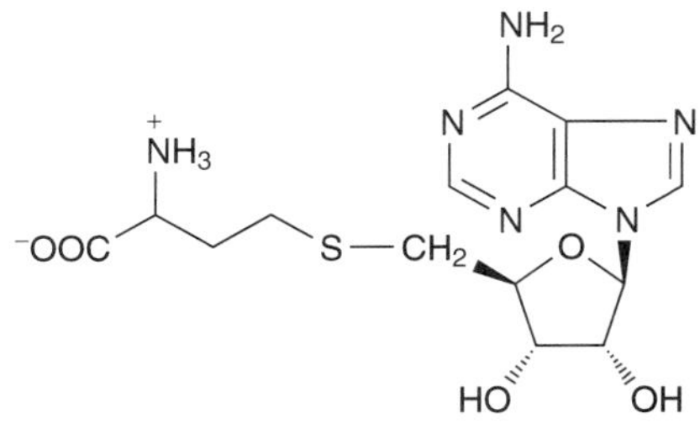

(10)

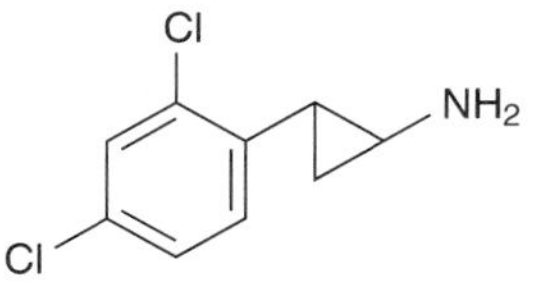

(11)

Figure 17.12. Inhibitors of phenylethanolamine *N*-methyltransferase.

binding mechanism were in operation, it would be expected that both inhibitors would be competitive with either substrate. More detail on similar uses of reversible inhibitors may be found elsewhere (76).

2.4 Slow-, Tight-, and Slow-Tight–Binding Inhibitors

Not all reversible inhibitors have an instantaneous effect on the rate of an enzymatic reaction. Some inhibitors, known as slow-binding enzyme inhibitors, can take a considerable time to establish the equilibrium between the free enzyme and inhibitor, and the enzyme-inhibitor complex. This time period may be on the scale of seconds, minutes, or even longer. The enzyme-inhibitor complexes have slow off (dissociation) rates, but the on (association) rates may be either slow or fast. Hence, the term slow binding does not necessarily indicate a slow binding of inhibitor to enzyme but rather the fact that reaching equilibrium is a

slow process. Other inhibitors, known as tight-binding inhibitors, bind their target enzyme with such high affinity that the population of free inhibitor molecules is significantly depleted when the enzyme-inhibitor complex is formed. Often, tight-binding inhibitors also have a slow onset of action, and are termed slow-tight-binding inhibitors. What these three types of inhibitors have in common is that, generally, the major assumptions of Michaelis-Menten kinetics do not hold true.

As with rapid reversible inhibition, for slow-binding inhibition to take place a significantly larger concentration of inhibitor than enzyme is required. However, reaching equilibrium slowly is incompatible with the assumption of Michaelis-Menten kinetics that inhibitors bind much more quickly than the enzyme turns over. Unlike rapid reversible and slow-binding inhibitors, both tight-binding and slow-tight-binding inhibitors are effective at concentrations comparable to that of the enzyme. At that point, the inhibitor concentration is no longer independent of the enzyme concentration, as assumed for Michaelis-Menten kinetics. A summary of the properties of reversible enzyme inhibitors is shown in Table 17.5. Although we give a brief overview of these types of inhibitors, excellent and more in-depth descriptions of slow-, tight-, and slow-tight-binding inhibitors have appeared elsewhere (71, 77–80).

2.4.1 Slow-Binding Inhibitors. Two different mechanisms have been suggested to rationalize the slow-binding behavior of competitive inhibitors (71, 78, 80). In the one-step mechanism A, the direct binding process of the inhibitor to the enzyme is slow (Equation 17.26); that is, the magnitude of $k_3[I]$ is small relative to $k_1[S]$ and k_2, the rate constants for the conversion of substrate to product.

Table 17.5 Classes of Reversible Inhibitors

Inhibitor Class	Ratio of Inhibitor to Enzyme Necessary for Inhibition	Rate at Which Equilibrium is Attained $E + I \rightleftharpoons E \cdot I$
Rapid, reversible	$I \gg E$	Fast
Tight binding	$I \approx E$	Fast
Slow binding	$I \gg E$	Slow
Slow-tight binding	$I \approx E$	Slow

$$E \underset{k_{-1}}{\overset{S,k_1}{\rightleftharpoons}} E \cdot S \overset{k_2}{\longrightarrow} E + P$$

$$k_{-3} \updownarrow I,k_3 \qquad\qquad (17.26)$$

$$E \cdot I$$

The slow on rate (k_3) has been attributed to the inhibitor encountering some barrier to binding at the active site. The inhibitor has to overcome this barrier by correct alignment. Once aligned properly, it binds so tightly that it is released very slowly from the enzyme, making the overall equilibrium process extremely slow. The equilibrium dissociation constant for the E $\cdot$ I complex K_i, derived from Equation 17.26, is given by Equation 17.27.

$$K_i = \frac{[E][I]}{[E \cdot I]} = \frac{k_{-3}}{k_3} \qquad (17.27)$$

This is the same equilibrium as that for a rapid reversible inhibitor (Equation 17.17). From Equation 17.27, it should be noted that, if K_i is very small (as with a tight-binding inhibitor) and [I] is varied in the region of K_i, even if the on rate (k_3) is diffusion controlled, both $k_3[I]$ and k_{-3} will be very small. Thus, the onset of inhibition for a tight-binding inhibitor can appear to be slow, even though k_3 is in the range expected for rapid reversible inhibitors (78). It is possible to carry out kinetic analyses of tight-binding inhibitors. This can be done either by including a preincubation step, to allow sufficient time for the enzyme and inhibitor to reach equilibrium, or by carrying out the reaction at very high concentrations of both substrate and inhibitor. More detailed discussion of these methods, with appropriate references, can be found in a recent volume by Copeland (71).

If the slow-binding inhibitor described by Equation 17.26 also binds very tightly, it is referred to as a slow-tight-binding inhibitor. For inhibitors of this type, K_i is given by Equation 17.28, where $[E_T]$ represents the total enzyme concentration (in all forms) present in solution.

$$K_i = \frac{([E_T] - [E \cdot I])([I] - [E \cdot I])}{[E \cdot I]} \qquad (17.28)$$

For slow-tight-binding inhibitors, k_{-3} is very small and formation of the E $\cdot$ I complex is essentially irreversible. Use of Equation 17.28 ensures that depletion of free enzyme and free inhibitor by formation of the E $\cdot$ I complex is taken into account.

In mechanism B, the more common mechanism for slow-binding inhibition (80), the initial equilibrium between the enzyme, inhibitor, and the E $\cdot$ I complex is fast. However, there is a subsequent slow rearrangement to form the final, more stable enzyme-inhibitor complex (E $\cdot$ I*) (Equation 17.29).

$$E \underset{k_{-1}}{\overset{S,k_1}{\rightleftharpoons}} E \cdot S \overset{k_2}{\longrightarrow} E + P$$

$$k_{-3} \updownarrow I,k_3 \qquad\qquad (17.29)$$

$$E \cdot I \underset{k_{-4}}{\overset{k_4}{\rightleftharpoons}} E \cdot I^*$$

Here the dissociation constant for the initial E $\cdot$ I complex is still k_{-3}/k_3, but there is also a dissociation constant for the formation of the E $\cdot$ I* complex. The second dissociation constant is given by Equation 17.30.

$$K_i^* = \frac{K_i k_{-4}}{k_4 + k_{-4}} = \frac{[E][I]}{[E \cdot I] + [E \cdot I^*]} \qquad (17.30)$$

To observe the slow onset of inhibition and the E $\cdot$ I complex, K_i^* must be smaller than K_i and k_{-4} smaller than k_4. However, if k_{-4} is considerably smaller than k_4, then the formation of the E $\cdot$ I* complex will be effectively irreversible (i.e., the inhibitor is of the slow-tight-binding variety). Under those circumstances it will again be necessary to take depletion of free enzyme and free inhibitor into account when determining K_i and K_i^* (78).

The slow rearrangement step has been correlated with conformational changes of the enzyme following initial binding of the inhibitor. It is possible that the enzyme in its transition state conformation may be better equipped to

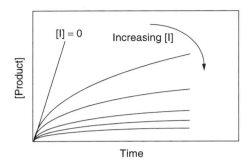

Figure 17.13. Reaction progress curves in the presence of increasing concentrations of a slow-binding inhibitor.

accommodate the inhibitor. A slow change to reach this optimal conformation will lead to tighter binding of the inhibitor and even slower release from the enzyme. An alternative suggestion is that the slow-binding process is linked to a requisite displacement of water molecules from the active site (81). Initially the inhibitor binds loosely to the enzyme, but upon release of water molecules the gain in entropy leads to a more stable $E \cdot I^*$ complex.

One way of quickly identifying a potential slow-binding inhibitor is to examine the progress of the reaction at increasing concentrations of inhibitor. Under initial velocity conditions (Section 2.2.1), an enzyme-catalyzed reaction will exhibit a linear increase in the amount of product formed over time. A reaction progress plot for a reaction carried out in the presence of a rapid reversible inhibitor will also be linear. However, a slow-binding inhibitor will initially show a linear relationship, although this will change as the inhibitor binds, resulting in a biphasic plot. Typical biphasic progress curves for a reaction in the presence of increasing concentrations of a slow-binding inhibitor are shown in Fig. 17.13. The initial burst of the reaction, the linear section of the graphs, can be described by competitive Michaelis-Menten kinetics. The higher the concentration of the inhibitor, the shorter the initial linear section of each curve and the slower the subsequent final steady-state rate, as observed in the asymptotes in Fig. 17.13. If the inhibitor concentration is small, the substrate might be too depleted to permit observation of steady-state rates.

A good comparison of rapid reversible and slow-binding inhibition can be found in a recent study on the inhibition of arginase, an enzyme that catalyzes the hydrolysis of L-arginine to yield L-ornithine and urea (Equation 17.31).

$$\text{L-arginine} \xrightarrow{\text{arginase}}$$

L-arginine

$$\text{L-ornithine} \qquad (17.31)$$

L-ornithine

$$+ \; H_2N \overset{O}{\underset{}{\big\Vert}} NH_2$$

Urea

Arginase competes with nitric oxide synthase (NOS) for arginine and, in doing so, helps regulate NOS. As a consequence, inhibitors of arginase may have therapeutic use in treating NO-dependent smooth muscle disorders, including erectile dysfunction (82). A series of arginine analogs were prepared and tested as inhibitors of arginase (83). Three examples are shown in Fig. 17.14. One of these, N^{ω}-hydroxy-L-arginine (**12**), is a competitive inhibitor of arginase at both pH 7.5 and pH 9.5 with K_i values of 2 and 1.6 μM, respectively. The two boronic acid derivatives, 2(S)-amino-6-boronohexanoic acid (**13**) and S-(2-borono-ethyl)-L-cysteine (**14**), were also competitive inhibitors at pH 7.5 with K_i values of 0.25 and 0.31 μM, respectively. However, at pH 9.5, the boronic acid derivatives both became slow-binding inhibitors, apparently binding by mechanism B and with lowered K_i values of 8.5 and 30 nM, respectively. It was suggested that, at low pH, the trigonal form of the boronic acid derivative predominates, and that

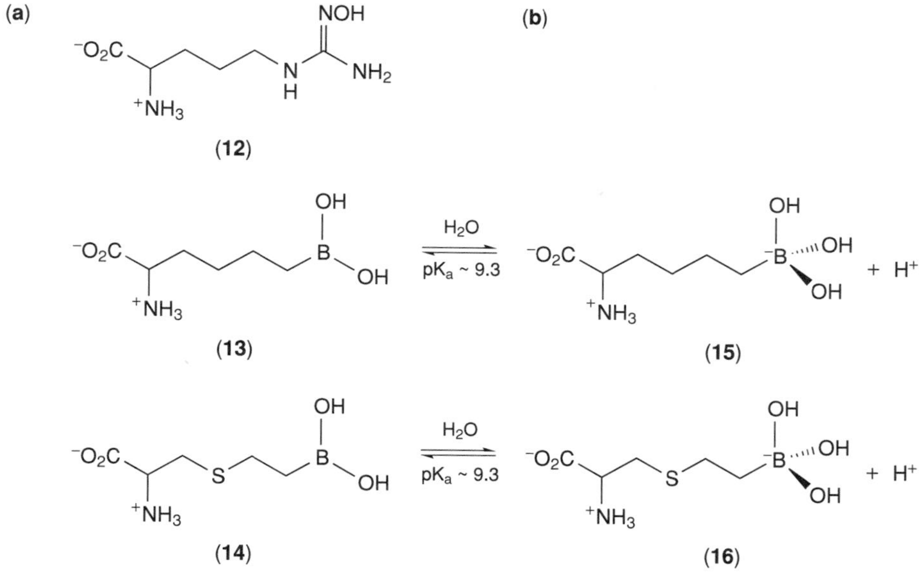

(a)

(b)

Figure 17.14. (a) Competitive and (b) slow-binding inhibitors of arginase.

this species binds with one hydroxyl, coordinating to one of the two requisite manganese ions. At pH 9.5 the tetrahedral species is the major form and this initially binds also with one hydroxyl coordinated to a manganese ion. Then, in a second, slower step, a water molecule that bridges the two active-site manganese ions is displaced by a second hydroxyl group on the boronic acid (83). Support for this mechanism is provided by crystal structures, showing both (**15**) and (**16**) are bound in

the active site of arginase as tetrahedral species at alkaline pH (82). Of course, compound (**12**) is unable to form the tetrahedral species and is a competitive inhibitor at all times.

Leucine aminopeptidase (LAP) is a metalloenzyme that has been inhibited in a slow-binding manner. This exopeptidase catalyzes the hydrolysis of N-terminal amino acids, particularly those with a leucine at the N-terminus, although it does have a broad specificity (Equation 17.32).

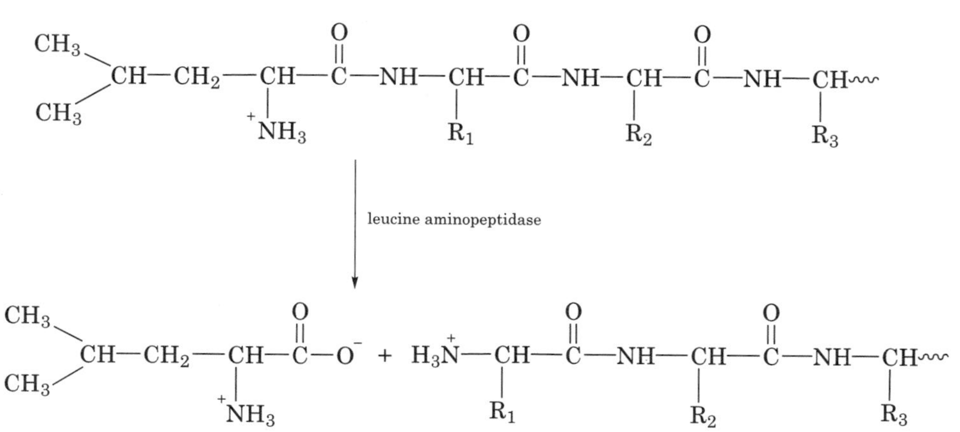

(17.32)

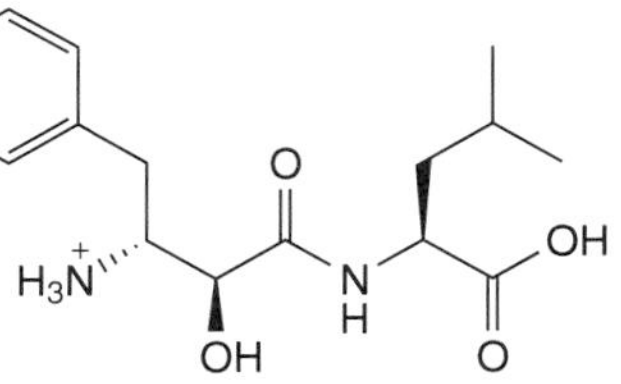

(17) 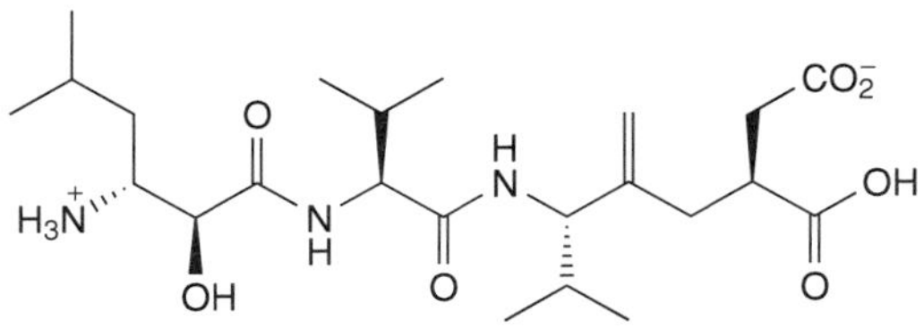   (18)

Figure 17.15. Slow-tight-binding inhibitors of leucine aminopeptidase.

Bestatin (**17**) (Fig. 17.15) and amastatin (**18**) have been identified as slow-tight-binding inhibitors of LAP from porcine kidney, with K_i values in the low nanomolar range (84). Later, bestatin was shown to be a slow-binding inhibitor of LAP employing mechanism B, with a K_i value of 0.11 μM and a K_i^* value of 1.3 nM. Values of 1.5×10^{-2} s^{-1} and 2×10^{-4} s^{-1} were obtained for k_4 and k_{-4} (Equation 17.29), respectively (85). It was assumed that the inhibition of bovine lens leucine aminopeptidase (blLAP) by amastatin would also proceed by mechanism B. This prediction was supported by an X-ray crystallography study of the amastatin-blLAP complex (86), which suggested that (**18**) (and, by analogy, **17**) initially binds to a Zn^{2+} atom in a groove in the active site. The slow step in binding was seen as a subsequent coordination to a second Zn^{2+} atom located deeper in the active site (86).

It is difficult to find clear-cut examples of slow-binding inhibition occurring by mechanism A. However, the inhibition of Factor Xa by a peptidyl-α-ketothiazole was found to be unusual because it appeared that the formation of E · I was partially rate limiting. Factor Xa is a trypsinlike protease found in the blood coagulation pathway, which cleaves prothrombin forming thrombin that, in turn, promotes blood clotting (Equation 17.33).

Inhibitors of Factor Xa activity offer potential as anticoagulants and several irreversible inhibitors of Factor Xa have been developed. One of the few tight-binding reversible inhibitors of Factor Xa is BnSO$_2$-D-Arg-Gly-Arg-ketothiazole (**19**).

The inhibitor could be displaced from Factor Xa by substrates and, based on steady-state assumptions, the dissociation constant for (**19**) was found to be 14 pM (87). However, the reaction progress curves indicated a slow-binding process, probably by mechanism B. Stopped-flow fluorescence studies, combined with kinetic analysis, showed that the isomerization step (E · I → E · I*) is unusually fast and that the formation of E · I is, at least, partially rate limiting.

In some instances the type of inhibition has been found to be isozyme specific. For example, inducibly expressed isozymes (iNOS) and constitutively expressed isozymes (cNOS) of nitric oxide synthase (NOS) all catalyze the conversion of L-arginine to L-citrulline and nitric oxide (Equation 17.34).

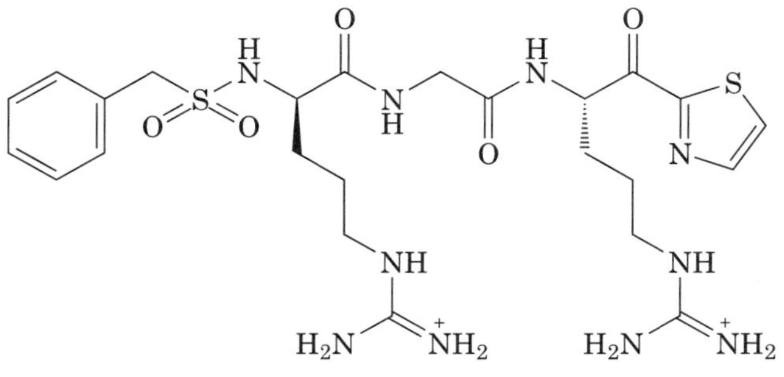

(19)

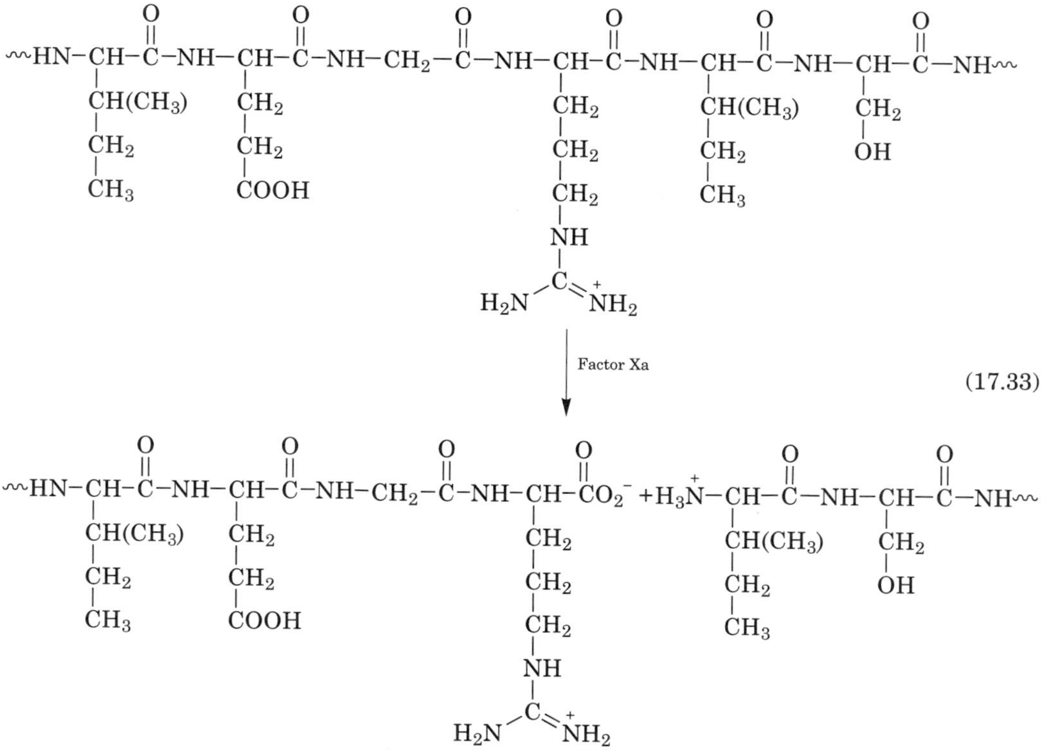

$$\text{Factor Xa}$$

$$(17.33)$$

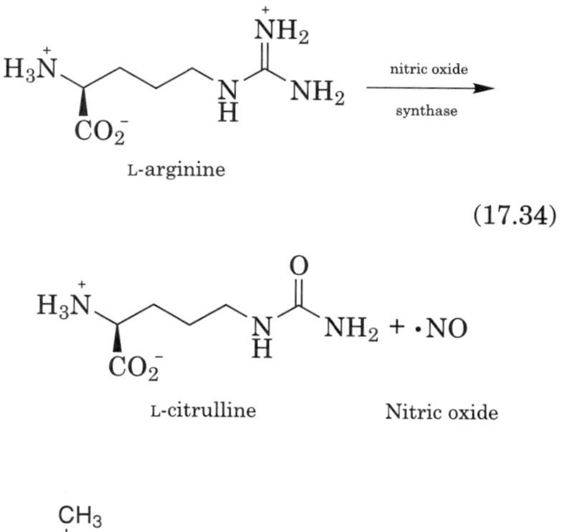

L-arginine

nitric oxide synthase

$$(17.34)$$

L-citrulline   Nitric oxide

The inhibition of human iNOS by N-(3-(aminomethyl)benzyl)acetamidine (**20**) (Fig. 17.16) was found to proceed by mechanism B, with an overall K_d of <7 nM. Conversely, inhibition of constitutive isoforms of the human enzyme was found to be rapidly reversible, with K_i values in the micromolar range (88). This is in contrast to results obtained for the arginine analog, L-N^G-nitroarginine (**21**), which was found to be a rapid reversible inhibitor of mouse macrophage iNOS, with a K_i of 4.4 μM, and a slow-binding inhibitor of brain cNOS with a K_d (assuming mechanism A) of 15 nM (89).

Many more examples of these types of inhibitors can be found in the review by Morrison and Walsh (78).

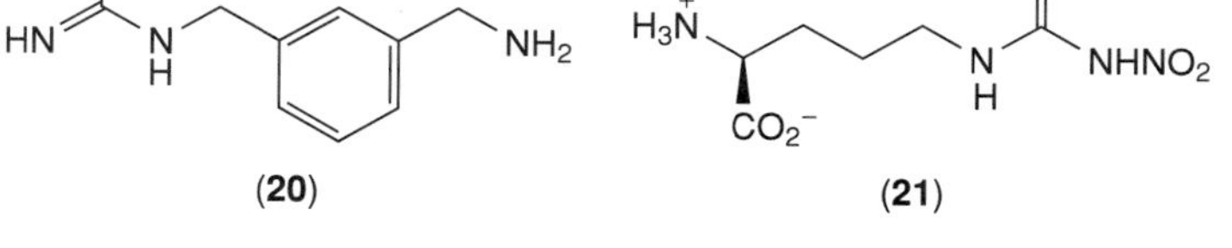

(20)   (21)

Figure 17.16. Inhibitors of nitric oxide synthase.

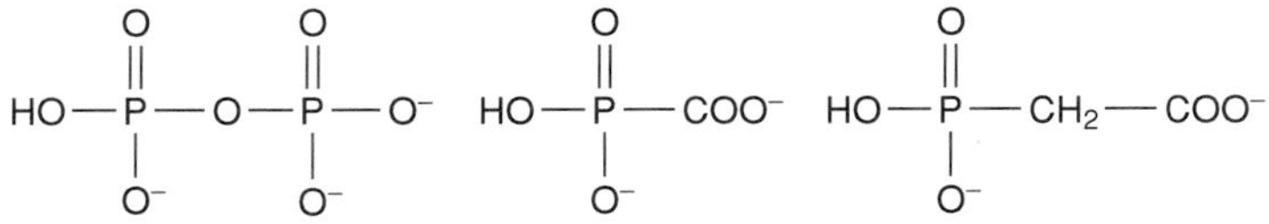

$$
\underset{\text{Pyrophosphate (PP}_i)}{\text{HO}-\overset{\overset{\displaystyle O}{\|}}{\underset{\underset{\displaystyle O^-}{|}}{P}}-O-\overset{\overset{\displaystyle O}{\|}}{\underset{\underset{\displaystyle O^-}{|}}{P}}-O^-} \qquad
\underset{(\mathbf{22})}{\text{HO}-\overset{\overset{\displaystyle O}{\|}}{\underset{\underset{\displaystyle O^-}{|}}{P}}-COO^-} \qquad
\underset{(\mathbf{23})}{\text{HO}-\overset{\overset{\displaystyle O}{\|}}{\underset{\underset{\displaystyle O^-}{|}}{P}}-CH_2-COO^-}
$$

Figure 17.17. Pyrophosphate analogs used to inhibit DNA polymerase.

2.5 Inhibitors Classified on the Basis of Structure/Mechanism

As with any reaction, an enzyme-catalyzed reaction must proceed from the ground state through a transition state before products are formed. In addition, there are often some high-energy intermediates along the pathway. Knowledge and understanding of an enzyme's mechanism permits the identification of the high-energy intermediates and the prediction of the structures of the transition states. Armed with that knowledge, it is possible to design enzyme inhibitors based on the structures of the various intermediates along the reaction pathway. Inhibitors designed in this manner are occasionally referred to as mechanism-based inhibitors. However, for purposes of this chapter, we will reserve that term for the covalently binding inhibitors described in Section 3.

2.5.1 Ground-State Analogs.
The ground state of an enzymatic reaction consists of the substrates and the products. Compounds that mimic the substrate of an enzymatic reaction have been examined earlier (Section 2.3) and are not discussed again here. There are many examples of enzymatic reactions that are inhibited by some or all of the reaction products. Both epinephrine and S-adenosyl-L-homocysteine, for example, are inhibitors of phenylethanolamine N-methyltransferase (Equation 17.35). In much the same way as described earlier for substrate analogs, product analogs can also be used to obtain information about the binding mechanism of enzymes (90).

Phosphonoformate (**22**) (Fig. 17.17) is an antiviral agent that is used clinically in the treatment of herpes simplex virus (HSV) and human cytomegalovirus (HCMV) (91). It acts as a product analog, blocking the pyrophosphate-binding site, in the reaction catalyzed by DNA polymerase (Equation 17.35). It is also effective, using the same mechanism, against HIV reverse transcriptase (91).

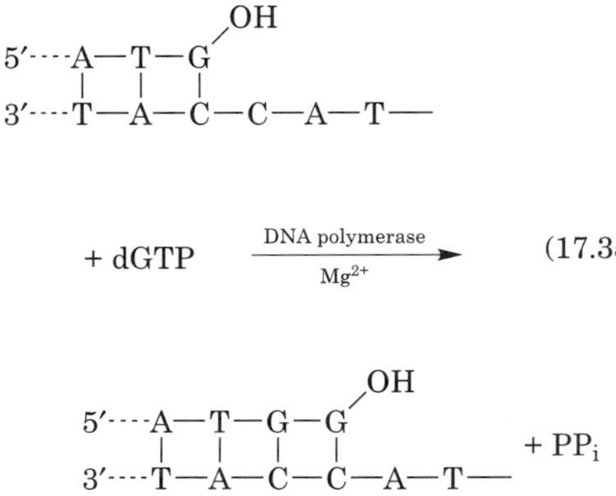

$$
\begin{array}{l}
5'\text{----A}-\text{T}-\text{G}^{\nearrow \text{OH}}\\
\quad\quad\;|\;\;\;\;|\;\;\;\;|\\
3'\text{----T}-\text{A}-\text{C}-\text{C}-\text{A}-\text{T}-
\end{array}
$$

$$
+ \text{dGTP} \xrightarrow[\text{Mg}^{2+}]{\text{DNA polymerase}} \tag{17.35}
$$

$$
\begin{array}{l}
5'\text{----A}-\text{T}-\text{G}-\text{G}^{\nearrow \text{OH}}\\
\quad\quad\;|\;\;\;\;|\;\;\;\;|\;\;\;\;|\\
3'\text{----T}-\text{A}-\text{C}-\text{C}-\text{A}-\text{T}-
\end{array} + \text{PP}_i
$$

DNA polymerase catalyzes the transfer of a complementary deoxynucleoside monophosphate moiety from its triphosphate (dNTP) to the 3' hydroxyl of the primer terminus, with subsequent release of pyrophosphate (PP$_i$, eq. 17.35). Initially, phosphonoformate (**22**) and phosphonoacetate (**23**) were identified as inhibitors of HSV DNA synthesis (92). Detailed kinetic studies (93), using DNA polymerase induced by avian herpes viruses, showed that phosphonoacetate (**23**) was a noncompetitive inhibitor of the four dNTPs. At low levels of dNTPs it was a noncompetitive inhibitor of the substrate DNA, becoming uncompetitive at saturating dNTP levels. It was also found that (**23**) was a competitive inhibitor of pyrophosphate, with a K_i value in the low micromolar range, in the dNTP-PP$_i$ exchange reaction catalyzed by a turkey virus DNA polymerase (93). The inhibition patterns were identical to those observed using pyrophosphate as an inhibitor. Therefore it was concluded that (**23**) acted as an analog of pyrophosphate and competed for the same binding site (93). Later, both (**22**) and (**23**) were confirmed as acting as pyrophosphate (i.e., product) analog inhibitors of isolated HSV DNA polymerase (94).

2.5.2 Multisubstrate Analogs. A large number of enzymatic reactions involve the simultaneous binding of two or more substrates at the active site. The bound substrates must be in close proximity to each other and positioned in such a way as to facilitate covalent bond formation or the transfer of a functional group from one substrate to another. Multisubstrate analog inhibitors mimic the simultaneous binding of two or more substrates at the active site of the enzyme. The advantage of this, for a bireactant system, is shown in Equation 17.36.

$$A + B \quad \begin{array}{c} \xrightarrow{K_{Bi}} A \cdot B \xrightarrow{k_N} P \\ \\ E \; \Big\| \; K_{MS} \\ \\ \xrightarrow[K_B]{K_A} \; E \cdot A \cdot B \xrightarrow{k_E} E + P \end{array} \qquad (17.36)$$

$$K_{MS} = \frac{1}{K_{Bi}} K_A K_B$$

There are two ways the two substrates, A and B, may bind to the enzyme to form an $E \cdot A \cdot B$ complex. First, and most likely, they bind individually (in either a random or an ordered fashion) with dissociation constants of K_A and K_B. Second, the substrates may come together, positioned in such a way as to facilitate their subsequent reaction with a dissociation constant of K_{Bi}. This reactive complex $A \cdot B$ then binds to the enzyme with a dissociation constant of K_{MS}. In general, the formation of $A \cdot B$ is entropically unfavorable. However, a bisubstrate analog, designed to mimic $A \cdot B$, can often be prepared by covalently connecting the corresponding substrates or substrate analogs with a suitable linker group. Linking the two groups effectively overcomes the unfavorable entropic barrier. It has been calculated that an ideal bisubstrate analog inhibitor can bind up to 10^8 times more tightly than the product of the substrate-binding constants (i.e., $1/K_{Bi}$ may be as high as 10^{-8} M). This figure is based on entropic considerations and also assumes a perfect fit of the bisubstrate analog inhibitor to the two binding sites on the enzyme (57).

Where does this high affinity come from? A multisubstrate analog inhibitor will bind more tightly than substrate analog inhibitor because it has (1) the entropic advantage of reduced molecularity and (2) an additive binding contribution from each of the substrates it mimics. For example, when two single-substrate analog inhibitors bind separately, but next to each other, two sets of translational and rotational entropies are lost. However, when a bisubstrate analog inhibitor binds it loses only a single set of translational and rotational entropies (57, 60). Further, let us assume that the bisubstrate analog binds to the same two sites as two single-substrate analog inhibitors. In that case there will be a gain in entropy from the release of water molecules from each substrate-binding site, as well as the favorable enthalpic contributions from the formation of hydrogen bonds, buried salt bridges, and so forth in each site. These favorable free-energy contributions will be the same for a bisubstrate analog as for the two individual inhibitors binding simultaneously. On the other hand, compared to the binding of a single-substrate analog, the multisubstrate analog inhibitor gains favorable binding enthalpies and entropies from the additional binding site(s), while still losing only one set of translational and rotational entropies. Thus the binding of a multisubstrate analog should be very tight, without needing any assistance from transition-state complementarity.

Inhibitors that combine two substrates are termed bisubstrate analogs, whereas those combining three substrates are termed trisubstrate analogs and so on, with the former being the most common. The design of a bisubstrate analog inhibitor ordinarily requires the development of two single-substrate analog inhibitors of reasonable affinity. The two single-substrate inhibitors are then connected by an appropriate linker, and the optimal length of the linker is determined experimentally. Under normal circumstances, the K_i value for a bisubstrate analog inhibitor can be expected to approximate the product of the K_i values of the two substrate analogs. A guide to a reasonably achievable K_i for a bisubstrate analog also may be obtained from the product of the K_M values of the individual substrates. For example, if two substrates of an enzymatic reaction have binding constants in the millimolar range, a bisubstrate analog would be expected

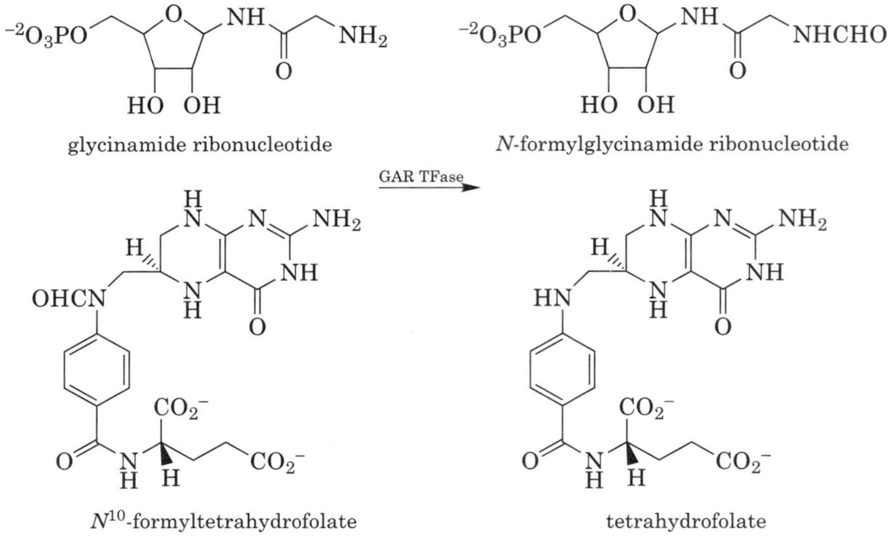

glycinamide ribonucleotide

N-formylglycinamide ribonucleotide

N^{10}-formyltetrahydrofolate

tetrahydrofolate

(17.37)

to have a K_i value in the micromolar range. Note also that, if the enzyme binds substrates in a random manner, then a multisubstrate inhibitor should exhibit competitive inhibition patterns with each substrate it mimics because the binding of the inhibitor should be mutually exclusive with that substrate. If the enzyme employs an ordered mechanism, then the inhibitor should be competitive with the first substrate to bind and uncompetitive with other substrates.

The multisubstrate analog approach to enzyme inhibition has the additional advantage in that it provides a high degree of specificity. The combination of two or more substrates will usually produce a unique structure, unlikely to bind to other enzymes that may utilize any one of the substrates. This approach has even been used to design isozyme-specific inhibitors (95). It should also be noted that the distinction between a transition-state analog (Section 2.5.3) and a multisubstrate analog inhibitor is often quite arbitrary. In fact many inhibitors described as transition-state analogs are often actually analogs of high-energy reaction intermediates that, in turn, may have structures somewhat akin to those of multisubstrate analog inhibitors. However, multisubstrate analog inhibitors are intended to mimic the combined substrates in their ground-state forms and do not require any contribution from transition-state stabiliza-

tion. Several general reviews on multisubstrate analog inhibitors have appeared (96–98), and multisubstrate analogs also receive some discussion in reviews on transition-state analogs (99–101).

Glycinamide ribonucleotide transformylase (GAR TFase) catalyzes the transfer of a formyl group from N^{10}-formyltetrahydrofolate to glycinamide ribonucleotide (Equation 17.37). This is a crucial step in de novo purine biosynthesis, which is essential for cell division, and GAR TFase has become a target enzyme for the development of antineoplastic agents.

Inglese et al. (102) were able to synthesize the bisubstrate inhibitor β-thioGARdideazafolate (β-TGDDF) (**24**) (Fig. 17.18). This compound combines nearly all the features of both substrates, linked by a stable thioether bridge, and was found to inhibit GAR TFase with a K_i value of 250 pM (102). β-TGDDF acted as a slow, tight-binding inhibitor (Section 2.4) and the K_i value was about three times lower than the product of the K_M values of the substrates. More recently, the crystal structure of the complex between BW1476U89 (**25**) and GAR TFase was obtained (103). BW1476U89 is another multisubstrate analog and has a K_i value of about 100 pM (104). The structure confirms that the inhibitor binds in those sites identified previously as substrate-

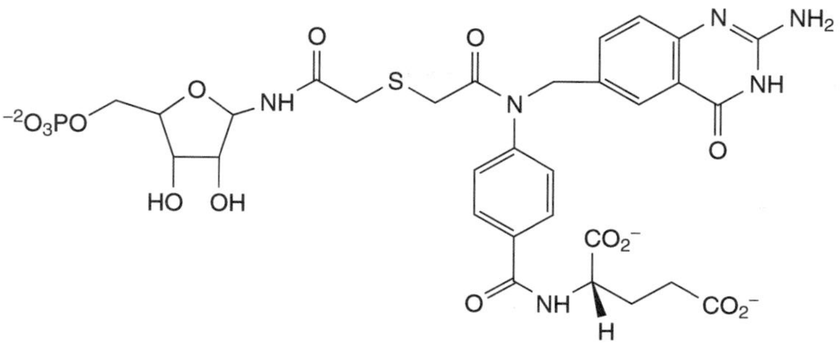

Figure 17.18. Bisubstrate analog inhibitors of GAR-TFase.

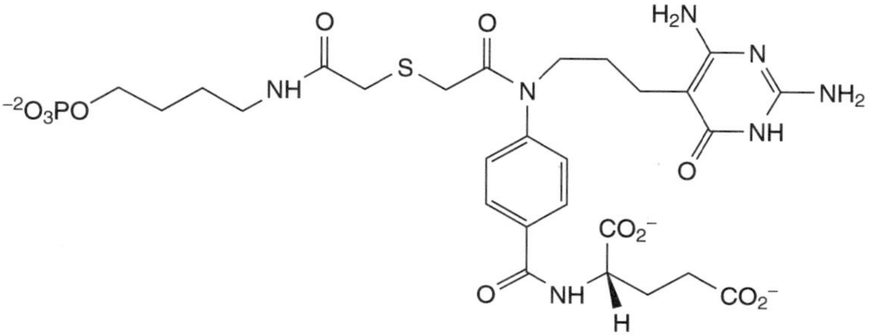

binding sites, and provides a starting point for development of even more potent transition-state analogs.

The condensation of carbamyl phosphate and L-aspartate, catalyzed by aspartate transcarbamoylase (ATCase), produces N-carbamyl-L-aspartate (Equation 17.38). This is one of the early steps in *de novo* pyrimidine biosynthesis, also a requirement for cell division,

making ATCase also a target for potential anticancer agents.

N-Phosphonoacetyl-L-aspartate (PALA) (**26**) (Fig. 17.19) was initially designed as a transition-state analog inhibitor of ATCase (105). It was found to have a K_i value of 27 nM, a value that is considerably lower than the K_M values of 27 μM and 17 mM for carbamyl phosphate and L-aspartate, respectively (105). PALA was

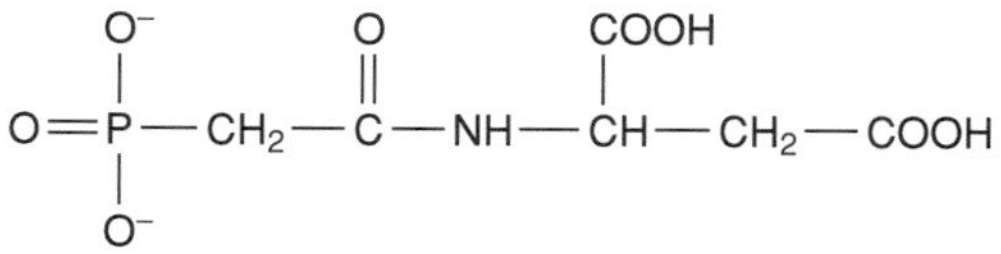

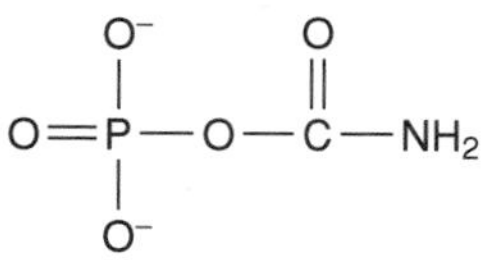

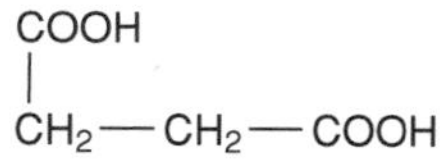

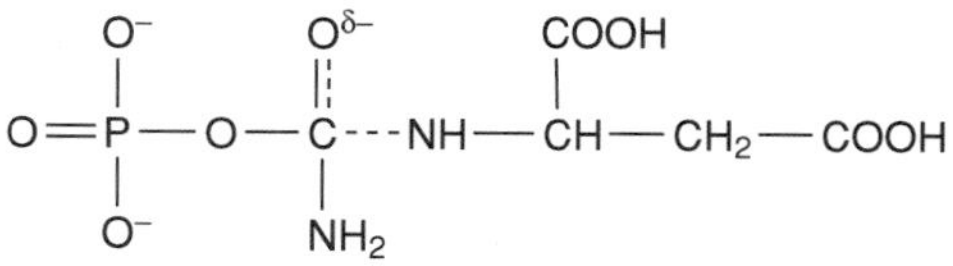

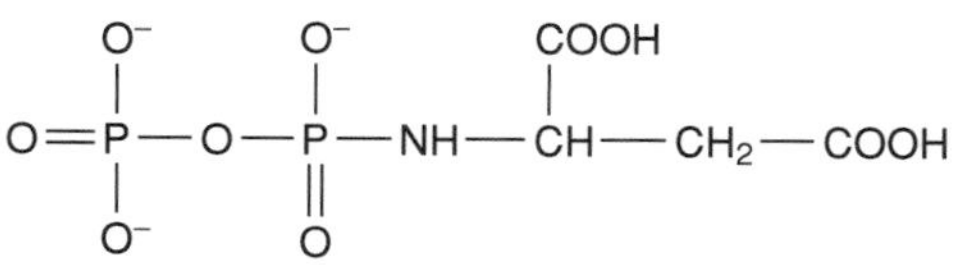

Figure 17.19. Putative transition state, substrate, and inhibitors of aspartate transcarbamylase.

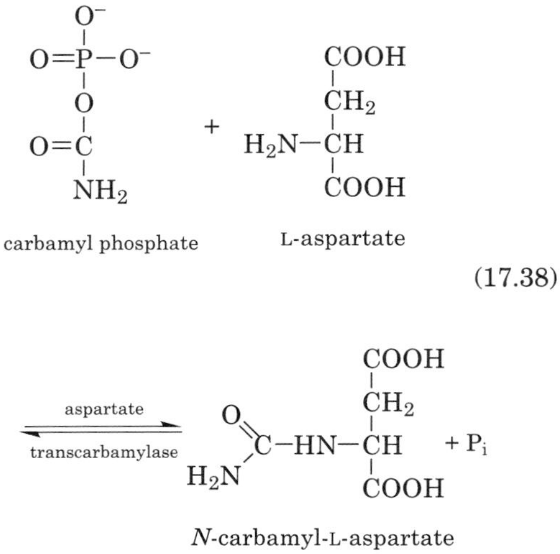

carbamyl phosphate   L-aspartate

$$(17.38)$$

N-carbamyl-L-aspartate

found to inhibit cell growth *in vivo* (106) and, eventually, underwent clinical trials as an anticancer agent (107).

PALA provides an example of the difficulties in distinguishing between a multisubstrate analog and a transition-state analog. As shown in Fig. 17.19, in effect PALA (**26**) combines two fragments, an analog of carbamyl phosphate (**27**) and succinate (**28**). The tight binding of PALA also suggested it was a potential transition-state analog. However, succinate has a K_i value of 90 μM, and the product of the K_i values of succinate and carbamyl phosphate is 24 nM, which is almost identical to the K_i value of PALA (105). As shown in Fig. 17.19, the transition-state structure (**29**) for the ATCase-catalyzed reaction is tetrahedral. The pyrophosphate analog (**30**) was expected to provide a much better mimic of the transition state, yet its K_i value of 0.24 μM was tenfold higher than that of PALA (108). It is not clear why there is this discrepancy, but a recent X-ray structure of the ATCase-PALA complex identified several groups that are positioned to bind to a tetrahedral transition state (109). Two of these, the side chain of Gln137 and the backbone carbonyl of Pro266, were positioned to interact with the amino group of the putative transition state (**29**). However, these groups would not be expected to interact so well to the analogous oxygen atom of the pyrophosphate transition-state

analog (**30**), perhaps leading to its weaker-than-expected binding.

The statins are a group of cholesterol-lowering agents that have become some of the largest selling drugs in the world. They lower serum cholesterol levels by competitively inhibiting 3-hydroxy-3-methylglutaryl-coenzyme A (HMG-CoA) reductase, a key enzyme in cholesterol biosynthesis (Equation 17.39).

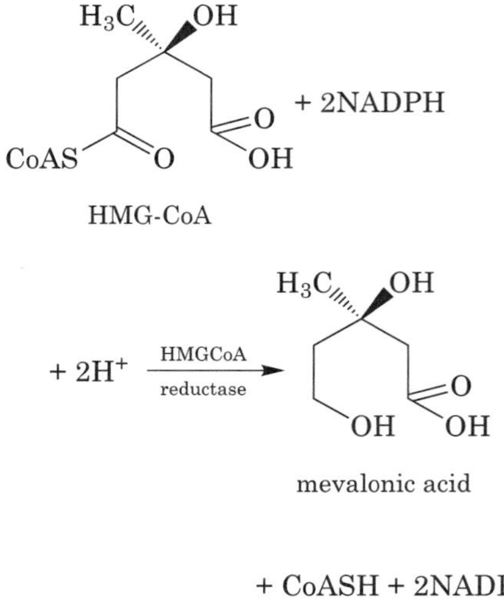

HMG-CoA

$$(17.39)$$

mevalonic acid

$$+ \text{CoASH} + 2\text{NADP}^+$$

Several statin inhibitors of HMG-CoA reductase are shown in Fig. 17.20. They consist of rigid, hydrophobic groups connected to an HMG-like group that, in inhibitors such as mevastatin (compactin) (**31**), simvastatin (**9**) (Fig. 17.1), and the dichlorophenol derivative (**32**) is present in the form of a lactone. *In vivo*, the lactone is converted to the free acid, as shown in Fig. 17.20 for mevastatin (**33**). More recently developed statins, such as fluvastatin (**34**) and atorvastatin (**35**), are prepared as the free acids. These inhibitors have K_i values in the low nanomolar range (110), significantly lower than the K_M value of the substrate HMG-CoA, which is in the micromolar range (110, 111). Given that these inhibitors did not appear to be transition-state analogs, Nakamura and Abeles (112, 113) conducted a number of experiments to determine the basis of the enhanced affinity of, in particular, (**31**) and (**32**).

Both mevastatin (**31**) and (**32**) were found to bind to the hydroxymethylglutarate portion

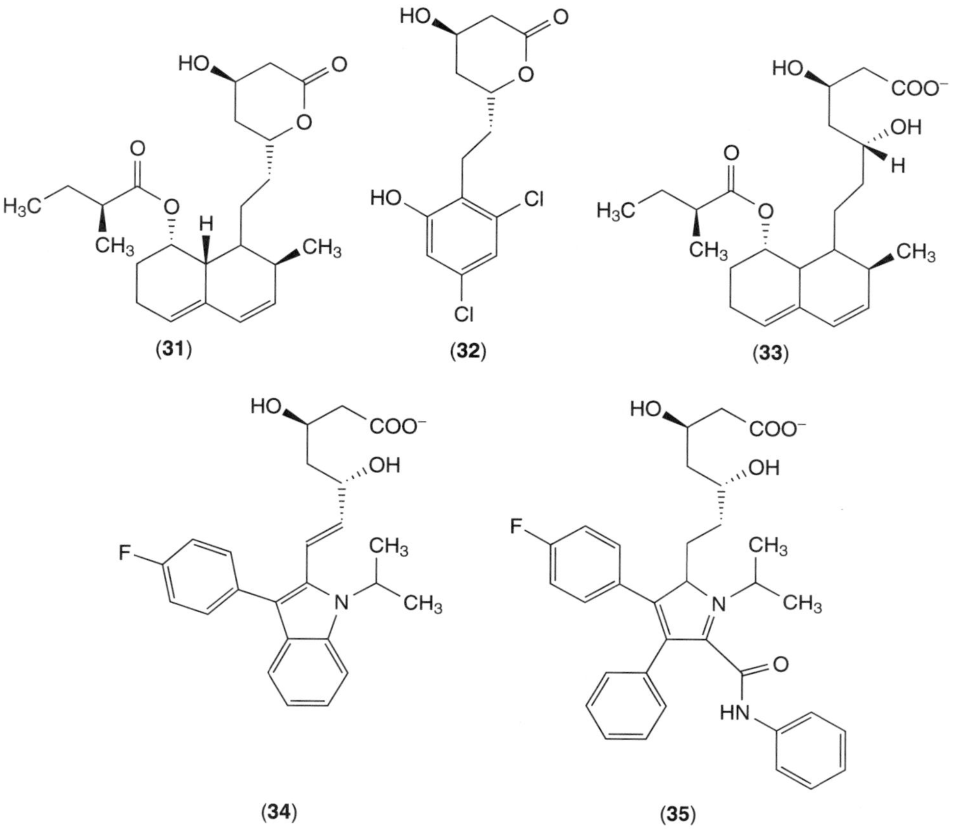

Figure 17.20. Statin inhibitors of HMGCoA-reductase.

of the active site, but not the NADPH region, whereas only (**31**) bound to the coenzyme A portion. D,L-Mevalonate and D,L-3,5-hydroxy-valerate, used as analogs of the upper portion of the statins, were both poor inhibitors, with K_i values in the millimolar range; however, analogs of the hydrophobic decalin region of mevastatin showed no inhibitory effect (112). Given that the K_i value for mevastatin is almost eight orders of magnitude lower than that of D,L-3,5-hydroxyvalerate, it is clear that the hydrophobic lower portion (and its covalent link) must play a significant role in the binding of (**31**) and, by implication, the binding of all the statins. Presumably, the upper portion of the inhibitor is necessary for specificity and the hydrophobic region for binding affinity. The hydrophobic region must be relatively nonspecific because a variety of hydrophobic groups (Fig. 17.20) are accepted. In

some cases (e.g., mevastatin), the hydrophobic group overlaps the CoASH site and in others, such as the dichlorophenol group of (**32**), it does not (112). The structure of the statin is analogous to that of a bisubstrate inhibitor, in that there is linked binding to two distinct binding sites on the enzyme, leading to greatly enhanced inhibition of the enzyme. For mevastatin, the entropic advantage provided by linking the mevalonate and decalin portions together is estimated to be approximately $5 \times 10^4\ M$ (113). This is quite a reasonable enhancement, given that the theoretical maximum is $10^8\ M$ (57), and it has been suggested that such a "hydrophobic anchor" is responsible for the enhanced binding of some inhibitors of alcohol dehydrogenase and adenosine deaminase (113).

Although this explanation appeared quite reasonable, it was thrown into doubt when X-

$$H_3\overset{+}{N}—Asp—Arg—Val—Tyr—Ile—His—Pro—Phe—His—Leu—COO^-$$

Angiotensin I

| angiotensin converting enzyme (ACE) | (17.40) |

$$H_3\overset{+}{N}—Asp—Arg—Val—Tyr—Ile—His—Pro—Phe—COO^- + H_3\overset{+}{N}—His—Leu—COO^-$$

Angiotensin II

ray structures of HMG-CoA reductase complexed with both substrates and products were obtained (114, 115). These structures showed that, if the statins bound so the HMG-like groups bound the HMG-binding pocket of the active site, the bulky hydrophobic groups of the statins would clash with the residues lining the narrow pocket into which part of the coenzyme A bound (115). However, recently, Istvan and Deisenhofer have obtained X-ray structures of HMG-CoA reductase bound to six individual statins, including (**9**), (**31**), (**34**), and (**35**) (116). This study showed the substrate-binding pocket rearranges to accommodate the statins, that the statins do bind to the HMG-binding region, that a shallow hydrophobic groove now accommodates the hydrophobic groups, and that none of the NADP(H)-binding pocket is occupied (116). *In toto*, the structural studies supported all interpretations made some 15 years earlier based on kinetic studies, and provided definitive evidence for a hydrophobic anchor enhancing the binding of the mevalonate portion of the statins.

The evolution of the angiotensin converting enzyme (ACE) inhibitors is an illuminating story in the development of enzyme inhibitors as therapeutic agents. As shown in Equation 17.40, ACE catalyzes the conversion of angiotensin I to angiotensin II.

Angiotensin II, itself a potent hypertensive agent, also stimulates the release of a second hypertensive agent, aldosterone. In addition, ACE catalyzes the cleavage of the nonapeptide vasodilating agent, bradykinin (not shown). Therefore an ACE inhibitor was seen to have the potential to limit three hypertensive actions. This premise was validated by *in vivo* results with teprotide, a peptide inhibitor of

ACE, which had been isolated from a South American pit viper (117).

At that time the structure of ACE was unknown, although it had been identified as a zinc metalloprotease. It was surmised that its mechanism and active site may resemble that of another metalloprotease, carboxypeptidase A, whose X-ray structure was known. (R)-2-Benzylsuccinic acid (**36**) (Fig. 17.21) had been identified as a potent inhibitor of carboxypeptidase A, and it was suggested that (**36**) resembled the collected products of the hydrolysis reaction (Fig. 17.21). In other words, (**36**) was a biproduct analog and, not unexpectedly, it was found to bind with an affinity resembling the combined affinities of the two products (118). Carboxypeptidase A appeared to have three main interactions with (**36**). Two substrate-binding sites bound the phenyl group and one carboxylate, and the Zn^{2+} ion, usually coordinated to the carbonyl of the amide bond being cleaved, was now bound to the second carboxylate. Combining those suggestions with studies with viper venom peptides, indicating that a C-terminal proline was effective in inhibiting ACE, a number of carboxyalkanoylproline derivatives were tested as ACE inhibitors (119). Of these, the succinyl-L-proline derivative (**37**) was found to be the most effective, with an IC_{50} value of 330 μM. Given that one carboxylate bound to the Zn^{2+} ion, a better zinc ligand, a thiol group, was substituted for this carboxylate, resulting in (**38**) with the IC_{50} value now reduced to 0.2 μM. Finally, after taking into account the differences between the active sites of ACE and carboxypeptidase A, captopril (**39**) was prepared. Captopril was found to be a competitive inhibitor of ACE, with a K_i value of 1.7 nM, and was

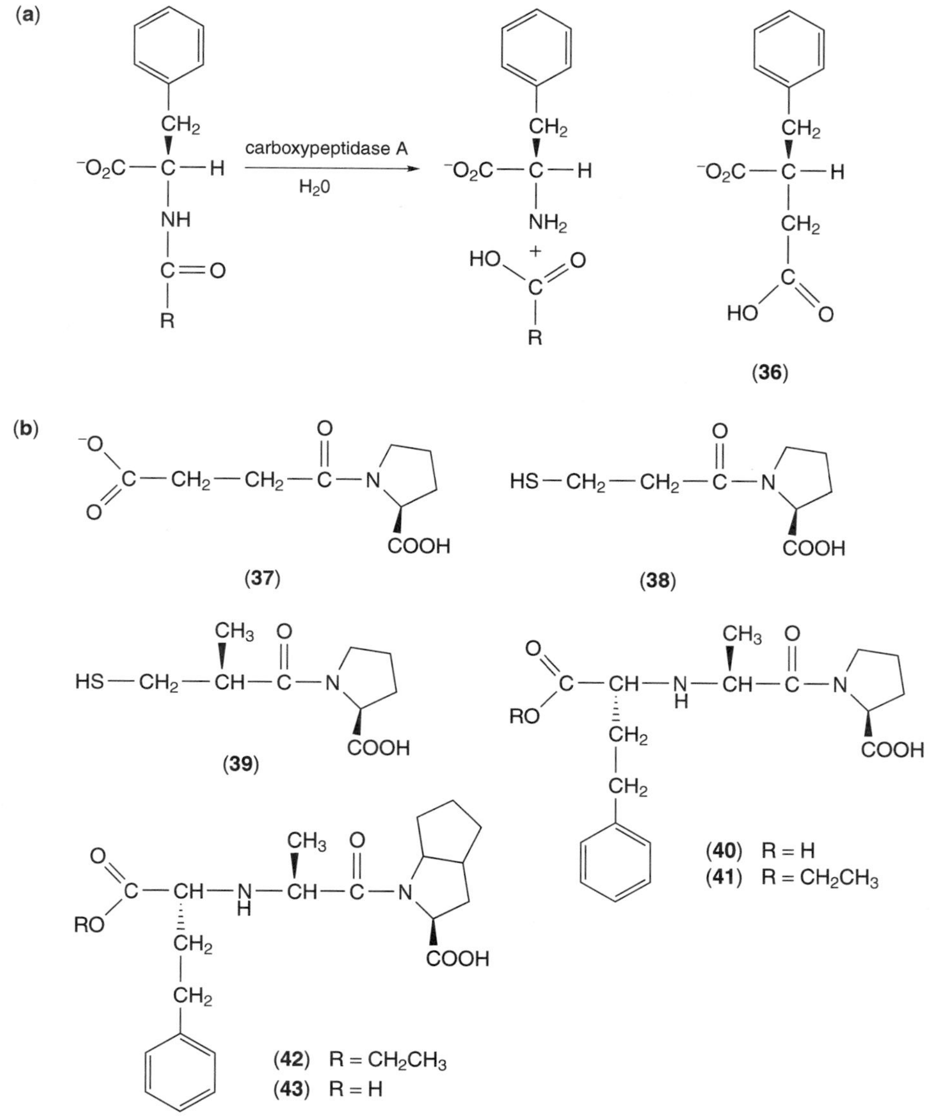

Figure 17.21. (a) Biproduct analog inhibitor of carboxypeptidase A and (b) several ACE inhibitors.

the first ACE inhibitor to be marketed. It was not long before attempts were made to make capropril more productlike, with the resultant development of enalaprilat (i.e. enalaprilat) (**40**). Enalaprilat was found to be a slow-tight-binding inhibitor (Section 2.4) of ACE with a K_i value below 1 nM, (120), but it was poorly absorbed orally. However, the ethyl ester (enalapril), (**41**) acted as a prodrug, had good oral activity, and was marketed. Note that it is

also possible that enalaprilat acts as a transition-state analog (Section 2.5.3), thereby accounting for its performing as a slow-tight-binding inhibitor (121). Following enalapril, many more ACE inhibitors have been developed mainly aimed at increasing oral bioavailability, removing side effects, or improving metabolism.These include ramipril (**42**), the ester prodrug of ramiprilat (**43**), with 10 times better bioavailability than that of enalapril.

Ramaprilat was also shown to be a slow-tight-binding inhibitor of ACE, operating by mechanism B, with K_i^* (Equation 17.30) of 7 pM (122). A more detailed discussion of the development of the ACE inhibitors is available (121).

2.5.3 Transition-State Analogs.

As a chemical reaction proceeds from substrates to products, it will pass through one or more transition states. The energy barrier imposed by the highest energy transition state controls the overall rate of the reaction. Enzymes bring about rate enhancements of 10^{10}–10^{15} (123) by lowering this energy barrier. They do this by having a greater affinity to the structure of the transition state than to the structures of either substrates or products. Although an enzyme may have good affinity for its substrate, as evidenced by a low dissociation constant (K_S, Equation 17.41), for the Michaelis (E · S) complex, the enzyme can further stabilize the inherently unstable transition state, for example, by forming extra electrostatic or hydrogen bonds, by providing more effective hydrophobic interactions, or by using structural rearrangements to exclude solvent, thereby strengthening existing electrostatic contacts.

$$
\begin{array}{ccccc}
\mathrm{E + S} & \xrightleftharpoons{K_N^{\ddagger}} & \mathrm{E + S^{\ddagger}} & \xrightarrow{k_N} & \mathrm{E + P} \\
\Big\updownarrow{\scriptstyle K_S} & & \Big\updownarrow{\scriptstyle K_T} & & \Big\updownarrow{} \\
\mathrm{E \cdot S} & \xrightleftharpoons{K_E^{\ddagger}} & \mathrm{E \cdot S^{\ddagger}} & \xrightleftharpoons{k_E} & \mathrm{E \cdot P}
\end{array} \quad (17.41)
$$

$$
\frac{K_T}{K_S} = \frac{K_E^{\ddagger}}{K_N^{\ddagger}} = \frac{k_E}{k_N}
$$

Simple transition-state theory states that the rate of an enzyme-catalyzed reaction is correlated with the rate of a noncatalyzed reaction by the same factor as the affinity of an enzyme for the transition state to the affinity of an enzyme for a substrate (Equation 17.41) (99).

Therefore, the magnitude of enzymatic catalysis (k_E/k_N) is related to the enhanced binding of the transition state to the enzyme (K_T/K_S). Compounds that can take advantage of this enhanced binding to the transition state can prove to be potent and selective enzyme inhibitors. Such compounds, referred to as transition-state analogs, can theoretically have ratios of the binding constants of inhibitor to substrate (K_i/K_S) on the order of 10^{-8} to 10^{-14}. In addition, transition-state analogs may have the further advantage of reduced molecularity, as outlined earlier (Section 2.5.2) for multisubstrate analog inhibitors. Several reviews on the theory and general aspects of transition-state analog inhibitors are available and are recommended for a more complete understanding of this topic (37, 96, 99, 100, 124, 125).

The design of a good transition-state mimic is quite challenging. It requires, at the least, sufficient knowledge of the mechanism of the target enzyme to predict transition-state structure(s). This is why transition-state analogs are sometimes (but not in this review) referred to as mechanism-based inhibitors. A detailed knowledge of the true energy profile, including details such as the existence of distinct chemical steps, high-energy intermediates, and their associated transition states, is also useful (126). Further, by definition, the transition state is unstable, often highly charged, and possesses partially broken/formed covalent bonds. Designing a *stable* compound that will closely mimic a transition state is impossible. However, the Hammond postulate states that the transition state between a reactant and a high-energy reaction intermediate will resemble the intermediate rather than the reactant. It is possible to design/synthesize an analog of a high-energy intermediate. Indeed, the majority of so-called transition-state analogs are actually analogs of high-energy reaction intermediates. Although a clear distinction exists, the design process is, for all practical purposes, the same.

It should also be noted that an enzyme is designed to initially recognize the features of its substrates. Often substrate binding brings about a conformational change in the enzyme that will then maximize the attractive forces between the enzyme and transition state. The transition-state analog may not possess those features of the substrate that facilitate rapid binding, even though its affinity for the enzyme is extremely high. Although some transition-state analogs bind rapidly to enzymes,

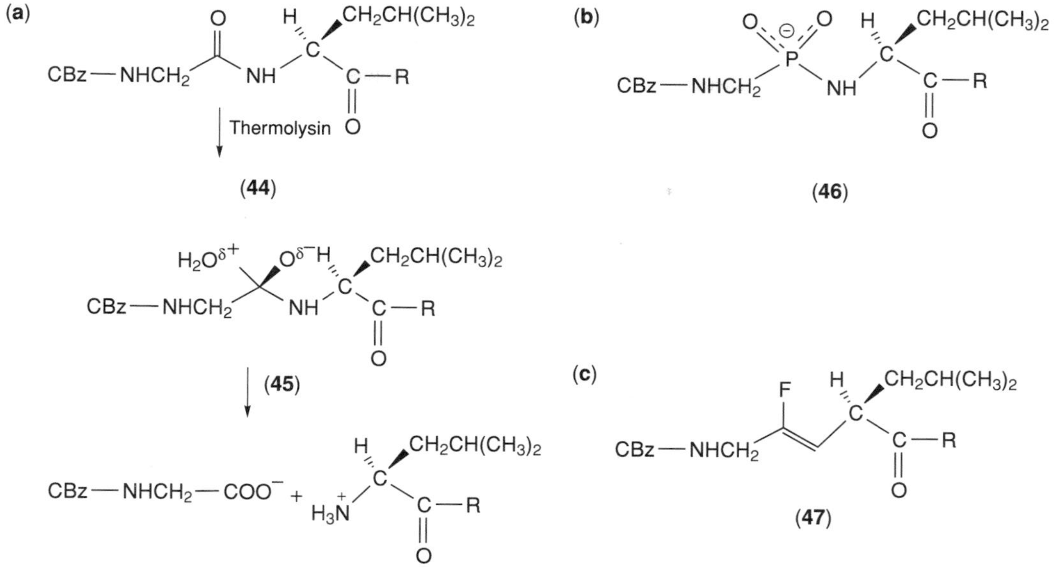

Figure 17.22. (a) Thermolysin-catalyzed hydrolysis of peptide analogs showing putative transition state, (b) phosphonamidate peptide analog, and (c) fluoroalkane peptide analog.

others bind slowly and show the properties of the slow-binding inhibitors described earlier in Section 2.4.

Slow binding, tight binding, or structural similarity to the assumed transition-state structure are not, in themselves, sufficient criteria to establish that an inhibitor is a true transition-state analog (127). Methotrexate (**7**), for example, is an extremely high-affinity (K_i = 58 pM), slow-binding inhibitor of dihydrofolate reductase (128). On the surface, it would appear that methotrexate could be classified as a transition-state analog. However, crystallographic studies have shown that methotrexate binds with its pterin ring in the opposite orientation to that of the substrate, dihydrofolate (129, 130). To distinguish between a high-affinity, ground-state analog and a putative transition-state analog requires a careful appraisal. There is a fundamental difference between the entropy change of a unimolecular enzymatic reaction and that of a multimolecular solution reaction (131). In addition, the appropriate rate constant for the nonenzymatic reaction is often either not available or hard to obtain (132, 133). These factors combine to make it difficult to evaluate

quantitatively the correlation between the enhanced rates of enzymatic reactions and the tight binding of transition-state analogs.

In an attempt to develop stringent criteria for the distinction between transition- state and ground-state analogs, Bartlett and Marlowe (127) overcame some of these inherent difficulties by comparing the binding affinities of a series of substrate analogs with those of the corresponding transition-state analogs. One consequence of Equation 17.41 is that, if there is a change in structure of a substrate that alters k_{cat}/K_M without altering the nonenzymatic rate of reaction, then an analogous structural change in the transition-state mimic should bring about a similar change in K_i. Put simply, there should be a linear relationship between the values of K_i for the transition-state analog and k_{cat}/K_M for the corresponding substrate. Bartlett and Marlow (127) designed a series of dipeptide analog substrates (**44**) (Fig. 17.22) for thermolysin in which the structural variation was remote from the reactive center and therefore unlikely to affect the nonenzymatic reaction rate. The reaction catalyzed by thermolysin is proposed to proceed by the tetrahedral transi-

Table 17.6   Correlation of K_i Values for Inhibitors of Thermolysin with K_M and K_M/k_{cat} Values for the Corresponding Substrates[a]

$$\text{CBz—NHCH}_2 \overset{O \overset{\ominus}{\underset{}{\diagup}} O}{\underset{P}{\diagdown}} \text{NH} \overset{H \quad CH_2CH(CH_3)_2}{\underset{\underset{O}{\parallel}}{\overset{C}{\diagup} \text{—R}}}$$

	Inhibitor Data	Corresponding Substrate Data	
	K_i (nM)	K_M (mM)	K_M/k_2 ($\mu M\ s^{-1}$)
R = D-Ala	1700	16.6	3200
R = NH$_2$	760	20.6	196
R = Gly	70	10.8	165
R = L-Phe	78	2.4	20
R = L-Ala	16.5	10.6	13.6
R = L-Leu	9.1	2.6	7.0

$$\text{CBz—NHCH}_2 \overset{F}{\diagup} \diagdown \overset{H \quad CH_2CH(CH_3)_2}{\underset{\underset{O}{\parallel}}{\overset{C}{\diagup} \overset{C}{\diagup}\text{—R}}}$$

	K_i (mM)		
R = Gly	1.80	10.8	165
R = L-Ala	1.48	10.6	13.6
R = L-Leu	0.32	2.6	7.0
R = L-Phe	0.19	2.4	20

[a]Data are from Ref. 127.

tion state (**45**), and with their long P—O bonds, the phosphonamidate compounds (**46**, Table 17.6) were expected to act as transition-state analog inhibitors. It was found that the K_i values for the putative transition-state analog inhibitors correlated linearly with the K_M/k_{cat} values of the corresponding substrates, although no correlation was found between K_i and K_M (Table 17.6). The fact that substrate binding (K_M) was relatively unaffected by a change at a remote site was not unexpected, but the observation that the binding of the phosphonamide inhibitors was greatly affected suggests that these inhibitors were, indeed, transition-state rather than ground-state analogs. Conversely, the K_i values for a series of fluoroalkene isosteres of the same substrates (**47**) (Fig. 17.22) correlated strongly with K_M but not K_M/k_{cat} (Table 17.6), indicating that the latter inhibitors were ground-state analogs (134). This approach has also been used to confirm that phosphonic acid peptides were transition-state analog inhibitors of pepsin (135).

One of the most popular targets for design of transition-state analogs is adenosine deami-

nase. Inhibitors of this enzyme have been used as immunosuppressants and are also potential antitumor agents, whereas lack of adenosine deaminase results in severe combined immunodeficiency disease (SCID).

Adenosine deaminase (ADA), which catalyzes the conversion of adenosine to inosine (Equation 17.42), is an extremely proficient enzyme, providing a rate enhancement of more than 12 orders of magnitude (123). The enzyme-catalyzed reaction is thought to pass through an unstable hydrated intermediate (**48**) (Fig. 17.23), with a K_T (Equation 17.41) in the region of $10^{-17}\ M$ (123). Clearly, even a crude analog of (**48**) would have the potential to be an extremely powerful inhibitor of ADA.

The structures of several inhibitors of ADA are shown in Fig. 17.23. Of these, the antibiotics coformycin (**49**) and (*R*)-deoxycoformycin (pentostatin, **50**) were found to be potent ADA inhibitors, with K_i values of $1 \times 10^{-11}\ M$ (136) and $2.5 \times 10^{-12}\ M$ (137), respectively. The K_M for adenosine is around 30 μM (138, 139), whereas the K_i of the product, inosine, is $10^{-4}\ M$. Thus, the two antibiotics show at least 10^6-fold greater affinity for ADA than for

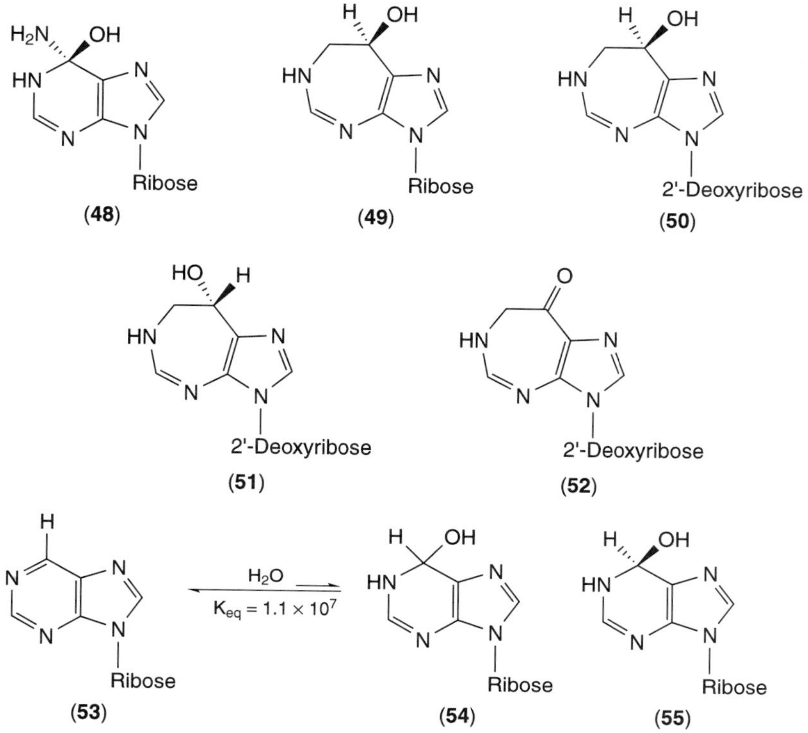

Figure 17.23. Reaction intermediate and inhibitors of the reaction catalyzed by adenosine deaminase.

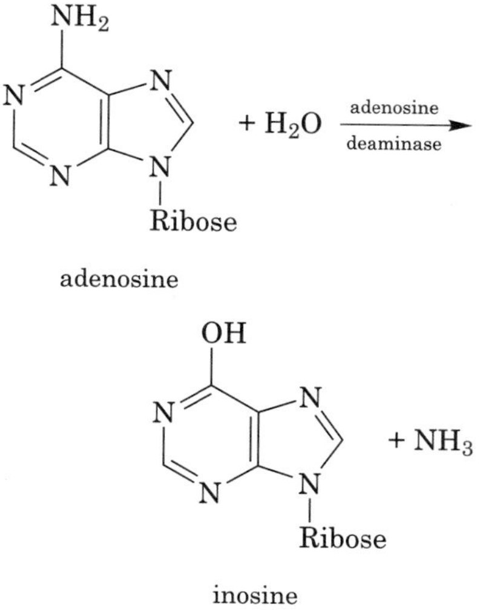

$$(17.42)$$

the substrate, suggesting that they are acting as transition-state analogs. By contrast, (S)-deoxycoformycin (**51**) and 8-ketodeoxycoformycin (**52**), with K_i values of 33 and 40 μM,

respectively, are clearly ground-state analogs. The differences in binding affinities of (R)- and (S)-deoxycoformycin translate to a difference in binding energy of almost 10 kcal/mol and provide an estimate of the energy that can be applied to substrate distortion in formation of the transition-state complex (139).

Purine ribonucleoside (**53**) was initially thought to bind to ADA as a ground-state analog, with an apparent K_i of 3 μM (138, 140). However, it was observed that the structure of the ligand and the enzyme were perturbed when purine riboside bound to ADA. ^{13}C-NMR spectroscopy showed that the ADA-bound purine riboside was sp^3 hybridized at C-6 (141). The NMR and UV spectra suggested that it was the hydrated form of purine riboside (**54**) that was binding to ADA and, using the unfavorable equilibrium constant for hydration in solution ($1.1 \times 10^{-7}\,M$, Fig. 17.23), a true K_i value of $3 \times 10^{-13}\,M$ could be calculated for (**54**) (142). Given the low concentration of the free hydrate in solution, and the

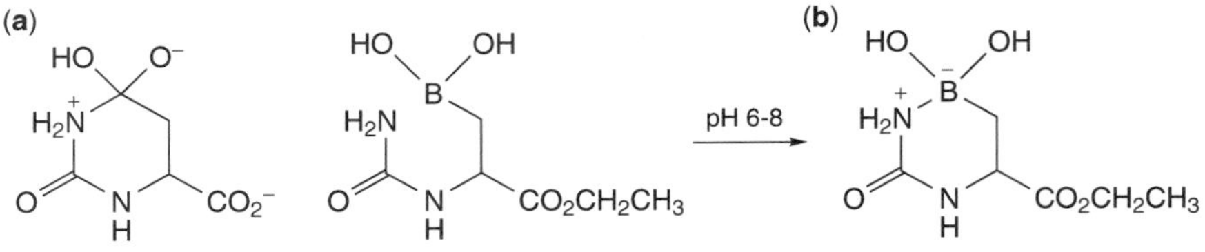

Figure 17.24. (a) Putative transition state for the dihydroorotase reaction, and (b) boronic acid transition-state analogs.

rapid onset of inhibition, it appears that purine riboside (53) itself initially binds and is then rapidly converted to (54) in the active site (141). This result, along with the high affinities of (R)-coformycin and (R)-deoxycoformycin, argues that the reaction proceeds by a stereospecific, direct attack of water rather than the double-displacement mechanism that also had been proposed (141). More recently, an X-ray study on adenosine deaminase, which had been crystallized in the presence of purine ribonucleoside, confirmed that it was the hydrated species of purine ribonucleoside that was present in the active site (143). Further, a triad of a zinc atom, a histidine residue, and an aspartic acid residue ensured that the binding was stereospecific, with the 6R isomer (55) being favored.

The adenosine deaminase story, in many ways, provides a perfect example of the general principles of enzymatic catalysis and the utility of enzyme inhibitors. ADA is an extremely efficient catalyst, producing a rate enhancement of 12 orders of magnitude. 6R-Hydroxy-1,6-dihydropurine riboside (55) has an affinity for ADA about 8 orders of magnitude greater than that for substrates or products; that is, it expresses a substantial fraction of the free energy of binding that separates the transition state from the ground state in an enzymatic reaction. Evidence of the extraordinary ability of an enzyme to discriminate between stereoisomers is provided by the 10^7-fold difference in binding affinities of the 8R-OH (50) and 8S-OH (51) stereoisomers of 2'-deoxycoformycin. Inhibitors were used to differentiate among several potential reaction mechanisms for ADA and, finally, an ADA in-

hibitor (pentostatin) has proved to be of therapeutic benefit.

Inhibitors of pyrimidine and purine biosynthesis are used as antineoplastic agents. As a consequence, dihydroorotase, which catalyzes the third step of de novo pyrimidine biosynthesis, the conversion of carbamyl aspartate to dihydroorotate (Equation 17.43), is a target for therapeutic intervention.

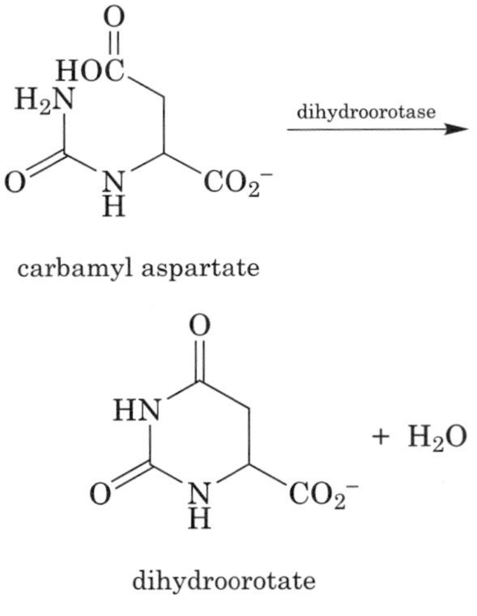

carbamyl aspartate

$$(17.43)$$

dihydroorotate

The reaction is thought to proceed through the tetrahedral-activated complex (56) (Fig. 17.24), which is a highly charged, unstable sp^3 carbon species (144, 145). At around neutral pH, compound (57), a boron-containing analog of carbamyl aspartate, rearranges to the stable, tetrahedral boronic acid derivative (58). The affinity of (58) for dihydroorotase ($K_i = 5$ μM) was found to be 10-fold greater than that of the carbamyl aspartate

$K_M = 50 \ \mu M$, indicating that (**58**) was probably acting as a transition-state analog (145). Tetrahedral boronic acid structures are stable, unlike the analogous sp^3 carbon species, and boronic acid derivatives of substrate peptides have proved to be quite potent inhibitors of a variety of proteases, particularly serine proteases (146).

Chorismate mutase catalyzes the conversion of chorismate to prephenate (Equation 17.44). This reaction is unusual, in that it is the only pericyclic [3,3] sigmatropic rearrangement (Claisen rearrangement) that is catalyzed by an enzyme.

chorismate

(17.44)

prephenate

Although chorismate mutase does provide a rate enhancement of 2×10^6 (147), this unimolecular reaction readily occurs without enzyme, under mild conditions. The reaction was expected to pass through a chairlike transition state (**59**) (Fig. 17.25) but early molecular orbital calculations indicated that the boatlike transition state (**60**) was not out of the question (147). In an attempt to define the transition-state structure, several compounds, each designed to mimic a putative transition state, were synthesized and tested as chorismate mutase inhibitors (147). The enzyme was found to be inhibited by the *exo*-carboxy nonane (**61**), with an apparent K_i value of $3.9 \times 10^{-4} \ M$. Conversely, the *endo*-carboxy nonane (**62**) did not inhibit the enzyme. The apparent K_i value of the adaman-

tane derivative (**63**), in which the chair-chair conformation is fixed, was about the same as that of (**61**). Taken together, this implied that the reaction proceeded through a chairlike transition state (147).

This approach was later refined by Bartlett and Johnson, who suggested that IC_{50}/K_M ratios of 7 for compound (**61**) and 12 for compound (**63**) indicated that these inhibitors were not particularly good transition-state analogs (148). In an attempt to improve potency, and to further define the stereochemistry of the transition state, they synthesized several compounds including the *exo*- and *endo*-carboxy unsaturated oxabicylic ethers, (**64**) and (**65**), respectively (148). The *exo*-compound (**64**) was not significantly better than its saturated carbocyclic analog (**61**), but the *endo*-derivative (**65**) bound chorismate mutase some 100-fold more tightly than did chorismic acid under the same conditions, with a K_i value of 120 nM (148, 149). Later, monoclonal antibodies elicited against (**65**) were found to be effective catalysts for the conversion of chorismate to prephenate, with rate enhancements of 200-fold in one case (150) and 10,000-fold in another (151). In both instances it was suggested that the rate enhancement was attributable to increased binding of the transition state by the antibody (150, 151).

X-ray structures are now available of the complexes of (**65**) with two chorismate mutase enzymes (152, 153), as well as with the less-efficient catalytic antibody (154). Although each active site was found to employ a different constellation of interactions with (**65**), the dissociation constants for the binding of (**65**) to the three proteins were strikingly similar, ranging from 0.6 to 3 μM (153, 154). However, the micromolar affinity of (**65**) for both enzyme and antibody is considerably weaker than might be expected for a good mimic of the transition state, and the antibody is not a particularly efficient catalyst. Wiest and Houk (155) have calculated that the bond lengths for the breaking and forming bonds in the transition state are considerably longer than those for (**65**), and the neutral inhibitor does not mimic the charge separation that builds up in the transition state. Although the two enzyme-active sites have evolved to complement the larger, polarized transition state, the anti-

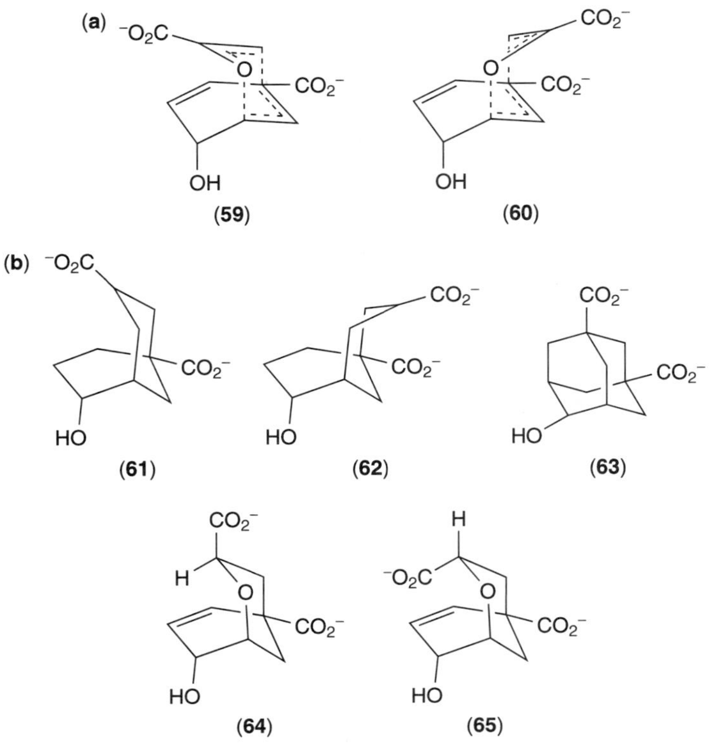

Figure 17.25. (a) Putative transition states for the chorismate-prephenate rearrangement and (b) structures of potential transition-state analogs.

body has no residues positioned to stabilize the polar-transition state. Further, the active site is smaller and makes more van der Waals contacts with the inhibitor, again features likely to impede catalysis. These features provide evidence for the innate difficulties associated with designing both good transition-state analogs and efficient catalytic antibodies.

3 RATIONAL DESIGN OF COVALENTLY BINDING ENZYME INHIBITORS

For the purposes of this chapter we have divided covalently binding enzyme inhibitors into categories according to Table 17.4. Pseudoirreversible inhibitors are discussed separately and the others are, in order of increasing specificity, chemical modifiers, affinity labels, and mechanism-based inhibitors. The targets for these inhibitors are the chemically reactive groups found within the enzyme's active site. These groups, in the majority of cases, are nucleophiles such as the —OH groups of serine, threonine, and tyrosine, the

—SH group of cysteine, and the —COOH groups of aspartic and glutamic acid residues. Other nucleophilic groups include the ϵ-amino group of lysine and the imidazole ring of histidine. In some cases the —NH$_2$ and —COOH groups of the enzyme's N- and C-termini, respectively, are also active-site nucleophiles, whereas enzymic cofactors may also provide targets for covalently binding inhibitors. Arginine is the only common amino acid that has an electrophilic side chain and it also can be modified with suitable nucleophilic agents. Kyte has recently provided an excellent overview of the general area of active-site modification and labeling (156).

The first group of covalently binding enzyme inhibitors, the chemical modifiers, are small organic molecules, generally electrophiles, that are used to modify the enzyme's side chains in such a way as to produce a stable covalent bond. These are often used to study enzyme inactivation and to identify residues potentially involved in binding and catalysis. Some of the commonly used reagents are

Table 17.7   Commonly Used Reagents for Chemical Modification

Residue Targeted	Reagent	Other Residues Labeled
Lysine	Acetic anhydride Isothiocyanates Trinitrobenzenesulfonate (TNBS) Cyanate	These reagents can also react with the N-terminal amino group
Histidine	Diethylpyrocarbonate (DEPC)	DEPC should be used at neutral pH to minimize reaction with lysines, cysteines, and tyrosines
Cysteine	Iodoacetamide, iodoacetate p-Hydroxymercuribenzoate Methyl methanethiosulfonate Ellman's reagent (DTNB) N-Ethylmaleimide	Iodoacetamide has the potential to modify histidines and lysines
Arginine	Phenylglyoxal Butanedione	Phenylglyoxal can react with lysine Butanedione should be used in the dark to prevent reaction with tryptophans, histidines, and tyrosines
Tyrosine	Tetranitromethane Chloramine T	Chloramine T also modifies histidines and methionines
Tryptophan	N-Bromosuccinimide 2-Hydroxy-5-nitrobenzyl bromide	
Serine	Diisopropylfluorophosphate Halomethyl ketones	
Aspartic Acid	Carbodiimides	
Glutamic Acid	Trimethyl oxonium fluoroborate Isoxazolium salts	

listed in Table 17.7. These compounds are chemically reactive and may lead to the modification of both catalytic and nonessential residues. As a consequence, experimental design (such as choice of reagent and reaction conditions, use of substrate protection, etc.) is of utmost importance in carrying out and interpreting chemical modification studies. Although inhibitors of this type are not the prime focus of this chapter (and are not discussed further), it should be noted that most of the kinetic equations that apply to affinity labels also apply to chemical modifiers, and there are a number of texts available that cover this topic (40, 157, 158).

Although the organic modifiers are usually not specific for a given enzyme, the second group, the affinity labels, have a degree of specificity built in. Sometimes described as active-site directed, irreversible inhibitors, affinity labels are usually substrate or product analogs that contain an additional chemically reactive moiety. They first bind to the en-

zyme's active site in a noncovalent fashion, like rapid reversible inhibitors. However, upon formation of the enzyme-inhibitor complex (E · I), they react by various mechanisms with one or more amino acid residues in close proximity in the enzyme's active site. This results in covalent bond formation between the enzyme and the inhibitor (E-I) (Equation 17.45).

$$\text{E} + \text{I} \underset{k_{-1}}{\overset{k_1}{\rightleftharpoons}} \text{E} \cdot \text{I} \overset{k_2}{\longrightarrow} \text{E} - \text{I} \quad (17.45)$$

Usually the inhibitor contains an electrophilic moiety that labels amino acids containing nucleophilic groups. However, in some cases, a nucleophilic species may be formed, which can react either with arginine or with any tightly bound organic or inorganic low molecular weight cofactors possessing electrophilic sites. Unlike the mechanism-based inhibitors described below, affinity labels do not require activation by catalysis at the enzyme's active site. Most often, the covalent bond formation occurs by an S_N2 alkylation-type mechanism, Schiff base formation, or acylation (156, 159).

Affinity labels, some of which have become successful therapeutic agents, are often used to identify catalytically important residues. In some cases, by examining the pH dependency of the rate of inactivation, it is possible to determine the pK_a of the labeled residue. Again, there are a number of excellent reviews on this topic (160–163), including a complete volume in the Methods in Enzymology series (159).

Recently, Pratt (164) and Krantz (165) have suggested that any inactivator that utilizes an enzyme's mechanism, in the broadest sense, should be described as a mechanism-based inhibitor. Although this is not unreasonable, we have, for the purposes of this chapter, adopted the more narrow view of Silverman (166). In this view, mechanism-based inhibitors (also called suicide substrates, Trojan horse inactivators, enzyme-induced inactivators, k_{cat} inhibitors, and latent inactivators) are described as unreactive compounds, the structure of which usually resembles that of a substrate or product of the target enzyme, and that undergo a catalytic transformation by the enzyme to species that, before release from the active site, inactivate the enzyme. Thus, these compounds usually contain a latent, reactive functional group that gets activated during the normal catalysis of the enzyme. Upon formation of the initial reversible enzyme-inhibitor complex E · I, the enzyme starts its normal catalytic cycle, leading in a usually rate-determining step to the formation of a highly reactive species, E · I' (Equation 17.46).

$$E + I \underset{k_{-1}}{\overset{k_1}{\rightleftharpoons}} E \cdot I \overset{k_2}{\longrightarrow} E \cdot I' \overset{k_3}{\underset{k_4}{\nearrow \atop \searrow}} \begin{matrix} E-I'' \\ \\ E + P \end{matrix} \quad (17.46)$$

The reactive species can either react with one of the enzyme active-site amino acid residues, to form a covalent bond between the enzyme and the inhibitor (E-I''), or be released into the medium to form product (P) and free

active enzyme (E). In some instances the reaction may occur between the reactive species and the enzyme's cofactor, again resulting in inactivation of the enzyme.

It should also be noted that the activation of a mechanism-based inhibitor by its target enzyme is, formally, an example of metabolic activation. However, there is a clear distinction between the activation of a mechanism-based inhibitor described above and the metabolic activation of a prodrug. In the latter case, an inactive precursor is metabolized in the body (either chemically or enzymatically) to metabolites that possess the desired activity. For example, Acyclovir (**3a**) must be metabolically converted to the triphosphate (**3b**) and released into the medium before it will inhibit viral DNA polymerase. Further discussion on prodrugs may be found in volume 2, chapter 14.

3.1 Evaluation of the Mechanism of Inactivation of Covalently Binding Enzyme Inhibitors

The inherent complexity of the inactivation mechanisms of covalently binding enzyme inhibitors makes it necessary to evaluate their proposed modes of action carefully. An overview of the criteria for the study of irreversible inhibitors is provided below.

3.1.1 Criteria for the Study of Affinity Labels. The evaluation of affinity labels is based on the fulfillment of the following criteria:

1. *Irreversible inactivation.* Inactivation by affinity labels leads to irreversible covalent bond formation between the enzyme and the inhibitor. Unlike the complex between and enzyme and a rapid, reversible inhibitor, the covalent enzyme-inhibitor complex is no longer in equilibrium with free enzyme and inhibitor. Therefore, exhaustive dialysis or gel filtration of the covalent enzyme-inhibitor complex cannot lead to the recovery of free, active enzyme. However, such experiments do not allow distinction among tight-binding, noncovalent inhibitors, affinity labels, and mechanism-based inactivators.

2. *Time- and concentration-dependent inactivation showing saturation kinetics.* The

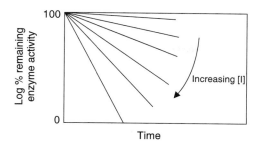

Figure 17.26. Pseudo first-order inactivation kinetics of an active-site directed irreversible inhibitor.

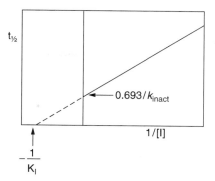

Figure 17.27. Kitz and Wilson plot.

scheme described by Equation 17.45 is analogous to that described for a simple enzyme-catalyzed reaction (Equation 17.9). This scheme can be described by Equation 17.47, which is analogous to the Michaelis-Menten equation (Equation 17.10).

$$-\frac{d[\mathrm{E}]}{dt} = \frac{k_2[\mathrm{I}]}{K_\mathrm{I} + [\mathrm{I}]} \quad \text{where} \quad K_\mathrm{I} = \frac{k_1}{k_{-1}}$$

$$(17.47)$$

According to Equation 17.47, an affinity label should exhibit time- and concentration-dependent inactivation. The rate of inactivation is proportional to low concentrations of inhibitor, whereas at high inhibitor concentrations saturation occurs and no further increase in the rate of inactivation is observed. A typical pseudo first-order plot of log enzyme activity vs. time is illustrated in Fig. 17.26. In some cases nonlinear plots may be obtained, particularly for mechanism-based inhibitors (166, 167).

3. *Saturation kinetics and determination of K_I and k_{inact}.* To distinguish the rate and binding constants of rapid reversible inhibitors (k_{cat} and K_i, respectively) from the rate and binding constants of irreversible inhibitors, the terms k_{inact} and K_I have been used. To determine k_{inact} and K_I, saturation kinetics must be obeyed. Saturation is reached when all of the free enzyme is converted to the reversible enzyme-inhibitor complex $\mathrm{E} \cdot \mathrm{I}$. At that point, the rate of inactivation is independent of k_1/k_- (Equation 17.45), assuming that the rate of

formation of the initial reversible enzyme-inhibitor complex (k_1) is significantly greater than the rate of formation of the covalent enzyme-inhibitor complex (k_2). Consequently, a higher concentration of inhibitor will not lead to an increased rate of inactivation. The K_I value represents the concentration of inhibitor leading to the half-maximum rate of inactivation (in analogy to a K_M value), and k_{inact} is the maximum rate of inactivation at the point of saturation (in analogy to k_{cat}). To determine the K_I and k_{inact} values, the enzyme is incubated at various subsaturating concentrations of the inhibitor, from which the half-life of inactivation at each inhibitor concentration is deduced. Using Kitz and Wilson plots (168), the half-life of inactivation at each inhibitor concentration is plotted against $1/[\mathrm{I}]$. A typical plot is illustrated in Fig. 17.27. The y-intercept represents the half-life of inhibition at infinite inhibitor concentration, with k_{inact} equal to $0.693/t_{1/2}$. K_I can be determined from the x-intercept, which is equal to $-1/K_\mathrm{I}$. If no saturation occurs with a tested inhibitor, the curve will intercept at the origin of the graph, implying that k_{inact} is much faster than the formation of the initial reversible enzyme-mechanism-based inhibitor complex. If this is observed, one might use a lower temperature or a different pH to lower k_{inact}. It should also be noted that, in general, if the affinity label is reacting with an ionizable group that is involved in catalysis, then the pH dependency of $k_{\mathrm{inact}}/K_\mathrm{I}$ should mirror that of $k_{\mathrm{cat}}/K_\mathrm{M}$.

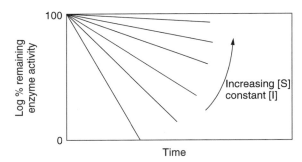

Figure 17.28. Using a substrate to protect an enzyme from inactivation by an active site-directed irreversible inhibitor.

4. *A binding stoichiometry of 1:1 of inhibitor to the enzyme's active site.* In general, complete inactivation of an enzyme requires the binding of one mole of inhibitor per mole of enzyme active site. Exceptions can be certain multimeric enzymes that are inactivated by binding of only one-half mole of inhibitor per mole of enzyme subunit, a phenomenon called half-site reactivity. The stoichiometry of binding is usually determined by incubating an excess of radiolabeled inhibitor with the enzyme to ensure complete irreversible inactivation, followed by either exhaustive dialysis or gel filtration. The binding stoichiometry of the obtained enzyme-inhibitor complex in the absence of free inhibitor is then examined for its radiolabel and protein content. More recently, developments in high-resolution mass spectrometry have allowed the determination of binding stoichiometry without the need for radiolabeled inhibitor.

5. *Substrate protection.* Ligands of the enzyme, either substrates or reversible inhibitors, should greatly decrease the rate of modification by the affinity label. Both affinity labels and mechanism-based inhibitors should be active-site directed, thereby competing with the substrate for the same binding site on the enzyme. This can be tested by incubating the enzyme with increasing amounts of substrate at constant inhibitor concentrations. As the substrate concentration is increased, the rate of inactivation will become slower because, under initial velocity conditions, a portion of the enzyme is protected as the E · S complex. A typical plot of the log of enzyme concentration vs. time at different substrate concentrations is shown in Fig. 17.28.

6. *Verification of covalent bond formation.* In many cases it can be difficult to differentiate between a covalently binding enzyme inhibitor and a very tight but noncovalently binding inhibitor. Although strongly denaturing conditions may not lead to the release of tight, noncovalently bound inhibitors, the covalent linkage between an enzyme and its inhibitor can sometimes be quite labile to nucleophiles and extremes of pH. A frequently used method to determine the covalently modified amino acid residue of an enzyme's active site is peptide mapping. Enzyme-inhibitor complexes, usually prepared from radioactive labeled inhibitor, are treated under mildly denaturing conditions with an appropriate protease. Subsequently, the peptide fragments obtained are usually resolved by high-pressure liquid chromatography and isolated. Analysis of the labeled peptides can be accomplished by Edman degradation and/or mass spectrometry. A good description and example of this method can be found elsewhere (169). Alternatively, electrospray ionization mass spectrometry has been used as a tool to determine the accurate mass of the proteins and enzyme-inhibitor complexes. In a study by Knight et al. (170), this method was successfully used to distinguish between covalent and noncovalent complexes because the latter did not survive the experimental conditions.

3.1.2 Criteria for the Study of Mechanism-Based Inactivators.

In addition to the requirements described above for an affinity label, a mechanism-based inhibitor should also demonstrate the following:

1. *Occurrence of a catalytic step.* The major difference between the mechanism of inactivation of mechanism-based inactivators vs. that of any other type of inhibitor is the obligate involvement of a catalytic step, that is, step 2 in Equation 17.46. Initially, the mechanism-based inhibitor binds reversibly to form the E · I complex. The enzyme then starts its normal catalytic cycle, resulting in the conversion of the inhibitor into a reactive species (I'). If the reactive species is electrophilic, it may react with an active-site nucleophile, much like an affinity label. If the reactive species is nucleophilic, it may react with an electrophilic species on the enzyme, probably an oxidized cofactor. Finally, a radical species may be generated that has the potential to react with an enzyme radical, or generate one by hydrogen atom abstraction. The experiments necessary to provide evidence for a catalytic step are obviously strongly dependent on the individual catalytic mechanism involved. The experiments may include spectrophotometric detection of oxidized or reduced cofactor, observing C—H bond cleavage by monitoring the release of tritium, or the detection of some component of cleaved inhibitor (such as fluoride ion as in some examples shown below).

2. *No release of the activated species before enzyme inactivation.* For a mechanism-based inactivator to retain its high specificity during inactivation, release of the reactive species from the active site must not be part of the normal mechanism of inactivation. A time-dependent increase in the rate of inactivation points to the release of an activated species before inactivation. This increase in the rate of inactivation is brought about by the accumulation of free reactive species in solution. Inhibitors generated in this manner have been termed metabolically activated affinity labels (171). In these cases, as with affinity labels, nonspecific covalent modification of residues other than those located in the active site cannot be excluded. A second test for a metabolically activated affinity label is to add an additional aliquot of fresh enzyme to the incubation buffer. The fresh enzyme should be inactivated at a higher rate than that of the first equivalent of enzyme because there is more reactive species present in solution. By contrast, the mechanism-based inhibitor should show no difference in rate until the concentration of inhibitor is depleted. It should also be noted that the observation of such rate increases necessitates that the reactive species is relatively stable and is not immediately quenched by the incubation buffer.

Additional tests such as the addition of nucleophilic scavengers (e.g., thiols such as dithiothreitol or β-mercaptoethanol) can provide further evidence for the presence of a free, reactive electrophilic species. The scavengers should quench all of the free reactive species, thereby protecting the enzyme from inhibition. Unfortunately, this method cannot exclude the possibility that a nucleophilic thiol may even attack the bound reactive species at the active site of the enzyme (which would also give rise to protection from inactivation). However, the use of a bulky thiol, such as reduced glutathione, should limit that possibility. An alternative scenario occurs wherein the released reactive species returns and reacts faster with an active-site nucleophile than with the added thiol. Clearly this is a complex problem and, consequently, it is advisable to use several different tests to avoid misleading conclusions.

3. *Partition ratio.* The partition ratio is the ratio of product release to enzyme inactivation and is a measure of the efficiency of the mechanism-based inhibitor. Formally, it refers to the ratio k_4/k_3 (Equation 17.46). The most efficient inactivators will have partition ratio of zero. In those cases, theoretically, every enzymatically processed inhibitor molecule will result in the inactivation of a molecule of enzyme. Even though the partition ratio is independent of

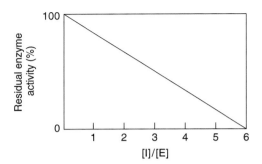

Figure 17.29. Determination of the partition ratio.

the initial concentration of inhibitor, it will depend on factors such as the rate of diffusion of the reactive species from the active site, its reactivity, and the proximity of the target for covalent bond formation. A number of different methods have been used to determine the partition ratio. For example, if, under the experimental conditions, the rate of inactivation is relatively fast compared to the chemical stability of the enzyme or the inhibitor, the partition ratio can be determined kinetically by titration of the enzyme activity. The titration measures the number of inhibitor molecules required to completely inactivate the enzyme. In an experiment of this type, increasing amounts of inhibitor are added to a known, fixed amount of enzyme, and the reaction is allowed to go to completion. After gel filtration or dialysis, a plot of the amount of inhibitor per enzyme active site and the remaining enzyme activity is drawn (Fig. 17.29). The intercept with the x-axis represents the minimum number of equivalents of inhibitor necessary to inactivate the enzyme completely (turnover number). A turnover number of 6, such as that shown in Fig. 17.29, indicates that on average 5 equivalents of inactivator are converted to product and only every sixth equivalent of inhibitor leads to irreversible covalent bond formation (i.e., the partition ratio equals the turnover ratio minus 1). Unfortunately, there are a number of factors associated with this method that may lead to misleading results (166). Another method for determining partition ratios is

equilibrium dialysis of the enzyme with radiolabeled inactivator, followed by determination of the amount of radiolabeled metabolites produced per radiolabeled enzyme. Perhaps the simplest method is for cases where the rate of product formation (i.e., $k_{cat} = k_4$ in Equation 17.46) can easily be measured. In this instance, both k_{cat} and k_{inact} are measured directly, with k_{cat}/k_{inact} being the partition ratio (166, 167).

A more detailed discussion of the requirements for mechanism-based inhibition can be found in a recent review by Silverman (166).

3.2 Affinity Labels

Affinity labels are potentially good drugs, although the presence of a reactive functional group can make them somewhat nonselective and prone toward reaction with other proteins and metabolites. If the affinity label is highly selective toward its target enzyme and has a great affinity for the enzyme's active site, this drawback can be overcome kinetically. Once the inhibitor is bound, the unimolecular reaction between the inhibitor and an amino acid residue in close proximity is entropically quite favorable compared to a bimolecular reaction between two free molecules in solution. This proximity effect has resulted in rate enhancements as great as 10^8 (172) and means that a reagent that is, in itself, only weakly active, may be highly reactive when it is reformulated as an affinity label. More in-depth discussion on this topic can be found elsewhere (39, 173, 174).

The design of a potent affinity label requires the study of the initial requirements for the inhibitor to bind to the active site. Next, regions of bulk tolerance are determined that are useful for the introduction of a reactive functional group. In some cases, it might be advantageous to place the reactive group at the end of a spacer arm, particularly if no nucleophilic amino acid residue is in close proximity to the reactive group. However, not only the location and orientation, but also the size and inherent reactivity of the reactive functional group are critical for its potential as an affinity label.

Perhaps the archetypical example of an affinity label is TPCK (**66**) (Fig. 17.30). This

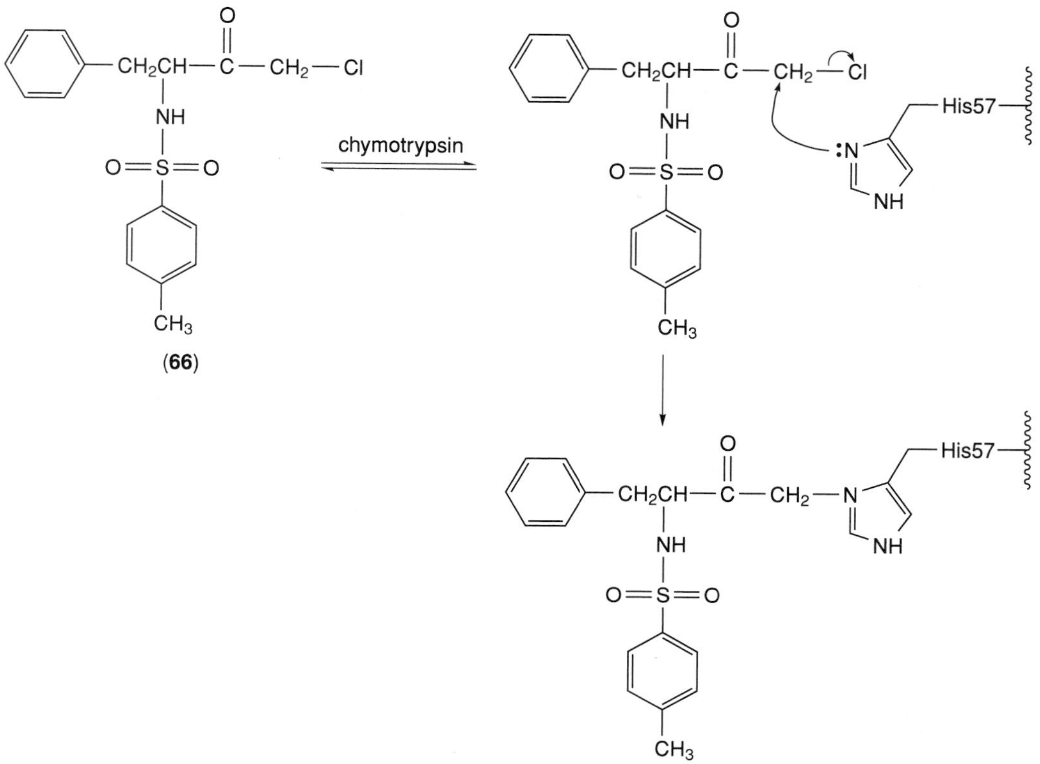

Figure 17.30. Inhibition of chymotrypsin by TPCK.

compound was designed to mimic substrates of chymotrypsin such as the tosyl-L-phenylalanine methyl ester (Equation 17.48), thereby providing a basis of affinity for the chymotrypsin-active site.

In addition to mimicking a substrate, it contains the halomethyl ketone moiety, to provide a point of covalent attachment (175). TPCK was shown to irreversibly inhibit chymotrypsin (it is still employed today to remove chymotrypsin from trypsin preparations) by specifically labeling a histidine residue (175), later identified as His57 (176). After the success of TPCK, chloromethyl ketones became

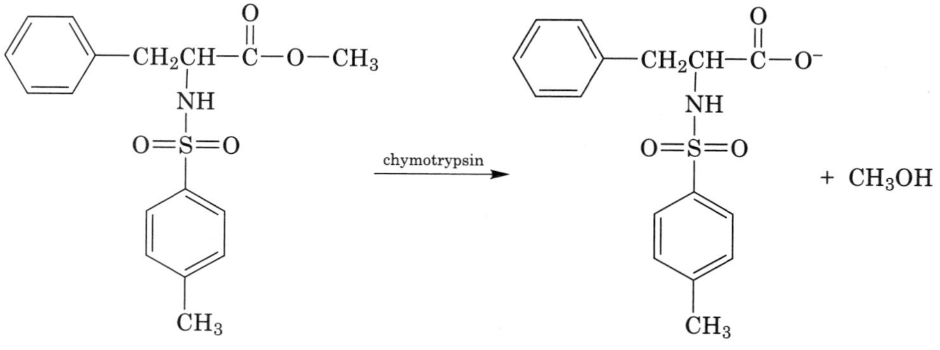

$$+ \ CH_3OH \tag{17.48}$$

methyl N-tosyl-L-phenylalanine

extremely popular for the inactivation of proteases. By incorporating part of the sequence of the physiological substrate into the halomethyl ketone, it was possible to obtain selective inactivation of individual proteases (177). This selective inactivation also meant that chloromethyl ketones became widely used as probes for the binding requirements and chemically reactive residues in the active sites of serine proteases, in particular. Replacement of the chloromethyl ketone moiety by a diazomethyl group provided a specific inactivation of cysteine proteases (172). The use of TPCK has not been restricted to chymotrypsin, as elegantly demonstrated in a recent report on the inhibition of human aldehyde dehydrogenase (178). As a group, proteases remain major targets for therapeutic intervention, and peptide-based affinity labels are still playing a major role in drug design (179).

The interconversion of (R)-mandelate and (S)-mandelate is catalyzed by mandelate racemase (Equation 17.49). The reaction can be reversibly inhibited by the substrate analog atrolactate (**67**), (Fig. 17.31). Because of its structural similarity to both (**67**) and the substrate and, given the reactivity of the epoxide group to nucleophiles, (R,S)-α-phenylglycidate (**68**) was synthesized as a potential affinity label of mandelate racemase. The compound was found to be an irreversible inhibitor, fitting all the criteria described in section 3.1.1 (180). Later it was established that (S)-α-phenylglycidate (S-αPGA) did not irreversibly inactivate the enzyme, binding noncovalently and with less affinity than R-αPGA (**69**). As shown in Figure 17.31, the epoxide ring of R-αPGA potentially is subject to attack at either of two carbons. Attack at the distal endocyclic carbon atom of (**69**) (path a) will result in the formation of (**70**), whereas attack at the α-carbon (path b) will yield (**71**). The crystal structure of the inactivated complex revealed that nucleophilic attack of the ε-amino group of Lys166 resulted in adduct (**72**), which is consistent with attack on the distal carbon of the epoxide ring (181). This structure confirmed the original design premise of Fee et al. (180), wherein it was thought that the distal oxirane carbon occupied the position similar to the α-proton in mandelate. Therefore, on binding of the

α-phenylglycidate to the enzyme, the electrophilic epoxide group would be subject to attack by the nucleophile responsible for α-proton abstraction in the normal catalytic cycle. Further confirmation is provided by the X-ray structure of (S)-atrolactate bound to the racemase (181), which reveals that Lys166 has been pushed away by the α-methyl group of (S)-atrolactate (which is positionally equivalent but much larger than the α-proton in (S)-mandelate). In both structures the positions of the remaining active-site residues are almost identical.

(R)-mandelate

(17.49)

(S)-mandelate

Perhaps the best-known affinity labeling reagent is aspirin (**73**) (Fig. 17.31), a member of the class of drugs known as the nonsteroidal anti-inflammatory drugs (NSAIDS), and whose activity was initially reported to result from its inhibition of prostaglandin biosynthesis (182, 183). Prostaglandins are involved in the inflammatory response and can cause headache and vascular pain in humans.

Prostaglandin synthase, which catalyzes the first step in the arachidonic acid cascade, is a heme protein and possesses two activities. As illustrated in Equation 17.50, a cyclooxygenase activity is used in the conversion of arachidonic acid to the bicyclic endoperoxide PGG_2, whereas a peroxidase activity catalyzes the subsequent reduction of PGG_2 to prostaglandin H_2. The latter serves as a branch point in the production of various prostaglandins as well as thromboxane A_2 and prostacyclin (PGI_2).

Aspirin (acetylsalicylic acid) was ultimately confirmed as an inhibitor of prostaglandin synthetase (184). Incubation of [*acetyl*-^{3}H] aspirin showed that one acetyl group was in-

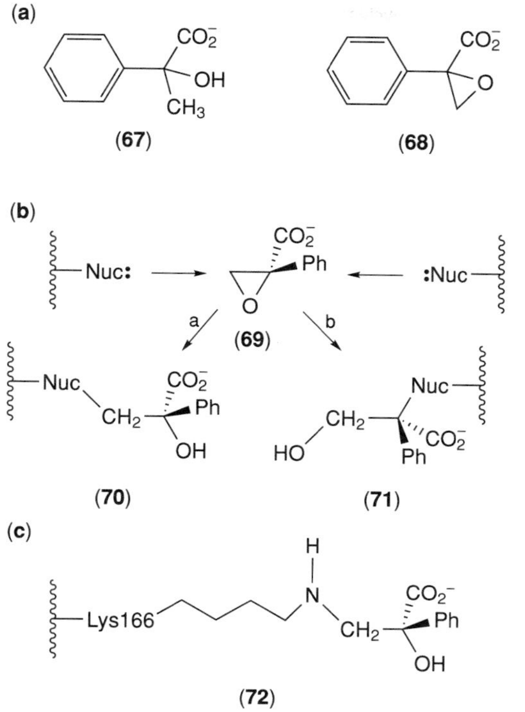
(a)
(67)
(68)
(b)
Nuc:
:Nuc
(69)
(70)
(71)
(c)
Lys166
(72)

Figure 17.31. (a) Inhibitors of mandelate racemase, (b) potential products from nucleophilic attack on the epoxide of R-α-phenylglycidate, and (c) adduct formed by attack of Lys166.

corporated per mole of enzyme (185, 186). Enzymatic digest of the labeled enzyme provided evidence that a serine residue, later identified as Ser530, was acetylated (186, 187), probably by the mechanism shown in Fig. 17.32. The X-ray structure of bromoaspirin (**74**) inactivated prostaglandin synthase has been solved, and the bromoacetylation of Ser530 confirmed (188). The structure was similar to that of flurbiprofen (**75**), complexed with prostaglandin synthase (189). Aspirin is the only NSAID known to inactivate prostaglandin synthase through covalent modification, and the brominated aspirin analog was also determined to be a potent irreversible inhibitor. Conversely, flurbiprofen (**75**), another NSAID, has been classified as a slow-tight-binding inhibitor (190) and was expected to induce a conformational change upon binding. However, there were no significant differences between the two X-ray structures, and it is yet to be determined whether the binding of aspirin also induces a conformational change in the enzyme.

Although affinity labels have played a major role in characterizing the active sites of a large number of proteases, they have also proved to be particularly useful in mapping nucleotide-binding sites (161, 163, 191). Numerous compounds, some of which are shown in Fig. 17.33, have been designed to be analogs of the various nucleosides and nucleotides. Perhaps the best known of these is 5′-p-fluorosulfonylbenzoyl adenosine (5′-FSBA) (**76**), which was designed to be an analog of ADP or ATP (**77**). It has both the adenosine and ribose moieties, as well as a carbonyl group adjacent to the 5′ position. The latter mimics the first phosphoryl group of the purine nucleotides. If

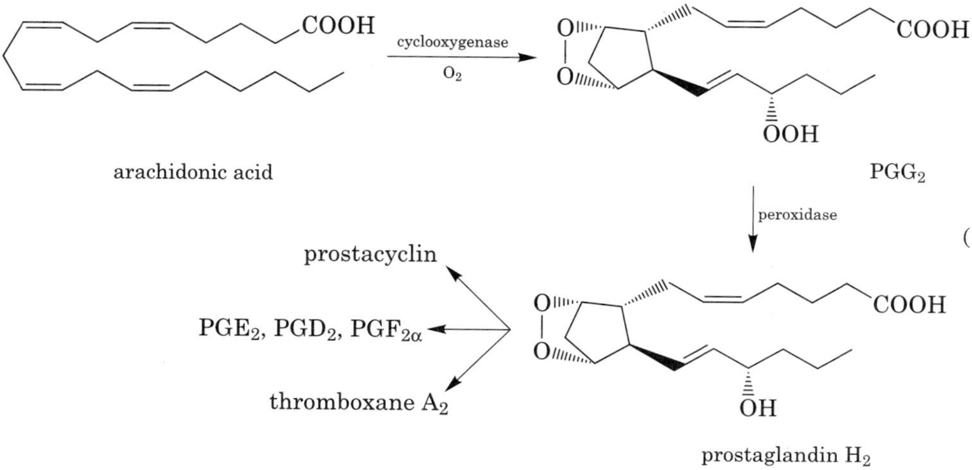
COOH
cyclooxygenase
O_2
arachidonic acid
COOH
OOH
PGG_2
peroxidase
prostacyclin
PGE_2, PGD_2, $PGF_{2\alpha}$
thromboxane A_2
COOH
OH
prostaglandin H_2

(17.50)

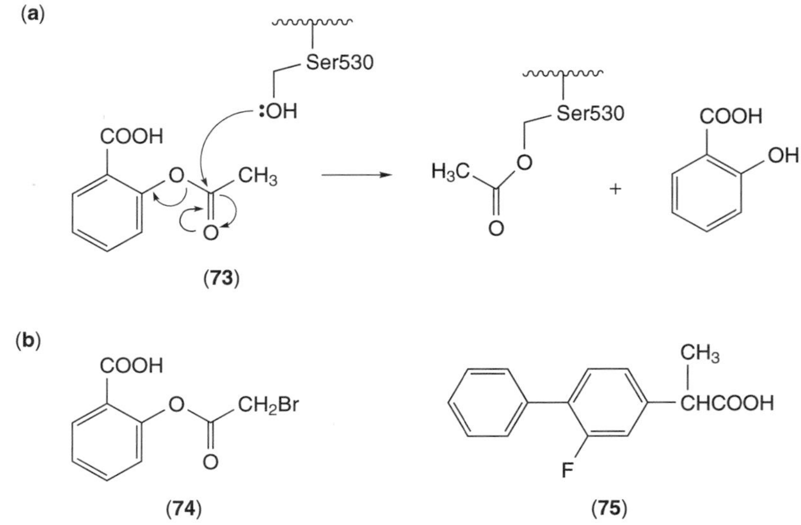

Figure 17.32. (a) Inactivation of prostaglandin H2 synthase by aspirin, and (b) inhibitors cocrystallized with prostaglandin synthase.

the molecule is arranged in an extended conformation, the reactive sulfonyl fluoride group would be found in a position analogous to that occupied by the γ-phosphoryl group of ATP. This was initially used to explore the regulatory site of glutamate dehydrogenase (192) and the active site of pyruvate kinase (193). It has now been employed to label the NAD and ATP sites of more than 50 proteins (163). Modifications to 5'-FSBA have provided the fluorescent probe (**78**) as well as the bifunctional affinity label (**79**), which has a photoactivatable azido group as well as the electrophilic fluorosulfonyl moiety (163). The bromodioxybutyl compound (**80**) contains the adenine, ribose, and 5'-monophosphate of adenosine monophosphate (AMP). It is also water soluble and negatively charged at neutral pH. As described above, a bromomethyl ketone group will react with a number of nucleophiles, whereas the dioxo group can potentially react with arginine residues. This reagent has a structural similarity to adenylosuccinate (**81**) and was used to identify a critical arginine residue in the active site of adenylosuccinate lyase, an enzyme whose deficiency in humans leads to severe mental retardation and autism (163).

3.3 Mechanism-Based Inhibitors

Mechanism-based inactivators have great potential as drugs because they are designed to be specific toward their target enzyme. Furthermore, because these compounds are unreactive until activated within their target enzyme, they are expected to show little or no cellular toxicity. The design of mechanism-based inhibitors requires an understanding of the binding specificity requirements for the ligand-recognition site of the enzyme, to promote the formation of the initial noncovalent enzyme-inhibitor complex E·I (Equation 17.46). In addition, the choice of an appropriate latent functional group requires knowledge of the catalytic mechanism of the target enzyme with its normal substrate. Finally, covalent bond formation by the activated inhibitor (I') will strongly depend on its inherent chemical reactivity, and its proximity to a susceptible amino acid residue or cofactor. A number of excellent reviews and monographs have appeared on the general design of mechanism-based inhibitors (166, 167, 194–201). The following examples have been chosen to emphasize both the potential for the use of

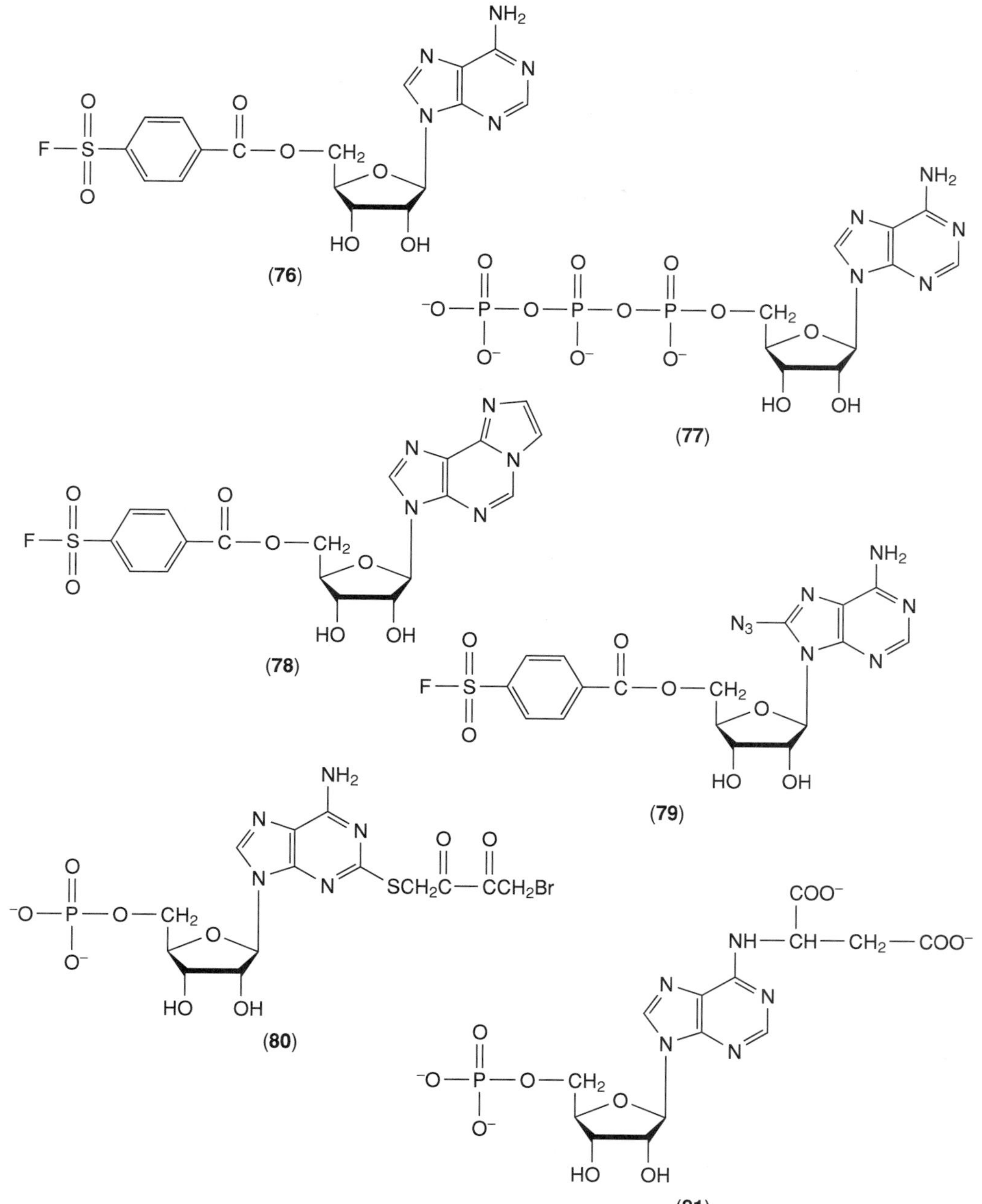

Figure 17.33. ATP (**77**), adenylosuccinate (**81**), and representative affinity label analogs.

mechanism-based inhibitors as drugs and the diversity of their mechanisms of inactivation.

Of all the classes of enzymes, the pyridoxal phosphate (PLP)-dependent enzymes have been found to be most susceptible to mechanism-based inhibitors (202). To some extent this is because the mechanism of catalysis by the PLP-dependent enzymes is extremely well characterized, making the design process somewhat easier. The initial steps in the mechanism for a PLP-dependent enzyme are shown in Equation 17.51.

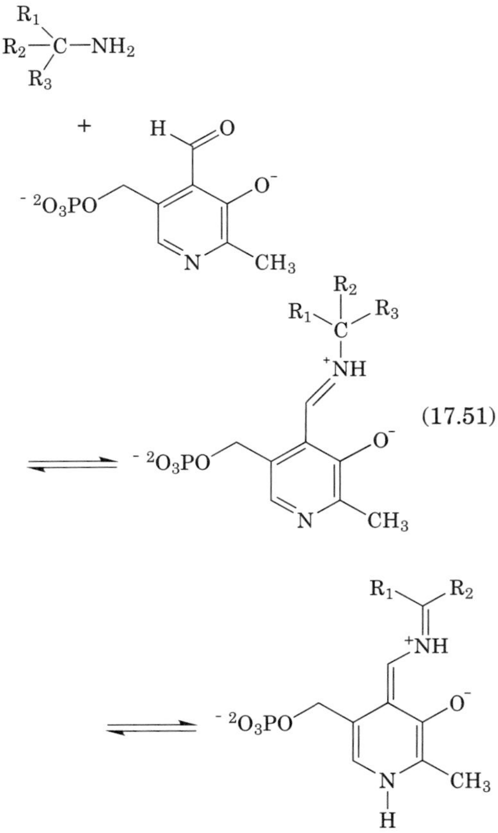

(17.51)

The first step involves Schiff base formation by the amino group of the substrate reacting with pyridoxal phosphate to form an aldimine. This is followed by loss of a functional group (R_3, Equation 17.51), usually by abstraction by an active-site base, to form a resonance-stabilized carbocation.

γ-Aminobutyric acid (GABA) is one of the major inhibitory neurotransmitters in the mammalian central nervous system. A decrease in the concentration of GABA had been shown to lead to convulsions. Therefore it was suggested that inhibitors of GABA transaminase, the enzyme responsible for the breakdown of GABA, may act as antiepileptic

agents, by providing an increase in the concentration of GABA in the brain.

GABA aminotransferase (GABA transaminase, GABA-T) catalyzes the conversion of γ-aminobutyric acid to succinic semialdehyde with the subsequent transfer of an amino group to pyruvate (Equation 17.52).

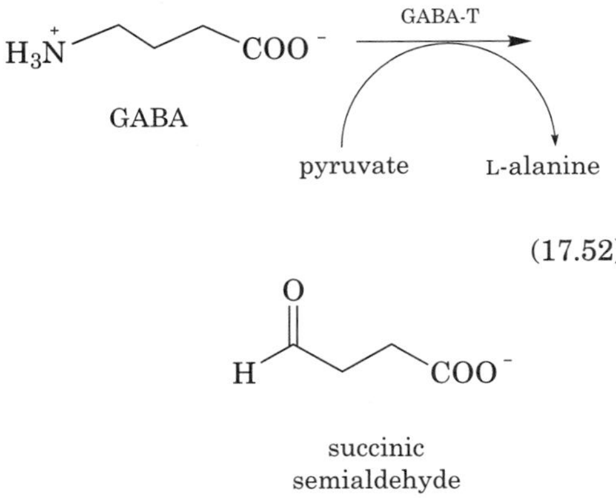

(17.52)

GABA-T is a PLP-dependent enzyme and a wide variety of mechanism-based inhibitors have been described for this enzyme (202, 203). These include inhibitors bearing an unsaturated moiety, a leaving group, as well as those forming a stable complex with the cofactor (202, 203). Vigabatrin (**8**), currently used as an antiepileptic drug, provides an excellent example of this approach. The proposed mechanism is shown in Fig. 17.34.

As with the normal mechanism of the enzyme, the inactivation starts with Schiff base formation with the enzyme-bound pyridoxal phosphate, followed by removal of an α-proton by an active-site base to form the reactive electrophilic intermediate (**82**). This then partitions between hydrolysis of the Schiff base linkage, resulting in the keto product (**83**) and enzyme reactivation, and Michael-type addition of an enzyme active-site nucleophile, resulting in a stable covalently bonded enzyme adduct (**84**).

Ornithine decarboxylase (ODC), another PLP-dependent enzyme, catalyzes the rate-limiting step in the biosynthesis of polyamines, i.e., the conversion of ornithine to putrescine (Equation 17.53).

The enzyme is a target for drugs against African sleeping sickness caused by *Trypano-*

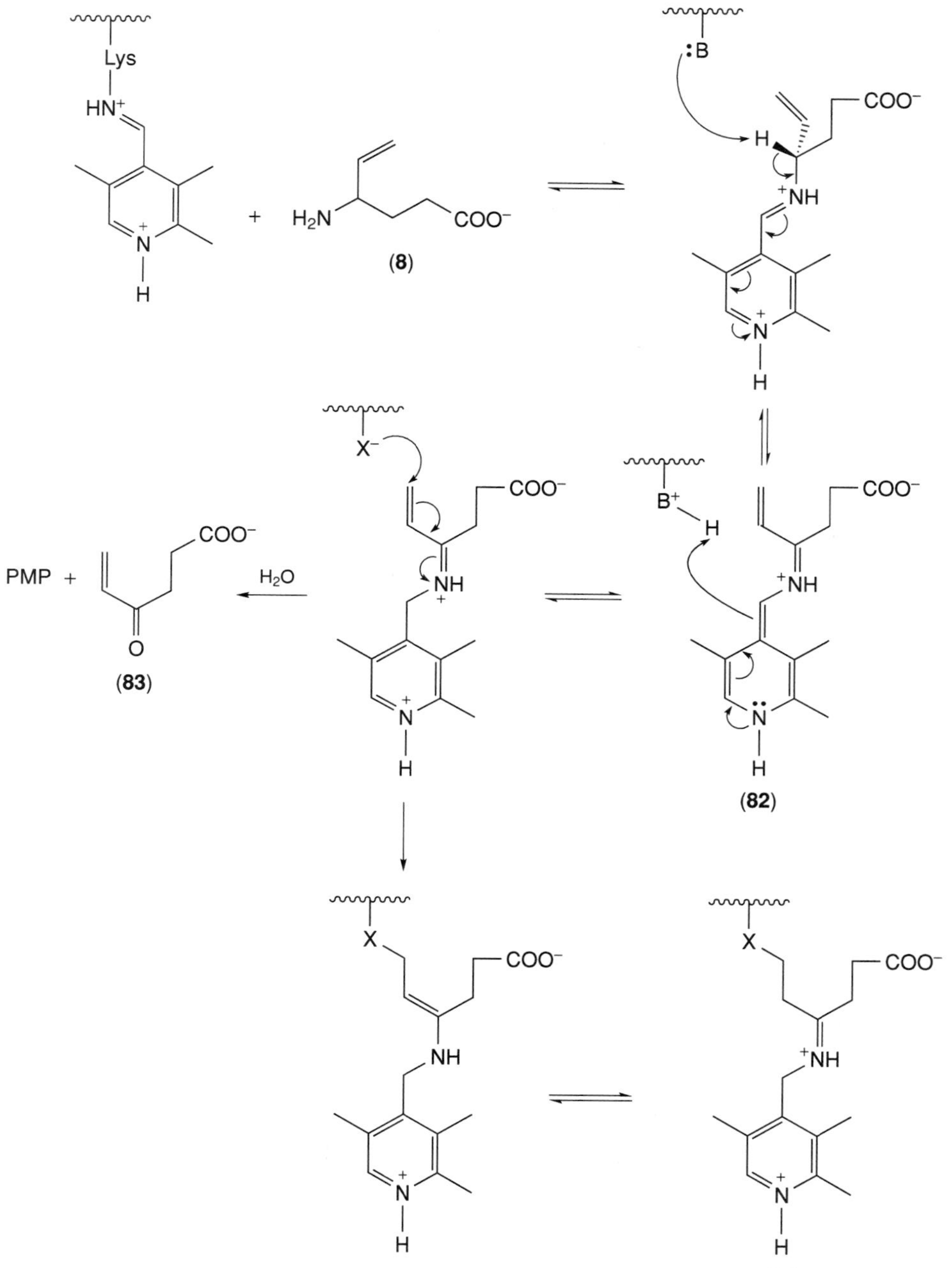

Figure 17.34. Inactivation of GABA transaminase by vigabatrin (PMP = pyridoxamine phosphate). For clarity, substituents on the pyridine ring are omitted.

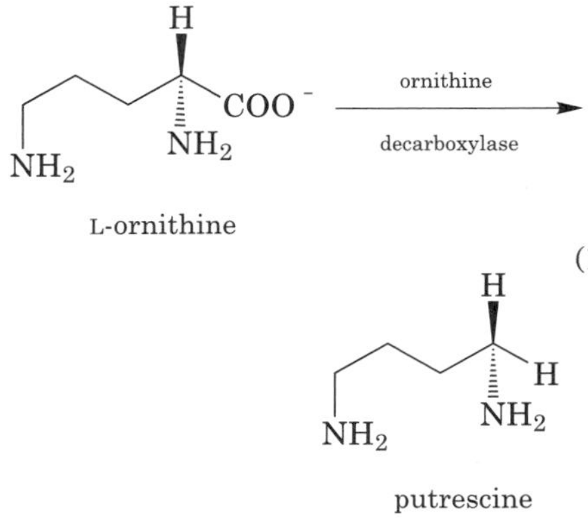

L-ornithine

ornithine
decarboxylase

(17.53)

putrescine

testosterone

5α-reductase
NADPH

(17.54)

dihydrotestosterone

soma brucei. One of the currently used drugs, eflornithine (α-difluoromethylornithine, DFMO) (**85**), is a mechanism-based inhibitor of ODC. The inactivation of ODC by DFMO involves the decarboxylation of DFMO by the enzyme, with subsequent stoichiometric binding of a reactive species to the enzyme (204). The proposed mechanism for inhibition of *T. brucei* ODC is outlined in Fig. 17.35.

DFMO initially forms a Schiff base with PLP (**86**), then, following decarboxylation, a fluoride ion is eliminated, thereby generating the electrophilic conjugated imine (**87**). Attack by the nucleophilic thiol group of Cys360 and subsequent elimination of a second fluoride anion yields a second conjugated imine (**88**). The second imine then undergoes a transaldimination reaction with the amino group of Lys69. The enamine formed in this reaction (**89**) may then undergo an internal cyclization to yield a cyclic imine (**90**). This is the main product formed by the alkylation of ODC by DFMO (204). Recently, the X-ray structure of the DFMO-inactivated *T. brucei* ODC K69A mutant was solved (205). This structure was of the second conjugated imine (**88**) complexed with ODC and showed that, within a single active site, the decarboxylated DFMO bridges two subunits, forming a Schiff base with the PLP on one monomer and a covalent bond to Cys360 on the second monomer.

The enzyme steroid 5α-reductase is an NADPH-dependent enzyme that catalyzes the conversion of testosterone to dihydrotestosterone, a more potent androgen (Equation 17.54).

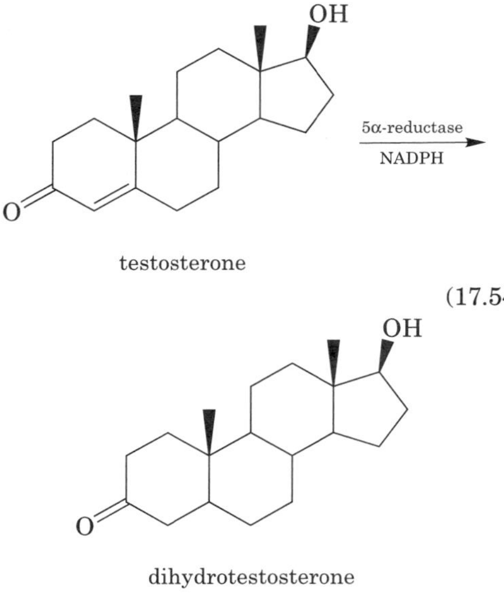

Dihydrotestosterone, rather than testosterone, had been implicated in endocrine disorders such as acne, enlargement of the prostate, and male pattern baldness, and it was suggested that 5α-reductase was an attractive therapeutic target. Initially, finasteride (**91**) (Fig. 17.36) was developed as a potent reversible inhibitor of 5α-reductase with a K_i in the low nanomolar range (206). Closer examination revealed that finasteride appeared to be a slow-binding, high-affinity inhibitor of the human prostate (type 2) 5α-reductase, with a K_i of less than 1 nM (207). Finasteride is currently the drug of choice in the treatment of benign prostatic hyperplasia, and it is now thought that finasteride is, in fact, a mechanism-based inhibitor (208, 209), which acts through an enzyme-bound NADP-dihydrofinasteride adduct (Fig. 17.36).

In this mechanism, reduction of finasteride (**91**) leads to the formation of an enolate (**92**) and the subsequent formation of an adduct with NADP$^+$ (**93**) (where PADPR = phosphoadenonsine diphosphoribose). The dissociation constant for the enzyme-inhibitor complex is less than 10^{-13} M, and the partition ratio for the enzyme-catalyzed formation of dihydrofinasteride (**94**) is less than 1.07 (208). Clearly, finasteride is an extremely efficient mechanism-based inhibitor. As shown in Fig. 17.36, the NADP$^+$-finasteride adduct will

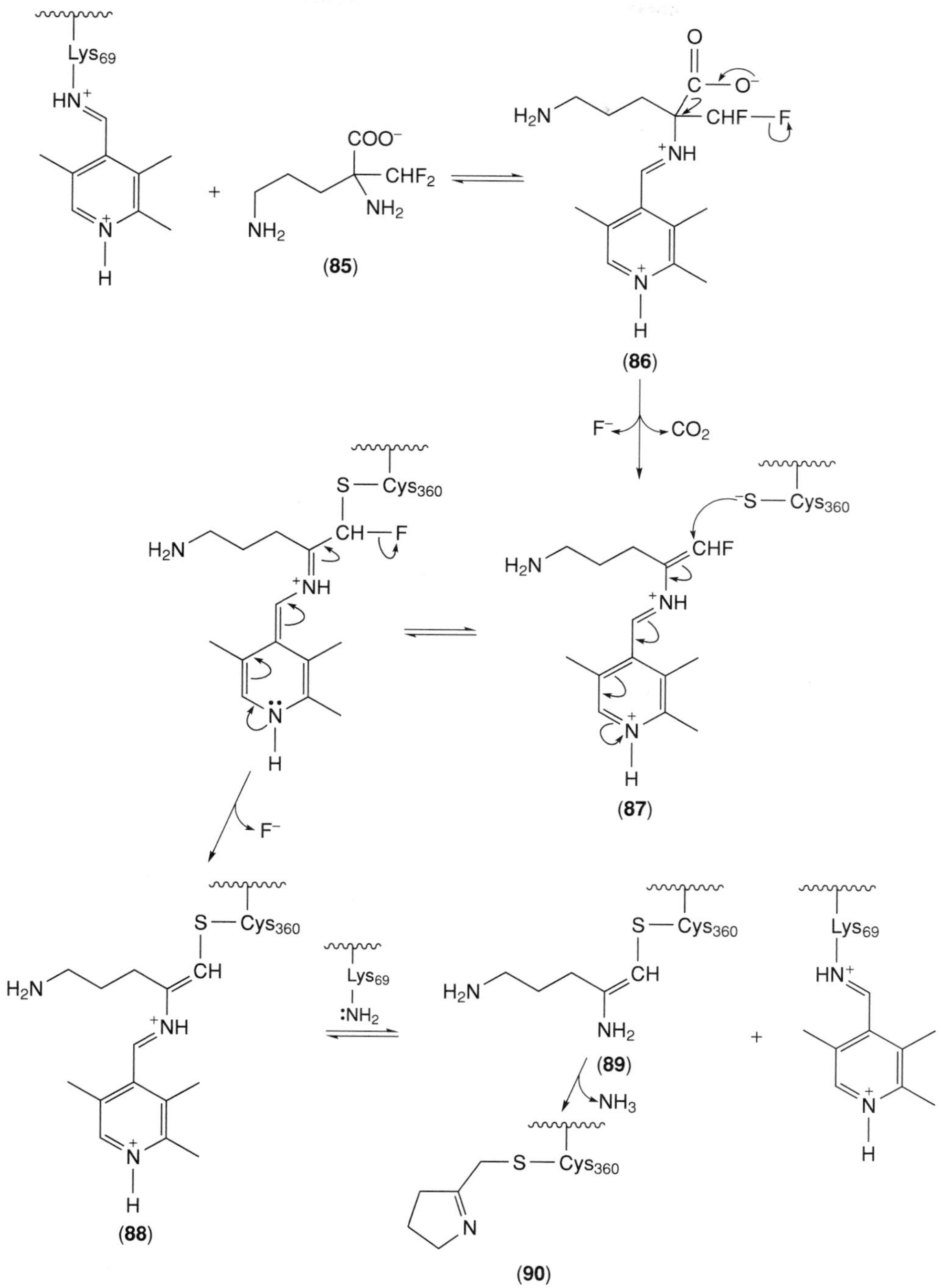

Figure 17.35. Inactivation of ornithine decarboxylase by eflornithine. For clarity, substituents on the pyridine ring are omitted.

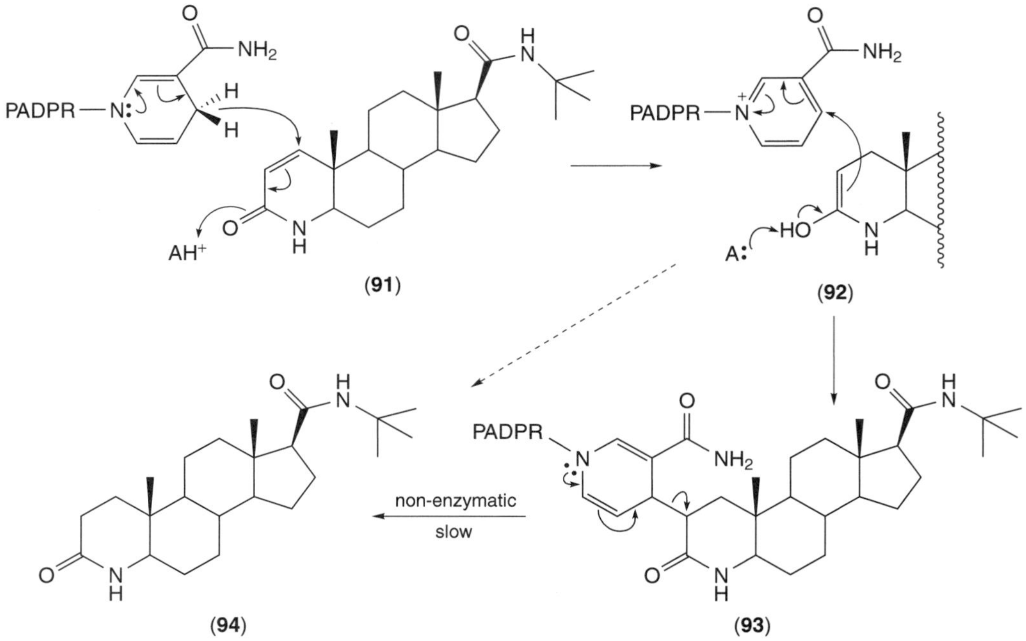

Figure 17.36. Inhibition of steroid 5α-reductase by finasteride (PADPR = phosphoadenosine diphosphoribose).

eventually dissociate and form dihydrofinasteride (**94**), although the half-life of 14 days also points to the effectiveness of finasteride as a steroid 5α-reductase inhibitor.

Enzymes involved in steroid biosynthesis have proved to be good targets, both for therapeutic intervention and for mechanism-based inactivators (2). Aromatase, for example, catalyzes the final, rate-limiting step in estrogen biosynthesis (Equation 17.55). Aromatase has proved susceptible to mechanism-based inhibitors such as formestane and exemestane. These are now both used in the treatment of breast cancer (210).

In the last decade there has been a considerable increase in the occurrence of antibiotic-resistant microbial pathogens. Vancomycin, one of the last resort antibiotics for treating some gram-positive bacterial infections, inhibits peptidoglycan synthesis by binding the terminal D-alanyl-D-alanine (D-Ala-D-Ala) dipeptide from pentapeptide precursors of *Enterococcus* cell walls. VanX is a zinc-dependent D-Ala-D-Ala dipeptidase (Equation 17.56), which has been implicated in high-level resis-

tance to vancomycin (211, 212). As a consequence, VanX has become a prime drug target for overcoming vancomycin resistance and a number of transition-state analogs have been prepared (213, 214).

The enzyme was also shown to process dipeptides with bulky C-terminal amino groups (213) and, using this knowledge, a novel mechanism-based inhibitor was recently developed (215). Its mechanism is shown in Fig. 17.37. D-Ala-D-Gly($S\Phi p$-CHF$_2$)-OH (**95**) is a dipeptide-like analog of D-Ala-D-Ala and is readily accepted by VanX. Cleavage of the peptide bond and elimination of D-alanine results in the formation of the metastable 2-p-difluoromethylthioglycine (**96**), which spontaneously decomposes, yielding ammonia, glyoxylic acid, and p-difluoromethyl thiophenol (**97**). Elimination of a fluoride ion results in the electrophilic 4-thioquinone fluoromethide (**98**), which irreversibly alkylates the enzyme (**99**). Interestingly, the turnover of the analog was faster than that of D-Ala-D-Ala itself. However, the partition ratio of 7500 in-

testosterone

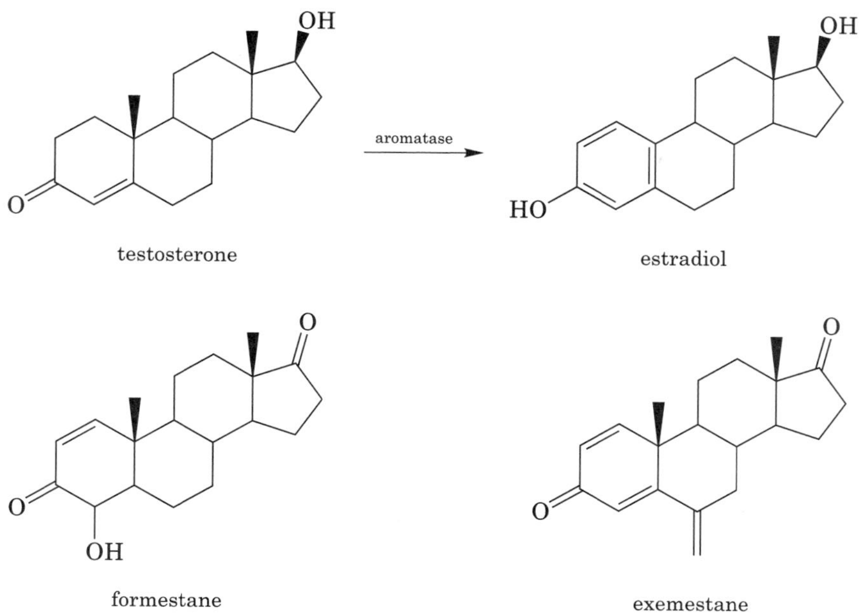

aromatase →

estradiol

(17.55)

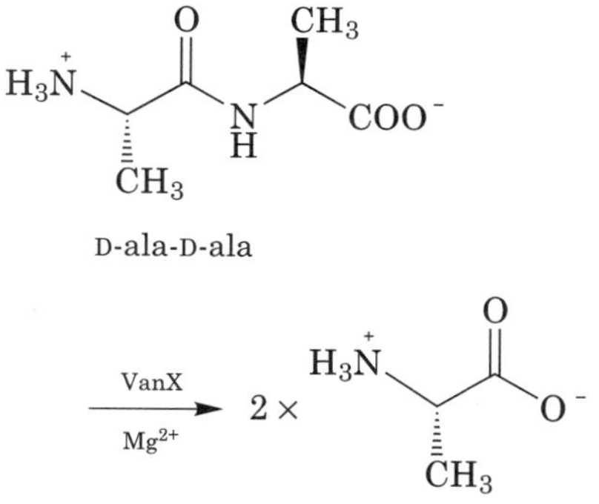

formestane

exemestane

D-ala-D-ala (17.56)

$$\xrightarrow[\text{Mg}^{2+}]{\text{VanX}} 2 \times$$

dicated that one of the reactive intermediates must be relatively long-lived (215).

More detailed examples of approaches used to design mechanism-based inhibitors may be found in excellent reviews by Silverman (166, 167) and by Ator and Ortiz de Montellano (216).

3.4 Pseudoirreversible Inhibitors

Pseudoirreversible inhibitors are the least common of the covalently binding enzyme inhibitors. They have some features in common with both affinity labels (Section 3.2) and mechanism-based inhibitors (Section 3.3) but they have one distinguishing feature; that is,

the covalent bond formed between the enzyme and the inhibitor is reversible (Equation 17.57). As with the affinity labels, initially they bind to the enzyme's active site in a non-covalent fashion to form an enzyme-inhibitor complex $E \cdot I$ but, unlike an affinity label, the pseudoirreversible inhibitor generally possesses unreactive functional groups. As with the mechanism-based inhibitor, the enzyme then starts the catalytic cycle and an active-site residue, usually one involved in covalent catalysis (60), reacts with the inhibitor, without producing a highly reactive species, and forms a covalent bond.

$$E + I \underset{k_{-1}}{\overset{k_1}{\rightleftharpoons}} E \cdot I \underset{k_{-2}}{\overset{k_2}{\rightleftharpoons}} E - I' \overset{k_3}{\longrightarrow} E + P \quad (17.57)$$

The covalently bound inhibitor mimics the normal covalent reaction intermediate occurring during the normal reaction mechanism. However, the covalent adduct is far more stable, with half-lives on the order of several hours to days. The free enzyme may then, depending on the lability of the E-I' bond, be regenerated by hydrolysis or reversal of the covalent bond. The utility of a pseudoirrevers-

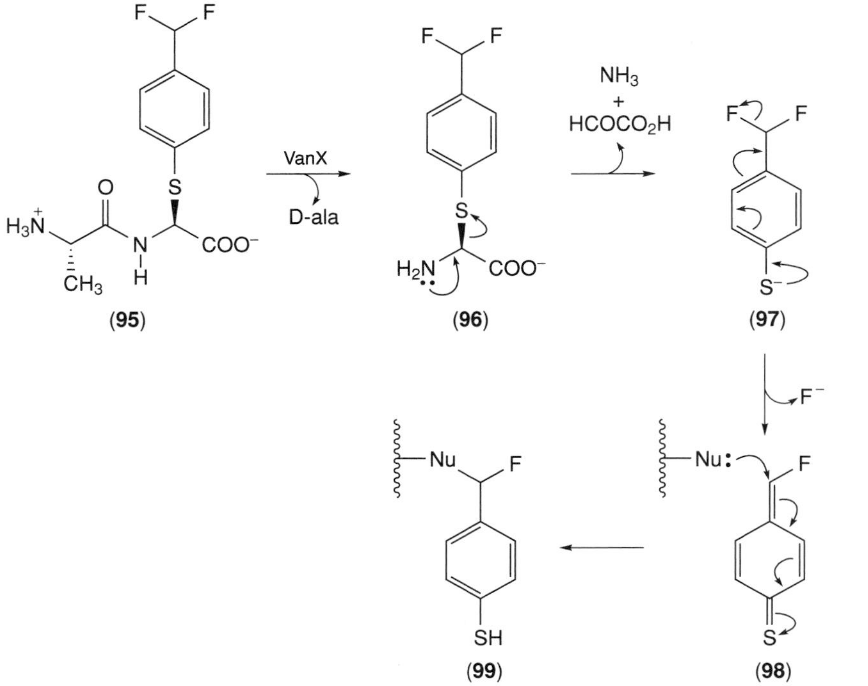

Figure 17.37. Mechanism-based inhibition of VanX.

ible inhibitor will be determined by a combination of the rate of formation of the covalent enzyme inhibitor adduct and the half-life for reactivation.

As may be expected, criteria for the study of pseudoirreversible inhibitors are very similar to those for both affinity labels and mechanism-based inhibitors. However, because of the inherent reversibility of pseudoirreversible inhibitors, it may be more difficult to obtain structural evidence for the covalent enzyme inhibitor adduct. Further, determination of the rate of reactivation and characterization of the products of the recovery process will also be of major importance in designating an inhibitor as pseudoirreversible.

Pseudoirreversible inhibitors can be broken into two classes, depending on how the active enzyme is regenerated. In the first class, exemplified by inhibitors of acetylcholinesterase, the enzyme is regenerated as the covalent E-I′ bond is hydrolyzed (i.e., $k_3 \gg k_{-2}$). As shown in Equation 17.58 , acetylcholinesterase catalyzes the hydrolysis of acetylcholine, yielding choline and acetate.

Acetylcholine is a neurotransmitter that relays nerve impulses across the neuromuscular junction. Acetylcholinesterase (AcChE) rapidly breaks down acetylcholine, thereby lowering its concentration in the synaptic cleft and ensuring that nerve impulses are of a finite length. As shown in Fig. 17.38, a nucleophilic serine residue reacts with the substrate to form an acetyl-serine intermediate (**100**) with concomitant release of choline. This intermediate is then rapidly hydrolyzed by wa-

$$\underset{\text{acetylcholine}}{H_3C-\overset{\overset{O}{\|}}{C}-O-CH_2CH_2\overset{+}{N}(CH_3)(CH_3)-CH_3} \xrightarrow[H_2O]{\text{acetylcholinesterase}} \underset{\text{acetate}}{H_3C-\overset{\overset{O}{\|}}{C}-O^-} + \underset{\text{choline}}{HO-CH_2CH_2\overset{+}{N}(CH_2)(CH_3)-CH_3} \quad (17.58)$$

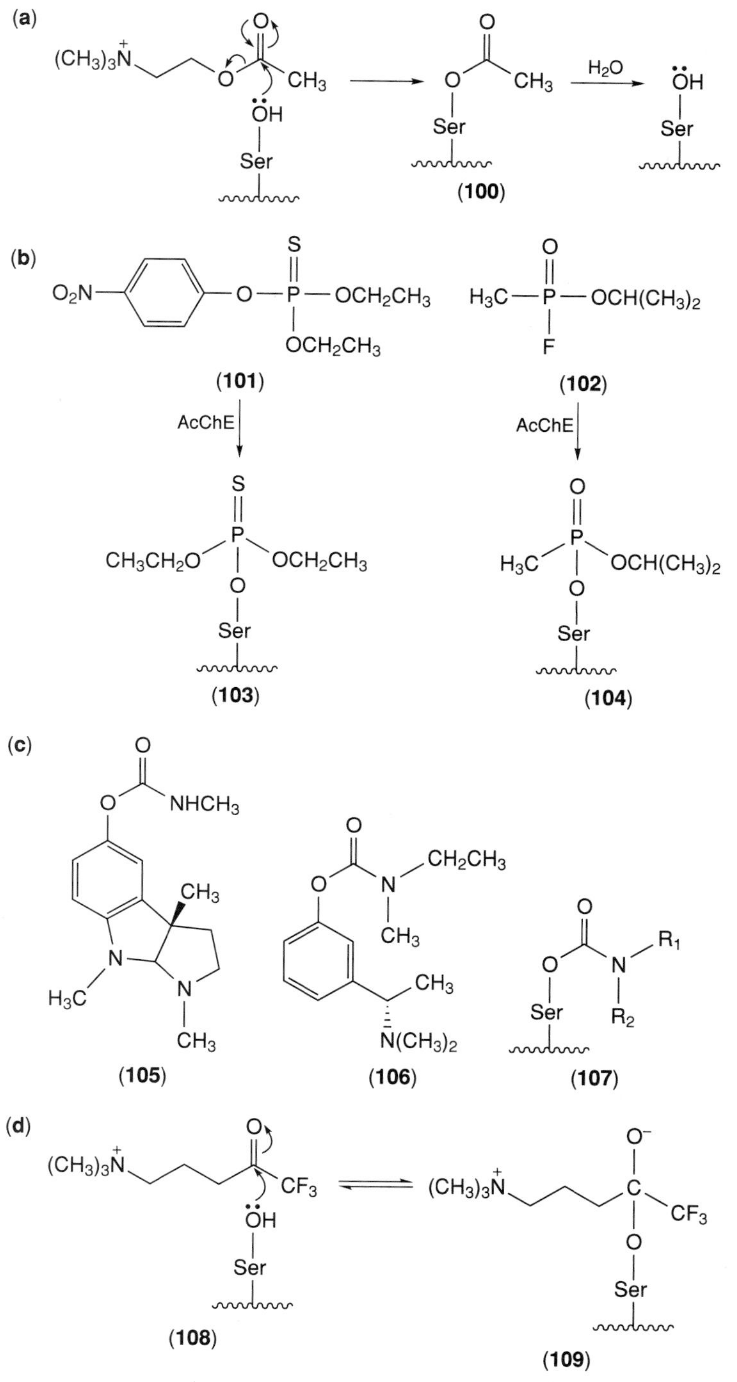

Figure 17.38. (a) Mechanism of reaction, (b) irreversible inhibitors, and (c,d) pseudoirreversible inhibitors of acetylcholinesterase.

ter, producing acetate and regenerated enzyme. Agents such as parathion (**101**) and sarin (**102**) have found utility as insecticides and nerve gases, respectively, because they react with the enzyme to form the active-site serine-phosphate esters, (**103**) and (**104**). These esters are hydrolyzed extremely slowly by water, making the inhibition effectively irreversible (i.e., both k_{-2} and k_3 are very small), although the inhibition can be overcome with high concentrations of strong nucleophiles such as hydroxylamine.

More recently, it has been established that inhibitors of acetylcholinesterase may play a role in the memory enhancement in patients with Alzheimer's disease (217). Unlike (**101**) and (**102**), carbamate inhibitors such as physostigmine (**105**) and rivastigmine (**106**) are classified as pseudoirreversible inhibitors because they react with AcChE to form a carbamylated serine (**107**). By comparison with the serine-phosphate ester, the carbamylated serine is rapidly hydrolyzed, thereby regenerating AcChE. For example, reactivation of the physostigmine-inactivated enzyme is rapid, with a $t_{1/2}$ of less than 40 min (218). Rivastigmine, a more useful therapeutic agent, is considerably longer acting, with a half-life of more than 10 h (217, 219). Overall, for pseudoirreversible inhibitors of this type, the effectiveness and duration of the "irreversible" inhibition will be controlled by the chemical nature of the groups transferred to the active-site nucleophile, making it readily amenable to manipulation.

In pseudoirreversible inhibitors of the second class, the enzyme is regenerated by the inhibitor simply dissociating from the enzyme; that is, the binding is covalent but reversible ($k_{-2} \gg k_3$). This class can also be exemplified by an AcChE inhibitor. For example, the trifluoromethyl ketone (**108**) binds to AcChE as a slow-binding inhibitor (Section 2.4.1) with a K_i value of 0.06 nM, and a k_{off} value of 6.7×10^{-6} s^{-1} (220). A linear correlation was observed between K_i values of a series of fluoromethyl ketones and the V_{max}/K_i value for the corresponding substrate (220). This suggests (127) that the tetrahedral adduct (**109**), in effect, mimics the transition state (or a high-energy intermediate), thereby accounting for the high affinity (Section

2.5.3). The affinity of the inhibitor for AcChE could be decreased (with a concomitant increase in the value of k_{off}), by sequentially reducing the number of fluorine atoms into the methyl group adjacent to the ketone (220). Finally, it should be noted that the two classes of pseudoirreversible inhibitor can be differentiated by examining the decomposition products of the inhibition reaction. When hydrolysis is required for enzyme regeneration, cleavage products, such as substituted carbamates, will be in evidence. Conversely, the trifluoromethyl ketones will not be broken down by AcChE and no decomposition products will be observed.

4 CONCLUSIONS

Enzyme inhibitors have long played an important role in medicine, pharmacology, and basic research. The advances in DNA technology have enabled cloning and overexpression of large numbers of enzymes, and the approaches described in this chapter have already led to the development of novel therapeutic agents. However, in the postgenomics era, large numbers of new targets have been identified. Although the drug discovery process moves toward structure-based drug design as its prime tool, even with high-throughput crystallography, not all target proteins will be readily accessible. The evolution of algorithms that can predict enzyme function and mechanism will ensure that the rational design of enzyme inhibitors not only complements structure-based approaches but continues to play a stand-alone role in the discovery of novel therapeutics.

REFERENCES

1. M. Sandler and H. J. Smith, Eds., *Design of Enzyme Inhibitors as Drugs*, Oxford University Press, Oxford, 1989.

2. M. Sandler and H. J. Smith, Eds., Design of Enzyme Inhibitors as Drugs, Vol. **2**, Oxford University Press, Oxford, 1994.

3. P. Krogsgaard-Larsen, T. Liljefors, and U. Madsen, Eds., *A Textbook of Drug Design and Development*, 2nd ed., Harwood Academic, Amsterdam, 1996.

4. H. J. Smith, Ed., *Smith and Williams' Introduction to the Principles of Drug Design and Action*, Harwood Academic, Amsterdam, 1998.

5. G. Gregoriadis, *Trends Biotechnol.*, **13**, 527–537 (1995).

6. I. A. Bakker-Woudenberg, G. Storm, and M. C. Woodle, *J. Drug Target.*, **2**, 363–371 (1994).

7. J. Kreuter, *Adv. Drug Del. Rev.*, **47**, 65–81 (2001).

8. F. C. Neuhaus and J. L. Lynch, *Biochemistry*, **3**, 471–480 (1964).

9. R. R. Rando, *Biochem. Pharmacol.*, **24**, 1153–1160 (1975).

10. J. N. Delgado and W. A. Remers, Eds., *Wilson and Gisvold's Textbook of Organic Medicinal and Pharmaceutical Chemistry*, 10th ed., Lippincott-Raven, Philadelphia, 1998.

11. N. Rastogi and H. L. David, *Res. Microbiol.*, **144**, 133–143 (1993).

12. A. G. Gilman, T. W. Rall, A. S. Nies, and P. Taylor, Eds., *Goodman and Gilman's the Pharmacological Basis of Therapeutics*, 8th ed., Pergamon, New York, 1990, p. 985.

13. H. J. Schaeffer, L. Beauchamp, P. de Miranda, G. B. Elion, D. J. Bauer, and P. Collins, *Nature*, **272**, 583–585 (1978).

14. P. A. Furman, M. H. St Clair, and T. Spector, *J. Biol. Chem.*, **259**, 9575–9579 (1984).

15. G. B. Elion, P. A. Furman, J. A. Fyfe, P. de Miranda, L. Beauchamp, and H. J. Schaeffer, *Proc. Natl. Acad. Sci. USA*, **74**, 5716–5720 (1977).

16. J. P. Durkin and T. Viswanatha, *J. Antibiot. (Tokyo)*, **31**, 1162–1169 (1978).

17. S. J. Cartwright and A. F. Coulson, *Nature*, **278**, 360–361 (1979).

18. R. Labia, V. Lelievre, and J. Peduzzi, *Biochim. Biophys. Acta*, **611**, 351–357 (1980).

19. C. Reading and T. Farmer, *Biochem. J.*, **199**, 779–787 (1981).

20. R. L. Charnas and J. R. Knowles, *Biochemistry*, **20**, 3214–3219 (1981).

21. A. R. English, J. A. Retsema, A. E. Girard, J. E. Lynch, and W. E. Barth, *Antimicrob. Agents Chemother.*, **14**, 414–419 (1978).

22. K. P. Fu and H. C. Neu, *Antimicrob. Agents Chemother.*, **15**, 171–176 (1979).

23. P. S. Mezes, A. J. Clarke, G. I. Dmitrienko, and T. Viswanatha, *FEBS Lett.*, **143**, 265–267 (1982).

24. D. G. Brenner and J. R. Knowles, *Biochemistry*, **23**, 5833–5839 (1984).

25. W. L. Washtien, *Mol. Pharmacol.*, **25**, 171–177 (1984).

26. D. R. Seeger, *J. Am. Chem. Soc.*, **71**, 1753 (1949).

27. D. A. Matthews, R. A. Alden, J. T. Bolin, S. T. Freer, R. Hamlin, N. Xuong, J. Kraut, M. Poe, M. Williams, and K. Hoogsteen, *Science*, **197**, 452–455 (1977).

28. B. Lippert, B. W. Metcalf, M. J. Jung, and P. Casara, *Eur. J. Biochem.*, **74**, 441–445 (1977).

29. M. Cziraky, *Pharmacoeconomics*, **14** (Suppl. 3), 29–38 (1998).

30. A. Endo, M. Kuroda, and K. Tanzawa, *FEBS Lett.*, **72**, 323–326 (1976).

31. V. W. Rodbell, *Adv. Lipid Res.*, **14**, 1 (1976).

32. A. W. Alberts, J. Chen, G. Kuron, V. Hunt, J. Huff, C. Hoffman, J. Rothrock, M. Lopez, H. Joshua, E. Harris, A. Patchett, R. Monaghan, S. Currie, E. Stapley, G. Albers-Schonberg, O. Hensens, J. Hirshfield, K. Hoogsteen, J. Liesch, and J. Springer, *Proc. Natl. Acad. Sci. USA*, **77**, 3957–3961 (1980).

33. A. P. Lea and D. McTavish, *Drugs*, **53**, 828–847 (1997).

34. H. S. Yee and N. T. Fong, *Ann. Pharmacother.*, **32**, 1030–1043 (1998).

35. S. Yu, K. Sugahara, K. Nakayama, S. Awata, and H. Kodama, *Metabolism*, **49**, 1025–1029 (2000).

36. A. Hestnes, O. Borud, H. Lunde, and L. Gjessing, *J. Ment. Defic. Res.*, **33**, 261–265 (1989).

37. G. R. Stark and P. A. Bartlett, *Pharmacol. Ther.*, **23**, 45–78 (1983).

38. I. H. Segel, *Enzyme Kinetics: Behavior and Analysis of Rapid Equilibrium and Seady-State Enzyme Systems*, Wiley-Interscience, New York, 1975.

39. E. Shaw in P. D. Boyer, Ed., *Chemical Modification by Active-Site Directed Reagents*, Academic Press, New York, 1970, pp. 91–147.

40. R. L. Lundblad, *Chemical Reagents for Protein Modification*, 2nd ed., CRC Press, Boca Raton, FL, 1991.

41. H. F. Hixson, Jr. and A. H. Nishikawa, *Arch. Biochem. Biophys.*, **154**, 501–509 (1973).

42. K. I. Skorey, N. A. Johnson, G. Huyer, and M. J. Gresser, *Protein Express. Purif.*, **15**, 178–187 (1999).

43. F. A. Norris and P. W. Majerus, *J. Biol. Chem.*, **269**, 8716–8720 (1994).

44. M. Knockaert, N. Gray, E. Damiens, Y. T. Chang, P. Grellier, K. Grant, D. Fergusson, J. Mottram, M. Soete, J. F. Dubremetz, K. Le

Roch, C. Doerig, P. Schultz, and L. Meijer, *Chem. Biol.*, **7**, 411–422 (2000).

45. J. S. Fowler, R. R. MacGregor, A. P. Wolf, C. D. Arnett, S. L. Dewey, D. Schlyer, D. Christman, J. Logan, M. Smith, H. Sachs, et al., *Science*, **235**, 481–485 (1987).

46. D. V. Santi and G. L. Kenyon in M. E. Wolff, Ed., *Burger's Medicinal Chemistry: Approaches to the Rational Design of Enzyme Inhibitors*, Wiley-Interscience, New York, 1980, pp. 349–391.

47. A. Muscate, C. L. Levinson, and G. L. Kenyon in M. Howe-Grant, Ed., *Kirk-Othmer Encyclopedia of Chemical Technology*, 4th. ed, John Wiley & Sons, New York, 1994, pp. 644–671.

48. A. Patel, H. J. Smith, and J. Stürzebecher, in ref. 4, pp. 261–330.

49. P. Veerapandian, Ed., *Structure-Based Drug Design*, Marcel Dekker, New York, 1997.

50. J. E. Ladbury and P. R. Connelly, Eds., *Structure-Based Drug Design: Thermodynamics, Modeling and Strategy*, Springer-Verlag, Berlin, 1997.

51. E. M. Gordon and J. F. Kerwin, Jr., Eds., *Combinatorial Chemistry and Molecular Diversity in Drug Discovery*, Wiley-Liss, New York, 1998.

52. K. Gubernator and H.-J. Böhm, Eds., *Structure-Based Ligand Design*, Wiley-VCH, Weinheim, Germany, 1998.

53. J. P. Dirlam, L. J. Czuba, B. W. Dominy, R. B. James, R. M. Pezzullo, J. E. Presslitz, and W. W. Windisch, *J. Med. Chem.*, **22**, 1118–1121 (1979).

54. J. C. Dearden and K. C. James in T. J. Perun and C. L. Propst Eds., *Computer-Aided Drug Design: Methods and Applications*, Marcel Dekker, New York, 1989, pp. 168–207.

55. T. Högberg and U. Norinder, in ref. 3, pp. 95–129.

56. Y. C. Martin, P. Willett, and S. R. Heller, Eds., *Designing Bioactive Molecules: Three Dimensional Techniques and Applications*, American Chemical Society, Washington, DC, 1998.

57. W. P. Jencks, *Adv. Enzymol. Relat. Areas Mol. Biol.*, **43**, 219–410 (1975).

58. W. P. Jencks, *Proc. Natl. Acad. Sci. USA*, **78**, 4046–4050 (1981).

59. J. Kyte, *Structure in Protein Chemistry*, Garland Publishing, New York, 1995, pp. 147–196.

60. A. R. Fersht, *Structure and Mechanism in Protein Science: A Guide to Enzyme Catalysis and Protein Folding*, Freeman, New York, 1999.

61. D. A. Dougherty, *Science*, **271**, 163–168 (1996).

62. N. S. Scrutton and A. R. Raine, *Biochem. J.*, **319**, 1–8 (1996).

63. S. K. Burley and G. A. Petsko, *Adv. Protein Chem.*, **39**, 125–189 (1988).

64. J. P. Gallivan and D. A. Dougherty, *Proc. Natl. Acad. Sci. USA*, **96**, 9459–9464 (1999).

65. J. L. Sussman, M. Harel, F. Frolow, C. Oefner, A. Goldman, L. Toker, and I. Silman, *Science*, **253**, 872–879 (1991).

66. A. Ordentlich, D. Barak, C. Kronman, N. Ariel, Y. Segall, B. Velan, and A. Shafferman, *J. Biol. Chem.*, **270**, 2082–2091 (1995).

67. A. Cornish-Bowden and C. W. Wharton, *Enzyme Kinetics*, IRL Press, Oxford, 1988.

68. W. W. Cleland in D. S. Sigman, and P. D. Boyer, Eds., The Enzymes, 3rd ed., Vol. **XIX**, Academic Press, San Diego, 1990, pp. 99–158.

69. T. Palmer, *Understanding Enzymes*, Prentice Hall/Ellis Horwood, London, 1995.

70. D. L. Purich, Ed., *Contemporary Enzyme Kinetics and Mechanism*, 2nd ed., Academic Press, San Diego, 1996.

71. R. A. Copeland, *Enzymes: A Practical Introduction to Structure, Mechanism and Data Analysis*, 2nd ed., Wiley-VCH, New York, 2000.

72. R. Eisenthal and A. Cornish-Bowden, *Biochem. J.*, **139**, 715–720 (1974).

73. C. L. Tsou, *Adv. Enzymol. Relat. Areas Mol. Biol.*, **61**, 381–436 (1988).

74. M. Dixon, *Biochem. J.*, **55**, 170–171 (1953).

75. R. G. Pendleton and I. B. Snow, *Mol. Pharmacol.*, **9**, 718–725 (1973).

76. H. J. Fromm, in ref. 70, pp. 207–227.

77. J. F. Morrison, *Trends Biochem. Sci.*, **7**, 102–105 (1982).

78. J. F. Morrison and C. T. Walsh, *Adv. Enzymol. Relat. Areas Mol. Biol.*, **61**, 201–301 (1988).

79. S. E. Szedlacsek and R. G. Duggleby in D. L. Purich, Ed., Methods in Enzymology, Vol. **249**, Academic Press, New York, 1995, pp. 144–180.

80. M. J. Sculley, J. F. Morrison, and W. W. Cleland, *Biochim. Biophys. Acta*, **1298**, 78–86 (1996).

81. D. H. Rich, *J. Med. Chem.* **28**, 263–273 (1985).

82. J. D. Cox, N. N. Kim, A. M. Traish, and D. W. Christianson, *Nat. Struct. Biol.*, **6**, 1043–1047 (1999).

83. D. M. Colleluori and D. E. Ash, *Biochemistry*, **40**, 9356–9362 (2001).

84. S. H. Wilkes and J. M. Prescott, *J. Biol. Chem.*, **260**, 13154–13162 (1985).

85. A. Taylor, C. Z. Peltier, F. J. Torre, and N. Hakamian, *Biochemistry*, **32**, 784–790 (1993).

86. H. Kim and W. N. Lipscomb, *Biochemistry*, **32**, 8465–8478 (1993).

87. A. Betz, P. W. Wong, and U. Sinha, *Biochemistry*, **38**, 14582–14591 (1999).

88. E. P. Garvey, J. A. Oplinger, E. S. Furfine, R. J. Kiff, F. Laszlo, B. J. Whittle, and R. G. Knowles, *J. Biol. Chem.*, **272**, 4959–4963 (1997).

89. E. S. Furfine, M. F. Harmon, J. E. Paith, and E. P. Garvey, *Biochemistry*, **32**, 8512–8517 (1993).

90. B. F. Cooper and F. B. Rudolph, in ref. 70, pp. 183–205.

91. B. Oberg, *Pharmacol. Ther.*, **40**, 213–285 (1989).

92. L. R. Overby, E. E. Robishaw, J. B. Schleicher, A. Reuter, N. L. Shipkowitz, and J. C.-H. Mao, *Antimicrob. Agents Chemother.*, **6**, 360–365 (1974).

93. S. S. Leinbach, J. M. Reno, L. F. Lee, A. F. Isbell, and J. A. Boezi, *Biochemistry*, **15**, 426–430 (1976).

94. B. Eriksson, A. Larsson, E. Helgstrand, N. G. Johansson, and B. Oberg, *Biochim. Biophys. Acta*, **607**, 53–64 (1980).

95. F. Kappler and A. Hampton, *J. Med. Chem.*, **33**, 2545–2551 (1990).

96. R. Wolfenden, *Annu. Rev. Biophys. Bioeng.*, **5**, 271–306 (1976).

97. A. D. Broom, *Fed. Proc.*, **45**, 2779–2783 (1986).

98. A. D. Broom, *J. Med. Chem.*, **32**, 2–7 (1989).

99. G. E. Lienhard, *Science*, **180**, 149–154 (1973).

100. A. Radzicka and R. Wolfenden, in ref. 70, pp. 229–257.

101. M. Mader and P. A. Bartlett, *Chem. Rev.*, **97**, 1281–1301 (1997).

102. J. Inglese, R. A. Blatchly, and S. J. Benkovic, *J. Med. Chem.*, **32**, 937–940 (1989).

103. C. Klein, P. Chen, J. H. Arevalo, E. A. Stura, A. Marolewski, M. S. Warren, S. J. Benkovic, and I. A. Wilson, *J. Mol. Biol.*, **249**, 153–175 (1995).

104. E. C. Bigham, W. R. Mallory, S. J. Hodson, D. S. Duch, R. Ferone, and G. K. Smith, *Heterocycles*, **35**, 1289–1307 (1993).

105. K. D. Collins and G. R. Stark, *J. Biol. Chem.*, **246**, 6599–6605 (1971).

106. E. A. Swyryd, S. S. Seaver, and G. R. Stark, *J. Biol. Chem.*, **249**, 6945–6950 (1974).

107. R. F. Morton, E. T. Creagan, S. A. Cullinan, J. A. Mailliard, L. Ebbert, M. H. Veeder, and M. Chang, *J. Clin. Oncol.*, **5**, 1078–1082 (1987).

108. N. M. Laing, W. W. Chan, D. W. Hutchinson, and B. Oberg, *FEBS Lett.*, **260**, 206–208 (1990).

109. L. Jin, B. Stec, W. N. Lipscomb, and E. R. Kantrowitz, *Proteins.*, **37**, 729–742 (1999).

110. A. Corsini, F. M. Maggi, and A. L. Catapano, *Pharmacol. Res.*, **31**, 9–27 (1995).

111. K. M. Bischoff and V. W. Rodwell, *Biochem. Med. Metab. Biol.*, **48**, 149–158 (1992).

112. C. E. Nakamura and R. H. Abeles, *Biochemistry*, **24**, 1364–1376 (1985).

113. R. H. Abeles, *Drug. Dev. Res.*, **10**, 221–234 (1987).

114. E. S. Istvan, M. Palnitkar, S. K. Buchanan, and J. Deisenhofer, *EMBO J.*, **19**, 819–830 (2000).

115. E. S. Istvan and J. Deisenhofer, *Biochim. Biophys. Acta*, **1529**, 9–18 (2000).

116. E. S. Istvan and J. Deisenhofer, *Science*, **292**, 1160–1164 (2001).

117. M. A. Ondetti and D. W. Cushman in R. L. Soffer, Ed., *Biochemical Regulation of Blood Pressure*, Wiley, New York, 1981, pp. 165–186.

118. L. D. Byers and R. Wolfenden, *Biochemistry*, **12**, 2070–2078 (1973).

119. D. W. Cushman, H. S. Cheung, E. F. Sabo, and M. A. Ondetti, *Biochemistry*, **16**, 5484–5491 (1977).

120. H. G. Bull, N. A. Thornberry, M. H. Cordes, A. A. Patchett, and E. H. Cordes, *J. Biol. Chem.*, **260**, 2952–2962 (1985).

121. R. B. Silverman, *The Organic Chemistry of Drug Design and Drug Action*, Academic Press Inc., San Diego, 1992, pp. 162–175.

122. P. Buenning, *J. Cardiovascular. Res.*, **10** (Suppl. 7): S31–S35, 1987.

123. A. Radzicka and R. Wolfenden, *Science*, **267**, 90–93 (1995).

124. R. Wolfenden, *Transition States of Biochemical Processes*, Plenum, New York, 1978.

125. V. L. Schramm, *Annu. Rev. Biochem.*, **67**, 693–720 (1998).

126. P. A. Bartlett, Y. Nakagawa, C. Johnson, S. Reich, and A. Luis, *J. Org. Chem.*, **53**, 3195–3210 (1988).

127. P. A. Bartlett and C. K. Marlowe, *Biochemistry*, **22**, 4618–4624 (1983).

128. J. W. Williams, J. F. Morrison, and R. G. Duggleby, *Biochemistry*, **18**, 2567–2573 (1979).

129. D. A. Matthews, R. A. Alden, J. T. Bolin, D. J. Filman, S. T. Freer, R. Hamlin, W. G. Hol, R. L. Kisliuk, E. J. Pastore, L. T. Plante, N. Xuong, and J. Kraut, *J. Biol. Chem.*, **253**, 6946–6954 (1978).

130. P. A. Charlton, D. W. Young, B. Birdsall, J. Feeney, and G. C. K. Roberts, *J. Chem. Soc. Chem. Commun.*, 922–924 (1979).

131. K. Schray and J. P. Klinman, *Biochem. Biophys. Res. Commun.*, **57**, 641–648 (1974).

132. L. Frick, R. Wolfenden, E. Smal, and D. C. Baker, *Biochemistry*, **25**, 1616–1621 (1986).

133. L. Frick, J. P. MacNeela, and R. Wolfenden, *Biorg. Chem.*, **15**, 100–108 (1987).

134. P. A. Bartlett and A. Otake, *J. Org. Chem.*, **60**, 3107–3111 (1995).

135. P. A. Bartlett and M. A. Giangiordano, *J. Org. Chem.*, **61**, 3433–3438 (1996).

136. S. Cha, R. P. Agarwal, and R. E. Parks, Jr., *Biochem. Pharmacol.*, **24**, 2187–2197 (1975).

137. R. P. Agarwal, T. Spector, and R. E. Parks, Jr., *Biochem. Pharmacol.*, **26**, 359–367 (1977).

138. R. Wolfenden, J. Kaufman, and J. B. Macon, *Biochemistry*, **8**, 2412–2415 (1969).

139. V. L. Schramm and D. C. Baker, *Biochemistry*, **24**, 641–646 (1985).

140. L. C. Kurz and C. Frieden, *Biochemistry*, **22**, 382–389 (1983).

141. L. C. Kurz and C. Frieden, *Biochemistry*, **26**, 8450–8457 (1987).

142. W. Jones, L. C. Kurz, and R. Wolfenden, *Biochemistry*, **28**, 1242–1247 (1989).

143. D. K. Wilson, F. B. Rudolph, and F. A. Quiocho, *Science*, **252**, 1278–1284 (1991).

144. R. I. Christopherson and M. E. Jones, *J. Biol. Chem.*, **255**, 3358–3370 (1980).

145. D. H. Kinder, S. K. Frank, and M. M. Ames, *J. Med. Chem.*, **33**, 819–823 (1990).

146. C. A. Kettner and A. B. Shenvi, *J. Biol. Chem.*, **259**, 15106–15114 (1984).

147. P. R. Andrews, E. N. Cain, E. Rizzardo, and G. D. Smith, *Biochemistry*, **16**, 4848–4852 (1977).

148. P. A. Bartlett and C. R. Johnson, *J. Am. Chem. Soc.*, **107**, 7793–7794 (1985).

149. P. A. Bartlett, Y. Nakagawa, C. R. Johnson, S. H. Reich, and A. Luis, *J. Org. Chem.*, **53**, 3195–3210 (1988).

150. D. Y. Jackson, J. W. Jacobs, R. Sugasuwara, S. H. Reich, P. A. Bartlett, and P. G. Schultz, *J. Am. Chem. Soc.*, **110**, 4841–4842 (1988).

151. D. Hilvert, S. H. Carpenter, K. D. Nared, and M. T. Auditor, *Proc. Natl. Acad. Sci. USA*, **85**, 4953–4955 (1988).

152. Y. M. Chook, H. Ke, and W. N. Lipscomb, *Proc. Natl. Acad. Sci. USA*, **90**, 8600–8603 (1993).

153. A. Y. Lee, P. A. Karplus, B. Ganem, and J. Clardy, *J. Am. Chem. Soc.*, **117**, 3627–3628 (1995).

154. M. R. Haynes, E. A. Stura, D. Hilvert, and I. A. Wilson, *Science*, **263**, 646–652 (1994).

155. O. Wiest and K. N. Houk, *J. Org. Chem.*, **59**, 7582–7584 (1994).

156. J. Kyte, *Mechanism in Protein Chemistry*, Garland Publishing, New York, 1995, pp. 314–344.

157. J. Eyzaguirre ed., *Chemical Modification of Enzymes: Active Site Studies*, Ellis Horwood Ltd., Chichester, 1987.

158. B. I. Kurganov, N. K. Nagradova, and O. I. Lavrik, Eds., *Chemical Modification of Enzymes*, Nova Science, New York, 1996.

159. W. B. Jakoby and M. Wilchek, Eds., Methods in Enzymology, Vol. **46**, Academic Press, New York, 1977.

160. T. Imoto and H. Yamada, in ref. 157, pp. 279–318.

161. R. F. Colman, *Subcell. Biochem.*, **24**, 177–205 (1995).

162. B. V. Plapp, in ref. 70, pp. 259–289.

163. R. F. Colman, *FASEB J.*, **11**, 217–226 (1997).

164. R. F. Pratt, *Bioorg. Med. Chem. Lett.*, **2**, 1323–1326 (1992).

165. A. Krantz, *Bioorg. Med. Chem. Lett.*, **2**, 1327–1334 (1992).

166. R. B. Silverman, in ref. 70, pp. 291–333.

167. R. B. Silverman, *Mechanism-Based Enzyme Inactivation: Chemistry and Enzymology*, CRC Press, Boca Raton, FL, 1988.

168. R. Kitz and I. B. Wilson, *J. Biol. Chem.*, **237**, 3245–3249 (1962).

169. D. D. Buechter, K. F. Medzihradszky, A. L. Burlingame, and G. L. Kenyon, *J. Biol. Chem.*, **267**, 2173–2178 (1992).

170. W. B. Knight, K. M. Swiderek, T. Sakuma, J. Calaycay, J. E. Shively, T. D. Lee, T. R. Covey, B. Shushan, B. G. Green, R. Chabin, et al., *Biochemistry*, **32**, 2031–2035 (1993).

171. S. D. Nelson, *J. Med. Chem.*, **25**, 753–765 (1982).

172. G. D. Green and E. Shaw, *J. Biol. Chem.*, **256**, 1923–1928 (1981).

173. R. B. Baker, *Design of Active-Site Directed Irreversible Enzyme Inhibitors*, Wiley-Interscience, New York, 1967.

174. J. A. Katzenellenbogen, *Ann. Rep. Med. Chem.*, 222–233 (1974).

175. G. Schoellmann and E. Shaw, *Biochemistry*, **2**, 252–255 (1963).

176. E. B. Ong, E. Shaw, and G. Schoellmann, *J. Biol. Chem.*, **240**, 694–698 (1965).

177. C. Kettner and E. Shaw in L. Lorand, Ed., Methods in Enzymology, Vol. **80**, Academic Press, New York, 1981, pp. 826–842.

178. M. Dryjanski, L. L. Kosley, and R. Pietruszko, *Biochemistry*, **37**, 14151–14156 (1998).

179. K. von der Helm, B. D. Korant, and J. C. Cheronis, Eds., *Proteases as Targets for Therapy*, Springer-Verlag, Berlin, 2000.

180. J. A. Fee, G. D. Hegeman, and G. L. Kenyon, *Biochemistry*, **13**, 2533–2538 (1974).

181. J. A. Landro, J. A. Gerlt, J. W. Kozarich, C. W. Koo, V. J. Shah, G. L. Kenyon, D. J. Neidhart, S. Fujita, and G. A. Petsko, *Biochemistry*, **33**, 635–643 (1994).

182. J. R. Vane, *Nat. New Biol.*, **231**, 232–235 (1971).

183. J. B. Smith and A. L. Willis, *Nat. New Biol.*, **231**, 235–237 (1971).

184. G. J. Roth, N. Stanford, and P. W. Majerus, *Proc. Natl. Acad. Sci. USA*, **72**, 3073–3076 (1975).

185. M. Hemler and W. E. Lands, *J. Biol. Chem.*, **251**, 5575–5579 (1976).

186. F. J. Van der Ouderaa, M. Buytenhek, D. H. Nugteren, and D. A. Van Dorp, *Eur. J. Biochem.*, **109**, 1–8 (1980).

187. G. J. Roth, E. T. Machuga, and J. Ozols, *Biochemistry*, **22**, 4672–4675 (1983).

188. P. J. Loll, D. Picot, and R. M. Garavito, *Nat. Struct. Biol.*, **2**, 637–643 (1995).

189. D. Picot, P. J. Loll, and R. M. Garavito, *Nature*, **367**, 243–249 (1994).

190. L. H. Rome and W. E. Lands, *Proc. Natl. Acad. Sci. USA*, **72**, 4863–4865 (1975).

191. R. F. Colman in T. E. Creighton, Ed., *Protein Function: A Practical Approach*, 2nd ed., Oxford University Press, Oxford, 1997, pp. 155–183.

192. P. K. Pal, W. J. Wechter, and R. F. Colman, *J. Biol. Chem.*, **250**, 8140–8147 (1975).

193. J. L. Wyatt and R. F. Colman, *Biochemistry*, **16**, 1333–1342 (1977).

194. R. H. Abeles, *Pure Appl. Chem.*, **53**, 149–160 (1980).

195. T. I. Kalman, *Drug. Dev. Res.*, **1**, 311–328 (1981).

196. C. T. Walsh, *Tetrahedron*, **38**, 871–909 (1982).

197. C. T. Walsh, *TIPS*, 254–257 (1983).

198. R. H. Abeles, *Chem. Eng. News.*, **61**, 48–56 (1983).

199. T. M. Penning, *TIPS*, 212–217 (1983).

200. C. T. Walsh, *Annu. Rev. Biochem.*, **53**, 493–535 (1984).

201. M. G. Palfreyman, P. Bey, and A. Sjoerdsma, *Essays Biochem.*, **23**, 28–81 (1987).

202. M. J. Jung and C. Danzin, in ref. 1, pp. 257–293.

203. B. Frølund, in ref. 3, pp. 264–266.

204. R. Poulin, L. Lu, B. Ackermann, P. Bey, and A. E. Pegg, *J. Biol. Chem.*, **267**, 150–158 (1992).

205. N. V. Grishin, A. L. Osterman, H. B. Brooks, M. A. Phillips, and E. J. Goldsmith, *Biochemistry*, **38**, 15174–15184 (1999).

206. T. Liang, M. A. Cascieri, A. H. Cheung, G. F. Reynolds, and G. H. Rasmusson, *Endocrinology*, **117**, 571–579 (1985).

207. B. Faller, D. Farley, and H. Nick, *Biochemistry*, **32**, 5705–5710 (1993).

208. H. G. Bull, M. Garcia-Calvo, S. Andersson, W. F. Baginski, H. K. Chan, D. E. Ellsworth, R. R. Miller, R. A. Stearns, R. K. Bakshi, G. H. Rasmusson, R. L. Tolman, R. W. Myers, J. W. Kozarich, and G. S. Harris, *J. Am. Chem. Soc.*, **118**, 2359–2365 (1996).

209. B. Azzolina, K. Ellsworth, S. Andersson, W. Geissler, H. G. Bull, and G. S. Harris, *J. Steroid Biochem. Mol. Biol.*, **61**, 55–64 (1997).

210. V. C. Njar and A. M. Brodie, *Drugs*, **58**, 233–255 (1999).

211. P. E. Reynolds, F. Depardieu, S. Dutka-Malen, M. Arthur, and P. Courvalin, *Mol. Microbiol.*, **13**, 1065–1070 (1994).

212. D. E. Bussiere, S. D. Pratt, L. Katz, J. M. Severin, T. Holzman, and C. H. Park, *Mol. Cell.*, **2**, 75–84 (1998).

213. Z. Wu, G. D. Wright, and C. T. Walsh, *Biochemistry*, **34**, 2455–2463 (1995).

214. Z. Wu and C. T. Walsh, *Proc. Natl. Acad. Sci. USA*, **92**, 11603–11607 (1995).

215. R. Araoz, E. Anhalt, L. Rene, M. A. Badet-Denisot, P. Courvalin, and B. Badet, *Biochemistry*, **39**, 15971–15979 (2000).

216. M. A. Ator and P. R. Ortiz de Montellano in D. S. Sigman and P. D. Boyer, Eds., The Enzymes, 3rd ed., Vol. XIX, Academic Press, San Diego, 1990, pp. 214–282.

217. J. Grutzendler and J. C. Morris, *Drugs*, **61**, 41–52 (2001).

218. E. Perola, L. Cellai, D. Lamba, L. Filocamo, and M. Brufani, *Biochim. Biophys. Acta*, **1343**, 41–50 (1997).

219. R. J. Polinsky, *Clin. Ther.*, **20**, 634–647 (1998).

220. K. N. Allen and R. H. Abeles, *Biochemistry*, **28**, 8466–8473 (1989).

CHAPTER EIGHTEEN

Chirality and Biological Activity

ALISTAIR G. DRAFFAN
GRAHAM R. EVANS
JAMES A. HENSHILWOOD
Celltech R&D Ltd.
Granta Park, Great Abington, Cambridge, United Kingdom

Contents

Burger's Medicinal Chemistry and Drug Discovery
Sixth Edition, Volume 1: Drug Discovery
Edited by Donald J. Abraham
ISBN 0-471-27090-3   © 2003 John Wiley & Sons, Inc.

1 GENERAL INTRODUCTION

1.1 Introduction

The subtle relationship between the efficacy and chirality of a drug is an area of research that has grown enormously over the past 20 years because of the recognition that single isomer drugs can be more potent and safer than their racemic mixtures. It is intended that this chapter will provide the reader with an appreciation of the area of chirality and biological activity. Indeed, consideration of chirality in drug design has become ubiquitous because of the greater understanding of stereoselective pharmacokinetics, pharmacodynamics, and receptor binding. The enabling technologies of chiral synthesis and analysis have provided the tools to drive these advances in the detailed understanding of the differential biological activity of stereoisomers. Isomers of identical constitution but differing in the arrangement of their atoms in space are defined as stereoisomers (enantiomers and diastereomers are subclasses). Enantiomers consist of a pair of molecular species that are mirror images of each other and are not superimposable.

The term chiralty is broadly used within chemistry and drug development; however, the terms chiralty or chiral are not always well understood by the reader. To clarify the matter, we can consider that two main situations exist. In the first case a sample consists of equal numbers of molecules having an opposite sense of chirality (heterochiral molecules). This sample is said to be chiral but racemic. The second case occurs when a sample is made up of molecules that all have the same sense of chiralty (homochiral molecules). In this case the sample is said to be chiral and nonracemic.

Recognition of the importance of chirality and biological activity has led to the position where the regulatory authorities will no longer consider the registration of a new racemic compound. Exceptions include cases where the stereoisomers interconvert *in vivo* or where there is a specific advantage or synergy associated with dosing both stereoisomers as a racemate. In addition to the benefit to the patient of a safer and more potent drug, there are numerous advantages to a company in developing a single isomer drug. For example, the cost and complexity of testing is simpler for single isomers, because the Food and Drug Administration (FDA), which regulates the approval and sale of drugs and other products in the United States, requires that both of the isomers within a racemate be tested. Thus, the overall development costs and time are greatly increased because of the requirement for information on all three, the racemate and both isomers separately. The use of a single isomer should also result in a lower dosage, at least one-half that of the racemate. Because of both the therapeutic advantages and the greater regulatory burden of proof associated with a racemate compared with a single isomer, sales of single isomer drugs have increased to over $120 billion dollars, representing more than 30% of the total market in 2000 compared with 3% in 1980 and 9% in 1990. When combined with the declining figures for new drug application approvals by the FDA, the efficient and rapid development of a single isomer drug is imperative (1).

The fundamental reason for the differential activity of stereoisomers is that the majority of molecules that make up living organisms are chiral, and moreover, exist in only one enantiomeric form. Thus, stereoisomers will be seen by the system as different molecules and will have different effects on the biological system. Typically therefore, one of the single enantiomers of a drug will demonstrate greater potency and/or less side effects than the corresponding racemate, and such examples are given within this chapter. For example, Vigabatrin, which is a selective GABA transamidase inhibitor, gained approval in 1997 as a racemate. While the two stereoisomers exhibit the same pharmacokinetics, only the S-enantiomer is active as an anti-epileptic. However, there are some limited cases, such as Tramadol, where a synergistic benefit associated with the dosing of the racemate is claimed in comparison with either single enantiomer (2).

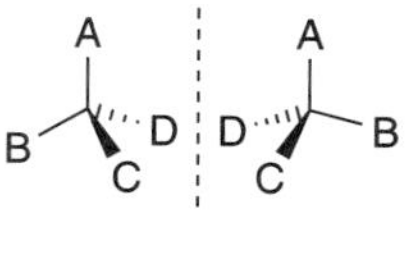

Figure 18.1.

1.2 Definition of Chirality

The definition of chirality and its measurement are described in great detail in a number of texts (3); however, a brief introduction to the key issues is given in this section. Specifically, chirality is a term referring to a property of a molecule that is nonsuperimposable on its mirror image as shown in Fig. 18.1, where such a molecule is chiral.

In the majority of cases, chirality results from the three dimensional orientation of four different substituents around a carbon atom forming the chiral center. In addition the orientation of atoms or groups around sulfur, phosphorus, and nitrogen atoms can sometimes form a chiral center. Examples of chiral drugs are numerous but include Certirizine (**1**), Rotigotine (**2**), and Ifosfamide (**3**).

When a molecule contains only one chiral center, the two stereoisomers are known as enantiomers. These may be referred to or labeled using the configurational descriptors as either *R* (*rectus* meaning righthanded) or *S* (*sinister* meaning left handed), or alternatively, D (dextrorotatory) or L (levorotatory). The D and L configurational descriptors are

also commonly used in classifying the configuration of sugars and amino acids (see below). In an achiral environment, enantiomers will behave identically, exhibiting for example, the same melting/boiling point, lipid solubility, nuclear magnetic resonance (NMR), infrared (IR), etc. However, in a chiral environment such as within the macromolecular components of a living system or a chiral high performance liquid chromatograpy (HPLC) system, the enantiomers display different properties, such as a different route and rate of metabolism, different biological activity, and different retention times in chiral HPLC. A frequently quoted analogy for the differing properties of enantiomers is the hand and glove example. The left and right hand are enantiomers of one another, that is they are nonsuperimposable mirror images. If the right hand or "enantiomer" is placed in the right hand glove, or "receptor", there is a good fit. Thus, in the case of a true drug and receptor there will be the desired effect. If the left hand, or "enantiomer" is placed in the right hand glove, there is either a poor fit or no fit.

As introduced previously, when a chiral compound is present as exactly a 1:1 mixture of its enantiomers, it is referred to as a racemate or racemic mixture. Thus, if a single enantiomer undergoes racemization, the pure enantiomer is converted to a 1:1 mixture of enantiomers. Perhaps the most famous example of a chiral drug is Thalidomide. In the early

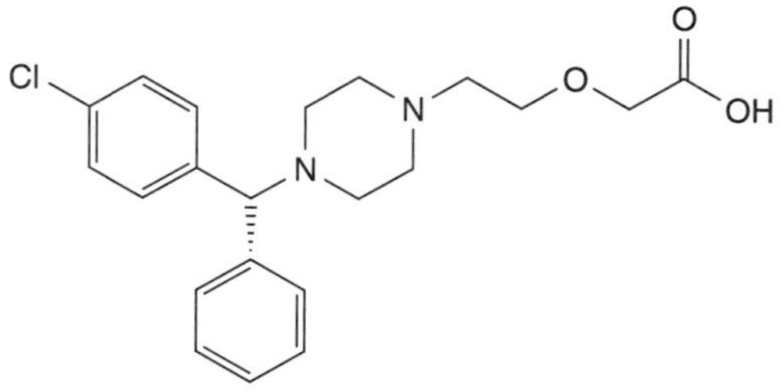

Levocertirizine (**1**)

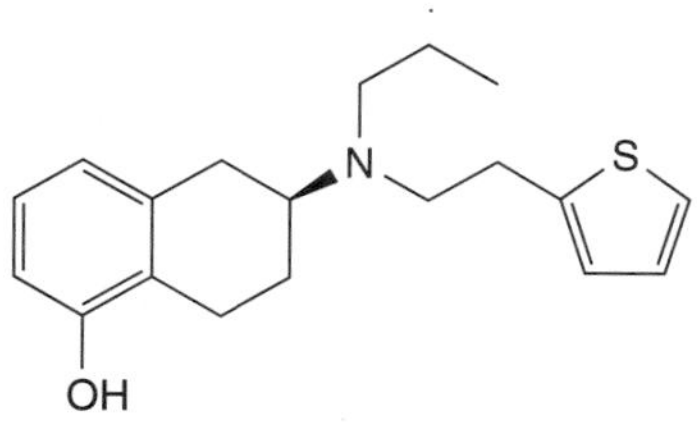

Retigotine (**2**)

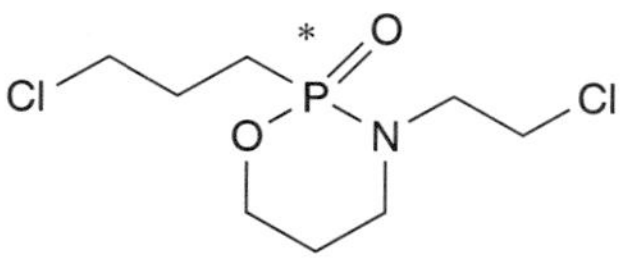

Ifosfamide (**3**)

Figure 18.2.

1960s, Thalidomide was widely prescribed as a sleeping pill and as a treatment for morning sickness, with claims that it was completely safe. We all know of the terrible birth defects suffered by children born to mothers who took the drug during pregnancy. The drug was taken as the racemate, and it has been shown that the *R*-enantiomer is responsible for the drug's anti-inflammatory activity, whereas the *S*-enantiomer causes the teratogenicity. Separation of the racemic mixture to give the patient only the *R*-enantiomer is not a simple answer to the problem. The liver contains an enzyme that converts the *R*- into the *S*-enatiomer, thus negating the benefit of giving the single enantiomer (4).

As described in this chapter, there are many reactions that can be performed by chemists to create new chiral centers. When these reactions are performed in such a way as to create one enantiomer in greater amounts than the other the process is called asymmetric or stereoselective synthesis. The term enantioselectivity refers to the efficiency with which the reaction produces one enantiomer. This efficiency is quantitatively described as the enantiomeric excess (ee) of the product, which is the percentage by which one enantiomer is produced in excess of the other. Thus a 45:8 mixture of two enantiomers will have an enantiomeric excess of $[(45 - 8)/(45 + 8)] \times 100$, which equals 70%. It should be noted that if neither the starting material or reaction system is chiral and non-racemic, then the product will be formed as an equal mixture of the enantiomers (i.e., a racemate).

Glucose is perhaps the most widely available chiral compound. It is a monosaccharide and part of the sugar group (carbohydrates) that occur naturally. Sugars, along with amino acids, constitute a special example and are commonly classified with a D- or L-configuration. In the case of sugars, the D-configuration is given when the hydroxyl group on the highest numbered chiral carbon atom is on the righthand side (with the structure drawn in the Fischer convention as shown in Fig. 18.3). Likewise for L-configured sugars, the hydroxyl group is on the lefthand side. In the case of the tetrose sugars there are two enantiomer pairs as illustrated in Fig. 18.3. Here, the enantiomer pairs of erythrose (4, 5), namely D-(−)-

Figure 18.3.

erythrose (4), and L-(+)-erythrose (5), and threose (6, 7), and L-(−)-threose (6) and D-(+)-threose (7), are shown, with each pair of enantiomers being diasteromeric with the other pair. Diastereomers can be simply defined as stereoisomers that are not enantiomers. The prefixes *erythro-* and *threo-* are applied to such systems that contain two asymmetric carbons where two of the groups are identical and the third is different. The *erythro* pair has the identical groups on the same side, whereas the *threo* pair has them on opposite sides.

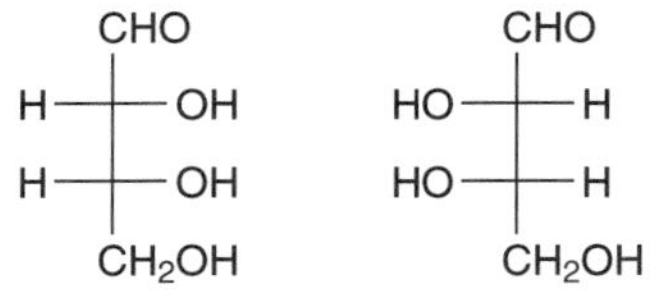

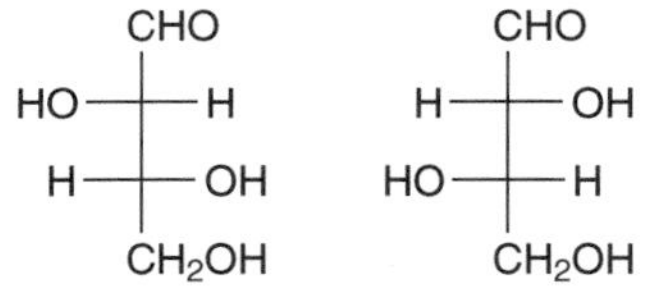

Finally, as further elucidation of this relationship where the molecule contains more than one chiral center the number of stereoisomers increases. In the case of the drug glycopyrrolate which contains two chiral centers, there are four possible stereoisomers as shown in Fig. 18.4. In general, the number of possible isomers can be calculated from the formula 2^n where n is the number of chiral centers.

The four stereoisomers can be divided as shown into two pairs of enantiomers, where the (*R*,*R*)-(8) and (*S*,*S*)-(9) stereoisomers are enantiomers of one another, and the (*S*,*R*)-(10) and (*R*,*S*)-(11) stereoisomers are also an enantiomeric pair. The stereoisomers that do not have an enantiomeric relationship to one another, such as (*R*,*R*)-(8) and (*R*,*S*)-(11) are known as diastereomers. Like enantiomers, these molecules are not superimposable on one another, but unlike enantiomers, they do not exhibit the same physical, chemical, and spectral characteristics. Thus, they have different melting/boiling points, lipid solubility,

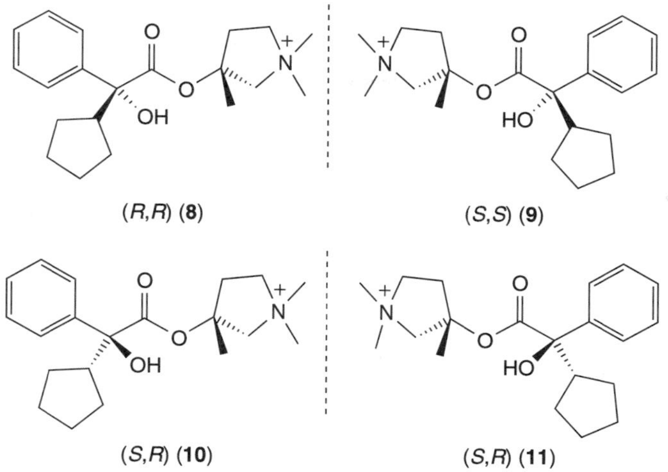

(R,R) (8)   (S,S) (9)

(S,R) (10)   (S,R) (11)   **Figure 18.4.**

NMR spectra, retention time in HPLC or thin layer chromatography (TLC), and can behave differently in chemical reactions with achiral reagents. The commercial glycopyrrolate product contains only the *threo* isomers (S,R)-(**10**) and (R,S)-(**11**).

1.3 Pharmacology

Biological systems are in the main constructed from homochiral molecules such as L-amino acids or D-sugars. Such systems give rise to a highly "chiral environment," and hence, it is not surprising that many drugs possessing asymmetric centers exhibit a high degree of steroselectivity in their interactions with biological macromolecules. In the past 20 years or so, pharmacological and toxicological investigations have clearly demonstrated significant differences in the biological activity of some isomeric pairs. Pharmacokinetic investigations have also led to a better understanding of racemic drug action.

It is important to introduce two other terms that compare the pharmacological activity of a pair of enantiomers. The isomer imparting the desired activity is called the eutomer (in the case of Thalidomide this is the R-enantiomer), whereas the isomer which is inactive or causes unwanted side effects is called the distomer (this is the S-enantiomer for Thalidomide). Comparison of the potencies of the two isomers comes from the eudismic ratio and this can be used *in vitro* or *in vivo*.

With the advancement in analytical and preparative technologies, the researcher is now more able to separate and study individual enantiomers. Pharmacological assessment of the behavior of chiral compounds in early phase research is imperative for selection of the correct isomer for development.

When a racemate is administered, the overall pharmacological effect may have one of three general outcomes described below.

1. All activity resides in one of the isomers, the other antipode being inactive.
2. Both isomers have equal activity.
3. Both isomers have the same activity but differ in potencies.

We will briefly highlight some examples that help to elucidate the above general classes with some pertinent examples. The antihypertensive agent α-methyldopa is an example where all the desired antihypertensive activity is confined to a single isomer (the L-enantiomer). It is noteworthy that L-(α)-methyldopa is a prodrug, being metabolized to the L-isomer of the active metabolite, and it is this metabolite that has the required activity (5). L-Dopa is marketed as the single enantiomer; during early development it was noted that the D-isomer exhibited serious side effects such as granulocytopenia (which is defined as a reduced number of blood granulocytes) (6).

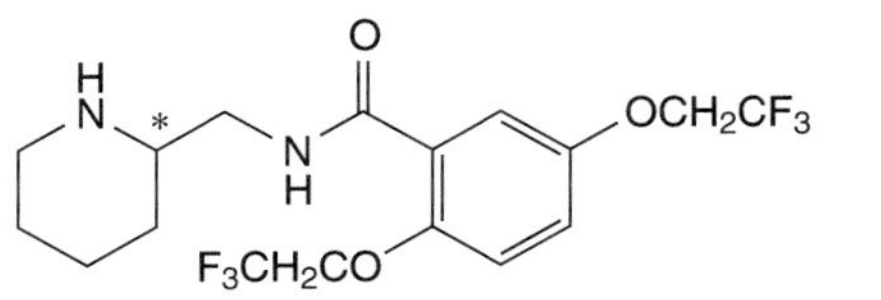

Flecainide (**12**)

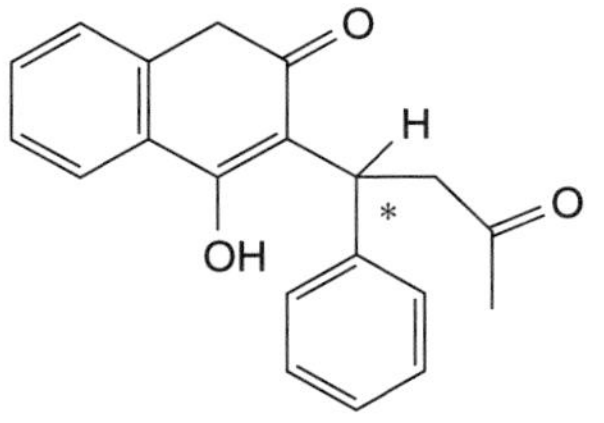

Warfarin (**13**)

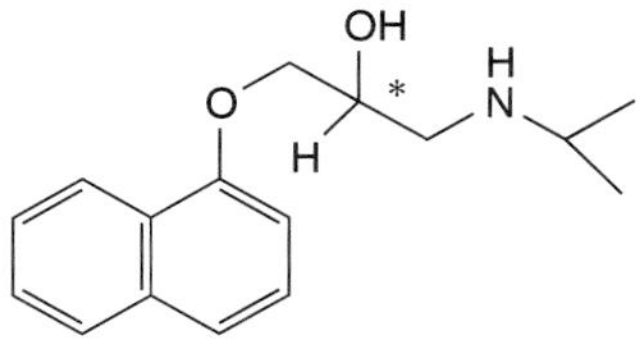

Propranolol (**14**)

Figure 18.5.

It has been reported that the single enantiomers of Flecainide (**12**) (Fig. 18.5) have similar *in vitro* pharmacological activity. Assessment of the effect of each enantiomer on the action potential characteristics in canine cardiac Purkinje fibers gave similar electrophysiological effects. The plasma concentration data of the enantiomers was only very moderately enantioselective. This gave rise to the authors concluding that there was no advantage in administering a single enantiomer of Flecainide over the racemate (7). It is, however, relatively more common to find examples where the enantiomers have similar qualitative pharmacological activity but differ in their potencies. Two classic examples are Warfarin (**13**) (see Fig. 18.5 for the structure and later on in the crystallization section for it's preparation) and Propranolol (**14**). The potency of S-(−)-Warfarin *in vivo* is two to five times greater than that of the R-enantiomer (8). However this difference in potency is offset by the two- to fivefold greater plasma clearance of S-(−)-Warfarin (9). These apparently offsetting properties are only part of the complex pharmacological story surrounding Warfarin, and to this day, it is still administered as the racemate.

β-blocking drugs such as Propranolol (**14**) (Fig, 18.5) have been shown to stereoselectively bind to β-receptors, and it is the S-(−)-enantiomer that exhibits the β-blocking activity. However, *in vitro*, the binding of the R-and S-enantiomers varies widely within this class of compound depending on the structure. Again a complex number of pharmacological actions come into play with the plasma binding of the R-enantiomer being much greater than it's antipode. The two enantiomers are also stereoselectively metabolized at different rates ($R > S$). Therefore, the pharmacological dynamic outcome can vary greatly between patients who have different P_{450} compositions (10). To add to this already complicated story, the bioavailability of S-(−)-Propranolol is reduced when given as the single enantiomer compared with the racemate. This suggests that the presence of R-(+)-propranolol has a beneficial effect on the availability of the S-(−)-enantiomer (11).

1.4 Protein Binding and Metabolism

The enantiomers of a specific drug can bind stereoselectively to plasma proteins. For example, acidic drugs bind in an enantioselective manner to human serum albumin (12). There are two binding sites on albumin, site I (Warfarin) and site II (indole) (13). The binding of L-tryptophan to site II is up to 100 times greater than that of D-tryptophan (14). It has been suggested that binding to albumin can be used as an indication of the extent of the binding to the drug receptor. For basic drugs, α_1-acid glycoprotein (AAG) is used and is relatively non-stereoselective (12, 15).

Enantioselective metabolism and clearance plays a prominent role in determining the pharmacological effect of a drug. For example, a highly potent rapidly cleared enantiomer may be of less benefit clinically than it's lower potency antipode, which is more slowly cleared. Returning to Warfarin, the S-enantiomer is eliminated mainly by 7-hydroxylation, whereas the R-enantiomer is eliminated by ketone reduction and oxidation to 6 and 8-hydroxywarfarin (16). Tramadol is a centrally acting analgesic with efficacy and potency

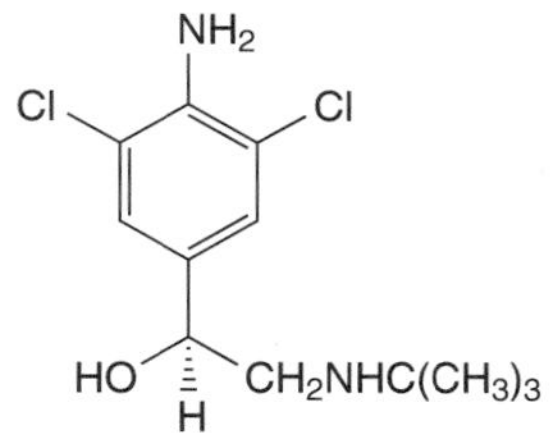

(R)-(−)-Clenbuterol (**15**)

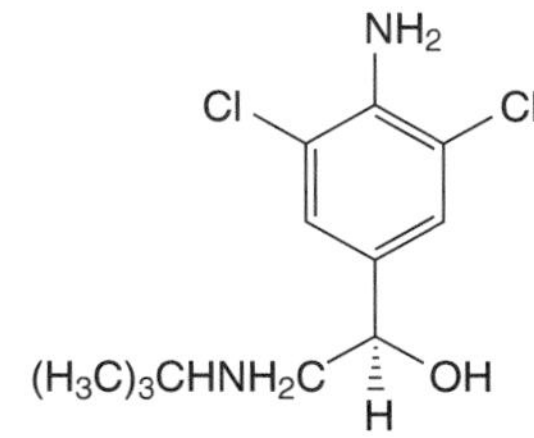

(S)-(+)-Clenbuterol (**16**)

Figure 18.6.

ranging between weak opioids and morphine, and it is currently used as a racemic mixture (Fig. 18.6). It is metabolized in the liver mainly to O-desmethyltramadol (ODT), mono-N-desmethyltramadol, and di-N,O-desmethyltramadol. Of the metabolites only the O-desmthyltramadol is pharmacologically active and the (+)-ODT has ~200 times greater activity for the μ opioid receptor (17). It is thought that this metabolite contributes largely to the analgesic properties of Tramadol, and for this reason numerous studies have been undertaken on the activity of the single enantiomers of Tramadol (17). The complex nature of the interaction of the single enantiomers of a drug with biological systems described in the introduction will have an effect on whether a single enantiomer or racemate is taken through to development.

The following sections describe the available methods for the separation of enantiomers or their preparation using asymmetric synthesis. The first sections involve separations of enantiomers using chromatography or crystallization technology. These are often considered to be the most expedient methods and should deliver the single enantiomers in the shortest timescale. Asymmetric synthesis has developed considerably over the last 20 years and now provides an alternative, and at

times more cost-effective, method of preparing the single enantiomers. It is noteworthy that the 2001 Nobel prize for chemistry was awarded to Sharpless, Knowles, and Noyori for their pioneering work in the area of asymmetric synthesis (18).

2 CHROMATOGRAPHIC SEPARATIONS

The separation of enantiomeric drugs and intermediates by chromatographic methods is a well-developed area that is broadly utilized from milligrams to tons (19). Initially, these chromatographic methods were used to determine the enantiopurity of the compound obtained from, for example, a separation process. With the continuing advancement of chiral stationary phases (CSP) and the development of chromatography, separation of enantiomers using chromatographic techniques is increasingly seen as the method of choice because of the speed at which the separation can be achieved (20).

A multitude of chromatographic separation methodologies exist, all of which can be applied to the separation of enantiomers; for example, liquid chromatography (LC) (21), gas chromatography (GC) (22), high performance liquid chromatography (HPLC) (23), capillary electrophoresis (CE) (24), super critical fluid chromatography (SFC) (25), simulated moving bed (SMB) (26), and membrane technologies (27). Once the appropriate technique has been chosen, at the analytical level, run time, sensitivity, and selectivity then need to be enhanced to improve the limits of detection and analysis time. At the preparative scale (milligram to gram), in addition to enantioselectivity, other factors such as high loading capacity, robustness, and chemical compatibility are essential requirements when selecting the CSP (28). In addition, the CSP needs to be readily available and provide a cheaper process when compared with the chemical/biological alternatives that are discussed in this chapter. However a significant advantage of the method is that almost any enantioseparation can be achieved with one of about 200 commercially available chiral selectors by the techniques described above (29). In this section, we will highlight a number of examples, from the small semi-preparative scale to large-potential manufacturing processes.

2.1 Small-Scale HPLC Examples

HPLC is now a widely available and user-friendly method employed for qualitative and quantitative analysis and is also one of the most expedient methods for providing the milligram quantities of stereochemically pure material required for initial testing. Often the identification of a suitable CSP to effect separation of a specific pair of enantiomers is seen as being labor intensive and requiring considerable experimentation. However, the availability of commercial databases that compile literature on LC enantioseparations makes this process significantly easier (30). The companies that supply CSPs also provide detailed information about a specific columns' suitability towards the separation of certain types of compounds (31). This helps to avoid a "trial and error" approach towards enantioseparations using chromatography. The use of column switchers to test a number of CSPs can also be of enormous assistance in a rational screening program.

One example of separation by HPLC is Clenbuterol, which is an orally active, sympathomimetic agent that has specificity for $\beta 2$-adrenoceptors. Owing to its bronchodilator properties, it has found use in the treatment of respiratory disorders in humans and animals (32). The two enantiomers of Clenbuterol have been separated using a chirobiotic column, which consists of a macrolide-type antibiotic stationary phase, using a mobile phase with composition of 70% MeOH, 30% acetonitrile, 0.3% acetic acid, and 0.2% triethylamine (33). The enantiomers eluted as follows: R-(−)-Clenbuterol (**15**) with a retention time of 8.35 minutes and S-(+)-Clenbuterol (**16**) with a retention time of 9.12 minutes. The single enantiomers obtained *through* chromatography were of >95% optical purity. It has been shown that (−)-Clenbuterol was 100–1000 times more potent than (+)-Clenbuterol in β-adrenergic agonist bioassays (34).

A number of 1,4-dihydropyridines (**17–20**), exhibiting axial chirality (chiralty stemming from the nonplanar arrangement of four groups about an axis), have been separated by small-scale HPLC methods. This is an impor-tant class of drugs that are potent blockers of calcium currents and have found use in the treatment of cardiac arrhythmias, peripheral vascular disorders, and hypertension (35). It has been shown that enantiomers of chiral DHP have opposite pharmacological profiles (35). One of the antipodes is a calcium entry activator, while the other is a calcium entry blocker. The analytical and semi-preparative separation using chiral HPLC for a number of DHPs of the structures (Fig. 18.7) has been described (36). Here a number of different CSP were utilized and their ability to separate the above DHPs determined.

2.2 Chromatographic Diastereoisomer Separation

Another approach to the separation of enantiomers by chromatography is to prepare a diastereoisomer of the enantiomer to be separated. As discussed in the introduction to this chapter, diastereomers exist if there is more than one chiral center, but are not enantiomers of one another. As such they do not have identical physical properties. In chromatography, formation of derivatives such as esters, amides, etc., often leads to better separation of the components. In the case of a racemate, if a chiral reagent (i.e., acid or amine) is employed, then a diastereomeric mixture results on treatment with such a derivatizing agent. One such example is the derivatization of Pirlindole, which is a racemic anti-depressant drug. Here the use of amino acid derivatives as chiral derivatizing agents (CDA) was shown to enable an effective and efficient separation (37). Preparation of the L-phenylalanine methyl ester (**21**) enabled separation of the Pirlindole enantiomers using a medium liquid pressure (MPLC) method. This is highlighted in Fig 18.8, after removal of the CDA the enantiomers of pirlindole were obtained in high optical purity. This gave several grams of each enantiomer, which permitted a study of the stereochemical influence at the pharmacological level. The interaction with monoamine oxidase A (MAO-A) and B (MAO-B) with Pirlindole racemate and single enantiomers using biochemical techniques (*in vitro* and *ex vivo* determination of rat brain MAO-A and MAO-B activity) was studied. *In vitro*, the MAO-A IC_{50} of (±)-Pirlindole, R-(−)-Pirlin-

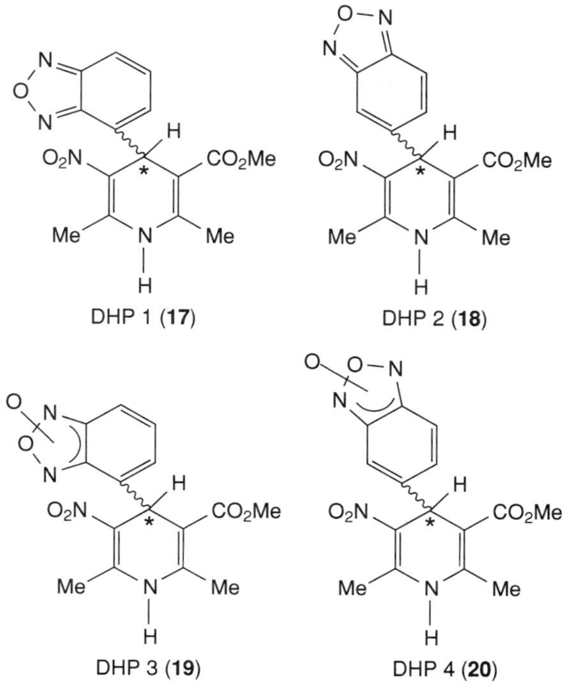

DHP 1 (**17**)   DHP 2 (**18**)

DHP 3 (**19**)   DHP 4 (**20**)   **Figure 18.7.**

dole (**22**), and S-(+)-Pirlindole (**23**) were 0.24, 0.43, and 0.18 μM, respectively. The differences between the three compounds were not significant, with a ratio between the two enantiomers R-(−)/S-(+) of 2.2 *in vitro* (38).

2.3 Preparative HPLC/SMB

In the initial discovery phase of drug research, time is the most important factor where a successful process must be rapidly identified, have a short run time, and have general applicability. As the phase of the project changes to full development, the process needs to be established and cost becomes a crucial factor. Thus, on scale up of an LC method to the preparative level (100 mg and above), a number of additional important aspects become relevant. The selection of a suitable CSP from the plethora available depends on the following factors: CSP availability, loading capacity and selectivity, throughput, and mobile phase.

The most successful and broadly applied chiral stationary phases comprise the cellulose-and amylose-based phases developed by Okamoto (Chiracel and Chiralpak) (39), brush-type phases developed by Pirkle (40),

some polyacrylamides (Chiraspher) (41), cross-linked diallyltartramide (42), and to a lesser extent, cyclo-dextrin based phases. Clearly for the larger scale separations, the availability of the CSP in larger quantities is a prerequisite. It should also be noted that at the preparative scale, it seems that up to 90% of racemic compounds tested have been resolved with just four different polysaccharide-based phases (43).

The degree of separation of the two enantiomers obviously plays an important part in the CSP selection. Another equally important parameter is the loading capacity of the stationary phase. The higher the loading capacity, the greater the amount of material that can be separated (44). For example the polysaccharide-based CSPs have a saturation capacity of 5–100 mg/g of CSP; this is clearly dependent on the type of racemate that is being resolved. On the other hand, protein-based CSPs have lower saturation capacities, of the order 0.1–0.2 mg/g of CSP.

For preparative chromatography, throughput can be defined as the amount of purified material obtained per unit of time and per unit

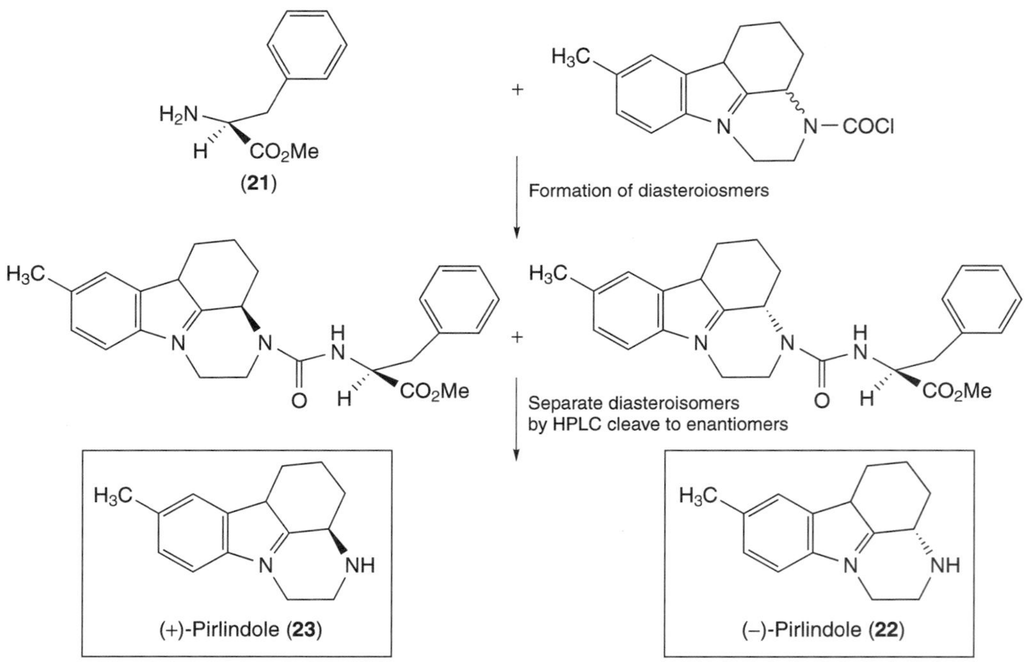

Figure 18.8.

mass of stationary phase. Several factors affect this including loading capacity, column efficiency, selectivity, column size, temperature, cycle time, flow rate, and the solubility of the racemate.

The mobile phase plays a crucial role in the separation process for at least three main reasons. The selectivity of the separation, retention time, and solubility of the racemate are directly affected by the mobile phase composition. Other parameters such as viscosity, solvent recovery, cost, and solvent handling properties also play a prominent role. This brief introduction is also applicable to the criteria for CSP selection for SMB.

An example of a drug separated by preparative HPLC is cetirizine dihydrochloride, a racemic drug that is a second generation antihistamine H_1 receptor antagonist. Studies on the effect of racemic and R (**25**) and S-Cetirizine (**26**) on nasal resistance indicated that both racemic and the R-enantiomer had similar activity. The racemate and R-enantiomer inhibit histamine and induced an increase in nasal resistance, thus indicating the antihistaminic properties of R-Cetirizine (45). The S-enantio-

mer was shown not to exhibit these antihistaminic effects. An asymmetric synthesis (46), and resolution of an intermediate have delivered the single enantiomer previously. However, for various reasons, the development of a preparative HPLC method seems to be the method of choice (47). The main reasons are the rapid scale up and the improved economics of this approach. Utilization of the amide (**24**) (Fig. 18.9) gave rise to a highly efficient separation using a Chiralpak AD column in a mixture of acetonitrile/iso-propanol 60:40. The efficiency of the separation can be measured by the α value (2.76) or the USP resolution (8.54). The α value and USP resolution numbers are measurements of how efficient the separation is; typically the higher the number, the better the separation. This enabled the production of 1.6 kg of both the (+) and (−) isomers of high purity.

Like all methods for separating chiral molecules, chromatographic separations do suffer from drawbacks: large quantities of expensive stationary phases are needed and large volumes of mobile phases are used, coupled with the resultant high dilution of separated prod-

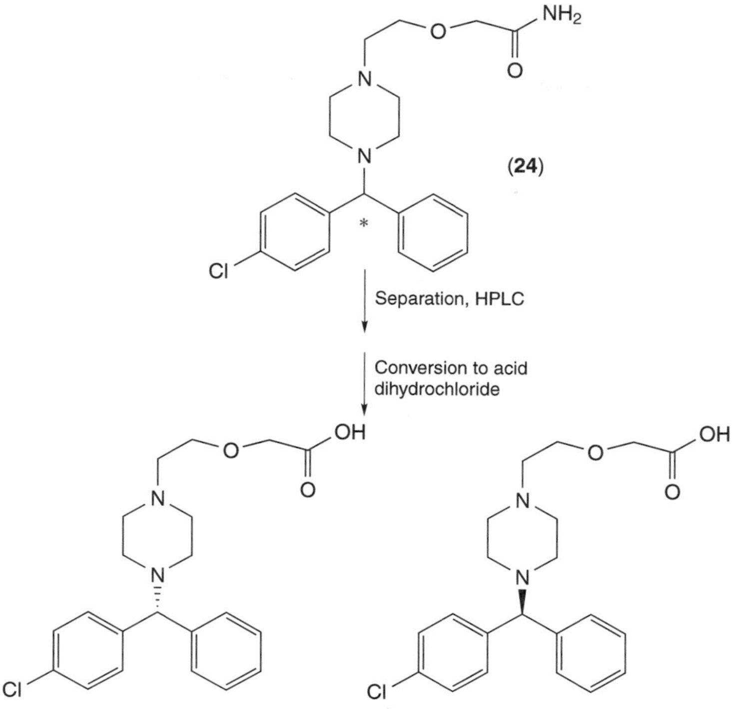

Figure 18.9.

ucts. A number of methods have been introduced in an attempt to improve on this technology, such as recycling (44). Perhaps the biggest advancement in recent times has been the introduction and application of SMB technology in the field of chiral separations (48). This technique was pioneered in the late 1950s by Universal Oil Products in the United States as a useful method for separation of oil derivatives and sugars (49). Initially SMB technology was applied to very large volumes of material. For example, xylene isomers are separated in thousands of ton quantities annually. The application of SMB to the separation of racemic mixtures has led to downsizing and modifications of this technology, but the main principles remain the same. The use of counter-current contact in SMB maximizes the driving force for mass transfer and the contact between the substrate and stationary phase. This provides a more efficient use of the adsorbent capacity than that of a simple batch system (50).

The separation of racemic mixtures is well suited to SMB technology, because these counter current systems can generally only perform two-component separations at a time (51). A detailed description of this technique is given in an excellent article by Guest (52). The SMB system generally consists of several columns, typically 6–12, which are connected in series. An arrangement of pumps and valves are set up to maximize the stationary phase utilization, allowing for better solvent efficiency and adsorbate concentration. This leads to two streams coming off the system in solution, one is termed the raffinate, which is enriched in the less adsorbed component, and the other termed extract, which is enriched in the more adsorbed component. The complex set of conditions and parameters that are required to optimize SMB chromatography has led to the design and process optimization being done by computer simulations (53). A number of examples will be discussed that highlights this growing area of chiral separa-

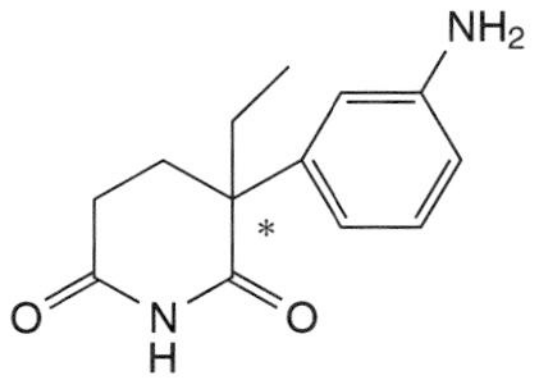

Aminoglutethimide (**27**)

Figure 18.10.

tions. It should be noted that the scale of operation is dependent on the column size and can lead to a range from tens of grams to tons of separated isomers. Clearly, the larger quantities separated imply that this technology has industrial applications.

The enantiomers of aminoglutethimide (**27**) (Fig. 18.10) have been separated using an SMB approach (48) (see also Section 3, Fig. 18.11 for more information on aminoglutethimide). A set of 16 columns (6 × 1.6 cm) containing Chiracel OJ were used. The feed concentration was 1.63% in a mixture of hexane:ethanol (15:85), which was used as the mobile phase. A feed rate of 0.45 ml/min and a mobile phase rate of 6 ml/min gave rise to a production of 5.27 g of each enantiomer per day. The S-(−)-enantiomer was obtained as the extract in solution, in a 99.8% purity, while the R-(+)-enantiomer also in solution as the raffinate, achieved a 99.9% purity. This would lead to a productivity of 59.9 g of each enantiomer per kilogram of CSP per day. It should be noted that one big advantage of SMB over preparative chromatography are the vast savings on mobile phase consumption; this is generally coupled to thin film evaporators that allow for very high levels of

recovery of the solvent. This becomes even more evident when a poorly soluble compound is used.

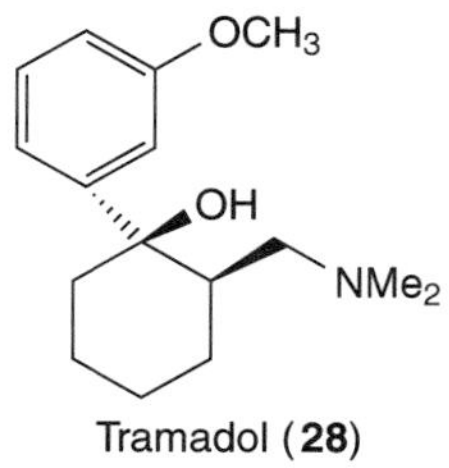

The two isomers of the racemic analgesic drug Tramadol (**28**) (Fig. 18.11) display differing affinities for various receptors. (−)-Tramadol mainly inhibits the reuptake of noradrenaline, whereas the (+)-isomer inhibits the reuptake of serotonin. In addition, the (+)-isomer and its primary metabolite, the O-desmethyl derivative, are selective agonists of μ opiate receptors (54). Tramadol has been efficiently separated using SMB; in addition, the resolution by crystallization is given in Section 3 of this chapter (55). Comparison between batch chromatography and SMB for the separation of tramadol was made. Use of 12 columns (100 × 21.2 mm ID), each packed with 20 g of Chiralpak AD 20-μm phase, and using a mobile phase composition of 2-propanol/light petroleum/diethylamine (5:95:0.1 v/v/v) with feed concentration of 20 g/L, obtained a very high productivity. Thus, 680 g of racemic tramadol could be separated per liter of stationary phase (which equates to 1.2 kg of racemate per kilogram of stationary phase per day). The solvent consumption of 144 L/kg of racemate should also be noted. This gives both (+)- and (−)-enantiomers of high optical purity, with the extract of 6.33 g/L and the raffinate of 7.69 g/L. Typically, the solvent (mobile phase) is readily recycled by the use of thin film evaporators, which further extends the economic practicality of the process.

2.4 Conclusions

It should be noted that all the techniques described in this chapter can be inter-linked. In other words, if one technique, i.e., asymmetric synthesis, failed to deliver enantiopure material, then another technique such as crystallization can be used to push through the product to the desired purity. As an example of this "double" approach, the application of SMB and crystallization to the separation of mandelic acid is noteworthy (56). When very high levels of enantiopurity are required, the efficiency and cost effectiveness of SMB may not be economical. However, if for example, a lower enantiomeric excess can be coupled with an enhancement by crystallization, then the

Tramadol (**28**)

Figure 18.11.

SMB approach becomes even more favorable. This can lead to substantial increases in the productivity of the SMB process. Further examples of coupling of two techniques will be given throughout this chapter.

In summary chromatographic separations offer an expedient method for the separation of enantiomers on a small scale. With the development of more efficient stationary phases and the application of SMB, this may become the method of choice for the separation of racemates. Each individual case deserves investigation by all of the techniques/approaches described in this section.

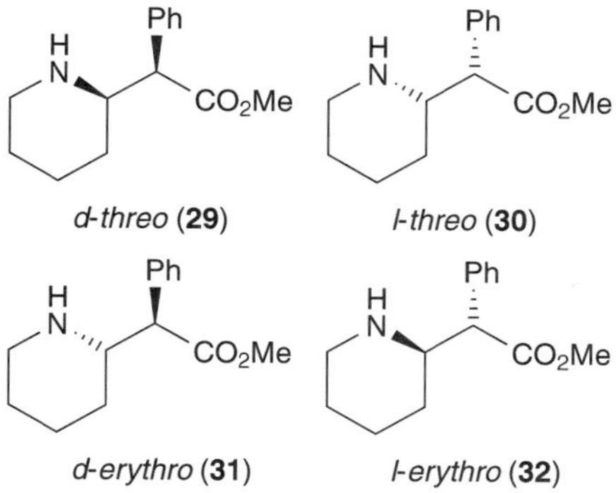

3 CLASSICAL RESOLUTION

Perhaps the most widely used method for the preparation of single enantiomers involves the classical resolution of a racemic mixture, which uses the formation of crystalline diastereomeric salts. As discussed in the introduction to this chapter, by converting a racemic mixture of enantiomers to two diastereomeric salts with differing physical properties, one being crystalline and the other remaining in solution, the molecules can be separated and simply converted back to the two separated enantiomers. With the advent of automation, the classical resolution approach offers a speedy and through racemate separation methodology. This enables the separation of small amounts of material (milligram to gram) and can be directly scaled up to provide an industrial process (kilogram to ton). A number of different approaches to this type of separation are highlighted in the following sections, where it should be noted that diastereomeric salt resolutions have mistakenly been considered to be a mysterious art. In fact, there is considerable information in the literature as to how to perform a resolution and the physical chemistry aspects associated with how to define and conduct the resolution to its optimum capability (57–59). We will not go into this in great detail but will highlight some pertinent points; for greater detail, the reader is directed to the monograph by Jacques et al. (57).

Figure 18.12.

3.1 Separation of the Active Pharmaceutical Ingredient

A number of single isomer switches, that is, where a drug that was previously sold as a racemate is developed and sold as a single isomer, have been isolated through classical resolution (60). This approach to a single isomer offers several advantages; first, the racemate is freely available and can be purchased to high levels of purity and quality. Second, the analytical methods will also be in place. Also, no new synthetic development chemistry is required, and hence this is the fastest route to the single enantiomers at the multigram scale. Generally, this is the first method to be tried. Some of the many available examples demonstrate the different nuances that can be applied in classical resolution to provide the single enantiomers in optimal yields and purities are given in this section.

An efficient and large scale resolution of methylphenidate (ritalin hydrochloride) using dibenzoyl-tartaric acid has been described (61). Ritalin is marketed for the treatment of children with attention deficient disorder (ADHD). Methylphenidate has two chiral centers and originally was marketed as a mixture of two racemates, 20% DL-*threo* (**29, 30**) and 80% DL-*erythro* (**31, 32**) (see Fig. 18.12 for the structures of all four isomers). As introduced previously, the *erythro*-isomer is defined as the case when the main chain of a molecule (drawn vertically in a Fischer projection) has identical or similar substituents at two adjacent non-identical chiral centers on the same side of the chain, whereas the *threo* isomer has

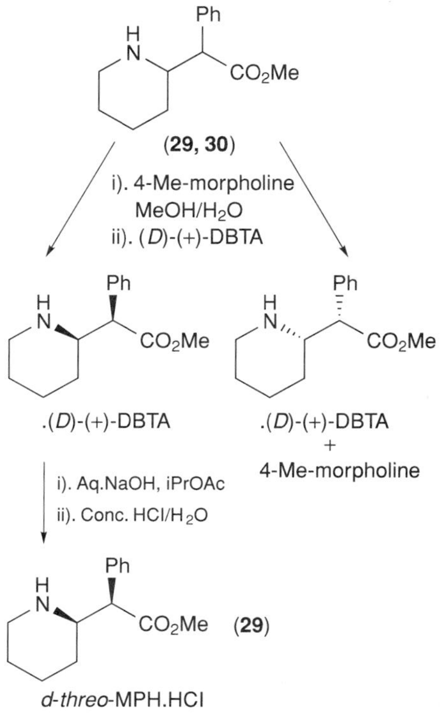

Figure 18.13.

the corresponding substituents on opposite sides. The racemic drug currently used in therapy comprises only the pair of *threo*-enantiomers (**29, 30**). The mode of action in humans is not completely understood, but methylphenidate presumably activates the brain stem arousal system and cortex to produce its stimulant effect. In addition, there is no specific evidence that clearly establishes the mechanism whereby methylphenidate produces its mental and behavioral effects in children or conclusive evidence regarding how these effects relate to the condition of the CNS. The D-*threo* (**29**) enantiomer has, however, been reported to be 5 to 38 times more active than the corresponding L-*threo* enantiomer (**30**) (62). The resolution shown in Fig. 18.13 uses the racemic hydrochloride salt as input material. The HCl salt is cracked to the free base *in situ* with 4-Me-morpholine, which then forms a salt with the resolving agent dibenzoyl-tartaric acid (DBTA). The required D-*threo*-methylphenidate (**29**) is removed as the crystalline salt of D-(+)-DBTA, leaving the L-

threo enantiomer (**30**) in solution with 4-methylmorpholine hydrochloride. The use of 4-methylmorpholine to effect base release *in situ* helps to streamline the process and to remove a costly free base isolation process. The D-*threo*-methylphenidate, (D)-(+)-DBTA, salt is readily converted into the hydrochloride salt. It is interesting to note that recently, Celgene and Norvatis received a FDA approvable letter for the use of dexmethylphenidate for use in ADHD. This consists of only the D-*threo* enantiomer (**29**), in comparison with the original product, which contained all four isomers (**29–32**).

Chemists at Chiroscience took an alternative approach to the D-*threo*-methylphenidate (**29**) single enantiomer (63). An efficient resolution using L-(−)-di-toluoyl-tartaric acid (DTTA) was developed. This left the required D-*threo* diastereoisomer in solution with a diastereomeric excess of 88% yield in 55% chemical yield. Conversion of this salt to the free base and subsequent crystallization of the hydrochloride salt gave >98% ee D-*threo* methylphenidate in high purity in an overall yield of 42%. The enhancement of the ee is caused by the eutectic point of methylphenidate hydrochloride, which is at 30% ee. A more detailed description of this phenomenon will be discussed later in this section.

(*S*)-Naproxen (**36**) is a non-steroidal anti-inflammatory drug that was introduced to market in 1976 by Syntex. The *S*-(+)-isomer is about 28 times more effective than the *R*-(−)-isomer (64). The annual sales in 1995 were about $1 billion; thus, a large amount of effort has been spent developing the synthesis of (*S*)-Naproxen (65). The resolution of racemic Naproxen (**33**), developed by Syntex, approaches the ideal case for a Pope Peachy resolution, that is, resolution using non-stoichiometric quantities of resolving agent (66). Here, a mixture of 1 equivalent (eq) of the racemic acid, 0.5 eq of an achiral amine base, and 0.5 eq of the chiral amine (*N*-alkylglucamine) are used (Fig. 18.13). This results in the formation of two salts: one is the insoluble (*S*)-Naproxen chiral amine (**34**), obtained in 45–47% yield and optical purity of 99%. The second salt that remains in solution contains (*R*)-Naproxen and the achiral amine (**35**). The insoluble salt of (*S*)-Naproxen (**34**) is removed

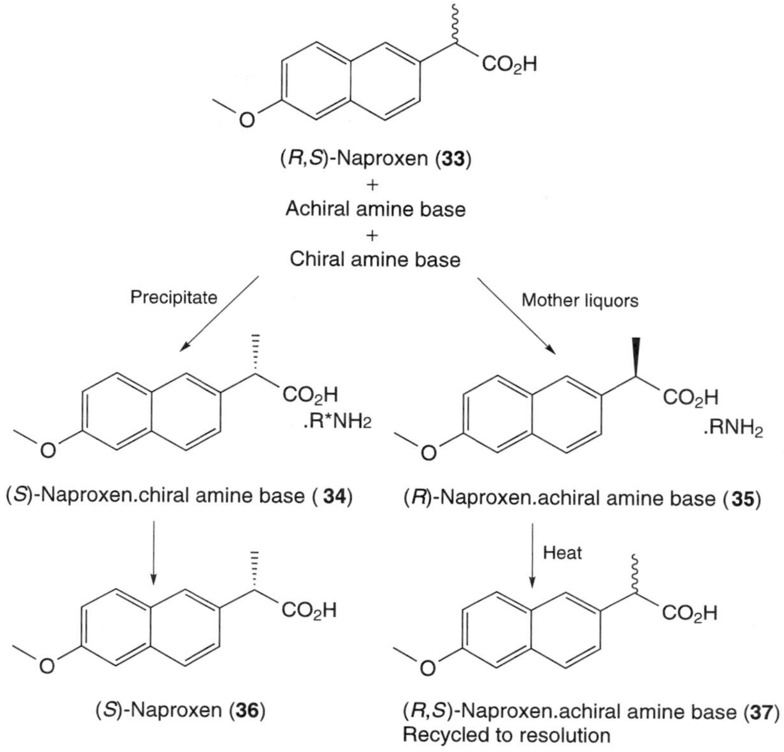

Figure 18.14.

by filtration. The mother liquors are then heated and the achiral amine base catalyzes racemization of the unwanted R-enantiomer. The resulting racemic mixture of the acid (R,S)-(**37**) can then be put back into the resolution loop. Using this process, the overall yield of (S)-Naproxen is >95%, based on the input of racemic acid. To further highlight the efficiency of this process, the N-alkylglucamine resolving agent is recovered in >98% per cycle.

Racemic bupivacaine hydrochloride (**38**, Marcaine) is currently used as an epidural anesthetic during labor and as a local anesthetic in minor operations. Clinical studies have shown that *levo*-bupivacaine (**41**) is less cardiotoxic in man, making it significantly safer than the racemate (67). Separation of the enantiomers was readily achieved using 0.25 eq of D-tartaric acid. This resulted in the isolation of a 2:1 (S)-bupivacaine D-tartaric acid salt (**39**) in 98% de, leaving the (R)-bupivacaine

free base (**40**) in solution. Conversion of the tartrate salt to (S)-bupivacaine hydrochloride (**39**) was obtained in 35–40% overall yield based on racemate input. To increase the economics of the process, a racemization of the unwanted R-enantiomer was required. Treatment of the liquors containing the enriched (R)-bupivacaine, tartaric acid, propanol, and propionic acid at reflux resulted in complete racemization in 2 h. By pertinent processing, the racemic free base thus obtained is isolated by crystallization and can be put back into the resolution cycle (68). Another fine example by chemists from Eli Lilly involves a clever resolution-racemization-recycle (R-R-R) process in the synthesis of Duloxetine (69).

As discussed in Section 2 of this chapter, Tramadol is a chiral drug substance that is currently used as a high potency analgesic agent. The preparation of Tramadol is shown in Fig 18.16, which results in the formation of all four possible stereoisomers from the Grig-

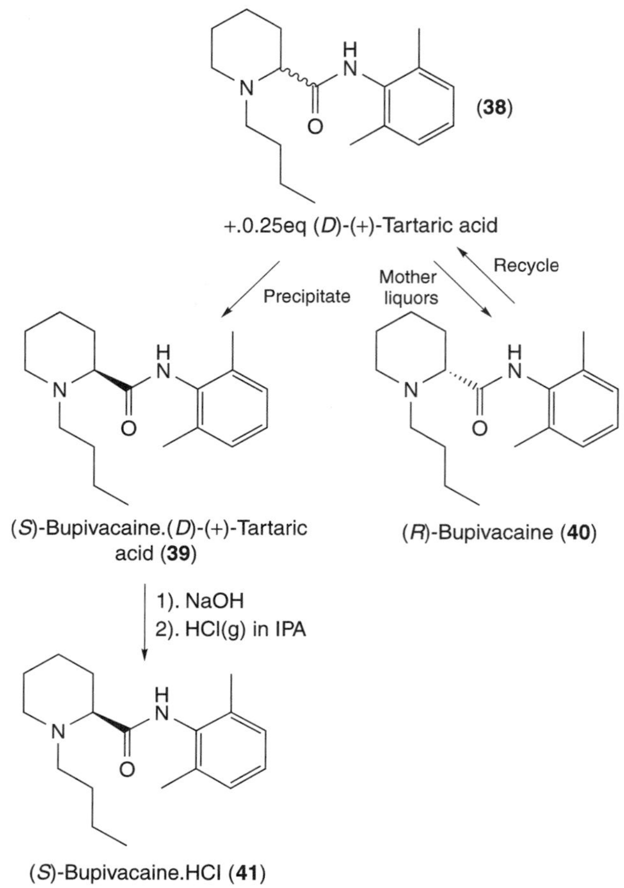

Figure 18.15.

nard reaction (70). The *trans* isomers (**42, 43**) form over the *cis* isomers (**44, 45**) in a ratio of ~8:2; the currently marketed racemate consists of only the *trans* isomers. It is possible to take this crude reaction mixture and selectively isolate either the (+)-*trans* isomer (**42**), by using di-*p*-toluoyl-D–tartaric acid [D-(+)-DTTA] resolving agent or the (−)-*trans* isomer (**43**) using L-(−)-DTTA. This highlights the high selectivity that can be achieved when using certain resolving agents. In the case of Tramadol, the *cis* isomers (**45, 46**) do not form crystalline salts with DTTA and therefore remain in solution. This results in a highly efficient process, where the chiral acid not only separates the single enantiomers (**42** or **43**) but also removes other impurities (i.e., *cis* isomers **44** and **45**) at the same time (71).

Another drug that is sold as a racemate is Etodolac (**46**), which is used as a non-steroidal anti-inflammatory agent (NSAID) that also has analgesic properties; it has the ability to retard the progression of skeletal changes in rheumatoid arthritis (72). It has been shown that the majority of therapeutic activity lies in the *S*-(+)-isomer (73). D-(−)-*N*-Methylglucamine (meglumine) is obtained by ring opening of D-glucose with methylamine, and hence it is readily available and inexpensive. Scientists at Chiroscience have described the use of meglumine to separate the enantiomers of Etodolac (74). It was shown that the meglumine salt possessed suitable properties to enable its use as a salt for pharmaceutical administration. Therefore, in the case of Etodolac, meglumine can not only be used to separate

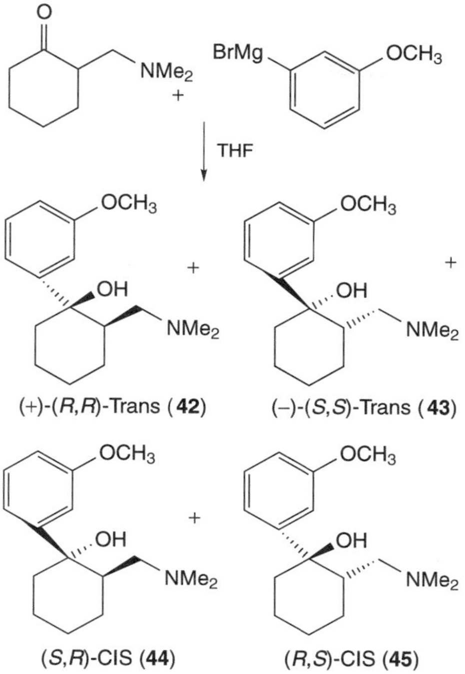

Figure 18.16.

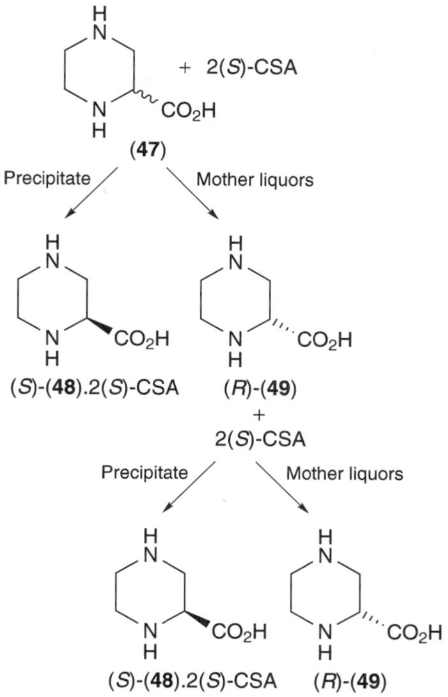

Figure 18.18.

the enantiomers, but it can also be used as the pharmaceutical salt form of choice.

In addition to the racemic drugs discussed in this section, resolutions are also used in the isolation of key building blocks for the pharmaceutical industry. An important class of these intermediates are amino acids, many of which are available as the single isomer from natural sources (see INTRODUCTION). The use of unnatural amino acids and D configured ones are expected to have a greater influence at the biological level. In the drive for molecular diversity and metabolic stability, a number of unnatural amino acids such as the non-proteinogenic piperazine carboxylic acid (**47**) (Fig. 18.18) have been developed. Specifically,

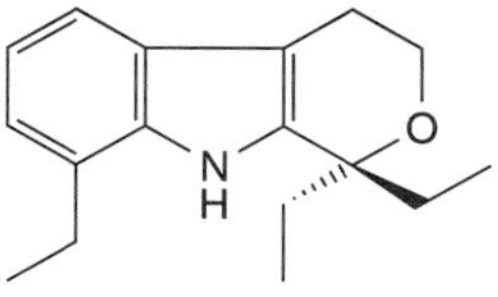

Etodolac (**46**)

Figure 18.17.

this amino acid has found use as an intermediate compound of the HIV proteinase inhibitor L-735,525 (75). The racemic cyclic amino acid (**47**) has been resolved with S-camphorsulfonic acid (CSA), which yields the S-isomer as the double CSA salt (**48**) as the precipitate (76). Retained in the mother liquors is the R-isomer (**49**). This can neatly be racemized to the S-isomer by mixing with S-CSA in a suitable solvent. On seeding with pure (S,S)-diastereomeric salt, a further quantity of the desired (S,S) product (**48**) is obtained, leaving the R-isomer (**49**) once more in the liquors. The whole cycle can be repeated and has been demonstrated with four complete cycles. To complete the whole process, the resolving agent is also readily recovered and recycled.

3.2 Separation of Intermediates to Single Enantiomer Active Pharmaceutical Ingredient

The previous examples given for diastereomeric salt resolution have all involved separation of the active pharmaceutical ingredient (API) or late stage intermediate. Whereas this

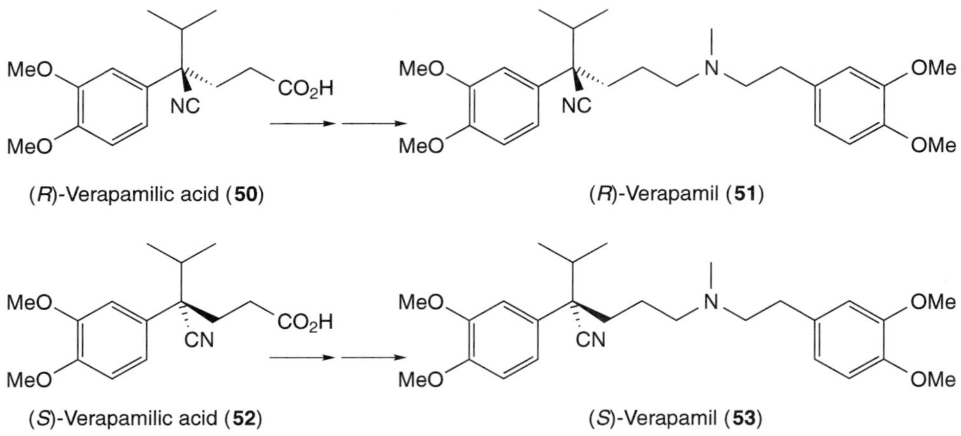

Figure 18.19.

does offer several advantages from the point of view of time and quality aspects, there are also a number of drawbacks. If, for example, a racemization of the unwanted isomer cannot be found, there would be a waste of 50% of material. Therefore, it can often be advantageous to conduct the separation at an earlier stage in the synthesis of the drug. This leads to better atom efficiency compared with resolution of the final product, resulting in a reduction of the overall amount of waste and cost.

One such example is Verapamil, which is a well-established treatment of cardiovascular ailments (77). S-(−)-Verapamil (51) has specific transmembrane calcium channel antagonist activity, whereas its antipode (53) influences a wider range of cell pump actions, including those for sodium ions (78). Verapamil has been separated into its single enantiomers by resolution with expensive resolving agents, which required multiple recrystallizations to effect complete separation (79). Looking into the synthetic sequence of Verapamil, several intermediates seemed to be attractive alternatives to Verapamil (80). The intermediate verapamilic acid (Fig. 18.19) was efficiently separated using α-methylbenzylamine (α-MBA), which is an extremely cheap resolving agent (81). Subsequent transformation of the easily obtained R- or S-verapamilic acid (50 or 52), required a further three to four synthetic steps to yield the active pharmaceutical ingredient.

The racemate aminoglutethimide (27) has been shown to be effective in the treatment of hormone-dependent breast cancer (Fig. 18.20). Further studies have shown that the R-enantiomer is more potent than its antipode as an aromatase inhibitor (82). The resolution of aminoglutethimide itself has been reported in the literature, using tartaric acid. This resolution suffers from the formation of solid solutions (83), which require endless crystallizations to deliver the single enantiomer (84). Use of a suitable precursor (54) enabled separation of the intermediate (55), by treatment with the alkaloid resolving agent (−)-cinchonidine. This chiral acid was then cyclized to nitroglutethimide, which on reduction, gave the desired R-aminoglutethimide (56) (85). It is noteworthy that in the case of aminoglutethimide, the amine functionality is an aniline moiety. Because of the low pK_a associated with this amine (2.5–4.6), the number of acidic resolving agents that can be employed are reduced, because they need to be of relatively high acidity to form a salt.

3.3 Crystallization-Induced Asymmetric Transformation

A number of amino acids have been separated by resolution, in certain cases the yield of the required diastereoisomer has been greater than 50% (86). p-Chlorophenylalanine is of considerable pharmacological interest, because of its ability to inhibit serotonin forma-

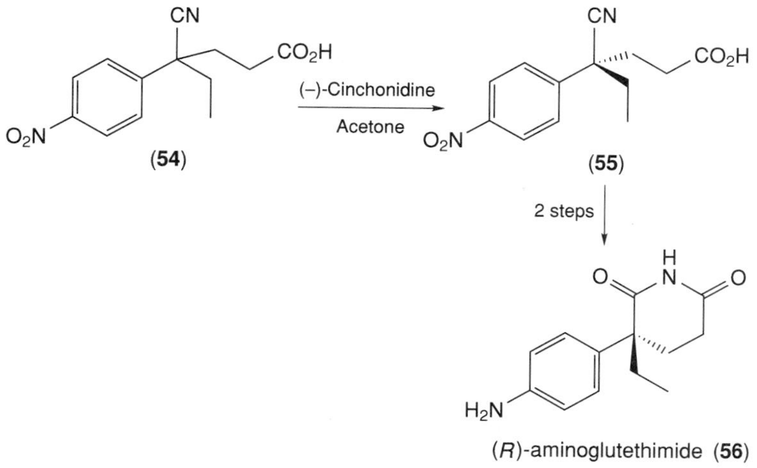

Figure 18.20.

tion in laboratory animals (87). Both the *R*- and *S*-enantiomers have also been used as building blocks in the synthesis of other drugs. An ingenious approach to *R-p*-chlorophenylalanine methyl ester, which is based on a one-pot resolution-racemization sequence, is highlighted in Fig. 18.21. Here, treatment of racemic *p*-chlorophenylalanine methyl ester (**57**) with 0.5 eq of D-tartaric acid and 0.1 eq of salicylaldehyde in methanol gave a 68% yield of 98% enantiomeric purity of the 2:1 *R-p*-chlorophenylalanine D-tartaric acid salt (**58**). The reason that the absolute yield is greater than 50% is caused by the *S*-enantiomer being racemized *in situ*. The 2:1 tartrate salt is crystalline and is therefore removed from the system by virtue of its insolubility. This drives the equilibrium further in favor of the 2:1 *R-p*-chlorophenylalanine D-tartrate salt (88).

While the common goal remains to be the rational design of resolving agents (89), it is clear that we are still away from this actually happening. An alternative "family" approach to classical resolution has been demonstrated by Vries et al. (90). A group of similar resolving agents are mixed simultaneously with the racemate. This was done to shorten the time required to complete the resolving agent screen. Note should be made that the families of resolving agents are very similar and that the crystalline species obtained by this method contained more than one of the resolv-

ing agents. As with all screens, analysis of the data is often time consuming and laborious. Bruggink et al. have shown that differential scanning calorimetry (DSC) of the isolated salts can help to quickly determine whether the isolated salt will provide a through resolution (91). However, with a methodical and precise screening protocol, it is nearly always possible to find a suitable resolving agent that effects separation of the enantiomers (92).

4 NONCLASSICAL RESOLUTION

4.1 Preferential Crystallization

A brief description of the type of "racemic" compounds is necessary for the reader to better understand the principles behind the application of crystallization methods to the separation of enantiomers. Three fundamental types of crystalline racemates exist. In the first, the crystalline racemate is a conglomerate, which exists as a mechanical mixture of crystals of two pure enantiomers. The second, which is the most common, consists of the two enantiomers in equal proportions in a well-defined arrangement within the crystal lattice; this is termed racemic compound. The third possibility occurs with the formation of a solid solution between the two enantiomers that coexist in an unordered manner in the crystal. This kind of racemate is called a pseu-

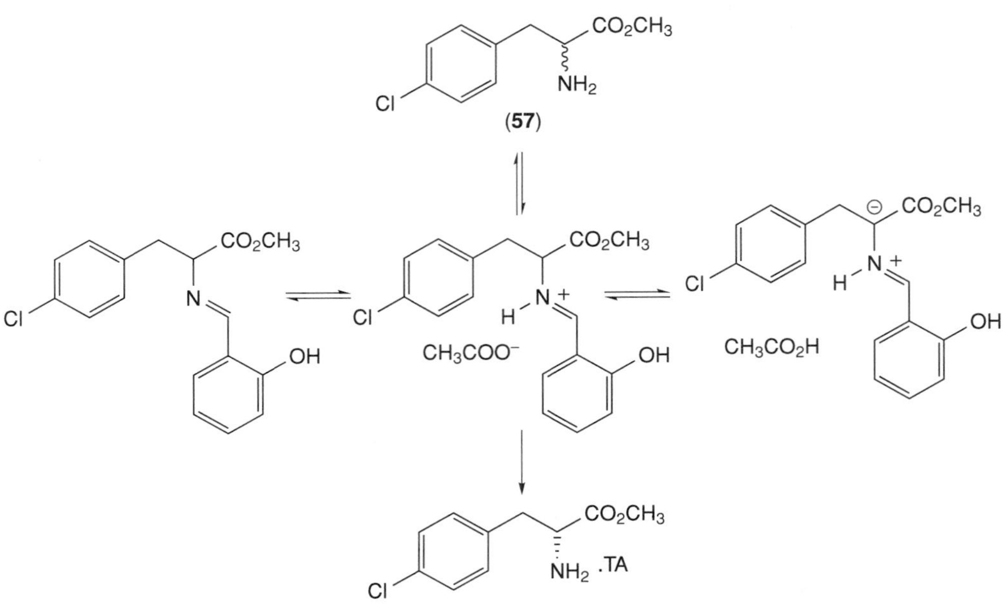

(R)-p-chlorophenylalanine.0.5eq(*D*)-tartaric acid **(58)**

Figure 18.21.

doracemate and is rather rare. Conglomerates have been estimated to be approximately 10% of all racemates (93). Diagrammatic representation of the first two types of racemate are shown in Fig. 18.22.

By understanding the appropriate phase diagrams, which describe the melting behav-

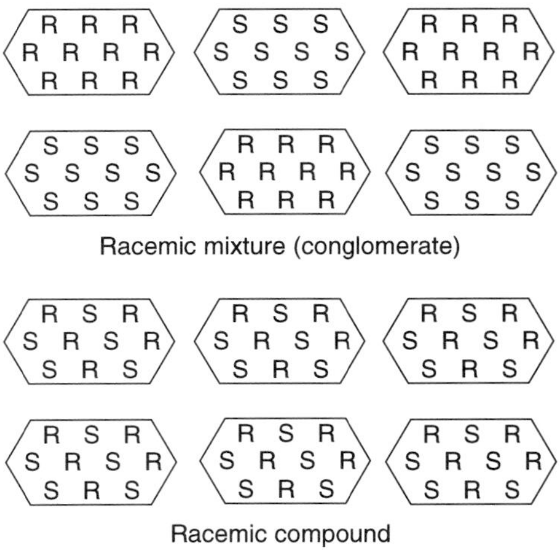

Racemic mixture (conglomerate)

Racemic compound

Figure 18.22.

ior of the two enantiomers (binary melting point phase diagram) or their solubility behavior in the presence of a solvent (ternary solubility phase diagram), separation of enantiomers can be reproduced. Phase diagrams for the three types of racemate are shown in Fig. 18.23. For a full and detailed explanation of this topic refer to the monograph of Jacques et al. (57).

4.2 Enrichment of Enantiomeric Excess by Crystallization

The attainment of high levels of enantiopurity is not always possible by enzymatic or diastereomeric resolutions or by asymmetric syntheses alone. It is however frequently possible to prepare a pure enantiomer from a partially resolved sample by simple recrystallization. For this process to proceed successfully it is necessary that the initial enantiopurity of the mixture is greater than that of the eutectic point in the phase diagram. By utilization of the phase diagram, the optimal quantity of solvent required can be calculated. It is also possible to calculate the maximum expected yield.

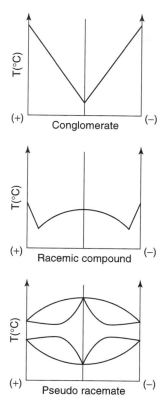

Figure 18.23.

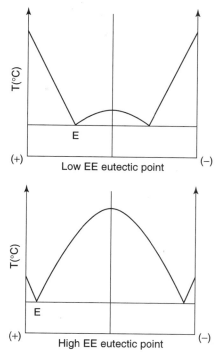

Figure 18.24.

Note should also be made that in some cases recrystallization reduces the enantiomeric excess, which can lead to crystallization of the racemate (94). In these cases the mother liquors contain moderately to highly enriched material. It is therefore important to plan the strategy at which point the enantiomer is recrystallized to optical purity. This may be from an enzymic resolution, or in the event that an asymmetric synthesis has failed, to deliver enantiopure product. As discussed in Section 3, the liquors from the diastereomeric resolution with DTTA of 88% de can be cleaved to the free base, and crystallization of the hydrochloride salt gives >98% ee. This is because of the fact that methylphenidate hydrochloride has a eutectic point of 30% ee. Davies et al. (95) and Winkler et al. (96) have prepared single enantiomer methylphenidate (**29**). Their approaches use an enantioselective synthesis; the enantiomeric excesses are 86% and 69%, respectively, thus requiring recrystallization

to deliver enantiopure product. Another example of this type of compound is Warfarin (**13**). Chemists at Dupont (97) developed an asymmetric hydrogenation approach, which gave Warfarin in ∼80% ee. Simple crystallization in an appropriate solvent yielded optically pure Warfarin, thus indicating that the eutectic point is below 80% ee. (See earlier section on the metabolism and binding properties of the Warfarin enantiomers).

The phase diagrams below highlight two typical cases, the first where the eutectic point E is close to the racemate, and the second where the eutectic approaches the single enantiomer as shown in Fig. 18.24. In the first case, it would be preferable to crystallize the enriched enantiomer to optical purity, e.g., methylphenidate. However, in the second case, a very stable racemic compound exists, giving rise to a high eutectic point. Here crystallization of enriched enantiomer mixture will only be successful at high ee. For example, verapamil hydrochloride requires that the ee be greater than 98% for crystallization to yield enantiopure product. Below this, the enantiopurity is reduced. In this case, it is advantageous to recrystallize the diastereomeric salt precursor to optical purity before proceeding to final product.

4.3 Resolution by Direct Crystallization

It is important to show how conglomerates are identified. We have already seen that they have specific phase diagrams as shown in Fig. 18.23. Other such data that support identification of a conglomerate are IR, X-ray data, and observation of a spontaneous resolution or resolution by entrainment. Note should be made that in 1848, Louis Pasteur separated the dextrorotatory and levorotatory crystals of sodium ammonium tartrate. This manual sorting of crystals is also known as triage, and by its very nature is time consuming and laborious. The readers are again directed towards the Jaques et al. monograph, which lists over 250 known examples of conglomerates (57). There are two possibilities for separation of enantiomers by direct crystallization. The first uses spontaneous resolution, which occurs when a conglomerate crystallizes. This crystallization may be followed by the mechanical separation of the crystals of the two enantiomers. Various techniques have been developed that aid this separation.

The second type of resolution by direct crystallization is known as entrainment. Here, the differences in the rate of crystallization of the enantiomers in a supersaturated solution give rise to a separation. Strict control of the conditions for the crystallization are required, with the system of crystals and solution not being allowed to come to equilibrium and time playing an important role. The occurrence of conglomerates has been estimated to be approximately 10% of all racemic compounds. We will now illustrate this phenomenon with some pertinent examples.

An example of use of the conglomerate Narwedine (**59**) in the synthesis of a natural product Galanthamine (**61**) which is an *Amaryllidaceae* alkaloid and has been used clinically for 30 years for neurological illnesses (98). More recently it has been approved for the use in the treatment of Alzheimer's disease (AD) (99). Galanthamine acts to inhibit acetylcholinesterase (AChE), thus increasing the levels

of acetylcholine. An increase in the level of acetylcholine in patients with AD has been shown to improve their cognitive performance. Galanthamine has been extracted from botanical sources; however, several tons of daffodil bulbs are needed to produce 1 kg of product. A synthetic route has been developed that uses a crystallization-induced chiral transformation (Fig. 18.25). This crystallization was first reported by Barton and Kirby (100) and further developed by Shieh and Carlson (101). The success of this transformation is based on two phenomena: narwedine (**59**), which crystallizes as a conglomerate, and (−)-narwedine (**60**), which equilibrates with (+)-narwedine through a retro-Michael intermediate. This process has now been developed so that (−)-narwedine (**60**) is routinely obtained in 80% yield from the racemate input, as shown in Fig. 18.25 (102).

Recently a number of potent 5-HT$_3$ receptor antagonists such as Ondansetron have been reported to be clinically effective for the blockade of chemotherapy-induced nausea and emesis (103). The structurally novel compound (**62**) has also been shown to be a highly potent 5-HT$_3$ antagonist (104); specifically, the R-(−)-(**62**) enantiomer was shown to be the most active. Comparison of the physical data of the racemate and single enantiomer indicated that this structure (**62**) exists as a conglomerate (104). By careful experimentation, the best concentration, temperature, and time for crystallization were discovered. Table 18.1 highlights the results obtained for the entrainment.

The initial concentration of the solution was 10.0 g of (±)-(**62**) in 50 g of acetone. In all runs, 10 mg of seed crystals were used. From the 10 runs highlighted in the 18.1, 21.0 g of R-(−)-(**62**) of >92.0% ee and 21.4 g of (S)-(+)-(**62**) of >90% ee are obtained from an input of 50.4 g of racemate. The table also nicely illustrates the continuous nature of the process, which coupled with the fact that no resolving agent, chiral auxiliary, enzyme, or catalyst is needed, underlines the economic advantages of this type of process.

The importance of amino acids as building blocks for asymmetric synthesis is well documented (105). A number of amino acids have been shown to exist as conglomerates. Shi-

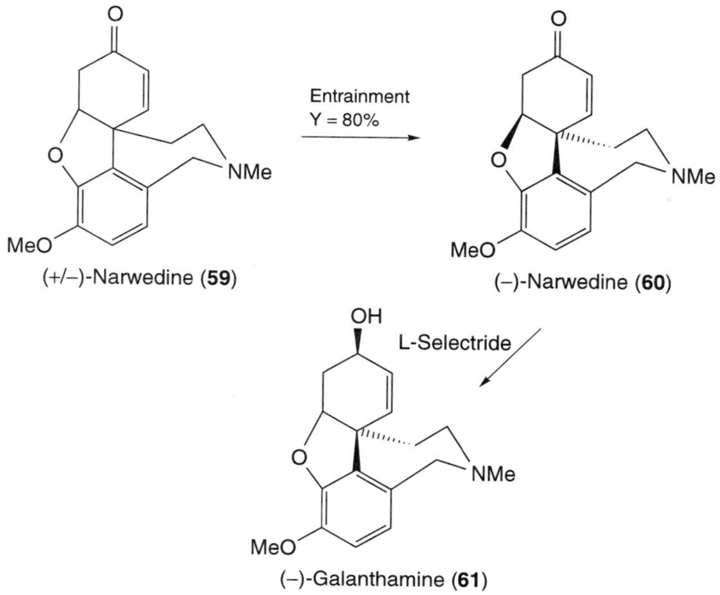

(+/−)-Narwedine (59)

Entrainment
Y = 80%

(−)-Narwedine (60)

L-Selectride

(−)-Galanthamine (61)

Figure 18.25.

raiwa et al. have described the preferential crystallization of racemic methionine hydrochloride (106). The obtained D- or L-methionine hydrochloride was, however, only ~75% optically pure, requiring a further recrystallization to furnish enantiopure product. Shiraiwa et al. have also recently disclosed the resolution of (2RS, 3SR)-2-amino-3-chlorobutanoic acid HCl again using entrainment (107). Here it was shown to be necessary to conduct the crystallization in an ethanol/5 M hydrochloric acid solvent mixture for optimal results. By careful control of the conditions, high levels of enantiomeric excess were obtained in the crystalline salt.

Chemists in Japan have developed an excellent approach to (+)-Diltiazem, which is a coronary vasodilator (108). An intermediate

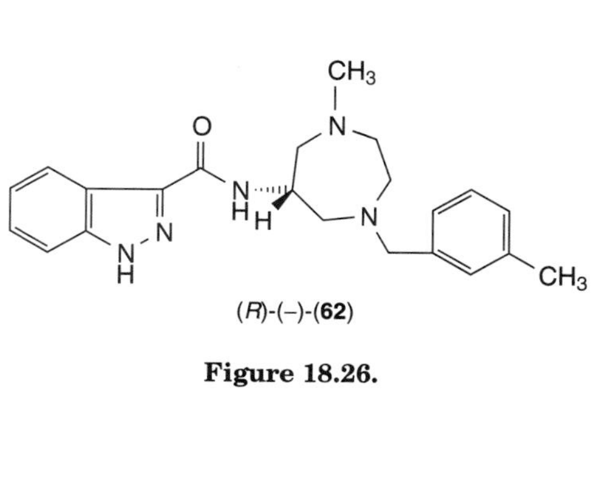

Figure 18.26.

(R)-(−)-(62)

is successfully resolved using preferential crystallization. The glycidic acid–substituted phenylesters were prepared; of the 30 synthesized, only one exhibited conglomerate properties (109). This was the 3-(4-methoxyphenyl)glycidic acid 4-chloro-3-methylphenyl ester (63). Table 18.2 summarizes the physical data collected, which is illustrative of the conglomerate nature of this compound.

The obtained single enantiomer (−)-epoxide (64) is then converted into the required (+)-isomer of Diltiazem (65) in several steps, as highlighted in Fig. 18.27.

Taxol is a natural product isolated in very low yield from *Taxus brevifolia* and is used in the treatment of cancer (110). The extreme chemical complexity of Taxol makes production by total synthesis uneconomical. However, a semisynthetic approach using the naturally derived 10-deacetylbaccatin III (66) condensation with N-benzoyl-(2R, 3S)-3-phenylisoserine (67) does provide an alternative and economic approach (111). N-benzoyl-(2R, 3S)-3-phenylisoserine (67) is also commonly known as the Taxol side-chain and has been prepared in optically active form using chiral auxiliaries or resolving agents (112). It has been shown that the Taxol side-chain is a conglomerate and can therefore be cheaply and

Table 18.1   Resolution of (62) by Preferential Crystallization

Run	(±) Added (g)	Seed	Time (minutes)	EE of Solution (%EE)	Amount of Crystals (g)	Rotation	%EE of Solid
1	—	(R)-(−)	220	22.7/+	2.0	−	95
2	4.0	(S)-(+)	210	19.4/−	4.6	+	92
3	4.6	(R)-(−)	190	21.2/+	3.8	−	95
4	3.8	(S)-(+)	205	21.0/−	3.9	+	95
5	3.9	(R)-(−)	210	21.5/+	3.8	−	95
6	3.8	(S)-(+)	220	16.4/−	5.0	+	90
7	5.0	(R)-(−)	200	21.3/+	3.4	−	92
8	3.4	(S)-(+)	215	21.2/−	3.9	+	94
9	3.9	(R)-(−)	190	22.2/+	4.0	−	93
10	4.0	(S)-(+)	220	21.4/−	4.0	+	95
11	4.0	(R)-(−)	200	20.8/+	4.0	−	95

Reprinted from H. Harada, *Tetrahedron Asymmetry*, vol. 8, T. Morie, Y. Hirokawa, and S. Kato, 1997, pp. 2367–2374. Reproduced with permission from Elsevier Science.

efficiently entrained to the single required enantiomer (113).

5   ENZYME-MEDIATED ASYMMETRIC SYNTHESIS

Enzymes have found frequent use in the synthesis of single isomer drugs from racemic or prochiral compounds at the larger manufacturing scales. The use of enzymes to effect chiral transformations in the medicinal chemistry laboratory has been far less frequent; however, the increasing availability of immobilized and stabilized forms of enzymes has made their use easier and the resultant transformations more predictable.

By virtue of their complex macromolecular structure, including a highly defined active site, enzymatic transformations generally proceed with a high degree of chemical selectivity and stereospecificity. Reactions are typically conducted under mild conditions of temperature, pressure, and pH, thus minimizing losses caused by unwanted side reactions or partial racemization. The use of extremophiles or cross-linked enzymes such as CLECs

do enable the use of higher temperatures, pressures, and organic solvents.

Enzymes can be utilized to affect a number of transformations; the broad spectrum of reactions, including amide bond formation, hydrolysis, esterification, reduction, oxidation, and carbon–carbon bond formation, has been reviewed elsewhere (114).

5.1   Amide Bond Formation

The use of enzymes to stereospecifically form amide bonds has been described in many texts (115); however, the commercial availability of cross-linked enzyme crystals (CLECs), for example, PeptiCLEC-TR, which is an immobilized form of *Thermolysin protease*, has been used in the synthesis of D2163 (68), a novel matrix metalloproteinase inhibitor (116). *In vitro* enzyme screening identified the all-natural *SSS*-isomer as the active product. The elegant CLEC (117) technology used in this example makes the enzyme stable to typical organic reaction conditions and enables facile removal of the enzyme at the end of the reaction by simple filtration. On this basis, it is

Table 18.2   Properties of (63) Indicating Conglomerate Nature

Compound	MP (°C)	Solubility (g/100 mL) THF	Solubility (g/100 mL) DMF	IR Spectrum
(±)-4.2	123–124	14.0	13.0	Identical
(−)-4.2	139–141	6.7	6.9	Identical

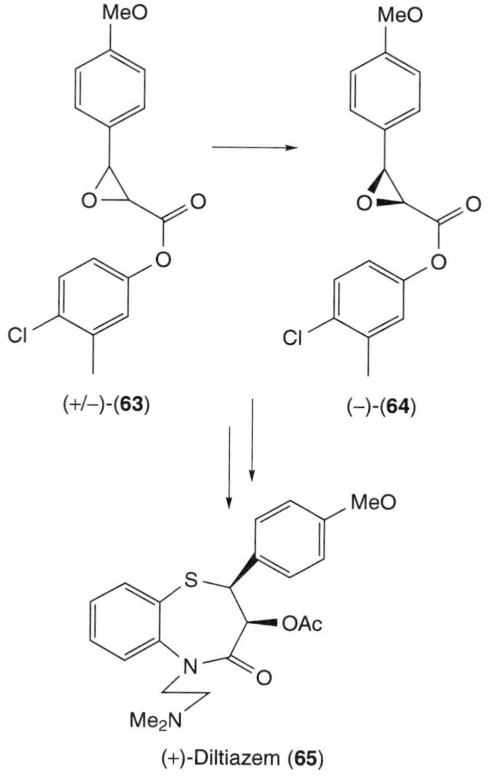

(+/−)-(63) → (−)-(64)

(+)-Diltiazem (65)

Figure 18.27.

anticipated that medicinal chemists will more commonly use these enzymes in the future.

The coupling of dipeptide (**69**) to the protected α-*thio* carboxylic acid (**70**) was conducted in organic solvent at high concentration with the desired product produced in a few hours with high enantiospecificity.

5.2 Transesterification and Hydrolysis

A widely used technique for separating racemic mixture is the use of enzyme mediated transesterification or hydrolysis. One important example is the separation of Naproxen (**33**), which is a member of the 2-arylpropionic acid class of profens that are broadly used as NSAIDs (see Section 2 for the separation of enantiomers using a crystallization approach). The important association between chirality and biological activity of this class of drugs has been extensively researched, where

the role of cyclooxygenase-independent properties of the *R*-enantiomers in the gastrointestinal toxicity of the racemates and the likelihood that the use of racemates increases the propensity of profens to alter the pharmacokinetics of other drugs has been described (118).

Whereas not all profens are sold as single isomers, Naproxen is sold as the single *S*-enantiomer (**36**) where various strategies including crystallization, chromatographic separation, asymmetric hydrogenation and enzymatic hydrolysis, and esterification have been used to prepare the single isomer (65). Specific examples include the use of *Candida cylindracea* lipase to enantioselectively prepare single isomer naproxen ester with trimethyl silyl methanol (119) and the use of *Candida rugosa* lipase in an enantioselective continuous hydrolysis of Naproxen methyl ester (120).

Pipecolic acid is a component of a number of active drugs, including bupivacaine (**38**) and thioridazine (**72**) (Fig. 18.30), which has been efficiently resolved as the racemic *n*-octyl pipecolate with *Aspergillus niger*. The *S*-isomer is obtained as the free acid in a 40% yield based on the available enantiomer with a 97% ee (121).

Propanolol (**14**) is a broadly used β-adrenergic receptor blocking agent that is sold as the racemate. However, the majority of the activity is associated with the *S*-enantiomer (**74**) (see Section 2) (122). The asymmetric

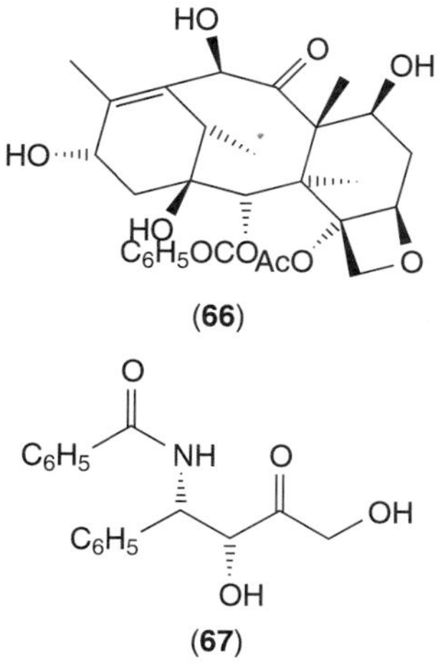

(**66**)

(**67**)

Figure 18.28.

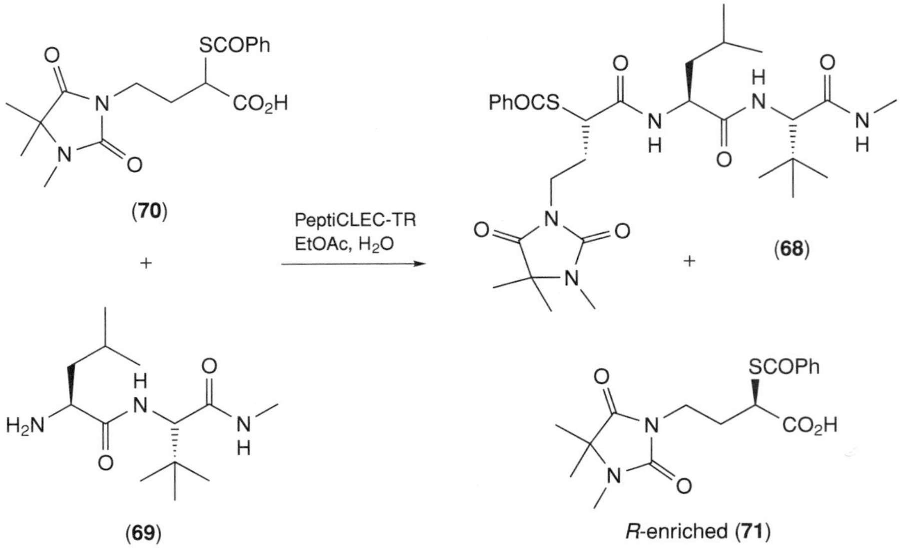

Figure 18.29.

synthesis of the desired S-enantiomer has been achieved by the selective acylation of the R-enantiomer of the key intermediate (**73**) as shown in Fig. 18.30.

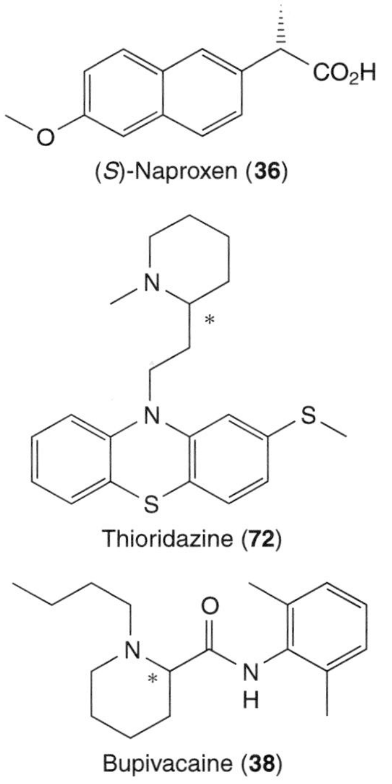

(S)-Naproxen (**36**)

Thioridazine (**72**)

Bupivacaine (**38**)

Figure 18.30.

5.3 Oxidation and Reduction

In addition to the widely reported techniques of amide bond formation, transesterification, and hydrolysis, enzymic enantioselective oxidation is also used in the synthesis of single isomer drugs. Patel described the efficient oxidation of benzopyran (**75**), an intermediate in the synthesis of potassium channel openers (123). The transformation was effected with a cell suspension of *Mortierella ramanniana* with glucose over a 48-h period; the isolated product (**77**) was obtained in a 76% yield with an optical purity of 97% and a chemical purity of 98%, as shown in Fig. 18.32.

Reduction with a variety of enzymes has been reported (114), including bakers yeast for the reduction of α-methyleneketones to the corresponding α-methylalcohol (124), a functionality that is present in a number of drugs. The reduction of an azidoketone (**78**) using *Pichia angusta* enzyme has been used in the synthesis of S-salmeterol (**79**) (125). Salmeterol (Serevent) is a potent, long-acting β2-adrenoreceptor used as a bronchodilitor in the treatment of asthma. Recently, Sepracor claimed that the S-enantiomer had a higher selectivity for β2 receptors and that it did not cause certain adverse effects associated with the administration of ($\pm$)- or (R)-salmeterol (126). The synthesis of (S)-salmeterol (**79**) is shown in Fig. 18.33.

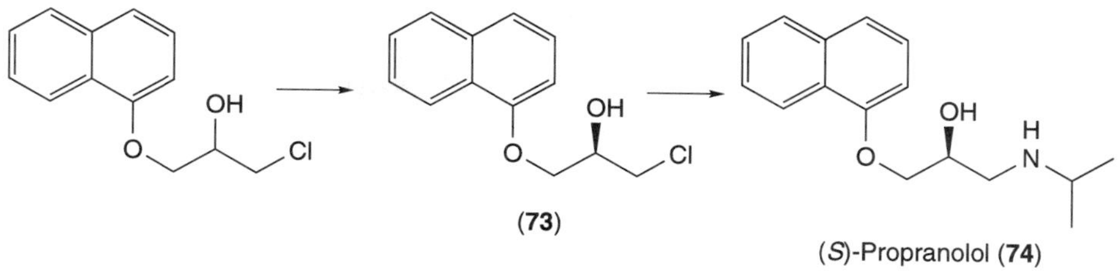

Figure 18.31.

6 ASYMMETRIC SYNTHESIS

Synthetic organic chemists have a vast array of tools at their disposal when faced with the challenge of preparing a chiral compound as a single enantiomer. The purpose of this section is to introduce the reader to some asymmetric approaches toward chiral drugs and medicinal compounds, highlighting examples where the stereoisomers behave differently in biological systems. There are many excellent books and reviews covering methods for asymmetric synthesis and their application to the preparation of pharmaceutical agents and complex natural products (127).

6.1 Chiral Pool

The use of enantiopure starting materials from nature in the synthesis of chiral drugs is not only of great historical significance but remains of critical importance to the pharmaceutical industry. Consideration of the current biggest-selling single enantiomer drugs shows how important this approach is (8 of the top 10 in 1996 were obtained from chiral pool

Figure 18.32.

starting materials or by synthetic manipulation of fermentation products) (127). The optically pure starting materials that have been used in drug synthesis include amino acids, hydroxy acids, terpenes, alkaloids, carbohydrates, and many more structurally diverse compounds. There are many syntheses involving clever manipulation of chiral pool starting materials and use of these chiral centers to induce further asymmetry (i.e., by diastereoselective reactions). We will briefly consider some examples in which all or most of the chiral centers in the target molecule originate directly from nature.

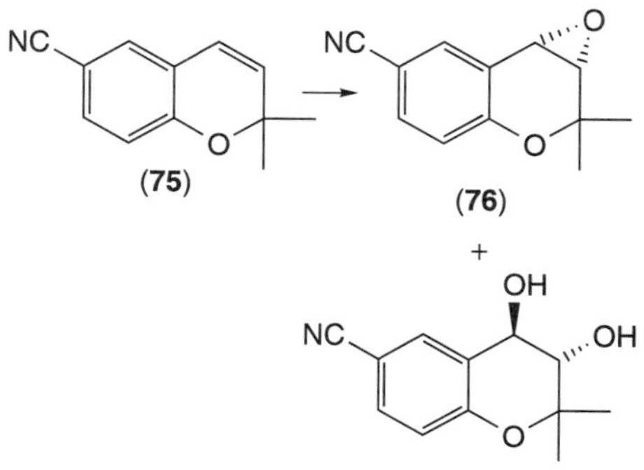

Angiotensin-converting enzymes (ACE) inhibitors are used mainly for the treatment of cardiovascular disorders and are among the biggest selling drugs worldwide (128). Enalapril (**80**) is synthesized from the natural amino acids L-alanine and L-proline (129). Lisinopril (**81**) incorporates a lysine derivative (130). One of the chiral centers in Captopril (**82**) is derived from proline, but the other is generated by chemical or enzymatic resolution (131). Cilazapril (**83**) is a conformationally restricted second generation ACE inhibitor developed by Roche, and the core is synthesized from a glutamic acid derivative and an amino acid–derived pyridazine (128, 132).

There are many other examples of drugs based on an amino acid backbone. Stoner et al. recently reported a synthesis of the HIV protease inhibitor ABT-378 (Lopinavir) (**84**) (133). In a similar synthesis to that of the related compound, Ritonavir, key intermediate (**85**) is prepared by stepwise diastereoselective reduction of enaminone (**86**). This means that the existing chiral center, derived from natural L-phenylalanine (protected to **87**), controls the formation of the two new stereocenters as

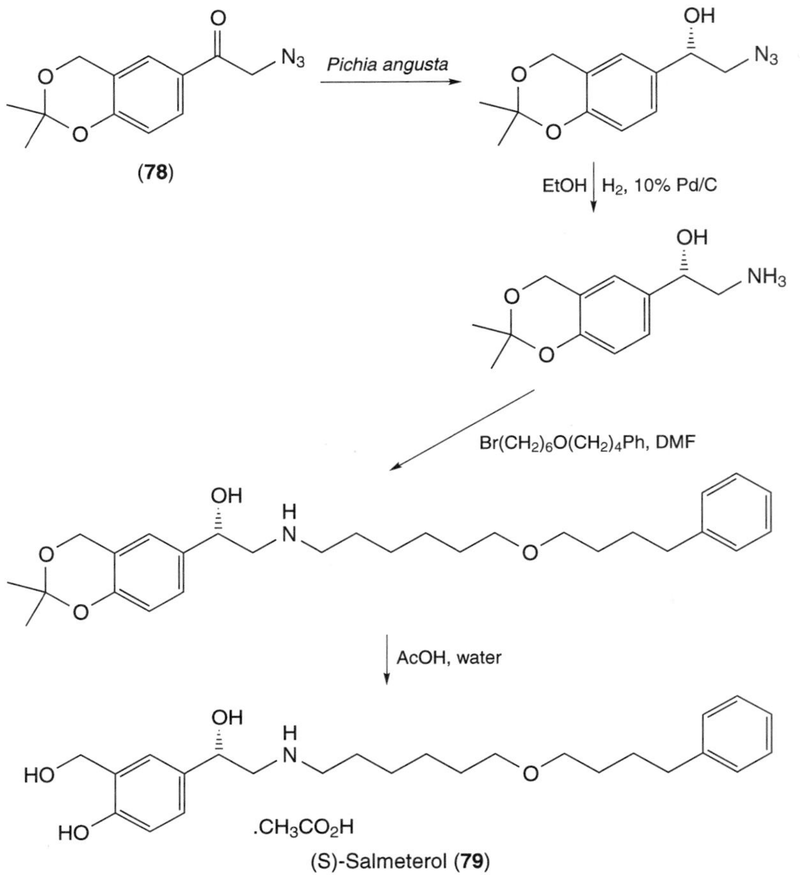

Figure 18.33.

discussed for chiral auxiliaries below. Two acylations then complete the synthesis, with the final chiral center clearly derived from L-valine.

The stereospecificity of binding at the histamine H_3-receptor was investigated by preparing a series of ligands from D- or L-histidine (**88**) (134). It was found that compounds such as (S)-(**89**) had greater affinity for the receptor than their R-enantiomers. In addition, replacing the aromatic moiety with a cyclohexyl group (e.g., **90**) switched the activity to agonism for compounds with an amino group in the chain.

Hydroxy acids are important chiral starting materials in the synthesis of many biologically active compounds (135). (S)-3-Hydroxy-γ-butyrolactone (**91**) is a very useful synthetic

unit available from D-pyranoses (136). Workers at Schering-Plough used this as the key starting material in a concise synthesis of Sch 57939 (**92**), a β-lactam–based cholesterol absorption inhibitor (137). The condensation between the dianion of (S)-3-hydroxy-γ-butyrolactone and an appropriate diaryl imine proceeded with very high diastereo- and enantioselectivity, generating azetidinone (**93**) with a *trans:cis* ratio of >95:5.

Researchers at Abbott have been investigating the use of pyrrolidinyl isoxazoles as nicotinic cholinergic channel activators (138). Until recently, ABT-418 (**97**) was undergoing clinical trials as a potential treatment for cognitive impairment and decline and for Alzheimer's disease. A short synthesis of ABT-418 was devised starting from commercially avail-

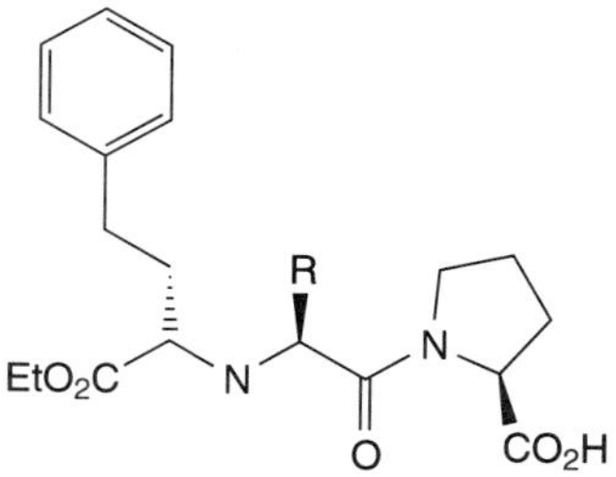

Enalapril R=CH₃ (**80**)
Lisinopril R=(CH₂)₄NH₂ (**81**)

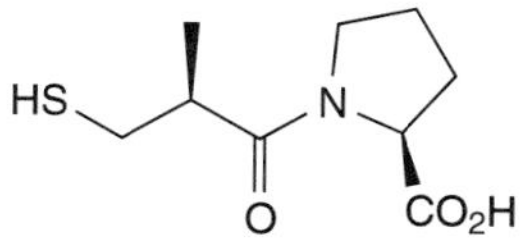

Captopril (**82**)

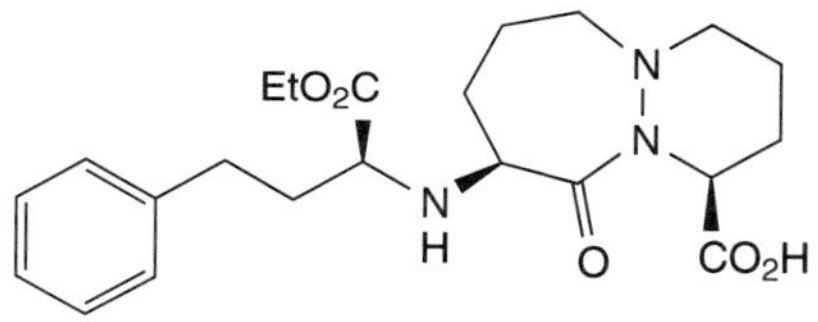

Cilazapril (**83**)

Figure 18.34.

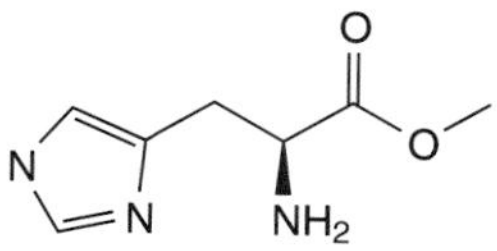

L-histidine methyl ester (**88**)

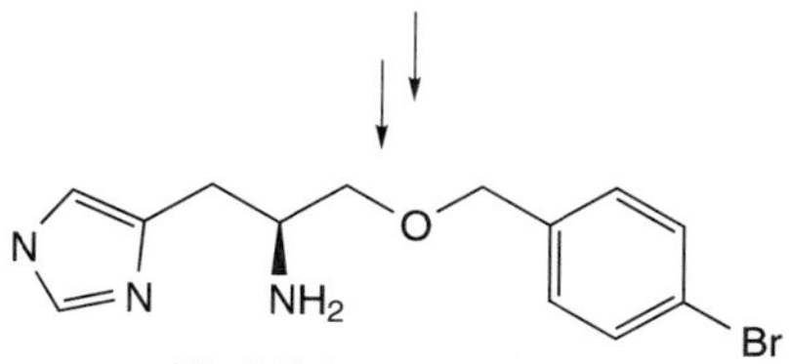

(S)-(**89**) (pK_i H₃ = 7.1)
c.f. (R)-(**89**) (pK_i H₃ = 6.7)

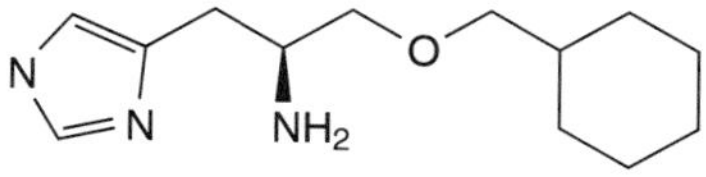

(S)-(**90**) (pK_i H₃ = 7.9)
c.f. (R)-(**90**) (pK_i H₃ = 6.8)

Figure 18.36.

able (S)-pyroglutamic acid methyl ester (**94**) (139). Acetone oxime dianion was added to the methyl ester (**94**) to generate an intermediate (**95**). Racemization of the chiral center was found to occur under basic conditions; however, this was avoided by immediate treat-

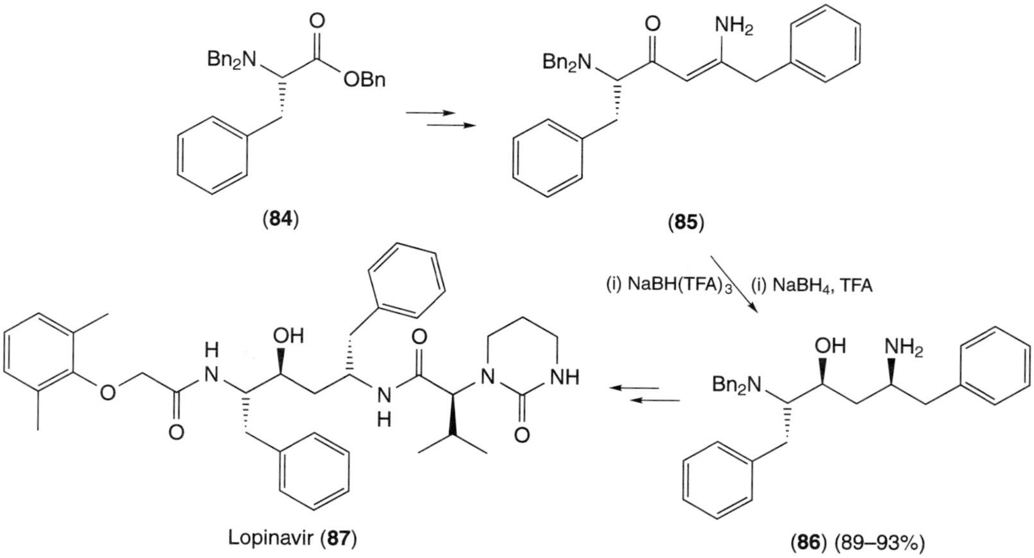

(**84**)

(**85**)

(i) NaBH(TFA)₃ (i) NaBH₄, TFA

Lopinavir (**87**)

(**86**) (89–93%)

Figure 18.35.

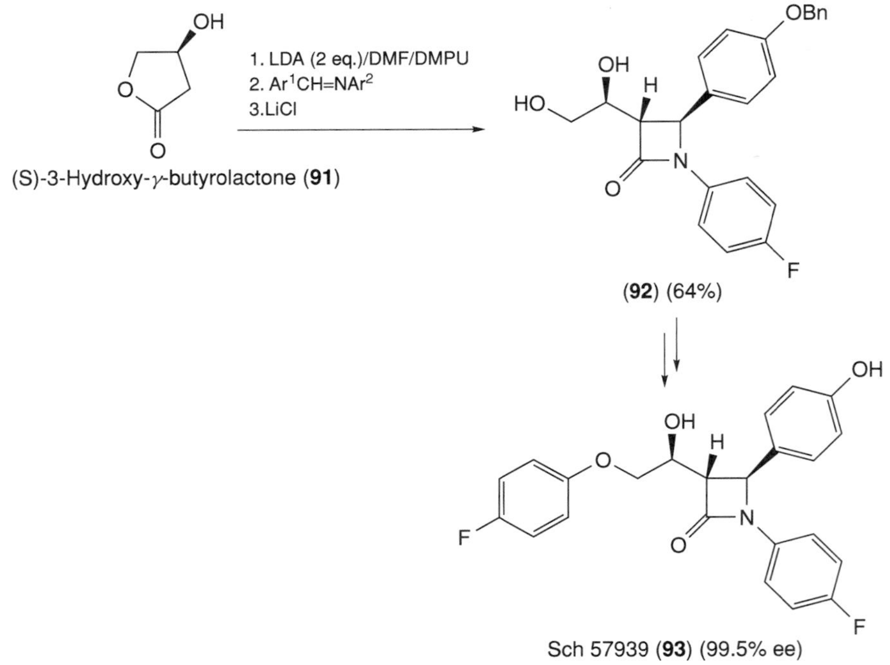

Figure 18.37.

ment with concentrated sulfuric acid resulting in cyclization and dehydration to amide isoxazole (**96**). Reduction and *N*-methylation yielded ABT-418 (**97**). The binding affinity of ABT-418 at neuronal cholinergic channel receptors was measured to be one order of magnitude greater than the corresponding *R*-enantiomer (K_i = 4.2 versus 44 n*M*) (138).

6.2 Chiral Auxiliary

In this approach the substrate is attached to a chiral, non-racemic unit that controls the formation of one or more new chiral groups. Reaction of the coupled unit with a reagent or prochiral substrate is designed to produce one diastereomeric product in excess. The auxil-

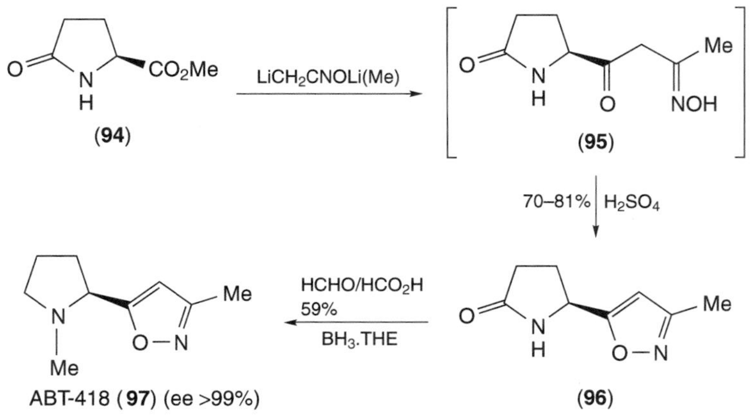

Figure 18.38.

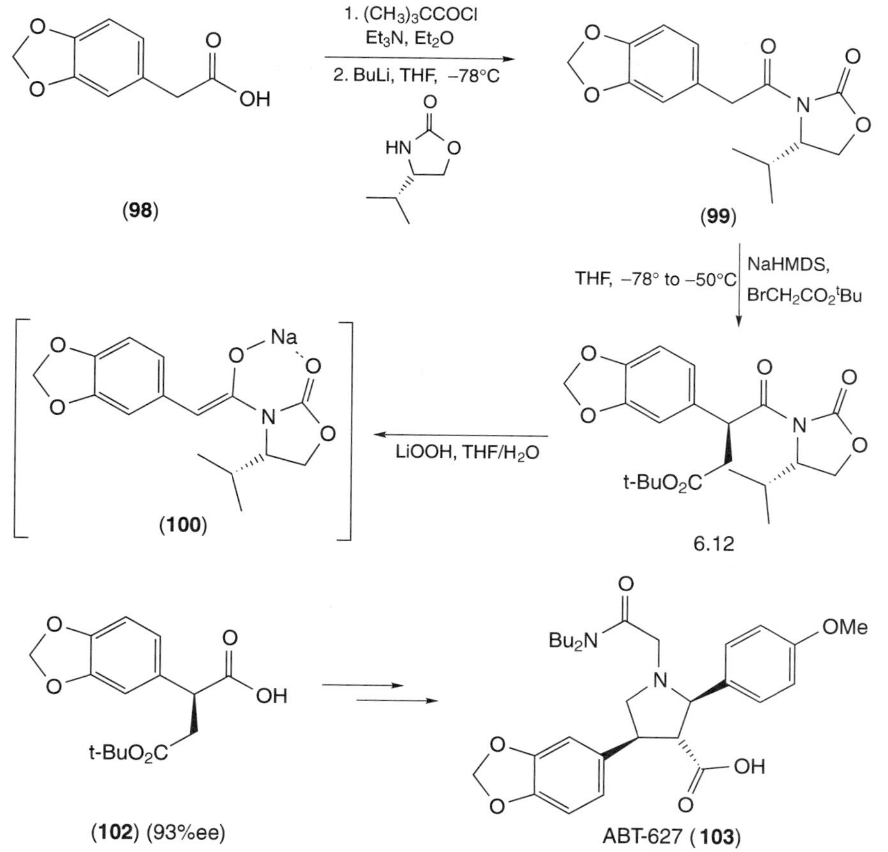

Figure 18.39.

iary is then removed (and preferably recovered), providing the product in high enantiomeric excess. This process is most attractive when both isomers of the auxiliary are readily available in enantiomerically pure form, and where the reaction leads to high levels of stereoselectivity in a predictable manner. Attaching and removing the auxiliary should be straightforward and proceed without loss of stereochemical integrity.

Many auxiliaries currently in use are derived from 1,2-amino alcohols (140). These are readily available from natural sources with little or no synthetic manipulation and can react in a variety of ways to form conformationally well-defined (usually cyclic) auxiliary systems. The use of oxazolidinones in asymmetric synthesis was developed by Evans et al., and these oxazolidinones have been used extensively in a

variety of different reactions (140, 141). The use of this chiral auxiliary in the preparation of pharmaceuticals is widespread, and there are several large-scale processes using such chemistry (142).

Abbott reported an improved synthesis of ABT-627 (**98**) involving an asymmetric alkylation of the valine-derived acyl oxazolidinone (**99**) (143). ABT-627 (Atrasentan) is a selective endothelin ET_A receptor antagonist under development for the treatment of cancer, particularly prostrate cancer. Acid (**100**) was activated as a mixed anhydride and treated with the lithium anion of the oxazolidinone to give (**101**). Both of the following deprotonation and alkylation steps must be controlled to give high levels of stereoselectivity. The (Z)-enolate (**102**) is favored, both kinetically and thermodynamically, by the bulky *iso*-propyl group

Table 18.3   Stereochemical Variation

$(3\alpha, 6)$	K_i (nM versus HIV-1)	IC$_{50}$ (μM, cell HIV-1)
R, R Tipranavir	0.008	0.03
R, S	0.018	0.14
S, R	0.032	0.41
S, S	0.22	1.7

and is held rigid by chelation to the carbonyl oxygen of the oxazolidinone. The major stereoisomer then results from alkylation of this chelated enolate anion from the least hindered "upper" face to yield (**103**) as the major product. There are many strategies for removal and recovery of an oxazolidinone auxiliary (141). In this case, hydrolysis with lithium peroxide provides the acid that is transformed into Atrasentan *through* a cyclization-ring contraction strategy controlled by the chirality present in (**103**).

Tipranavir (PNU-140690) is a potent third-generation HIV protease inhibitor in clinical development by Boehringer Ingelheim (under license from Pharmacia). The biological activity of such 5,6-dihydro-4-hydroxy-2-pyrone sulfonamides shows considerable stereochemical variation (Table 18.3) (144). The R-configuration is preferred at both chiral centers (3α and 6), and Tripanavir is more than 50 times as potent as its enantiomer in a cell culture assay using HIV-1$_{IIIB}$–infected H9 cells. An asymmetric synthesis (145) begins with the Michael addition of an aryl cuprate (derived from commercially available Grignard reagent **105**) to the unsaturated oxazolidinone imide (**104**), yielding the adduct as a single diastereomer (**106**). The nitrogen protecting group was changed and an acetyl group introduced to give ketone (**107**), which undergoes a stereoselective aldol reaction with an acetylenic ketone (**108**). The highest diastereoselectivity was obtained for this reaction using Ti(O^{n}Bu)Cl$_3$ as the Lewis acid. Both of the critical asymmetric steps to form new chiral centers are controlled by the (R)-phenyl oxazolidinone. The chiral auxiliary is removed when (**109**) is treated with base to form the lactone ring. This is followed by two further steps that generate PNU-140690 (**110**) as a single enantiomer.

The enantioselective synthesis of dopaminergic benzyltetrahydroisoquinolines and their binding to D$_1$ and D$_2$ dopamine receptors was investigated by Cabedo et al. (146). The synthetic route, illustrated by the preparation of the (1S)-isomer involves stereoselective reduction of the isoquinolinium salt (**114**) with (R)-phenylglycinol (introduced in protected form as **112**) as the chiral auxiliary. The (1R)-enantiomer of (**115**), prepared in an analogous fashion using (S)-phenylglycinol, binds to dopamine receptors with considerably less affinity (>100 μM versus D$_1$ and 61.2 μM versus D$_2$). In contrast, stereochemical differentiation was not observed at the dopamine uptake site for these compounds.

Two different chiral auxiliary approaches have been applied to the synthesis of NPS 1407 and it's enantiomer (**119**) (147). NPS 1407 is an antagonist of the glutamate NMDA receptor that has *in vivo* activity in neuroprotection and anti-convulsant assays. The R-enantiomer was synthesized in four steps from (**116**) with the chiral center introduced by a completely stereoselective alkylation of hydrazone (**117**). The chiral auxiliary, S-(−)-1-amino-2-(methoxylmethyl)pyrrolidine (SAMP), was introduced by condensation with aldehyde (**116**) and removed by catalytic hydrogenolysis. In the second method, the S-enantiomer was formed in a four-step sequence with the chiral center installed by the Michael addition of chiral amine (**121**) (formed in one step from the readily available α-methylbenzylamine) to benzyl crotonate (**120**). NPS 1407 (**123**) was found to be 12 times more potent than it's enantiomer (**119**) at the NMDA receptor in an *in vitro* assay.

An example of the use of a terpene as a chiral auxiliary is provided by the synthesis of the anti-viral reverse transcriptase inhibitor Lamivudine (148). The nucleoside analog is marketed by Biochem Pharma (now Shire Pharmaceuticals) and Glaxo Wellcome (now GlaxoSmithKline) for the treatment of HIV and hepatitis B virus infection. In the

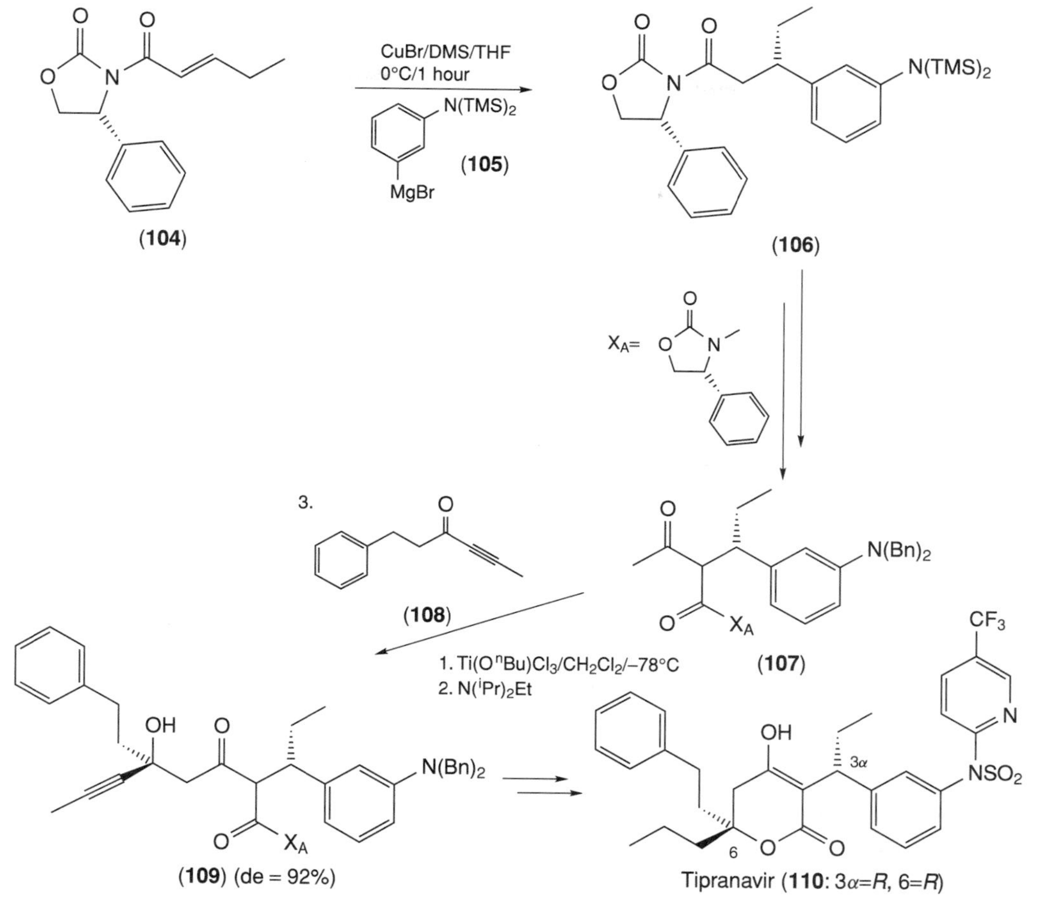

Figure 18.40.

production route, the glycolate derived from (−)-menthol (**124**) is coupled with thioacetyl dimer (**125**). The chiral auxiliary directs reaction to install the desired (2R)-stereochemistry in (**126**). *In situ* formation of chloro compound (**127**) is followed by a stereoselective coupling reaction with trimethylsilyl cytosine again directed by the (−)-menthyl carboxylate. Reductive removal of the auxiliary yields Lamivudine (**129**) as a single isomer that was found to have favorable toxicological and pharmacokinetic properties to the racemate.

6.3 Chiral Reagent

In this approach, asymmetry is induced in a prochiral molecule or functional group by reaction with a stoichiometric amount of an en-

antio-enriched reagent system. The reaction proceeds through diastereomeric transition states, resulting in the preferential formation of one enantiomer or diastereomer. Current reagents can lack generality and may be difficult to prepare in both chiral forms. At least one equivalent of the chiral component is required, which can present economical and practical difficulties. Many examples are provided by the reduction of double bonds, especially ketones. Ketone (**130**) was reduced enantioselectively using either (+) or (−)-*b*-chlorodiisopinocampheylborane (**149**). Reduction with (−)-*b*-chlorodiisopinocampheylborane generated the alcohol (S)-(**131**), which was transformed into the (1R,3S)-isochroman compound, (1R,3S)-(**132**), through a ste-

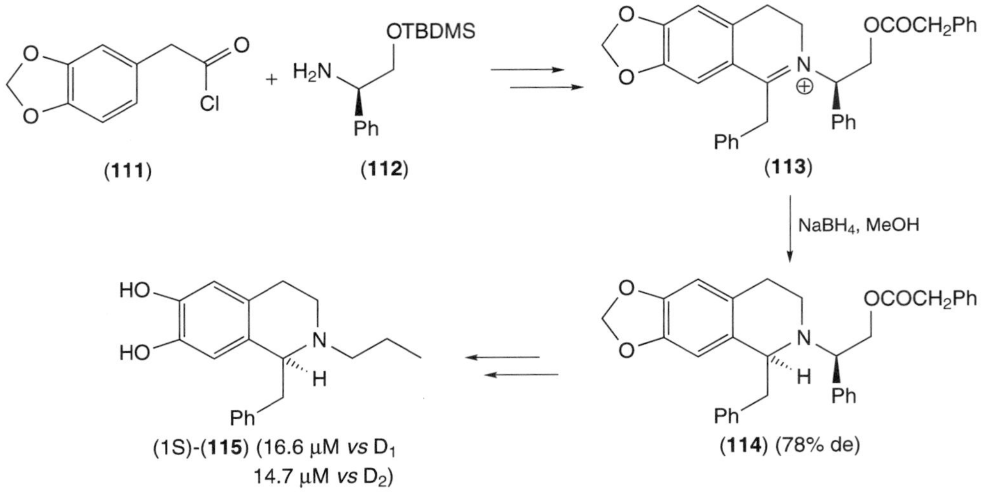

Figure 18.41.

reospecific cyclization to form the *cis* stereochemistry across the ring. The enantiomer, (1*S*,3*R*)-(**132**), was prepared in a directly analogous manner by reduction with (+)-*b*-chlorodiisopinocampheylborane. This study represents another example of stereodifferentiation at the dopamine receptors, with nearly a 5000-fold difference in D_1 potency observed between the two isomers in an *in vitro* assay.

A recent paper by scientists at Bristol-Myers Squibb reports the synthesis of a new class of calcium-activated potassium channel modulators (150). The compounds were investigated for their ability to increase channel opening at large conductance (BK or maxi-K) channels and showed a limited degree of stereospecificity. The key step in the synthesis is the direct oxidation of the enolate derived from (**133**) with either isomer of camphorsulfonyl oxaziridine, a reagent developed by Davis (151). Both enantiomers of the 3-aryl-3-hydroxyindol-2-ones were prepared with very high enantiopurity (>95% ee) using opposite enantiomers of the chiral oxaziridine. (−)-(**134**) was found to be a better activator of a cloned BK channel than the (+)-isomer at 20 μM, generating a current increase of 141% compared with 124% for (+)-(**134**).

6.4 Chiral Catalyst

The use of a chiral catalyst represents the ideal method for asymmetric synthesis because only small amounts of the chiral mediator are required and no modifications of the prochiral substrate are necessary. In many systems both enantiomers of the product can be prepared in a predictable and reproducible manner. The pharmaceutical industry is particularly interested in the capability of new catalyst systems to operate as reliable manufacturing processes on a large scale (127, 152). Substantial effort continues to be expended by the synthetic organic community with the goal of extending the number of efficient and broadly applicable catalyst systems capable of generating high levels of enantiomeric excess in a wide range of substrates (127).

The reduction of ketones by borane catalyzed by chiral oxazaborolidines such as (**136**), derived from the enantiomeric amino alcohols, has been applied to the synthesis of several drug candidates (127). This system is known as the CBS (Corey, Bakshi, Shibata) reduction (153), and Corey himself has applied it to the synthesis of pharmaceutical compounds (154). A further example is provided by the synthesis of MK-499 (**137**), a

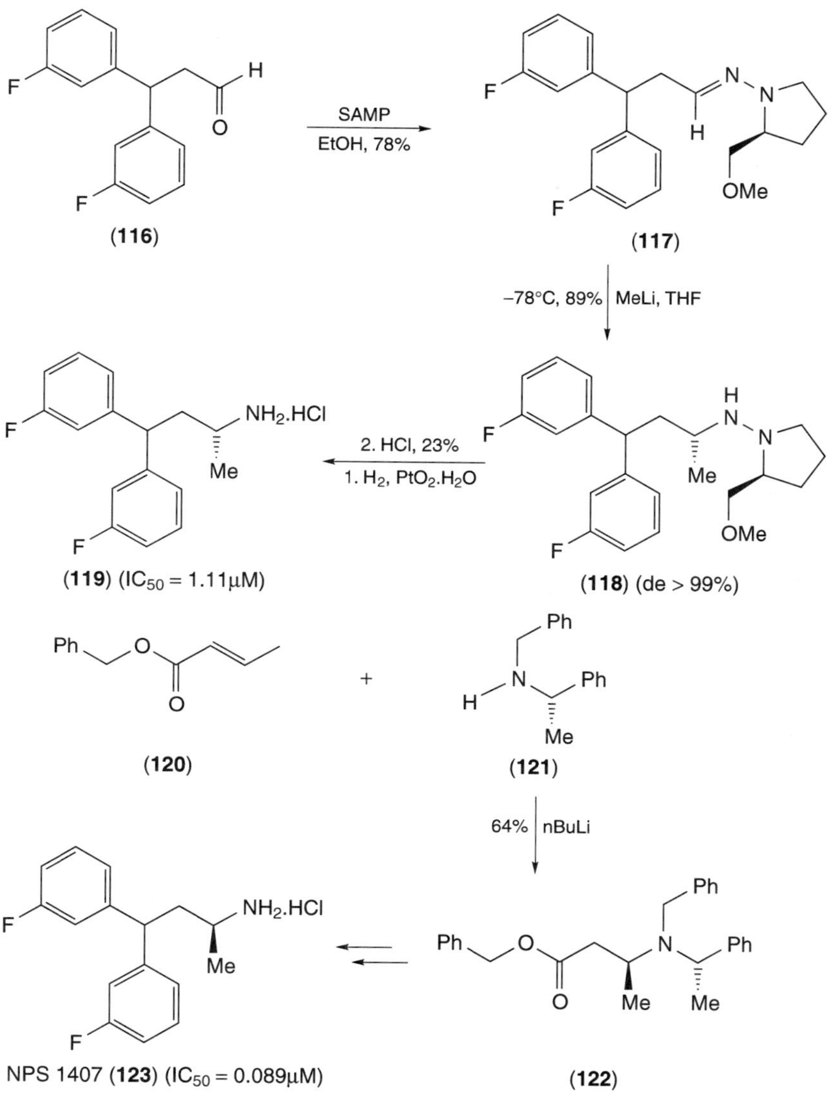

Figure 18.42.

potassium channel blocker that was developed for the treatment of cardiac arrhythmia by Merck (155).

Asymmetric hydrogenations with transition metal catalysts have been applied to single enantiomer synthesis in the pharmaceutical industry with considerable success. ChiroTech and Pfizer developed an improved synthesis of glutarate derivative (139), an intermediate required for the synthesis of Candoxatril (140) (156). The drug, a neutral endopeptidase inhib-

itor, was in clinical development for the treatment of hypertension and congestive heart failure, and its enantiomer does not possess the same biological activity. Several catalysts and conditions were screened before arriving at optimized conditions using cationic rhodium-(R,R)-MeDuPHOS (141), which provided the product with complete enantioselectivity and avoided previously observed problems associated with isomerization of the enone starting material. The reaction could be conducted at a

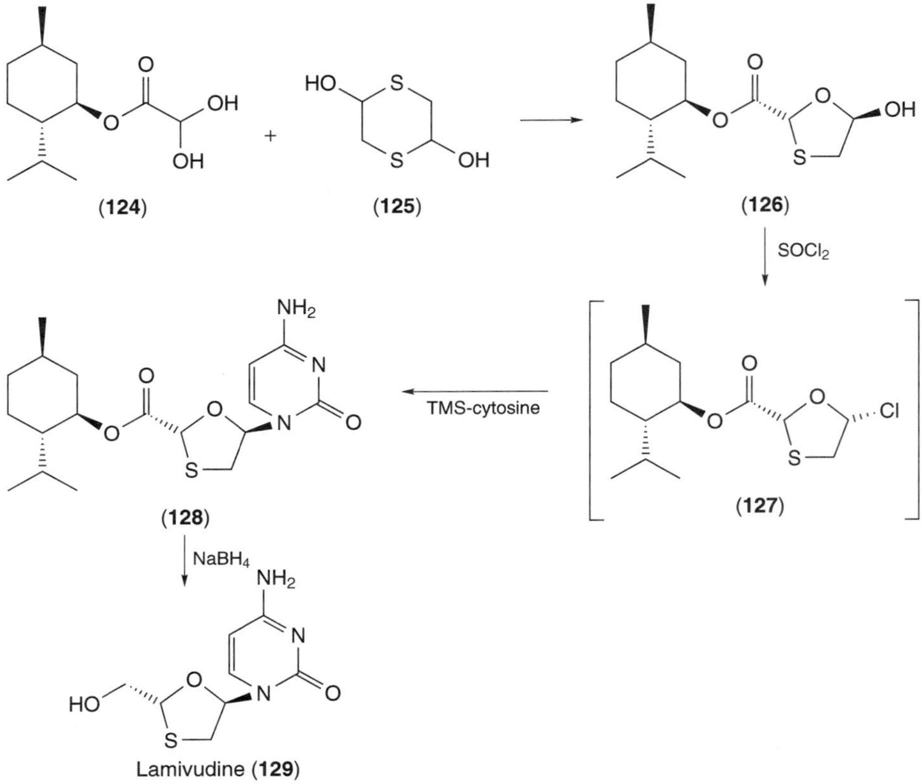

Figure 18.43.

high substrate-to-catalyst ratio of 3500:1 without a detrimental effect on enantiomeric excess or reaction rate. In catalytic asymmetric reactions, it is clearly economically advantageous to minimize the amount of catalyst that may comprise expensive chiral material and transition metals.

A method for the asymmetric dihydroxylation of alkenes to yield *cis*-diols was developed by the research group of Sharpless using chiral ligands derived from the cinchona alkaloids dihydroquinidine (DHQD) and dihydroquinine (DHQ) with a catalytic amount of osmium tetroxide (157). Although they are diastereomers, the phthalazine ligands act as "pseudo-enantiomeric" ligands, i.e., they give opposite asymmetric induction in a predictable manner. This procedure was recently used to prepare both isomers of combretadioxolane (**144**), a chiral analog of the natural product Combretastatin A-4 (**146**) (158). Com-

bretastatin A-4 displays antitubulin activity and cytotoxicity to tumor cells and is therefore an interesting lead structure for new anticancer drugs. The asymmetric synthesis of (*S,S*)-combretadioxolane (**144**) involved treatment of the *trans*-stilbene (**142**) with "AD-mix-α" [containing (DHQ)$_2$-PHAL] (**145**), whereas the enantiomer (*R,R*)-combretadioxolane resulted from use of AD-mix-β, which contains (DHQD)$_2$-PHAL as the chiral ligand. The tubulin polymerization-inhibitory activity of (*S,S*)-combretadioxolane was comparable with combretastatin A-4 (IC$_{50}$ = 4–6 μM) in an *in vitro* assay, whereas (*R,R*)-combretadioxolane was essentially inactive (IC$_{50}$ > 50 μM). In addition, (*S,S*)-combretadioxolane was 20 times more potent than vincristine as an *in vitro* growth inhibitor of the multidrug-resistant cell line PC-12.

Workers at SmithKline Beecham reported the stereoselective synthesis of inhibitors of

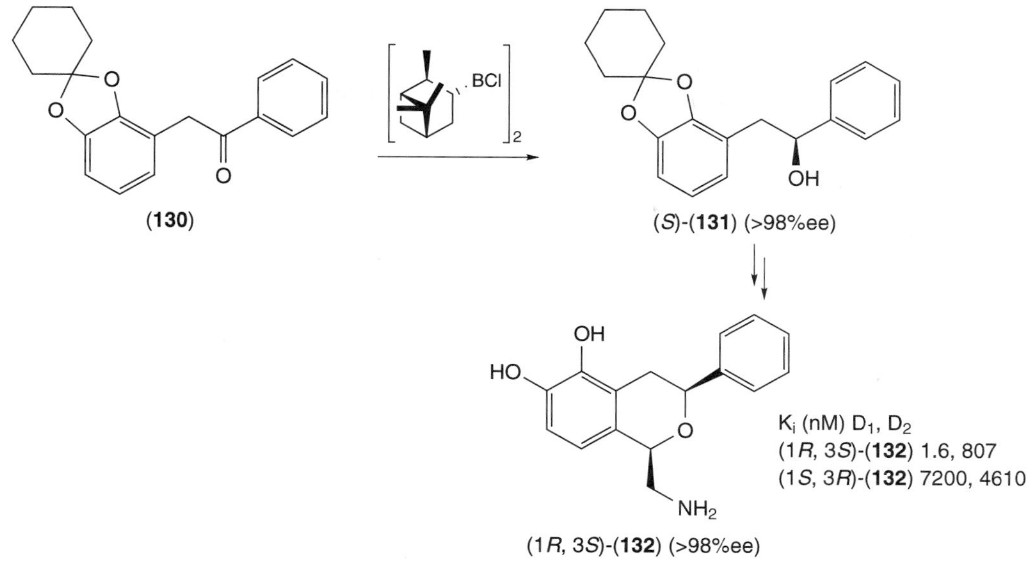

(130)

(S)-(131) (>98%ee)

K_i (nM) D_1, D_2
(1R, 3S)-**(132)** 1.6, 807
(1S, 3R)-**(132)** 7200, 4610

(1R, 3S)-**(132)** (>98%ee)

Figure 18.44.

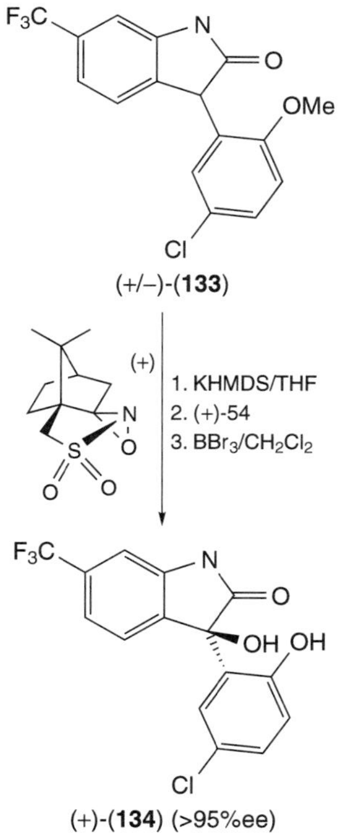

(+/−)-**(133)**

1. KHMDS/THF
2. (+)-54
3. BBr$_3$/CH$_2$Cl$_2$

(+)-**(134)** (>95%ee)

Figure 18.45.

the cysteine protease cathepsin K (159). A procedure was sought to allow preparation of either enantiomer of azido alcohol (**148**). This was readily achieved by Jacobsen asymmetric desymmetrization of the *meso*-epoxide (**147**) using azidotrimethylsilane catalyzed by chromium salen complex (**149**) (160). Use of the (R,R)-salen catalyst shown generated (3S, 4R)-(**148**), whereas the (S,S)-catalyst provided the (3R,4S)-azido silyl alcohol, both with very high enantioselectivity. After removal of the silyl group and reduction of the azido moiety, the resultant enantiomeric amino alcohols were transformed into diastereomers (4S)- and (4R)-(**150**) by reaction with leucine, amide formation, and oxidation. The cathepsin K activity for the diastereomers showed the (4S)-isomers to be up to 40-fold more potent than the corresponding (4R)-(**150**) in an enzyme assay.

A large scale synthesis of the drug Nelfinavir, an HIV protease inhibitor developed by Agouron (now Pfizer) was reported with the amino alcohol derived from (**148**), prepared using the Jacobsen procedure described above (161).

A similar approach uses the chromium-Salen complex (**149**) to open racemic terminal epoxides in a highly efficient resolution pro-

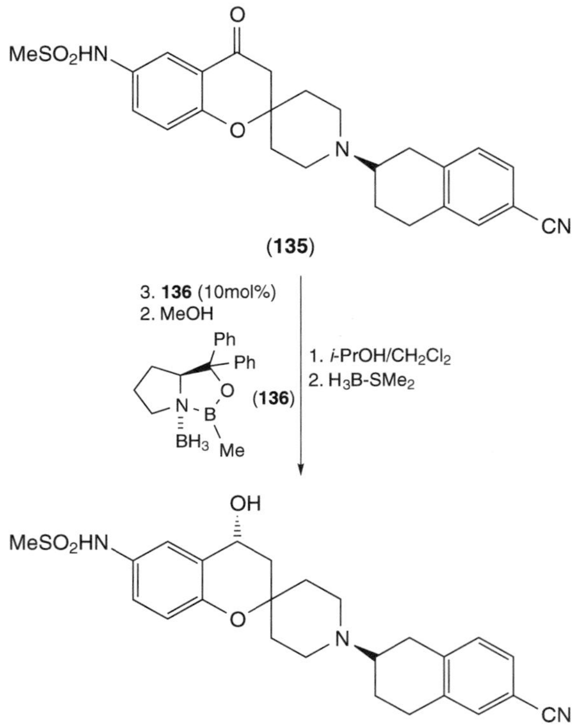

(135)

3. **136** (10mol%)
2. MeOH

1. *i*-PrOH/CH₂Cl₂
2. H₃B-SMe₂

(136)

MK 499 (**137**) (98%de; 92% yield)

Figure 18.46.

cess that has been applied to the synthesis of biologically active compounds (162). As with any such resolution process, the maximum yield of enantiopure material is 50% based on starting material. Terminal epoxides are easy

to prepare in racemic form, and conversely, difficult to prepare as single enantiomers by epoxidation of the corresponding alkene. (*R*)-9-[2-(phosphonomethoxy)propyl]adenine (*R*-PMPA) is a nucleotide reverse transcriptase

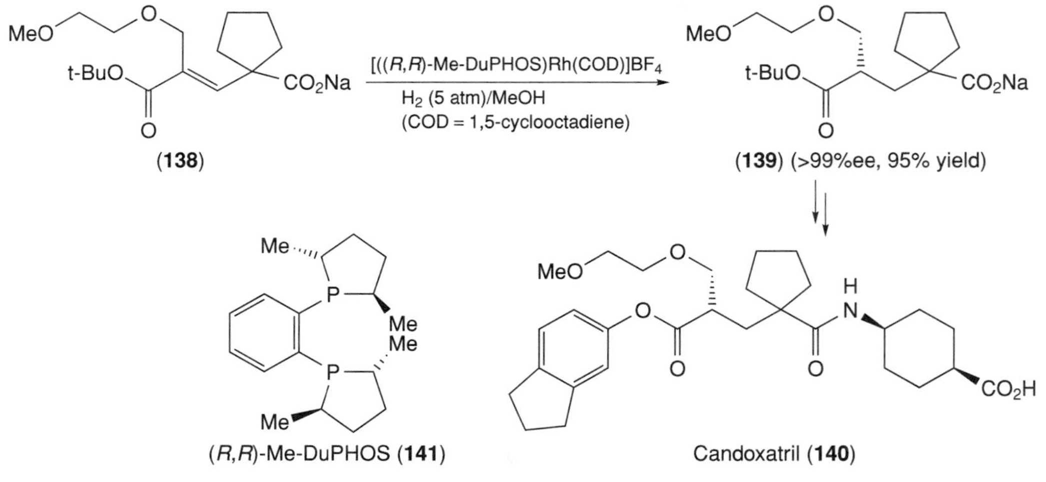

(138)

[((*R*,*R*)-Me-DuPHOS)Rh(COD)]BF₄

H₂ (5 atm)/MeOH
(COD = 1,5-cyclooctadiene)

(139) (>99%ee, 95% yield)

(*R*,*R*)-Me-DuPHOS (**141**)

Candoxatril (**140**)

Figure 18.47.

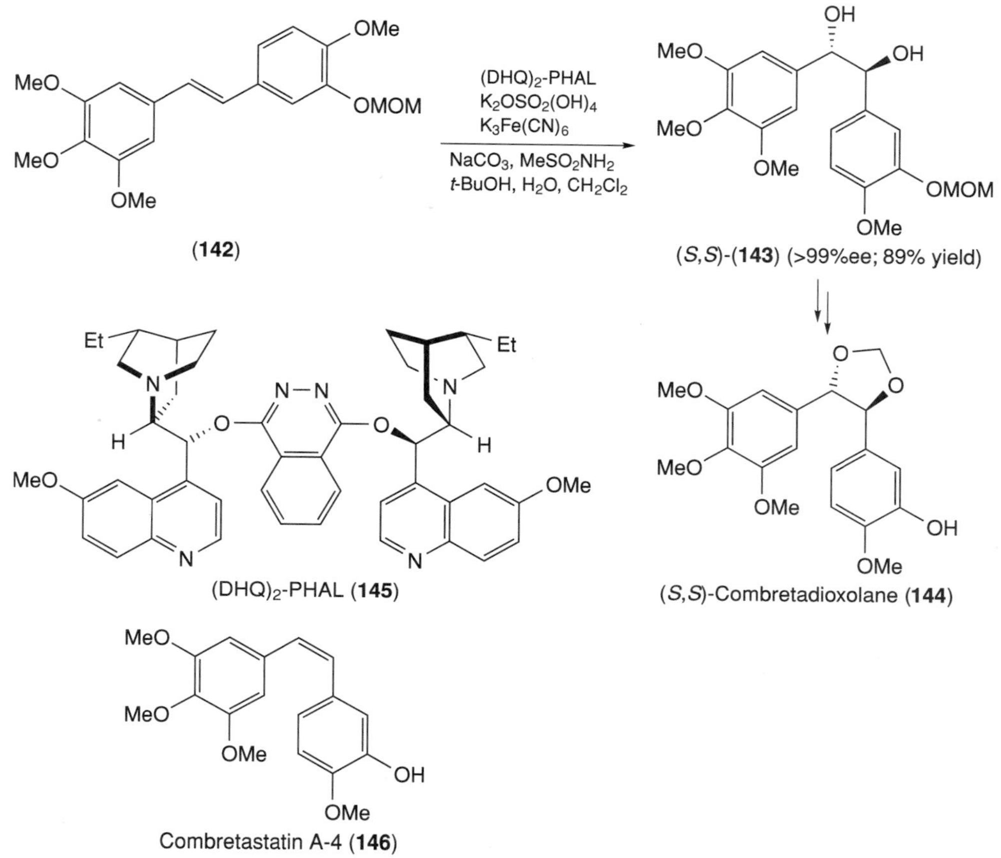

Figure 18.48.

inhibitor being developed by Gilead Sciences and a collaborative group from the University of Washington for the treatment and prevention of HIV infection (163). The compound can be prepared through kinetic resolution of propylene oxide using (*S*,*S*)-(**149**) and the resultant (*R*)-1-amino-2-propanol (**153**) was transformed into (*R*)-PMPA (**154**) in five steps (162).

In 1997, Tokunaga et al. reported the hydrolytic kinetic resolution of racemic terminal epoxides using a Co(III)-Salen catalyst (164). This remarkably general process uses only water as the nucleophile and provides the synthetically useful chiral epoxides and diols in highly enantioenriched form. The catalyst can be recycled and the reactions conducted under solvent-free conditions.

The process has been used by academic and industrial groups and is operated by Rhodia ChiRex on a large scale (165).

A wide variety of synthetic processes have been rendered asymmetric through the use of a chiral catalyst. In addition to the types of reaction described above, chiral transition metal catalysts have been used to influence the stereochemical course of isomerization, cyclization, and coupling reactions. As an example, an approach towards the natural product (−)-epibatidine (**158**) was recently reported by Namyslo and Kaufmann (166). Epibatidine is a potent analgesic and a nicotinic receptor agonist. The synthesis involves an asymmetric Heck-type hydroarylation between the bicyclic alkene (**155**) and pyridyl iodide (**156**). A number of bidentate chiral li-

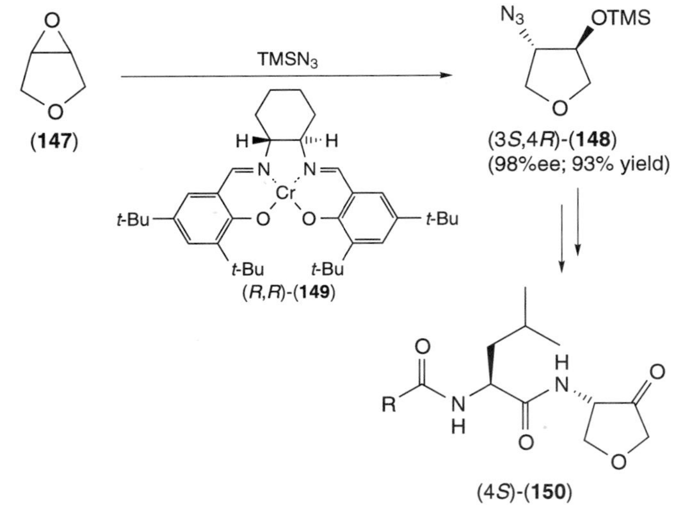

Figure 18.49.

 gands were investigated with BINAP (**159**), which were observed to give the highest enantioselectivity. By using the (*R*)- or (*S*)-BINAP ligand, both enantiomers of (**157**) were accessible with about the same level of enantioselection.

The continuing development of efficient and practical asymmetric processes will be one of the major driving forces in the future of drug discovery and development. In particular, the design of new general and practical catalytic processes will help explore the link between chirality and biological activity.

7   CONCLUSIONS

The ultimate focus of the endeavors of medicinal chemists is to develop a successful drug that will cure patients. However, with the increased regulatory requirements within the competitive biotechnology and pharmaceutical industry, the initial research to achieve this objective must be conducted in a rapid and thorough manner. During the drug research and development process, the important and subtle relationship between chirality and biological activity should be carefully considered.

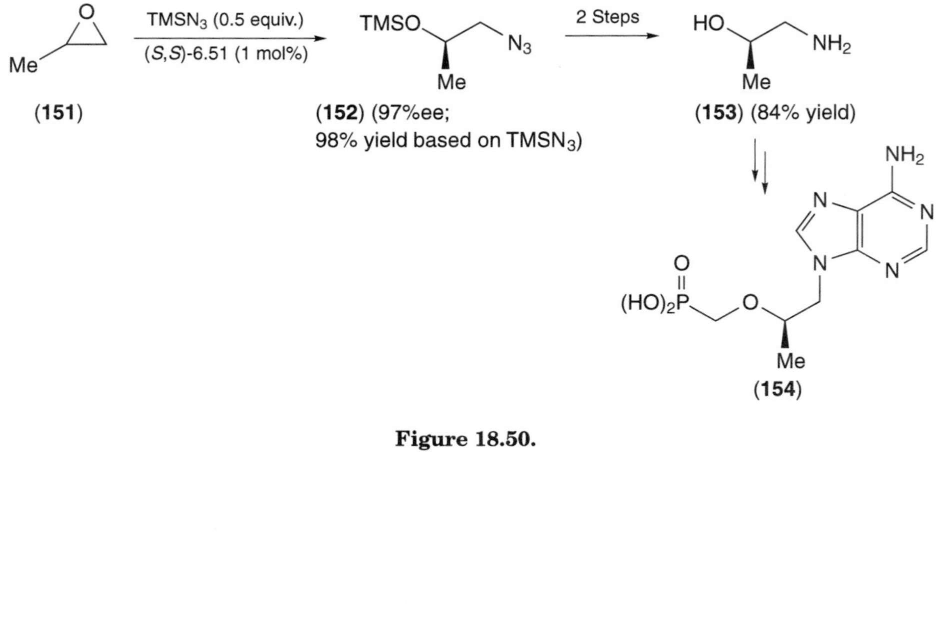

Figure 18.50.

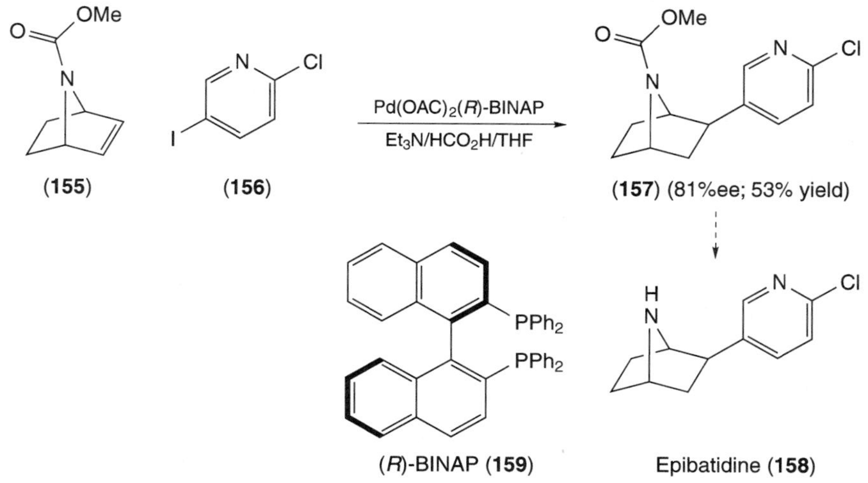

Figure 18.51.

Enantiomers frequently display markedly different biological activity; however, the fact that a large and adaptable toolbox of chemical and biological techniques to obtain single isomers are available allows the medicinal chemist to avoid working with mixtures of stereoisomers.

As reviewed in this chapter, there are numerous synthetic strategies available to the medicinal chemist that offer their own particular drawbacks and advantages. In the early stages of research it may be preferable to separate isomers by chromatography, thus providing both single enantiomers for biological testing. It should be noted that all the techniques described in this chapter can be used in conjunction with one another. That is to say, if one technique such as asymmetric synthesis failed to deliver enantiopure material, then another technique such as crystallization can be used to push through the product to the desired purity. As an example of this "double" approach, the use of SMB and crystallization in the separation of mandelic acid is worthy of note (56). The use of asymmetric hydrogenation followed by asymmetric enzymic transformation to obtain single isomer products has also been described by Taylor et al. at Chiro-Tech (167).

In conclusion, if a chiral center is present in a molecule designed and synthesized by a medicinal chemist, there are a broad number of methods available to prepare or isolate either isomer. From the examples given in this chapter, stereoisomers frequently display markedly different biological properties where the desirable properties associated with one isomer may not be apparent when the corresponding racemic mixture is tested either *in vivo* or *in vitro*.

REFERENCES

1. A. Michaels and J. Fuller, *Financial Times (Lond.)*, **23**, 21 (2001).

2. R. B. Raffa, E. Friderichs, W. Reimann, R. P. Shank, E. E. Codd, J. L. Vaught, H. I. Jacoby, and N. Selve, *J. Pharmacol. Exp. Ther.*, **267**, 331–340 (1993).

3. M. B. Smith and J. March, *March's Advanced Organic Chemistry*, 5th ed., Wiley, New York, 2001, pp. 125–217; E. L. Eliel, S. H. Wilen, and L. N. Mander, *Stereochemistry of Organic Compounds*, Wiley, New York, 1994; E. L. Eliel, S. H. Wilen, and M. P. Doyle, *Basic Organic Stereochemistry*, Wiley, New York, 2001.

4. T. D. Stephens, *Chem. Br.*, **37**, 38 (2001).

5. J. J. Baldwin and W. B. Abrams in I. W. Wainer and D. E. Drayer, Eds., *Stereochemically Pure Drugs: An Industrial Perspective*, Marcel Dekker, New York, 1988, p. 311.

6. G. C. Cotzias, P. S. Papavasiliou, R. Gellene, *N. Engl. J. Med.*, **280**, 337 (1969).

7. H. K. Kroemer, J. Turgeon, R. A. Parker, and D. M. Roden, *Clin. Pharmacol. Ther.*, **46**, 584 (1989).

8. R. A. O'Reily, *Clin. Pharmacol. Ther.*, **16**, 348 (1974).

9. A. Breckenbridge, M. Orme, H. Wesseling, R. J. Lewis, and R. Gibbons, *Clin. Pharmacol. Ther.*, **15**, 424 (1974).

10. T. Walle, J. G. Webb, E. E. Bagwell, U. K. Walle, H. B. Daniell, and T. E. Gaffney, *Biochem. Pharamacol.*, **37**, 115 (1988).

11. W. Lindner, M. Rath, K. Stoschitzky, and H. J. Semmelrock, *Chirality*, **1**, 10 (1989).

12. D. E. Drayer in I. W. Wainer and D. E. Drayer, Eds., *Drug Sterochemistry-Analytical Methods and Pharmacology*, Marcel Dekker, New York, 1988, p. 209.

13. K. J. Fehske, W. E. Muller, and U. Wollert, *Biochem. Pharmacol.*, **30**, 687 (1981).

14. R. H. McMenamy and J. L. Oncley, *J. Biol. Chem.*, **233**, 1436 (1958).

15. W. E. Muller in I. W. Wainer and D. E. Drayer, Eds., *Drug Sterochemistry-Analytical Methods and Pharmacology*, Marcel Dekker, New York, 1988, p. 227.

16. S. Toon, L. K. Low, M. Gibaldi, W. F. Trager, R. A. O'Reily, C. H. Motley, and D. A. Goulart, *Clin. Pharmacol. Ther.*, **39**, 15 (1986).

17. M. A. Campanero, B. Calahorra, M. Valle, I. F. Troconiz, and J. Honorato, *Chirality*, **11**, 272 (1999).

18. R. Stevenson, *Chem. Br.*, **37**, 24 (2001).

19. N. M. Maier, P. Franco, and W. Lindner, *J. Chromatogr. A*, **906**, 3 (2001).

20. L. Miller, C. Orihuela, R. Fronek, D. Honda, and O. Dapremont, *J. Chromatogr. A*, **849**, 309 (1999).

21. V. M. Meyer, *Chirality*, **7**, 567 (1995); O. P. Kleidernigg, M. Lammerhofer, and W. Lindner, *Enantiomer*, **1**, 387 (1996).

22. V. Schurig, *J. Chromatogr.* **441**, 135 (1988); K. Watabe, S. C. Chang, E. Gil-Av, and B. Koppenhofer, *Synthesis*, **3**, 225 (1987).

23. E. R. Francotte in S. Ahuja, Ed., *Chiral Separations, Applications and Technology*, chap. 10, American Chemical Society, Washington DC, 1997, p. 271.

24. K. D. Altria, N. W. Smith, and C. H. Turnbull, *Chromatographia*, **46**, 664 (1997); K. D. Altria, M. A. Kelly, and B. J. Clark, *Trends Anal. Chem.*, **17**, 214 (1998).

25. K. L. Williams, L. C. Sander, and S. A. Wise, *J. Pharm. Biomed. Anal.*, **15**, 1789 (1997);N. Bargmann-Leyder, A. Tambute, and M. Claude, *Chirality*, **7**, 311 (1995).

26. M. Juza, M. Mazzotti, and M. Morbidelli, *Trends Biotechnol.*, **18**, 108 (2000).

27. J. T. F. Keurentjes and F. J. M. Voermans in A. N. Collins, G. N. Sheldrake, and J. Crosby, Eds., *Chirality and Industry. II. Developments in the Manufacture of Optically Active Compounds*, chap. 8, Wiley, New York, 1997, p. 157.

28. S. C. Stinson, *Chem. Eng. News*, **73**, 44 (1995).

29. E. Francotte, *J. Chromatogr. A*, **666**, 565 (1994).

30. CHIRBASE, available online at http://chirbase.u-3mrs.fr, accessed on July 29, 2002.

31. Daicel Chemical Industries, Ltd., available online at http://www.daicel.co.jp/chiral, accessed on July 29, 2002. NOVASEP, available online at http://www.novasep.com, accessed on July 29, 2002. Astec, available online at http://www.astecusa.com, accessed on July 29, 2002.

32. D. Boyd, M. O'Keeffe, and M. R. Smyth, *Analyst*, **119**, 1467 (1994).

33. D. A von Deutsch, I. K. Abukhalaf, L. E. Wineski, H. Y. Aboul-Enein, S. A. Pitts, B. A. Parks, R. A. Oster, D. F. Paulsen, and D. E. Potter, *Chirality*, **12**, 637 (2000).

34. B. Waldeck, E. Widmark, *Acta Pharmacol. Toxicol.*, **56**, 221–227 (1985).

35. D. J. Triggle, D. A. Langs, and R. A. Janis, *Med. Res. Rev.*, **9**, 123 (1989); V. C. O. Njar and A. M. H. Brodie, *Drugs*, **58**, 233 (1999).

36. S. Visentin, P. Amiel, A. Gasco, B. Bonnet, C. Suteu, and C. Roussel, *Chirality*, **11**, 602 (1999).

37. P. Tullio, A. Ceccato, J-F. Liegeois, B. Pirotte, A. Felikidis, M. Stachow, P. Hubert, J. Crommen, J. Geczy, and J. Delarge, *Chirality*, **11**, 261 (1999).

38. J. Bruhwyler, J. F. Liegeois, J. Gerardy, J. Damas, E. Chleide, C. Lejeuns, E. Decamp, P. De Tullio, J. Delarge, A. Dresse, and J. Geczy, *Behav. Pharmacol.*, **9**, 731 (1998).

39. T. Shibata, I. Okamoto, and K. Ishii, *J. Liq. Chromatogr.*, **9**, 313 (1986); E. Yashima and Y. Okamoto, *Bull. Chem. Soc. Jpn.*, **68**, 3289 (1995).

40. W. H. Pirkle, D. W. House, and J. H. Finn, *J. Chromatogr.*, **192**, 143 (1980).

41. J. N. Kinkel in A. Subramanian, Ed., *A Practical Approach to Chiral Separations by Liquid Chromatography*, VHC, New York, 1994.

42. S. G. Allenmark, S. Andersson, P. Moller, and D. Sanchez, *Chirality*, **7**, 248 (1995).

43. M. Meurer, U. Altenhöner, J. Straube, and H. Schmidt-Traub, *J. Chromatogr. A*, **769**, 71–79 (1997).

44. M. Schulte, R. Ditz, R. M. Devant, J. N. Kinkel, and F. Charton, *J. Chromatogr. A*, **769**, 93 (1997).

45. D. Y. Wang, F. Hanotte, C De Vos, and P. Clement, *Eur. J. Allerg. and Clin. Immunol.*, **56**, 339 (2001).

46. E. J. Corey and C. J. Helal, *Tetrahedron Lett.*, **37**, 4837 (1996).

47. D. A. Pflum, H. Scot Wilkinson, G. J. Tanoury, D. W. Kessler, H. B. Kraus, C. H. Senanayake, and S. A. Wald, *Org. Process Res. Dev.*, **5**, 110 (2001).

48. E. R. Francotte and P. Richert, *J. Chromatogr. A*, **769**, 101 (1997); M. Negawa and F. Shoji, *J. Chromatogr.*, **590**, 113 (1992).

49. G. Ganetsos, P. E. Barker, J. A. Johnson, R. G. Kabza, K. Hashimoto, S. Adachi, Y. Shirai, M. Morishita, B. Balannec, G. Hotier, and H. Makai in G. Ganetsos and P. E. Barker, Eds., *Preparative and Production Scale Chromatography*, chaps. 11–15, Marcel Dekker, New York, 1993, p. 233; D. B. Broughton and C. G. Gerhold, inventors, Universal Oil Prod. Co., assignee, US patent 2,985,589, May 23, 1961.

50. D. M. Ruthven, *Principles of Adsorption and Adsorption Processes*, chap. 12, Wiley, New York, 1984, p. 380.

51. E. R. Francotte, *Chim. Nouvelle*, **53**, 1541 (1996).

52. D. W. Guest, *J. Chromatogr. A*, **760**, 159 (1997).

53. M. Schulte and J. Strube, *J. Chromatgr. A*, **906**, 399 (2001).

54. K. E. Goeringer, B. K. Logan, and G. D. Christian, *J. Anal. Toxicol.* **21**, 529 (1997).

55. E. Cavoy, M.-F. Deltent, S. Lehoucq, and D. Miggiano, *J. Chromatogr. A*, **769**, 49 (1997).

56. H. Lorenz, P. Sheehan, and A. Seidel-Morgenstern, *J. Chromatogr. A*, **908**, 201 (2001).

57. J. Jacques, A. Collet, and S. H. Wilen, *Enantiomers, Racemates and Resolutions*, Krieger: Malabar, Florida, 1994.

58. A. Collet, M. J. Brienne, and J. Jacques, *Chem. Rev.*, **80**, 215 (1980).

59. S. H. Wilen in E. L. Eliel, Ed., *Tables of Resolving Agents and Optical Resolutions*, University of Notre Dame Press, Notre Dame, IN, 1972; P. Newman, *Optical Resolution Procedures for Chemical Compounds*, vol. **1–3**, Optical Resolution Center, Manhattan College, New York, 1978–1984.

60. M. J. Cannarsa, *Chimica Oggi*, **17**, 28 (1999).

61. M. Prashad, D. Har, O. Repic, T. J. Blacklock, and P. Giannousis, *Tetrahedron Asymmetry*, **10**, 3111 (1999).

62. R. A. Maxwell, E. Chaplin, S. B. Eckhardt, J. R. Soares, and G. Hite, *J. Pharmacol. Exp. Ther*, **173**, 158 (1970).

63. S. Faulconbridge, H. S. Zavareh, G. R. Evans, and M. Langston, inventors, Medeva Europe Ltd. (GB), assignee, World patent WO98/25902, June 18, 1998.

64. A. S. C. Chan, *Chemtech.*, **3**, 46 (1993).

65. P. J. Harrington and E. Loderwijk, *Org. Process Res. Dev.*, **1**, 72 (1997).

66. W. J. Pope and S. J. Peachey, *J. Chem. Soc.*, **75**, 1066 (1899).

67. R. Gristwood, H. Bardsley, H. Baker, and J. Dickens, *J. Exp. Opin. Invest. Drugs*, **3**, 1209 (1994).

68. M. Langston, U. C. Dyer, G. A. C. Frampton, G. Hutton, C. J. Lock, B. M. Skead, M. Woods, and H. Zavareh, *Org. Process Res. Dev.*, **4**, 530 (2000).

69. B. A. Astleford and L. O. Weigel in A. N. Collins, G. N. Sheldrake, and J. Crosby, Eds., *Chiralty in Industry I*, chap. 6, Wiley, New York, 1997, p. 99.

70. K. Flick and E. Frankus, inventors, Gruenenthal Chemie, assignee, US patent 3,652,589, March 28, 1972.

71. G. R. Evans, inventor, Darwin Discovery Ltd., (US) assignee, World Patent WO00/32554, June 8, 2000; G. R. Evans, J. A. Henshilwood, and J. O'Rourke, *Tetrahedron Asymmetry*, **12**, 1663 (2001).

72. L. G. Humber, *Med. Res. Rev.*, **7**, 1 (1987).

73. L. G. Humber, *J. Med. Chem.*, **29**, 871 (1986).

74. M. Woods, U. C. Dyer, J. F. Andrews, C. N. Morfitt, R. Valentine, and J. Sanderson, *Org. Process Res. Dev.*, **4**, 418 (2000).

75. D. Askin, K. K. Eng, R. M. Purick, K. M. Wells, R. P. Volante, and P. J. Reider, *Tetrahedron Lett.*, **35**, 673 (1994).

76. M. Kottenhahn, K. Stingl, and K. Drauz, inventors, Degussa (DE), assignee, US patent 6,093,823, July 25, 2000.

77. M. Eichelbaum, *Federation. Proc.*, **43**, 2298 (1984); M. Eichelbaum, *Biochem. Pharmacol.*, **37**, 93 (1988).

78. H. Echizen, T. Brecht, S. Neidergesass, B. Volgelgesang, and M. Eichelbaum, *Am. Heart J.* **109**, 210 (1985).

79. O. Ehrmann, H. Nagel, and W. Karau, inventors, Knoll Aq (DE), assignee, US patent 5,457,224, October 10, 1995 and World patent WO94/08950, April 14, 1994.

80. E. J. Trieber, M. Raschack, and F. Dengel, inventors, Knoll Aq (DE), assignee, German patent 2059923, 1972.

81. R. M. Bannister, M. H. Brookes, G. R. Evans, R. B. Katz, and N. D. Tyrrell, *Org. Process Res. Dev.*, **4**, 467 (2000).

82. P. E. Graves, H. A. Salhanick, *Endocrinology*, **105**, 52 (1979).

83. See ref. 57, pp. 299–301, 382–383.

84. N. Finch, R. Dziemian, J. Cohen, and B. G. Steinetz, *Experientia*, **31**, 1002 (1975).

85. M. J. Bunegar, U. C. Dyer, G. R. Evans, R. P. Hewitt, S. W. Jones, N. Henderson, C. J. Richard, S. Sivaprasad, B. M. Skead, M. A. Stark, and E. Teale, *Org. Process Res. Dev.*, **3**, 442 (1999).

86. See ref. 57, pp. 374–375.

87. K. P. Datla and G. Curzon, *Neuropharmacology*, **35**, 315 (1996); S. Salvadori, C. Bianchi, L. H. Lazurus, V. Scaranari, M. Attila, and R. Tomatis, *J. Med. Chem.*, **35**, 4651 (1992).

88. C. A. Maryanoff, L. Scott, R. D. Shah, and F. J. Villani Jr., *Tetrahedron Asymmetry*, **9**, 3247 (1998).

89. A. Bruggink in A. N. Collins, G. N. Sheldrake, and J. Crosby, Eds., *Chirality in Industry II*, chap. 5, Wiley, New York, 1997, p. 81.

90. T. Vries, H. Wynberg, E. van Echten, J. Koek, W. ten Hoeve, R. M. Kellogg, Q. B. Broxterman, A. Minnaard, B. Kaptien, S. van der Sluis, L. Hulshof, and J. Kooistra, *Angew. Chem. Int. Ed.*, **37**, 2349 (1998).

91. E. Ebbers, G. J. A. Arianns, B. Zwanenburg, and A. Bruggink, *Tetrahedron Asymmetry*, **9**, 2745 (1998); U. C. Dyer, D. A. Henderson, and M. B. Mitchell, *Org. Process Res. Dev.*, **3**, 161 (1999).

92. S. H. Wilen, A. Collet, and J. Jacques, *Tetrahedron*, **33**, 2725 (1977).

93. Z. J. Li, M. T. Zell, E. J. Munson, and D. J. W. Grant, *J. Pharm. Sci.*, **88**, 337 (1999).

94. R. Tamura, T. Ushio, H. Takahashi, K. Nakamura, N. Azuma, F. Toda, and K. Endo, *Chirality*, **9**, 220 (1997).

95. H. M. L. Davies, T. Hansen, D. W. Hopper, and S. A. Panaro, *J. Am. Chem. Soc.*, **121**, 6509 (1999).

96. J. M. Axten, R. Ivy, L. Krim, and J. D. Winkler, *J. Am. Chem. Soc.*, **121**, 6511 (1999).

97. A. Robinson, H. Y. Li, and J. Feaster, *Tetrahedron Lett.*, **37**, 8321 (1996).

98. W. Goppel, W. Betram, *Psychiatr. Neurol. Med. Psychol.*, **23**, 712 (1971).

99. H. A. M. Mucke, *Drugs Future*, **33**, 259 (1997).

100. D. H. R. Barton and G. W. Kirby, *J. Chem. Soc.*, 806 (1962).

101. W. Shieh and J. A. Carlson, *J. Org. Chem.*, **59**, 5463 (1994).

102. B. Kuenburg, L. Czollner, J. Frohlich, and U. Jordis, *Org. Process Res. Dev.*, **3**, 425 (1999).

103. G. L. Plosker and K. L. Goa, *Drugs*, **42**, 805 (1991).

104. H. Harada, T. Morie, Y. Hirokawa, and S. Kato, *Tetrahedron Asymmetry*, **8**, 2367 (1997).

105. R. M. Williams in J. E. Baldwin and P. D. Magnus, Eds., *Synthesis of Optically Active '61-Amino Acids*, vol. **7**, Pergamon Press, Oxford, UK 1989; R. O. Duthaler, *Tetrahedron*, **50**, 1539 (1994).

106. T. Shiraiwa, H. Miyazaki, T. Watanabe, and H. Kurokawa, *Chirality*, **9**, 48 (1997).

107. T. Shiraiwa, H. Miyazaki, A. Ohta, K. Motonaka, E. Kobayashi, M. Kubo, and H. Kurokawa, *Chirality*, **9**, 656 (1997).

108. K. Abe, H. Inoue, and T. Nagao, *Yakugaku Zasshi*, **108**, 716 (1988).

109. S. Yamada, K. Morimatsu, R. Yoshioka, Y. Ozaki, and H. Seko, *Tetrahedron Asymmetry*, **9**, 1713 (1998).

110. E. K. Rowinsky, L. A. Cazenav, and R. C. Donehower, *J. Natl. Cancer Inst.*, **82**, 1247 (1990).

111. J. N. Denis, A. E. Green, D. Guenard, F. Gueritte-Voegelein, L. Mangatal, and P. Potier, *J. Am. Chem. Soc.*, **110**, 5917 (1988).

112. J. Kearns and M. M. Kayser, *Tetrahedron Lett.*, **35**, 2845 (1994).

113. R. P. Srivastava, J. K. Zjawiony, J. R. Peterson, and J. D. McChesney, *Tetrahedron Asymmetry*, **5**, 1683 (1994).

114. R. N. Patel, *Adv. Appl. Microbiol.*, **47**, 33 (2000); V. Gotor, *Biocat and Biotrans* **18**, 87 (2000); W. A. Loughlin, *Bioresource Technology*, **74**, 49 (2000); S. M. Roberts, *J. Chem. Soc., Perkin Trans.*, **1**, 1 (1999).

115. C.-H. Wong and G. M. Whitesides, *Enzymes in Synthetic Organic Chemistry*, Pergamon, New York, 1994.

116. A. D. Baxter, J. B. Bird, R. Bannister, R. Bhogal, D. T. Manallack, R. W. Watson, D. A. Owen, J. Montana, J. Henshilwood, and R. C. Jackson in N. Clendeninn and K. Appelt, Eds., *Matrix Metalloproteinase Inhibitors in Cancer Therapy*, Humana Press, Totawa, NJ, 2000, pp. 193–221.

117. N. L. St. Clair and N. L. Nathrough, *J. Am. Chem. Soc.*, **114**, 7314 (1992); N. Khalaf, C. P. Govardan, J. J. Lalonde, R. A. Persichetti, Y. F. Wang, and A. L. Margolin, *J. Am. Chem. Soc.*, **118**, 5495 (1996).

118. A. M. Evans, *J. Clin. Pharmacol.* **36** (Suppl 12), 7S (1996).

119. S.-W. Tsai and H.-J. Wei, *Enzyme Microb. Technol.*, **16**, 328 (1994).

120. J.-Y. Xin, S.-B. Li, Y. Xy, J.-R. Chui; and C.-G. Xi, *J. Chem. Tech. Biotechnol.*, **76**, 579 (2001).

121. M. E. Swarbrick, F. Gosselin, and W. D. Lubell, *J. Org. Chem.*, **64**, 1993–2002 (1999).

122. W. L. Nelson and T. R. Burke, *J. Org. Chem.*, **43**, 3641 (1978).

123. R. M. Patel, *Stereosel. Biocatal.*, Marcel Dekker, Inc., New York, 2000, pp. 87–130.

124. E. P. Siqueira Fihlo, J. A. R. Rodrigues, and P. J. S. Moran, *Tetrahedron Asymmetry*, **12**, 847 (2001).

125. P. A. Procopiou, G. E. Morton, M. Todd, and G. Webb, *Tetrahedron Asymmetry*, **12**, 2005 (2001).

126. T. P. Jerusi, inventor, Sepracor Inc. (US), assignee, World patent WO99/13867, March 25, 1999.

127. J. D. Morrison, Ed., *Asymmetric Synthesis*, Academic Press, San Diego, CA, 1983; G. Procter, *Asymmetric Synthesis*, Oxford University Press, Oxford, UK, 1996; M. Nógrádi, *Stereoselective Synthesis*, 2nd ed., VCH, Weinheim, Germany, 1995; D. J. Ager and M. L. East, Eds., *Asymmetric Synthetic Methodology*, CRC Press, Boca Raton, FL, 1995; D. J. Ager, Ed., *Handbook of Chiral Chemicals*, Marcel Dekker, New York, 1999; P. O'Brien, *J. Chem. Soc.*, **1**, 95–113 (2001); H. Tye and P. J. Comino, *J. Chem. Soc.*, **1**, 1729–1747 (2001); P. I. Dalko and L. Moisan, *Chem. Int. Ed.*, **40**, 3726–3748 (2001); K. C. Nicolaou and E. J. Sorensen, *Classics in Total Synthesis*, VCH, Basel, Switzerland, 1996.

128. S. Redshaw in C. R. Ganellin and S. M. Roberts, Eds., *Medicinal Chemistry*, 2nd ed., Academic Press, San Diego, 1993, pp. 163–186.

129. A. A. Patchett, E. Harris, E. W. Tristram, M. J. Wyvratt, M. T. Wu, D. Taub, E. R. Peterson, T. J. Ikeler, J. ten Broeke, L. G. Payne, D. L. Ondeyka, E. D. Thorsett, W. J. Greenlee, N. S. Lohr, R. D. Hoffsommer, H. Joshua, W. V. Ruyle, J. W. Rothrock, S. D. Aster, A. L. Maycock, F. M. Robinson, R. Hirschmann, C. S. Sweet, E. H. Ulm, D. M. Gross, T. C. Vassil, and C. A. Stone, *Nature*, **288**, 280 (1980).

130. T. J. Blacklock, R. F. Shuman, J. W. Butcher, W. E. Shearin, J. Budavari, and V. J. J. Grenda, *J. Org. Chem.*, **53**, 836 (1988).

131. T. Ohashi and J. Hasegawa in A. N. Collins, G. N. Sheldrake, and J. Crosby, Eds., *Chirality in Industry II: Developments in the Commercial Manufacture and Applications of Optically Active Compounds*, Wiley, Chichester, UK, 1997, p. 269.

132. M. R. Attwood, C. H. Hassall, A. Krohn, G. Lawton, and S. Redshaw, *J. Chem. Soc.*, **1**, 1011–1019 (1986).

133. E. J. Stoner, A. J. Cooper, D. A. Dickman, L. Kolaczkowski, J. E. Lallaman, J.-H. Liu, P. A. Oliver-Shaffer, K. M. Patel, J. B. Paterson Jr., D. J. Plata, D. A. Riley, H. L. Sham, P. J. Stengel, and J.-H. J. Tien, *Org. Process Res. Dev.*, **4**, 264–269 (2000).

134. J. T. Kovalainen, J. A. M. Christiaans, S. Kotisaari, J. T. Laitinen, P. T. Männistö, L. Tuomisto, and J. Gynther, *J. Med. Chem.*, **42**, 1193–1202 (1999).

135. G. M. Coppola and H. F. Schuster, *Chiral α-Hydroxy Acids in Enantioselective Syntheses*, VCH, Weinheim, Germany, 1997.

136. G. Wang and R. Hollinsworth, *J. Org. Chem.*, **64**, 1036–1038 (1999).

137. G. G. Wu, *Org. Process Res. Dev.*, **4**, 298–300 (2000); G. Wu, Y. S. Wong, X. Chen, and Z. Ding, *J. Org. Chem.*, **64**, 3714–3718 (1999).

138. D. S. Garvey, J. T. Wasicak, M. W. Decker, J. D. Brioni, M. J. Buckley, J. P. Sullivan, G. M. Carrera, M. W. Holladay, S. P. Arneric, and M. Williams, *J. Med. Chem.*, **37**, 1055–1059 (1994).

139. N.-H. Ling, Y. He, and H. Kopecka, *Tetrahedron Lett.* **36**, 2563–2566 (1995).

140. D. J. Ager, I. Prakash, and D. R. Schaad, *Chem. Rev.*, **96**, 835–875 (1996).

141. D. A. Evans, J. M. Takacs, L. R. McGee, D. J. Mathre, and J. Bartroli, *Pure Appl. Chem.*, **53**, 1109 (1981); D. A. Evans, M. D. Ennis, and D. J. Mathre, *J. Am. Chem. Soc.*, **104**, 1737 (1982); D. A. Evans, *Aldrichimica Acta*, **15**, 23 (1982).

142. D. R. Schaad in ref. 127, pp. 287–300.

143. S. J. Wittenberger and M. A. McLaughlin, *Tetrahedron Lett.*, **40**, 7175–7178 (1999).

144. S. R. Turner, J. W. Strohbach, R. A. Tommasi, P. A. Aristoff, P. D. Johnson, H. I. Skulnick, L. A. Dolak, E. P. Seest, P. K. Tomich, M. J. Bohanon, M.-M. Horng, J. C. Lynn, K. T. Chong, R. R. Hinshaw, K. D. Watenpaugh, M. N. Janakiraman, and S. Thaisrivongs, *J. Med. Chem.*, **41**, 3467–3476 (1998).

145. T. M. Judge, G. Phillips, J. K. Morris, K. D. Lovasz, K. R. Romines, G. P. Luke, J. Tulinsky, J. M. Tustin, R. A. Chrusciel, L. A. Dolak, S. A. Mizsak, W. Watt, J. Morris, S. L. Vander Velde, J. W. Strohbach, and R. B. Gammill, *J. Am. Chem. Soc.*, **119**, 3627–3628 (1997).

146. N. Cabedo, I. Andreu, M. C. Ramírez de Arellano, A. Chagraoui, A. Serrano, A. Bermejo, P. Protais, and D. Cortes, *J. Med. Chem.*, **44**, 1794–1801 (2001).

147. S. T. Moe, D. L. Smith, E. G. DelMar, S. M. Shimizu, B. C. Van Wagenen, M. F. Balandrin, Y. Chien, J. L. Raszkiewicz, L. D. Artman, H. S. White, and A. L. Mueller, *Bioorg. Med. Chem. Lett.*, **10**, 2411–2415 (2000).

148. M. D. Goodyear, P. O. Dwyer, M. L. Hill, A. J. Whitehead, R. Hornby, and P. Hallett, inventors, Glaxo Group Ltd. (GB), assignee, World patent WO-09529174, April 18, 1995.

149. M. P. DeNinno, R. Schoenleber, R. J. Perner, L. Lijewski, K. E. Asin, D. R. Britton, R. MacKenzie, and J. W. Kebabian, *J. Med. Chem.*, **34**, 2561–2569 (1991).

150. P. Hewawasam, N. A. Meanwell, V. K. Gribkoff, S. I. Dworetzky, and C. G. Boissard, *Bioorg. Med. Chem. Lett.*, **7**, 1255–1260 (1997).

151. F. A. Davis and B.-C. Chen, *Chem. Rev.*, **92**, 919 (1992).

152. M. K. O'Brien and B. Vanase, *Curr. Opin. Drug Discovery Dev.*, **3**, 793-806 (2000); R. A. Sheldon, Ed., *Chirotechnology: Industrial Synthesis of Optically Active Compounds*, Marcel Dekker, New York, 1993.

153. E. J. Corey, R. K. Bakshi, and S. Shibita, *J. Am. Chem. Soc.* **109**, 5551 (1987); E. J. Corey, R. K. Bakshi, S. Shibita, C.-P. Chen, and V. K. Singh, *J. Am. Chem. Soc.*, **109**, 7925 (1987).

154. E. J. Corey and J. O. Link, *Tetrahedron Lett.*, **31**, 601 (1990); E. J. Corey and J. O. Link, *J. Org. Chem.*, **56**, 442 (1991).

155. Y.-J. Shi, D. Cai, U.-H. Dolling, A. W. Douglas, D. M. Tschaen, and T. R. Verhoeven, *Tetrahedron Lett.*, **35**, 6409–6412 (1994).

156. M. J. Burk, F. Bienewald, S. Challenger, A. Derrick, and J. A. Ramsden, *J. Org. Chem.*, **64**, 3290–3298 (1999).

157. H. C. Kolb, M. S. VanNieuwenhze, and K. B. Sharpless, *Chem. Rev.*, **94**, 2483 (1994).

158. R. Shirai, H. Takayama, A. Nishikawa, Y. Koiso, and Y. Hashimoto, *Bioorg. Med. Chem. Lett.*, **8**, 1997–2000 (1998).

159. A. E. Fenwick, A. D. Gribble, R. J. Ife, N. Stevens, and J. Witherington, *Biorg. Med. Chem. Lett.*, **11**, 199–202 (2001).

160. L. E. Martinez, J. L. Leighton, D. H. Carsten, and E. N. Jacobsen, *J. Am. Chem. Soc.*, **117**, 5897 (1995); S. E. Schaus, J. F. Larrow, and E. N. Jacobsen, *J. Org. Chem.*, **62**, 4197–4199 (1997).

161. S. E. Zook, J. K. Busse, and B. C. Borer, *Tetrahedron Lett.*, **41**, 7017–7021 (2000).

162. J. F. Larrow, S. E. Schaus, and E. N. Jacobsen, *J. Am. Chem. Soc.*, **118**, 7420–7421 (1996); S. E. Schaus and E. N. Jacobsen, *Tetrahedron Lett.*, **37**, 7937–7940 (1996).

163. C.-C. Tsai, K. E. Follis, A. Sabo, T. W. Beck, R. F. Grant, N. Bischofberger, R. E. Benveniste, and R. Black, *Science*, **270**, 1197–1199 (1995).

164. M. Tokunaga, J. F. Larrow, F. Kakiuchi, and E. N. Jacobsen, *Science*, **277**, 936 (1997).

165. J. M. Keith, J. F. Larrow, and E. N. Jacobsen, *Adv. Synth. Catal.*, **343**, 5–26 (2001).

166. J. H. Namyslo and D. E. Kaufmann, *Synlett.*, 804–806 (1999).

167. S. J. C. Taylor, K. E. Holt, R. C. Brown, P. A. Keene, and I. N. Taylor in R. N. Patel, Ed., *Stereoselective Biocatalysis*, Marcel Dekker, New York, 2000, p. 397.

CHAPTER NINETEEN

Structural Concepts in the Prediction of the Toxicity of Therapeutical Agents

HERBERT S. ROSENKRANZ
Department of Biomedical Sciences
Florida Atlantic University
Boca Raton, Florida

Contents

Burger's Medicinal Chemistry and Drug Discovery
Sixth Edition, Volume 1: Drug Discovery
Edited by Donald J. Abraham
ISBN 0-471-27090-3 © 2003 John Wiley & Sons, Inc.

1 INTRODUCTION

The increased acceptance and availability of various structure-activity relationship (SAR) approaches in health hazard identification (1, 2) is accompanied by many opportunities and some pitfalls. The latter are derived from the availability of various computer-based SAR platforms whose basis and performance characteristics are not transparent to the user. Such programs, in the hand of the non-expert, may be misused. SAR models and associated technologies on the other hand, while not crystal balls, provide the expert toxicologist with meaningful information regarding the putative toxicological profile of candidate agents. They can guide in the design of agents with decreased or without unwanted side effects and yet retain or even enhance therapeutic effectiveness. Finally, the SAR technology can provide insights into the mechanism whereby a chemical exerts its toxic effects and thereby provide a better understanding of the risk that the agent poses to humans.

However, to achieve these aims, it is essential that the performance characteristics and the basis of the SAR model be known. This involves several critical steps in the SAR model development. These are listed below, each of which will be amplified:

1. Development of database
2. Model building
3. Model characterization
4. Model validation
5. Model application to individual agents and for mechanistic evaluations

Although the present review focuses on the MULTICASE SAR methodology (3–5), the concepts discussed herein apply to all generally available SAR techniques used to study toxicological phenomena. Basically SAR approaches that have been used fall into two categories: those that are based on statistical automated algorithms not dependent on prior expert judgment and those that are *a priori*

Table 19.1 Some SAR Approaches Used in Toxicology

Designation	Approach[a]	References
MULTICASE	I	3–5
TOPKAT	I	6
COMPACT	I	7
DEREK	II	8–10
ONCOLOGIC	II	11, 12
Hazard Expert	II	13
PROGOL	I	14
Structural Alerts	II	15

[a]Approach I indicates statistical automated algorithms not dependent on prior expert judgment. Approach II indicates a rule-based technique that requires prior expert judgment.

rule-based requiring prior expert knowledge (Table 19.1). However, as will be stressed herein, even the approach not requiring prior expert input is very much dependent on human expertise at various stages of the model development and interpretation process.

Reviews and assessments of the various SAR methodologies used to analyze toxicological phenomena are available (16–21).

1.1 Development of Database

Most experimental data compilations of toxicological effects both in the public domain as well as in proprietary databases were not developed for SAR purposes. Thus, with respect to some toxicological phenomena, the database may be rich with certain chemical classes such as chloroarene and lacking in data relating to others, e.g., aminoarenes. Yet, unlike the SAR models developed for drug discovery, SAR models of toxicological phenomena must be able to handle databases composed of noncongeneric chemicals. Additionally, as a consequence of how toxicological data are generated, there may be a paucity of data altogether. Yet, for optimal SAR models of toxicological phenomena, the "learning set" should include at least 300 non-congeneric chemicals (3, 22).

Accordingly, the human expert may suggest, that for certain purposes, the results of certain assays be pooled, e.g., rat and mouse carcinogenicity or the results of the *Salmonella* and *E. coli* WP *uvrB* mutagenicity assays (23, 24). Obviously, such data pooling must be based on a sound scientific basis as well as data that show extensive concordance between the experimental results of the systems to be pooled, i.e., that a substantial number of chemicals must give identical results in the two systems, thereby indicating that results obtained with one system can be amalgamated with those obtained in the other (25). For example, when the same chemicals were tested for their ability to induce sister chromated exchanges and chromosomal aberrations in cultured Chinese hamster ovary (CHO) cells, they showed divergent results (26). Hence, the results of the two assays cannot be amalgamated into a single database to develop an SAR model of cytogenetic effects. Similarly, even using the same indicator system, results cannot be merged if different criteria are used to interpret the significance of the results. That situation prevails with respect to the induction of mutations at the thymidine kinase locus of mouse lymphoma cells *vis-a-vis* the criteria used by the U.S. National Toxicology Program versus those employed by the U.S. Environmental Protection Agency's GeneTox Program. In fact, each data set gives rise to a distinct SAR model (27–29).

On the other hand, the consensus database of potential developmental toxicity in humans, based on experimental results in animals, observations in exposed humans, and expert judgment, yields a coherent SAR model of developmental risks to humans (30). That model is distinct from SAR models of developmental toxicity to individual rodent species (31).

1.2 Model Building

Once a "learning set" (i.e., database) satisfying preset criteria for acceptance (3) has been developed, the model building phase can begin. In general, this is a straightforward process that is specific for the SAR method employed.

Here, I will exemplify the various stages with the MULTICASE SAR system (3–5, 32). Thus, once the structures of the chemicals and an indication of their potency (i.e., either active, marginally active, and inactive, or a continuous scale of potencies) are entered, the program identifies the chemical substructures significantly associated with the toxicological phenomenon under investigation (Table 19.2). Each of these structural determinants ("toxicophore") is associated with a base potency and a probability of activity (see Fig. 19.1). The latter is derived from the distribution of active and inactive molecules that contain the toxicophore. The program also identifies the chemicals that give rise to the toxicophore (Table 19.3 and Fig. 19.7). This enables the human expert (see below) to ascertain whether the structures of the chemicals giving rise to the toxicophores are germane to the test chemical whose toxicity is predicted.

In addition to the toxicophores, the program also identifies modulators for specific toxicophores (Table 19.4). These are substructures or physicochemical parameters that determine whether the specific toxicity inherent in the toxicophore will be expressed or whether it is augmented further.

Thus, when faced with a chemical of unknown activity, the program uses the presence or absence of toxicophores and of modulators to predict its toxicity (Figs. 19.1–19.3). Thus, the presence of the toxicophore OH—C= (a phenol) endows a chemical with an 87.5% probability of being a contact allergen and a potency of 51 (moderate activity, see Table 19.3). That basal activity is modulated by $-25.8 \times$ electronegativity (see Table 19.3). For the example in Fig. 19.1, this results in a further increase in potency. The total potency of 55 units corresponds to a moderately strong activity (Table 19.3). A chemical with that toxicophore may also contain a structural modulator that augments the basal activity further (Fig. 19.2). On the other hand, the chemical may contain a modulator which completely abolishes a chemical's potential to be an allergen (Fig. 19.3). Additionally, the MULTICASE SAR program will identify substructures that are absent from the learning set and therefore may introduce an element of uncertainty in the prediction, i.e., the "unknown" substructure could represent either potential toxicophore or a modulator that alters a rec-

Table 19.2 Major Toxicophores Associated with Allergic Contact Dermatitis in Humans

Fragment							N*	Inactives*	Marginals*	Actives*	Toxicophore No.
1	2	3	4	5	6	7					
N	—CH₂—						50	1	0	49	1
N	—CH₃—						26	1	0	25	2
Cl	—CH₂—						8	0	0	8	3
OH	—c=						86	10	0	76	4
SH	—CH₂—						19	0	1	18	5
NH	—CH₂—						26	1	1	24	6
NH₂	—CH₂—						7	0	0	7	7
cH	=cH	—c	=cH—			⟨3—NH₂⟩	18	0	0	18	8
CH″	—CO	—CH=					15	0	0	15	9
O	—CO	—C	—CH₂—				21	1	5	15	10
NH	—c	=cH—					17	0	1	16	11
CO	—CH	—CH₂—					9	0	0	9	12
N	=C—	—c=					24	1	0	23	13
CO	—N—						14	0	0	14	14
OH	—CO						8	1	0	7	15
cH	—c	—cH	=c	—cH=			25	1	0	24	16
cH	—cH	—cH	—cH	—cH	=c⟩—		17	4	0	13	17
S	—c:=						10	0	0	10	18
CO	—CH₂	—CH₂—					5	5	0	0	19

The database and derivation of the SAR model have been described (33).

*N indicates the number of chemicals in the database that contain that toxicophore. "Inactives," "marginals," and "actives" indicate the distribution of that toxicophore among activity groups.

Toxicophore No. 4 is shown embedded in chemicals in Figs. 19.1–19.3 and No. 5 is shown in Fig. 19.5.

C indicates a carbon atom shared by two rings; (3—NH₂) indicates an amino group attached to the third non-hydrogen atom from the left. In toxicophore No. 17, the last carbon to the right is shown as unsubstituted. This means that it can be substituted with any atom except a hydrogen. On the other hand, in toxicophore No. 8, the penultimate carbon is shown unsubstituted; it can only be substituted by an amino group (i.e., (3—NH₂)). However, the last carbon of that toxicophore is shown with an attached hydrogen. It cannot be substituted by any other atom.

Table 19.3   Derivation of Toxicophore: The 19 Molecules Containing Fragment SH—CH₂

Chemicals	Potency[a]
2,3-Dimercapto-1-propanol	55
2-Mercaptoethanesulfonic acid	55
2-Mercaptoethyl methyl sulfone	45
2-Mercaptoethyl urea	35
2-Methoxyethyl mercaptoacetate	35
N-(1,1-dimethylolethyl) mercaptoacetamide	35
N,N-dimethyl mercaptoacetamide	35
N-(2-mercaptoethyl) acetamide	25
N-(2-mercaptoethyl) pyrolidone	55
N-(mercaptoacetyl) urea	35
N-(mercaptoacetyl) glycine	35
N-(mercaptoethyl) morpholine	35
N-methyl mercaptoacetamide	35
N-trimethylolmethyl mercaptoacetamide	45
Cysteine	45
Mercaptoacetamide	55
Mercaptoacethydrazide	45
Mercaptoacetic acid	45
Thioglycerol	55

The program identifies the chemicals that are responsible for toxicophore No. 5 of Table 19.1 (see also Fig. 19.7). The toxicophore is shown embedded in a molecule in Fig. 19.5.

[a]The allergenic potencies were defined based on the percent responders in the human maximization test as follows (33): 10, Non-sensitizer; 25, "marginal" (4–7% responders); 39, "weak" (8–23% responders); 49, "moderate": (24–55% responders); 59, "strong" (56–83% responders); 69, "extreme" (84–100% responders).

ognized toxicophore or a noninformative structure unrelated to toxicity (Fig. 19.4).

It should be stressed that not every experimental data set gives rise to a coherent SAR model. Failure to construct a model may be caused by the fact that the experimental data are invalid or that they do not reflect a specific toxicological phenomenon. Additionally, the phenomenon under investigation may be so complex or be the result of so many different mechanisms that the experimental database is not sufficiently large to describe it. With this in mind, it should be stressed that the predictivity of the SAR model will be a reflection of the complexity of the phenomenon, the size of the database (i.e., the number of chemicals for which experimental data are available), and the ratio of actives/inactives in the data set (3, 22).

In view of the above considerations, once an SAR model has been developed, it requires extensive validation and characterization.

1.3   Model Characterization

As mentioned above, the nature of the SAR model that is derived is a reflection of the complexity of the toxicological phenomenon that it describes, as well as of the size of the learning and the extent to which it includes chemical classes and/or substructures that are representative of the chemical species to which it will be applied. Thus, the chemical substructures present among therapeutics are much greater and diverse than, for example, those used or generated in the chemical or agricultural industries. This means that SAR models used to examine pharmacologically active substance must contain a greater variety of chemical substructures. This may well translate into a requirement for a larger experimental data set (i.e., one containing an increased number of chemicals).

In evaluating the SAR model, it is of importance to determine the relationship between its predictivity and the size of the database to determine whether the model is optimal. This can be ascertained by first determining the model's predictivity (see below), and then systematically decreasing the size of the database by random deletion of chemicals to determine the predictive parameters of the model derived from the reduced data set. Doing this iteratively will allow a determination of the relationship between database size and concordance between predicted and experimentally derived results (22). If the relationship, including the value for the SAR model derived from the total database is linear, then the model will not be optimally predictive and consideration should be given to obtaining additional experimental data and deriving a further model. On the other hand, if the relationship including the data for the SAR model derived from the total database is no longer linear, the size of the data set may be satisfactory. Incremental data may not yield a correspondingly significant increase in the model's performance. Thus, the predictivity of the SAR model of mutagenicity in *Salmonella* improves linearity until a database size of 350 chemicals is reached, and then it plateaus (22).

Table 19.4 List of MODULATORS Related to Toxicophore OH —c=

Fragment								Constant = 51.0 QSAR	Toxicophore No.
1	2	3	4	5	6	7	8		
2D	[N—]⟨——6.5A——⟩[NH—]							16.7	1
CO	—CH₂	—CH₂—						−53.3	2
cH	≡c	⟩—c	=cH—				⟨3—CO⟩		−38.8	3
OH	≡c	≡c	—CO	—c≡					25.2	4
cH	=cH	≡cH	=cH	=c≡	=cH—		⟨2—OH⟩		8.3	5
CH₂	—CH	≡c	=cH	—cH	=cH—	—CH—	⟨3—cH=⟩		13.4	6
cH	=cH	=cH	=c	=cH	=cH—	≡cH—	⟨4—OH⟩		−10.3	7
cH	=cH	=cH	=cH	—c	=cH—	—CH₂	⟨5—OH⟩		−10.3	8
OH	≡c	c	—O	—CO	—CH₂	—CH—			−20.3	9
OH	—c	=cH	=cH	=cH	=cH	≡cH—			−20.6	10
OH	—c	≡c	=cH	≡c	—CH₂	—CH₂	—CH₃		−22.4	11
(HOMO + LUMO)/2								−25.8	12

Modulators associated with toxicophore No. 4 of Table 19.2. Each of the modulators augments or decreases the activity inherent in the toxicophore (i.e., 51.0 units; see Figs. 19.1–19.3). (HOMO + LUMO)/2 describes the electronegativity of the molecule. That value is multiplied by −25.8. Modulator No. 5 and No. 2 are shown embedded in chemicals in Figs. 19.2 and 19.3, respectively. Modulator No. 1 describes a 2D distance descriptor of 6.5 Å between two atoms. For interpretation of the structures see legend to Table 19.2.

```
The molecule contains the Toxicophore    (nr.occ.= 2):

        OH  -c"

    *** 76 out of the known  86 molecules ( 88%) containing such a

Toxicophore are Contact Allergens with an average activity of 49.

(conf.level=100%)

        *** QSAR Contribution :                    Constant is    51.04

     ** The following Modulator is also present:

        Electronegativity   = -0.15 ;    Its contribution is    3.83
                                                              ------
     ** Total projected QSAR activity                           54.87

    *** The probability that this molecule is a Contact Allergen is
87.5%  **
     ** The projected  Allergic Potency is 54.9 CASE units **
```

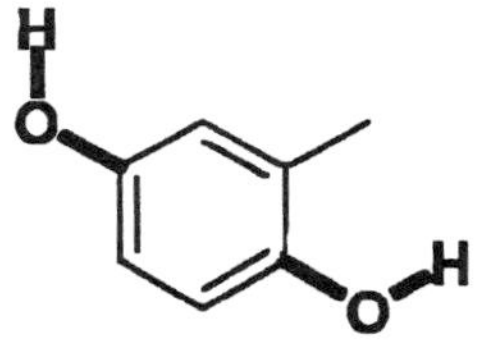

Figure 19.1. Prediction of the contact allergenicity of 2-methyl-1,4-benzenediol. The prediction is based on the presence of the toxicophore (shown in bold). The potency is modulated further by the electronegativity (see Table 19.4). A potency of 55 units indicates a moderately strong allergen (see Table 19.3).

Another concern relates to the effect of the ratio of active to inactive chemicals in the data set. Some SAR models are most predictive when that ratio is unity (3, 22). Hence, for a model that will be widely used for hazard identification and risk assessment purposes, it would be of importance to determine whether its performance is optimal. Thus, if the number of inactives exceeds the number of actives, the number of inactives can be decreased by randomly removing the appropriate number of inactives and determining the performance of the resulting SAR model. The random deletion of inactives and the model derivation should be repeated several times to ascertain that a robust model has been derived. We found that because the nature of the toxicophores is determined primarily by the actives and because the "quality" of the toxicophores

is a function of the size of the database (22, 34, 35), it follows that if the number of actives exceeds the number of inactives that removal of actives to achieve a ratio of unity is not the optimal solution. Rather, we have found that supplementing the database with randomly selected chemicals from a "pool" of normal physiological chemicals (amino acids, sugar, lipids, purines, pyrimidines, etc., but excluding hormones, prostaglandins, and vitamins), assuming these chemicals to be inactive, is a viable alternative (36, 37). This is based on the recognition that the biological and/or toxicological phenomena being modeled occur in a milieu that is rich in these physiological chemicals.

Finally, the "informational content" of an SAR model determines its coverage. Thus, if a test molecule contains a substructure un-

```
The molecule contains the Toxicophore   (nr.occ.= 2):

        OH  -c"

  *** 76 out of the known  86 molecules ( 88%) containing such a

Toxicophore are contact allergies with an average activity of  49.

(conf.level=100%)

        *** QSAR Contribution :                    Constant is    51.04

     ** The following Modulators are also present:

     ( 1) cH =c  -cH =cH -c  =              <2-OH >   Activating   8.33

          Electronegativity   = -0.17 ;    Its contribution is     4.29
                                                               ------
      ** Total projected QSAR activity                            63.65

   *** The probability that this molecule is a Contact Allergen is

87.5%  **

      ** The projected Allergic Potency is 63.7 CASE units **
```

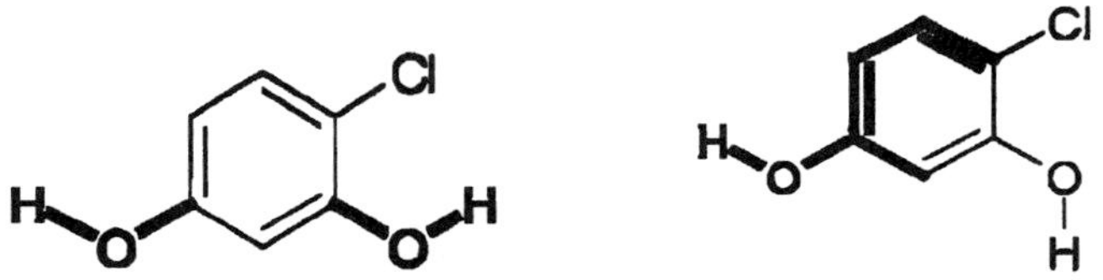

Figure 19.2. Prediction of the contact allergenicity of 4-chloro-1,3-benzenediol. In addition to the probability of activity and the basal potency derived from the toxicophore (shown in bold in A), the chemical also contains an activating modulator (shown in bold in B), which further augments the potency. A potency of 64 units indicates a very strong potency (Table 19.3).

known to the model, this introduces a measure of uncertainty into the SAR prediction. In the MULTICASE SAR program, such an "unknown" moiety is flagged (Fig. 19.4). We have found that a satisfactory approach to determining informational content is to challenge an SAR model with a panel of 10,000 chemicals representative of the "universe of chemicals" and determining the frequency with which the SAR predictions are accompanied by a "warning" of the presence of "unknown" substructures. An enumeration of the frequency with which the individual unknown moieties are present will allow a determination of their importance and thereby identifies chemicals that should be tested and the results included in the model to improve the predictive performance. This is based on the observation that the greater the informational content (i.e., the fewer warnings of "unknown" moieties), the greater the model's predictivity (22, 34, 35).

1.4 Model Validation

In its application to toxicology, SAR can serve two functions: (1) to predict a specific toxico-

The molecule contains the Toxicophore   (nr.occ.= 1):

        OH -c"

*** 76 out of the known 86 molecules (88%) containing such a

Toxicophore are Contact Allergens with an average activity of 49.

(conf.level=100%)

    *** QSAR Contribution :   Constant is   51.04

   ** The following Modulators are also present:

(1) CO -CH2-CH2-   Inactivating -53.33

      Electronegativity   = -0.10 ;   Its contribution is   2.51

  ** Total projected QSAR activity   0.22

   ** The molecule contains the following DEACTIVATING Fragment:

      CO   -CH2
            \\
          CH2 -c"

*** The probability that this molecule is a Contact Allergen is 63.6%   **

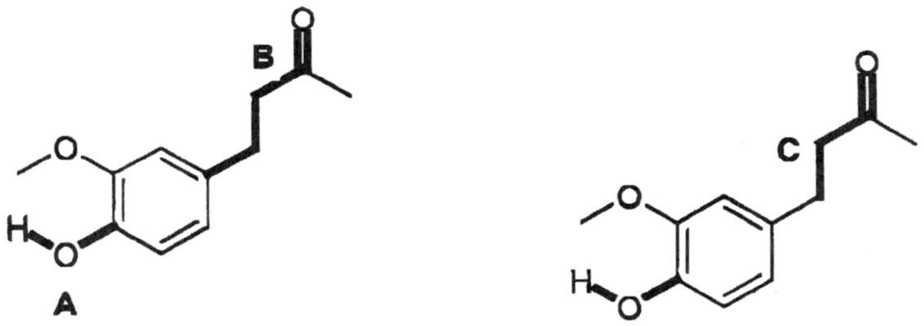

Figure 19.3. Prediction of the lack of contact allergenicity of zingerone. Whereas the presence of the toxicophore (A) is associated with a probability of activity and a potency, the presence of the inactivating modulator (B) abolishes the potency. Moreover, the presence of a deactivating moiety (C), which is present in five chemicals in the database that are devoid of allergenicity (Table 19.2, No. 19), further decreases the likelihood that the zigerone is a contact allergen.

logical effect based on the identification of substructures significantly associated with that activity and (2) to gain insight into the mechanistic basis of that effect.

To be useful in its predictive mode, the performance of a model does not need to be perfect, but it must be known. The predictivity of an SAR model is defined by the concordance between the predictions of chemicals external to the SAR model and the experimentally determined toxicities. The predictivity is governed by the sensitivity (number of correct positive predictions/total number of positive chemicals) and the specificity (number of correct negative predictions/total number of negative chemicals) (22). Moreover, because the basic function of SAR applied to toxicological phenomena is the prevention, reduction, or elimination of harmful chemicals from the home, the environment, and the workplace, risk averse prediction models are preferred. That is achieved by the development of SAR models that yield a low frequency of false negative predictions, i.e., high specificity. Obviously, ideally the model should have high sensitivity as well as high specificity (38).

```
0*** WARNING *** The following functionalities are UNKNOWN to me:

  *** O  -C. =C. -

  *** CO -O  -C. =
```

** The molecule does not contain any known Biophore   **

it is therefore presumed to be INACTIVE

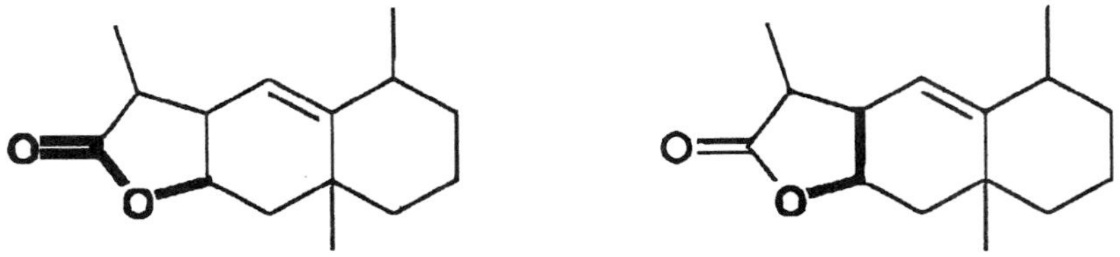

Figure 19.4. Prediction of the lack of contact allergenicity of of dehydroalantolactone. The chemical contains no toxicophore; therefore, it is presumed to be inactive. However, it contains two structures (shown in bold) that are "unknown" to the model. That introduces an element of uncertainity in the prediction.

The simplest way to determine predictivity parameters is to remove initially from the data set a random representative sample (e.g., 5%) to be used as a "tester set," to develop the SAR model on the remaining chemicals (i.e., 95%), and then challenge the model with the "tester set" and ascertain the predictivity. However, as has been demonstrated on a number of occasions, the predictivity of an SAR model is determined by the size of the database (22), and as in most instances, the size of the available data set is not optimal, therefore, further decreasing the size of the learning set by sequestering the "tester set" is not optimal.

To overcome this limitation, a cross-validation approach has been used (39). In that procedure, a portion of the database (e.g., 5%) is randomly selected and removed, and a model is developed from the remaining 95%. That model is challenged with the "tester set" (5%). That procedure is repeated 20 times, and the cumulative predictivity is determined. The final SAR model includes the complete database (i.e., 100%). Because the predictive performance is a function of the size of the database, the performance of the final model will be better than that based on 95% of the data. When,

however, the learning set consists of less than 150 chemicals, a more tedious procedure may be required, wherein one to two chemicals (i.e., n-1 or n-2) are removed at a time to serve as the "tester set" and the process is repeated n or n/2 times.

1.5 Applications and Mechanistic Studies

As has been mentioned earlier (Table 19.1), SAR methodologies can be divided into two general non-mutually exclusive approaches: (1) hypothesis driven and (2) knowledge based. The former is rule driven, wherein specific properties or chemical substructures are looked for, e.g., mutagens are electrophiles and hence one would look for electrophilic or proelectrophilic moieties. This approach assumes that mutations are caused solely by covalent binding of electrophiles to DNA. Agents that induce mutations by a nonelectrophilic (i.e., non-DNA damaging) mechanism will not be detected. Thus, agents that mutagenize purely as a result of intercalation between DNA base pairs (e.g., acridine orange, ethidium bromide) will not be identified. Such rules are based on prior knowledge and/or in-

```
The molecule contains the Biophore

            SH   -CH2

*** 18 out of the known 19 molecules ( 95%) containing such a

toxicophore are Contact Allergens with an average activity of 42.

(conf.level=100%)


      *** QSAR Contribution :                         Constant is    52.50

      ** The following Modulators are also present:

      (2D) [CO -] <-- 5.2A --> [SH -]                 Inactivating   -7.6

           Electronegativity   = 0.10 ;     Its contribution is   -0.67

                                                                  -----
       ** Total projected QSAR activity                           44.23

    *** The probability that this molecule is a Contact

            Allergen is 95.0%  **

    ** The projected Allergic Cont activity is 44.3 CASE units **
```

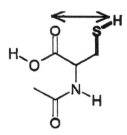

Figure 19.5. Prediction of the contact allergenicity of *N*-acetyl-L-cysteine. The prediction is based on the presence of the toxicophore (shown in bold), which is present in 19 chemicals in the database (18 allergens and 1 marginal allergen; see Table 19.3). The arrow indicates the 5.2 Å distance described by the inactivating modulator.

tuition and do not necessarily require adherence to strict statistical criteria.

The approach illustrated herein, exemplified by MULTICASE (3), is knowledge based. The input consists of the structures and toxicological activities of the chemicals in the learning set. The program then identifies structural descriptors (toxicophores) that are significantly associated with activity (see Table 19.3). The human expert participates in setting criteria for the inclusion of experimen-

tal results in the database (3) as well as in examining the plausibility of the final model based on exact knowledge of the toxicological phenomenon under investigation. The human expert again also determines the acceptability of individual predictions (see below).

Once an SAR model has been developed and validated, it can be applied in a number of fashions. SAR methodologies, such as MULTICASE (3–5), which document predictions (Table 19.2), are obviously preferable to those

The molecule contains the Toxicophore

```
        S    -C
               \ \
                 N
               /
             C"
```

*** 5 out of the known 5 molecules (100%) containing such Biophore

    are Mouse carcinogens with an average activity of 62.

    (conf.level=97%)

  ** This Biophore exists in a significantly different environment

    than in the data base (i.e. 5.45); It may not be relevant

*** QSAR Contribution :   Constant is   64.00

  ** Total projected QSAR activity   64.00

*** The probability that this molecule is a Mouse carcinogen is 85.7%

  ** The projected Mouse carcinogenic potency is 64.0 CASE units **

Figure 19.6. Prediction of the carcinogenicity in mice of epitholone A. The structure of epitholone A (toxicophore shown in bold) is given in Fig. 19.7.

that operate like a "black box." The latter simply provides a likelihood that a test chemical is active or inactive. When, however, the SAR prediction is accompanied by documentation of the basis of that forecast, the human expert can determine whether it is justified and whether it applies to the specific test chemical.

Thus, the mucolytic agent N-acetyl-L-cysteine is predicted to have a potential to induce allergic contact dermatitis by virtue of the biophore SH—CH$_2$ (Fig. 19.5). Moreover, examination of the chemicals that contribute to that toxicophore reveals that indeed they all have the substructure in an environment that is similar to the one found in N-acetyl-L-cysteine (Table 19.3). On the other hand, the tubulin polymerization perturber (and potential antineoplastic agent) epitholone A (Fig. 19.6) is predicted to be a mouse carcinogen by virtue of the toxicophore units shown in bold. That toxicophore is present in five molecules in the learning set. The presence of that toxicophore

is associated with an 89% probability of carcinogenicity and a potency of 63 units, which corresponds to a TD$_{50}$ value of 0.039 mmol/kg per day (40). However, the program flags the toxicophore because its environment in epitholone A is significantly different from that of the molecules in the learning set (Fig. 19.6). In fact, examination of the structures of the molecules that contribute to the biophore (Fig. 19.7) indicates that indeed the molecules are quite different from that of epitholone A, and hence, the prediction of carcinogenicity can be disregarded (however, also see below).

Moreover the molecules that contributed to this toxicophore (Fig. 19.7), even though they contain the S—C≡N—C≡ moiety (Fig. 19.6), also contain functionalities (i.e., "structural alerts") that are associated with carcinogenicity/genotoxicity such as nitro, amino, and hydrazino groups. In fact, these could be responsible for the murine carcinogenicity of these chemicals. Obviously, these latter functionalities are absent in epitholone A.

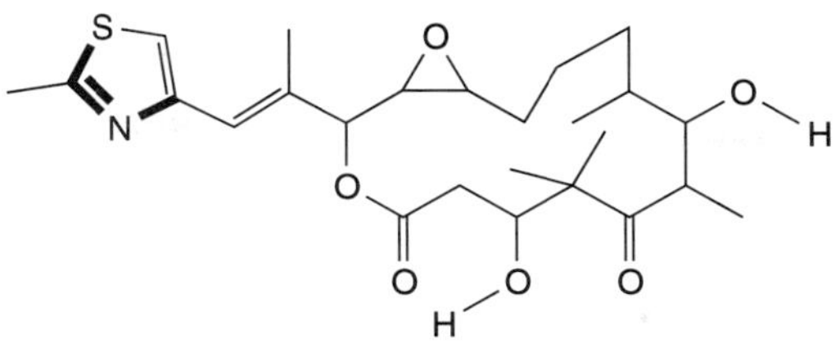

Epitholone A

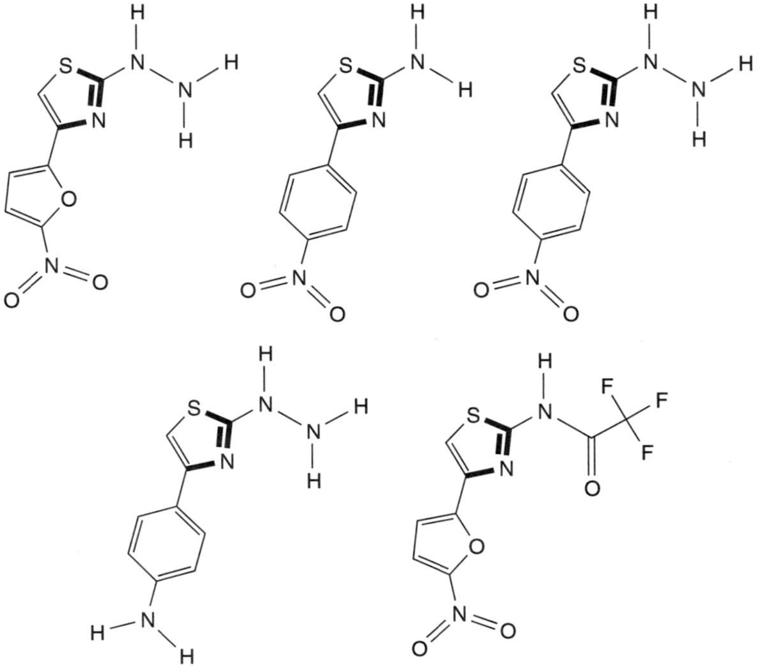

Figure 19.7. Structures of epitholone A and of chemicals that contain the toxicophore. The toxico-phore (Fig. 19.6) is shown in bold.

Table 19.5   SAR Predictions Related to the Potential Carcinogenicity of Epitholone A

SAR Model	Prediction	References
Mutagenicity: *Salmonella*	Negative	22, 47
Error-prone DNA repair	Negative	48
Unscheduled DNA synthesis	Negative	49
Mouse MTD	Positive	50
Rat LD_{50}	Positive	SAR model based on RTECS
Cell toxicity	Positive	51
Inhibition GJIC	Negative	52

A positive response indicates a potential for maximum tolerated dose of less than 0.9 mmol/kg; an LD_{50} value of less than 7.2 mmol/kg or a toxicity (IC_{50}) for cultured BALB/3T3 cells of less than 1 μM.

GJIC, gap junctional intercellular communication; RTECS, Registry of Toxic Effects of Chemical Substances.

Table 19.6 Predicted Toxicological Profile of *N*-Acetylcysteine

	Multicase	
SAR Model	Probability (%)	Potency (units)
Structure alerts	0	0
Salmonella mutagenicity	0	0
SOS chromotest	0	0
umu/SOS repair	0	0
Carcinogenicity: rodents-NTP	0	0
Carcinogenicity: mice-NTP	0	0
Carcinogenicity: rats-NTP	0	0
Carcinogenicity: rodent-CPDB	0	0
Carcinogenicity: mice-CPDB	0	0
Carcinogenicity: rats-CPDB	0	0
Inhibition gap junction intercell comm	0	0
Binding to Ah receptor	0	0
Mutations in mouse lymphoma (NTP)	0	0
Mutations in mouse lymphoma (GenTox)	0	0
Sister chromatic exchanges *in vitro*	0	0
Chromosomal aberrations *in vitro*	0	0
Unscheduled DNA synthesis *in vitro*	0	0
Cell transformation	0	0
Drosophila somatic mutations	0	0
Sister chromatic exchanges *in vivo*	0	0
Induction of micronuclei *in vivo*	0	0
Yeast malsegregation	0	0
Inhibition of tubulin polymerization	0	0
Sensory irritation	89	72
Eye irritation	72	52
Respiratory hypersensitivity	0	0
Allergic contact dermatitis	95	44
Rat lethality (LD50)	0	0
Mouse MTD	0	0
Rat MTD	0	0
Cellular toxicity (3T3)	0	0
Cellular toxicity (HeLa)	0	0
Nephrotoxicity: male rats ($\alpha 2\mu$globulin)	0	0
Inhibition human cyt. P4502D	0	0
Developmental toxicity: hamster	0	0
Developmental toxicity: human	0	0
Aquatic toxicity (minnows)	0	0
Water solubility: 3.88	$\log P$ (Octanol: water): -1.79	
Electronegativity: 0.10		

NTP and CPDB refer to the U.S. National Toxicology Program carcinogenicity assays (45) and to the Carcinogenic Potency Data Bases (46), respectively.

Based on all of these considerations, the "human expert" would overrule the prediction of rodent carcinogenicity. Additionally, in overriding the computer-based prediction, cognisance was also taken of the understanding that the vast majority of recognized human carcinogens are genotoxicants, i.e., "genotoxic carcinogens" (41–44). Epitholone A, on the other hand, was not predicted to be genotoxic (i.e., a DNA-damaging agent), evidenced by its lack of potential to induce mutations in *Salmonella*, error-prone DNA repair, or unscheduled DNA synthesis in rat hepatocytes (Table 19.5). Thus, even if the potential for murine carcinogenicity were accepted, in view of the fact that the vast majority of rec-

The molecule contains the Biophore   (nr.occ.= 1):

```
        NH   -CH
                 \
                   C."
```

*** 38 out of the known 41 molecules (93%) containing such a Biophore

        are perturbers of Tubulin Polymerization

*** QSAR Contribution :   Constant is   85.89

  ** Total projected QSAR activity   85.89

  ** The probability that this molecule inhibits Tubulin

       Polymerization is 93% **

  ** The projected Tubulin Polymerization Inhibitory activity is 85.9

       CASE units **

Figure 19.8. Prediction of the ability of colchicine to inhibit tubulin polymerization. The structure of colchicine is shown in Fig. 19.9. The biophore is shown in bold (a) in Fig. 19.9.

ognized human carcinogens are mutagens/genotoxicants or are hormones and epitholone A is neither, it would not represent a human risk.

If, based on the above, it were accepted that epitholone A is not genotoxic, and if the human expert examining the documentation wished not to override the prediction of carci-

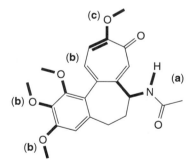

Figure 19.9. Structure of colchicine. The biophore A (bold, see Fig. 19.8) is responsible for the therapeutic effectiveness. Toxicophore B (see Fig. 19.10; shown in bold) is responsible for the induction of sister chromatid exchanges (SCE) *in vivo*. Removal of toxicophore B or its replacement be isopropoxy groups abolishes the induction of SCEs without affecting the therapeutic potential.

nogenicity in mice based on the differences in chemical environments between epitholone A and the molecules responsible for the toxicophore (Figs. 19.6 and 19.7), he could examine mechanisms of non-genotoxic carcinogenicity, even though its relevance to human may not be applicable. One of the mechanisms of non-genotoxic carcinogenicity is inhibition of intercellular communication (53). Epitholone A does not possess such a potential (Table 19.5). Another mechanism for non-genotoxic rodent carcinogenesis may involve systemic or cell toxicity followed by mitogenesis (54–56). This may occur as a consequence of including the maximum tolerated dose (MTD) in the cancer bioassay protocol. When this is done, up to 50% of chemicals tested are found to be rodent carcinogens (54). Obviously, this MTD situation rarely, if ever, applies to humans. Still, epitholone A has the potential for inducing cellular as well as systemic toxicity (Table 19.5), which may explain its potential carcinogenicity in mice, were we to discount the difference in chemical environment.

Obviously, the availability of a number of characterized and validated SAR models allows the development of a putative toxicologi-

The molecule contains the toxicoophore (nr.occ.= 3):

```
        CH3 -O
               \
                 C"
```

*** 8 out of the known 8 molecules (100%) containing such a
    toxicophore are Mouse SCE inducers with an average activity of
    57.

*** QSAR Contribution :   Constant is   73.17

  ** The following Modulators are also present:
     (3) CH3-O -c =   Inactivating   -7.41

     (1) CH3-O -c =cH -   Inactivating   -7.41

          Log partition coeff.= 3.19 ;   LogP contribution is   -7.65

  ** Total projected QSAR activity   50.70

  ** The probability that this molecule induces Mouse SCEs is 90.0% **

  ** The projected Mouse SCE inducing activity is 50.7   CASE units **

Figure 19.10. The potential of colchicine to induce sister chromatic exchanges *in vivo*. The structure of colchicine and of the toxicophore B is given in Fig. 19.9. One of the inactivating modulators (c) is also shown in bold in Fig. 19.9.

cal profile (Table 19.6). This can be used as a guideline in the product developmental phase to select lead compounds least likely to induce unwanted side effects. However, the SAR approach can also be used to optimize beneficial effects and decrease or eliminate unwanted toxic effects.

Thus, let us examine colchicine (CH), an anti-inflammatory agent that has been in use for several centuries for the treatment of gout. The anti-inflammatory potential of CH is understood to derive from its ability to inhibit tubulin polymerization (iTP) (57). That is also the basis of the anticancer activity of paclitaxel (Taxol) (58–60). The structural basis of that activity derives from the presence in CH of the NH—CH—C.= moiety (Figs. 19.8 and 19.9), which endows the molecule with a 93% probability of activity. However, colchicine also has the potential for inducing sister chromatid exchanges (SCEs) *in vivo* (Fig. 19.10). This SCE-inducing ability may endow it with genotoxic and developmental toxicity poten-

tials. However, the potential for inducing SCEs *in vivo* is associated with the methoxy moiety (Figs. 19.9 and 19.10). Removal of that moiety or replacing it with an isopropoxy group abolishes the SCE-inducing ability of CH without affecting its potential for iTP (i.e., the basis of its anti-inflammatory action).

Finally, SAR approaches can also be used to provide a basis for making intelligent risk assessments. Thus, it has been shown that the similarity in biophores/toxicophores present in different SAR models of toxicological phenomena provides a measure of mechanistic similarity (3). The SAR models of mutagenicity in *Salmonella* and of error-prone DNA repair (SOS Chromotest) show significant overlaps (Table 19.7). This is not unexpected because DNA is the target of both phenomena, and the tester strain used for the *Salmonella* mutagenicity assays contains a plasmid that codes for error-prone DNA repair (61). In fact there is a substantial (though not complete) overlap among chemicals that cause the two

Table 19.7   Structural Commonalities among SAR Models

SAR Models	Percent
Salmonella mutagenicity and SOS chromotest	57
Salmonella mutagenicity and iGJIC	10
Salmonella mutagenicity and iTP	9
Salmonella mutagenicity and Mnt	53
Mnt and iTP	71

iGJIC, inhibition of gap functional intercellular communications; iTP, inhibition of tubulin polymerization; Mnt, induction of bone marrow micronuclei *in vivo*.

phenomena (48, 62). On the other hand, there is little overlap between *Salmonella* mutagenicity and inhibition of gap junctional intercellular communication (Table 19.7), which is considered the epigenetic (non-genotoxic) phenomenon *par excellence* (53). Nor do the SAR models for *Salmonella* mutagenicity and inhibition of tubulin polymerization overlap significantly (Table 19.7), which is further support for the fact that genotoxicity and inhibition of tubulin polymerization can be dissociated (see above).

With respect to the *in vivo* induction of micronuclei (Mnt), a different situation prevails. There is considerable overlap between the toxicophores associated with Mnt and those with the induction of mutation is *Salmonella* (Table 19.7). This is not surprising, because the induction of Mnt is known to involve a genotoxic mechanism (63, 64). Indeed, when attempting to identify potential genotoxic carcinogens, when a chemical is found to induce mutations in *Salmonella*, this result is frequently followed by a Mnt test to determine

whether the chemical is genotoxic *in vivo* as well (43, 65) and thus represent a risk to humans.

However, it was found that there is also substantial overlap between Mnt and iTP, the latter being a non-genotoxic phenomenon (Table 19.7) (66). This finding suggests that Mnt can occur by genotoxic as well as non-genotoxic mechanisms. Thus, a positive Mnt response by a chemical that does not induce mutations in *Salmonella* does not necessarily represent a carcinogenic risk to humans.

Discodermolide (Fig. 19.11) is a promising antineoplastic agent, which like paclitaxel, inhibits tubulin polymerization (67), but being considerably more water-soluble than paclitaxel, discodermolide may present certain therapeutic advantages while also being effective against paclitaxel-resistant cells (67). Neither discodermolide nor paclitaxel are mutagenic in *Salmonella* (and in fact neither is predicted to be a rodent carcinogen). However, both of these agents have a potential (determined by SAR) to induce Mnt *in vivo*. In fact, for paclitaxel that potential has been determined experimentally. This has led to the suggestion that paclitaxel, because of its ability to induce Mnt, presented a carcinogenic risk (68). However, based on the above findings (Table 19.7), it can be assumed that the ability of discodermolide and of paclitaxel to induce Mnt is independent of genotoxicity, and in fact, derives from iTP. Thus, it does not represent an unreasonable risk to humans who are treated with those antineoplastic agents. In fact, the biophores in discodermolide responsible for the induction of Mnt and iTP are identical (Fig. 19.11).

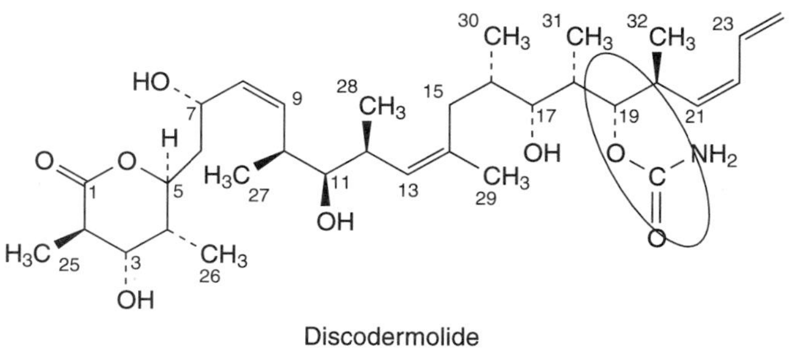

Discodermolide

Figure 19.11. Structure of discodermolide. The circled biophore is responsible for the inhibition of tubulin polymerization.

2 CONCLUSIONS

SAR methodologies, in their present state, coupled with human expertise, can be used to determine and to understand the potential toxicity of therapeutic agents. In fact, this approach can be used to engineer molecules devoid of the moieties associated with these unwanted side effects. It must be understood, however, that while SAR techniques can be used to accelerate the identification and development of safe therapeutic agents, it is to be used as an adjunct to experimental determinations.

3 ACKNOWLEDGMENTS

The support of the Vira Heinz Endowment is gratefully acknowledged.

REFERENCES

1. National Research Council, *Science and Judgment in Risk Assessment*, National Academy Press, Washington, DC, 1994.
2. I. D. McKinney, A. Richard, C. Waller, M. C. Newman, and F. Gerberich, *Toxicol. Sci.*, **56**, 8–17 (2000).
3. H. S. Rosenkranz, A. R. Cunningham, Y. P. Zhang, H. G. Claycamp, O. T. Macina, N. B. Sussman, S. G. Grant, and G. Klopman, *SAR QSAR Environ. Res.*, **10**, 277–298 (1999).
4. G. Klopman and H. S. Rosenkranz, *Mutat. Res.*, **305**, 33–46 (1994).
5. G. Klopman and H. S. Rosenkranz, *Toxicol. Lett.*, **79**, 145–155 (1995).
6. K. Einslein, V. K. Gombar, and B. W. Blake, *Mutat. Res.*, **305**, 47–61 (1994).
7. D. F. V. Lewis, C. Ioannides, and D. V. Parke, *Environ. Health Perspect.*, **104**, 1011–1016 (1996).
8. D. M. Sanderson and C. G. Earnshaw, *Human Exp. Toxicol.*, **10**, 261–273 (1991).
9. N. Greene, *J. Chem. Inf. Comput. Sci.*, **37**, 148–150 (1996).
10. J. E. Ridings, M. D. Barratt, R. Cary, C. G. Earnshaw, C. E. Eggington, M. K. Ellis, P. N. Judson, J. J. Langowski, C. A. Marchant, M. P. Payne, W. P. Watson, and T. D. Yih, *Toxicology*, **106**, 267–279 (1996).
11. Y. T. Woo, D. Y. Lai, M. F. Argus, and J. C. Arcos, *Toxicol. Lett.*, **79**, 219–228 (1995).
12. Y. T. Woo, D. Y. Lai, M. F. Argus, and J. C. Arcos, *Environ. Carcin. Ecotoxicol. Rev. C.*, **16**, 101–102 (1998).
13. F. Darvas, A. Papp, A. Allerdyce, E. Benfenati, G. Fini, M. Tichy, N. Sobb, and A. Citti, AAAI Spring Symposium on Predictive Toxicology of Chemicals: Experiences and Impact of AI Tools, Technical Report SS-99-01, AAAI Press, Mento Park, CA, 1999.
14. R. D. King, S. H. Muggleton, A. Srinivasan, and M. J. E. Sternberg, *Proc. Natl. Acad. Sci.*, **93**, 438–442 (1996).
15. J. Ashby, *Environ. Mutagen.*, **7**, 919–921 (1985).
16. A. M. Richard, *Mutat. Res.*, **305**, 73–77 (1994).
17. A. M. Richard, *Knowledge Engineer. Rev.*, **14**, 307–317 (1999).
18. R. D. Combes and P. Judson, *Pestic. Sci.*, **45**, 179–194 (1995).
19. A. M. Richard, *Toxicol. Lett.* **102–103**, 611–616 (1998).
20. C. Helma, E. Gottmann, and S. Kramer, *Stat. Methods. Med. Res.*, **9**, 1–30 (2000).
21. A. M. Richard and R. Benigni, *SAR QSAR Environ. Res.*, **13**, 1–19 (2002).
22. M. Liu, N. Sussman, G. Klopman, and H. S. Rosenkranz, *Mutat. Res.*, **358**, 63–72 (1996).
23. D. M. Maron and B. N. Ames, *Mutat. Res.*, **113**, 173–215 (1983).
24. D. Brusick, V. F. Simmon, H. S. Rosenkranz, V. Ray, and R. S. Stafford, *Mutat. Res.*, **76**, 169–190 (1980).
25. J. Pet-Edwards, H. S. Rosenkranz, V. Chankong, and Y. Y. Haimes, *Mutat. Res.*, **153**, 167–185 (1985).
26. H. S. Rosenkranz, F. K. Ennever, M. Dimayuga, and G. Klopman, *Environ. Mol. Mutagen.*, **16**, 149–177 (1990).
27. A. D. Mitchell, A. E. Auletta, D. Clive, P. E. Kirby, M. M. Moore, and B. C. Myhr, *Mutat. Res.*, **394**, 177–303 (1997).
28. B. Henry, S. G. Grant, G. Klopman, and H. S. Rosenkranz, *Mutat. Res.*, **397**, 313–335 (1998).
29. S. G. Grant, Y. P. Zhang, G. Klopman, and H. S. Rosenkranz, *Mutat. Res.*, **465**, 201–229 (2000).
30. M. Ghanooni, D. R. Mattison, Y. P. Zhang, O. T. Macina, H. S. Rosenkranz, and G. Klopman, *Am. J. Obstet. Gynecol.*, **176**, 799–806 (1997).
31. J. Gómez, O. T. Macina, D. R. Mattison, Y. P. Zhang, G. Klopman, and H. S. Rosenkranz, *Teratology*, **60**, 190–205 (1999).
32. H. S. Rosenkranz, Y. P. Zhang, and G. Klopman, *Altern. Anim. Test.*, **26**, 779–809 (1998).

33. C. Graham, R. Gealy, O. T. Macina, M. H. Karol, and H. S. Rosenkranz, *Quant. Struct. Activ. Relat.*, **15**, 224–229 (1996).

34. N. Takihi, Y. P. Zhang, G. Klopman, and H. S. Rosenkranz, *Mutagenesis*, **8**, 257–264 (1993).

35. N. Takihi, Y. P. Zhang, G. Klopman, and H. S. Rosenkranz, *Qual. Assur. Good Pract. Regul. Law*, **2**, 255–264 (1993).

36. H. S. Rosenkranz and A. R. Cunningham, *Mutat. Res.*, **476**, 133–137 (2001).

37. H. S. Rosenkranz and A. R. Cunningham, *SAR QSAR Environ. Res.*, **12**, 267–274 (2001).

38. V. Chankong, Y. Y. Haimes, H. S. Rosenkranz, and J. Pet-Edwards, *Mutat. Res.*, **153**, 135–166 (1985).

39. Y. P. Zhang, N. Sussman, G. Klopman, and H. S. Rosenkranz, *Quant. Struct. Activ. Relat.*, **16**, 290–295 (1997).

40. A. R. Cunningham, H. S. Rosenkranz, Y. P. Zhang, and G. Klopman, *Mutat. Res.*, **398**, 1–17 (1998).

41. F. K. Ennever, T. J. Noonan, and H. S. Rosenkranz, *Mutagenesis*, **2**, 73–78 (1987).

42. H. Bartsch and C. Malaveille, *Cell Biol. Toxicol.*, **5**, 115–127 (1989).

43. J. Ashby and R. S. Morrod, *Nature*, **352**, 185–186 (1991).

44. M. D. Shelby, *Mutat. Res.*, **204**, 3–15 (1988).

45. J. Ashby and R. W. Tennant, *Mutat. Res.*, **257**, 229–306 (1991).

46. L. S. Gold, N. B. Manley, T. H. Slone, G. B. Garfinkel, L. Rohrbach, and B. N. Ames, *Environ. Health Perspect.*, **100**, 65–135 (1993).

47. H. S. Rosenkranz and G. Klopman, *Mutat. Res.*, **228**, 51–80 (1990).

48. V. Mersch-Sundermann, G. Klopman, and H. S. Rosenkranz, *Mutat. Res.*, **340**, 81–91 (1996).

49. Y. P. Zhang, A. van Praagh, G. Klopman, and H. S. Rosenkranz, *Mutagenesis*, **9**, 141–149 (1994).

50. H. S. Rosenkranz and G. Klopman, *Environ. Mol. Mutagen.*, **21**, 193–206 (1993).

51. H. S. Rosenkranz, E. J. Matthews, and G. Klopman, *Altern. Anim. Test.*, **20**, 549–562 (1992).

52. M. Rosenkranz, H. S. Rosenkranz, and G. Klopman, *Mutat. Res.*, **381**, 171–188 (1997).

53. J. E. Trosko and C. C. Chang in R. W. Hoerger and F. D. Hoerger, Eds., *Banbury Report 31: Carcinogen Risk Assessment: New Directions in Qualitative and Quantitative Aspects*, Cold Spring Harbor Laboratory Press, Cold Spring Harbor, NY, 1988, pp. 139–174.

54. B. N. Ames and L. S. Gold, *Proc. Natl. Acad. Sci. USA*, **87**, 7772–7776 (1990).

55. S. M. Cohen and L. B. Ellwein, *Science*, **249**, 1007–1011 (1990).

56. S. Preston-Martin, M. C. Pike, R. K. Ross, P. A. Jones, and B. E. Henderson, *Cancer Res.*, **50**, 7415–7421 (1990).

57. E. ter Haar, H. S. Rosenkranz, E. Hamel, and B. W. Day, *Bioorg. Med. Chem.*, **4**, 1659–1671 (1996).

58. E. Hamel, *Med. Res. Rev.*, **16**, 207–231 (1996).

59. P. B. Schiff and S. B. Horwitz, *Proc. Natl. Acad. Sci. USA*, **77**, 1561–1565 (1980).

60. P. B. Schiff, J. Fant, and S. B. Horwitz, *Nature (Lond.)*, **277**, 665–667 (1979).

61. J. McCann, N. E. Spingarn, J. Kobori, and B. N. Ames, *Proc. Natl. Acad. Sci. USA*, **72**, 979–983 (1975).

62. V. Mersch-Sundermann, U. Schneider, G. Klopman, and H. S. Rosenkranz, *Mutagenesis*, **9**, 205–224. (1994).

63. J. A. Heddle, M. C. Cimino, M. Hayashi, F. Romagna, M. D. Shelby, J. D. Tucker, Ph. Vanparys, and J. T. MacGregor, *Environ. Mol. Mutagen.*, **18**, 277–291 (1991).

64. K. H. Mavournin, D. H. Blakey, M. C. Cimino, M. F. Salamone, and J. A. Heddle, *Mutat. Res.*, **239**, 29–80 (1990).

65. H. Tinwell and J. Ashby, *Environ. Health Perspect.*, **102**, 758–762 (1994).

66. E. ter Haar, B. W. Day, and H. S. Rosenkranz, *Mutat. Res.*, **350**, 331–337 (1996).

67. E. ter Haar, R. J. Kowalski, E. Hamel, C. M. Lin, R. E. Longley, S. P. Gunasekera, H. S. Rosenkranz, and B. W. Day, *Biochemistry*, **35**, 243–250 (1996).

68. H. Tinwell and J. Ashby, *Carcinogenesis*, **15**, 1499–1501 (1994).

CHAPTER TWENTY

Natural Products as Leads for New Pharmaceuticals

A. D. Buss
MerLion Pharmaceuticals
Singapore Science Park, Singapore

B. Cox
Medicinal Chemistry
Respiratory Diseases Therapeutic Area
Novartis Pharma Research Centre
Horsham, United Kingdom

R. D. Waigh
Department of Pharmaceutical Sciences
University of Strathclyde
Glasgow, Scotland

Contents

Burger's Medicinal Chemistry and Drug Discovery
Sixth Edition, Volume 1: Drug Discovery
Edited by Donald J. Abraham
ISBN 0-471-27090-3 © 2003 John Wiley & Sons, Inc.

1 INTRODUCTION

Of the 520 new pharmaceuticals approved between 1983 and 1994, 39% were derived from natural products, the proportion of antibacterials and anticancer agents of which was over 60% (1). Between 1990 and 2000, a total of 41 drugs derived from natural products were launched on the market by major pharmaceutical companies (Table 20.1), including azithromycin, orlistat, paclitaxel, sirolimus (rapamycin), Synercid, tacrolimus, and topotecan. In 2000, one-half of the top-selling pharmaceuticals were derived from natural products, having combined sales of more than US $40 billion. These included the biggest selling anticancer drug paclitaxel, the "statin" family of hypolipidemics, and the immunosuppressant cyclosporin. During 2001 we have seen the launch of caspofungin from Merck and galantamine from Johnson & Johnson, with rosuvastatin, telithromycin, daptomycin, and ecteinascidin-743 due to follow in 2002.

Despite the figures, the popularity of natural products, particularly those from higher plants as leads for new pharmaceuticals, tends to fluctuate. At the time of writing, several of the world's biggest pharmaceutical companies have reined back their natural product drug discovery programs and have placed great faith in combinatorial chemistry, coupled to very high throughput screening. Time will tell whether this is a wise stratagem, or whether the unique features of compounds that are themselves derived from living organisms will once again see renewed acceptance.

The abundance of plant and microbial secondary metabolites and their value in medicine are undisputed, but one question that is only partly answered concerns the reasons for this abundance of complex chemical substances. In the past, the production of what we

would now call "bioactive" substances was a mystery. A modern view is that these compounds have a role in protecting the otherwise defenseless, stationary plant from attack by mammals, insects, fungi, bacteria, and viruses. Taking morphine as an example of a secondary metabolite whose value to the plant is not entirely obvious, 14 steps are required from available amino acids, including at least one step that is highly substrate specific (2). The presence of morphine in the tissues of *Papaver somniferum* must therefore confer a selectional advantage on the plant (3): genetic code is required for each of the enzymes involved in the biosynthesis, valuable amino acids are utilized in forming the enzymes, and a relatively scarce nutrient (nitrogen) is locked up in the compounds produced. If the morphine did not continue to have value for the plant, mutants would have arisen with the advantage of not having a drain on their metabolic resources.

We can only guess at the ecological functions of morphine. Perhaps a mammalian herbivore that consumed too many poppies would become drowsy and itself fall prey to a carnivore. It may be significant that the cannabinoids, produced in greatest abundance in the nutritious growing tips of the plant, also induce mental effects that would compromise a herbivore's ability to escape a predator. Whatever their natural protective functions, natural products are a rich source of biologically active compounds that have arisen as the result of natural selection, over perhaps 300 million years. The challenge to the medicinal chemist is to exploit this unique chemical diversity. The following account illustrates how natural products have been used as what are called *lead compounds*, or templates for the development of important medicines.

Table 20.1 Drugs Derived from Natural Products (1990–2000)

Name	Originator	Indication/Use
Acarbose	Bayer	Diabetes
Artemisinin	Kunming & Guilin	Malaria
Azithromycin	Pliva	Antibiotic
Carbenin	Sankyo	Antibiotic
Cefetamet pivoxil	Takeda	Antibiotic
Cefozopran	Takeda	Antibiotic
Cefpimizole	Ajinomoto	Antibiotic
Cefsulodin	Takeda	Antibiotic
Clarithromycin	Taisho	Antibiotic
Colforsin daropate	Nippon Kayaku	Asthma
Docetaxel	Aventis	Cancer
Dronabinol	Solvay	Alzheimer's disease
Galantamine	Intelligen	Alzheimer's disease, arthritis
Gusperimus	Nippon Kayaku	Arthritis
Irinotecan	Yakult Honsha	Cancer
Ivermectin	Merck & Co	Parasiticide
Lentinan	Ajinomoto	Cancer
LW-50020	Sankyo	Immunomodulation
Masoprocol	Access	Cancer
Mepartricin	SPA	Benign prostatic hyperplasia
Miglitol	Bayer	Diabetes
Mizoribine	Asahi Chemical	Arthritis
Mycophenolate mofetil	Hoffman-LaRoche	Arthritis
Orlistat	Hoffman-LaRoche	Obesity
Paclitaxel	Bristol-Myers Squibb	Cancer
Pentostatin	Warner-Lambert	Leukemia
Podophyllotoxin	Nycomed Pharma	Human papillomavirus
Policosanol	Dalmer	Hyperlipidaemia
Everolimus	Novartis	Immunomodulation
Sirolimus	American Home Products	Immunomodulation
Sizofilan	Taito	Cancer, hepatitis-B virus
Subreum	OM Pharma	Arthritis
Synercid	Novartis	Antibiotic
Tacrolimus	Fujisawa	Immunomodulation
Teicoplanin	Aventis	Antibiotic
Tirilazad mesylate	Pharmacia & Upjohn	Subarachnoid haemorrhage
Topotecan	GlaxoSmithKline	Diabetes
Ukrain	Nowicky Pharma	Cancer, HIV/AIDS
Vinorelbine	Pierre Fabre	Cancer
Voglibose	Takeda	Diabetes, obesity
Z-100	Zeria	Immunomodulation

2 DRUGS AFFECTING THE CENTRAL NERVOUS SYSTEM

2.1 Morphine Alkaloids

The history of the opium alkaloids is too well known to warrant repetition here, but the analgesics based on morphine (**1**) are too important to be left out of an account of natural products as leads. Thus we shall summarize the clinically more important developments that have occurred since the isolation of morphine in 1803. Codeine (**2**) continues to be used widely for the treatment of moderate pain and, although present in the opium poppy (*Papaver somniferum*), it is normally synthesized in higher yield from morphine (**4**).

Other than codeine, the earliest significant semisynthetic derivative of morphine is the diacetate heroin (**3**), which is still widely used in terminal cancer where its addictiveness is ir-

(1) morphine R₁ = R₂ = H
(2) codeine R₁ = CH₃, R₂ = H
(3) heroin R₁ = R₂ = COCH₃

(1) morphine $R_1 = R_2 = H$
(2) codeine $R_1 = CH_3$, $R_2 = H$
(3) heroin $R_1 = R_2 = COCH_3$

relevant. Acetylation masks the polar hydroxy groups, so that penetration into the central nervous system (CNS) is enhanced; hydrolysis then occurs to liberate the phenolic hydroxyl, giving an active analgesic, and ultimately regenerates morphine (5). Heroin was thus one of the first prodrugs.

Modifications to the C-ring of morphine are legion, but none of the derivatives is free from addictive liability, though many have been used clinically. *N*-Demethylation and realkylation yield more interesting analogs, notably *N*-allylnormorphine and nalorphine (4), which is a morphine antagonist (6). Further modification leads to naloxone (5), which unlike nalorphine has very little agonist activity (7) and has retained a place in therapy for treatment of opiate-induced respiratory depression. Naloxone will also precipitate withdrawal symptoms in opiate addicts, thereby facilitating diagnosis.

(4) nalorphine

Total synthesis of morphine is difficult, but analogs lacking the dihydrofuran ring are accessible (8) from 1-benzylisoquinolines, in analogy with the biosynthesis of morphine, to

(5) naloxone

give the morphinans (**6**). The system may be simplified even further (9), to give the benzomorphans (**7**), although neither these nor the morphinans have provided the long-sought analgesic without addictive properties.

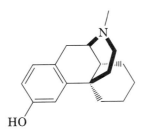

(6) morphinan

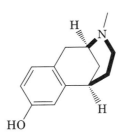

(7) benzomorphan

A semisynthetic route to morphine analogs was found (10) from thebaine (**8**) using Diels–Alder reactions in the C-ring. Adducts such as (**9**) have the distinction of enormous potency (11), sufficient to immobilize rhinoceroses at moderate dose levels! Unfortunately, the addictive liability runs parallel to the increase in analgesic potency, a tendency that was partly overcome (12) in the analog buprenorphine (**10**).

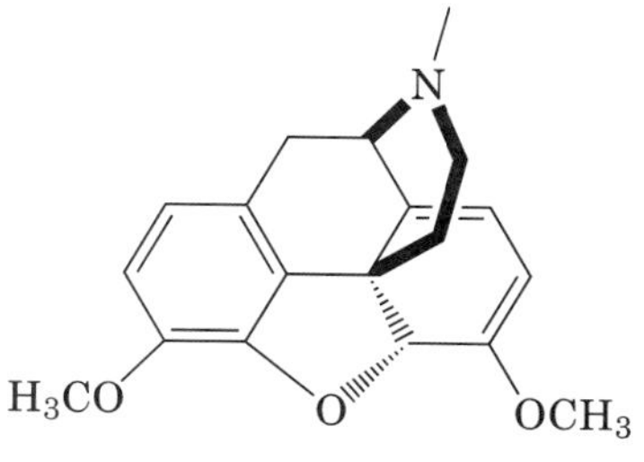

(8) thebaine

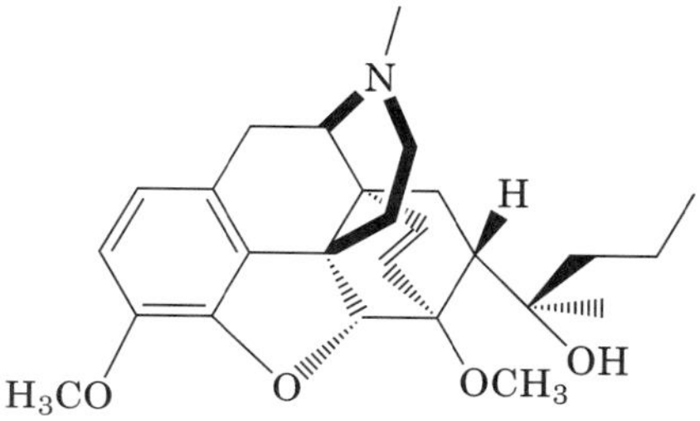

(9) etorphine

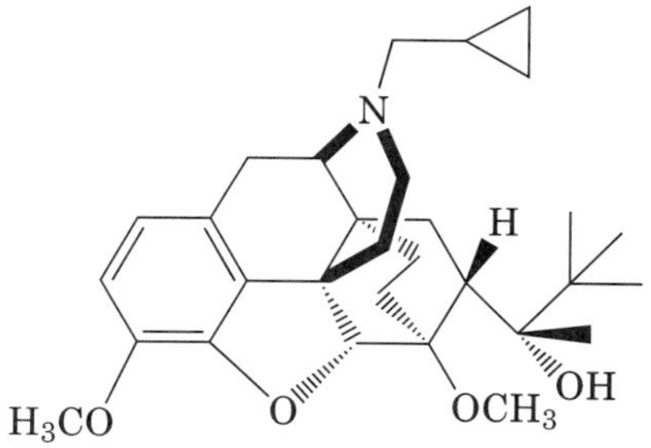

(10) buprenorphine

vation that meperidine (pethidine) (**12**) unexpectedly produced a reaction in mice known as Straub tail, normally characteristic of the morphine series (15). Meperidine itself is still used widely in childbirth in the belief that there is a lower incidence of respiratory depression in the fetus. The realization that 4-phenylpiperidines, which are not obvious structural analogs of morphine, could give rise to useful analgesic effects, led to the synthesis of many thousands of derivatives (16), many with far greater potency than that of meperidine. Unfortunately, as potency increases so do addiction liability and respiratory depression.

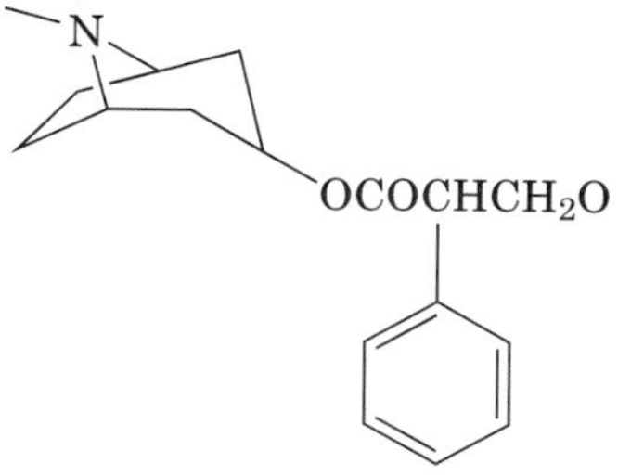

(11) atropine

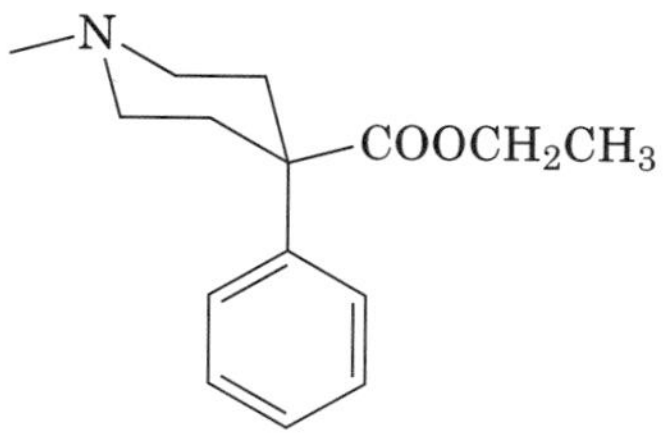

(12) pethidine

All this work was carried out in ignorance of the nature of the natural transmitter(s), which subsequently proved to be the peptides known as endorphins and their pentapeptide fragments, the enkephalins (13). It is perhaps significant that vastly improved understanding of the biochemical basis for analgesia and the characterization of a family of related receptors (14), known as δ, κ, and μ, have so far failed to yield any better drugs for the treatment of pain.

A series of analgesics that were discovered initially in an attempt to obtain smooth muscle relaxants based on another natural product, atropine (**11**), started with the obser-

2.2 Conotoxins

Elan Pharmaceuticals is developing SNX-111 (Ziconotide), the synthetic equivalent of ω-Conopeptide-MVIIA, found in the venom of the predatory marine snail *Conus magus*, for the treatment of severe pain and ischemia by the intrathecal or intravenous routes. The peptide has the structure H-^{1}Cys-Lys-Gly-Lys-Gly-Ala-Lys-^{8}Cys-Ser-Arg-Leu-Met-Try-Asp-^{15}Cys-^{16}Cys-Thr-Gly-Ser-^{20}Cys-Arg-Ser-Gly-Lys-^{25}Cys-NH$_2$cyclic(1–16),(8–20),(15–25)-tris(disulfide), which does not make it an

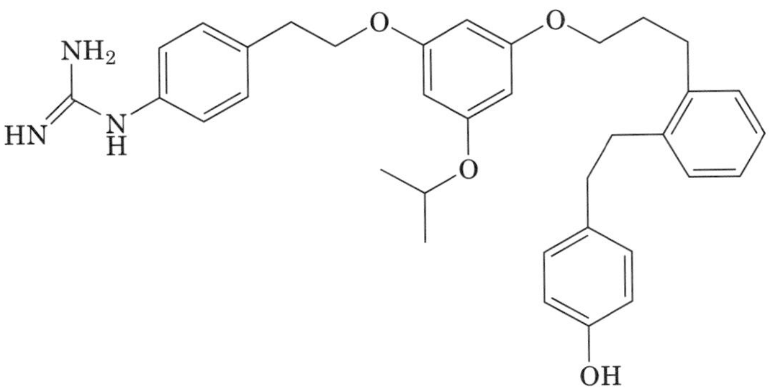

(13) conotoxin analog

easy target for synthesis and gives it poor distribution properties *in vivo* (17).

SNX-111 blocks N-type calcium channels, which are located throughout the CNS on neuronal somata, dendrites, dendritic spines, and axon terminals, where they play a major role in the regulation of the neurotransmitters associated with pain transmission and stroke. The drive is to discover an orally active, selective, small-molecule modulator of N-type calcium channels to overcome the disadvantages of administration of SNX-111.

High-throughput screening campaigns have resulted in a number of leads being identified; whereas others have chosen to modify known drugs shown to block N-type channels. Workers at Parke-Davis, however, employed a ligand-based approach using the three-dimensional solution structure of the peptide (18). Compounds such as (**13**) were designed where key binding motifs are attached to an alkylphenyl ether scaffold. The compound had an IC$_{50}$ value of 3.3 μM in a human N-type channel assay but showed no selectivity over the L-type channel. Structure-activity work on the conotoxins has shown that other regions of the peptide, absent in these synthetic ligands, are responsible for channel family selectivity (17, 18).

2.3 Cannabinoids

The plant *Cannabis sativa* has been used by humans for thousands of years, both for the effects when ingested and for making rope from the fibers in the stem. The major constituent of pharmacological interest is Δ_9-tetrahy-

drocannabinol (**14**) (THC), which has a multiplicity of actions. In animals the effects include sedation and apparent hallucinations (19), which are similar to the major effects in the CNS in humans. There are also cardiovascular effects, notably tachycardia and postural hypotension, that can be separated from the CNS action, as in the synthetic analog $\Delta_{6a,10a}$-dimethylheptylTHC (**15**), which has minimal CNS activity (20).

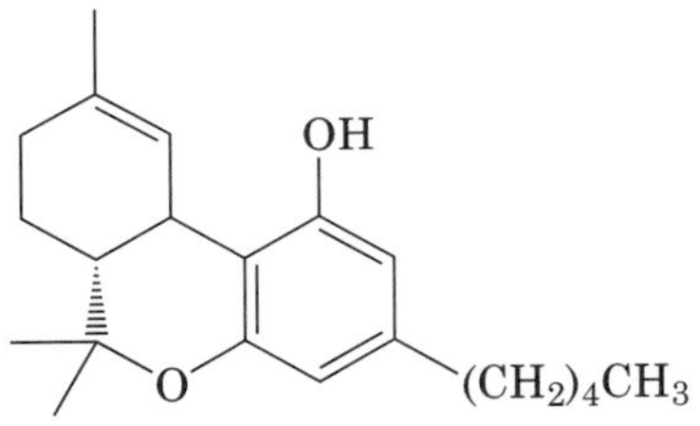

(14) THC

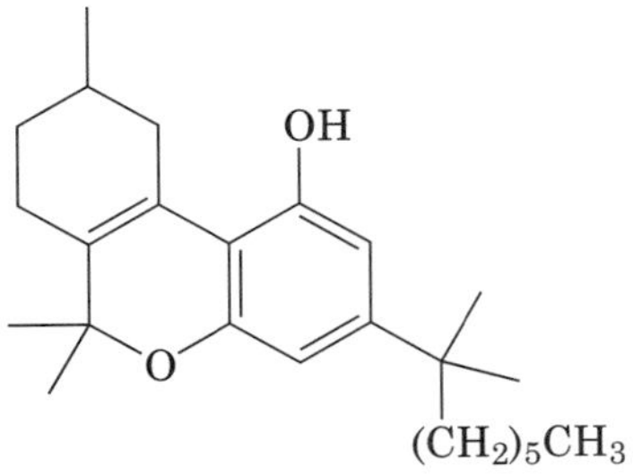

(15)

Given the widespread illicit use of *C. sativa*, it was perhaps inevitable that eventually one or two cancer patients receiving chemotherapy would dose themselves with their own sedative in the form of marijuana. An unexpected blessing from this uncontrolled combination was a reduction in the nausea experienced during chemotherapy. A variety of anticancer agents cause severe nausea and vomiting, including nitrogen mustard, adriamycin, 5-azacytidine, cyclophosphamide, and methotrexate: the unique situation arose in which the remedy was discovered by the patients themselves (21). Although smoking reefers gives rapid absorption and close control of the effects, smoking is itself carcinogenic and cannot be recommended to those who are unaccustomed to it; thus, when the physicians in charge were made aware of their patients' discovery, they devised a controlled clinical trial in which measured doses of THC were dissolved in sesame oil and administered in gelatin capsules. A placebo was similarly prepared for use in a randomized, double-blind, crossover experiment (21). The results left no doubt that a majority of patients benefited from THC pretreatment, even those who had previously been refractory to the effects of the standard antiemetics such as prochlorperazine. There remained the problem of tachycardia associated with THC treatment. The multiplicity of effects of THC have led to the synthesis of large numbers of analogs (22), particularly in the hope of finding non-morphine-like analgesics without addictiveness and without the other CNS effects of THC. The analog nabilone (**16**) had been shown to exert less effect than that of THC on the cardiovascular system, while retaining the mixture of CNS actions, including analgesic, antianxiety, and antipsychotic properties (23). When tested as an antiemetic, nabilone proved to be superior to THC (24) and has been used for this purpose for more than 30 years. The first 10 years of clinical experience was reviewed (25).

After the demonstration of THC binding sites in the CNS (26), a search for an endogenous ligand produced the long-chain ethanolamine derivative (**17**) of arachidonic acid, known as anandamide (27). Subsequently, the glycerol ester of arachidonic acid (**18**), known as 2-AG, was shown to be a more abundant endogenous ligand in the brain than anandamide (28). Further development has tended

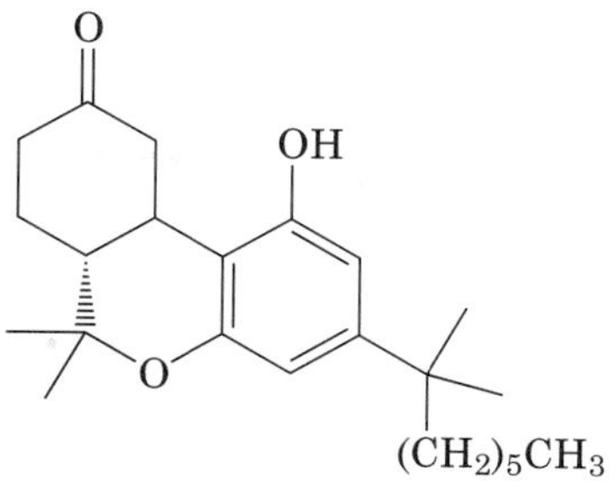

(16) nabilone

to concentrate on analogs of the natural ligands, notably the methyl derivative of anandamide (**19**), which is resistant to the amide hydrolase that terminates the action of anandamide itself and the dimethylheptyl analog (**20**) that is traceable to the earlier modifications to THC (29). Such analogs tend to have activity similar to that of THC.

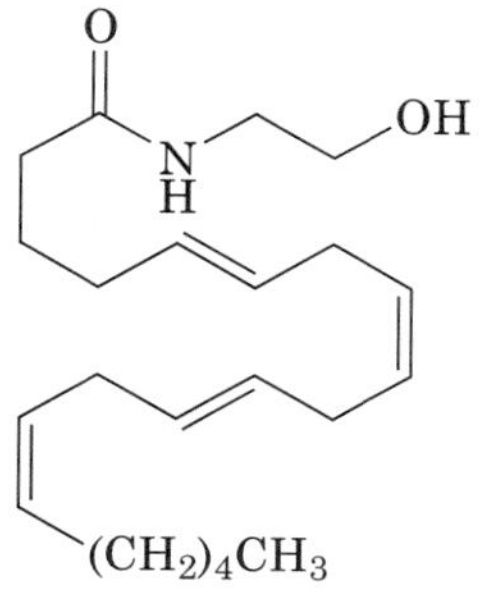

(17) anandamide

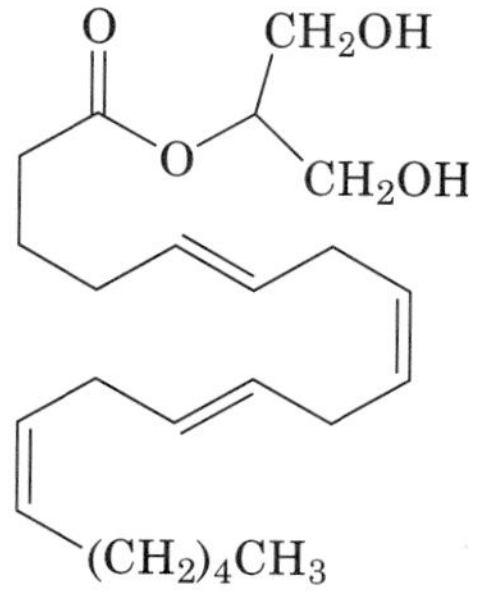

(18) 2-AG

An interesting twist in the tail is provided by the observation that anandamide is also a ligand for the so-called enigmatic vanilloid re-

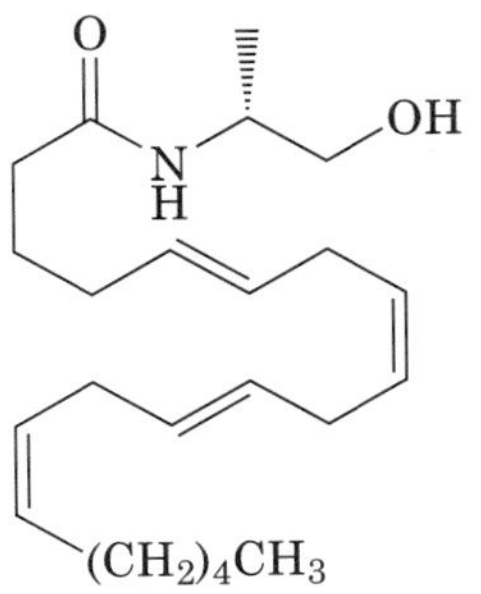

(19) R-methanandamide

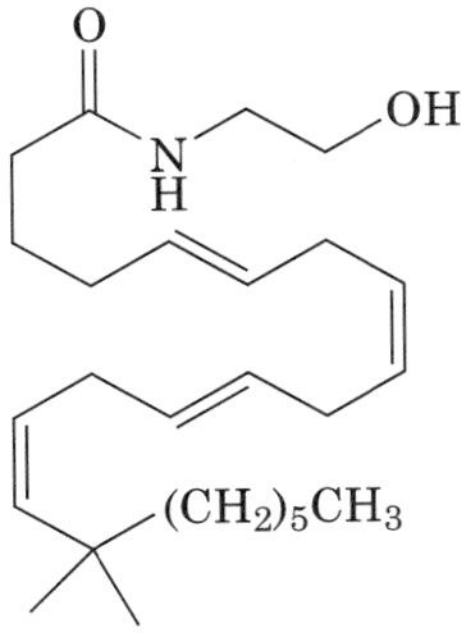

(20)

plus a hydrolase and a transport protein, interference with any or all of which might provide new drugs.

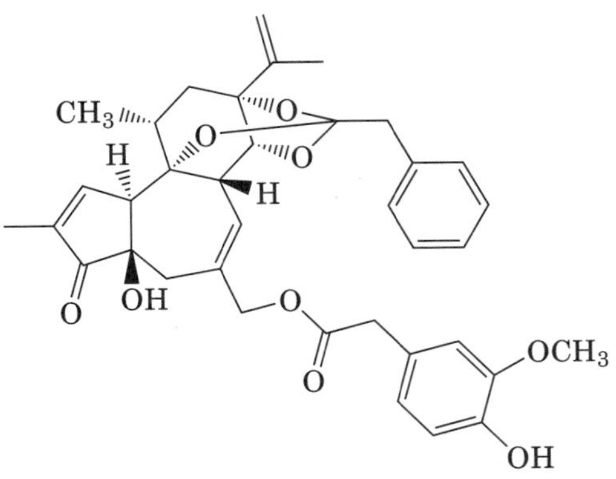

(22) resiniferatoxin

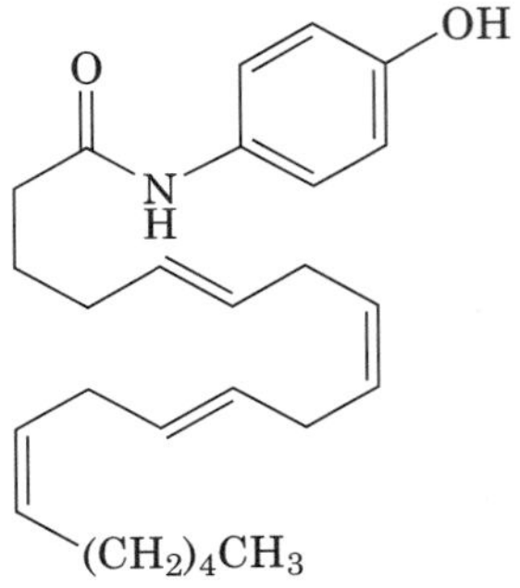

(23) AM404

ceptors, previously characterized through their interactions with two other natural products, capsaicin (**21**) and resiniferatoxin (**22**) (30) and responsible for the "hot" sensation caused by compounds in, for example, chilies. A functional vanilloid receptor was cloned in 1997 and is activated by heat and acid as well as the chemical ligands (30). A combination of the anandamide structure with a vanilloid motif, as in AM404 (**23**), enhances the anandamide transport inhibitory properties (29). The situation is complex from the viewpoint of drug design, not least because there are two cannabinoid (CB) receptors,

The cannabinoid acids, which are devoid of psychotropic activity, are promising anti-inflammatory agents (31) and it is possible that the next useful therapeutic agent will come from this direction, rather than the sought-after analgesic.

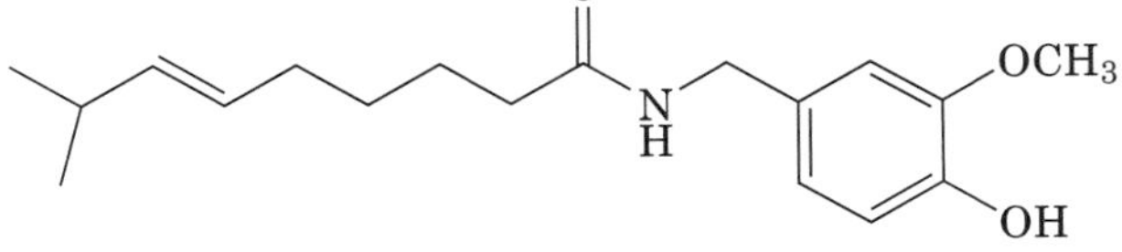

(21) capsaicin

2.4 Asperlicin

Cholecystokinin (CCK) is a peptide hormone, present in the gut and CNS; it is one of the most abundant peptides in the brain (32, 33). The whole peptide is composed of 33 amino acids, but the C-terminal octapeptide H-Asp-Tyr(SO$_3$H)-Met-Gly-Trp-Met-Asp-Phe-NH$_2$ possesses the full range of activities, sufficient for it to be classed as a neurotransmitter (34). Specific, high-affinity binding sites have been found on mammalian CNS cell membranes and in other organs such as pancreas, gall bladder, and colon (35). The latter have been classed as CCK-A receptors, but the majority of CNS receptors were classed as CCK-B, based on affinity differences for various agonists and antagonists (36). To confuse the issue slightly, the gastrin receptor in the stomach is closely related to the CCK-B (now known as the CCK$_2$) receptor (37) and is stimulated by the C-terminal tetrapeptide of CCK: in the periphery, gastrin receptors are the same as CCK$_2$ receptors (38).

The effects of CCK on intestinal smooth muscle and pancreas are easy to demonstrate pharmacologically, unlike the role in the CNS, which is a matter for conjecture. It was assumed that the CNS activity must be significant, given the abundance of the peptide in the brain, and that the discovery of antagonists might lead to new drug treatments, as yet unspecified (39).

Fishing in microbial broths, using radioreceptors as bait, produced asperlicin (24), the first potent, competitive and selective CCK-A (CCK$_1$) antagonist, from a culture medium of *Aspergillus alliaceus* (40).

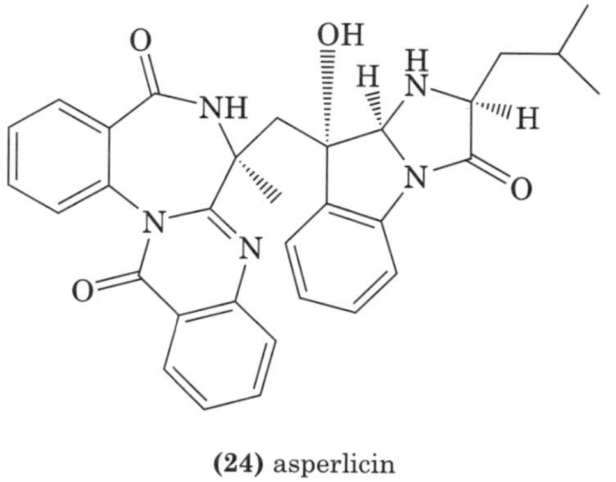

Asperlicin is moderately potent, poorly soluble in water, and not bioavailable by the oral route (41). When discovered it was also, with morphine, one of the very few nonpeptides with affinity for a peptide receptor (peptoids are discounted in this assessment). It was an interesting target for synthetic modification, particularly viewed as a benzodiazepine derivative with potential CNS activity.

Based on the benzodiazepine nucleus, and an overt mimic of diazepam, one of the first successful synthetic analogs was L-364,286 (25), which had potency on CCK-A receptors similar to that of asperlicin. Better receptor affinity was achieved with 3-amide–substituted benzazepines: the 2-indolyl derivative L-364,718, also known as MK-329 (26), is five orders of magnitude more potent than asperlicin (42) at CCK-A receptors and is a valuable pharmacological tool.

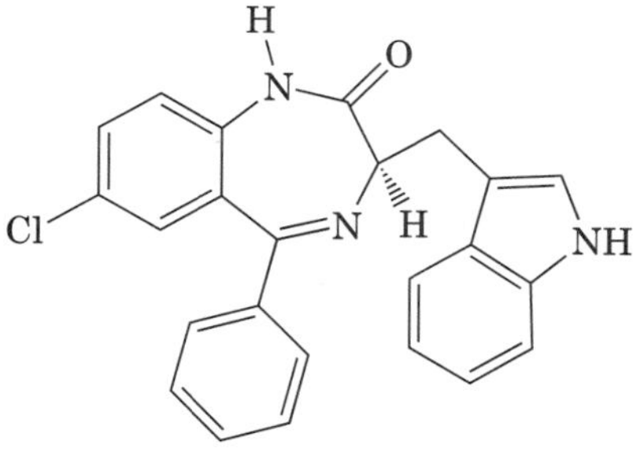

(25)

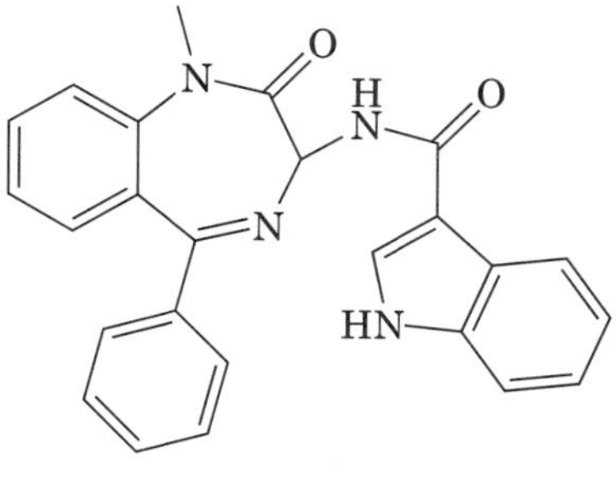

(26)

Modification of the 3-amide to give a urea linkage as in (27) led to a reduction in CCK-A receptor affinity. Importantly, discrimination between CCK-A and CCK-B receptors by (27)

(24) asperlicin

is governed by the stereochemistry at C3, the (S)-enantiomer showing greater affinity for CCK-A receptors. The (R)-enantiomer, known as L-365,260, prefers CCK-B receptors, antagonizes gastrin-stimulated acid secretion in animal models, and, among other CNS effects, induces analgesia in primates and displays anxiolytic properties (32).

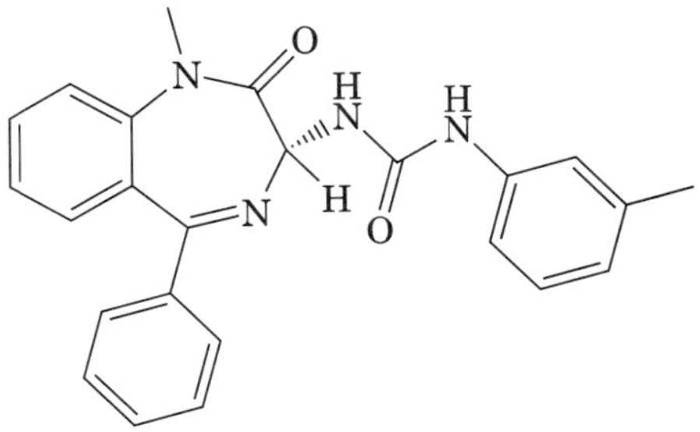

(27)

Further development in this series has very substantially improved receptor affinity: YM-022 (28) has IC_{50} 0.05 nM/kg (38). Clinical trials of compounds in this series have been disappointing because of poor bioavailability, but the general concept of finding a therapeutic agent through antagonism of CCK_2 receptors is still viable and it is reported that the number of patents in this area has increased in the last 5 years (43).

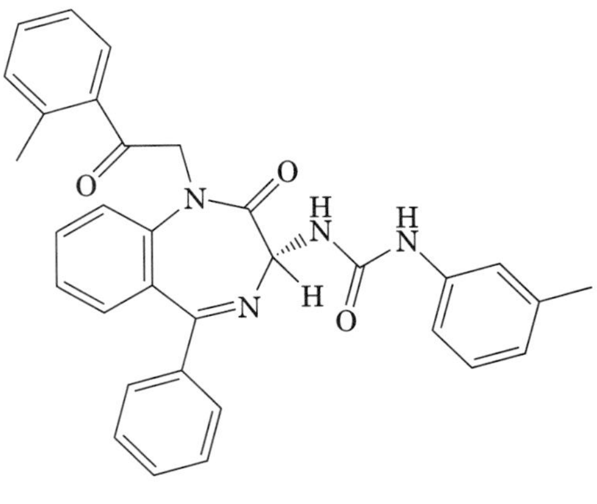

(28) YM-022

3 NEUROMUSCULAR BLOCKING DRUGS

3.1 Curare, Decamethonium, and Atracurium

The development and use of muscle relaxants, to allow a reduction in the level of anesthesia during surgery, follows entirely from studies of South American arrow poisons (44) and particularly from the isolation by King (45) of pure D-tubocurarine (29) in the 1930s, from tube curare. Another of the South American blowpipe poisons, calabash curare, was used for similar purposes and developed (46, 47), to give alcuronium (30) from the alkaloid C-toxiferine 1 (31). Both types of curare paralyze skeletal muscle by a similar mechanism, antagonizing the effect of acetylcholine at the neuromuscular junction (48).

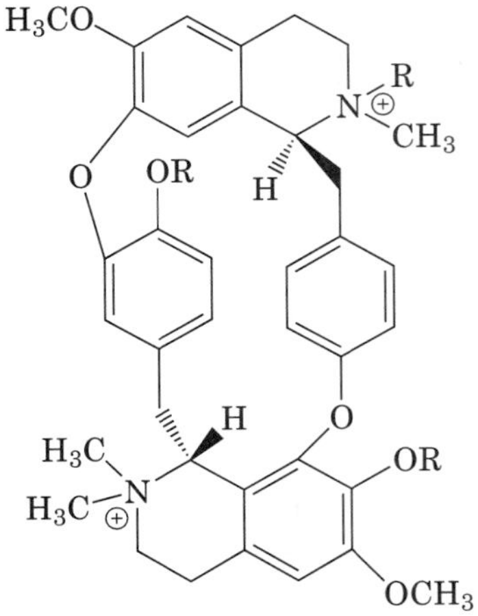

(29) tubocurarine R = H
(32) metocurine R = CH_3

The muscle-paralyzing curare alkaloids are quaternary salts that are not absorbed when taken orally. For surgical procedures they must be administered by intravenous injection, which results in onset of paralysis in at most a few minutes: anesthesia is normally induced before administration of the muscle relaxant (44), which is followed by artificial respiration. Although the neuromuscular blocking agents are potentially lethal when administered alone, in the environment of an operating theater they are truly life-saving

(31) C-toxiferine 1 R = CH$_3$
(30) alcuronium R =CH$_2$CH=CH$_2$

drugs that have made a major impact on survival rates during surgery.

At the time of King's work in the 1930s there were no spectroscopic aids to structure elucidation, and it is not surprising that he made a small error in the structure assigned to D-tubocurarine, believing it to have two quaternary nitrogens, a mistake that was not corrected (49) until 1970. The methylation product of D-tubocurarine, known as metocurine **(32)** is a more potent muscle relaxant. It was known for a long time as dimethyltubocurarine because of the error in the structure allocated to compound **(29)**. King's error, in assigning a bisquaternary structure to a molecule with one quaternary and one protonated tertiary nitrogen, led to a large number of highly active synthetic bisquaternaries. The simplest of these was decamethonium **(33)**, which was nothing more than two trimethylammonium end groups connected with a decamethylene chain. As one of a series with different chain lengths (50), decamethonium became the prototype for many more complex structures with 10 atoms between the quaternary centers, which appeared to be optimal for

binding to the acetylcholine receptor at the neuromuscular junction.

Unlike tubocurarine, decamethonium depolarizes the muscle endplate, rendering the membrane insensitive to acetylcholine (48). The action of tubocurarine is competitive and can be overcome with increased concentrations of acetylcholine, brought about by administration of an anticholinesterase: the latter is thus an antidote to tubocurarine, but not to decamethonium. Despite the lack of an antidote, decamethonium was used very widely for over two decades. One of its disadvantages is an overlong duration of action, during which time the patient has to be maintained on artificial respiration, because the muscle of the diaphragm is also susceptible to the actions of the drug. An early and highly successful attempt (51) to shorten the action of decamethonium gave suxamethonium **(34)**, a diester formed between succinic acid and two molecules of choline, which hydrolyzes rapidly in the presence of pseudocholinesterase.

Tubocurarine suffers from cardiovascular side effects induced by direct interactions with ganglionic acetylcholine receptors and from stimulation of histamine release, so analogs have been well worth pursuing. The macrocyclic structure of tubocurarine is a difficult synthetic target, but fortunately ring-opened analogs, such as laudexium **(35)**, have high potency and relatively few side effects (52). The main problem with **(35)** is the duration of action, which at about 40 min is too long for many operations. Two approaches have been used to shorten the duration of action. The concept of pH-controlled Hofmann elimination was employed successfully (53) in the design of atracurium **(36)**, which in clinical use (54) has the big advantage that the drug disappears at a constant rate, irrespective of liver or kidney function. Some ester hydrolysis contributes to the destruction of atracurium *in vivo*, as might be expected. A slightly later development (55) centered on an empirical search for structures that would undergo ester hydrolysis more rapidly, resulting in mivacurium **(37)**, which has a slightly shorter duration of action than that of atracurium, the latter being about 15–20 min.

(33) decamethonium

decomposition of suxamethonium (34)

(35) laudexium

4 ANTICANCER DRUGS

4.1 Catharanthus (Vinca) Alkaloids

In 1949 Canadian researchers at the University of Western Ontario began investigating the medicinal properties of the rosy periwinkle (*Catharanthus roseus*), a plant that had been used for many years to treat diabetes mellitus in the West Indies. Despite finding that the plant extract when given orally had no effect on blood sugar levels in rats or rabbits, the researchers noted that when given intravenously, the extract caused the animals to succumb to bacterial infection and die. This curious observation prompted further studies, which showed that the plant extract reduced levels of white blood cells, causing granulocytopenia and bone marrow damage, toxic effects that are encountered with many antitumor drugs (56). These findings led the Canadian group to isolate an alkaloid fraction with potent cytotoxic activity. The active principle was eventually purified and became known as vinblastine (38), a dimeric indoledihydroindole alkaloid.

Concurrently, researchers at the Lilly Research Laboratories had been investigating extracts of *C. roseus* and they too had detected cytotoxic activity, specifically against acute lymphocytic leukemia (57, 58). The U.S. group isolated several alkaloids, including vinblastine and another closely related alkaloid, vincristine (39).

Although many other alkaloids have been isolated from *C. roseus*, only vinblastine and vincristine have been developed for clinical use. The antiproliferative activity of the two compounds is related to their specific interaction with tubulin, thus preventing assembly of tubulin into microtubules and arresting cell division (59). However, despite this apparent identical mechanism of action and their clear chemical similarities, vinblastine and vincristine display very different clinical effects. Vinblastine, for example, is used to treat Hodgkin's disease and metastatic testicular tumors, whereas vincristine is used mainly in combination with other anticancer drugs for the treatment of acute lymphocytic leukemia in children. Toxicity profiles are also different, in that vinblastine causes bone-marrow depression, whereas peripheral neuropathy often proves to be dose-limiting in vincristine therapy.

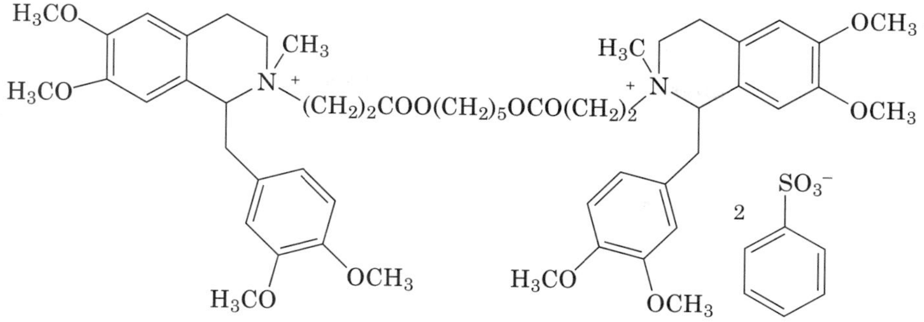

(36) atracurium

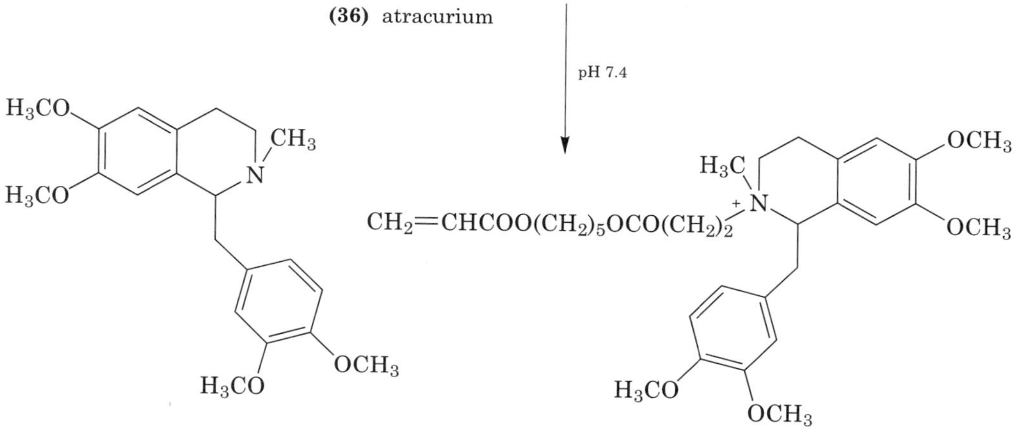

pH 7.4

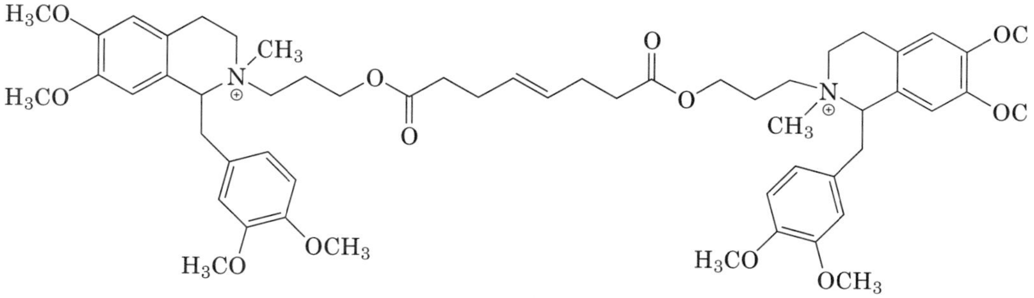

$$CH_2{=}CHCOO(CH_2)_5OCO(CH_2)_2^+$$

(37) mivacurium

Lilly introduced vinblastine and vincristine into the clinic in 1960 and 1963, respectively, but this did not preclude the search for improved derivatives. A chemical modification program aimed at improving antitumor activity and reducing toxicity was initiated in 1972 (60). Concern about the neurotoxicity displayed by vincristine, its chemical instability, and low natural abundance (0.03 g/kg dried plant material) led to vinblastine's being chosen as a template for semisynthetic modifica-

tion. Selective ammonolysis of the ester function at C-3 and hydrolysis of the adjacent acetyl group yielded the desacetyl vinblastine amide, vindesine (**40**). Better yields of vindesine were obtained from the hydrazide (**41**) on treatment with nitrous acid and reacting the resultant azide (**42**) with ammonia. The azide (**42**) proved to be a useful intermediate for the preparation of a range of substituted amides, although vindesine proved to be the derivative of choice, with significant differ-

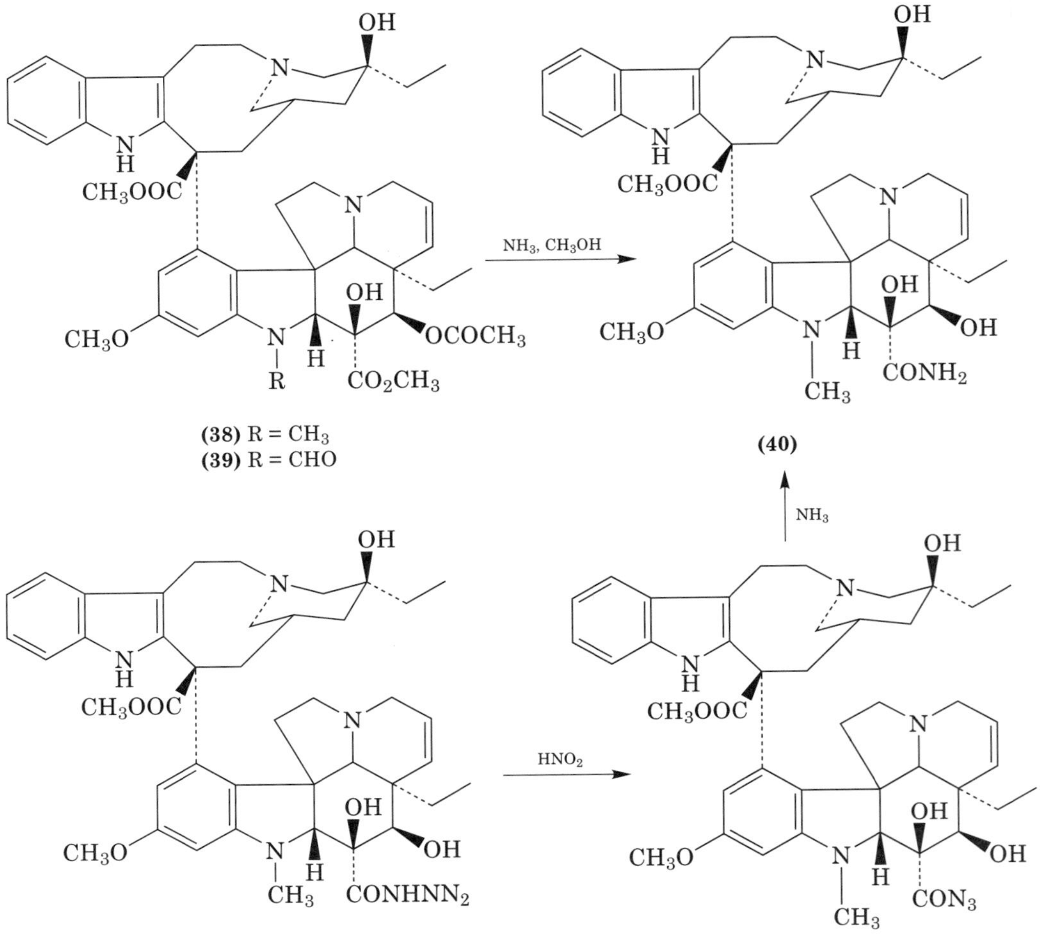

(38) R = CH$_3$
(39) R = CHO

NH$_3$, CH$_3$OH

(40)

NH$_3$

CH$_3$OOC ... CONHNN$_2$

(41)

HNO$_2$

CH$_3$OOC ... CON$_3$

(42)

ences in the spectrum of antitumor activity and toxicity compared to that of the naturally occurring alkaloids. Phase I clinical trials commenced in 1977 and vindesine has been used for the treatment of non-small cell lung cancer, lymphoblastic leukemia, and non-Hodgkin's lymphomas. In combination with cisplatin, vindesine ranks among the foremost treatments for non-small cell lung cancer with respect to response rate and survival (61). Back in the 1950s, the U.S. researchers could not have guessed that 30 years on, the demand for Catharanthus alkaloids would necessitate the processing of around 8000 kg of plant material per year (62)!

4.2 Camptothecin

Camptothecin (**43**) was first isolated by Monroe Wall and Mansukh Wani in 1966, after ethanolic extracts of *Camptotheca acuminata*, a tree native to China, showed unusual and potent antitumor activity (63). Starting with 19 kg of dried wood and bark, Wall and Wani painstakingly purified the principal active component with a combination of hot solvent extraction, an 11-stage Craig countercurrent partition process, silica gel chromatography, and crystallization. Camptothecin was characterized as a novel pentacyclic alkaloid, present as just 0.01% w/w of the stem bark of *C. acumi-*

nata. Of particular note was the unusual activity that camptothecin displayed in L1210 and P388 mouse leukemia life-prolongation assays. The compound also inhibited the growth of solid tumors *in vivo* and the water-soluble sodium salt was progressed to phase II clinical trials before being withdrawn because of severe bladder toxicity.

(43) Camptothecin: $R^1 = R^2 = R^3 = H$
(44) 10-hydroxycamptothecin: $R^1 = R^2 = H$, $R^3 = OH$
(48) 7-ethyl-10-hydroxycamptothecin: $R^1 = C_2H_5$, $R^2 = H$, $R^3 = OH$
(46) Topotecan: $R^1 = H$, $R^2 = CH_2—N(CH_3)_2$, $R^3 = OH$

Interest in camptothecin gained new impetus in 1985, when it was discovered that the compound exerts its antitumor activity through a novel mechanism of action (64). Camptothecin binds to the covalent complex formed by topoisomerase I and DNA, which initiates DNA replication and thus stabilizes the enzyme–DNA complex and prevents cell proliferation. The elucidation of the mechanism of action provided a means of evaluating camptothecin analogs as topoisomerase inhibitors *in vitro* and efforts then focused on synthesizing water-soluble analogs with broad-spectrum antitumor activities. The α-hydroxy lactone (ring E) and, in particular, the 20(*S*)-form proved essential for maintaining biolog-

ical activity, but the 10-hydroxy analog (44) showed greater activity than that of (43) (65). Wall and Wani successfully deployed the Friedlander reaction between substituted 2-aminobenzaldehydes and the tricyclic intermediate (45), to synthesize a variety of ring-A–substituted analogs. These studies may have prompted SmithKline Beecham (now GlaxoSmithKline) to synthesize the water-soluble 10-hydroxycamptothecin analog topotecan (46) that was first approved in 1996 for the treatment of recurrent ovarian cancer and, 2 years later, for small cell lung cancer (66). Irinotecan (47), developed by Daiichi and Yakult Honsha in Japan and marketed by Pharmacia, was also approved in 1996 for the treatment of advanced colorectal cancer. Irinotecan is inactive as a topoisomerase I inhibitor, but acts as a prodrug of the active 7-ethyl-10-hydroxy-camptothecin (48) (67).

(47) Irinotecan: $R^1 = C_2H_5$, $R^2 = H$,

$R^3 = O—C—N \qquad N$
$\qquad \quad \| \\ \qquad \quad O$

4.3 Paclitaxel and Docetaxel

Regarded as the tree of death by the Greeks and used to prepare arrow poison by the Celts, the yew tree has been associated with death and poisoning for centuries (68, 69). The English yew, *Taxus baccata*, was used to make funeral wreaths and it was believed that one could die by merely standing beneath the boughs of the tree.

Yew certainly contains highly toxic metabolites and their potency and fast duration of action has often made extracts of yew the poison of choice for numerous murders and suicide attempts. It is thus ironic that extracts from the Pacific yew, *T. brevifolia*, after being

(45)

tested in the National Cancer Institute's (NCI) screening program during the 1960s, yielded what was described (70) as the most exciting anticancer compound discovered in the previous 20 years; that is, paclitaxel (**49**) (originally given the name taxol by Wall and Wani).

(49)

The initial isolation and characterization of paclitaxel proved particularly difficult because of (*1*) its very low natural abundance in *T. brevifolia* bark (although this was the best known source, the isolated yield was only 0.02% w/w, equivalent to 650 mg per tree), (*2*) the poor analytical data obtained from the purified compound, and (*3*) the failure of paclitaxel to give crystals that were suitable for X-ray analysis (71). The structure of paclitaxel was published in 1971 (72), but further biological testing continued to be troubled by difficulties. The compound showed only modest *in vivo* activity in various leukemia assays, which was no better than that displayed by a number of other new compounds at the time. In addition to the limited supplies of paclitaxel (the complexity of the molecule precluded chemical synthesis), the compound was very poorly soluble in water, which made formulation difficult. However, various new assays were developed in the 1970s, including the murine B16 melanoma model, in which paclitaxel showed very good activity, and another boost came when Horwitz et al. (73) discovered that the compound prevented cell division by a unique mode of action. In contrast to the antimitotic vinblastine and podophyllotoxin analogs (q.v.), which prevent microtubule assembly, paclitaxel inhibits cell division by promoting assembly of stable microtubule bundles, which leads to cell death.

Phase I clinical trials were initiated in 1983, but these were to proceed at a slow and tortuous pace and proved all but disastrous when the high levels of oil-based adjuvant used to formulate paclitaxel caused severe allergic reactions in many volunteers. Undaunted by the formulation problem and spurred on by paclitaxel's novel mechanism of action, clinicians were able eventually to minimize the allergic events and demonstrate useful activity. Phase II clinical trials began in 1985 despite continuing supply problems, and 4 years later the program received a significant boost when McGuire et al. (74) reported good responses from patients suffering from refractory ovarian cancer, a disease that kills some 12,500 women a year in the United States alone.

In many ways, the development of paclitaxel mirrored that of the camptothecin analogs, both being dogged for many years by supply issues, poor pharmacokinetics, and toxicity, but the subsequent uncovering of novel mechanisms of action fueled renewed efforts to develop these leads into important new anticancer agents (75).

In 1991 Bristol-Myers Squibb in conjunction with the NCI agreed to manage the supplies of paclitaxel and were granted a licence to further develop the compound. The following year the U.S. Federal Drug Administration approved paclitaxel for the treatment of ovarian cancer in patients unresponsive to standard treatments, and in December 1993 approval was given for the treatment of metastatic breast cancer.

The sourcing of paclitaxel from *T. brevifolia* was a major problem (76) because to treat just the groups of patients suffering ovarian cancer in the United States would require about 25 kg of compound per year, necessitating the felling of some 38,000 trees (70)! Although the Pacific yew is not a rare tree, it is extremely slow growing and such harvesting could not be sustained indefinitely. It has been estimated that there were enough trees available to maintain a supply of paclitaxel for only 2–7 years (77). The isolation of paclitaxel from other Taxus species has been investigated at length and reasonable quantities have been obtained from the needles of several species including *T. baccata*. Using the needles has alleviated the supply problem because they can be harvested without damaging the tree. However, the needles contain much higher quantities of several biosynthetic precursors of paclitaxel and two of these, baccatin III (50) and 10-desacetylbaccatin III (51) have been used to prepare paclitaxel semisynthetically. One approach, developed by Potier et al. (78), involved acylation of the sterically hindered C-13 position of baccatin III with cinnamic acid and subsequent double-bond functionalization through hydroxyamination, to give paclitaxel together with various regio- and stereoisomers. A better approach involved protection of 10-desacetylbaccatin III as the triethylsilyl ether, followed by direct acylation with the phenylisoserine derivative (52), giving paclitaxel in 38% overall yield (79). Further improvements were made using less sterically demanding acylating reagents; for example, acylation with the β-lactam (53) gave paclitaxel in up to 90% yield (80) and this may be the preferred method for commercial production in the future.

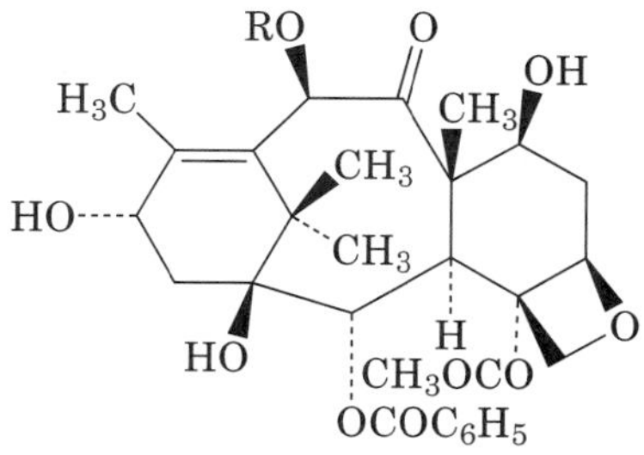

the C-13 ester side-chain can be tolerated. Thus, the N-t-(butoxycarbonyl) derivative, docetaxel (54), which appears to be more potent than paclitaxel (81) and has better solubility characteristics, has been developed and launched by Aventis for the treatment of ovarian, breast, and lung cancers.

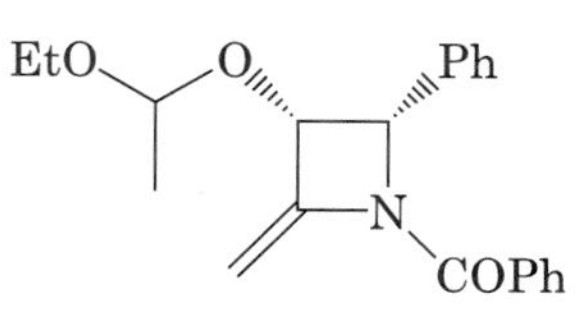

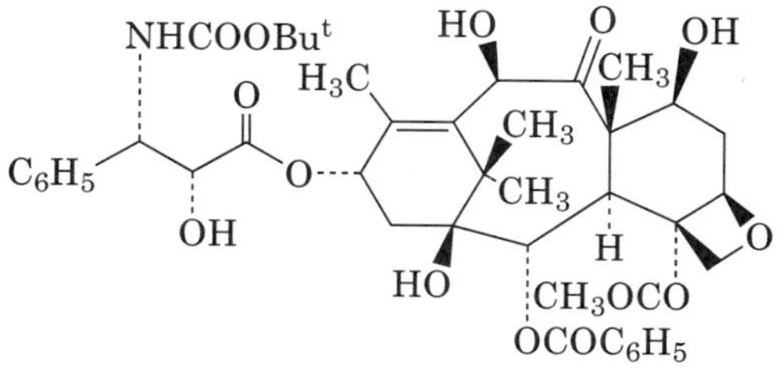

(54)

Various "protaxols," designed to release paclitaxel in situ under physiological conditions, have been prepared by acylating the C-2′ hydroxyl group. Nicolaou et al. (82) reported the synthesis of the sulfone (55), which is soluble and stable in aqueous media, but is able to release paclitaxel rapidly in human blood plasma.

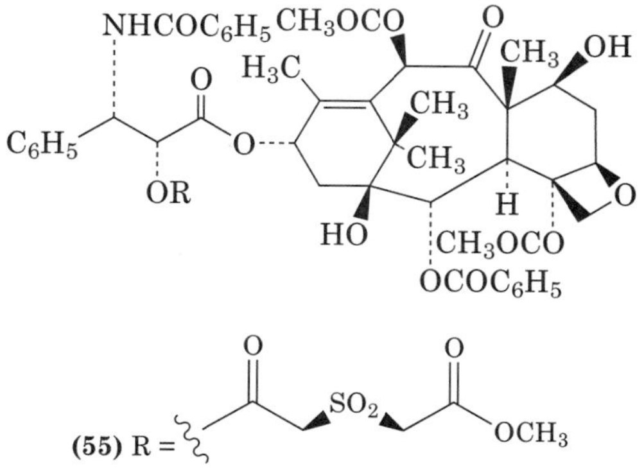

(50) R = COCH₃
(51) R = H

(53)

(55) R =

These semisynthetic approaches also provide access to analogs with potential advantages over paclitaxel itself. Structure-activity studies have shown that, although the oxetane ring appears to be essential for activity, wide variation in the nature and stereochemistry of

Plant tissue culture (70), microbial fermentation (83), and total synthesis (84, 85) provide other possibilities for the production of paclitaxel and its derivatives, although it is far from certain whether any of them will be commercially viable.

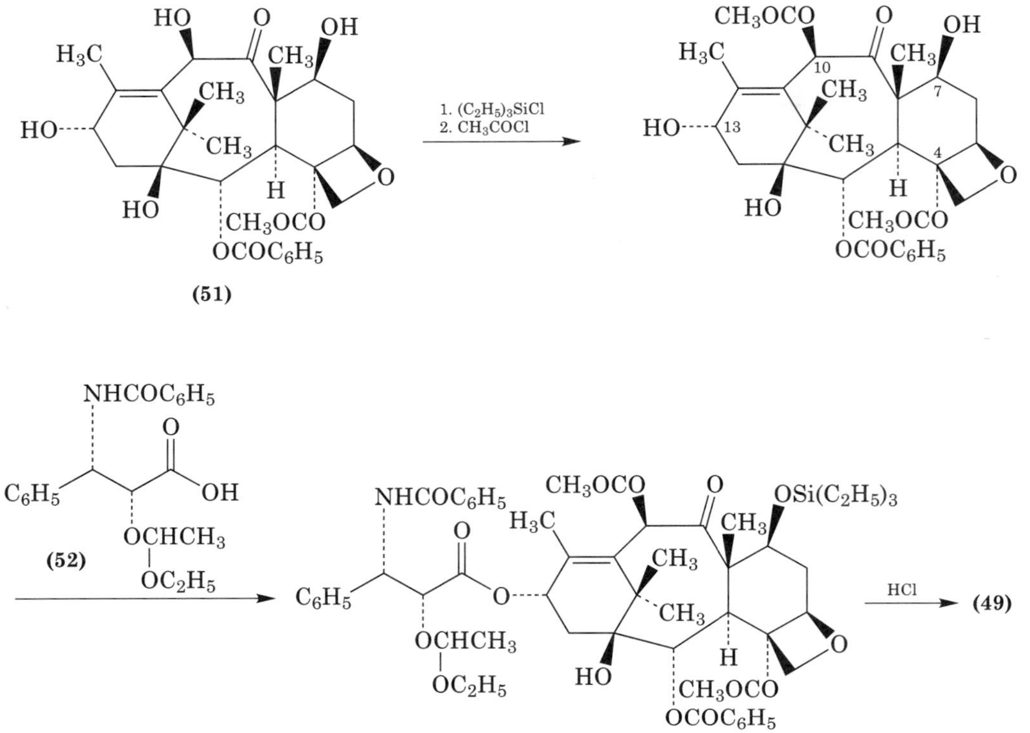

(51)

(52)

4.4 Epothilones

Epothilones A (56) and B (57), 16-membered macrocyclic polyketide lactones, were first isolated from the cellulose-degrading myxobacterium *Sorangium cellulosum* by Hoefle, Reichenbach, and coworkers (86) as narrow-spectrum antifungal and cytotoxic metabolites. The compounds were then tested by the National Cancer Institute in the United States and found to be highly active against breast and colon cancer cell lines (87). Subsequently,

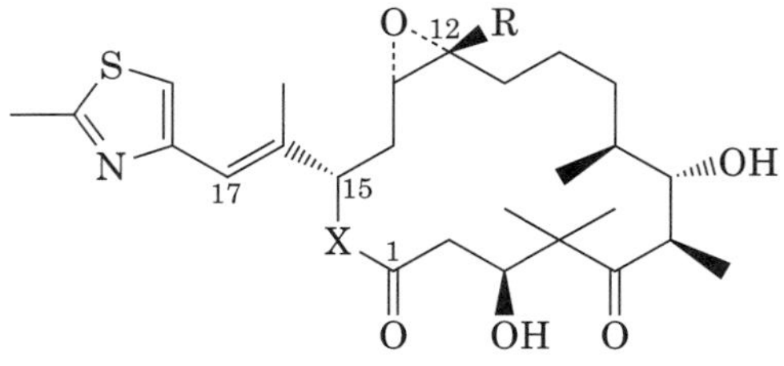

(56) epothilone A: X = O, R = H
(57) epothilone B: X = O, R = CH₃
(59) BMS 247550: X = NH, R = CH₃

Bollag et al. (88) at the Merck Research Laboratories discovered that the epothilones stabilize microtubule assembly and thus inhibit cell division by the same mechanism as that of paclitaxel (see above). This observation, together with their less complex chemical structure, increased water solubility, more rapid action *in vitro*, and effectiveness against multidrug-resistant tumor cells, has prompted significant interest in the epothilones as anticancer agents.

On learning the absolute stereochemistry of (56) and (57), three academic research groups embarked on the total synthesis of the epothilones. Nicolaou, Danishefsky, and Schinzer independently adopted successful, elegant synthetic approaches involving olefin metathesis, macrolactonization, Suzuki coupling, or ester–enolate–aldehyde condensation (89). Within 3 years of the disclosure of their absolute stereochemistry, 17 different total syntheses of the natural products were reported. These syntheses paved the way for the generation of a large number of epothilone analogs for biological evaluation, including the use of solid-phase combinatorial approaches.

The academic groups focused on modifications around the core macrocyclic lactone, establishing important structure–activity relationships, but not improving on the *in vitro* biological activity of the most active natural product, epothilone B (**57**). *In vivo* biological data were comparatively scarce and, although one group reported that epothilones B (**57**) and D (**58**) showed activity in murine tumor models, researchers at Bristol-Myers Squibb have shown that (**58**) lacks *in vivo* activity as a result of rapid metabolic inactivation (90). It was postulated that esterase-mediated hydrolysis of the macrocyclic lactone formed an inactive ring-opened species and, therefore, efforts were focused on replacing the lactone with a more stable macrocyclic lactam moiety. Several macrocyclic lactam derivatives were synthesized from (**57**) and (**58**). Of note was the preparation of BMS-247550 (**59**) in a three-step synthesis from epothilone B (**57**), utilizing a novel Pd(0)-catalyzed ring-opening reaction followed by reduction and macrolactamization. BMS-247550 (**59**), which is in phase I clinical trials, retains its activity against human cancer cells that are naturally insensitive to paclitaxel or that have developed resistance to paclitaxel, both *in vitro* and *in vivo* (91).

(**58**) epothilone D

4.5 Podophyllotoxin, Etoposide, and Teniposide

The development of the natural constituents of Podophyllum Resin into effective semisynthetic and, ultimately, totally synthetic compounds for the treatment of various kinds of cancer provides one of the most sustained and intriguing stories of drug discovery (92, 93).

The story has all the classic ingredients, starting with observation and reasoning, extending through chance into new areas, and characterized throughout by persistence and determination, particularly when biological activity had to be traced to very minor constituents in the crude plant extract.

Podophyllum peltatum (may apple, or American mandrake) and *P. emodi* are, respectively, American and Himalayan plants, widely separated geographically but used in both places as cathartics in folk medicine (94). An alcoholic extract of the rhizome known as podophyllin was included in many pharmacopoeias for its gastrointestinal effects; it was included in the U.S.P., for example, from 1820 to 1942. At about this time the beneficial effect of podophyllin, applied topically to benign tumors known as condylomata acuminata, was demonstrated clinically (95). This usage was not inspirational, given that there are records of topical application in the treatment of cancer by the Penobscot Indians of Maine and, subsequently, by various medical practitioners in the United States from the 19th century (96). The crude resinous podophyllin is an irritant and unpleasant mixture unsuited to systemic administration.

The first chemical constituent was isolated from podophyllin in 1880 and named podophyllotoxin (97). A structure was proposed in 1932 and after some fine-tuning (98) was shown to be the lignan (**60**). As might be expected, the crude resin contains a variety of chemical types, including the flavonols quercetin and kaempferol (99). Although these other constituents undoubtedly have biological activity, it is the lignans that have received most attention and to which we shall devote the remainder of this section.

Chemists at Sandoz in the early 1950s reasoned that crude podophyllin might contain lignan glycosides with anticancer activity, which might be more water soluble and less toxic than podophyllotoxin (92). The reasoning for the latter is not entirely clear, but in the event they proved to be correct in both respects. Careful isolation gave podophyllotoxin β-D-glucopyranoside (**61**) its 4'-desmethyl analog (**62**) and some less important lignans lacking the B-ring hydroxy group (100–102). Unfortunately, the sugar deriva-

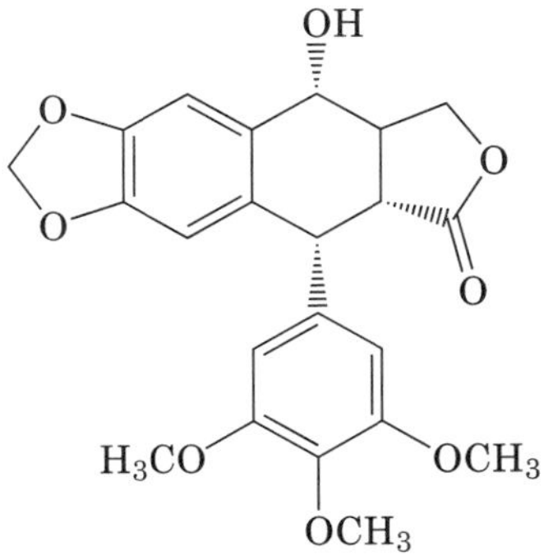

(60) podophyllotoxin

tives were less active as inhibitors of cell proliferation than were the aglycones, as well as less toxic; however, as expected, they were much more water soluble (92). While continuing work to isolate more natural lignans, a substantial program of structural modification of the known compounds was undertaken, with a view to protecting the glucosides from hydrolytic enzymes and also to improve cellular uptake. Most of these changes were ineffective: the per-acylated derivatives, for example, were insoluble in water and had inferior cytostatic effects (103).

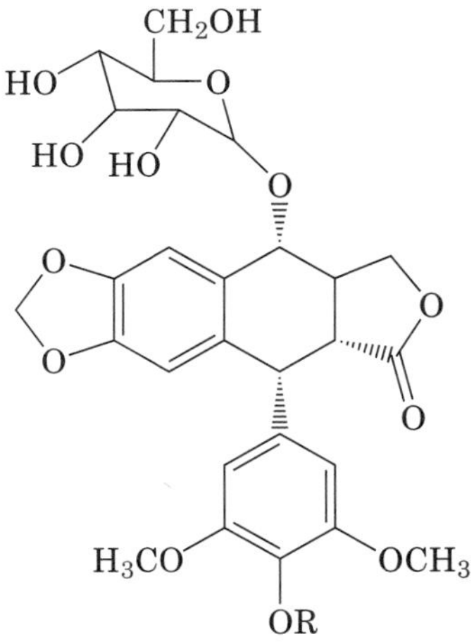

(61) R = CH$_3$
(62) R = H

Condensation of the glucosides with a variety of aldehydes was more useful, in that not all the hydroxy groups were blocked. Despite

this, water solubility was a problem with the podophyllotoxin derivatives (63). Gastrointestinal absorption was greatly improved, however, as was chemical stability (104), and positive effects were observed in a few cancer patients with the benzylidene derivative (64). It was at this point that luck played a hand, backed up by a good deal of determination. A crude podophyllin fraction, which was simpler and cheaper to prepare than pure podophyllin glucoside, was also treated with benzaldehyde to give a mixture of benzylidene derivatives, about 80% of which was compound (64). The crude product was found to be more potent than compound (64) and subsequently to possess a different mode of action from that of the lead compounds: rather than arresting cells in metaphase, cells were prevented from entering mitosis altogether (105). The crude mixture was marketed for cancer treatment as Proresid.

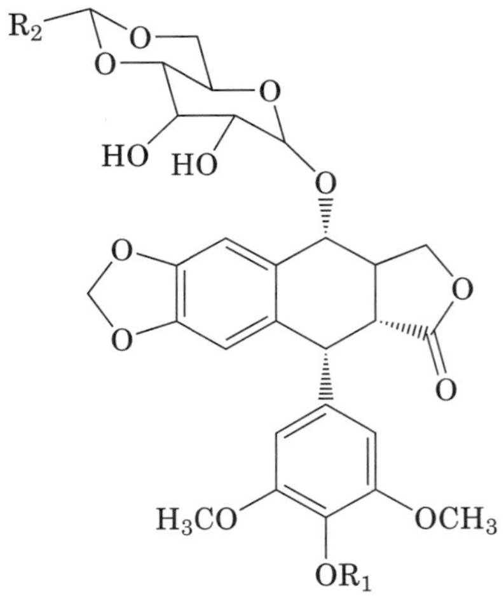

(63) R$_1$ = H,CH$_3$ R$_2$ = various alkyl, aryl
(64) R$_1$ = CH$_3$ R$_2$ = C$_6$H$_5$

Improved biological assay methods (106) indicated the presence of an unknown, highly active constituent of Proresid. For example, Proresid prolonged the life of mice inoculated with L1210 leukemia cells (93), an effect that was not observed with the known major constituent. In the early 1960s chromatographic and spectroscopic techniques were not as highly developed as they are now and more than 2 years' work was required to isolate and identify the unknown component of the mixture, which proved to be the 4′-desmethoxy-1-epi analog (**65**) of the podophyllotoxin glucoside adduct (92). Present only in very small amounts in the derivatized extract, it was necessary to devise a synthesis from readily available materials. It was fortunate that the desired 1β configuration was readily secured from 1α-hydroxy-4′-desmethylpodophyllotoxin, itself obtained by selective demethylation of podophyllotoxin: the remainder of the synthesis would now be considered fairly routine (107).

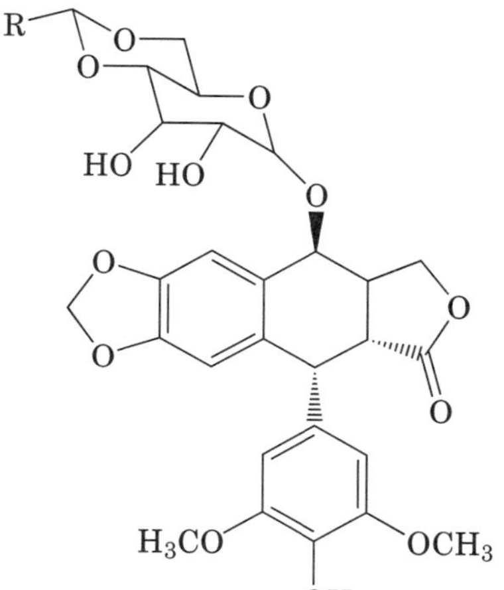

(65) R = C₆H₅
(67) R = 2-thienyl
(68) R = CH₃

Given a large supply of the key intermediate (**66**), it was straightforward to prepare a number of aldehyde derivatives, resulting in analogs with up to a 1000-fold increase in potency (108). The selected adducts were those prepared from thiophen-2-aldehyde, giving teniposide (**67**), and from acetaldehyde, giving etoposide (**68**). Both drugs are of value, etoposide in the treatment of small-cell lung cancer and testicular cancer, teniposide in the treatment of lymphomas and leukemias. The thiophene derivative is also of use in the treatment of brain tumors (93).

The natural products, podophyllotoxin and its congeners, are "spindle poisons" that inhibit cell proliferation by binding to tubulin

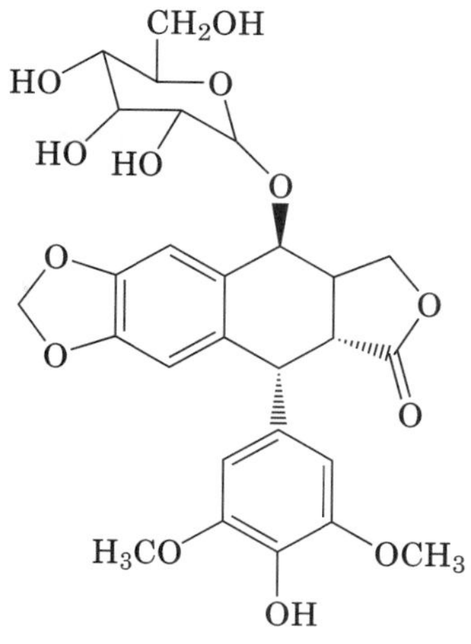

(66)

and preventing formation of microtubules (105). Presumably this effect is sufficient to account for the success of podophyllin in the treatment of condylomata acuminata, although the crude extract contains many other candidates for a contribution to the biological activity. As has been described, a very minor component of the natural mixture, missing the 4′ hydroxy group, having the 1β- instead of the 1α-hydroxy configuration and with this hydroxy group conjugated with β-D-glucose, must be treated with an aldehyde to produce the highly active and most important derivatives. These derivatives do not bind to tubulin, but have been shown to be inhibitors of topoisomerase II, which may account for most of the observed biological effects, including DNA strand breaks, that lead to anticancer activity (109).

4.6 Marine Sources

Cytosine arabinoside (**69**), a synthetic analog of the C-nucleosides spongouridine (**70**) and spongothymidine (**71**) from the sea sponge *Cryptotheca cripta*, was the first and, so far, the only marine-derived compound used routinely as an anticancer agent (110). However, a number of chemically diverse natural products from marine sources have been progressed to clinical trials. The three most advanced compounds are in phase II trials; ecteinascidin-743 (**72**), a tetrahydroisoquinoline alkaloid isolated from the mangrove ascidian *Ecteinascidia turbinata*, bryostatin-1 (**73**), a macrolide isolated from the bryozoan *Bugula neritina*, and dolastatin-10 (**74**), a linear peptide from the sea mollusk *Dolabella auricularia*. Ecteinascidin-743 (**72**) was first isolated by Rinehart's group at the University of Illinois (111) and has been licensed to PharmaMar (Zeltia), which plans, together with Ortho Biotech, to launch the compound in late 2002 initially for the treatment of soft tissue sarcoma (112).

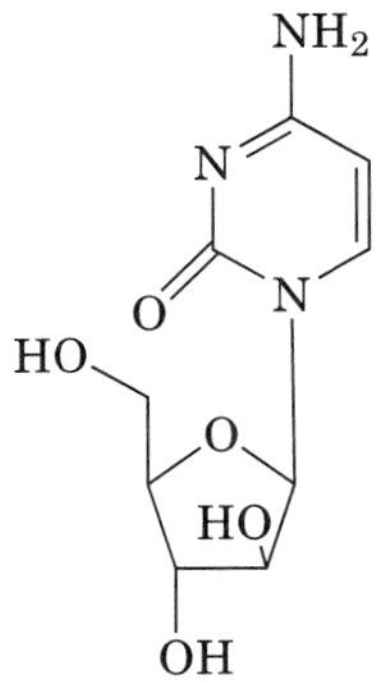

(69) Cytosine arabinoside

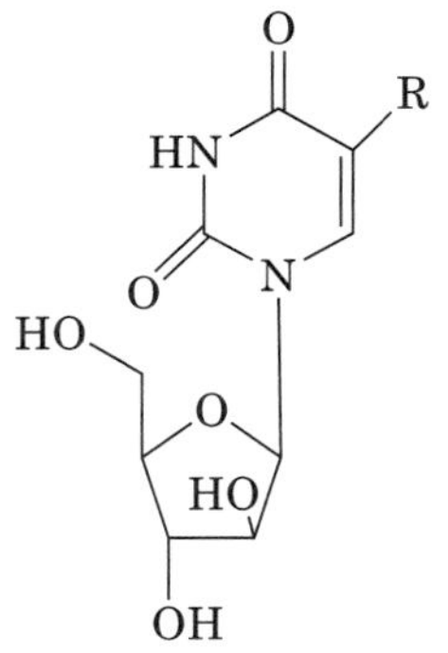

(70) Spongouridine: R = H
(71) Spongothymidine: R = CH$_3$

Yields of marine-derived natural products are invariably low and supply problems have delayed their development as useful pharmaceutical agents. For example, over 3000 kg of the sea squirt *E. turbinata* is required to produce 3 g of ecteinascidin-743, sufficient for just one cycle of treatment (113) and 1000 kg of *B. neritina* yields 1.5 g of bryostatin-1 (114). For-

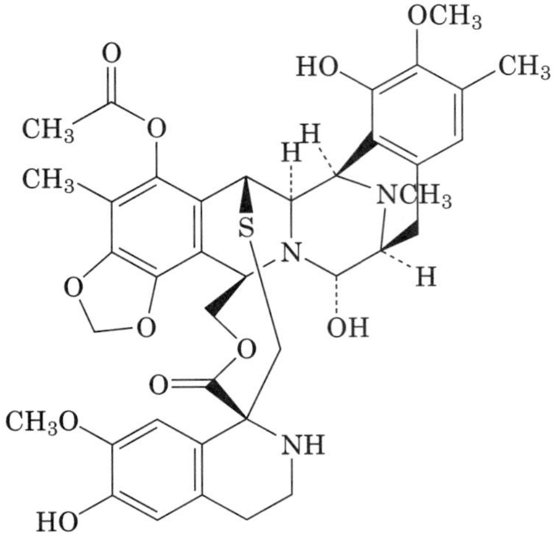

(72) Ecteinascidin 743

tunately, supplies of ecteinascidin-743 (**72**) and also bryostatin-1 (**73**) have been met by aquaculture techniques (115), and more viable synthetic routes are now available for (**73**) (116) and dolastatin-10 (**74**) (117).

5 ANTIBIOTICS

5.1 β-Lactams

In 1929 Alexander Fleming published the results of his chance finding that a Penicillium mold caused lysis of staphylococcal colonies on an agar plate (118). He also showed that the culture filtrate, named penicillin, possessed activity against important pathogens including Gram-positive bacteria and Gram-negative cocci. However, it was not until 1940 that the true therapeutic efficacy of penicillin was revealed, when Chain et al. (119) successfully tested the material in mice that had been previously infected with a lethal dose of streptococci. Several years later the precise chemical structure of the main active component, benzylpenicillin (**75**), was determined and efforts to synthesize the compound were initiated (120). Benzylpencillin proved to be an elusive target because of the instability of the β-lactam ring: it was unstable under acid conditions and was deactivated by β-lactamase enzymes produced by various Gram-positive and Gram-negative bacteria.

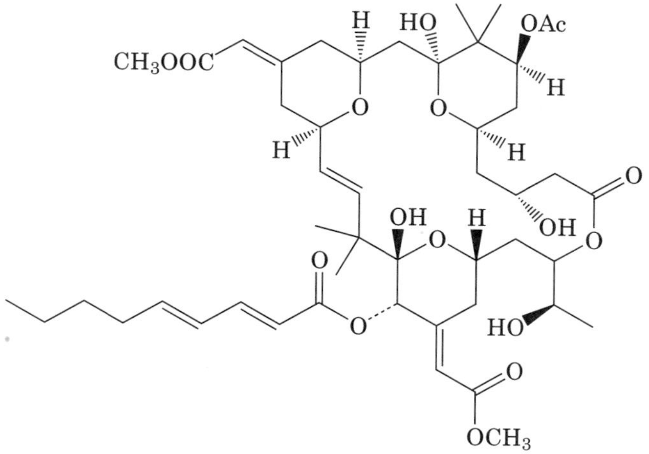

(73) Bryostatin 1

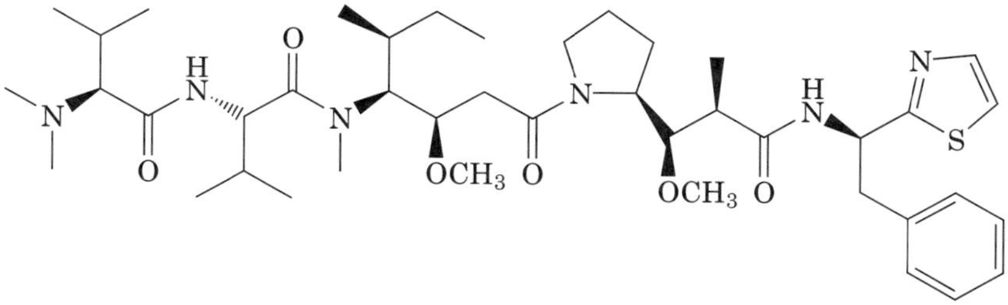

(74) Dolastatin 10

The discovery that the fused β-lactam nucleus, 6-aminopenicillanic acid (6-APA) (76), could be obtained from cultures of *Penicillium chrysogenum* led to the preparation of new, semisynthetic derivatives with improved stability to gastric acid and β-lactamases, and with activity against a wider range of pathogenic organisms (121). Sheehan (122) showed that compound (76) would react readily with acid chlorides to form new penicillin derivatives with novel substituents at the 6-position. Methicillin (77), with a sterically demanding 2,6-dimethoxybenzamide side-chain, was the first semisynthetic penicillin to show resistance to staphylococcal β-lactamases, although the compound was still acid labile. Ampicillin (78) has an α-aminophenylacetamido side-chain and displays good activity against Gram-negative organisms, it is stable to acid and thus can be administered orally, although it is susceptible to degradation by β-lactamases. Amoxycillin (79) differs from ampicillin by the addition of a single

hydroxy group, but the compound is better absorbed by the gastrointestinal tract.

Clavulanic acid (80), isolated from *Streptomyces clavuligerus*, is similar in structure to the penicillins, except oxygen replaces sulfur in the five-membered ring (123). Clavulanic acid has weak antibacterial activity, but is a potent inhibitor of β-lactamases (124). A mixture of clavulanic acid and the β-lactamase–sensitive amoxycillin was introduced in 1981 as Augmentin and has proved to be an effective combination to combat β-lactamase–producing bacteria (125). In 2001, 20 years after its launch, Augmentin is the best-selling antibacterial worldwide.

The clinical introduction of the penicillin group of antibiotics prompted an intensive search for novel antibiotic-producing organisms and Selman Waksman demonstrated the value of actinomycetes in this role, discovering the aminoglycoside streptomycin (81) from *Streptomyces griseus* in 1943 (126). Pharma-

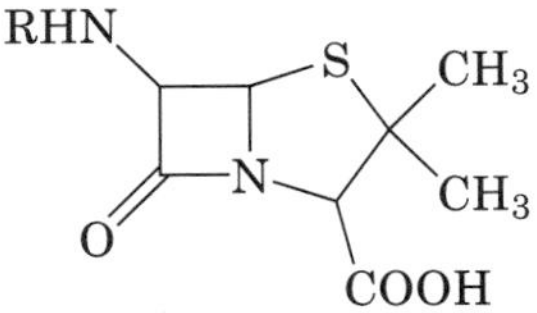

(75) R = COCH$_2$Ph

(76) R = H

(77) R = CO—[3,5-dimethoxyphenyl]

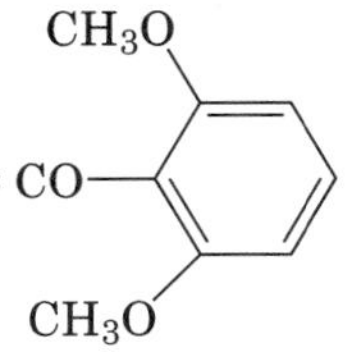

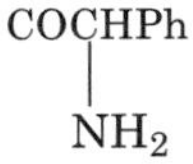

(78) R = COCHPh
              |
              NH$_2$

(79) R = COCH—[4-hydroxyphenyl]—OH
              |
              NH$_2$

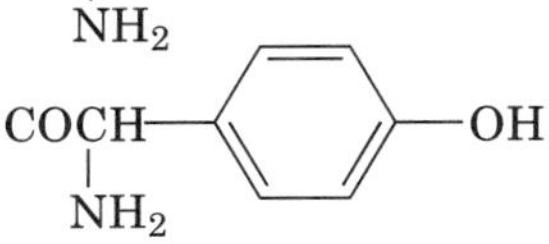

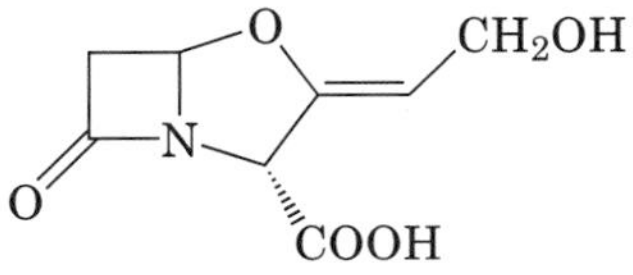

(80)

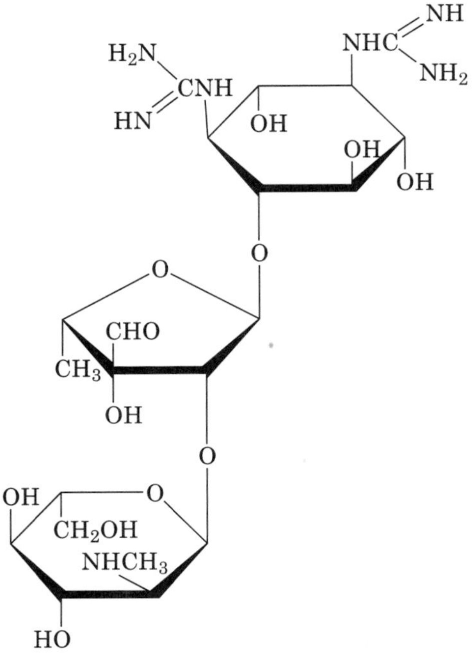

(81)

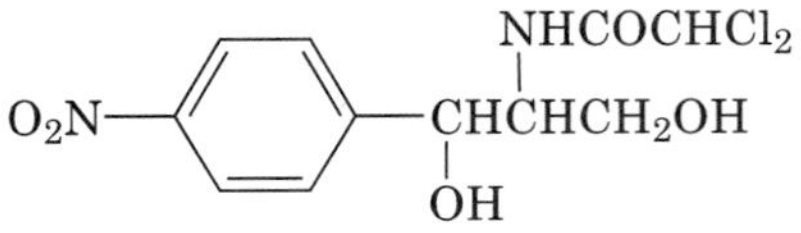

(82)

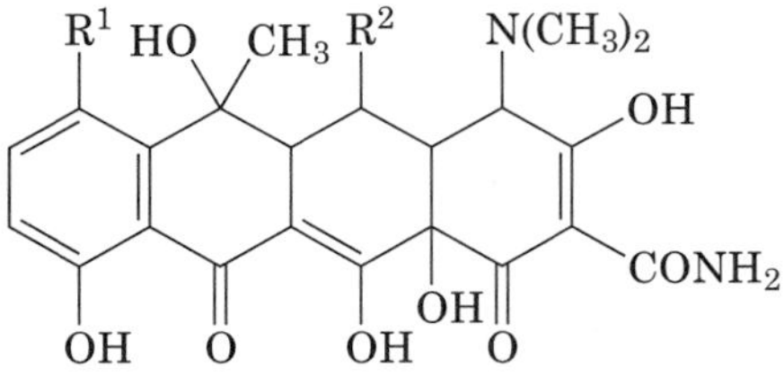

(83) R^1 = -Cl, R^2 = -H
(85) R^1 = -H, R^2 = -OH

ceutical companies also embarked on large programs of screening soil samples for antibiotic-producing microorganisms (127). Chloramphenicol (**82**) was isolated from *Streptomyces venezuelae* in 1948 and other clinically important antibiotics followed: chlortetracycline (**83**), neomycin (**84**), oxytetracyclin (**85**), erythromycin (**86**), oleandomycin (**87**), kanamycin (**88**), and rifamycin (**89**).

In 1948 Giuseppe Brotzu isolated the fungus *Cephalosporium acremonium* from a water sample collected off the coast of Sardinia. The culture showed significant antimicrobial activity, but Brotzu could not interest the Italian authorities in his discovery. He then turned to a friend in England for help, who

arranged for Howard Florey at Oxford to receive a sample of the producing culture. Eventually, an antibacterial substance was isolated and named cephalosporin C (**90**) (128). The compound, which had a structure similar to that of the penicillins, except it had a dihydrothiazine ring fused to the β-lactam core,

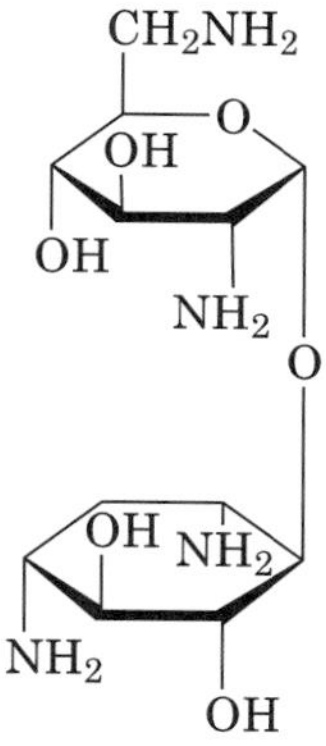

(84)

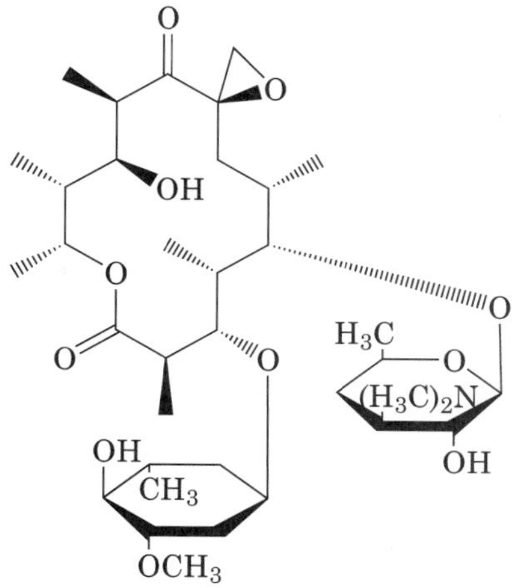

(87)

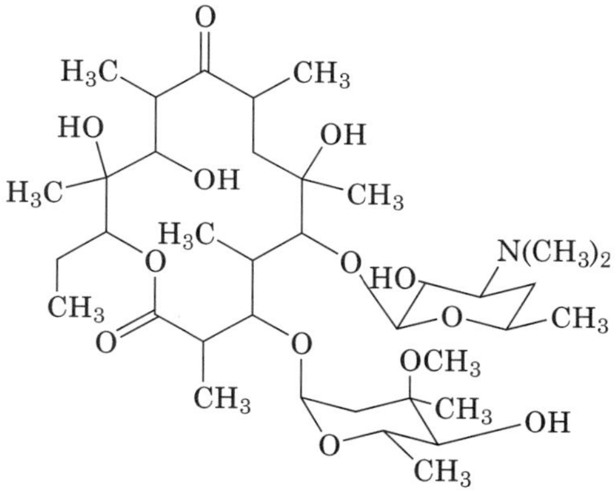

(86)

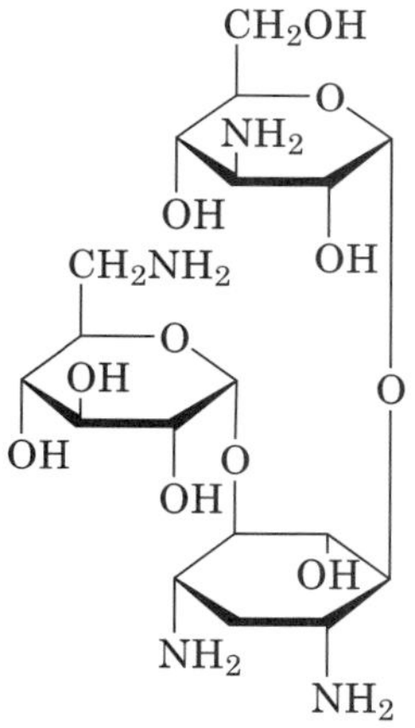

(88)

showed good resistance to β-lactamases and was less toxic than benzylpenicillin. However, plans to market the compound were terminated with the introduction of methicillin (see above).

The discovery that the basic structural building block of cephalosporin C, that is, 7-aminocephalosporanic acid (7-ACA) (**91**), could be synthesized led to the preparation of numerous cephalosporin derivatives in a similar way to the synthesis of penicillins from 6-aminopenicillanic acid (129, 130). Modification of the substituent at the 7-position, while retaining the 3-acetoxymethyl group, gave cephalothin (**92**), cephacetrile (**93**), and cephapirin (**94**), so-called first-generation cephalosporins with good activity against Gram-posi-

tive bacteria, although the acetyl ester was susceptible to degradation by esterases and thus limited the duration of action. Replacement of the acetoxy group by other substituents rendered the products less prone to esterase attack. For example, the pyridinium derivative, cephaloridine (**95**), has a longer duration of action than that of cephalothin.

The first orally active cephalosporin was cephaloglycin (**96**), which possessed a phenylglycine substituent in the C-7 side-chain, although the labile 3-acetoxymethyl group was retained. Replacing the acetoxy group with a proton or chlorine, for example, cephalexin

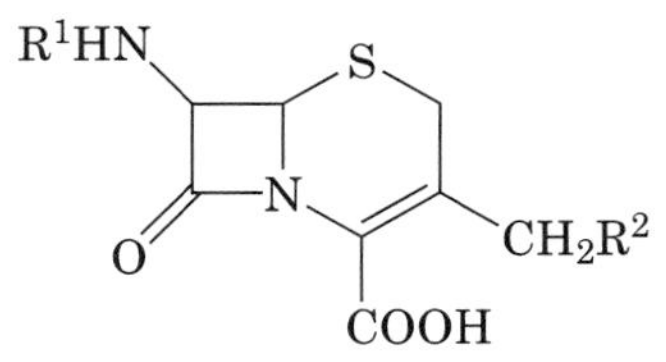

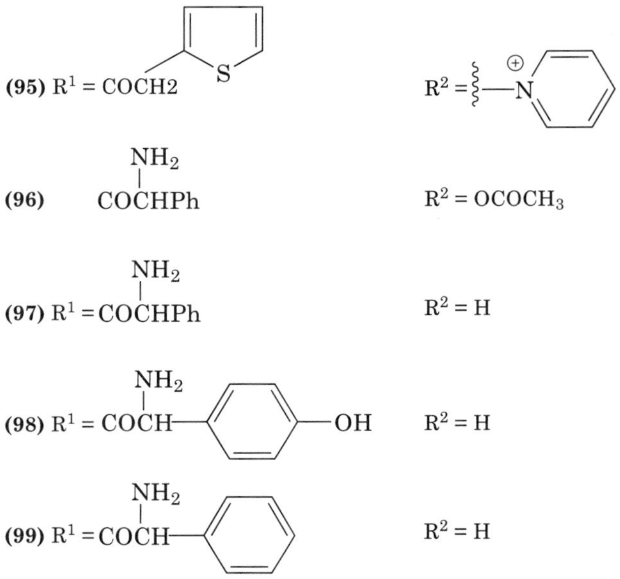

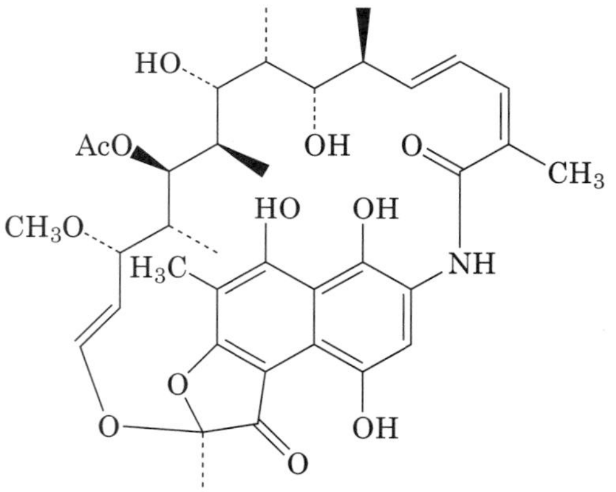

(95) R^1 = COCH2 [thiophene]

R^2 = —N⁺[pyridinium]

(96)   COCHPh, NH₂

R^2 = OCOCH₃

(97) R^1 = COCHPh, NH₂

R^2 = H

(98) R^1 = COCH(NH₂)—[phenyl]—OH

R^2 = H

(99) R^1 = COCH(NH₂)—[phenyl]

R^2 = H

(89)

(97), cefadroxil (98), cephradine (99), and cefaclor (100), extended the duration of action of
these orally active products. Cefaclor has been
classified as a second-generation cephalospo-

rin because it has a wider spectrum of activity,
which includes Gram-negative bacteria such
as *Haemophilus influenzae.* Cephamandole
(101) and cefuroxime (102) are parenterally
administered cephalosporins with similar activities against clinically important Gram-
negative bacteria and are also resistant to
many types of β-lactamases.

The newer third-generation cephalosporins, including ceftazidime (103), ceftizoxime
(104), and ceftriaxone (105), which all contain
an α-aminothiazolyl group in the C-7 side-
chain, have been developed for treating specific pathogens such as *Pseudomonas aerugi-
nosa.* Thienamycin (106), isolated from
Streptomyces cattleya in 1976, represented a
new class of β-lactam antibiotics produced by
bacteria where the sulfur of the penicillin nucleus was replaced by a methylene group
(131). An *N*-formylimidoyl derivative, imipenem (107), was the first example from this

(100) R¹ = COCHPh (with NH₂)   R² = Cl

(101) R¹ = COCHPh (with NH₂)   R² = CH₂S—(1-methyltetrazol-5-yl)

(102) R¹ = COC(=NOCH₃)(thiophene)   R² = CH₂OCONH₂

(103) R¹ = COC(=NOC(CH₃)₂COOH)(aminothiazole)   R² = CH₂–N⁺(pyridine)

(104) R¹ = COC(=NOCH₃)(aminothiazole)   R² = H

(105) R¹ = COC(=NOCH₃)(aminothiazole)   R² = CH₂S—(triazine)

new class of carbapenem antibiotics to become available for clinical use (132). Imipenem has a very broad spectrum of activity against most Gram-positive and Gram-negative aerobic and anaerobic bacteria.

Screening bacteria such as *Pseudomonas acidophila* and *Chromobacterium violacium* for production of β-lactam antibiotics resulted in the discovery of naturally occurring monobactams, which had moderate antimicrobial activity (133–135). Side-chain varia-

tions, as developed for the penicillins and cephalosporins, led to compounds with improved activity against both Gram-positive and Gram-negative bacteria. A derivative containing the α-aminothiazoyl group, a side-chain component common to the third-generation cephalosporins (see above), showed specific activity against Gram-negative aerobic bacteria, including *Pseudomonas* spp., and was stable to most types of β-lactamases. The compound aztreonam (**108**) became the first

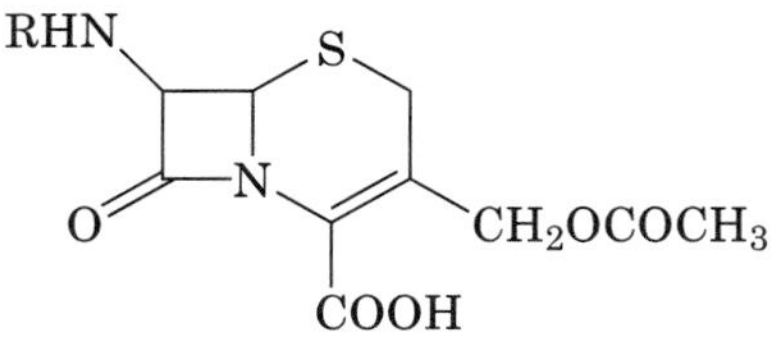

(90) R = COCH₂CH₂CH₂CHNH₂
COOH

(91) R = H

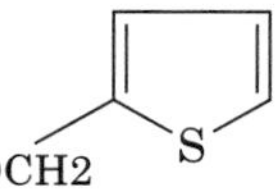

(92) R = COCH2

(93) R = COCH₂CN

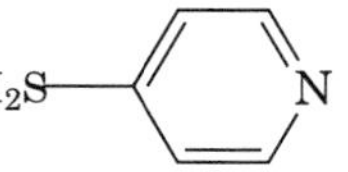

(94) R = COCH₂S—

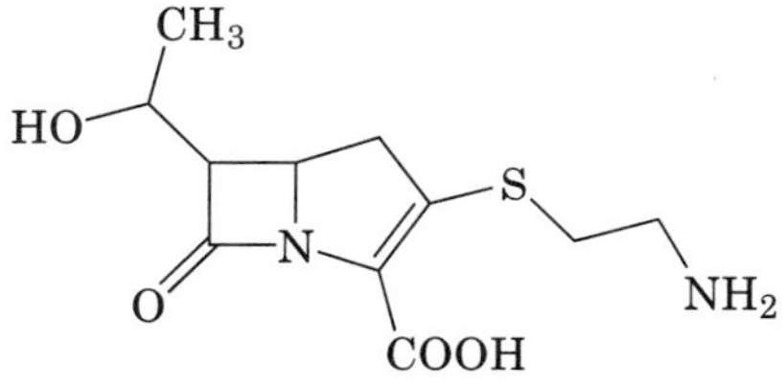

(106)

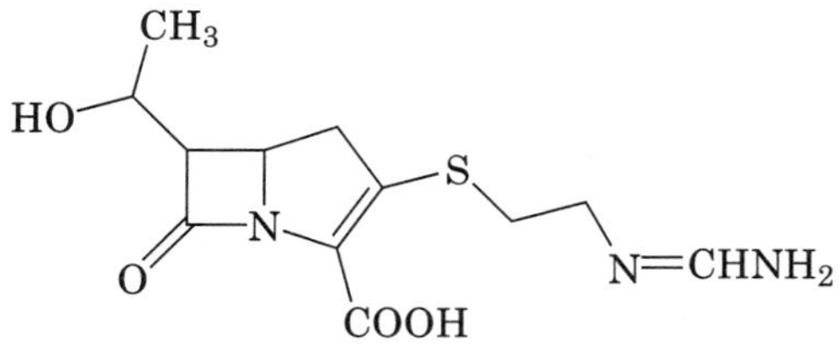

(107)

commercially available monobactam and showed a mode of action similar to that of the other β-lactam antibiotics by blocking bacterial cell wall synthesis (136).

5.2 Erythromycin Macrolides

Erythromycin (**109**) was isolated, in 1952, from a strain of *Saccharopolyspora erythraea*

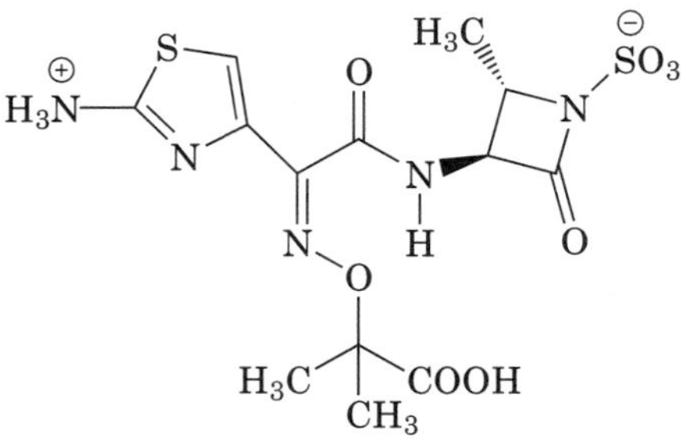

(108)

(formerly *Streptomyces erythreus*). As a broad-spectrum antibiotic erythromycin has proved invaluable for the treatment of bacterial infections in patients with β-lactam hypersensitivity and is also the drug of choice in the treatment of infections caused by species of *Legionella*, *Mycoplasma*, *Campylobacter*, and *Bordetella* (137).

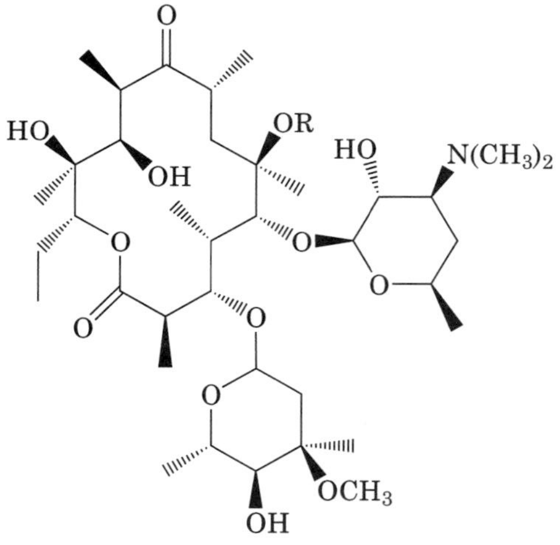

(109) Erythromycin A, R = H
(114) Clarithromycin, R = CH₃

Although safe and effective, erythromycin is not a perfect antibacterial. The presence of hydroxy groups suitably disposed with respect to the keto function at C-9 leads to the formation of a tautomeric mixture of hemiketals (138). The 6,9-hemiketal (**110**) may be dehydrated in stomach acid to give the inactive Δ₈ analog (**111**), which may undergo further ring closure to give the 9,12-tetrahydrofuran (**112**) that is also inactive (139). The Δ₈ derivative

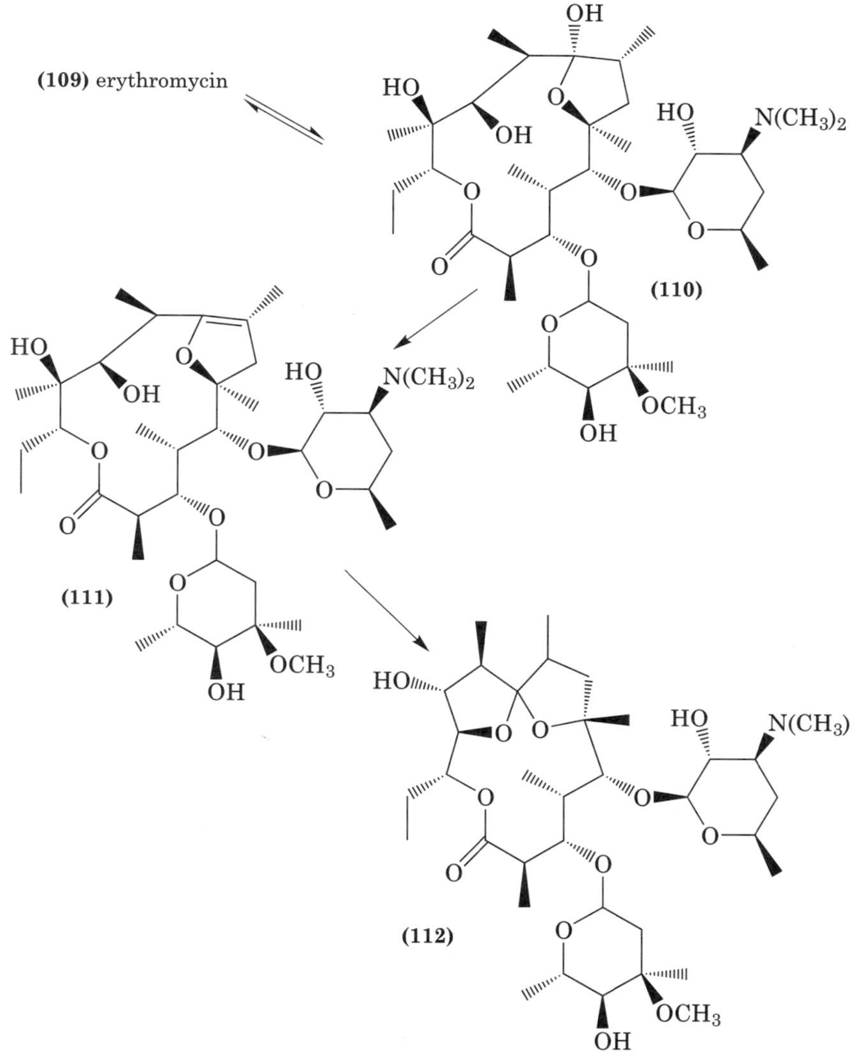

(109) erythromycin

(110)

(111)

(112)

(111) may be responsible for some gastrointestinal disturbance (140). To avoid these problems by increasing the stability to acid, the 2′-stearate, estolate, and ethylsuccinate esters have been prepared (141), but even when the tablets are enteric-coated the bioavailability is erratic and relatively frequent dosing is required (137).

An understanding of the acid-catalyzed decomposition of erythromycin has led to a variety of semisynthetic derivatives with improved oral bioavailability (142). Reductive amination of the 9-keto function gives erythromycylamine,

which reacts with (2-methoxyethoxy)acetaldehyde (143) to give dithromycin. Beckmann rearrangement of the 9-oxime followed by reduction and methylation (144) gives azithromycin (**113**), which shows good activity against Gram-negative bacteria, including *Haemophilus influenzae*. An alternative for prevention of cyclization between the 9-keto and 6-hydroxy is to mask the 6-hydroxy group. If the 6-hydroxy is methylated (145), the result is clarithromycin (**114**), which like (**113**), has an improved pharmacokinetic profile compared with that of the parent molecule.

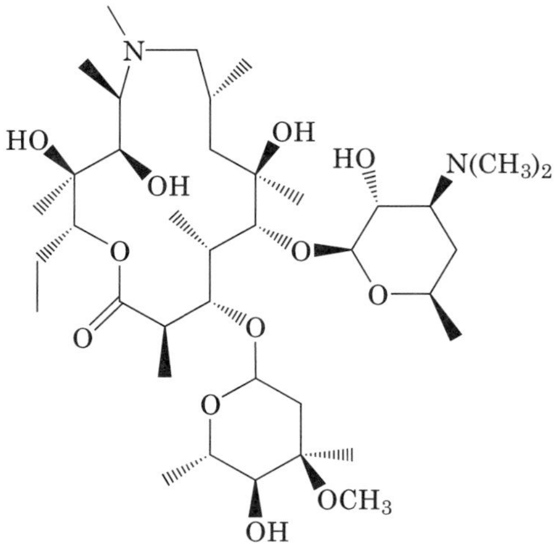

(113) Azithromycin

Both azithromycin and clarithromycin have been used for various bacterial infections for a number of years. Within the last decade, resistance has emerged to a range of antibacterials, including the macrolides, arising from methylation of an adenine in the 23S ribosomal RNA target site, which prevents binding (146). The invention of the ketolides [e.g., telithromycin (**115**)] overcomes MLS_B resistance by removing the L-cladinose moiety at position 3: the exposed hydroxyl is also oxi-

dised to a ketone (147). The loss of potency that would ensue is compensated by two further modifications, which improve binding, formation of a carbamate at positions 11/12, and extension with a heterocycle-substituted side-chain. In ABT 773 a similar side-chain is placed at position 6, with comparable results (147).

5.3 Streptogramins

The streptogramins are produced by *Streptomyces* species and have been classified into two groups: Group A are polyunsaturated macrocyclic lactones and Group B are cyclic hexadepsipeptides. Both groups bind bacterial ribosomes and inhibit protein synthesis at the elongation step and they act synergistically against many Gram-positive microorganisms. However, the naturally occurring streptogramins are poorly soluble in water and this, until recently, has limited their use to treat bacterial infections. New, water-soluble derivatives have been developed and the semisynthetic dalfopristin (**116**) and quinupristin (**117**) mixture (Synercid) has been approved for the treatment of Gram-positive infections, including multidrug-resistant strains of *Enterococcus faecium, Staphylococcus aureus,* and *S. pneumoniae* (148).

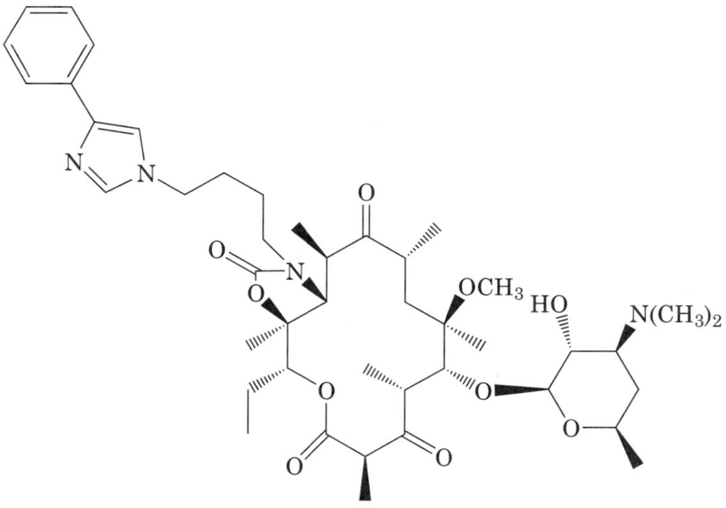

(115) Telithromycin

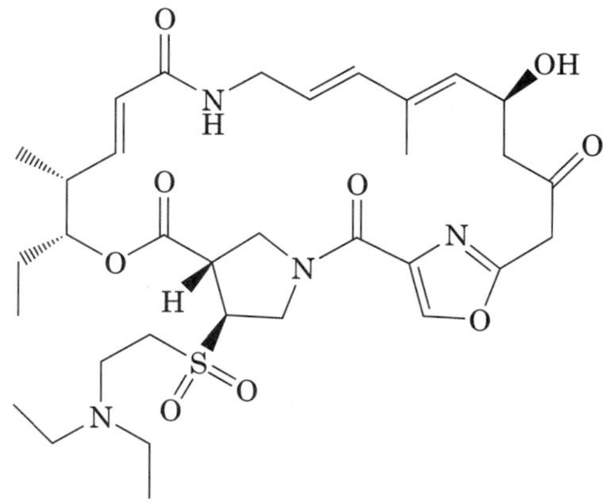

(116) Dalfopristin

water soluble (150), despite the hydrogen-bonding ability of the polyhydroxylated hexapeptide.

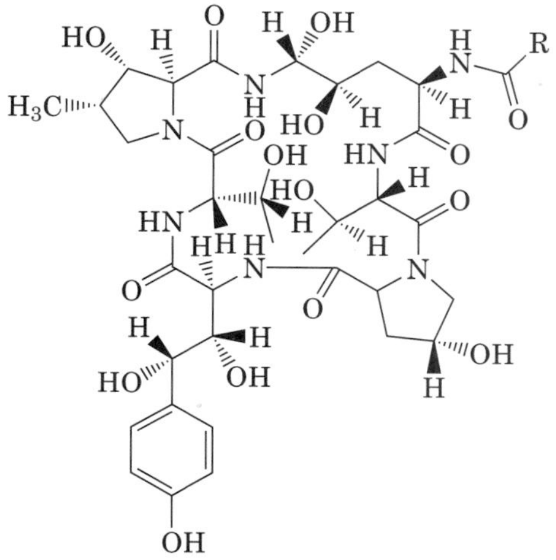

(118) echinocandin B, R = linoleyl

5.4 Echinocandins

The fungal metabolite echinocandin B (**118**) is one of the lipopeptides, in which a cyclic hexapeptide is combined with a long-chain fatty acid. Echinocandin B inhibits β-1,3-glucan synthesis and as a result has anti-Candida and anti-*Pneumocystis carinii* activity (149). As a group, the echinocandins are not orally bioavailable, are haemolytic, and are not very

Synthesis of the cyclic hexapeptide is unattractive for the purpose of securing analogs with improved biological activity because of the unusual nature of the amino acids used

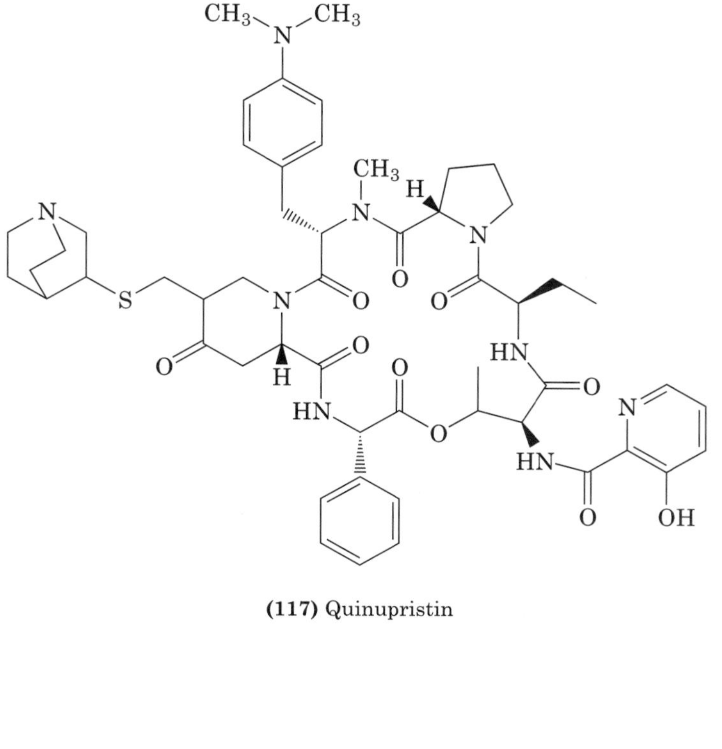

(117) Quinupristin

and the complex stereochemistry generated by the high degree of hydroxylation. However, echinocandin B can be produced efficiently by fermentation of a culture of *Aspergillus nidulans* and then deacylated by fermentation with *Actinoplanes utahensis* (151). The free amino group thus exposed can be derivatized with a number of active esters. Synthesis of the amide from 4-octylbenzoic acid gives cilofungin (**119**), which has specifically high potency against *Candida albicans* and some other Candida species (151).

(**119**) cilofungin, R = ![benzene ring]—O(CH$_2$)$_7$CH$_3$

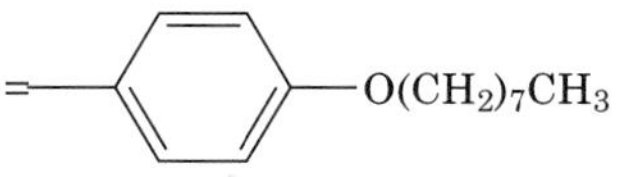

For systemic use cilofungin had to be given intravenously and unfortunately ran into problems associated with the cosolvent (PEG) (152). A better derivative, LY-303366 (**120**) is

now in clinical trial and has the major advantage of oral bioavailability (153). Many other antifungal peptides are under investigation (152). The member of this series that is furthest advanced is caspofungin (MK-991, L-743,872) (**121**), following its approval by the FDA, early in 2001, for the treatment of aspergillosis. The two analogs, LY-303366 and caspofungin, have been compared against clinical fungal isolates *in vitro* (154) and the latter has been evaluated in immunosuppressed mice (155).

6 CARDIOVASCULAR DRUGS

6.1 Lovastatin, Simvastatin, and Pravastatin

One of the most significant natural product discoveries in the last 25 years has been a fungal secondary metabolite called lovastatin (**122**). Heralded as a major breakthrough in

LY-303366, R = ![three benzene rings]—OC$_5$H$_{11}$

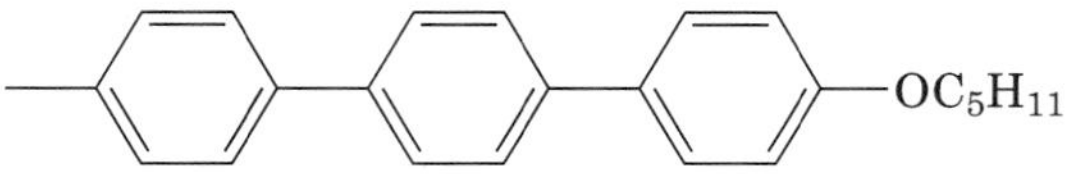

(**120**)

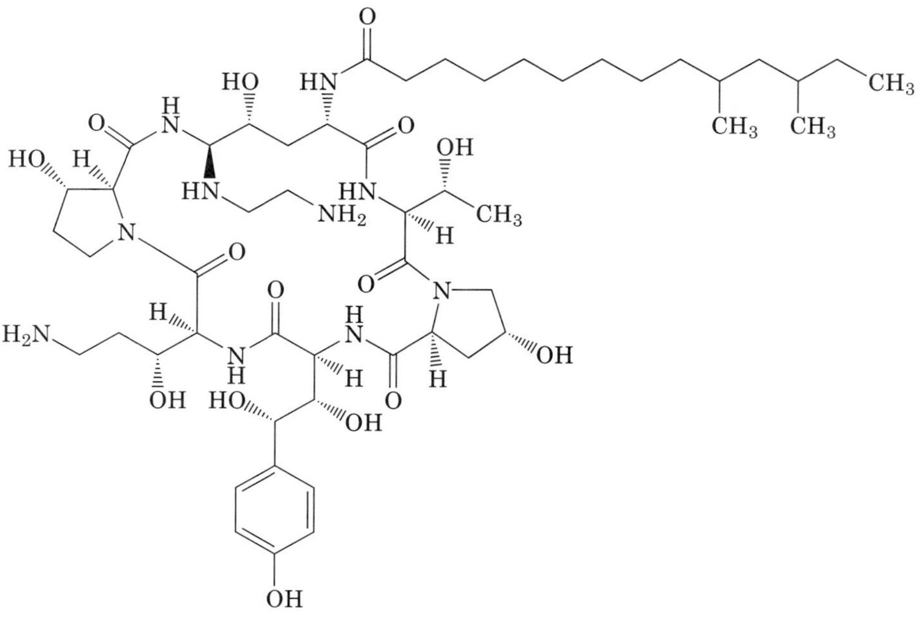

(**121**) caspofungin

the treatment of coronary heart disease (156), lovastatin was introduced onto the market by Merck in 1987 for the treatment of hypercholesterolemia, a condition marked by elevated levels of cholesterol in the blood.

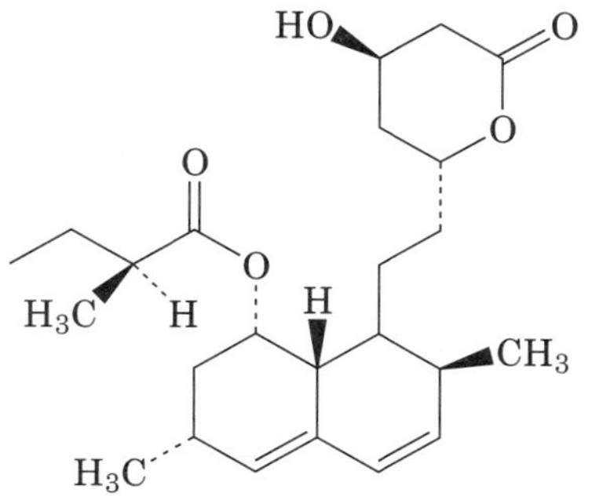

(122)

Lovastatin works by inhibiting 3-hydroxy-3-methylglutaryl coenzyme A (HMG-CoA) reductase, a key rate-limiting enzyme in the cholesterol biosynthetic pathway. However, the first specific inhibitors of this enzyme were discovered several years earlier by Endo et al. at Sankyo (157). The compounds, which are structurally related to lovastatin, were isolated from *Penicillium citrinum* and shown to block cholesterol synthesis in rats and lower cholesterol levels in the blood. Development of the most active compound, designated ML-236B (123), is believed to have been curtailed because of toxicity problems (158).

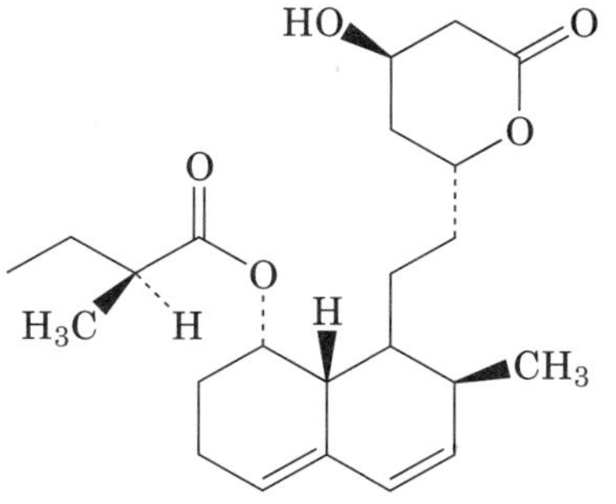

(123)

Brown et al. at Beechams also reported the isolation of (123), but as a metabolite from *Penicillium brevicompactum* (159). The

group, naming the compound compactin, reported its antifungal activity but failed to reveal its mode of action as an inhibitor of HMG-CoA reductase. The search for naturally occurring inhibitors of HMG-CoA reductase gained pace and after spending several years developing appropriate screens, Merck found during only the second week of testing a culture of *Aspergillus terreus* that displayed interesting inhibitory activity (160). In February 1979 the active component, lovastatin (mevinolin), was isolated and characterized (161), and in November the following year Merck was granted patent protection in the United States. Although lovastatin proved to be identical to monocolin K, a metabolite isolated earlier from *Monasus ruber* (162), the chemical structure of the latter compound had not been reported, whereas Merck filed for patent protection giving complete structural details for lovastatin.

The discovery of compactin and lovastatin prompted efforts to develop derivatives with improved biological properties (163, 164). Modification of the methylbutyryl side chain of lovastatin led to a series of new ester derivatives with varying potency and, in particular, introduction of an additional methyl group α to the carbonyl gave a compound with 2.5 times the intrinsic enzyme activity of lovastatin (165). The new derivative, named simvastatin (124), was the second HMG-CoA reductase inhibitor to be marketed by Merck. Both lovastatin and simvastatin are prodrugs and are hydrolyzed to their active open-chain dihydroxy acid forms in the liver (166). A third compound, pravastatin (125), launched by Sankyo and Squibb in 1989, is the open hydroxyacid form of compactin that was first identified as a urinary metabolite in dogs. Pravastatin is produced by microbial biotransformation of compactin.

The HMG-CoA reductase inhibitors described above bind to two active sites on the enzyme: the hydroxymethylglutaryl binding domain and an adjacent hydrophobic pocket to which the decalin moiety binds (167). The recognition that the ring-opened hydroxy acids resemble mevalonic acid and that the decalin moiety could be replaced by 4-fluorophenyl–substituted heterocycles led to the launch of several new products including fluvastatin

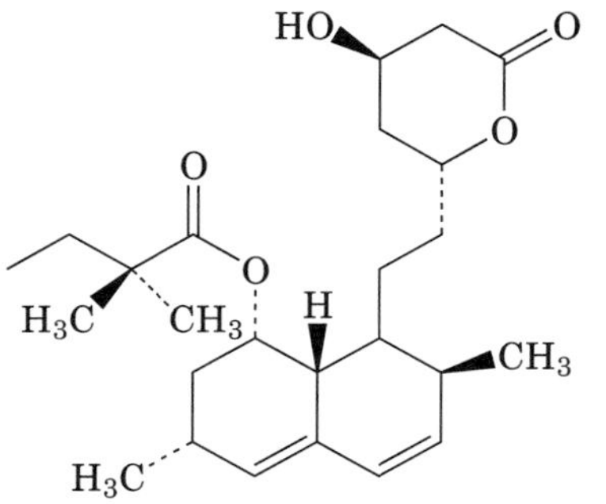

(124)

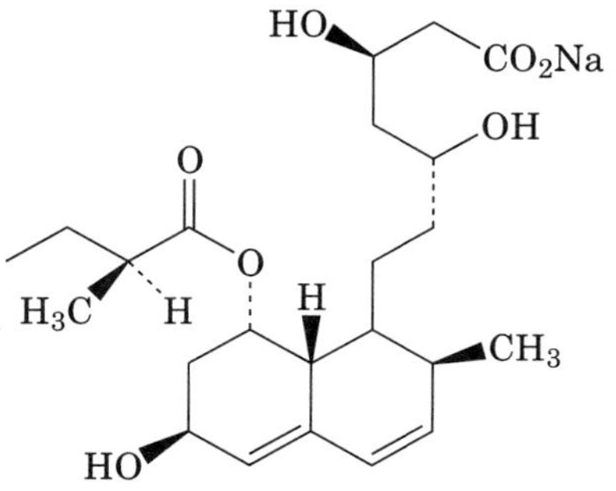

(125)

(**126**), the ill-fated cerivastatin (**127**), and the so-called turbostatin atorvastatin (**128**). Although cerivastatin was withdrawn from the market in 2001 because of fatal adverse drug–drug interactions, the "statins" remain one of the fastest growing segments of the pharmaceutical industry. The latest member of this group of cholesterol-lowering drugs, Astra-

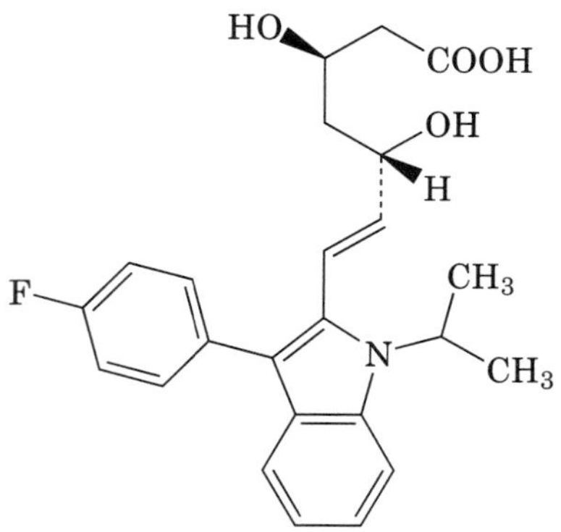

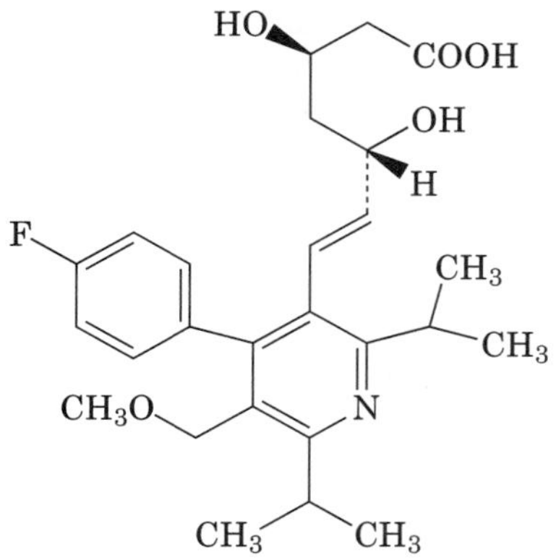

(**127**) Cerivastatin

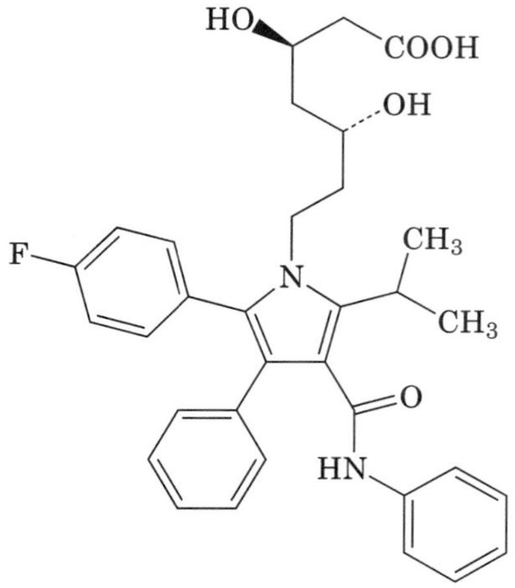

(**128**) Atorvastatin

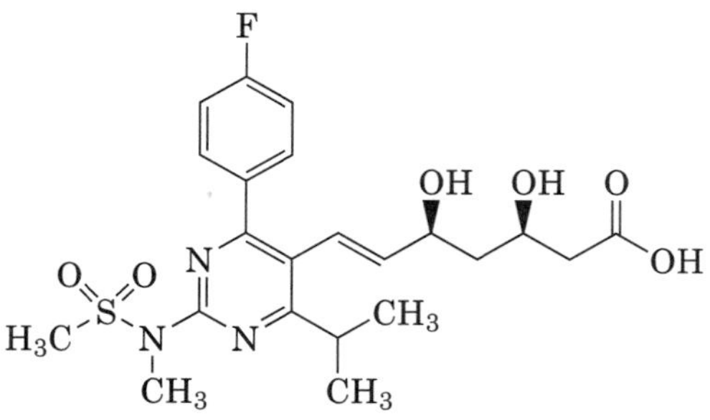

(**126**) Fluvastatin

(**129**) Rosuvastatin

Zeneca's rosuvastatin (**129**), is due to be launched in 2002 and is forecast to achieve sales of US $2.8 billion by 2005 (168).

6.2 Teprotide and Captopril

While studying the physiological effects of snake poisoning, Ferreira (169) discovered that specific components in the venom of the pit viper *Bothrops jararaca* inhibited degradation of the peptide bradykinin and potentiated its hypotensive action. The "potentiating factors" proved to be a family of peptides that worked by inhibiting the dipeptidyl carboxypeptidase, angiotensin-converting enzyme (ACE) (170, 171). In addition to catalyzing the degradation of bradykinin, ACE also catalyzes the conversion of human prohormone, angiotensin 1, to the potent vasoconstrictor octapeptide, angiotensin II. However, the significance of ACE in the pathogenesis of hypertension was not fully appreciated until the 1970s after Ondetti et al. (172) had first isolated and then synthesized the naturally occurring nonapeptide, teprotide (**130**). The compound proved to be a specific potent inhibitor of ACE and showed excellent antihypertensive properties in clinical trials, although its use was limited by the lack of oral activity.

Pyr–Trp–Pro–Arg–Pro–Gln–Ile–Pro–Pro

(**130**)

The discovery of teprotide led to a search for new, specific, orally active ACE inhibitors. Ondetti et al. (172) proposed a hypothetical model of the active site of ACE, based on analogy with pancreatic carboxypeptidase A, and used it to predict and design compounds that would occupy the carboxy-terminal binding site of the enzyme. Carboxyalkanoyl and mercaptoalkanoyl derivatives of proline were found to act as potent, specific inhibitors of ACE and 2-D-methyl-3-mercaptopropanoyl-L-proline (**131**) (captopril) was developed and launched in 1981 as an orally active treatment for patients with severe or advanced hypertension. Captopril, modeled on the biologically active peptides found in the venom of the pit viper, made an important contribution to the understanding of hypertension and paved the

way for other ACE inhibitors, such as enalapril (**132**) and lisinopril, which have had a major impact on the treatment of cardiovascular disease (173).

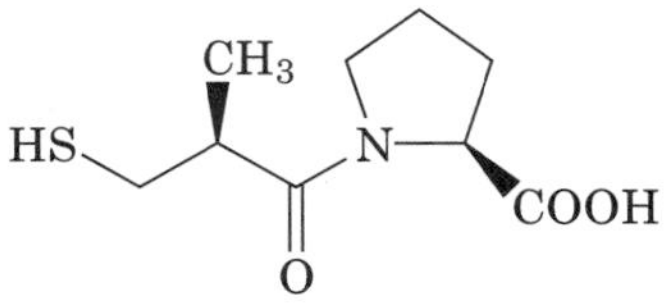

(**131**)

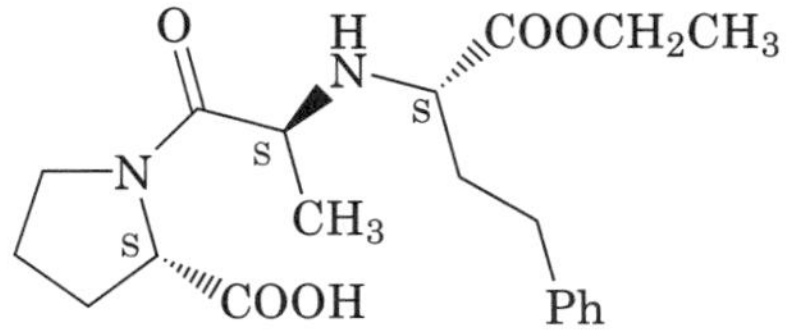

(**132**)

6.3 Adrenaline, Propranolol, and Atenolol

The true clinical potential of β-adrenoceptor blocking agents for treating angina, atrial fibrillation, and tachycardias was first recognized by James Black and colleagues at ICI (174). Black noted a report from Neil Moran of Emory University in 1958, showing that dichloroisoprenaline antagonized the effects of adrenaline on heart rate and muscle tension. The first effective β-adrenoceptor blocker, pronethalol (**133**), was synthesized 2 years later by the ICI group and marketed for limited use in 1963. Toxicity problems soon led pronethalol to be replaced by the 1-naphthyl analog, propranolol (**134**), which became the first β-adrenoceptor antagonist approved for general use, being more potent and yet devoid of the partial agonist or intrinsic sympathomimetic activity shown by many other analogs. Compounds with improved selectivity for the β-adrenoceptor of cardiac muscle (β-1-adreno-

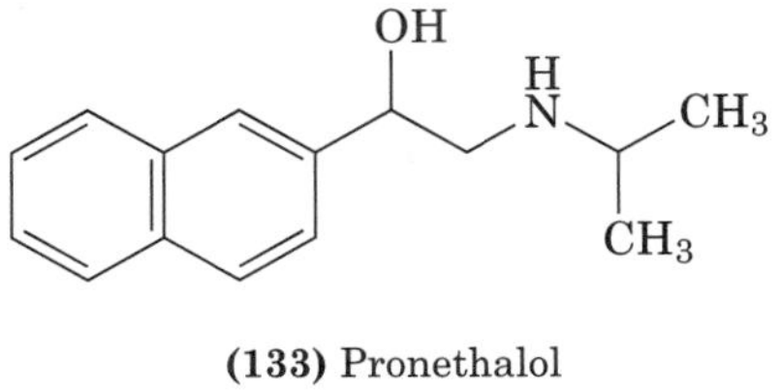

(**133**) Pronethalol

ceptor blockers) were to follow, including atenolol (**135**), which became the most frequently prescribed β-blocker and one of the best-selling drugs of the time.

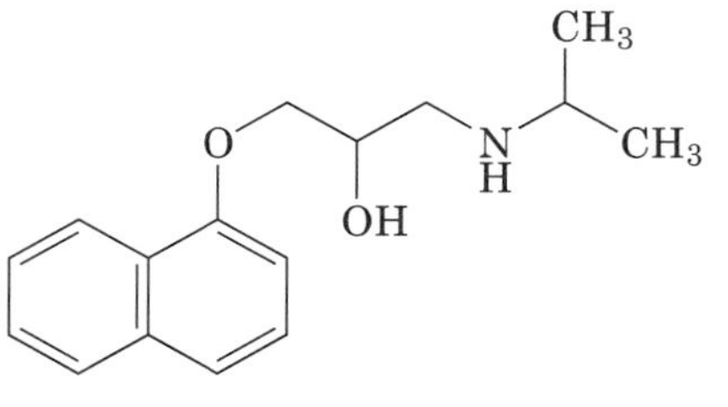

(134) Propranolol

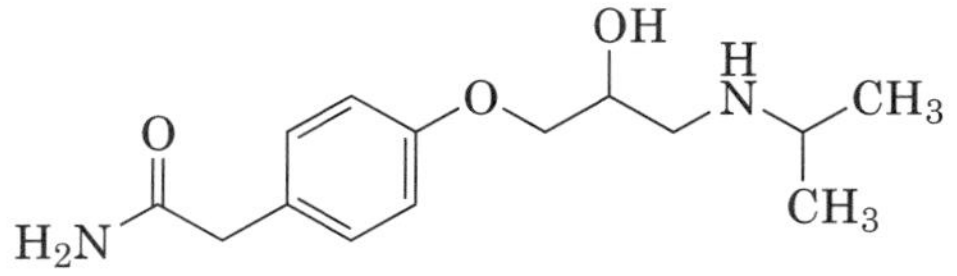

(135) Atenolol

6.4 Dicoumarol and Warfarin

Sweet clover has a long history of medicinal use, often as an antiflammatory or analgesic preparation in the form of ointments and poultices. *Melilotus officinalis* (yellow sweet clover, or ribbed melilot) was reputed to have been a favorite herbal treatment used by King Henry VIII of England and the plant is still referred to as King's Clover in some publications (175).

The plant flourishes in poor soil and was cultivated extensively in Europe for cattle fodder and for soil improvement. In the early 1920s *M. officinalis* was planted on the prairies of North Dakota and Alberta, Canada, but with disastrous consequences. Soon cattle and sheep throughout these regions began literally bleeding to death. The mysterious hemorrhagic disease was traced to clover fodder that had not been stored properly and had become "spoiled," or moldy. However, the insolubility of the anticoagulant component and the difficulty of assaying extracts for biological activity made the task of isolating the active principal component intractable (176). It took almost 20 years before the compound was identified as 3,3′-methylenebis(4-hydroxycoumarin) (**136**), an oxidative degradation metabolite of coumarin (**137**), itself a common component of *Melilotus* sp. Soon after the compound had been identified, trials were initiated that confirmed the oral anticoagulant activity in humans and in 1942 it was marketed under the name dicoumarol (177). The compound had a slow, erratic onset of action and efforts were initiated to prepare synthetic analogs that acted faster and had longer duration of action. A 4-hydroxycoumarin residue, substituted at the 3-position, proved essential for biological activity and in 1948, after synthesizing over 150 compounds, a 4-hydroxycoumarin derivative that was longer acting and more potent than dicoumarol was selected not for clinical use, but as a rodenticide for development by the Wisconsin Alumni Research Foundation! The compound (**138**),

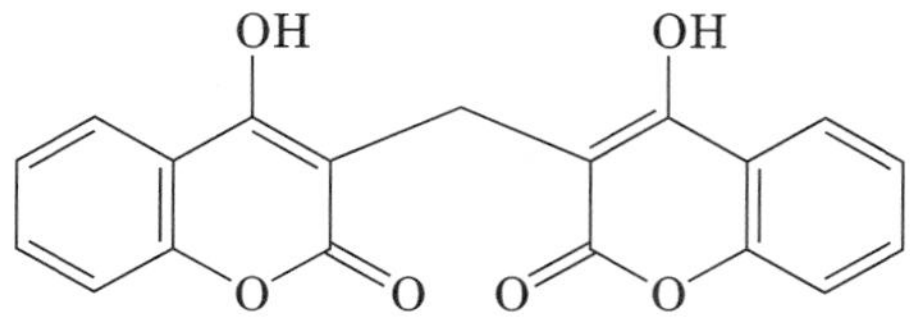

(136)

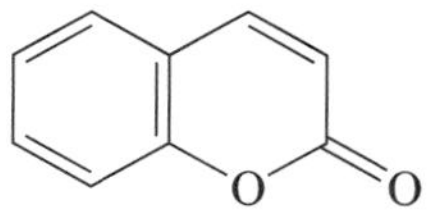

(137)

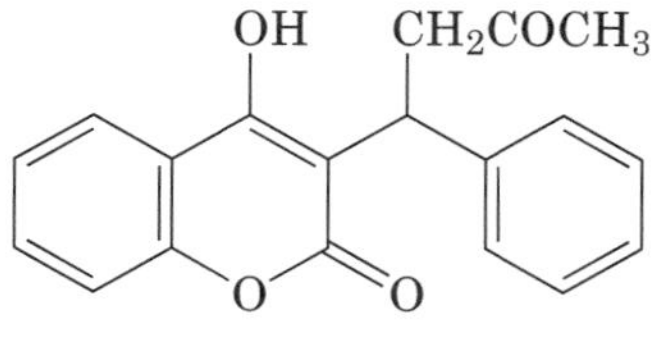

(138)

named warfarin (an acronym derived from the name of the institute coupled with "arin" from coumarin), became a household name for rat poison. Concern over the use of oral anticoagulants and the inherent risk of hemorrhage inhibited the development of warfarin as a therapeutic agent. However, in 1951, a

U.S. Army cadet unsuccessfully attempted to commit suicide by taking massive doses of the compound. The incident prompted further clinical trials that resulted in warfarin being used as the anticoagulant of choice for prevention of thromboembolic disease (177).

The mode of action of the coumarin anticoagulants involves blocking the regeneration of reduced vitamin K and induces a state of functional vitamin K deficiency, thus interfering with the blood-clotting mechanism (178).

7 ANTIASTHMA DRUGS

7.1 Khellin and Sodium Cromoglycate

The toothpick plant, *Ammi visnaga*, had been used for centuries in Egypt as an antispasmodic agent to treat renal colic and ureteral spasm. In 1879 one of the plant's main constituents was isolated, crystallized, and named khellin (139) (179). Subsequently, the pure compound was shown to relax smooth muscle and in 1938 the chemical structure was characterized as a chromone derivative (180). In 1945 a medical technician took khellin to treat renal colic and found instead that it acted as a potent coronary vasodilator and relieved his angina (181). This chance discovery, together with earlier observations, led to khellin being used as a coronary artery vasodilator and for treating bronchial asthma (182). However, its clinical use was severely limited by some unpleasant gastrointestinal side effects.

(139)

Five years later, a small British pharmaceutical company, called Benger Laboratories, initiated a program to synthesize khellin analogs as potential bronchodilators for treating asthma, and had prepared a series of compounds that relaxed guinea pig bronchial smooth muscle and protected the animals against allergen-induced bronchospasm (183).

A clinical pharmacologist on Benger's staff, who suffered from chronic asthma, questioned the validity of the animal model and decided instead to test the compounds on himself. He then prepared a "soup" of guinea pig fur, inhaled the vapors to induce a reproducible asthma attack, and assessed the effects of the synthesized khellin derivatives. Many of the compounds first prepared were insoluble in water and caused nausea and other unpleasant side effects when taken orally. This led to the test compounds being formulated as aerosol sprays and in 1958, an aerosol preparation of a chromone-2-carboxylic acid derivative (140) was found to exert a protectant effect, albeit short lived, against bronchial allergen challenge without showing the bronchodilator activity seen with other compounds. The compound was completely inactive in the guinea pig asthma model and afforded its protectant effect in humans only when inhaled as an aerosol.

(140)

About two new compounds were tested each week and in 1965, after synthesizing some 670 analogs, a bischromone was prepared that gave good protection, even when inhaled up to 6 h before bronchial allergen challenge (184). The compound sodium cromoglycate (141) was obtained by condensing diethyl oxalate with the bis(hydroxy acetophenone) (142) and cyclizing the resultant bis(2,4-dioxobutyric acid) ester (143) under acidic conditions (185). The essential chemical features required for activity appeared to be the coplanarity of the chromone nuclei, the flexible dioxyalkyl link, and the carboxyl groups in the 2-positions. It is believed to act by stabilizing tissue mast cells against degranulation, thereby preventing release of inflammatory mediators (186).

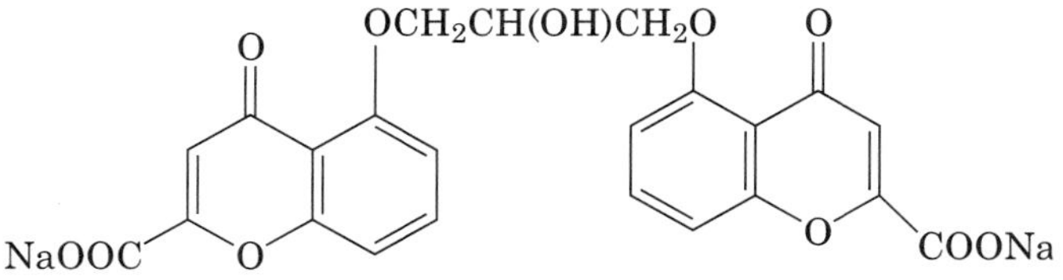

(141)

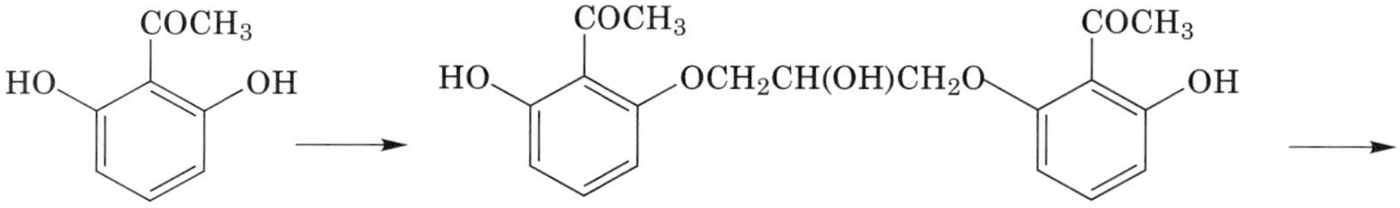

(142)

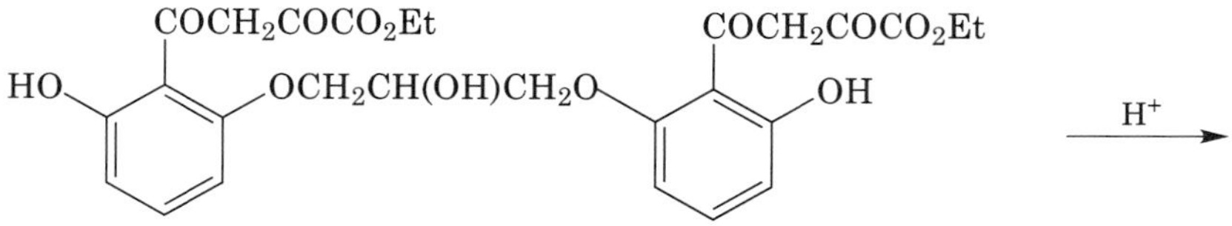

(143)

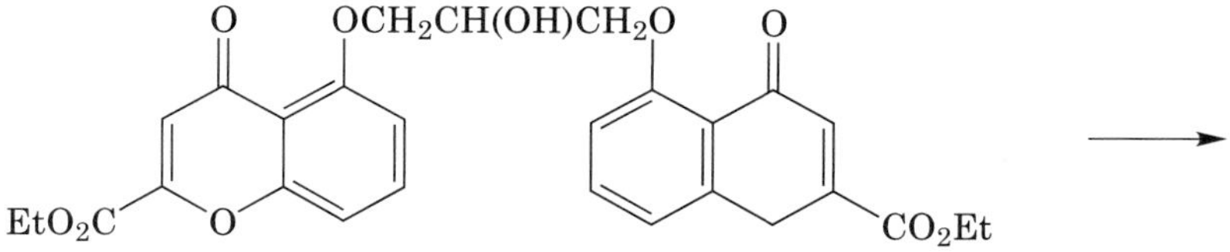

⟶ (141)

Sodium cromoglycate entered clinical trials in 1967 and emerged to become a first-line prophylactic treatment for bronchial asthma.

The coronary dilator properties of khellin have not been ignored and at least one successful program was initiated to prepare analogs for testing as potential antiangina drugs (187, 188). Benziodarone (144) was the first useful compound to emerge from the Labaz laboratories in Belgium based on the benzofuran ring system. However, the compound caused hepatotoxicity in man and was soon superseded by amiodarone (145), a more potent coronary dilator for treating angina. In 1970 the first report of antiarrhythmic activity in the clinic was published (189) and amiodarone became established for prophylactic

(144) benziodarone

control of supraventricular and ventricular arrhythmias during the 1980s (188).

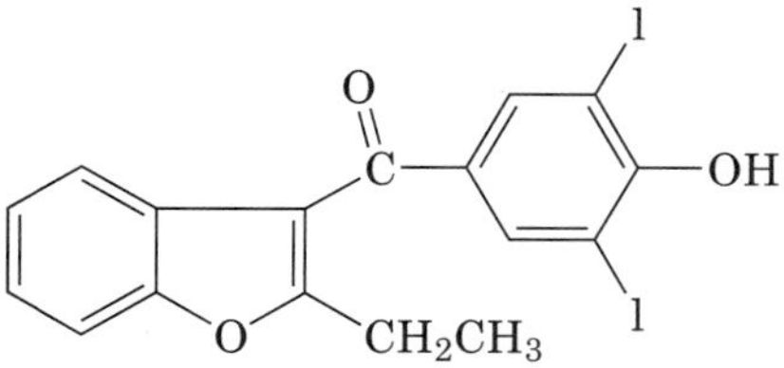

7.2 Ephedrine, Isoprenaline, and Salbutamol

The Chinese have been using a plant extract known as *ma huang* to treat asthma and hay

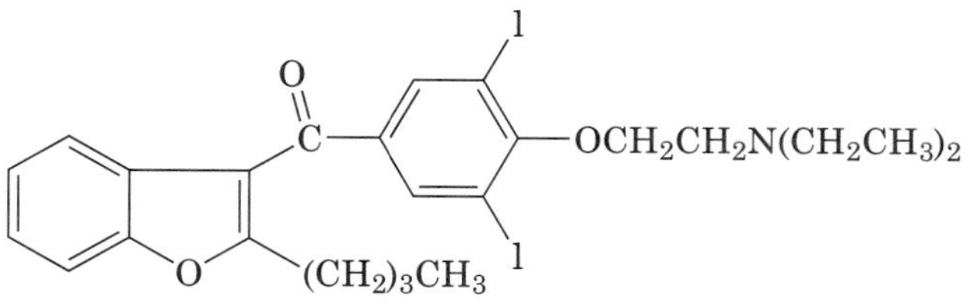

(145) amiodarone

fever for thousands of years. The extract is prepared from several species of *Ephedra*, a small leafless shrub found in China. Following experiments at the Peking Union Medical College and then at the University of Pennsylvania and the Mayo Clinic in the United States, the active ingredient, ephedrine (146), was introduced into Western medicine in 1926 as an orally active bronchodilator for the treatment of acute asthma (190, 191).

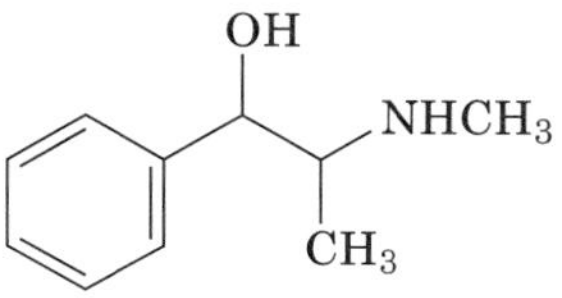

(146)

Ephedrine is related to another natural product that has been used to treat asthma, that is, the adrenal hormone adrenaline (147) (epinephrine). Adrenaline is a potent agonist of both α- and β-adrenoceptors and thus produces arterial hypertension as an undesirable side effect. In 1951 a synthetic alternative, isoprenaline (148), was introduced and for almost 20 years it was considered the drug of choice for treating bronchospasm associated with acute asthmatic attack (191). Isoprenaline is a specific β-adrenoceptor agonist and, although it has no vasoconstrictor activity, the compound does have marked cardiac stimulant properties and a short duration of action. Ahlquist's concept (192) of two types of adrenoceptor was developed further by Lands et al. (193), who established the existence of β_1- and β_2-adrenoceptor subtypes. Clear structure-activity relationships emerged with the preparation of compounds related to adrenaline and ephedrine; the basic requirement for β-adrenoceptor agonist activity was an aromatic ring

substituted by an ethanolamine side-chain. The branched methyl substituent on the side-chain was associated with prolonged duration of action (i.e., ephedrine), whereas aromatic hydroxylation (in isoprenaline) prevented penetration across the blood–brain barrier and thus prevented stimulation of the CNS (191). However, 1,2-dihydroxy substituents were found to promote enzymic degradation, and replacement of the 3-hydroxy group by a hydroxymethyl substituent was required to extend the duration of action. In 1969 salbutamol (149) was launched by Glaxo as a longer-lasting, selective β_2-adrenoceptor agonist for the treatment of bronchial asthma (194) and, recently, a lipophilic ether analog, salmeterol (150), was introduced with an even longer duration of action that has potential advantage in the prevention of nocturnal asthma.

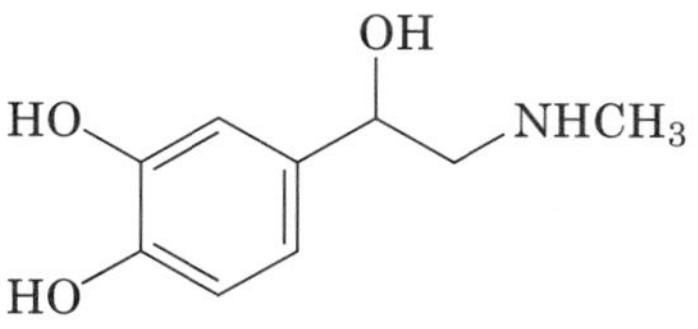

(147)

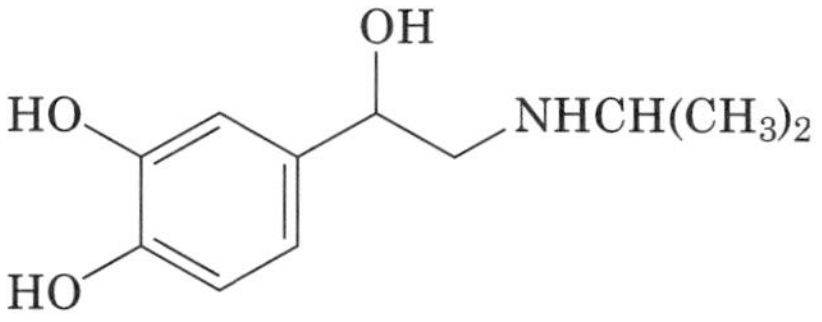

(148)

Despite the many chemical alterations that have been carried out on the phenylethanolamine "template," the key chemical features associated with modern β-agonists can be seen

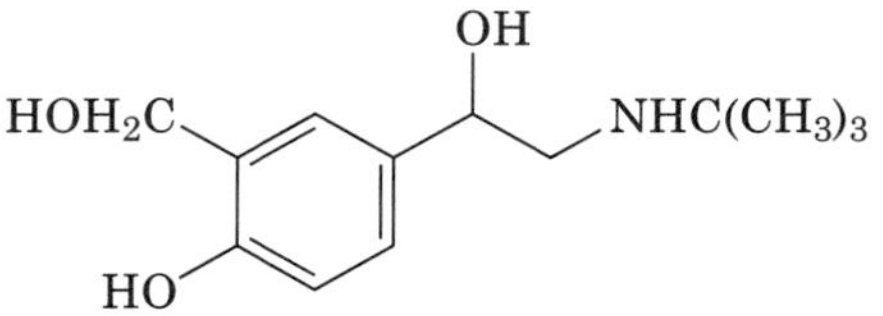

(149)

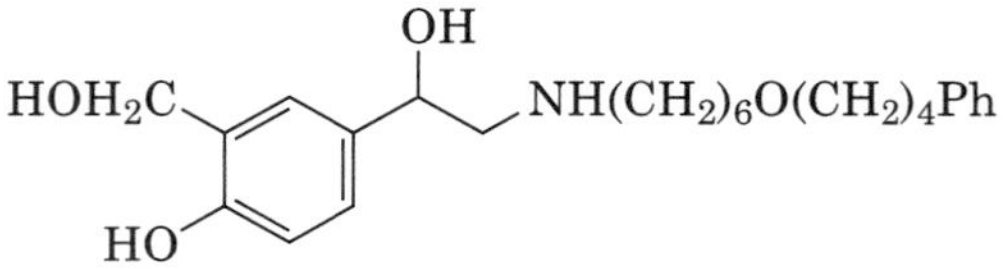

(150)

to have originated from the naturally occurring compounds, adrenaline and ephedrine.

7.3 Contignasterol

The use of inhaled corticosteroids such as fluticasone propionate to treat asthma and rhinitis has been well documented and will not be repeated here. Less well known is an unusual, highly oxygenated marine-derived steroid isolated from the sponge *Petrosia contignata* that possesses a unique cyclic hemiacyl sidechain (**151**). The compound was isolated by Andersen and coworkers (195) at the University of British Columbia and found to possess anti-inflammatory properties *in vivo*. Contignasterol is being developed by Inflazyme, in collaboration with Aventis, for the treatment of asthma and other inflammatory diseases and has progressed to phase II clinical trials.

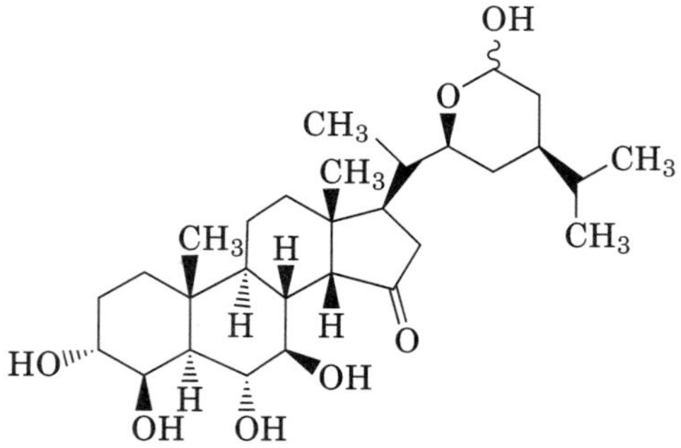

8 ANTIPARASITIC DRUGS

8.1 Artemisinin, Artemether, and Arteether

Artemisia annua (sweet wormwood, qing hao) has been used in Chinese medicine for well over 1000 years. The earliest recommendation is for the treatment of hemorrhoids, but there is a written record of use in fevers dated 340 A.D. Modern development dates from the isolation of a highly active antimalarial, artemisinin (qinghaosu), in 1972, and has been carried out almost entirely in China. Much of the original literature is therefore in Chinese, but there is an excellent review on qinghaosu by Trigg (196) and an account of the uses of *A. annua* (197). This section is largely a summary of these two articles.

Artemisinin (**152**) is a sesquiterpene lactone with an unusual peroxide bridge. One of the earliest modifications involved catalytic reduction of the peroxide, resulting in loss of one oxygen and total loss of antimalarial activity (196) in the adduct (**153**). The role of the peroxide bridge in producing antimalarial effects was not fully understood, but it appeared essential for activity, so much of the early work on analogs conserved this structural feature as an empirical finding. The mechanism

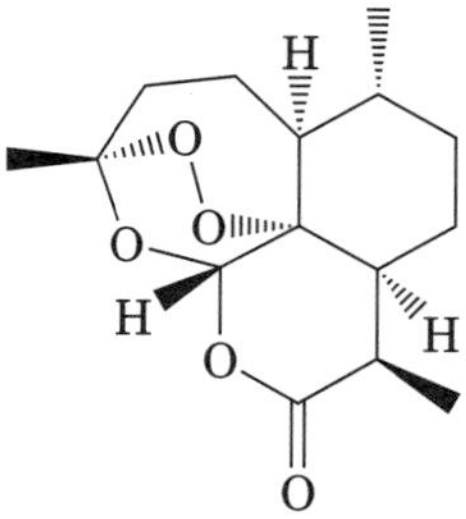

(152) artemisinin

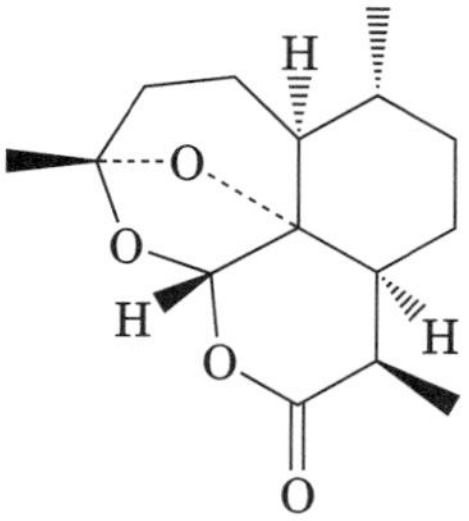

(151) Contignasterol

(153)

of action of artemisinin has since been elucidated (198, 199), although it is not without controversy (200, 201). The drug has a high affinity for hemozoin, a storage form of hemin that is retained by the parasite after digestion of hemoglobin, leading to a highly selective accumulation of the drug by the parasite. Artemisinin then decomposes in the presence of iron, probably from the hemozoin, and releases free radicals, which kill the parasite. The peroxide bridge is therefore a crucial part of the drug molecule, as was suspected from structure-activity studies. Elucidation of the mechanism of action has led to the synthesis of a range of simple analogs capable of iron-catalyzed decomposition, some of which have good antimalarial activity (202).

In retrospect, it is not surprising that the peroxide-bridged compound (154), isolated from *Artabotrys uncinatus*, also has antimalarial activity (197). Because peroxides of this kind are likely to be formed from a variety of precursors in dried plant material (see below), there may well be many more antimalarials of this kind to be found.

(154)

Artemisinin is an excellent antimalarial, approximately equal in potency to chloroquine, with a good therapeutic index except on the fetus. The preparation of semisynthetic derivatives has been stimulated primarily by a requirement for improved solubility because artemisinin is relatively insoluble in both water and oil.

Reduction of (152) with sodium borohydride occurs at the lactone carbonyl, leaving the peroxide intact (196, 197). The resulting cyclic hemiacetal, dihydroartemisinin (155), which is a more potent antimalarial than the parent compound, shows typical acetal reac-

tivity. In the presence of acid, a highly reactive carbocation intermediate allows S_N1-type substitution with a variety of nucleophiles. For example, boron trifluoride catalyzes reactions with methanol and ethanol to give artemether (156) and arteether (157), respectively, two of the most important derivatives (196). Both are more potent than the parent compound and have improved solubility in oil. Artemether has been chosen for development in the West under the name Paluther.

(155) R = H
(156) R = CH$_3$ artemether
(157) R = CH$_2$CH$_3$ arteether
(158) R = COCH$_2$CH$_2$COONa sodium artesunate

Water solubility can be greatly improved by the standard ploy of esterification with succinic acid and conversion to the sodium salt. Applied to compound (155), this technique gives sodium artesunate (158), a water-soluble prodrug that may be given intravenously (196). It may be assumed that hydrolysis occurs *in vivo* to give back (155) as the active antimalarial because (156) has been shown to be unstable in aqueous solution and because analogous carboxylic acids with a nonhydrolyzable ether link are relatively inactive.

There are two reasons for the great interest being shown in artemisinin and its derivatives. First, there is little cross resistance with *Plasmodium falciparum* between the members of this series and the quinoline-based antimalarials like chloroquine (203). On the contrary, significant potentiation of effect is observed in combination with chloroquine analogs such as mefloquine (204). Second, the high lipid solubility of, for example, artemether ensures rapid penetration into the CNS, so these sesquiterpene lactones are first-line drugs for the treatment of cerebral malaria caused by *P. falciparum* (197), which is otherwise fatal.

It seems highly likely (205) that most of the artemisinin found in dried plant material is formed by autoxidation after the death of the plant. From the medicinal chemist's point of view this is unimportant, but some plant biochemists might have doubts about the description of artemisinin as a "natural product." In our view, air drying in sunlight is a natural, although not a botanical, process. It is probable that many other plant-derived peroxides are formed in a similar way.

Whole plant extracts often show promising activity that may not be traceable to single components. This is obviously not true of *Artemisia annua* extracts, but it is interesting to note that other constituents, notably methoxylated flavones, have potentiating effects on the antimalarial activity of artemisinin (206).

The reported effect of artemisinin on systemic lupus erythematosus (196) is intriguing, given the history of use of quinine-type antimalarials in this disease.

8.2 Quinine, Chloroquine, and Mefloquine

The use of Cinchona bark (e.g., *Cinchona succirubra*) by South American indians to treat fevers and the subsequent importation of the bark into Europe by Jesuit priests in the 17th century is well known (207). At that time malaria was widespread, even as far north as eastern Scotland, and there was no effective treatment for "the ague." Although quinine (**159**) is not very potent or long acting, a good sample of Cinchona bark contains about 5% of the alkaloid (208). This high concentration permitted genuinely therapeutic doses of bark to be given and allowed the pure alkaloid to be isolated (209) as early as 1820. During the next 100 years quinine was the only effective treatment for malaria known to Europeans. Without quinine, life in the tropics was impossible for those without natural immunity to malaria. "One thing that was compulsory was the taking of five grains of quinine a day. . . . And if you didn't take it and got ill your salary was liable to be stopped" (210). Supplies of quinine to Europe were threatened

during World War I, stimulating a major program of research into synthetic analogs.

(**159**) quinine

The chemical techniques available to chemists in the period 1820–1920, although improving rapidly, did not allow a structure to be proposed for quinine with any confidence: the first completely correct proposal (211) came in 1922 and was finally confirmed by total synthesis (212) as late as 1945. However, part structures were known, such as the 6-methoxyquinoline moiety, from long before, and were sufficient to allow the synthesis of mimics. The first clinically successful mimics were the 8-aminoquinolines.

In the early years of the 20th century, synthetic organic chemistry was a young discipline, largely governed by empirical rules. Progress toward synthetic analogs of complex natural structures was governed as much by synthetic feasibility as by a desire for close mimicry. The first quinine analogs were, therefore, a combination of the accessible 6-methoxyquinoline part of the quinine structure, with elements of the first successful antimicrobial agents, such as 9-aminoacridine. Nitration followed by reduction could be used to generate a number of new molecules from a variety of parent heterocycles. It is recorded (213) that 4-, 6-, and 8-aminoquinolines have antimalarial properties and, quite extraordinarily, two of these chemical classes are still used today, have quite different uses as antimalarials, and quite possibly have different modes of action.

The first of the 8-aminoquinolines to be introduced into medicine was pamaquine (**160**), not long after World War I (214). Despite greater toxicity than that of quinine, this class of drugs was found to have radical curative ability against the relapsing malarias. Several hundred analogs were tested during World War II and of these, primaquine (**161**) survives to the present day for short-term use as a radical curative (215).

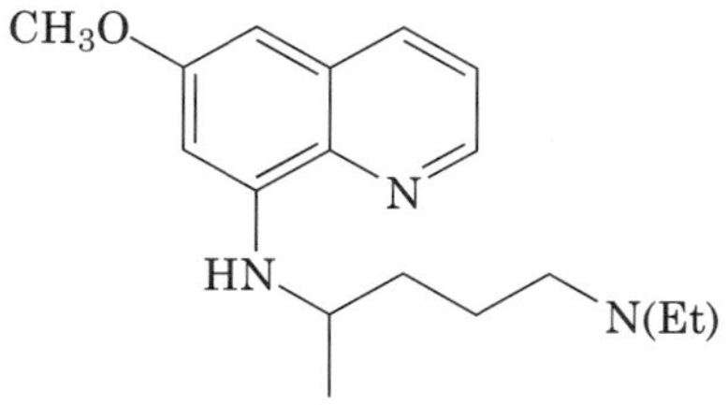

(**160**) pamaquine

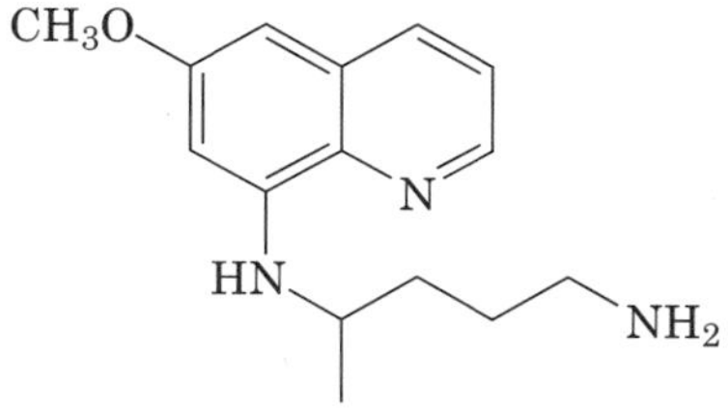

(**161**) primaquine

Quinacrine (**162**) is an obvious embodiment of the principle outlined above; as a derivative of both quinine and 9-aminoacridine it combined a known antimalarial with a known antimicrobial. The result was a useful, relatively nontoxic antimalarial, although it stained the skin and eyeballs yellow (216). Despite this side effect and a high incidence of gastrointestinal disturbance, quinacrine was widely used during World War II by European troops in East Asia. The availability of the results of medicinal chemistry research to both sides in wartime is a curious feature of antimalarial development, highlighted below.

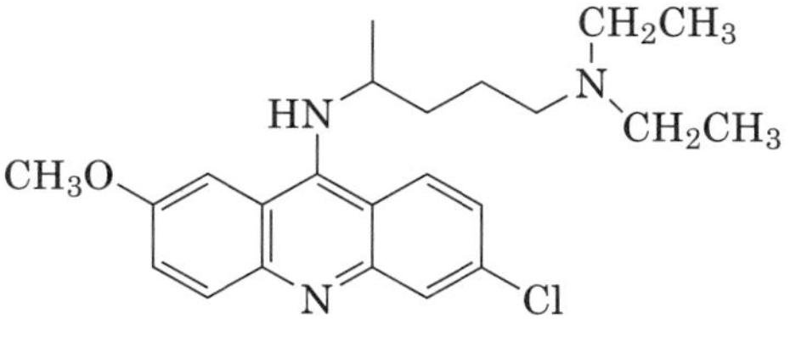

(**162**) quinacrine (mepacrine)

As has been explained, the major stimulus for research into synthetic antimalarials was not so much the therapeutic inadequacy of quinine as the potential lack of availability in times of social upheaval. During World War II, the United States encouraged the planting of Cinchona in Costa Rica, Peru, and Ecuador (216). The total synthesis of quinine was too difficult in the 1940s and is unlikely to become economically viable even in the new millennium. This problem was partly overcome with quinacrine, which was used widely in World War II, although quinacrine has the defects described above. The conceptual derivation of chloroquine (**163**) from quinacrine is obvious and apparently happened twice, in Germany and the United States, the latter about 10 years after the Germans had discarded the drug as being too toxic! The story of the rediscovery of chloroquine is fascinating, as an account of human muddle and misjudgment, finally leading to an extraordinarily valuable drug (216).

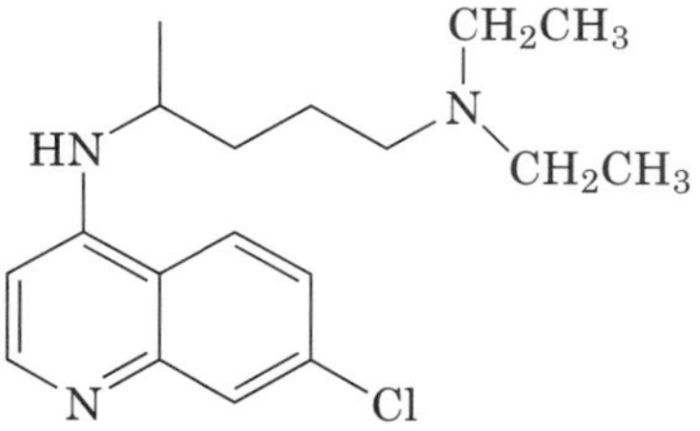

(**163**) chloroquine

Over decades of sublethal exposure the resistance of all types of malaria has increased to a point where chloroquine no longer offers certain protection (217). With the partial exception of quinine and dihydroquinine (218), resistance to antimalarials had reached the stage at the time of the Vietnam war where more research was required. The development of mefloquine (**164**) was a continuation of the World War II effort, with a gap of about 20 years. Resistance to chloroquine had developed widely during that period, but surprisingly less so to quinine, given the obvious similarities in structure. This observation stimulated a reappraisal of quinolines, known as quinoline methanols, which bear a hydroxy group on the α-carbon of a substituent attached to the 4-position (219). Up to 1944, a total of 177 quinoline methanols had been synthesized and tested, resulting in one compound (165) with activity superior to that of quinine. In human volunteers there was a high incidence of phototoxicity associated with (165), so research on quinoline methanols in 1944 had ceased in favor of the 4-amino series, which included chloroquine. Reappraisal of about 100 of the World War II compounds confirmed the high activity and phototoxicity of (165) and also showed the high potency of an analog (166), which had reduced phototoxicity (219). These data, together with results from about 200 newer compounds, fostered the belief that phototoxicity was separable from antimalarial activity. Extensive evaluation of (166) in humans with chloroquine-resistant *Plasmodium falciparum* infections showed promise, but with a significant incidence of toxic reactions; the dose required was also inconveniently large.

Two hypotheses concerning the effect of the 2-phenyl substituent were proposed. One was that metabolic oxidation was blocked at

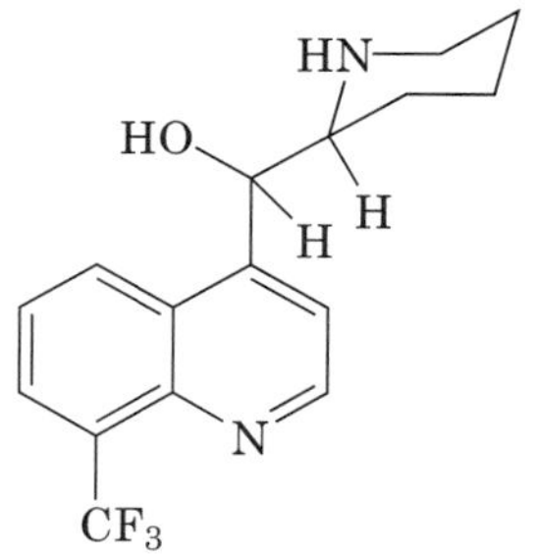

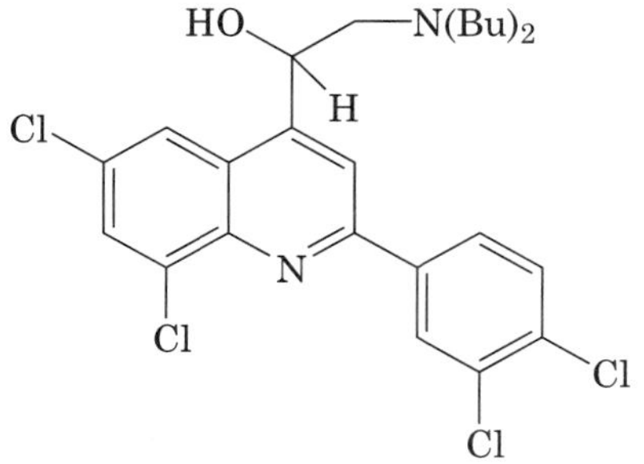

(166)

this position, so that duration of action was prolonged, which was considered desirable. Second, the UV chromophore was enlarged, which would increase the likelihood of drug-induced photosensitivity. The phenyl substituent was thus replaced by trifluoromethyl in the 2-position (220). Before the first such derivatives were tested, further analogs were prepared with an additional trifluoromethyl group on the benzene ring. This was serendipitous because the first series of 2-trifluoromethyl analogs had low potency and were also photosensitizing. The series with two trifluoromethyl groups, one at position 2 and another in the 6-, 7-, or 8-position were all potent and free from phototoxicity (221). The most potent was mefloquine (164), a very successful drug but one that produces unacceptable CNS effects in a small proportion of users (222); parasite resistance has also been observed in parts of Southeast Asia (217). There is now a serious attempt by the World Health Organization to find new antimalarials.

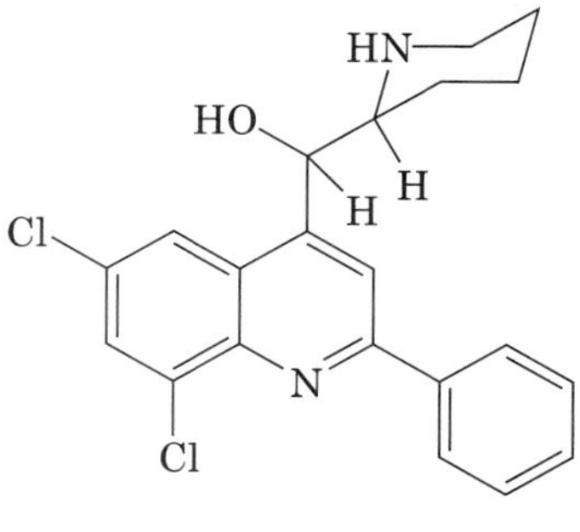

Physicians are pragmatic when choosing therapy for patients whose suffering is not alleviated by accepted methods. A drug that has been shown to be toxicologically safe may be utilized in a new area for the flimsiest of reasons. Thus Page (223) described his use of quinacrine in two cases of lupus erythematosus as being based on "[a] chance observation . . . ," although he did not describe the observation that led to his decision. He did, however, record that quinine had been tried previously and "prevented extension of the lesions," so this may have been the basis for his rationale. In any event, the beneficial effects of quinacrine were remarkable and ap-

(164) mefloquine

(165)

peared to be related to the degree of yellowing of the skin that, as described earlier, is a common side effect of the use of quinacrine in malaria.

Among Page's group of patients with lupus erythematosus were two with rheumatoid arthritis, whose symptoms also responded to treatment with quinacrine. The following year, other physicians (224) conducted a trial of quinacrine on a larger group of patients with rheumatoid arthritis; the results encouraged Haydu (225) to test chloroquine on similar patients, again with positive results. A year later, two more physicians (226) compared quinacrine with chloroquine and found the latter to be better tolerated, the majority of patients gaining some benefit. Both quinacrine and chloroquine caused gastrointestinal disturbances, which led to a trial (227) of hydroxychloroquine (**167**), an unsuccessful antimalarial but with less effect on the gut, thus allowing larger doses to be given. Hydroxychloroquine has remained part of the standard drug therapy for rheumatoid arthritis ever since.

(**167**) hydroxychloroquine

So far, the choice of quinine-like drugs to treat rheumatoid arthritis has been based on preliminary selection as antimalarials. Because the two types of action are presumably unconnected, there might be some value in a screening program aimed directly at rheumatoid disease.

8.3 Avermectins and Milbemycins

There is no major distinction between the avermectins and milbemycins, which are based on the same complex polyketide macrocycle (**168**): the avermectins are oxygenated at C-13 and bear a disaccharide on this oxygen. They have been isolated from cultures of a number of *Streptomyces* species, obtained from all over the world (228).

The avermectins, particularly, have been the subject of intense commercial interest because they possess potent activity against both nematode and arthropod parasites of livestock (229). A full discussion of structure-activity relationships would be out of place here, not least because the data are voluminous, so we shall concentrate on the development of ivermectin, which has been a major success.

Structural designation of avermectins is quaintly based on three series: A, B; a, b; and 1, 2. These are illustrated diagrammatically. Greater activity resides in the B series, with a free OH at position 5. There is little difference in potency between the a and b series. In the more potent B series there are important differences between the 1 series and the 2 series; B_1 is the more active orally, whereas B_2 is the more potent by injection. There are also differences in their spectrum of activity (230). The spectrum of activity was kept as broad as possible by hydrogenation of a mixture of avermectins B_1a and B_1b to give ivermectin (**169**), which contains at least 80% of 22,23-dihydroavermectin B_1a and not more than 20% of 22,23-dihydroavermectin B_1b.

Ivermectin was developed for, and has been highly successful in, the treatment and control of parasites in cattle, horses, sheep, pigs, and dogs. Following studies in humans with river blindness (onchocerciasis) (231–233), the developers of ivermectin (Mectizan) have participated in a major program aimed at eradication of the disease. The sufferers inhabit some of the poorest parts of Africa and cannot pay for their treatment, so the drug has been donated by Merck and Co. Since 1996 more than 20 million treatments have been given (234). The drug does not kill the adult worms that cause onchocerciasis (235), but is useful in interrupting the life cycle (236). Ivermectin is also of value in treatment of scabies (237). A great deal of information on the biological aspects of the use of ivermectin has recently been summarized (238).

9 CONCLUSION

Natural product research has been the single most successful strategy for discovering new

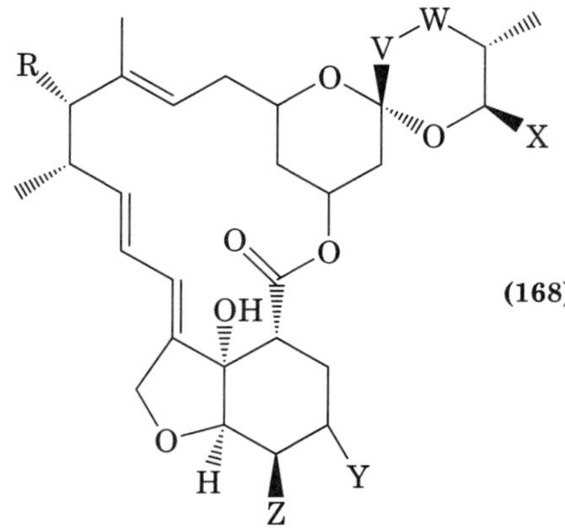

(168)

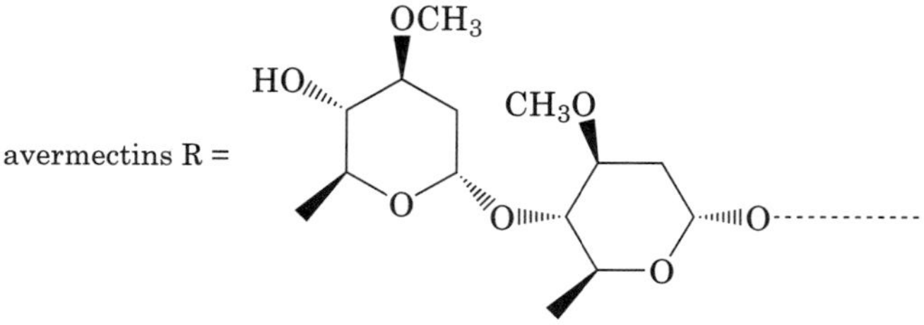

avermectins R =

milbemycins R = H

In the avermectins the series are designated as follows (Y = CH$_3$):

> A, Z = CH$_3$
> B, Z = H
> a, X = CH(CH$_3$)CH$_2$CH$_3$
> b, X = CH(CH$_3$)$_2$
> 1, V-W = CH=CH
> 2, V-W = CH$_2$CH(OH)

For further details of these descriptors, in the milbemycins, see Ref. 228.

In ivermectin **(169)**, V-W = CH$_2$CH$_2$, X = CH(CH$_3$)CH$_2$CH$_3$ (major) or CH(CH$_3$)$_2$ (minor), Y = CH$_3$ and Z = H

pharmaceuticals and has contributed dramatically to extending human life and improving clinical practice. As long as Nature continues to yield novel, diverse chemical entities possessing selective biological activities, natural products will play an important role as leads for new pharmaceuticals. An interesting recent example is the alkaloid galantamine (Ni-

valin, Reminyl) **(170)**, originally isolated from the bulbs of the *Amaryllidaceae* family (snowdrops, daffodils, etc.), which has found use in the symptomatic treatment of Alzheimer's Disease (239). It is a reversible and competitive inhibitor of acetylcholinesterase that also interacts allosterically with nicotinic acetylcholine receptors to potentiate the action of

(169) X = CH(CH$_3$)CH$_2$CH$_3$ (major) or CH(CH$_3$)$_2$ (minor)

agonists. By acting to enhance the reduced central cholinergic function associated with this disease, significant improvements in cognition and behavioral symptoms have been observed in patients. In this case it is the alkaloid itself that is used as the active compound and it will be interesting to see whether development leads to better drugs. There are as yet relatively few publications in this area, although Sanochemia is interested (240, 241).

(170)

Over 90% of bacterial, fungal, and plant species are still waiting to be investigated (242). High throughput screening methods will allow even greater numbers of samples to

be tested against more biological targets (243, 244), although this approach sometimes produces more data than can be conveniently integrated into a research program. An alternative view is that the elucidation of the biological effects of chosen compounds, in some detail, will yield insight into biological processes that may open avenues for medicinal chemistry research that is not based on pure chance. This view is based on the recognition that secondary metabolites have been produced and ruthlessly selected, by evolution, over a long period of time. Either way, the medicinal chemist has a wonderful opportunity to continue utilizing the rich chemical diversity offered by nature, as is shown in two recent reviews that explore this topic in some detail (245, 246).

The best approach for the identification of natural product leads is a matter of debate. Some very inventive techniques have been used in the bioassay-guided method; for example, by spraying TLC plates with reactive media that respond by producing a color change in the presence of an active compound. An alternative is to use an ethnobotanical or ethnopharmacological technique, whereby the accumulated wisdom of many generations of native plant users may be harnessed in the search for better medicines for all. These two techniques may be combined, so that the native people describe the uses to which they put the plant and the researchers devise a bioassay that is used to find the active components. The problem with any bioassay-guided technique, however, is that the inactive constituents are not identified. This represents a considerable waste, given that the plant has had to be collected, preserved, and identified. An alternative view is that it is best to extract all the constituents, with a view to screening in whichever way is appropriate, at that time or in the future. With modern high-performance liquid chromatography facilities it is possible to reduce a plant to its secondary metabolites, as single compounds, in a few days: the products are then able to be screened in a high throughput manner in an equally short time and the compounds can be reevaluated when new screens become available. One thing is certain: the variety of natural product structures, after perhaps 300 million years of natural selection, far exceeds the bounds of human imagination, unlike the typical output from combinatorial chemistry!

REFERENCES

1. G. M. Cragg, D. J. Newman, and K. M. Snader, *J. Nat. Prod.*, **60**, 52 (1997).

2. R. Gerardy and M. H. Zenk, *Phytochemistry*, **32**, 79–86 (1993).

3. M. J. Stone and D. H. Williams, *Mol. Microbiol.*, **6**, 29–34 (1992).

4. R. J. Bryant, *Chem. Ind.*, 146–153 (1988).

5. C. E. Inturissi, M. Schultz, S. Shin, J. G. Umans, L. Angel, and E. J. Simm, *Life Sci.*, **33** (Suppl. 1), 773 (1983).

6. W. Sneader, *Drug Discovery: The Evolution of Modern Medicine*, John Wiley & Sons, Inc., New York, 1985, pp. 78–80 summarizes the confusion surrounding the early work.

7. A. F. Casy and R. T. Parfitt, *Opioid Analgesics*, Plenum, New York, 1986, p. 407.

8. R. Grewe and A. Mondon, *Chem. Ber.*, **81**, 279 (1948).

9. See Ref. 7, p. 153.

10. K. W. Bentley and D. G. Hardy, *Proc. Chem. Soc.*, 220 (1963).

11. G. F. Blane, A. L. A. Boura, A. E. Fitzgerald, and R. E. Lister, *Br. J. Pharmacol.*, **30**, 11 (1967).

12. J. W. Lewis, *Adv. Biochem. Psychopharmacol.*, **8**, 123 (1974).

13. J. Hughes, T. W. Smith, H. W. Kosterlitz, L. A. Fothergill, B. A. Morgan, and H. R. Morris, *Nature*, **258**, 577–579 (1975).

14. J. A. H. Lord, A. A. Wakerfield, J. Hughes, and H. W. Kosterlitz, *Nature*, **267**, 495–499 (1977).

15. O. Schaumann, *Arch. Exp. Pathol. Pharmacol.*, **196**, 109–136 (1940).

16. See Ref. 7, pp. 209–301.

17. B. Cox, *Curr. Rev. Pain*, **4**, 448–498 (2000).

18. B. Cox and J. C. Denyer, *Expert Opin. Ther. Pat.*, **8**, 1237–1250 (1998).

19. A. G. Gilman, T. W. Rail, A. S. Nies, and P. Taylor, *Goodman and Gilman's The Pharmacological Basis of Therapeutics*, 8th ed., Pergamon Press, New York, 1990, p. 550.

20. L. Lemberger, *Clin. Pharmacol. Therap.*, **39**, 1–4 (1986).

21. S. E. Sallan, N. E. Zinberg, and E. Frei, *N. Engl. J. Med.*, **293**, 795–797 (1975).

22. R. K. Razdan, in P. Krogsgaard-Larsen, S. Brogger Christensen, and H. Kofod, Eds., *Natural Products and Drug Development*, Munksgaard, Copenhagen, 1984, pp. 486–499.

23. L. Lemberger and H. Rowe, *Clin. Pharmacol. Ther.*, **18**, 720–726 (1976).

24. T. S. Herman, L. E. Einhorn, S. E. Jones, C. Nagy, A. B. Chester, J. C. Dean, B. Furnas, S. D. Williams, S. A. Leigh, R. T. Dorr, and T. E. Moon, *N. Engl. J. Med.*, **300**, 1295 (1979).

25. A. Ward and B. Holmes, *Drugs*, **30**, 127–144 (1985).

26. W. A. Devane, F. A. Dysarz, R. M. Johnson, L. S. Melvin, and A. C. Howlett, *Mol. Pharmacol.*, **34**, 605–613 (1988).

27. W. A. Devane, L. Hanus, A. Breuer, R. G. Pertwee, L. A. Stevenson, G. Griffin, D. Gibson, A. Mandelbaum, A. Etinger, and R. Mechoulam, *Science*, **258**, 1946–1949 (1992).

28. N. Stella, P. Schweitzer, and D. Piomelli, *Nature*, **388**, 773–778 (1997).

29. A. D. Khanolkar and A. Makryannis, *Life Sci.*, **65**, 607–616 (1999).

30. A. Szallasi and V. Di Marzo, *Trends Neurosci.*, **23**, 491–497 (2000).

31. S. H. Burstein, *Pharmacol. Ther.*, **82**, 87–96 (1999).

32. M. G. Bock, *Drugs of the Future*, **16**, 631–640 (1991) provides a succinct summary.

33. R. S. L. Chang, V. J. Lotti, R. L. Monaghan, J. Birnbaum, E. O. Stapley, M. A. Goetz, G. Al-

bers-Schonberg, A. A. Patchett, J. M. Liesch, O. D. Hensens, and J. P. Springer, *Science*, **230**, 177–179 (1985).

34. P. R. Dodd, J. A. Edwardson, and G. J. Dockray, *Regul. Pept.*, **1**, 17 (1980).

35. R. B. Innis and S. H. Snyder, *Proc. Natl. Acad. Sci. USA*, **77**, 6917–6921 (1980).

36. D. R. Hill, N. J. Campbell, T. M. Shaw, and G. N. Woodruff, *J. Neurosci.*, **7**, 2967–2976 (1987).

37. R. A. Gregory, *Bioorg. Chem.*, **8**, 497–511 (1979).

38. J. Dunlop, *Gen. Pharmacol.*, **31**, 519–524 (1998).

39. P. S. Anderson, R. M. Freidinger, B. E. Evans, M. G. Bock, K. E. Rittle, R. M. Dipardo, W. L. Whitter, D. F. Veber, R. S. L. Chang, and V. J. Lotti, *Int. Cong. Ser. Excerpta Med.*, **766** (Gastrin Cholecystokinin), 235–242 (1987).

40. M. A. Goetz, M. Lopez, R. L. Monaghan, R. S. L. Chang, V. J. Lotti, and T. B. Chen, *J. Antibiot.*, **38**, 1633–1637 (1985).

41. B. E. Evans, *Z. Gastroenterol. Verh.*, **26**, 269–271 (1991).

42. B. E. Evans, K. E. Rittle, M. G. Bock, R. M. Dipardo, R. M. Freidinger, W. L. Whitter, G. F. Lundell, D. F. Veber, and P. S. Anderson, *J. Med. Chem.*, **31**, 2235–2246 (1988).

43. I. M. McDonald, *Expert Opin. Therap. Pat.*, **11**, 445–462 (2001).

44. I. G. Marshall and R. D. Waigh in A. L. Harvey, Ed., *Drugs from Natural Products*, Ellis Horwood, Chichester, UK, 1993, pp. 131–151.

45. H. King, *J. Chem. Soc.*, 1381 (1935).

46. P. Karrer in D. Bovet, F. Bovet-Nitti, and G. B. Marini-Bettolo, Eds., *Curare and Curare-like Agents*, Elsevier, Amsterdam, 1959, pp. 125–136.

47. P. G. Waser, *Helv. Physiol. Pharmacol. Acta II* (Suppl. VIII) (1953).

48. W. C. Bowman, *Pharmacology of Neuromuscular Function*, J. Wright, Bristol, 1990.

49. A. J. Everett, L. A. Lowe, and S. Wilkinson, *J. Chem. Soc. Chem. Commun.*, 1020–1021 (1970).

50. R. B. Barlow and H. R. Ing, *Br. J. Pharmacol. Chemother.*, **3**, 298 (1948).

51. D. Bovet, F. Bovet-Nitti, S. Guarini, V. Longo, and R. Fusco, *Arch. Intern. Pharmacol. Ther.*, **88**, 1–50 (1951).

52. E. P. Taylor and H. O. J. Collier, *Nature*, **167**, 692 (1951).

53. J. B. Stenlake, R. D. Waigh, G. H. Dewar, R. Hughes, D. J. Chapple, and G. G. Coker, *Eur. J. Med. Chem.*, **16**, 515–524 (1981).

54. R. Hughes in J. Norman, Ed., *Clinics in Anaesthesiology*, Vol. **3**, W. B. Saunders, London, 1985, pp. 331–345.

55. J. E. Caldwell, T. Heier, J. B. Kitts, D. P. Lynam, M. R. Fahey, and R. D. Miller, *Br. J. Anaesth.*, **63**, 393–399 (1989).

56. R. L. Noble, C. T. Beer, and J. H. Cutts, *Ann. N. Y. Acad. Sci.*, 882–894 (1958).

57. I. S. Johnson, H. F. Wright, and G. H. Svoboda, *J. Lab. Clin. Med.*, **54**, 830 (1959).

58. G. H. Svoboda, *Lloydia*, **24**, 173 (1961).

59. A. C. Sartorelli and W. A. Creasey, *Annu. Rev. Pharmacol.*, **9**, 51 (1969).

60. K. Gerzon, "Dimeric Catharanthus Alkaloids," in J. M. Cassady and J. D. Douros, Eds., *Anticancer Agents Based on Natural Product Models*, Academic Press, New York, 1980, pp. 271–317.

61. J. B. Sorensen and H. H. Hansen, *Investigational New Drugs*, **11**, 103–133 (1993).

62. J. Mann, *Murder Magic and Medicine*, Oxford University Press, Oxford, 1992, pp. 213–214.

63. M. E. Wall, M. C. Wani, C. E. Cook, K. H. Palmer, H. T. McPhail, and G. A. Sim, *J. Am. Chem. Soc.*, **88**, 3888 (1966).

64. Y.-H. Hsiang, R. Hertzberg, S. Hecht, and L. Lui, *J. Biol. Chem.*, **260**, 14873–14878 (1985).

65. M. E. Wall and M. C. Wani, "Camptothecin and Analogues" in *Human Medicinal Agents from Plants*, ACS Symposium Series **534**, American Chemical Society, Washington, DC, 1993, pp. 149–169 and references therein.

66. W. D. Kingsbury, J. C. Boehm, D. R. Jakas, K. G. Holden, S. M. Hecht, G. Gallagher, M. J. Caranfa, F. L. McCabe, L. F. Faucette, R. K. Johnson, and R. P. Hertzberg, *J. Med. Chem.*, **34**, 98–107 (1991).

67. S. Sawada, S. Okajima, R. Aiyama, K. Nokata, T. Furuta, T. Yokokura, E. Sugino, K. Yamaguchi, and T. Miyasaka, *Chem. Pharm. Bull.*, **39**, 1446–1450 (1991).

68. J. Caesar, *The Battle for Gaul, Book 6, Section 31*, A. Wiseman and P. Wiseman translators, Chatto and Windus, London, 1980, p. 126.

69. T. Bryan-Brown, *Q. J. Pharm. Pharmacol.*, **5**, 205–219 (1932).

70. D. G. I. Kingston, "Taxol and Other Anticancer Agents from Plants" in J. D. Coombes, Ed., *New Drugs from Natural Sources*, IBC Technical Services, 1992, pp. 101–108.

71. M. Suffness, "Taxol: From Discovery to Therapeutic Use" in J. A. Bristol, Ed., *Annual Report of Medicinal Chemistry*, Vol. **28**, Academic Press, New York, 1993, pp. 305–314, provides a good review of the discovery and development of taxol and related derivatives.

72. M. C. Wani, H. L. Taylor, M. E. Wall, P. Coggon, and A. T. McPhail, *J. Am. Chem. Soc.*, **93**, 2325–2327 (1971).

73. P. B. Schiff, J. Fant, and S. B. Horwitz, *Nature*, **277**, 665–667 (1979).

74. W. P. McGuire, E. K. Rowinsky, N. B. Rosenhein, F. C. Grunbine, D. S. Ettinger, D. K. Armstrong, and R. C. Donehower, *Ann. Intern. Med.*, **111**, 273–279 (1989).

75. M. E. Wall and M. C. Wani, *Cancer Res.*, **55**, 753–760 (1995).

76. G. M. Cragg, S. A. Schepartz, M. Suffness, and M. R. Grever, *J. Nat. Prod.*, **56**, 1657–1668 (1993).

77. D. G. I. Kingston, *Pharmacol. Ther.*, **52**, 1–34 (1991).

78. L. Mangatal, M.-T. Adeline, D. Guenard, F. Gueritte-Voegelein, and P. Potier, *Tetrahedron*, **45**, 4177–4190 (1989).

79. J. N. Denis, A. E. Greene, D. Guenard, F. Gueritte-Voegelein, L. Mangatal, and P. Potier, *J. Am. Chem. Soc.*, **110**, 5917–5919 (1988).

80. C. Palomo, A. Arrieta, F. Cossio, J. M. Aizpurua, A. Mielgo, and N. Aurrekoetxea, *Tetrahedron Lett.*, **31**, 6429–6432 (1990).

81. F. Gueritte-Voegelein, D. Guenard, F. Lavelle, M.-T. Le Goff, L. Mangatal, and P. Potier, *J. Med. Chem.*, **34**, 992–998 (1991).

82. K. C. Nicolaou, C. Riemer, M. A. Kerr, D. Rideout, and W. Wrasidlo, *Nature*, **364**, 464–466 (1993).

83. A. Stierle, G. Strobel, and D. Stierle, *Science*, **260**, 214–217 (1993).

84. K. C. Nicolaou, Z. Yang, J. J. Liu, H. Ueno, P. G. Nantermet, R. K. Guy, C. F. Claibome, J. Renaud, E. A. Couladouros, K. Paulvannan, and E. J. Sorensen, *Nature*, **367**, 630–634 (1994).

85. R. A. Holton, H. B. Kim, C. Somoza, F. Liang, R. J. Biediger, P. D. Boatman, M. Shindo, C. C. Smith, and S. Kim, *J. Am. Chem. Soc.*, **116**, 1597–1600 (1994).

86. G. Höfle, N. Bedorf, K. Gerth, and H. Reichenbach, Ger. Pat. DE 91-4138042 (1993); *Chem Abstr.*, **120**, 52841 (1993).

87. M. R. Grever, S. A. Schepartz, and B. A. Chabner, *Semin. Oncol.*, **19**, 622–638 (1992).

88. D. M. Bollag, P. A. McQueney, J. Zhu, O. Hensens, L. Koupal, J. Liesch, M. Goetz, E. Lazarides, and C. M. Woods, *Cancer Res.*, **55**, 2325–2333 (1995).

89. For an excellent review of the "Chemical Biology of Epothilones," see K. C. Nicolaou, F. Roschangar, and V. Dionisios, *Angew. Chem. Int. Ed. Engl.*, **37**, 2014–2045 (1998) and references therein.

90. R. M. Borzilleri, X. Zheng, R. J. Schmidt, J. A. Johnson, S.-H. Kim, J. D. DiMarco, C. R. Fairchild, J. Z. Gougoutas, F. Y. F. Lee, B. H. Long, and G. D. Vite, *J. Am. Chem. Soc.*, **122**, 8890–8897 (2000).

91. F. Y. F. Lee, R. Borzilleri, C. R. Fairchild, S.-H. Kim, B. H. Long, C. Reventos-Suarez, G. D. Vite, W. C. Rose, and R. A. Kramer, *Clin. Cancer Res.*, **7**, 1429–1437 (2001).

92. H. Stahelin and A. von Wartburg in E. Jucker, Ed., *Progress in Drug Research*, Birkhauser-Verlag, Basel, Vol. **33**, 1989, pp. 169–266.

93. H. Stahelin and A. von Wartburg, *Cancer Res.*, **51**, 5–15 (1991) present a shorter and more readable account.

94. M. G. Kelly and J. L. Hartwell, *J. Natl. Cancer Inst.*, **14**, 967–986 (1954).

95. I. W. Kaplan, *New Orleans Med. Surg. J.*, **94**, 388 (1942).

96. J. L. Hartwell and A. W. Schrecker in L. Zechmeister, Ed., *Progress in the Chemistry of Organic Natural Products*, 1958, pp. 83–166 provide a detailed review of the earlier developments and background.

97. V. Podwyssotzki, *Arch. Exp. Pathol. Pharmacol.*, **13**, 29 (1880).

98. J. L. Hartwell and A. W. Schrecker, *J. Am. Chem. Soc.*, **73**, 2909–2916 (1951).

99. K. S. Pankajarnani and T. R. Seshadri, *Proc. Ind. Acad. Sci.*, **36A**, 157 (1952) through *Chem. Abstr.*, **48**, 2702 (1954). See Ref. 77 for a wider discussion.

100. A. Stoll, J. Renz, and A. von Wartburg, *J. Am. Chem. Soc.*, **76**, 3103–3104 (1954).

101. A. Stoll, A. von Wartburg, E. Angliker, and J. Renz, *J. Am. Chem. Soc.*, **76**, 6413–6414 (1954).

102. A. von Wartburg, E. Angliker, and J. Renz, *Helv. Chim. Acta*, **40**, 1331–1357 (1957).

103. I. Jardine in J. M. Cassady and J. D. Douros, Eds., *Anticancer Agents Based on Natural Product Models*, Academic Press, New York, Vol. **16**, 1980, pp. 319–351 provides a useful review of the middle years.

104. H. Emmenegger, H. Stahelin, J. Rutschmann, J. Rertz, and A. von Wartburg, *Drug Res.*, **11**, 327–333, 459–469 (1961).

105. H. Stahelin, *Planta Med.*, **22**, 336–347 (1972).

106. H. Stahelin, *Med. Exp. (Basel)*, **7**, 92–102 (1962).

107. M. Kulm and A. von Wartburg, *Helv. Chim. Acta*, **52**, 948–955 (1969).

108. C. Keller-Juslen, M. Kuhn, A. von Wartburg, and H. Stahelin, *J. Med. Chem.*, **14**, 936–940 (1971).

109. B. H. Long and A. Minocha, *Proc. Am. Assoc. Cancer Res.*, **24**, 321 (1983).

110. O. J. McConnell, R. E. Longley, and F. E. Koehn, in V. P. Gullo, Ed., *The Discovery of Natural Products with Therapeutic Potential*, Butterworth-Heinemann, Boston, p. 109 (1994).

111. K. L. Rinehart, T. G. Holt, N. L. Fregeau, J. G. Stroh, P. A. Keifer, F. Sun, L. H. Li, and D. G. Martin, *J. Org. Chem.*, **55**, 4512–4515 (1990).

112. Reuters News Service, 11 July (2001).

113. J. Adams and P. J. Elliott, *Oncogene*, **19**, 6687–6692 (2000).

114. G. R. Pettit, F. Gao, D. Sengupta, J. C. Coll, C. L. Herald, D. L. Doubek, J. M. Schmidt, J. R. Camp, J. J. Rudloe, and R. A. Nieman, *Tetrahedron*, **47**, 3601–3610 (1991).

115. M. Jaspars, *Chem. Ind.*, 51–55 (1999).

116. E. J. Corey, D. Y. Gin, and R. Kania, *J. Am. Chem. Soc.*, **118**, 9202–9203 (1996).

117. Y. Hamada, K. Hayashi, and T. Shiori, *Tetrahedron Lett.*, **32**, 931–934 (1991).

118. A. Fleming, *Br. J. Exp. Med.*, **10**, 226–236 (1929).

119. E. Chain, H. W. Florey, A. D. Gardner, N. G. Heatley, M. A. Jennings, J. Orr-Ewing, and A. G. Sanders, *Lancet*, **2**, 226–228 (1940).

120. Ref. 6, pp. 298–315, provides a good review of the discovery and development of penicillin antibiotics.

121. Ref. 19, pp. 1065–1085, summarizes the pharmacological properties of the more important commercial penicillins.

122. J. C. Sheehan, "Molecular Modification in Drug Design," in *Advances in Chemistry Series, No. 45*, American Chemical Society, Washington, DC, 1964, pp. 15–24.

123. T. F. Howarth, A. G. Brown, and T. J. King, *J. Chem. Soc. Chem. Commun.*, 266–267 (1976).

124. C. Reading and P. Hepburn, *Biochem. J.*, **179**, 67–76 (1979).

125. A. P. Ball, A. M. Geddes, P. G. Davey, I. D. Farrell, and G. R. Brookes, *Lancet*, **1**, 620–623 (1980).

126. Ref. 6, pp. 321–324, provides an interesting account of this discovery.

127. See Ref. 6, pp. 324–300.

128. G. G. F. Newton and E. P. Abraham, *Nature*, **175**, 548 (1955).

129. H. J. Smith, "Design of Antimicrobial Chemotherapeutic Agents," in *Smith and William's Introduction to the Principles of Drug Design*, 2nd ed., Wright, London, 1988, pp. 285–288.

130. E. H. Flynn, Ed., *Cephalosporins and Penicillins Chemistry and Biology*, Academic Press, New York, 1972.

131. G. Albers-Schonberg et al., *J. Am. Chem. Soc.*, **100**, 6491–6499 (1978).

132. W. J. Leanza, K. J. Wildonger, T. W. Miller, and B. G. Christensen, *J. Med. Chem.*, **22**, 1435–1436 (1979).

133. A. Imada, K. Kitano, K. Kintaka, M. Muroi, and M. Asai, *Nature*, **289**, 590–591 (1981).

134. R. B. Sykes, C. M. Cimarusti, D. P. Bonner, K. Bush, D. M. Floyd, N. H. Georgopapadakou, W. H. Koster, W. C. Liu, W. L. Parker, P. A. Principe, M. L. Rathnum, W. A. Slusarchyk, W. H. Trejo, and J. S. Wells, *Nature*, **291**, 489–491 (1981).

135. R. B. Sykes, D. P. Bonner, K. Bush, N. H. Georgopapadakou, and J. S. Wells, *J. Antimicrob. Chemother.*, **8** (Suppl. E), 1–16 (1981).

136. R. B. Sykes and D. P. Bonner, "Monobactam Antibiotics: History and Development," in J. D. Williams and P. Woods, Eds., *Aztreonam, The Antibiotic Discovery for Gram-negative Infections*, Royal Society Medicine International Congress Symposium Series No. 89, Royal Society Medicine, London, 1985, pp. 3–24.

137. N. Bahal and M. C. Nahata, *Ann. Pharmacother.*, **26**, 46–55 (1992).

138. J. Barber, J. I. Gyi, G. A. Morris, D. A. Pye, and J. K. Sutherland, *J. Chem. Soc. Chem. Commun.*, 1040–1041 (1990).

139. P. Kurath, P. H. Jones, R. S. Egan, and T. J. Perun, *Experientia*, **27**, 362 (1971).

140. S. Omura, K. Tsuzuki, T. Sunazuka, S. Marui, H. Toyoda, N. Inatomi, and Z. Itoh, *J. Med. Chem.*, **30**, 1941–1943 (1987).

141. L. D. Bechtol, V. C. Stephens, C. T. Pugh, M. B. Perkal, and P. A. Coletta, *Curr. Ther. Res.*, **20**, 610 (1976).

142. H. A. Kirst and G. D. Sides, *Antimicrob. Agents Chemother.*, **33**, 1413–1418 (1989), provide a useful, brief review.

143. P. Luger and R. Maier, *J. Cryst. Mol. Struct.*, **9**, 329 (1979).

144. G. M. Bright, A. A. Nagel, J. Bordner, K. A. Desai, J. N. Dibrino, J. Nowakowska, L. Vincent, R. M. Watrous, F. C. Sciavolino, A. R. English, J. A. Retsema, M. R. Anderson, L. A. Brennan, R. J. Borovoy, C. R. Cimochowski, J. A. Faiella, A. E. Girard, D. Girard, C. Herbert, M. Manousos, and R. Mason, *J. Antibiot.*, **41**, 1029–1047 (1988).

145. S. Moromoto, Y. Takahashi, Y. Watanabe, and S. Omura, *J. Antibiot.*, **37**, 187–189 (1984).

146. For an overview, see R. Leclercq, *J. Antimicrob. Chemother.*, **48**, 9–23 (2001).

147. S. Douthwaite and W. S. Champney, *J. Antimicrob. Chemother.*, **48**, 1–8 (2001).

148. G. Bonfiglio and P. M. Furneri, *Expert Opin. Invest. Drugs*, **10**, 185–198 (2001).

149. J. S. Tkacz in J. Sutcliffe and N. H. Georgopapadakou, Eds., *Emerging Targets in Antibacterial and Antifungal Chemotherapy*, Chapman and Hall, New York, 1992, pp. 504–508.

150. J. M. Balkovec, R. M. Black, M. L. Hammond, J. V. Heck, R. A. Zambias, G. Abruzzo, K. Bartizal, H. Kropp, C. Trainor, R. E. Schwartz, D. C. McFadden, K. H. Nollstadt, L. A. Pittarelli, M. A. Powles, and D. M. Schatz, *J. Med. Chem.*, **35**, 194–198 (1992).

151. R. Gordee and M. Debono, *Drugs of the Future*, **14**, 939 (1989).

152. A. J. De Lucca, *Expert Opin. Invest. Drugs*, **9**, 273–299 (2000).

153. S. Y. Ablordeppey, P. Fan, J. H. Ablordeppey, and L. Mardenborough, *Curr. Med. Chem.*, **6**, 1151–1195 (1999).

154. M. A. Pfaller, F. Marco, S. A. Messer, and R. N. Jones, *Diagn. Microbiol. Infect. Dis.*, **30**, 251–255 (1998).

155. G. K. Abruzzo, C. J. Gill, A. M. Flattery, L. Kong, C. Leighton, J. G. Smith, V. B. Pikounis, K. Bartizal, and H. Rosen, *Antimicrob. Agents Chemother.*, **44**, 2310–2318 (2000).

156. E. E. Slater and J. S. McDonald, *Drugs*, Suppl. 3, 72–82 (1988).

157. A. Endo, M. Kuroda, and Y. Tsujita, *J. Antibiot.*, **29**, 1346–1348 (1976).

158. D. J. Gordon and B. M. Riffind, *Ann. Int. Med.*, **107**, 759–761 (1987).

159. A. G. Brown, T. C. Smale, T. J. King, R. Hasenkamp, and R. H. Thompson, *J. Chem. Soc. Perkin Trans. 1*, 1165–1170 (1976).

160. P. R. Vagelos, *Science*, **252**, 1080–1084 (1991), gives a brief, chronological account of the discovery of lovastatin.

161. A. W. Alberts et al., *Proc. Natl. Acad. Sci. USA*, **77**, 3957–3961 (1980).

162. A. Endo, *J. Antibiot.*, **32**, 852–854 (1979).

163. A. W. Alberts, *Am. J. Cardiol.*, **62**, 10J–15J (1988), and references therein.

164. S. M. Grundy, "HMG Co A Reductase Inhibitors: Clinical Applications and Therapeutic Potential," in B. M. Rifkind, Ed., *Drug Treatment of Hyperlipidemia*, Marcel Dekker, New York, 1991, pp. 139–167.

165. W. F. Hoffrnann, A. W. Alberts, P. S. Anderson, J. S. Chen, R. L. Smith, and A. K. Willard, *J. Med. Chem.*, **29**, 849–852 (1986).

166. E. E. Slater and J. S. MacDonald, *Drugs*, **36** (Suppl. 3), 72–82 (1988).

167. C. E. Nakamura and R. H. Abeles, *Biochemistry*, **24**, 1364–1376 (1985).

168. Reuters News Service, 24 Sept (2001).

169. S. H. Ferreira, *Br. J. Pharmacol.*, **24**, 163–169 (1965).

170. S. H. Ferreira, L. J. Greene, V. A. Alabaster, Y. S. Bakhle, and J. R. Vane, *Nature*, **225**, 379–380 (1970).

171. S. H. Ferreira, D. C. Bartelt, and L. J. Greene, *Biochemistry*, **9**, 2583–2593 (1970).

172. M. A. Ondetti, B. Rubin, and D. W. Cushman, *Science*, **196**, 441–444 (1977).

173. R. A. Maxwell and S. B. Eckhardt, *Drug Discovery: A Casebook and Analysis*, Humana Press, Clifton, NJ, 1990, pp. 19–34.

174. Ref. 6, pp. 105–114, chronicles the development of the β-adrenoceptor blocking agents.

175. D. Potterton, Ed., *Culpeper's Colour Herbal*, Foulsham, London, 1983, p. 123.

176. K. P. Link, *Harvey Lectures*, Series 39, 1944, pp. 162–216.

177. K. P. Link, *Circulation*, **19**, 97–107 (1959).

178. See Ref. 19, pp. 1317–1322.

179. Mustafa, *C. R. Acad. Sci. Paris*, **89**, 442 (1879).

180. E. Spath and W. Gruber, *Ber. Dtsch. Chem. Ges.*, **71**, 106 (1938).

181. G. V. Anrep and G. Misrahy, *Gaz. Fac. Med. Cairo*, **13**, 33 (1945).

182. G. V Anrep, G. S. Barsourn, M. R. Kenawy, and G. Misrahy, *Lancet*, 557–558 (1947).

183. G. B. Kauffman, *Educ. Chem.*, **21**, 42–45 (1984).

184. See Ref. 62, p. 192.

185. H. Cairns, C. Fitzmaurice, D. Hunter, P. B. Johnson, J. King, T. B. Lee, G. H. Lord, R. Minshull, and J. S. G. Cox, *J. Med. Chem.*, **15**, 583–589 (1972).

186. See Ref. 19, pp. 630–632.

187. B. N. Singh, *Am. Heart J.*, **106**, 788–797 (1983).

188. B. N. Singh, N. Venkatesh, K. Nademanee, M. A. Josephson, and R. Karman, *Prog. Cardiovasc. Dis.*, **31**, 249–280 (1989).

189. J. van Schepdael and H. Solvay, *Presse Med.*, **78**, 1849–1855 (1970).

190. See Ref. 62, pp. 189–191.

191. See Ref. 6, pp. 98–105.

192. R. P. Ahlquist, *Am. J. Physiol.*, **153**, 586–600 (1948).

193. A. M. Lands, F. P. Luduena, and H. J. Buzzo, *Life Sci.*, **6**, 2241–2249 (1967).

194. See Ref. 173, pp. 333–348.

195. D. L. Burgoyne, R. J. Anderson, and T. M. Allen, *J. Org. Chem.*, **57**, 525–528 (1992).

196. P. I. Trigg, in H. Wagner, H. Hikino, and N. R. Farnsworth, Eds., *Economic and Medicinal Plant Research*, Academic Press, London, Vol. **3**, 1989, pp. 19–55.

197. W. Tang and G. Eisenbrand, Eds., *Chinese Drugs of Plant Origin*, Springer-Verlag, Berlin, 1992, pp. 161–175.

198. S. R. Meshnick, A. Thomas, A. Ranz, C.-M. Xu, and H.-Z. Pan, *Mol. Biochem. Parasitol.*, **49**, 181–190 (1991).

199. S. R. Meshnick, Y.-Z. Yang, V. Lima, F. Kuypers, S. Kamchonwongpaisan, and Y. Yuthavong, *Antimicrob. Agents Chemother.*, **37**, 1108–1114 (1993).

200. P. L. Olliaro et al., *Trends Parasitol.*, **17**, 122–126 (2001).

201. G. H. Posner and S. R. Meshnick, *Trends Parasitol.*, **17**, 266–267 (2001).

202. For example: J. Cazelles et al., *J. Chem. Soc. Perkin Trans. 1*, 1265–1270 (2000); M. D. Bachi et al., *Bioorg. Med. Chem. Lett.*, **8**, 903–908 (1998).

203. J. Karbwang, K. N. Bangchang, A. Thanavibul, D. Bunnag, T. Chongsuphajaisiddhi, and T. Harinasuta, *Lancet*, **340**, 1245 (1992), report some clinical experience to support the data in Refs. 196 and 197.

204. A. N. Chawira, D. C. Warhurst, B. L. Robinson, and W. Peters, *Trans. R. Soc. Trop. Med. Hyg.*, **81**, 554–558 (1987).

205. G. D. Brown, personal communication.

206. B. C. Elford, M. F. Roberts, J. D. Phillipson, and R. J. M. Wilson, *Trans. R. Soc. Trop. Med. Hyg.*, **81**, 434–436 (1987).

207. A. I. White in C. O. Wilson, O. Gisvold, and R. F. Doerge, Eds., *Textbook of Organic, Medic-*

inal and Pharmaceutical Chemistry, 7th ed., J. B. Lippincott, Philadelphia, 1977, pp. 247–268.

208. F. A. Fluckiger and D. Hanbury, *Pharmacographia, A History of the Principal Drugs of Vegetable Origin, Met With in Great Britain and British India*, Macmillan, London, 1879, pp. 361–362.

209. J. Pelletier and J. Caventou, *Ann. Chim. Phys.*, **XV**, 292 (1820).

210. Anonymous, quoted by C. Allen, *Tales from the Dark Continent*, Warner, London, 1992, p. 30.

211. P. Rabe, *Berichte*, **55**, 522 (1922).

212. R. B. Woodward and W. E. Doering, *J. Am. Chem. Soc.*, **67**, 860 (1945).

213. F. Schonhofer et al., *Z. Physiol. Chem.*, **274**, 1 (1942).

214. P. Miffilens, *Naturwissenschaften*, **14**, 1162–1166 (1926).

215. See Ref. 19, pp. 988–991.

216. G. R. Coatney, *Am. J. Trop. Med. Hyg.*, **12**, 121–128 (1963).

217. P. Winstanley and P. Olliario, *Expert Opin. Invest. Drugs*, **7**, 261–271 (1998).

218. Anonymous, *Bull. World Health Org.*, **61**, 169–178 (1983).

219. L. H. Schmidt, R. Crosby, J. Rasco, and D. Vaughan, *Antimicrob. Agents Chemother.*, **13**, 1011–1030 (1978).

220. R. M. Pinder and A. Burger, *J. Med. Chem.*, **11**, 267 (1968).

221. C. J. Ohnmacht, A. R. Patel, and R. E. Lutz, *J. Med. Chem.*, **14**, 926 (1971).

222. J. E. Rosenblatt, *Mayo Clin. Proc.*, **74**, 1161–1175 (1999).

223. F. Page, *Lancet*, 755 (1951).

224. A. Freedman and F. Bach, *Lancet*, 321 (1952).

225. G. O. Haydu, *Am. J. Med. Sci.*, **225**, 71 (1953).

226. J. Forestier and A. Certonciny, *Rev. Rhum. Mal. Osteoartic.*, **21**, 395 (1954).

227. A. L. Scherbel, S. L. Schuchter, and J. W. Harrison, *Cleve. Clin. Q.*, **24**, 98 (1957); see also A. L. Scherbel, *Am. J. Med.*, **75**, 1 (1983).

228. H. G. Davies and R. H. Green, *Chem. Soc. Rev.*, **20**, 211–269 (1991), provide structural details of a large number of analogs.

229. H. G. Davies and R. H. Green, *Nat. Prod. Rep.*, **3**, 87–121 (1986).

230. W. C. Campbell, M. H. Fisher, E. O. Stapley, G. Albers-Schonberg, and T. A. Jacob, *Science*, **221**, 823–828 (1983).

231. K. Awadzi, K. Y. Dadzie, H. Schulzkey, D. R. W. Haddock, H. M. Gillies, and M. A. Aziz, *Ann. Trop. Med. Parasitol.*, **79**, 63 (1985).

232. B. M. Greene, H. R. Taylor, E. W. Cupp, R. P. Murphy, A. T. White, M. A. Aziz, H. Schulzkey, S. A. Danna, H. S. Newland, L. P. Goldschmidt, C. Auer, A. P. Hanson, S. V. Freeman, E. W. Reber, and P. N. Williams, *N. Engl. J. Med.*, **313**, 133–138 (1985).

233. F. A. Drobniewski, *Microbiology Europe*, 24–28 (1993).

234. F. O. Richards, E. S. Miri, M. Katabarwa, A. Eyamba, M. Sauerbrey, G. Zea-Flores, K. Korve, W. Mathai, M. A. Homeida, I. Mueller, E. Hilyer, and D. R. Hopkins, *Am. J. Trop. Med. Hyg.*, **65**, 108–114 (2001).

235. K. Awadzi, S. K. Attah, E. T. Addy, N. O. Opoku, and B. T. Quartey, *Trans. R. Soc. Trop. Med. Hyg.*, **93**, 189–194 (1999).

236. B. A. Boatin, J. M. Hougard, E. S. Alley, L. K. B. Akpoboua, L. Yameogo, N. Dembele, A. Seketeli, and K. Y. Dadzie, *Ann. Trop. Med. Parasitol.*, **92**, S47–S60 (1998).

237. B. Leppard and A. E. Naburi, *Br. J. Dermatol.*, **143**, 520–523 (2000).

238. C. N. Burkhart, *Vet. Hum. Toxicol.*, **42**, 30–35 (2000).

239. L. J. Scott and K. L. Goa, *Drugs*, **60**, 1095–1122 (2000).

240. M. A. H. Mucke, J. Froehlich, and U. Jordis, WO 0032199 (2000).

241. U. Jordis, J. Froehlich, M. Treu, M. Hirnschall, L. Czollner, B. Kaelz, and S. Welzig, WO 0174820 (2001).

242. J. D. Coombes, Ed., *New Drugs from Natural Sources*, IBC Technical Services, London, 1992, pp. 59–62, 93–100.

243. G. G. Yarbrough, D. P. Taylor, R. T. Rowlands, M. S. Crawford, and L. Lasure, *J. Antibiot.*, **46**, 535–544 (1993).

244. W. H. Moos, G. D. Green, and M. R. Pavia, "Recent Advances in the Generation of Molecular Diversity," in J. A. Bristol, Ed., *Annual Report of Medicinal Chemistry*, Vol. **28**, Academic Press, New York, 1993, pp. 315–324.

245. Y.-Z. Shu, *J. Nat. Prod.*, **61**, 1053–1071 (1998).

246. D. J. Newman, G. M. Cragg, and K. M. Snader, *Nat. Prod. Rep.*, **17**, 215–234 (2000).

Index

Terms that begin with numbers are indexed as if the number were spelled out; e.g., "3D models" is located as if it were spelled "ThreeD models."

A-77003
  structure-based design, 437–438
A-80987
  structure-based design, 438, 439
A-306552, 675
Absolv program, 389
Absorption, distribution, metabolism, and excretion (ADME)
  as bottleneck in drug discovery, 592
  estimation systems, 389–390
  molecular modeling, 154–155
Absorption, distribution, metabolism, excretion, and toxicity (ADMET), 389
  and druglikeness screening, 245
  and virtual screening filter cascade, 267
ABT-418, 808–810
ABT-627, 811–812
ABT-773, 876
Academic databases, 387–388
Acarbose, 849
Accelrys databases, 384–385
Accord, 377, 385
Accord Database Explorer, 385
Acebutol
  renal clearance, 38
ACE inhibitors, 718, 881
  asymmetric synthesis, 807, 809
  comparative molecular field analysis, 151–153
  conformationally restricted peptidomimetics, 640–641
  molecular modeling, 131, 132–133, 145
  multisubstrate analogs, 746–748
  receptor-relevant subspace, 204
  structure-based design, 432–433

transition state analogs, 650–651
Acetic acid
  CML representation, 372
Acetylcholine, 772
  cation-π bonding, 724–725
  conformationally restricted analogs, 697–698
  muscarinic agonist analogs, 143–144
Acetylcholinesterase inhibitors, 718
  CoMFA study, 58–59
  pseudoirreversible, 772–774
  substrates from acetylcholine analogs, 697–698
  target of structure-based drug design, 449–450
  volume mapping, 140
  X-ray crystallographic studies, 482
N-Acetylcysteine
  toxicological profile prediction, 838, 840
Acetylsalicylic acid (aspirin), 762–763
Acquire database, 386
Actimomycin D
  thermodynamics of binding to DNA, 183
Actinoplanes utahensis, 878
Active Analog Approach, 58, 60, 639
  flow of information in, 146
  and molecular modeling, 134, 151
  and systematic search, 144–145
Active-site directed, irreversible inhibitors, 755
Activity
  binding affinity contrasted, 131–135
Activity-guided fractionation, 597
Activity similarity, 255
Acyclovir, 717, 719, 756
Acyl halides
  filtering from virtual screens, 246

ADAM
  geometric/combinatorial search, 295
ADAMs, 652
ADAPT, 53
Adenosine deaminase inhibitors, 717
  mass-spectrometric screening for ligands to, 604
  transition-state analogs, 750–752
  X-ray crystallographic studies, 482
S-Adenosyl-L-homocysteine, 733, 740
S-Adenosyl-L-methionine, 733
ADEPT suite, 225, 226, 237
ADME studies, *See* Absorption, distribution, metabolism, and excretion (ADME)
ADMET studies, *See* Absorption, distribution, metabolism, excretion, and toxicity (ADMET)
ADP analogs, 763–764
Adrenaline, 885, 886
β-Adrenoreceptor antagonists, *See* β-Blockers
Afferent, 387
Affinity calculation, 118–122
Affinity capillary electrophoresis-mass spectrometry, 599–600
Affinity chromatography-mass spectrometry, 598–599
Affinity grids, 292–293
Affinity labels, 756–759, 760–764
Affinity NMR, 571
2-AG, 853
AG31
  structure-based design, 428–429
AG85
  structure-based design, 428
AG331
  structure-based design, 428–429
AG337
  structure-based design, 428